T0320550

SEMICONDUCTOR QUANTUM OPTICS

The emerging field of semiconductor quantum optics combines semiconductor physics and quantum optics, with the aim of developing quantum devices with unprecedented performance. In this book researchers and graduate students alike will reach a new level of understanding to begin conducting state-of-the-art investigations.

The book combines theoretical methods from quantum optics and solid-state physics to give a consistent microscopic description of light–matter- and many-body-interaction effects in low-dimensional semiconductor nanostructures. It develops the systematic theory needed to treat semiconductor quantum-optical effects, such as strong light–matter coupling, light–matter entanglement, squeezing, as well as quantum-optical semiconductor spectroscopy. Detailed derivations of key equations help readers learn the techniques and nearly 300 exercises help test their understanding of the materials covered.

The book is accompanied by a website hosted by the authors, containing further discussions on topical issues, latest trends, and publications on the field. The link can be found at www.cambridge.org/9780521875097.

MACKILLO KIRA and STEPHAN W. KOCH are Professors of Theoretical Physics in the Department of Physics, Philipps-Universität Marburg.

SEMICONDUCTOR QUANTUM OPTICS

MACKILLO KIRA
Philipps-Universität Marburg

STEPHAN W. KOCH
Philipps-Universität Marburg

CAMBRIDGE
UNIVERSITY PRESS

University Printing House, Cambridge CB2 8BS, United Kingdom

One Liberty Plaza, 20th Floor, New York, NY 10006, USA

477 Williamstown Road, Port Melbourne, VIC 3207, Australia

314-321, 3rd Floor, Plot 3, Splendor Forum, Jasola District Centre, New Delhi - 110025, India

79 Anson Road, #06-04/06, Singapore 079906

Cambridge University Press is part of the University of Cambridge.

It furthers the University's mission by disseminating knowledge in the pursuit of education, learning and research at the highest international levels of excellence.

www.cambridge.org
Information on this title: www.cambridge.org/9780521875097

First published 2012

A catalogue record for this publication is available from the British Library

Library of Congress Cataloging in Publication data
Kira, Mackillo, 1969–
Semiconductor quantum optics / Mackillo Kira and Stephan W. Koch.
p. cm.
ISBN 978-0-521-87509-7 (hardback)
1. Semiconductors. 2. Quantum optix. 3. Quantum electrodynamics.
I. Koch, S. W. (Stephan W.) II. Title.
QC611.6.Q36K57 2011
621.3815′2–dc23
2011025901

ISBN 978-0-521-87509-7 Hardback

Additional resources for this publication at www.cambridge.org/9780521875097

Contents

Preface *page* xi

1 Central concepts in classical mechanics 1
- 1.1 Classical description 1
- 1.2 Statistical description of particles 13
- Exercises 23
- Further reading 25

2 Central concepts in classical electromagnetism 26
- 2.1 Classical description of electromagnetic fields 26
- 2.2 Particle aspects of electromagnetic waves 32
- 2.3 Generalized wave and Helmholtz equations 38
- Exercises 45
- Further reading 46

3 Central concepts in quantum mechanics 48
- 3.1 Schrödinger equation 49
- 3.2 Expectation values in quantum mechanics 54
- Exercises 63
- Further reading 64

4 Central concepts in stationary quantum theory 65
- 4.1 Stationary Schrödinger equation 65
- 4.2 One-dimensional Schrödinger equation 67
- 4.3 Classification of stationary eigenstates 71
- 4.4 Generic classification of energy eigenstates 82
- Exercises 84
- Further reading 85

5 Central concepts in measurement theory 86
- 5.1 Hermitian operators 86
- 5.2 Eigenvalue problems 87
- 5.3 Born's theorem 93

Exercises 98
Further reading 99

6 Wigner's phase-space representation 101
6.1 Wigner function 101
6.2 Wigner-function dynamics 111
6.3 Density matrix 115
6.4 Feasibility of quantum-dynamical computations 116
Exercises 118
Further reading 119

7 Hamiltonian formulation of classical electrodynamics 121
7.1 Basic concepts 122
7.2 Hamiltonian for classical electrodynamics 124
7.3 Hamilton equations for light–matter system 129
7.4 Generalized system Hamiltonian 136
Exercises 139
Further reading 140

8 System Hamiltonian of classical electrodynamics 141
8.1 Elimination of the scalar potential 141
8.2 Coulomb and Lorentz gauge 143
8.3 Transversal and longitudinal fields 148
8.4 Mode expansion of the electromagnetic field 153
Exercises 159
Further reading 160

9 System Hamiltonian in the generalized Coulomb gauge 162
9.1 Separation of electronic and ionic motion 162
9.2 Inclusion of the ionic polarizability 165
9.3 Generalized Coulomb potential 170
9.4 Generalized light-mode functions 177
Exercises 190
Further reading 191

10 Quantization of light and matter 193
10.1 Canonical quantization 193
10.2 Second quantization of light 198
10.3 Eigenstates of quantized modes 203
10.4 Elementary properties of Fock states 211
Exercises 216
Further reading 217

11 Quasiparticles in semiconductors 218
11.1 Second-quantization formalism 218

11.2 System Hamiltonian of solids 229
Exercises 236
Further reading 238

12 Band structure of solids 240
12.1 Electrons in the periodic lattice potential 240
12.2 Systems with reduced effective dimensionality 246
Exercises 251
Further reading 252

13 Interactions in semiconductors 253
13.1 Many-body Hamiltonian 253
13.2 Light–matter interaction 254
13.3 Phonon–carrier interaction 264
13.4 Coulomb interaction 267
13.5 Complete system Hamiltonian in different dimensions 269
Exercises 276
Further reading 277

14 Generic quantum dynamics 279
14.1 Dynamics of elementary operators 279
14.2 Formal properties of light 288
14.3 Formal properties of general operators 295
Exercises 300
Further reading 302

15 Cluster-expansion representation of the quantum dynamics 304
15.1 Singlet factorization 305
15.2 Cluster expansion 310
15.3 Quantum dynamics of expectation values 315
15.4 Quantum dynamics of correlations 316
15.5 Scattering in terms of correlations 318
Exercises 321
Further reading 322

16 Simple many-body systems 324
16.1 Single pair state 324
16.2 Hydrogen-like eigenstates 328
16.3 Optical dipole 335
Exercises 342
Further reading 343

17 Hierarchy problem for dipole systems 345
17.1 Quantum dynamics in the $\hat{\mathbf{A}} \cdot \hat{\mathbf{p}}$ picture 345
17.2 Light–matter coupling 351

17.3 Dipole emission 351
17.4 Quantum dynamics in the $\hat{\mathbf{E}} \cdot \hat{\mathbf{x}}$ picture 356
Exercises 361
Further reading 363

18 Two-level approximation for optical transitions 365
18.1 Classical optics in atomic systems 365
18.2 Two-level system solutions 377
Exercises 386
Further reading 387

19 Self-consistent extension of the two-level approach 388
19.1 Spatial coupling between light and two-level system 388
19.2 Maxwell-optical Bloch equations 394
19.3 Optical Bloch equations with radiative coupling 399
Exercises 403
Further reading 403

20 Dissipative extension of the two-level approach 405
20.1 Spin representation of optical excitations 405
20.2 Dynamics of Pauli spin matrices 406
20.3 Phenomenological dephasing 408
20.4 Coupling between reservoir and two-level system 412
Exercises 418
Further reading 418

21 Quantum-optical extension of the two-level approach 420
21.1 Quantum-optical system Hamiltonian 420
21.2 Jaynes–Cummings model 427
Exercises 436
Further reading 437

22 Quantum dynamics of two-level system 438
22.1 Formal quantum dynamics 438
22.2 Quantum Rabi flopping 441
22.3 Coherent states 445
22.4 Quantum-optical response to superposition states 449
Exercises 454
Further reading 456

23 Spectroscopy and quantum-optical correlations 457
23.1 Quantum-optical spectroscopy 457
23.2 Quantum-statistical representations 459
23.3 Thermal state 464
23.4 Cluster-expansion dynamics 467

23.5 Quantum optics at the singlet–doublet level 473
Exercises 476
Further reading 478

24 General aspects of semiconductor optics 480
24.1 Semiconductor nanostructures 480
24.2 Operator dynamics of solids in optical regime 486
24.3 Cluster-expansion dynamics 490
24.4 Relevant singlets and doublets 491
24.5 Dynamics of singlets 492
Exercises 496
Further reading 497

25 Introductory semiconductor optics 499
25.1 Optical Bloch equations 499
25.2 Linear response 502
25.3 Coherent vs. incoherent quantities 507
25.4 Temporal aspects in semiconductor excitations 512
Exercises 518
Further reading 519

26 Maxwell-semiconductor Bloch equations 521
26.1 Semiconductor Bloch equations 521
26.2 Excitonic states 526
26.3 Semiconductor Bloch equations in the exciton basis 529
26.4 Linear optical response 532
26.5 Excitation-induced dephasing 541
Exercises 547
Further reading 548

27 Coherent vs. incoherent excitons 550
27.1 General singlet excitations 550
27.2 Incoherent excitons 556
27.3 Electron–hole correlations in the exciton basis 563
Exercises 568
Further reading 570

28 Semiconductor luminescence equations 572
28.1 Incoherent photon emission 572
28.2 Dynamics of photon-assisted correlations 577
28.3 Analytic investigation of the semiconductor luminescence 582
28.4 Excitonic signatures in the semiconductor luminescence 588
Exercises 590
Further reading 591

29 Many-body aspects of excitonic luminescence 593
- 29.1 Origin of excitonic plasma luminescence 593
- 29.2 Excitonic plasma luminescence 597
- 29.3 Direct detection of excitons 602
- Exercises 604
- Further reading 606

30 Advanced semiconductor quantum optics 608
- 30.1 General singlet–doublet dynamics 609
- 30.2 Advanced quantum optics in the incoherent regime 614
- 30.3 Advanced quantum optics in the coherent regime 616
- Exercises 622
- Further reading 622

Appendix Conservation laws for the transfer matrix 627
- A.1 Wronskian-induced constraints 627
- A.2 Current-induced constraints 628
- A.3 Explicit conservation laws 630
- Further reading 632

Index 633

Preface

A wide variety of quantum-optical effects can be understood by analyzing atomic model systems interacting with the quantized light field. Often, one can fully calculate and even measure the quantum-mechanical wave function and its dependence on both the atomic and the light degrees of freedom. By elaborating on and extending this approach, researchers perpetually generate intriguing results and new insights allowing for the exploration and utilization of effects encountered only in the realm of quantum phenomena.

By now, quantum-optical investigations have evolved from atoms all the way to complex systems, such as solids, in particular semiconductors. As a profound conceptual challenge, the optical transitions in semiconductors typically involve an extremely large number of electronic states. Due to their electric charge, the optically active electrons experience strong Coulomb interaction effects. Furthermore, they are coupled to the lattice vibrations of the solid crystal. For such an interacting many-body system, the overwhelmingly large number of degrees of freedom makes it inconceivable to measure the full wave function; we obviously need new strategies to approach semiconductor quantum optics. The combination of quantum-optical and many-body interactions not only leads to prominent modifications of the effects known from atomic systems but also causes new phenomena without atomic counterparts.

In this book, we develop a detailed microscopic theory for the analysis of semiconductor quantum optics. As central themes, we discuss how the quantum-optical approach can be systematically formulated for solids, which new aspects and prospects arise, and which conceptual modifications have to be implemented. The presented material is largely based on our own research and teaching endeavors on various topics in quantum mechanics, many-body theory, solid-state physics, optics, laser theory, quantum-optics, and semiconductor quantum optics. Our experience shows that one needs a systematic combination of optical and many-body theory to truly understand and predict quantum-optical effects in semiconductors. Therefore, we have implemented a multifaceted approach where we first discuss the basic quantum-theoretical techniques and concepts. We then present the central steps to quantize the light field and the many-body system. Altogether, we develop a systematic theory for semiconductor quantum optics and present its main consequences.

One of our major goals is to provide a bridge between "traditional" quantum optics and many-body theory. We naturally cannot present the final conclusion on this topic because the combination of quantum optics and many-body quantum dynamics actually poses one of the most difficult problems in contemporary physics, which does not allow for general exact results. Therefore, the material in this book is designed such that it can be applied to generate new systematic approximations to the full many-body/quantum optics. We believe that this work will contribute to an expansion of the general knowledge base needed to diversify our understanding of quantum mechanics in complex many-body systems.

This book introduces all the central concepts and develops the main steps of the theory needed for a precise formulation and analysis of many relevant phenomena. We thoroughly discuss the emerging effects as we cross the boundary from the classical to the quantum-optical features of semiconductor systems. Even though we present a research outlook beyond the basic investigations in the last chapter of this book, the more detailed applications and many-body extensions of the presented theory are covered in our second book, *Semiconductor Quantum Optics: Advanced Many-Body Aspects* (Cambridge University Press, to be published). Whereas this first book develops a working knowledge up to a level where one can start doing research on semiconductor quantum optics, our second book deepens the analysis to an advanced level and examines intriguing new phenomena and details.

This first book has been designed in such a way that it can be used for self-study as well as classroom teaching for advanced undergraduate or regular postgraduate courses with a different emphasis on the topics. To follow many details of the theory development and to deepen the basic understanding, we recommend reflecting upon the presented material through the exercises given at the end of each chapter. Chapters 1–6 can serve as supplementary material in teaching quantum mechanics, especially, for keen beginners or to provide a complementary view besides the standard books. Chapters 7–10 are well suited for introducing light quantization and quantum field theory while chapters 11–15 present the elements of solid-state and many-body theory. Especially, the cluster-expansion method (Chapter 15) provides a common starting point to bridge the "traditional" and the semiconductor quantum optics.

After these foundations are carefully laid, one can design a pure quantum-optics course based on Chapters 16–24. Here, we have paid special attention to study atomic phenomena with the goal to provide a connection to semiconductor quantum optics. The remaining chapters, 25–30, present the material for lectures discussing the essence of semiconductor quantum optics. Naturally, we recommend combining the material from all the chapters for the full learning experience. Further information, figure downloads, and comments on the book can be found at http://sqo.physik.uni-marburg.de.

This series of two books has been written during an extended period from 2006–2011. Most of this work has been done at the Philipps-Universität Marburg – we truly appreciate the research-oriented infrastructure provided here. We also thank the members of our department, our collaborators, and students for the inspirational research and teaching interactions we have had during our efforts to reveal and explain new semiconductor quantum phenomena. We also have enjoyed the hospitality of several collaborating institutes. In particular, we want to thank Professor Steven Cundiff (M.K., JILA visiting fellow program,

University of Colorado), Professor Ilkka Tittonen (M.K., Aalto University (Finland) visitor program), and Professor Jerome V. Moloney (S.W.K, University of Arizona) for allowing us to expand our book as well as our collaborations during the extended visits there. We are also very grateful to Renate Schmid who has helped us tremendously in coordinating the LaTeX manuscript and who provided us with prolific language and consistency checks. Many of our students have been instrumental in creating this book. Especially, the members of the cluster-crunching-club (CCC) are thanked dearly for their careful reading of the first draft. Our main collaborators as well as the current CCC member list can be found under the web link mentioned above.

1

Central concepts in classical mechanics

Historically, the scientific exploration of new phenomena has often been guided by systematic studies of observations, i.e., experimentally verifiable facts, which can be used as the basis to construct the underlying physical laws. As the apex of the investigations, one tries to identify the minimal set of fundamental assumptions – referred to as the *axioms* – needed to describe correctly the experimental observations. Even though the axioms form the basis to predict the system's behavior completely, they themselves have no rigorous derivation or interpretation. Thus, the axioms must be viewed as the elementary postulates that allow us to formulate a systematic description of the studied system based on well-defined logical reasoning. Even though it might seem unsatisfactory that axioms cannot be "derived," one has to acknowledge the paramount power of well-postulated axioms to predict even the most exotic effects. As is well known, the theory of classical mechanics can be constructed using only the three Newtonian axioms. On this basis, an infinite variety of phenomena can be explained, ranging from the cyclic planetary motion all the way to the classical chaos.

In this book, we are mainly interested in understanding how the axioms of classical and quantum mechanics can be applied to obtain a systematic description for the phenomena of interest. Especially, we want to understand how many-particle systems can be modeled, how quantum features of light emerge, and how these two aspects can be combined and utilized to explore new intriguing phenomena in semiconductor quantum optics. These investigations can always be traced back to fundamental equations that define the axioms in a mathematical form. In particular, these equations can be applied to develop the systematic theory for semiconductor quantum-optical phenomena. As background for our quantum-mechanical studies, we summarize the basic equations needed to formulate the theory of classical mechanics in this first chapter.

1.1 Classical description

The motion of a classical particle can be analyzed when its position x is known as function of time t, yielding a well-defined trajectory $x(t)$. For example, the velocity of the particle, $v(t) = \frac{d}{dt}x(t) = \dot{x}(t)$, follows from the first-order time derivative of the trajectory.

To define $x(t)$ and $v(t)$ under arbitrary conditions, one also needs to know the mass m of the particle which defines the particle's inertia resisting changes of $v(t)$ resulting from external forces $F(x, t)$ that can depend on both position and time. For simplicity, and since we want to focus on the underlying concepts, we use the notation for simple one-dimensional motion. More generally, position, velocity, force etc. are all three-dimensional vectorial quantities.

The fundamental description of the classical motion can always be expressed using Hamilton's formulation of mechanics. As a first step, we identify the system Hamiltonian

$$H = \frac{p^2}{2m} + U(x), \tag{1.1}$$

where p is the momentum of the particle and $U(x)$ is the potential of the external forces. In the Hamilton formalism, one describes x and p as independent variables such that the Hamiltonian is actually a function $H(x, p)$ of x and p. In general, (x, p) define the canonical phase-space coordinates and $(x(t), p(t))$ describes a phase-space trajectory. Figure 1.1 (solid and dashed lines) presents several trajectories connecting the same initial (x_i, p_i) and final (x_f, p_f) phase-space points. In principle, we can find infinitely many possible trajectories when the time evolves from t_i to t_f. However, depending on its initial conditions, a given classical particle follows one particular trajectory (solid line).

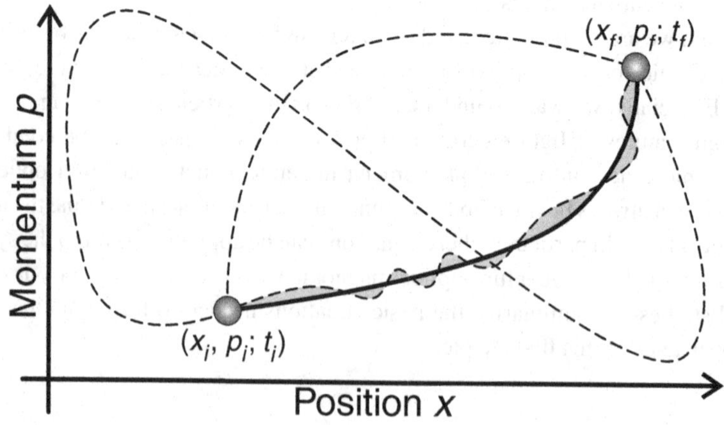

Figure 1.1 Phase-space trajectories of a one-dimensional particle. The particle is at (x_i, p_i) at the initial time $t = t_i$ and has moved to (x_f, p_f) by the final time $t = t_f$. Several arbitrary trajectories (dashed lines) are shown together with the classical trajectory (solid line). The shaded curve illustrates one possible infinitesimal variation around the classical path.

In order to determine classical trajectories, one often identifies the action

$$S(x_i, p_i, t_i;\; x_f, p_f; t_f) = \int_{t_i}^{t_f} dt' \left[p(t')\,\dot{x}(t') - H\left(x(t'),\, p(t')\right)\right]$$
$$= \int_{t_i}^{t_f} dt'\, L\left(x(t'),\, \dot{x}(t')\right), \tag{1.2}$$

which consists of the time integral from the initial (t_i) to the final time (t_f). The integrand is the product of momentum and velocity minus the Hamiltonian, i.e., the Lagrangian L. In principle, the action can be evaluated for any phase-space trajectory connecting the initial and final positions and momenta. The classical path is defined to be that solution which minimizes the action with the constraints $x(t_i) = x_i$, $x(t_f) = x_f$, $p(t_i) = p_i$, and $p(t_f) = p_f$. This condition guarantees that any arbitrary path (dashed line in Fig. 1.1) connects the investigated initial (x_i, p_i) and final (x_f, p_f) phase-space points.

We now see that S depends on the $x(t)$ and $p(t)$ values at every time point $t \in [t_i, t_f]$. Mathematically, S is thus functionally dependent on $x(t)$ and $p(t)$. To distinguish this from ordinary functions, one often denotes the functional dependence using brackets, i.e., $S \equiv S[x, p]$. Thus, the minimization of S with respect to the functions $x(t)$ and $p(t)$ yields the classical path. Practically, we use the variational principle to perform the minimization. If we assume that $x_{\rm cl}(t)$ and $p_{\rm cl}(t)$ are the correct classical solutions, we know that they minimize $S[x, p]$. In other words, $S[x, p]$ does not change if we introduce arbitrary infinitesimal variations $\delta x(t)$ and $\delta p(t)$ around $[x_{\rm cl}(t), p_{\rm cl}(t)]$. Due to the constraints stated above, the variations must satisfy

$$\delta x(t_i) = \delta x(t_f) = 0, \qquad \delta p(t_i) = \delta p(t_f) = 0 \tag{1.3}$$

for any arbitrary path.

We may now compute the changed S caused by the variations:

$$S\left[x_{\rm cl}(t) + \delta x(t),\, p_{\rm cl}(t) + \delta p(t)\right] = \int_{t_i}^{t_f} dt' \left[\left(p_{\rm cl}(t') + \delta p(t')\right) \frac{d}{dt'}\left(x_{\rm cl}(t') + \delta x(t')\right)\right.$$
$$\left. - H\left(x_{\rm cl}(t') + \delta x(t'),\, p_{\rm cl}(t') + \delta p(t')\right)\right]. \tag{1.4}$$

Since the variations are infinitesimal, we only need to consider the linear changes

$$S\left[x_{\rm cl}(t) + \delta x(t),\, p_{\rm cl}(t) + \delta p(t)\right] = \int_{t_i}^{t_f} dt' \left[p_{\rm cl}(t')\,\dot{x}_{\rm cl}(t') - H\left(x_{\rm cl}(t'),\, p_{\rm cl}(t')\right)\right.$$
$$+ p_{\rm cl}(t')\frac{d}{dt'}\delta x(t') + \delta p(t')\frac{d}{dt'}x_{\rm cl}(t')$$
$$\left. - \frac{\partial H}{\partial x_{\rm cl}(t')}\,\delta x(t') - \frac{\partial H}{\partial p_{\rm cl}(t')}\,\delta p(t')\right]. \tag{1.5}$$

Here, we used the Taylor expansion,

$$f(x+\delta x) = f(x) + \delta x \frac{df(x)}{dx} + \sum_{j=2}^{\infty} \frac{\delta x^j}{j!} \frac{d^j f(x)}{dx^j}, \tag{1.6}$$

and stopped after the first-order terms to include the infinitesimal changes in H.

We notice that the first two terms of Eq. (1.5) constitute the action of the classical trajectory. Integrating by parts and using the condition (1.3), we can move the time derivative, i.e.,

$$\begin{aligned}\int_{t_i}^{t_f} dt'\, p_{\rm cl}(t') \frac{d}{dt'} \delta x(t') &= \Big|_{t_i}^{t_f} p_{\rm cl}(t')\, \delta x(t') - \int_{t_i}^{t_f} dt'\, \delta x(t') \frac{d}{dt'} p_{\rm cl}(t') \\ &= -\int_{t_i}^{t_f} dt'\, \delta x(t') \frac{d\, p_{\rm cl}(t')}{dt'}.\end{aligned} \tag{1.7}$$

Using the result (1.7) in Eq. (1.5), we find

$$\begin{aligned} S\left[x_{\rm cl}(t) + \delta x(t),\ p_{\rm cl}(t) + \delta p(t)\right] = S\left[x_{\rm cl}(t),\ p_{\rm cl}(t)\right] \\ - \int_{t_i}^{t_f} dt'\, \delta x(t') \left[\frac{d\, p_{\rm cl}(t')}{dt'} + \frac{\partial H}{\partial x_{\rm cl}(t')}\right] \\ + \int_{t_i}^{t_f} dt'\, \delta p(t') \left[\frac{d\, x_{\rm cl}(t')}{dt'} - \frac{\partial H}{\partial p_{\rm cl}(t')}\right], \end{aligned} \tag{1.8}$$

after having organized the position and momentum variations separately.

We can now directly read off the action changes resulting from the infinitesimal displacements,

$$\begin{aligned} \delta S &\equiv S\left[x_{\rm cl}(t) + \delta x(t),\ p_{\rm cl}(t) + \delta p(t)\right] - S\left[x_{\rm cl}(t),\ p_{\rm cl}(t)\right] \\ &= -\int_{t_i}^{t_f} dt'\, \delta x(t') \left[\frac{d\, p_{\rm cl}(t')}{dt'} + \frac{\partial H}{\partial x_{\rm cl}(t')}\right] + \int_{t_i}^{t_f} dt'\, \delta p(t') \left[\frac{d\, x_{\rm cl}(t')}{dt'} - \frac{\partial H}{\partial p_{\rm cl}(t')}\right]. \end{aligned} \tag{1.9}$$

The action is minimized by the classical trajectory only if δS vanishes for arbitrary infinitesimal variations. Since $\delta x(t')$ and $\delta p(t')$ have mutually independent functional forms satisfying Eq. (1.3), the terms within the brackets must vanish separately for all times, producing the classical Hamilton equations

$$\begin{cases} \dfrac{d\, p_{\rm cl}(t)}{dt} = -\dfrac{\partial H}{\partial x_{\rm cl}(t)} \\ \dfrac{d\, x_{\rm cl}(t)}{dt} = \dfrac{\partial H}{\partial p_{\rm cl}(t)} \end{cases} \tag{1.10}$$

that uniquely determine the classical trajectory.

To identify Newton's laws, i.e., the principal axioms of classical mechanics, we insert Eq. (1.1) into Eq. (1.10), yielding

$$\begin{cases} \dot{p}_{\rm cl} = -\frac{\partial U(x_{\rm cl})}{\partial x_{\rm cl}} \\ \dot{x}_{\rm cl} = \frac{p_{\rm cl}}{m} \end{cases} . \tag{1.11}$$

We see that the momentum is directly defined by the product of velocity and mass, i.e., $p = mv$, while any temporal changes are caused by the spatial derivative of the potential that can be identified as the external force

$$F(x) \equiv -\frac{\partial U(x)}{\partial x} \tag{1.12}$$

acting upon the particle. The absence of an external force, i.e. $F = 0$, has the consequence that p remains constant based on Eq. (1.11). Hence, we find the linear motion

$$x_{\rm cl}(t) = \frac{p}{m}(t - t_i) + x_i, \quad \text{if } F = 0 \tag{1.13}$$

as the solution for the classical trajectory. In other words, the Hamilton equations (1.10) imply Newton's first law: *If no force is acting upon a particle, it continues its motion linearly or stays at rest.*

In cases where an external force is present, the functional form of the motion can become arbitrarily complicated. In this case, one can take the second derivative of the trajectory to produce

$$\frac{d}{dt}(m\dot{x}_{\rm cl}) = \dot{p}_{\rm cl} = F(x_{\rm cl}) \tag{1.14}$$

directly from Eq. (1.11). We see that the time derivative of the momentum $p_{\rm cl}$ equals the force acting on the particle, which is Newton's second law.

To find Newton's third law, we have to generalize the description to many-particle systems. For this purpose, we extend the Hamilton formalism to three dimensions and N particles. In this situation, position and momenta become three-dimensional vectors and each particle must be labeled by a discrete index j. Thus, we can rewrite the Hamiltonian (1.1) using

$$H = \sum_{j=1}^{N} \frac{\mathbf{p}_j^2}{2m_j} + U(\mathbf{x}_1, \mathbf{x}_2, \cdots, \mathbf{x}_N), \tag{1.15}$$

where U describes a generic potential acting among the N particles. In the most general form, each particle can have a different mass m_j. The classical Hamilton equations can now be derived with steps identical to those producing Eq. (1.10), see Exercise 1.2. We only need to keep track of the particle index and convert the differentiation into three dimensions (3D). Thus, the classical Hamilton equations follow from

$$\begin{cases} \frac{d\,\mathbf{p}_j(t)}{dt} = -\frac{\partial H}{\partial \mathbf{x}_j(t)} \\ \frac{d\,\mathbf{x}_j}{dt} = \frac{\partial H}{\partial \mathbf{p}_j(t)} \end{cases} . \tag{1.16}$$

The emerging vectorial differentiation follows from the convention

$$\frac{\partial f(\mathbf{x}_j)}{\partial \mathbf{x}_j} = \sum_{\alpha=1}^{3} \mathbf{e}_\alpha \frac{\partial f(x_{j,1}, x_{j,2}, x_{j,3})}{\partial x_{j,\alpha}} = \nabla_{\mathbf{x}_j} f(\mathbf{x}_j), \tag{1.17}$$

where $\alpha = 1, 2, 3$ refers to the different Cartesian coordinates and $\mathbf{e}_\alpha$ denotes the corresponding unit vectors. The same convention is applied to the momenta, which eventually simplifies the classical Hamilton equations into

$$\begin{cases} \frac{d\,\mathbf{p}_j(t)}{dt} = -\nabla_{\mathbf{x}_j} U(\mathbf{x}_1, \mathbf{x}_2, \cdots, \mathbf{x}_N) \\ \frac{d\,\mathbf{x}_j}{dt} = \frac{\mathbf{p}_j}{m_j} \end{cases} \tag{1.18}$$

for a 3D system with N particles.

We may now identify the force acting upon particle j via

$$\mathbf{F}_j \equiv -\nabla_{\mathbf{x}_j} U(\mathbf{x}_1, \mathbf{x}_2, \cdots, \mathbf{x}_N) \tag{1.19}$$

which converts Eq. (1.18) into

$$\begin{cases} \frac{d\,\mathbf{p}_j(t)}{dt} = \mathbf{F}_j \\ \frac{d\,\mathbf{x}_j}{dt} = \frac{\mathbf{p}_j}{m_j} \end{cases}. \tag{1.20}$$

The total momentum of the system, $\mathbf{p}_{\mathrm{tot}} \equiv \sum_{j=1}^{N} \mathbf{p}_j$, satisfies then

$$\frac{d\,\mathbf{p}_{\mathrm{tot}}}{dt} = \sum_{j=1}^{N} \mathbf{F}_j \equiv \mathbf{F}_{\mathrm{tot}}, \tag{1.21}$$

where $\mathbf{F}_{\mathrm{tot}}$ is just the total external force. If there are no external forces, $\mathbf{F}_{\mathrm{tot}}$ vanishes and $\mathbf{p}_{\mathrm{tot}}$ remains constant, producing a linear motion for the center-of-mass coordinate $\mathbf{R} \equiv \sum_{j=1}^{N} \frac{m_j}{M} \mathbf{x}_j$ with the total mass $M = \sum_{j=1}^{N} m_j$. For a two-particle system, we find

$$\mathbf{F}_1 + \mathbf{F}_2 = 0, \quad \Leftrightarrow \quad \mathbf{F}_1 = -\mathbf{F}_2 \tag{1.22}$$

if no external forces are present. In this situation, F_1 (F_2) is the force that particle $j = 2$ ($j = 1$) enacts upon particle $j = 1$ ($j = 2$). Thus, Eq. (1.22) predicts that two interacting particles produce forces on each other that have equal magnitude but opposite directions. This constitutes Newton's third law.

Consequently, Eqs. (1.15)–(1.16) provide the fundamental description for the classical mechanics of particles, i.e., they directly imply Newton's laws. Even though they cannot be derived from a set of more fundamental concepts, they provide systematic rules to predict the actual behavior of classical particles. Thus, Eqs. (1.15)–(1.16) can be used as a general starting point to investigate, e.g., classical many-body systems with or without external force fields.

1.1.1 Conservation laws

The use of Hamilton's equations becomes particularly efficient once the system has specific conservation laws. For example, Eq. (1.21) directly predicts that the total momentum,

$$\mathbf{p}_{\text{tot}} = \sum_{j=1}^{N} \mathbf{p}_j, \tag{1.23}$$

remains constant if the external forces, $\mathbf{F}_{\text{tot}}$, vanish. In other words, the total momentum is always a conserved quantity in an isolated system regardless of how complicated the interactions among the particles are. This constitutes the conservation law for the momentum.

The Hamiltonian (1.16) itself contains a sum over the kinetic energies,

$$E_{\text{kin},j} = \frac{\mathbf{p}_j^2}{2m_j}, \tag{1.24}$$

of individual particles and the potential. Assuming particles with time-independent masses, it is straightforward to evaluate the dynamics of H,

$$\begin{aligned}\dot{H} &= \sum_{j=1}^{N} \frac{d}{dt}\frac{\mathbf{p}_j^2}{2m_j} + \frac{d}{dt}U(\mathbf{x}_1, \mathbf{x}_2, \cdots, \mathbf{x}_N) \\ &= \sum_{j=1}^{N} \frac{\mathbf{p}_j}{m_j} \cdot \dot{\mathbf{p}}_j + \sum_{j=1}^{N} \dot{\mathbf{x}}_j \cdot \nabla_{\mathbf{x}_j} U(\mathbf{x}_1, \mathbf{x}_2, \cdots, \mathbf{x}_N)\end{aligned} \tag{1.25}$$

that follows directly from the chain rule of differentiation. We can now apply the relation (1.18) to express $\dot{\mathbf{x}}_j$ in terms of the momentum. This identification yields

$$\dot{H} = \sum_{j=1}^{N} \frac{\mathbf{p}_j}{m_j} \cdot \left[\dot{\mathbf{p}}_j + \nabla_{\mathbf{x}_j} U(\mathbf{x}_1, \mathbf{x}_2, \cdots, \mathbf{x}_N)\right] = 0, \tag{1.26}$$

where we used the $\dot{\mathbf{p}}_j$ equation of (1.18). The result shows that H is constant. Since H is the sum of kinetic and potential energies, it actually defines the total energy. In other words, also the total energy is a conserved quantity in many-body systems which are not subjected to external forces.

Looking at the angular momentum,

$$\mathbf{L}_j \equiv \mathbf{x}_j \times \mathbf{p}_j, \tag{1.27}$$

of particle j, we see that the total angular momentum satisfies

$$\dot{\mathbf{L}}_{\text{tot}} = \sum_{j=1}^{N} \dot{\mathbf{L}}_j = \sum_{j=1}^{N} \left(\dot{\mathbf{x}}_j \times \mathbf{p}_j + \mathbf{x}_j \times \dot{\mathbf{p}}_j\right). \tag{1.28}$$

With the help of Eq. (1.20), we find

$$\dot{\mathbf{L}}_{\text{tot}} = \sum_{j=1}^{N} \left(\frac{\mathbf{p}_j}{m_j} \times \mathbf{p}_j + \mathbf{x}_j \times \mathbf{F}_j\right) = \sum_{j=1}^{N} \mathbf{x}_j \times \mathbf{F}_j \tag{1.29}$$

because the cross-product of parallel vectors vanishes. At the same time, we recognize that $\boldsymbol{\tau}_j \equiv \mathbf{x}_j \times \mathbf{F}_j$ identifies the torque acting on particle j. Thus, the temporal changes of $\mathbf{L}_{\text{tot}}$ are defined by the total external torque,

$$\dot{\mathbf{L}}_{\text{tot}} = \sum_{j=1}^{N} \boldsymbol{\tau}_j \equiv \boldsymbol{\tau}_{\text{tot}}. \tag{1.30}$$

If the system is isolated, we have no external torque such that the total angular momentum is a conserved quantity, which establishes the third central conservation law.

The conservation laws not only constrain the processes that are possible in isolated systems but they can also be used to simplify significantly many calculations. For example, one can analyze a one-dimensional problem by solving the second-order differential equation (1.14). However, the conservation of energy can be directly used to find the simple relation

$$E = \frac{p^2}{2m} + U(x) = \frac{1}{2}m\dot{x}^2 + U(x), \tag{1.31}$$

where we have expressed the momentum in terms of the velocity. Since the total energy is constant, we can determine the velocity directly from

$$\dot{x} = \pm\sqrt{\frac{2}{m}\left[E - U(x)\right]}, \tag{1.32}$$

which is a first-order differential equation and, thus, simpler to solve than the original Eq. (1.14). Besides this, the relation (1.32) can directly be applied to identify constraints for the solutions of classical problems. For example, we see directly that a particle cannot penetrate regions where the potential energy exceeds E. These classically forbidden regions will be studied further in connection with the quantum description of particles.

1.1.2 Single-particle motion for simple potentials

Even the simplest one-dimensional (1D) single-particle motion is governed by differential equations (1.11) that cannot be solved analytically for a generic potential $U(x)$. Thus, one often must resort to numerical approaches to obtain the trajectories for the classical motion of particles. In order to gain some fundamental insight, we first take a look at a few of the analytically solvable cases.

The simplest situation is found when the external potential is constant producing a vanishing force, $F = -dU(x)/dx = 0$. In this case, Eq. (1.11) reduces into

$$\begin{cases} \dot{p}_{\text{cl}} = 0 \\ \dot{x}_{\text{cl}} = \frac{p_{\text{cl}}}{m} \end{cases} \tag{1.33}$$

with the solution

$$p_{\text{cl}}(t) = p_0, \qquad x_{\text{cl}}(t) = x_0 + \frac{p_0}{m}t, \qquad t \geq 0, \tag{1.34}$$

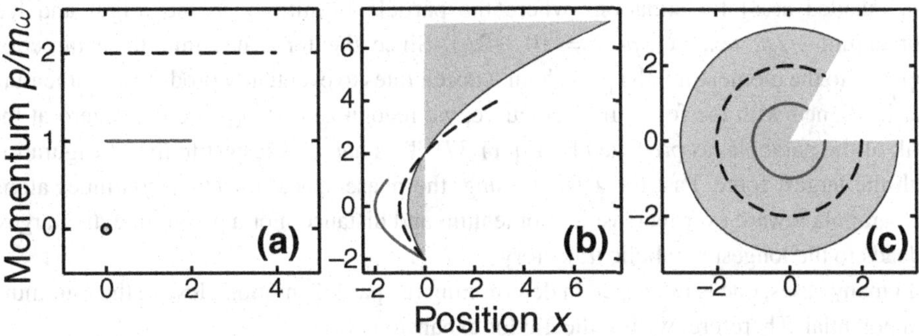

Figure 1.2 Classical phase-space trajectories, $[x_{\rm cl}(t), p_{\rm cl}(t)]$, of a one-dimensional particle. **(a)** The free-particle trajectories are shown for three initial conditions $(x_0 = 0, p_0 = 0)$ (circle), $(x_0 = 0, p_0 = p)$ (solid line), and $(x_0 = 0, p_0 = 2p)$ (dashed line). **(b)** The particle is initially at $(x_0 = 0, p_0 = -2p)$ and is subject to a linear potential $U(x) = F_0\, x$ with $F_0 = +|F|$ (solid line), $F_0 = +2|F|$ (dashed line), and $F_0 = +3|F|$ (shaded area). **(c)** The particle is initially at $(x_0 = x, p_0 = 0)$ (solid line), $(x_0 = 2x, p_0 = 0)$ (dashed line), and $(x_0 = 3x, p_0 = 0)$ (shaded area) while it is affected by a harmonic potential with the characteristic frequency ω. In all cases, the three trajectories are evolved for identical time intervals.

where x_0 and p_0 denote the initial values at $t = 0$. The resulting trajectories can be presented in phase space as $[x_{\rm cl}(t), p_{\rm cl}(t)]$ curves.

Figure 1.2(a) presents the solution (1.34) for three different initial position and momentum conditions: $(x_0, p_0) = (0, 0)$ (circle), $(x_0, p_0) = (0, p)$ (solid line), and $(x_0, p_0) = (0, 2p)$ (dashed line). In phase space, the $p_0 = 0$ case produces a trajectory that remains at a point because the particle is at rest for all times. For nonvanishing p_0, the trajectories are always horizontal lines. If we evolve the different cases for an equal time interval, the length of the line is proportional to p_0.

If the particle is influenced by a linear potential, $U(x) = F_0 x$, it is subject to a constant force $F(x) = dU(x)/dx = F_0 \neq 0$. As a result, Eq. (1.11) yields

$$\begin{cases} \dot{p}_{\rm cl} = F_0 \\ \dot{x}_{\rm cl} = \frac{p_{\rm cl}}{m}, \end{cases} \tag{1.35}$$

which has the general solution

$$p_{\rm cl}(t) = p_0 + F_0 t, \qquad x_{\rm cl}(t) = x_0 + \frac{p_0}{m} t + \frac{1}{2m} F_0 t^2, \qquad t \geq 0. \tag{1.36}$$

Substituting $t = \frac{p_{\rm cl}(t) - p_0}{F_0}$ into $x_{\rm cl}(t)$ produces

$$x_{\rm cl}(t) = x_0 + \frac{p_{\rm cl}(t)^2 - p_0^2}{2m F_0}, \tag{1.37}$$

which describes a parabolic phase-space trajectory. Figure 1.2(b) presents particle trajectories for three different magnitudes of the linear force: F_0 (solid line), $2F_0$ (dashed line), and

$3F_0$ (shaded area) for situations where the particle is initially at the origin and has momentum $-2p$, i.e., $(x_0, p_0) = (0, -2p)$. Since the force has initially a direction opposite to the momentum, the particle first decelerates to eventually produce a momentum that is aligned with the force. In each curve, we recognize that $p_{cl}(t)$ crosses zero at the peak of the parabola, as predicted by Eq. (1.37). This happens fastest in the configuration with the largest force. Past the zero crossing, the phase-space trajectory continues along the parabola toward ever increasing momentum and distance. For a given time, the largest F leads to the longest parabolic trajectory.

In many cases, one is interested in determining the particle motion close to the minimum of a potential. Therefore, we use the Taylor expansion (1.6),

$$\begin{aligned} U(x) &= \sum_{j=0}^{\infty} \frac{(x - x_{min})^j}{j!} \frac{d^j\, U(x_{min})}{dx_{min}^j} \\ &= U(x_{min}) + \frac{(x - x_{min})^2}{2} \frac{d^2 U(x_{min})}{dx_{min}^2} + \mathcal{O}\left([x - x_{min}]^3\right), \end{aligned} \tag{1.38}$$

where the first-order derivative $d\, U(x_{min})/dx_{min}$ vanishes and the second-order differentiation must be positive at the minimum. Thus, we may replace its value by $m\omega^2 \geq 0$. Quite often we can choose the coordinate system freely such that we can set $U(x_{min})$ and x_{min} to zero, without loss of generality. Thus, we can express the potential via a parabolic approximation, yielding the harmonic potential

$$U(x) = \frac{1}{2} m\omega^2 x^2. \tag{1.39}$$

We will see that the quantity ω defines the characteristic frequency of a periodic motion.

As we insert Eq. (1.39) into Eq. (1.11), we find

$$\left\{ \begin{array}{l} \dot{p}_{cl} = -m\omega^2 x_{cl} \\ \dot{x}_{cl} = \frac{p_{cl}}{m} \end{array} \right. \quad \Leftrightarrow \quad \frac{d}{dt} \begin{pmatrix} \frac{p_{cl}}{m\omega} \\ x_{cl} \end{pmatrix} = \begin{pmatrix} 0 & -\omega \\ \omega & 0 \end{pmatrix} \begin{pmatrix} \frac{p_{cl}}{m\omega} \\ x_{cl} \end{pmatrix} \tag{1.40}$$

that can equivalently be expressed in a matrix form. Since the resulting equation is a simple linear equation, it can be solved to produce

$$\begin{aligned} \begin{pmatrix} \frac{p_{cl}(t)}{m\omega} \\ x_{cl}(t) \end{pmatrix} &= \begin{pmatrix} \cos\omega t & -\sin\omega t \\ \sin\omega t & \cos\omega t \end{pmatrix} \begin{pmatrix} \frac{p_0}{m\omega} \\ x_0 \end{pmatrix} \\ &= \begin{pmatrix} \frac{p_0}{m\omega}\cos\omega t - x_0 \sin\omega t \\ \frac{p_0}{m\omega}\sin\omega t + x_0 \cos\omega t \end{pmatrix}. \end{aligned} \tag{1.41}$$

This result shows that the trajectory $\left[x_{cl}(t), \frac{p_{cl}(t)}{m\omega}\right]$ follows a circular path in the phase space. Independent of the initial conditions $\left(x_0, \frac{p_0}{m\omega_0}\right)$, the oscillation frequency is uniquely given by ω. Figure 1.2(c) presents three trajectories corresponding to the initial conditions

$(x, 0)$ (solid line), $(2x, 0)$ (dashed line), and $(3x, 0)$ (shaded area). We see that the initial condition just defines the radius of the circular trajectory while the angular velocity remains unchanged.

It is interesting to notice that changing the potential from constant to linear leads to a change of the phase-space trajectories from linear to parabolic. Thus, it might seem that the power-law dependence of the potential trivially defines the functional class of the phase-space trajectory. However, the use of a quadratic potential dramatically modifies the trajectories into purely oscillatory ones such that no simple power-law relation is found between $U(x)$ and the trajectories. In fact, the classical trajectories become more and more complicated as the nonlinear character of $U(x)$ is increased further. For example, a general $U(x) \propto x^{2n}$ also produces an oscillatory motion, however, the oscillation frequency depends on the initial conditions. This situation is referred to as *anharmonic oscillations* because the classical trajectory cannot be characterized with a single frequency.

To investigate the anharmonic behavior further, we consider the potential

$$U(x) = mR^2\omega^2 \left(1 - \cos\frac{x}{R}\right). \tag{1.42}$$

Physically, this potential is realized in a rigid pendulum where gravitation provides the restoring force. The geometry and the force directions are illustrated in Fig. 1.3(a) where R is the length of the pendulum and x measures the deviation from the equilibrium position along the arc. Thus, $x/R \equiv \theta$ directly defines the displacement angle. The strength of the force is defined by the constant $F \equiv m\omega^2 R$ and the angle of displacement relative to the resting position. The identification of the harmonic frequency ω follows from the expansion (1.38) close to the minimum $x = 0$.

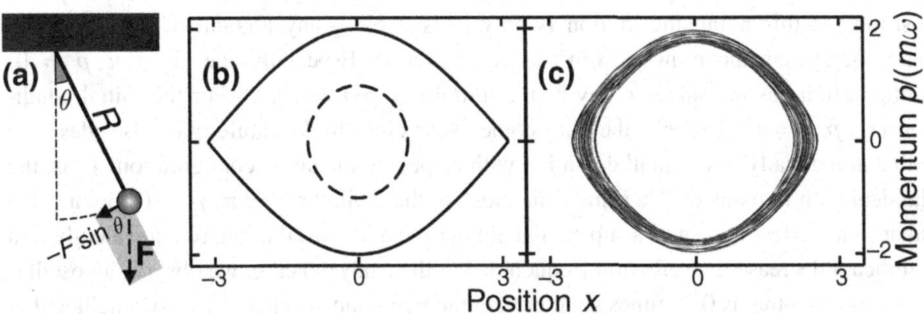

Figure 1.3 Classical phase-space trajectories for an anharmonic oscillator. **(a)** Schematic picture of a pendulum where the mass m is attached rigidly at the distance R from the origin. The force **F** (gravitation force) acts always downwards such that the restoring force along the circular path is given by $-F\sin\frac{x}{R}$ where x defines the displacement. **(b)** Classical phase-space trajectories $[\bar{x}_{cl}(t), \bar{p}_{cl}(t)]$ for the initial conditions $(\bar{x}_0 = 0, \bar{p}_0 = 1)$ (dashed line) and $(\bar{x}_0 = 0, \bar{p}_0 = 1.9999)$ (solid line). **(c)** Solution for a driven anharmonic oscillator with external force $0.1 \sin 2\omega t$. The initial conditions are $(\bar{x}_0 = 0, \bar{p}_0 = 1.76)$. All quantities are scaled according to Eq. (1.44).

Using the potential (1.42), the corresponding Hamilton equations become

$$\begin{cases} \dot{p}_{\mathrm{cl}} = -m\omega^2 R \sin \frac{x_{\mathrm{cl}}(t)}{R} \\ \dot{x}_{\mathrm{cl}} = \frac{p_{\mathrm{cl}}}{m} \end{cases}. \tag{1.43}$$

In the limit of small deviations, we can use the first-order Taylor expansion $\sin x_{\mathrm{cl}}(t)/R = x_{\mathrm{cl}}(t)/R$ to recover Eq. (1.40) of the fully harmonic system. However, for larger values of x_{cl}, Eq. (1.43) has strong anharmonic corrections whose consequences are most easily discussed with the help of numeric solutions. For this purpose, we introduce scaled units for both position and momentum

$$\bar{x}(t) \equiv \frac{x_{\mathrm{cl}}(t)}{R}, \qquad \bar{p}(t) \equiv \frac{p_{\mathrm{cl}}(t)}{m\omega R}, \tag{1.44}$$

respectively. Substitution into Eq. (1.43) produces

$$\begin{cases} \dot{\bar{p}}(t) = -\omega \sin \bar{x}(t) \\ \dot{\bar{x}}(t) = \omega\, \bar{p}(t) \end{cases}. \tag{1.45}$$

To obtain representative results, we numerically integrate Eq. (1.45) up to $t = 50\frac{2\pi}{\omega}$, i.e., we compute the classical trajectory during the first 50 oscillations as if the motion were harmonic.

The computed phase-space trajectories are shown in Fig.1.3(b) for the initial conditions, $(\bar{x}_0, \bar{p}_0) = (0, 1)$ (dashed line) and $(\bar{x}_0, \bar{p}_0) = (0, 1.9999)$ (solid line). For the low-momentum case, the phase-space trajectory is nearly circular. The anharmonic corrections merely reduce the frequency of oscillation to 0.93 times that of the purely harmonic motion. As we increase $\bar{p}_0$ to 1.9999, the phase-space trajectory is distorted far from a circle. In particular, we find discontinuous apexes at the phase-space points $(\pm\pi, 0)$. Physically, $(\bar{x} = \pm\pi, \bar{p} = 0)$ corresponds to a configuration where the pendulum is near its upright position. At this point, the motion is only quasistable – any amount of deviation will make the system move away from this configuration. Especially, the $(\bar{x} = \pi, \bar{p} = 0)$ configuration has the same energy as the initial $(\bar{x}_0, \bar{p}_0) = (0, 2)$. For the initial condition $(\bar{x}_0, \bar{p}_0) = (0, 1.9999)$, the path comes very close to the labile point. Besides, this path automatically has a small deviation with respect to the labile configuration. Thus, the classical path remains only a finite time close to the labile $(\bar{x} = \pm\pi, \bar{p} = 0)$ point. This anharmonic effect is seen as a substantial elongation of the oscillation, or equivalently as a significant decrease in oscillation frequency. For the analyzed case, we observe an oscillation frequency that is 0.26 times smaller than the harmonic ω value. We also conclude that the anharmonic modifications to the classical paths strongly depend on the initial condition.

Even though the Hamilton equations are structurally simple, they describe surprisingly complicated situations. To illustrate this aspect, we add an external sinusoidal force to Eq. (1.45) yielding

$$\begin{cases} \dot{\bar{p}}(t) = -\omega \sin \bar{x}(t) + K \cos \Omega t \\ \dot{\bar{x}}(t) = \omega\, \bar{p}(t) \end{cases}. \tag{1.46}$$

Physically, this can be understood as an external force that is applied to periodically alter the motion of the pendulum. By properly adjusting this force, we can generate transitions from one anharmonic path to another. Since the different anharmonic trajectories can be qualitatively very different, we may anticipate that the disturbed paths display a nontrivial combination of very different forms. For illustration, we solve the equations for a case with a weak external force using $K = 0.1$ and $\Omega = 2\omega$ with the initial condition $(x_0, p_0) = (0,\ 1.76)$.

Figure 1.3(c) follows the classical trajectory for a time corresponding to the first 50 oscillations in case the motion is harmonic. We find that the trajectory never repeats the same oscillatory path due to the external disturbance. Instead, the trajectory seems to randomly jump from one solution branch to another. Consequently, the classical trajectory displays all possible forms of anharmonic solutions such that the overall motion becomes quite chaotic. In fact, the simple pendulum with external force is known to be a simple prototype system that displays classical chaos. A similar chaotic behavior is observed in a double pendulum without the external force. The appearance of chaotic trajectories demonstrates that even simple classical systems can display extremely complicated paths whose properties depend even on infinitesimal changes in the initial conditions. Hence, again we see that the simple set of Newtonian axioms provides the basis to explore a vast spectrum of nontrivial effects.

1.2 Statistical description of particles

In principle, the Hamilton equations (1.18) deterministically predict the motion of a set of classical particles at any time after the initial configuration. Even though one often needs to apply numerical schemes to evolve the system in time, one can rather easily simulate systems with millions of classical particles. As an obvious complication, the solutions may depend sensitively on initial conditions due to the potentially chaotic behavior of the classical trajectories. In cases where one wants to model realistic experimental situations, the determination of the actual initial conditions may be influenced by the finite accuracy with which they can be measured. Thus, one often needs to investigate the influence of random small changes of the initial conditions on the solutions, especially when chaotic behavior is expected.

To describe mathematically the uncertainty, one can introduce a probability distribution of initial conditions $\rho(x, p\,; t = 0)$. For simplicity, we start the discussion here for a one-dimensional system. The time-evolved $\rho(x, p\,; t)$ determines the probability density of finding the system in a given phase-space state (x, p) at the time t. Due to its probabilistic nature, the phase-space distribution must be normalized and positive for all times,

$$\rho(x, p\,; t) \geq 0, \qquad \int_{-\infty}^{\infty} dx \int_{-\infty}^{\infty} dp\, \rho(x, p\,; t) = 1. \tag{1.47}$$

The actual probability of finding the particle within a phase-space region, $|x - x_0| \leq \Delta x/2$ and $|p - p_0| \leq \Delta p/2$ centered at (x_0, p_0), is then given by

$$P(x_0, p_0\,;\Delta x, \Delta p) \equiv \int_{x_0-\frac{\Delta x}{2}}^{x_0+\frac{\Delta x}{2}} dx \int_{p_0-\frac{\Delta p}{2}}^{p_0+\frac{\Delta p}{2}} dp\; \rho(x, p\,; t). \tag{1.48}$$

The condition (1.47) guarantees that the probability always varies in the range $0 \leq P \leq 1$.

The concept of a probability distribution can be directly generalized for a system consisting of N 3D particles. Here, the probability density for a phase-space configuration $(x_1, \cdots, x_N, p_1, \cdots, p_N)$ is given by $\rho(\mathbf{x}_1, \cdots, \mathbf{x}_N, \mathbf{p}_1, \cdots, \mathbf{p}_N\,; t)$ that satisfies positivity and normalizability

$$\begin{aligned} &\rho(\mathbf{x}_1, \cdots, \mathbf{x}_N, \mathbf{p}_1, \cdots, \mathbf{p}_N\,; t) \geq 0, \\ &\int d^3x_1 \cdots d^3x_N d^3p_1 \cdots d^3p_N\; \rho(\mathbf{x}_1, \cdots, \mathbf{x}_N, p_1, \cdots, \mathbf{p}_N\,; t) = 1, \end{aligned} \tag{1.49}$$

in analogy to Eq. (1.47). A perfect knowledge of the N-particle state implies that ρ becomes arbitrarily accurate. In other words, ρ should then identify $(\mathbf{x}_{\mathrm{cl},1}, \cdots, \mathbf{x}_{\mathrm{cl},N}, \mathbf{p}_{\mathrm{cl},1}, \cdots, \mathbf{p}_{\mathrm{cl},N})$ as the only possible phase-space configuration. For single-particle systems ($N = 1$) this means that the probability becomes $P(x_0, p_0\,; \Delta x, \Delta p) = 1$ only if the phase-space interval includes the classical values, i.e. $|x_0 - x_{\mathrm{cl}}| \leq \Delta x/2$ and $|p_0 - p_{\mathrm{cl}}| \leq \Delta p/2$, for all nonvanishing Δx and Δp. In any other case, the probability $P(x_0, p_0\,; \Delta x, \Delta p)$ should vanish.

Mathematically, such an accurate distribution can be described with the help of the δ-function that has the general property

$$\int_{-\infty}^{\infty} dx\; \delta(x - x_0)\; f(x) = f(x_0) \tag{1.50}$$

for any analytic function $f(x)$. In general, the δ-function can be expressed in several equivalent ways, see Exercises 1.7–1.9. For example,

$$\delta(x) \equiv \lim_{\epsilon \to 0} \frac{1}{\sqrt{2\pi\epsilon}}\, e^{-\frac{x^2}{\epsilon^2}}, \qquad \delta(\mathbf{r}) = \delta(x)\,\delta(y)\,\delta(z) \tag{1.51}$$

identifies one possible representation. We see that the used Gaussian is extremely narrowed down to concentrate the distribution around a single value. At the same time, the height of the Gaussian peak increases because the integral over x must remain unchanged. The δ-function can directly be generalized into three dimensions.

With the help of the δ-function, the classical distribution becomes

$$\rho_{\mathrm{accu}}\left(\mathbf{x}_1, \cdots, \mathbf{x}_N, \mathbf{p}_1, \cdots, \mathbf{p}_N\,; t\right) \equiv \prod_{j=1}^{N} \delta\left(\mathbf{x}_j - \mathbf{x}_{\mathrm{cl,j}}(t)\right) \delta\left(\mathbf{p}_j - \mathbf{p}_{\mathrm{cl,j}}(t)\right). \tag{1.52}$$

This mathematical expression implies that the probability density is completely concentrated on the classical phase-space point, that is in general time dependent. By using the property (1.50), it is straightforward to show that the choice (1.52) implies the probability of one only for those arbitrarily small phase-space regions that contain the classical path. Besides this, the perfectly sharp classical probability distributions are fully characterized by the classical trajectory $\left[x_{\text{cl},1}(t), \cdots, x_{\text{cl},N}(t),\ \ p_{\text{cl},1}(t), \cdots, p_{\text{cl},N}(t)\right]$. Thus, the numerical description of the accurate phase-space distributions requires

$$\text{Dim}_{\text{num}}\left[(x_{\text{cl},1}, \cdots, x_{\text{cl},N}, p_{\text{cl},1}, \cdots, p_{\text{cl},N})\right] = 6N \tag{1.53}$$

real-valued elements to be stored in the computer memory at any given time.

We need a much larger number of elements to describe the general ρ because each phase-space coordinate within ρ is a continuous variable in real space. Thus, the dimension of ρ is given by

$$\text{Dim}\left[\rho(\mathbf{x}_1, \cdots, \mathbf{x}_N, \mathbf{p}_1, \cdots, \mathbf{p}_N\,; t)\right] = \mathbb{R}^{6N}. \tag{1.54}$$

This means that in a numerical analysis of ρ, one must, in principle, store the probability distribution at

$$\text{Dim}_{\text{num}}[\rho] = \mathcal{N}^{6N} \tag{1.55}$$

values if each continuous phase-space coordinate is discretized into $\mathcal{N}$ intervals. Thus, the numerical evolution and the storage of ρ is much more demanding than the classical evolution of individual phase-space trajectories $[\mathbf{r}_j(t), \mathbf{p}_j(t)]$ that requires only $6N$ real numbers for each time step.

If we consider computers that are capable of handling 10^9 numbers within a reasonable computation time, a straightforward numerical algorithm can evolve roughly $N = 10^8$ particles along their classical paths. At the same time, the statistical calculation is limited to only a few particles due to an exponentially growing number of elements, according to Eq. (1.55). For example, already $\mathcal{N} = 10$ yields $\text{Dim}_{\text{num}} = 10^{12}$ for two three-dimensional particles. Thus, one needs to consider and develop appropriate and efficient numerical schemes to handle the solutions for many-body systems with nontrivial statistics.

1.2.1 Liouville equation

Before we address the numerical solutions for ρ, we need to know how to actually evolve ρ in time. Since this problem is dimension independent, we first consider how $\rho(x, p\,; t)$ can be computed for a one-dimensional system. For any given time t, $\rho(x, p\,; t)$ can be viewed as a distribution that is a function of the initial conditions. For each initial value, the particle follows a classical trajectory $[x_{\text{cl}}(x, p, t+t'),\ \ p_{\text{cl}}(x, p, t+t')]$ where t' denotes the elapsed time. Since $\rho(x, p\,; t+t')$ defines the probability distribution of classical paths, its dynamics should directly contain all possible paths. Looking at the evolution over an infinitesimal time step Δt, we can make the ansatz

$$\rho(x, p\,; t + \Delta t) = \rho(x_{\text{cl}}(x, p, -\Delta t),\ p_{\text{cl}}(x, p, -\Delta t);\ t). \tag{1.56}$$

According to this ansatz, the phase-space point $[x_{\text{cl}}(x, p, -\Delta t),\ \ p_{\text{cl}}(x, p, -\Delta t)]$ is moved along the classical trajectory to be at (x, p) when the time Δt has elapsed. To test the ansatz, we set $\Delta t = 0$ producing

$$\rho(x,\ p,\ t) = \rho(x_{\text{cl}}(x, p, 0),\ p_{\text{cl}}(x, p, 0);\ t). \tag{1.57}$$

This is an identity because of the initial conditions $x_{\text{cl}}(x, p, 0) = x$ and $p_{\text{cl}}(x, p, 0) = p$. Thus, the ansatz (1.56) correctly describes the ρ dynamics over the time Δt, which in principle can also be macroscopic.

The form of Eq. (1.56) is particularly useful if we consider infinitesimal Δt because we can then use the Taylor expansion (1.6), yielding

$$\begin{aligned}
x_{\text{cl}}(x,\ p,\ -\Delta t) &= x - \dot{x}\Delta t + \mathcal{O}(\Delta t^2) = x - \frac{\partial H}{\partial p}\Delta t + \mathcal{O}(\Delta t^2) \\
p_{\text{cl}}(x,\ p,\ -\Delta t) &= p - \dot{p}\,\Delta t + \mathcal{O}(\Delta t^2) = p + \frac{\partial H}{\partial x}\Delta t + \mathcal{O}(\Delta t^2).
\end{aligned} \tag{1.58}$$

Here, we used Hamilton's equations (1.10) to express the $\dot{x}$ and $\dot{p}$ terms. We may now substitute the result (1.58) into Eq. (1.56) to obtain

$$\begin{aligned}
\rho(x,\ p\,;t + \Delta t) &= \rho\left(x - \frac{\partial H}{\partial p}\Delta t,\ p + \frac{\partial H}{\partial x}\Delta t,\ t\right) + \mathcal{O}(\Delta t^2) \\
&= \rho\,(x,\ p\,;t) - \frac{\partial H}{\partial p}\frac{\partial}{\partial x}\rho\,(x,\ p\,;t)\,\Delta t \\
&\quad + \frac{\partial H}{\partial x}\frac{\partial}{\partial p}\rho\,(x,\ p\,;t)\,\Delta t + \mathcal{O}(\Delta t^2).
\end{aligned} \tag{1.59}$$

In the second step, we Taylor expanded the phase-space distribution. Rearranging the terms, dividing both sides of the equation by Δt and taking the limit $\Delta t \to 0$, we can identify the time derivative of the probability distribution function as

$$\begin{aligned}
\frac{\partial}{\partial t}\rho\,(x,\ p\,;t) &= \lim_{\Delta t\to 0}\frac{\rho\,(x,\ p\,;t + \Delta t) - \rho\,(x,\ p\,;t)}{\Delta t} \\
&= -\frac{\partial H}{\partial p}\frac{\partial}{\partial x}\rho\,(x,\ p\,;t) + \frac{\partial H}{\partial x}\frac{\partial}{\partial p}\rho\,(x,\ p\,;t)\,.
\end{aligned} \tag{1.60}$$

This differential equation is known as the classical Liouville equation

$$\begin{aligned}
&\frac{\partial}{\partial t}\rho\,(x,\ p\,;t) = -i\hat{L}\,\rho\,(x,\ p\,;t)\,, \\
&\qquad i\hat{L} \equiv \frac{\partial H}{\partial p}\frac{\partial}{\partial x} - \frac{\partial H}{\partial x}\frac{\partial}{\partial p} = \frac{\partial H}{\partial p}\frac{\partial}{\partial x} + F(x)\frac{\partial}{\partial p}.
\end{aligned} \tag{1.61}$$

Here, we identified the classical Liouville operator $\hat{L}$ and the force $F(x)$ which acts on the particle. We introduced an imaginary unit in order to write the equation in analogy to the full quantum-mechanical formulation, discussed in Chapter 3.

The Liouville equation has the formal solution

$$\rho\,(x,\ p\,;t) = e^{-i\hat{L}\,t}\ \rho\,(x,\ p\,;0)\,. \tag{1.62}$$

Here, as well as later in this book, we use the convention that any function of an operator has to be interpreted as its Taylor-expanded form, i.e., $e^{-i\hat{L}\,t} = \sum_{j=0}^{\infty}\frac{[-i\hat{L}]^j}{j!}$.

Obviously, the j-th power of $[\hat{L}]^j$ introduces a j-th-order differentiation with respect to the phase-space coordinates. The result (1.62) can also be used for 3D multi-particle systems where

$$i\hat{L} \equiv \sum_{j=1}^{\mathcal{N}} \left[\frac{\partial H}{\partial \mathbf{p}_j} \cdot \frac{\partial}{\partial \mathbf{x}_j} - \frac{\partial H}{\partial \mathbf{x}_j} \cdot \frac{\partial}{\partial \mathbf{p}_j} \right] \tag{1.63}$$

follows from a straightforward generalization of the result (1.61).

To illustrate how the solution (1.62) can be evaluated in practice, we solve the Liouville Eq. (1.61) for a free particle with $H = p^2/2m$. In this situation, the formal solution (1.62) produces

$$\begin{aligned} \rho\,(x,p\,;t) &= e^{-\frac{pt}{m}\frac{\partial}{\partial x}}\,\rho\,(x,p\,;0) \\ &= \sum_{j=0}^{\infty} \frac{1}{j!} \left(-\frac{pt}{m}\right)^j \frac{\partial^j \rho\,(x,p\,;0)}{\partial x^j}. \end{aligned} \tag{1.64}$$

We notice that the series of differentiations produces a Taylor expansion of $\rho\,(x,p\,;0)$. Since $\rho\,(x,p\,;0)$ is assumed to be analytic, the Taylor expansion can be summed up to produce

$$\rho\,(x,p\,;t) = \rho\left(x - \frac{pt}{m}, p\,;0\right). \tag{1.65}$$

Hence, the free-particle case produces

$$\rho\,(x,p\,;t) = \rho\,(x_{\rm cl}(x,p\,;-t), p_{\rm cl}(x,p\,;-t); 0)\,. \tag{1.66}$$

This result is in agreement with Eq. (1.56), as we can verify by first setting $t \to 0$ and then $\Delta t \to t$ in Eq. (1.56). In fact, one can show that Eq. (1.66) is a general solution of the Liouville Eq. (1.61). These aspects are the central topics of Exercise 1.10. Thus, one can solve the classical $\rho(t)$ by evolving only a set of initial values along their classical path. Since each trajectory can be computed with only $6N$ elements – according to Eq. (1.53) – the numerical task is significantly reduced. In particular, we find that the numerical effort is $M \times 6N$ for M classical paths instead of $\mathcal{N}^{2N}$ for the brute force approach according to Eq. (1.55).

If the initial phase-space distribution has a Gaussian shape,

$$\rho\,(x,p\,;0) = \frac{1}{2\pi\,\Delta x\,\Delta p}\, e^{-\frac{(x-x_0)^2}{2\Delta x^2}}\, e^{-\frac{(p-p_0)^2}{2\Delta p^2}}, \tag{1.67}$$

it is centered at (x_0, p_0) and has the width Δx and Δp in position and momentum, respectively. We may then apply Eq. (1.66) to determine the phase-space distribution at any later time,

$$\rho\,(x,p\,;t) = \frac{1}{2\pi\,\Delta x\,\Delta p}\, e^{-\frac{(x_{\rm cl}(x,p,;-t)-x_0)^2}{2\Delta x^2}}\, e^{-\frac{(p_{\rm cl}(x,p,;-t)-p_0)^2}{2\Delta p^2}}. \tag{1.68}$$

The classical trajectories of vanishing, linear, and harmonic potential can be obtained from Eqs. (1.34), (1.36), and (1.41), respectively. Especially, we find that the center of the distribution, $(x, p) \equiv [x_{\text{cent}}(t),\ p_{\text{cent}}(t)]$, must satisfy

$$\begin{cases} x_{\text{cl}}(x_{\text{cent}}(t), p_{\text{cent}}(t)\ ;\ -t) = x_0 \\ p_{\text{cl}}(x_{\text{cent}}(t), p_{\text{cent}}(t)\ ;\ -t) = p_0 \end{cases} \Leftrightarrow \begin{cases} x_{\text{cent}}(t) = x_{\text{cl}}(x_0, p_0\ ;\ t) \\ p_{\text{cent}}(t) = p_{\text{cl}}(x_0, p_0\ ;\ t) \end{cases} \tag{1.69}$$

because the distribution center has to follow the classical phase-space trajectory starting from (x_0, p_0).

Figure 1.4 presents the time evolution of the classical phase-space distribution when the system is subject to: (a) no force, (b) a constant force, and (c) a harmonic force. More specifically, we have plotted the contour lines of $\rho(x, p\ ; t)$ for values corresponding to 0.5, 0.1, 0.01, 0.001, and 0.0001 of the peak value of ρ. In all calculations, we use the scaled units

$$\bar{x} \equiv \frac{x}{a_0}, \qquad \bar{p} \equiv \frac{p}{m\omega a_0} \tag{1.70}$$

motivated by (1.44). In the general case, a_0 identifies a typical length scale while $\frac{2\pi}{\omega}$ sets the typical time scale. We choose the initial distribution to be a Gaussian with the widths $\Delta x = \frac{a_0}{2\sqrt{2}}$ and $\Delta p = m\omega a_0$. In other words, we choose the parameters such that the scaled momentum initially has a $2\sqrt{2} \approx 2.8284$ times wider spread than the $\bar{x}$ does in order to make it easier to follow how the shape of the distribution changes with time.

The free-particle case (Fig. 1.4(a)) yields $\rho(x, p\ ; t)$ where $[x_{\text{cent}}(t),\ p_{\text{cent}}(t)]$ has a constant momentum while the position moves linearly in time, as predicted by Eqs. (1.34) and (1.69). We see that the overall shape of $\rho(x, p\ ; t)$ becomes strongly distorted because the low-momentum points move more slowly in the x-direction than the high-momentum

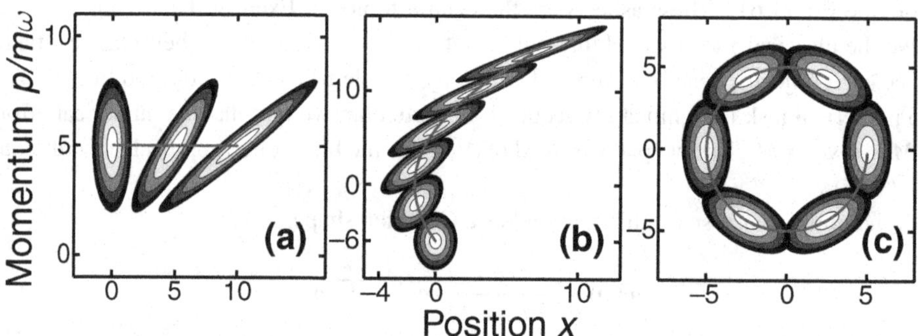

Figure 1.4 Dynamics of classical statistical distributions. **(a)** The free-particle solution is shown for three representative times. The initial distribution is a Gaussian centered at $(\bar{x}_0 = 0, \bar{p}_0 = 5)$. **(b)** Motion of $\rho(t)$ under the influence of a constant force. The initial Gaussian is centered at $(\bar{x}_0 = 0, \bar{p}_0 = -6)$ and the linear potential is taken as $U(x) = -12\, m\omega a_0 \bar{x}$. **(c)** Motion of $\rho(t)$ under the influence of a harmonic force. The initial Gaussian is centered at $(\bar{x}_0 = 0, \bar{p}_0 = 5)$ and six snapshots corresponding to the times $t = 0,\ \pi/3\omega,\ 2\pi/3\omega\ 3\pi/3\omega,\ 4\pi/3\omega$, and $5\pi/3\omega$ are shown. In all cases, the solid line indicates the classical $[x_{\text{cent}}(t),\ p_{\text{cent}}(t)]$ trajectory. The contour lines correspond to $1/2$, 10^{-1}, 10^{-2}, and 10^{-3} of the peak value.

ones. As a result, $\rho(x, p\ ; t)$ becomes tilted and stretched as time elapses. However, the more elliptic distribution has a narrower minor axis such that the area under $\rho(x, p\ ; t)$ remains unchanged.

The linear potential $U(x) = -Fx$ (Fig. 1.4(b)) produces a $\rho(x, p\ ; t)$ whose center follows a parabolic trajectory (solid line) while the elliptical shape of $\rho(x, p\ ; t)$ is both stretched and rotated in time. The parabolic $[x_{\text{cent}}(t),\ p_{\text{cent}}(t)]$ trajectory (line) follows the classical motion affected by a constant force. As for the free particle, each phase-space point of $\rho(x, p\ ; t)$ follows a different trajectory. In this case, all trajectories are parabolic but they have different maxima and curvatures, which distorts the shape of the original distribution.

For the harmonic potential, $U(x) = \frac{1}{2}m\omega^2 x^2$, the classical trajectories are circles in the phase space, as predicted by Eq. (1.41). In this case, $\rho(x, p\ ; t)$ is simply rotated in clockwise direction without distorting the shape. As shown in Fig. 1.4(c), the elliptical shape of $\rho(x, p\ ; t)$ is tilted to match the rotation angle at any given moment. At the same time, the center of the distribution follows a circular path.

The classical paths can become significantly more complicated for a general potential. As an example, we consider $\rho(x, p\ ; t)$ for the anharmonic oscillator already analyzed in Fig. 1.3(a). Here, we assume $\bar{x}_0 = 0$ and $\bar{p}_0 = 1.76 m\omega a_0$. The resulting $\rho(x, p\ ; t)$ is shown in Fig. 1.5(a) for several t. Since each phase-space point propagates now very differently, the original Gaussian shape of the distribution becomes strongly distorted in shape, even though we recognize some form of rotation due to the anharmonic oscillations. In fact, the classical trajectories depend so sensitively on the starting point (x, p) that a distribution of them will eventually spread across a very broad range of phase space. As a general tendency, the spreading of $\rho(x, p\ ; t)$ becomes larger as the system is evolved longer even though the area under $\rho(x, p\ ; t)$ remains constant. Thus, $\rho(x, p\ ; t)$ starts to resemble more and more a fractal structure. This example shows that one can detect potentially chaotic behavior by following phase-space distributions in time.

To compare the obtained result with a truly chaotic evolution, we add the same oscillatory disturbance to the anharmonic oscillator that we already used to produce Fig. 1.3(c). The resulting $\rho(x, p\ ; t)$ is presented in Fig. 1.5(f) for otherwise the same conditions as in Fig. 1.5(e). We observe that the truly chaotic configuration yields a $\rho(x, p\ ; t)$ that is very similar to that of the anharmonic distribution. This is a strong hint that the appearance of fractal phase-space signatures can indeed be related to chaotic behavior. The oscillatory disturbance causes the classical trajectory to jump from one anharmonic solution to another. As this happens, a single trajectory never repeats the same path but jumps chaotically between infinitely many solution branches that cover very different regions in the phase space.

Regardless of the numerical approach used, solving $\rho(x, p\ ; t)$ is considerably more difficult than evaluating individual classical paths. However, the probability distribution allows us to describe classical systems whose properties are known only probabilistically. By following $\rho(x, p\ ; t)$, we can deduce how accurately the position and momentum of the system can be determined at any later time. As typical features, both the center of

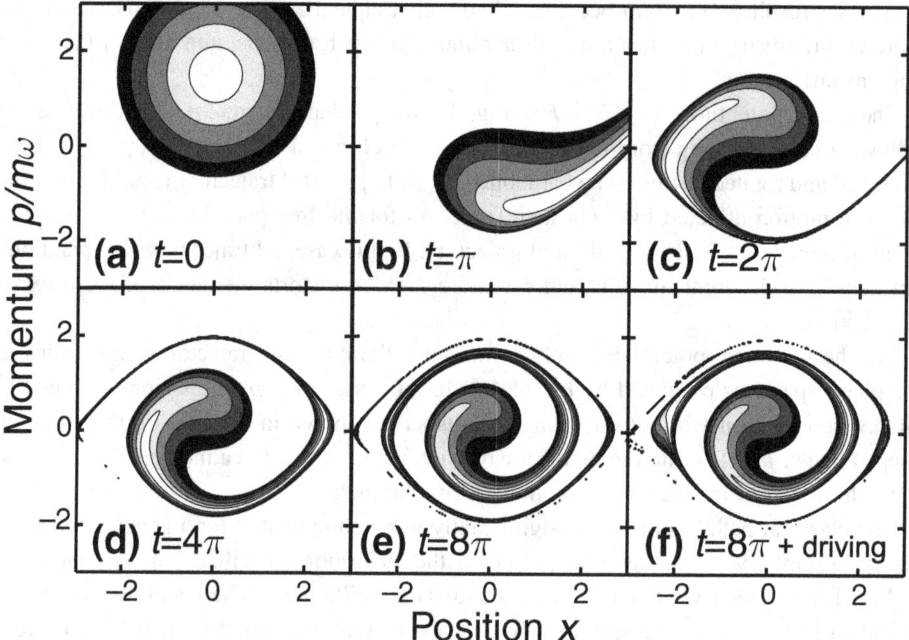

Figure 1.5 Dynamics of classical statistical distributions for the anharmonic oscillator. The system is the same as in Figs. 1.3(b) and (c) and the Gaussian initial state is centered at $(\bar{x}_0 = 0, \bar{p}_0 = 1.76)$ with the width $\Delta\bar{x} = \Delta\bar{p} = 1/2$. The computation without external force: contour plots of ρ are shown at **(a)** $t = 0$, **(b)** $t = \pi/\omega$, **(c)** $t = 2\pi/\omega$, **(d)** $t = 4\pi/\omega$, and **(e)** $t = 8\pi/\omega$. **(f)** The final $\rho(t)$ at $t = 8\pi/\omega$ is shown when the external driving force $0.1 \times \sin 2\omega t$ is added. The contours are evaluated at the values $1/2,\ 10^{-1},\ 10^{-2}$, and 10^{-3}.

the distribution and its shape can change drastically even for relatively simple classical systems. Especially, systems with strong sensitivity to the initial conditions can produce fractal patterns for $\rho(x, p\,; t)$. Furthermore, as we will see in the later chapters of this book, $\rho(x, p\,; t)$ is a useful concept to connect classical mechanics to the quantum-mechanical description of particles that fully incorporates the inherent inaccuracy in particle position and momentum.

1.2.2 Classical averaging of statistical distributions

If we have a set of phase-space distributions $\{\rho_\lambda(x, p\,; t)\}$, we may always construct a new distribution,

$$\rho(x, p\,; t) \equiv \sum\nolimits_\lambda P_\lambda\, \rho_\lambda(x, p\,; t), \tag{1.71}$$

as a superposition of the original ones. The resulting $\rho(x, p\,; t)$ is a classical phase-space distribution if the coefficients P_λ are positive definite and normalized to unity, i.e.,

$$P_\lambda \geq 0, \qquad \sum_\lambda P_\lambda = 1. \tag{1.72}$$

Since each individual $\rho_\lambda(x, p\,; t)$ is assumed to be a genuine classical phase-space distribution, all of them must satisfy the Liouville equation (1.61). Due to its linearity, the Liouville equation produces the general property

$$\begin{aligned}
&\frac{\partial}{\partial t}\rho_\lambda(x, p\,; t) = -i\hat{L}\rho_\lambda(x, p\,; t) \\
&\qquad \Rightarrow \frac{\partial}{\partial t}\sum_\lambda P_\lambda\,\rho_\lambda(x, p\,; t) = -i\hat{L}\sum_\lambda P_\lambda\,\rho_\lambda(x, p\,; t) \\
&\qquad \Rightarrow \frac{\partial}{\partial t}\rho(x, p\,; t) = -i\hat{L}\rho(x, p\,; t).
\end{aligned} \tag{1.73}$$

In other words, if we find different solutions to the Liouville equation (1.61), any additive combination of them will also be a solution.

The summed $\rho = \sum_\lambda P_\lambda\,\rho_\lambda$ has a relatively straightforward interpretation. By combining different distributions, we simply increase the statistical uncertainty to find the system at a given phase-space point. The procedure outlined in Eqs. (1.71)–(1.72) introduces *classical averaging* of probabilities that essentially redistributes the realizations across a wider portion of the phase space. In general, the classical averaging either increases or at least maintains the statistical scatter of events described by the individual probability distributions.

Quite often we are not interested in the full phase-space distribution but only in the probability distribution of position or momentum. These quantities can be directly accessed via the *marginal distributions*

$$\rho_p(x\,; t) \equiv \int dp\;\rho(x, p\,; t), \qquad \rho_x(p\,; t) \equiv \int dx\;\rho(x, p\,; t), \tag{1.74}$$

where $\rho_p(x\,; t)$ and $\rho_x(p\,; t)$ respectively determine the probability distributions for the particle position and momentum alone. These definitions can be directly generalized for N particles having the full 3D dependence.

To get some feeling for the marginal distributions, we look at

$$\rho_\pm(x, p\,; t) = \frac{1}{2\pi\,\Delta x\,\Delta p}\,e^{-\frac{\left[x-(\pm x_0+\frac{p}{m}t)\right]^2}{2\Delta x^2}}\,e^{-\frac{(p\pm p_0)^2}{2\Delta p^2}} \tag{1.75}$$

that follows from Eq. (1.68) by using the classical free-particle trajectory $x_{\text{cl}}(x, p; -t) = x - \frac{p}{m}t$. The different signs for (x_0, p_0) determine the central position and momentum direction at $t = 0$. We combine two distributions with $(-x_0, +p_0)$ and $(+x_0, -p_0)$ such that the partial distributions are initially centered at the opposite sides of the origin and propagate toward each other. The combined distribution uses equal probability, $P_+ = P_- = \frac{1}{2}$, for both parts. A straightforward integration produces a marginal distribution

$$\rho_p(x\ ;t) = \frac{1}{2}\sqrt{\frac{1}{2\pi(\Delta x^2 + \Delta v^2 t^2)}}\left[e^{-\frac{(x+x_0-v_0t)^2}{2(\Delta x^2+\Delta v^2 t^2)}} + e^{-\frac{(x-x_0+v_0t)^2}{2(\Delta x^2+\Delta v^2 t^2)}}\right], \tag{1.76}$$

where we have identified the central velocity $v_0 \equiv p_0/m$ and fluctuations in the velocity $\Delta v \equiv \Delta p/m$. From Eq. (1.76), we see that the position distribution broadens as

$$\Delta x(t) \equiv \sqrt{\Delta x^2 + \Delta v^2 t^2}. \tag{1.77}$$

In other words, the momentum uncertainty of the combined distributions produces an increasing spread of the positions, as seen earlier in Fig. 1.4(a).

According to Eq. (1.76), the initial form of $\rho_p(x\ ;t=0)$ corresponds to a distribution where the phase-space probability is concentrated at two separate Gaussian peaks centered at $x = -a_0$ and $x = a_0$. At times $0 < t < a_0/v_0$ the Gaussians propagate toward each other, they meet, and then move apart for the times $t > a_0/v_0$. Thus, the setup (1.76) defines the classic scenario where two independent probability distributions "collide." Figure 1.6 demonstrates how the corresponding total marginal distribution $\rho_p(x\ ;t)$ (shaded area) evolves in time if $\bar{x}_0 = 1$, $\bar{p}_0 \equiv \bar{v}_0 = 10$, and $\Delta\bar{x} = \Delta\bar{p} = 1/2$. The solid lines indicate the individual $\rho_\pm$ components within ρ_p.

In Fig. 1.6, we see clearly that the separate probability distributions first approach each other until they merge at $t = x_0/v_0$. At this time, the central region of $\rho_p(x\ ;t)$ (dark area) is twice that of the individual distributions (solid line). Just before and after the "collision," $\rho_p(x\ ;t)$ is clearly non-Gaussian due to the addition of the distributions. After the collision, ($t \gg x_0/v_0$), $\rho_p(x\ ;t)$ separates again into two distinct Gaussians, unaltered by the collision.

The chosen example has a relatively small momentum spread such that the shown Gaussians remain almost unchanged in width. Figure 1.6(b) presents the Gaussian width for up to $t = 5x_0/v_0$, based on Eq. (1.77). The filled circles indicate the $\Delta x(t)$ width for the five first snapshots presented in Fig. 1.6(a). We see that $\Delta x(t)$ is almost unchanged

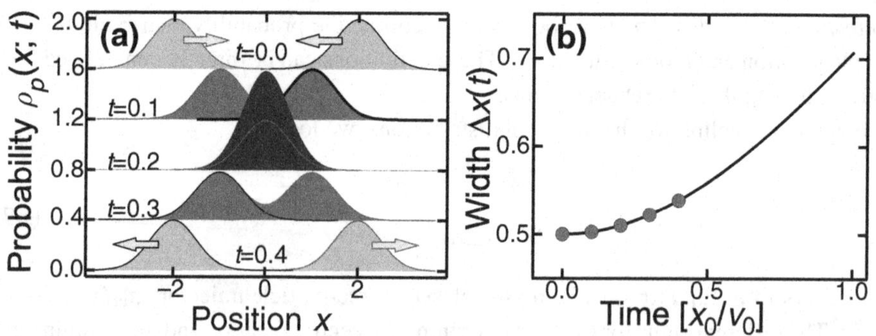

Figure 1.6 Classical averaging of two statistical distributions. **(a)** The time evolution of the marginal distributions is shown for the situation where the individual Gaussian components propagate into opposite directions. **(b)** The Δx width of the individual Gaussians is plotted as function of time. The circles indicate the times at which the actual snapshots in frame (a) are taken.

at the early times and broadens significantly for large t. These broadening features will become important in Chapter 3 where we investigate the quantum aspects of particles.

Exercises

1.1 Instead of using Hamilton's formalism, the theory of classical mechanics can also be formulated using the Lagrangian $\mathcal{L}(x, \dot{x}, t) = \frac{1}{2}m\dot{x}^2 - U(x)$ that contains position and velocity. The action is then $S(f, i; t) = \int_{t_i}^{t_f} dt' L(x(t'), \dot{x}(t'), t')$. Using the variation principle and constraints $\delta x(t_i) = \delta x(t_f) = 0$, show that S is minimized by a classical path that satisfies the Lagrange equation

$$\frac{d}{dt}\frac{\partial \mathcal{L}}{\partial \dot{q}} = \frac{\partial \mathcal{L}}{\partial q}.$$

Show also that its solution is identical to that of Hamilton Eq. (1.11).

1.2 Derive Hamilton's equations for N particles starting from the action

$$S(f, i; t) = \int_{\theta_i}^{t_f} dt' \left[\sum_j \mathbf{p}_j(t') \cdot \mathbf{x}_j(t) - H\left(\{\mathbf{x}(t)\}, \{\mathbf{p}(t)\}; t\right)\right]$$

where $\{\mathbf{x}(t)\} = \mathbf{x}_1(t), \mathbf{x}_2(t), \ldots, \mathbf{x}_N(t)$ and $\{\mathbf{p}(t)\} = \mathbf{p}_1(t), \ldots, \mathbf{p}_N(t)$ denote the positions and momenta of all three-dimensional particles.

1.3 Show that Hamilton equations (1.18) imply that $\dot{\mathbf{R}}$ vanishes in the absence of external forces. Here, $\mathbf{R} = \sum_{j=1}^{N} \frac{m_j}{M}\mathbf{x}_j$ refers to the center-of-mass coordinate, $M = \sum_{j=1} m_j$ is the total mass, and the external force is given by $\mathbf{F}_{ext} = \sum_{j=1}^{N} \nabla_{\mathbf{x}_j} U(\{\mathbf{x}\})$.

1.4 Assume that a one-dimensional particle propagates toward a potential barrier. The classical turning point x_{tp} is defined by $E = U(x_{\text{tp}})$. Assume that the potential can be written as $U(x) = E + \alpha(x - x_{\text{tp}})$ in the vicinity of x_{tp}. **(a)** Show that the particle's trajectory is $x(t) = x_{\text{tp}} - \frac{\alpha}{2m}t^2$ in the vicinity of x_{tp}. **(b)** Show that a classical particle can penetrate into the barrier only if we allow for the analytic continuation of time into the complex plane. What is the corresponding $x(t)$ in the vicinity of x_{tp}?

1.5 Show that Eqs. (1.36) and (1.41) are general solutions for particles moving in the linear potential $U(x) = F_0 x$ and the harmonic potential $U(x) = \frac{1}{2}mx^2$, respectively.

1.6 Show that scaling (1.44) converts Eq. (1.43) into Eq. (1.45). Show that the pendulum's ω is independent of its mass in the gravitational field. Show also that the pendulum motion is harmonic for small displacements.

1.7 **(a)** The delta function can be defined through

$$\delta(x) = \lim_{\varepsilon \to 0} \delta_\varepsilon^{\text{box}}(x), \text{ where } \delta_\varepsilon^{\text{box}}(x) = \begin{cases} \frac{1}{\varepsilon}, & |x| \le \frac{\varepsilon}{2} \\ 0, & \text{otherwise} \end{cases}.$$

Evaluate $\lim_{\varepsilon\to 0}\int dx\, \delta_\varepsilon^{\rm box}(x)\, f(x)$ where $f(x)$ is an analytic function. **Hint:** *Taylor expand $f(x)$ around $x = 0$.*
(b) Show that also $\delta_\varepsilon^L(x) = \frac{1}{\pi}\frac{\varepsilon}{x^2+\varepsilon^2}$ identifies a δ function in the limit $\varepsilon \to 0$.

1.8 Start from two functions

$$\delta_\varepsilon^{\rm tri}(x) \equiv \begin{cases} \frac{|\varepsilon-x|}{\varepsilon^2}, & |x| \le \varepsilon \\ 0, & \text{otherwise} \end{cases} \quad \text{and} \quad \theta(x) = \begin{cases} 0, & x < 0 \\ 1, & x \ge 0 \end{cases}$$

and show that **(a)** $\int_{-\infty}^{\infty} dx\, f(x)\, \frac{d\delta(x)}{dx} = \lim_{\varepsilon\to 0}\int_{-\infty}^{\infty} dx\, f(x)\, \frac{d\delta_\varepsilon^{\rm tri}(x)}{dx} = -f'(0)$ for any analytic function $f(x)$. **(b)** Show also that $\frac{d}{dx}\theta(x) = \delta(x)$. **(c)** Compare $\frac{d}{dx}\delta_\varepsilon(x)$ computed using $\delta_\varepsilon^{\rm box}(x)$, $\delta_\varepsilon^{\rm tri}(x)$, and the Gaussian form of Eq. (1.51).

1.9 Show that the δ function satisfies the following properties **(a)** $\delta(x) = \delta(-x)$, **(b)** $\delta(ax) = \frac{1}{|a|}\delta(x)$, **(c)** $\delta(g(x)) = \sum_j \frac{\delta(x-x_j)}{|g'(x_j)|}$ where $g(x)$ is an analytic function with roots $g(x_j) = 0$. **Hint:** *Start from* (1.50).

1.10 Assume that $x_{\rm cl}(x, p; t)$ and $p_{\rm cl}(x, p; t)$ define generic particle trajectories when the particle is found at (x, p) for $t = 0$. Show that any analytic $\rho(x, p; t) \equiv \rho(x_{\rm cl}(x, p; -t), p_{\rm cl}(x, p; -t); 0)$ satisfies

$$\frac{\partial \rho(x, p; t)}{\partial t} = \left(-\frac{\partial H}{\partial p}\frac{\partial}{\partial x} + \frac{\partial H}{\partial x}\frac{\partial}{\partial p} \right) \rho(x, p; t).$$

Hint: *Show that* $\frac{\partial p_{\rm cl}}{\partial p}\frac{\partial x_{\rm cl}}{\partial x} - \frac{\partial p_{\rm cl}}{\partial x}\frac{\partial x_{\rm cl}}{\partial p}$ *is constant in time.*

1.11 Compute $\rho(x, p; t)$ for a harmonic force using the result (1.66). Show that $\rho(x, p; t)$ is simply rotated in phase space.

1.12 Identify

$$[x(t)] \equiv \int dx \int dp\, x\, \rho(x, p, t) \text{ and } [p(t)] \equiv \int dx \int dp\, p\, \rho(x, p, t)$$

as the average values of (x, p). **(a)** Derive the equation of motion for $[(x(t)]$ and $[p(t)]$ and show that $\frac{d}{dt}[x(t)] = \int dx \int dp \frac{\partial H}{\partial p}\rho(x, p, t)$ and $\frac{d}{dt}[p(t)] = \int dx \int dp \left(-\frac{\partial H}{\partial x}\right) \rho(x, p, t)$. **(b)** The fluctuations of position are defined by

$$\Delta x_\lambda^2 \equiv \int dx \int dp\, x^2 \rho_\lambda(x, p, t) - [x(t)]^2.$$

Show that the classical averaging (1.71) always increases the total fluctuations, i.e., $\Delta x^2 \ge \Delta x_\lambda^2$ for all λ.

1.13 A perfectly sharp phase-space distribution follows from

$$\rho_s(x, p, x_0, p_0; t) = \delta\left(x - x_{\rm cl}(x_0, p_0; t)\right) \delta\left(p - p_{\rm cl}(x_0, p_0; t)\right),$$

where (x_0, p_0) is the phase space at the initial moment. **(a)** Show that a general distribution can always be presented through classical averaging

$$\rho(x, p; t) = \int dx_0 \int dp_0\, P(x_0, p_0)\, \rho_s(x, p, x_0, p_0; t).$$

What is the interpretation of $P(x_0, p_0)$?
(b) Show that ρ_s can alternatively be expressed as

$$\rho_s(x, p, x_0, p_o; t) = \delta\left(x_0 - x_{\text{cl}}(x, p; -t)\right) \delta\left(p_0 - p_{\text{cl}}(x, p; -t)\right).$$

Using the properties of the δ function, show that the generic ρ follows from

$$\rho(x, p; t) = \rho(x_{\text{cl}}\,(x, p; -t), p_{\text{cl}}(x, p; -t); 0)\,.$$

1.14 **(a)** Verify that the marginal distribution (1.76) is obtained from a classical averaging of Eq. (1.75) with the weights $P_\pm = 1/2$. **(b)** Which total width in position is obtained from Eq. (1.76)?

Further reading

For further reading and many more details of the Lagrange and Hamilton formulation of classical mechanics, we recommend:

L. D. Landau and E. M. Lifschitz (1976). *Mechanics*, 3rd edition, Reading, Pergamon Press.

A. L. Fetter and J. D. Walecka (1980). *Theoretical Mechanics of Particles and Continua*, New York, McGraw-Hill.

H. Goldstein (1980). *Classical Mechanics*, 2nd edition, Reading, Addison-Wesley.

M. G. Calkin (2005). *Lagrangian and Hamiltonian Mechanics*, Singapore, World Scientific.

F. Scheck (2007). *Mechanics: From Newton's Laws to Deterministic Chaos*, 5th edition, Heidelberg, Springer.

C. Gignoux and B. Silvestre-Brac (2009). *Solved Problems in Lagrangian and Hamiltonian Mechanics*, New York, Springer.

Excellent discussions of classical and quantum chaos can be found in:

M. C. Gutzwiller (1990). *Chaos in Classical and Quantum Mechanics*, New York, Springer.

H.-J. Stöckmann (2000). *Quantum Chaos: An Introduction*, Cambridge, Cambridge University Press.

F. Haake (2010). *Quantum Signatures of Chaos*, 3rd edition, Berlin, Springer.

2
Central concepts in classical electromagnetism

The mass of a classical particle defines its inertia against force-induced motion changes. The resulting dynamics can be completely described on the basis of Newton's three laws, see Chapter 1. As an additional feature, some particles are electrically charged. These charges do not only lead to forces between the particles but they also determine how strongly they interact with an electromagnetic force field. Furthermore, charged particles act as sources of electromagnetic fields themselves.

If we want to model these effects, we have to describe the interplay between the charges and the electromagnetic fields. This analysis involves a simultaneous treatment of particles, their charges, and the fields induced by them. As a first step, we present in this chapter the fundamental axioms leading to the theory of classical electromagnetism.

2.1 Classical description of electromagnetic fields

The classical electromagnetism follows axiomatically from Maxwell's equations:

$$\begin{aligned} \nabla\cdot\mathbf{E}(\mathbf{r}) &= \frac{1}{\varepsilon_0}\rho_Q(\mathbf{r}), & \nabla\cdot\mathbf{B}(\mathbf{r}) &= 0, \\ \nabla\times\mathbf{E}(\mathbf{r}) &= -\frac{\partial}{\partial t}\mathbf{B}(\mathbf{r}), & \nabla\times\mathbf{B}(\mathbf{r}) &= \frac{1}{c^2}\frac{\partial}{\partial t}\mathbf{E}(\mathbf{r}) + \mu_0\mathbf{j}(\mathbf{r}). \end{aligned} \tag{2.1}$$

Here, $\mathbf{E}(\mathbf{r})$ is the electric field, $\mathbf{B}(\mathbf{r})$ denotes the magnetic field, $\rho_Q(\mathbf{r})$ is the charge density, and $\mathbf{j}(\mathbf{r})$ the current density. The constants appearing in Eq. (2.1) define the speed of light $c = 1/\sqrt{\epsilon_0\mu_0}$ via the dielectric constant ϵ_0 and the vacuum permeability μ_0:

$$\begin{aligned} c &= 299\,792\,458\,\mathrm{m\,s^{-1}}, \\ \epsilon_0 &= 8.854\,187\,817\times 10^{-12}\,\mathrm{F\,m^{-1}}, \qquad \mu_0 = 4\pi\,10^{-7}\,\mathrm{NA^{-2}}, \end{aligned} \tag{2.2}$$

expressed in SI units.

If the particles are described classically, the definition of the charge and current distribution follows from

$$\rho_Q(\mathbf{r}) \equiv \int d^3p\, Q\, \rho(\mathbf{r}, \mathbf{p}\,; t) = Q\rho_p(\mathbf{r}\,; t),$$

$$\mathbf{j}(\mathbf{r}) \equiv \int d^3p\, Q\mathbf{v}\, \rho(\mathbf{r}, \mathbf{p}\,; t), \tag{2.3}$$

respectively. Here, Q is the charge, $\mathbf{v}$ is the velocity, and $\rho(\mathbf{r}, \mathbf{p}\,; t)$ is the phase-space distribution representing the state of the particle. Furthermore, $\rho_p(\mathbf{r}\,; t) \equiv \int d^3p\, \rho(\mathbf{r}, \mathbf{p}t)$ is the marginal distribution where the momentum degree of freedom is integrated over, see Eq. (1.74). In cases where the particle's position and momentum are known precisely, we may replace $\rho(\mathbf{r}, \mathbf{p}\,; t)$ by delta functions $\delta(\mathbf{r} - \mathbf{x}_j)\delta(\mathbf{p} - \mathbf{p}_j)$.

In multi-particle systems, we label the individual particles by j, such that

$$\rho_Q(\mathbf{r}) = \sum_{j=1}^{\mathcal{N}} Q_j \delta(\mathbf{r} - \mathbf{x}_j(t)) \quad \text{and}$$

$$\mathbf{j}(\mathbf{r}) = \sum_{j=1}^{\mathcal{N}} Q_j \mathbf{v}_j(t) \delta(\mathbf{r} - \mathbf{x}_j(t)). \tag{2.4}$$

As we will see later, the joint description of light and matter requires a separation of *canonical momentum* $\mathbf{p}_j$ and *kinetic momentum* $m_j\mathbf{v}_j$. In most other cases, we can simply use the replacement $\mathbf{v}_j \rightarrow \mathbf{p}_j/m_j$.

2.1.1 Wave equation

To analyze the basic properties of electromagnetic fields, we first treat a system without external charges. In this case, Maxwell's equations (2.1) eventually produce

$$\begin{aligned} \nabla \times \nabla \times \mathbf{E}(\mathbf{r}) &= -\frac{\partial}{\partial t} \nabla \times \mathbf{B}(\mathbf{r}) \\ &= -\frac{1}{c^2} \frac{\partial^2}{\partial t^2} \mathbf{E}(\mathbf{r}) - \mu_0 \frac{\partial}{\partial t} \mathbf{j}(\mathbf{r}). \end{aligned} \tag{2.5}$$

For the double *curl* operation, we use the general relation

$$\nabla \times \nabla \times \mathbf{E}(\mathbf{r}) = \nabla\left[\nabla \cdot \mathbf{E}(\mathbf{r})\right] - \nabla^2 \mathbf{E}(\mathbf{r}). \tag{2.6}$$

This expression can be simplified because Maxwell's first equation in the absence of external charges reduces to

$$\nabla \cdot \mathbf{E}(\mathbf{r}) = 0, \tag{2.7}$$

showing that the field is divergence free. With this result, we obtain

$$\nabla \times \nabla \times \mathbf{E}(\mathbf{r}) = -\nabla^2 \mathbf{E}(\mathbf{r}) \tag{2.8}$$

allowing us to convert Eq. (2.5) into the *wave equation*

$$\nabla^2 \mathbf{E}(\mathbf{r}) - \frac{1}{c^2} \frac{\partial^2}{\partial t^2} \mathbf{E}(\mathbf{r}) = \mu_0 \frac{\partial}{\partial t} \mathbf{j}(\mathbf{r}). \tag{2.9}$$

Sometimes, the LHS is expressed with the d'Alambert operator $\Box \equiv \nabla^2 - \frac{1}{C^2}\frac{\partial^2}{\partial t^2}$.

In the absence of external charges, one often expresses the current

$$\mathbf{j}(\mathbf{r}) \equiv \frac{\partial}{\partial t}\mathbf{P}(\mathbf{r}), \tag{2.10}$$

via the polarization $\mathbf{P}(\mathbf{r})$ inside the matter. If we further assume that the matter has a dispersionless and linear response, i.e., $\mathbf{P}(\mathbf{r})$ is directly proportional to the exciting $\mathbf{E}(\mathbf{r})$, we can use the linear-response relation

$$\mathbf{P}(\mathbf{r}, t) \equiv \epsilon_0 \, \chi(\mathbf{r}) \, \mathbf{E}(\mathbf{r}, t). \tag{2.11}$$

As a consequence of the assumed instantaneous proportionality between $\mathbf{E}(\mathbf{r})$ and $\mathbf{P}(\mathbf{r})$, the linear susceptibility χ does not depend on time. Under these conditions, we can transform the wave equation into its customary form

$$\nabla^2\mathbf{E}(\mathbf{r}) - \frac{n^2(\mathbf{r})}{c^2}\frac{\partial^2}{\partial t^2}\mathbf{E}(\mathbf{r}) = 0. \tag{2.12}$$

Here, we used $\mu_0\epsilon_0 = \frac{1}{c^2}$ and introduced the refractive index

$$n^2(\mathbf{r}) \equiv 1 + \chi(\mathbf{r}). \tag{2.13}$$

To explore the properties of the wave equation (2.12), we consider the special case with constant $n^2(\mathbf{r}) = n^2$ and analyze electromagnetic fields in the form

$$\mathbf{E}(\mathbf{r}, t) = E(z, t) \, \mathbf{e}_x. \tag{2.14}$$

By inserting the ansatz (2.14) into Eq. (2.12), we obtain the scalar wave equation

$$\frac{\partial^2}{\partial z^2}E(z, t) - \frac{n^2}{c^2}\frac{\partial^2}{\partial t^2}E(z, t) = 0. \tag{2.15}$$

Structurally, this resembles the Liouville equation (1.61) for particles, with the exception that the wave equation is a second-order partial differential equation in time. It is straightforward to show that

$$E(z, t) = f_\nu\left(z \mp \frac{ct}{n}\right) \tag{2.16}$$

is a solution of the wave equation for any analytical function f_ν.

The expression $f_\nu(z - ct/n)$ describes a field that propagates from left to right without temporal shape changes. An example of such a solution is illustrated in Fig. 2.1 showing a pulsed electrical field for three different times. The solution with the plus sign, $f_\nu(z+ct/n)$ propagates from right to left. These electromagnetic fields show all the hallmarks of waves, i.e., they propagate through space with the speed of light without any form of distortions.

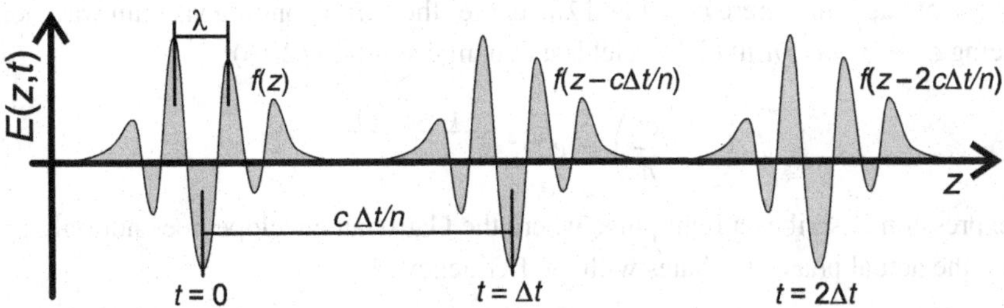

Figure 2.1 One-dimensional propagation of electromagnetic waves. An electrical field in the form of a pulse propagates undeformed according to Eq. (2.16). The electric field is presented at three different times. The spatial oscillation period identifies an average wave length λ while the propagation length $\frac{c}{n}\Delta t$ is defined by the speed of light c and the refractive index n.

2.1.2 Superposition of waves

Since the wave equation (2.15) is linear, any combination of the individual solutions

$$E(z,t) = \sum_\nu \left[B_{\nu,\rightarrow}\, f_\nu\left(z - \frac{ct}{n}\right) + B_{\nu,\leftarrow}\, f_\nu\left(z + \frac{ct}{n}\right)\right], \tag{2.17}$$

is also a solution such that the superposition principle applies. The constants $B_{\nu,\rightarrow}$ and $B_{\nu,\leftarrow}$ that appear are associated with modes propagating from right to left and left to right, respectively. The superposition principle expresses the fact that any linear combination of the individual solutions yields a new solution. In the current context, Eq. (2.17) can be regarded as a *superposition-based averaging of waves* in contrast to the particle averaging discussed in Section 1.2.2. However, $E(z,t)$ does not have a probabilistic interpretation because an electric field can have both positive and negative values. In fact, the mathematical description of electromagnetic phenomena becomes much more flexible if we use complex-valued functions f_ν as basis states.

In the following, we study Gaussian solutions

$$f_\nu(z) = e^{inqz}\, e^{-\frac{(z-z_0)^2}{2\Delta z^2}}, \tag{2.18}$$

where q identifies the wave vector while z_0 and Δz denote the center and width, respectively. In general, q determines how fast the phase of the field is changed. As a typical definition, the wave length λ identifies the distance at which the phase oscillates by 2π, as shown in Fig. 2.1. Thus, we define

$$nq \equiv \frac{2\pi}{\lambda}, \tag{2.19}$$

which simplifies to

$$q \equiv \frac{2\pi}{\lambda_0} \tag{2.20}$$

in the case of vacuum, where $n = 1$ and λ_0 denotes the corresponding vacuum wave length. Replacing $z \to z - ct/n$ in (2.18) yields a dynamic solution (2.16),

$$f_\nu\left(z - \frac{ct}{n}\right) = e^{inqz}\, e^{-\frac{(z-z_0-\frac{ct}{n})^2}{2\Delta z^2}}\, e^{-i\frac{q}{c}t}. \tag{2.21}$$

This expression describes a light pulse where the Gaussian envelope does not only propagate but the actual phase oscillates with the frequency

$$\omega_q \equiv \frac{q}{c}, \tag{2.22}$$

which establishes the dispersion relation between ω and q.

To study the spatial distribution of an electromagnetic field, one often looks at its intensity distribution

$$I(\mathbf{r}\,;t) \equiv |\mathbf{E}(\mathbf{r},t)|^2 = \mathbf{E}^\star(\mathbf{r},t)\cdot\mathbf{E}(\mathbf{r},t). \tag{2.23}$$

Since the intensity is a positive definite quantity, it can be viewed as a measure for the probability of detecting light at a given position. Similar to our analysis of the classical superposition of two statistical distributions in Section 1.2.2, we now study how $I(\mathbf{r}\,;t)$ behaves if we consider two counterpropagating pulses,

$$E(z,t) = \frac{1}{\sqrt{2}}\left[e^{inqz}\, e^{-\frac{(z+z_0-\frac{ct}{n})^2}{2\Delta z^2}} + e^{-inqz}\, e^{-\frac{(z-z_0+\frac{ct}{n})^2}{2\Delta z^2}}\right] e^{-i\omega_q t}. \tag{2.24}$$

The intensity corresponding to this superposition state is

$$\begin{aligned} I(z,t) = {} & \frac{1}{2}\left[e^{-\frac{(z+z_0-\frac{ct}{n})^2}{\Delta z^2}} + e^{-\frac{(z-z_0+\frac{ct}{n})^2}{\Delta z^2}}\right] \\ & + e^{-\frac{(z+z_0-\frac{ct}{n})^2}{2\Delta z^2}}\, e^{-\frac{(z-z_0+\frac{ct}{n})^2}{2\Delta z^2}} \cos 2nqz. \end{aligned} \tag{2.25}$$

Here, the terms in the first line correspond to the intensities of the individual pulses. This part is fully analogous to the classical averaging of particles according to Eqs. (1.71)–(1.72). However, the second line in Eq. (2.25) produces an additional new term which contributes during the interval of spatio-temporal pulse overlap.

The consequences of the new term are best discussed with the help of a graphical analysis. For this purpose, we investigate a situation analogous to the classical averaging example in Fig. 1.6. In particular, we consider a field $E(z, t = 0)$ that originally consists of two counterpropagating Gaussians centered at $z_0 = \pm 2\,a_0$. The $\cos 2nqz$-dependent contribution in Eq. (2.25) has a nontrivial form only for nonvanishing q, i.e., if the phases of the Gaussian components are genuinely different. For simplicity, we set the refractive index to $n = 1$ and assume a wave length $\lambda = 0.4\,a_0$ defining the wave vector $q = 2\pi/\lambda$ as well as the frequency $\omega_q = q/c$. Figure 2.2(a) presents the computed intensity (2.25)

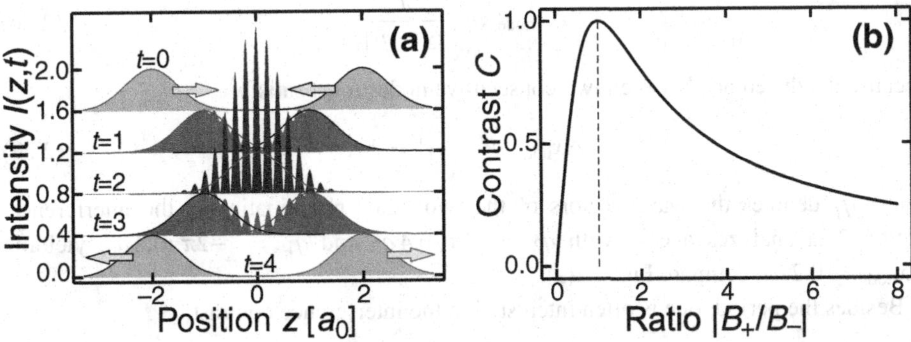

Figure 2.2 Interference of two wave packets. **(a)** The spatial intensity $I(z,t)$ is presented for the time sequence $t = 0,\ 0.5\frac{z_0}{c},\ \frac{z_0}{c},\ 1.5\frac{z_0}{c},\ 2\frac{z_0}{c}$, from top to bottom. The consecutive times are offset by 0.4. The results are computed using Eq. (2.25). **(b)** The interference contrast is plotted as function of the amplitude ratio of the individual field components, compare Eq. (2.29).

(shaded area) for five different times. The solid lines show the intensities of the individual field components at the corresponding times.

We observe at the early times that the total intensity $I(z,t)$ is just the sum of two separate Gaussians. As these Gaussians approach each other, $I(z,t)$ deviates more and more from the simple summation result. In particular, the last term of Eq. (2.25) yields a strongly oscillating *interference pattern* within the region where the pulses overlap. We see from Eq. (2.25) that the oscillation period is defined by nq which is proportional to the phase of the individual pulses. The interference pattern displays several remarkable features: (i) if we have maximum overlap ($t = z_0/c$), $I(z,t)$ peaks at a value that is four times larger than the intensity of an individual pulse. This peak value exceeds the result of the simple intensity addition by a factor of two. (ii) However, the area under $I(z,t)$ is not changed because $I(z,t)$ displays pronounced minima between the peaks, constituting the overall interference pattern. (iii) For the analyzed case, $I(z,t)$ dips all the way down to zero, producing several nodes where no light is detected.

Since the interference produces a strongly oscillating intensity, one can anticipate that different effects may happen at the node and peak positions. There is no field present at the nodes in contrast to the interference maxima, where the field intensity is enhanced. Hence, the energy transfer from the electromagnetic field to a particle is strongly position dependent. Furthermore, the interference pattern also changes the overall functional form of the $I(z,t)$ distribution, which may critically alter symmetry-dependent effects. These interference features are genuine wave properties and do not appear in the description of classical particle motion.

In general, any two overlapping waves with different phases yield an interference pattern that depends directly on the phase difference. By constructing $I(z,t)$ for two fields with wave vectors $\mathbf{q}_1$ and $\mathbf{q}_2$ in analogy to Eq. (2.25), we find that the emerging interference pattern has the period

$$\Delta x_{\text{node}} \equiv \frac{2\pi}{|q_2 - q_1|} \tag{2.26}$$

because the difference between two consecutive nodes must satisfy

$$\Delta x_{\text{node}} |q_2 - q_1| = 2\pi, \tag{2.27}$$

where q_j denotes the wave vectors of the two fields participating in the interference. Figure 2.2a analyzes a case with $q_2 = 2\pi/0.4\,a_0$ and $q_1 = -2\pi/0.4\,a_0$, yielding $\Delta x_{\text{node}} = 0.2\,a_0$, compare Fig. 2.2(a).

Besides the period, one is often interested in the interference contrast,

$$C \equiv \frac{I_{\text{max}} - I_{\text{min}}}{I_{\text{max}} + I_{\text{min}}}, \tag{2.28}$$

where I_{max} (I_{min}) denotes the peak (dip) intensity of the interference pattern. A vanishing contrast indicates that no interference pattern is present while $C = 1$ implies maximal interference with perfectly vanishing intensity at the nodes.

If we analyze $I(z, t)$ at $z = 0$, Eq. (2.25) tells us that the contrast is $C = 1$ for the central region. However, as soon as we use unequal $B_\pm$ coefficients in the superposition, the contrast is no longer perfect. Instead, we find

$$C = \frac{2r}{1 + r^2}, \tag{2.29}$$

where $r = |B_+|/|B_-|$ is the component ratio at $z = 0$. This result can be directly generalized to arbitrary z if we take r as the amplitude ratio of the interfered components at a given position. Figure 2.2(b) presents C as a function of r. We see that C peaks if the components have equal amplitudes, i.e., $r = 1$ (vertical line).

2.2 Particle aspects of electromagnetic waves

Even though classical electromagnetic fields predominantly exhibit wave properties, we can also identify some analogies to particle-like features. For example, outside the region of interference the propagation of the wave packets resembles the time evolution of a classical probability distribution. Since the study of classical particle propagation is much easier than solving either the wave or the Liouville equation, it is often meaningful to identify and utilize the pertinent particle aspects within waves to analyze their propagation properties at a simplified level.

To start this analysis, we Fourier transform both sides of Eq. (2.12), yielding the Helmholtz equation

$$\nabla^2 \mathbf{E}(\mathbf{r}, \omega) + n^2(\mathbf{r}) \frac{\omega^2}{c^2} \mathbf{E}(\mathbf{r}, \omega) = 0. \tag{2.30}$$

Here, we used the conventions,

$$\mathbf{E}(\mathbf{r}, \omega) = \int_{-\infty}^{\infty} dt\ \mathbf{E}(\mathbf{r}, t)\ e^{+i\omega t},$$

$$\mathbf{E}(\mathbf{r}, t) = \frac{1}{2\pi}\int_{-\infty}^{\infty} d\omega\ \mathbf{E}(\mathbf{r}, \omega)\ e^{-i\omega t}, \tag{2.31}$$

where the chosen sign for $e^{\pm i\omega t}$ differs from the standard mathematical format because we want to match the definition of the Fourier transformation with the temporal oscillations of the electromagnetic field, compare Eq. (2.21). Inserting Eq. (2.22) brings Eq. (2.30) into the form

$$\nabla^2 \mathbf{E}(\mathbf{r}, \omega) + n^2(\mathbf{r})\, \mathbf{q}^2\, \mathbf{E}(\mathbf{r}, \omega) = 0. \tag{2.32}$$

In the case of a constant refractive index $n(\mathbf{r}) = n$, we find the plane-wave solutions

$$\mathbf{E}_{\mathbf{q}}(\mathbf{r}) = e^{in\mathbf{q}\cdot\mathbf{r}}\mathbf{e}_{\mathbf{q}}, \tag{2.33}$$

where $\mathbf{e}_{\mathbf{q}}$ denotes the vectorial direction of the electric field.

2.2.1 Eikonal equation

In many situations, the field $\mathbf{E}(\mathbf{r}, \omega)$ has a slowly varying envelope, $\mathbf{E}_0(\mathbf{r}, \omega)$, motivating the ansatz

$$\mathbf{E}(\mathbf{r}, \omega) \equiv \mathbf{E}_0(\mathbf{r}, \omega)\ e^{i|\mathbf{q}|S(\mathbf{r})}. \tag{2.34}$$

Here, we assume that both $\mathbf{E}_0(\mathbf{r}, \omega)$ and $S(\mathbf{r})$ are only weakly position dependent. To insert the ansatz into Eq. (2.32), we use

$$\nabla \mathbf{E}(\mathbf{r}, \omega) = [\nabla \mathbf{E}_0(\mathbf{r}, \omega)]\ e^{i|\mathbf{q}|S(\mathbf{r})} + i|\mathbf{q}|\,[\nabla S(\mathbf{r})]\,\mathbf{E}(\mathbf{r}, \omega),$$

$$\nabla^2 \mathbf{E}(\mathbf{r}, \omega) = \left[\nabla^2 \mathbf{E}_0(\mathbf{r}, \omega)\right]\ e^{i|\mathbf{q}|S(\mathbf{r})} + 2i|\mathbf{q}|\,[\nabla \mathbf{E}_0(\mathbf{r}, \omega)\cdot \nabla S(\mathbf{r})]\,e^{i|\mathbf{q}|S(\mathbf{r})}$$
$$+\ i|\mathbf{q}|\left[\nabla^2 S(\mathbf{r})\right]\mathbf{E}(\mathbf{r}, \omega) - \mathbf{q}^2\,[\nabla S(\mathbf{r})]^2\,\mathbf{E}(\mathbf{r}, \omega), \tag{2.35}$$

and obtain

$$\left[\nabla^2 \mathbf{E}_0(\mathbf{r}, \omega)\right]\ e^{i|\mathbf{q}|S(\mathbf{r})} + i|\mathbf{q}|\left[2\nabla \mathbf{E}_0(\mathbf{r}, \omega)\cdot \nabla S(\mathbf{r}) + \mathbf{E}_0(\mathbf{r}, \omega)\cdot \nabla^2 S(\mathbf{r})\right]$$
$$+\,\mathbf{q}^2\left[-\,[\nabla S(\mathbf{r})]^2 + n(\mathbf{r})^2\right]\mathbf{E}(\mathbf{r}, \omega) = 0. \tag{2.36}$$

This is a nonlinear differential equation that cannot be solved analytically for a general $n(\mathbf{r})$. However, the equation significantly simplifies in the *geometrical optics* which is the limit of vanishing wave length, i.e., $\lambda_0 = 2\pi/q \to 0$. Since a vanishing wave length implies that q becomes very large, the last term proportional to $\mathbf{q}^2$ clearly dominates in Eq. (2.36). Ignoring all the smaller contributions, we see that Eq. (2.36) implies that $\left[-\,[\nabla S(\mathbf{r})]^2 + n(\mathbf{r})^2\right]$ vanishes. This yields the so-called *eikonal equation*

$$[\nabla S(\mathbf{r})]^2 - n(\mathbf{r})^2 = 0. \tag{2.37}$$

Taking the square root on both sides leads to the first-order differential equation,

$$|\nabla S(\mathbf{r})| = \pm n(\mathbf{r}). \tag{2.38}$$

The solution of this equation defines $S(\mathbf{r})$ telling us how the phase fronts of the electromagnetic wave change as a function of position.

2.2.2 Classical trajectories within wave equation

The eikonal equation (2.37) has an interesting connection to the description of particle motion, as we can see by making the following formal identifications,

$$\mathbf{p}(\mathbf{r}) \equiv \sqrt{m}\,\nabla S(\mathbf{r}), \quad 2\,[E - U(\mathbf{r})] \equiv n^2(\mathbf{r}). \tag{2.39}$$

At this stage, the constants m, E, and $U(\mathbf{r})$ have no physical meaning but we will show later that they can be associated with the particle mass, energy, and potential, respectively. The vector field $\mathbf{p}(\mathbf{r})$ is curl free, i.e., $\nabla \times \mathbf{p}(\mathbf{r}) = 0$, because the *curl* of a gradient field always vanishes, $\nabla \times [\nabla S(\mathbf{r})] = 0$.

Inserting the substitutions (2.39) into the eikonal equation (2.37), we obtain

$$\frac{\mathbf{p}(\mathbf{r})^2}{m} - 2\,[E - U(\mathbf{r})] = 0 \quad \Leftrightarrow \quad \frac{\mathbf{p}(\mathbf{r})^2}{2m} + U(\mathbf{r}) = E. \tag{2.40}$$

This result merely states the fact that the sum of kinetic and potential energy yield the total energy of a classical particle. Thus, we may solve $\mathbf{p}(\mathbf{r})$ from the Hamilton equations (1.10) that produce a classical trajectory $\left[\mathbf{r}(t), \mathbf{p}(t)\right]$ also defining the function $\mathbf{p}(\mathbf{r})$. Due to the conservation of energy, Eqs. (1.31)–(1.32), the classical trajectory $\mathbf{p}(\mathbf{r})$ always satisfies Eq. (2.40). The actual wave phase can then be obtained from

$$\frac{\partial}{\partial r_j} S(\mathbf{r}) = \frac{p_j(r_1, r_2, r_3)}{\sqrt{m}}, \tag{2.41}$$

with the generic constraint $\nabla \times \mathbf{p}(\mathbf{r}) = 0$ for any classical field, see Exercise 2.5.

Even though we do not want to elaborate on this point, we would like to mention here that the eikonal equation (2.37) can be taken as the starting point of the so-called Hamilton–Jacobi theory. In this approach, one computes $S(\mathbf{r})$ with the goal to describe the propagation of classical particles. Hence, the phase of the electromagnetic field can be related to particle-like properties and the electromagnetic field in the geometrical-optics limit propagates along the classical trajectories, which in this context are known as the *light rays*.

Even though ray optics and ray tracing are important concepts in classical optics and optical design, there are several important effects that cannot be understood at this level. In particular, the classical path does not allow for interference effects. Moreover, the full solution of the wave equation also describes diffraction of the field where the electromagnetic wave scatters into multiple directions if it encounters an obstacle. Furthermore, as we will see in Chapter 4, light can also penetrate into or even through a thin object as an exponentially decaying, evanescent wave. Obviously, a real-valued classical $S(\mathbf{r})$ defining the phase in $e^{i|\mathbf{q}|S(\mathbf{r})}$ cannot describe such an exponentially decaying field.

The eikonal approximation is, however, very useful to determine the path of the electromagnetic field if diffraction and evanescent effects can be ruled out. As illustration, we analyze a situation where light is propagating in two dimensions through an interface where the refractive index changes according to

$$n(x) = \begin{cases} n_1, & x < 0 \\ \frac{n_1+n_2}{2}, & x = 0 \\ n_2, & x > 0 \end{cases} \quad \Leftrightarrow \quad V(x) = \begin{cases} V_1, & x < 0 \\ \frac{V_1+V_2}{2}, & x = 0 \\ V_2, & x > 0 \end{cases}. \tag{2.42}$$

For the sake of completeness, we have also identified the potential $V_j \equiv E - n(x_j)^2/2$, based on Eq. (2.39). In the case of Eq. (2.42), the refractive index and the potential do not depend on the y coordinate. Hence, the path in this direction simply follows from the Hamilton equations (1.10), $\dot{p}_y = 0$ and $\dot{y} = p_y/m$, with the solution

$$p_y(t) = p_y, \quad y(t) = y_0 + \frac{p_y\, t}{m} \tag{2.43}$$

in analogy to Eq. (1.34).

The dynamics is more difficult in the x direction because $\frac{\partial}{\partial x}V(x)$ diverges at the origin due to the discontinuity in $V(x)$. However, we can easily solve this problem by using the energy conservation, Eq. (1.31), to formulate the strict connection

$$\begin{aligned} p_{x,1}^2 + p_y^2 &= 2m[E - V_1] = m n_1^2, \quad x < 0 \\ p_{x,2}^2 + p_y^2 &= 2m[E - V_2] = m n_2^2, \quad x > 0. \end{aligned} \tag{2.44}$$

It turns out to be beneficial to parametrize the problem via a propagation angle,

$$\begin{aligned} p_{x,1} &= \sqrt{m}\, n_1 \cos\theta_1, \quad p_y = \sqrt{m}\, n_1 \sin\theta_1 \\ p_{x,2} &= \sqrt{m}\, n_2 \cos\theta_2, \quad p_y = \sqrt{m}\, n_2 \sin\theta_2, \end{aligned} \tag{2.45}$$

as identified in Fig. 2.3.

Since the momenta are piece-wise constant, Eq. (2.41) yields a linear space dependence of the slowly varying phase,

$$S(x, y) = \begin{cases} S_0 + \dfrac{p_{x,1}x + p_y y}{\sqrt{m}} = S_0 + n_1 (x\cos\theta_1 + y \sin\theta_1), & x < 0 \\ S_0 + \dfrac{p_{x,2}x + p_y y}{\sqrt{m}} = S_0 + n_2 (x\cos\theta_2 + y \sin\theta_2), & x > 0 \end{cases}. \tag{2.46}$$

Substituting this result into (2.34) yields

$$e^{i|\mathbf{q}|S(\mathbf{r})} = \begin{cases} e^{i|\mathbf{q}|S_0 + n_1|\mathbf{q}|(x\cos\theta_1 + y\sin\theta_1)} \equiv e^{i\phi_0 + i\mathbf{q}_1\cdot\mathbf{r}}, & x < 0 \\ e^{i|\mathbf{q}|S_0 + n_2|\mathbf{q}|(x\cos\theta_2 + y\sin\theta_2)} \equiv e^{i\phi_0 + i\mathbf{q}_2\cdot\mathbf{r}}, & x > 0 \end{cases}. \tag{2.47}$$

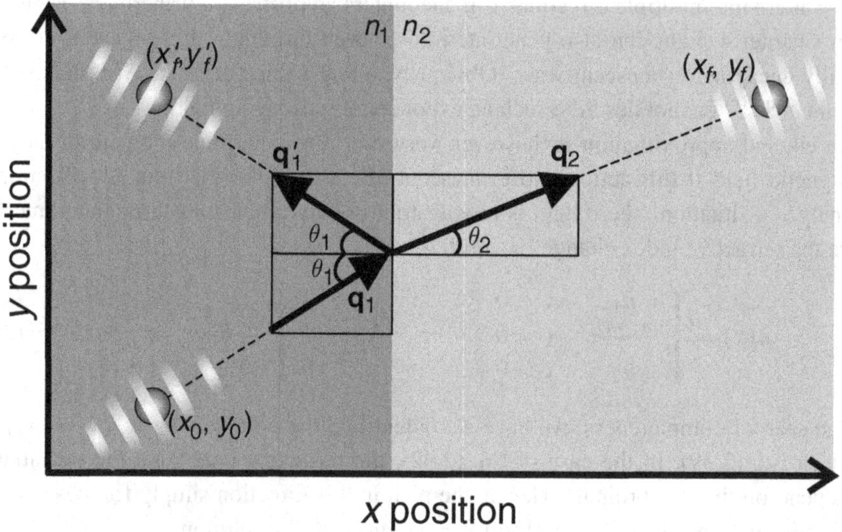

Figure 2.3 Propagation of a light ray through an $n_1 - n_2$ index step located at $z = 0$. The particle-like trajectories (dashed lines) are aligned with the wave vector (arrows), according to Eq. (2.53). Furthermore, the beginning and ending points of the propagation are indicated. The wave vectors are perpendicular to the wave fronts (fading areas) and their propagation directions satisfy Snell's law (2.49), i.e., the in-plane components of the wave vectors are conserved.

Here, ϕ_0 is the arbitrary overall phase and the wave vector is given as

$$\mathbf{q}_j \equiv n_j |\mathbf{q}| \left(\mathbf{e}_x \cos\theta_j + \mathbf{e}_y \sin\theta_j \right), \tag{2.48}$$

inside the regions $x < 0\,(j = 1)$ and $x > 0\,(j = 2)$.

Equation (2.47) shows that the eikonal equation (2.37) yields the correct plane-wave solution within each individual layer. In addition, since p_y has the same value on either side of the interface, Eq. (2.45) directly implies Snell's law,

$$n_1 \sin\theta_1 = n_2 \sin\theta_2, \tag{2.49}$$

which defines the propagation direction of waves when they are refracted.

The analysis can be extended by choosing a solution branch that is reflected from the $x = 0$ boundary. A purely classical particle would naturally follow only one path but the actual wave aspects allow for this diffractive feature. The path of the reflected part can be associated with the wave vector

$$\mathbf{q}'_1 \equiv n_1 |\mathbf{q}| \left(-\mathbf{e}_x \cos\theta_1 + \mathbf{e}_y \sin\theta_1 \right), \tag{2.50}$$

compare Fig. 2.3. If a light ray is split into multiple paths, one also needs to define the probability amplitude for each path, i.e., one must determine the transmission and reflection coefficients. This analysis is beyond the ray analysis presented here and requires a closer inspection of genuine wave properties, see Chapter 3.

Even though the determined $\mathbf{r}(t)$ describes a purely fictitious "path," it is instructive to follow it in different regions. In particular, the solutions (2.45) allow us to evaluate Hamilton's equation, $\dot{x} = p_x/m$. If we assume that the particle starts out on the LHS of the $x = 0$ step, we find the following solution branches

$$x(t) = \begin{cases} \frac{p_{x,1}t}{m}(t - t_0) = q_{x,1}\frac{t-t_0}{\sqrt{m}\,|\mathbf{q}|}, & t < t_0 \\ \frac{p_{x,2}t}{m}(t - t_0) = q_{x,2}\frac{t-t_0}{\sqrt{m}\,|\mathbf{q}|}, & t > t_0 \text{ transmitted part} \\ -\frac{p_{x,1}t}{m}(t - t_0) = -q_{x,1}\frac{t-t_0}{\sqrt{m}\,|\mathbf{q}|}, & t > t_0 \text{ reflected part} \end{cases}, \tag{2.51}$$

where we used (2.45) and expressed the momentum via Eqs. (2.48)–(2.50). The identified $t_0 \equiv -mx_0/p_{x,1}$ defines the time when the "particle" arrives at the refractive-index boundary after it started from its initial position $x_0 < 0$. We can rewrite the solution (2.43) as

$$\begin{aligned} y(t) &= y_0 + \frac{p_y\, t}{m} = y_0 + \frac{p_y\, t_0}{m} + \frac{p_y}{m}(t - t_0) \\ &= y_0 + \frac{p_y t_0}{m} + q_{y,j}\frac{t - t_0}{\sqrt{m}|\mathbf{q}|}, \end{aligned} \tag{2.52}$$

and combine it with Eq. (2.51) to express the classical trajectory as

$$\mathbf{r}(t) = \mathbf{r}_0 + \begin{cases} \mathbf{q}_1\,\tau, & t < t_0 \\ \mathbf{q}_2\,\tau, & t > t_0 \text{ transmitted part} \\ \mathbf{q}_1'\,\tau, & t > t_0 \text{ reflected part} \end{cases}. \tag{2.53}$$

Here, we defined $\mathbf{r}_0 \equiv \left(0, y_0 + \frac{p_y t_0}{m}\right)$ and $\tau \equiv \frac{t-t_0}{\sqrt{m}|\mathbf{q}|}$. We now see that the fictitious path leads into the same directions as the wave vectors in the phase $e^{i|\mathbf{q}|S(\mathbf{r})}$. Thus, it is convenient to associate certain paths, i.e., rays with the propagation of the light. Figure 2.3 presents examples of such paths together with contours of constant phase $e^{i|\mathbf{q}|S(\mathbf{r})}$ (fading areas).

We already know that the phase structure of a wave propagates unchanged in the direction perpendicular to the phase fronts. Thus, the classical paths, i.e., rays do capture this aspect of the electromagnetic fields. However, as mentioned before, the particle description cannot account for diffraction, evanescent propagation, or interference aspects of the full wave propagation. In fact, even the concept of reflection needs some generalization in order to produce the correct path for the reflected light. These aspects will be addressed more thoroughly in Chapter 4 where we construct the full numerical solution to the Helmholtz equation (2.30). Nonetheless, the presented analysis allows us to identify particle-like properties even within waves if we follow the approximative propagation of the wave fronts. This interesting result will help us to define how the concept of a classical particle can be generalized.

2.3 Generalized wave and Helmholtz equations

Whereas the wave dynamics naturally originates from the wave equation (2.15), the association of wave-front propagation with light rays and temporally moving particles is made through the static Helmholtz equation. Since there can be more than one partial differential equation yielding the same Helmholtz equation, it is interesting to study how particle aspects emerge from a more general class of "wave" equations.

For this purpose, we consider the generalized equation

$$\frac{\partial^2}{\partial z^2}\,\psi(z,t) + \left(\frac{i}{\beta}\frac{\partial}{\partial t}\right)^M \psi(z,t) = 0, \tag{2.54}$$

which also yields the Helmholtz equation (2.30). To check this, we use the Fourier transformation similar to Eq. (2.31) and obtain

$$\frac{\partial^2}{\partial z^2}\,\psi(z,\omega) + \left(\frac{\omega}{\beta}\right)^M \psi(z,\omega) = 0. \tag{2.55}$$

The physical meaning of the constant β, the power $M = 1, 2, \cdots$, and the wave function $\psi(z,t)$ depends on the particular physical case under consideration. For example, Eq. (2.54) reproduces the wave equation (2.15) for the choice $\beta = c/n$ and $M = 2$. In this case, $\psi(z,t)$ can be identified with the scalar electric field. For any other combination of M and β, Eq. (2.54) describes generalized waves where $\psi(z,t)$ is the respective wave function.

2.3.1 Formal eigenvalue problem

Before we attempt to solve Eq. (2.55), we identify a general dispersion relation

$$q^2 \equiv \left(\frac{\omega}{\beta}\right)^M, \tag{2.56}$$

that casts Eq. (2.55) into the form of an eigenvalue problem

$$\frac{\partial^2}{\partial z^2}\,\phi_q(z) = -q^2\phi_q(z). \tag{2.57}$$

We recognize that q is real valued only if $\left(\frac{\omega}{\beta}\right)^M$ is positive definite.

Generalizing to the three-dimensional case, we have

$$\frac{1}{n^2(\mathbf{r})}\nabla^2\,\phi_{\mathbf{q}}(\mathbf{r}) = -\mathbf{q}^2\phi_{\mathbf{q}}(\mathbf{r}). \tag{2.58}$$

Here, we also allow $\mathbf{q}$ to depend on position via the substitution $\mathbf{q}^2 \to \mathbf{q}^2\,n^2(\mathbf{r})$ where, in analogy to Eq. (2.30), $n^2(\mathbf{r})$ is a positive-definite, nonvanishing function. The eigenvalue problem (2.58) has the form of an ordinary linear differential equation. In vacuum $n^2(\mathbf{r}) = 1$, we find the simple plane-wave solution

$$\phi_q(z) = \frac{1}{\sqrt{2\pi}} e^{iqz}, \qquad \phi_{\mathbf{q}}(\mathbf{r}) = \left(\frac{1}{2\pi}\right)^{\frac{3}{2}} e^{i\mathbf{q}\cdot\mathbf{r}} \tag{2.59}$$

for one- and three-dimensional systems, respectively.

Mathematically, Eqs. (2.57)–(2.58) constitute eigenvalue problems for a Hermitian operator, $\frac{1}{n^2(\mathbf{r})}\nabla^2$ or $\partial^2/\partial^2 z$. Consequently, the different eigenfunctions are orthogonal and form a complete set of functions. For the plane-wave solutions, the orthogonality relation is

$$\int_{-\infty}^{\infty} dz\, \phi_q^\star(z)\, \phi_{q'}(z) = \frac{1}{2\pi}\int_{-\infty}^{\infty} dz\, e^{i(q'-q)z} = \delta(q'-q),$$
$$\int d^3r\, \phi_{\mathbf{q}}^\star(\mathbf{r})\, \phi_{\mathbf{q}'}(\mathbf{r}) = \frac{1}{(2\pi)^3}\int d^3r\, e^{i(\mathbf{q}'-\mathbf{q})\cdot\mathbf{r}} = \delta(\mathbf{q}'-\mathbf{q}), \tag{2.60}$$

which follows directly from the definition of the delta function, see Eq. (1.50) and Exercise 2.7. The completeness relation is then given by

$$\int_{-\infty}^{\infty} dq\, \phi_q^\star(z)\, \phi_q(z') = \delta(z'-z), \quad \int d^3q\, \phi_{\mathbf{q}}^\star(\mathbf{r})\, \phi_{\mathbf{q}}(\mathbf{r}') = \delta(\mathbf{r}'-\mathbf{r}). \tag{2.61}$$

For more general eigenvalue problems (2.58), we might encounter situations where the solutions must satisfy boundary conditions. In such cases, **q** is not continuous but discrete with the consequence that the normalization involves the Kronecker delta. More precisely, $\delta(\mathbf{q}'-\mathbf{q})$ is replaced by $\delta_{\mathbf{q}',\mathbf{q}}$ in Eq. (2.60) and the integral, $\int d^3\mathbf{q}$, is replaced by a sum, $\sum_{\mathbf{q}}$. Further properties are studied in Chapter 3.

2.3.2 Temporal properties of generalized waves

For simplicity, we now discuss continuous plane-wave solutions in one dimension. Due to their completeness and orthogonality, any function can be expressed as a linear superposition of such plane waves,

$$\psi(z,t) = \int_{-\infty}^{\infty} dq\, c_q(t)\, \phi_q(z) = \frac{1}{\sqrt{2\pi}}\int_{-\infty}^{\infty} dq\, c_q(t)\, e^{iqz}, \tag{2.62}$$

$$c_q(t) = \int_{-\infty}^{\infty} dz\, \psi(z,t)\phi_q^\star(z) = \frac{1}{\sqrt{2\pi}}\int_{-\infty}^{\infty} dz\, \psi(z,t)\, e^{-iqz}. \tag{2.63}$$

The constants $c_q(t)$ determine the weights at each given time t. We also recognize that the integral (2.62) generalizes the wave-averaging (2.17) to superpositions consisting of infinitely many wave components. Due to this specific decomposition, we refer to Eq. (2.62) as the *mode expansion* in terms of the eigenstates $\phi_q(z)$. For plane waves, the mapping relations define the ordinary one-dimensional Fourier transformation.

In particular, the property (2.63) can be derived on the basis of the orthogonality and completeness relations (2.60)–(2.61), see Exercise 2.7.

We now consider a case where $\psi(z,t)$ vanishes at $z \to \pm\infty$. We also assume that $\psi(z,t)$ is normalizable in the usual quadratic form,

$$\int_{-\infty}^{\infty} dz\, \psi^\star(z,t)\, \psi(z,t) = \int_{-\infty}^{\infty} dz\, |\psi^\star(z,t)|^2 = 1. \tag{2.64}$$

Clearly,

$$\rho(z,t) \equiv |\psi(z,t)|^2 \tag{2.65}$$

corresponds to the intensity distribution of the electromagnetic field (2.23), defining the field strength at a given position. While $\rho(z,t)$ is the probability distribution,

$$P(z_0, \Delta z\,;t) \equiv \int_{z_0-\frac{\Delta z}{2}}^{z_0+\frac{\Delta z}{2}} dz\, |\psi(z,t)|^2 \tag{2.66}$$

defines the probability of finding waves within the interval $|z - z_0| \leq \Delta z/2$.

Next, we want to take a look at the coefficient $c_q(t)$, Eq. (2.63). For the calculation, we need to evaluate two integrals. The first one is

$$\begin{aligned} I_{\text{1st}} &= \int_{-\infty}^{\infty} dz\, \phi_q^\star(z)\, \frac{\partial^2}{\partial z^2}\psi(z,t) = \int_{-\infty}^{\infty} dz\, \psi(z,t)\, \frac{\partial^2}{\partial z^2}\, \phi_q^\star(z) \\ &= -q^2 \int_{-\infty}^{\infty} dz\, \psi(z,t)\;\; \phi_q^\star(z) = -q^2 c_q(t), \end{aligned} \tag{2.67}$$

where we twice integrate by parts assuming vanishing $\psi(z,t)$ at the borders and use the relation (2.57) to identify the eigenvalue. In the last step, we apply the projection property (2.63). The second integral that appears can be evaluated as

$$\begin{aligned} I_{\text{2nd}} &= \int_{-\infty}^{\infty} dz\, \phi_q^\star(z) \left(\frac{i}{\beta}\frac{\partial}{\partial t}\right)^M \psi(z,t) = \left(\frac{i}{\beta}\frac{\partial}{\partial t}\right)^M \int_{-\infty}^{\infty} dz\, \phi_q^\star(z)\, \psi(z,t) \\ &= \left(\frac{i}{\beta}\frac{\partial}{\partial t}\right)^M c_q(t), \end{aligned} \tag{2.68}$$

where we interchanged the order of integration and differentiation. Thus, the generalized wave equation (2.54) yields

$$-q^2 c_q(t) + \left(\frac{i}{\beta}\frac{\partial}{\partial t}\right)^M c_q(t) = 0 \tag{2.69}$$

for the individual plane-wave components.

The exponential form

$$c_q(t) = c_q(0)\, e^{-i\omega_q t} \tag{2.70}$$

constitutes a proper solution of Eq. (2.69) whenever the dispersion relation

$$\left(\frac{\omega_q}{\beta}\right)^M = q^2 \quad \Rightarrow \quad \omega_q = \beta\, |q|^{\frac{2}{M}} \tag{2.71}$$

is satisfied. For the wave equation with $M = 2$ and $\beta = c$, we recover the linear dispersion relation (2.22). For any other M, the connection between wave vector and frequency is nonlinear.

Since the q integral within the mode expansion (2.62) contains both positive and negative q, we can restrict the analysis to positive-valued ω_q without any loss of generality. Thus, we may insert the result (2.70) into (2.62), producing

$$\psi(z,t) = \frac{1}{\sqrt{2\pi}} \int_{-\infty}^{\infty} dq\; c_q(0)\, e^{i(qz-\omega_q t)}. \tag{2.72}$$

The initial values for the mode coefficients follow from Eq. (2.63),

$$c_q(0) = \frac{1}{\sqrt{2\pi}} \int_{-\infty}^{\infty} dz\; \psi(z,0)\, e^{-iqz}, \tag{2.73}$$

such that the full evolution can always be obtained as long as we know the wave function $\psi(z,t)$ at time $t = 0$. It is straightforward to show that the norm of $\psi(z,t)$ remains constant, i.e., $\int dz\; |\psi(z,t)|^2 = 1$ for all times, simply by applying the completeness relation (2.61) together with the result (2.72).

For the case $M = 2$ and $\beta = c$, Eq. (2.71) implies the dispersion $\omega_q = c|q|$. Inserting this into Eq. (2.72) yields

$$\begin{aligned}
\psi_{M=2}(z,t) &= \frac{1}{\sqrt{2\pi}} \int_{-\infty}^{\infty} dq\; c_q(0)\, e^{i(qz-|q|ct)} \\
&= \int_{-\infty}^{0} dq\; \frac{c_q(0)}{\sqrt{2\pi}}\, e^{i(qz+qct)} + \int_{0}^{\infty} dq\; \frac{c_q(0)}{\sqrt{2\pi}}\, e^{i(qz-qct)} \\
&= \int_{0}^{\infty} dq\; \frac{c_{-q}(0)}{\sqrt{2\pi}}\, e^{-iq(z+ct)} + \int_{0}^{\infty} dq\; \frac{c_q(0)}{\sqrt{2\pi}}\, e^{iq(z-ct)},
\end{aligned} \tag{2.74}$$

where we separated the integral into the parts with positive and negative q. In the last step, we substituted $q \to -q$ for the integration variable. We now recognize that the integrations produce the Fourier transformations of a complex-valued function $f(z)$ and its complex conjugate $f^{\star}(z)$. In particular, the $z \pm ct$ arguments yield propagating and counterpropagating solutions,

$$\psi_{M=2}(z,t) = f^{\star}_{\leftarrow}(z+ct) + f_{\rightarrow}(z-ct), \tag{2.75}$$

in full analogy to the result (2.17). In other words, according to Eq. (2.54), wave fronts propagate undistorted with the speed $\beta = c$ for $M = 2$.

To see whether similarly simple solutions can be found for other values of M, we consider the case $M = 1$, yielding

$$\psi_{M=1}(z,t) = \frac{1}{\sqrt{2\pi}} \int_{-\infty}^{\infty} dq\; c_q(0)\; e^{i(qz-\beta q^2 t)}, \tag{2.76}$$

based on Eqs. (2.56) and (2.72). However, due to the quadratic term that appears, this expression cannot be simplified further for a generic $c_q(0)$.

2.3.3 Generalized waves and particles

Even though $\psi_M(z,t)$ cannot be solved analytically for an arbitrary M and $\psi_M(z,0)$, we can still analyze the overall behavior by assuming Gaussian waves for the cases $M = 1$ and $M = 2$. In particular, we consider here

$$\psi(z,0) = \left(\frac{1}{2\pi\;\Delta z^2}\right)^{\frac{1}{4}} e^{-\frac{z^2}{4\Delta z^2}}\; e^{iq_0 z}, \tag{2.77}$$

to deduce the principal features of Eq. (2.76). The wave function (2.77) has an envelope width Δz whereas q_0 determines how fast the phase changes in space. This Gaussian is normalized according to Eq. (2.64) and it defines the $c_q(0)$ coefficients via the Fourier transformation (2.73),

$$\begin{aligned} c_q(0) &= \left(\frac{1}{2\pi}\right)^{\frac{1}{4}} \frac{1}{\sqrt{2\pi\,\Delta z}} \int_{-\infty}^{\infty} dz\; e^{-\frac{z^2}{4\Delta z^2}}\; e^{iq_0 z}\; e^{-iqz} \\ &= \left(\frac{1}{2\pi}\right)^{\frac{1}{4}} e^{-\Delta z^2 (q-q_0)^2} \frac{1}{\sqrt{2\pi\,\Delta z}} \int_{-\infty}^{\infty} dz\; e^{-\frac{\left[z - i\Delta z^2 (q-q_0)\right]^2}{4\Delta z^2}}. \end{aligned} \tag{2.78}$$

The emerging integral can be evaluated with the help of the general relation

$$\int_{-\infty}^{\infty} dx\; e^{-\frac{(x-\alpha)^2}{\epsilon^2}} = \sqrt{\pi\,\epsilon^2}, \quad \forall\; \alpha \in \mathbb{C} \text{ and } \mathrm{Re}\left[\epsilon^2\right] > 0. \tag{2.79}$$

Thus, we eventually find

$$c_q(0) = \left(\frac{2\Delta z^2}{\pi}\right)^{\frac{1}{4}} e^{-\Delta z^2 (q-q_0)^2}, \tag{2.80}$$

defining the decomposition of the Gaussian (2.77) into plane waves.

Since we know $c_q(0)$, we can now use Eq. (2.76) to express the dynamics,

$$\begin{aligned} \psi_{M=1}(z,t) &= \left(\frac{2\Delta z^2}{\pi}\right)^{\frac{1}{4}} \frac{1}{\sqrt{2\pi}} \int_{-\infty}^{\infty} dq\; e^{-\Delta z^2 (q-q_0)^2}\; e^{i(qz-\beta q^2 t)} \\ &= \left(\frac{2\Delta z^2}{\pi}\right)^{\frac{1}{4}} \frac{e^{iq_0 z - i\beta q_0^2 t}}{\sqrt{2\pi}} \int_{-\infty}^{\infty} dq\; e^{-(\Delta z^2 + i\beta t) q^2 + i(z - 2\beta q_0 t) q}, \end{aligned} \tag{2.81}$$

where we changed the integration variable $q \to q + q_0$ and reorganized the terms within the exponent. The remaining Gaussian integral can be evaluated by completing the square and applying the relation (2.79). This calculation produces the freely propagating wave function,

$$\psi_{M=1}(z,t) = \left[\frac{\Delta z^2}{2\pi(\Delta z^2 + i\beta t)^2}\right]^{\frac{1}{4}} e^{-\frac{(z-2\beta q_0 t)^2}{4(\Delta z^2+i\beta t)}} e^{iq_0 z - i\beta q_0^2 t}. \tag{2.82}$$

From this result, we obtain the time-dependent probability distribution

$$P_{M=1}(z,t) = |\psi_{M=1}(z,t)|^2 = \left[\frac{1}{\pi\Delta z(t)^2}\right]^{\frac{1}{2}} e^{-\frac{(z-2\beta q_0 t)^2}{2\Delta z(t)^2}}, \tag{2.83}$$

where we identified the temporal width,

$$\Delta z(t)^2 \equiv \Delta z^2 + \frac{\beta^2 t^2}{\Delta z^2}, \tag{2.84}$$

for the Gaussian pulse. The center of $P_{M=1}(z,t)$ moves with the velocity

$$v_0 \equiv 2\beta q_0. \tag{2.85}$$

Interestingly, this velocity depends on the initial condition, i.e., q_0, whereas the ordinary wave equation yields the same propagation velocity $c = \beta$ for any initial state. Moreover, according to the ordinary wave equation the wave fronts move without distorting their shape, see Eq. (2.75). In contrast to this, Eq. (2.84) shows that the $M = 1$ case yields a Gaussian width that increases with time. Additionally, the wave function experiences phase distortions according to (2.82).

Just as the generalized wave equation with $M = 2$ and $\beta = c$ describes the physics of propagating electromagnetic fields, also the case $M = 1$ has intriguing connections with actual physical systems. To clearly see these connections, we compare $P_{M=1}(z,t)$ with the marginal distribution $\rho_p(x,t)$ of a classical particle, see Section 1.2.2. Based on Eqs. (2.83) and (1.76), we can make the following associations

$$v_0 \equiv 2\beta q_0, \qquad \Delta v(t) = \Delta v \equiv \frac{\beta}{\Delta z},$$
$$\Delta z(t) \equiv \Delta x(t) = \sqrt{\Delta z^2 + \Delta v^2 t^2}. \tag{2.86}$$

In other words, the generalized wave equation with $M = 1$ produces solutions whose center propagates like a particle whose velocity is determined by the initial conditions. Furthermore, the resulting wave spreads like an ensemble of classical particles with a statistical momentum distribution.

This information will become extremely valuable when we quantize the particles in the next chapter. As the major new feature of the wave aspects, the association (2.86) predicts that the fluctuations of position and momentum become connected. In particular, we find that the particle-like Gaussian waves must satisfy a relation

$$\Delta z\,\Delta v = \beta, \qquad \Delta z(t)\,\Delta v(t) \geq \beta, \tag{2.87}$$

such that the uncertainties of position Δz and momentum, $\Delta p \equiv m\Delta v$, cannot simultaneously become arbitrarily small. This is a genuine wave property not observed for classical particles. Moreover, $\psi(z,t)$ can display other wave aspects such as interference, diffraction, and evanescent propagation, which eventually generalize the particle properties into the quantum world.

Before we discuss the full extent of the quantum theory of particles, we briefly summarize the similarities and differences between electromagnetic and particle waves. For this purpose, we analyze the wave function $\psi_M(z,t)$ for $M=2$ and $M=1$. With the help of (2.86), the particle wave (2.82) can be expressed via

$$\psi_{M=1}(z,t) = \left[\frac{1}{2\pi(\Delta z + i\Delta v_0 t)^2}\right]^{\frac{1}{4}} e^{-\frac{(z-v_0 t)^2}{4\Delta z(\Delta z + i\Delta v t)}}\, e^{iq_0\left(z-\frac{v_0}{2}t\right)}. \tag{2.88}$$

If we choose the same Gaussian initial condition (2.77) for the electromagnetic waves, the wave-equation solution (2.75) yields

$$\psi_{M=2}(z,t) = \left[\frac{1}{2\pi\Delta z^2}\right]^{\frac{1}{4}} e^{-\frac{(z-ct)^2}{4\Delta z^2}}\, e^{iq_0(z-ct)}. \tag{2.89}$$

We see that the center of the Gaussians moves with constant velocity if we choose $v_0 = c$. Besides this, the propagation of the electromagnetic field is defined by the actual values for c, q_0, and Δz whereas the particle wave has an additional dependence on Δv. However this dependence is redundant due to the conditions (2.86) that also determine

$$\Delta v = \frac{v_0}{2q_0\Delta z} \tag{2.90}$$

for a known combination of $v_0 = c$, q_0, and Δz.

Figure 2.4 presents the real part of the wave function $\psi_M(z,t)$ as a function of position z. For this illustration, we used $\Delta z = 0.4\,a_0$, $q_0 = \frac{2\pi}{a_0}0.8$, and $v_0 = c$, where a_0 denotes the unit of the length scale. The presented electromagnetic (solid line) and particle (shaded area) waves are shown at times (a) $t=0$, (b) $t = 2\,a_0/c$, and (c) $t = 4\,a_0/c$ when the center of the Gaussian is at $z=0$, $z = 2\,a_0$, and $z = 4\,a_0$, respectively. We observe that the electromagnetic wave $\psi_{M=2}(z,t)$ propagates without shape changes while the propagating particle wave $\psi_{M=1}(z,t)$ is significantly distorted. Only the envelope $|\psi_{M=2}(z,t)|$ (dashed line) allows us to recognize the Gaussian shape which propagates and broadens as predicted by Eq. (2.84). Besides the broadening, we observe that the pulse phase is distorted in a way that the oscillations are faster before than after the center. These distortions of the particle-like phase fronts are known as the phenomenon of chirp.

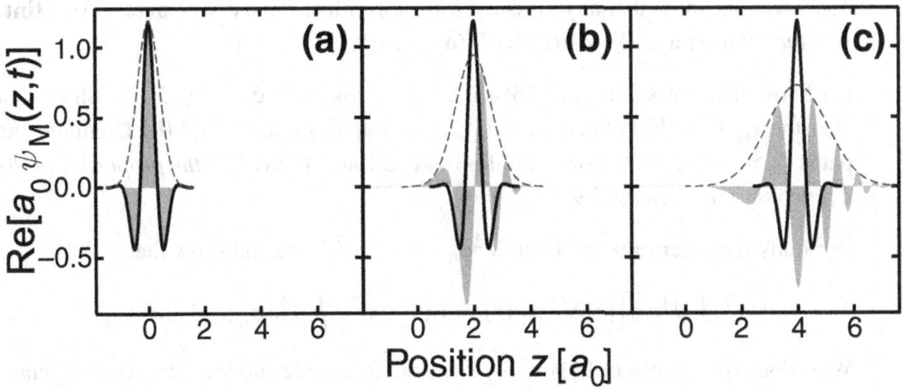

Figure 2.4 Propagation of electromagnetic vs. particle waves. The wave function is shown after the propagation time **(a)** $t = 0$, **(b)** $t = 2a_0/v_0$, and **(c)** $t = 4a_0/v_0$. The time-evolution of the electromagnetic (solid line) and particle (shaded area) waves is compared. The dashed line indicates the $|\psi_{M=1}(z,t)|$ envelope of the particle waves. The results are computed by solving Eqs. (2.88)–(2.89) with $\Delta z = 0.4\,a_0$, $q_0 = \frac{2\pi}{a_0}0.8$, and $v_0 = c$.

Exercises

2.1 **(a)** Show that $\int_{-\infty}^{\infty} dx\, x^{2j+1} e^{-ax^2} = 0$ and $\int_{-\infty}^{\infty} dx\, x^{2j} e^{-a^2} = \frac{\Gamma\left(j+\frac{1}{2}\right)}{q^{j+\frac{1}{2}}} = \frac{(2j)!\sqrt{\pi}}{j!4^j a^{j+\frac{1}{2}}}$ (see also Exercise 2.2). **(b)** Show that $\int_{-\infty}^{\infty} dx\, e^{-ax^2+bx} = \sqrt{\frac{\pi}{a}}\, e^{\frac{b^2}{4a}}$. Do this first by completing the square and then by Taylor expanding e^{bx}. Show also that this relation can be applied for $a \in \mathbb{C}$ as long as Re[a] < 0. **Hint:** *See Exercise 2.2.*

2.2 Gaussian integrals can be conveniently computed with the help of the gamma function: $\Gamma(z) \equiv \int_0^{\infty} t^{z-1} e^{-t} dt$: **(a)** Show that $\int_{-\infty}^{\infty} dx\, x^{2n} e^{-x^2} = \Gamma\left(n + \frac{1}{2}\right)$. **(b)** Show that $\Gamma(x+1) = x\Gamma(x)$. **(c)** Show that $\left[\int_{-\infty}^{\infty} dx\, e^{-x^2}\right]^2 = \pi$. **Hint:** *Convert this into a radial integral in two dimensions.* **(d)** Show that $\Gamma(n+1) = n!$ and $\Gamma\left(n + \frac{1}{2}\right) = \frac{(2n)!}{n!4^n}\sqrt{\pi}$.

2.3 **(a)** Show that $E_{\pm}(z,t) = fv(z \mp ct/n)$ is a solution of the wave equation (2.15). **(b)** The temporal and spatial Fourier transformation of $E(z,t)$ follows from Eqs. (2.31) and (2.63), respectively. How are they connected based on the generic solution $E_{\pm}(z,t)$?

2.4 By using the Gaussian wave packets $E_q(z,t) = e^{i(qz-\omega_q t)}\, e^{-(z-ct)^2/4\Delta z^2}$, show that the interference pattern has a period $\Delta x_{\text{node}} = 2\pi/|q - q'|$ in intensity when we consider a superposition state of E_q and $E_{q'}$.

2.5 Any classical particle trajectory can be described in phase space $(\mathbf{r}, \mathbf{p})$ such that $\mathbf{p} = \mathbf{p}(\mathbf{r})$ becomes a function of $\mathbf{r}$. Starting from Hamilton's equations and energy conservation, show that any classical trajectory must satisfy $\nabla \times \mathbf{p}(\mathbf{r}) = 0$. **Hint:** *Use the relation* $\mathbf{a} \times (\nabla \times \mathbf{b}(\mathbf{r})) = \nabla(\mathbf{a} \cdot \mathbf{b}(\mathbf{r})) - (\mathbf{a} \cdot \nabla)\mathbf{b}(\mathbf{r})$.

2.6 **(a)** Show that the solution (2.46) satisfies the eikonal equation (2.37). Show that also the $\mathbf{q}_1'$ branch, defined by Eq. (2.50), satisfies Eq. (2.37). **(b)** Compute the paths (2.50) by starting from Hamilton's equations. **Hint:** *Use the properties of the* δ *function, see Exercise 2.9.*

2.7 **(a)** Study the eigenvalue problem $\nabla^2 \phi_{\mathbf{q}}(\mathbf{r}) = -\mathbf{q}^2 \phi_{\mathbf{q}}(\mathbf{r})$ and show that

$$\int d^3r\, \phi_{\mathbf{q}}^*(r)\, \nabla^2\, \phi_{\mathbf{q}'}(r) = (\mathbf{q}'^2 - \mathbf{q}^2) \int d^3r\, \phi_{\mathbf{q}}^*(r) \phi_{\mathbf{q}'}(r).$$

Why does this imply orthogonality for nondegenerate modes? **(b)** Also degenerate solutions can be orthogonalized, e.g., via the Gram–Schmidt method. Assume that a generic function $\psi(r) = \sum_{\mathbf{q}} c_{\mathbf{q}}\, \phi_{\mathbf{q}}(r)$ is expressed in terms of the modes $\phi_{\mathbf{q}}(r)$. Show that the coefficients follow from $c_{\mathbf{q}} = \int d^3r\, \phi_{\mathbf{q}}^*(r)\, \psi(r)$ and that this eventually produces the completeness relation

$$\sum\nolimits_{\mathbf{q}} \phi_{\mathbf{q}}^*(\mathbf{r})\, \phi_{\mathbf{q}}(\mathbf{r}') = \delta(\mathbf{r} - \mathbf{r}')\,.$$

2.8 Evaluate (2.82) by starting from (2.77) and using the relations derived in Exercise 2.1.

2.9 Construct the probability distribution (2.83) out of the solution (2.82). Show that this generalized wave spreads like the statistical particle distribution (1.76). Verify that the associations (2.86) connect the generalized wave and particle distributions.

2.10 For light interacting with matter, the wave equation is also coupled to the matter polarization $\left(\frac{\partial^2}{\partial z^3} - \frac{1}{c^2}\frac{\partial^2}{\partial t^2}\right) E(z,t) = \mu_0 \delta(z) \frac{\partial^2 P(t)}{\partial t^2}$, where the polarization is assumed to be in a plane. **(a)** Show that $E(z,t) = E_0(z-t) - \frac{1}{2}\mu_0 \frac{\partial}{\partial u} P(u)\Big|_{u=t-\frac{|z|}{c}}$ is a solution. **(b)** Assume that the polarization response follows from $P(\omega) = \varepsilon_0 \chi(\omega) E(0,\omega)$ and compute the transmitted field $E_T(z,t)$ at $z > 0$ as a function of E_0 and $\chi(\omega)$.

Further reading

The most recent list of natural constants can, e.g., be found at:
http://physics.nist.gov/cuu/Constants/index.html

Reference books for special functions and vector algebra are:

G. B. Arfken and H. J. Weber (2005). *Mathematical Methods for Physicists*, 6th edition, New York, Elsevier.

P. Dennery and A. Krzywicki (1967). *Mathematics for Physicists*, New York, Harper & Row.

F. W. Byron, Jr. and R. W. Fuller (1992). *Mathematics of Classical and Quantum Physics*, New York, Dover.

For further reading on classical electromagnetism, we recommend:

J. D. Jackson (1999). *Classical Electrodynamics*, 3rd edition, New York, Wiley & Sons.
J. R. Reitz, F. J. Milford, and R. W. Christy (2008). *Foundations of Electromagnetic Theory*, 4th edition, New York, Addison-Wesley.
D. J. Griffiths (1995). *Introduction to Electrodynamics*, 2nd edition, Saddle River, NJ, Prentice-Hall.
L. D. Landau and E. M. Lifschitz (2003). *The Classical Theory of Fields*, 4th edition, Amsterdam, Butterworth–Heinemann.

3

Central concepts in quantum mechanics

In the early 1900s more and more experimental evidence was accumulated indicating that microscopic particles show wave-like properties in certain situations. These particle-wave features are very evident, e.g., in measurements where electrons are diffracted from a double slit to propagate toward a screen where they are detected. Based on the classical averaging of particles discussed in Section 1.2.2, one expects that the double slit only modulates the overall intensity, not the spatial distribution. Experimentally, however, one observes a nearly perfect interference pattern at the screen implying that the electrons exhibit wave averaging features such as discussed in Section 2.1.2. This behavior, originally unexpected for particle beams, persists even if the experiment is repeated such that only one electron at a time passes the double slit before it propagates to the detection screen. Thus, the wave aspect must be an inherent property of individual electrons and not an ensemble effect.

Another, independent argument for the failure of classical physics is that the electromagnetic analysis of atoms leads to the conclusion that the negatively charged electron(s) should collapse into the positively charged ion because the electron–ion system loses its energy due to the emission of radiation. As we will see, this problem can be solved by including the wave aspects of particles into the analysis. In particular, as discussed in Section 2.3.3, waves can never be localized to a point without increasing their momentum and energy beyond bounds. Hence, a wave-like electron in an atom occupies an energetically stable standing-wave state that is static and distributed around the ion. This distribution implies that the electron undergoes no classical motion with the consequence that the electron–ion system does not radiate light, making it completely stable. The configuration where the electron wave function exhibits minimal fluctuations both in position and momentum defines the so-called ground state, which sets a lower bound to the energy of the electronic states.

Since atoms are building blocks of molecules and solids, it is obvious that the classical description also fails for these systems. Thus, we most certainly need to include the wave aspects of electrons in order to have a systematic description of semiconductors. As a fundamental property, the wave nature of microscopic particles must be postulated via an axiom. This was first done in the year 1924 when Louis de Broglie formulated the wave–particle dualism stating that any particle simultaneously exhibits wave and particle properties. This feature constitutes one of the foundations of quantum mechanics.

3.1 Schrödinger equation

The wave–particle dualism can be formulated mathematically by extending the treatment of particle waves presented in Chapter 2. For this, we make use of the fact that the generalized particle waves produce exactly the same behavior as a statistical distribution of free particles, compare Section 2.3.3. Furthermore, we know that the movement of the slowly varying phase fronts follows the dynamics of classical particles, as discussed in Section 2.2.2. Thus, the wave–particle dualism can be fully described if we identify a generalized wave equation whose slowly varying phase is described by the Hamilton equation (1.10). When this connection is made, both the wave function $\psi(x, t)$ and the particle paths describe different aspects of reality, one assigned to the wave properties and the other one to the particle features.

The starting point of quantum mechanics can be formulated axiomatically via the time-dependent Schrödinger equation

$$i\hbar\frac{\partial}{\partial t}\psi(\mathbf{r},t) = \left[-\frac{\hbar^2}{2m}\nabla^2 + U(\mathbf{r})\right]\psi(\mathbf{r},t). \tag{3.1}$$

This equation has the form of the generalized wave equation (2.54) that is a first-order partial differential equation with respect to time. As a new element, the Schrödinger equation contains the potential $U(\mathbf{r})$. The constant $\beta \equiv \hbar/2m$ is explicitly defined through the particle mass m and Planck's constant $\hbar$. In general, $\hbar = 1.054\,571\,628 \times 10^{-34}$ J s defines an extremely small quantity from the macroscopic point of view. As we will see later, it is the tiny value of $\hbar$ that is the reason why quantum features are dominant only for microscopic particles.

The wave function $\psi(\mathbf{r}, t)$ in (3.1) is complex valued. Therefore, it is sometimes useful to solve the respective Schrödinger equations both for $\psi(\mathbf{r}, t)$ and for its complex conjugate $\psi^\star(\mathbf{r}, t)$. By taking the complex conjugate of both sides of Eq. (3.1) and multiplying by -1, we find

$$i\hbar\frac{\partial}{\partial t}\psi^\star(\mathbf{r},t) = -\left[-\frac{\hbar^2}{2m}\nabla^2 + U(\mathbf{r})\right]\psi^\star(\mathbf{r},t) \tag{3.2}$$

for the complex conjugated wave function.

To verify that the Schrödinger equation describes classical particles as a limiting case, we Fourier transform both sides of Eq. (3.1) with the help of Eq. (2.31),

$$\begin{aligned}\hbar\omega\,\psi(\mathbf{r},\omega) &= -\frac{\hbar^2}{2m}\nabla^2\psi(\mathbf{r},\omega) + U(\mathbf{r})\psi(\mathbf{r},\omega)\\ \Leftrightarrow \nabla^2\psi(\mathbf{r},\omega) &+ \frac{2m}{\hbar^2}\left[\hbar\omega - U(\mathbf{r})\right]\psi(\mathbf{r},\omega) = 0.\end{aligned} \tag{3.3}$$

We see that this expression has exactly the form of the Helmholtz equation (2.30) by making the identification $\frac{2m}{\hbar^2}[\hbar\omega - U(\mathbf{r})] \equiv \mathbf{q}^2 n^2(\mathbf{r})$. The dispersion relation (2.56) yields $\mathbf{q}^2 \equiv \omega/\beta = 2m\omega/\hbar$ and we formally identify the square of the refractive index, $n^2(\mathbf{r}) \equiv 1 - U(\mathbf{r})/\hbar\omega\beta$. As we discuss in Section 2.2.1, the Helmholtz equation (3.3) describes particle aspects via $\psi \propto e^{i|\mathbf{q}|\tilde{S}(\mathbf{r})}$ in the limit where $|\mathbf{q}|$ becomes large while $\tilde{S}(\mathbf{r})$ remains slowly varying. In the present case, the identified wave vector is proportional to $1/\sqrt{\hbar}$ such that we can reach the large $|\mathbf{q}|$ limit by formally letting $\hbar \to 0$.

In detail, we look for classical aspects of the Schrödinger equation via the ansatz

$$\psi(\mathbf{r}) \equiv \psi_0(\mathbf{r})\, e^{\frac{i}{\hbar}S(\mathbf{r})}. \tag{3.4}$$

We assume that both the envelope $\psi_0(\mathbf{r})$ and the phase $S(\mathbf{r})$ are only weakly $\mathbf{r}$ dependent on the scale set by $\hbar^{-1}$. Inserting the ansatz (3.4) into Eq. (3.3), we can follow steps identical to those performed in Eqs. (2.35)–(2.36). The leading term of order $\hbar^{-2}$ then yields the eikonal equation

$$-\frac{1}{\hbar^2}[\nabla S(\mathbf{r})]^2\,\psi_0(\mathbf{r})\,e^{\frac{i}{\hbar}S(\mathbf{r})} + \frac{2m}{\hbar^2}[\hbar\omega - U(\mathbf{r})]\,\psi_0(\mathbf{r})\,e^{\frac{i}{\hbar}S(\mathbf{r})} = 0$$

$$\Rightarrow \frac{[\nabla S(\mathbf{r})]^2}{2m} + U(\mathbf{r}) = \hbar\omega. \tag{3.5}$$

At this stage, we make the direct identifications

$$\mathbf{p}(\mathbf{r}) \equiv \nabla S(\mathbf{r}), \qquad E \equiv \hbar\omega \tag{3.6}$$

which convert Eq. (3.5) into

$$\frac{\mathbf{p}^2(\mathbf{r})}{2m} + U(\mathbf{r}) = E. \tag{3.7}$$

This is just the classical equation for a particle with mass m. The solutions yield the phase-space trajectories $\mathbf{p}(\mathbf{r})$.

According to the axiom of the wave–particle dualism, both the wave-function and classical-particle properties are genuine – not fictitious – and simultaneous aspects of any particle. It is very satisfying to see that the wave–particle dualism can be formulated mathematically via the Schrödinger equation. In classical mechanics, one only discusses the motion of the phase fronts while the wave aspects are ignored. In contrast, the quantum mechanical treatment incorporates wave and particle aspects via the wave functions.

3.1.1 Free quantum-mechanical propagation

The elementary properties of quantized particles can be nicely illustrated by following how the wave function of free particles evolves in time. As an example, we analyze the one-dimensional propagation of a Gaussian wave packet. Based on the results of

Section 2.3.3, we can directly determine $\psi(z,t)$ using Eq. (2.82) with the replacement $\beta \to \hbar/2m$,

$$\psi(z,t) = \left[\frac{1}{2\pi\,\Delta z^2(t)}\right]^{\frac{1}{4}} e^{-\frac{(z-z_0 v_0 t)^2}{4\Delta z_0 \Delta z(t)}}\, e^{iq_0(z-z_0)-i\frac{\hbar q_0^2}{2m}t}, \tag{3.8}$$

where q_0 denotes the wave vector, z_0 is the central position, and $v_0 \equiv \hbar q_0/m$ defines the propagation velocity of the Gaussian wave packet. The initial width Δz_0 of the Gaussian distribution evolves according to

$$\Delta z(t) \equiv \Delta z_0 + i\frac{\hbar t}{2m\Delta z_0} \tag{3.9}$$

in analogy to Eq. (2.84).

Associating $e^{\frac{i}{\hbar}S(z)} \equiv e^{iq_0 z}$ in Eq. (3.8), we find for the slowly varying phase $S(z) = \hbar q_0 z$. Furthermore, the exponent in $e^{-i\frac{\hbar q_0^2}{2m}t}$ determines the dispersion relation for the frequency $\omega \equiv \hbar q_0^2/2m$. As we combine this information with Eqs. (3.6)–(3.7), we conclude that the slowly varying parts of the wave function identify the particle's momentum and energy as

$$p = \frac{\partial S(z)}{\partial z} = \hbar q_0, \qquad E = \hbar\omega = \frac{\hbar^2 q_0^2}{2m_0} = \frac{p^2}{2m}, \tag{3.10}$$

respectively.

To incorporate the nontrivial wave aspects, we also need to consider the temporal changes of the envelope. For this purpose, we look at the probability distribution to observe the particle at the position z,

$$\rho_p(z,t) \equiv |\psi(z,t)|^2 = \frac{1}{\sqrt{2\pi}\,|\Delta z(t)|}\, e^{-\frac{(z-z_{\mathrm{cl}}(t))^2}{2|\Delta z(t)|^2}}. \tag{3.11}$$

We see that the center of the wave packet follows the classical trajectory

$$z_{\mathrm{cl}}(t) = v_0 t = \frac{p_0 t}{m}. \tag{3.12}$$

Besides the purely classical propagation of its center, the Gaussian wave packet spreads as can be seen from the increasing $|\Delta z(t)|$. This additional dynamical feature is a consequence of the fact that waves cannot be localized in contrast to classical particles, compare Section 2.3.3. Thus, the degree of spreading can be used as a phenomenological measure to determine when wave aspects of particles become important.

If Planck's constant was really zero, Eq. (3.9) would predict that no temporal spreading occurs implying a purely classical motion. To see how the finite $\hbar$ influences the particle's motion, we determine the spreading time

$$t_{\mathrm{spread}} \equiv \frac{2m\Delta z_0^2}{\hbar}, \tag{3.13}$$

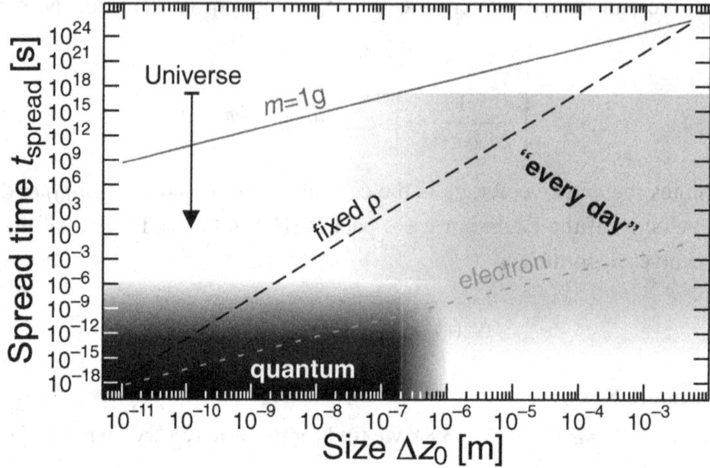

Figure 3.1 Spreading time of freely propagating particles. Equation (3.13) is used to determine the time a Gaussian wave packet spreads by a factor of $\sqrt{2}$ as a function of its initial size Δz_0. The particle is assumed to have $m = 1$ g (solid line), constant density $\rho_m = 1000\,\text{kg}\,\text{m}^{-3}$ (dashed line), and electron mass (dotted line). The light- and dark-shaded areas discriminate between the typical classical and quantum realm, respectively.

i.e., the time at which $|\Delta z(t)|$ has increased by $\sqrt{2}$ compared to its original width Δz_0. We decrease Δz_0 to localize the particle position within an ever smaller region and investigate t_{spread} for the two different conditions where we keep either the mass m or the mass density ρ_m constant as a function of Δz_0. For the constant ρ_m, we assume the body to be spherical with the radius Δz_0 such that $m = \frac{4\pi}{3}\Delta z^3 \rho_m$ is the volume times ρ_m. Figure 3.1 shows t_{spread} for a macroscopic particle with $m = 1$ g (dashed line) and for a macroscopic body like water with $\rho_m = 1000\,\text{kg}\,\text{m}^{-3}$ (solid line), respectively. We see that any system larger than 10^{-4} exhibits a spreading time t_{spread} that is greater than $t = 10^{17}$ s, which is comparable with the life time of the Universe. Thus, any macroscopic system truly constitutes a classical entity.

If we take a 1 μs spreading time as the rough boundary between quantum and classical objects, we see in Fig. 3.1 that Δz_0 must be smaller than 2 nm for objects to be in the quantum world. This spreading size can be realized only for extremely small masses of the order of $m = \frac{4}{3}\pi\rho_m[2\,\text{nm}]^3 = 3.35 \times 10^{-23}$ kg, which implies that the corresponding particle is microscopic. Such nanoscale particles exhibit true quantum features due to their prominent wave-like features. Thus, the properties of elementary particles, such as electrons, typically require a full quantum treatment. Even though there is a fuzzy transition region between the purely classical and the purely quantum-mechanical world, we can get a rough estimate for the extent of the quantum realm (dark shaded area in Fig. 3.1) by defining that times $t \leq 1$ μs and sizes smaller than 1 μm yield true quantum-mechanical behavior. Due to its very small mass $m_e = 9.109\,382\,15 \times 10^{-31}$ kg, an electron clearly is a microscopic entity. The corresponding spreading time is plotted in Fig. 3.1 as a dotted line. Assuming initial localization accuracies of Δz_0 up to 15 μm produces dominant wave aspects within $t_{\text{spread}} = 1\ \mu\text{s}$ such that electrons basically always behave quantum mechanically.

3.1.2 Interpretation of the wave function

Based on the analysis above, a completely classical description of particles is possible only if the wave function $\psi(x,t) = \psi_0(\mathbf{r},t)\, e^{\frac{i}{\hbar}S(\mathbf{r},t)}$ can be accurately expressed invoking only the slowly varying phase $S(\mathbf{r},t)$. The classical description fails for microscopic particles because the particle waves spread too rapidly and thus cannot be localized. In other words, a rapidly changing $\psi_0(\mathbf{r},t)$ implies the dominance of wave aspects over particle properties. Thus, one must replace the concept of localized particles by that of delocalized waves in order to explain microscopic phenomena. This replacement considerably changes the interpretation of the physics involved.

In quantum systems, the wave function $\psi(\mathbf{r},t)$ is the fundamental quantity which we can compute by solving the full Schrödinger equation (3.1). Since the classical concepts must be extended considerably to incorporate genuine wave aspects, we clearly need a new interpretation that allows us to extract and predict specific quantum-mechanical properties from $\psi(\mathbf{r},t)$ directly.

Since classical electromagnetic fields are always describable as waves, we can develop useful concepts by extending the treatment of the electric field $\mathbf{E}(\mathbf{r})$ to particle waves. In connection with the wave averaging, see Section 2.1.2, we have already identified the intensity $I(\mathbf{r}) = |\mathbf{E}(\mathbf{r})|^2$ as the quantity measuring the strength of the electromagnetic wave as a function of position $\mathbf{r}$. Since waves cannot be localized, it is only meaningful to define the probability of detecting them at a given position. Hence, when properly normalized, $I(\mathbf{r})$ can also be interpreted as the $\mathbf{r}$-dependent probability distribution. For example, when $E(\mathbf{r})$ has a node, there is no field to be detected at that position while a maximum of $I(\mathbf{r})$ implies the largest probability of detecting the wave there.

This probabilistic interpretation can be generalized directly for particle waves: a properly normalized $|\psi(\mathbf{r},t)|^2$ can be understood as the probability distribution to detect the particle at a given position. In order for the probability to be normalized to unity, the wave function has to be square normalizable

$$\int d^3r\, |\psi(\mathbf{r},t)|^2 = 1. \tag{3.14}$$

Square normalizability of a function implies that it must decay sufficiently fast toward zero at infinity to guarantee convergence under an integral. Based on Eq. (2.64), the Schrödinger equation (3.1) conserves this property such that the wave function is square normalizable at all times. Thus, one can understand $|\psi(\mathbf{r},t)|^2$ as a measure describing how "intense" the particle wave is at any given position. In other words, $|\psi(\mathbf{r},t)|^2$ can be viewed as a distribution function that determines the probability

$$P(\mathbf{r}, \Delta r, t) \equiv \int_{|\mathbf{r}'-\mathbf{r}|<\Delta r} d^3r' \, |\psi(\mathbf{r}', t)|^2, \tag{3.15}$$

of finding a particle within the volume of a Δr-sized sphere centered at $\mathbf{r}$. Since microscopic particles follow no deterministic path, their dynamics can be described only via a probability distribution $|\psi(\mathbf{r}, t)|^2$ in full analogy to the marginal distribution $\rho_p(z)$ defining the statistical description of classical particles, see Section 1.2. Thus, we can use all the tools of probability theory to determine the statistical properties of quantized particles.

Using the wave-packet result (3.11), we can compute the average position of a free particle via

$$\begin{aligned}\langle \hat{z} \rangle &\equiv \int_{-\infty}^{\infty} dz \, z \, |\psi(z, t)|^2 = \frac{1}{\sqrt{2\pi}|\Delta z(t)|} \int_{-\infty}^{\infty} dz \, z \, e^{-\frac{(z-z_{\text{cl}}(t))^2}{2|\Delta z(t)|^2}} \\ &= \frac{1}{\sqrt{2\pi}|\Delta z(t)|} \int_{-\infty}^{\infty} dz \, (z_{\text{cl}}(t) + [z - z_{\text{cl}}(t)]) \, e^{-\frac{(z-z_{\text{cl}}(t))^2}{2|\Delta z(t)|^2}} = z_{\text{cl}}(t).\end{aligned} \tag{3.16}$$

The second evaluation step adds and subtracts $z_{\text{cl}}(t)$ to the integrand. At this point, we realize that the integrand within the brackets is odd, yielding a vanishing contribution for this term. The remaining integrand produces $z_{\text{cl}}(t) \int_{-\infty}^{\infty} dz \, |\psi(z, t)|^2 = z_{\text{cl}}(t)$, which can be evaluated by performing the Gaussian integral with the help of Eq. (2.79). We see that the average position of the freely propagating Gaussian wave packet follows the classical path $z_{\text{cl}}(t)$ even though $\psi(z, t)$ itself behaves nonclassically. Thus, we can conclude that classical mechanics detects only the average position of the particle while it ignores all the fluctuations.

3.2 Expectation values in quantum mechanics

The result (3.16) introduces the important concept of how to compute physically relevant quantities from a known wave function. If we want to determine the average value of a generic $\mathbf{r}$-dependent property, i.e., $F(\mathbf{r})$, we evaluate it via the expectation value

$$\langle \hat{F} \rangle \equiv \int_{-\infty}^{\infty} d^3r \, \psi^\star(\mathbf{r}, t) \, F(\mathbf{r}) \, \psi(\mathbf{r}, t) \tag{3.17}$$

which defines the average of the quantity $F(\mathbf{r})$ when the three-dimensional position distribution is given by $|\psi(\mathbf{r}, t)|^2$. Equation (3.17) also shows that we may include an arbitrary constant phase $\psi(\mathbf{r}, t) \rightarrow e^{i\varphi} \, \psi(\mathbf{r}, t)$ without altering either the probability distribution or any expectation value. Thus, a pure phase φ can be chosen arbitrarily in the description of quantum-mechanical states. We adopt here a commonly used notation where the argument $\hat{F}$ of the expectation value $\langle \cdots \rangle$ is expressed with a hat to stress that $\hat{F}$ is actually a quantum-mechanical operator. More general forms of $\hat{F}$ are discussed in Section 5.1.

In general, expectation values are additive, i.e.,

$$\langle[\hat{F}+\hat{G}]\rangle \equiv \int_{-\infty}^{\infty} d^3r\ \psi^\star(\mathbf{r},t)\ [F(\mathbf{r})+G(\mathbf{r})]\ \psi(\mathbf{r},t) = \langle\hat{F}\rangle + \langle\hat{G}\rangle \tag{3.18}$$

based on the definition (3.17). Furthermore,

$$\langle C\,\hat{F}\rangle \equiv C\,\langle\hat{F}\rangle, \qquad \langle C\rangle = C, \tag{3.19}$$

where C is not an operator but a generic complex-valued number. However, there is no simple relation allowing us to evaluate the product $\langle\hat{F}\hat{G}\rangle$ in terms of $\langle\hat{F}\rangle$ and $\langle\hat{G}\rangle$. This aspect is elaborated further in Section 3.2.2.

As discussed above, the average $\langle\hat{z}\rangle = z_{\text{cl}}(t)$ produces the classical path for a freely propagating one-dimensional particle. To access its quantum fluctuations, we must analyze more complicated expectation values. Like in any probabilistic description, the simplest characterization of fluctuations follows from the variance

$$\begin{aligned}\langle[\hat{z}-\langle\hat{z}\rangle]^2\rangle &= \langle\hat{z}^2 - 2\langle\hat{z}\rangle\hat{z} + \langle\hat{z}\rangle^2\rangle = \langle\hat{z}^2\rangle - 2\langle\hat{z}\rangle\langle\hat{z}\rangle + \langle\hat{z}\rangle^2 \\ &= \langle\hat{z}^2\rangle - \langle\hat{z}\rangle^2.\end{aligned} \tag{3.20}$$

Here, we have used the fact that $\langle\hat{z}\rangle$ constitutes a real-valued number and thus can be taken out of any expectation value. The final resulting (3.20) is obtained if we also use the relations (3.18)–(3.19).

For the Gaussian wave packet (3.11), the evaluation of its variance follows very similar steps as in Eq. (3.16), eventually producing

$$\langle[\hat{z}-\langle\hat{z}\rangle]^2\rangle = \Delta z_0^2 + \left(\frac{\hbar t}{2m\Delta z_0}\right)^2 \equiv |\Delta z(t)|^2 \tag{3.21}$$

in agreement with Eq. (3.9). Thus, the variance directly identifies the extent of fluctuations around the classical path. Since this connection also holds for generic wave packets, we may adopt the notation where the variance is expressed via

$$\langle[\hat{z}-\langle\hat{z}\rangle]^2\rangle \equiv |\Delta z(t)|^2. \tag{3.22}$$

Unlike a classical system, a quantum-mechanical particle wave always spreads in time due to its inherent fluctuations in momentum, as discussed in Section 3.1.1.

3.2.1 Particle momentum

To characterize the momentum of a particle, we need to identify the corresponding momentum operator. Since $\langle\hat{\mathbf{r}}\rangle$ defines the average position of either a classical or a quantum particle, we may use $m\frac{d}{dt}\langle\hat{\mathbf{r}}\rangle$ to define the average momentum $\langle\hat{\mathbf{p}}\rangle$ of the particle, based on the classical analogy. Thus, we evaluate the momentum indirectly via

$$\langle \hat{\mathbf{p}} \rangle = m\frac{d}{dt}\langle \hat{\mathbf{r}} \rangle = m\frac{d}{dt}\int_{-\infty}^{\infty} d^3r\ \psi^\star(\mathbf{r},t)\,\mathbf{r}\,\psi(\mathbf{r},t)$$

$$= \int_{-\infty}^{\infty} d^3r\ m\left[\frac{\partial\,\psi^\star(\mathbf{r},t)}{\partial t}\,\mathbf{r}\,\psi(\mathbf{r},t) + \psi^\star(\mathbf{r},t)\,\mathbf{r}\,\frac{\partial\,\psi(\mathbf{r},t)}{\partial t}\right], \tag{3.23}$$

after having applied the product rule of differentiation. The temporal derivatives of the wave function that appear can now be evaluated by substituting the Schrödinger equation (3.1)–(3.2) into Eq. (3.23), producing

$$\langle \hat{\mathbf{p}} \rangle = \int_{-\infty}^{\infty} d^3r \left\{\frac{m}{i\hbar}\left[+\frac{\hbar^2}{2m}\nabla^2\psi^\star(\mathbf{r},t) - U(\mathbf{r})\psi^\star(\mathbf{r},t)\right]\mathbf{r}\,\psi(\mathbf{r},t)\right.$$

$$\left. +\ \psi^\star(\mathbf{r},t)\,\mathbf{r}\,\frac{m}{i\hbar}\left[-\frac{\hbar^2}{2m}\nabla^2\psi(\mathbf{r},t) + U(\mathbf{r})\psi(\mathbf{r},t)\right]\right\}$$

$$= \frac{\hbar}{2i}\int_{-\infty}^{\infty} d^3r\ \left\{\left[\nabla^2\psi^\star(\mathbf{r},t)\right]\mathbf{r}\,\psi(\mathbf{r},t) - \psi^\star(\mathbf{r},t)\,\mathbf{r}\left[\nabla^2\psi(\mathbf{r},t)\right]\right\}, \tag{3.24}$$

where the potential terms cancel each other leaving only the second-order differentiations with respect to the space coordinate. So far, the expectation value $\langle \hat{\mathbf{p}} \rangle$ is not yet in the standard form of Eq. (3.17). Thus, we need to manipulate the differentiations to sandwich them between $\psi^\star$ and ψ. This conversion can be performed by applying

$$\left[\nabla^2\psi^\star(\mathbf{r},t)\right]\mathbf{r}\,\psi(\mathbf{r},t) - \psi^\star(\mathbf{r},t)\,\mathbf{r}\left[\nabla^2\psi(\mathbf{r},t)\right] = 2\psi^\star(\mathbf{r},t)\left[\nabla\psi(\mathbf{r},t)\right]$$

$$+\nabla\cdot\left\{\left[\nabla\psi^\star(\mathbf{r},t)\right]\mathbf{r}\,\psi(\mathbf{r},t) - \psi^\star(\mathbf{r},t)\,\mathbf{r}\left[\nabla\psi(\mathbf{r},t)\right]\right\} - \nabla|\psi(\mathbf{r},t)|^2, \tag{3.25}$$

which follows from the basic properties of ∇, see Exercise 3.3.

Replacing the integrand within Eq. (3.24) by the expression on the RHS of Eq. (3.25), we conclude that only the first term $2\psi^\star(\mathbf{r})\left[\nabla\psi(\mathbf{r})\right]$ is in the standard form of the expectation value (3.17). However, the two last terms can be shown to vanish by expressing them in the format $\nabla \odot \mathbf{F}(\mathbf{r})$, where $\odot$ refers to ordinary, dot, or cross product. The resulting integrals can be evaluated with the help of the Gauss theorem

$$\int_V d^3r\ \nabla \odot \mathbf{F}(\mathbf{r}) = \oint_{\delta V} d\mathbf{S} \odot \mathbf{F}(\mathbf{r}), \tag{3.26}$$

where the integration is evaluated over the volume V while $\oint_{\delta V}$ is evaluated over the surface δV of V. The infinitesimal surface element $d\mathbf{S}$ points in the direction normal to the surface. In the integration (3.24), the δV is at infinity where the wave function must vanish due to the square normalizability (3.14). Consequently, the $\nabla \odot \mathbf{F}(\mathbf{r})$ contributions in Eq. (3.25) vanish after the integration (3.24), simplifying the expression for the momentum expectation value to

$$\langle \hat{\mathbf{p}} \rangle = \int_{-\infty}^{\infty} d^3r\ \psi^\star(\mathbf{r},t)\left[-i\hbar\nabla\right]\psi(\mathbf{r},t). \tag{3.27}$$

Since this has exactly the form of the expectation value (3.17), we can identify

$$\hat{\mathbf{p}} \equiv -i\hbar\nabla \tag{3.28}$$

as the particle momentum operator.

Considering the example of the one-dimensional Gaussian wave function (3.8), the momentum operator reduces to the $\hat{p}_z = -i\hbar\frac{\partial}{\partial z}$ component of the full three-dimensional $\hat{\mathbf{p}}$. A straightforward integration then yields the expectation value and variance of the momentum as

$$\langle \hat{p}_z \rangle = \hbar q_0, \qquad \Delta p_z^2 \equiv \langle \left[\hat{p}_z - \langle \hat{p}_z \rangle\right]^2 \rangle = \frac{\hbar^2}{4\Delta z_0^2}. \tag{3.29}$$

This result confirms the connection between the wave vector q_0 and the momentum, in agreement with Eq. (3.10). We can also express the relation (3.21) as

$$|\Delta z(t)|^2 = \Delta z_0^2 + \left(\frac{\Delta p_z}{m} t\right)^2. \tag{3.30}$$

In full analogy to the statistical distributions of classical particles, the momentum spread is constant in time while Δp yields spatial broadening, as discussed in Section 1.2.1. Thus, the spreading of freely propagating particle waves has partially a classical origin.

In spite of similarities between the wave function analysis and the classical statistical description, we see that the spread of the quantum-mechanical momentum is related to the position spreading. In particular, we find that the Gaussian wave packet always produces

$$\Delta z(t)^2 \Delta p(t)^2 \geq \frac{\hbar}{4}, \tag{3.31}$$

which states that the accuracy in position and momentum cannot simultaneously be made arbitrarily small. Once we develop the operator description further in Chapter 5, we will show that Eq. (3.31) constitutes one version of the Heisenberg uncertainty principle that any physical particle must always satisfy. Thus, particles cannot be reduced to precise δ-function-like distributions in phase space, in contrast to the classical description. Instead, the quantum-mechanical description associates intrinsic fluctuations to the particle that originate from its wave-like features, which cannot be explained classically.

Figure 3.2(a) presents $\langle \hat{z} \rangle$ (solid line) as a function of time for an electron that is initially at rest. The particle has $q_0 = 2\pi 10^9\ \mathrm{m}^{-1}$ corresponding to the velocity $v_0 \equiv p_0/m_e = \hbar q_0/m_e = 0.727\ \mathrm{nm\ fs}^{-1}$ for an electron. Whereas the resulting average trajectory is independent of the Gaussian distribution, the average extent of the fluctuations around the trajectory grows with an increasing width. The corresponding average $\langle \hat{p}_z \rangle$ is shown in Fig. 3.2(b) as a solid line. The average extent of the momentum fluctuations is presented for the cases with $\Delta z_0 = 0.2$ nm (shaded area) and $\Delta z_0 = 0.4$ nm (dashed lines). The average particle position moves linearly while the fluctuations grow nonlinearly. At

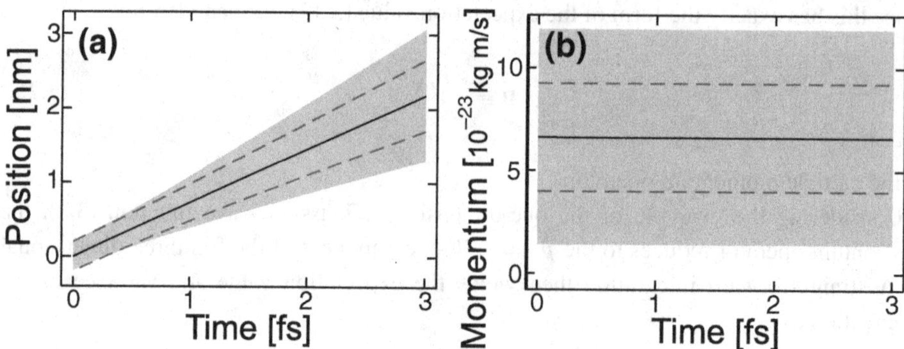

Figure 3.2 Average position, momentum, and fluctuations of freely propagating particle waves. **(a)** The average position (solid line) is shown together with the fluctuations for the situations where the initial width is $\Delta z_0 = 0.2$ nm (shaded area) and $\Delta z_0 = 0.4$ nm (dashed lines). **(b)** The corresponding average momentum and fluctuations are plotted as a function of time. The wave packet is Gaussian with $q_0 = 2\pi 10^9$ m^{-1} and $z_0 = 0$ at the initial time. The trajectory and fluctuations are computed from Eqs. (3.12), (3.29), and (3.30).

the same time, both the average momentum and the momentum fluctuations remain constant in time, as they would for a classical particle with a statistical distribution. However, the value of the momentum fluctuations increases with a decrease of the position fluctuations, as predicted by the Heisenberg uncertainty principle, which makes the classical and quantum descriptions distinctly different.

3.2.2 Commutation relations

For the expectation value of the product between the momentum operator $\hat{\mathbf{p}}$ and an arbitrary $\hat{\mathbf{r}}$-dependent $\hat{F}$ operator, we obtain

$$\begin{aligned} \langle \hat{\mathbf{p}} \hat{F} \rangle &= \int d^3r\, \psi^\star(\mathbf{r},t)\,[-i\hbar\nabla F(\mathbf{r})\psi(\mathbf{r},t)] \\ &= \int d^3r \left\{\psi^\star(\mathbf{r},t)\,[-i\hbar F(\mathbf{r})\nabla\psi(\mathbf{r},t)] + \psi^\star(\mathbf{r},t)\,[-i\hbar\nabla F(\mathbf{r})]\,\psi(\mathbf{r},t)\right\} \\ &= \langle \hat{F}\hat{\mathbf{p}}\rangle + \int d^3r\, \psi^\star(\mathbf{r},t)\,[-i\hbar\nabla F(\mathbf{r})]\,\psi(\mathbf{r},t) \end{aligned} \tag{3.32}$$

where we used (3.17) and the product rule of differentiation. The last step identifies the expectation value $\langle \hat{F}\hat{\mathbf{p}}\rangle$ of the operator product appearing in the reversed order plus an additional integral expression. Since the emerging additional integral does not necessarily vanish, we have to conclude that the expectation value of an operator product depends on the order of the operators in the product.

Such order-related effects can be analyzed by defining the commutator

$$\left[\hat{A},\,\hat{B}\right]_- \equiv \hat{A}\hat{B} - \hat{B}\hat{A} \tag{3.33}$$

between two generic operators. If the commutator vanishes, i.e. if the operators $\hat{A}$ and $\hat{B}$ commute, we can arbitrarily change their ordering just as in a product of classical functions. For the case of noncommuting operators, the commutator can be evaluated by always keeping in mind that the operators act on an arbitrary wave function on the RHS of the expression. For example, the result (3.32) can be written as the operator expression

$$\langle \hat{\mathbf{p}} \hat{F} \rangle = \langle \hat{F} \hat{\mathbf{p}} \rangle + \langle \left[\hat{\mathbf{p}}, \hat{F} \right]_- \rangle \Rightarrow \left[\hat{\mathbf{p}}, \hat{F} \right]_- \equiv -i\hbar \nabla F(\mathbf{r}), \tag{3.34}$$

which follows when we move the first term from the RHS to the LHS. This way, we have the definition of the commutator on the LHS and identify the left-over term on the RHS as the evaluated commutator.

We now use this result to compute the commutator between position and momentum. For this purpose, we simply consider the case $F(\mathbf{r}) = \mathbf{r} = r_1\mathbf{e}_1 + r_2\mathbf{e}_2 + r_3\mathbf{e}_3$, yielding

$$\left[\hat{r}_j, \hat{p}_k\right]_- = -\left[\hat{p}_k, \hat{r}_j\right]_- = i\hbar \frac{\partial}{\partial r_k} r_j = i\hbar\, \delta_{j,k} \tag{3.35}$$

because the differentiation is nonzero only for matching vectorial components j and k. This relation is often referred to as the canonical commutation relation. The noncommuting nature of $\hat{\mathbf{r}}$ and $\hat{\mathbf{p}}$ stems from the wave aspects of the particles, which prevents us from completely separating position and momentum properties, as discussed in connection with Eq. (3.31). To evaluate more general commutation relations, we can often use

$$\begin{aligned} &\left[\hat{A}, \hat{B}\right]_- = -\left[\hat{B}, \hat{A}\right]_-, \quad \left[\hat{A}, \hat{B} + \hat{C}\right]_- = \left[\hat{A}, \hat{B}\right]_- + \left[\hat{A}, \hat{C}\right]_-, \\ &\left[\hat{A}, \hat{B}\hat{C}\right]_- = \left[\hat{A}, \hat{B}\right]_- \hat{C} + \hat{B}\left[\hat{A}, \hat{C}\right]_- \end{aligned} \tag{3.36}$$

which introduce the exchange, additivity, and the product rule, respectively.

3.2.3 Canonical quantization

Looking again at the Schrödinger equation (3.1), we recognize that we may directly replace $-\hbar^2\nabla^2$ by $\hat{\mathbf{p}}^2$ based on Eq. (3.28), such that

$$i\hbar \frac{\partial}{\partial t} \psi(\mathbf{r}, t) = \left[\frac{\hat{\mathbf{p}}^2}{2m} + U(\hat{\mathbf{r}}) \right] \psi(\mathbf{r}, t). \tag{3.37}$$

From this form, we can now identify the Hamilton operator

$$H(\hat{\mathbf{r}}, \hat{\mathbf{p}}) \equiv \frac{\hat{\mathbf{p}}^2}{2m} + U(\hat{\mathbf{r}}) \tag{3.38}$$

such that the Schrödinger equation (3.1) is cast into the form

$$i\hbar \frac{\partial}{\partial t} \psi(\mathbf{r}, t) = H(\hat{\mathbf{r}}, \hat{\mathbf{p}})\, \psi(\mathbf{r}, t). \tag{3.39}$$

We notice that the Hamilton operator (3.38) has exactly the same functional form as the classical Hamilton function (1.1). By going from the classical to the quantum expression, we replace the canonical space and momentum coordinates by operators and introduce the nontrivial commutation relation (3.35). Here, the numerically small but nonzero factor $\hbar$ appears, which is a hallmark of all quantum features in the system. The Hamilton operator, often referred to as the Hamiltonian, $H(\hat{\mathbf{r}}, \hat{\mathbf{p}})$ acts upon the wave function – producing the Schrödinger equation. All functions of the classical canonical variables have to be replaced by operators and one can define their properties only probabilistically. The axiomatic steps from the classical $H(\mathbf{r}, \mathbf{p})$ to the quantized $H(\hat{\mathbf{r}}, \hat{\mathbf{p}})$-based dynamics are often referred to as the canonical quantization. Interestingly, the form of classical mechanics is mostly conserved except for potentially small $\hbar$-dependent corrections.

In general, the classical and quantum properties are identical only in situations where $\hbar$ can be approximated to be zero. This fact yields the correspondence principle: *The quantum description approaches the classical one if the $\hbar$-dependent contributions are negligibly small.* In particular, if one is interested in $\langle\hat{\mathbf{p}}\rangle$, this average value should be much larger than the fluctuation limit set by the Heisenberg uncertainty relation (3.31). Nevertheless, the value of $\hbar$ is appreciable for microscopic particles as discussed in Section 3.1.1. Thus, in the description of microscopic phenomena, the sharp classical values are replaced by expectation values and an intrinsic fuzziness due to the unavoidable fluctuations. Mathematically, these fluctuations follow from the probability distribution that can be determined from the solution of the Schrödinger equation. The quantum-mechanical theory simultaneously contains wave and particle aspects, as discussed in Section 3.1.

3.2.4 Representations of position and momentum operators

The properties of the momentum and position operators become clearer if we investigate a spatial Fourier transformation and its inverse

$$\psi(\mathbf{k}, t) \equiv \frac{1}{(2\pi)^3} \int d^3r\, \psi(\mathbf{r}, t)\, e^{-i\mathbf{k}\cdot\mathbf{r}},$$
$$\psi(\mathbf{r}, t) \equiv \int d^3k\, \psi(\mathbf{k}, t)\, e^{i\mathbf{k}\cdot\mathbf{r}}. \tag{3.40}$$

Using

$$-i\hbar\nabla\psi(\mathbf{r}, t) \equiv \int d^3k\, \hbar\mathbf{k}\, \psi(\mathbf{k}, t)\, e^{i\mathbf{k}\cdot\mathbf{r}} \tag{3.41}$$

and the basic properties of the Fourier transformation, we obtain

$$\langle\hat{\mathbf{p}}\rangle = \int d^3k\, \psi^\star(\mathbf{k}, t)\hbar\mathbf{k}\, \psi(\mathbf{k}, t) = \int d^3k\, \hbar\mathbf{k}\, |\psi(\mathbf{k}, t)|^2. \tag{3.42}$$

Hence, we see that we can equally well evaluate the momentum in $\mathbf{k}$ space where the momentum operator simply becomes multiplicative, $\hat{\mathbf{p}} \equiv \hbar\mathbf{k}$. If we compare this with $\hat{\mathbf{p}} \equiv -i\hbar\nabla$, Eq. (3.28), we realize that the representation of operators depends on the basis in which the wave function is defined. We discuss the general representation of operators in Section 5.1 where we show how to specify operator properties without defining the actual basis of the wave functions.

If we consider expectation values of momentum-dependent functions, $G(\mathbf{p})$, we can compute them via

$$\langle \hat{G} \rangle = \int d^3k \; \psi^\star(\mathbf{k}, t) \, G(\hbar \mathbf{k}) \; \psi(\mathbf{k}, t) = \int d^3k \; G(\hbar \mathbf{k}) \, |\psi(\mathbf{k}, t)|^2$$
$$= \int d^3r \; \psi^\star(\mathbf{r}, t) \, G(-i\hbar\nabla) \, \psi(\mathbf{r}, t), \tag{3.43}$$

where only the $\mathbf{k}$-space expression can be written without explicitly keeping the order among $\psi^\star$, $\hat{G}$, and ψ. Similarly, the expectation value of any analytic $\mathbf{r}$-dependent function $F(\mathbf{r})$ can be evaluated using

$$\langle \hat{F} \rangle = \int d^3r \; \psi^\star(\mathbf{r}, t) F(\mathbf{r}) \, \psi(\mathbf{r}, t) = \int d^3k \; F(\mathbf{r}) \, |\psi(\mathbf{r}, t)|^2$$
$$= \int d^3k \; \psi^\star(\mathbf{k}, t) + F(i\nabla_{\mathbf{k}}) \; \psi(\mathbf{k}, t). \tag{3.44}$$

From this equation, we can deduce for the position operator $\hat{\mathbf{r}} \equiv i\nabla_{\mathbf{k}}$ in the $\mathbf{k}$ space where $\nabla_{\mathbf{k}}$ denotes that the differentiation is performed with respect to the wave vector. In the derivation of Eq. (3.44), we use the property

$$\mathbf{r}^J \psi(\mathbf{r}, t) = \int d^3k \, \psi(\mathbf{k}, t) \left[(-i\hbar\nabla_{\mathbf{k}})^J e^{+i\mathbf{k}\cdot\mathbf{r}} \right]$$
$$= \int d^3k \left[(+i\hbar\nabla_{\mathbf{k}})^J \psi(\mathbf{k}, t) \right] e^{+i\mathbf{k}\cdot\mathbf{r}}. \tag{3.45}$$

Here, $\nabla_{\mathbf{k}}$ has been converted to act upon $\psi(\mathbf{k}, t)$ via integrating by parts and using the Gauss theorem (3.26) as well as the square integrability of ψ.

To illustrate the nature of momentum-space and real-space representation, we compare the one-dimensional Gaussian wave packet (3.8) (upper row) and its Fourier transformation (lower row)

$$\psi(k_z, t) = \frac{1}{\sqrt{2\pi}} \int_{-\infty}^{\infty} dz \, \psi(z, t) \, e^{-ik_z z}$$
$$= \left[\frac{2\Delta z_0^2}{\pi} \right]^{\frac{1}{4}} e^{-\Delta z_0^2 (k_z - q_0)^2} \; e^{-ik_z z_0 - i\frac{\hbar k_z^2}{2m} t} \tag{3.46}$$

in Fig. 3.3. Here, we used the electron mass $m = 9.109\,382\,15 \times 10^{-31}$ kg, initial position $z_0 = 0$, $\Delta z_0 = 0.2$ nm, and $q_0 = 2\pi\, 10^9\,\mathrm{m}^{-1}$. Based on Eq. (3.13), we expect $|\psi(z, t)|$ to become spatially broader by a factor of $\sqrt{5} \approx 2.24$ after propagating for $2t_{\text{spread}} = 1.382$ fs. Figure 3.3(a), (b), and (c) present $|\psi(z, t)|$ (shaded area) and $\mathrm{Re}\,[\psi(z, t)]$ (solid line) for the times $t = 0$, $t = 2t_{\text{spread}}$, and $t = 4t_{\text{spread}}$. The vertical line indicates the computed $\langle \hat{z} \rangle$ for each time. We see that the center of the wave packet is moving linearly while the spatial width of the wave packet spreads, in agreement with the results in Fig. 3.2.

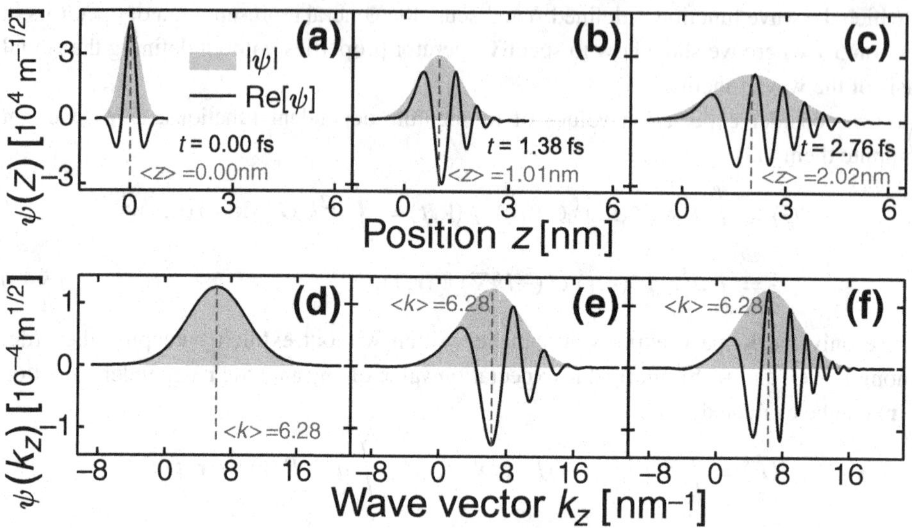

Figure 3.3 Freely propagating particle wave (top row). The real part of the wave function (solid line) and $|\psi(z,t)|$ (shaded area) are shown in real space for the time sequence **(a)** $t = 0$ fs, **(b)** $t = 1.382$ fs, **(c)** $t = 2.764$ fs. The corresponding momentum-space representations of $\psi(k_z,t)$ (solid line) and $|\psi(k_z,t)|$ are plotted in the bottom row. The vertical lines indicate the positions of the first-order momenta, i.e., $\langle \hat{z} \rangle$ and $\langle \hat{k}_z \rangle$. The time evolution of the initially Gaussian wave packet is computed from Eqs. (3.8) and (3.46) where we have set $q_0 = 2\pi 10^9\,\mathrm{m}^{-1}$, $z_0 = 0$, and $\Delta z_0 = 0.2$ nm.

As time elapses, $\mathrm{Re}\,[\psi(z,t)]$ displays faster oscillations for positions on the RHS of the center, i.e., strong chirp is observed. As discussed in Section 2.3.3, the spreading of the freely propagating wave packet in real space is directly related to the chirp.

The corresponding momentum representation is shown in Fig. 3.3(d), (e), and (f) where $\mathrm{Re}\left[\psi(k_z,t)\right]$ (solid line) is compared with $|\psi(k_z,t)|$ (shaded area). The vertical line indicates the computed average value of $\langle \hat{p}_z \rangle / \hbar$ that defines the average wave vector $\langle \hat{k}_z \rangle$. We see that even though significant chirp develops into $\mathrm{Re}\left[\psi(k_z,t)\right]$, $|\psi(k_z,t)|$ remains constant in time. As a result, $\langle \hat{k}_z \rangle$ and any fluctuation in momentum also remain constants. These observations can be confirmed by computing the probability distribution

$$|\psi(k_z,t)|^2 = \left[\frac{2\Delta z_0^2}{\pi}\right]^{\frac{1}{2}} e^{-2\Delta z_0^2 (k_z - q_0)^2} \tag{3.47}$$

from Eq. (3.46). Using $p_z = \hbar k_z$, we can express the probability distribution

$$|\psi(p_z,t)|^2 \equiv \frac{1}{\hbar}\left|\psi\left(\frac{p_z}{\hbar},t\right)\right|^2 = \left[\frac{2\Delta z_0^2}{\hbar\pi}\right]^{\frac{1}{2}} e^{-2\left(\frac{\Delta z_0}{\hbar}\right)^2 (p_z - \hbar q_0)^2} \tag{3.48}$$

in terms of the momentum. The $1/\hbar$ in front of the probability distribution is a consequence of the normalization requirement, $\int dp_z |\psi(p_z, t)|^2 = 1$. In this format, we not only verify the stationarity of the free-space momentum distribution but we also see that the fluctuations of the momentum are given by the exponent, yielding $\Delta p_z^2 = \frac{\hbar^2}{4\Delta z_0^2}$, in agreement with Eq. (3.29).

Exercises

3.1 (a) Show that Eq. (2.54) produces the Schrödinger equation $i\hbar\frac{\partial}{\partial t}\psi(z,t) = -\frac{\hbar^2}{2m}\nabla^2\,\psi(z,t)$ of free particles for $M = 1$. What is the value of β? **(b)** Derive Eq. (3.8) from Eq. (2.82). What is the probability distribution in this specific case? **(c)** Evaluate $\psi(k,t) = \frac{1}{\sqrt{2\pi}}\int dx\,\psi(z,t)\,e^{-ikz}$ by starting from Eq. (3.8).

3.2 (a) Show that $\frac{\partial}{\partial t}\int d^3r|\psi(r,t)|^2 = 0$ by starting from the Schrödinger equation (3.1). **(b)** Show that the Schrödinger equation produces the continuity equation $\frac{\partial}{\partial t}\rho_P(r,t)+\nabla\cdot j(r,t) = 0$, where $\rho_P(r,t) = |\psi(r,t)|^2$ is the probability density and $\mathbf{j}(r,t) \equiv \frac{\hbar}{2mi}\left[\psi^\star(r,t)\nabla\psi(r,t) - \psi(r,t)\nabla\psi^\star(r,t)\right]$. **(c)** Evaluate $\mathbf{j}(t) = \int d^3r\,\mathbf{j}(r,t)$ for a Gaussian wave packet (3.8).

3.3 (a) Show that $\nabla A(\mathbf{r})B(\mathbf{r}) = B(\mathbf{r})\nabla A(\mathbf{r}) + A(\mathbf{r})\nabla B(\mathbf{r})$,
$\nabla\cdot A(\mathbf{r})\mathbf{B}(\mathbf{r}) = \mathbf{B}(\mathbf{r})\cdot\nabla\mathbf{A}(\mathbf{r}) + \mathbf{A}(\mathbf{r})\nabla\cdot\mathbf{B}(\mathbf{r})$, and
$\nabla\cdot[\mathbf{A}(\mathbf{r})\mathbf{B}(\mathbf{r})] = \mathbf{B}(\mathbf{r})\nabla\cdot\mathbf{A}(\mathbf{r}) + \mathbf{A}(\mathbf{r})\cdot\nabla\mathbf{B}(\mathbf{r})$. **(b)** Verify that the relation (3.25) follows from the properties derived in part **(a)**.

3.4 Evaluate the average momentum $\langle\hat{p}\rangle$ of the Gaussian wave packet (3.8) by starting from **(a)** Eq. (3.24) and **(b)** Eq. (3.27). Verify that the results of **(a)** and **(b)** are identical. **(c)** Evaluate $\langle\hat{p}\rangle$ completely in momentum representation. **Hint:** *Start from the result of Exercise 3.1(c).*

3.5 Verify the exchange, additivity, and product rules of commutators, i.e., Eq. (3.36).

3.6 (a) Show that $e^{\hat{A}}\,\hat{B} = \hat{C} \Rightarrow \hat{B} = e^{-\hat{A}}\,\hat{C}$. **Hint:** *Taylor expand the exponential operators.* **(b)** Show that

$$\begin{aligned}\hat{F}(\lambda) &\equiv e^{\lambda\hat{S}}\,\hat{A}\,e^{-\lambda\hat{S}}\\ &= \hat{A} + \lambda[\hat{S},\hat{A}]_- + \frac{\lambda^2}{2!}[\hat{S},[\hat{S},\hat{A}]_-]_- + \frac{\lambda^3}{3!}[\hat{S},[\hat{S},[\hat{S},\hat{A}]_-]_-]_- \ldots .\end{aligned}$$

Hint: *Taylor expand* $\hat{F}(\lambda)$.

3.7 (a) Show that the expectation value $\langle\hat{F}\rangle \equiv \int d^3r\,\psi^\star(\mathbf{r},t)\,F(\mathbf{r})\,\psi(r,t)$ becomes $\langle\hat{F}\rangle = \int d^3k\,\psi^\star(\mathbf{k},t)\,F(i\nabla_k)\,\psi(k,t)$ in momentum space. **(b)** Assume that $F(\mathbf{r}) = e^{i\mathbf{k}_0\cdot\mathbf{r}}$. Show that $\langle\hat{F}\rangle = \int d^3k\,\psi^\star(\mathbf{k},t)\,\psi(\mathbf{k}-\mathbf{k}_0,t)$ for arbitrary $\psi(r,t)$, which defines the convolution theorem of the Fourier transformation. Verify this result also by Taylor expanding $F(i\nabla_{\mathbf{k}})\psi(\mathbf{k},\, t)$.

3.8 Assume that we need to express the system in new coordinates $\mathbf{R} = \mathbf{R}(\mathbf{r})$ where $\mathbf{r}$ is the old coordinate. How must we choose the new wave function $\tilde{\psi}(\mathbf{R}, t)$ such that the generic expectation value $\langle \hat{F} \rangle = \int d^3R\, \tilde{\psi}^\star(\mathbf{R}, t)\, F(\mathbf{r}(\mathbf{R}))\, \tilde{\psi}(\mathbf{R}, t) = \int d^3r\, \psi^\star(\mathbf{r}, t)\, F(\mathbf{r})\, \psi(r, t)$ remains invariant? Assume that $\psi(\mathbf{r}, t)$ and $F(\mathbf{r})$ are known analytic functions. Confirm that Eq. (3.48) satisfies the transformation.

3.9 Assume that a Gaussian momentum distribution (3.46) is split into two parts $\psi_1(k_z, t) = \mathcal{N}_1 e^{-\alpha^2(k-q_0-\Delta k)^2}\, \psi(k_z, t)$ and $\psi_2(k_z, t) = \mathcal{N}_2 \left(1 - e^{-\alpha^2(k-q_0-\Delta k)^2}\right)$ $\psi(k_z, t)$. **(a)** Normalize $\psi_j(k_z, t)$. **(b)** Compute $\psi_j(z, t)$ and show that ψ_1 and ψ_2 propagate with different speeds.

Further reading

Excellent introductions to the theory of quantum mechanics and much relevant background information on the material presented in this chapter can be found in the following books:

L. I. Schiff (1955). *Quantum Mechanics*, 2nd edition, New York, McGraw-Hill.
A. S. Davydov (1976). *Quantum Mechanics*, 2nd edition, Oxford, Pergamon Press.
A. Messiah (1999). *Quantum Mechanics*, New York, Dover.
L. D. Landau and E. M. Lifschitz (2003). *Quantum Mechanics: Nonrelativistic Theory*, 3rd edition, Amsterdam, Butterworth–Heinemann.
E. Merzbacher (1998). *Quantum Mechanics*, 3rd edition, New York, Wiley & Sons.
S. Gasiorowicz (2003). *Quantum Physics*, 3rd edition, Phoenix, Wiley & Sons.

4

Central concepts in stationary quantum theory

Whereas the time-dependent Schrödinger equation (3.1) uniquely determines the quantum dynamics of particles, it is often sufficient to use a much simpler static description. In particular, the stationary Schrödinger equation, $\hat{H}\phi_E(\mathbf{r}) = E\,\phi_E(\mathbf{r})$, defines an eigenvalue problem whose solutions can be used to obtain a relatively simple interpretation of the pertinent quantum features. For example, if the system is initially in a particular eigenstate $\phi_E(\mathbf{r})$, it stays there for all times such that all expectation values remain constant. Clearly, such a system does not have time-dependent currents that could radiate energy in the form of electromagnetic fields. Thus, quantum mechanics predicts that an atom is stable if it is in an eigenstate $\phi_E(\mathbf{r})$. This stationarity explains why quantized atoms do not collapse despite the classical description's complete failure to explain this elementary aspect.

In this chapter, we summarize how stationary quantum-mechanical properties emerge from the dynamic Schrödinger equation. Especially, we formulate the stationary Schrödinger equation in terms of a generic Helmholtz equation. We develop a transfer-matrix technique that can be applied to calculate the stationary wave functions for arbitrary one-dimensional problems. These solutions show how interference effects, tunneling phenomena, and bound states with a discrete energy spectrum relate to the wave aspects of the particles.

4.1 Stationary Schrödinger equation

Fourier transforming the Schrödinger Eq. (3.39) with respect to time yields the eigenvalue problem

$$\hbar\omega\,\psi(\mathbf{r},\omega) = H(\hat{\mathbf{r}},\hat{\mathbf{p}})\,\psi(\mathbf{r},\omega), \quad \Leftrightarrow \quad H(\hat{\mathbf{r}},\hat{\mathbf{p}})\,\phi_E(\mathbf{r}) = E\,\phi_E(\mathbf{r}). \tag{4.1}$$

Associating $\hbar\omega$ with the energy E defines the stationary Schrödinger equation with the eigenfunction $\psi(\mathbf{r},\omega)$ via $\psi_E(\mathbf{r})$. The energy eigenvalue E can be used to identify the state and thus plays the role of an energy quantum number. Using the explicit form of the Hamiltonian, Eq. (3.38), allows us to write Eq. (4.1) in the form of the Helmholtz equation (3.3),

$$\nabla^2 \psi_E(\mathbf{r}) + \mathbf{k}^2(\mathbf{r})\, \psi_E(\mathbf{r}) = 0, \quad \mathbf{k}^2(\mathbf{r}) \equiv \frac{2m}{\hbar^2}\left[E - U(\mathbf{r})\right], \tag{4.2}$$

where we identify a position- and E-dependent wave vector $\mathbf{k}(\mathbf{r})$. Since $\mathbf{k}(\mathbf{r})$ determines E, it is often convenient to define the quantum number in terms of the wave vector $\mathbf{k}_0 = \mathbf{k}(\mathbf{r}_0)$ at a fixed position $\mathbf{r}_0$. To connect Eq. (4.2) more closely to the formal eigenvalue problem discussed in Section 2.3.1, we rewrite it in the form

$$\nabla^2 \phi_{\mathbf{k}_0}(\mathbf{r}) = -\mathbf{k}^2(\mathbf{r})\, \phi_{\mathbf{k}_0}(\mathbf{r}) \tag{4.3}$$

which is a direct generalization of Eq. (2.57).

Due to the generic $\mathbf{r}$ dependence of the wave vector, this eigenvalue problem may have solutions only for certain discrete $\mathbf{k}_0$ values, as shown explicitly in Section 4.3.3. When we are dealing with particles, $\mathbf{k}_0 = \mathbf{k}(\mathbf{r}_0)$ uniquely identifies the eigenstates (see Exercise 4.1) and the orthogonality relation

$$\int d^3r\, \phi^\star_{\mathbf{k}_0}(\mathbf{r})\, \phi_{\mathbf{k}'_0}(\mathbf{r}) = \begin{cases} \delta\left(\mathbf{k}'_0 - \mathbf{k}_0\right), & \mathbf{k}_0, \mathbf{k}'_0 \text{ continuous} \\ \delta_{\mathbf{k}'_0,\mathbf{k}_0}, & \text{discrete} \end{cases} \tag{4.4}$$

of Eq. (2.60). As a minor modification of our earlier discussion, we also allow here for the possibility of discrete eigenstates.

The completeness relation (2.61) is

$$\sum_{\mathbf{k}_0} \phi^\star_{\mathbf{k}_0}(\mathbf{r})\, \phi_{\mathbf{k}_0}(\mathbf{r}') = \delta(\mathbf{r}' - \mathbf{r}). \tag{4.5}$$

Here and in the following, we always use the convention that $\sum_{\mathbf{k}_0}$ has to be replaced by the integral $\int d^3\mathbf{k}_0$ in the case of continuous states. The proof of the central orthogonality and completeness relations is the topic of Exercise 4.1. In the case where multiple eigenvalues exist for a given $\mathbf{k}_0$, we are dealing with a degenerate system. The degenerate and nondegenerate cases can be treated analogously if we assume that, e.g., the Gram–Schmidt orthogonalization procedure is used to guarantee the applicability of Eqs. (4.4)–(4.5).

The orthogonality and completeness of the states $\phi_{\mathbf{k}_0}(\mathbf{r})$ allow us to solve the full quantum dynamics, following steps analogous to those producing Eqs. (2.72)–(2.73) starting from Eqs. (2.62)–(2.63). This way, the resulting quantum dynamics can be expressed via the mode expansion

$$\psi(\mathbf{r}, t) = \sum_{\mathbf{k}_0} c_{\mathbf{k}_0}\, \phi_{\mathbf{k}_0}(\mathbf{r})\, e^{-i\omega(\mathbf{k}_0)t}, \quad \omega(\mathbf{k}_0) \equiv \frac{\hbar \mathbf{k}_0^2}{2m} + \frac{U(\mathbf{r}_0)}{\hbar}, \tag{4.6}$$

$$c_{\mathbf{k}_0} = \int d^3r\, \phi^\star_{\mathbf{k}_0}(\mathbf{r})\, \psi(\mathbf{r}, 0). \tag{4.7}$$

Again, $\sum_{\mathbf{k}_0}$ has to be converted into an integral for continuous quantum numbers. For example, the $\phi_{\mathbf{k}_0}$ solutions are ordinary plane waves with continuous $\mathbf{k}_0$ values in the absence of a potential. The corresponding $\omega(\mathbf{k}_0)$ dispersion is identical to Eq. (2.71), converting Eqs. (4.6)–(4.7) into Eqs. (2.62)–(2.63) for this special case.

If we know the wave function at the initial time and use a complete and orthogonal set of $\phi_{\mathbf{k}_0}(\mathbf{r})$ states, Eqs. (4.6)–(4.7) yield the dynamical solutions at any time. Thus, one can investigate many system aspects simply by analyzing the stationary Schrödinger equation. For example, if $\phi_{\mathbf{k}_0}(\mathbf{r})$ exists only for discrete energies $E = \hbar\omega(\mathbf{k}_0)$, we know that the wave function can oscillate only with the corresponding discrete frequencies $\omega(\mathbf{k}_0)$.

Technically, the stationary Schrödinger equation (4.1) is easier to solve than the dynamic equation (3.39) because it is an ordinary differential equation in contrast to the partial differential equation (3.39). Nonetheless, $\phi_{\mathbf{k}_0}(\mathbf{r})$ can be determined analytically only in very few exemplary cases. Solutions exist, e.g., for $U(\mathbf{r}) \propto \mathbf{r}^2$ or for $U(\mathbf{r}) \propto 1/|\mathbf{r}|$. For more complicated $U(\mathbf{r})$, one usually has to use approximative methods or invoke numerical treatments.

4.2 One-dimensional Schrödinger equation

As an illustration, we discuss the generic one-dimensional Helmholtz equation

$$\frac{\partial^2}{\partial z^2}\phi_E(z) + k^2(z)\,\phi_E(z) = 0, \qquad k(z) \equiv \sqrt{\frac{2m}{\hbar^2}\left[E - U(z)\right]}. \tag{4.8}$$

For arbitrary $k(z)$, we can only find numerical solutions. One possible method involves discretizing the z axis at the points $z_1, z_2, \cdots, z_N$, as shown in Fig. 4.1. We can then approximate the wave vector by a set of discrete values

$$k_j \equiv k\left(z'_j\right), \qquad \text{with } z_j \leq z'_j \leq z_{j+1}, \tag{4.9}$$

where z'_j is a point within the discretization interval z_j and z_{j+1}. In Fig. 4.1, $z'_j = \frac{z_j + z_{j+1}}{2}$ corresponds to the middle point of the interval. In some other cases, it is convenient to choose $z'_j = z_j$ to be the first point of the interval. The length of each interval is given by

$$l_j \equiv z_{j+1} - z_j. \tag{4.10}$$

It is clear that the set of k_j values approaches the actual $k(z)$ as the discretization is made small enough. In the limit $l_j \to 0$, the Helmholtz solution with the discretized and the actual $k(z)$ must become identical.

The convergence is typically very fast such that we may replace $k(z)$ by k_j in Eq. (4.8) to get

$$\frac{\partial^2}{\partial z^2}\phi_{k_0}(z) + k_j^2\,\phi_{k_0}(z) = 0, \quad \text{for } z_j \leq z < z_{j+1}. \tag{4.11}$$

This equation has analytic plane-wave solutions,

$$\phi_{k_0}(z) \equiv \phi_j(z) = A_j\, e^{ik_j(z-z_j)} + B_j\, e^{-ik_j(z-z_j)}, \quad \text{for } z_j \leq z < z_{j+1}, \tag{4.12}$$

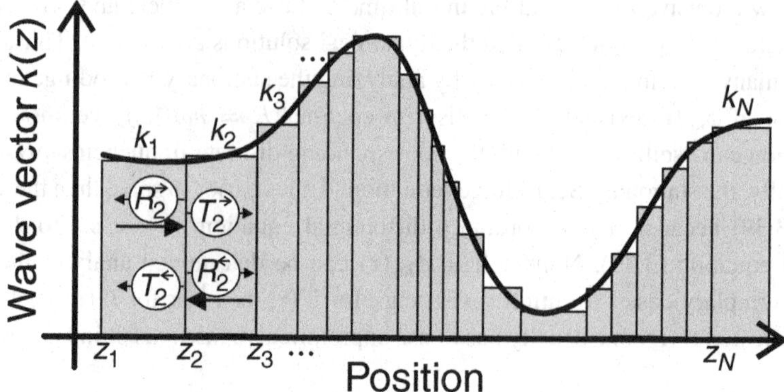

Figure 4.1 Transfer-matrix approach together with discretized one-dimensional potential. The discretization intervals are defined by z_j $(j = 1, 2, \cdots)$ and the corresponding wave vectors are given by k_j. At each boundary, the incoming field is transmitted and reflected to the next layer. The propagation direction of the incoming wave is assigned with $\rightarrow$ if it propagates from left to right; the opposite direction has the arrow $\leftarrow$. The thick arrow indicates the incoming part whereas the transmission and reflection are shown as thin arrows. The single-boundary transmission $T_j^{\rightarrow/\leftarrow}$ and reflection $R_j^{\rightarrow/\leftarrow}$ coefficients are indicated symbolically.

within each discrete interval. To obtain the complete solution, we need a rule to connect the coefficients A_j and B_j of the different intervals. The A_j coefficient defines the amplitude of the plane-wave component $e^{ik_j(z-z_j)}$ that propagates from left to right in Fig. 4.1. Thus, we associate it with the propagation direction $\rightarrow$. The opposite propagation direction, $\leftarrow$, is then connected with the B_j component. If all k_j values are equal, we recover the plane-wave solutions as the eigenstates. In this case, the application of Eqs. (4.6)–(4.7) produces the dynamic solution (3.8) for the freely propagating Gaussian wave packet.

To connect the A_j and B_j coefficients, we start from the general properties of the Helmholtz equation (4.8). We first check how the differentiation must be performed at the $z = z_{j+1}$ boundary. In this situation, the wave function can be expressed as

$$\phi_{k_0}(z) \equiv \phi_j(z)\,\theta(z_{j+1} - z) + \phi_{j+1}(z)\,\theta(z - z_{j+1}), \quad z_j \leq z < z_{j+2}, \tag{4.13}$$

where the Heaviside step function is defined to be

$$\theta(z) \equiv \int_{-\infty}^{z} dz'\,\delta(z') = \begin{cases} 0, & z < 0 \\ 1, & z \geq 0 \end{cases}$$
$$\Leftrightarrow \quad \frac{\partial}{\partial z}\theta(z - z_j) = \delta(z - z_j), \quad \frac{\partial}{\partial z}\theta(z_j - z) = -\delta(z - z_j). \tag{4.14}$$

This can also be expressed with the δ function, introduced in connection with Eq. (1.50). If we now apply $\partial/\partial z$ to Eq. (4.13) and use the product rule of differentiation, we obtain

$$\begin{aligned}\frac{\partial\,\phi_{k_0}(z)}{\partial z} &= \frac{\partial\,\phi_j(z)}{\partial z}\,\theta(z_{j+1}-z) + \phi_j(z)\,\frac{\partial\,\theta(z_{j+1}-z)}{\partial z} \\ &+ \frac{\partial\,\phi_{j+1}(z)}{\partial z}\,\theta(z-z_{j+1}) + \phi_{j+1}(z)\frac{\partial\,\theta(z-z_{j+1})}{\partial z} \\ &= \frac{\partial\,\phi_j(z)}{\partial z}\,\theta(z_{j+1}-z) + \frac{\partial\,\phi_{j+1}(z)}{\partial z}\,\theta(z-z_{j+1}) \\ &+ \left[\phi_{j+1}(z)-\phi_j(z)\right]\,\delta(z-z_{j+1}),\end{aligned} \tag{4.15}$$

where the last step follows by applying the differentiation rule (4.14) for the $\theta(z)$ function. Since the original Helmholtz equation does not contain the divergent and nonanalytic δ function, it is necessary that the prefactor of the δ-function term vanishes in the limit $z \to z_{j+1}$.

In the next step, we differentiate Eq. (4.15), producing a potentially diverging and nonanalytic contribution $\left[\frac{\partial\,\phi_{j+1}(z)}{\partial z} - \frac{\partial\,\phi_j(z)}{\partial z}\right]\,\delta(z-z_{j+1})$. Obviously, this term must also vanish for $z \to z_{j+1}$, based on the same argument as for the calculation (4.15). Thus, the Helmholtz equation (4.11) produces the two conditions

$$\phi_{j+1}(z_{j+1}) - \phi_j(z_{j+1}) = 0, \qquad \left[\frac{\partial\,\phi_{j+1}(z)}{\partial z} - \frac{\partial\,\phi_j(z)}{\partial z}\right]_{z=z_{j+1}} = 0 \tag{4.16}$$

for the wave function at the z_{j+1} interface. These relations are intuitively obvious because the second-order differentiation in Eq. (4.11) produces an analytic result only if the wave function and its first-order derivative are continuous.

By inserting the explicit plane-wave components (4.12) into Eq. (4.16), we find

$$\begin{cases} \left[A_{j+1}+B_{j+1}\right] - \left[A_j\,e^{ik_jl_j} + B_j\,e^{-ik_jl_j}\right] = 0 \\ ik_{j+1}\left[A_{j+1}-B_{j+1}\right] - ik_j\left[A_j\,e^{ik_jl_j} - B_j\,e^{-ik_jl_j}\right] = 0, \end{cases} \tag{4.17}$$

which forms a pair of linear equations with four unknowns. Thus, we can always express two of the coefficients in terms of the other two. For identification reasons, it is convenient to organize Eq. (4.17) into a matrix format

$$\begin{pmatrix} B_j \\ A_{j+1} \end{pmatrix} = \begin{pmatrix} R^{\rightarrow}_{j+1} e^{2ik_jl_j} & T^{\leftarrow}_{j+1} e^{ik_jl_j} \\ T^{\rightarrow}_{j+1} e^{ik_jl_j} & R^{\leftarrow}_{j+1} \end{pmatrix} \begin{pmatrix} A_j \\ B_{j+1} \end{pmatrix}. \tag{4.18}$$

Using Eq. (4.17), we identify the transmission and the reflection coefficients

$$\begin{aligned} T^{\rightarrow}_{j+1} &= \frac{2k_j}{k_j+k_{j+1}}, & R^{\rightarrow}_{j+1} &= \frac{k_j-k_{j+1}}{k_j+k_{j+1}} \\ T^{\leftarrow}_{j+1} &= \frac{2k_{j+1}}{k_j+k_{j+1}}, & R^{\leftarrow}_{j+1} &= \frac{k_{j+1}-k_j}{k_j+k_{j+1}}. \end{aligned} \tag{4.19}$$

The physical interpretation follows if we analyze a situation where the region $z_{j+1} \leq z < z_{j+2}$ has no component B_{j+1} propagating toward the z_{j+1} boundary, see Fig. 4.1 for $j = 1$. Thus, only the A_j component within the region $z_j \leq z < z_{j+1}$ propagates to the z_{j+1} interface, allowing us to identify this as the incoming field. Equation (4.18) then predicts that the incoming light is partially reflected from the z_{j+1} interface with the amplitude $B_j = R^{\rightarrow}_{j+1} e^{2ik_j l_j} A_j$ to the region $z < z_{j+1}$ while it is partially transmitted to the region $z > z_{j+1}$ with the amplitude $A_{j+1} = T^{\rightarrow}_{j+1} e^{ik_j l_j} A_j$. If the incoming field has the opposite propagation direction, i.e., $\leftarrow$, we must set A_j to zero while B_{j+1} defines the amplitude. In this context, the transmission and reflection coefficients follow from $T^{\leftarrow}_{j+1}$ and $R^{\leftarrow}_{j+1}$, respectively. Thus, the identified T and R coefficients define simple transmission and reflection coefficients through an interface. The arrows $\rightarrow$ and $\leftarrow$ indicate the direction of the incoming field. These two transmission-reflection scenarios are indicated in Fig. 4.1 at the boundary z_2.

The bifurcation of waves into transmitted and reflected branches was introduced phenomenologically in Section 2.2.2 where the classical paths corresponding to light rays were studied. In our current analysis, we generalize this treatment by including the full wave aspects instead of just the slowly varying $e^{\pm i \int dz\, k(z)}$ path in Eq. (2.47). Thus, the pure wave aspects produce transmission and reflection coefficients for each path. In classical optics, these are known as the Fresnel coefficients.

When we actually solve Eq. (4.11), it is convenient to rearrange Eq. (4.18) into

$$\begin{pmatrix} A_{j+1} \\ B_{j+1} \end{pmatrix} = \frac{1}{T^{\leftarrow}_{j+1} e^{ik_j l_j}} \begin{pmatrix} D_{j+1} e^{2ik_j l_j} & R^{\leftarrow}_{j+1} \\ -R^{\rightarrow}_{j+1} e^{2ik_j l_j} & 1 \end{pmatrix} \begin{pmatrix} A_j \\ B_j \end{pmatrix},$$
$$\text{with } D_j \equiv T^{\leftarrow}_j T^{\rightarrow}_j - R^{\leftarrow}_j R^{\rightarrow}_j, \tag{4.20}$$

because one can then determine the (A_{j+1}, B_{j+1}) coefficients based on the knowledge of (A_j, B_j) in the previous layer. If we insert the relations (4.19) into D_{j+1} we always find $D_j = 1$ within all layers. To simplify the notation further, we introduce a vector and a matrix

$$\mathbf{s}_j \equiv \begin{pmatrix} A_j \\ B_j \end{pmatrix}, \quad \mathbf{M}_{j+1} \equiv \frac{1}{T^{\leftarrow}_{j+1} e^{ik_j l_j}} \begin{pmatrix} D_{j+1} e^{2ik_j l_j} & R^{\leftarrow}_{j+1} \\ -R^{\rightarrow}_{j+1} e^{2ik_j l_j} & 1 \end{pmatrix}, \tag{4.21}$$

respectively. Clearly, $\mathbf{s}_j$ defines the plane-wave components in a given layer and $\mathbf{M}_{j+1}$ determines how the coefficients between two adjacent layers are connected. This way, we can present Eq. (4.20) compactly via

$$\mathbf{s}_{j+1} = \mathbf{M}_{j+1}\, \mathbf{s}_j, \tag{4.22}$$

which determines the fundamental step yielding the transfer-matrix approach.

We start from the coefficient $\mathbf{s}_0$ of layer $j = 0$ and connect it to $\mathbf{s}_j$ of any arbitrary layer j, by applying Eq. (4.22) j times. The transfer-matrix approach becomes then

$$\begin{aligned}\mathbf{s}_j &= \mathbf{M}_j\,\mathbf{s}_{j-1} = \mathbf{M}_j\mathbf{M}_{j-1}\,\mathbf{s}_{j-2} = \cdots \\ &= \mathbf{M}_j\mathbf{M}_{j-1}\cdots\mathbf{M}_1\,\mathbf{s}_0 \equiv \mathcal{M}_{j,0}\,\mathbf{s}_0.\end{aligned} \tag{4.23}$$

Here, we identified the product of the matrices to be the transfer matrix $\mathcal{M}_{j,0}$ connecting the coefficient of layers 0 and j. Obviously, the plane-wave coefficients in one layer, e.g., $\mathbf{s}_0$, define $\mathbf{s}_j$ in all other layers. Thus, we can determine the eigenfunction $\phi_j(z)$ in Eq. (4.12) simply by multiplying the $\mathbf{M}_j$ matrices multiple times. The corresponding numerical task involves N multiplications of complex 2×2 matrices, which makes the transfer-matrix approach very efficient. Once the relevant $\mathcal{M}_{j,0}$ matrices are known, we can directly generate the eigenstate by using Eq. (4.12) defined uniquely by $\mathbf{s}_0$.

We can find the physical interpretation of the transfer matrix using

$$\begin{aligned}&\mathcal{M}_{j,l} = \mathbf{M}_j\mathbf{M}_{j-1}\cdots\mathbf{M}_{l+1} \equiv \frac{1}{\mathcal{T}^{\leftarrow}_{j,l}}\begin{pmatrix} \mathcal{D}_{j,l} & \mathcal{R}^{\leftarrow}_{j,l} \\ -\mathcal{R}^{\rightarrow}_{j,l} & 1 \end{pmatrix}, \qquad j > l, \\ &\mathcal{M}_{j,l} = \left[\mathcal{M}_{l,j}\right]^{-1}, \qquad j < l, \\ &\mathcal{M}_{j,j} = \mathbb{I}_{2\times 2}, \qquad \mathcal{D}_{j,l} \equiv \mathcal{T}^{\leftarrow}_{j,l}\mathcal{T}^{\rightarrow}_{j,l} - \mathcal{R}^{\leftarrow}_{j,l}\mathcal{R}^{\rightarrow}_{j,l}\end{aligned} \tag{4.24}$$

which is a direct generalization of Eq. (4.21). This identification is always possible because the complex-valued 2×2 matrix is still expressed with four independent complex-valued variables. By using a similar analysis as in Eqs. (4.18)–(4.20), we find that $\mathcal{T}^{\rightarrow}_{j,l}$ and $\mathcal{R}^{\rightarrow}_{j,l}$ [$\mathcal{T}^{\leftarrow}_{j,l}$ and $\mathcal{R}^{\leftarrow}_{j,l}$] determine the transmission and reflection coefficients between the regions limited by z_l and z_j if the incoming field has the propagation direction $\rightarrow$ [$\leftarrow$]. Equation (4.24) fully defines $\mathcal{M}_{j,l}$ such that we can choose the indices j and l completely arbitrarily.

4.3 Classification of stationary eigenstates

In many relevant problems, the particle has an energy E that exceeds the potential energy. This situation is the only possible scenario for a moving particle in classical mechanics. Classical paths involve a positive definite kinetic energy, which is possible only within the regions where $E > U(z)$. Equation (4.12) shows that the corresponding quantum-mechanical solution consists of propagating plane waves because the wave vectors that appear are real valued. Thus, we can make a general classification that quantum particles are described by propagating waves within a given region whenever classical trajectories are allowed there. However, we should keep in mind that the quantum description leads to significant modifications of the classical results. For example, the wave aspects can produce a bifurcation of paths at each z_j interface and the counterpropagating plane waves can lead to an interference pattern due to the built-in wave averaging in the solution (4.12).

Due to the intrinsic wave aspects, the stationary Schrödinger equation (4.3) also has solutions in regions where E is lower than the local potential energy. Since these solutions exist within classically forbidden regions, this branch has no classical analogy. Nonetheless, the developed transfer-matrix formalism can be directly applied by replacing the real-value wave vector by a purely imaginary one,

$$k_j = k(z_j) \equiv \sqrt{\frac{2m}{\hbar^2}\left[E - U(z_j)\right]} = i\sqrt{\frac{2m}{\hbar^2}\left[U(z_j) - E\right]} \equiv i\kappa_j, \tag{4.25}$$

which follows directly from Eq. (4.8). If the condition $E < U(z_j)$ is satisfied, the identified κ_j is real valued and the solution (4.12) is converted into

$$\phi_j(z) = A_j\, e^{-\kappa_j(z-z_j)} + B_j\, e^{\kappa_j l_j} e^{-\kappa_j(z_{j+1}-z)}, \quad \text{for } z_j \le z < z_{j+1}. \tag{4.26}$$

This expression shows that the solution decays exponentially from the z_j to z_{j+1} interface for the A_j component and from the z_{j+1} to z_j boundary for the B_j component. Thus, a quantum-mechanical particle can penetrate into a classically forbidden region as an exponentially decaying tail. The effect is known as particle tunneling. Under suitable conditions, the Schrödinger equation (4.3) produces such tunneling solutions within classically forbidden regions. A similar scenario also exists for electromagnetic waves. Here, the decaying contributions are often referred to as evanescent solutions.

In a situation where classically allowed regions are surrounded only by classically forbidden regions, the condition of square normalizability eliminates the mathematically possible, exponentially growing contributions from the true asymptotic solution. We show in Section 4.3.3 that then only states with a discrete set of energies can satisfy the boundary conditions. Thus, the existing boundary conditions lead to a discretization of the energy and confine the solutions mostly within the classically allowed regions. This solution type yields the bound states for both particle and electromagnetic waves.

4.3.1 Propagating solutions

Next, we elaborate on the properties of propagating solutions by considering a situation where, in the regions $z_j \le z < z_{j+1}$ and $z_l < z < z_{l+1}$, we have energies that exceed the respective local potential. In this case, the corresponding wave vectors k_j and k_l are real valued, see Eq. (4.8). The investigation in Appendix A provides us with the general relations,

$$\begin{aligned} &\mathcal{T}^{\leftarrow}_{j,l} = \frac{k_j}{k_l}\mathcal{T}^{\rightarrow}_{j,l}, \quad \left[\mathcal{T}^{\leftarrow}_{j,l}\right]^{\star}\mathcal{R}^{\rightarrow}_{j,l} + \mathcal{T}^{\leftarrow}_{j,l}\left[\mathcal{R}^{\leftarrow}_{j,l}\right]^{\star} = 0, \\ &\frac{k_j}{k_l}\left|\mathcal{T}^{\rightarrow}_{j,l}\right|^2 + \left|\mathcal{R}^{\rightarrow}_{j,l}\right|^2 = 1, \quad \frac{k_l}{k_j}\left|\mathcal{T}^{\leftarrow}_{j,l}\right|^2 + \left|\mathcal{R}^{\leftarrow}_{j,l}\right|^2 = 1, \quad \text{if } k_j, k_l \in \mathbb{R}, \end{aligned} \tag{4.27}$$

among the coefficients defining the transfer matrix.

If the system is characterized by the real-valued k_0 in the asymptotic region $z < z_1$ and by k_N in the other asymptotic region $z > z_N$, we can construct the solution

$$\phi_{k_0}^{\rightarrow}(z) = \frac{1}{\sqrt{2\pi}} \begin{cases} e^{+ik_0(z-z_0)} + \mathcal{R}_{N,0}^{\rightarrow}\, e^{-ik_0(z-z_0)}, & z < z_0 \\ \phi_j^{\rightarrow}(z), \;\; j = 0, \cdots, N-1 & z_j \le z < z_{j+1} \\ \mathcal{T}_{N,0}^{\rightarrow}\, e^{+ik_N(z-z_N)}, & z \ge z_N \end{cases} \tag{4.28}$$

for the wave propagating in the direction $\rightarrow$. The counterpropagating solution is

$$\phi_{k_0}^{\leftarrow}(z) = \sqrt{\frac{k_N}{2\pi\, k_0}} \begin{cases} \mathcal{T}_{N,0}^{\leftarrow}\, e^{-ik_0(z-z_N)}, & z < z_0 \\ \phi_j^{\leftarrow}(z), \;\; j = 0, \cdots, N-1 & z_j \le z < z_{j+1} \, . \\ e^{-ik_N(z-z_0)} + \mathcal{R}_{N,0}^{\leftarrow}\, e^{+ik_N(z-z_0)}, & z \ge z_N \end{cases} \tag{4.29}$$

One can show (see Exercise 4.5) that these solutions are orthogonalized via

$$\int d^3r \left[\phi_{\mathbf{k}_0}^{\lambda}(\mathbf{r})\right]^{\star} \phi_{\mathbf{k}_0'}^{\lambda'}(\mathbf{r}) = \delta_{\lambda',\lambda}\, \delta\left(\mathbf{k}_0' - \mathbf{k}_0\right), \tag{4.30}$$

where λ and λ' refer to either the propagating $\rightarrow$ or counterpropagating $\leftarrow$ state.

Interestingly, many properties of $\phi_{k_0}^{\rightarrow}(z)$ can be deduced from its behavior in the asymptotic regions $z < z_0$ and $z > z_N$. For example, the probability distribution

$$\left|\phi_{k_0}^{\rightarrow}(z)\right|^2 = \frac{1}{2\pi} \begin{cases} 1 + \left|\mathcal{R}_{N,0}^{\rightarrow}\right|^2 + 2\left|\mathcal{R}_{N,0}^{\rightarrow}\right| \cos 2\left[k_0 z - \varphi_{0,R}\right], & z < z_0 \\ \left|\phi_j^{\rightarrow}(z)\right|^2, \;\; j = 0, \cdots, N-1 & \text{otherwise} \\ \left|\mathcal{T}_{N,0}^{\rightarrow}\right|^2, & z \ge z_N \end{cases} \tag{4.31}$$

has significant contributions in the asymptotic regions. The phase factor $\varphi_{0,R} \equiv k_0 z_0 + \varphi_R$ contains the phase of the reflected wave $\mathcal{R}_{N,0}^{\rightarrow} = |\mathcal{R}_{N,0}^{\rightarrow}| e^{\varphi_R}$.

In the region $z < z_0$, the probability distribution consists of a constant $1 + |\mathcal{R}_{N,0}^{\rightarrow}|^2$ and a $\cos 2k_0 z$-dependent interference contribution, in full analogy to the wave-averaging investigation in Section 2.1.2. In the present case, we have the interference between the incoming particle wave with wave vector $q_1 = k_0$, and its reflected part with wave vector $q_2 = -k_0$. In full agreement with Eq. (2.26), the $\cos 2k_0 z$-dependent interference pattern has a period, $\Delta z_{\text{dip}} = 2\pi / |q_2 - q_1| = \pi / k_0$, that defines the distance between two consecutive dips in the interference pattern. Based on the definition (2.28), the contrast of the interference

$$C = \frac{\left(1 + \left|\mathcal{R}_{N,0}^{\rightarrow}\right|\right)^2 - \left(1 - \left|\mathcal{R}_{N,0}^{\rightarrow}\right|\right)^2}{\left(1 + \left|\mathcal{R}_{N,0}^{\rightarrow}\right|\right)^2 + \left(1 - |\mathcal{R}_{N,0}^{\rightarrow}|\right)^2} = \frac{2\left|\mathcal{R}_{N,0}^{\rightarrow}\right|}{1 + \left|\mathcal{R}_{N,0}^{\rightarrow}\right|^2} \tag{4.32}$$

depends only on the magnitude of $|\mathcal{R}_{N,0}^{\rightarrow}|$. This result is the same as the generic Eq. (2.29) where the ratio r is replaced by the reflection coefficient $|\mathcal{R}_{N,0}^{\rightarrow}|$. As shown in Fig. 2.2(b), the interference pattern is most pronounced if $|\mathcal{R}_{\vec{N},0}|$ is equal to one. These aspects are typical hallmarks of the wave averaging that is indeed an intrinsic feature of the stationary wave functions. In the analyzed case, the incoming and reflected components produce a pronounced interference pattern whenever their relative amplitudes are close to equal.

Even though, $\phi_{\vec{k_0}}(z)$ is not normalized to one, the absolute square $|\phi_{\vec{k_0}}(z)|^2$ still resembles a probability-density distribution. Thus, the integral

$$\int_{z_0 - n\Delta z_{\text{dip}}}^{z_0} dz \, \left|\phi_{\vec{k_0}}(z)\right|^2 = \left(1 + |\mathcal{R}_{\vec{N},0}|^2\right) n\Delta z_{\text{dip}} \tag{4.33}$$

tells us that the interference pattern does not alter the overall probability of finding the particle within the asymptotic region $z < z_0$ because n can be made an arbitrarily large integer. Hence, the probabilities of finding an incoming or reflected component can be treated additively, as discussed in Section 1.2.2. In particular, the state has an incoming component with probability one and it is reflected with probability $|\mathcal{R}_{\vec{N},0}|^2$.

Since the probability of detecting the transmitted and reflected components must add up to one, the transmission probability must be $1 - |\mathcal{R}_{\vec{N},0}|^2$. The relation (4.27) produces the transmission probability of $\frac{k_N}{k_0}|\mathcal{T}_{\vec{N},0}|^2$. This follows directly from

$$\int_{z_N}^{z_N + \Delta z_T} dz \, \left|\phi_{\vec{k_0}}(z)\right|^2 = \left|\mathcal{T}_{\vec{N},0}\right|^2 \Delta z_T \tag{4.34}$$

if we set $\Delta z_T = \frac{k_N}{k_0} n\Delta z_{\text{dip}}$ and compare it with the overall probability in the region $z_0 - n\Delta z_{\text{dip}} < z < z_0$. The matching of the different regions can also be written via $\Delta z_T / v_N = n\Delta z_{\text{dip}} / v_0$ if we associate a classical velocity $v_j \equiv \hbar k_j / m$ with each plane-wave component. Thus, the probabilities in the regions $z < z_0$ and $z > z_N$ can be compared only if a classical particle passes them within a matching propagation time $t_j = \Delta z_j / v_j$. This property is transmitted to the quantum realm as we can see from the fact that the center of the wave packet propagates classically in free space, see Eq. (3.11). In other words, the transmitted or reflected particle must spend the same amount of time within the region of interest (ROI) before different $\int_{\text{ROI}} dz \, |\phi_{\vec{k_0}}(z)|^2$ can be compared with each other in a probabilistic sense.

The analysis above can also be made for the counterpropagating component. Again, we see that the reflection produces an interference pattern that enhances $|\phi_{\vec{k_0}}(z)|^2$ locally up to a factor of four or decreases it maximally down to zero at the dips. Therefore, the overall transmission and reflection probabilities are identified without the interference pattern,

$$\begin{aligned} T^{\text{prob}}_{k_0,\rightarrow} &= \frac{k_N}{k_0}\left|\mathcal{T}_{\vec{N},0}\right|^2, \quad R^{\text{prob}}_{k_0,\rightarrow} = \left|\mathcal{R}_{\vec{j},l}\right|^2, \\ T^{\text{prob}}_{k_0,\leftarrow} &= \frac{k_0}{k_N}\left|\mathcal{T}_{\overleftarrow{N},0}\right|^2, \quad R^{\text{prob}}_{k_0,\leftarrow} = \left|\mathcal{R}_{\overleftarrow{N},0}\right|^2, \quad k_0, k_N \in \mathbb{R} \end{aligned} \tag{4.35}$$

i.e., they always add up to one according to Eq. (4.27).

4.3.2 Tunneling solutions

When we have a situation where a tunneling solution exists in only one of the asymptotic regions, the total wave function (4.28) can be written as

$$\phi_{k_0}^{\rightarrow}(z) = \frac{1}{\sqrt{2\pi}} \begin{cases} e^{+ik_0(z-z_0)} + \mathcal{R}_{N,0}^{\rightarrow} e^{-ik_0(z-z_0)}, & z < z_0 \\ \phi_j^{\rightarrow}(z), \; j = 0, \cdots, N-1 & z_j \le z < z_{j+1} \\ \mathcal{T}_{N,0}^{\rightarrow} e^{-\kappa_N(z-z_N)}, & z \ge z_N \end{cases} \tag{4.36}$$

if the region $z > z_N$ is classically forbidden. In this region, the propagating plane-wave solution is converted into the exponentially decaying tunneling solution. If we also consider the counterpropagating wave (4.29), we find

$$\phi_{k_0}^{\leftarrow}(z) = \sqrt{\frac{k_N}{2\pi k_0}} \begin{cases} \mathcal{T}_{N,0}^{\leftarrow} e^{-ik_0(z-z_N)}, & z < z_0 \\ \phi_j^{\leftarrow}(z), \; j = 0, \cdots, N-1 & z_j \le z < z_{j+1} \, . \\ e^{+\kappa_N(z-z_0)} + \mathcal{R}_{N,0}^{\leftarrow} e^{-\kappa_N(z-z_0)}, & z \ge z_N \end{cases} \tag{4.37}$$

This branch contains an exponentially growing "incoming" component which grows without bound and thus cannot be normalized. Therefore, it must be excluded from the solution. Generally, we have to impose the boundary condition

$$\lim_{\mathbf{r}\to\infty} \phi_{\mathbf{r}_0}(\mathbf{r})\Big|_{\mathbf{r}\in V_{\text{bound}}} = 0 \tag{4.38}$$

if $\mathbf{r} \in V_{\text{bound}}$ denotes the forbidden region where no classical solution exists.

To see that the remaining tunneling solution is indeed orthogonal, we need to consider the general conservation laws,

$$\mathcal{T}_{j,l}^{\leftarrow} = \frac{i\kappa_j}{k_l}\mathcal{T}_{j,l}^{\rightarrow}, \quad \left[\mathcal{T}_{j,l}^{\leftarrow}\right]^\star + \mathcal{T}_{j,l}^{\leftarrow}\left[\mathcal{R}_{j,l}^{\leftarrow}\right]^\star = 0,$$
$$\left[\mathcal{R}_{j,l}^{\leftarrow}\right]^\star - \left[\mathcal{R}_{j,l}^{\leftarrow}\right]^\star = \frac{k_l}{i\kappa_j}\mathcal{T}_{j,l}^{\rightarrow}, \quad \text{if } \kappa_j, k_l \in \mathbb{R}, \tag{4.39}$$

derived in Appendix A. With the help of these relations, it is straightforward to show that

$$\int d^3r \left[\phi_{\mathbf{k}_0}^{\rightarrow}(\mathbf{r})\right]^\star \phi_{\mathbf{k}_0'}^{\rightarrow}(\mathbf{r}) = \delta\left(\mathbf{k}_0' - \mathbf{k}_0\right) \tag{4.40}$$

is also satisfied if tunneling is included in the description.

To illustrate the basic features of tunneling, we consider a step-like potential

$$U(z) = \theta(z)\, U_0. \tag{4.41}$$

In this situation, we have a single boundary at $z_0 = 0$ such that the propagating solution, Eq. (4.36), reduces to

$$\begin{aligned} \phi_k^{\rightarrow}(z) &= \frac{\theta(-z)}{\sqrt{2\pi}}\left[e^{+ikz} + R_0^{\rightarrow} e^{-ikz}\right] + \frac{\theta(-z)}{\sqrt{2\pi}} T_0^{\rightarrow} e^{-\kappa z} \\ &= \frac{\theta(-z)}{\sqrt{2\pi}}\left[e^{+ikz} + \frac{k-i\kappa}{k+i\kappa} e^{-ikz}\right] + \frac{\theta(z)}{\sqrt{2\pi}} \frac{2k}{k+i\kappa} e^{-\kappa z} \end{aligned} \tag{4.42}$$

because we can directly apply the single-boundary relation (4.19). The corresponding tunneling component has

$$\kappa = \sqrt{\frac{2m}{\hbar^2}U_0 - k^2} \tag{4.43}$$

based on Eq. (4.25). It is interesting to see that the reflection probability is $|R_0^{\rightarrow}|^2 = 1$, which implies total reflection.

To find a dynamical solution, we assume that the initial momentum distribution is a Gaussian

$$c_k^{\rightarrow} = \left[\frac{2\Delta z_0^2}{\pi}\right]^{\frac{1}{4}} e^{-\Delta z_0^2(k-q_0)^2}\, e^{-ikZ} \tag{4.44}$$

in analogy to Eq. (3.46). The counterpropagating component vanishes because of the condition (4.38). We choose the initial value for the average position Z to be far away from the tunneling region relative to the width Δz_0 of the wave packet. The wave packet then approaches the tunneling region with the velocity $v_0 = \hbar q_0/m$.

To study the tunneling solutions, we assume that the average wave vector q_0 corresponds to the energy $E = \hbar^2 q_0^2/2m$ much smaller than U_0 such that the incoming wave packet can only penetrate into the region $z > 0$ via tunneling. After inserting Eq. (4.44) into (4.6), we find

$$\begin{aligned}
\psi(z,t) &= \int dk\, c_k^{\rightarrow}\, \phi_{\mathbf{k}}^{\rightarrow}(z)\, e^{-i\frac{\hbar k^2}{2m}t} \\
&= \theta(-z)\left[\frac{1}{2\pi\,\Delta z^2(t)}\right]^{\frac{1}{4}} e^{-\frac{(z-Z-v_0t)^2}{4\Delta z_0 \Delta z(t)}}\, e^{iq_0(z-Z)-i\frac{\hbar q_0^2}{2m}t} \\
&\quad + \theta(-z)\left[\frac{2\Delta z_0^2}{\pi}\right]^{\frac{1}{4}} \int dk\, \frac{k-i\kappa}{k+i\kappa}\, e^{-\Delta z^2(k_z-q_0)^2}\, e^{-ik(z+Z)-i\frac{\hbar k^2}{2m}t} \\
&\quad + \theta(z)\left[\frac{2\Delta z_0^2}{\pi}\right]^{\frac{1}{4}} \int dk\, \frac{2k}{k+i\kappa}\, e^{-\Delta z^2(k_z-q_0)^2}\, e^{ik(z-Z)-i\frac{\hbar k^2}{2m}t},
\end{aligned} \tag{4.45}$$

where the incoming field produces the free propagation result (3.8) while the transmission and reflection integrals have to be solved numerically.

Figure 4.2 displays an electron wave packet penetrating a tunneling barrier at $z_0 = 0$. For this example, we used the average wave vector $q_0 = 8.98\,\text{nm}^{-1}$, a region with the potential $U_0 = 7.67\,\text{eV}$, and an electron that is initially centered at $Z = -1.29\,\text{nm}$. Since the wave packet has an initial width $\Delta z_0 = 0.3\,\text{nm}$, it is sufficiently far from the tunneling region. The average electron energy $\hbar^2 q_0^2/2m = 3.07\,\text{eV}$ is set to be 2.5 times lower than U_0 such that we may expect tunneling to appear in the barrier region (light shaded area). With these numbers, the center of the Gaussian reaches the $z_0 = 0$ step in 1.24 fs. Since $\kappa_0 = 10.99\,\text{nm}^{-1}$, we expect a tunneling depth of a few hundred Ångströms.

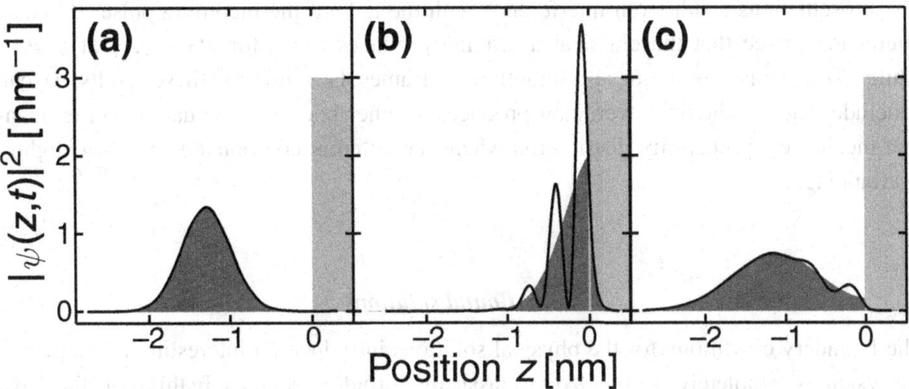

Figure 4.2 Tunneling of a wave packet into a classically forbidden barrier. The wave packet (solid line) is shown together with the tunneling barrier (light area). The dark area presents the artificial result of classical averaging. The time evolution is shown **(a)** for the initial time, **(b)** at $t = 1.24$ fs if the wave packet arrives at the barrier, and **(c)** at $t = 2.49$ fs if the wave packet exits the barrier. The results are computed using Eq. (4.45) with $Z = -1.29$ nm, $\Delta z_0 = 0.3$ nm, $q_0 = 8.98\,\text{nm}^{-1}$, and $U_0 = 7.67$ eV for an electron.

The quantum dynamics of the wave function is obtained by numerically integrating the transmitted $\phi_T(z,t)$ and reflected $\phi_R(z,t)$ components of Eq. (4.45). These are added together with the incoming part using the wave averaging, as indicated in Eq. (4.45). The solid lines in Fig. 4.2 present the actual probability distribution $|\psi(z,t)|^2$. The dark shaded area shows the corresponding "classical" probability distribution

$$P_{\text{class}}(z,t) \equiv \theta(-z)\left[|\phi_{\text{in}}(z,t)|^2 + |\phi_R(z,t)|^2\right] + \theta(-z)|\phi_T(z,t)|^2 \tag{4.46}$$

that is obtained by adding the probabilities of incoming, reflected, and transmitted parts via the classical averaging, based on the discussion in Section 1.2.2.

As we see in Fig. 4.2(a), the initial $t = 0$ probability distribution is a well-localized Gaussian positioned left from the potential wall. At the time $t = 1.24$ fs, the center of the incoming pulse is exactly at the $z = 0$ boundary of the potential wall. Figure 4.2(b) shows that a significant portion of the incoming field is already reflected back. In particular, the incoming and reflected parts produce a pronounced interference pattern, as discussed in connection with Eq. (4.32). Since the reflection is nearly perfect, we observe the maximum contrast of $C = 1$. Comparing this result with $P_{\text{class}}(z,t)$, we realize that the peaks of the $|\psi(z,t)|^2$ interference pattern are two times higher than the results obtained via classical averaging. Besides the interference, $|\psi(z,t)|^2$ displays an exponentially decaying tail in the barrier region, indicating that tunneling occurs. With the current choice of $1/\kappa_0$, the tail penetrates a few hundred Ångströms into the barrier.

At the later time $t = 2.49$ fs, we mostly see the reflected contribution in Fig. 4.2(c). There are no propagating components within the potential, implying that the electron penetrated it only for a short time. We also see mild oscillations in the reflected part.

These oscillations result from interference with the tails of the incoming pulse. It is also interesting to see that the classical averaging produces kinks for $P_{\text{class}}(z,t)$ at $z=0$ while $|\psi(z,t)|^2$ is continuous and smooth in all frames. Generalizing these results, we can conclude that the classical averaging produces a rather poor approximation to the quantum mechanical probability distribution whenever different components are even slightly overlapping.

4.3.3 Bound solutions

The boundary conditions for the physical solutions introduce an interesting new effect if the system is completely surrounded by classically forbidden regions. In this case, the wave function must decay to zero, i.e., $\lim_{|\mathbf{r}|\to\infty}\phi_{\mathbf{k}_0}(\mathbf{r})=0$. This boundary condition induces more restrictions than the wave function has degrees of freedom. As a general consequence, the boundary condition can be satisfied only by a discrete set of quantum numbers $\mathbf{k}_0$ with the consequence that the energy of the system becomes quantized.

To see the quantization explicitly, we consider a one-dimensional system with classically forbidden regions on both sides, $z\to\pm\infty$. We can still analyze this configuration as a general transfer-matrix problem where the $\mathbf{s}_0$ part in the region $z_0<z<z_1$ is connected with the $\mathbf{s}_{-\infty}$ and $\mathbf{s}_{+\infty}$ parts. Equation (4.23) tells us that these pieces are connected via

$$\begin{aligned}\mathbf{s}_{+\infty}&=\mathcal{M}_{+\infty,0}\,\mathbf{s}_0,\quad\text{and}\quad \mathbf{s}_0=\mathcal{M}_{0,-\infty}\,\mathbf{s}_{-\infty}\\ &\Rightarrow\ \mathbf{s}_{+\infty}=\mathcal{M}_{+\infty,0}\,\mathcal{M}_{0,-\infty}\,\mathbf{s}_{-\infty}.\end{aligned}\tag{4.47}$$

All physical solutions in the asymptotic regions can only have exponentially decaying components, as shown in connection with Eqs. (4.36)–(4.37). As a result, Eq. (4.38) produces two boundary conditions

$$\mathbf{s}_{+\infty}\equiv\begin{pmatrix}A_{+\infty}\\0\end{pmatrix},\qquad \mathbf{s}_{-\infty}\equiv\begin{pmatrix}0\\B_{-\infty}\end{pmatrix}.\tag{4.48}$$

Inserting these into Eq. (4.23) produces

$$\begin{aligned}\begin{pmatrix}A_{+\infty}\\0\end{pmatrix}&=\frac{1}{\mathcal{T}^{\leftarrow}_{\infty,0}\mathcal{T}^{\leftarrow}_{0,-\infty}}\begin{pmatrix}\mathcal{D}_{\infty,0}&\mathcal{R}^{\leftarrow}_{\infty,0}\\-\mathcal{R}^{\rightarrow}_{\infty,0}&1\end{pmatrix}\\ &\quad\times\begin{pmatrix}\mathcal{D}_{0,-\infty}&\mathcal{R}^{\leftarrow}_{0,-\infty}\\-\mathcal{R}^{\rightarrow}_{0,-\infty}&1\end{pmatrix}\begin{pmatrix}0\\B_{-\infty}\end{pmatrix},\end{aligned}\tag{4.49}$$

where we have expressed $\mathcal{M}$ via Eq. (4.24). Evaluating the products yields

$$\begin{pmatrix}A_{+\infty}\\0\end{pmatrix}=\frac{B_{-\infty}}{\mathcal{T}^{\leftarrow}_{\infty,0}\mathcal{T}^{\leftarrow}_{0,-\infty}}\begin{pmatrix}\mathcal{D}_{\infty,0}\mathcal{R}^{\leftarrow}_{0,-\infty}-\mathcal{R}^{\leftarrow}_{\infty,0}\\1-\mathcal{R}^{\rightarrow}_{\infty,0}\mathcal{R}^{\leftarrow}_{0,-\infty}\end{pmatrix}.\tag{4.50}$$

The only nontrivial solution is found if the product of the reflection coefficients

$$\mathcal{R}^{\leftarrow}_{0,-\infty}\mathcal{R}^{\rightarrow}_{\infty,0} = 1, \tag{4.51}$$

is equal to 1. This condition cannot be satisfied for all k_0 values. Instead, we find only discrete eigenvalues verifying that the boundary condition $\lim_{|\mathbf{r}|\to 0}\phi_{\mathbf{k}_0}(\mathbf{r}) = 0$ indeed enforces energy quantization.

In fact, Eq. (4.51) has a rather natural explanation. If the wave function vanishes at all boundaries, any propagating solution must be reflected back to its original position. If we now consider a loop where a wave starts from z_0, it is first reflected from $+\infty$ and then from $-\infty$ to return back to z_0. This loop formally has a quantum-mechanical amplitude $\mathcal{R}^{\leftarrow}_{0,-\infty}\mathcal{R}^{\rightarrow}_{\infty,0}$. For any bound state, this process must have probability 1 and the overall phase must be unchanged. This aspect is enforced by the condition (4.51). Otherwise, each additional closed loop would produce a different phase, which would lead to a wave averaging of ϕ_{k_0} to zero.

To determine the basic features of bound states, we consider a box-like potential

$$U(z) = \begin{cases} 0, & |z| \leq \frac{L}{2} \\ U_0, & |z| > \frac{L}{2} \end{cases}, \tag{4.52}$$

where U_0 defines the depth of the potential box while L is its spatial extent. This potential has classically forbidden solutions for any positive U_0 because the asymptotic $z > L/2$ regions can have $E < U(z)$. Since the potential box is symmetric with respect to $z = 0$, it is beneficial to choose $z_0 = 0$ and $z_{\pm 1} = \pm L/2$. From Eq. (4.20), we can then read off the reflection coefficient between z_0 and z_1,

$$\mathcal{R}^{\rightarrow}_{1,0} = R^{\rightarrow}_{1}\, e^{2ik_0\frac{L}{2}} = \frac{k_0 - i\kappa_1}{k_0 + i\kappa_1}\, e^{ik_0 L}. \tag{4.53}$$

Here, the single-boundary reflection coefficient is expressed with the help of Eq. (4.19) and $k_1 = i\kappa_1 = i\sqrt{\frac{2m}{\hbar^2}U_0 - k_0^2}$. Since there is no additional reflection in region $z > z_1 = L/2$, the total reflection coefficient from the asymptotic region is also given by $\mathcal{R}^{\rightarrow}_{\infty,0} = \mathcal{R}^{\rightarrow}_{1,0}$. Due to the symmetry of the system, the reflection from the other asymptotic regime is the same producing

$$\mathcal{R}^{\rightarrow}_{\infty,0} = \mathcal{R}^{\leftarrow}_{0,-\infty} = \frac{k_0 - i\kappa_1}{k_0 + i\kappa_1}\, e^{ik_0 L}. \tag{4.54}$$

Inserting this result into Eq. (4.51) yields the strict condition

$$\left(\frac{k_0 - i\kappa_1}{k_0 + i\kappa_1}\right)^2 e^{2ik_0 L} = 1 \tag{4.55}$$

to find bound states. This relation can be evaluated analytically only for strongly confined particles, implying $\kappa_1 \gg k_0$ because $U_0 \gg E$. In this limit, we must only solve $e^{2ik_0 L} = 1$ producing the usual particle-in-a-box solutions,

$$k_0 = \frac{\pi}{L} n, \quad n = 1, 2, \cdots \quad \Rightarrow E_n = \frac{\hbar^2 k_0^2}{2m} = \frac{\hbar^2 \pi^2}{2mL^2} n^2 \tag{4.56}$$

that exist only for discrete wave vectors and energies. If U_0 is finite, we still find only discrete solutions but need a numerical iteration to find the roots of Eq. (4.55).

In practice, one can often reduce the general one-dimensional problem to a finite number of intervals because one can use $U(z) = const$ in the asymptotic regions $z < z_{-N}$ and $z > z_{+N}$. In this situation, we find exact relations $\mathcal{R}^{\rightarrow}_{\infty,0} = \mathcal{R}^{\rightarrow}_{+N,0}$ and $\mathcal{R}^{\leftarrow}_{0,-\infty} = \mathcal{R}^{\leftarrow}_{0,-N}$ such that we may solve the roots of Eq. (4.51) with a finite number of transfer-matrix multiplications. Once the discrete set of k_0 is found, we can also construct the corresponding wave function. By applying the boundary condition (4.48) in the transfer-matrix relation (4.23), we eventually find

$$\mathbf{s}_0 = \begin{pmatrix} 1 \\ \mathcal{R}^{\rightarrow}_{+N,0} \end{pmatrix} \quad \text{with } k_0 \Rightarrow \mathcal{R}^{\rightarrow}_{+N,0} \mathcal{R}^{\leftarrow}_{0,-N} = 1 \tag{4.57}$$

for the reference interval. We also see that $|A_0| = 1$ and $|B_0| = 1$ such that the corresponding components of $\mathbf{s}_0$ must have an equal magnitude for any bound state. Thus, the resulting wave averaging produces $\phi_0(z)$ with a maximal interference contrast between the propagating A_0 and counterpropagating B_0 components. In other words, the formation of bound states can also be considered as the case where a perfect interference pattern is formed between the asymptotic tunneling regions.

The computed $\mathbf{s}_0$ can also be used to determine the A_j and B_j components at every interval $j = -N, -N+1, \cdots, +N$. We can thus express the generic bound-state wave function via

$$\phi_{k_0}(z) = \mathcal{N} \begin{cases} \mathcal{T}^{\leftarrow}_{0,-N} \mathcal{R}^{\rightarrow}_{+N,0} e^{+\kappa_{-N}(z - z_{-N})}, & z < z_{-N} \\ \phi_j(z), & z_{-N} \le z \le z_{+N} \\ \mathcal{T}^{\rightarrow}_{+N,0} e^{-\kappa_{+N}(z - z_{+N})}, & z < z_{+N} \end{cases}. \tag{4.58}$$

For bound solutions, the integral $\int dz \; |\phi_{k_0}(z)|^2$ is finite because the solution decays exponentially in the asymptotic regions. Since the explicit structure of $\phi_j(z)$ contributes significantly to $\int dz \; |\phi_{k_0}(z)|^2$, the normalization $\mathcal{N}$ generally can only be evaluated numerically. Once the norm is settled, the different discrete states are orthogonal via the $\delta_{k_0,k_0'}$ condition in Eq. (4.4).

We now determine the eigenenergies $E = \hbar^2 k_0^2 / 2m$ and the bound-state wave functions assuming a finite potential height U_0 in Eq. (4.52). Again, we consider the case of an electron and take a potential width $L = 5$ nm. Figure 4.3(a) presents the lowest E (bright area), the second (dim area), and the third (dark area) root. The thick solid lines indicate energies of the corresponding infinite U_0 solutions, provided by Eq. (4.56). The dashed line indicates the case where $E = U_0$.

As a general trend, the $E_n(U_0)$ approaches the result (4.56) as U_0 is made larger. Thus, we may identify n as a suitable quantum number to label the solutions. The lowest

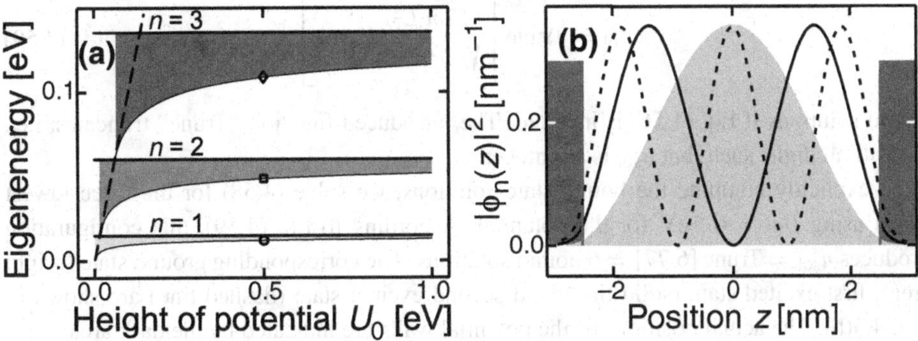

Figure 4.3 Bound-state solutions for an electron in a finite potential well. **(a)** The bound-state energies are shown as a function of the potential height for the ground state (light area), the first (dim area), and the second (dark area) excited state. The horizontal thick lines indicate the bound states for the infinite-barrier solution (4.56) and the dashed line shows $E = U_0$. **(b)** The corresponding probability distributions for $U_0 = 0.5\,\text{eV}$. The ground state (light area) is compared with the first (solid line) and the second (dashed line) excited state. The dark area indicates the extent of the walls.

allowed energy state, $n = 1$, is often referred to as the ground state because the quantum system cannot have an energy below this value. The other discrete solutions are called excited states. Figure 4.3(a) shows that the ground-state energy approaches zero only if U_0 becomes zero as well, which actually implies freely propagating solutions. Thus, the ground state has always a higher energy than classically allowed ($E > 0$) for a particle within the bounding walls. This distinct quantum feature can be traced back to the particle's wave aspects because a quantum-mechanical wave must be bent within the box for the bound solution. This creates $\phi_{E_1}(z)$ that has nonvanishing momentum, kinetic energy, and eventually total energy due to the imminent spatial variations. The energy associated with the ground state is often referred to as the zero-point energy because the quantum fluctuations elevate the energy above the corresponding classical situation.

When we compare the ground state with the first ($n = 2$) and the second ($n = 3$) excited states, we see that the ground state approaches the perfect-box limit (4.56) most rapidly. It is also interesting to note that all the shown bound states have a significantly lower energy than U_0 if U_0 becomes sufficiently large (dashed line in Fig. 4.3(a)). Thus, we can conclude that the system has an appreciable number of bound states if U_0 is large enough. On the other hand, we see that, e.g., bound states beyond the third solution (dark shaded area) no longer exist if the $E_{n=3}(U_0)$ curve crosses the perfect-box energy for the second state (thick solid line). In the same way, all solutions beyond the second one vanish if the $E_{n=2}(U_0)$ curve crosses the perfect-box energy for the ground state. Thus, we find a simple estimate for the number of bound states by defining when U_0 crosses the perfect-box energies (4.56). In other words, we have

$$n_{\mathrm{sol}} = \mathrm{Trunc}\left[\sqrt{\frac{2m\,L^2}{\hbar^2\pi^2}U_0} + 1\right] \tag{4.59}$$

bound solutions if Eq. (4.56) is inverted. The introduced function "Trunc" truncates, i.e., cuts off all digits such that n_{sol} is an integer.

To explicitly compute the bound-state solutions, we solve (4.58) for the three lowest states using $U_0 = 0.5\,\mathrm{eV}$ for the potential. According to Eq. (4.59), this configuration produces $n_{\mathrm{sol}} = \mathrm{Trunc}\,[6.77] = 6$ bound solutions. The corresponding ground state (bright area), first excited state (solid line), and second excited state (dashed line) are shown in Fig. 4.3(b). The actual positions of the potential walls are indicated by the dark area.

We see that all bound-state solutions penetrate into the bounding walls as exponentially decaying tunneling tails. As a general trend, the degree of tunneling increases for elevated E_n. We also see that the bound states are described by standing waves within the walls, as discussed in connection with Eq. (4.57). The ground state has a single maximum and no nodes. The second state has two maxima and one node and the third state has three maxima and two nodes. It is obvious that the energy of the state must increase with the number of nodes because more nodes in the standing wave means faster variation for $\psi_E(z)$ and thus larger kinetic and overall energy.

4.4 Generic classification of energy eigenstates

Our one-dimensional analysis shows that we obtain physically very different configurations for the respective regions where classical solutions do or do not exist. If no classical solutions exist, we find exponentially decaying tunneling states. This observation also holds for a general d-dimensional eigenvalue problem. Thus, we may classify the solution types qualitatively without solving the stationary Schrödinger equation (4.1). We only need to investigate the properties of the wave vector $\mathbf{k}^2(\mathbf{r})$ as a function of position for the energy range of interest. If regions with negative $\mathbf{k}^2(\mathbf{r})$ are found, those have only tunneling solutions while regions with positive $\mathbf{k}^2(\mathbf{r})$ have also classical solutions.

The classification of the solutions is shown in Fig. 4.4. There, the regions where classical solutions exist are indicated as white areas while the tunneling regions are shown as dark areas. The related propagating and tunneling solutions are denoted by a wave symbol or an exponentially decaying curve, respectively. The symbolic presentation of the solutions covers the whole space including the asymptotic regions distinguished by the sign in $\pm\infty$. If all asymptotic regions have tunneling solutions, we find only bound states due to the boundary condition (4.38). If one of the asymptotic regions allows for classical solutions while the others have tunneling solutions, the quantum-mechanical solution still has a continuous spectrum. In this situation, the particles can penetrate into the tunneling regions as decaying tails, as discussed in Section 4.3.2. If all asymptotic regions allow for classical paths, the quantum-mechanical solutions have no propagation limitations.

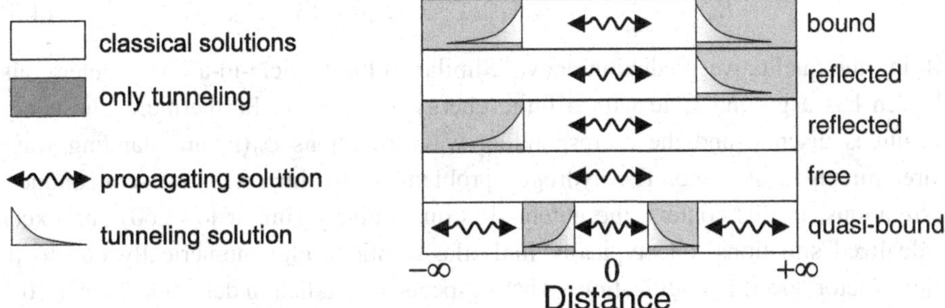

Figure 4.4 Qualitative classification of energy eigenstates. The solutions can be classified into bound, reflected, propagating, or quasibound solutions based on where tunneling (dark shaded) or classically allowed (white) regions exist. The propagating solutions are symbolized by their wave character while the tunneling solutions are shown as an exponentially decaying symbol.

In another interesting situation, one has an internal region surrounded by tunneling regions of a finite extent, as shown in the bottom part of Fig. 4.4. Obviously, this configuration does not impose strictly confining conditions because the classically forbidden regions do not extend to infinity. Nevertheless, the classically forbidden regions can act like mirrors that confine waves for a finite time to within the classically allowed regions bounded by the tunneling regions. This scenario has a full analogy with a Fabry Pérot cavity where the electromagnetic wave is confined between highly reflecting mirrors. Similarly, the particle wave may dwell a very long time within the "cavity" confined by the tunneling regions. In this situation, the particle becomes quasibound if its energy matches the discrete energetic structure of the "cavity." In this situation, the particle eventually propagates through the tunneling regions with probability 1 but it is captured between the tunneling regions for a substantially long time. This phenomenon is referred to as resonant tunneling. It can have a very similar resonance structure for the transmission probability as true bound states have.

To give an example of how one can make qualitative predictions for solution types in nontrivial problems, we consider the hydrogen problem. In this case, the negatively charged electron is attracted by the Coulomb potential,

$$U(\mathbf{r}) = -\frac{e^2}{4\pi\epsilon_0|\mathbf{r}|}, \tag{4.60}$$

created by the positively charged proton which is considered to be a point-like object at the origin. We see that the eigenenergy E must be negative in order to produce tunneling solutions $E < U(\mathbf{r})$. Since the corresponding asymptotic tunneling regions, $|\mathbf{r}| > -e^2/4\pi\epsilon_0 E$, surround the propagation regions, the qualitative arguments of Fig. 4.4 predict that the hydrogen atom has a set of discrete bound states for negative energies.

In fact, the stationary Schrödinger equation $\hat{H}\phi_E(\mathbf{r}) = E\phi_E(\mathbf{r})$ of the hydrogen atom constitutes one of the few three-dimensional eigenvalue problems that can be solved analytically. The actual energy eigenstates of hydrogen are quantized according to

$$E_n = -\frac{E_B}{n^2}, \quad n = 1, 2, \cdots, \quad E_B \approx 13.6\,\text{eV}, \tag{4.61}$$

verifying the qualitative prediction above. Similar to the particle-in-a-box problem, also hydrogen has a ground state with a finite energy $E_1 = E_B$. Furthermore, the energy spectrum is discrete and the corresponding wave functions $\phi_n(\mathbf{r})$ are standing waves in three-dimensional space. The hydrogen problem is also important for semiconductor investigations. In this context, the potential is qualitatively similar to (4.60) but, except for idealized situations, one typically finds the solutions only numerically due to the semiconductor-specific modifications. These aspects are studied in detail in Chapter 16.

Exercises

4.1 **(a)** Show that the solutions of the Helmholtz equation $[\nabla^2 + q^2(\mathbf{r})]\phi_{\mathbf{q}_0}(\mathbf{r}) = 0$ are orthogonal satisfying $\int d^3r\, \phi^*_{\mathbf{q}_0}(\mathbf{r})\phi(\mathbf{r})\phi_{\mathbf{q}'_0}(\mathbf{r}) = \delta_{\mathbf{q}_0,\mathbf{q}'_0}$ where $\mathbf{q}_0 = \mathbf{q}(\mathbf{r_0})$ and $\mathbf{q}^2(\mathbf{r}) = \mathbf{q}_0^2\, Q(\mathbf{r})$ if $Q(\mathbf{r})$ is independent of $\mathbf{q_0}$. Show that this property follows from the wave equation (2.12). **(b)** Show that the same orthogonality relation $\int d^3r\, \phi^*_{\mathbf{q}_0}(\mathbf{r})\, \phi_{\mathbf{q}'_0}(\mathbf{r}) = \delta_{\mathbf{q},\mathbf{q}'_0}$ also follows from the stationary Schrödinger equation (4.2). **Hint:** *Study why the integration quadrature $Q(\mathbf{r})$ is missing for particles.* **(c)** Compare the completeness relations for particles and for electromagnetic waves (use the scalar form).

4.2 **(a)** Differentiate $\phi_{k_0}(z) = \phi_j(z)\theta(z_{j+1} - z) + \phi_{j+1}(z - z_{j+1})$ with $\partial/\partial z$ and $\partial^2/\partial z^2$. **(b)** Construct the Helmholtz equation for $\phi_j(z)$ and $\phi_{j+1}(z)$ and show that the Helmholtz equation (4.11) produces the condition (4.16).

4.3 **(a)** Construct the condition (4.17) for the plane-wave decomposition (4.12) and identify the transmission and reflection coefficients (4.19). **(b)** Show that $D_j \equiv T_j^{\leftarrow} T_j^{\rightarrow} - R_j^{\leftarrow} R_j^{\rightarrow} = 1$ and $T_j^{\lambda} = 1 + R_j^{\lambda}$. **(c)** Show that T_j^{λ} and R_j^{λ} define the Fresnel coefficients for light. **(d)** Show that Eq. (4.18) can be rearranged to get Eq. (4.20).

4.4 **(a)** Express the generic transmission and reflection coefficients $\mathcal{T}^{\lambda}_{j+1,0}$ and $\mathcal{R}^{\lambda}_{j+1,0}$ recursively in terms of $\mathcal{T}^{\lambda}_{j,0}$, $\mathcal{R}^{\lambda}_{j,0}$, T^{λ}_{j+1}, and $R^{\lambda+1}_{j+1}$. **(b)** Show that the solution $\mathcal{T}^{\rightarrow}_{j+1,0} = \frac{\mathcal{T}^{\rightarrow}_{j,0} T^{\rightarrow}_{j+1}\, e^{ik_j l_j}}{1 - R^{\rightarrow}_{j+1}\mathcal{R}^{\leftarrow}_{j+1,0}\, e^{2ik_j l_j}}$ implies that $\mathcal{T}^{\rightarrow}_{j+1,0}$ can be constructed out of the sum over all paths that bifurcate at z_{j+1} and the interval $z \in [z_0, z_j]$.

4.5 The integrals of plane waves follow from $\int_0^\infty dx\, e^{iqx} = \pi\delta(q) + iP(1/q)$ and $\int_{-\infty}^0 dx\, e^{-iqx} = \pi\delta(q) - iP(1/q)$ where $P(1/q)$ refers to the Cauchy principal-value integral $\int_{-\varepsilon}^{\varepsilon} dx \frac{1}{x}\, f(x)$. This integral vanishes for symmetric functions $f(x)$. Use this information and the conservation laws (4.27) to show that the solutions (4.28)–(4.29) are orthogonal according to
$\int dz \left[\phi^{\lambda}_{k_0}(z)\right]^* \left[\phi^{\lambda'}_{k'_0}(z)\right] = \delta_{\lambda',\lambda}\, \delta\left(k'_0 - k_0\right)$. **Hint:** *Use properties of the δ function, derived in Exercise 1.9.*

4.6 Show that also the tunneling solution (4.36) is orthogonal via $\int d^3r \left[\phi_{\mathbf{k}_0}^{\rightarrow}(r)\right]^* \phi_{\mathbf{k}'_0}^{\rightarrow}(r) = \delta_{\mathbf{k}'_0,\mathbf{k}_0}$. Why are tunneling solutions always orthogonal with respect to both propagating and bound solutions?

4.7 Show that the bound-state solutions for the potential (4.52) are orthogonal.

4.8 (**a**) Show that the tunneling probability through a potential barrier $U(z) = \begin{cases} U_0, & z \leq L/2 \\ 0, & \text{otherwise} \end{cases}$ is $|T^\lambda|^2 = \left|\frac{T_1^{\rightarrow} T_2^{\leftarrow} e^{-\kappa L}}{1-R_2^{\rightarrow} R_1^{\leftarrow} e^{-2\kappa L}}\right|^2$ where T_j^λ and R_j^λ define the transmission and reflection coefficients through the $z = -\frac{L}{2}$ $(j = 1)$ and $z = +\frac{L}{2}$ $(j = 2)$ boundary. (**b**) Assume that a single tunneling barrier j has the transmission and reflection coefficients $\mathcal{T}_{B,j}^\lambda$ and $\mathcal{R}_{B,j}^\lambda$, respectively, and that two such barriers are separated by distance D. Show that the total transmission coefficient is $\mathcal{T}_{2\text{bar}}^{\rightarrow} = \frac{\mathcal{T}_{B,1}^{\rightarrow} \mathcal{T}_{B,2}^{\rightarrow} e^{ik_0 D}}{1-\mathcal{R}_{B,2}^{\rightarrow} \mathcal{R}_{B,1}^{\leftarrow} e^{2ik_0 D}}$. (**c**) The system above allows for a new phenomenon called resonant tunneling that yields 100 percent tunneling probability under certain discrete k_0 values. What is the explicit condition for this? **Hint:** *Analyze the result above and apply the relations* (4.27) *in the asymptotic regions.*

Further reading

The transfer-matrix for electromagnetic and particle waves is discussed e.g. in:

M. Born and E. Wolf (2003). *Principles of Optics: Electromagnetic Theory of Propagation, Interference and Diffraction of Light*, 7th edition, Cambridge, Cambridge University Press.

E. Merzbacher (1998). *Quantum Mechanics*, 3rd edition, New York, Wiley & Sons.

Tunneling and evanescent light is discussed in:

M. Razavy (2003). *Quantum Theory of Tunneling*, Singapore, World Scientific.

C. J. Chen (1993). *Introduction to Scanning Tunneling Microscopy*, Oxford, Oxford University Press.

F. de Fornel (1997). *Evanescent Waves from Newtonian Optics to Atom Optics*, Heidelberg, Springer.

L. Novotny and B. Hecht (2006). *Principles of Nano-Optics*, Cambridge, Cambridge University Press.

Detailed discussions of the harmonic oscillator and the hydrogen problem can be found in the quantum-mechanics books listed in Chapter 3.

5

Central concepts in measurement theory

The full quantum-mechanical treatment systematically describes both the wave and particle properties of microscopic systems. The wave aspect automatically implies several nontrivial effects such as interference, tunneling, and the existence of bound states. All these features have been demonstrated in experiments but cannot be described classically.

However, the increased generality of a quantum theory, as compared with the traditional classical treatment, comes with a price. For example, particle movements now are interpreted probabilistically and measurable classical quantities are replaced by operators and their expectation values. Thus, it is not immediately clear how actual measurements, which after all are performed in the macroscopic world, are connected with the quantum-mechanical theory.

It turns out that the connection between quantum theory and measurements can be established only axiomatically, e.g., via Born's theorem that defines the generalized probabilistic interpretation. With the help of this theorem, the relation of different operators and eigenstates to measurements can be formulated precisely. In this chapter, we first introduce a more abstract operator formalism and analyze the related quantum dynamics with the goal to connect them to actual measurements.

5.1 Hermitian operators

Before we determine the explicit connection between quantum theory and measurements, we need to develop the formal aspects of operators and expectation values somewhat further. For this purpose, we start by evaluating

$$\hat{\mathbf{p}}\,|\psi(\mathbf{r},t)|^2 = \left[\hat{\mathbf{p}}\,\psi^\star(\mathbf{r},t)\right]\psi(\mathbf{r},t) + \psi^\star(\mathbf{r},t)\,\hat{\mathbf{p}}\,\psi(\mathbf{r},t). \tag{5.1}$$

From Eq. (3.17), we know that the expectation values have to be evaluated in the properly sandwiched format to produce the correct $\psi^\star(\mathbf{r},t)\,\hat{\mathbf{p}}\,\psi(\mathbf{r},t)$ argument for the $\langle\hat{\mathbf{p}}\rangle$ integral. To generate this form, we use $\hat{\mathbf{p}} \equiv -i\hbar\nabla$ and the conversion,

$$\psi^\star(\mathbf{r},t)\,\hat{\mathbf{p}}\,\psi(\mathbf{r},t) = -i\hbar\nabla\,|\psi(\mathbf{r},t)|^2 + \left[i\hbar\nabla\,\psi^\star(\mathbf{r},t)\right]\psi(\mathbf{r},t), \tag{5.2}$$

together with Gauss' theorem (3.26) to produce

$$\begin{aligned}\langle \hat{\mathbf{p}} \rangle &= -\int d^3r \; i\hbar\nabla\, |\psi(\mathbf{r},t)|^2 + \int_{-\infty}^{\infty} d^3r \; \left[i\hbar\nabla\, \psi^\star(\mathbf{r},t) \right] \psi(\mathbf{r},t) \\ &= i\hbar \oint d\mathbf{S}\, |\psi(\mathbf{r},t)|^2 + \int d^3r \; [-i\hbar\nabla\, \psi(\mathbf{r},t)]^\star \, \psi(\mathbf{r},t) \\ &= \int d^3r \; \left[\hat{\mathbf{p}}\, \psi(\mathbf{r},t) \right]^\star \psi(\mathbf{r},t). \end{aligned} \tag{5.3}$$

Here, the integral over the surface at infinity vanishes because $|\psi(\mathbf{r},t)|^2$ must decay to zero if r approaches infinity due to the square integrability, Eq. (3.14).

We generalize the result, Eq. (5.3) using the definition of the Hermitian adjoint operator $\hat{F}^\dagger$,

$$\langle \hat{F} \rangle \equiv \int_{-\infty}^{\infty} d^3r \; \left[F^\dagger(\mathbf{r}) \psi(\mathbf{r},t) \right]^\star \psi(\mathbf{r},t), \tag{5.4}$$

which introduces an alternative way to compute the expectation value $\langle \hat{F} \rangle$. The exponent $\dagger$ symbolizes the Hermitian conjugation which is also referred to as the Hermitian adjoint of a given operator. It is straightforward to show that Hermitian conjugation satisfies the following relations

$$\left[\hat{G}^\dagger\right]^\dagger = \hat{G}, \quad \left[\hat{F} + \hat{G}\right]^\dagger = \hat{F}^\dagger + \hat{G}^\dagger, \quad \text{and} \quad \left[\hat{F}\,\hat{G}\right]^\dagger = \hat{G}^\dagger \hat{F}^\dagger, \tag{5.5}$$

such that applying the Hermitian conjugation twice produces an identity for an arbitrary operator $\hat{G}$. At the same time, the conjugation of a sum of operators can be performed individually while one must reverse the order of operators in conjugated operator products.

Even though Eq. (5.3) shows that $\hat{\mathbf{p}}^\dagger = \hat{\mathbf{p}}$, this is not valid for a generic operator $\hat{F}^\dagger \neq \hat{F}$. In the special case of $\hat{F}^\dagger = \hat{F}$, the operator $\hat{F}$ is denoted as self-adjoint or Hermitian. Hermitian operators have a special importance in measurements because they construct the very important class of operators that define observables, related to experimentally measurable properties such as position, intensity, momentum, etc. Since measurements must produce real-valued outcomes, expectation values corresponding to observables must also be real valued even though the wave functions are generally complex. One can easily convince oneself that any Hermitian operator with $\hat{F}^\dagger = \hat{F}$, produces strictly real-valued expectation values – and vice versa – based on Eqs. (3.17) and (5.4) for arbitrary wave functions, see Exercise 5.1. Thus, any observable physical quantity must be represented by a Hermitian operator.

5.2 Eigenvalue problems

The treatment of Hermitian operators $\hat{O}$ becomes particularly efficient if we consider the eigenvalue problem

$$O(\mathbf{r})\, \phi_\lambda(\mathbf{r}) = \lambda\; \phi_\lambda(\mathbf{r}), \tag{5.6}$$

where λ denotes a generic eigenvalue. Due to the diagonality of $O(\mathbf{r})$, Eq. (5.6) is not the most general form, however, it already covers all the examples in Chapters 3–4. For example, the stationary Schrödinger equation (4.1) can be written in this format. Mathematically, Eq. (5.6) defines a linear problem and we may apply several powerful relations known from linear algebra. To explore the implications of quantum mechanics, one may view Eq. (5.6) as a typical matrix eigenvalue problem. However, the dimension of the matrices may approach infinity if all quantum-mechanically possible cases are included. In many situations, a single λ has multiple eigensolutions implying a certain level of degeneracy. The degenerate eigenstates $\phi_{\nu,l}$ can always be identified by introducing an additional label $l = 1, 2, \cdots, N_\lambda$, where N_λ defines the degree of degeneracy of the eigenvalue λ. In quantum mechanics, the labels λ and l are commonly referred to as the quantum numbers because they uniquely define the eigenstates related to the quantum-mechanical operator $\hat{O}$.

One can easily show (see Exercise 5.2) that λ is real valued and the relation

$$(\lambda - \nu) \int d^3r\ \phi^\star_\lambda(\mathbf{r})\phi_\nu(\mathbf{r}) = 0 \tag{5.7}$$

is always satisfied for any solution of a Hermitian eigenvalue problem. This means that two eigenfunctions are orthogonal whenever their eigenvalues are different. Even if degeneracy exists, we may apply the standard Gram–Schmidt orthogonalization to provide orthogonality also among the degenerate states. If the eigenvalues are discrete, it is beneficial to orthogonalize the eigenstates via

$$\int d^3r\ \phi^\star_{\lambda,j}(\mathbf{r})\ \phi_{\nu,k}(\mathbf{r}) = \delta_{j,k}\ \delta_{\lambda,\nu}. \tag{5.8}$$

It is also possible that we find infinitely many continuous values for λ. A very good example is provided by the plane-wave solution $\phi_\mathbf{q}(\mathbf{r}) = \frac{2}{[2\pi]^{\frac{3}{2}}} e^{i\mathbf{q}\cdot\mathbf{r}}$, discussed in Section 2.3.1. In this situation, the normalization follows via

$$\int d^3r\ \phi^\star_{\lambda,j}(\mathbf{r})\ \phi_{\nu,k}(\mathbf{r}) = \delta_{j,k}\ \delta(\lambda - \nu) \tag{5.9}$$

if ν and λ are continuous. Linear algebra yields yet another strong statement expressing that all eigenstates of Hermitian operators define a complete set of functions. This can be expressed with the help of the completeness relation

$$\sum\nolimits_\lambda \phi^\star_\lambda(\mathbf{r})\phi_\lambda(\mathbf{r}') = \delta(\mathbf{r} - \mathbf{r}'), \tag{5.10}$$

in analogy to Eq. (2.61).

The completeness of ϕ_λ means that any wave function can be expressed as a linear superposition

$$\psi(\mathbf{r},t) = \sum_{\lambda} c_\lambda(t)\,\phi_\lambda(\mathbf{r})$$
$$= \sum_{\lambda,j=1} c_{\lambda,j}(t)\,\phi_{\lambda,j}(\mathbf{r}) + \sum_{j=1} \int d\lambda\, c_{\lambda,j}\,\phi_{\lambda,j}(\mathbf{r}). \quad (5.11)$$

In the second line, we explicitly allow for both discrete and continuous λ states appearing in the sum and the integral, respectively, as well as for a possible degeneracy denoted by the sums over j whose upper limit is the degree of degeneracy. By applying the orthogonality relations (5.8)–(5.9), we can always deduce the expansion coefficients,

$$c_\lambda(t) = \int d^3r\,\phi^\star_\lambda(\mathbf{r})\,\psi(\mathbf{r},t), \quad (5.12)$$

whenever the functions ϕ_λ are known, compare Eq. (2.63). Since the $\phi_\lambda(\mathbf{r})$ are stationary, the time dependence of $c_\lambda(t)$ completely stems from the wave function $\psi(\mathbf{r},t)$.

5.2.1 Dirac notation

There is always a certain degree of arbitrariness in choosing the most appropriate representation of the wave function. For example, one can represent ϕ_λ in either real or momentum space, as discussed in Section 3.2.4. To focus the analysis entirely on the generic features, it is useful to introduce Dirac's abstract notation,

$$\psi(\mathbf{r},t) \to |\psi(t)\rangle \equiv \langle\psi(t)|^\dagger, \qquad \psi^\star(\mathbf{r},t) \to \langle\psi(t)| \equiv |\psi(t)\rangle^\dagger,$$
$$\phi_\lambda(\mathbf{r}) \to |\phi_\lambda\rangle \equiv \langle\phi_\lambda|^\dagger, \qquad \phi^\star_\lambda(\mathbf{r}) \to \langle\phi_\lambda| \equiv |\phi_\lambda\rangle^\dagger, \quad (5.13)$$

where one associates a state with a so-called "bra" $\langle\cdots|$ and a "ket" $|\cdots\rangle$ form of the wave function or eigenstate in abstract space. In Eq. (5.13), we have also defined the effect of Hermitian conjugation converting a bra into a ket and vice versa. The "bra–ket" notation becomes particularly useful when we want to evaluate matrix elements,

$$\langle\phi_\lambda|\hat{O}|\phi_\nu\rangle \equiv \int d^3r\,\phi^\star_\lambda(\hat{\mathbf{r}})\,O(\mathbf{r})\,\phi_\nu(\mathbf{r})$$
$$\equiv \int d^3p\,\phi^\star_\lambda(\hat{\mathbf{p}})\,O(\mathbf{p})\,\phi_\nu(\mathbf{p}), \quad (5.14)$$

because we can write them compactly without the need to define the explicit representation. If we choose $\hat{O}$ as the identity operator $\hat{\mathbb{I}}$, we may express the orthogonality relations (5.8) and (5.9) in the form

$$\langle\phi_\lambda|\phi_\nu\rangle = \begin{cases} \delta_{\lambda,\nu}, & \forall \text{ discrete } \lambda \text{ and } \nu \\ \delta(\lambda-\nu), & \forall \text{ continuous } \lambda \text{ and } \nu \end{cases}. \quad (5.15)$$

Furthermore, we can reformulate Eqs. (5.12) and (5.11) as

$$c_\lambda(t) = \langle\phi_\lambda|\psi(t)\rangle \quad \Leftrightarrow \quad |\psi(t)\rangle = \sum_\lambda c_\lambda(t)|\phi_\lambda\rangle, \tag{5.16}$$

respectively. The Dirac notation also produces a simple expression for the completeness relation (5.10)

$$\sum_\lambda |\phi_\lambda\rangle\langle\phi_\lambda| = \hat{\mathbb{I}}, \tag{5.17}$$

where the δ function of real space is replaced by the identity operator.

Since $\hat{\mathbf{r}}$ is a Hermitian operator, it has a complete and orthogonal set of eigenstates $|\mathbf{r}\rangle$,

$$\hat{\mathbf{r}}|\mathbf{r}\rangle = \mathbf{r}|\mathbf{r}\rangle, \quad \langle\mathbf{r}|\mathbf{r}'\rangle = \delta(\mathbf{r}-\mathbf{r}'), \quad \sum_{\mathbf{r}} |\mathbf{r}\rangle\langle\mathbf{r}| \equiv \int d^3r\, |\mathbf{r}\rangle\langle\mathbf{r}| = \hat{\mathbb{I}}, \tag{5.18}$$

just like any other operator with continuous eigenvalues. The inner products,

$$\langle\mathbf{r}|\phi_\lambda\rangle = \phi_\lambda(\mathbf{r}), \qquad \langle\mathbf{r}|\psi(t)\rangle = \psi(\mathbf{r},t), \tag{5.19}$$

present any given set of states in real space. More generally, for $|\beta\rangle$ and $|\phi_\lambda\rangle$, the inner product

$$\langle\beta|\phi_\lambda\rangle \equiv \phi_\lambda(\beta), \quad \langle\beta|\phi_\lambda\rangle^\dagger = \langle\phi_\lambda|\beta\rangle \equiv \phi_\lambda^\star(\beta) = \langle\beta|\phi_\lambda\rangle^\star \tag{5.20}$$

defines the β representation of the state $|\phi_\lambda\rangle$. The Hermitian conjugation of an inner product is nothing but the usual complex conjugation, as one can easily verify using (5.13). Since the set $|\beta\rangle$ can be chosen arbitrarily as long as it is complete and orthogonal, it is clear that the explicit representation of the state $|\phi_\lambda\rangle$ can be performed in many different ways while its abstract form is unique. A completely analogous situation appears in linear algebra where any vector or matrix can be represented equivalently with arbitrarily many different choices for the basis vectors.

An even clearer connection to linear algebra is obtained by considering the eigenvalue problem (5.6) of a Hermitian operator $\hat{O}$,

$$\hat{O}|\phi_\lambda\rangle = \lambda|\phi_\lambda\rangle, \tag{5.21}$$

in the abstract notation. By projecting both sides of the equation onto $\langle\beta|$, we get

$$\langle\beta|\hat{O}|\phi_\lambda\rangle = \lambda\langle\beta|\phi_\lambda\rangle = \lambda\phi_\lambda(\beta) \tag{5.22}$$

if the association (5.20) is made. Inserting the completeness relation $\sum_{\beta'}|\beta'\rangle\langle\beta'|$ into $\langle\beta|\hat{O}|\phi_\lambda\rangle$, we find

$$\langle\beta|\hat{O}|\phi_\lambda\rangle = \sum_{\beta'}\langle\beta|\hat{O}|\beta'\rangle\langle\beta'|\phi_\lambda\rangle = \sum_{\beta'} O_{\beta,\beta'}\,\phi_\lambda(\beta'), \tag{5.23}$$

where we defined the Hermitian matrix $O_{\beta,\beta'} \equiv \langle\beta|\hat{O}|\beta'\rangle$. Consequently, the quantum-mechanical eigenvalue problem (5.22) can be expressed as

$$\sum_{\beta'} O_{\beta,\beta'}\,\phi_\lambda(\beta') = \lambda\,\phi_\lambda(\beta), \tag{5.24}$$

which has exactly the typical matrix form encountered in linear algebra. Thus, quantum-mechanical eigenstates and eigenvalues can always be obtained by solving a matrix problem.

5.2.2 Central eigenvalue problems

For the purposes of an explicit eigenvalue calculation, one eventually needs to determine the specific representation. The real- and momentum-space representations are the most commonly used. Thus, we first choose the $|\beta\rangle$ states to be $|\mathbf{r}\rangle$ states, such that Eq. (5.24) casts into the form

$$\int d^3r'\, O_{\mathbf{r},\mathbf{r}'}\,\phi_\lambda(\mathbf{r}') = \int d^3r'\,\langle\mathbf{r}|\hat{O}|\mathbf{r}'\rangle\,\phi_\lambda(\mathbf{r}') = \lambda\,\phi_\lambda(\mathbf{r}), \tag{5.25}$$

where we have used $O_{\mathbf{r},\mathbf{r}'} = \langle\mathbf{r}|\hat{O}|\mathbf{r}'\rangle$ in the second step. Since the operator $\hat{O}$ is arbitrary, the eigenvalue problem (5.25) reduces to the diagonal Eq. (5.6) only if

$$\langle\mathbf{r}|\,\hat{O}\,|\mathbf{r}'\rangle = \delta(\mathbf{r}-\mathbf{r}')O(\mathbf{r}'). \tag{5.26}$$

In cases where $\hat{O}$ is not diagonal in $\mathbf{r}$, we automatically end up with the linear matrix problem (5.24) and nonvanishing off-diagonal matrix elements $\langle\mathbf{r}|\hat{O}|\mathbf{r}'\rangle \equiv O(\mathbf{r},\mathbf{r}')$ for $\mathbf{r} \neq \mathbf{r}'$. Since $\mathbf{r}$ is continuous, the corresponding Eq. (5.24) can be viewed as an integral equation. Even though many relevant quantum-mechanical operators can be expressed via diagonal matrix elements in real space, it is sometimes convenient formally to keep the more general form $\langle\mathbf{r}|\hat{O}|\mathbf{r}'\rangle$ in connection with $\phi_\lambda(\mathbf{r}')$. The resulting integral equations can then be used, e.g., as the basis for the path-integral interpretation of quantum mechanics or the Green's function approach.

While the operators are defined in abstract space, the matrix elements $\langle\mathbf{r}|\hat{O}|\mathbf{r}'\rangle$ are either simple numbers or differentiations. In the following, we deal with $\hat{O}$ in its purely abstract form while the earlier Chapters 3–4 explicitly used the diagonal $O(\mathbf{r})$. As an illustration, we consider the simple case $\hat{O} = \hat{p}$. By combining Eqs. (3.28) and (5.26), we find

$$\langle\mathbf{r}|\,\hat{\mathbf{p}}\,|\mathbf{r}'\rangle = \delta(\mathbf{r}-\mathbf{r}')\,[-i\hbar\nabla_{\mathbf{r}'}], \tag{5.27}$$

where the differentiation is performed with respect to $\mathbf{r}'$. Together with Eq. (5.25), this identification produces

$$\int d^3r'\,\delta(\mathbf{r}-\mathbf{r}')\left[-i\hbar\nabla_{\mathbf{r}'}\,\phi_\lambda(\mathbf{r}')\right] = \lambda\,\phi_\lambda(\mathbf{r})$$
$$\Leftrightarrow \quad -i\hbar\nabla\,\phi_\lambda(\mathbf{r}) = \lambda\,\phi_\lambda(\mathbf{r}). \tag{5.28}$$

The solutions are plane waves

$$\langle \mathbf{r}|\mathbf{p}\rangle = \phi_{\mathbf{p}}(\mathbf{r}) = \left[\frac{1}{2\pi\hbar}\right]^{\frac{3}{2}} e^{i\frac{\mathbf{p}}{\hbar}\cdot\mathbf{r}}, \quad \langle \mathbf{p}|\mathbf{r}\rangle = \phi^{\star}_{\mathbf{p}}(\mathbf{r}) = \left[\frac{1}{2\pi\hbar}\right]^{\frac{3}{2}} e^{-i\frac{\mathbf{p}}{\hbar}\cdot\mathbf{r}} \tag{5.29}$$

and the eigenvalue is $\lambda = \mathbf{p}$. By setting $\mathbf{p} \equiv \hbar\mathbf{k}$, we find the same complete set of orthogonal plane waves as discussed in connection with Eq. (2.59). We can represent any wave function either in real or in momentum space

$$\psi(\mathbf{r},t) = \int d^3p\, \langle \mathbf{r}|\mathbf{p}\rangle\langle \mathbf{p}|\psi(t)\rangle = \left[\frac{1}{2\pi\hbar}\right]^{\frac{3}{2}} \int d^3p\, \psi(\mathbf{p},t)\, e^{i\frac{\mathbf{p}}{\hbar}\cdot\mathbf{r}}$$

$$\psi(\mathbf{p},t) = \int d^3r\, \langle \mathbf{p}|\mathbf{r}\rangle\langle \mathbf{r}|\psi(t)\rangle = \left[\frac{1}{2\pi\hbar}\right]^{\frac{3}{2}} \int d^3r\, \psi(\mathbf{r},t)\, e^{-i\frac{\mathbf{p}}{\hbar}\cdot\mathbf{r}}, \tag{5.30}$$

where we used the completeness relation and Eq. (5.29). Introducing the simple conversion

$$\psi(\mathbf{k},t) \equiv \left[\frac{\hbar}{2\pi}\right]^{\frac{3}{2}} \psi(\mathbf{p} = \hbar\mathbf{k}, t), \tag{5.31}$$

we see that Eq. (5.30) is nothing but the spatial Fourier transformation (3.40) with a properly adjusted norm.

The system Hamiltonian (3.38) can also be expressed completely in the abstract format

$$\hat{H} = \frac{\hat{\mathbf{p}}^2}{2m} + U(\hat{\mathbf{r}}). \tag{5.32}$$

Since both position and momentum operators are Hermitian and they appear separately in $\hat{H}$, the Hamiltonian is also Hermitian. The corresponding Schrödinger equation (3.39) can also be expressed in its abstract form

$$i\hbar\frac{\partial}{\partial t}|\psi(t)\rangle = \hat{H}\,|\psi(t)\rangle. \tag{5.33}$$

Since this clearly is a linear equation, all quantum-mechanical properties follow from linear algebra. In particular, the sum of two solutions is also a solution such that the superposition principle must be generally valid.

The RHS of the Schrödinger equation produces an important Hermitian eigenvalue problem

$$\hat{H}\,|\phi_E\rangle = E|\phi_E\rangle, \tag{5.34}$$

which defines the stationary Schrödinger equation introduced in Section 4.1. The eigenvalue E corresponds to the eigenenergy of the state. In many cases, E is degenerate such that the quantum number E is often appended to include an implicit degeneracy index. In addition, the resulting states are orthogonalized. For free particles, the potential part is missing from $\hat{H}$. It is therefore easy to evaluate

$$\hat{H}\,|\mathbf{p}\rangle = \frac{\hat{p}^2}{2m}|\mathbf{p}\rangle = \frac{\mathbf{p}^2}{2m}|\mathbf{p}\rangle \;\;\Rightarrow\;\; E \equiv E_{\mathbf{p}} = \frac{\mathbf{p}^2}{2m}. \tag{5.35}$$

Clearly, the same E is produced by all the states with the same $|\mathbf{p}|$. Nevertheless, it is convenient to keep the full vectorial expression $\mathbf{p}$ to label the state since the $|\mathbf{p}\rangle$ states are complete and orthogonal.

Since Eq. (5.34) constitutes a Hermitian eigenvalue problem, its solutions always form a complete set of orthogonal states. Consequently, any wave function can be expressed as a linear superposition

$$|\psi(t)\rangle = \sum\nolimits_E |\phi_E\rangle\langle\phi_E|\psi(t)\rangle \equiv \sum\nolimits_E c_E(t)\,|\phi_E\rangle, \tag{5.36}$$

where we identified the expansion coefficients $c_E(t) \equiv \langle\phi_E|\psi(t)\rangle$. Inserting this expansion into the Schrödinger equation (5.33), we find

$$i\hbar\frac{\partial}{\partial t}c_E(t) = E\,c_E(t), \tag{5.37}$$

where we projected both sides of Eq. (5.33) onto $\langle\phi_E|$ and used the property (5.34). The resulting dynamics follows from a simple first-order differential equation with the solution $c_E(t) = c_E(0)e^{-i\frac{Et}{\hbar}}$. Thus, we can explicitly write Eq. (5.36) as

$$|\psi(t)\rangle = \sum\nolimits_E c_E(0)e^{-i\frac{Et}{\hbar}}\,|\phi_E\rangle, \quad c_E(0) = \langle\phi_E|\psi(0)\rangle, \tag{5.38}$$

in analogy to the mode expansion (4.6)–(4.7). The quantum dynamics can always be solved formally if $|\psi(0)\rangle$ and the states $|\phi_E\rangle$ are known. However, in most realistic situations the solutions of the stationary Schrödinger equation are unknown and even the numerical solutions are tedious. Nonetheless, even the formal knowledge of the eigenvalue problem and the related quantum dynamics can help us to develop suitable approaches to systematically approximate the solution of the full quantum problem.

5.3 Born's theorem

The mode-expansion approach (5.36) also works if we use a basis $|\phi_\lambda\rangle$ provided by the generic Hermitian operator $\hat{O}$ that defines an observable of a given measurement. In other words, we can expand the wave function into

$$|\psi(t)\rangle = \sum\nolimits_\lambda c_\lambda(t)\,|\phi_\lambda\rangle, \quad \text{with } c_\lambda(t) \equiv \langle\phi_\lambda|\psi(t)\rangle, \tag{5.39}$$

with the expansion coefficient $c_\lambda(t)$ that can be time dependent. Additionally, the λ sum implicitly includes the possible degenerate states. Since any observable is Hermitian, this expansion can be performed with respect to any measurable quantity.

Equation (5.39) has an intimate connection to measurements that can be established axiomatically. The resulting Born's theorem can be formulated through three statements: (i) A single measurement will produce an outcome λ_{out} with the value $\lambda_{\text{out}} = \lambda$ corresponding to one of the possible eigenvalues of the corresponding observable $\hat{O}$. (ii) The selection process is completely random such that one generally cannot predict which λ value is observed in an individual measurement. (iii) However, quantum mechanics determines the probability distribution to detect λ. If the measurement is repeated $\mathcal{N}$ times under identical conditions, we can determine the number of times, n_λ, the measurement outcome is λ. Even though quantum mechanics cannot predict the outcome of individual measurements, it yields a strict connection between n_λ and the quantum-mechanical expansion coefficient

$$|c_\lambda(t)|^2 = \lim_{\mathcal{N}\to\infty} \frac{n_\lambda}{\mathcal{N}}, \tag{5.40}$$

where t refers to the time of the measurement after the initial state preparation. Clearly, all the measurements must be performed under identical conditions requiring that the initial state is always reset. Equation (5.40) identifies $|c_\lambda(t)|^2$ as the probability of detecting the value λ in a measurement. The implications of Born's theorem are illustrated schematically in Fig. 5.1.

The analysis above must be altered only slightly if λ is continuous. For example, the measurement of $\hat{\mathbf{r}}$ produces a continuous spectrum of $\mathbf{r}$ outcomes. The corresponding $c_\mathbf{r} = \langle \mathbf{r}|\psi(t)\rangle = \psi(\mathbf{r}, t)$ defines the wave function while $|c_\mathbf{r}|^2 = |\psi(\mathbf{r}, t)|^2$ is the corresponding probability distribution. Since $\lambda = \mathbf{r}$ is continuous, one always finds zero counts to detect exactly the value $\mathbf{r}$ if $\mathcal{N}$ is finite. To avoid this problem, we modify Eq. (5.40) such that $n_{\lambda=\mathbf{r}}$ defines the number of measured counts within the infinitesimal volume Δr^3

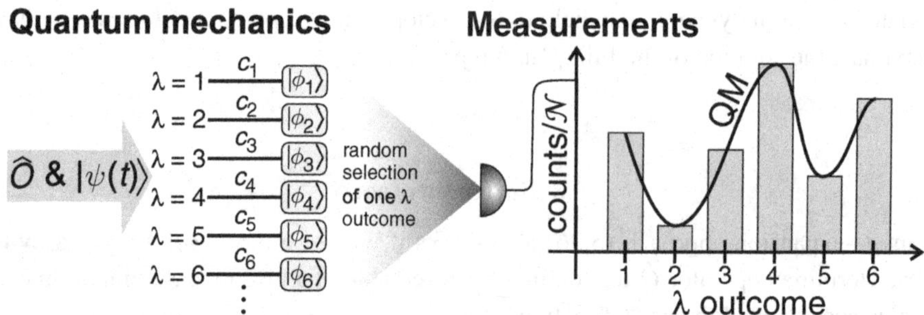

Figure 5.1 Schematic presentation of Born's theorem for a measurement corresponding to the observable $\hat{O}$. The quantum description defines the wave function $|\psi(t)\rangle = \sum_\lambda c_\lambda|\phi_\lambda\rangle$ and the coefficients c_λ define the mode expansion in terms of the eigenstates to the operator $\hat{O}$. The measurement randomly selects an outcome to be one of the possible eigenvalues λ. The quantum-mechanical distribution $|c_\lambda|^2$ (solid line) determines how many times an individual λ value appears (bar chart) if the measurement is repeated $\mathcal{N}$ times.

surrounding $\mathbf{r}$. As a result, even a finite number of measurements produces a nonvanishing $n_{\mathbf{r}}$. Born's theorem establishes the connection

$$P(\mathbf{r}, \Delta r, t) = \lim_{\mathcal{N}\to\infty} \frac{n_{\mathbf{r}}}{\mathcal{N}}, \tag{5.41}$$

based on Eqs. (3.15) and (5.40).

In a way, Born's theorem can be viewed as the most general formulation of the probabilistic interpretation of quantum mechanics. After the probability distributions are accepted conceptually, the analysis of the expectation values follows as discussed in Section 3.2. At a general level, Born's theorem can be interpreted to imply that individual measurements produce random outcomes whose overall distribution is governed by quantum mechanics while the individual outcomes cannot be predicted. Born's theorem is the corner stone of the Copenhagen interpretation of quantum mechanics. There are also other alternative schools of thought, however, the Copenhagen interpretation is the most widely tested and accepted one. Thus, this book analyzes all quantum-mechanical phenomena following this perspective.

5.3.1 Heisenberg uncertainty principle

Based on Born's theorem, a measurement can produce exact results only if the wave function occupies a single eigenstate, i.e., $|\psi(t)\rangle = |\phi_\lambda\rangle$, at the moment of the measurement. This case is very rare because one always has infinitely many possibilities to distribute the wave function among the infinitely many eigenstates. For example, a precise position measurement follows only if the wave function is $|\psi(t)\rangle = |\mathbf{r}_0\rangle$ producing $\psi(\mathbf{r}, t) = \langle\mathbf{r}|\mathbf{r}_0\rangle = \delta(\mathbf{r} - \mathbf{r}_0)$, implying a δ-function precision. However, as shown in Section 3.1.1, such a wave function will spread to a finite width $|\psi(t)\rangle$ even after an infinitesimally short time evolution. For a δ-function space accuracy, the probability distribution of the momentum, $|\langle\mathbf{p}|\mathbf{r}_0\rangle|^2 = 1/[2\pi\hbar]^3$ according to Eq. (5.29) is distributed evenly over all momenta, implying infinite average energy. Consequently, it is not physically meaningful to assume absolute accuracy in position measurements. Similar arguments can also be made for a wide range of other measurements such that one should always be concerned with the probabilistic interpretation.

The accuracy of two (or more) different measurements can be correlated. In order to study this on a general level, we consider measurements of the two observables $\hat{A}$ and $\hat{B}$. Both of them should yield the average outcomes $\langle\hat{A}\rangle$ and $\langle\hat{B}\rangle$ as well as the fluctuations

$$\Delta A^2 \equiv \langle\left[\hat{A} - \langle\hat{A}\rangle\right]^2\rangle, \qquad \Delta B^2 \equiv \langle\left[\hat{B} - \langle\hat{B}\rangle\right]^2\rangle. \tag{5.42}$$

Since the fluctuations follow from a positive-definite $|c_\lambda(t)|^2$ distribution, the variance of any observable must be positive definite. Using the relations (5.5), it is easy to show that the commutator of any two Hermitian operators satisfies

$$\left[\hat{A}, \hat{B}\right]_- = i\hat{M}, \tag{5.43}$$

where $\hat{M}$ is also a Hermitian operator. For example, $\left[\hat{\mathbf{r}}_j, \hat{\mathbf{p}}_j\right]_- = i\hbar$ fulfills this relation based on Eq. (3.35).

We may now introduce a general operator

$$\hat{C}_x = \hat{A} - \langle\hat{A}\rangle + ix\left(\hat{B} - \langle\hat{B}\rangle\right), \tag{5.44}$$

and an expectation value

$$\begin{aligned} f(x) \equiv \left\langle \hat{C}_x^\dagger \hat{C}_x \right\rangle &= \Delta A^2 + ix\langle\left[\hat{A}, \hat{B}\right]_-\rangle + x^2 \Delta B^2 \\ &= \Delta A^2 + x^2 \Delta B^2 - x\langle\hat{M}\rangle. \end{aligned} \tag{5.45}$$

We see that $f(x)$ is a real-valued parabola that has a minimum because ΔB^2 is positive definite. As for any such parabola, $f'(x_{\min}) = 0$ is satisfied at the minimum. Using this relation, we find $x_{\min} = \langle\hat{M}\rangle/2\Delta B^2$ and

$$f(x_{\min}) = \Delta A^2 - \frac{\langle\hat{M}\rangle^2}{4\Delta B^2}, \tag{5.46}$$

defining the smallest possible value for $f(x)$.

On a more general level, $f(x)$ can also be expressed via

$$f(x) = \left\langle \hat{C}_x^\dagger \hat{C}_x \right\rangle = \langle\psi|\hat{C}_x^\dagger\, \hat{C}_x|\psi\rangle = |N|^2 \langle\phi|\phi\rangle \tag{5.47}$$

after we have realized that $\hat{C}_x|\psi\rangle \equiv N|\phi\rangle$ can be understood to define the wave function $|\phi\rangle$ times a normalization factor N. Hence, we verify the positivity,

$$f(x) = |N|^2 \langle\phi|\phi\rangle \geq 0, \qquad \forall x. \tag{5.48}$$

Combining the results (5.46)–(5.48), we find

$$\Delta A^2 - \frac{\langle\hat{M}\rangle^2}{4\Delta B^2} \geq 0 \quad \Leftrightarrow \quad \Delta A\, \Delta B \geq \frac{|\langle\hat{M}\rangle|}{2}. \tag{5.49}$$

This is intimately related to the Cauchy–Schwarz inequality, see Exercise 5.6.

In quantum mechanics, Eq. (5.49) defines the general form of the Heisenberg uncertainty relation. It tells us that if two observables do not commute, we cannot simultaneously measure both of them with absolute accuracy. Instead, the fluctuations in the $\hat{A}$ and $\hat{B}$ measurements are connected via the Heisenberg uncertainty relation:

$$\Delta A\, \Delta B \geq \frac{\left|\left\langle\left[\hat{A}, \hat{B}\right]_-\right\rangle\right|}{2}. \tag{5.50}$$

In the special case where the equal sign is realized in the Heisenberg uncertainty relation, the system is in a minimum-uncertainty state with respect to $\hat{A}$ and $\hat{B}$ measurements. By choosing $\hat{\mathbf{r}}_j$ and $\hat{\mathbf{p}}_j$, we find the position–momentum uncertainty

$$\Delta \mathbf{r}_j \, \Delta \mathbf{p}_j \geq \frac{\hbar}{2}, \tag{5.51}$$

which is exactly the same as Eq. (3.31) constructed for the Gaussian wave packets. However, the Gaussian wave packet is a minimum-uncertainty state of position and momentum only if it does not contain chirp, i.e., at the time $t = 0$ before it starts to spread.

5.3.2 Schrödinger and Heisenberg picture

In many investigations, one wants to determine the time evolution of specific observables while the detailed knowledge of the full wave-function dynamics is not needed. To analyze how one best describes such a situation and to reveal the generic features of the quantum dynamics, we start by considering some formal properties of the time-dependent Schrödinger equation (5.33). As long as $\hat{H}$ is not explicitly time dependent, one can write the simple formal solution,

$$i\hbar \frac{\partial}{\partial t} |\psi(t)\rangle = \hat{H} \, |\psi(t)\rangle \qquad \Leftrightarrow \qquad |\psi(t)\rangle = e^{-i\frac{\hat{H}t}{\hbar}} |\psi(0)\rangle. \tag{5.52}$$

This result can be used to express the time evolution of generic operator expectation values via

$$\langle \hat{O} \rangle \equiv \langle \psi(t) | \hat{O} | \psi(t) \rangle = \langle \psi(0) | e^{+i\frac{\hat{H}t}{\hbar}} \, \hat{O} e^{-i\frac{\hat{H}t}{\hbar}} | \psi(0) \rangle, \tag{5.53}$$

where the time dependence of the conjugated wave function is given by $\langle \psi(t)| = \langle \psi(0)| e^{+i\frac{\hat{H}t}{\hbar}}$.

The two ways to write the expectation value according to Eq. (5.53) allow for two different interpretations. First, in the so-called Schrödinger picture, one considers a static operator $\hat{O}$ that is sandwiched between two time-dependent wave functions. Alternatively, we can look at the last expression in Eq. (5.53) and interpret it to have the generic time-dependent operator

$$\hat{O}(t) \equiv e^{+i\frac{\hat{H}t}{\hbar}} \, \hat{O} \, e^{-i\frac{\hat{H}t}{\hbar}}, \tag{5.54}$$

sandwiched between two static wave functions. This yields the so-called Heisenberg picture where all dynamical features are contained in the time dependence of the operators. The formal evolution of the operators then follows from

$$\begin{aligned} i\hbar \frac{\partial}{\partial t} \hat{O}(t) &= \left[i\hbar \frac{\partial}{\partial t} e^{+i\frac{\hat{H}t}{\hbar}} \right] \hat{O} e^{-i\frac{\hat{H}t}{\hbar}} + e^{+i\frac{\hat{H}t}{\hbar}} \hat{O} \left[i\hbar \frac{\partial}{\partial t} e^{-i\frac{\hat{H}t}{\hbar}} \right] \\ &= -\hat{H} \, e^{+i\frac{\hat{H}t}{\hbar}} \hat{O} e^{-i\frac{\hat{H}t}{\hbar}} + e^{+i\frac{\hat{H}t}{\hbar}} \hat{O} e^{-i\frac{\hat{H}t}{\hbar}} \, \hat{H} \\ &= -\hat{H} \, \hat{O}(t) + \hat{O}(t) \, \hat{H}, \end{aligned} \tag{5.55}$$

where we applied the generic relation

$$i\hbar\frac{\partial}{\partial t}e^{\pm i\frac{\hat{H}t}{\hbar}} = \mp\hat{H}e^{\pm i\frac{\hat{H}t}{\hbar}} = \mp e^{\pm i\frac{\hat{H}t}{\hbar}}\hat{H}, \tag{5.56}$$

which is valid because $\hat{H}$ commutes with the operator in the exponent. We recognize that the final expression on the RHS of (5.55) is nothing but a commutation relation. This is known as the Heisenberg equation of motion,

$$i\hbar\frac{\partial}{\partial t}\hat{O}(t) = \left[\hat{O}(t),\, \hat{H}\right]_- \tag{5.57}$$

showing that the full quantum dynamics can completely be expressed in terms of operator equations. In many cases, $\left[\hat{O}(t),\, \hat{H}\right]_-$ produces a new operator which differs significantly from the known $\hat{O}(t)$. Evaluating the Heisenberg equation also for these new operators introduces even more different operators such that one ends up with an infinite hierarchy of operator equations. We discuss in Chapter 15 how this hierarchy can be systematically truncated to produce approximate solutions for the quantum dynamics.

As a simple example, we consider a free particle and the dynamics of $\hat{\mathbf{r}}(t)$ and $\hat{\mathbf{p}}(t)$. Since $\hat{H} = \hat{p}^2/2m$, we find

$$\begin{cases} i\hbar\frac{\partial}{\partial t}\hat{r}(t) = \left[\hat{\mathbf{r}}(t),\, \hat{H}\right]_- = i\hbar\frac{\hat{p}(t)}{m} \\ i\hbar\frac{\partial}{\partial t}\hat{p}(t) = \left[\hat{\mathbf{p}}(t),\, \hat{H}\right]_- = 0 \end{cases} \Leftrightarrow \begin{cases} \frac{\partial}{\partial t}\hat{r}(t) = \frac{\hat{p}(t)}{m} \\ \frac{\partial}{\partial t}\hat{p}(t) = 0 \end{cases}. \tag{5.58}$$

The resulting equations have the same form as the classical Hamilton equations (1.33). Thus, the expectation value of $\langle\hat{\mathbf{r}}(t)\rangle$ and $\langle\hat{\mathbf{p}}(t)\rangle$ follow a classical path as shown in Fig. 3.2. However, Eq. (5.58) must be understood as an operator equation since $\hat{\mathbf{r}}(t)$ and $\hat{\mathbf{p}}(t)$ satisfy the canonical commutation relation (3.35) at all times. Consequently, the operator order does matter when we consider quantities like $\hat{\mathbf{r}}^n(t)\hat{\mathbf{p}}^m(t)$, which induces true quantum features such as the Heisenberg uncertainty relation (5.51).

Exercises

5.1 **(a)** Show that the Hermitian conjugation produces the relations (5.5) based on the definitions (3.17) and (5.4). **(b)** Show that $\langle\hat{F}\rangle^* = \langle\hat{F}^\dagger\rangle$. **(c)** Show that a real-valued $\langle\hat{F}\rangle$ for arbitrary quantum states implies that $\hat{F}$ must be self-adjoint, i.e., $\hat{F} = \hat{F}^\dagger$. **(d)** Show that the operator $\hat{O} = \partial/\partial z$ is not Hermitian.

5.2 **(a)** Assume that the operator $\hat{O}$ is Hermitian. Show that any eigenstate $\phi_\lambda(\mathbf{r})$ of $\hat{O}$ must produce a real-valued eigenvalue. **(b)** Derive Eq. (5.7) and show that the orthogonality (5.8) and the completeness relation (5.9) follow directly. **(c)** Show that the completeness relation is given by Eq. (5.17) in Dirac's abstract notation.

5.3 **(a)** Represent the stationary Schrödinger equation $\hat{H}|\phi_E\rangle = E|\phi_E\rangle$ in momentum space. Which aspects make this form relatively difficult to solve? **(b)** Assume a vanishing potential $\hat{V} = 0$ and solve $\langle p|\phi_E\rangle$ and E. **Hint:** *Perform the calculation in momentum space.* Show that the choice $|\phi_E\rangle = |p\rangle$ works. Which other choices are possible?

5.4 **(a)** Compute $|\psi_1\rangle \equiv e^{i\mathbf{R}\cdot\hat{\mathbf{p}}}|\psi\rangle$ and $|\psi_2\rangle \equiv e^{+i\mathbf{P}\cdot\hat{\mathbf{r}}}|\psi\rangle$ in real and momentum space when $\langle\mathbf{r}|\psi\rangle$ and $\langle\mathbf{p}|\psi\rangle$ are known. **(b)** Compute $\langle\psi_1|\psi\rangle$ and $\langle\psi_2|\psi\rangle$.

5.5 **(a)** Assume that a measurement determines the observable $\hat{A}$. Use Born's theorem to show that $\Delta A^2 = \langle[\hat{A} - \langle\hat{A}\rangle^2]^2\rangle$ is positive definite. **(b)** Show that any Hermitian operator must satisfy $[\hat{A}, \hat{B}]_- = i\hat{M}$, where $\hat{M}$ is Hermitian. **(c)** Show that $\left\langle\hat{C}_x^\dagger\hat{C}_x\right\rangle = \Delta A^2 + x^2\Delta B^2 - x\langle\hat{M}\rangle$ when $\hat{C}_x = \hat{A} - \langle\hat{A}\rangle + ix(\hat{B} - \langle\hat{B}\rangle)$.

5.6 One defines a generic complex-valued function $G^*(\mathbf{r}, x) = f(\mathbf{r}) + ix\,g(\mathbf{r})$, where both $f(\mathbf{r})$ and $g(\mathbf{r})$ are square integrable. **(a)** Minimize $F(x) = \int d^3r|G(r, x)|^2$ with respect to x and show that this yields the Cauchy–Schwarz inequality

$$\int d^3r\,|f^*(r)g(r)|^2 \leq \int d^3rf^*(r)f(r)\int d^3rg^*(r)g(r)\,.$$

(b) Choose $|f\rangle = (\hat{A} - \langle\hat{A}\rangle)|\psi\rangle$ and $|g\rangle = (\hat{B} - \langle\hat{B}\rangle)|\psi\rangle$ and show that the Cauchy–Schwarz inequality implies the Heisenberg uncertainty relation (5.50).

5.7 **(a)** Derive the Heisenberg equation of motion for the harmonic oscillator $\hat{H} = \frac{1}{2}m\hat{p}^2 + \frac{1}{2}m\omega^2\hat{x}^2$. **(b)** Show that the general solution follows from

$$\hat{x}(t) = \hat{x}(0)\cos\omega t + \frac{\hat{p}(0)}{m\omega}\sin\omega t \quad \hat{p}(t) = \hat{p}(0)\cos\omega t - \hat{x}(0)\,m\omega\sin\omega t.$$

5.8 **(a)** Derive the Heisenberg equation of motion for the anharmonic oscillator $\hat{H} = \frac{1}{2}m\hat{p}^2 + \frac{1}{2}m\omega^2\hat{x}^2 + Q^4\hat{x}^4$ and show that this produces an infinite hierarchy of equations. **(b)** Use the approximation $\frac{1}{4}\langle[\hat{x}^3\hat{p} + \hat{x}^2\hat{p}\hat{x} + \hat{x}\hat{p}\hat{x}^2 + \hat{p}\hat{x}^3]\rangle = \langle\hat{p}\rangle\langle\hat{x}\rangle^3$ to close the hierarchy. Assuming that anharmonic contributions are weak, how is the harmonic solution modified by Q?

Further reading

Discussions of the central relations in linear algebra, including the Gram–Schmidt orthogonalization, can be found in:

G. Strang (2005). *Introduction to Linear Algebra*, 3rd edition, Wellesley, MA, Wellesley-Cambridge Press.

G. E. Dhilov (1977). *Linear Algebra*, Mineola, NY, Dover.

D. Poole (2006). *Linear Algebra: A Modern Introduction*, 2nd edition, Belmont, CA, Thomson-Brooks/Cole.

Many details concerning the presentation, interpretation, and philosophy of quantum mechanics are presented in the following books:

P. A. M. Dirac (1999). *The Principles of Quantum Mechanics*, 4th edition, Oxford, Clarendon Press.
W. Heisenberg (1983). *The Physical Principles of the Quantum Theory*, Princeton, NJ, Princeton University Press.
M. Born (1965). *Natural Philosophy of Cause and Chance*, Oxford, Oxford University Press.
R. P. Feynman and A. R. Hibbs (1965). *Quantum Mechanics and Path Integrals*, New York, McGraw-Hill.
R. P. Feynman (1967). *The Character of Physical Law*, Cambridge, MA, MIT Press.
J. A. Wheeler and W. H. Zurek (1983). *Quantum Theory and Measurement*, Princeton, NJ, Princeton University Press.
D. Bohm (1951). *Quantum Mechanics*, New York, Prentice-Hall.
J. S. Bell (2004). *Speakable and Unspeakable in Quantum Mechanics*, Cambridge, Cambridge University Press.
D. Bohm and B. J. Hiley (1993). *The Undivided Universe: An Ontological Interpretation of Quantum Theory*, London, Routledge.
J. T. Cushing (1994). *Quantum Mechanics: Historical Contingency and the Copenhagen Hegemony*, Chicago, University of Chicago Press.

6

Wigner's phase-space representation

So far, we have described the quantum dynamics either in the Schrödinger picture using wave functions or via time-dependent operators in the Heisenberg picture. The wave function itself can be expressed in arbitrarily many different representations. As a common feature, any quantum description replaces the exact classical quantities by probabilistic distributions incorporating the inherent wave characteristics of quantum objects. We refer to any representation that determines *all* quantum features as the quantum statistics of the system. Due to the uniqueness of the quantum statistics, any specific representation can be uniquely converted into any other one.

In this context, it is interesting to search for the quantum-statistical representation that most closely resembles the classical description. In classical mechanics, position and momentum appear as independent variables, e.g., as arguments of the phase-space distribution $\rho(x, p\,; t)$ and the Liouville equation (1.61), derived in Section 1.2.1. As a generalization of the classical approach, we discuss in this chapter the quantum statistics via a phase-space distribution in the Schrödinger picture where simultaneously momentum and position appear as arguments.

6.1 Wigner function

We start the analysis by investigating the functions

$$\rho(\mathbf{r}', \mathbf{r};\ t) \equiv \psi(\mathbf{r}', t)\, \psi^{\star}(\mathbf{r}, t), \qquad \rho(\mathbf{k}', \mathbf{k};\ t) \equiv \frac{\psi(\mathbf{k}', t)\, \psi^{\star}(\mathbf{k}, t)}{[2\pi]^3} \tag{6.1}$$

instead of the probability distributions discussed so far. Both ρ functions can be used to describe the quantum statistics because they have a simple one-to-one mapping with the wave function. For example, $\tilde{\psi}(\mathbf{r}, t) \equiv \rho(\mathbf{r}, \mathbf{r}_0;\, t)\, [\rho(\mathbf{r}_0, \mathbf{r}_0;\, t)]^{-\frac{1}{2}}$ produces $\psi(\mathbf{r}, t)$ times a constant phase if we choose the reference point $\mathbf{r}_0$ such that $\rho(\mathbf{r}_0, \mathbf{r}_0;\, t) \neq 0$. Since a constant phase does not alter an expectation value, $\tilde{\psi}(\mathbf{r}, t)$ is basically equivalent to the original wave function. A similar unique mapping exists between $\rho(\mathbf{k}', \mathbf{k};\, t)$ and $\tilde{\psi}(\mathbf{k}, t)$. Furthermore, the ρ functions uniquely define the probability distributions

$$P(\mathbf{r},t)=\rho(\mathbf{r},\mathbf{r};t)=|\psi(\mathbf{r},t)|^2,$$
$$P(\mathbf{p},t)=\left[\frac{2\pi}{\hbar}\right]^3\rho\left(\mathbf{k}=\frac{\mathbf{p}}{\hbar},\mathbf{k}=\frac{\mathbf{p}}{\hbar};t\right)=|\psi(\mathbf{p},t)|^2, \tag{6.2}$$

through their diagonal elements.

The representations in (6.1) directly produce either a position or a momentum distribution but not both simultaneously. In principle, one could also consider quantum-statistical quantities such as $\psi^\star(\mathbf{r},t)\psi(\mathbf{p},t)$ to describe position and momentum simultaneously. However, this expression is generally complex valued such that it cannot be interpreted as a usual phase-space distribution.

We can easily establish a connection between $\rho(\mathbf{r}',\mathbf{r};t)$ and $\rho(\mathbf{k}',\mathbf{k};t)$ via Fourier transformation (3.40),

$$\rho(\mathbf{k}',\mathbf{k};t)=\frac{1}{[2\pi]^6}\int d^3r\,d^3r'\,\rho(\mathbf{r}',\mathbf{r};t)\,e^{i(\mathbf{k}\cdot\mathbf{r}-\mathbf{k}'\cdot\mathbf{r}')}, \tag{6.3}$$

and similarly for the inverse Fourier transformation. Since the Fourier transformation and its inverse define unique one-to-one mappings, we have verified again that $\rho(\mathbf{r}',\mathbf{r};t)$ and $\rho(\mathbf{k}',\mathbf{k};t)$ describe the very same quantum statistics.

As an alternative, we can also perform the Fourier transformation with respect to the center-of-mass coordinates

$$\begin{cases}\Delta\mathbf{x}\equiv\mathbf{r}-\mathbf{r}'\\ \mathbf{X}\equiv\frac{\mathbf{r}+\mathbf{r}'}{2}\end{cases}\Leftrightarrow\begin{cases}\mathbf{r}=\mathbf{X}+\frac{\Delta\mathbf{x}}{2}\\ \mathbf{r}'=\mathbf{X}-\frac{\Delta\mathbf{x}}{2}\end{cases} \tag{6.4}$$

converting Eq. (6.3) into

$$\rho(\mathbf{k}',\mathbf{k};t)=\int\frac{d^3X\,d^3\Delta x}{[2\pi]^6}\,\rho\left(\mathbf{X}-\frac{\Delta\mathbf{x}}{2},\mathbf{X}+\frac{\Delta\mathbf{x}}{2};t\right)\,e^{i(\mathbf{k}-\mathbf{k}')\cdot\mathbf{X}+i\frac{\mathbf{k}+\mathbf{k}'}{2}\cdot\Delta\mathbf{x}}. \tag{6.5}$$

We recognize that the exponential contains the center-of-mass and relative values of the wave vectors,

$$\begin{cases}\Delta\mathbf{q}\equiv\mathbf{k}-\mathbf{k}'\\ \mathbf{Q}\equiv\frac{\mathbf{k}+\mathbf{k}'}{2}\end{cases}\Leftrightarrow\begin{cases}\mathbf{k}=\mathbf{Q}+\frac{\Delta\mathbf{q}}{2}\\ \mathbf{k}'=\mathbf{Q}-\frac{\Delta\mathbf{q}}{2}\end{cases}. \tag{6.6}$$

With this identification, Eq. (6.5) becomes

$$\rho\left(\mathbf{Q}-\frac{\Delta\mathbf{q}}{2},\mathbf{Q}+\frac{\Delta\mathbf{q}}{2};t\right)=\int\frac{d^3X\,d^3\Delta x}{[2\pi]^6}\,\rho\left(\mathbf{X}-\frac{\Delta\mathbf{x}}{2},\mathbf{X}+\frac{\Delta\mathbf{x}}{2};t\right)\,e^{i\Delta\mathbf{q}\cdot\mathbf{X}+i\mathbf{Q}\cdot\Delta\mathbf{x}}. \tag{6.7}$$

For later reference, we note that this function produces the momentum distribution for $\Delta\mathbf{q}=\mathbf{0}$, based on the relation (6.2).

Since the Fourier transformation is defined by the relative and center-of-mass coordinates, we may equally well identify it through

$$W(\Delta\mathbf{q}, \mathbf{Q}; t) \equiv \rho\left(\mathbf{Q} - \frac{\Delta\mathbf{q}}{2}, \mathbf{Q} + \frac{\Delta\mathbf{q}}{2}; t\right)$$
$$= \int d^3X\, d^3\Delta x\, \frac{\rho\left(\mathbf{X} - \frac{\Delta\mathbf{x}}{2}, \mathbf{X} + \frac{\Delta\mathbf{x}}{2}; t\right)}{[2\pi]^6}\, e^{i\Delta\mathbf{q}\cdot\mathbf{X} + i\mathbf{Q}\cdot\Delta\mathbf{x}}. \tag{6.8}$$

Obviously, $W(\Delta\mathbf{q}, \mathbf{Q}; t)$ is still connected to ρ by a one-to-one mapping, i.e., W also defines the quantum statistics because we have merely exchanged the momentum indices. If we Fourier transform only the $\Delta\mathbf{q}$ coordinate of $W(\Delta\mathbf{q}, \mathbf{Q}; t)$ back into real space, we obtain the so-called Wigner function

$$W(\mathbf{R}, \mathbf{Q}; t) \equiv \int d^3\Delta q\, W(\Delta\mathbf{q}, \mathbf{Q}; t)\, e^{-i\mathbf{R}\cdot\Delta\mathbf{q}}$$
$$= \int d^3X\, d^3\Delta x\, \frac{\rho\left(\mathbf{X} - \frac{\Delta\mathbf{x}}{2}, \mathbf{X} + \frac{\Delta\mathbf{x}}{2}; t\right)}{[2\pi]^6}\, e^{i\mathbf{Q}\cdot\Delta\mathbf{x}} \int d^3\Delta q\, e^{i(\mathbf{X}-\mathbf{R})\cdot\Delta\mathbf{q}}$$
$$= \frac{1}{[2\pi]^3} \int d^3\Delta x\, \rho\left(\mathbf{R} - \frac{\Delta\mathbf{x}}{2}, \mathbf{R} + \frac{\Delta\mathbf{x}}{2}; t\right) e^{i\mathbf{Q}\cdot\Delta\mathbf{x}}. \tag{6.9}$$

Here, we used

$$\int d^3\Delta q\, e^{i(\mathbf{X}-\mathbf{R})\cdot\Delta\mathbf{q}} = [2\pi]^3 \delta(\mathbf{X} - \mathbf{R}) \tag{6.10}$$

and the integration rule (1.50).

The Wigner function has several interesting properties. First of all, its complex conjugate produces

$$W^\star(\mathbf{R}, \mathbf{Q}; t) = \frac{1}{[2\pi]^3} \int d^3\Delta x\, \rho\left(\mathbf{R} + \frac{\Delta\mathbf{x}}{2}, \mathbf{R} - \frac{\Delta\mathbf{x}}{2}; t\right) e^{-i\mathbf{Q}\cdot\Delta\mathbf{x}}$$
$$= \frac{1}{[2\pi]^3} \int d^3\Delta x\, \rho\left(\mathbf{R} - \frac{\Delta\mathbf{x}}{2}, \mathbf{R} + \frac{\Delta\mathbf{x}}{2}; t\right) e^{i\mathbf{Q}\cdot\Delta\mathbf{x}}$$
$$= W(\mathbf{R}, \mathbf{Q}; t), \tag{6.11}$$

where we have exchanged the integration variable $\Delta\mathbf{x} \leftrightarrow -\Delta\mathbf{x}$ to get the final result. Since the complex conjugation of the Wigner function leaves it unchanged, it is obviously a real-valued distribution.

The first coordinate of $W(\mathbf{R}, \mathbf{Q}; t)$ refers to the position while $\hbar\mathbf{Q} \equiv \mathbf{p}$ has the character of a momentum such that $W(\mathbf{R}, \mathbf{Q}; t)$ resembles a classical phase-space distribution. Hence, it is interesting to evaluate the marginal distributions

$$W_{\mathbf{R}}(\mathbf{Q}; t) \equiv \int d^3R\, W(\mathbf{R}, \mathbf{Q}; t) = \int d^3\Delta q\, W(\Delta\mathbf{q}, \mathbf{Q}; t) \int d^3R\, e^{-i\mathbf{R}\cdot\Delta\mathbf{q}}$$
$$= [2\pi]^3 W(\Delta\mathbf{q} = \mathbf{0}, \mathbf{Q}; t) = [2\pi]^3 \rho(\mathbf{Q}, \mathbf{Q}; t) = |\psi(\mathbf{Q})|^2, \tag{6.12}$$

where we have used (6.1) and (6.8). Clearly, this marginal distribution defines the momentum distribution (6.2) of the quantized system. The other marginal distribution can also be computed from Eq. (6.9) with the result

$$\begin{aligned}
W_{\mathbf{Q}}(\mathbf{R};t) &\equiv \int d^3Q\; W(\mathbf{R},\mathbf{Q};t)\\
&= \int d^3\Delta x\, \rho\left(\mathbf{R}-\frac{\Delta\mathbf{x}}{2},\mathbf{R}+\frac{\Delta\mathbf{x}}{2};t\right)\frac{1}{[2\pi]^3}\int d^3Q\; e^{i\mathbf{Q}\cdot\Delta\mathbf{x}}\\
&= \int d^3\Delta x\, \rho\left(\mathbf{R}-\frac{\Delta\mathbf{x}}{2},\mathbf{R}+\frac{\Delta\mathbf{x}}{2};t\right)\delta(\Delta\mathbf{x})\\
&= \rho(\mathbf{R},\mathbf{R};t) = P(\mathbf{r};t).
\end{aligned} \tag{6.13}$$

This marginal distribution corresponds to the probability distribution for the position.

It is also easy to prove that the Wigner function is normalized

$$\int d^3R\, d^3Q\; W(\mathbf{R},\mathbf{Q};t) = \int d^3R\; W_{\mathbf{Q}}(\mathbf{R};t) = \int d^3Q\; W_{\mathbf{R}}(\mathbf{Q};t) = 1, \tag{6.14}$$

based on Eqs. (6.12)–(6.13). Consequently, the Wigner function resembles a classical phase-space distribution in several ways: it is real valued, normalized to one, and its marginal distributions are true probability distributions for either momentum or position. Furthermore, the Wigner function determines any purely position- or momentum-dependent quantity via a simple averaging procedure. Combining the results (6.12)–(6.13) with Eqs. (3.17) and (3.43), we find

$$\begin{aligned}
\langle F(\hat{\mathbf{r}})\rangle &= \int d^3R\, d^3Q\; F(\mathbf{R})W(\mathbf{R},\mathbf{Q};t),\\
\langle G(\hat{\mathbf{p}})\rangle &= \int d^3R\, d^3Q\; G(\hbar\mathbf{Q})W(\mathbf{R},\mathbf{Q};t).
\end{aligned} \tag{6.15}$$

Note that even though many aspects of the Wigner function resemble those of a classical phase-space distribution, there are also important differences.

6.1.1 Averages in phase space

We start looking at the genuine quantum aspects of the Wigner function by investigating more complicated averages in phase space. At first sight, the arguments **R** and **Q** seem to appear as ordinary classical variables parametrizing the Wigner function. Thus, the particular order should not be relevant if we evaluate phase-space averages of the type

$$\begin{aligned}
\langle \mathbf{R}_j\hbar\mathbf{Q}_k\rangle_{\mathrm{ps}} &\equiv \int d^3R\, d^3Q\; \mathbf{R}_j\hbar\mathbf{Q}_k W(\mathbf{R},\mathbf{Q};t)\\
&= \int d^3R\, d^3Q\; \hbar\mathbf{Q}_k\mathbf{R}_j\; W(\mathbf{R},\mathbf{Q};t)\\
\Leftrightarrow \langle \mathbf{R}_j\hbar\mathbf{Q}_k\rangle_{\mathrm{ps}} &= \langle \hbar\mathbf{Q}_k\mathbf{R}_j\rangle_{\mathrm{ps}}.
\end{aligned} \tag{6.16}$$

However, we find a different result $\langle \hat{\mathbf{r}}_j \hat{\mathbf{p}}_j \rangle \neq \langle \hat{\mathbf{p}}_j \hat{\mathbf{r}}_j \rangle$ for the quantum-mechanical expectation values. Thus, it is clear that we cannot use the simple substitutions $\mathbf{R}_j \to \hat{\mathbf{r}}_j$ and $\hbar\mathbf{Q}_j \to \hat{\mathbf{p}}_j$ for mixed operator products, such as $\langle \mathbf{R}_j \hbar\mathbf{Q}_k \rangle_{\text{ps}}$. To clearly mark the different definitions, we use the notation $\langle \hat{O} \rangle_{\text{ps}}$ to distinguish the phase-space average from the actual quantum-mechanical $\langle \hat{O} \rangle$.

To find the correct connection between the generic $\langle \cdots \rangle_{\text{ps}}$ and $\langle \cdots \rangle$, we consider the simplest nontrivial case

$$
\begin{aligned}
\langle \mathbf{R}_j \hbar\mathbf{Q}_k \rangle_{\text{ps}} &\equiv \int d^3R\, d^3Q\, d^3\Delta x\; \mathbf{R}_j\, \hbar\mathbf{Q}_k \frac{\psi\left(\mathbf{R} - \frac{\Delta\mathbf{x}}{2}\right)\, \psi^\star\left(\mathbf{R} + \frac{\Delta\mathbf{x}}{2}\right)}{(2\pi)^3} e^{i\mathbf{Q}\cdot\Delta\mathbf{x}} \\
&= \int d^3R\, d^3Q\, d^3\Delta x\; \mathbf{R}_j \frac{\psi\left(\mathbf{R} - \frac{\Delta\mathbf{x}}{2}\right)\, \psi^\star\left(\mathbf{R} + \frac{\Delta\mathbf{x}}{2}\right)}{(2\pi)^3} \left[-i\hbar \frac{\partial}{\partial \Delta\mathbf{x}_k} e^{i\mathbf{Q}\cdot\Delta\mathbf{x}} \right] \\
&= \int d^3R\, d^3Q\, d^3\Delta x\; \mathbf{R}_j \left[i\hbar \frac{\partial}{\partial \Delta\mathbf{x}_k} \frac{\psi^\star\left(\mathbf{R} + \frac{\Delta\mathbf{x}}{2}\right)\, \psi\left(\mathbf{R} - \frac{\Delta\mathbf{x}}{2}\right)}{(2\pi)^3} \right] e^{i\mathbf{Q}\cdot\Delta\mathbf{x}}.
\end{aligned}
\tag{6.17}
$$

Here, we have first inserted the definitions (6.1) and (6.9) to express the Wigner function in terms of the quantum-mechanical wave functions. The second step follows from the generic relation $\hbar\mathbf{Q}_j\, e^{i\mathbf{Q}\cdot\Delta\mathbf{x}} = -i\hbar \frac{\partial}{\partial \Delta\mathbf{x}_j} e^{i\mathbf{Q}\cdot\Delta\mathbf{x}}$, and in the third step, we applied integration by parts using the fact that the wave function must vanish at infinity.

The differentiation in the last line of (6.17) is evaluated using

$$
\begin{aligned}
D &\equiv \frac{\partial}{\partial \Delta\mathbf{x}_k} \psi^\star\left(\mathbf{R} + \frac{\Delta\mathbf{x}}{2}\right) \psi\left(\mathbf{R} - \frac{\Delta\mathbf{x}}{2}\right) \\
&= \left[\frac{\partial}{\partial \Delta\mathbf{x}_k} \psi\left(\mathbf{R} - \frac{\Delta\mathbf{x}}{2}\right) \right] \psi^\star\left(\mathbf{R} + \frac{\Delta\mathbf{x}}{2}\right) + \psi\left(\mathbf{R} - \frac{\Delta\mathbf{x}}{2}\right) \left[\frac{\partial}{\partial \Delta\mathbf{x}_k} \psi^\star\left(\mathbf{R} + \frac{\Delta\mathbf{x}}{2}\right) \right] \\
&= \psi^\star\left(\mathbf{R} + \frac{\Delta\mathbf{x}}{2}\right) \left[-\frac{1}{2}\frac{\partial}{\partial \mathbf{R}_k} \psi\left(\mathbf{R} - \frac{\Delta\mathbf{x}}{2}\right) \right] \\
&\quad + \left[\frac{1}{2}\frac{\partial}{\partial \mathbf{R}_k} \psi^\star\left(\mathbf{R} + \frac{\Delta\mathbf{x}}{2}\right) \right] \psi\left(\mathbf{R} - \frac{\Delta\mathbf{x}}{2}\right).
\end{aligned}
\tag{6.18}
$$

Substituting this result into Eq. (6.17), we obtain

$$
\begin{aligned}
\langle \mathbf{R}_j \hbar\mathbf{Q}_k \rangle_{\text{ps}} &= \int d^3R\, d^3\Delta x \left\{ \psi^\star\left(\mathbf{R} + \frac{\Delta\mathbf{x}}{2}\right) \mathbf{R}_j \left[-\frac{i\hbar}{2}\frac{\partial}{\partial \mathbf{R}_k} \psi\left(\mathbf{R} - \frac{\Delta\mathbf{x}}{2}\right) \right] \right. \\
&\quad \left. + \left[\frac{i\hbar}{2}\frac{\partial}{\partial \mathbf{R}_k} \psi^\star\left(\mathbf{R} + \frac{\Delta\mathbf{x}}{2}\right) \right] \mathbf{R}_j\, \psi\left(\mathbf{R} - \frac{\Delta\mathbf{x}}{2}\right) \right\} \int \frac{d^3Q}{(2\pi)^3} e^{i\mathbf{Q}\cdot\Delta\mathbf{x}} \\
&= \int d^3R\, d^3\Delta x \left\{ \psi^\star\left(\mathbf{R} + \frac{\Delta\mathbf{x}}{2}\right) \mathbf{R}_j \left[-\frac{i\hbar}{2}\frac{\partial}{\partial \mathbf{R}_k} \psi\left(\mathbf{R} - \frac{\Delta\mathbf{x}}{2}\right) \right] \right. \\
&\quad \left. + \left[-\frac{i\hbar}{2}\frac{\partial}{\partial \mathbf{R}_k} \psi\left(\mathbf{R} + \frac{\Delta\mathbf{x}}{2}\right) \right]^\star \mathbf{R}_j\, \psi\left(\mathbf{R} - \frac{\Delta\mathbf{x}}{2}\right) \right\} \delta(\Delta\mathbf{x}),
\end{aligned}
\tag{6.19}
$$

which follows after we have first evaluated the **Q**-integral, producing the δ function. Using this δ function to evaluate the $\Delta\mathbf{x}$-integral yields

$$\langle \mathbf{R}_j \hbar \mathbf{Q}_k \rangle_{\text{ps}} = \int d^3R \left\{ \psi^\star(\mathbf{R})\, \mathbf{R}_j \left[-\frac{i\hbar}{2} \frac{\partial}{\partial \mathbf{R}_k} \psi(\mathbf{R}) \right] + \left[-\frac{i\hbar}{2} \frac{\partial}{\partial \mathbf{R}_k} \psi(\mathbf{R}) \right]^\star \mathbf{R}_j\, \psi(\mathbf{R}) \right\}, \tag{6.20}$$

which defines the phase-space averages if the wave function of the system is known.

In Eq. (6.20), we recognize that the first term is the quantum-mechanical expectation value, Eq. (3.17), and the second term is the adjoint expectation value, Eq. (5.4), of the product of position $\mathbf{R}_j \to \hat{\mathbf{r}}_j$ and momentum operators $-i\hbar \frac{\partial}{\partial \mathbf{R}_k} \to \hat{\mathbf{p}}_k$. Clearly, the ordering of $\hat{\mathbf{r}}_j$ and $\hat{\mathbf{p}}_j$ is important because of the operator nature of the canonical pair.

Summarizing our calculations, we see that the phase-space average can be expressed in terms of a quantum-mechanical expectation value,

$$\begin{aligned} \langle \mathbf{R}_j \hbar \mathbf{Q}_k \rangle_{\text{ps}} &\equiv \int d^3R\, d^3Q\; \mathbf{R}_j\, \hbar \mathbf{Q}_j\; W(\mathbf{R}, \mathbf{Q}; t) \\ &= \int d^3R\, d^3Q\; \hbar \mathbf{Q}_j\, \mathbf{R}_j\; W(\mathbf{R}, \mathbf{Q}; t) \\ &= \frac{1}{2} \langle \hat{\mathbf{r}}_j \hat{\mathbf{p}}_k + \hat{\mathbf{p}}_j \hat{\mathbf{r}}_k \rangle. \end{aligned} \tag{6.21}$$

Clearly, the phase-space average is neither $\langle \hat{\mathbf{p}}_j \hat{\mathbf{r}}_j \rangle$ nor $\langle \hat{\mathbf{r}}_j \hat{\mathbf{p}}_j \rangle$ but a symmetrized version of them. This observation holds also more generally: the phase-space average $\langle \hat{O} \rangle$ always produces symmetrized expectation values of the actual quantum-mechanical position and momentum operators within $\hat{O}$. Only if $\hat{O}$ consists of pure position or momentum operators, do $\langle \hat{O} \rangle_{\text{ps}}$ and $\langle \hat{O} \rangle$ become identical.

Obviously, the quantum statistics has significant consequences if the operator order is important such as in $\langle \hat{\mathbf{r}}_j \hat{\mathbf{p}}_j \rangle$ and $\langle \hat{\mathbf{p}}_j \hat{\mathbf{r}}_j \rangle$. For example, the canonical commutation relation (3.35) produces

$$\begin{aligned} &\langle \hat{\mathbf{p}}_j \hat{\mathbf{r}}_j \rangle = \langle \hat{\mathbf{r}}_j \hat{\mathbf{p}}_j \rangle - i\hbar \\ \Rightarrow \quad &\langle \hat{\mathbf{p}}_j \hat{\mathbf{r}}_j \rangle_{\text{ps}} = \left\langle \frac{\hat{\mathbf{p}}_j \hat{\mathbf{r}}_j + \hat{\mathbf{r}}_j \hat{\mathbf{p}}_j}{2} \right\rangle = \langle \hat{\mathbf{r}}_j \hat{\mathbf{p}}_j \rangle - \frac{i\hbar}{2} = \langle \hat{\mathbf{p}}_j \hat{\mathbf{r}}_j \rangle + \frac{i\hbar}{2}, \end{aligned} \tag{6.22}$$

where the additional $\pm i\hbar/2$ connects the phase-space and the quantum-mechanical averages. It is also clear that the relation between a generic phase-space average $\left\langle \hat{\mathbf{r}}_j^N \hat{\mathbf{p}}_j^M \right\rangle_{\text{ps}}$ and the corresponding $\left\langle \hat{\mathbf{r}}_j^N \hat{\mathbf{p}}_j^M \right\rangle$ becomes more involved for arbitrary N and M, see Exercise 6.3.

6.1.2 Quantum properties of the Wigner function

Another important difference between the classical and the quantum-statistical distribution is that the classical $\rho(x, p; t)$ can have absolute accuracy, i.e., it can simultaneously be a δ function both in position and momentum, whereas the Wigner function must satisfy the Heisenberg uncertainty relation (5.51). Thus, the Wigner function cannot become arbitrarily narrow in both variables.

An interesting situation arises if the marginal distributions exhibit an interference pattern. In this situation, the Wigner function becomes negative valued. As a consequence, the Wigner function has the nature of a quasiprobability distribution which deviates from a genuine phase-space probability. Whenever the Wigner function has negative-valued regions, there is no corresponding classical state. In practice, one often takes the partial negativity of $W(\mathbf{R}, \mathbf{Q}; t)$ as an indicator that the corresponding state exhibits strong quantum features.

We show in Exercise 6.4 that the values of the Wigner function are bounded within

$$-\frac{1}{\pi} \leq W(Z, Q; t) \leq \frac{1}{\pi}. \tag{6.23}$$

The upper limit can also be understood as one manifestation of the Heisenberg uncertainty relation. For example, we cannot localize the Wigner distribution inside a region smaller than $\pi \Delta Z \Delta Q = \pi/2$ where $\pi \Delta Z \Delta Q$ is the area of an ellipse with the axes ΔZ and ΔQ, and the RHS follows if we use the uncertainty relation, Eq. (5.51). As a consequence, the height of $W(Z, Q; t)$ must be limited by Eq. (6.23) due to the normalization (6.14). At the Heisenberg uncertainty limit, a Gaussian Wigner function has $\Delta Z = \Delta Z_0(t)$ and $\Delta Q = 1/2\Delta Z_0$ producing

$$W_{\text{Gauss}}(Z, Q; t) = \frac{1}{\pi} e^{-\frac{[Z-Z_0(t)]^2}{2\Delta Z_0^2(t)}} e^{-2\Delta Z_0^2[Q-Q_0(t)]^2}. \tag{6.24}$$

This function has the peak value $1/\pi$ at $(Z, Q) = (Z_0(t), Q_0(t))$ matching the upper limit of Eq. (6.23).

6.1.3 Negativity of the Wigner function

To get a better feeling for the negativity of the Wigner function and its relation to true quantum features, we investigate wave-averaging properties which do not exist in classical particle mechanics. For this purpose, we consider a superposition state of two Gaussian wave packets that propagate toward one another and eventually "collide." More specifically, we wave average Eq. (3.8) where the propagating (counterpropagating) component is centered at $-z_0$ $(+z_0)$ and propagates to the positive (negative) direction with momentum $p_0 = \hbar q_0$ $(p_0 = -\hbar q_0)$. The corresponding superposition state is

$$\psi(z, t) = \left[\frac{N^2}{2\pi \Delta z^2(t)}\right]^{\frac{1}{4}} e^{iq_0 z_0 - i\frac{\hbar q_0^2 t}{2m}} \left[C_1 e^{-\frac{(z-z_0-v_0 t)^2}{4\Delta z_0 \Delta z(t)}} e^{-iq_0 z} + C_2 e^{-\frac{(z+z_0-v_0 t)^2}{4\Delta z_0 \Delta z(t)}} e^{+iq_0 z}\right], \tag{6.25}$$

where $v_0 \equiv \hbar q_0/m$ determines the velocity associated with the wave vector q_0. The prefactors C_1 and C_2 determine the weighting of the respective contributions and N is the normalization constant,

$$N = \frac{1}{|C_1|^2 + |C_2|^2 + 2\mathrm{Re}\left[C_1^\star C_2\right] e^{-2q_0^2 \Delta z_0^2} e^{-z_0^2/2\Delta z_0^2}}. \tag{6.26}$$

This norm approaches $\frac{1}{|C_1|^2+|C_2|^2}$ if $|q_0 \Delta z_0|$ becomes large.

Since we consider a one-dimensional example, we use the one-dimensional Wigner function,

$$\begin{aligned} W(Z, Q; t) &\equiv \frac{1}{2\pi} \int_{-\infty}^{\infty} d\Delta x\, \rho\left(Z - \frac{\Delta x}{2}, Z + \frac{\Delta x}{2}; t\right) e^{iQ\cdot\Delta x} \\ &= \frac{1}{2\pi} \int_{-\infty}^{\infty} d\Delta x\, \psi\left(Z - \frac{\Delta x}{2}, t\right) \psi^\star\left(Z + \frac{\Delta x}{2}, t\right) e^{iQ\cdot\Delta x}. \end{aligned} \tag{6.27}$$

Inserting the Gaussian superposition state, we have to perform a lengthy but straightforward calculation involving a series of Gaussian integrations, see Exercise 2.6 and Exercise 6.6. We finally obtain the Wigner function in the form

$$\begin{aligned} W(Z, Q; t) &= \frac{N|C_1|^2}{\pi} e^{-\frac{\left[Z - \frac{\hbar Q}{m}t - z_0\right]^2}{2\Delta z_0^2}} e^{-2\Delta z_0^2 [Q+q_0]^2} \\ &+ \frac{N|C_2|^2}{\pi} e^{-\frac{\left[Z - \frac{\hbar Q}{m}t + z_0\right]^2}{2\Delta z_0^2}} e^{-2\Delta z_0^2 [Q-q_0]^2} \\ &+ \frac{N|C_1 C_2|}{\pi} e^{-\frac{\left[Z - \frac{\hbar Q}{m}t\right]^2}{2\Delta z_0^2}} e^{-2\Delta z_0^2 Q^2} 2\cos\left(\varphi_{12} + 2\left[Q z_0 + q_0\left(Z - \frac{\hbar Q}{m}t\right)\right]\right), \end{aligned} \tag{6.28}$$

where φ_{12} defines the phase between the coefficients C_1 and C_2. In other words, $C_1^\star C_2 = |C_1 C_2| e^{i\varphi_{12}}$. Taking the cross section along $Q = 0$, we find

$$\begin{aligned} W(Z, 0; t) &= \frac{N e^{-2\Delta z_0^2 q_0^2}}{\pi} \left(|C_1|^2 e^{-\frac{[Z-z_0]^2}{2\Delta z_0^2}} + |C_2|^2 e^{-\frac{[Z+z_0]^2}{2\Delta z_0^2}} \right) \\ &+ \frac{N|C_1 C_2|}{\pi} 2e^{-\frac{Z^2}{2\Delta z_0^2}} \cos(\varphi_{12} + 2q_0 Z). \end{aligned} \tag{6.29}$$

If $\Delta z_0^2 q_0^2$ becomes large, $W(Z, 0; t)$ is dominated by the expression in the last line, i.e., it oscillates between $-\frac{N|C_1 C_2|}{\pi}$ and $+\frac{N|C_1 C_2|}{\pi}$. The minimum value can reach $-1/\pi$ for sufficiently large $\Delta z_0^2 q_0^2$ and $|C_1| = |C_2|$.

This little exercise shows that we can find interference-pattern-like oscillations (6.28) in certain parts of phase space simply by constructing the wave function via wave averaging.

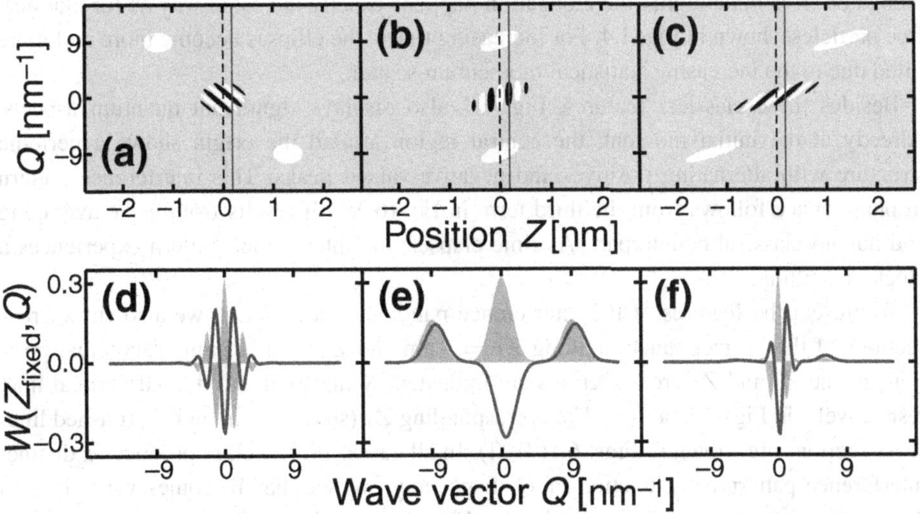

Figure 6.1 Time evolution of the Wigner function for a two-Gaussian superposition state. The contour lines of $W(Z, Q; t)$ are shown at **(a)** $t = 0\,\text{fs}$, **(b)** $t = 1.24\,\text{fs}$, and **(c)** $t = 2.49\,\text{fs}$. The color code is chosen to show $W(Z, Q; t) \geq 0.1$ (white), $W(Z, Q; t) \leq -0.1$ (black), and $|W(Z, Q; t)| \leq 0.1$ (gray). The vertical lines indicate the cross sections at $Z \equiv Z_1 = 0$ (dashed line) and $Z \equiv Z_2 = 0.18\,\text{nm}$ (solid line). The corresponding $W(Z_1, Q; t)$ (shaded area) and $W(Z_2, Q; t)$ (solid line) are plotted for the times **(d)** $t = 0\,\text{fs}$, **(e)** $t = 1.24\,\text{fs}$, and **(f)** $t = 2.49\,\text{fs}$. In the calculations, we used Eq. (6.28) with the free-electron mass, $z_0 = 1.29\,\text{nm}$, $\Delta z_0 = 0.3\,\text{nm}$, and $q_0 = 8.98\,\text{nm}^{-1}$.

The cosine term in (6.28) resembles the typical form of an interference pattern. However, this time the interference happens in phase space.

In connection with Eq. (2.24), we discussed that propagating waves in real space display an interference pattern only if the pulses overlap. In phase space, however, the interference pattern is always present, regardless of the spatial overlap of the particle-wave components. Thus, the Wigner function can easily reveal whether the system has wave-averaging related quantum features.

Figure 6.1 presents the Wigner function (6.28) in phase space for the times (a) $t = 0\,\text{fs}$, (b) $t = 1.24\,\text{fs}$, and (c) $t = 2.49\,\text{fs}$. The Wigner function is constructed for the superposition of two freely propagating Gaussians. The Gaussian is initially placed at $z_0 = 1.29\,\text{nm}$ away from the origin, it has the width $\Delta z_0 = 0.30\,\text{nm}$, the wave vector is $q_0 = 8.98\,\text{nm}^{-1}$, and the particle has the mass of an electron. The counterpropagating components are mixed with an equal weight $C_1 = C_2 = 1/\sqrt{2}$. The regions with $W \geq 0.1$ are shown as white while the negative $W \leq -0.1$ regions are black. The gray shading indicates $|W| \leq 0.1$.

The large white ellipses in Fig. 6.1 show that the Wigner function is initially peaked around $(+z_0, -q_0)$ and $(-z_0, +q_0)$. The two parts correspond to separate wave packets propagating toward each other from the opposite sides of the origin. As time elapses, the phase-space distributions move along the Z axis and the shape of the ellipses becomes elongated. It is notable that the elongation happens exactly the same way as for classical free particles, shown in Fig. 1.4. For increasing times, the ellipses become more and more tilted due to the increasing statistical momentum scatter.

Besides these classical features, Fig. 6.1 also displays significant quantum aspects. Already at the initial moment, the central region around the origin shows a periodic structure with alternating positive- and negative-valued peaks. This interference pattern in phase-space follows from the third term in Eq. (6.28). It results from wave averaging and has no classical counterpart. As time elapses, the interference pattern experiences a clockwise tilting.

To present the features of the interference pattern in more detail, we also show cross sections of the Wigner function along $Z_1 = 0\,\text{nm}$ and $Z_2 = 0.182\,\text{nm}$. The actual positions of the Z_1 and Z_2 cross sections are indicated by the solid and dashed vertical line, respectively, in Figs. 6.1(a)–(c). The corresponding Z_1 (shaded area) and Z_2 (dashed line) cross sections are shown in Figs. 6.1(d)–(f). In all cases, $W(Z_j, Q; t)$ produces a distinct interference pattern as a function of Q. Especially, we see that W comes very close to the extreme values $\pm 1/\pi$, indicated by the thin horizontal lines. Thus, we clearly see that true quantum features, such as the phase-space interference pattern, induce pronounced negativity of the Wigner function.

The top row in Fig. 6.1 shows that the time-development of $W(Z_j, Q; t)$ manifests itself in a tilting of the entire interference pattern. At $t = 1.24\,\text{ps}$, the two components of the wave packet are aligned on top of each other along the Z axis. The $Z_1 = 0\,\text{nm}$ slice (shaded area in Fig. 6.1(e)) shows two positive bumps at $Q = \pm q_0$ and one large peak at $Q = 0$. At the same time, the $Z_2 = 0.18\,\text{nm}$ slice (solid line in Fig. 6.1(e)) shows the two positive bumps close to $\pm q_0$ and a strongly negative dip close to $Q = 0$. If we now integrate $W_Q(Z_j) = \int dQ\ W(Z_j, Q)$, the Z_1 slice averages to a large value whereas $W_Q(Z_2)$ is more or less zero. Since $W_Q(Z)$ is a probability distribution for the position, we conclude that a significant spatial interference pattern is observed at the time $t = 1.24\,\text{ps}$ and that the period is roughly $0.36\,\text{nm}$, twice the observed peak-dip separation in $W_Q(Z_2)$. This is in full agreement with the period-estimate $\Delta z_{\text{node}} = \pi/q_0 = 0.36\,\text{nm}$ provided by Eq. (2.26). For the times $t = 0\,\text{fs}$ and $t = 2.49\,\text{fs}$, the corresponding $\int dQ\ W(Z_j, Q)$ approaches zero, verifying that the spatial interference pattern is detected only if the wave-function components overlap.

To identify the actual shape of the spatial interference pattern, we compute the marginal distributions $W_P(Z) = \int dQ W(Q, Z; t)$ for the Wigner functions shown in Fig. 6.1. These position probability distributions are plotted in Fig. 6.2 at the times $t = 0\,\text{fs}$ (frame (d)), $t = 1.24\,\text{fs}$ (frame (e)), and $t = 2.49\,\text{fs}$ (frame (f)). The upper row repeats the corresponding Wigner functions already shown in Fig. 6.1.

At the initial time, Fig. 6.2(d), $W_P(Z) = |\psi(Z, t)|^2$ consists of two distinctly separated Gaussians. Interestingly, the interference pattern in the corresponding Wigner function $W(Z, Q; t)$ around the origin averages to zero in $W_P(Z)$. The phase-space interference pattern is tilted so much that the integral $\int dQ W(Z, Q; t)$ practically vanishes close to $Z = 0$.

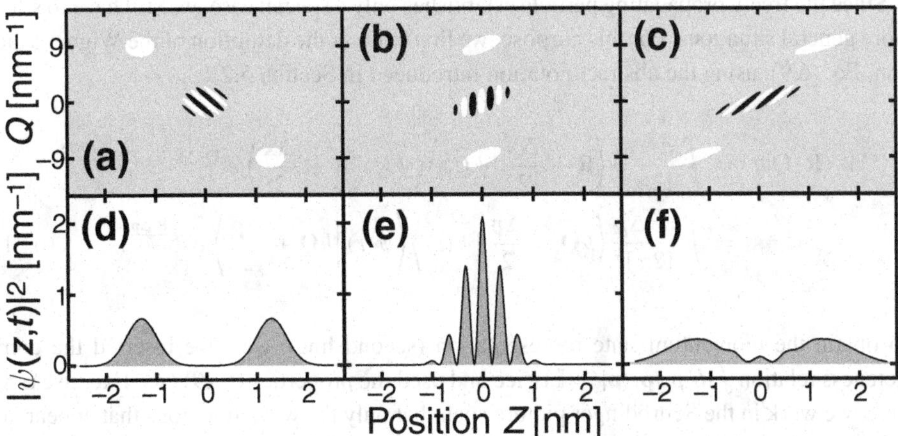

Figure 6.2 Time evolution of the Z-dependent marginal distribution for a two-Gaussian superposition state. Contour plots of the Wigner function are shown at **(a)** $t = 0\,\text{fs}$, **(b)** $t = 1.24\,\text{fs}$, and **(c)** $t = 2.49\,\text{fs}$. The contours are defined as in Fig. 6.1. The corresponding marginal distributions $W_Q(Z) = \int dQ W(Z, Q)$ are shown at the times **(d)** $t = 0\,\text{fs}$, **(e)** $t = 1.24\,\text{fs}$, and **(f)** $t = 2.49\,\text{fs}$. For the calculations, we used the same parameters as in Fig. 6.1.

However, at the times where the originally separated Gaussians overlap, a strong interference occurs also in the marginal distribution, Fig. 6.2(d). This is consistent with the fact that the corresponding phase-space interference pattern, in Fig. 6.2(b), is now vertically aligned such that the integral $\int dQ W(Z, Q; t)$ remains finite. With increasing time, the continuing tilting of the phase-space interference again leads to a vanishing interference in the marginal distribution, see Fig. 6.1(f) where clearly separated Gaussian pulses begin to emerge. Due to the spread in momentum, the late-time Gaussians are broader than the early-time ones.

A close inspection shows that the interference peaks in Fig. 6.2(e) are 3.12 times higher than the separate Gaussian peaks in Fig. 6.2(d). Since this factor clearly is larger than two, the increase cannot be explained by a simple addition of probabilities.

6.2 Wigner-function dynamics

The dynamics of the free-particle Wigner function shows the same functional evolution as the classical phase-space distribution. The connection becomes even clearer if we identify a classical path for free propagation, $Z_{\text{cl}}(Z, P = \hbar Q; t) \equiv Z - Pt/m = Z - \hbar Q t/m$ and $P_{\text{cl}}(Z, P = \hbar Q; t) = P = \hbar Q$ in analogy to Eq. (1.34). This allows us to represent the Wigner function (6.28) via

$$W(Z, Q; t) = W(Z_{\text{cl}}(Z, P = \hbar Q; -t), P_{\text{cl}}(Z, P = \hbar Q; -t)/\hbar; 0). \qquad (6.30)$$

This equation is identical to Eq. (1.66) showing that the Wigner function follows exactly the dynamics of the statistical phase-space distribution.

Since the freely propagating particle establishes only a special case, we still have to study more general situations. For this purpose, we first express the definition of the Wigner function, Eq. (6.9), using the abstract notation introduced in Section 5.2.1,

$$\begin{aligned} W(\mathbf{R}, \mathbf{Q}; t) &= \int \frac{d^3\Delta x}{[2\pi]^3} \left\langle \mathbf{R} - \frac{\Delta \mathbf{x}}{2} \middle| \psi(t) \right\rangle \left\langle \psi(t) \middle| \mathbf{R} + \frac{\Delta \mathbf{x}}{2} \right\rangle e^{i\mathbf{Q}\cdot\Delta\mathbf{x}} \\ &= \int \frac{d^3\Delta p}{[2\pi]^3} \left\langle \hbar\mathbf{Q} - \frac{\Delta \mathbf{p}}{2} \middle| \psi(t) \right\rangle \left\langle \psi(t) \middle| \hbar\mathbf{Q} + \frac{\Delta \mathbf{p}}{2} \right\rangle e^{-i\frac{\mathbf{R}\cdot\Delta\mathbf{p}}{\hbar}}. \end{aligned} \tag{6.31}$$

To obtain the momentum-state representation (second line), we have inserted the completeness relation $\int d^3p\, |\mathbf{p}\rangle\langle\mathbf{p}| = \mathbb{I}$ twice and used the properties (5.29) (see Exercise 6.8). Here, we work in the Schrödinger picture such that only the wave functions that appear are time dependent. Differentiating Eq. (6.31) with respect to time, we obtain

$$\begin{aligned} i\hbar\frac{\partial\, W(\mathbf{R}, \mathbf{Q}; t)}{\partial t} = \int \frac{d^3\Delta x}{[2\pi]^3} \Bigg[&\left\langle \mathbf{R} - \frac{\Delta \mathbf{x}}{2} \middle| \hat{H} \middle| \psi(t) \right\rangle \left\langle \psi(t) \middle| \mathbf{R} + \frac{\Delta \mathbf{x}}{2} \right\rangle \\ &- \left\langle \mathbf{R} - \frac{\Delta \mathbf{x}}{2} \middle| \psi(t) \right\rangle \left\langle \psi(t) \middle| \hat{H} \middle| \mathbf{R} + \frac{\Delta \mathbf{x}}{2} \right\rangle \Bigg] e^{i\mathbf{Q}\cdot\Delta\mathbf{x}} \end{aligned} \tag{6.32}$$

if we apply the product rule of differentiation and use Eq. (5.52). The total Hamiltonian $\hat{H} = \hat{\mathbf{p}}^2/2m + \hat{U}$ consists of the sum of kinetic- and potential-energy operators. Since $\hat{\mathbf{p}}$ is purely multiplicative in momentum space while $\hat{U}$ is multiplicative in real space, it is beneficial to use the respective representations for the different components within Eq. (6.32). Implementing this division, Eq. (6.32) casts into the form

$$\begin{aligned} i\hbar\frac{\partial\, W(\mathbf{R}, \mathbf{Q}; t)}{\partial t} = \int \frac{d^3\Delta p}{[2\pi]^3} \Bigg[&\left\langle \hbar\mathbf{Q} - \frac{\Delta \mathbf{p}}{2} \middle| \frac{\hat{\mathbf{p}}^2}{2m} \middle| \psi(t) \right\rangle \left\langle \psi(t) \middle| \hbar\mathbf{Q} + \frac{\Delta \mathbf{p}}{2} \right\rangle \\ &- \left\langle \hbar\mathbf{Q} - \frac{\Delta \mathbf{p}}{2} \middle| \psi(t) \right\rangle \left\langle \psi(t) \middle| \frac{\hat{\mathbf{p}}^2}{2m} \middle| \hbar\mathbf{Q} + \frac{\Delta \mathbf{p}}{2} \right\rangle \Bigg] e^{-i\frac{\mathbf{R}\cdot\Delta\mathbf{p}}{\hbar}} \\ + \int \frac{d^3\Delta x}{[2\pi]^3} \Bigg[&\left\langle \mathbf{R} - \frac{\Delta \mathbf{x}}{2} \middle| \hat{U} \middle| \psi(t) \right\rangle \left\langle \psi(t) \middle| \mathbf{R} + \frac{\Delta \mathbf{x}}{2} \right\rangle \\ &- \left\langle \mathbf{R} - \frac{\Delta \mathbf{x}}{2} \middle| \psi(t) \right\rangle \left\langle \psi(t) \middle| \hat{U} \middle| \mathbf{R} + \frac{\Delta \mathbf{x}}{2} \right\rangle \Bigg] e^{i\mathbf{Q}\cdot\Delta\mathbf{x}}. \end{aligned} \tag{6.33}$$

For the purely momentum- and position-dependent operators, we can use

$$\begin{aligned} \frac{\hat{\mathbf{p}}^2}{2m} \left| \hbar\mathbf{Q} \pm \frac{\Delta \mathbf{p}}{2} \right\rangle &= \frac{\left[\hbar\mathbf{Q} \pm \frac{\Delta \mathbf{p}}{2}\right]^2}{2m} \left| \hbar\mathbf{Q} \pm \frac{\Delta \mathbf{p}}{2} \right\rangle \\ \hat{U} \left| \mathbf{R} \pm \frac{\Delta \mathbf{x}}{2} \right\rangle &= U\left(\mathbf{R} \pm \frac{\Delta \mathbf{x}}{2}\right) \left| \mathbf{R} \pm \frac{\Delta \mathbf{x}}{2} \right\rangle, \end{aligned} \tag{6.34}$$

respectively, which allows us to reduce the kinetic-energy part of Eq. (6.33) into

$$
\begin{aligned}
i\hbar \frac{\partial\, W(\mathbf{R},\mathbf{Q};t)}{\partial t}\bigg|_{\mathrm{kin}} &= \int \frac{d^3\Delta p}{[2\pi]^3} \left\langle \hbar\mathbf{Q} - \frac{\Delta\mathbf{p}}{2} \,|\psi(t)\right\rangle \left\langle \psi(t) \middle| \hbar\mathbf{Q} + \frac{\Delta\mathbf{p}}{2} \right\rangle \\
&\quad \times \left[\frac{\left[\hbar\mathbf{Q} - \frac{\Delta\mathbf{p}}{2}\right]^2}{2m} - \frac{\left[\hbar\mathbf{Q} + \frac{\Delta\mathbf{p}}{2}\right]^2}{2m} \right] e^{-i\frac{\mathbf{R}\cdot\Delta\mathbf{p}}{\hbar}} \\
&= -\int \frac{d^3\Delta p}{[2\pi]^3} \left\langle \hbar\mathbf{Q} - \frac{\Delta\mathbf{p}}{2} \,|\psi(t)\right\rangle \left\langle \psi(t) \middle| \hbar\mathbf{Q} + \frac{\Delta\mathbf{p}}{2} \right\rangle \\
&\quad \times \frac{\hbar\mathbf{Q}}{m} \cdot \Delta\mathbf{p}\, e^{-i\frac{\mathbf{R}\cdot\Delta\mathbf{p}}{\hbar}}.
\end{aligned} \tag{6.35}
$$

Using $\Delta\mathbf{p}\, e^{-i\frac{\mathbf{R}\cdot\Delta\mathbf{p}}{\hbar}} = i\hbar\nabla_{\mathbf{R}}\, e^{-i\frac{\mathbf{R}\cdot\Delta\mathbf{p}}{\hbar}}$, we obtain

$$
\begin{aligned}
i\hbar \frac{\partial\, W(\mathbf{R},\mathbf{Q};t)}{\partial t}\bigg|_{\mathrm{kin}} &= -\int \frac{d^3\Delta p}{[2\pi]^3} \left\langle \hbar\mathbf{Q} - \frac{\Delta\mathbf{p}}{2} \,|\psi(t)\right\rangle \left\langle \psi(t) \middle| \hbar\mathbf{Q} + \frac{\Delta\mathbf{p}}{2} \right\rangle \\
&\quad \times i\hbar \frac{\hbar\mathbf{Q}}{m} \cdot \nabla_{\mathbf{R}} e^{-i\frac{\mathbf{R}\cdot\Delta\mathbf{p}}{\hbar}} \\
&= -i\hbar \frac{\hbar\mathbf{Q}}{m} \cdot \nabla_{\mathbf{R}}\, W(\mathbf{R},\mathbf{Q};t)
\end{aligned} \tag{6.36}
$$

because we can exchange differentiation and integration for the analytic Wigner functions. After that, the remaining integral is identical to the original definition (6.31) of the Wigner function.

The same sequence of manipulations can be applied to the potential-energy part of Eq. (6.37), producing

$$
\begin{aligned}
i\hbar \frac{\partial\, W(\mathbf{R},\mathbf{Q};t)}{\partial t}\bigg|_{\mathrm{pot}} &= \int \frac{d^3\Delta x}{[2\pi]^3} \left\langle \mathbf{R} - \frac{\Delta\mathbf{x}}{2} \,|\psi(t)\right\rangle \left\langle \psi(t) \middle| \mathbf{R} + \frac{\Delta\mathbf{x}}{2} \right\rangle \\
&\quad \times \left[U\left(\mathbf{R} - \frac{\Delta\mathbf{x}}{2}\right) - U\left(\mathbf{R} + \frac{\Delta\mathbf{x}}{2}\right) \right] e^{i\mathbf{Q}\cdot\Delta\mathbf{x}}.
\end{aligned} \tag{6.37}
$$

It is instructive to Taylor expand the potential-energy terms that appear via

$$
U\left(\mathbf{R} \pm \frac{\Delta\mathbf{x}}{2}\right) = \sum_{j=0}^{\infty} \frac{1}{j!} \left[\pm \frac{\Delta\mathbf{x}}{2} \cdot \nabla_{\mathbf{R}} \right]^j U(\mathbf{R}). \tag{6.38}
$$

Whereas the terms with an even value of j cancel in (6.37), we obtain from the odd-valued contributions

$$
\begin{aligned}
U\left(\mathbf{R} - \frac{\Delta\mathbf{x}}{2}\right) - U\left(\mathbf{R} + \frac{\Delta\mathbf{x}}{2}\right) &= -\Delta\mathbf{x}\cdot\nabla_{\mathbf{R}} U(\mathbf{R}) \\
&\quad - \sum_{j=1}^{\infty} \frac{2}{(2j+1)!} \left[\frac{\Delta\mathbf{x}}{2} \cdot \nabla_{\mathbf{R}} \right]^{2j+1} U(\mathbf{R}) \\
&\equiv -\Delta\mathbf{x}\cdot\nabla_{\mathbf{R}} U(\mathbf{R}) + i\hbar U_{\mathrm{qm}}(\mathbf{R}, \Delta\mathbf{x}),
\end{aligned} \tag{6.39}
$$

where we lumped the remaining Taylor-expansion terms into the quantum potential U_{qm}. This term vanishes for any constant, linear, or quadratic potential.

Inserting (6.39) into Eq. (6.37) yields

$$
\begin{aligned}
i\hbar \left.\frac{\partial W(\mathbf{R},\mathbf{Q};t)}{\partial t}\right|_{\text{pot}} &= \int \frac{d^3\Delta x}{[2\pi]^3} \left\langle \mathbf{R} - \frac{\Delta\mathbf{x}}{2} \,|\psi(t)\right\rangle \left\langle \psi(t) \,\middle|\, \mathbf{R} + \frac{\Delta\mathbf{x}}{2}\right\rangle \\
&\quad \times \left[-\Delta\mathbf{x}\cdot\nabla_{\mathbf{R}}U(\mathbf{R}) + i\hbar U_{\text{qm}}(\mathbf{R},\Delta\mathbf{x})\right] e^{i\mathbf{Q}\cdot\Delta\mathbf{x}} \\
&= i\hbar \int \frac{d^3\Delta x}{[2\pi]^3} \left\langle \mathbf{R} - \frac{\Delta\mathbf{x}}{2} \,|\psi(t)\right\rangle \left\langle \psi(t) \,\middle|\, \mathbf{R} + \frac{\Delta\mathbf{x}}{2}\right\rangle \\
&\quad \times \left[\nabla_{\mathbf{R}}U(\mathbf{R})\cdot\nabla_{\hbar\mathbf{Q}} + U_{\text{qm}}(\mathbf{R},\Delta\mathbf{x})\right] e^{i\mathbf{Q}\cdot\Delta\mathbf{x}}.
\end{aligned} \tag{6.40}
$$

As before, the $\nabla_{\hbar\mathbf{Q}}$ differentiation can be moved outside the integral and the integral itself defines the Wigner function such that

$$
\begin{aligned}
i\hbar \left.\frac{\partial W(\mathbf{R},\mathbf{Q};t)}{\partial t}\right|_{\text{pot}} &= i\hbar\nabla_{\mathbf{R}}U(\mathbf{R})\cdot\nabla_{\hbar\mathbf{Q}}W(\mathbf{R},\mathbf{Q};t) \\
&\quad + i\hbar \left.\frac{\partial W(\mathbf{R},\mathbf{Q};t)}{\partial t}\right|_{\text{qm}}.
\end{aligned} \tag{6.41}
$$

The quantum correction that appears can be evaluated more explicitly if we insert U_{qm} of Eq. (6.39) into Eq. (6.37) and apply the relation

$$
[\Delta\mathbf{x}\cdot\nabla_{\mathbf{R}}]^{2j+1}\, e^{i\mathbf{Q}\cdot\Delta\mathbf{x}} = -i(-1)^j \left[\nabla_{\mathbf{Q}}\cdot\nabla_{\mathbf{R}}\right]^{2j+1} e^{i\mathbf{Q}\cdot\Delta\mathbf{x}}. \tag{6.42}
$$

These calculation steps eventually produce the nonlinear quantum contributions

$$
\left.\frac{\partial W(\mathbf{R},\mathbf{Q};t)}{\partial t}\right|_{\text{qm}} \equiv \sum_{j=1}^{\infty} \frac{2}{\hbar}\frac{(-1)^j}{(2j+1)!} \left[\frac{\nabla_{\mathbf{Q}}\cdot\nabla_{\mathbf{R}'}}{2}\right]^{2j+1} U(\mathbf{R}')W(\mathbf{R},\mathbf{Q};t)\Bigg|_{\mathbf{R}'=\mathbf{R}} \tag{6.43}
$$

that follow after we have identified the Wigner function as in the derivation of the linear terms in Eq. (6.41). Combining Eqs. (6.36) and (6.41) with Eq. (6.33), we obtain

$$
\begin{aligned}
i\hbar\frac{\partial W(\mathbf{R},\mathbf{Q};t)}{\partial t} &= i\hbar \left.\frac{\partial W(\mathbf{R},\mathbf{Q};t)}{\partial t}\right|_{\text{kin}} + i\hbar \left.\frac{\partial W(\mathbf{R},\mathbf{Q};t)}{\partial t}\right|_{\text{pot}} \\
&= i\hbar\left[-\frac{\hbar\mathbf{Q}}{m}\cdot\nabla_{\mathbf{R}} + \nabla_{\mathbf{R}}U(\mathbf{R})\cdot\nabla_{\hbar\mathbf{Q}}\right] W(\mathbf{R},\mathbf{Q};t) \\
&\quad + i\hbar \left.\frac{\partial W(\mathbf{R},\mathbf{Q};t)}{\partial t}\right|_{\text{qm}}.
\end{aligned} \tag{6.44}
$$

Identifying the momentum coordinate $\mathbf{P} = \hbar\mathbf{Q}$ and the classical force $\mathbf{F}(\mathbf{R}) \equiv -\nabla_{\mathbf{R}}U(\mathbf{R})$ yields

$$\frac{\partial\, W(\mathbf{R},\mathbf{Q};t)}{\partial t} = -\left[\frac{\mathbf{P}}{m}\cdot\nabla_{\mathbf{R}} - \mathbf{F}(\mathbf{R})\cdot\nabla_{\mathbf{P}}\right] W(\mathbf{R},\mathbf{Q};t) + \left.\frac{\partial\, W(\mathbf{R},\mathbf{Q};t)}{\partial t}\right|_{\mathrm{qm}}$$
$$\equiv -i\hat{L}\, W(\mathbf{R},\mathbf{Q};t) + \left.\frac{\partial\, W(\mathbf{R},\mathbf{Q};t)}{\partial t}\right|_{\mathrm{qm}}. \tag{6.45}$$

In the second line, we have combined the respective terms into the classical Liouville operator, encountered earlier in connection with Eq. (1.73).

We now have the intriguing result that the dynamics of the Wigner function contains a Liouville contribution that evolves $W(\mathbf{R},\mathbf{Q};t)$ just like a classical phase-space distribution. Additionally, we identified the quantum contribution that vanishes only for simple potentials like

$$U_{\mathrm{quad}}(\mathbf{r}) \equiv U_0 + \mathbf{U}_1\cdot\mathbf{r} + \mathbf{r}\cdot\mathbf{U}_{2,a}\mathbf{U}_{2,b}\cdot\mathbf{r}. \tag{6.46}$$

Consequently, only the anharmonic part of the potential can produce a phase-space evolution which differs from the classical dynamics. The fact that for certain potentials $W(\mathbf{R},\mathbf{Q};t)$ follows a classical time evolution should not be understood to imply classical behavior of quantum particles because $W(\mathbf{R},\mathbf{Q};t)$ itself can display strong nonclassical features.

6.3 Density matrix

The quantum-dynamical Eq. (6.45) shows that any time evolution is linear. If we now define a set of Wigner-function solutions $W_\lambda(\mathbf{R},\mathbf{Q};t)$, also a linear superposition of them,

$$W_{\mathrm{class}}(\mathbf{R},\mathbf{Q};t) = \sum\nolimits_\lambda p_\lambda\, W_\lambda(\mathbf{R},\mathbf{Q};t), \tag{6.47}$$

is a solution. For any physical system, the coefficients p_λ must be positive-definite real-valued numbers. If the p_λ coefficients add up to $\sum_\lambda p_\lambda = 1$, the relation (6.47) corresponds to the classical averaging (1.71), introduced in Section 1.2.2. Since the classical averaging cannot induce quantum features in the system, the wave-averaging aspects cannot be described via Eq. (6.47).

Taking the Fourier transformation of Eq. (6.31)

$$\begin{aligned}\int d^3Q\; W(\mathbf{R},\mathbf{Q};t)\, e^{-i\Delta\mathbf{r}\cdot\mathbf{Q}} &= \int \frac{d^3\Delta x}{[2\pi]^3}\left\langle \mathbf{R}-\frac{\Delta\mathbf{x}}{2}\,|\psi(t)\right\rangle\left\langle\psi(t)\,\middle|\mathbf{R}+\frac{\Delta\mathbf{x}}{2}\right\rangle \\ &\quad\times\int\frac{d^3Q}{[2\pi]^3}\, e^{i[\Delta\mathbf{r}-\Delta\mathbf{r}]\cdot\mathbf{Q}} \\ &= \left\langle \mathbf{R}-\frac{\Delta\mathbf{r}}{2}\,|\psi(t)\right\rangle\left\langle\psi(t)\,\middle|\mathbf{R}+\frac{\Delta\mathbf{r}}{2}\right\rangle \\ &= \left\langle \mathbf{R}-\frac{\Delta\mathbf{r}}{2}\middle|\hat{\rho}(t)\middle|\mathbf{R}+\frac{\Delta\mathbf{r}}{2}\right\rangle,\end{aligned} \tag{6.48}$$

we see that

$$\hat{\rho}(t) = |\psi(t)\rangle\langle\psi(t)| \tag{6.49}$$

uniquely defines the Wigner function and vice versa. In particular, the Wigner function also defines $\hat{\rho}(t)$

$$\begin{aligned}\left\langle \mathbf{R} - \frac{\Delta\mathbf{r}}{2} \middle| \hat{\rho}(t) \middle| \mathbf{R} + \frac{\Delta\mathbf{r}}{2} \right\rangle &= \sum_\lambda p_\lambda \int d^3Q\, W_\lambda(\mathbf{R}, \mathbf{Q}; t)\, e^{-i\Delta\mathbf{r}\cdot\mathbf{Q}} \\ &= \left\langle \mathbf{R} - \frac{\Delta\mathbf{r}}{2} \middle| \sum_\lambda p_\lambda |\psi_\lambda(t)\rangle\langle\psi_\lambda(t)| \middle| \mathbf{R} + \frac{\Delta\mathbf{r}}{2} \right\rangle \\ &\Leftrightarrow \quad \hat{\rho}(t) = \sum_\lambda p_\lambda |\psi_\lambda(t)\rangle\langle\psi_\lambda(t)|. \end{aligned} \tag{6.50}$$

The operator $\hat{\rho}(t)$ is often referred to as the density matrix. Since $\hat{\rho}(t)$ is a Hermitian operator, i.e., $\hat{\rho}(t) = \hat{\rho}^\dagger(t)$, it has real-valued eigenvalues. From Eq. (6.50), we see that the eigenstates are $|\psi_\lambda\rangle \equiv |\psi_\lambda(t)\rangle$ while $\lambda = p_\lambda$ determines the eigenvalues. Physically, the p_λ can be interpreted as the probability of finding the system in the state $|\psi_\lambda\rangle$. Thus, the p_λ must be positive definite for any physical system and satisfy the overall normalization $\sum_\lambda p_\lambda = 1$.

In most cases, we do not know the basis $|\psi\rangle_\lambda$. In this situation, the density matrix casts into the form

$$\hat{\rho} = \sum\nolimits_{\lambda,\nu} |\phi_\lambda\rangle c_{\lambda,\nu} \langle\phi_\nu|, \tag{6.51}$$

where $|\phi_\lambda\rangle$ is an arbitrary basis of orthogonal functions. The coefficient $c_{\lambda,\nu} = \langle\phi_\nu|\hat{\rho}|\phi_\lambda\rangle$ is then a complex-valued Hermitian matrix defining the representation of $\hat{\rho}$ in terms of $|\phi_\lambda\rangle$. The matrix diagonalization of $c_{\lambda,\nu}$ yields

$$\sum\nolimits_\nu c_{\nu,\nu'}\, \beta_\lambda(\nu') = p_\lambda\, \beta_\lambda(\nu), \tag{6.52}$$

where $\beta_\lambda(\nu)$ is an eigenvector and p_λ is the eigenvalue. Here, $\beta_\lambda(\nu)$ defines the eigenstate via

$$|\psi_\lambda\rangle \equiv \sum\nolimits_\nu \beta_\lambda(\nu)|\phi_\nu\rangle, \quad \langle\psi_\lambda| \equiv \sum\nolimits_\nu \beta^\star_\lambda(\nu)\langle\phi_\nu|. \tag{6.53}$$

Thus, we can always express an arbitrary density matrix in the form of Eq. (6.50) using the series of steps outlined above.

6.4 Feasibility of quantum-dynamical computations

We can use the strong analogies between the Wigner function and classical probability distributions to estimate for which systems numerical quantum-dynamical computations are feasible. As discussed in Section 1.2, a classical problem with D degrees of freedom can be solved by following $2D$ real-valued phase-space coordinates. If the precise classical paths are replaced by a statistical distribution, one needs to follow an $\mathbb{R}^{2D}$-dimensional phase-space distribution, according to Eq. (1.55). Hence, the quantum statistics is also represented by an $\mathbb{R}^{2D}$-dimensional object. Since we are primarily interested in developing a quantum-optical theory for semiconductors in this book, we have many degrees of freedom such that we have to deal with a very large value for D.

The large dimension of the quantum-statistical description clearly becomes a major concern when we attempt to develop numerically feasible solution schemes to evaluate the time evaluation. For example, a direct discretization of each phase-space coordinate into $n_{\#}$ elements needs

$$\mathrm{Qs}(D) = [n_{\#}]^{2D} \tag{6.54}$$

elements if we try to store quantum-statistical distributions numerically. We identify $\mathrm{Qs}(D)$ as the quantum-statistical might that defines how many elements are needed to describe the quantum statistics of a system with D degrees of freedom. Since the straightforward discretization yields an exponentially growing $\mathrm{Qs}(D)$ as function of D, one often needs to develop more subtle methods to numerically evaluate the quantum properties of many-body systems.

To gain some insight into what kind of problems are feasible if we use the brute force discretization, Fig. 6.3(a) presents $\mathrm{Qs}(D)$ as a function of D if $n_{\#} = 2$ (circles), $n_{\#} = 10$ (squares), and $n_{\#} = 100$ (diamonds) discretization points are used. The shaded area shows the D dependence of the classical problem with one perfect path in the phase space. The horizontal line indicates where more than 10^{12} elements are needed – this number is close

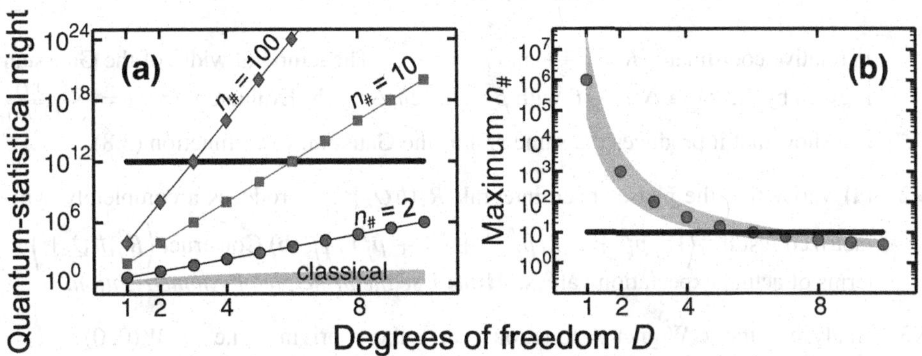

Figure 6.3 The quantum-statistical might $\mathrm{Qs}(D)$ of numerical solutions. The estimate is based on the direct discretization of D degrees of freedom, i.e., Eq. (6.54). **(a)** The computed $\mathrm{Qs}(D)$ is shown as a function of D. The system is assumed to have $n_{\#} = 2$ (circles), $n_{\#} = 10$ (squares), and $n_{\#} = 100$ (diamonds) discretization points. The shaded area indicates the number of elements needed in the corresponding classical computations. The horizontal line marks the 10^{12} limit. **(b)** The largest number of discretization points $n_{\#}$ possible if the numerical treatment allows for $\mathrm{Qs}(D) = 10^{12}$ elements (circles). The shaded area indicates the range of $n_{\#}$ obtained for $10^{9} \leq \mathrm{Qs}(D) \leq 10^{15}$. The horizontal line indicates the limit of $n_{\#} = 10$.

to the near-future numerical capabilities of computers such that it provides a good estimate for the numerical feasibility of this approach. We see that for $n_{\#} = 10$ discretization points, we exceed this limit for $D = 6$ while $n_{\#} = 100$ only allows for the treatment of $D = 3$ degrees of freedom. Thus, a straightforward discretization of the many-body dynamics is feasible only for systems with very few particles.

In Fig. 6.3(b) (circles), we plot the number $n_{\#}$ of discretization points that produces $\text{Qs}(D) = 10^{12}$ as a function of D. The shaded area indicates those $n_{\#}$ values obtained if the numerics allows for $10^9 \leq \text{Qs}(D) \leq 10^{12}$. The horizontal line identifies the limit of $n_{\#} = 10$ discretization points. Again, we see that the numerical feasibility is limited to rather small values of $n_{\#}$ as D is elevated. Furthermore, the curve for constant $\text{Qs}(D)$ decays very fast such that the $n_{\#} = 10$ limit is already crossed at $D = 6$ (solid line). We also see that decreasing the $\text{Qs}(D)$ value does not help much for elevated D, which means that realistic many-body computations always become tedious if the quantum dynamics is treated via direct discretization. Thus, we need to develop the quantum theory further in order to find a more feasible approach to evaluate the many-body dynamics of realistic systems.

Exercises

6.1 **(a)** Show that a Gaussian wave packet, Eq. (3.8), produces the density matrix

$$\rho(z, z'; t) = \frac{1}{\sqrt{2\pi|\Delta z(t)|^2}} e^{-\frac{|R-v_0 t|^2}{2|\Delta z(t)|^2}} e^{i\frac{\beta(R-v_0 t)}{2\Delta z_0|\Delta z|^2}r} e^{-\frac{r^2}{8|\Delta z(t)|^2}}$$

in relative coordinates $R = \frac{z+z'}{2}$ and $r = z - z'$. The temporal width of the Gaussian is given by $\Delta z(t) \equiv \Delta z_0 + i\beta$ with $\beta \equiv \hbar t/2m\Delta z_0$. **(b)** Evaluate $\tilde{\psi}(z, t) = \frac{\rho(z,0;t)}{\sqrt{\rho(0,0;t)}}$ and show that it produces the same $\langle \hat{F} \rangle$ as the Gaussian wave function (3.8).

6.2 **(a)** Verify that the phase-space integral $\langle R_j [\hbar Q_k]^J \rangle_{ps}$ produces a completely symmetrized result $\frac{1}{J}\left\langle \left(\hat{r}_j \hat{p}_k^J + \hat{p}_k \hat{r}_j \hat{p}_k^{J-1} + \cdots + \hat{p}_k^J \hat{r}_j \right) \right\rangle$. **(b)** Construct $\left\langle R_j^2 [\hbar Q_k]^2 \right\rangle$ in terms of actual expectation values. **Hint:** *Use the principle of symmetrization.*

6.3 Analyze the Wigner function at the origin, i.e., $W(0, 0) = \frac{1}{2\pi} \int d\Delta x\, \psi^\star \left(-\frac{\Delta x}{2}\right) \psi \left(\frac{\Delta x}{2}\right)$. **(a)** Construct the condition for $\psi(z)$ and $\psi^\star(z)$ that produces the extremum value for $W(0, 0)$ with the constraint $\int dz |\psi(z)|^2 = 1$. **Hint:** *Use the variational principle and apply the Lagrange multiplier.* **b)** Show that $-1/\pi \leq W(0, 0) \leq 1/\pi$. Why is this result valid for any phase-space point? **(c)** Show that this limit also holds when classical averaging (6.47) is implemented.

6.4 Show that the wave function (6.25) is normalized if we choose N according to Eq. (6.26).

6.5 The width of a freely propagating Gaussian is $\Delta z(t) = \Delta z_0 + i\beta \quad \beta = \hbar t/2m\Delta z_0$.
(a) Show that

$$\Delta z(t) + \Delta^* z^*(t) = 2\Delta z_0\,, \qquad \frac{1}{\Delta z(t)} + \frac{1}{\Delta z^*(t)} = \frac{2\Delta z_0}{|\Delta z(t)|^2}$$

$$\Delta z(t) - \Delta z^*(t) = 2i\beta\,, \qquad \frac{1}{\Delta z(t)} - \frac{1}{\Delta z^*(t)} = -\frac{2i\beta}{|\Delta z(t)|^2}.$$

(b) Verify the result (6.28). **Hint:** *Make use of the relations above and use the results for Gaussian integration discussed in Exercise 2.6.* **(c)** Identify the contributions in ρ which produce the cosine interference pattern.

6.6 **(a)** Verify that Eq. (6.30) reproduces the result (6.28). **(b)** Calculate the marginal distribution $W_Q(z) = \int dQ\, W(z, Q; t)$ by starting from Eq. (6.28).

6.7 Show that the Wigner function can be constructed from either the real- or the momentum-space representation, as given by Eq. (6.31). **Hint:** *Start from the real-space form and use the properties* $\mathbb{I} = \int d^3p\, |p\rangle\langle p|$ *and* $\langle r|p\rangle = \left(\frac{1}{2\pi\hbar}\right)^{\frac{3}{2}} e^{i\frac{\mathbf{p}}{\hbar}\cdot\mathbf{r}}$.

6.8 Show that a generic potential $U(R)$ results in the dynamics described by Eqs. (6.41) and (6.43).

6.9 Assume that the density matrix is given by $\hat{\rho} = \sum_{\lambda,v} |\phi_\lambda\rangle C_{\lambda,v}\langle\phi_v|$, where $|\phi_\lambda\rangle$ forms a complete set of orthogonal states.
(a) Show that $C_{\lambda,v}$ is a Hermitian matrix. **(b)** Assume that $C_{\lambda,v}$ is diagonalized, i.e., $\sum_{v'} C_{v,v'}\beta_\lambda(v') = p_\lambda\,\beta_\lambda(v)$. Show that $\hat{\rho} = \sum_\lambda |\psi_\lambda\rangle p_\lambda\langle\psi_\lambda|$, where $|\psi_\lambda\rangle = \sum_v \beta_\lambda(v)|\phi_v\rangle$. Show that $\beta_\lambda(v) = \langle\phi_v|\psi_\lambda\rangle$.

Further reading

The Wigner function was originally introduced in:

E. P. Wigner (1932). On the quantum correction for thermodynamic equilibrium, *Phys. Rev.*, **40**, 749.

The following references present a few examples where Wigner functions have been measured in different systems:

D. T. Smithey, M. Beck, M. G. Raymer, and A. Faridani (1993). Measurement of the Wigner distribution and the density matrix of a light mode using optical homodyne tomography: Application to squeezed states and the vacuum, *Phys. Rev. Lett.*, **70**, 1244.

D. Leibfried, D. M. Meekhof, B. E. King, *et al.* (1996). Experimental determination of the motional quantum state of a trapped atom, *Phys. Rev. Lett.*, **77**, 4281.

C. Kurtsiefer, T. Pfau, and J. Mlynek (1997). Measurement of the Wigner function of an ensemble of Helium atoms, *Nature*, **363**, 150.
A. I. Lvovsky, H. Hansen, T. Aichele, *et al.* (2001). Quantum state reconstruction of the single-photon Fock state, *Phys. Rev. Lett.*, **87**, 05042.
G. Breitenbach, S. Schiller, and J. Mlynek, (1997). Measurement of the quantum states of squeezed light, *Nature*, **387**, 471.
A. Ourjoumtsev, R. Tualle-Brouri, J. Laurat, and P. Grangier (2006). Generating optical Schrödinger kittens for quantum information processing, *Science*, **312**, 83.

The Wigner function is a central concept in quantum optics. These books discuss different aspects of Wigner functions:

Y. S. Kim and M. E. Noz (1991). *Phase Space Picture of Quantum Mechanics*, Singapore, World Scientific.
U. Leonhardt (1997). *Measuring the Quantum State of Light*, Cambridge, Cambridge University Press.
W. P. Schleich (2001). *Quantum Optics in Phase Space*, Darmstadt, Wiley & Sons.
M. O. Scully and M. S. Zubairy (2002). *Quantum Optics*, Cambridge, Cambridge University Press.
C. C. Gerry and P. L. Knight (2005). *Introductory Quantum Optics*, Cambridge, Cambridge University Press.
C. Zachos, D. Fairlie, and T. Curtright (2005). *Quantum Mechanics in Phase Space*, Singapore, World Scientific.
W. Vogel and D.-G. Welsch (2006). *Quantum Optics*, 3rd edition, Darmstadt, Wiley & Sons.
D. F. Walls and G. J. Milburn (2008). *Quantum Optics*, 2nd edition, Berlin, Springer.

7
Hamiltonian formulation of classical electrodynamics

At the conceptual level, the quantization procedure replaces the pure particle description of classical mechanics by a particle-wave treatment satisfying the requirements of the fundamental wave–particle dualism. Technically, one can always start from a generalized wave equation and use it as a basis to discuss wave-front propagation in the form of rays following classical particle trajectories, see Sections 2.2.1 and 3.1. Consequently, any wave theory inherently contains that aspect of the wave–particle dualism.

Especially, Maxwell's equations of electrodynamic theory include both wave and particle aspects of electromagnetic fields. Hence, it could be that there are no additional "quantum," i.e., wave–particle dualistic aspects. However, as we discuss in this chapter, Maxwell's equations can be expressed in the framework of Hamilton's formulation of particle mechanics. This then suggests that in addition to the approximate ray-like behavior, electromagnetic fields do have a supplementary level of particle-like properties.

Due to the fundamental axiom of wave–particle dualism, the additional particle-like aspects of electromagnetic fields must then also have a wave counterpart that can be axiomatically introduced via the canonical quantization scheme described in Section 3.2.3. Since the spatial dependency in Maxwell's equation contains wave–particle dualism already at the classical level, it is clear that the new level of quantization cannot be a simple real-space feature but has to involve some other space describing the structure of the electromagnetic fields. In fact, this quantization yields several very concrete implications such as quantization of the light energy, elementary fluctuations in the light-field amplitude and intensity, and more. These aspects are the basic elements of quantum optics, vitally important to explain how electromagnetic fields interact with matter.

For example, a purely classical analysis of black-body radiation yields the ultraviolet catastrophe where the emitted energy density approaches infinity for elevated light frequencies. After countless attempts, this problem was eventually solved by Max Planck, who proposed in 1900 "as the last resort" the hypothesis of the quantized light energy, which correctly suppresses the thermal emission of high-frequency light. Retrospectively, one can understand the quantization as a fundamental necessity because all aspects of light must satisfy the wave–particle dualism. It is interesting to notice that Planck's quantization hypothesis provides one of the founding ideas for the development of quantum mechanics. Yet, we now know that the quantization of light is much more involved than the quantization of particles because already the classical Maxwell's equations are dualistic. It took several decades and pioneering contributions from many researchers to develop a satisfactory scheme for the simultaneous quantization of light and matter.

In this chapter, we discuss the relevant theoretical steps needed to perform the consistent quantization of interacting light–matter systems. We start by analyzing how particle aspects can be identified at the level of Maxwell's equations. The major task is then to convert these equations into the Hamilton formalism that we have used exclusively for the description of particles up to now. Since we want to find a joint description for real particles and electromagnetic fields, we need a generic Hamiltonian and Lagrangian including the particles, the fields, and the electromagnetic interactions. The results can then be used as a general basis for the description of many-body systems coupled to electromagnetic fields. The eventual quantization of this theory provides us with a fully consistent platform to investigate semiconductor quantum optics.

7.1 Basic concepts

At the simplest level, we start from the one-dimensional electromagnetic wave equation without sources,

$$\frac{\partial^2}{\partial z^2}E(z,t) - \frac{1}{c^2}\frac{\partial^2}{\partial t^2}E(z,t) = 0, \tag{7.1}$$

which has the generic plane-wave solution

$$E_{\text{pl}}(z,t) = E_0\, e^{iq_z z - i\omega_{q_z} t}, \qquad \omega_q = c|q|, \tag{7.2}$$

where the speed of light c connects the frequency ω_q and the wave vector q. The prefactor E_0 determines the electrical field strength.

Assuming that we measure the complex-valued $E_{\text{pl}}(z,t)$ at the position z_0, we determine either the real or the imaginary part of $E_{\text{pl}}(z,t)$,

$$\begin{aligned} X(t) &\equiv \text{Re}\left[E_{\text{pl}}(z_0,t)\right] = E_0\cos(\omega_{q_z}t - q_z z_0), \\ Y(t) &\equiv \text{Im}\left[E_{\text{pl}}(z_0,t)\right] = -E_0\sin(\omega_{q_z}t - q_z z_0), \end{aligned} \tag{7.3}$$

respectively. As shown in Fig. 7.1(a), this solution identifies a point in the XY-plane. Alternatively, one can view $[X(t), Y(t)]$ as a vector (arrow) that moves with time on a circle in a clockwise direction. The trajectory of $[X(t), Y(t)]$ has an exact analogy to the phase-space trajectory of a particle in a harmonic potential shown in Fig. 1.2. Thus, the identified $[X(t), Y(t)]$ can potentially be associated with a particle-like $[x_{\text{cl}}(t), p_{\text{cl}}(t)]$-structure within Maxwell's equations.

If this particle-field association is generally correct, we can always make the connection

$$x_j(t) \leftrightarrow \text{Re}\left[E_j(\mathbf{r}_0,t)\right], \qquad p_j(t) \leftrightarrow \text{Im}\left[E_j(\mathbf{r}_0,t)\right] \tag{7.4}$$

between a particle's position–momentum coordinate pair and the pair formed by the real and imaginary part of the electric field defined at the reference point $\mathbf{r}_0$. Here, j labels

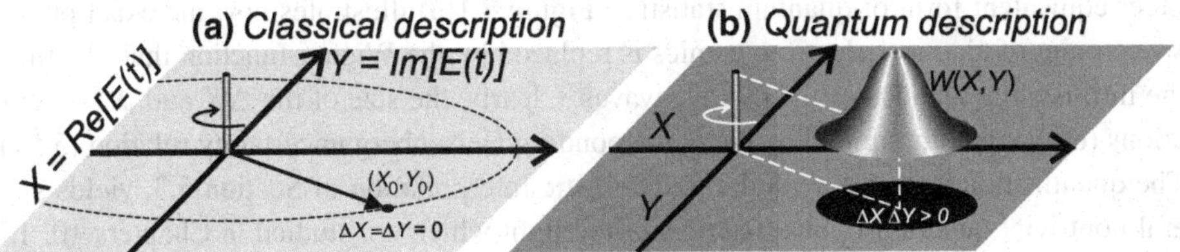

Figure 7.1 Principal idea of field quantization. **(a)** Classical description: The real and imaginary part of a single-mode electric field can be presented as a single point ($X_0 = \mathrm{Re}[E(t)]$, $Y_0 = \mathrm{Im}[E(t)]$) moving clockwise on a circular path (dashed line) in the XY-plane. The representation has no uncertainty, i.e., the fluctuations ΔX and ΔY vanish such that one can represent the phase-space path as a rotating vector (arrow). **(b)** Quantum description: The canonical quantization replaces the precise vector by a phase-space distribution (3D contour) that rotates in time. Due to its intrinsic fuzziness, both X and Y direction exhibit fluctuations limited by the Heisenberg uncertainty principle. The typical range of the fluctuations is indicated by the dark area.

the different independent components of **E**. Remarkably, the analogy (7.4) involves the electrical field's magnitude showing that this, and not the field position, is associated with the particle position $x_j(t)$. In other words, the associated variable pair is not defined in real space but in the space of complex field amplitudes. Thus, Eq. (7.4) implies a new level of electromagnetic particle features that is completely different from the eikonal approximation leading to real-space ray trajectories, as discussed in Section 2.2.1.

Assuming that the association (7.4) is valid, we can describe the evolution of the electric-field magnitude via a classical equation of motion similar to that for (x_j, p_j). In analogy to the harmonic oscillator, we anticipate a system Hamiltonian in the form

$$H = \sum_j (T_j + U_j), \quad T_j \equiv \frac{p_j^2}{2m}, \quad U_j \equiv \frac{1}{2} m\omega^2 x_j^2, \tag{7.5}$$

where we identify the kinetic and the potential energy as in Eq. (1.39). The corresponding Hamilton equations (1.10) are

$$\begin{cases} \dot{p}_j = -\frac{\partial H}{\partial x_j} \\ \dot{x}_j = \frac{\partial H}{\partial p_j} \end{cases} \Rightarrow \begin{cases} \dot{p}_j = -m\omega^2 x_j \\ \dot{x}_j = \frac{p_j}{m} \end{cases}. \tag{7.6}$$

Their solution (1.41) can be cast into the form of Eq. (7.3) if we set $\omega = \omega_{q_z}$ and choose the initial conditions properly. We will see later that the choice of the "particle" mass is completely arbitrary such that it can be set to any desired value. In fact, m is only needed to produce a suitable scaling for the X and Y axes in the **E**-amplitude space.

Since the identified canonical pair (x_j, p_j) has the properties of classical space-momentum variables, wave–particle dualism implies the existence of the corresponding wave aspects. With the help of the canonical quantization presented in Section 3.2.3, we can obtain a fully quantum-mechanical description of the electromagnetic field whose precise amplitude is replaced, e.g., by a quantum-mechanical phase-space distribution or any other equivalent form of quantum statistics. Figure 7.1(a) illustrates how the exact phase-space point of classical electrodynamics is replaced by the Wigner function that identifies the intrinsic fluctuations of the particle waves. Clearly, the size of the ΔX and ΔY fluctuations (dark area) are limited by the corresponding Heisenberg uncertainty relation (5.51). The quantization, together with the probabilistic interpretation of Section 5.3, yields several nontrivial quantum-optical features of the light which are studied in Chapters 10, 15, 17, 21–24, and 28–30.

Before we quantitatively perform the identification (7.4) on the basis of Maxwell's equations, we briefly recapitulate the Lagrange formulation of classical mechanics because this will help us to identify the proper canonical variable pairs for the electromagnetic fields. The starting point is the Lagrange function

$$L(x,\dot{x},t) = T(\dot{x}) - U(x,t), \qquad T(\dot{x}) \equiv \frac{1}{2}m\dot{x}^2, \tag{7.7}$$

where the kinetic energy T is expressed via the velocity $\dot{x} \equiv p/m$ and U is the potential energy. The Lagrange function identifies the canonical momentum via

$$p \equiv \frac{\partial L}{\partial \dot{x}} \tag{7.8}$$

which is needed for the Legendre transformation leading to the Hamiltonian

$$H(x,p,t) = p\dot{x} - L(x,\dot{x},t). \tag{7.9}$$

Using Hamilton's equation (7.6), we can derive the Lagrange equation

$$\frac{d}{dt}\left[\frac{\partial L}{\partial \dot{x}}\right] = \frac{\partial L}{\partial x}. \tag{7.10}$$

This equation can also be obtained from a variational principle. In Lagrange's theory of mechanics, the Lagrange equation constitutes the basic axiom, providing an alternative approach to Hamilton's or Newton's formulation of classical mechanics.

7.2 Hamiltonian for classical electrodynamics

We look for the particle aspects of Maxwell's equations assuming a general configuration where $\mathcal{N}$ charged particles couple to the electromagnetic field. As we will verify later, Maxwell's equations (2.1) follow from the Lagrangian

$$\begin{aligned} L = & \sum_{j=1}^{\mathcal{N}} \left[\frac{1}{2}m_j\dot{\mathbf{r}}_j^2 + Q_j\dot{\mathbf{r}}_j \cdot \mathbf{A}(\mathbf{r}_j,t) - Q_j\phi(\mathbf{r}_j,t)\right] \\ & + \frac{\varepsilon_0}{2}\int d^3r \left[\mathbf{E}^2(\mathbf{r},t) - c^2\mathbf{B}^2(\mathbf{r},t)\right]. \end{aligned} \tag{7.11}$$

Here, m_j is the mass and Q_j is the charge of particle j that is found at the position $\mathbf{r}_j$ propagating with the velocity $\dot{\mathbf{r}}_j$. The vector potential $\mathbf{A}(\mathbf{r},t)$ and the scalar potential $\phi(\mathbf{r},t)$ define the electric and magnetic fields via

$$\mathbf{E}(\mathbf{r},t) = -\dot{\mathbf{A}}(\mathbf{r},t) - \nabla\phi(\mathbf{r},t), \quad \mathbf{B}(\mathbf{r},t) = \nabla \times \mathbf{A}(\mathbf{r},t), \tag{7.12}$$

respectively. As before, the dot above the quantity denotes differentiation with respect to time, i.e., $\dot{\mathbf{A}} \equiv \partial\mathbf{A}/\partial t$.

From Eq. (7.11), we see that the expression $Q_j \dot{\mathbf{r}}_j \cdot \mathbf{A}(\mathbf{r}_j,t) - Q_j\phi(\mathbf{r}_j,t)$ provides the coupling between the particles and the electromagnetic field, i.e., it describes the system's light–matter interaction. The explicit format of the interactions can be modified via the gauge transformation

$$\mathbf{A}(\mathbf{r},t) \to \mathbf{A}(\mathbf{r},t) + \nabla F(\mathbf{r},t), \qquad \phi(\mathbf{r},t) \to \phi(\mathbf{r},t) - \frac{\partial}{\partial t}F(\mathbf{r},t). \tag{7.13}$$

Even though the scalar and vector potentials are altered by the gauge, the electric and magnetic fields as well as the Lagrangian (7.11) are invariant under the $F(\mathbf{r},t)$-dependent gauge, see Exercise 7.1.

To quantize the system, it is convenient to proceed from the Lagrangian to the Hamiltonian following the standard procedure introduced in connection with Eqs. (7.8)–(7.9). In general, the particle system is described by a set of $(\mathbf{r}_j, \dot{\mathbf{r}}_j)$ values. In the same manner, the electromagnetic field is defined by $\mathbf{A}(\mathbf{r})$, $\dot{\mathbf{A}}(\mathbf{r})$, and $\phi(\mathbf{r})$. In full analogy with the particle system, $(\mathbf{A}(\mathbf{r}), \dot{\mathbf{A}}(\mathbf{r}))$ can be treated as a pair of independent variables. Formally, the same is true for the pair $(\phi(\mathbf{r}), \dot{\phi}(\mathbf{r}))$, however, the Lagrangian does not contain any $\dot{\phi}(\mathbf{r})$ dependence. Note that here and in the following, we suppress the time argument of the variables and fields whenever it is not needed explicitly.

The independent pair $(\mathbf{r}_k, \dot{\mathbf{r}}_k)$ can also be presented in terms of the position and the canonical momentum

$$\mathbf{p}_k \equiv \frac{\partial L}{\partial \dot{\mathbf{r}}_k} = \sum_{\alpha=1}^{3} \mathbf{e}_\alpha \frac{\partial L}{\partial \dot{r}_{k,\alpha}} \tag{7.14}$$

based on Eq. (7.8). In other words, we treat $(\mathbf{r}_k, \mathbf{p}_k)$ as independent variables when we convert the Lagrangian into the Hamiltonian with the help of the Legendre transformation (7.9). Using Cartesian coordinates with the basis vectors $\mathbf{e}_1 = \mathbf{e}_x$, $\mathbf{e}_2 = \mathbf{e}_y$, and $\mathbf{e}_3 = \mathbf{e}_z$ allows us to decompose position, velocity, and momentum into $r_{j,\alpha}$, $\dot{r}_{j,\alpha}$, and $p_{j,\alpha}$, respectively, where α labels the three independent Cartesian components.

Evaluating (7.14) with L given by Eq. (7.11), we find

$$\begin{aligned}
\mathbf{p}_k &= \sum_{\alpha=1}^{3} \mathbf{e}_\alpha \frac{\partial}{\partial \dot{r}_{k,\alpha}} \sum_{\beta=1}^{3} \left[\frac{1}{2} m_k \dot{r}_{k,\beta}^2 + Q_k \dot{r}_{k,\beta} A_\beta(\mathbf{r}_k)\right] \\
&= \sum_{\alpha=1}^{3} \mathbf{e}_\alpha \sum_{\beta=1}^{3} \left[m_k \delta_{\alpha,\beta} \dot{r}_{k,\beta} + Q_k \delta_{\alpha,\beta} A_\beta(\mathbf{r}_k)\right] \\
&= \sum_{\alpha=1}^{3} \left[m_k \mathbf{e}_\alpha \dot{r}_{k,\beta} + Q_k \mathbf{e}_\alpha A_\alpha(\mathbf{r}_k)\right] = m\dot{\mathbf{r}}_k + Q_k \mathbf{A}(\mathbf{r}_k).
\end{aligned} \tag{7.15}$$

Here, the second line follows from a straightforward differentiation if we take into account that the different Cartesian components are mutually independent. We notice from (7.15) that the canonical momentum is not equal to $m\dot{\mathbf{r}}$ that is often referred to as the kinetic momentum, i.e.,

$$\mathbf{p}_j^{\text{kin}} \equiv m\dot{\mathbf{r}}_j, \quad j = 1, 2, \cdots, \mathcal{N},$$
$$\mathbf{p}_j = \mathbf{p}_j^{\text{kin}} + Q_j \mathbf{A}(\mathbf{r}_j) = m_j \dot{\mathbf{r}}_k + Q_j \mathbf{A}(\mathbf{r}_j). \tag{7.16}$$

In other words, the most general form of the light–matter interaction inevitably leads to the separation of canonical and kinetic momentum. Sometimes, the difference $Q_k \mathbf{A}(\mathbf{r}_k)$ is referred to as the potential momentum.

7.2.1 Functional derivative

Before we investigate the connection between the Lagrangian (7.11) and Maxwell's equations, it is beneficial to analyze the underlying mathematical structure. We start by identifying a *functional*. As a general definition, F is a functional of $f(\mathbf{r})$ if a continuous set of $f(\mathbf{r})$-values assigns only one value to F. In other words, F is defined by the mapping $f(\mathbf{r}) \mapsto F$. This mapping is very different from the usual operation that maps equal amounts of elements between two chosen sets. Especially, the functional does not provide a one-to-one correspondence because one can never identify an inverse transformation from F to $f(\mathbf{r})$. The functional dependence is denoted by brackets, i.e., $F[f]$, and it can often be written in the form of an integral

$$G[f] = \int_{\mathcal{V}} d^3r \; g(f(\mathbf{r}),\, f'(\mathbf{r}), \cdots), \tag{7.17}$$

where $\mathcal{V}$ is the integration volume and g identifies the integration kernel. In the most general case, f can also be a vectorial field. The functional mapping from $f(\mathbf{r})$ (solid line) and a set of $f(\mathbf{r}) + \delta f(\mathbf{r})$ variations (shaded area) into a single F value is schematically illustrated in Fig. 7.2.

The Lagrangian L given by Eq. (7.11) is a *functional* of $\mathbf{A}(\mathbf{r})$, $\dot{\mathbf{A}}(\mathbf{r})$, and $\phi(\mathbf{r})$, which can explicitly be denoted by $L[\mathbf{A}, \dot{\mathbf{A}}, \phi]$. At the same time, L has a normal dependence on the particle positions $\mathbf{r}_j$, which allows us to identify the canonical momenta via the usual differentiation of the L function, see Eq. (7.15). Also the canonical momenta that correspond to the vector and scalar potentials follow from an analogous differentiation procedure, after we generalize the operation of differentiation to include also the so-called functional derivative.

If the function $f(\mathbf{r})$ is altered by an infinitesimal amount $\delta f(\mathbf{r})$, i.e., $f(\mathbf{r}) \to f(\mathbf{r}) + \delta f(\mathbf{r})$, it is clear that the altered functional $G[f(\mathbf{r}) + \delta f(\mathbf{r})]$ should stay in the vicinity of the original functional $G[f(\mathbf{r})]$, as symbolically shown by the fuzzy area in Fig. 7.2(b). In physical problems, it is reasonable to assume that the actual functional mapping is analytic, e.g., the integration kernel $g(f(\mathbf{r}),\, f'(\mathbf{r}), \cdots)$ in Eq. (7.17) is a well-behaved function.

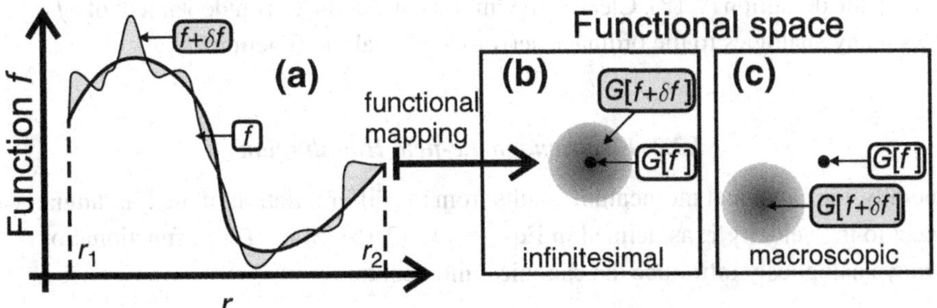

Figure 7.2 Schematic representation of functional dependency. **(a)** A function $f(r)$ (solid line) assumes unique values producing an r-dependent "trajectory." The function is varied by $\delta f(r)$ (shaded area) to produce a new trajectory $f(r)+\delta f(r)$. The entire trajectory is mapped into a single value $G[f(r)]$ via functional mapping. The results of $G[f(r)]$ (black dot) and a set of $G[f(r)+\delta f(r)]$ results (shaded area) are shown for **(b)** infinitesimal and **(c)** macroscopic $\delta f(r)$.

In this situation, infinitesimal variations of $f(\mathbf{r})$ can be expected to alter the functional linearly such that

$$G[f+\delta f] = G[f] + \int d^3r \, \frac{\delta G[f]}{\delta f(\mathbf{r})} \, \delta f(\mathbf{r}) + \mathcal{O}(\delta f^2), \tag{7.18}$$

where $\delta G[f]/\delta f(\mathbf{r})$ is the *functional derivative* defining how the linear but arbitrary $\delta f(\mathbf{r})$ alters the functional. The corresponding $G[f+\delta f]$ is shown schematically in Fig. 7.2(b) as the fuzzy area around $G[f]$. If δf becomes macroscopic, $G[f+\delta f]$ can differ considerably from the original $G[f]$, as shown in Fig. 7.2(c), so that one needs to include δf to all orders in Eq. (7.18).

To illustrate how the functional differentiation works, we evaluate it for

$$G[f] \equiv \int d^3r \; g\left(f(\mathbf{r})\right), \tag{7.19}$$

where $g(x)$ is an arbitrary analytic function. We can write,

$$\begin{aligned} G[f+\delta f] &= \int d^3r \; g\left(f(\mathbf{r}) + \delta f(\mathbf{r})\right) \\ &= \int d^3r \; \left\{ g\left(f(\mathbf{r})\right) + g'\left(f(\mathbf{r})\right) \delta f(\mathbf{r}) \right\} + \mathcal{O}(\delta f^2) \\ &= G[f] + \int d^3r \; g'\left(f(\mathbf{r})\right) \delta f(\mathbf{r}) + \mathcal{O}(\delta f^2), \end{aligned} \tag{7.20}$$

where lines two and three follow by Taylor expanding G. We recognize that the functional derivative of G must be

$$\frac{\delta G[f]}{\delta f(\mathbf{r})} \equiv g'\left(f(\mathbf{r})\right) \quad \text{if} \quad G[f] = \int d^3r \; g\left(f(\mathbf{r})\right) \tag{7.21}$$

based on the definition (7.18). Clearly, the functional derivative is independent of $\delta f(\mathbf{r})$ and shows many analogies to the ordinary derivative of analytic functions.

7.2.2 Electromagnetic-field Hamiltonian

Generally, the canonical momentum results from the differentiation of the Lagrangian with respect to the velocity $\dot{\mathbf{r}}$, as defined in Eqs. (7.14)–(7.15). Since L is a functional of $\dot{\mathbf{A}}(\mathbf{r})$, we may analogously introduce the canonical momentum

$$\Pi_A(\mathbf{r}) \equiv \frac{\delta L}{\delta \dot{\mathbf{A}}(\mathbf{r})} \equiv \sum_{\alpha=1}^{3} \mathbf{e}_\alpha \frac{\delta\, L}{\delta \dot{A}_\alpha(\mathbf{r})} \tag{7.22}$$

for the vector potential. Here, we define the functional derivative with respect to a vector field as the sum over the functional derivatives with respect to its Cartesian components in the corresponding direction $\mathbf{e}_\alpha$.

When we evaluate Eq. (7.22) for the Lagrangian (7.11), we notice that only the $\mathbf{E}(\mathbf{r}) = -\dot{\mathbf{A}}(\mathbf{r}) - \nabla\phi(\mathbf{r})$ part of L depends on $\dot{\mathbf{A}}(\mathbf{r})$. We thus obtain

$$\begin{aligned}\Pi_A &= \frac{\delta}{\delta \dot{\mathbf{A}}(\mathbf{r})} \frac{\varepsilon_0}{2} \int d^3r \, \left\{\dot{\mathbf{A}}(\mathbf{r}) + \nabla\phi(\mathbf{r})\right\}^2 \\ &= \frac{\varepsilon_0}{2} \sum_\alpha \mathbf{e}_\alpha \frac{\delta}{\delta \dot{A}_\alpha(\mathbf{r})} \int d^3r \sum_\beta \left\{\dot{A}_\beta(\mathbf{r}) + \nabla_\beta\phi(\mathbf{r})\right\}^2,\end{aligned} \tag{7.23}$$

where we expressed both the differentiation and the squared vector field as sums over the Cartesian components involved. The functional derivative that appears has exactly the form of Eq. (7.19) allowing us to apply the relation (7.21) with the result

$$\Pi_A = \varepsilon_0 \sum_\alpha \mathbf{e}_\alpha (\dot{A}_\alpha(\mathbf{r}) + \nabla_\alpha \phi(r)) = \varepsilon_0(\dot{\mathbf{A}}(\mathbf{r}) + \nabla\phi(\mathbf{r})) = -\varepsilon_0 \mathbf{E}(\mathbf{r}). \tag{7.24}$$

Here, we identified the electric field using Eq. (7.12). From a purely formal point of view, we may also identify the canonical momentum

$$\Pi_\phi(\mathbf{r}) = \frac{\delta L}{\delta \dot{\phi}(\mathbf{r})} = 0 \tag{7.25}$$

for the scalar potential. It strictly vanishes because L has no $\dot{\phi}$ dependence.

Knowing the relevant canonical momenta, we can now determine the standard Hamiltonian via the Legendre transformation

$$\begin{aligned}H &= \sum\nolimits_j \mathbf{p}_j \cdot \dot{\mathbf{r}}_j + \int d^3r \left[\Pi_A(\mathbf{r}) \cdot \dot{\mathbf{A}}(\mathbf{r}) + \Pi_\phi(\mathbf{r})\,\dot{\phi}(\mathbf{r})\right] - L \\ &= \sum\nolimits_j \frac{\mathbf{p}_j \cdot \left[\mathbf{p}_j - Q_j \mathbf{A}(\mathbf{r}_j)\right]}{m_j} + \int d^3r \left[\Pi_A(\mathbf{r}) \cdot \left(\frac{\Pi_A(r)}{\varepsilon_0} - \nabla\phi\right)\right] - L,\end{aligned} \tag{7.26}$$

where we properly converted the $\dot{\mathbf{r}}_j$, $\dot{\mathbf{A}}(\mathbf{r})$, and $\dot{\phi}(\mathbf{r})$ that appear into the corresponding canonical momenta using Eqs. (7.15), (7.24), and (7.25). Inserting the Lagrangian, Eq. (7.11), into Eq. (7.26) produces the Hamiltonian

$$H = \sum_j \left(\frac{(\mathbf{p}_j - Q_j \mathbf{A}(\mathbf{r}_j))^2}{2m_j} + Q_j \phi(\mathbf{r}_j) \right) + \int d^3r \left[\frac{\Pi_A^2(\mathbf{r})}{2\varepsilon_0} + \frac{\varepsilon_0 c^2}{2} (\nabla \times \mathbf{A}(\mathbf{r}))^2 - \Pi_A(\mathbf{r}) \cdot \nabla\phi(\mathbf{r}) \right]. \tag{7.27}$$

We can also use $\Pi_A = -\varepsilon_0 \mathbf{E}$ and $\nabla \times \mathbf{A} = \mathbf{B}$ to simplify the Hamiltonian into

$$H = \sum_j \left[\frac{(\mathbf{p}_j - Q_j \mathbf{A}(\mathbf{r}_j))^2}{2m_j} + Q_j \phi(\mathbf{r}_j) \right] + \varepsilon_0 \int d^3r\, \mathbf{E}(\mathbf{r}) \cdot \nabla\phi(\mathbf{r}) + \frac{\varepsilon_0}{2} \int d^3r \left[\mathbf{E}(\mathbf{r})^2 + c^2 \mathbf{B}(\mathbf{r})^2 \right] \tag{7.28}$$

in terms of **E**, **B**, and ϕ. We see that the usual kinetic energy is transformed via the so-called minimal substitution

$$\mathbf{p}_k \to \mathbf{p}_k - Q_k \mathbf{A}(\mathbf{r}_k). \tag{7.29}$$

In Section 7.3.2, we show that the minimal substitution produces the general light–matter interaction fulfilling the condition of gauge invariance.

7.3 Hamilton equations for light–matter system

At this stage, we have two equivalent forms of the Hamiltonian, Eqs. (7.27) and (7.28). As a consistency check, we first verify that they produce the correct classical dynamics where the particle trajectories should follow from Newton's equation,

$$m_j \ddot{\mathbf{r}}_j = \mathbf{F}_{\text{Lor}}(\mathbf{r}_j, \dot{\mathbf{r}}_j), \qquad \mathbf{F}_{\text{Lor}}(\mathbf{r}_j, \dot{\mathbf{r}}_j) \equiv Q_j \mathbf{E}(\mathbf{r}_j) + Q_j \dot{\mathbf{r}}_j \times \mathbf{B}(\mathbf{r}_j). \tag{7.30}$$

Here, $\mathbf{F}_{\text{Lor}}(\mathbf{r}_j, \dot{\mathbf{r}}_j)$ is the Lorentz force that acts on a charged particle in the presence of electromagnetic fields. Besides the particle dynamics at the level of Newton's equation, the derived Hamiltonian should produce Maxwell's equations (2.1) for the electromagnetic field. If H satisfies both of these requirements, the developed formalism can indeed be applied to simultaneously describe charged particles, electromagnetic fields, and all possible interactions among them.

In Section 1.1 of this book, we investigate the classical particle dynamics on the basis of a variational principle. While the system Hamiltonian defines the action (1.2) as a functional of the particle trajectories, the classical trajectories minimize this action. The minimization procedure yields Hamilton's equations (1.10) that can serve as the fundamental basis for the theory of classical mechanics.

We now use this framework for the coupled system of particles and electromagnetic fields. As a first step, we generalize the action (1.2) to also include the electromagnetic fields,

$$S \equiv \int_{t_1}^{t_2} dt \left\{ \sum_j \mathbf{p}_j \cdot \dot{\mathbf{r}}_j + \int d^3r \left[\Pi_A(\mathbf{r}) \cdot \dot{\mathbf{A}}(\mathbf{r}) + \Pi_\phi(\mathbf{r}) \dot{\phi}(\mathbf{r}) \right] - H \right\}. \tag{7.31}$$

This action is a functional of all the identified pairs of canonical variables, $S \equiv S[\mathbf{r}_j,\ \mathbf{p}_j,\ \mathbf{A},\ \Pi_A,\ \phi,\ \Pi_\phi]$. As an abbreviation, we denote the group of all canonical pairs by the symbol $\{\odot\}$.

If H and the canonical momenta are identified correctly, the minimization of the action must yield the appropriate classical equations. If $S[\{\odot\}]$ assumes its minimum, the change $\delta S[\{\odot\}]$ must vanish with respect to all infinitesimal variations $\mathbf{r}_j \to \mathbf{r}_j + \delta\mathbf{r}_j$, $\mathbf{p}_j \to \mathbf{p}_j + \delta\mathbf{p}_j$, $\mathbf{A}(\mathbf{r}) \to \mathbf{A}(\mathbf{r}) + \delta\mathbf{A}(\mathbf{r})$, $\Pi_A(\mathbf{r}) \to \Pi_A(\mathbf{r}) + \delta\Pi_A(\mathbf{r})$, $\phi \to \phi + \delta\phi$, and $\Pi_\phi \to \Pi_\phi + \delta\Pi_\phi$. Here, all canonical quantities have to be treated as independent variables and the infinitesimal variations must vanish at the initial t_1 and the final t_2 giving, e.g., $\delta\mathbf{r}_j(t_1) = \delta\mathbf{r}_j(t_2) = 0$. These variations produce

$$\begin{aligned}
\delta S \equiv S[\{\odot + \delta\odot\}] - S[\{\odot\}] = \int_{t_1}^{t_2} dt \Big[& \sum_j \delta\mathbf{p}_j \cdot \dot{\mathbf{r}}_j + \mathbf{p}_j \cdot \delta\dot{\mathbf{r}}_j \\
& + \int d^3r \left(\delta\Pi_A(\mathbf{r}) \cdot \dot{\mathbf{A}}(\mathbf{r}) + \Pi_A(\mathbf{r}) \cdot \delta\dot{\mathbf{A}}(\mathbf{r}) \right) \\
& + \int d^3r \left(\delta\Pi_\phi(r)\, \dot{\phi}(\mathbf{r}) + \Pi_\phi(\mathbf{r})\, \delta\dot{\phi}(\mathbf{r}) \right) \\
& - \sum_j \left(\frac{\partial H}{\partial \mathbf{p}_j} \cdot \delta\mathbf{p}_j + \frac{\partial H}{\partial \mathbf{r}_j} \cdot \delta\mathbf{r}_j \right) \\
& - \int d^3r \left(\frac{\delta H}{\delta \Pi_A(\mathbf{r})} \cdot \delta\Pi_A(\mathbf{r}) + \frac{\delta H}{\delta \mathbf{A}(\mathbf{r})} \cdot \delta\mathbf{A}(\mathbf{r}) \right) \\
& - \int d^3r \left(\frac{\delta H}{\delta \Pi_\phi(\mathbf{r})} \delta\Pi_\phi(\mathbf{r}) + \frac{\delta H}{\delta \phi(\mathbf{r})} \delta\phi(\mathbf{r}) \right) \Big].
\end{aligned} \tag{7.32}$$

The variations with respect to the vector and scalar potentials and their canonical momentum fields produce functional derivatives of H. Using the constraint $\delta\mathbf{r}_j(t_1) = \delta\mathbf{r}_j(t_2) = 0$, we perform the partial integration

$$\int_{t_1}^{t_2} dt\, \mathbf{p}_j \cdot \frac{d}{dt}\delta\mathbf{r}_j = \Big|_{t_1}^{t_2} \mathbf{p}_j \cdot \delta\mathbf{r}_j - \int_{t_1}^{t_2} dt\, \dot{\mathbf{p}}_j \delta\mathbf{r}_j = -\int_{t_1}^{t_2} dt\, \dot{\mathbf{p}}_j \cdot \delta\mathbf{r}_j, \tag{7.33}$$

which allows us to move the time derivative away from the variations. An analogous partial integration has to be performed for the parts containing $\delta\dot{\mathbf{A}}(\mathbf{r}) = \frac{d}{dt}\delta\mathbf{A}$, and $\delta\dot{\phi} = \frac{d}{dt}\delta\phi$. As we collect all variations together, the change in the action (7.32) yields

$$\delta S = \int_{t_1}^{t_2} dt \Big\{ \sum_j \left(\dot{\mathbf{r}}_j - \frac{\partial H}{\partial \mathbf{p}_j} \right) \cdot \delta \mathbf{p}_j - \sum_j \left(\dot{\mathbf{p}}_j + \frac{\partial H}{\delta \mathbf{r}_j} \right) \cdot \delta \mathbf{r}_j$$
$$+ \int d^3r \left[\left(\dot{\mathbf{A}}(\mathbf{r}) - \frac{\delta H}{\delta \Pi_A(\mathbf{r})} \right) \cdot \delta \Pi_A(\mathbf{r}) - \left(\dot{\Pi}_A(\mathbf{r}) + \frac{\delta H}{\delta \mathbf{A}(\mathbf{r})} \right) \cdot \delta \mathbf{A}(\mathbf{r}) \right]$$
$$+ \int d^3r \left[\left(\dot{\phi}(\mathbf{r}) - \frac{\delta H}{\delta \Pi_\phi(\mathbf{r})} \right) \delta \Pi_\phi(\mathbf{r}) - \left(\dot{\Pi}_\phi(\mathbf{r}) + \frac{\delta H}{\delta \phi(\mathbf{r})} \right) \delta \phi(\mathbf{r}) \right] \Big\}. \quad (7.34)$$

Since the different variations are mutually independent, each term multiplying a variation must vanish separately. The only exception is the first term in the last line of Eq. (7.34) because the corresponding canonical momentum $\Pi_\phi = 0$, as shown by Eq. (7.25). Therefore, we have to set $\delta \Pi_\phi$ to zero such that the first term in the last line of Eq. (7.34) does not produce an additional constraint for the classical dynamics. With these considerations, Eq. (7.34) yields Hamilton's equations in the form,

$$\begin{cases} \dot{\mathbf{r}}_j = +\frac{\partial H}{\partial \mathbf{p}_j} \\ \dot{\mathbf{p}}_j = -\frac{\partial H}{\partial \mathbf{r}_j} \end{cases}, \quad \begin{cases} \dot{\mathbf{A}}(\mathbf{r}) = +\frac{\delta H}{\delta \Pi_A(\mathbf{r})} \\ \dot{\Pi}_A(\mathbf{r}) = -\frac{\delta H}{\delta \mathbf{A}(\mathbf{r})} \end{cases}, \quad \begin{cases} \Pi_\phi = 0 \\ \dot{\Pi}_\phi = 0 = -\frac{\delta H}{\delta \phi(\mathbf{r})} \end{cases} \quad (7.35)$$

because each $\delta \odot$ represents an independent infinitesimal variation.

7.3.1 Classical particle equations

We start to evaluate the Hamilton equations (7.35) by inserting the Hamiltonian (7.27) into the first one, yielding

$$\dot{\mathbf{r}}_j = \frac{\mathbf{p}_j - Q_j \mathbf{A}(\mathbf{r}_j)}{m_j} = \frac{\mathbf{p}_j^{kin}}{m_j} \equiv \mathbf{v}_j. \quad (7.36)$$

We see that the particle velocity is connected with the kinetic momentum and not the canonical momentum, as discussed in Section 7.2. To complete the particle dynamics, we evaluate

$$\frac{\partial H}{\partial \mathbf{r}_j} \equiv \nabla_j H = \sum_{\alpha=1}^{3} \mathbf{e}_\alpha \frac{\partial H}{\partial r_{j,\alpha}}$$
$$= \frac{1}{2m_j} \nabla_j (\mathbf{p}_j - Q_j \mathbf{A}(\mathbf{r}_j))^2 + Q_j \nabla_j \phi(\mathbf{r}_j)$$
$$= \frac{m_j}{2} \nabla (\mathbf{v}_j \cdot \mathbf{v}_j) + Q_j \nabla \phi(\mathbf{r}_j), \quad (7.37)$$

where the last step follows after we identify the velocity using Eq. (7.36). Here, we have to keep in mind that the velocity has an implicit $\mathbf{r}_j$ dependence because it also contains $\mathbf{A}(\mathbf{r}_j)$.

To simplify Eq. (7.37), we apply the general relation,

$$\mathbf{v} \times (\nabla \times \mathbf{v}) = \frac{1}{2}\nabla(\mathbf{v}\cdot\mathbf{v}) - (\mathbf{v}\cdot\nabla)\mathbf{v}$$
$$\Leftrightarrow \quad \nabla(\mathbf{v}\cdot\mathbf{v}) = 2\mathbf{v}\times(\nabla\times\mathbf{v}) + 2(\mathbf{v}\cdot\nabla)\mathbf{v}. \tag{7.38}$$

Using Eq. (7.36), the differentiation of the velocity can be evaluated explicitly to produce

$$\nabla_j \cdot m_j\mathbf{v}_j = -Q_j\nabla_j\cdot\mathbf{A}(\mathbf{r}_j)$$
$$\nabla_j \times m_j\mathbf{v}_j = -Q_j\nabla_j\times\mathbf{A}(\mathbf{r}_j). \tag{7.39}$$

With the help of Eqs (7.38)–(7.39), we can write Eq. (7.37) as

$$\frac{\partial H}{\partial \mathbf{r}_j} = m_j\mathbf{v}_j\times(\nabla_j\times\mathbf{v}_j) + m_j(\mathbf{v}_j\cdot\nabla_j)\mathbf{v}_j + Q_j\nabla_j\phi(\mathbf{r}_j)$$
$$= -Q_j\mathbf{v}_j\times(\nabla_j\times A(\mathbf{r}_j)) - Q_j(\mathbf{v}_j\cdot\nabla_j)\mathbf{A}(\mathbf{r}_j) + Q_j\nabla_j\phi(\mathbf{r}_j), \tag{7.40}$$

such that

$$\dot{\mathbf{p}}_j = -\frac{\partial H}{\partial \mathbf{r}_j} = Q_j\mathbf{v}_j\times\mathbf{B}(\mathbf{r}_j) + Q_j(\mathbf{v}_j\cdot\nabla_j)\mathbf{A}(\mathbf{r}_j) - Q_j\nabla\phi(\mathbf{r}_j), \tag{7.41}$$

after we have identified the magnetic field according to Eq. (7.12).

To obtain Newton's equation for the particle, we use Eq. (7.41) to determine the product of mass and acceleration,

$$m_j\dot{\mathbf{v}}_j = \frac{\partial}{\partial t}(\mathbf{p}_j - Q_j\mathbf{A}(\mathbf{r}_j)) = \dot{\mathbf{p}}_j - Q_j\dot{\mathbf{A}}(\mathbf{r}_j) - Q_j(\dot{\mathbf{r}}_j\cdot\nabla_j)\mathbf{A}(\mathbf{r}_j). \tag{7.42}$$

Here, the $\mathbf{r}_j$ argument of the vector potential must also be treated as a dynamic variable yielding the ∇_j contribution. Substituting (7.41) into Eq. (7.42) leads to

$$m_j\ddot{\mathbf{r}}_j = m_j\dot{\mathbf{v}}_j$$
$$= Q_j\mathbf{v}_j\times\mathbf{B}(\mathbf{r}_j) + Q_j\left(\left[\mathbf{v}_j - \dot{\mathbf{r}}_j\right]\cdot\nabla_j\right)\mathbf{A}(\mathbf{r}_j) - Q_j(\dot{\mathbf{A}}(\mathbf{r}_j) + \nabla_j\phi(\mathbf{r}_j))$$
$$= Q_j\mathbf{v}_j\times\mathbf{B}(\mathbf{r}_j) + Q_j\mathbf{E}(\mathbf{r}_j) \equiv \mathbf{F}_{\mathrm{Lor}}(\mathbf{r}_j, \dot{\mathbf{r}}_j), \tag{7.43}$$

where we used (7.12) to identify the electric field and Eq. (7.36) to set $\mathbf{v}_j - \dot{\mathbf{r}}_j = 0$. Equation (7.43) is identical to Eq. (7.30) verifying that the Lorentz force describes how a classical field influences a classical charged particle. Thus, we have shown that the Hamiltonian (7.27) indeed yields the correct electromagnetically induced particle motion. It is interesting to note that Eq. (7.43) is gauge invariant because it contains only the gauge-invariant electric and magnetic fields.

7.3.2 Classical equations for the electromagnetic field

We now continue to derive the classical dynamics of the electromagnetic field by explicitly evaluating the functional derivatives in Eq. (7.35). The simplest derivative produces

$$\begin{aligned}\frac{\delta H}{\delta \Pi_A(\mathbf{r})} &= \sum\nolimits_\alpha \mathbf{e}_\alpha \frac{\delta H}{\delta \Pi_A(\mathbf{r})_\alpha} \\ &= \sum\nolimits_\alpha \mathbf{e}_\alpha \frac{\delta}{\delta \Pi_{A,\alpha}(\mathbf{r})} \int d^3r \left[\frac{\Pi_A^2(\mathbf{r})}{2\varepsilon_0} - \Pi_A(\mathbf{r}) \cdot \nabla\phi(\mathbf{r}) \right],\end{aligned} \tag{7.44}$$

where only the Π_A-dependent terms in Eq. (7.27) contribute. The integrand contains both a linear and a quadratic term such that the explicit functional derivatives can be evaluated with the help of Eq. (7.21), yielding

$$\frac{\delta H}{\delta \Pi_A(\mathbf{r})} = \sum_\alpha \mathbf{e}_\alpha \left[\frac{1}{\varepsilon_0} \Pi_{A,\alpha}(\mathbf{r}) - \nabla_\alpha \phi(\mathbf{r}) \right] = \frac{\Pi_A(\mathbf{r})}{\varepsilon_0} - \nabla\phi(\mathbf{r}). \tag{7.45}$$

Thus, we obtain

$$\begin{aligned} \dot{\mathbf{A}}(\mathbf{r}) &= \frac{\delta H}{\delta \Pi_A(\mathbf{r})} = \frac{\Pi_A(\mathbf{r})}{\varepsilon_0} - \nabla\phi(\mathbf{r}) \\ \Leftrightarrow \quad & \Pi_A(\mathbf{r}) = \varepsilon_0(\dot{\mathbf{A}}(\mathbf{r}) + \nabla\phi(\mathbf{r})),\end{aligned} \tag{7.46}$$

which is identical to (7.24) and therefore yields no new information.

At this stage, only the Hamilton equations for $\dot{\Pi}_A$ and $\dot{\Pi}_\phi$ are left to produce new results. We first evaluate

$$\begin{aligned}\frac{\delta H}{\delta \mathbf{A}(\mathbf{r})} &= \sum\nolimits_j \frac{\delta}{\delta \mathbf{A}(\mathbf{r})} \frac{(\mathbf{p}_j - Q_j \mathbf{A}(\mathbf{r}_j))^2}{2m_j} \\ &\quad + \frac{\varepsilon_0 c^2}{2} \frac{\delta}{\delta \mathbf{A}(\mathbf{r})} \int d^3r (\nabla \times \mathbf{A}(\mathbf{r}))^2,\end{aligned} \tag{7.47}$$

where we included only the truly **A**-dependent contributions of the Hamiltonian (7.27). The first part of Eq. (7.47) can also be written as

$$\begin{aligned}\frac{\delta H^{1\text{st}}}{\delta \mathbf{A}(\mathbf{r})} &= \sum\nolimits_j \frac{\delta}{\delta \mathbf{A}(\mathbf{r})} \int d^3r \frac{[\mathbf{p}_j - Q_j \mathbf{A}(\mathbf{r})]^2}{2m_j} \delta(\mathbf{r} - \mathbf{r}_j) \\ &= -\sum\nolimits_j \frac{Q_j(\mathbf{p}_j - Q_j\mathbf{A}(\mathbf{r}))}{m_j} \delta(\mathbf{r} - \mathbf{r}_j) \\ &= -\sum\nolimits_j Q_j \mathbf{v}_j \delta(\mathbf{r} - \mathbf{r}_j).\end{aligned} \tag{7.48}$$

Here, we first expressed the $\mathbf{r}_j$-dependent part via an integral allowing us to replace the argument of the vector potential. Once we have this integral form, the functional derivative in the second step can be taken directly following (7.21). The last step in (7.48) simply identifies the particle velocity according to Eq. (7.36).

The second part of Eq. (7.47) can be evaluated using

$$\begin{aligned} I[A+\delta A] &\equiv \int d^3r(\nabla\times[\mathbf{A}(\mathbf{r})+\delta\mathbf{A}(\mathbf{r})])^2 \\ &= \int d^3r\left[(\nabla\times\mathbf{A}(\mathbf{r}))^2+2\,(\nabla\times\mathbf{A}(\mathbf{r}))\cdot(\nabla\times\delta\mathbf{A}(\mathbf{r}))\right] \\ &= \int d^3r\left[\mathbf{B}(\mathbf{r})^2+2\mathbf{B}(\mathbf{r})\cdot(\nabla\times\delta\mathbf{A}(\mathbf{r}))\right], \end{aligned} \tag{7.49}$$

where we used $\mathbf{B}=\nabla\times\mathbf{A}$ and kept only the contributions linear in the variation because they determine the functional derivative.

The last term in Eq. (7.49) can be simplified using

$$\int d^3r\,\mathbf{B}(\mathbf{r})\cdot[\nabla\times\delta\mathbf{A}(\mathbf{r})] = \int d^3r\,\{[\nabla\times\mathbf{B}(\mathbf{r})]\cdot\delta\mathbf{A}(\mathbf{r}) \\ -\nabla\cdot[\mathbf{B}(\mathbf{r})\times\delta\mathbf{A}(\mathbf{r})]\}\,. \tag{7.50}$$

This relation is generally valid as can be verified by applying the standard differentiation rules and the vector relation $\mathbf{a}\cdot(\mathbf{b}\times\mathbf{c}) = (\mathbf{a}\times\mathbf{b})\cdot\mathbf{c}$. With the help of Gauss' theorem (3.26), the integral over the term $\nabla\cdot[\cdots]$ in (7.50) can be converted into the surface integral, $\oint d\mathbf{S}\cdot\mathbf{B}(\mathbf{r})\times\delta\mathbf{A}(\mathbf{r})$, which vanishes because the surface integral is performed at infinity where the electromagnetic fields are zero. Using this result in Eq. (7.49), we are left with

$$I[A+\delta A] = I[A]+2\int d^3r\,\delta\mathbf{A}(\mathbf{r})\cdot(\nabla\times\mathbf{B}(\mathbf{r}))\,, \tag{7.51}$$

which produces

$$\frac{\delta}{\delta\mathbf{A}(\mathbf{r})}\int d^3r\,(\nabla\times\mathbf{A}(\mathbf{r}))^2 = 2\nabla\times B(\mathbf{r}). \tag{7.52}$$

Combining this expression with (7.48), we eventually find

$$\frac{\delta H}{\delta\mathbf{A}(\mathbf{r})} = -\sum\nolimits_j Q_j\mathbf{v}_j\delta(\mathbf{r}-\mathbf{r}_j)+\varepsilon_0c^2\nabla\times\mathbf{B}(\mathbf{r}). \tag{7.53}$$

Inserting this into Hamilton's equation (7.35) yields

$$\begin{aligned} \dot{\Pi}_A(\mathbf{r}) = -\frac{\delta H}{\delta\mathbf{A}(\mathbf{r})} &= +\sum_j Q_j\mathbf{v}_j\delta(\mathbf{r}-\mathbf{r}_j)-\varepsilon_0c^2\nabla\times\mathbf{B}(\mathbf{r}) \\ \Leftrightarrow -\varepsilon_0\frac{\partial}{\partial t}\mathbf{E}(\mathbf{r}) &= +\sum\nolimits_j Q_j\mathbf{v}_j\delta(\mathbf{r}-\mathbf{r}_j)-\varepsilon_0c^2\nabla\times\mathbf{B}(\mathbf{r}), \end{aligned} \tag{7.54}$$

where we used (7.24) to convert the canonical momentum into the electric field. Reorganizing the terms, we can write

$$\nabla\times\mathbf{B}(\mathbf{r}) = \frac{1}{c^2}\frac{\partial}{\partial t}\mathbf{E}(\mathbf{r})+\mu_0\sum\nolimits_j Q_j\mathbf{v}_j\delta(\mathbf{r}-\mathbf{r}_j), \tag{7.55}$$

where we implemented $c^2 = 1/\varepsilon_0\mu_0$ with μ_0 being the permittivity of vacuum.

To complete the evaluation of Hamilton's equations, we still need to calculate $\delta H/\delta\phi(\mathbf{r})$. To get a smooth derivation, it is beneficial to express the second term in the Hamiltonian (7.27) using the identity

$$\sum_j Q_j\phi(\mathbf{r}_j) = \sum_j \int d^3r\, Q_j\delta(\mathbf{r}-\mathbf{r}_j)\phi(\mathbf{r}). \tag{7.56}$$

After this substitution, the functional derivative of the Hamiltonian (7.27) yields

$$\frac{\delta H}{\delta\phi(\mathbf{r})} = \sum_j Q_j\delta(\mathbf{r}-\mathbf{r}_j) + \nabla\cdot\Pi_A(\mathbf{r}) = \sum_j Q_j\delta(\mathbf{r}-\mathbf{r}_j) - \varepsilon_0\nabla\cdot\mathbf{E}(\mathbf{r}), \tag{7.57}$$

where Π_A is expressed via $\mathbf{E}$ using (7.24). Since $\delta H/\delta\phi(\mathbf{r}) = 0$, see (7.35), Eq. (7.57) implies

$$\nabla\cdot\mathbf{E}(\mathbf{r}) = \frac{1}{\varepsilon_0}\sum\nolimits_j Q_j\delta(\mathbf{r}-\mathbf{r}_j), \tag{7.58}$$

which completes the evaluation of Hamilton's equations.

In addition, we directly obtain from Eq. (7.12)

$$\begin{aligned} &\nabla\times\mathbf{E}(\mathbf{r}) = -\nabla\times\frac{\partial\mathbf{A}(\mathbf{r})}{\partial t} - \nabla\times(\nabla\phi(\mathbf{r})) = -\frac{\partial}{\partial t}\nabla\times\mathbf{A}(\mathbf{r}) = -\frac{\partial}{\partial t}\mathbf{B}(\mathbf{r}) \\ &\nabla\cdot\mathbf{B}(\mathbf{r}) = \nabla\cdot(\nabla\times\mathbf{A}(\mathbf{r})) = 0, \end{aligned} \tag{7.59}$$

because $\nabla\times\nabla f$ and $\nabla\cdot\nabla\times f$ must vanish for any function f.

Combining Eqs. (7.55), (7.58), and (7.59) yields all four *Maxwell equations*

$$\begin{aligned} &\nabla\times\mathbf{E}(\mathbf{r}) = -\frac{\partial}{\partial t}\mathbf{B}(\mathbf{r}), \quad \nabla\times\mathbf{B}(\mathbf{r}) = \frac{1}{c^2}\frac{\partial}{\partial t}\mathbf{E}(\mathbf{r}) + \mu_0\sum_j Q_j\mathbf{v}_j\delta(\mathbf{r}-\mathbf{r}_j) \\ &\nabla\cdot\mathbf{E}(\mathbf{r}) = \frac{1}{\varepsilon_0}\sum_j Q_j\delta(\mathbf{r}-\mathbf{r}_j), \quad \text{and} \quad \nabla\cdot\mathbf{B}(\mathbf{r}) = 0 \end{aligned} \tag{7.60}$$

that describe the dynamics of classical light fields in the presence of charged particles.

The quantity

$$n_Q(\mathbf{r}) \equiv \sum\nolimits_j Q_j\delta(\mathbf{r}-\mathbf{r}_j). \tag{7.61}$$

defines the charge distribution of point-like particles. Furthermore, we identify

$$\mathbf{j}(\mathbf{r}) = \sum\nolimits_j Q_j\mathbf{v}_j\delta(\mathbf{r}-\mathbf{r}) \tag{7.62}$$

as the total current distribution. Both $n_Q(\mathbf{r})$ and $\mathbf{j}(\mathbf{r})$ are known once we have determined the classical trajectories from

$$m\ddot{\mathbf{r}}_j = Q_j\dot{\mathbf{r}}_j\times\mathbf{B}(\mathbf{r}_j) + Q_j\mathbf{E}(\mathbf{r}_j), \tag{7.63}$$

see (7.43). Jointly, Eqs. (7.60) and (7.63) form the closed set of Maxwell–Lorentz equations that describe how classical particles interact with classical electromagnetic fields. The Maxwell–Lorentz equations are gauge invariant because only gauge-invariant quantities appear. Thus, we have verified at this stage that the Hamiltonian (7.27) indeed provides the general description of charged particles in the presence of electromagnetic fields.

It is interesting to notice that the Hamilton formulation identifies the vector potential as the "position variable" and $\Pi_A = -\varepsilon_0 \mathbf{E}$ as the corresponding "canonical momentum." Thus, $\mathbf{A}$ and Π_A have a functional structure that can formally be related to particle-like properties. This is the very information we need in order to quantize the electromagnetic fields. Since the quantization of particles involves the replacement of the canonical variable pair by noncommuting operators, we may anticipate that the quantization of $\mathbf{A}$ and Π_A converts them into operators and the quantum features follow from the respective Schrödinger equation. This field quantization is done in Chapter 10. Unfortunately, the quantization of the scalar potential is much more difficult because Π_ϕ vanishes. Nonetheless, many important quantum-optical effects can already be understood on the basis of the quantized vector potential.

7.4 Generalized system Hamiltonian

In solid-state systems, one quite often has the situation where only a small fraction of the charged particles can move appreciable distances. Figure 7.3(a) schematically presents a scenario where electrons (large shaded spheres) move through a lattice of polarizable ions (black and white circles). Since the ions in solids are confined to their lattice sites and just oscillate around their equilibrium positions, it is not convenient to include their motion in the description of the interacting light–matter systems, even though that would be the fundamentally correct approach.

As an alternative, one often includes ionic contributions via the polarization $\mathbf{P}$ that reacts to electromagnetic fields entering the solid. Quite often, this part of the light–matter interaction can be approximated by a real-valued susceptibility

$$\mathbf{P}(\mathbf{r}) = \epsilon_0\, \chi_{\text{ion}}(\mathbf{r})\, \mathbf{E}(\mathbf{r}), \quad n(\mathbf{r})^2 \equiv 1 + \chi_{\text{ion}}(\mathbf{r}), \tag{7.64}$$

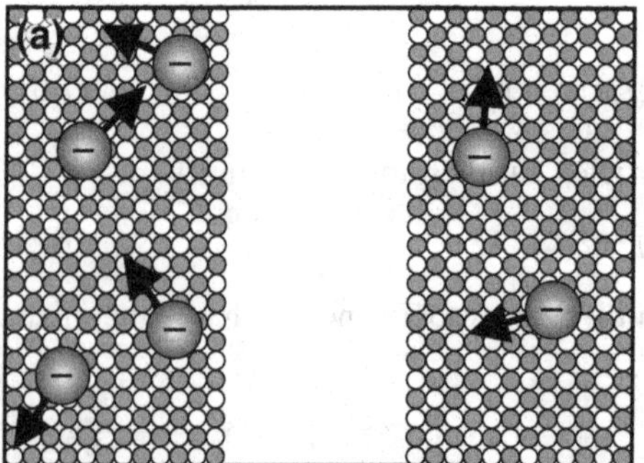

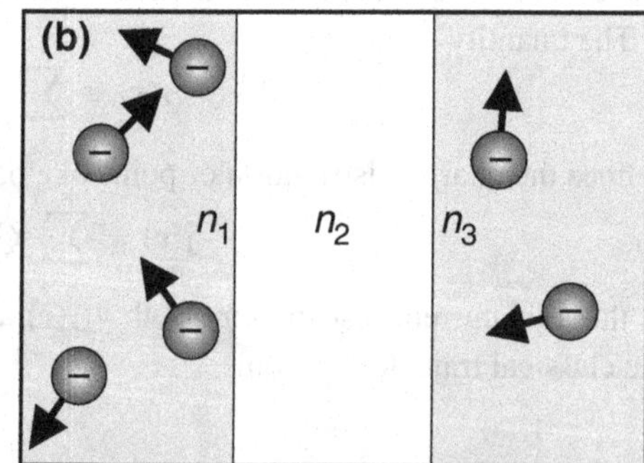

Figure 7.3 Schematic representation of the dielectric model for solid-state systems. **(a)** Electrons (large shaded spheres) move, as indicated by arrows, within the nearly rigid lattice of the polarizable ions (white and dark circles). **(b)** The approximative description replaces the polarizable atoms by refractive indices n_1, n_2, and n_3 within the respective regions.

which describes the strength of the induced ionic polarization at a given position. Typically, $\chi_{\text{ion}}(\mathbf{r})$ can be taken as a real-valued constant that is characteristic for a given solid. In a composite material, the magnitude of $\chi_{\text{ion}}(\mathbf{r})$ may be different in the different material regions. Equivalently, the effects of the background polarization can be treated via a refractive index $n(\mathbf{r})$ that slows down the speed of light to $c/n(\mathbf{r})$.

The main advantage of the approximate description is the significant reduction of the degrees of freedom needed in the actual many-body treatment. The corresponding approximation is schematically depicted in Fig. 7.3(b) where the electrons move within the solid (shaded area) that has a background refractive index n_1 or n_3 while the electrons do not enter the middle region having a different refractive index n_2.

To systematically incorporate the approximation of the ions as dielectric background into the Hamilton formulation of particle–light coupling, we have to implement the appropriate modifications. We start by introducing the generalized Lagrangian

$$L = \sum_{j=1}^{\mathcal{N}} \left[\frac{1}{2} m \dot{\mathbf{r}}_j^2 + Q_j \dot{\mathbf{r}}_j \cdot \mathbf{A}(\mathbf{r}_j) - Q_j \phi(\mathbf{r}_j) \right] + \frac{\varepsilon_0}{2} \int d^3r \left[n^2(\mathbf{r}) \mathbf{E}^2(\mathbf{r}) - c^2 \mathbf{B}^2(\mathbf{r}) \right],$$
$$\mathbf{E}(\mathbf{r}) = -\dot{\mathbf{A}}(\mathbf{r}) - \nabla\phi(\mathbf{r}), \qquad \mathbf{B}(\mathbf{r}) = \nabla \times \mathbf{A}(\mathbf{r}), \tag{7.65}$$

where we have also included the definitions (7.12), for the sake of completeness. As a difference to the original L in Eq. (7.11), the energy related to $\mathbf{E}^2$ is modified by the refractive-index profile, $n^2(\mathbf{r})$, representing the background polarization. With an analogous derivation as in Section 7.2.2, we find the canonical momenta

$$\Pi_A(\mathbf{r}) = \frac{\delta L}{\delta \dot{\mathbf{A}}(\mathbf{r})} = \varepsilon_0 n^2(\mathbf{r}) \left[\dot{\mathbf{A}}(\mathbf{r}) + \nabla\phi(\mathbf{r}) \right] = -\varepsilon_0 n^2(\mathbf{r})\, \mathbf{E}(\mathbf{r}),$$
$$\Pi_\phi(\mathbf{r}) = \frac{\delta L}{\delta \dot{\phi}(\mathbf{r})} = 0, \tag{7.66}$$

where $\Pi_A(\mathbf{r})$ is modified by $n^2(\mathbf{r})$ relative to its original format (7.24). The canonical particle momentum remains unchanged, i.e., Eq. (7.15) is still valid. Consequently, we may directly construct the system Hamiltonian

$$H = \sum_{j=1}^{\mathcal{N}} \left(\frac{\left[\mathbf{p}_j - Q_j \mathbf{A}(\mathbf{r}_j)\right]^2}{2m_j} + Q_j \phi(\mathbf{r}_j) \right) + \int d^3r \left[\frac{\Pi_A^2(\mathbf{r})}{2\varepsilon_0 n^2(\mathbf{r})} + \frac{\varepsilon_0 c^2}{2} (\nabla \times \mathbf{A}(\mathbf{r}))^2 - \Pi_A(\mathbf{r}) \cdot \nabla\phi(\mathbf{r}) \right] \tag{7.67}$$

that can be derived following steps similar to those leading to Eq. (7.28).

Evaluating Hamilton's equations (7.35), we find that the classical particle trajectory is still determined by Eq. (7.43). Thus, the presence of the dielectric background only modifies Maxwell's equations. By repeating the derivation steps of Section 7.3.2, we eventually find

$$\begin{aligned}\nabla \cdot [n^2(\mathbf{r})\,\mathbf{E}(\mathbf{r})] &= \frac{1}{\varepsilon_0}\sum\nolimits_j Q_j\delta(\mathbf{r}-\mathbf{r}_j)\\ \nabla \times \mathbf{B}(\mathbf{r}) &= \frac{1}{\varepsilon_0 c^2}\sum\nolimits_j Q_j\mathbf{v}_j\delta(\mathbf{r}-\mathbf{r}_j) + \frac{n^2(\mathbf{r})}{c^2}\dot{\mathbf{E}}(\mathbf{r})\end{aligned} \tag{7.68}$$

while $\nabla \times \mathbf{E} = \dot{\mathbf{B}}$ and $\nabla \cdot \mathbf{B} = 0$ remain unchanged.

As the next step, we check how the presence of the refractive index modifies the solutions of Maxwell's equations. In cases where we have a constant $n(\mathbf{r}) = n$, we can express Eq. (7.68) as

$$\begin{aligned}\nabla \cdot \mathbf{E}(\mathbf{r}) &= \frac{1}{\varepsilon_0}\sum\nolimits_j \frac{Q_j}{n^2}\delta(\mathbf{r}-\mathbf{r}_j)\\ \nabla \times \mathbf{B}(\mathbf{r}) &= \frac{1}{\varepsilon_0 c^2}\sum\nolimits_j Q_j\mathbf{v}_j\delta(\mathbf{r}-\mathbf{r}_j) + \frac{n^2}{c^2}\dot{\mathbf{E}}(\mathbf{r}),\end{aligned} \tag{7.69}$$

where we see that the refractive index effectively reduces the influence of the charge density on $\mathbf{E}(\mathbf{r})$ by a factor of n^2. This reduction is a consequence of the fact that the polarization of matter leads to a partial screening of any test charge put inside the polarizable matter.

To consider another limit, we assume a situation where no charges exist, i.e., $Q_j = 0$. Equation (7.69) then produces $\nabla \cdot \mathbf{E}(\mathbf{r}) = 0$ such that

$$\begin{aligned}&\nabla \times [\nabla \times \mathbf{E}(\mathbf{r})] = \nabla[\nabla \cdot \mathbf{E}(\mathbf{r})] - \nabla^2\mathbf{E}(\mathbf{r}) = -\nabla^2\mathbf{E}(\mathbf{r})\\ &\nabla \times [\nabla \times \mathbf{E}(\mathbf{r})] = \nabla \times \dot{\mathbf{B}}(\mathbf{r}) = \frac{\partial}{\partial t}\nabla \times \mathbf{B}(\mathbf{r}) = \frac{n^2}{c^2}\ddot{\mathbf{E}}(\mathbf{r})\\ \Rightarrow \quad &\nabla^2\mathbf{E}(\mathbf{r}) + \frac{n^2}{c^2}\frac{\partial^2}{\partial t^2}\mathbf{E}(\mathbf{r}) = 0.\end{aligned} \tag{7.70}$$

Here, the second relation follows from Eq. (7.69) for a vanishing current. The final form of (7.70) is nothing but a wave equation where the propagation speed of light is reduced to c/n.

Thus, on the basis of the Hamiltonian (7.67), we recover the expected effects resulting from the presence of a background refractive index. In this book, we use this approach as the systematic starting point to incorporate the consequences of a dielectric background without the need to describe each ion in the material separately. On this basis, we develop the pragmatic description of semiconductor quantum optics using the canonical formulation of the electromagnetic field also if the trivial part of the matter response is treated phenomenologically via $n(\mathbf{r})$. The effect of nonuniform $n(\mathbf{r})$ on charge-generated fields as well as on light propagation is investigated in Chapter 9 in more detail. This knowledge is then applied to quantize the light–matter system in structures with an arbitrary dielectric profile.

Exercises

7.1 (**a**) Show that one can add an arbitrary function $\frac{d}{dt}G(\mathbf{r},t)$ to the Lagrangian without changing the classical trajectories resulting from Eq. (7.31). (**b**) Show that $\mathbf{E}(r,t)$, $B(r,t)$, and the Lagrangian (7.11) remain invariant under the gauge transformation (7.13). **Hint:** *Use the insight obtained from part* (**a**) . (**c**) Construct a constant field $\mathbf{E}(r,t) = E_0\hat{\mathbf{e}}_z$, $\mathbf{B}(r,t) = 0$ in terms of either $A(r,t)$ or $\phi(r,t)$. What is the gauge transformation connecting these two forms?

7.2 Use the vectorial relations $\mathbf{A}\times(\mathbf{B}\times\mathbf{C}) = \mathbf{B}(\mathbf{A}\cdot\mathbf{C}) - (\mathbf{C}\cdot\mathbf{A})\mathbf{B}$, $\mathbf{A}\times\mathbf{B} = -\mathbf{B}\times\mathbf{A}$, and $\mathbf{A}\cdot(\mathbf{B}\times\mathbf{C}) = (\mathbf{A}\times\mathbf{B})\cdot\mathbf{C}$ to verify the following relations:
(**a**) $\nabla\times(\mathbf{A}\times\mathbf{B}) = \mathbf{A}(\nabla\cdot\mathbf{B}) + \mathbf{B}(\nabla\cdot\mathbf{A}) - (\mathbf{A}\cdot\nabla)\mathbf{B} - (\mathbf{B}\cdot\nabla)\mathbf{A}$
(**b**) $\mathbf{A}\times(\nabla\times\mathbf{B}) + \mathbf{B}\times(\nabla\times\mathbf{A}) = \nabla\cdot(\mathbf{A}\cdot\mathbf{B}) - (\mathbf{A}\cdot\nabla)\mathbf{B} - (\mathbf{B}\cdot\nabla)\mathbf{A}$ and

$$\nabla\times(\mathbf{A}\times\mathbf{B}) = \mathbf{A}\times(\nabla\times\mathbf{B}) + \mathbf{B}\times(\nabla\times\mathbf{A}) + \mathbf{A}(\nabla\cdot\mathbf{B}) + \mathbf{B}(\nabla\cdot\mathbf{A}) - \nabla(\mathbf{A}\cdot\mathbf{B})$$

(**c**) $\nabla(\mathbf{A}\cdot\mathbf{A}) = 2\mathbf{A}\times(\nabla\times\mathbf{A}) + 2\mathbf{A}(\nabla\cdot\mathbf{A})$
(**d**) $\nabla\times(\nabla\times\mathbf{A}) = \nabla(\nabla\cdot\mathbf{A}) - \nabla^2\mathbf{A}$.

7.3 (**a**) Assume that a functional has the form $G[f] = \int d^3r\, g(\mathbf{r}, f(\mathbf{r}), \nabla f(\mathbf{r}))$. Use the relation (7.18) and show that the functional derivative is $\frac{\delta G[f]}{\delta f(\mathbf{r})} = \frac{\partial g}{\partial f(\mathbf{r})} - \nabla\cdot\frac{\partial g}{\partial\nabla f(\mathbf{r})}$. (**b**) Assume that a functional has the form $G[f] = \sum_j g(f(\mathbf{r}_j))$. Show that its functional derivative is $\frac{\delta G[f]}{\delta f(\mathbf{r})} = \sum_j \delta(\mathbf{r}-\mathbf{r}_j)g'(f(\mathbf{r}))$.

7.4 Construct the variation of action, i.e., Eq. (7.34), by starting from the definition (7.31).

7.5 Use the generalized system Hamiltonian (7.67) and show that the classical trajectory follows from $m_j\ddot{\mathbf{r}}_j = Q_j\mathbf{v}_j\times\mathbf{B}(\mathbf{r}_j) + Q_j\mathbf{E}(\mathbf{r}_j)$, in analogy to Eq. (7.43).

7.6 Using the generalized system Hamiltonian (7.67), derive the explicit form of Hamilton's equation $\dot{A} = \frac{\delta H}{\delta\Pi_A(\mathbf{r})}$. Use the original definition (7.18) of variations and compare it to the result obtained via the relation (7.21). At the final step, express $\Pi_A(\mathbf{r})$ in terms of gauge-invariant fields.

7.7 Show that the generalized Hamiltonian (7.67) produces Maxwell's equations through (**a**) $\dot{\Pi}_A = -\delta H/\delta\mathbf{A}(\mathbf{r})$ and (**b**) $\dot{\Pi}_\phi = 0 = -\delta H/\delta\phi(\mathbf{r})$.

7.8 Break the gauge symmetry by introducing an artificial $\dot{\phi}$-dependent contribution into the Lagrangian (7.11), i.e., use

$$L' = \sum_{j=1}^{N} \left[\frac{1}{2} m_j \dot{\mathbf{r}}_j^2 + Q_j \dot{\mathbf{r}}_j \cdot \mathbf{A}(\mathbf{r}_j, t) - Q_j \phi(\mathbf{r}_j, t) \right]$$
$$+ \frac{\epsilon_0}{2} \int d^3r \left[\dot{\mathbf{A}}^2(\mathbf{r}) + 2\dot{\mathbf{A}}(\mathbf{r}) \cdot \nabla\phi(\mathbf{r}) + [\nabla\phi(\mathbf{r})]^2 \right.$$
$$\left. -c^2 [\nabla \times \mathbf{A}(\mathbf{r})]^2 + \frac{1}{K^2} [\dot{\phi}(\mathbf{r})]^2 \right].$$

(a) Define $\Pi_\phi = \delta L' / \delta \dot{\phi}(\mathbf{r})$ and construct the modified system Hamiltonian H' based on Eq. (7.26). **(b)** Which dynamics follows from Hamilton's equations $\dot{\phi}(\mathbf{r}) = \delta H / \delta \Pi_\phi(\mathbf{r})$ and $\dot{\Pi}_\phi = -\delta H / \delta \phi(\mathbf{r})$? Set $K \to \infty$ and show that the choice (7.35) is the correct one. **(c)** Assume that we set $\mathbf{A}(\mathbf{r}) = 0$ and $Q_j \to 0$. Show that $\phi(\mathbf{r})$ then follows from a wave equation. What is the speed of "light"and what happens in the limit $K \to \infty$?

Further reading

More background information on the Lagrange formulation of electromagnetism is found, e.g., in:

C. Cohen-Tannoudji, J. Dupont-Roc, and G. Grynberg (1997). *Photons & Atoms: Introduction to Quantum Electrodynamics*, Wiley & Sons.
F. Melia (2001). *Electrodynamics*, Chicago, University of Chicago Press.
B. di Bartolo (2004). *Classical Theory of Electromagnetism*, 2nd edition, Singapore, World Scientific.

The Lagrange formulation of the electromagnetic field is often discussed at the level of a relativistic description. Such treatments can be found in:

M. Schwartz (1987). *Principles of Electrodynamics*, New York, Dover.
F. W. Hehl and Y. N. Obukhov (2003). *Foundations of Classical Electrodynamics: Charge, Flux, and Metric*, Boston, Birkhäuser.
L. D. Landau and E. M. Lifshitz (2003). *The Classical Theory of Fields*, 4th edition, Amsterdam, Butterworth–Heinemann.

Functional derivatives are a standard tool in variation and density matrix theory. For more background information, we recommend:

I. M. Gelfand and S. V. Fomin (1991). *Calculus of Variations*, Mineola, NY, Dover.
D. S. Sholl and J. A. Steckel (2009). *Density Functional Theory: A Practical Introduction*, Hoboken, NJ, Wiley & Sons.

8

System Hamiltonian of classical electrodynamics

The analysis in Chapter 7 shows how Maxwell's equations can be expressed in the framework of Hamilton's formalism, one of the fundamental descriptions of classical physics. Since we want to use canonical field quantization to proceed from the classical to the full quantum-mechanical theory, it is important to identify the proper pairs of canonical variables that will be converted into noncommuting operators. So far, we know that $\Pi_A(\mathbf{r})$ is the canonical "momentum" corresponding to the vector field $\mathbf{A}(\mathbf{r})$. However, Maxwell's equations also contain the scalar potential $\phi(\mathbf{r})$ that has a vanishing canonical momentum, as shown in Section 7.2.2. Thus, the quantization of the scalar potential seems to be, at the same time, less relevant and more complicated than the quantization of the vector potential.

To deal with the scalar potential, we adopt an approach where it is treated classically while the vector potential is fully quantized. We determine the exact result for $\phi(\mathbf{r})$ from Maxwell's equations and substitute it into the classical Hamiltonian. In this procedure, it is important to choose the gauge properly. Therefore, we compare in this chapter results produced in the Coulomb gauge with those in the Lorentz gauge.

8.1 Elimination of the scalar potential

Since we do not quantize the scalar potential $\phi(\mathbf{r})$, we want to eliminate it as an independent variable from the Hamiltonian. For this purpose, we derive an explicit expression for $\phi(\mathbf{r})$ as a solution of Maxwell's equations, expressing it in terms of other truly independent variables such as particle positions.

Explicitly, the scalar potential appears in two places in the Hamiltonian (7.27). We first consider the last term of Eq. (7.27),

$$H_{\text{last}} = -\int d^3r\, \Pi_A(\mathbf{r}) \cdot \nabla\phi(\mathbf{r}) = \varepsilon_0 \int d^3r\, \mathbf{E}(\mathbf{r}) \cdot \nabla\phi(\mathbf{r}), \tag{8.1}$$

where we used Eq. (7.24) to express $\Pi_A(\mathbf{r})$ in terms of $\mathbf{E}(\mathbf{r})$. We integrate by parts and use Gauss' theorem (3.26) to obtain

$$H_{\text{last}} = \varepsilon_0 \oint d\mathbf{S} \cdot \mathbf{E}(\mathbf{r})\, \phi(\mathbf{r}) - \int d^3r\, \varepsilon_0 \left[\nabla \cdot \mathbf{E}(\mathbf{r})\right] \phi(\mathbf{r})$$
$$= 0 - \int d^3r \sum_j Q_j \delta(\mathbf{r} - \mathbf{r}_j) \phi(\mathbf{r}) = -\sum_j Q_j \phi(\mathbf{r}_j). \tag{8.2}$$

Here, the surface integral vanishes because $\mathbf{E}(\mathbf{r})$ decays to zero at the boundaries of the system. Furthermore, we have used the appropriate Maxwell equation (7.60) to express $\nabla \cdot \mathbf{E}(\mathbf{r})$ in terms of the charge density.

Inserting (8.2) into the Hamiltonian (7.27) shows that this contribution exactly cancels the scalar-potential term within the particle sum $\sum_j$, leaving us with

$$H = \sum_j \frac{\left[\mathbf{p}_j - Q_j \mathbf{A}(\mathbf{r}_j)\right]^2}{2m_j} + \int d^3r \left[\frac{\Pi_A^2(\mathbf{r})}{2\varepsilon_0} + \frac{\varepsilon_0 c^2}{2} (\nabla \times \mathbf{A}(\mathbf{r}))^2\right]. \tag{8.3}$$

Even though this form appears to be simpler than the original Hamiltonian, it still contains an implicit dependence on ϕ within the $\Pi_A = \varepsilon_0 \left(\dot{\mathbf{A}} + \nabla\phi\right)$ term, according to Eq. (7.24). Thus, we must work further on this contribution,

$$H_\Pi \equiv \int d^3r \frac{\Pi_A^2(\mathbf{r})}{2\varepsilon_0} = \frac{\varepsilon_0}{2} \int d^3r\, \mathbf{E}(\mathbf{r}) \cdot \mathbf{E}(\mathbf{r})$$
$$= -\frac{\varepsilon_0}{2} \int d^3r\, \mathbf{E}(\mathbf{r}) \cdot \left[\dot{\mathbf{A}}(\mathbf{r}) + \nabla\phi(\mathbf{r})\right]$$
$$= -\frac{\varepsilon_0}{2} \int d^3r\, \mathbf{E}(\mathbf{r}) \cdot \dot{\mathbf{A}}(\mathbf{r}) - \frac{\varepsilon_0}{2} \int d^3r \mathbf{E}(\mathbf{r}) \cdot \nabla\phi(\mathbf{r}). \tag{8.4}$$

Here, the two first steps result from the substitutions $\Pi_A(\mathbf{r}) = -\varepsilon_0 \mathbf{E}(\mathbf{r})$ and $\mathbf{E}(\mathbf{r}) = -\frac{\partial}{\partial t}\mathbf{A} - \nabla\phi$ given by Eqs. (7.24) and (7.12), respectively. The last term has the form of Eq. (8.1) such that we can use (8.2) to replace it by $\frac{1}{2} \sum_j Q_j \phi(\mathbf{r}_j)$. This leads to

$$H_\Pi = -\frac{\varepsilon_0}{2} \int d^3r\, \mathbf{E}(\mathbf{r}) \cdot \dot{\mathbf{A}}(\mathbf{r}) + \frac{1}{2} \sum_j Q_j \phi(\mathbf{r}_j). \tag{8.5}$$

Using again $\mathbf{E} = -\dot{\mathbf{A}} - \nabla\phi$, we obtain

$$H_\Pi = \frac{1}{2} \sum_j Q_j \phi(\mathbf{r}_j) + \frac{\varepsilon_0}{2} \int d^3r\, \nabla\phi(\mathbf{r}) \cdot \dot{\mathbf{A}}(\mathbf{r}) + \frac{\varepsilon_0}{2} \int d^3r\, \dot{\mathbf{A}}^2(\mathbf{r})$$
$$= \frac{1}{2} \sum_j Q_j \phi(\mathbf{r}_j) - \frac{\varepsilon_0}{2} \int d^3r\, \phi(\mathbf{r}) \frac{\partial}{\partial t} \left[\nabla \cdot \mathbf{A}(\mathbf{r})\right] + \frac{\varepsilon_0}{2} \int d^3r\, \dot{\mathbf{A}}^2(\mathbf{r}), \tag{8.6}$$

where the middle term has been converted via partial integration as before. Inserting (8.6) into Eq. (8.3) brings the system Hamiltonian into the form

$$H = \sum_j \frac{\left[\mathbf{p}_j - Q_j \mathbf{A}(\mathbf{r}_j)\right]^2}{2m_j} + \frac{1}{2} \sum_j Q_j \phi(\mathbf{r}_j) - \frac{\varepsilon_0}{2} \int d^3r\, \phi(\mathbf{r}) \frac{\partial}{\partial t} \left[\nabla \cdot \mathbf{A}(\mathbf{r})\right]$$
$$+ \frac{\varepsilon_0}{2} \int d^3r \left[\dot{\mathbf{A}}^2(\mathbf{r}) + c^2 (\nabla \times \mathbf{A}(\mathbf{r}))^2\right]. \tag{8.7}$$

No matter which formal rearrangements we perform, we have to realize at this stage that we cannot completely eliminate the scalar potential without knowing the relation between ϕ and $\mathbf{A}$.

8.2 Coulomb and Lorentz gauge

To simplify the analysis as much as possible, we can make use of the fact that we have a free choice of the electromagnetic gauge. For this purpose, we first assume arbitrary vector and scalar potentials $\tilde{\mathbf{A}}(\mathbf{r})$ and $\tilde{\phi}(\mathbf{r})$, respectively, and then try to find simpler ones $\mathbf{A}(\mathbf{r})$ and $\phi(\mathbf{r})$. For this, we utilize the fact that we can always introduce new vector and scalar potentials $\tilde{\mathbf{A}} = \mathbf{A} + \nabla F$ and $\tilde{\phi} = \phi - \dot{F}$, respectively, without any change of the underlying physics, compare Eq. (7.13). One often chooses F such that it satisfies

$$\nabla \cdot \tilde{\mathbf{A}}(\mathbf{r}) = \nabla^2 F(\mathbf{r}) \qquad \Leftrightarrow \qquad \nabla \cdot \mathbf{A}(\mathbf{r}) = 0. \tag{8.8}$$

Hence, $\mathbf{A}$ becomes a divergence-free vector field. This choice is known as the *Coulomb gauge*.

Alternatively, we may proceed from the generic connection,

$$\begin{cases} \nabla \cdot \tilde{\mathbf{A}} = \nabla \cdot \mathbf{A} + \nabla^2 F \\ \dot{\tilde{\phi}} = \dot{\phi} - \ddot{F} \end{cases}$$

$$\Rightarrow \qquad \nabla \cdot \tilde{\mathbf{A}} + \frac{1}{c^2}\dot{\tilde{\phi}} = \nabla \cdot \mathbf{A} + \frac{1}{c^2}\dot{\phi} + \left[\nabla^2 - \frac{1}{c^2}\frac{\partial^2}{\partial t^2}\right] F. \tag{8.9}$$

We may now choose F to satisfy

$$\left[\nabla^2 - \frac{1}{c^2}\frac{\partial^2}{\partial t^2}\right] F(\mathbf{r}, t) = \left[\nabla \cdot \tilde{\mathbf{A}}(\mathbf{r}, t) + \frac{1}{c^2}\dot{\tilde{\phi}}(\mathbf{r}, t)\right]$$
$$\Leftrightarrow \qquad \nabla \cdot \mathbf{A}(\mathbf{r}, t) + \frac{1}{c^2}\dot{\phi}(\mathbf{r}, t) = 0. \tag{8.10}$$

This choice is known as the *Lorentz gauge*.

8.2.1 Scalar-potential elimination in the Coulomb gauge

The Coulomb gauge directly simplifies the Hamiltonian (8.7) because it eliminates the $\nabla \cdot \mathbf{A}$-dependent contribution that appears. As the remaining task, we only need to find a way to treat the $\sum_j Q_j \phi(\mathbf{r}_j)$ term. For this purpose, we consider Maxwell's equation (7.60)

$$\nabla \cdot \mathbf{E}(\mathbf{r}) = \nabla \cdot \left[-\dot{\mathbf{A}}(\mathbf{r}) - \nabla\phi(\mathbf{r})\right] = -\frac{\partial}{\partial t}\nabla \cdot \mathbf{A}(\mathbf{r}) - \nabla^2\phi(\mathbf{r}) = -\nabla^2\phi(\mathbf{r}), \tag{8.11}$$

where we have used (7.12) and the Coulomb gauge (8.8). By substituting this into Eq. (7.60), we obtain the Poisson equation

$$\nabla^2\phi(\mathbf{r}) = -\frac{1}{\varepsilon_0}\sum\nolimits_j Q_j\delta(\mathbf{r}-\mathbf{r}_j), \tag{8.12}$$

that determines the potential created by a set of point-like charges. As shown in many textbooks on electromagnetism (see Exercise 8.2), the corresponding scalar potential is

$$\phi(\mathbf{r}) = \sum\nolimits_k \frac{Q_k}{4\pi\varepsilon_0|\mathbf{r}-\mathbf{r}_k|}. \tag{8.13}$$

As this is inserted into Eq. (8.7), we obtain

$$H = \sum\nolimits_j \frac{\left[\mathbf{p}_j - Q_j\mathbf{A}(\mathbf{r}_j)\right]^2}{2m_j} + \frac{1}{2}\sum\nolimits_{j,k}\frac{Q_jQ_k}{4\pi\varepsilon_0|\mathbf{r}_j-\mathbf{r}_k|} + \frac{\varepsilon_0}{2}\int d^3r\left[\dot{\mathbf{A}}^2(\mathbf{r}) + c^2\mathbf{B}^2(\mathbf{r})\right] \tag{8.14}$$

showing that the scalar potential is replaced by an explicit pairwise interaction among the charged point-like particles. This many-body interaction can be expressed via the standard Coulomb potential

$$V_{j,k}(\mathbf{r}) = \frac{Q_jQ_k}{4\pi\varepsilon_0|\mathbf{r}|}, \tag{8.15}$$

where $\mathbf{r} = \mathbf{r}_j - \mathbf{r}_k$. In principle, the Coulomb interaction also involves the self-interaction contributions $V_{j,j}(0)$. However, these produce a position $\mathbf{r}_j$ independent additive constant in H such that the $j = k$ terms do not contribute to the particle dynamics described by Hamilton's equations (7.35). Even though the constant self-interaction diverges, we may always remove it by choosing the zero level of energy accordingly. Thus, we may simply omit the $j = k$ terms from Eq. (8.14) replacing $\sum_{j,k}$ by $\sum_{j,k\neq j}$.

To express H completely in terms of $\mathbf{A}$, we can apply the vectorial relation

$$\begin{aligned}\nabla\cdot(\mathbf{A}(\mathbf{r})\times[\nabla\times\mathbf{A}(\mathbf{r})]) &= [\nabla\times\mathbf{A}(\mathbf{r})]\cdot[\nabla\times\mathbf{A}(\mathbf{r})] \\ &\quad - \mathbf{A}(\mathbf{r})\times\nabla\cdot[\nabla\times\mathbf{A}(\mathbf{r})] \\ &= \mathbf{B}^2(\mathbf{r}) - \mathbf{A}(\mathbf{r})\cdot[\nabla\times(\nabla\times\mathbf{A}(\mathbf{r}))] \\ &= \mathbf{B}^2(\mathbf{r}) - \mathbf{A}(\mathbf{r})\cdot\left[\nabla[\nabla\cdot\mathbf{A}(\mathbf{r})] - \nabla^2\mathbf{A}(\mathbf{r})\right]\end{aligned} \tag{8.16}$$

together with the identification $\mathbf{B} = \nabla\times\mathbf{A}$. In Coulomb gauge, $\nabla\cdot\mathbf{A} = 0$ such that (8.16) simplifies to

$$\mathbf{B}^2(\mathbf{r}) = \nabla\cdot[\mathbf{A}(\mathbf{r})\times\mathbf{B}(\mathbf{r})] - \mathbf{A}(\mathbf{r})\cdot\nabla^2\mathbf{A}(\mathbf{r}). \tag{8.17}$$

Inserting this expression into the Hamiltonian (8.14), we notice that the spatial integral over the contribution $\nabla\cdot[\mathbf{A}(\mathbf{r})\times\mathbf{B}(\mathbf{r})]$ vanishes based on Gauss' theorem (3.26) and the vanishing of $\mathbf{B}$ at infinity. Hence, we can write the system Hamiltonian in its final form

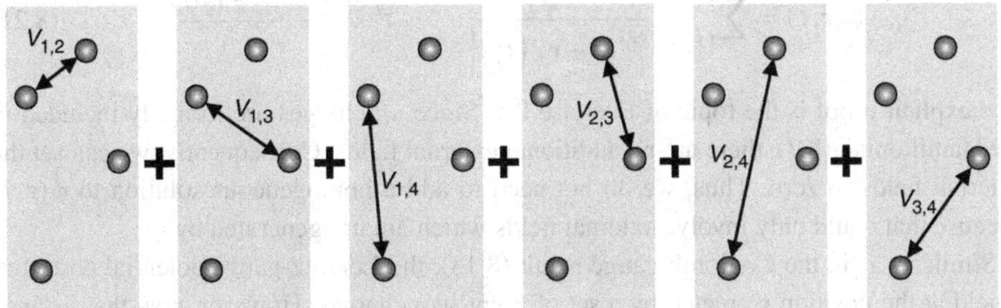

Figure 8.1 Schematic presentation of pairwise Coulomb interaction among charged particles. The pairwise interaction configurations are presented among four charged particles (spheres). Each interaction possibility is included only once, according to Eq. (8.19).

$$H = \sum_j \frac{\left[\mathbf{p}_j - Q_j\mathbf{A}(\mathbf{r}_j)\right]^2}{2m_j} + \frac{1}{2}\sum_{j,k\neq j} V_{j,k}(\mathbf{r}_j - \mathbf{r}_k) + \frac{\varepsilon_0}{2}\int d^3r \left[\dot{\mathbf{A}}^2(\mathbf{r}) - c^2\mathbf{A}(\mathbf{r})\cdot\nabla^2\mathbf{A}(\mathbf{r})\right], \tag{8.18}$$

which will be the starting point of our many-body investigations within the Coulomb gauge. We notice that each pairwise interaction is included only once because the factor $1/2$ removes the double counting from $\sum_{j,k\neq j} V_{j,k}(\mathbf{r}_j - \mathbf{r}_k)$.

The Coulomb sum in Eq. (8.18) can also be written as

$$\frac{1}{2}\sum_{j,k\neq j} V_{j,k} = \sum_{j=1}^{\mathcal{N}-1}\sum_{k=j+1}^{\mathcal{N}} V_{j,k} = \sum_{k=1}^{\mathcal{N}-1}\sum_{j=k+1}^{\mathcal{N}} V_{j,k}, \tag{8.19}$$

which follows from a simple reorganization of terms. The summation limits are chosen for a system with exactly $\mathcal{N}$ particles. Figure 8.1 symbolically presents the Coulomb sum over $\mathcal{N} = 4$ particles (circles). In general, we always find $\mathcal{N}(\mathcal{N}-1)/2$ different interacting pairs. In the four-particle example of Fig. 8.1, we have six different interaction configurations.

8.2.2 Scalar-potential elimination in the Lorentz gauge

We may eliminate the scalar potential also in the Lorentz gauge (8.10). In this case, Maxwell's equations (7.60) produce

$$\left[\nabla^2 - \frac{1}{c^2}\frac{\partial^2}{\partial t^2}\right]\phi(\mathbf{r},t) = -\frac{1}{\varepsilon_0}\sum_j Q_j\,\delta(\mathbf{r}-\mathbf{r}_j) \tag{8.20}$$

for the scalar potential. As in the Coulomb gauge, this relation can be applied to express $\phi(\mathbf{r})$ in terms of interactions among pure particle quantities. Structurally, Eq. (8.20) is a one-dimensional wave equation whose inhomogeneous solution is given by

$$\phi(\mathbf{r}, t) = \sum_j \frac{1}{4\pi\varepsilon_0} \frac{Q_j}{\left|\mathbf{r} - \mathbf{r}_j\left(t'_j\right)\right|}, \qquad t'_j = t - \frac{|\mathbf{r} - \mathbf{r}_j(t)|}{c}. \tag{8.21}$$

The explicit proof is the topic of Exercise 8.4. Since *all* charges are formally included in the Hamiltonian (8.7), there are no additional external fields. Consequently, we can set the external fields to zero. Thus, we do not need to add a homogeneous solution to $\phi(\mathbf{r}, t)$ because that could only involve external fields which are not generated by Q_j.

Similarly as in the Coulomb-gauge result (8.13), the Lorentz-gauge potential describes a field at the position $\mathbf{r}$ created by a set of point-like charges. However, now the distance is evaluated via $|\mathbf{r} - \mathbf{r}_j(t'_j)|$ at the retarded time $t'_j = t - \frac{|\mathbf{r}-\mathbf{r}_j(t)|}{c}$. This retardation can produce significant complications for quantum investigations because it is not a priori clear, e.g., how a pair $(\mathbf{r}(t), \mathbf{p}(t'))$ containing different times should be treated in the canonical quantization. Nevertheless, we can complete the simplification of the classical Hamiltonian following a derivation similar to Eq. (8.18). This calculation eventually yields

$$\begin{aligned} H = &\sum_j \frac{\left[\mathbf{p}_j - Q_j\mathbf{A}(\mathbf{r}_j)\right]^2}{2m_j} + \frac{1}{2}\sum_{j,k\neq j} V_{j,k}\left(\mathbf{r}_j(t) - \mathbf{r}_k(t_{j,k})\right) \\ &+ \frac{\varepsilon_0}{2}\int d^3r \left[\dot{\mathbf{A}}^2(\mathbf{r}) - c^2\mathbf{A}(\mathbf{r})\cdot\nabla^2\mathbf{A}(\mathbf{r})\right] \\ &+ \frac{\varepsilon_0}{2}\int d^3r \frac{1}{c^2}\left[\phi(\mathbf{r}, t)\ddot{\phi}(\mathbf{r}, t) - \dot{\phi}(\mathbf{r}, t)\dot{\phi}(\mathbf{r}, t)\right] \\ \text{with} \quad & t_{j,k} \equiv t - \frac{|\mathbf{r}_j(t) - \mathbf{r}_k(t)|}{c}. \end{aligned} \tag{8.22}$$

This Hamiltonian describes a more difficult dynamics than its Coulomb-gauge counterpart because of the $t_{j,k}$ retardation in the pairwise interactions and in $\phi(\mathbf{r}, t)$.

If the retardation effects are important, $\phi(\mathbf{r}, t)$ cannot be approximated as instantaneous. Fundamentally, this produces several complications for solid-state investigations where rapidly moving electrons are propagating through a slowly varying potential landscape created by the nearly rigid ions. If the retardation in the electron–ion interaction cannot be eliminated, we cannot approximate ions as an effective potential or as background refractive index, as was proposed in Section 7.4, because ionic contributions to Eq. (8.20) would be subject to retardation effects. In this situation, also the ion motion must be computed at the same level as the electronic motion. Thus, the exact Lorentz-gauge formulation prevents us from identifying several central concepts in solid-state physics such as the band structure because the ion-lattice induced potential follows from a nonintegrable dynamics within Eq. (8.20).

To get a simple estimate under which conditions retardation effects are critical in solids, we consider typical interparticle distances and related retardation times. In solids, the electronic particle density n_{3D} ranges roughly from dilute $10^9\,\text{cm}^{-3}$ up to dense $3 \times 10^{22}\,\text{cm}^{-3}$ that also defines the typical density of ions within the lattice. For quasi one- and two-dimensional systems, the density scales with $n_{2D} = [n_{3D}]^{\frac{2}{3}}$ and $n_{1D} = [n_{3D}]^{\frac{1}{3}}$, respectively. If we know the particle density, we can directly deduce the typical interparticle

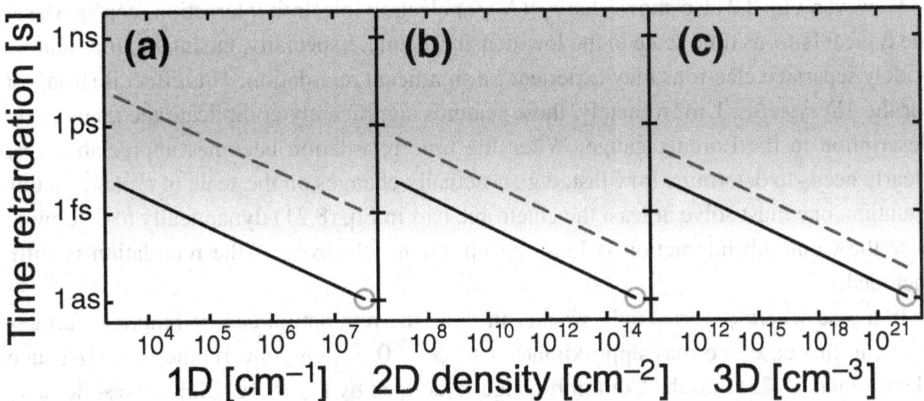

Figure 8.2 Retardation time in the Coulomb interaction as a function of particle density. The retardation time between nearest neighbors (solid line) is compared with $t_d^{\mathrm{ret}}(N)$ (dashed line) for the $n = 1000$-th closest neighbor. The fuzzy area identifies the typical time scales for quantum-optical solid-state investigations. The retardation time is presented for **(a)** one-, **(b)** two-, and **(c)** three-dimensional systems. The curves are computed using Eq. (8.23).

distance to be $|\mathbf{r}_d| \equiv [n_{d\mathrm{D}}]^{-\frac{1}{d}}$, where d denotes the effective dimensionality. The distance to the N-th closest particle follows from $|\mathbf{r}_d(N)| \equiv N^{\frac{1}{d}}[n_{d\mathrm{D}}]^{-\frac{1}{d}}$. With these length-scale estimates, the typical retardation time between a particle and its N-th closest partner is

$$t_d^{\mathrm{ret}}(N) \equiv \frac{|\mathbf{r}_d(N)|}{c} = \frac{1}{c}\left[\frac{N}{n_{d\mathrm{D}}}\right]^{\frac{1}{d}}, \quad d = 1, 2, 3. \tag{8.23}$$

Figure 8.2 presents the computed retardation time for (a) one-, (b) two-, and (c) three-dimensional systems. The solid line indicates $t_d^{\mathrm{ret}}(1)$ for the closest pair as a function of the particle density n_d. The dashed line defines $t_d^{\mathrm{ret}}(N)$ for the $N = 1000$th partner. For a charge-neutral system, the maximum electron density matches the ion density. The corresponding retardation is indicated by the circle, which also defines the typical ion–electron separation. Since the electron–ion separation is roughly dimension independent, i.e., $|\mathbf{r}_d| = 3$Å using the density ranges mentioned above, we find a rough estimate for $t^{\mathrm{red}}(1)$ describing the retardation of the interactions between an electron and the closest ion.

Many relevant solid-state processes involve time scales ranging from 1 fs up to 1 ns. This time regime is indicated by the fuzzy area in Fig. 8.2. We notice that the typical many-body interaction times are three to nine orders of magnitude larger than the electron–ion retardation such that the electron–ion interactions can very accurately be approximated as instantaneous. Hence, even in Lorentz gauge, we may ignore the retardation among electrons and ions in the Hamiltonian (8.22) such that we are still allowed to introduce simplifications such as the electronic band structure or the effective dielectric profile resulting from rigid ions.

However, Fig. 8.2 also shows that $t_d^{\mathrm{ret}}(N)$ for electron–electron interactions can approach the typical fs-to-ns time scale in the low-density regime. Especially, the interactions among widely separated electrons may experience a significant retardation. This effect is strongest for the 1D systems. Unfortunately, these features significantly complicate the many-body description in the Lorentz gauge. When the time retardation becomes appreciable, one clearly needs to determine how fast, e.g., ϕ actually changes on the scale of $t_d^{\mathrm{ret}}(N)$. In this situation, one must solve at least the electronic part in Eq. (8.21) dynamically to determine how the Coulomb interaction is building up among electrons if the retardation is fully included.

In a case where ϕ varies only weakly in time, the retardation can be ignored. Furthermore, in this case, we may approximate $\dot{\phi} \approx \ddot{\phi} \approx 0$, which converts the Lorentz-gauge Hamiltonian (8.22) into the Coulomb-gauge H defined by Eq. (8.19). Since the solid-state description can be formulated exactly and much more simply in the Coulomb gauge, it is the obvious choice for our semiconductor quantum-optics investigations and we will use it from now on.

8.3 Transversal and longitudinal fields

Based on Eq. (8.8), the Coulomb gauge is implemented by demanding a divergence-free vector potential, $\nabla \cdot \mathbf{A}(\mathbf{r}) = 0$. To find out how this can be realized in practice, we start from the elementary decomposition of the vectorial field,

$$\mathbf{F}(\mathbf{r}) \equiv \mathbf{F}_L(\mathbf{r}) + \mathbf{F}_T(\mathbf{r}), \tag{8.24}$$

into its longitudinal $\mathbf{F}_L$ and transversal $\mathbf{F}_T$ components. These components must satisfy the relations

$$\nabla \times \mathbf{F}_L(\mathbf{r}) = \mathbf{0}, \qquad \nabla \cdot \mathbf{F}_T(\mathbf{r}) = 0. \tag{8.25}$$

Consequently, the $\nabla \cdot \mathbf{A}(\mathbf{r}) = 0$ condition of the Coulomb gauge demands that the vector potential is a transversal field. As we construct the electric field, $\mathbf{E}(\mathbf{r}) = -\dot{\mathbf{A}}(\mathbf{r}) - \nabla\phi(\mathbf{r})$, we realize that its transversal component is $\mathbf{E}_T(\mathbf{r}) = -\dot{\mathbf{A}}(\mathbf{r})$ while its longitudinal part is $\mathbf{E}_L(\mathbf{r}) = -\nabla\phi(\mathbf{r})$.

The physical meaning of transversal and longitudinal components becomes transparent when we use the Fourier transformation

$$\mathbf{F}(\mathbf{q}) \equiv \int d^3r\, \mathbf{F}(\mathbf{r})\, e^{-i\mathbf{q}\cdot(\mathbf{r})}, \qquad \mathbf{F}(\mathbf{r}) = \frac{1}{(2\pi)^3} \int d^3q\, \mathbf{F}(\mathbf{q})\, e^{i\mathbf{q}\cdot\mathbf{r}}. \tag{8.26}$$

With this, we can easily evaluate

$$\begin{aligned}\nabla \odot \mathbf{F}(\mathbf{r}) &= \frac{1}{(2\pi)^3} \int d^3q\, \nabla \odot e^{i\mathbf{q}\cdot\mathbf{r}}\, \mathbf{F}(\mathbf{q}) \\ &= \frac{1}{(2\pi)^3} \int d^3q\, i\mathbf{q} \odot \mathbf{F}(\mathbf{q})\, e^{i\mathbf{q}\cdot\mathbf{r}},\end{aligned} \tag{8.27}$$

where $\odot$ symbolizes either the cross or the dot product between two vectors. From Eq. (8.27) we see that $\nabla \cdot \mathbf{F}_T(\mathbf{r}) = 0$ implies $\mathbf{q} \cdot \mathbf{F}_T(\mathbf{q}) = 0$ and $\nabla \times \mathbf{F}_L(\mathbf{r}) = 0$ is true if $\mathbf{q} \times \mathbf{F}_L(\mathbf{q})$ vanishes. In other words, the longitudinal part of $\mathbf{F}(\mathbf{q})$ is parallel to $\mathbf{q}$ while the transversal part of $\mathbf{F}(\mathbf{q})$ is perpendicular to $\mathbf{q}$.

The transversal–longitudinal decomposition expresses the vector field $\mathbf{F}(\mathbf{q})$ in terms of three perpendicular vectors in the Fourier space, which gives these components a simple physical interpretation. Based on the vectorial character of the decomposition, we can always express the three components of $\mathbf{F}(\mathbf{q})$ as

$$\mathbf{F}(\mathbf{q}) = F_{T,1}(\mathbf{q})\,\mathbf{e}_{\mathbf{q}}^{[1]} + F_{T,2}(\mathbf{q})\,\mathbf{e}_{\mathbf{q}}^{[2]} + F_{L,3}(\mathbf{q})\,\mathbf{e}_{\mathbf{q}}^{[3]}, \tag{8.28}$$

where $\mathbf{e}_{\mathbf{q}}^{[j]}$ designates three orthogonal vectors. They have the following properties

$$\mathbf{e}_{\mathbf{q}}^{[j]} \cdot \mathbf{e}_{\mathbf{q}}^{[k]} = \delta_{j,k} \quad \text{and} \quad \mathbf{e}_{\mathbf{q}}^{[i]} \times \mathbf{e}_{\mathbf{q}}^{[j]} = \varepsilon_{i,j,k}\,\mathbf{e}_{\mathbf{q}}^{[k]}, \quad \{i,j,k\} = \{1,2,3\}, \tag{8.29}$$

where the Levi–Civita symbol has the properties

$$\varepsilon_{i,j,k} = \begin{cases} +1, & \text{for even permutations of } \varepsilon_{1,2,3} \\ -1, & \text{for odd permutations of } \varepsilon_{1,2,3} \\ 0, & \text{otherwise} \end{cases}. \tag{8.30}$$

We defined $\mathbf{e}_{j,\mathbf{q}}$ such that its two first components are transversal and the last one is longitudinal.

Based on Eqs. (8.25), (8.27), and (8.28), any transversal component must satisfy

$$\begin{aligned} &\nabla \cdot \mathbf{F}_T(\mathbf{r}) = 0 \quad \Leftrightarrow \quad i\mathbf{q} \cdot \mathbf{F}_T(\mathbf{q}) = 0, \\ &\mathbf{F}_T(\mathbf{q}) = \sum\nolimits_{j=1}^{2} F_{T,j}(\mathbf{q})\,\mathbf{e}_{\mathbf{q}}^{[j]}, \end{aligned} \tag{8.31}$$

whereas a longitudinal field produces $\mathbf{q} \cdot F_L(q) \neq 0$. We see that the longitudinal basis vector is aligned with $\mathbf{q}$ whereas the transversal components $\mathbf{e}_{\mathbf{q}}^{[j]}$ can be chosen as any two orthogonal directions perpendicular to $\mathbf{q}$, see Fig 8.3. If we present $\mathbf{q}$ in a Cartesian basis,

$$\begin{aligned} &\mathbf{q} = q_x\,\mathbf{e}_x + q_y\,\mathbf{e}_y + q_z\,\mathbf{e}_z \equiv \mathbf{q}_\parallel + q_z\mathbf{u}_z, \\ &q = \sqrt{q_x^2 + q_y^2 + q_z^2}\,, \qquad q_\parallel = \sqrt{q_x^2 + q_y^2}, \end{aligned} \tag{8.32}$$

we can determine the longitudinal and transversal components uniquely via

$$\begin{aligned} &\mathbf{e}_{\mathbf{q}}^{[3]} \equiv \begin{cases} \frac{\mathbf{q}}{q}, & q \neq 0 \\ \mathbf{e}_z, & q = 0 \end{cases} \\ &\mathbf{e}_{\mathbf{q}}^{[2]} = \begin{cases} \frac{\mathbf{u}_z \times \mathbf{q}}{q_\parallel}, & q_\parallel \neq 0 \\ \mathbf{e}_y, & q_\parallel = 0 \end{cases}, \qquad \mathbf{e}_{\mathbf{q}}^{[1]} = \begin{cases} \frac{\mathbf{q} \times (\mathbf{q} \times \mathbf{u}_z)}{q_\parallel^2}, & q_\parallel \neq 0 \\ \mathbf{e}_x, & q_\parallel = 0 \end{cases}. \end{aligned} \tag{8.33}$$

It is straightforward to show that this choice satisfies (8.29). Figure 8.3 presents the decomposition of $\mathbf{F}(\mathbf{q})$ into transversal and longitudinal components for three different values of $\mathbf{q}$. According to Eq. (8.33), the basis vectors depend on $\mathbf{q}$. The unit vectors in Fig. 8.3 are

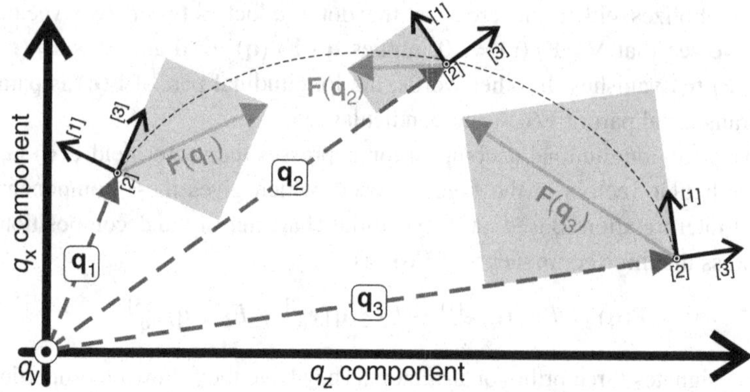

Figure 8.3 Decomposition of a generic vector field into transversal and longitudinal components in Fourier space. The three arrows present a generic vector field $\mathbf{F}(\mathbf{q}_j)$ evaluated at three different $\mathbf{q}_j$ values (dashed arrows). The corresponding directions of transversal ([1] and [2]) and longitudinal ([3]) basis vectors are shown in connection with each $\mathbf{q}_j$, as defined by Eq. (8.33). The component [2] points out of the paper. The shaded rectangles indicate the decomposition of $\mathbf{F}(\mathbf{q}_j)$ into transversal (side along [1]) and longitudinal (side along [3]) components.

symbolized by $[j] \equiv \mathbf{e}_{j,\mathbf{q}}$ such that [3] defines the longitudinal component while [1] and [2] (directed out of the page) determine the linearly independent transversal components.

The explicit knowledge of the basis (8.33) allows us to construct an algorithm to identify the longitudinal component

$$\mathbf{F}_L(\mathbf{q}) = \left(\mathbf{e}_{\mathbf{q}}^{[3]} \cdot \mathbf{F}(\mathbf{q})\right) \mathbf{e}_{\mathbf{q}}^{[3]} = \frac{\mathbf{q} \cdot \mathbf{F}(\mathbf{q})}{q^2}\, \mathbf{q} \equiv \frac{\mathbf{q}}{q^2}\, \mathbf{q} \cdot \mathbf{F}(\mathbf{q})$$
$$\mathbf{F}_L(\mathbf{r}) = \frac{1}{(2\pi)^3} \int d^3q\, \frac{\mathbf{q}}{q^2}\, \mathbf{q} \cdot \mathbf{F}(\mathbf{q})\, e^{i\mathbf{q}\cdot\mathbf{r}} \tag{8.34}$$

based on simple vector algebra. In the same way, we can extract the transversal part of any vector field by applying the transformation

$$\mathbf{F}_T(\mathbf{q}) = \mathbf{F}(\mathbf{q}) - \mathbf{F}_L(\mathbf{q}) = \left(1 - \frac{\mathbf{q}}{q^2}\mathbf{q}\cdot\right) \mathbf{F}(\mathbf{q})$$
$$\mathbf{F}_T(\mathbf{r}) = \frac{1}{(2\pi)^3} \int d^3q \left(1 - \frac{\mathbf{q}}{q^2}\mathbf{q}\cdot\right) \mathbf{F}(\mathbf{q})\, e^{i\mathbf{q}\cdot(\mathbf{r})}, \tag{8.35}$$

which again follows from simple vector algebra. It is straightforward to see that the decomposition $\mathbf{F}(\mathbf{r}) = \mathbf{F}_T(\mathbf{r}) + \mathbf{F}_L(\mathbf{r})$ is unique.

In real space, the general orthogonality $\mathbf{F}_T(\mathbf{q}) \cdot \mathbf{F}_L(\mathbf{q}) = -\mathbf{F}_T(\mathbf{q}) \cdot \mathbf{F}_L(-\mathbf{q}) = 0$ produces a vanishing integral equation

$$\int d^3r\, \mathbf{F}_T(\mathbf{r}) \cdot \mathbf{F}_L(\mathbf{r}) = \frac{1}{(2\pi)^3} \int d^3q\, \mathbf{F}_T(\mathbf{q}) \cdot \mathbf{F}_L(-\mathbf{q}) = 0 \tag{8.36}$$

due to the convolution theorem, see Exercise 8.7. This relation does not necessarily imply that $\mathbf{F}_T(\mathbf{r}) \cdot \mathbf{F}_L(\mathbf{r})$ vanishes.

8.3.1 Poisson equation

To decompose the electric field into its transversal and longitudinal parts, we Fourier transform Eq. (7.12) to produce

$$\mathbf{E}(\mathbf{q},t) = -\dot{\mathbf{A}}(\mathbf{q},t) - i\mathbf{q}\,\phi(\mathbf{q},t). \tag{8.37}$$

Clearly, the ϕ-dependent component is aligned with $\mathbf{q}$ identifying it as purely longitudinal. At the same time, $\nabla \cdot \mathbf{A}(\mathbf{r})$ vanishes in the Coulomb gauge such that the $\mathbf{A}$-dependent component is transversal. In other words, we find the automatic separation

$$\mathbf{E}_T(\mathbf{r}) = -\dot{\mathbf{A}}(\mathbf{r}), \qquad \mathbf{E}_L(\mathbf{r}) = -\nabla\phi(\mathbf{r}). \tag{8.38}$$

Similarly, the magnetic field follows from

$$\mathbf{B}(\mathbf{r},t) = \nabla \times \mathbf{A}(\mathbf{r},t) \quad \Leftrightarrow \quad \mathbf{B}(\mathbf{q},t) = i\mathbf{q} \times \mathbf{A}(\mathbf{q},t) \tag{8.39}$$

based on (7.12). Since $\mathbf{B}(\mathbf{q},t)$ is perpendicular to $\mathbf{q}$, it is purely transversal independent of the gauge.

The longitudinal electric field can be computed starting from the Fourier transformed Poisson equation (8.12),

$$-q^2\phi(\mathbf{q}) = -\frac{1}{\varepsilon_0}\sum\nolimits_j Q_j\, e^{-i\mathbf{q}\cdot\mathbf{r}_j} \quad \Leftrightarrow \quad \phi(\mathbf{q}) = \sum\nolimits_j \frac{Q_j}{\varepsilon_0 q^2}\, e^{-i\mathbf{q}\cdot\mathbf{r}_j}. \tag{8.40}$$

Fourier transformation into real space produces

$$\begin{aligned}\phi(\mathbf{r}) &= \frac{1}{(2\pi)^3}\sum_j \int d^3q\, \frac{Q_j}{\varepsilon_0 q^2}\, e^{i\mathbf{q}\cdot(\mathbf{r}-\mathbf{r}_j)} \\ &= \frac{1}{(2\pi)^2}\sum_j \int_0^{2\pi} d\varphi \int_0^{\infty} dq\, q^2 \frac{Q_j}{\varepsilon_0 q^2} \int_0^{\pi} d\theta\, \sin\theta\, e^{iq|\mathbf{r}-\mathbf{r}_j|\cos\theta},\end{aligned} \tag{8.41}$$

where we have chosen the integration axes such that the q_z direction is aligned with $(\mathbf{r}-\mathbf{r}_j)$. All the integrals can be performed analytically yielding

$$\phi(\mathbf{r}) = \sum\nolimits_j \frac{Q_j}{4\pi\varepsilon_0|\mathbf{r}-\mathbf{r}_j|}, \tag{8.42}$$

see Exercise 8.8. This result has been used in Eq. (8.13) leading to the Coulomb potential (8.15). For later reference, we write the Coulomb potential and its Fourier transformation as

$$V_{i,j}(\mathbf{r}_i - \mathbf{r}_j) = \frac{Q_i Q_j}{4\pi\varepsilon_0|\mathbf{r}_i - \mathbf{r}_j|} \quad \Leftrightarrow \quad V_{i,j}(\mathbf{q}) = \frac{Q_i Q_j}{\varepsilon_0|\mathbf{q}|^2}\, e^{-i\mathbf{q}\cdot\mathbf{r}_j}. \tag{8.43}$$

8.3.2 Wave equation

The Coulomb gauge also simplifies the dynamics of the transversal fields. By substituting Eqs. (8.38)–(8.39) into Maxwell's equations (7.60), we find

$$\nabla \times (\nabla \times \mathbf{A}(\mathbf{r})) = -\frac{1}{c^2}\frac{\partial^2}{\partial t^2}\mathbf{A}(\mathbf{r}, t) \\ -\frac{1}{c^2}\frac{\partial}{\partial t}\nabla\phi(\mathbf{r}) + \mu_0 \sum\nolimits_j Q_j \mathbf{v}_j \delta(\mathbf{r} - \mathbf{r}_j). \tag{8.44}$$

The LHS of Eq. (8.44) can be simplified using

$$\nabla \times (\nabla \times \mathbf{A}(\mathbf{r})) = \nabla(\nabla \cdot \mathbf{A}(\mathbf{r})) - \nabla^2 \mathbf{A}(\mathbf{r}). \tag{8.45}$$

Since the first term vanishes in Coulomb gauge, Eq. (8.44) reduces to the wave equation

$$\left(\nabla^2 - \frac{1}{c^2}\frac{\partial^2}{\partial t^2}\right)\mathbf{A}(\mathbf{r}, t) = -\mu_0 \left[\sum\nolimits_j Q_j \mathbf{v}_j \delta(\mathbf{r} - \mathbf{r}_j) - \varepsilon_0 \nabla \frac{\partial \phi(\mathbf{r})}{\partial t}\right], \tag{8.46}$$

where we used $c^2 = 1/\varepsilon_0\mu_0$ and organized the source terms to the RHS.

Since $\mathbf{A}$ is transversal, also the vector field

$$\mathbf{J}_T(\mathbf{r}) \equiv \sum\nolimits_j Q_j \mathbf{v}_j \delta(\mathbf{r} - \mathbf{r}_j) - \varepsilon_0 \nabla \frac{\partial \phi(\mathbf{r})}{\partial t} \tag{8.47}$$

must be transversal. The source term containing the scalar potential can be evaluated with the help of Eq. (8.42), yielding

$$\frac{\partial}{\partial t}\phi(\mathbf{r}) = \sum\nolimits_j \frac{\partial}{\partial t}\frac{Q_j}{4\pi\varepsilon_0|\mathbf{r} - \mathbf{r}_j(t)|} = -\sum\nolimits_j Q_j \dot{\mathbf{r}}_j \cdot \nabla \frac{1}{4\pi\varepsilon_0|\mathbf{r} - \mathbf{r}_j|} \\ = -\sum\nolimits_j Q_j \mathbf{v}_j \cdot \nabla \frac{1}{4\pi\varepsilon_0|\mathbf{r} - \mathbf{r}_j|}. \tag{8.48}$$

With this result, we see that

$$\nabla \frac{\partial}{\partial t}\phi(\mathbf{r}) = -\sum\nolimits_j Q_j \nabla(\mathbf{v}_j \cdot \nabla)\frac{1}{4\pi\varepsilon_0|\mathbf{r} - \mathbf{r}_j|} \equiv \mathbf{F}_\phi(\mathbf{r}), \tag{8.49}$$

which has the Fourier transform,

$$F_\phi(\mathbf{q}) \equiv \sum\nolimits_j Q_j\, \mathbf{q}(\mathbf{v}_j \cdot \mathbf{q})\frac{1}{\varepsilon_0 q^2} e^{-i\mathbf{q}\cdot\mathbf{r}_j} = \frac{1}{\varepsilon_0}\frac{\mathbf{q}}{q^2}\mathbf{q} \cdot \sum\nolimits_j Q_j \mathbf{v}_j\, e^{-i\mathbf{q}\cdot\mathbf{r}_j}. \tag{8.50}$$

The Fourier transform of $J_T(\mathbf{r})$ thus becomes

$$J_T(\mathbf{q}) = \sum\nolimits_j Q_j \mathbf{v}_j e^{-i\mathbf{q}\cdot\mathbf{r}_j} - \frac{\mathbf{q}}{q^2}\mathbf{q} \cdot \sum\nolimits_j Q_j \mathbf{v}_j e^{-i\mathbf{q}\cdot\mathbf{r}_j} \\ = \left(1 - \frac{\mathbf{q}}{q^2}\mathbf{q}\cdot\right)\sum\nolimits_j Q_j \mathbf{v}_j\, e^{-i\mathbf{q}\cdot\mathbf{r}_j} = \left(1 - \frac{\mathbf{q}}{q^2}\,\mathbf{q}\cdot\right)\mathbf{J}(\mathbf{q}). \tag{8.51}$$

The result is identical to (8.35) where the charge current $\mathbf{J}$ is converted into its transversal form. Thus, we can write the transversal wave equation (8.46) as

$$\left(\nabla^2 - \frac{1}{c^2}\frac{\partial^2}{\partial t^2}\right)\mathbf{A}(\mathbf{r},t) = -\mu_0 \mathbf{J}_T(\mathbf{r},t), \tag{8.52}$$

where the source is the purely transversal current.

8.4 Mode expansion of the electromagnetic field

To prepare the equations for the canonical quantization in Chapter 10, we start from the homogeneous part of Eq. (8.52)

$$\left(\nabla^2 - \frac{1}{c^2}\frac{\partial^2}{\partial t^2}\right)\mathbf{A}_0(\mathbf{r},t) = 0 \quad \text{with} \quad \nabla\cdot\mathbf{A}_0(\mathbf{r},t) = 0 \tag{8.53}$$

and implement the temporal Fourier transformation,

$$\begin{aligned} \mathbf{A}_0(\mathbf{r},\omega) &= \int_{-\infty}^{\infty} dt\, \mathbf{A}_0(\mathbf{r},t)\, e^{+i\omega t}, \\ \mathbf{A}_0(\mathbf{r},t) &= \frac{1}{2\pi}\int_{-\infty}^{\infty} d\omega\, \mathbf{A}_0(\mathbf{r},\omega)\, e^{-i\omega t} \end{aligned} \tag{8.54}$$

leading to

$$\left(\nabla^2 + \left(\frac{\omega}{c}\right)^2\right)\mathbf{A}_0(\mathbf{r},\omega) = 0, \qquad \nabla\cdot\mathbf{A}_0(\mathbf{r},\omega) = 0. \tag{8.55}$$

This equation can be brought into the form of the Helmholtz equation (2.30) after we identify the wave vector $\mathbf{q}$, the dispersion relation, and the mode function,

$$\omega_{\mathbf{q}} \equiv c|\mathbf{q}|, \quad \mathbf{u}_{\mathbf{q}}(\mathbf{r}) \equiv A_0(, \omega_{\mathbf{q}}), \tag{8.56}$$

respectively. With these identifications, Eq. (8.55) becomes

$$\left(\nabla^2 + \mathbf{q}^2\right)\mathbf{u}_{\mathbf{q}}(\mathbf{r}) = 0, \quad \nabla\cdot\mathbf{u}_{\mathbf{q}}(\mathbf{r}) = 0. \tag{8.57}$$

This eigenvalue problem can be treated on the basis of the formal eigenvalue-problem analysis in Section 2.3.1. Thus, we already know that Eq. (8.57) produces a complete set of transversal modes which can be used as the basis for a mode expansion of the vector potential.

8.4.1 Modes with periodic boundary conditions

Before we actually express $\mathbf{A}(\mathbf{r},t)$ in terms of the mode functions, we have to extend the treatment of Sections 2.3.1–2.3.2 to cover also transversal vectorial fields. For this purpose, we generalize the plane-wave solution (2.59) directly for transversal modes,

$$\mathbf{u_q(r)} = \mathcal{N}_\mathbf{q} \begin{cases} e^{i\mathbf{q}\cdot\mathbf{r}}\,\mathbf{e}_\mathbf{q}^{[j]}, & \text{for} \quad \mathbf{u_q} \in \mathbb{C} \\ \mathbf{e}_\mathbf{q}^{[j]} \cos(\mathbf{q}\cdot\mathbf{r}) \;\&\; \mathbf{e}_\mathbf{q}^{[j]} \sin(\mathbf{q}\cdot\mathbf{r}), & \text{for} \quad \mathbf{u_q} \in \mathbb{R} \end{cases}, \tag{8.58}$$

where $\mathcal{N}_\mathbf{q}$ defines the norm for modes pointing in the direction of the transversal unit vectors $\mathbf{e}_\mathbf{q}^{[j]}$, $j = 1, 2$. The basis states can be chosen to be either real- or complex-valued vectorial fields. In the following, we assume that the $\mathbf{q}$ index implicitly includes the reference to the transversal directions $\mathbf{e}_\mathbf{q}^{[1]}$ or $\mathbf{e}_\mathbf{q}^{[2]}$ as well as to the $\cos(\mathbf{q}\cdot\mathbf{r})$ and $\sin(\mathbf{q}\cdot\mathbf{r})$ branches if we choose a real-valued mode basis.

In the next step, we determine the norm in a situation where the light is confined within a periodic box whose sides have the length $\mathcal{L}$. This choice is very natural in the analysis of lattice-periodic solid-state systems, as will be discussed later in Section 12.1. The periodicity implies the condition

$$\mathbf{u_q}(\mathbf{r} + j\,\mathcal{L}\,\mathbf{e}_\lambda) = \mathbf{u_q(r)}, \qquad j = 0, \pm1, \pm2, \cdots, \qquad \lambda = x,\ y,\ z. \tag{8.59}$$

The solution (8.58) can satisfy (8.59) only for discrete values of the wave vector,

$$\mathbf{q} = \sum_{\lambda=x,y,z} \Delta q\, l_\lambda \mathbf{e}_\lambda, \quad \Delta q = \frac{2\pi}{\mathcal{L}}, \quad l_\lambda = 0, \pm1, \pm2, \cdots, \tag{8.60}$$

where Δq defines the discretization interval.

As the quantization box is made larger, i.e., $\lim_{\mathcal{L}\to\infty} \Delta q = 0$, the wave vector eventually becomes continuous. In other words, we can convert the results obtained for discrete momenta (8.60) into results for continuous ones by applying the limit $\mathcal{L} \to \infty$. Thus, the chosen approach is applicable for both periodic and infinite systems. In particular, the physically relevant quantities often follow from sums of the type

$$\lim_{\mathcal{L}\to\infty} \frac{1}{\mathcal{L}^3} \sum_\mathbf{q} \mathbf{F_q} = \lim_{\mathcal{L}\to\infty} \frac{1}{(2\pi)^3} \sum_\mathbf{q} \Delta q^3 \mathbf{F_q} = \frac{1}{(2\pi)^3} \lim_{\Delta q\to 0} \sum_\mathbf{q} \Delta q^3 \mathbf{F_q}$$
$$= \frac{1}{(2\pi)^3} \int d^3q\, \mathbf{F_q}, \tag{8.61}$$

where $\mathbf{F_q}$ is a general function. For large enough $\mathcal{L}$, it is beneficial to identify Δq using the relation (8.60). Then, the large-$\mathcal{L}$ limit produces an integral after we identify a Riemann sum for the infinitesimal wave-vector spacing Δq, which converts the expression into an integral.

Since the physical properties are repeated with the period $\mathcal{L}$, it is meaningful to normalize the mode functions with respect to the box volume $\mathcal{L}^3$. Thus, the orthogonality relation (2.60) is slightly modified into

$$\int_{\mathcal{L}^3} d^3r\, \mathbf{u}_\mathbf{q}^\star(\mathbf{r}) \cdot \mathbf{u}_{\mathbf{q}'}(\mathbf{r}) = \delta_{\mathbf{q},\mathbf{q}'} \tag{8.62}$$

and we define the norm as

$$\mathcal{N}_{\mathbf{q}} = \begin{cases} \frac{1}{\sqrt{\mathcal{L}^3}} = \sqrt{\frac{\Delta q^3}{(2\pi)^3}}, & \text{for } \mathbf{u_q} \in \mathbb{C} \\ \sqrt{\frac{2}{\mathcal{L}^3}} = \sqrt{\frac{\Delta q^3}{\pi^3}}, & \text{for } \mathbf{u_q} \in \mathbb{R} \end{cases}. \tag{8.63}$$

In comparison with the continuous basis (2.59), we have chosen here to include $\Delta q^{\frac{3}{2}}$ in the basis states. If we take the large $\mathcal{L}$ limit of Eq. (8.62), we end up with

$$\lim_{\mathcal{L}\to\infty} \frac{1}{\Delta q^3} \int_{\mathcal{L}^3} d^3r\, \mathbf{u}^\star_{\mathbf{q}}(\mathbf{r}) \cdot \mathbf{u}_{\mathbf{q}'}(\mathbf{r}) = \lim_{\Delta q\to 0} \frac{\delta_{\mathbf{q},\mathbf{q}'}}{\Delta q^3} = \delta(\mathbf{q} - \mathbf{q}'). \tag{8.64}$$

Here, the last equality holds because this is one possible definition of the delta function. We also see that if we identify the continuous mode function as

$$\mathbf{u}_{\mathbf{q},\mathrm{cont}}(\mathbf{r}) \equiv \frac{\mathbf{u_q}(\mathbf{r})}{\sqrt{\Delta q^3}}, \tag{8.65}$$

the orthogonality relation (8.64) is identical to that introduced in Eq. (2.60). Furthermore, the norm of the complex-valued $\mathbf{u}_{\mathbf{q},\mathrm{cont}}$ is given by $(2\pi)^{-\frac{3}{2}}$ just as in Eq. (2.59).

In full analogy to Eq. (2.61), the transversal modes satisfy the completeness relation

$$\mathbf{e}_\lambda \cdot \sum_{\mathbf{q}} \mathbf{u}^\star_{\mathbf{q}}(\mathbf{r})\, \mathbf{u_q}(\mathbf{r}') \cdot \mathbf{e}_{\lambda'} = \delta^T_{\lambda,\lambda'}(\mathbf{r} - \mathbf{r}'), \qquad \lambda, \lambda' = x, y, z. \tag{8.66}$$

The $\sum_{\mathbf{q}} \mathbf{u}^\star_{\mathbf{q}}(\mathbf{r})\, \mathbf{u_q}(\mathbf{r}')$ that appears is a dyadic tensor whose projection into Cartesian basis vectors is diagonal and the δ^T-function operates within the functional space of transversal vector fields. In other words, we find that

$$\int d^3r'\, \mathbf{F}(\mathbf{r}') \cdot \sum_{\mathbf{q}} \mathbf{u}^\star_{\mathbf{q}}(\mathbf{r})\, \mathbf{u_q}(\mathbf{r}') = \mathbf{F}_T(\mathbf{r}) \tag{8.67}$$

projects the transversal part of a generic vector field $\mathbf{F}(\mathbf{r})$ in the same way as Eq. (8.35). If $\mathbf{F}(\mathbf{r})$ is transversal to begin with, $\sum_{\mathbf{q}} \mathbf{u}^\star_{\mathbf{q}}(\mathbf{r})\, \mathbf{u_q}(\mathbf{r}')$ acts like an ordinary delta function. For continuous modes, the $\mathbf{q}$ sum can be converted into an integral using the relation (8.61). This way, we can write

$$\mathbf{e}_\lambda \cdot \frac{1}{(2\pi)^3} \int d^3q\, \mathbf{u}^\star_{\mathbf{q},\mathrm{cont}}(\mathbf{r})\, \mathbf{u}_{\mathbf{q},\mathrm{cont}}(\mathbf{r}') \cdot \mathbf{e}_{\lambda'} = \delta^T_{\lambda,\lambda'}(\mathbf{r} - \mathbf{r}'), \tag{8.68}$$

which generalizes (2.61) for transversal fields.

8.4.2 Real-valued mode expansion

Without loss of generality, we can choose real-valued basis states to express any transversal vector field in terms of the mode expansion

$$\mathbf{A}(\mathbf{r},t) = \sum_{\mathbf{q}} \mathcal{A}_{\mathbf{q}}\, C_{\mathbf{q}}(t)\, \mathbf{u}_{\mathbf{q}}(\mathbf{r}), \quad \mathbf{u}_{\mathbf{q}} \in \mathbb{R}$$

$$\mathcal{A}_{\mathbf{q}} C_{\mathbf{q}}(t) = \int_{\mathcal{L}^3} d^3r\, \mathbf{u}_{\mathbf{q}}(\mathbf{r}) \cdot \mathbf{A}(\mathbf{r},t), \quad C_{\mathbf{q}}, \mathcal{A}_{\mathbf{q}} \in \mathbb{R}, \tag{8.69}$$

in analogy to Eqs. (2.62)–(2.63). The expansion coefficient $C_{\mathbf{q}}(t)$ defines the temporal dynamics of the vector potential and the time-independent normalization $\mathcal{A}_q$ will be determined in Section 8.4.3.

Inserting the mode expansion into Eq. (8.52) yields

$$\sum_{\mathbf{q}} \mathcal{A}_{\mathbf{q}} \left[C_{\mathbf{q}} \nabla^2 \mathbf{u}_{\mathbf{q}}(\mathbf{r}) - \frac{1}{c^2} \ddot{C}_{\mathbf{q}}\, \mathbf{u}_{\mathbf{q}}(\mathbf{r}) \right] = -\mu_0 \mathbf{J}_T(\mathbf{r},t). \tag{8.70}$$

Using the Helmholtz equation (8.57) to express the spatial derivatives, we obtain

$$-\sum_{\mathbf{q}} \mathcal{A}_{\mathbf{q}} \left[\mathbf{q}^2 C_{\mathbf{q}} + \frac{1}{c^2} \ddot{C}_{\mathbf{q}} \right] \mathbf{u}_{\mathbf{q}}(\mathbf{r}) = -\mu_0 \mathbf{J}_T(\mathbf{r},t). \tag{8.71}$$

We now extract the modal component $\mathbf{q}$ by multiplying both sides by $c^2 \mathbf{u}^\star_{\mathbf{q}}(\mathbf{r})$. Integration over the volume $\mathcal{L}^3$ leads to

$$c^2 \mathbf{q}^2 C_{\mathbf{q}} + \ddot{C}_{\mathbf{q}} = \frac{c^2 \mu_0}{\mathcal{A}_{\mathbf{q}}} J_{T,\mathbf{q}}(t), \qquad J_{T,\mathbf{q}}(t) \equiv \int d^3r\, \mathbf{u}^\star_{\mathbf{q}}(\mathbf{r}) \cdot \mathbf{J}_T(\mathbf{r},t). \tag{8.72}$$

We may simplify this further using $c^2\mu_0 = 1/\varepsilon_0$ and the dispersion relation (8.56), resulting in

$$\ddot{C}_{\mathbf{q}} = -\omega_{\mathbf{q}}^2 C_{\mathbf{q}} + \frac{J_{T,\mathbf{q}}(t)}{\varepsilon_0 \mathcal{A}_{\mathbf{q}}}, \tag{8.73}$$

which defines how each modal component of the electromagnetic field evolves.

8.4.3 Particle aspects

Equation (8.73) clearly resembles the classical dynamics of a harmonic oscillator that is driven by a source. In other words, we have found a direct connection to the principal idea presented in Fig. 7.1, identifying the particle aspects within electromagnetic waves by virtue of the harmonic-oscillator model. To make this identification mathematically more concrete, we define a canonical momentum corresponding to $C_{\mathbf{q}}$

$$\begin{cases} \Pi_{\mathbf{q}} \equiv \dot{C}_{\mathbf{q}} \\ \dot{\Pi}_{\mathbf{q}} = -\omega_{\mathbf{q}}^2 C_{\mathbf{q}} + \frac{J_{T,\mathbf{q}}(t)}{\varepsilon_0 \mathcal{A}_{\mathbf{q}}} \end{cases}. \tag{8.74}$$

Combining Eqs. (8.38) and (8.69), we find

$$\mathbf{E}_T(\mathbf{r}) = -\dot{\mathbf{A}}(\mathbf{r}) = -\sum_{\mathbf{q}} \mathcal{A}_{\mathbf{q}}\, \Pi_{\mathbf{q}}(t)\, \mathbf{u}_{\mathbf{q}}(\mathbf{r}). \tag{8.75}$$

This expression connects the canonical momentum and the transversal electric field via a mode expansion, which will considerably simplify the description of the particle aspects. We can now evaluate $J_{T,\mathbf{q}}(t)$ using (8.47)

$$\begin{aligned} J_{T,\mathbf{q}}(t) &= \int d^3r\, \mathbf{u}^\star_{\mathbf{q}}(\mathbf{r}) \cdot \left[\sum_j Q_j \mathbf{v}_j \delta(\mathbf{r}-\mathbf{r}_j) - \varepsilon_0 \nabla\dot{\phi}(\mathbf{r})\right] \\ &= \int d^3r\, \mathbf{u}^\star_{\mathbf{q}}(\mathbf{r}) \cdot \sum_j Q_j \mathbf{v}_j\, \delta(\mathbf{r}-\mathbf{r}_j) \\ &= \sum_j Q_j\, \mathbf{u}^\star_{\mathbf{q}}(\mathbf{r}_j) \cdot \mathbf{v}_j \end{aligned} \tag{8.76}$$

because the $\nabla\dot{\phi}$ contribution vanishes under the integral on the basis of (8.36).

Next, we use the mode expansions (8.69) and (8.75) to express the system Hamiltonian (8.18),

$$H = \sum_j \frac{\left[\mathbf{p}_j - Q_j \sum_{\mathbf{q}} \mathcal{A}_{\mathbf{q}} C_{\mathbf{q}}\, \mathbf{u}_{\mathbf{q}}(\mathbf{r}_j)\right]^2}{2m_j} + \frac{1}{2}\sum_{j,k\neq j} V_{j,k}(\mathbf{r}_j - \mathbf{r}_k) + H_0. \tag{8.77}$$

The purely field-dependent contribution H_0 is given by

$$\begin{aligned} H_0 = \frac{\varepsilon_0}{2}\int d^3r \Big[&\sum_{\mathbf{q},\mathbf{q}'} \mathcal{A}_{\mathbf{q}} \mathcal{A}_{\mathbf{q}'}\, \Pi_{\mathbf{q}}\, \Pi_{\mathbf{q}'}\, \mathbf{u}_{\mathbf{q}}(\mathbf{r}) \cdot \mathbf{u}_{\mathbf{q}'}(\mathbf{r}) \\ &-\sum_{\mathbf{q},\mathbf{q}'} \mathcal{A}_{\mathbf{q}} \mathcal{A}_{\mathbf{q}'}\, C_{\mathbf{q}}\, C_{\mathbf{q}'}\, \mathbf{u}_{\mathbf{q}}(\mathbf{r}) \cdot c^2\nabla^2 \mathbf{u}_{\mathbf{q}'}(\mathbf{r})\Big]. \end{aligned} \tag{8.78}$$

Combining Eqs. (8.56) and (8.57) into $c^2\nabla^2\mathbf{u}_{\mathbf{q}'}(\mathbf{r}) = -\omega^2_{\mathbf{q}'}\mathbf{u}_{\mathbf{q}'}(\mathbf{r})$ and using this in H_0 yields

$$H_0 = \frac{\varepsilon_0}{2}\sum_{\mathbf{q},\mathbf{q}'} \mathcal{A}_{\mathbf{q}} \mathcal{A}_{\mathbf{q}'} \left(\Pi_{\mathbf{q}}\, \Pi_{\mathbf{q}'} + \omega^2_{\mathbf{q}'}\, C_{\mathbf{q}} C_{\mathbf{q}'}\right) \int d^3r\, \mathbf{u}_{\mathbf{q}}(\mathbf{r}) \cdot \mathbf{u}_{\mathbf{q}'}(\mathbf{r}). \tag{8.79}$$

Since we are dealing with real-valued $\mathbf{u}_{\mathbf{q}}$, the integral produces a Kronecker delta $\delta_{\mathbf{q},\mathbf{q}'}$ such that

$$H_0 = \sum_{\mathbf{q}} \frac{\varepsilon_0}{2} \mathcal{A}^2_{\mathbf{q}} \left[\Pi^2_{\mathbf{q}} + \omega^2_{\mathbf{q}} C^2_{\mathbf{q}}\right]. \tag{8.80}$$

This form resembles the Hamiltonian for a harmonic oscillator if we set $\mathcal{A}^2_{\mathbf{q}}\varepsilon_0 = 1$. Thus, we identify

$$\mathcal{A}_{\mathbf{q}} = \frac{1}{\sqrt{\varepsilon_0}} \tag{8.81}$$

as the mode-independent normalization constant.

To check whether the results (8.77) and (8.80) identify particle-like aspects, we test if the usual Hamilton equations

$$\dot{C}_{\mathbf{q}} = \frac{\partial H}{\partial \Pi_{\mathbf{q}}} \quad \text{and} \quad \dot{\Pi}_{\mathbf{q}} = -\frac{\partial H}{\partial C_{\mathbf{q}}} \tag{8.82}$$

produce the correct equations of motion. Since only H_0 depends on $\Pi_{\mathbf{q}}$, we directly obtain

$$\frac{\partial H}{\partial \Pi_{\mathbf{q}}} = \frac{\partial H_0}{\partial \Pi_{\mathbf{q}}} = \Pi_{\mathbf{q}} \quad \Rightarrow \quad \dot{C}_{\mathbf{q}} = \Pi_{\mathbf{q}}, \tag{8.83}$$

which agrees with Eq. (8.74). Furthermore,

$$\begin{aligned}\frac{\partial H}{\partial C_{\mathbf{q}}} &= \sum_j \frac{\left[\mathbf{p}_j - Q_j \mathbf{A}(\mathbf{r}_j)\right]}{m_j} \cdot (-Q_j) \frac{\partial \mathbf{A}(\mathbf{r}_j)}{\partial C_{\mathbf{q}}} + \omega_{\mathbf{q}}^2 C_{\mathbf{q}} \\ &= -\sum_j Q_j \mathbf{v}_j \cdot \mathcal{A}_{\mathbf{q}}\, \mathbf{u}_{\mathbf{q}}(\mathbf{r}_j) + \omega_{\mathbf{q}}^2\, C_{\mathbf{q}} \\ &= -\mathcal{A}_{\mathbf{q}} J_{T,\mathbf{q}} + \omega_{\mathbf{q}}^2 C_{\mathbf{q}} = -\frac{J_{T,\mathbf{q}}}{\varepsilon_0 \mathcal{A}_{\mathbf{q}}} + \omega_{\mathbf{q}}^2 C_{\mathbf{q}},\end{aligned} \tag{8.84}$$

where we used (8.76), (8.81), and (7.36). This way, we obtain

$$\dot{\Pi}_{\mathbf{q}} = -\frac{\partial H}{\partial C_{\mathbf{q}}} = -\omega_{\mathbf{q}}^2 C_{\mathbf{q}} + \frac{J_{T,\mathbf{q}}}{\varepsilon_0 \mathcal{A}_{\mathbf{q}}} \tag{8.85}$$

in full agreement with (8.73). Consequently, the mode-expansion coefficients $C_{\mathbf{q}}$ and $\Pi_{\mathbf{q}}$ have a one-to-one correspondence to the particle position and canonical momentum, respectively.

In summary, we have found that the system Hamiltonian in Coulomb gauge is given by

$$H = \sum_j \frac{\left[\mathbf{p}_j - Q_j \mathbf{A}(\mathbf{r}_j)\right]^2}{2m_j} + \frac{1}{2} \sum_{j,k \neq j} V_{j,k}(\mathbf{r}_j - \mathbf{r}_k) - \sum_{\mathbf{q}} \frac{1}{2}\left[\Pi_{\mathbf{q}}^2 + \omega_{\mathbf{q}}^2 C_{\mathbf{q}}^2\right], \tag{8.86}$$

where the canonical pair, $(C_{\mathbf{q}}, \Pi_{\mathbf{q}})$, determines the transversal electromagnetic field through the mode expansions

$$\mathbf{A}(\mathbf{r}) = \sum_{\mathbf{q}} \frac{1}{\sqrt{\varepsilon_0}} C_{\mathbf{q}}\, \mathbf{u}_{\mathbf{q}}(\mathbf{r}) \quad \text{and} \quad \mathbf{E}(\mathbf{r}) = -\sum_{\mathbf{q}} \frac{1}{\sqrt{\varepsilon_0}} \Pi_{\mathbf{q}}\, \mathbf{u}_{\mathbf{q}}(\mathbf{r}). \tag{8.87}$$

In these expressions, the mode functions are real-valued transversal solutions of the Helmholtz equation (8.57). This basis choice does not limit the applicability of the approach because Eqs. (8.86)–(8.87) describe arbitrary electromagnetic fields in the Coulomb gauge. Nevertheless, it is sometimes more convenient to consider a complex-valued basis. We investigate this possibility further in Chapter 10, after having quantized the particle properties of light.

In general, the Hamiltonian (8.86) and the mode expansion (8.87) represent classical electromagnetism coupled with particles in a format where canonical positions and momenta are defined analogously for particles and fields. Thus, we have a formulation that provides a possibility to quantize both the transversal light field and the particles following equivalent steps. The introduction of the Coulomb gauge and the use of the mode expansion allow us to identify the particle properties of electromagnetic waves and to determine the explicit form of the Coulomb-induced many-body contributions. Both aspects are vitally important for our goal to develop a computationally feasible approach describing semiconductor quantum optics at a microscopic level.

Exercises

8.1 The new and old vector and scalar potentials are connected via $\tilde{\mathbf{A}} = \mathbf{A} + \nabla F$ and $\tilde{\phi} = \phi - \dot{F}$. Show that **(a)** choice $\nabla^2 F = \nabla \cdot \mathbf{A}(r)$ implies $\nabla \cdot \mathbf{A} = 0$ and **(b)** choice $\left[\nabla^2 - \frac{1}{c^2}\frac{\partial^2}{\partial t^2}\right] F = \left[\nabla \cdot \tilde{A} + \frac{1}{c^2}\dot{\tilde{\phi}}\right]$ yields $\nabla \cdot \mathbf{A} + \frac{1}{c^2}\dot{\phi} = 0$.

8.2 **(a)** Show that $\nabla_r \cdot \mathbf{r} = 3$, $\nabla_{\mathbf{r}}|\mathbf{r} - \mathbf{r}'| = \frac{\mathbf{r}-\mathbf{r}'}{|\mathbf{r}-\mathbf{r}'|}$, and $\nabla_r f(|\mathbf{r} - \mathbf{r}'|) = \frac{\mathbf{r}-\mathbf{r}'}{|\mathbf{r}-\mathbf{r}'|} f'(|\mathbf{r} - \mathbf{r}'|)$ where $f(\mathbf{r})$ is an analytic function. **(b)** Show that $\nabla_r^2 \frac{1}{|\mathbf{r}-\mathbf{r}'|} = 0$ for $\mathbf{r} \neq \mathbf{r}'$. Show also that $\int_V d^3r' \nabla_r^2 \frac{1}{|\mathbf{r}-\mathbf{r}'|} = \begin{cases} 4\pi\,, & \text{if } \mathbf{r} \in V \\ 0\,, & \text{otherwise} \end{cases}$ **Hint:** *Isolate sphere V_R with radius R around* $\mathbf{r}' = \mathbf{r}$. **(c)** Conclude that $\nabla_{\mathbf{r}}^2 \frac{1}{|\mathbf{r}-\mathbf{r}'|} = 4\pi\delta(\mathbf{r} - \mathbf{r}')$ and construct the solution (8.13) based on this information.

8.3 **(a)** Show that the summation $\sum_{j=1}^{\mathcal{N}} \sum_{k=1}^{\mathcal{N}} V_{j,k}$ yields $\mathcal{N}^2$ terms. Assume that $V_{j,j=0}$ and $V_{j,k} = V_{k,j}$ to show that this sum has $\mathcal{N}(\mathcal{N} - 1)/2$ different terms. **(b)** Show that these unique contributions follow from $\sum_{j=1}^{\mathcal{N}-1} \sum_{k=j+1}^{\mathcal{N}} V_{j,k}$. Set $V_{j,k} = 1$ to evaluate the total number of pairs. Show that $\sum_{j=1}^{\mathcal{N}-1} j = \mathcal{N}(\mathcal{N} - 1)/2$. **(c)** Show that $\sum_{j_1=1}^{\mathcal{N}-K+1} \sum_{j_2=j+1}^{\mathcal{N}-K+2} \cdots \sum_{j_K=j_{k-1}+1}^{\mathcal{N}} V_{j_1 j_2 \cdots j_K}$ yields $\binom{\mathcal{N}}{K} = \frac{\mathcal{N}!}{K!(\mathcal{N}-K)!}$ different terms. **Hint:** *This includes all ordered* $\{j_1 < j_2 < \cdots < j_K\}$ *choices among the* $\mathcal{N}$ *possible choices.* **(d)** Use the result 8.3(c) to prove $\sum_{j=1}^{N} j^2 = N(1 + N)(1 + 2N)/6$ and $\sum_{j=1}^{N} j^3 = N^2(1 + N)^2/4$.

8.4 **(a)** Confirm that $\phi(r, t) = \sum_j \frac{1}{4\pi\epsilon_0} \frac{Q_j}{\left|r - r_j\left(t_j'\right)\right|}$ satisfies Eq. (8.20), provided that $t_j' = t - \frac{|r - r_j(t)|}{c}$ is the retarded time. **(b)** Assume particles that are moving linearly in time, i.e., $\mathbf{r}_j(t) = \mathbf{r}_{j,0} + \boldsymbol{v}_j t$. Evaluate $\phi(r, t)$ with and without (instantaneous ϕ_{inst}) time retardation. Expand $\phi(r, t)$ up to the first power of v_j/c. What are the first-order corrections to the instantaneous ϕ_{inst}?

8.5 Eliminate the scalar potential from the Hamiltonian (8.7) by using the Lorentz gauge (8.10) and (8.21). Verify that the Hamiltonian (8.22) reduces to (8.18) for slowly varying ϕ if retardation effects are negligible.

8.6 **(a)** Show that the transversal electric field is $\mathbf{E}_T(\mathbf{r}) = -\dot{\mathbf{A}}(\mathbf{r})$ while the longitudinal part follows from $\mathbf{E}_L(\mathbf{r}) = -\nabla\phi(\mathbf{r})$, in the Coulomb gauge. Show that the magnetic field is always transversal. **(b)** Show that (8.22) produces the usual dot and cross product rules among two vectors $\mathbf{a} = \sum_j a_j \mathbf{e}_{\mathbf{q}}^{[j]}$ and $\mathbf{b} = \sum_j b_j \mathbf{e}_{\mathbf{q}}^{[j]}$. **(c)** Show that the basis choice (8.33) satisfies (8.22).

8.7 **(a)** Assume that the vector field has a form $\mathbf{F}(\mathbf{r}) = \mathbf{e}_x\, e^{ikz}\, e^{-\frac{\mathbf{r}^2}{R^2}}$. Identify its transversal and longitudinal parts. What are $\mathbf{F}_T(\mathbf{q}) \cdot \mathbf{F}_L(\pm\mathbf{q})$ and $\mathbf{F}_T(r) \cdot \mathbf{F}_L(\mathbf{r})$? **(b)** Show that any vectorial field produces $\int d^3r\, \mathbf{F}_T(\mathbf{r}) \cdot \mathbf{F}_L(\mathbf{r}) = 0$. **Hint:** *Convert everything into Fourier space.* Verify that the explicit field of part (a) indeed satisfies this relation.

8.8 **(a)** Fourier transform the Poisson Eq. (8.12) and show that it yields Eq. (8.40). **(b)** Assume that a scalar function has a Fourier transformation $f(\mathbf{q}) = 1/q^2$. Fourier transform this into real space and show that $f(\mathbf{r}) = \frac{1}{4\pi^2 r}\int_{-\infty}^{\infty} dx \frac{\sin x}{x}$. Show that $\int_{-\infty}^{\infty} dx \frac{\sin x}{x+i\varepsilon} = \pi$. **Hint:** *Use the Cauchy integral theorem.* Conclude that $f(\mathbf{r}) = 1/4\pi r$ and construct the potential (8.42) based on this information.

8.9 **(a)** Verify that Eq. (8.56) is one possible solution of the Helmholtz equation (8.57). Show that the norm is given by Eq. (8.63). **(b)** Show that the projection of generic vector fields can be constructed using the relation (8.67).

8.10 **(a)** Verify that the mode expansion (8.87) converts the Hamiltonian (8.18) into (8.86). **(b)** Start from Hamilton's equations $\dot{C}_{\mathbf{q}} = \Pi_{\mathbf{q}}$ and $\dot{\Pi}_{\mathbf{q}} = -\omega_{\mathbf{q}}^2 C_{\mathbf{q}} + \frac{J_{T,\mathbf{q}}}{\varepsilon_0 \mathcal{A}_{\mathbf{q}}}$ and show that the mode expansion (8.87) eventually produces the wave equation (8.44).

Further reading

The Coulomb and Lorentz gauge and the resulting Poisson and wave equations are thoroughly discussed in several classical electromagnetism books, see further reading in Chapter 2. The mode expansion is a standard step in the quantization procedure of light. See e.g.:

R. Loudon (2000). *The Theory of Quantum Light*, 3rd edition, Oxford, Oxford University Press.

M. O. Scully and M. S. Zubairy (2002). *Quantum Optics*, Cambridge, Cambridge University Press.

P. Meystre and M. Sargent III (2007). *Elements of Quantum Optics*, 4th edition, Berlin, Springer.

D. F. Walls and G. J. Milburn (2008). *Quantum Optics*, 2nd edition, Berlin, Springer.

G. Grynberg, A. Aspect, and C. Fabre (2010). *Introduction to Quantum Optics: From the Semi-classical Approach to Quantized Light*, Cambridge, Cambridge University Press.

The properties of the transversal delta function are discussed, e.g., in:

C. Cohen-Tannoudji, J. Dupont-Roc, and G. Grynberg (1997). *Photons & Atoms: Introduction to Quantum Electrodynamics*, Wiley & Sons.

9

System Hamiltonian in the generalized Coulomb gauge

As discussed in the previous chapter, we adopt the Coulomb gauge for all our further investigations starting from the many-body Hamiltonian (8.86) and the mode expansion (8.87). Before we proceed to quantize the Hamiltonian, we want to make sure that our analysis is focused on the nontrivial quantum phenomena. Thus, we first have to identify and efficiently deal with the trivial parts of the problem.

Often, the experimental conditions are chosen such that only a subset of all the electrons in a solid interacts strongly with the transversal electromagnetic fields while the remaining electrons and the ions are mostly passive. To describe theoretically such a situation in an efficient way, it is desirable to separate the dynamics of reactive electrons from the almost inert particles that merely produce a background contribution. This background can often be modeled as an optically passive response that is frequency independent and does not lead to light absorption.

In this chapter, we show how the passive background contributions can be systematically identified and included in the description. As the first step, we introduce the generalized Coulomb gauge to eliminate the scalar potential and to express the mode functions and the canonical variables. This leads us to a new Hamiltonian with altered Coulomb potential and mode functions. This generalized Hamiltonian allows us to efficiently describe optically active many-body systems in the presence of an optically passive background.

9.1 Separation of electronic and ionic motion

Solids, molecules, and atoms consist of negatively charged electrons that move around positively charged ions. In solids, the ions are located at the lattice sites of a rigid crystal structure. Since the ions are extremely massive compared with the mobile electrons, their movement is significantly slower. Thus, one can often separate the electronic from the ion motion focusing in the first step on the electronic motion by assuming the ions to be at rest. Corrections are included at a later stage where thermal ionic oscillations around the lattice sites are treated via the electron–phonon interactions. This approach is related to the Born–Oppenheimer approximation that significantly simplifies investigations of quantized many-body systems.

To distinguish electrons from ions, we use capital indices and labels ($j \to J$, $\mathbf{r} \to \mathbf{R}$, $\mathbf{p} \to \mathbf{P}$, $m \to M$) for the ions and keep the lower-case notation for the electrons. This way, the Hamiltonian (8.86) casts into the form

$$H = \sum_j \frac{\left[\mathbf{p}_j - Q_j\mathbf{A}(\mathbf{r}_j)\right]^2}{2m_j} + \sum_J \frac{[\mathbf{P}_J - Q_J\mathbf{A}(\mathbf{R}_J)]^2}{2M_J} + \frac{1}{2}\sum_{j,k\neq j} V_{j,k}(\mathbf{r}_j - \mathbf{r}_k)$$
$$+ \sum_{j,J} V_{j,J}(\mathbf{r}_j - \mathbf{R}_J) + \frac{1}{2}\sum_{J,K\neq J} V_{J,K}(\mathbf{R}_J - \mathbf{R}_K)$$
$$+ \sum_{\mathbf{q}} \frac{\Pi_{\mathbf{q}}^2 + \omega_{\mathbf{q}}^2 C_{\mathbf{q}}^2}{2}. \tag{9.1}$$

We see that the electron–ion Coulomb interaction term is summed over all (j, J) combinations because $\mathbf{r}_j$ and $\mathbf{R}_J$ automatically refer to different particles, unlike the indices within the electron–electron and ion–ion contributions. If the system has $\mathcal{N}$ electrons and $\mathcal{M}$ ions, we find $\mathcal{N}\mathcal{M}$, $\mathcal{N}(\mathcal{N}-1)/2$, and $\mathcal{M}(\mathcal{M}-1)/2$ different terms for the Coulomb interacting electron–ion, electron–electron, and ion–ion pairs, respectively. As in Fig. 8.1, where we show the interaction configurations between $\mathcal{N} = 4$ equally charged particles, we schematically illustrate in Fig. 9.1 the interaction possibilities for $\mathcal{N} = 2$ electrons (shaded circles) and $\mathcal{M} = 3$ (dark circles) ions.

To take advantage of the massive and nearly motionless nature of the ions, we formally set the ion mass to infinity. As a result, the Hamiltonian simplifies to

$$H = \sum_j \frac{\left[\mathbf{p}_j - Q_j\mathbf{A}(\mathbf{r}_j)\right]^2}{2m_j} + \sum_{j,J} V_{j,J}(\mathbf{r}_j - \mathbf{R}_J) + \frac{1}{2}\sum_{j,k\neq j} V_{j,k}(\mathbf{r}_j - \mathbf{r}_k)$$
$$+ \frac{1}{2}\sum_{J,K\neq J} V_{J,K}(\mathbf{R}_J - \mathbf{R}_K) + \sum_{\mathbf{q}} \frac{\Pi_{\mathbf{q}}^2 + \omega_{\mathbf{q}}^2 C_{\mathbf{q}}^2}{2}, \tag{9.2}$$

where the canonical momentum of the ions no longer appears. Consequently, we do not need to keep $\mathbf{R}_J$ as an independent variable allowing us to lump the effects of the ions into formally known functions. This treatment follows the same logic as the introduction of the

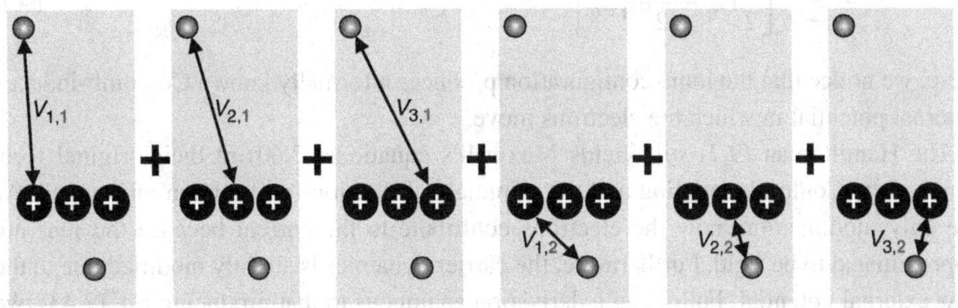

Figure 9.1 Schematic presentation of pairwise Coulomb interaction among electrons and ions. The interaction configurations are constructed for two electrons (shaded spheres) and three ions (black circles). Each interaction possibility is included only once, according to Eq. (9.1).

scalar potential which was possible because the canonical momentum Π_ϕ did not appear as dynamical variable. For example, we may write

$$Q_j U(\mathbf{r}_j) \equiv \sum_J V_{j,J}(\mathbf{r}_j - \mathbf{R}_J) = \sum_J \frac{Q_j Q_J}{4\pi\varepsilon_0 |\mathbf{r}_j - \mathbf{R}_J|}$$
$$= Q_j \int d^3r \sum_J Q_J\, \delta(\mathbf{r} - \mathbf{R}_J)\, \frac{1}{4\pi\varepsilon_0 |\mathbf{r}_j - \mathbf{r}|}, \tag{9.3}$$

which follows directly from the definition (8.15) of the Coulomb interaction. Identifying the ionic charge density,

$$n_{\text{ion}}(\mathbf{r}) = \sum_J Q_J\, \delta(\mathbf{r} - \mathbf{R}_J) \tag{9.4}$$

simplifies the potential U into

$$U(\mathbf{r}) = \int d^3r' \, \frac{n_{\text{ion}}(\mathbf{r}')}{4\pi\varepsilon_0 |\mathbf{r} - \mathbf{r}'|}. \tag{9.5}$$

A similar derivation produces the ion–ion interaction

$$V_{\text{ion}} \equiv \frac{1}{2} \sum_{J,K\neq J} V_{J,K}(\mathbf{R}_J - \mathbf{R}_K)$$
$$= \frac{1}{2} \int d^3r \int d^3r' \, \frac{n_{\text{ion}}(\mathbf{r}')\, n_{\text{ion}}(\mathbf{r})}{4\pi\varepsilon_0 |\mathbf{r} - \mathbf{r}'|} \left[1 - \delta(\mathbf{r} - \mathbf{r}')\right], \tag{9.6}$$

where the δ function eliminates the self-interaction. Even though V_{ion} is a constant, in fact infinite, we keep it for later reference. Based on the identifications (9.3) and (9.6), the Hamiltonian (9.2) reduces to

$$H = \sum_j \left(\frac{\left[\mathbf{p}_j - Q_j \mathbf{A}(\mathbf{r}_j)\right]^2}{2m_j} + Q_j U(\mathbf{r}_j) \right) + \frac{1}{2} \sum_{j,k\neq j} V_{j,k}(\mathbf{r}_j - \mathbf{r}_k) + V_{\text{ion}}$$
$$+ \sum_{\mathbf{q}} \left[\frac{1}{2} \Pi_{\mathbf{q}}^2 + \frac{1}{2} \omega_{\mathbf{q}} C_{\mathbf{q}}^2 \right]. \tag{9.7}$$

Here, we notice that the ionic configuration produces a formally known Coulomb-induced external potential in which the electrons move.

The Hamiltonian (9.7) still yields Maxwell's equations (7.60) in their original form because the Coulomb coupling and the minimal substitution form remain unchanged. As the only modification, now the electrons contribute to the current because the ions are approximated to be rigid. Furthermore, the carrier dynamics is slightly modified due to the new external potential. Following a derivation analogous to that producing Eq. (7.43), we obtain

$$m\dot{\mathbf{v}}_j = Q_j \mathbf{v}_j \times \mathbf{B}(\mathbf{r}_j) + Q_j \mathbf{E}(\mathbf{r}_j) - Q_j \nabla U(\mathbf{r}_j). \tag{9.8}$$

This can be brought into the form of Eq. (7.43), when we identify the external longitudinal electric field to be $\mathbf{E}_{\text{ext}}(\mathbf{r}) \equiv -\nabla U(\mathbf{r})$. Consequently, the Hamiltonian (9.7) provides a fully consistent system description even though we have approximated the dynamics of the ions.

The presented elimination of the ions is an example of a more general principle: the appearance of external fields and potentials depends on whether or not the dynamics of all entities are included in the Hamiltonian. If only part of the dynamics is described by H, the missing contributions appear as external potentials or fields.

9.2 Inclusion of the ionic polarizability

In general, atoms and ions in the solid-state lattice consist of nuclei and optically inert core electrons. Despite their nearly rigid nature, the lattice entities can adiabatically react to the electromagnetic field because, e.g., $\mathbf{E}(\mathbf{r})$ can move the core electrons with respect to the nuclei, which creates a dipole polarization of the matter. Thus, the approximation of completely rigid ions is not entirely satisfactory. However, the ionic polarization aspect can be incorporated phenomenologically because this polarizability is almost frequency independent and does not produce absorption of light.

In solids, we can usually treat the $\mathbf{E}(\mathbf{r})$-generated ionic/atomic polarization as a linear, frequency-independent, and absorption-free, response. Thus, we can approximate this ionic aspect via the phenomenological linear response,

$$\begin{aligned} \boldsymbol{P}_{\text{ion}}(\mathbf{r}, t) &\equiv \varepsilon_0\, \chi(\mathbf{r})\, \mathbf{E}(\mathbf{r}, t), \\ \boldsymbol{J}_{\text{ion}}(\mathbf{r}, t) &\equiv \frac{\partial\, \boldsymbol{P}_{\text{ion}}(\mathbf{r}, t)}{\partial t} = \varepsilon_0\, \chi(\mathbf{r})\, \frac{\partial\, \mathbf{E}(\mathbf{r}, t)}{\partial t}, \end{aligned} \tag{9.9}$$

where $\chi(\mathbf{r})$ defines the background susceptibility produced by the lattice of massive particles. In principle, the solid-state system can consist of multiple different material compositions producing a spatial dependence of $\chi(\mathbf{r})$ but no frequency dependence. Usually, $\chi(\mathbf{r})$ can be replaced by a constant within each homogeneous material layer of the system. In Eq. (9.9), we use the standard connection between polarization and current. Since the linear response is frequency independent, $\chi(\mathbf{r})$ cannot be time dependent such that we can also express the polarization current in terms of the susceptibility.

To phenomenologically incorporate the polarizability, we simply add the polarization current into Maxwell's equation (7.60),

$$\nabla \times \mathbf{B} = \frac{1 + \chi(\mathbf{r})}{c^2} \frac{\partial}{\partial t} \mathbf{E}(\mathbf{r}) + \mu_0 \sum\nolimits_j Q_j \mathbf{v}_j\, \delta(\mathbf{r} - \mathbf{r}_j), \tag{9.10}$$

where the effects of the individual electrons are still fully included. As usual, we identify the refractive index,

$$n^2(\mathbf{r}) \equiv 1 + \chi(\mathbf{r}) \geq 1. \tag{9.11}$$

Since we assume a situation where the lattice does not absorb light, the refractive index is real valued and must be at least $n(\mathbf{r}) \geq 1$. We can now write Eq. (9.10) as

$$\nabla \times \mathbf{B} = \frac{n^2(\mathbf{r})}{c^2}\frac{\partial}{\partial t}\mathbf{E}(\mathbf{r}) + \mu_0 \sum\nolimits_j Q_j \mathbf{v}_j \delta(\mathbf{r} - \mathbf{r}_j). \tag{9.12}$$

This contains only the currents of the individual electrons because the polarization current of the lattice is already included via the refractive index. Using

$$\mathbf{E}(\mathbf{r}) = -\dot{\mathbf{A}}(\mathbf{r}) - \nabla\phi(\mathbf{r}), \quad \mathbf{B}(\mathbf{r}) = \nabla \times \mathbf{A}(\mathbf{r}), \tag{9.13}$$

we find

$$\begin{aligned} -\nabla \times (\nabla \times \mathbf{A}(\mathbf{r}, t)) &- \frac{n^2(\mathbf{r})}{c^2}\frac{\partial^2}{\partial t^2}\mathbf{A}(\mathbf{r}, t) \\ &= -\mu_0 \sum\nolimits_j Q_j \mathbf{v}_j\, \delta(\mathbf{r} - \mathbf{r}_j) + \frac{\partial}{\partial t}\frac{n^2(\mathbf{r})\nabla\phi(\mathbf{r})}{c^2}. \end{aligned} \tag{9.14}$$

This shows that the polarizable ions reduce the speed of light to $c/n(\mathbf{r})$.

We already know that the generalized system Hamiltonian, introduced in Section 7.4, produces the correct dynamics since Eqs. (9.12) and (7.68) are identical. With this Hamiltonian, we find an additional $n(\mathbf{r})$-dependent modification of the Maxwell equation,

$$\nabla \cdot [n^2(\mathbf{r})\,\mathbf{E}(\mathbf{r})] = \frac{1}{\varepsilon_0}\left[\sum\nolimits_j Q_j \delta(\mathbf{r} - \mathbf{r}_j) + \sum\nolimits_J Q_J \delta(\mathbf{r} - \mathbf{R}_J)\right], \tag{9.15}$$

based on Eq. (7.68) and the separation of the electronic and ionic charges. If $n(\mathbf{r}) = n_0$, we can divide both sides of (9.15) by n_0^2; the emerging effective charge, Q_j/n_0^2, is smaller than its free-space Q_j due to screening by the surrounding polarization.

9.2.1 Generalized Coulomb gauge

The results above show that the generalized system Hamiltonian, introduced in Section 7.4, already includes the effect of the polarizability of the nearly rigid ion lattice. We also know that the Coulomb gauge is the suitable choice when *all* charged particles are fully incorporated in the Hamiltonian and we aim for the eventual quantization of both light and matter. Since it is clearly beneficial to describe the nearly rigid ion lattice phenomenologically, we need to consider how the Coulomb-gauge condition $\nabla \cdot \mathbf{A}(\mathbf{r}) = 0$ (8.8) must be modified in order to produce a Hamiltonian to which a straightforward canonical quantization can be applied.

As we examine Eq. (9.14), we realize that the $\nabla \times (\nabla \times \mathbf{A}(\mathbf{r}, t))$ contribution is always transversal. We then introduce the *generalized Coulomb gauge*

$$\nabla \cdot [n^2(\mathbf{r})\mathbf{A}(\mathbf{r})] = 0, \tag{9.16}$$

which reduces into the usual Coulomb gauge (8.8) if the refractive index is constant. The generalized Coulomb gauge forces the $n^2(\mathbf{r})\mathbf{A}(\mathbf{r})$ contribution within Eq. (9.14) always to be transversal. Thus, $n^2(\mathbf{r})\mathbf{A}(\mathbf{r})$ is now a transversal field while $\mathbf{A}(\mathbf{r})$ is not necessarily transversal. Using this gauge together with (9.13), allows us to convert Eq. (9.15) into the generalized Poisson equation

$$\nabla \cdot [n^2(\mathbf{r})\,\nabla\phi(\mathbf{r})] = -\frac{1}{\varepsilon_0}\Big[\sum\nolimits_j Q_j\delta(\mathbf{r}-\mathbf{r}_j) + \sum\nolimits_J Q_J\delta(\mathbf{r}-\mathbf{R}_J)\Big]. \tag{9.17}$$

Thus, we can formally express $n^2(\mathbf{r})\,\nabla\phi(\mathbf{r})$ as a function of the electronic positions after having solved the generalized Poisson equation, which will be studied in Sections 9.3–9.3.1. There, we show that the general solution has the form

$$\begin{aligned}
Q_j\phi(\mathbf{r}_j) &= \sum\nolimits_k V^{\text{gen}}_{j,k}(\mathbf{r}_j,\mathbf{r}_k) + U^{\text{gen}}(\mathbf{r}_j)\\
U^{\text{gen}}(\mathbf{r}_j) &\equiv \sum\nolimits_J V^{\text{gen}}_{j,J}(\mathbf{r}_j,\mathbf{R}_J)\\
Q_J\phi(\mathbf{R}_J) &= \sum\nolimits_j V^{\text{gen}}_{J,j}(\mathbf{R}_J,\mathbf{r}_j) + \sum\nolimits_K V^{\text{gen}}_{J,K}(\mathbf{R}_J,\mathbf{R}_K)\\
U^{\text{gen}}_{\text{ion}} &\equiv \frac{1}{2}\sum\nolimits_{J,K\neq J} V^{\text{gen}}_{J,K}(\mathbf{R}_J,\mathbf{R}_K)
\end{aligned} \tag{9.18}$$

in analogy to the identification made in connection with Eqs. (8.42)–(8.43). Here, we identified the effective lattice potential in the second line, as was done in Eq. (9.3), and the total potential energy contribution of all ions in the fourth line, corresponding to Eq. (9.6). The generalized Coulomb potential $V^{\text{gen}}_{j,k}(\mathbf{r}_j,\mathbf{r}_k)$ that appears has a full vectorial dependence on the two position coordinates of the interacting particles that cannot be reduced into the simple dependence on the distance $|\mathbf{r}_j - \mathbf{r}_k|$ as a consequence of the generalized Coulomb gauge.

The generalized Coulomb gauge converts Eq. (9.14) into the generalized wave equation

$$-\nabla\times[\nabla\times\mathbf{A}(\mathbf{r})] - \frac{1}{c^2}\frac{\partial^2}{\partial t^2}n^2(\mathbf{r})\mathbf{A}(\mathbf{r}) = -\mu_0\Big[\sum\nolimits_j Q_j\mathbf{v}_j\delta(\mathbf{r}-\mathbf{r}_j)\Big]_T. \tag{9.19}$$

Altogether, the generalized Poisson and wave equations plus the generic relations (9.13) define Maxwell's equations within the generalized Coulomb gauge (9.16). Thus, they can serve as a general starting point to describe how light propagates through a solid consisting of individual electrons in the presence of the nearly rigid, polarizable lattice. In other words, we have managed to separate the description of the optically reactive electrons from the optically passive response of the ions. This will considerably simplify further investigations because we only need to quantize the electronic part once the trivial response can be described through $n(\mathbf{r})$ and the ionic potential $U(\mathbf{r})$.

9.2.2 System Hamiltonian

We can now combine the results obtained in Sections 9.1 and 9.2 to include systematically the ionic contribution via the potential and the refractive index profile it creates. We start from the generalized system Hamiltonian (7.67) and apply the generalized Coulomb gauge (9.16) to eliminate the scalar potential. By following a derivation analogous to that yielding the Hamiltonian (8.18), we obtain

$$H = \sum_j \frac{\left[\mathbf{p}_j - Q_j \mathbf{A}(\mathbf{r}_j)\right]^2}{2m_j} + \frac{1}{2}\sum_j Q_j\,\phi(\mathbf{r}_j) + \frac{1}{2}\sum_J Q_J\,\phi(\mathbf{R}_J)$$
$$+ \frac{\varepsilon_0}{2}\int d^3r \left[n^2(\mathbf{r})\dot{\mathbf{A}}^2(\mathbf{r}) + c^2\mathbf{A}(\mathbf{r})\cdot\nabla\times[\nabla\times\mathbf{A}(\mathbf{r})]\right]. \tag{9.20}$$

Here, we have again separated the electron and the rigid-ion contributions as in Eq. (9.2).

We can now eliminate the scalar potential following the same steps as those resulting in Eqs. (8.13)–(8.14). Next, we use (9.18) to identify the generalized pairwise interactions. This way, we can express the system Hamiltonian as

$$H = \sum_j \left[\frac{\left[\mathbf{p}_j - Q_j \mathbf{A}(\mathbf{r}_j)\right]^2}{2m_j} + Q_j U^{\mathrm{gen}}(\mathbf{r}_j)\right] + \frac{1}{2}\sum_{j,k\neq j} V^{\mathrm{gen}}_{j,k}(\mathbf{r}_j,\,\mathbf{r}_k) + U^{\mathrm{gen}}_{\mathrm{ion}}$$
$$+ \frac{\varepsilon_0}{2}\int d^3r \left[n^2(\mathbf{r})\dot{\mathbf{A}}^2(\mathbf{r}) + c^2\mathbf{A}(\mathbf{r})\cdot\nabla\times[\nabla\times\mathbf{A}(\mathbf{r})]\right], \tag{9.21}$$

where the self-interactions are removed as in Eq. (9.1). The explicit form of the generalized Coulomb interaction depends on $n(\mathbf{r})$ and has to be obtained by solving Eq. (9.17).

In order to find a suitable mode expansion for the vector potential, we look at the homogeneous part of Eq. (9.19). By extending the analysis of Section 8.4, we find the generalized Helmholtz equation

$$\mathbf{q}^2 n^2(\mathbf{r})\,\mathbf{u}_{\mathbf{q}}(\mathbf{r}) - \nabla\times\left[\nabla\times\mathbf{u}_{\mathbf{q}}(\mathbf{r})\right] = 0, \quad \nabla\cdot\left[n^2(\mathbf{r})\,\mathbf{u}_{\mathbf{q}}(\mathbf{r})\right] = 0, \tag{9.22}$$

which reduces to Eq. (8.57) if we set $n(\mathbf{r}) = 1$. Again, the resulting modes form a complete set but the normalization (8.62) has to be modified,

$$\int d^3r\; n^2(\mathbf{r})\,\mathbf{u}^{\star}_{\mathbf{q}}(\mathbf{r})\cdot\mathbf{u}_{\mathbf{q}'}(\mathbf{r}) = \delta_{\mathbf{q},\mathbf{q}'}, \tag{9.23}$$

due to the generalized Coulomb gauge. Also the completeness relation (8.66) is slightly changed,

$$\mathbf{e}_\lambda\cdot\sum_{\mathbf{q}}\mathbf{u}^{\star}_{\mathbf{q}}(\mathbf{r}')\,\mathbf{u}_{\mathbf{q}}(\mathbf{r})\cdot\mathbf{e}_{\lambda'} = \frac{1}{n^2(\mathbf{r}')}\,\delta^T_{\lambda,\lambda'}(\mathbf{r}-\mathbf{r}'), \quad \lambda,\lambda' = x,\,y,\,z, \tag{9.24}$$

where the δ^T-function operates within the functional space of transversal vector fields, in analogy to Eq. (8.67).

Altogether, we find

$$\int d^3r'\, \mathbf{F}(\mathbf{r}') \cdot \sum_{\mathbf{q}} \mathbf{u}^{\star}_{\mathbf{q}}(\mathbf{r})\, \mathbf{u}_{\mathbf{q}}(\mathbf{r}') = \frac{1}{n^2(\mathbf{r})} \mathbf{F}_T(\mathbf{r}), \tag{9.25}$$

which reduces into Eq. (8.67) if the refractive index becomes unity. If the function $\mathbf{F}_{\mathrm{T,G}}(\mathbf{r})$ satisfies the generalized Coulomb gauge (9.16), the relation (9.25) produces the projection

$$\int d^3r'\, \mathbf{F}_{\mathrm{T,G}}(\mathbf{r}') \cdot n^2(\mathbf{r}') \sum_{\mathbf{q}} \mathbf{u}^{\star}_{\mathbf{q}}(\mathbf{r}')\, \mathbf{u}_{\mathbf{q}}(\mathbf{r}) = \mathbf{F}_{\mathrm{T,G}}(\mathbf{r}). \tag{9.26}$$

The explicit proof of relations (9.23)–(9.26) is the topic of Exercise 9.3.

Since the mode functions and the vector potential satisfy identical gauges, we may define the vector potential through a mode expansion

$$\mathbf{A}(\mathbf{r}) = \sum_{\mathbf{q}} \frac{C_{\mathbf{q}}\, \mathbf{u}_{\mathbf{q}}(\mathbf{r})}{\sqrt{\varepsilon_0}}, \tag{9.27}$$

where we assume a real-valued basis as before in connection with Eq. (8.87). At the same time, $\mathbf{A}(\mathbf{r})$ also defines that part of the electrical field that satisfies the generalized Coulomb gauge

$$\mathbf{E}(\mathbf{r}) \equiv \mathbf{E}_{\mathrm{T,G}}(\mathbf{r}) - \nabla\phi(\mathbf{r}), \qquad \mathbf{E}_{\mathrm{T,G}}(\mathbf{r}) \equiv -\dot{\mathbf{A}}(\mathbf{r}),$$
$$\nabla \cdot \left[n^2(\mathbf{r})\, \mathbf{E}_{\mathrm{T,G}}(\mathbf{r})\right] = 0. \tag{9.28}$$

In other words, $\mathbf{E}_{\mathrm{T,G}}(\mathbf{r})$ is transversal in the generalized sense and establishes the canonical momentum corresponding to $\dot{C}_{\mathbf{q}}$:

$$\mathbf{E}_{\mathrm{T,G}}(\mathbf{r}) \equiv -\dot{\mathbf{A}}(\mathbf{r}) = -\sum_{\mathbf{q}} \frac{\Pi_{\mathbf{q}}\, \mathbf{u}_{\mathbf{q}}(\mathbf{r})}{\sqrt{\varepsilon_0}}, \qquad \Pi_{\mathbf{q}} \equiv \dot{C}_{\mathbf{q}} \tag{9.29}$$

in full analogy to Eq. (8.87). If we insert the mode expansions (9.27) and (9.29) into Eq. (9.21), we find

$$H = \sum_j \left(\frac{\left[\mathbf{p}_j - \frac{Q_j}{\sqrt{\varepsilon_0}} \sum_{\mathbf{q}} C_{\mathbf{q}}\, \mathbf{u}_{\mathbf{q}}(\mathbf{r})\right]^2}{2m_j} + Q_j U^{\mathrm{gen}}(\mathbf{r}_j) \right)$$
$$+ \frac{1}{2} \sum_{j,k\neq j} V^{\mathrm{gen}}_{j,k}(\mathbf{r}_j - \mathbf{r}_k) + V^{\mathrm{gen}}_{\mathrm{ion}} + \sum_{\mathbf{q}} \frac{\left[\Pi^2_{\mathbf{q}} + \omega^2_{\mathbf{q}} C^2_{\mathbf{q}}\right]}{2}, \tag{9.30}$$

which defines the system Hamiltonian in the generalized Coulomb gauge.

It is interesting to notice that the phenomenological treatment of the nearly rigid ion lattice does not alter the structure of the standard Coulomb-gauge Hamiltonian (9.7). However, the polarizability, i.e. a nontrivial refractive-index profile, alters the explicit form of both the Coulomb interaction $V^{\mathrm{gen}}_{j,k}(\mathbf{r})$ and the mode functions because they must be obtained as solutions of the generalized Poisson and wave equations, respectively. Due to the refractive-index profile that appears, the solutions $\mathbf{u_q}$ can potentially display intriguing wave features such as interference, diffraction, tunneling, i.e., evanescent propagation, and refraction. We analyze these aspects in Section 9.4.3. After both $V_{j,k}^{\text{gen}}(\mathbf{r})$ and $\mathbf{u_q}(\mathbf{r})$ are known, the usual Hamilton equations,

$$\begin{cases} \dot{\mathbf{p}}_j = -\frac{\partial H}{\partial \mathbf{r}_j} \\ \dot{\mathbf{r}}_j = \frac{\partial H}{\partial \mathbf{p}_j} \end{cases} \qquad \begin{cases} \dot{\Pi}_{\mathbf{q}} = -\frac{\partial H}{\partial C_{\mathbf{q}}} \\ \dot{C}_{\mathbf{q}} = \frac{\partial H}{\partial \Pi_{\mathbf{q}}} \end{cases}, \tag{9.31}$$

produce the correct Maxwell–Lorentz equations. The particle trajectory follows from (9.8), where U is replaced by its generalized form U^{gen}. The electromagnetic field is defined by the generalized Poisson and wave equations (9.17) and (9.19), respectively. This derivation is analogous to the one in Section 8.4.3 and is the subject of Exercise 9.5.

9.3 Generalized Coulomb potential

The modifications of the Coulomb potential due to a dielectric background can be investigated directly when we determine the scalar potential of a single test charge embedded within the specific dielectric structure. In this situation, Eq. (9.17) casts into the form

$$\nabla \cdot [n^2(\mathbf{r})\, \nabla\phi_j(\mathbf{r})] = -\frac{Q_j}{\varepsilon_0}\, \delta(\mathbf{r} - \mathbf{r}_j), \tag{9.32}$$

where Q_j and $\mathbf{r}_j$ determine the charge and position of the test charge, respectively. The total solution to (9.17) follows from the superposition of the test charge solutions, yielding

$$\phi(\mathbf{r}) = \sum\nolimits_j \phi_j(\mathbf{r}) + \sum\nolimits_J \phi_J(\mathbf{r}). \tag{9.33}$$

In the homogeneous case where the system consists only of a single material, the refractive index is position independent, $n(\mathbf{r}) = n_0$, such that Eq. (9.32) reduces to

$$\nabla^2\phi_j^0(\mathbf{r}) = -\frac{Q_j}{\varepsilon_0\, n_0^2}\, \delta(\mathbf{r} - \mathbf{r}_j). \tag{9.34}$$

This mathematical problem has already been solved in Chapter 8, leading to the Coulomb potential

$$\phi_j^0(\mathbf{r}) = \frac{Q_j}{4\pi\varepsilon_0 n_0^2 |\mathbf{r} - \mathbf{r}_j|}, \tag{9.35}$$

where the interaction strength is reduced by the factor n_0^2, as predicted in connection with Eq. (9.15). Obviously, this produces the generalized Coulomb interaction $V_{j,k}^{\text{gen}}(\mathbf{r})$ that is also reduced by the factor n_0^2 with respect to its free-space value.

To determine the generalized Coulomb potential in a nontrivial situation, we calculate the potential near a surface between two different dielectric materials. More specifically, we assume that the dielectric constant makes an abrupt jump at the interface in the z direction and does not depend on x, y,

$$n(z) = n_-\, \theta(-z) + n_+\, \theta(z), \tag{9.36}$$

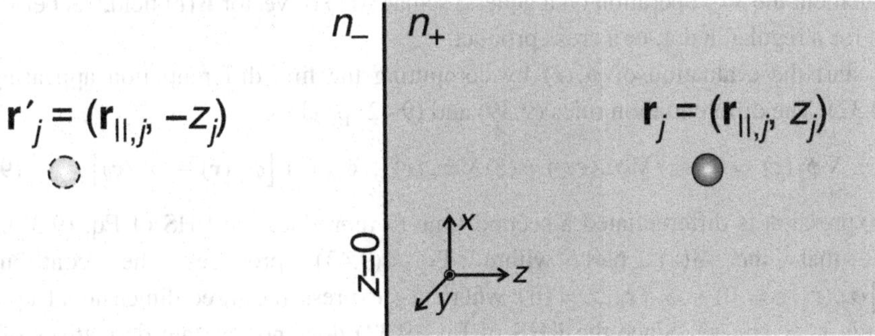

Figure 9.2 Charge and image charge appearing in a dielectric environment. The charge (sphere) is positioned within a region with dielectric constant n_+. The created potential can be constructed by assuming that an image charge (dashed circle) is induced in the region with dielectric constant n_-. At the interface $z = 0$, the dielectric constant changes abruptly from n_- to n_+.

see Fig. 9.2. We locate the charge (sphere) in the region $z > 0$. Since the refractive index is constant within the regions $z < 0$ and $z > 0$, it is meaningful to present the scalar potential,

$$\phi_j(\mathbf{r}) = \theta(-z)\,\phi_-(\mathbf{r}) + \theta(z)\,\phi_+(\mathbf{r}), \tag{9.37}$$

through two independent $\phi_\pm(\mathbf{r})$ functions that will be determined below.

Before we enter the actual computation of $\phi_j(\mathbf{r})$, we repeat a few of the known relations for the Heaviside and delta function. The Heaviside function is defined by

$$\theta(z) = \begin{cases} 0, & z < 0 \\ \frac{1}{2}, & z = 0 \\ 1, & z > 0 \end{cases} = \int_{-\infty}^{z} dz'\,\delta(z'). \tag{9.38}$$

It is straightforward to show that it satisfies the following relations

$$\theta(-z) = 1 - \theta(z), \quad \nabla\theta(z) = \delta(z)\,\mathbf{e}_z, \quad \nabla\theta(-z) = -\delta(z)\,\mathbf{e}_z. \tag{9.39}$$

The δ function that appears has the properties

$$\int_{-|a|}^{|b|} dz\,F(z)\,\delta(z) = \frac{F_- + F_+}{2}$$
$$F_\pm \equiv \lim_{z\to 0^\pm} F(z), \quad \text{for all} \quad |a| > 0, \quad |b| > 0. \tag{9.40}$$

Thus, we can write

$$F(z)\,\delta(z) = \frac{F_- + F_+}{2}\,\delta(z) \tag{9.41}$$

if F remains finite. We also frequently apply the generic relation

$$\nabla \odot [A(\mathbf{r})\mathbf{B}(\mathbf{r})] = [\nabla A(\mathbf{r})] \odot \mathbf{B}(\mathbf{r}) + A(\mathbf{r})\nabla \odot \mathbf{B}(\mathbf{r}) \tag{9.42}$$

to implement the $\nabla\odot$ operation on a general scalar $A(\mathbf{r})$ or vector $\mathbf{B}(\mathbf{r})$ field. As before, $\odot$ stands for a regular, a dot, or a cross product.

We start the evaluation of $\phi_j(\mathbf{r})$ by computing the first differentiation appearing in Eq. (9.32). The differentiation rules (9.39) and (9.42) produce

$$\nabla\phi_j(\mathbf{r}) = \theta(-z)\,\nabla\phi_-(\mathbf{r}) + \theta(z)\,\nabla\phi_+(\mathbf{r}) + \mathbf{e}_z\,\delta(z)\left[\phi_+(\mathbf{r}) - \phi_-(\mathbf{r})\right]. \tag{9.43}$$

This expression is differentiated a second time to reproduce the LHS of Eq. (9.32). We notice that the $\delta(z)$ part within Eq. (9.43) produces the contribution $\frac{\partial\delta(z)}{\partial z}\left[\phi_+(\mathbf{r}_\parallel, z=0) - \phi_-(\mathbf{r}_\parallel, z=0)\right]$ where we express the three-dimensional coordinates via $\mathbf{r} = (\mathbf{r}_\parallel,\, z)$. Since the RHS of Eq. (9.32) does not contain derivatives of the delta function, we conclude that the term within the brackets must vanish, such that

$$\begin{aligned} \phi_+(\mathbf{r}_\parallel, 0) &= \phi_-(\mathbf{r}_\parallel, 0) \\ \nabla\phi_j(\mathbf{r}) &= \theta(-z)\,\nabla\phi_-(\mathbf{r}) + \theta(z)\,\nabla\phi_+(\mathbf{r}). \end{aligned} \tag{9.44}$$

We can now continue with the second differentiation

$$\begin{aligned} \nabla\cdot\left[n^2(z)\,\nabla\phi_j(\mathbf{r})\right] &= \delta(z)\left[n_+^2\,\nabla\phi_+(\mathbf{r}_\parallel, 0) - n_-^2\,\nabla\phi_-(\mathbf{r}_\parallel, 0)\right]\cdot\mathbf{e}_z \\ &\quad + \theta(-z)\,n_-^2\,\nabla^2\phi_-(\mathbf{r}) + \theta(z)\,n_+^2\,\nabla^2\phi_+(\mathbf{r}). \end{aligned} \tag{9.45}$$

Here, we have performed steps similar to those made in connection with Eq. (9.43).

Equation (9.45) is the LHS of Eq. (9.32) that does not have a δ function at $z = 0$. Consequently, the expression within the bracket in (9.45) must vanish such that the generalized Poisson equation produces

$$\theta(-z)\,n_-^2\,\nabla^2\phi_-(\mathbf{r}) + \theta(z)\,n_+^2\,\nabla^2\phi_+(\mathbf{r}) = \frac{Q_j}{\epsilon_0}\,\delta(\mathbf{r}-\mathbf{r}_j). \tag{9.46}$$

Additionally, the components must satisfy the conditions

$$\begin{aligned} \phi_+(\mathbf{r}_\parallel, 0) &= \phi_-(\mathbf{r}_\parallel, 0) \\ n_-^2\,\mathbf{e}_z\cdot\nabla\phi_-(\mathbf{r}_\parallel, 0) &= n_+^2\,\mathbf{e}_z\cdot\nabla\phi_+(\mathbf{r}_\parallel, 0) \end{aligned} \tag{9.47}$$

demanding that the scalar potential and its divergence along the normal of the $z = 0$ interface must be continuous.

9.3.1 Image potentials

According to Eq. (9.46), only the $\phi_+(\mathbf{r})$ branch of the solution is directly affected by the charge because $\mathbf{r}_j$ is assumed to be within the region $z > 0$. Thus, we can express Eq. (9.46) through

$$\begin{cases} n_-^2\,\nabla^2\phi_-(\mathbf{r}) = 0, & z < 0 \\ n_+^2\,\nabla^2\phi_+(\mathbf{r}) = \frac{Q_j}{\epsilon_0}\,\delta(\mathbf{r}-\mathbf{r}_j), & z > 0 \end{cases}. \tag{9.48}$$

We may now make use of the results (9.34)–(9.35), which determine the potential created by a point charge embedded in a dielectric. Here, Q_j is simply suppressed by n_+^2. The $\phi_+(\mathbf{r})$ branch contains the direct Coulomb term created by Q_j,

$$\phi_+(\mathbf{r}) = \frac{Q_j}{4\pi\varepsilon_0 n_+^2|\mathbf{r}-\mathbf{r}_j|} + f(\mathbf{r}). \tag{9.49}$$

Here, we have added an arbitrary function that must satisfy $\nabla^2 f(\mathbf{r}) = 0$ in the region $z > 0$.

Clearly, the charge Q_j should induce a similar potential in the region $z < 0$, i.e.,

$$\phi_-(\mathbf{r}) = \frac{C\,Q_j}{4\pi\varepsilon_0|\mathbf{r}-\mathbf{r}_j|}. \tag{9.50}$$

This expression has no poles for $z < 0$ because the charge is situated in the region $z > 0$. This means that $\nabla^2\phi_-(\mathbf{r})$ vanishes for $z < 0$ such that Eq. (9.48) is satisfied. The constant C that appears and the function $f(\mathbf{r})$ must be chosen such that the solution satisfies the boundary conditions (9.47).

The condition (9.47) implies that $\phi_\pm(\mathbf{r})$ is continuous at the $z = 0$ interface, i.e., $\mathbf{r} = \mathbf{r}_\parallel$. Thus, Eqs. (9.49)–(9.50) impose

$$\frac{Q_j}{4\pi\varepsilon_0 n_0^2|\mathbf{r}_\parallel-\mathbf{r}_j|} + f(\mathbf{r}_\parallel) = \frac{C\,Q_j}{4\pi\varepsilon_0|\mathbf{r}_\parallel-\mathbf{r}_j|}, \tag{9.51}$$

which implies that $f(\mathbf{r})$ must have the same $\mathbf{r}_\parallel$ dependence as the direct charge term. Such a field can be created only by a point charge located at $\mathbf{r}_j = \mathbf{r}_{\parallel,j} + z_j\,\mathbf{e}_z$, i.e. by a so-called *image charge* situated at the position

$$\mathbf{r}'_j \equiv \mathbf{r}_{\parallel,j} - z_j\,\mathbf{e}_z \tag{9.52}$$

on the opposite side of the $z = 0$ interface. The positioning of the image charge (dashed sphere) with respect to the actual charge (sphere) is shown in Fig. 9.2. Since the image charge produces a vanishing $\nabla^2 f(\mathbf{r})$ for $z > 0$, we find that

$$f(\mathbf{r}) = \frac{A\,Q_j}{4\pi\varepsilon_0|\mathbf{r}-\mathbf{r}'_j|} \tag{9.53}$$

is the only possible configuration.

Collecting the results (9.49)–(9.50) and (9.53), we find the explicit form of the potential

$$\begin{cases} \phi_-(\mathbf{r}) = \frac{C\,Q_j}{4\pi\varepsilon_0|\mathbf{r}-\mathbf{r}_j|}, & z < 0 \\ \phi_+(\mathbf{r}) = \frac{Q_j}{4\pi\varepsilon_0}\left[\frac{1}{n_+^2|\mathbf{r}-\mathbf{r}_j|} + \frac{A}{|\mathbf{r}-\mathbf{r}'_j|}\right], & z > 0 \end{cases}. \tag{9.54}$$

A straightforward differentiation produces

$$\begin{cases} \nabla\phi_-(\mathbf{r}) = \frac{C\,Q_j}{4\pi\varepsilon_0}\frac{\mathbf{r}_j-\mathbf{r}}{|\mathbf{r}-\mathbf{r}_j|^3}, & z < 0 \\ \nabla\phi_+(\mathbf{r}) = \frac{Q_j}{4\pi\varepsilon_0}\left[\frac{\mathbf{r}_j-\mathbf{r}}{n_+^2|\mathbf{r}-\mathbf{r}_j|^2} + \frac{A\left(\mathbf{r}'_j-\mathbf{r}\right)}{|\mathbf{r}-\mathbf{r}'_j|^3}\right] & z > 0 \end{cases} \tag{9.55}$$

when we use the property $\nabla\frac{1}{|\mathbf{r}|} = -\frac{\mathbf{r}}{|\mathbf{r}|^3}$. The A and C coefficients become fixed through the two boundary conditions (9.47), eventually yielding the explicit form

$$\phi_j(\mathbf{r}) = \frac{Q_j}{4\pi n_+^2 \varepsilon_0} \begin{cases} \frac{2n_+^2}{n_-^2+n_+^2}\frac{1}{|\mathbf{r}-\mathbf{r}_j|}, & z<0 \\ \frac{1}{|\mathbf{r}-\mathbf{r}_j|} + \frac{n_+^2-n_-^2}{n_+^2+n_-^2}\frac{1}{|\mathbf{r}-\mathbf{r}_j'|}, & z>0 \end{cases}. \tag{9.56}$$

The image-charge technique is generally used if one wants to compute the electrostatic potential created by a charge in a geometrically well-defined conducting surrounding. Here, we have applied this technique for a simple dielectric surrounding and we can obviously generalize it to handle multiple dielectric layers as well. We also see from Eq. (9.56) that the image charge contribution is multiplied by a prefactor $\frac{n_+^2-n_-^2}{n_+^2+n_-^2}$, which is either positive ($n_+ > n_-$) or negative ($n_+ < n_-$). Thus, the image charge contributes either constructively or destructively, depending on the relative magnitudes of the dielectric constants on either side of the interface.

9.3.2 Generalized Coulomb potential

We now use the solution (9.56) to construct the total Coulomb potential. For simplicity, we assume that all the charges are located in the region $z > 0$ such that the combination of Eqs. (9.33) and (9.56) produces

$$\phi(\mathbf{r}) = \sum_k \phi_k(\mathbf{r}) = \sum_k \frac{Q_j}{4\pi\, n_+^2\, \varepsilon_0} \begin{cases} \frac{2n_+^2}{n_-^2+n_+^2}\frac{1}{|\mathbf{r}-\mathbf{r}_k|}, & z<0 \\ \frac{1}{|\mathbf{r}-\mathbf{r}_k|} + \frac{n_+^2-n_-^2}{n_+^2+n_-^2}\frac{1}{|\mathbf{r}-\mathbf{r}_k'|}, & z>0 \end{cases}, \tag{9.57}$$

where the k sum implicitly also includes the sum over the solid-state ions. We can easily verify that the relations (9.18) are fulfilled by expressing $Q_j\phi(\mathbf{r}_j) = \sum_k V_{j,k}^{\text{gen}}(\mathbf{r}_j, \mathbf{r}_k) + U^{\text{gen}}(\mathbf{r}_j)$ in terms of the generalized Coulomb potential

$$V_{j,k}^{\text{gen}}(\mathbf{r}, \mathbf{r}_k) \equiv \frac{Q_j\, Q_k}{4\pi\, n_+^2\, \varepsilon_0} \begin{cases} \frac{2n_+^2}{n_-^2+n_+^2}\frac{1}{|\mathbf{r}-\mathbf{r}_k|}, & z<0 \\ \frac{1}{|\mathbf{r}-\mathbf{r}_k|} + \frac{n_+^2-n_-^2}{n_+^2+n_-^2}\frac{1}{|\mathbf{r}-\mathbf{r}_k'|}, & z>0 \end{cases}, \tag{9.58}$$

where all charges are in the region $z > 0$. We see that $V_{j,k}^{\text{gen}}(\mathbf{r}, \mathbf{r}_k)$ has the full vectorial dependence on the positions of the charge and its image. However, we find that the in-plane dependence of $V_{j,k}^{\text{gen}}(\mathbf{r}, \mathbf{r}_k)$ involves only the relative distance, i.e.,

$$V_{j,k}^{\text{gen}}(\mathbf{r}, \mathbf{r}_k) = V_{j,k}^{\text{gen}}(\mathbf{r}_\parallel - \mathbf{r}_{\parallel,k}, z, z_k) = V_{j,k}^{\text{gen}}(|\mathbf{r}_\parallel - \mathbf{r}_{\parallel,k}|, z, z_k) \tag{9.59}$$

because the analyzed system is planar.

To determine the effect of the dielectric step on the Coulomb interaction, we analyze $V^{\text{gen}}_{j,k}(\mathbf{r}, \mathbf{r}_k)$ at the scale of a typical interaction length a_0. Then the characteristic magnitude of the interaction is defined by

$$E_0 \equiv \frac{Q_j\, Q_k}{4\pi\, n_+^2\, \varepsilon_0\, a_0}, \tag{9.60}$$

which is the strength of the Coulomb interaction at the distance $|\mathbf{r} - \mathbf{r}_k| = a_0$ without a dielectric step, i.e., for $n(\mathbf{r}) = n_+$. In this situation, Eq. (9.58) reduces to

$$V^{\text{gen}}_k(\mathbf{r}) \equiv V^{\text{gen}}_{j,k}(\mathbf{r}, \mathbf{r}_k) = E_0 \begin{cases} \frac{2n_+^2}{n_-^2+n_+^2}\frac{a_0}{|\mathbf{r}-\mathbf{r}_k|}, & z < 0 \\ \frac{a_0}{|\mathbf{r}-\mathbf{r}_k|} + \frac{n_+^2-n_-^2}{n_+^2+n_-^2}\frac{a_0}{|\mathbf{r}-\mathbf{r}'_k|}, & z > 0 \end{cases}. \tag{9.61}$$

Figure 9.3(a) presents $V^{\text{gen}}_k(\mathbf{r})$ as a function of z if the in-plane separation is $|\mathbf{r}_\parallel - \mathbf{r}_{\parallel,k}| = a_0$ and the charge is positioned at $z_k = 0.5\,a_0$ (dotted line), $z_k = 2\,a_0$ (dashed line), and $z_k = 10\,a_0$ (solid line) from the $z = 0$ interface. The actual positions of the test charges are indicated by the filled circles. The refractive index is chosen to be $n_- = 1$ corresponding to vacuum and $n_+ = 3.6$, which is roughly the dielectric background constant of GaAs. As a comparison, we have also plotted the Coulomb potential in bulk GaAs (thin dashed line) by setting $n_- = n_+ = 3.6$ and $z_k = 10\,a_0$. In this situation, the Coulomb potential reduces to its regular form $V^{\text{gen}}_k(\mathbf{r}) \to \frac{E_0\, a_0}{|\mathbf{r}-\mathbf{r}_k|}$ because there is no image charge.

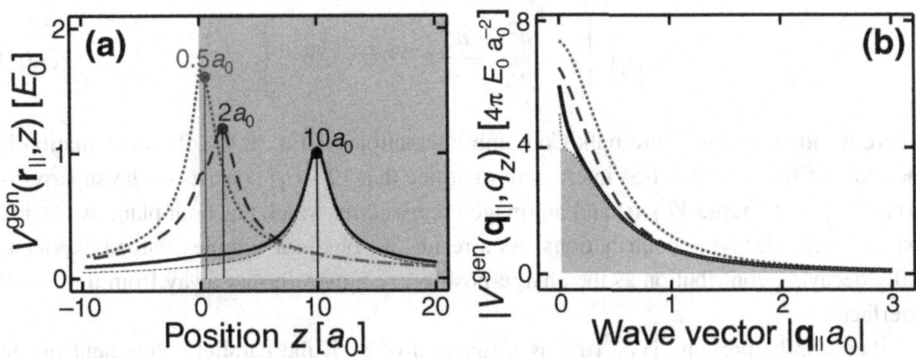

Figure 9.3 Generalized Coulomb potential close to the interface of two different dielectric materials. (**a**) The computed $V^{\text{gen}}_k(\mathbf{r}_\parallel, z)$ is shown when the test charge is positioned at $z_k = 0.5\,a_0$ (dotted line), $z_k = 2\,a_0$ (dashed line), and $z_k = 10\,a_0$ (solid line) from the $z = 0$ interface. The typical interaction length of the Coulomb potential is defined by a_0 and the in-plane separation is chosen $|\mathbf{r}_\parallel| = a_0$. The energy scale E_0 is given by Eq. (9.60). The circles indicate the position of the test charge and the thin dashed line (with transparent shading) presents the Coulomb potential in bulk. The dielectric constant changes from $n_- = 1$ to $n_+ = 3.6$ (shaded area). (**b**) The corresponding Fourier transformed $V^{\text{gen}}_k(\mathbf{q}_\parallel, q_z)$ as a function of $\mathbf{q}_\parallel$ for a fixed $q_z a_0 = 1$. The shaded area presents the bulk result and the thin dotted line has $z_k = 47\,a_0$; otherwise, the line styles correspond to frame (**a**).

The presence of the dielectric interface is important only if the charge is relatively close to it. Already by moving the charge to $z_k = 10\,a_0$ away from the surface, the homogeneous GaAs (thin dashed line) and $V_k^{\rm gen}(\mathbf{r})$ (solid line) become basically identical. However, $V_k^{\rm gen}(\mathbf{r})$ significantly differs from the bulk Coulomb potential if the test charge is closer to the $z = 0$ boundary. We clearly see that the dielectric interface acts as a partial mirror for the scalar potential. It can significantly enhance the Coulombic effects if interactions take place near the interface. In the limit of large n_+, the Coulombic effects can even become twice as large because the mirror effectively "doubles" the charge-dependent effects at short distances. In the analyzed system, the smallest $z_k = 0.5\,a_0$ (dotted line) already produces 1.66 times the homogeneous GaAs case such that the image-charge enhancement can become significant close to the GaAs surface. In GaAs, the typical length scale for interactions is roughly $a_0 = 10\,\mathrm{nm}$ such that these surface effects are limited to nanoscale distances.

In many-body investigations, it is often beneficial to analyze the Coulomb interactions via the Fourier transformed potential,

$$V_k^{\rm gen}(\mathbf{q}) \equiv \int d^3r\, V_k^{\rm gen}(\mathbf{r})\, e^{-i\mathbf{q}\cdot\mathbf{r}}. \tag{9.62}$$

The integral can be evaluated analytically,

$$\begin{aligned} V_k^{\rm gen}(\mathbf{q}) &= \frac{Q_j\, Q_k}{\varepsilon_0\, n_+^2\, \mathbf{q}^2}\, e^{-i\mathbf{q}\cdot\mathbf{r}_{\parallel,k}} \left[1 + \frac{n_+^2 - n_-^2}{n_-^2 + n_+^2} e^{-i\mathbf{q}_\parallel\cdot\mathbf{r}_{\parallel,k}}\, e^{-|\mathbf{q}_\parallel z_k|}\right] \\ &= V_{i,j}(\mathbf{q}) \left[1 + \frac{n_+^2 - n_-^2}{n_-^2 + n_+^2} e^{-i\mathbf{q}_z\, z_k}\, e^{-|\mathbf{q}_\parallel z_k|}\right], \end{aligned} \tag{9.63}$$

where we identified the pure-bulk Coulomb interaction via Eq. (8.43). This calculation is the topic of Exercise 9.7. It is interesting to notice that $V_k^{\rm gen}(\mathbf{q})$ consists of the superposition of a direct-charge $V_{i,j}(\mathbf{q})$ and an image-charge term, which has both plane-wave and exponentially decaying contributions. As a result, we obtain a complex-valued exponentially decaying contribution as the charge-position z_k moves further away from the $z = 0$ interface.

Figure 9.3(b) presents $\left|V_k^{\rm gen}(\mathbf{q})\right|$ as a function of q_z if the parallel component of the wave vector is fixed to $|\mathbf{q}_\parallel a_0| = 1/2$. The test charge is positioned at $z_k = 0.5\,a_0$ (dashed line), $z_k = 2\,a_0$ (dotted line), $z_k = 10\,a_0$ (solid line), and $z_k = 47\,a_0$ (thin dashed line). For comparison, we have also plotted the bulk GaAs $V_{i,j}(\mathbf{q})$ as the shaded area. We see that $\left|V_k^{\rm gen}(\mathbf{q})\right|$ is enhanced relative to $V_{i,j}(\mathbf{q})$ as z_k becomes smaller. We also observe that the image-charge effect weakens as we move away from the interface such that Coulomb interaction effects in regions away from dielectric interfaces can be described well via the homogeneous $V_{i,j}(\mathbf{q})$. We mostly concentrate on this situation in this book because we want to explore how many-body and quantum-optical effects become mingled without the complications resulting from surface and/or interface effects.

9.4 Generalized light-mode functions

To implement the generalized Coulomb gauge in practice, we also need to determine the generalized light modes explicitly from Eq. (9.22). In many relevant cases, the refractive-index profile $n(\mathbf{r})$ is piecewise constant, defined by the material compositions in the different regions. If $n(\mathbf{r}) = n(z)$ depends only on the z-coordinate, the dielectric structure is planar and we may look for the analytic eigenmodes $\mathbf{u_q}(\mathbf{r})$ from the generalized Helmholtz equation (9.22) in its simplified form

$$\mathbf{q}^2 n^2(z)\,\mathbf{u_q}(\mathbf{r}) - \nabla \times \left[\nabla \times \mathbf{u_q}(\mathbf{r})\right] = 0, \quad \nabla \cdot [n^2(z)\,\mathbf{u_q}(\mathbf{r})] = 0. \tag{9.64}$$

For more general functional dependencies of $n(\mathbf{r})$, one typically has to solve $\mathbf{u_q}(\mathbf{r})$ numerically.

To illustrate the basic properties of $\mathbf{u_q}(\mathbf{r})$, we consider the mode functions for an index step at $z = 0$ where the refractive index changes from n_- to n_+ when the $z = 0$ plane is crossed, Eq. (9.36). Due to the planar symmetry, we look for the solution of Eq. (9.64) in the form

$$\mathbf{u_q}(\mathbf{r}) = \mathbf{u_q^-}(\mathbf{r})\,\theta(-z) + \mathbf{u_q^+}(\mathbf{r})\,\theta(z) \tag{9.65}$$

such that the mode function is described by $\mathbf{u_q}^{-(+)}(\mathbf{r})$ for $z < 0$ ($z > 0$). Applying $\nabla \times \mathbf{u}(\mathbf{r})$ and $\nabla \times [\nabla \times \mathbf{u}(\mathbf{r})]$ to (9.65), we can convert (9.64) into the ordinary Helmholtz equation

$$\left[\nabla^2 + \mathbf{q}^2 n_\pm^2\right] \mathbf{u_q^\pm}(\mathbf{r}) = 0, \qquad \nabla \cdot \mathbf{u_q^\pm}(\mathbf{r}) = 0 \tag{9.66}$$

for transversal fields. To obtain these results, we proceeded in a way analogous to that producing Eqs. (9.46)–(9.47). In the process, we find the conditions

$$\begin{aligned} \left[\mathbf{e}_z \times \mathbf{u_q^-}\right]_{z=0} &= \left[\mathbf{e}_z \times \mathbf{u_q^+}\right]_{z=0}, \\ \mathbf{e}_z \times \left[\nabla \times \mathbf{u_q^-}\right]_{z=0} &= \mathbf{e}_z \times \left[\nabla \times \mathbf{u_q^+}\right]_{z=0}. \end{aligned} \tag{9.67}$$

Consequently, the in-plane component of $\mathbf{u_k}(\mathbf{r})$ and $\nabla \times \mathbf{u_k}(\mathbf{r})$ must be continuous at the $z = 0$ boundary.

Since both $\mathbf{u_q^+}$ and $\mathbf{u_q^-}$ must be transversal and satisfy the Helmholtz equation (9.66) with constant $n(\mathbf{r}) = n_\pm$, we may express them as linear combinations of transversal plane-wave solutions

$$\mathbf{u}^{\mathrm{pl}}_{\sigma,\mathbf{k}}(\mathbf{r}) = e^{i\mathbf{k}\cdot\mathbf{r}}\,\mathbf{e}^{[\sigma]}_{\mathbf{k}}, \qquad \sigma = 1, 2, \tag{9.68}$$

in full analogy with Eqs. (8.57)–(8.58). The unit vectors, $\mathbf{e}^{[1]}_{\mathbf{q}}$ and $\mathbf{e}^{[2]}_{\mathbf{q}}$, point in the two possible transversal directions explicitly defined by Eq. (8.33) such that $\mathbf{u}^{\mathrm{pl}}_{j,\mathbf{q}}$ is transversal, i.e., $\nabla \cdot \mathbf{u}^{\mathrm{pl}}_{\mathbf{q}}(\mathbf{r}) = 0$. To satisfy the Helmholtz equation (9.66), the wave vector must have the magnitude

$$\mathbf{k} = \mathbf{q}\,n_\pm \quad \text{for mode} \quad \mathbf{u}^\pm_q(\mathbf{r}). \tag{9.69}$$

Moreover, the plane-wave solutions produce the simple differentiation relation

$$\nabla \odot \mathbf{u}^{\mathrm{pl}}_{\sigma,\mathbf{k}}(\mathbf{r}) = i\mathbf{k} \odot e^{i\mathbf{k}\cdot\mathbf{r}}\,\mathbf{e}^{[\sigma]}_{\mathbf{k}} = i\,|\mathbf{k}|\,\mathbf{e}^{[3]}_{\mathbf{k}} \odot \mathbf{e}^{[\sigma]}_{\mathbf{k}}\,e^{i\mathbf{k}\cdot\mathbf{r}}$$
$$= i\,|\mathbf{k}|\,\mathbf{e}^{[3]}_{\mathbf{k}} \odot \mathbf{u}^{\mathrm{pl}}_{\sigma,\mathbf{k}}(\mathbf{r}), \qquad \sigma = 1, 2, \tag{9.70}$$

where $\mathbf{e}^{[3]}_{\mathbf{k}} = \frac{\mathbf{k}}{|\mathbf{k}|} = \frac{\mathbf{q}}{|\mathbf{q}|}$ defines the longitudinal basis vector.

9.4.1 Transmission and reflection of light modes

As the remaining task, we need to identify the vectorial directions $\mathbf{q}$ and suitable linear combinations of plane waves such that the two boundary conditions (9.67) are satisfied. For this purpose, we consider the situation in Fig. 9.4 where an incoming plane wave with $\mathbf{k} = \mathbf{q}^{\mathrm{in}}$ is transmitted and reflected in the directions $\mathbf{q}^{\mathrm{T}}$ and $\mathbf{q}^{R}$. Hence, for $z < 0$, we have the two plane-wave components, $\mathbf{q}^{\mathrm{in}}$ and $\mathbf{q}^{R}$, while for $z > 0$ we only have the transmitted $\mathbf{q}^{\mathrm{T}}$ contribution, i.e.,

$$\mathbf{u}^{-}_{\sigma,\mathbf{q}}(\mathbf{r}) = \mathbf{u}^{\mathrm{pl}}_{\sigma,\mathbf{q}^{\mathrm{in}}}(\mathbf{r}) + R^{[\sigma]}_{\mathbf{q}}\,\mathbf{u}^{\mathrm{pl}}_{\sigma,\mathbf{q}^{\mathrm{R}}}(\mathbf{r})$$
$$\mathbf{u}^{+}_{\sigma,\mathbf{q}}(\mathbf{r}) = T^{[\sigma]}_{\mathbf{q}}\,\mathbf{u}^{\mathrm{pl}}_{\sigma,\mathbf{q}^{\mathrm{T}}}(\mathbf{r}), \qquad \sigma = 1, 2. \tag{9.71}$$

The transmission $T^{[\sigma]}_{\mathbf{q}}$ and reflection $R^{[\sigma]}_{\mathbf{q}}$ coefficients that appear are computed below.

The actual directions of $\mathbf{q}^{\mathrm{T}}$, $\mathbf{q}^{\mathrm{R}}$, and $\mathbf{q}^{\mathrm{in}}$ are connected by the boundary conditions (9.67). The plane-wave component at $z = 0$,

$$\mathbf{u}^{\mathrm{pl}}_{\sigma,\mathbf{k}}(\mathbf{r}_{\|}, z = 0) = e^{i\mathbf{k}_{\|}\cdot\mathbf{r}_{\|}}\,\mathbf{e}^{[\sigma]}_{\mathbf{k}}, \qquad \sigma = 1, 2, \tag{9.72}$$

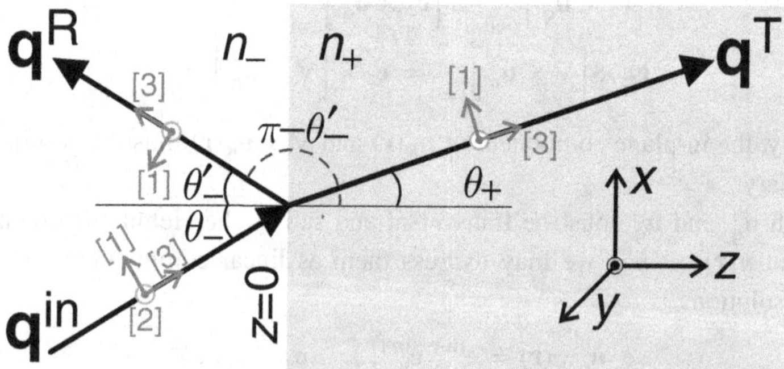

Figure 9.4 Schematic presentation of the plane-wave propagation through a boundary between two dielectric materials. The dielectric constant changes from n_- (white region) to n_+ (shaded region). The incoming ($\mathbf{q}^{\mathrm{in}}$), reflected ($\mathbf{q}^{\mathrm{R}}$), and transmitted ($\mathbf{q}^{\mathrm{T}}$) wave vectors are indicated by the large arrows. The plane of incidence is set to coincide with the xz-plane. The angles of incidences are θ_-, $\pi - \theta'_-$, and θ_+ for the incoming, reflected, and transmitted components, respectively. The corresponding transversal ([1] and [2]) as well as longitudinal ([3]) unit vectors are shown as small arrows. The [2] component points out of the paper toward the reader.

depends only on the in-plane coordinate $\mathbf{r}_{\parallel}$ and the in-plane wave vector $\mathbf{k}_{\parallel}$ that follow from the decompositions $\mathbf{r} = (\mathbf{r}_{\parallel}, z)$ and $\mathbf{k} = (\mathbf{k}_{\parallel}, k_z)$. The boundary conditions (9.67) connect the modes at $z = 0$ for all values of $\mathbf{r}_{\parallel}$. This property can be satisfied only if all in-plane components have the identical

$$\mathbf{q}_{\parallel}^{\text{in}} = \mathbf{q}_{\parallel}^{\text{R}} = \mathbf{q}_{\parallel}^{\text{T}}. \tag{9.73}$$

In other words, all wave vectors are confined to the same plane of incidence spanned by the incident wave vector and the normal of the $z = 0$ interface. Since the direction of Cartesian coordinates can be assigned arbitrarily, we can choose the wave vectors to be within the xz-plane that also defines the plane of incidence, as shown in Fig. 9.4.

However, one can also adopt a more general approach where the angle of incidence θ of all components is expressed via the angle between the normal to the $z = 0$ interface and the wave vector $\mathbf{q}^{\text{T}}$, $\mathbf{q}^{\text{R}}$, or $\mathbf{q}^{\text{in}}$, see Fig. 9.4. To simplify the notation, we express the transversal and longitudinal basis vectors compactly via

$$\mathbf{e}_{\mathbf{k}}^{[\sigma]} = \mathbf{e}_{\theta}^{[\sigma]}, \qquad \sigma = 1, 2, 3, \tag{9.74}$$

where $\mathbf{k}$ is either $\mathbf{q}^{\text{in}}$, $\mathbf{q}^{\text{R}}$, or $\mathbf{q}^{\text{R}}$. As before, we have chosen the $[\sigma]$ index such that the transversal–longitudinal division follows from the general relation (8.33), i.e., $\sigma = 3$ refers to the longitudinal direction while $j = 1, 2$ are related to the transversal components. Consequently, we can use the generic properties

$$\mathbf{e}_{\theta}^{[j]} \cdot \mathbf{e}_{\theta}^{[k]} = \delta_{j,k}, \quad \text{and} \quad \mathbf{e}_{\theta}^{[i]} \times \mathbf{e}_{\theta}^{[j]} = \varepsilon_{i,j,k}\, \mathbf{e}_{\theta}^{[k]}, \tag{9.75}$$

based on Eq. (8.29).

To obtain explicit expressions for $\mathbf{e}_{\theta}^{[\sigma]}$, we choose $\mathbf{q}^{\text{in}}$ in the xz-plane. Based on Fig. 9.4, the incoming light (from left-to-right direction) has the wave vector

$$\begin{aligned} \mathbf{q}^{\text{in}} &= |\mathbf{q}|\, n_-\, \mathbf{e}_{\theta_-}^{[3]} \equiv q_x^{\text{in}}\, \mathbf{e}_x + q_z^{\text{in}}\, \mathbf{e}_z \\ q_x^{\text{in}} &= |\mathbf{q}|\, n_-\, \sin\theta_-, \qquad q_z^{\text{in}} = |\mathbf{q}|\, n_-\, \cos\theta_- \end{aligned} \tag{9.76}$$

with the magnitude $|\mathbf{q}^{\text{in}}| = |\mathbf{q}|\, n_-$, based on relations (9.68)–(9.69). The corresponding longitudinal and transversal vector components are computed using the relations (8.33),

$$\mathbf{e}_{\mathbf{q}^{\text{in}}}^{[\sigma]} = \mathbf{e}_{\theta_-}^{[\sigma]} = \begin{cases} \sin\theta_-\, \mathbf{e}_x + \cos\theta_-\, \mathbf{e}_z, & \sigma = 3 \\ \mathbf{e}_y, & \sigma = 2 \\ \cos\theta_-\, \mathbf{e}_x - \sin\theta_-\, \mathbf{e}_z, & \sigma = 1 \end{cases}. \tag{9.77}$$

If the plane of incidence has an arbitrary direction, we must express $\mathbf{e}_{\theta_-}^{[\sigma]}$ in their implicit form provided by Eq. (8.33).

The reflected and the transmitted component can be expressed via a similar analysis. The explicit choice in Fig. 9.4 leads to

$$\begin{aligned}
&\mathbf{q}^{R} = |\mathbf{q}|\, n_-\, \mathbf{e}^{[3]}_{\pi-\theta'_-} \equiv q^{\mathrm{R}}_x e_x + q^{\mathrm{R}}_z e_z, \\
&q^{\mathrm{R}}_x = |\mathbf{q}|\, n_- \, \sin\theta'_-, \qquad q^{\mathrm{R}}_z = |\mathbf{q}|\, n_- \, \cos\theta'_- \\
&\mathbf{e}^{[\sigma]}_{\mathbf{q}^{\mathrm{R}}} = \mathbf{e}^{[\sigma]}_{\pi-\theta'_-} = \begin{cases} \sin\theta'_-\, \mathbf{e}_x - \cos\theta'_-\, \mathbf{e}_z, & \sigma = 3 \\ \mathbf{e}_y, & \sigma = 2 \\ -\cos\theta'_-\, \mathbf{e}_x - \sin\theta'_-\, \mathbf{e}_z, & \sigma = 1 \end{cases}.
\end{aligned} \tag{9.78}$$

The overall angle of the reflection $\pi - \theta'_-$ corresponds to the actual angle between the z axis and $\mathbf{q}^{\mathrm{R}}$, just like for the incoming parts. The transmitted branch resembles very much the incoming branch but can have a different propagation direction θ_+, as shown in Fig. 9.4. More explicitly, we find

$$\begin{aligned}
&\mathbf{q}^{\mathrm{T}} \equiv |\mathbf{q}|\, n_+\, \mathbf{e}^{[3]}_{\theta_+,T} \equiv q^{\mathrm{T}}_x \mathbf{e}_x + q^{\mathrm{T}}_z \mathbf{e}_z, \\
&q^{\mathrm{T}}_x = |\mathbf{q}|\, n_+ \, \sin\theta_+, \qquad q^{\mathrm{T}}_z = |\mathbf{q}|\, n_+ \, \cos\theta_+ \\
&\mathbf{e}^{[\sigma]}_{\mathbf{q}^{\mathrm{T}}} = \mathbf{e}^{[\sigma]}_{\theta_+} = \begin{cases} \sin\theta_+\, \mathbf{e}_x + \cos\theta_+\, \mathbf{e}_z, & \sigma = 3 \\ \mathbf{e}_y, & \sigma = 2 \\ \cos\theta_+\, \mathbf{e}_x - \sin\theta_+\, \mathbf{e}_z, & \sigma = 1 \end{cases},
\end{aligned} \tag{9.79}$$

which identifies the mode functions (9.71). Figure 9.4 also shows the three basis vectors corresponding to one longitudinal ($\sigma = 3$) and two transversal ($\sigma = 1, 2$) directions.

Combining (9.76) and (9.78)–(9.79) with (9.73), we find

$$\begin{aligned}
\mathbf{q}^{\mathrm{in}}_{\parallel} = \mathbf{q}^{\mathrm{R}}_{\parallel} \quad &\Leftrightarrow \quad \theta'_- = \theta_- \\
\mathbf{q}^{\mathrm{in}}_{\parallel} = \mathbf{q}^{\mathrm{T}}_{\parallel} \quad &\Leftrightarrow \quad n_- \sin\theta_- = n_+ \sin\theta_+.
\end{aligned} \tag{9.80}$$

Hence, the incoming θ_- and the reflected θ'_- are identical. At the same time, the transmission direction must satisfy Snell's law, discussed earlier in connection with Eq. (2.49).

9.4.2 Boundary conditions

At this stage, we still need to determine the magnitudes of the transmission and reflection coefficients such that the boundary conditions (9.67) are fulfilled. A direct substitution of the ansatz (9.71) casts the first boundary condition into the form

$$\begin{aligned}
&\left(\mathbf{e}_z \times \mathbf{e}^{[\sigma]}_{\theta_-}\right) e^{i\mathbf{q}^{\mathrm{in}}_{\parallel}\cdot\mathbf{r}_{\parallel}} + R^{[\sigma]}_{\mathbf{q}} \left(\mathbf{e}_z \times \mathbf{e}^{[\sigma]}_{\pi-\theta_-}\right) e^{i\mathbf{q}^{\mathrm{R}}_{\parallel}\cdot\mathbf{r}_{\parallel}} = T^{[\sigma]}_{\mathbf{q}} \left(\mathbf{e}_z \times \mathbf{e}^{[\sigma]}_{\theta_+}\right) e^{i\mathbf{q}^{\mathrm{T}}_{\parallel}\cdot\mathbf{r}_{\parallel}} \\
\Leftrightarrow \quad &\mathbf{e}_z \times \left(\mathbf{e}^{[\sigma]}_{\theta_-} + R^{[\sigma]}_{\mathbf{q}}\, \mathbf{e}^{[\sigma]}_{\pi-\theta_-} - T^{[\sigma]}_{\mathbf{q}}\, \mathbf{e}^{[\sigma]}_{\theta_+}\right) = 0,
\end{aligned} \tag{9.81}$$

where the dependence on $\mathbf{q}_\parallel \cdot \mathbf{r}_\parallel$ can be eliminated using (9.73). Together with (9.70), the second boundary condition yields

$$\begin{aligned}
&\mathbf{e}_z \times \left(i|\mathbf{q}^{\rm in}|\mathbf{e}^{[3]}_{\theta_-} \times \mathbf{e}^{[\sigma]}_{\theta_-}\right) e^{i\mathbf{q}^{\rm in}_\parallel\cdot\mathbf{r}_\parallel} + R^{[\sigma]}_{\mathbf{q}}\,\mathbf{e}_z \times \left(i|\mathbf{q}^{\rm R}|\mathbf{e}^{[3]}_{\pi-\theta_-} \times \mathbf{e}^{[\sigma]}_{\pi-\theta_-}\right) e^{i\mathbf{q}^{\rm R}_\parallel\cdot\mathbf{r}_\parallel} \\
&\quad = T^{[\sigma]}_{\mathbf{q}}\,\mathbf{e}_z \times \left(i|\mathbf{q}^{\rm T}|\mathbf{e}^{[3]}_{\theta_+} \times \mathbf{e}^{[\sigma]}_{\theta_+}\right) e^{i\mathbf{q}^{\rm T}_\parallel\cdot\mathbf{r}_\parallel} \\
&\Leftrightarrow \quad \mathbf{e}_z \times \left(n_-\,\mathbf{e}^{[\bar\sigma]}_{\theta_-} + n_-\,R^{[\sigma]}_{\mathbf{q}}\,\mathbf{e}^{[\bar\sigma]}_{\pi-\theta_-} - n_+\,T^{[\sigma]}_{\mathbf{q}}\,\mathbf{e}^{[\bar\sigma]}_{\theta_+}\right) = 0.
\end{aligned} \tag{9.82}$$

Here, we used the relations $|\mathbf{q}^{\rm in}| = |\mathbf{q}^{\rm R}| = n_-\,|\mathbf{q}|$ and $|\mathbf{q}^{\rm T}| = n_+\,|\mathbf{q}|$ as well as $\mathbf{e}^{[3]}_\theta \times \mathbf{e}^{[\sigma]}_\theta = \epsilon^{3,\sigma,\bar\sigma}\mathbf{e}^{[\bar\sigma]}_\theta$, where

$$\bar\sigma = \begin{cases} 2, & \sigma = 1, \\ 1, & \sigma = 2 \end{cases} \tag{9.83}$$

converts one transversal index into the other one. Since the Levi–Civita symbol is identical in all the terms that appear, it can be eliminated yielding the final expression in Eq. (9.82). In summary, we find the two boundary conditions

$$\begin{cases} \mathbf{e}_z \times \left(\mathbf{e}^{[\sigma]}_{\theta_-} + R^{[\sigma]}_{\mathbf{q}}\,\mathbf{e}^{[\sigma]}_{\pi-\theta_-} - T^{[\sigma]}_{\mathbf{q}}\,\mathbf{e}^{[\sigma]}_{\theta_+}\right) = 0 \\ \mathbf{e}_z \times \left(n_-\,\mathbf{e}^{[\bar\sigma]}_{\theta_-} + n_-\,R^{[\sigma]}_{\mathbf{q}}\,\mathbf{e}^{[\bar\sigma]}_{\pi-\theta_-} - n_+\,T^{[\sigma]}_{\mathbf{q}}\,\mathbf{e}^{[\bar\sigma]}_{\theta_+}\right) = 0 \end{cases} \tag{9.84}$$

that uniquely fix the transmission and reflection coefficients.

To solve Eq. (9.84), we first need to determine $\mathbf{e}_z \times \mathbf{e}^{[\sigma]}_\theta$ for the different transversal components. With the help of (9.77)–(9.79), we find

$$\begin{aligned}
&\mathbf{e}_z \times \mathbf{e}^{[1]}_{\theta_-} = \cos\theta_-\,\mathbf{e}_y, \quad \mathbf{e}_z \times \mathbf{e}^{[1]}_{\pi-\theta_-} = -\cos\theta_-\,\mathbf{e}_y, \\
&\mathbf{e}_z \times \mathbf{e}^{[1]}_{\theta_+} = \cos\theta_+\,\mathbf{e}_y, \quad \mathbf{e}_z \times \mathbf{e}^{[2]}_{\theta} = -\mathbf{e}_x, \\
&\theta = \theta_-,\ \pi-\theta_-,\ \text{or}\ \theta_+.
\end{aligned} \tag{9.85}$$

When these relations are used in (9.84), we obtain

$$\begin{cases} \cos\theta_- - \cos\theta_-\,R^{[1]}_{\mathbf{q}} - \cos\theta_+\,T^{[1]}_{\mathbf{q}} = 0 \\ n_- + n_-\,R^{[1]}_{\mathbf{q}} - n_+\,T^{[1]}_{\mathbf{q}} = 0 \end{cases} \tag{9.86}$$

for the $\sigma = 1$ modes and

$$\begin{cases} 1 + R^{[2]}_{\mathbf{q}} - T^{[2]}_{\mathbf{q}} = 0 \\ n_-\,\cos\theta_- - n_-\,\cos\theta_-\,R^{[2]}_{\mathbf{q}} - n_+\,\cos\theta_+ T^{[2]}_{\mathbf{q}} = 0 \end{cases} \tag{9.87}$$

for the $\sigma = 2$ modes. Thus, the transversal $\sigma = 1$ and $\sigma = 2$ components produce different $T^{[\sigma]}_{\mathbf{q}}$ and $R^{[\sigma]}_{\mathbf{q}}$ coefficients. Qualitatively, all $\sigma = 2$ modes have the direction $\mathbf{e}^{[\sigma=2]}_\theta$ perpendicular to the plane of incidence. The corresponding mode functions are often referred to as s-polarized plane waves; the letter s stems from the German word "senkrecht" meaning perpendicular. The remaining $\mathbf{e}^{[\sigma=1]}_\theta$ modes reside within the plane of incidence. This branch of solutions is often referred to as p-polarized plane waves originating from the word "parallel."

9.4.3 Fresnel coefficients for s- and p-polarized modes

When we solve Eq. (9.87), we find the transmission and reflection coefficients

$$T_{\mathbf{q}}^{[2]} = \frac{2n_- \cos\theta_-}{n_- \cos\theta_- + n_+ \cos\theta_+}, \qquad R_{\mathbf{q}}^{[2]} = \frac{n_- \cos\theta_- - n_+ \cos\theta_+}{n_- \cos\theta_- + n_+ \cos\theta_+}, \tag{9.88}$$

which constitute the Fresnel coefficients for the s-polarized modes. The same analysis can be repeated for the p-polarized modes, producing

$$T_{\mathbf{q}}^{[1]} = \frac{2n_- \cos\theta_-}{n_+ \cos\theta_- + n_- \cos\theta_+}, \qquad R_{\mathbf{q}}^{[1]} = \frac{n_+ \cos\theta_- - n_- \cos\theta_+}{n_+ \cos\theta_- + n_- \cos\theta_+}. \tag{9.89}$$

Both s and p-polarized modes connect θ_- and θ_+ via Snell's law (9.80)

$$n_- \sin\theta_- = n_+ \sin\theta_+. \tag{9.90}$$

Altogether, the generalized Coulomb gauge yields a vector potential that can be expressed in terms of modes which follow from the usual classical analysis of light propagation through dielectric materials.

If n_- is greater than n_+, Snell's law (9.90) predicts two distinctly different solution branches,

$$\theta_+ = \begin{cases} \arcsin \frac{n_- \sin\theta_-}{n_+}, & \theta_- \leq \arcsin \frac{n_+}{n_-} \\ \frac{\pi}{2} + i\beta_+, & \theta_- > \arcsin \frac{n_+}{n_-} \end{cases},$$

$$\beta_+ \equiv \operatorname{arccosh} y = \ln\left[y + \sqrt{y^2 - 1} \right], \qquad y \equiv \frac{n_- \sin\theta_-}{n_+}, \tag{9.91}$$

depending on the magnitude of the incident angle θ_- for the incoming field, within $\mathbf{u}_{\sigma,\mathbf{q}}^-$. Since it is physically possible that θ_- of the incoming wave can exceed the critical angle $\theta_{\text{crit}} \equiv \arcsin \frac{n_+}{n_-}$, the transmitted wave can propagate with an angle of incidence θ_+ that has an imaginary part β_+. To deal with this feature, we must generalize Snell's law to also include complex-valued angles.

If θ_- becomes greater than θ_{crit}, the transmitted solution branch $\mathbf{u}_{\sigma,\mathbf{q}}^+$ has a wave vector

$$\begin{aligned} \mathbf{q}^T &= |\mathbf{q}|\, n_+ \sin\left(\frac{\pi}{2} + i\beta_+\right) \mathbf{e}_x + |\mathbf{q}|\, n_+ \cos\left(\frac{\pi}{2} + i\beta_+\right) \mathbf{e}_z \\ &= |\mathbf{q}|\, n_+ \cosh\beta_+\, \mathbf{e}_x + i|\mathbf{q}|\, n_+ \sinh\beta_+\, \mathbf{e}_z \\ &= |\mathbf{q}|\, n_- \sin\theta_-\, \mathbf{e}_x + i|\mathbf{q}|\, n_+ \sinh\beta_+\, \mathbf{e}_z, \qquad \theta_- > \theta_{\text{crit}} \end{aligned} \tag{9.92}$$

which follows directly from Eqs. (9.79) and (9.91). The last result implies that the in-plane, i.e., $\mathbf{e}_x$ component of $\mathbf{q}^T$ equals that for the incoming and reflected components within $\mathbf{u}_{\sigma,\mathbf{q}}^-$. Furthermore, the z component of the transmitted wave vector is purely imaginary such that the z dependency of $\mathbf{u}_{\sigma,\mathbf{q}}^+$ describes an exponential decay instead of the usual plane-wave propagation. This scenario is very much analogous to the tunneling phenomenon of particles discussed in Section 4.3.2. For light, this phenomenon is often referred to as evanescent decay instead of light tunneling.

Based on the result (9.92), we see that the wave vector has the properties

$$\begin{aligned} |\mathbf{q}^T| &= |\mathbf{q}|\, n_+ \sqrt{\cosh^2 \beta_+ - \sinh^2 \beta_+} = |\mathbf{q}|\, n_+ \\ |\mathbf{q}^T_{\parallel}| &= \cosh \beta_+ |\mathbf{q}|\, n_+ = \cosh \beta_+ |\mathbf{q}^T| \geq |\mathbf{q}^T|, \qquad \theta_- > \theta_{\text{crit}} \end{aligned} \tag{9.93}$$

if the critical angle is exceeded. In particular, we observe that the in-plane component $\mathbf{q}^T_{\parallel} = \mathbf{q}^T_x$ becomes longer than the wave vector itself. Obviously, the total length can be conserved only with the help of an imaginary perpendicular component because then $(\mathbf{q}^T_{\perp})^2 = (\mathbf{q}^T_z)^2 < 0$ is negatively valued such that it subtracts the appropriate amount from $|\mathbf{q}^T|^2 = \left(\mathbf{q}^T_{\parallel}\right)^2 + \left(\mathbf{q}^T_{\perp}\right)^2$ in order to conserve the overall length of the wave vector. In other words, the generalized wave equation produces evanescent decay whenever one of the wave-vector components exceeds the magnitude of the wave vector itself.

To illustrate evanescent decay, Fig. 9.5(a) shows the computed real (shaded area) and imaginary (solid line) parts of the transmission angle θ_+ as a function of the angle of incidence when we choose $n_+ = 3.6$ and $n_- = 1$. This scenario is realized if the incoming plane wave propagates from GaAs-type material into vacuum or air. Under these conditions, we find the critical angle $\theta_{\text{crit}} = 0.2815$, which corresponds to an angle of incidence of 16.1°. We see that θ_+ is purely real below the critical angle (vertical line) but obtains a significant imaginary part when the critical angle is exceeded. We also see that the real part of θ_+ clamps to the value $\pi/2$ above the critical angle.

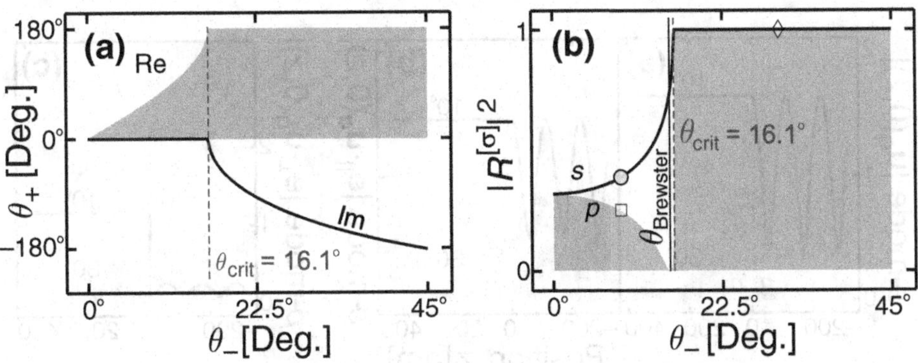

Figure 9.5 Transition to evanescent decay at a dielectric boundary. The system is schematically shown in Fig. 9.4. The dielectric constants are taken as $n = 3.6$ and $n_- = 1$. **(a)** The real (shaded area) and imaginary (solid line) parts of the angle of incidence θ_+ for the transmitted light are plotted as a function of the incoming angle θ_-. The vertical line indicates the critical angle θ_{crit}. **(b)** The corresponding reflection probability for s- (solid line) and p-polarized (shaded area) modes is plotted as a function of the incoming angle θ. The vertical line indicates the Brewster angle and the dashed vertical line denotes θ_{crit}. The circle (square) indicates the reflection probability at $\theta_- = 10°$ for the s- (p-) polarized mode. The diamond indicates $\theta_- = 30°$.

We have also computed the corresponding reflection coefficient as a function of the incidence angle. Figure 9.5(b) presents the reflection probability $|R_{\mathbf{q}}^{[\sigma]}|^2$ for the s-polarized (solid line) and p-polarized (shaded area) modes computed as a function of θ_- using Eqs. (9.88)–(9.90). We see that, depending on the polarization, the modes have a very different behavior below the critical angle (dashed, vertical line). While the s-polarized mode yields an $\left|R_{\mathbf{q}}^{[2]}\right|^2$ that increases monotonously from the value 0.319 up to perfect reflection (unity), the reflectivity for p polarization drops all the way to zero and then reaches unity above the critical angle. The angle of vanishing $\left|R_{\mathbf{q}}^{[1]}\right|^2$ follows directly from Eqs. (9.89)–(9.90) as

$$\theta_B \equiv \arctan \frac{n_+}{n_-}. \tag{9.94}$$

This special direction is called the Brewster angle, which yields completely vanishing reflection for the p-polarized mode. For the parameters used, we find $\theta_B = 0.2709$ corresponding to a 15.5° angle of incidence. The corresponding $\theta_- = \theta_B$ is indicated in Fig. 9.5 by a vertical solid line. For the studied situation, the Brewster angle is only slightly smaller than the critical angle.

Once the transmission and reflection coefficients are determined using Eqs. (9.88)–(9.90), we can use Eqs. (9.68) and (9.71) to construct the actual mode function

$$\mathbf{u}_{\sigma,\mathbf{q}}(\mathbf{r}) = \theta(-z)\left[e^{i\mathbf{q}^{\text{in}}\cdot\mathbf{r}}\,\mathbf{e}_{\theta_-}^{[\sigma]} + R_{\mathbf{q}}^{[\sigma]}\,e^{i\mathbf{q}^{\text{R}}\cdot\mathbf{r}}\,\mathbf{e}_{\pi-\theta_-}^{[\sigma]}\right] + \theta(z)\,T_{\mathbf{q}}^{[\sigma]}\,e^{i\mathbf{q}^{\text{T}}\cdot\mathbf{r}}\,\mathbf{e}_{\theta_+}^{[\sigma]}. \tag{9.95}$$

Figure 9.6 presents the in-plane and perpendicular components of $\mathbf{u}_{\sigma,\mathbf{q}}(\mathbf{r})$ computed for the same system as in Fig. 9.5. We have used $\theta_- = \pi/18 \equiv 10°$ (solid line) and

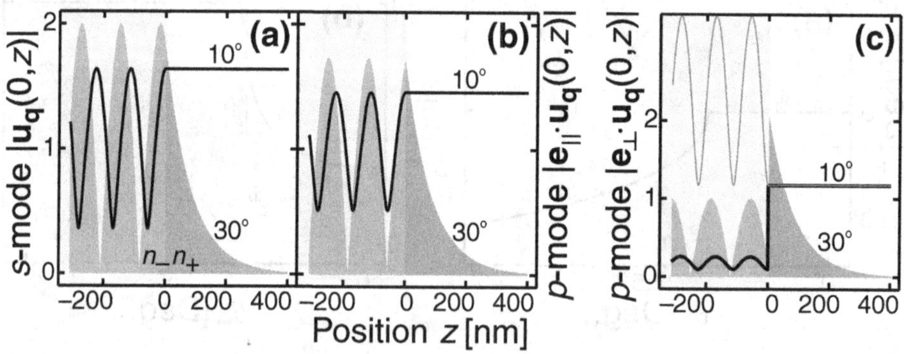

Figure 9.6 Generalized s- and p-polarized modes propagating through a single dielectric interface. Angles of incidence are $\theta_- = \pi/18$ (solid line) and $\theta_- = \pi/6$ (shaded area) corresponding to the cases analyzed in Fig. 9.5(b). The vertical line indicates the $z = 0$ interface where the dielectric constant changes from $n_- = 3.6$ to $n_+ = 1$. **(a)** The magnitude of the s-polarized mode is shown together with the **(b)** in-plane and **(c)** perpendicular component of the p-polarized mode. In frame **(c)**, the thin solid line defines the perpendicular component of $n^2(z)\,\mathbf{u}_{1,\mathbf{q}}(0,z)$ computed for the $\theta_- = \pi/18$ angle of incidence. All calculations assume the vacuum wave vector to be $|\mathbf{q}| = \frac{2\pi}{800\,\text{nm}^{-1}}$.

$\theta_- = \pi/6 \equiv 30°$ (shaded area) to analyze a situation below and above the critical angle of 16.1°, respectively. The corresponding reflection probabilities can be read off from Fig. 9.5(b). We see that the s- and the p-polarized modes have a reflection probability of 0.39 and 0.25, respectively, for a 10° angle of incidence while both of them exhibit the reflection probability of one (total reflection) for 30°. To construct the mode functions, we have assumed that the light mode has the wave length $\lambda = 800\,\text{nm}$ in vacuum, which produces $|\mathbf{q}| = 2\pi/\lambda = 7.85\,\mu\text{m}^{-1}$.

Since the s-polarized mode has only an in-plane component, according to Eqs. (9.77)–(9.79), we have plotted $|\mathbf{u}_{2,\mathbf{q}}(0, z)|$ as a function of z in Fig. 9.6(a). The corresponding p-polarized mode has both in-plane and perpendicular components. Therefore, we plot the in-plane $|\mathbf{e}_x \cdot \mathbf{u}_{1,\mathbf{q}}(0, z)|$ and the perpendicular $|\mathbf{e}_z \cdot \mathbf{u}_{1,\mathbf{q}}(0, z)|$ projections separately in Figs. 9.6(b) and 9.6(c), respectively. We see that the in-plane components are always continuous and smooth functions even at the $z = 0$ interface, as demanded by the boundary conditions (9.67).

The generalized Coulomb gauge allows for a discontinuity of the perpendicular component. Only the p-polarized mode has a component perpendicular to the $z = 0$ interface, as seen from the schematic Fig. 9.4. Figure 9.6(c) shows the perpendicular component for a 10° (solid line) and a 30° angle of incidence. In both cases, we observe that the mode is discontinuous at the $z = 0$ interface (vertical line). In fact, it is straightforward to prove that the solution (9.95) satisfies the generalized Coulomb gauge regardless of the discontinuity. To check how the multiplication of the mode function by $n^2(z)$ alters the discontinuity, we have also plotted $n^2(z)|\mathbf{e}_z \cdot \mathbf{u}_{1,\mathbf{q}}(0, z)|$ (thin solid line) for the 10° case. This product produces fully continuous perpendicular components suggesting that the discontinuity results from the generalized Coulomb gauge (9.64).

As we compare the solutions below and above the critical angle, we note a significant qualitative change in the behavior of the transmitted part. In all cases, the transmitted $|\mathbf{u}_{2,\mathbf{q}}(0, z)|$, i.e., the solution within the region $z > 0$ has a constant magnitude when the angle of incidence is below the critical value 16.128°. For larger angles, the transmitted mode exponentially decreases due to evanescent decay. For the parameters used, the z component of the transmitted wave vector is $\mathbf{q}_z^{\text{T}} = 11.8i\,\mu\text{m}^{-1}$ identifying the decay constant $\Delta z_{\text{dec}} \equiv \frac{1}{\text{Im}[\mathbf{q}_z^{\text{T}}]} = 85\,\text{nm}$. Figures 9.6(a)–(c) present evanescently decaying modes whose magnitude is reduced by $e^{-1} \approx 0.368$ within an 85 nm decay distance. It is also interesting to see that an interference pattern is formed for $z < 0$ due to the simultaneous presence of the incoming and reflected components.

9.4.4 Transfer-matrix solutions for generalized modes

The mode functions can also be constructed for generic planar structures when we introduce a few minor modifications to the investigation in Sections 9.4–9.4.3. To generalize the description, we make use of the fact that the s-polarized solution (9.88) greatly resembles the transmission and reflection coefficients (4.19) obtained for the discretized one-dimensional Helmholtz equation (4.11). In particular, when we identify a scalar wave vector

$$k_{\pm}^{[2]} \equiv n_{\pm}|\mathbf{q}| \cos\theta_{\pm}, \tag{9.96}$$

the transmission and reflection coefficients cast into the form

$$T_{\mathbf{q}}^{[2]} = \frac{2\,k_{-}^{[2]}}{k_{-}^{[2]} + k_{+}^{[2]}}, \qquad R_{\mathbf{q}}^{[2]} = \frac{k_{-}^{[2]} - k_{+}^{[2]}}{k_{-}^{[2]} + k_{+}^{[2]}}, \tag{9.97}$$

which is identical to the $\rightarrow$ branch in Eq. (4.19). This agreement suggests that the vectorial mode function problem can be mapped into a one-dimensional scalar problem once the $k_{\pm}$ are properly identified.

Before we attempt to prove this, we want to connect the result (9.89) for p-polarization with Eq. (9.97). For this purpose, we reorganize Eq. (9.86) into

$$\left\{ \begin{array}{l} \{n_-\} + \left\{n_-\,R_{\mathbf{q}}^{[1]}\right\} - \left\{n_+\,T_{\mathbf{q}}^{[1]}\right\} = 0 \\ \frac{\cos\theta_-}{n_-}\,\{n_-\} - \frac{\cos\theta_-}{n_-}\,\left\{n_-\,R_{\mathbf{q}}^{[1]}\right\} - \frac{\cos\theta_+}{n_+}\,\left\{n_+ T_{\mathbf{q}}^{[1]}\right\} = 0 \end{array} \right. . \tag{9.98}$$

We identify the effective amplitudes

$$i_{\mathbf{q}}^{[1]} = n_-, \qquad r_{\mathbf{q}}^{[1]} = n_-\,R_{\mathbf{q}}^{[1]}, \quad \text{and} \quad t_{\mathbf{q}}^{[1]} = n_+\,T_{\mathbf{q}}^{[1]} \tag{9.99}$$

for the incoming, reflected, and transmitted components, respectively. Like for the s-polarized case, the $\frac{\cos\theta_{\pm}}{n_{\pm}}$ terms appearing in (9.98) identify effective scalar wave vectors

$$k_{\pm}^{[1]} \equiv |\mathbf{q}|\,\frac{\cos\theta_{\pm}}{n_{\pm}}. \tag{9.100}$$

With the help of these, the effective transmission and reflection coefficients become

$$\left\{ \begin{array}{l} i_{\mathbf{q}}^{[1]} + r_{\mathbf{q}}^{[1]} - t_{\mathbf{q}}^{[1]} = 0 \\ k_-^{[1]}\,i_{\mathbf{q}}^{[1]} - k_-^{[1]}\,r_{\mathbf{q}}^{[1]} - k_+^{[1]}\,t_{\mathbf{q}}^{[1]} = 0 \end{array} \right. . \tag{9.101}$$

These equations yield

$$t_{\mathbf{q}}^{[1]} = \frac{2\,k_-^{[1]}}{k_-^{[1]} + k_+^{[1]}}\,i_{\mathbf{q}}^{[1]}, \qquad r_{\mathbf{q}}^{[1]} = \frac{k_-^{[1]} - k_+^{[1]}}{k_-^{[1]} + k_+^{[1]}}\,i_{\mathbf{q}}^{[1]}. \tag{9.102}$$

Thus, the effective transmission and reflection coefficients have a form that is identical to that of the s-mode solution (9.97) and transfer-matrix coefficients (4.19). We can thus introduce the effective quantities

$$i_{\mathbf{q}}^{[2]} = 1, \qquad r_{\mathbf{q}}^{[2]} = R_{\mathbf{q}}^{[2]}, \quad \text{and} \quad t_{\mathbf{q}}^{[2]} = T_{\mathbf{q}}^{[2]} \tag{9.103}$$

for the s-polarized modes, allowing us to solve for the corresponding transmission and reflection coefficients via the same analysis as for the p-polarized modes.

To connect the general s- and p-polarized solution branches to the transfer-matrix method, we look for a simple transformation such as Eqs. (9.99) and (9.103). More specifically, we attempt to solve the generic Eq. (9.64) by segmenting it into piecewise constant parts

$$\mathbf{q}^2 n_j^2\, \mathbf{u_q}(\mathbf{r}) - \nabla \times \left[\nabla \times \mathbf{u_q}(\mathbf{r})\right] = 0, \quad \text{for } z_j \le z < z_{j+1},$$
$$\nabla \cdot \left[n^2(z)\, \mathbf{u_q}(\mathbf{r})\right] = 0, \tag{9.104}$$

where the z-dependent discretization can be performed as in Fig. 4.1 if $n(z)$ is continuous. For each interval, we then construct a superposition of propagating and counterpropagating solutions using the ansatz

$$\mathbf{u}_{\sigma,\mathbf{q}}(\mathbf{r}) \equiv \mathbf{u}^{(j)}_{\sigma,\mathbf{q}}(\mathbf{r}), \qquad z_j \le z < z_{j+1},$$
$$\mathbf{u}^{(j)}_{\sigma,\mathbf{q}}(\mathbf{r}) = A^{[\sigma]}_j\, e^{i\mathbf{q}_{\theta_j}\cdot[\mathbf{r}-z_j\,\mathbf{e}_z]} + B^{[\sigma]}_j\, e^{i\mathbf{q}_{\pi-\theta_j}\cdot[\mathbf{r}-z_j\,\mathbf{e}_z]}$$
$$\mathbf{q}_{\theta_j} = n_j|\mathbf{q}|\,\mathbf{e}^{[3]}_{\theta_j} = n_j|\mathbf{q}|\,\sin\theta_j\,\mathbf{e}_\parallel + n_j|\mathbf{q}|\,\cos\theta_j\,\mathbf{e}_z. \tag{9.105}$$

These expressions generalize Eqs. (9.71)–(9.72) and (9.76)–(9.79) to be applicable within a given layer j, in analogy to Eq. (4.12). The propagation direction of the plane wave is defined by the angle of incidence θ_j associated with the propagation from left to right. The direction $\pi - \theta_j$ implies propagation into the direction mirrored with respect to the xy-plane, i.e. the plane wave has the same in-plane wave vector but its q_z component is reversed such that it corresponds to propagation from right to left. Thus, a given θ_j uniquely identifies the related transversal unit vectors through the relation (9.77).

Since we eventually want to solve Eq. (9.104) using the transfer-matrix method, we identify the mapping into an effective mode,

$$\bar{\mathbf{u}}^{(j)}_{\sigma,\mathbf{q}}(\mathbf{r}) \equiv \begin{cases} n_j\,\mathbf{u}^{(j)}_{1,\mathbf{q}}(\mathbf{r}), & \sigma = 1 \\ \mathbf{u}^{(j)}_{1,\mathbf{q}}(\mathbf{r}), & \sigma = 2 \end{cases}$$
$$\equiv a^{[\sigma]}_j\, e^{i\mathbf{q}_{\theta_j}\cdot[\mathbf{r}-z_j\,\mathbf{e}_z]} + b^{[\sigma]}_j\, e^{i\mathbf{q}_{\pi-\theta_j}\cdot[\mathbf{r}-z_j\,\mathbf{e}_z]},$$
$$\left\{a^{[\sigma]}_j,\, b^{[\sigma]}_j\right\} = \begin{cases} \left\{n_j\,A^{[1]}_j,\, n_j\,B^{[1]}_j\right\} & \sigma = 1 \\ \left\{A^{[2]}_j,\, B^{[2]}_j\right\}, & \sigma = 2 \end{cases}, \tag{9.106}$$

following the steps taken in connection with Eqs. (9.99) and (9.103). For later reference, we also identify an effective wave vector,

$$k^{[\sigma]}_j \equiv \begin{cases} |\mathbf{q}|\,\frac{\cos\theta_j}{n_j}, & \sigma = 1 \\ n_j|\mathbf{q}|\cos\theta_j, & \sigma = 2 \end{cases}, \tag{9.107}$$

which generalizes Eqs. (9.100) for arbitrary layers. Like for a single step in the dielectric profile, the mode function must now satisfy the boundary conditions

$$\left[\mathbf{e}_z \times \mathbf{u}_{\mathbf{q}}^{(j)}\right]_{z=z_{j+1}} = \left[\mathbf{e}_z \times \mathbf{u}_{\mathbf{q}}^{(j+1)}\right]_{z=z_{j+1}},$$
$$\mathbf{e}_z \times \left[\nabla \times \mathbf{u}_{\mathbf{q}}^{(j)}\right]_{z=z_{j+1}} = \mathbf{e}_z \times \left[\nabla \times \mathbf{u}_{\mathbf{q}}^{(j+1)}\right]_{z=z_{j+1}}, \tag{9.108}$$

at each $z = z_{j+1}$ interface.

In analogy to the derivation resulting in (9.73), the generic n_j system can satisfy the boundary condition (9.108) at all values of $\mathbf{r}_{\|}$ only if the in-plane momentum is conserved from layer to layer. Thus, we again find Snell's law,

$$n_j \sin\theta_j = n_{j+1} \sin\theta_{j+1}, \tag{9.109}$$

which must be fulfilled at every planar interface. All propagation directions θ_j can now be determined, once we know the n_j-layer structure and θ_k inside one of the layers. With this input, we can calculate θ_j from Eq. (9.109) using a straightforward recursive procedure. To find solutions in all situations, we must also allow for the possibility to have complex-valued θ_j based on the generalization of Eq. (9.91). The imaginary part of θ_j signifies the physical fact that the light propagation is converted into an evanescent decay within that particular layer.

To settle the connection between the plane-wave coefficients, we follow the same steps as those producing (9.86). After having implemented the mappings (9.106) and (9.107), we eventually find

$$\begin{cases} a_j^{[\sigma]} e^{ik_j l_j} + b_j^{[\sigma]} e^{-ik_j^{[\sigma]} l_j^{[\sigma]}} = a_{j+1}^{[\sigma]} + b_{j+1}^{[\sigma]} \\ k_j a_j^{[\sigma]} e^{ik_j l_j} - k_j b_j^{[\sigma]} e^{-ik_j^{[\sigma]} l_j^{[\sigma]}} = k_{j+1} a_{j+1}^{[\sigma]} - k_{j+1} b_{j+1}^{[\sigma]} \end{cases}, \tag{9.110}$$

which is identical to Eq. (4.17). The effective layer thickness is

$$l_j^{[\sigma]} \equiv \begin{cases} n_j^2 \left(z_{j+1} - z_j\right), & \sigma = 1 \\ z_{j+1} - z_j, & \sigma = 2 \end{cases}. \tag{9.111}$$

The result (9.110) shows that the construction of planar s- and p-polarized solutions can indeed be converted into a scalar problem

$$\frac{\partial^2}{\partial z^2} \phi_{\mathbf{q}}^{[\sigma]}(z) + \left[k_j^{[\sigma]}\right]^2 \phi_{\mathbf{q}}^{[\sigma]}(z) = 0, \quad \text{for } j\text{-th layer}. \tag{9.112}$$

The boundary conditions (9.110) are a consequence of the requirement that $\phi_{\mathbf{q}}^{[\sigma]}(z)$ and $\frac{\partial}{\partial z}\phi_{\mathbf{q}}^{[\sigma]}(z)$ have to be continuous at every interface. Thus, both s- and p-polarized modes have a one-to-one relation to the scalar transfer-matrix calculation discussed in Section 4.3.1. This mapping is made possible through the introduction of the effective quantities (9.106), (9.107), and (9.111). After we have identified the effective wave vector and layer thicknesses, the other constants can be directly obtained via

$$\begin{pmatrix} a_j^{[\sigma]} \\ b_j^{[\sigma]} \end{pmatrix} = \mathcal{M}_{j,l}^{[\sigma]} \begin{pmatrix} a_j^{[\sigma]} \\ b_j^{[\sigma]} \end{pmatrix}, \tag{9.113}$$

where the transfer-matrix $\mathcal{M}_{j,l}^{[\sigma]}$ follows from the same recursive multiplication of elementary matrices as in Eq. (4.24). In particular, $\mathcal{M}_{j,l}^{[\sigma]}$ is uniquely defined by the $k_j^{[\sigma]}$ and $l_j^{[\sigma]}$ values defining the planar structure. Thus, we can always solve the light modes in an arbitrary planar structure using a one-dimensional Helmholtz equation.

We may not only apply the transfer-matrix method directly to describe the propagation of waves through arbitrary planar structures but we can also classify the modes based on the identification made in Section 4.3. If the border regions of the system, i.e., $\lim_{|z|\to\infty}$, have a real-valued $k_j^{[\sigma]}$, the generalized mode has propagating solutions for continuous $\mathbf{q}$. If we have an imaginary $k_j^{[\sigma]}$ for only one of the $z \to \infty$ or $z \to -\infty$ limits, we still have a continuous $\mathbf{q}$ but the light mode decays as an evanescent wave in the particular asymptotic region. We only find solutions for discrete values of $\mathbf{q}$ if $k_j^{[\sigma]}$ is imaginary in both asymptotic regions. In the case of planar structures, these light modes never escape the innermost dielectric layers such that they are often referred to as guided modes that are analogous to the bound states of the stationary Schrödinger equation.

Figure 9.7(a) presents the result of a transfer-matrix calculation, where the first layer is air ($n_1 = 1$, white area), the second layer has the refractive index $n_2 = 3.63$ (dark area) and the third layer has the refractive index $n_3 = 2.97$ (dim area). The n_2 and n_3 layers are altered such that we have a periodic structure of $\lambda_n/4$ layers. This is followed by a $3\lambda_{n_2}/2$ thick layer with a subsequent periodic lattice of $\lambda_n/4$ layers. The dielectric structure ends with an infinitely thick n_2 layer. The actual dielectric profile is presented as a dashed line in Fig. 9.7(a). The wave length of interest is chosen to be $\lambda = 785$ nm in vacuum, which produces $\lambda_n = \lambda/n$ within the dielectric medium. We use here the incidence angle, $\theta_j = 0$,

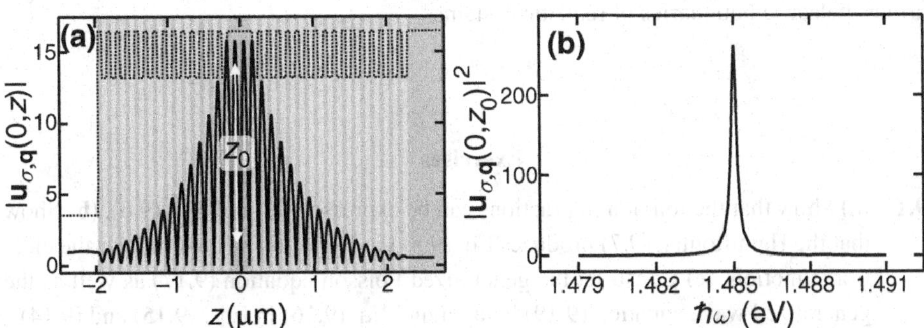

Figure 9.7 Light modes in a cavity sandwiched between two distributed Bragg reflectors (DBRs). **(a)** The actual dielectric structure consists of air (white area) and several alternating layers that have the refractive index of either $n_2 = 3.63$ (dark area) or $n_3 = 2.97$ (dim area). The dashed line presents the actual $n(z)$ profile. The DBRs consist of multiple quarter wave length layers and the cavity is three half-wave lengths wide. The cavity is designed to sustain a light mode with wave length $\lambda = 785$ nm for the $\theta_- = 0$ angle of incidence. The corresponding $|\mathbf{u}_{\sigma,\mathbf{q}}(0,z)|$ is plotted as a solid line. The z_0 position of the mode maximum is indicated by the white arrow. **(b)** The strength of the cavity mode, $|\mathbf{u}_{\sigma,\mathbf{q}}(0,z_0)|^2$ at z_0, is shown as a function of energy. The conversion to energy follows from $\hbar\omega = \hbar c|\mathbf{q}|$.

such that the s- and p-polarized results produce the same magnitude for the mode. The resulting $|\mathbf{u}_{\sigma,\mathbf{q}}(0,\ z)|$ (solid line) is plotted as a function of z.

We observe that $|\mathbf{u}_{\sigma,\mathbf{q}}(0,\ z)|$ displays a complicated interference structure where the light is strongly localized within the $3\lambda_{n_2}/2$ wide central region. We see nearly exponential decay within the stack of $\lambda_n/4$ layers such that only a small field fraction leaks into the asymptotic regions. Altogether, the dielectric structure defines a cavity for the light mode. In fact, the periodic $\lambda_n/4$ layers construct the so-called distributed Bragg reflector (DBR) while the $3\lambda_{n_2}/2$ is the cavity. This is one of the standard configurations where one can place, e.g., low-dimensional semiconductors at the position of the field maximum, indicated by z_0. To verify that the DBRs produce a cavity for selected frequencies, we have also computed $|\mathbf{u}_{\sigma,\mathbf{q}}(0,\ z_0)|^2$ as a function of energy $\hbar\omega = \hbar c|\mathbf{q}|$ in Fig. 9.7(b). We see that $|\mathbf{u}_{\sigma,\mathbf{q}}(0,\ z_0)|^2$ peaks strongly around energy 1.485 eV, which corresponds to the $\lambda = 785$ nm assumed to construct the dielectric cavity.

A semiconductor material that is positioned close to z_0 is hundreds of nanometers away from the closest dielectric layer while the typical Coulomb-interaction length is of the order of 10 nm in GaAs type systems. Hence, based on the analysis in Section 9.3, we conclude that these dielectric layers hardly influence the potential, except by reducing the charge by n_+^2. Thus, one can often use the bulk Coulomb potential $V_{i,j}/n_+^2$ instead of the fully self-consistent solution of the generalized Poisson equation (9.32). However, a similar simplification clearly does not apply to the transversal modes because the bulk modes are simple plane waves. In particular, even distant dielectric layers can strongly modify the generalized light modes because transversal modes can propagate over long distances. Thus, one must typically solve the fully self-consistent $\mathbf{u}_{\sigma,\mathbf{r}}(\mathbf{r})$ to explain how the modes are modified via interference between components reflected and diffracted from various dielectric boundaries of realistic systems.

Exercises

9.1 **(a)** Show that the ion–ion interactions can be expressed through Eq. (9.6). **(b)** Show that the Hamiltonian (9.7) produces Eq. (9.8) via Hamilton's equations for the electronic motion. **(c)** Construct the generalized Poisson equation (9.17) as well as the generalized wave equation (9.19) by applying Eq. (9.16) in Eqs. (9.15) and (9.14).

9.2 Verify that the Hamiltonian (7.67) yields Eq. (9.20) after we formally eliminate the scalar potential. Which explicit equation identifies $\phi(\mathbf{r})$? Show that the identifications (9.18) yield Eq. (9.21).

9.3 **(a)** Show that the solutions to the generalized Helmholtz equation (9.22) are normalized through Eq. (9.23). **(b)** Show that $\int d^3r\, \mathbf{u}_q^\star(\mathbf{r}) \cdot n^2(\mathbf{r})\, \mathbf{F}(\mathbf{r}) = \int d^3r\, \mathbf{u}^\star(\mathbf{r}) \cdot [n^2(\mathbf{r})\mathbf{F}(\mathbf{r})]$. **(c)** Define $C_\mathbf{q} \equiv \int d^3r'\, \mathbf{u}_\mathbf{q}^\star(\mathbf{r}) \cdot [\mathbf{n}^2(\mathbf{r})\, \mathbf{F}(\mathbf{r})]$ and construct $\mathbf{F}_G(\mathbf{r}) = \sum_\mathbf{q} C_\mathbf{q}\, \mathbf{u}_\mathbf{q}(\mathbf{r})$. Choose then $\mathbf{F}_T(\mathbf{r}) = n^2(\mathbf{r})\mathbf{F}_G(\mathbf{r})$ and show that $\mathbf{e}_\lambda \cdot \sum_\mathbf{q} \mathbf{u}_\mathbf{q}^\star(\mathbf{r})\mathbf{u}_\mathbf{q}(\mathbf{r}') \cdot$

$\mathbf{e}_{\lambda'} = \frac{1}{n^2(\mathbf{r}')}\delta^T_{\lambda,\lambda'}(\mathbf{r}-\mathbf{r}')$ because $\mathbf{F}_T(\mathbf{r})$ is an arbitrary transversal vector field. **(d)** Show that $\mathbf{F}_G(\mathbf{r}) = \frac{1}{n^2(\mathbf{r})}[n^2(\mathbf{r})\mathbf{F}(\mathbf{r})]_T$ and verify result (9.25).

9.4 Assume that $\mathbf{F}(\mathbf{r}) = \mathbf{e}_x\, e^{ikz}\, e^{-\frac{r^2}{R^2}}$ and $n^2(z) = 1 + \theta(z)$. **(a)** Construct $\mathbf{F}_G(\mathbf{r}) = \frac{1}{n^2(z)}[n^2(z)\mathbf{F}(\mathbf{r})]_T$. **(b)** Compute $\nabla \cdot F_G(\mathbf{r})$.

9.5 **(a)** Show that the mode expansions (9.27) and (9.29) convert the Hamiltonian (9.21) into Eq. (9.30). **(b)** Verify that Hamilton's equations (9.31) yield the correct Maxwell–Lorentz equations.

9.6 **(a)**Assuming $n^2(z) = n_-^2\,\theta(-z) + n_+^2\,\theta(z)$ and $\phi(\mathbf{r}) = \phi_-(\mathbf{r})\,\theta(-z) + \phi_+(\mathbf{r})\,\theta(z)$, show that the generalized Poisson equation (9.32) produces Eqs. (9.46)–(9.47). **(b)** Define the coefficients A and C based on the boundary conditions (9.47).

9.7 Fourier transform the Coulomb potential (9.58) to produce Eq. (9.63). **Hint:** *Express* $1/|\mathbf{r}|$ *in its three-dimensional form and perform the* $\int d^2 r_{\|}$ *integral first.*

9.8 **(a)** Show that the ansatz (9.65) produces Eqs. (9.66)–(9.67) when inserted into Eq. (9.64) having $n(z) = n_-\,\theta(-z) + n_+\,\theta(z)$. **(b)** Show that the plane-wave solution (9.71) can be expressed using Eqs. (9.76)–(9.79). Verify Snell's law (9.80).

9.9 Assume that $n_+ > n_-$ and that θ_- exceeds the critical angle such that the propagation in the region $z > 0$ is evanescent. **(a)** Rewrite $e^{[3]}_{\theta_+}$, $\mathbf{q}_T$ and the solution (9.71) for this case. **(b)** Use the boundary conditions (9.67) to produce $T^{[\sigma]}_{\mathbf{q}}$ and $R^{[\sigma]}_{\mathbf{q}}$. Show that one can apply the analytic continuation (9.91) in connection with Eqs. (9.88)–(9.89). **(c)** Show that the solution $n^2(z)u_{\sigma,\mathbf{q}}(\mathbf{r})$ is transversal also for the evanescent parts. **Hint:** *Compute* $\nabla \cdot [n^2(z)\, u_{\sigma,\mathbf{q}}(\mathbf{r})]$.

9.10 Assume that $n(z)$ is piece-wise constant. **(a)** Using the ansatz (9.105), show that it must satisfy the boundary conditions (9.108). **(b)** Introduce an effective mode (9.106) and an effective wave vector (9.107). Show that they imply (9.109)–(9.111), which can always be mapped into a scalar problem. **(c)** Show that the Brewster-angle condition (9.94) means $k^{[1]}_j = k^{[1]}_{j+1}$, i.e., no change in the effective wave vector. Why is there no reflection in this case?

Further reading

The Born–Oppenheimer approximation is widely used to investigate the quantum dynamics of atoms, molecules, and solids. The following books contain an introduction to this topic:

W. Barford (2005). *Electronic and Optical Properties of Conjugate Polymers*, Oxford, Oxford University Press.

A. Groß (2009). *Theoretical Surface Science*, Berlin, Springer.
M. Baer (2006). *Beyond Born–Oppenheimer: Canonical Intersections and Electronic Nonadiabatic Coupling Terms*, Hoboken, NJ, Wiley & Sons.

Calculus with Heaviside and delta functions is discussed, e.g., in the following books:

R. P. Kanwal (2006). *Generalized Functions: Theory and Technique*, 2nd edition, Boston, Birkhäuser.
G. B. Arfken and H. J. Weber (2005). *Mathematical Methods for Physicists*, 6th edition, New York, Elsevier.

10
Quantization of light and matter

Chapters 7–9 present the classical description of many-body systems in a way that allows us to identify the canonical variables for the coupled system of matter and electromagnetic fields. Thus, we are now in the position to apply the canonical quantization scheme outlined in Section 3.2.3. We already know that the quantization extends the particle concept to include also wave aspects such that the overall description satisfies the wave–particle duality. Once both matter and light are quantized, we have a full theory which can be applied to treat many interesting phenomena in the field of semiconductor quantum optics.

The quantization is conceptually more challenging for light than for particles because Maxwell's equations already describe classical waves. However, the mode expansion for the vector potential and the generalized transversal electric field allows us to identify the particle aspects associated with light waves. This approach presents the system dynamics in the form of classical Hamilton equations for the mode-expansion coefficients. Thus, the canonical quantization deals with these coefficients and supplements an additional wave character to them. In other words, the light quantization introduces complementarity at several levels: classical light is already fundamentally a wave while its dualistic particle aspects emerge in ray-like propagation, as discussed in Chapter 2. At the same time, the mode expansion identifies additional particle aspects and the quantization of the mode-expansion coefficients creates a new level of wave–particle dualism. In this chapter, we apply the canonical quantization scheme to derive the quantized system Hamiltonian. To simplify the notation, we introduce the technique of second quantization and analyze the elementary properties of quantized light eigenstates.

10.1 Canonical quantization

The canonical quantization of Section 3.2.3 can be summarized into the following four pragmatic steps:

(i) We first need to identify the classical system Hamiltonian $H_{\text{cl}}(\{q\}, \{p\})$ in terms of general position $\{q\}$ and momentum $\{p\}$ coordinates where $\{q\}, \{p\}$ should be understood as the most general set of canonical pairs. The corresponding Hamilton equations produce the classical dynamics for the given system.

(ii) The dualistic wave aspects are introduced axiomatically. In practice, we simply replace each q_j, p_j by the operators $\hat{q}_j, \hat{p}_j$ that satisfy the canonical commutation relation

$$\left[\hat{q}_j, \hat{p}_k\right]_- \equiv \hat{q}_j \hat{p}_k - \hat{p}_k \hat{q}_j = i\hbar\, \delta_{j,k}. \tag{10.1}$$

We can then use various representations for these operators, e.g.,

$$\left(\hat{q}_j \hat{=} \mathbf{q}_j,\ \hat{p}_j \hat{=} -i\hbar\frac{\partial}{\partial \mathbf{q}_j}\right), \quad \text{or} \quad \left(\hat{q}_j \hat{=} +i\hbar\frac{\partial}{\partial \mathbf{p}_j},\ \hat{p}_j \hat{=} p_j\right), \tag{10.2}$$

in real or momentum space, respectively. This procedure introduces the quantized Hamiltonian

$$\hat{H} \equiv H_{\text{cl}}\left(\{\hat{q}_1, \ldots, \hat{q}_N\}, \{\hat{p}_1, \ldots \hat{p}_N\}\right). \tag{10.3}$$

(iii) The properties of quantized particles can be computed using the time-dependent Schrödinger equation

$$i\hbar\frac{\partial}{\partial t}\Psi(q_1, \ldots, q_N; t) = \hat{H}\left(\{\mathbf{q}_j\}, \left\{-i\hbar\frac{\partial}{\partial \mathbf{q}_j}\right\}\right)\Psi(q_1, \ldots, q_N; t) \tag{10.4}$$

given here in the real-space representation.

(iv) Following the probabilistic interpretation of quantum mechanics, the distribution

$$P(q_1, \ldots, q_N) = |\Psi(q_1, \ldots, q_N; t)|^2 \tag{10.5}$$

is normalized, i.e., $\int dq_1 \ldots dq_N P(q_1, \ldots, q_N) = 1$ such that it defines the probability density of finding generalized particles at the positions $q_1, \ldots q_N$.

These four steps define the so-called first quantization where the Schrödinger equation is the particle-wave equation in analogy to the wave equation for the electromagnetic fields, see the discussion in Chapter 3.

10.1.1 Toward semiconductor quantum optics

To apply the canonical quantization for solid-state systems, we start from the classical system Hamiltonian in the generalized Coulomb gauge,

$$H = \sum_j \left(\frac{\left[\mathbf{p}_j - Q_j A(\mathbf{r}_j)\right]^2}{2m_j} + Q_j\, U^{\text{gen}}(\mathbf{r}_j)\right) + \frac{1}{2}\sum_{j,k\neq j} V^{\text{gen}}_{j,k}(\mathbf{r}_j, \mathbf{r}_k) + V^{\text{gen}}_{\text{ion}} + \sum_{\mathbf{q}}\frac{1}{2}\left[\Pi^2_{\mathbf{q}} + \omega^2_{\mathbf{q}} C^2_{\mathbf{q}}\right], \tag{10.6}$$

see Eq. (9.30). This form only describes the optically active charges Q_j while the nearly rigid ionic crystal produces the lattice potential $U^{\text{gen}}(\mathbf{r}_j)$, a constant energy contribution $V^{\text{gen}}_{\text{ion}}$ due to the ion–ion interaction, and the dielectric profile $n(\mathbf{r})$. The polarizability of the lattice, in principle, modifies the pairwise interactions $V^{\text{gen}}_{j,k}(\mathbf{r}_j, \mathbf{r}_k)$ among all optically active electrons. When all charges are far apart from the dielectric interfaces, the derivations in Section 9.3 show that the interaction can be approximated by the usual Coulomb potential

$$V^{\text{gen}}_{j,k}(\mathbf{r}_j, \mathbf{r}_k) = V^{\text{gen}}_{j,k}(\mathbf{r}_j - \mathbf{r}_k) = \frac{Q_j\, Q_k}{4\pi\varepsilon_0\, n^2 |\mathbf{r}_k - \mathbf{r}_j|}, \tag{10.7}$$

where $n = n(\mathbf{r})$ is a constant within the specific semiconductor material.

The index profile $n(\mathbf{r})$ can significantly modify the transversal light modes that must simultaneously satisfy the generalized Helmholtz equation and the generalized Coulomb gauge

$$\mathbf{q}^2 n^2(\mathbf{r})\, \mathbf{u_q}(\mathbf{r}) - \nabla \times \left[\nabla \times \mathbf{u_q}(\mathbf{r})\right] = 0, \quad \nabla \cdot [n^2(\mathbf{r})\, \mathbf{u_q}(\mathbf{r})] = 0. \tag{10.8}$$

Since this eigenvalue problem is Hermitian and only real-valued quantities appear in Eq. (10.8), we may formally choose $\mathbf{u_q}(\mathbf{r})$ to be real valued as well. This constraint is released in Section 10.2.2 where we aim for a more flexible description of the quantized light. The specific solution strategy for planar semiconductor structures is discussed in Sections 9.4–9.4.4.

With the help of the modes, the generalized transverse vector potential and electric field can be written as

$$\mathbf{A}(\mathbf{r}) = \sum\nolimits_{\mathbf{q}} \frac{1}{\sqrt{\varepsilon_0}} C_{\mathbf{q}}\, \mathbf{u_q}(\mathbf{r}), \qquad \mathbf{E}(\mathbf{r}) = -\frac{\partial}{\partial t}\mathbf{A}(\mathbf{r}) = -\sum\nolimits_{\mathbf{q}} \frac{1}{\sqrt{\varepsilon_0}} \Pi_{\mathbf{q}}\, \mathbf{u_q}(\mathbf{r}),$$
$$\mathbf{u_q}(\mathbf{r}) \in \mathbb{R}, \quad \nabla \cdot \left[n^2(\mathbf{r})\, \mathbf{A}(\mathbf{r})\right] = 0, \qquad \nabla \cdot \left[n^2(\mathbf{r})\, \mathbf{E}(\mathbf{r})\right] = 0. \tag{10.9}$$

The longitudinal part of the electrical field is included in the Coulomb interaction and in $U^{\text{gen}}(\mathbf{r}_j)$. These contributions are not quantized in contrast to the real-valued $C_{\mathbf{q}}$ and $\Pi_{\mathbf{q}}$. The canonical pairs $\{\mathbf{q}_j, \mathbf{p}_j\}$ and $\{C_{\mathbf{q}}, \Pi_{\mathbf{q}}\}$ satisfy the Hamilton equations

$$\dot{\mathbf{r}}_j = \frac{\partial H}{\partial \mathbf{p}_j}, \qquad \dot{\mathbf{p}}_j = -\frac{\partial H}{\partial \mathbf{r}_j},$$
$$\dot{C}_{\mathbf{q}} = \frac{\partial H}{\partial \Pi_{\mathbf{q}}}, \qquad \dot{\Pi}_{\mathbf{q}} = -\frac{\partial H}{\partial C_{\mathbf{q}}}, \tag{10.10}$$

which produce the classical Lorentz equation (9.8) and Maxwell's equations (7.69)–(7.70).

At this stage, we introduce the canonical quantization

$$\left[\hat{q}_j, \hat{p}_k\right]_- \equiv \hat{q}_j\, \hat{p}_k - \hat{p}_k\, \hat{q}_j = i\hbar\, \delta_{j,k},$$
$$\left[\hat{C}_{\mathbf{q}}, \hat{\Pi}_{\mathbf{q}'}\right]_- = \hat{C}_{\mathbf{q}}\, \hat{\Pi}_{\mathbf{q}'} - \hat{\Pi}_{\mathbf{q}'}\, \hat{C}_{\mathbf{q}} = i\hbar\, \delta_{\mathbf{q},\mathbf{q}'} \tag{10.11}$$

for both matter and light. Introducing the operators into Eq. (10.6), we obtain the quantized system Hamiltonian

$$\hat{H} = \sum_j \left(\frac{\left[-i\hbar\nabla_{\mathbf{r}_j} - Q_j\hat{\mathbf{A}}(\mathbf{r}_j)\right]^2}{2m_j} + Q_j\, U^{\text{gen}}(\mathbf{r}_j) \right)$$

$$+ \frac{1}{2}\sum_{j,k\neq j} V^{\text{gen}}_{j,k}(\mathbf{r}_j - \mathbf{r}_k) + V^{\text{gen}}_{\text{ion}} + \sum_{\mathbf{q}} \frac{1}{2}\left[\hat{\Pi}^2_{\mathbf{q}} + \omega^2_{\mathbf{q}}\hat{C}^2_{\mathbf{q}}\right], \quad (10.12)$$

where we have expressed the particle momentum $\hat{\mathbf{p}}_j = -i\hbar\nabla\mathbf{r}_j$ in real space while $\hat{\Pi}_{\mathbf{q}}$ and $\hat{C}_{\mathbf{q}}$ are given in an abstract space, i.e., their representation is not yet determined.

If we now choose the $C_{\mathbf{q}}$-representation, we have $\hat{\Pi}_{\mathbf{q}} \hat{=} -i\hbar\frac{\partial}{\partial C_{\mathbf{q}}}$. In this formulation, we have to deal with the quantum properties of the quantized light–matter system by solving the time-dependent Schrödinger equation

$$i\hbar\frac{\partial}{\partial t}\Psi\left(\{\mathbf{r}_j\}, \{C_{\mathbf{q}}\}; t\right) = \hat{H}\,\Psi\left(\{\mathbf{r}_j\}, \{C_{\mathbf{q}}\}; t\right), \quad (10.13)$$

where $\{\mathbf{r}_j\}$ and $\{C_{\mathbf{q}}\}$ refer to the group of particle positions and mode-expansion coefficients and Ψ is a many-body wave function. In this book, we typically use capital Greek letters, such as Ψ and Φ, to assign many-body states whereas single-particle wave functions are denoted by lower-case Greek letters such as ψ and ϕ.

As we have seen in Sections 3.1.2 and 5.3,

$$P\left(\{\mathbf{r}_j\}, \{C_{\mathbf{q}}\}; t\right) = \left|\Psi\left(\{\mathbf{r}_j\}, \{C_{\mathbf{q}}\}; t\right)\right|^2 \quad (10.14)$$

determines the probability distribution of finding the particles at the positions $\mathbf{r}_1, \ldots, \mathbf{r}_N$ and the mode-expansion coefficients at the values $C_{\mathbf{q}_1}, \ldots, C_{\mathbf{q}_M}$. This probability distribution accounts for a certain level of intrinsic fuzziness as an inherent consequence of the wave aspects in both $C_{\mathbf{q}}$ and $\mathbf{r}_j$. For particles, the wave function exists within the same *real* coordinate space in which also the classical particle motion takes place. For light, $C_{\mathbf{q}}$ is defined in an *auxiliary* space different from the real space where the mode functions exist. Consequently, the light quantization supplements each mode with a new auxiliary space in which the additional, quantum-mechanics based, wave aspects are defined.

10.1.2 Real vs. auxiliary quantization space

As usual, the properties of the quantized system can be deduced from the expectation values

$$\langle\hat{F}\rangle \equiv \int d^3\{\mathbf{r}_j\} \int d\{C_{\mathbf{q}}\}$$
$$\times\, \Psi^\star\left(\{\mathbf{r}_j\}, \{C_{\mathbf{q}}\}; t\right)\, F(\{\mathbf{r}_j\}, \{C_{\mathbf{q}}\})\,\Psi\left(\{\mathbf{r}_j\}, \{C_{\mathbf{q}}\}; t\right), \quad (10.15)$$

which generalizes Eq. (3.17) for generic light–matter operators $\hat{F}$. For example, the knowledge of $\langle\hat{\Pi}_{\mathbf{q}}\rangle$ allows us to construct the average electric field $\langle\hat{\mathbf{E}}(\mathbf{r})\rangle = -\sum_{\mathbf{q}} \frac{\mathbf{u}_{\mathbf{q}}(\mathbf{r})\,\langle\hat{\Pi}_{\mathbf{q}}\rangle}{\sqrt{\varepsilon_0}}$ as a function of $\mathbf{r}$, based on Eq. (10.9). Hence, the real-space properties of light are described by the original classical modes $\mathbf{u_q}(\mathbf{r})$ by virtue of the mode expansion

$$\hat{\mathbf{A}}(\mathbf{r}) = \sum_{\mathbf{q}} \frac{\hat{C}_{\mathbf{q}}\, \mathbf{u_q}(\mathbf{r})}{\sqrt{\varepsilon_0}}, \quad \hat{\mathbf{E}}(\mathbf{r}) = -\sum_{\mathbf{q}} \frac{\hat{\Pi}_{\mathbf{q}}\, \mathbf{u_q}(\mathbf{r})}{\sqrt{\varepsilon_0}}, \quad \mathbf{u_q} \in \mathbb{R}. \tag{10.16}$$

Thus, the wave aspects of $\hat{\mathbf{A}}(\mathbf{r})$ and $\hat{\mathbf{E}}(\mathbf{r})$ in $\mathbf{r}$ space are still defined by the classical Maxwell equation while the quantum-optical aspects stem from the operators $\hat{C}_{\mathbf{q}}$ and $\hat{\Pi}_{\mathbf{q}}$ defined in the auxiliary space. As a consequence, e.g., the quantization does not produce new real-space interference effects.

Figure 10.1 schematically explains the distinction between the representation of the electric field (a) in $\mathbf{r}$ space and (b) in the auxiliary $(C_{\mathbf{q}}, \Pi_{\mathbf{q}})$ space of the light quantization. At this point, the phase-space representation of quantized modes is purely illustrative – a more thorough discussion follows in Chapter 22. In real space, $\langle \hat{\mathbf{E}}(\mathbf{r}) \rangle$ (solid line) displays the same functional form as a classical field constructed out of modes with weights $\langle \hat{\Pi}_{\mathbf{q}_j} \rangle$. At the same time, a quantum-mechanical distribution in the auxiliary space corresponds to each $\mathbf{u_q}$ component within $\hat{\mathbf{E}}(\mathbf{r})$. For the four representative modes $\mathbf{q}_j$, $j = 1, 2, 3, 4$, these distributions are shown as fuzzy regions in Fig. 10.1(b). The points indicate the corresponding expectation values $(\langle \hat{C}_{\mathbf{q}} \rangle, \langle \hat{\Pi}_{\mathbf{q}} \rangle)$. In the chosen example, the wave functions of the modes $\mathbf{q}_3$ and $\mathbf{q}_4$ overlap partially, which produces an interference pattern (periodic bright-dark stripes) due to the intrinsic wave averaging within the wave function Ψ. This interference is not detectable in the real-space $\langle \hat{\mathbf{E}}(\mathbf{r}) \rangle$ (solid line in Fig. 10.1(a)) because any Ψ that produces the same value for $\langle \hat{\Pi}_{\mathbf{q}_j} \rangle$ yields the same $\langle \hat{\mathbf{E}}(\mathbf{r}) \rangle$ regardless of whether or not Ψ contains an interference pattern. Especially, the $\mathbf{r}$-dependent oscillations of $\langle \hat{\mathbf{E}}(\mathbf{r}) \rangle$ result from averaging of "classical" waves defined by the superposition $\sum_j \langle \hat{\Pi}_{\mathbf{q}_j} \rangle\, \mathbf{u}_{\mathbf{q}_j}(\mathbf{r})$,

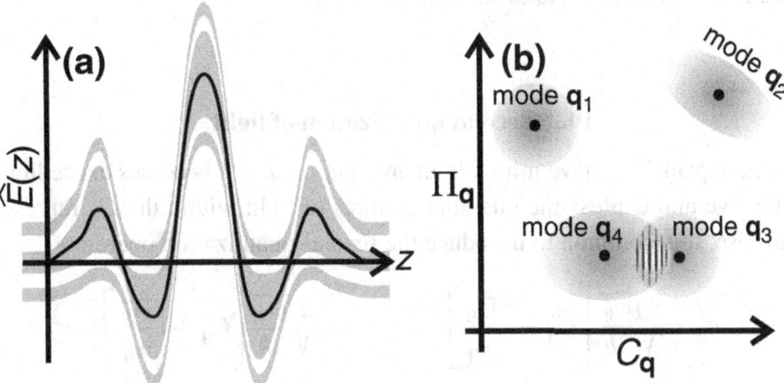

Figure 10.1 Schematic representation of quantized light. **(a)** The probability distribution $P(z,\ E)$ of possible E values is shown as a function of the real-space position z. The average $\langle \hat{E}(z) \rangle$ (solid line) is plotted together with the $P(z,\ E)$ regions having high probability (shaded areas). **(b)** The quantized field can also be decomposed into the $\mathbf{q}_j$ modes. The corresponding phase-space distributions (shaded areas) and average values of $\left(\langle \hat{C}_{\mathbf{q}_j} \rangle,\ \langle \hat{\Pi}_{\mathbf{q}_j} \rangle\right)$ (points) are symbolized for four representative modes.

where the quantum-state indiscriminate factors $\langle \hat{\Pi}_{\mathbf{q}_j} \rangle$ appear as weights. In other words, interference in the quantum world has a completely different origin from the real-space interference of classical waves.

The mode-expansion form (10.16) indicates that the quantum fluctuations can also induce nontrivial spatial effects. For example, we can always analyze how strongly the fluctuations of light change as a function of real-space position,

$$\Delta\langle\hat{\mathbf{E}}(\mathbf{r})\rangle^2 = \sum_{\mathbf{q},\mathbf{q}'} \frac{\mathbf{u}_{\mathbf{q}}(\mathbf{r})\,\mathbf{u}_{\mathbf{q}'}(\mathbf{r})}{\varepsilon_0} \left[\langle \hat{C}_{\mathbf{q}}\,\hat{C}_{\mathbf{q}'}\rangle - \langle \hat{C}_{\mathbf{q}}\rangle \left\langle \hat{C}_{\mathbf{q}'}\right\rangle\right]. \tag{10.17}$$

From this expression, we see now that $\Delta\langle\hat{\mathbf{E}}(\mathbf{r})\rangle$ vanishes only if the fluctuations

$$\Delta\left\langle \hat{C}_{\mathbf{q}}\,\hat{C}_{\mathbf{q}'}\right\rangle \equiv \left\langle \hat{C}_{\mathbf{q}}\,\hat{C}_{\mathbf{q}'}\right\rangle - \langle \hat{C}_{\mathbf{q}}\rangle \left\langle \hat{C}_{\mathbf{q}'}\right\rangle \tag{10.18}$$

are strictly zero. Since the functional form of $\Delta\left\langle \hat{C}_{\mathbf{q}}\,\hat{C}_{\mathbf{q}'}\right\rangle$ reflects the width and extent of the quantum distribution (fuzzy areas in Fig. 10.1(b)), the intrinsic fuzziness in auxiliary space defines the magnitude and distribution of the E fluctuations. According to Born's theorem, any measurement of $\hat{\mathbf{E}}$ produces a scatter of outcomes that on average follows a distribution $P(E, \mathbf{r})$, as discussed in Section 5.3.

The shaded areas in Fig. 10.1 schematically show the highly probable $(E, \mathbf{r})$ combinations. The average value of these regions produces $\langle\hat{\mathbf{E}}(\mathbf{r})\rangle$ (solid line) with the classical interference pattern. Additionally, the cross sections of $P(E, \mathbf{r})$ along the E direction also exhibit an interference structure related to the interference pattern between the modes $\mathbf{q}_3$ and $\mathbf{q}_4$ in the auxiliary space. In other words, the quantized field can be viewed as a formal distribution $P(E, \mathbf{r})$, where the E extension stems from the quantization while the $\mathbf{r}$ dependence follows from the classical wave properties.

10.2 Second quantization of light

Since we can, in principle, have infinitely many light modes, it becomes exceedingly cumbersome to solve and express the quantum features of light within the $C_{\mathbf{q}}$ representation. For this purpose, it is common to introduce the second-quantization operators

$$b^\dagger_{\mathbf{q}} = \sqrt{\frac{\omega_{\mathbf{q}}}{2\hbar}}\left[\hat{C}_{\mathbf{q}} - i\frac{\hat{\Pi}_{\mathbf{q}}}{\omega_{\mathbf{q}}}\right], \qquad b_{\mathbf{q}} = \sqrt{\frac{\omega_{\mathbf{q}}}{2\hbar}}\left[\hat{C}_{\mathbf{q}} + i\frac{\hat{\Pi}_{\mathbf{q}}}{\omega_{\mathbf{q}}}\right]. \tag{10.19}$$

Despite its name, the second quantization does not produce an additional level of quantization but expresses the system properties with carefully chosen abstract-space operators. We will extensively use the operators $b_{\mathbf{q}}$ and $b^\dagger_{\mathbf{q}}$ in our further investigations. We do not use the operator notation (hat above the operator quantity) because the operator nature is obvious from the context. The second-quantization operators are connected to the canonical operator pair via the simple linear transformation (10.19), which has the inverse

$$\hat{C}_{\mathbf{q}} = \sqrt{\frac{\hbar}{2\omega_{\mathbf{q}}}}\left(b_{\mathbf{q}} + b_{\mathbf{q}}^{\dagger}\right), \qquad \hat{\Pi}_{\mathbf{q}} = -i\sqrt{\frac{\hbar\omega_{\mathbf{q}}}{2}}\left(b_{\mathbf{q}} - b_{\mathbf{q}}^{\dagger}\right). \tag{10.20}$$

Inserting this into Eq. (10.11), we find the Bosonic commutation relations,

$$\left[b_{\mathbf{q}}, b_{\mathbf{q}'}^{\dagger}\right]_{-} = \delta_{\mathbf{q},\mathbf{q}'}, \qquad \left[b_{\mathbf{q}}, b_{\mathbf{q}'}\right]_{-} = 0 = \left[b_{\mathbf{q}}^{\dagger}, b_{\mathbf{q}'}^{\dagger}\right]_{-}. \tag{10.21}$$

One obvious benefit of the transformation (10.19) is that $b_{\mathbf{q}}$ and $b_{\mathbf{q}}^{\dagger}$ are unitless and satisfy simple commutation relations.

Next, we attempt to express the quantized light entirely in terms of $b_{\mathbf{q}}$ and $b_{\mathbf{q}}^{\dagger}$. A direct substitution of Eq. (10.20) into (10.16) produces

$$\begin{aligned}
\hat{\mathbf{A}}(\mathbf{r}) &= \sum_{\mathbf{q}} \sqrt{\frac{\hbar}{2\varepsilon_0\omega_{\mathbf{q}}}}\left(b_{\mathbf{q}} + b_{\mathbf{q}}^{\dagger}\right)\mathbf{u}_{\mathbf{q}}(\mathbf{r}) \equiv \sum_{\mathbf{q}} \frac{\mathcal{E}_{\mathbf{q}}}{\omega_{\mathbf{q}}}\left(b_{\mathbf{q}} + b_{\mathbf{q}}^{\dagger}\right)\mathbf{u}_{\mathbf{q}}(\mathbf{r}), \quad \mathbf{u}_{\mathbf{q}} \in \mathbb{R} \\
\hat{\mathbf{E}}(\mathbf{r}) &= \sum_{\mathbf{q}} i\sqrt{\frac{\hbar\omega_{\mathbf{q}}}{2\varepsilon_0}}\left(b_{\mathbf{q}} - b_{\mathbf{q}}^{\dagger}\right)\mathbf{u}_{\mathbf{q}}(\mathbf{r}) \equiv \sum_{\mathbf{q}} i\mathcal{E}_{\mathbf{q}}\left(b_{\mathbf{q}} - b_{\mathbf{q}}^{\dagger}\right)\mathbf{u}_{\mathbf{q}}(\mathbf{r})
\end{aligned} \tag{10.22}$$

for real-valued mode functions. Here, we have identified the so-called vacuum-field amplitude

$$\mathcal{E}_{\mathbf{q}} \equiv \sqrt{\frac{\hbar\omega_{\mathbf{q}}}{2\varepsilon_0}}. \tag{10.23}$$

The combination $\mathcal{E}_{\mathbf{q}}\,\mathbf{u}_{\mathbf{q}}(\mathbf{r})$ determines the strength of the electric field at a given position. As we show in Section 10.4.2, this product also determines the typical magnitude of the quantum fluctuations.

We now express the last term of the Hamiltonian (10.12) in terms of $b_{\mathbf{q}}$ and $b_{\mathbf{q}}^{\dagger}$ using the straightforward substitution of (10.20),

$$\begin{aligned}
\hat{H}_0 &= \sum_{\mathbf{q}} \frac{1}{2}\left[\hat{\Pi}_{\mathbf{q}}^2 + \omega_{\mathbf{q}}^2 C_{\mathbf{q}}^2\right] \\
&= \sum_{\mathbf{q}} \frac{1}{2}\left[-\frac{\hbar\omega_{\mathbf{q}}}{2}\left(b_{\mathbf{q}} - b_{\mathbf{q}}^{\dagger}\right)\left(b_{\mathbf{q}} - b_{\mathbf{q}}^{\dagger}\right) + \frac{\hbar\omega_{\mathbf{q}}}{2}\left(b_{\mathbf{q}} + b_{\mathbf{q}}^{\dagger}\right)\left(b_{\mathbf{q}} + b_{\mathbf{q}}^{\dagger}\right)\right] \\
&= \sum_{\mathbf{q}} \frac{\hbar\omega_{\mathbf{q}}}{4}\left[-\cancel{b_{\mathbf{q}}\,b_{\mathbf{q}}} + b_{\mathbf{q}}\,b_{\mathbf{q}}^{\dagger} + b_{\mathbf{q}}^{\dagger}\,b_{\mathbf{q}} - \cancel{b_{\mathbf{q}}^{\dagger}\,b_{\mathbf{q}}^{\dagger}}\right. \\
&\qquad \left. + \cancel{b_{\mathbf{q}}\,b_{\mathbf{q}}} + b_{\mathbf{q}}\,b_{\mathbf{q}}^{\dagger} + b_{\mathbf{q}}^{\dagger}\,b_{\mathbf{q}} + \cancel{b_{\mathbf{q}}^{\dagger}\,b_{\mathbf{q}}^{\dagger}}\right].
\end{aligned} \tag{10.24}$$

Here, the $b_{\mathbf{q}}\,b_{\mathbf{q}}$ and $b_{\mathbf{q}}^{\dagger}\,b_{\mathbf{q}}^{\dagger}$ terms cancel each other. Using $b_{\mathbf{q}}\,b_{\mathbf{q}}^{\dagger} = 1 + b_{\mathbf{q}}^{\dagger}\,b_{\mathbf{q}}$, following from Eq. (10.21), we find

$$\hat{H}_0 = \sum_{\mathbf{q}} \frac{\hbar\omega_{\mathbf{q}}}{2}\left[b_{\mathbf{q}} b^{\dagger}_{\mathbf{q}} + b^{\dagger}_{\mathbf{q}} b_{\mathbf{q}}\right] = \sum_{\mathbf{q}} \hbar\omega_{\mathbf{q}} \left(b^{\dagger}_{\mathbf{q}} b_{\mathbf{q}} + \frac{1}{2}\right). \tag{10.25}$$

With this result, the system Hamiltonian (10.12) becomes

$$\hat{H} = \sum_j \left(\frac{\left[\hat{\mathbf{p}}_j - Q_j \hat{\mathbf{A}}(\mathbf{r}_j)\right]^2}{2m_j} + Q_j\, U^{\text{gen}}(\mathbf{r}_j) \right)$$
$$+\frac{1}{2}\sum_{j,k\neq j} V^{\text{gen}}_{j,k}(\mathbf{r}_j,\, \mathbf{r}_k) + V^{\text{gen}}_{\text{ion}} + \sum_{\mathbf{q}} \hbar\omega_{\mathbf{q}} \left[b^{\dagger}_{\mathbf{q}}\, b_{\mathbf{q}} + \frac{1}{2}\right]. \tag{10.26}$$

10.2.1 Unitary transformations

Since the quantization is performed in auxiliary space, we can have many alternative representations. One useful scheme is to introduce a unitary transformation:

$$B_{\mathbf{q}} = \sum_{\mathbf{q}'} \mathbf{U}_{\mathbf{q},\,\mathbf{q}'}\, b_{\mathbf{q}'}, \qquad B^{\dagger}_{\mathbf{q}} = \sum_{\mathbf{q}'} \mathbf{U}^{\star}_{\mathbf{q},\mathbf{q}'}\, b^{\dagger}_{\mathbf{q}'} \equiv \sum_{\mathbf{q}'} b^{\dagger}_{\mathbf{q}'}\, \mathbf{U}^{H}_{\mathbf{q}',\mathbf{q}}. \tag{10.27}$$

Here, we impose the condition that the indices of $B_{\mathbf{q}}$ and $b_{\mathbf{q}'}$ have equal magnitude $|\mathbf{q}| = |\mathbf{q}'|$, which implies equal energies $\hbar\omega_{\mathbf{q}} = \hbar\omega_{\mathbf{q}'}$. The transformation (10.27) is unitary if the introduced matrix satisfies

$$\mathbf{U}\mathbf{U}^{H} = \mathbb{I} \;\Leftrightarrow\; \sum_{\mathbf{q}''} \mathbf{U}_{\mathbf{q},\,\mathbf{q}''} \mathbf{U}^{H}_{\mathbf{q}'',\,\mathbf{q}'} = \sum_{\mathbf{q}''} \mathbf{U}_{\mathbf{q},\,\mathbf{q}''} \mathbf{U}^{\star}_{\mathbf{q}',\,\mathbf{q}''} = \delta_{\mathbf{q},\,\mathbf{q}'}$$
$$\mathbf{U}^{H}\,\mathbf{U} = \mathbb{I} \Leftrightarrow \sum_{\mathbf{q}''} \mathbf{U}^{H}_{\mathbf{q},\,\mathbf{q}''}\, \mathbf{U}_{\mathbf{q}'',\,\mathbf{q}'} = \sum_{\mathbf{q}''} \mathbf{U}^{\star}_{\mathbf{q}'',\,\mathbf{q}}\, \mathbf{U}_{\mathbf{q}'',\,\mathbf{q}'} = \delta_{\mathbf{q},\,\mathbf{q}'}. \tag{10.28}$$

For the sake of brevity, we introduce the summation convention

$$\mathbf{M}_{\mathbf{q},\,\mathbf{q}'}\, \mathbf{y}_{\mathbf{q}'} \equiv \sum_{\mathbf{q}'} \mathbf{M}_{\mathbf{q},\,\mathbf{q}'}\, \mathbf{y}_{\mathbf{q}'},$$
$$\mathbf{M}_{\mathbf{q},\,\mathbf{q}''}\, \mathbf{N}_{\mathbf{q}',\,\mathbf{q}''} \equiv \sum_{\mathbf{q}''} \mathbf{M}_{\mathbf{q},\,\mathbf{q}''}\, \mathbf{N}_{\mathbf{q}',\,\mathbf{q}''}, \tag{10.29}$$

where two or more identical indices in an expression imply the summation of the identical indices over all values. This way, we can write the unitary transformations in the compact form

$$B_{\mathbf{q}} = \mathbf{U}_{\mathbf{q},\,\mathbf{q}'}\, b_{\mathbf{q}'}, \qquad B^{\dagger}_{\mathbf{q}} = \mathbf{U}^{\star}_{\mathbf{q},\,\mathbf{q}'}\, b^{\dagger}_{\mathbf{q}'},$$
$$b_{\mathbf{q}} = \mathbf{U}^{\star}_{\mathbf{q}',\mathbf{q}}\, B_{\mathbf{q}'}, \qquad b^{\dagger}_{\mathbf{q}} = \mathbf{U}_{\mathbf{q}',\mathbf{q}}\, B^{\dagger}_{\mathbf{q}'}, \tag{10.30}$$

where the last two relations introduce the inverse transformations. To check whether the Bosonic commutation still applies, we evaluate

$$\left[B_{\mathbf{q}}, B^{\dagger}_{\mathbf{q}'}\right]_{-} = \mathbf{U}_{\mathbf{q},\mathbf{q}''}\,\mathbf{U}^{\star}_{\mathbf{q}',\mathbf{q}'''}\left[b_{\mathbf{q}''}, b^{\dagger}_{\mathbf{q}'''}\right]_{-} = \mathbf{U}_{\mathbf{q},\mathbf{q}''}\,\mathbf{U}_{\mathbf{q}',\mathbf{q}'''}\,\delta_{\mathbf{q}'',\mathbf{q}'''}$$
$$= \mathbf{U}_{\mathbf{q},\mathbf{q}''}\,\mathbf{U}^{\star}_{\mathbf{q}',\mathbf{q}''} = \sum\nolimits_{\mathbf{q}''}\mathbf{U}_{\mathbf{q},\mathbf{q}''}\mathbf{U}^{\star}_{\mathbf{q}',\mathbf{q}''} = \delta_{\mathbf{q},\mathbf{q}'} \tag{10.31}$$

based on the unitarity (10.28). A very similar calculation produces

$$\left[B_{\mathbf{q}}, B_{\mathbf{q}'}\right]_{-} = 0 = \left[B^{\dagger}_{\mathbf{q}}, B^{\dagger}_{\mathbf{q}'}\right]_{-} \tag{10.32}$$

showing that also the new operators satisfy Bosonic commutation relations.

The transformations (10.30) can now be inserted into Eq. (10.25), yielding

$$H_0 = \sum\nolimits_{\mathbf{q}} \hbar\omega_{\mathbf{q}}\left(\mathbf{U}_{\mathbf{q}',\mathbf{q}}\,B^{\dagger}_{\mathbf{q}'}\,B_{\mathbf{q}''}\,\mathbf{U}^{\star}_{\mathbf{q}'',\mathbf{q}} + \frac{1}{2}\right)$$
$$= \sum\nolimits_{\mathbf{q}} \hbar\omega_{\mathbf{q}}\left(\mathbf{U}^{\star}_{\mathbf{q}'',\mathbf{q}}\,\mathbf{U}_{\mathbf{q}',\mathbf{q}}\,B^{\dagger}_{\mathbf{q}'}\,B_{\mathbf{q}''} + \frac{1}{2}\right). \tag{10.33}$$

The $\mathbf{q}$ sum over all states that appears can be separated into sums over magnitude and directions, i.e., $\sum_{\mathbf{q}} = \sum_{|\mathbf{q}|}\sum_{\mathbf{q}\in|\mathbf{q}|}$, converting Eq. (10.33) into

$$H_0 = \sum\nolimits_{|\mathbf{q}|} \hbar\omega_{|\mathbf{q}|}\left(\sum\nolimits_{\mathbf{q}\in|\mathbf{q}|}\mathbf{U}_{\mathbf{q}',\mathbf{q}}\,\mathbf{U}^{\star}_{\mathbf{q}'',\mathbf{q}}\right)B^{\dagger}_{\mathbf{q}'}\,B_{\mathbf{q}''} + \sum\nolimits_{\mathbf{q}}\frac{1}{2}\hbar\omega_{\mathbf{q}}$$
$$= \sum\nolimits_{|\mathbf{q}|} \hbar\omega_{|\mathbf{q}|}\left(\sum\nolimits_{\mathbf{q}',\mathbf{q}''\in|\mathbf{q}|}\delta_{\mathbf{q}',\mathbf{q}''}\,B^{\dagger}_{\mathbf{q}'}B_{\mathbf{q}''}\right) + \sum\nolimits_{\mathbf{q}}\frac{1}{2}\hbar\omega_{\mathbf{q}}$$
$$= \sum_{|\mathbf{q}|}\hbar\omega_{|\mathbf{q}|}\sum_{\mathbf{q}\in|\mathbf{q}|}B^{\dagger}_{\mathbf{q}}\,B_{\mathbf{q}} + \sum_{\mathbf{q}}\frac{1}{2}\hbar\omega_{\mathbf{q}} = \sum\nolimits_{\mathbf{q}}\hbar\omega_{\mathbf{q}}\left(B^{\dagger}_{\mathbf{q}}B_{\mathbf{q}} + \frac{1}{2}\right), \tag{10.34}$$

where the second step follows from the unitarity condition (10.28). We also have used the property $\omega_{\mathbf{q}} = \omega_{|\mathbf{q}|}$ and expressed the $\mathbf{q}'$ and $\mathbf{q}''$ sums explicitly with the constraint that the unitary transformations are defined among states with identical $|\mathbf{q}|$. As a result, we find an unaltered format for the energy of the transversal fields, such that we may express the quantization using arbitrary basis states that are connected via unitary transformations.

10.2.2 Complex-valued modes

To see how the quantized field is modified, we substitute Eq. (10.27) into Eq. (10.22)

$$\hat{\mathbf{A}}(\mathbf{r}) = \sum_{|\mathbf{q}|}\frac{\mathcal{E}_{|\mathbf{q}|}}{\omega_{|\mathbf{q}|}}\sum_{\mathbf{q}\in|\mathbf{q}|,\,\mathbf{q}'\in|\mathbf{q}|}\left(\mathbf{U}^{\star}_{\mathbf{q}',\mathbf{q}}\,\mathbf{u}_{\mathbf{q}}(\mathbf{r})\,B_{\mathbf{q}'} + \mathbf{U}_{\mathbf{q}',\mathbf{q}}\mathbf{u}_{\mathbf{q}}(\mathbf{r})\,B^{\dagger}_{\mathbf{q}'}\right). \tag{10.35}$$

The emerging terms $\mathbf{U}_{\mathbf{q}',\mathbf{q}}\,\mathbf{u}_{\mathbf{q}}(\mathbf{r})$ introduce new complex-valued basis states,

$$\mathbf{u}'_{\mathbf{q}}(\mathbf{r}) \equiv \sum_{\mathbf{q}'\in|\mathbf{q}|}\mathbf{U}^{\star}_{\mathbf{q},\mathbf{q}'}\,\mathbf{u}_{\mathbf{q}}(\mathbf{r}), \qquad \left[\mathbf{u}'_{\mathbf{q}}(\mathbf{r})\right]^{\star} \equiv \sum_{\mathbf{q}'\in|\mathbf{q}|}\mathbf{U}_{\mathbf{q},\mathbf{q}'}\,\mathbf{u}_{\mathbf{q}}(\mathbf{r}), \tag{10.36}$$

containing the original real-valued coefficients $\mathbf{u_q}(\mathbf{r}) = \mathbf{u_q^\star}(\mathbf{r})$. Using (10.36) in (10.35) yields

$$\begin{aligned}\hat{\mathbf{A}}(\mathbf{r}) &= \sum_{|\mathbf{q}|} \frac{\mathcal{E}_{|\mathbf{q}|}}{\omega_{|\mathbf{q}|}} \sum_{\mathbf{q}' \in |\mathbf{q}|} \left(\mathbf{u}'_{\mathbf{q}'}(\mathbf{r})\, B_{\mathbf{q}'} + \left[\mathbf{u}'_{\mathbf{q}'}(\mathbf{r})\right]^\star B^\dagger_{\mathbf{q}'} \right) \\ &= \sum_{\mathbf{q}} \frac{\mathcal{E}_{\mathbf{q}}}{\omega_{\mathbf{q}}} \left(\mathbf{u}'_{\mathbf{q}}(\mathbf{r})\, B_{\mathbf{q}} + \left[\mathbf{u}'_{\mathbf{q}}(\mathbf{r})\right]^\star B^\dagger_{\mathbf{q}} \right).\end{aligned} \tag{10.37}$$

An analogous derivation converts the electric field into

$$\hat{\mathbf{E}}(\mathbf{r}) = \sum_{|\mathbf{q}|} i\mathcal{E}_{\mathbf{q}} \left(\mathbf{u}'_{\mathbf{q}}(\mathbf{r})\, B_{\mathbf{q}} - \left[\mathbf{u}'_{\mathbf{q}}(\mathbf{r})\right]^\star B^\dagger_{\mathbf{q}} \right). \tag{10.38}$$

In both cases, the mode expansion remains structurally unaltered with the exception that one needs to account for the complex-valued nature of the new basis states. These states are solutions of the generalized Helmholtz equation (10.8),

$$\mathbf{q}^2 n^2(\mathbf{r})\, \mathbf{u}'_{\mathbf{q}}(\mathbf{r}) - \nabla \times \left[\nabla \times \mathbf{u}'_{\mathbf{q}}(\mathbf{r})\right] = 0, \quad \nabla \cdot \left[n^2(\mathbf{r})\, \mathbf{u}'_{\mathbf{q}}(\mathbf{r})\right] = 0, \tag{10.39}$$

because this equation is separately satisfied by each mode-expansion element of Eq. (10.36).

We may now evaluate the integral

$$\begin{aligned}\left\langle \mathbf{u_q} | \mathbf{u}'_{\mathbf{q}'} \right\rangle_n &\equiv \int d^3r\, \mathbf{u}^\star_{\mathbf{q}}(\mathbf{r}) \cdot n^2(\mathbf{r})\, \mathbf{u}'_{\mathbf{q}'}(\mathbf{r}) = \int d^3r\, \mathbf{u_q}(\mathbf{r}) \cdot n^2(\mathbf{r})\, \mathbf{u}'_{\mathbf{q}'}(\mathbf{r}) \\ &= \sum_{\mathbf{q}'' \in |\mathbf{q}'|} \mathbf{U}^\star_{\mathbf{q}',\mathbf{q}''} \int d^3r\, \mathbf{u_q}(\mathbf{r}) \cdot n^2(\mathbf{r})\, \mathbf{u}_{\mathbf{q}''}(\mathbf{r}) \\ &= \sum_{\mathbf{q}'' \in |\mathbf{q}'|} \mathbf{U}^\star_{\mathbf{q}',\mathbf{q}''}\, \delta_{\mathbf{q}'',\mathbf{q}'} = \mathbf{U}^\star_{\mathbf{q}',\mathbf{q}}\end{aligned} \tag{10.40}$$

after having implemented the orthogonality relation (9.23) for the original mode functions. Since $\mathbf{U}^\star_{\mathbf{q}',\mathbf{q}}$ provides a transformation between modes with equal energy, the complete matrix element vanishes for $|\mathbf{q}| \neq |\mathbf{q}'|$. For equal $|\mathbf{q}'| = |\mathbf{q}|$, the unitary transformation can be deduced from the matrix element between two solution groups. Following a very similar calculation, it is straightforward to show that the new complex-valued basis is orthogonal and produces the same completeness relation (9.24) as the original real-valued basis. This proof is the content of Exercise 10.3.

Since all the central relations remain unchanged, we do not actually need to explicitly keep track of the mode functions. We simply make a choice of a specific basis $\mathbf{u_q}(\mathbf{r})$, which in general may be complex valued, i.e., $\mathbf{u}^\star_{\mathbf{q}}(\mathbf{r}) \neq \mathbf{u_q}(\mathbf{r})$. Since these modes satisfy the generalized Helmholtz equation (10.39), derivations similar to those in Section 9.2.1 produce the general orthogonality

$$\int d^3r\, \left[\mathbf{u}_{\mathbf{q}'}(\mathbf{r})\right]^\star \cdot n^2(\mathbf{r})\, \mathbf{u_q}(\mathbf{r}) = \delta_{\mathbf{q},\mathbf{q}'} \tag{10.41}$$

as well as the completeness relation

$$\mathbf{e}_\lambda \cdot \sum_{\mathbf{q}} \mathbf{u}^\star_{\mathbf{q}}(\mathbf{r}')\, \mathbf{u}_{\mathbf{q}}(\mathbf{r}) \cdot \mathbf{e}_{\lambda'} = \frac{1}{n^2(\mathbf{r}')} \delta^T_{\lambda,\lambda'}(\mathbf{r}-\mathbf{r}') \quad \lambda, \lambda' = x,\, y,\, z \tag{10.42}$$

among the chosen transversal modes. Thus, we may present the quantized light in any orthogonal and complete basis of the generalized transversal modes. For complex-valued $\mathbf{u}_{\mathbf{q}}$, the mode expansions are

$$\begin{aligned} \hat{\mathbf{A}}(\mathbf{r}) &= \sum\nolimits_{\mathbf{q}} \frac{\mathcal{E}_{\mathbf{q}}}{\omega_{\mathbf{q}}} \left[\mathbf{u}_{\mathbf{q}}(\mathbf{r})\, B_{\mathbf{q}} + \mathbf{u}^\star_{\mathbf{q}}(\mathbf{r})\, B^\dagger_{\mathbf{q}} \right] \\ \hat{\mathbf{E}}(\mathbf{r}) &= \sum\nolimits_{\mathbf{q}} i\mathcal{E}_{\mathbf{q}} \left[\mathbf{u}_{\mathbf{q}}(\mathbf{r})\, B_{\mathbf{q}} - \mathbf{u}^\star_{\mathbf{q}}(\mathbf{r})\, B^\dagger_{\mathbf{q}} \right] \end{aligned} \tag{10.43}$$

according to Eqs. (10.37)–(10.38). From now on, we use the notation that the Bosonic photon operators are denoted by capital $B_{\mathbf{q}}$ and $B^\dagger_{\mathbf{q}}$ in connection with complex-valued modes. The resulting system Hamiltonian is then

$$\begin{aligned} \hat{H} &= \sum\nolimits_j \left(\frac{\left[\hat{\mathbf{p}}_j - Q_j \hat{\mathbf{A}}(\mathbf{r}_j) \right]^2}{2m_j} + Q_j\, U^{\text{gen}}(\mathbf{r}_j) \right) \\ &\quad + \frac{1}{2} \sum_{j,k \neq j} V^{\text{gen}}_{j,k}(\mathbf{r}_j,\, \mathbf{r}_k) + V^{\text{gen}}_{\text{ion}} + \sum_{\mathbf{q}} \hbar\omega_{\mathbf{q}} \left[B^\dagger_{\mathbf{q}} B_{\mathbf{q}} + \frac{1}{2} \right], \end{aligned} \tag{10.44}$$

where we have combined Eqs. (10.25)–(10.26) and (10.34).

10.3 Eigenstates of quantized modes

To identify some properties of the quantized field, we consider the eigenstates of an individual mode

$$\hat{H}_1 = \hbar\omega \left[B^\dagger B + \frac{1}{2} \right], \tag{10.45}$$

where $B = B_{\mathbf{q}}$ refers to $\hbar\omega_{\mathbf{q}} = \hbar\omega$. The Hamiltonian (10.45) is that of a simple harmonic oscillator. The corresponding stationary Schrödinger equation is then

$$\hat{H}_1 |\psi_n\rangle = E_n |\psi_n\rangle, \tag{10.46}$$

where we denote the single-mode wave function in abstract space using Dirac's notation introduced in Section 5.2.1. To obtain a particular representation, we implement the projections

$$\psi_n(x) = \langle x | \psi_n \rangle, \qquad \psi_n(y) = \langle y | \psi_n \rangle. \tag{10.47}$$

To gain further insight, we let the Hamiltonian act on the state $B|\psi_n\rangle$,

$$\begin{aligned}\hat{H}_1 B|\psi_n\rangle &= \hbar\omega\left(B^\dagger BB + \frac{1}{2}B\right)|\psi_n\rangle \\ &= \hbar\omega\left(-B + BB^\dagger B + B\frac{1}{2}\right)|\psi_n\rangle = -\hbar\omega B|\psi_n\rangle + B\hat{H}_1|\psi_n\rangle \\ &= [-\hbar\omega + E_n]\, B|\psi_n\rangle, \end{aligned} \tag{10.48}$$

where the second step follows from the commutation relation (10.31) stating $B^\dagger B = BB^\dagger - 1$. Equation (10.48) shows that not only $|\psi_n\rangle$ but also $B|\psi_n\rangle$ is an eigenstate of Eq. (10.46). The energy eigenvalue corresponding to $B|\psi_n\rangle$ is decreased by $\hbar\omega$ relative to that of $|\psi_n\rangle$. Formally, the energy quantum $\hbar\omega$ is annihilated such that $B_{\mathbf{q}}$ is often referred to as the *annihilation operator*. Following similar derivation steps, we can show that $B^\dagger|\psi_n\rangle$ is also an eigenstate because

$$\hat{H}_1 B^\dagger|\psi_n\rangle = (E_n + \hbar\omega)B^\dagger|\psi_n\rangle \tag{10.49}$$

just has its energy increased by $\hbar\omega$, i.e. the energy quantum $\hbar\omega$ is created. Thus, $B_{\mathbf{q}}^\dagger$ is often referred to as the *creation operator*.

If we now apply the annihilation and creation operators N times, we obtain

$$\begin{aligned} \left|\psi_{n,N}^-\right\rangle &\equiv B^N|\psi_n\rangle, \qquad \hat{H}_1\left|\psi_{n,N}^-\right\rangle = [E_n - N\hbar\omega]\left|\psi_{n,N}^-\right\rangle, \\ \left|\psi_{n,N}^+\right\rangle &\equiv [B^\dagger]^N|\psi_n\rangle, \qquad \hat{H}_1\left|\psi_{n,N}^+\right\rangle = [E_n + N\hbar\omega]\left|\psi_{n,N}^+\right\rangle. \end{aligned} \tag{10.50}$$

These relations seem to imply that the energy can be reduced without a bound by successive applications of the annihilation operator. However, we already know that the harmonic oscillator has a ground state which has the lowest possible energy of the system, see Section 4.3.3. Thus, the ground state must satisfy

$$B\,|\psi_0\rangle = 0, \tag{10.51}$$

which limits the number of physically relevant annihilations in $\left|\psi_{n,N}^-\right\rangle$ to $N = n$. The ground state of the quantized light is often referred to as the vacuum.

10.3.1 Explicit representation of operators

To determine the ground-state configuration based on Eq. (10.51), we need to consider the explicit representation instead of the abstract one. For this purpose, we introduce the scaled position and momentum operators via

$$\begin{aligned} \hat{x} &\equiv \frac{B + B^\dagger}{2}, \qquad \hat{y} \equiv \frac{B - B^\dagger}{2i} \\ \Leftrightarrow \quad B &= \hat{x} + i\hat{y}, \qquad B^\dagger = \hat{x} - i\hat{y}, \end{aligned} \tag{10.52}$$

respectively. Equation (10.20) provides a simple connection $\hat{C} = \sqrt{\frac{2\hbar}{\omega}}\,\hat{x}$ and $\hat{\Pi} = \sqrt{2\hbar\omega}\,\hat{y}$ between the scaled and the original canonical operator pairs. We see that any $\hat{x}$- or $\hat{y}$-related coordinate exists in the auxiliary space of quantization, compare Section 10.1.2. To avoid any confusion with the actual $\hat{r}$ and $\hat{p}$ space of true particles, the operators $\hat{x}$ and $\hat{y}$ are often referred to as *quadrature operators*. These quadrature operators satisfy the commutation relation

$$[\hat{x}, \hat{y}]_- = \frac{i}{2}, \tag{10.53}$$

where instead of the $\hbar$ in the canonical commutator, we have the factor $1/2$. Thus, $\hat{x}$ and $\hat{y}$ behave like a canonical operator pair, however, they are dimensionless.

The x-space representation of $\hat{x}$ and $\hat{y}$ follows from

$$(\hat{x}, \hat{y}) \to \left(x, -\frac{i}{2}\frac{\partial}{\partial x}\right) \tag{10.54}$$

based on the identification (10.2). More rigorously, the conversion from abstract to x-space representation can be formulated via

$$\langle x|\,\hat{x}\,|x'\rangle = x'\,\delta(x'-x), \qquad \langle x|\,\hat{y}\,|x'\rangle = \delta(x'-x)\left[-\frac{i}{2}\frac{\partial}{\partial x'}\right]$$
$$\langle x|\,\hat{O}\,|x'\rangle = \delta(x'-x)O(x), \tag{10.55}$$

where the last format converts the typical x-dependent function into x space. These conversions can also be applied to represent the Bosonic operators (10.52)

$$\langle x|\,B\,|x'\rangle = \delta(x-x')\left[x + \frac{1}{2}\frac{\partial}{\partial x}\right] = \delta(x'-x)\left[x' + \frac{1}{2}\frac{\partial}{\partial x'}\right],$$
$$\langle x|\,B^\dagger\,|x'\rangle = \delta(x-x')\left[x - \frac{1}{2}\frac{\partial}{\partial x}\right] = \delta(x'-x)\left[x' - \frac{1}{2}\frac{\partial}{\partial x'}\right] \tag{10.56}$$

in the x representation. The $|x\rangle$ that appears defines the eigenstate of the position operator

$$\hat{x}|x\rangle = x\,|x\rangle, \qquad \langle x|\hat{x} = \langle x|x, \qquad \langle x|x'\rangle = \delta(x-x') \tag{10.57}$$

and satisfies the completeness relation

$$\int dx\,|x\rangle\langle x| = \mathbb{I}, \tag{10.58}$$

as shown in Section 5.2.1.

The conversion of the abstract-space expressions into x space is obtained via

$$\langle x|\,\hat{O}\,|\psi\rangle = \langle x|\,\hat{O}\,\mathbb{I}\,|\psi\rangle = \int dx'\,\langle x|\,\hat{O}\,|x'\rangle\langle x'|\psi\rangle$$
$$= \int dx'\,O(x)\,\delta(x-x')\,\psi(x') = O(x)\,\psi(x). \tag{10.59}$$

The first expression simply inserts the identity $\mathbb{I} = \int dx' \,|x'\rangle\langle x'|$ between the operator and $|\psi\rangle$. The next step then implements the identification (10.55) to represent the operator in x space. These two steps allow us to convert any abstract-notation quantity into x space.

10.3.2 Properties of creation and annihilation operators

To get familiar with the creation and annihilation operators, we verify that $B^\dagger$ is really the Hermitian adjoint of B. For this purpose, we use

$$\begin{aligned}\langle\psi|\,B\,|\psi\rangle &= \langle\psi|\,\mathbb{I}\,B\,\mathbb{I}\,|\psi\rangle = \int dx \int dx' \,\langle\psi|x\rangle\,\langle x|B|x'\rangle\,\langle x'|\psi\rangle \\ &= \int dx \int dx' \,\psi^\star(x)\,\delta(x'-x)\left[\left(x' + \frac{1}{2}\frac{\partial}{\partial x'}\right)\psi(x')\right] \\ &= \int dx\,\psi^\star(x)\left[\left(x + \frac{1}{2}\frac{\partial}{\partial x}\right)\psi(x)\right],\end{aligned} \tag{10.60}$$

where $\psi(x)$ is an arbitrary wave function. Via integration by parts, we can convert the expression such that the operator acts on $\psi^\star(x)$,

$$\begin{aligned}\langle\psi|\,B\,|\psi\rangle &= \int_{-\infty}^{\infty} dx\,[x\,\psi(x)]^\star\,\psi(x) + \int_{-\infty}^{\infty} dx\,\psi^\star(x)\frac{1}{2}\left(\frac{\partial}{\partial x}\psi(x)\right) \\ &= \int_{-\infty}^{\infty} dx\,[x\,\psi(x)]^\star\,\psi(x) + \left|_{-\infty}^{\infty}\frac{|\psi(x)|^2}{2} - \int_{-\infty}^{\infty} dx\left[\frac{1}{2}\frac{\partial}{\partial x}\psi(x)\right]^\star\psi(x)\right. \\ &= \int_{-\infty}^{\infty} dx\left[\left(x - \frac{1}{2}\frac{\partial}{\partial x}\right)\psi(x)\right]^\star\psi(x) \\ &= \int_{-\infty}^{\infty} dx\left[\langle x|B^\dagger|\psi\rangle\right]^\star\langle x|\psi\rangle.\end{aligned} \tag{10.61}$$

Since any proper wave function is square integrable (i.e., $\int_{-\infty}^{\infty} dx\,|\psi(x)|^2$ is finite), it must vanish at the integration borders. Additionally, we use the fact that x and $\partial/\partial x$ are real-valued operators that are not influenced by the complex conjugation. From Eq. (10.56), we see that the resulting expression $x - \frac{1}{2}\frac{\partial}{\partial x}$ is directly related to $B^\dagger$.

A similar derivation can be done for $\langle\psi|B^\dagger|\psi\rangle = \int dx\langle\psi|B^\dagger|x\rangle\langle x|\psi\rangle$ yielding $\int dx\,[\langle x|B|\psi\rangle]^\star\,\langle x|\psi\rangle$. Since $|\psi\rangle$ is an arbitrary wave function, we have thus verified that B and $B^\dagger$ are a pair of adjunct operators with the properties

$$\begin{aligned}\langle\psi|\,B &= \left(B^\dagger\,|\psi\rangle\right)^\dagger, \qquad \langle\psi|\,B^\dagger = (B\,|\psi\rangle)^\dagger \\ B\,|\psi\rangle &= \left(\langle\psi|\,B^\dagger\right)^\dagger, \qquad B^\dagger\,|\psi\rangle = (\langle\psi|\,B)^\dagger\,.\end{aligned} \tag{10.62}$$

10.3.3 Fock states

By following the formal steps of Section 10.3.1, the ground-state condition (10.51) can be converted into the explicit x-space format,

$$\langle x|B|\psi_0\rangle = \langle x|(\hat{x}+i\hat{y})|\psi_0\rangle = \left(x+\frac{1}{2}\frac{\partial}{\partial x}\right)\psi_0(x) = 0. \tag{10.63}$$

The resulting differential equation has the normalized solution

$$\psi_0(x) = \left(\frac{2}{\pi}\right)^{\frac{1}{4}} e^{-x^2} \tag{10.64}$$

that defines the ground state of the quantized light mode. The corresponding eigenenergy follows from

$$\begin{aligned} E_0 = E_0\langle\psi_0|\psi_0\rangle &= \langle\psi_0|\hat{H}_1|\psi_0\rangle = \langle\psi_0|\hbar\omega\left(B^\dagger B+\frac{1}{2}\right)|\psi_0\rangle \\ &= \hbar\omega\left[\langle\psi_0|B^\dagger B|\psi_0\rangle + \frac{1}{2}\langle\psi_0|\psi_0\rangle\right] = \frac{1}{2}\hbar\omega_0 \end{aligned} \tag{10.65}$$

because $B^\dagger B|\psi_0\rangle$ vanishes due to Eq. (10.51).

Since the ground state is now explicitly known, we may generate all the excited states with the help of Eq. (10.50). In particular, the n-th excited state is obtained by operating n times with the creation operator on the ground state,

$$|\psi_n\rangle = N_n[B^\dagger]^n|\psi_0\rangle, \qquad \hat{H}_1|\psi_n\rangle = \hbar\omega\left(n+\frac{1}{2}\right)|\psi_n\rangle. \tag{10.66}$$

We see that the eigenenergy $E_n = \hbar\omega\left(n+\frac{1}{2}\right)$ increases via equidistantly spaced steps of $\hbar\omega$.

We normalize $|\psi_n\rangle$ by demanding $\langle\psi_n|\psi_n\rangle = 1$,

$$I_n = \langle\psi_n|\psi_n\rangle = N_n^2\langle\psi_0|B^n(B^\dagger)^n|\psi_0\rangle, \tag{10.67}$$

based on Eqs. (10.66) and (10.62). For the explicit evaluation, we adopt the strategy where we commute the annihilation operators one by one toward $|\psi_0\rangle$. In the first step, we obtain

$$\begin{aligned} B(B^\dagger)^n &= BB^\dagger(B^\dagger)^{n-1} = (B^\dagger)^{n-1} + B^\dagger B(B^\dagger)^{n-1} \\ &= n(B^\dagger)^{n-1} + (B^\dagger)^n B, \end{aligned} \tag{10.68}$$

where we used the commutator relation, $BB^\dagger = 1 + B^\dagger B$, n times. Repeating the pertinent steps for the second B yields

$$\begin{aligned} B\,B\,(B^\dagger)^n|\psi_0\rangle &= n\,B\,(B^\dagger)^{n-1}|\psi_0\rangle + B(B^\dagger)^n \cancel{B|\psi_0\rangle} \\ &= n(n-1)\,(B^\dagger)^{n-2}|\psi_0\rangle + n(B^\dagger)^{n-1}\cancel{B|\psi_0\rangle} \\ &= n(n-1)\,(B^\dagger)^{n-2}|\psi_0\rangle, \end{aligned} \tag{10.69}$$

where the terms with $B|\psi_0\rangle$ vanish due to Eq. (10.51). As we go through this procedure k-times, we eventually find

$$B^k(B^\dagger)^n|\psi_0\rangle = \frac{n!}{(n-k)!}(B^\dagger)^{n-k}|\psi_0\rangle. \tag{10.70}$$

Inserting this result into Eq. (10.67), we get

$$I_n = N_n^2\,\frac{n!}{0!}\,\langle\psi_0|\psi_0\rangle = N_n^2\,n! = 1 \quad\Rightarrow\quad N_n = \frac{1}{\sqrt{n!}}, \tag{10.71}$$

which determines the norm.

The eigenstates found are often referred to as the *Fock number states,*

$$|n\rangle \equiv |\psi_n\rangle = \frac{1}{\sqrt{n!}}(B^\dagger)^n|0\rangle \quad \text{with} \quad B|0\rangle = 0. \tag{10.72}$$

It is often convenient to label the Fock states via the quantum number n identifying each state uniquely. Acting with either the creation or annihilation operator upon the Fock state k times, and using Eqs. (10.70) and (10.72), we find

$$(B^\dagger)^k|n\rangle = \sqrt{\frac{(n+k)!}{n!}}|n+k\rangle, \qquad (B)^k|n\rangle = \sqrt{\frac{n!}{(n-k)!}}|n-k\rangle. \tag{10.73}$$

This result can be applied for all values of k and n because the factorial can be understood as a special limit of the gamma function

$$\Gamma(x) = \int_0^\infty dt\,t^{x-1}\,e^{-t}, \qquad n! = \Gamma(n+1). \tag{10.74}$$

In particular, $(n-k)! = \Gamma(n-k+1)$ diverges whenever k is greater than n. As a result, if we act with B for $k > n$ times on the $|n\rangle$ state, we correctly obtain a vanishing result due to the $1/\sqrt{\Gamma(n-k+1)} = 0$ factor in (10.73). Thus, we adopt this Γ-function representation of the factorials because then we do not have to explicitly impose limitations for the k values in Eq. (10.73).

10.3.4 Fock states in x space

We may also express the Fock states in the x-space representation by combining Eqs. (10.72) and (10.59) to produce

$$\psi_n(x) = \langle x|n\rangle = \frac{1}{\sqrt{n!}}\langle x|(B^\dagger)^n|0\rangle = \frac{1}{\sqrt{n!}}\left(x - \frac{1}{2}\frac{\partial}{\partial x}\right)^n \psi_0(x). \tag{10.75}$$

To evaluate this expression, we use the general transformation

$$e^{x^2}\left(-\frac{1}{2}\frac{\partial}{\partial x}\right)e^{-x^2} = x - \frac{1}{2}\frac{\partial}{\partial x}, \tag{10.76}$$

which implies

$$\begin{aligned}\left(x-\frac{1}{2}\frac{\partial}{\partial x}\right)^n &= e^{x^2}\left(-\frac{1}{2}\frac{\partial}{\partial x}\right)e^{-x^2}e^{x^2}\left(-\frac{1}{2}\frac{\partial}{\partial x}\right)e^{-x^2}\dots e^{x^2}\left(-\frac{1}{2}\frac{\partial}{\partial x}\right)e^{-x^2}\\ &= e^{x^2}\left(-\frac{1}{2}\frac{\partial}{\partial x}\right)^n e^{-x^2}.\end{aligned} \tag{10.77}$$

Thus, the Fock state $|n\rangle$ can be expressed via

$$\begin{aligned}\psi_n(x) &= \frac{1}{\sqrt{n!}}e^{x^2}\left(-\frac{1}{2}\frac{\partial}{\partial x}\right)^n e^{-x^2}\left(\frac{2}{\pi}\right)^{\frac{1}{4}}e^{-x^2}\\ &= \left(\frac{2}{\pi}\right)^{\frac{1}{4}}\frac{1}{\sqrt{n!2^n}}(-1)^n e^{-x^2}e^{2x^2}\left(\frac{1}{\sqrt{2}}\frac{\partial}{\partial x}\right)^n e^{-2x^2}\\ &= \left(\frac{2}{\pi}\right)^{\frac{1}{4}}\frac{(-1)^n}{\sqrt{n!2^n}}e^{-x^2}e^{(\sqrt{2}x)^2}\left(\frac{\partial}{\partial\sqrt{2}x}\right)^n e^{-(\sqrt{2}x)^2}.\end{aligned} \tag{10.78}$$

The factor

$$H_n(s) \equiv (-1)^n e^{+s^2}\left(\frac{\partial}{\partial s}\right)^n e^{-s^2} \tag{10.79}$$

identifies one possible representation of the Hermite polynomials which have the properties

$$\begin{aligned}H_n(s) &= \sum_{j=0}\frac{(-1)^j n!}{j!(n-2j)!}(2s)^{n-2j}\\ H_0(x) &= 1, \qquad H_1(x) = 2x, \qquad H_2(x) = 4x^2-2.\end{aligned} \tag{10.80}$$

Thus, the Fock state can be written as

$$\psi_n(x) = \langle x|n\rangle = \left(\frac{2}{\pi}\right)^{\frac{1}{4}}\frac{1}{\sqrt{n!\,2^n}}e^{-x^2}H_n(\sqrt{2}x). \tag{10.81}$$

This form expresses the eigenstates of the quantized light mode in x space.

In general, the Hermite polynomials belong to the class of special functions whose analytic properties are well known. For example, the Hermite polynomials are orthogonalized via

$$\int_{-\infty}^{\infty}dx\, H_n(\sqrt{2}x)\, H_m(\sqrt{2}x)\, e^{-2x^2} = 2^n n!\sqrt{\frac{\pi}{2}}\,\delta_{n,m} \tag{10.82}$$

and they form a complete set. This result implies that also the Fock states are orthogonal and complete.

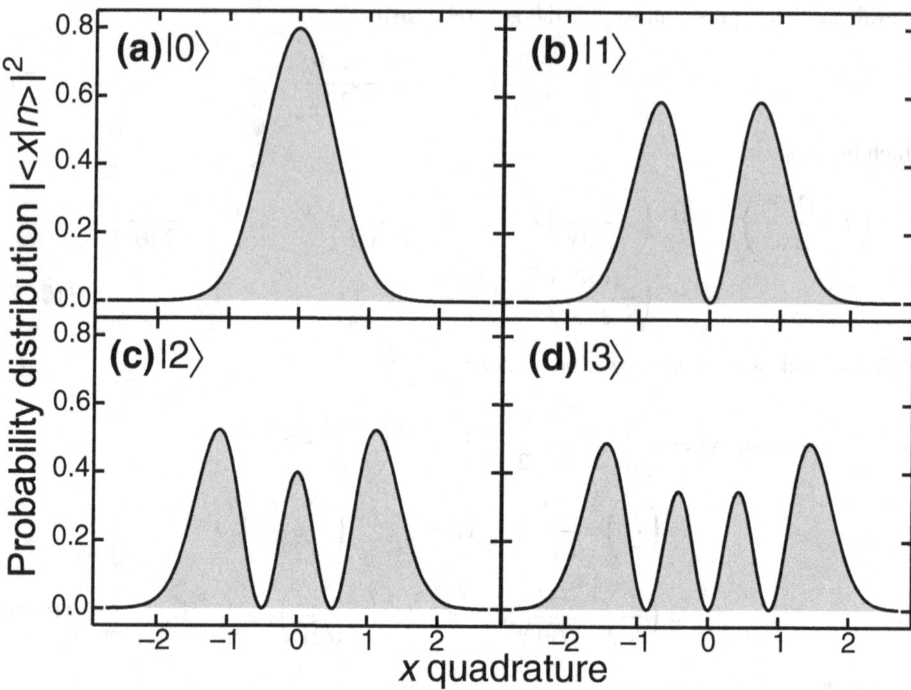

Figure 10.2 Probability distributions $P(x)$ for different Fock states. The distribution $P(x) = |\psi_n(x)|^2$, evaluated using Eq. (10.81), is plotted as a function of the x quadrature for **(a)** the vacuum $n = 0$, **(b)** the one-photon $n = 1$, **(c)** the two-photon $n = 2$, and **(d)** the three-photon $n = 3$ state.

The properties of Fock states are visualized in Fig. 10.2 where the probability distribution $P_n(x) = |\psi_n(x)|^2$ is plotted as a function of the x quadrature for (a) the ground state ($n = 0$) and for three excited states with quantum numbers (b) $n = 1$, (c) $n = 2$, and (d) $n = 3$. As a general feature, the probability distribution $P_n(x)$ shows similarities to a standing wave around $x = 0$.

It is important to note that the Fock states are also eigenstates of the harmonic-oscillator problem having the quadratic potential $V(x) \propto x^2$, see Exercise 10.6. Since this oscillator potential grows without bound for $|x| \to \infty$, the corresponding Schrödinger equation has only decaying tunneling-type solutions in the asymptotic regions. Thus, the set of $\psi_n(x)$ must represent bound states with a discrete energy spectrum. Outside the asymptotic regions, $\psi_n(x)$ consists of counterpropagating wave components that produce an interference pattern, as discussed in Section 4.3.3. We see from Fig. 10.2 that only the vacuum state, $|0\rangle$, is a Gaussian. For all the cases with $n \geq 1$, we find that the corresponding $|\psi_n(x)|^2$ vanishes at one or more points. The number of these nodes is equal to n. We always find a node at $x = 0$ for odd n whereas a local maximum at $x = 0$ appears for even n. As we increase n, the probability distribution becomes wider and has more nodes.

10.4 Elementary properties of Fock states

By now, we know that the Fock states are the eigenstates of the harmonic-oscillator problem,

$$\hbar\omega\left(B^{\dagger}B+\frac{1}{2}\right)|n\rangle = E_n|n\rangle, \qquad E_n=\hbar\omega\left(n+\frac{1}{2}\right), \tag{10.83}$$

with the ground state $|0\rangle$ having the energy $E_0=\frac{1}{2}\hbar\omega$. The excited states, $|n>0\rangle$, have a discrete spectrum where the energies are equidistantly spaced with the separation $\hbar\omega$. The excitation from $|0\rangle$ to the n-th excited state takes as much energy as collectively exciting n independent harmonic oscillators from the ground state to the first excited state. Thus, the energy spectrum of the harmonic oscillator can be associated with indistinguishable particles, where each excited state $|n\rangle$ corresponds to n existing particles while the vacuum refers to the complete absence of particles. This association is schematically made via the bubble next to the specific energy (filled circle) in Fig. 10.3. Hence, we establish a formal analogy between Fock states and a set of identical particles. We will show later that the corresponding many-body state is symmetric such that the harmonic oscillator model describes the mathematical structure of Bosons.

The possible connection between particle excitations and Fock states is often used to introduce the concept of a photon. More precisely, one may say that $|n\rangle$ is an n-photon state. However, the concept of the photon as a quantum-mechanical particle should not be taken too literally. Even though a discussion in terms of photons is sometimes useful for intuitive reasons, $|n\rangle$ does not represent a real particle for light. First of all, the light quantization is performed in the auxiliary space of mode-expansion coefficients while the real-space aspects are governed by the classical Maxwell equations. Thus, a photon cannot

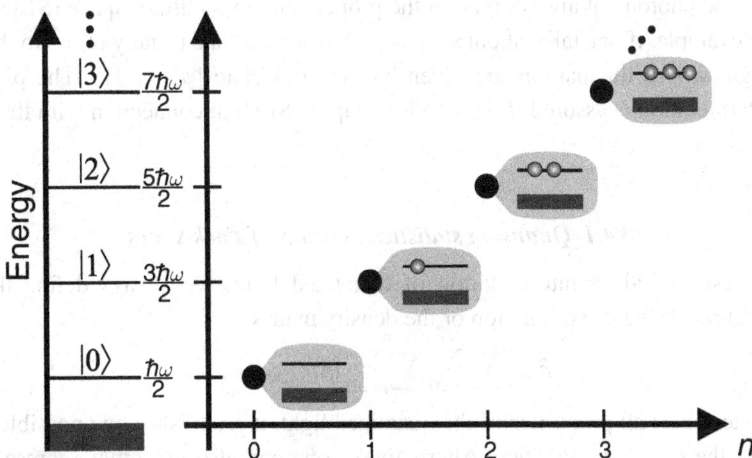

Figure 10.3 Fock states and photons. The energetic structure of Fock states is shown as a function of the quantum-number n (ladder structure as well as filled circles). The excited state $|n\rangle$ corresponds to n existing particles, here denoted as photons (spheres within the bubbles).

be a particle-like object that moves in real space, even though one can form a propagating wave packet whose *fluctuations* have the character of a specific Fock state.

In addition, $|n\rangle$ just presents a complete set of basis states that can be applied to determine the generic state of light,

$$|\psi\rangle = \sum_{n=0}^{\infty} c_n |n\rangle \quad P_n \equiv |c_n|^2, \tag{10.84}$$

where P_n defines the probability of finding the system in the Fock-state $|n\rangle$. Thus, one can have arbitrarily many wave-function configurations whose average energy and photon number,

$$\langle \hat{H} \rangle = \langle \psi | \hbar\omega \left(B^\dagger B + \frac{1}{2} \right) | \psi \rangle = \hbar\omega \left(\langle B^\dagger B \rangle + \frac{1}{2} \right)$$
$$\langle B^\dagger B \rangle = \langle \psi | B^\dagger B | \psi \rangle = \sum_{n=0}^{\infty} n \, P_n, \tag{10.85}$$

remain unchanged if the coefficients c_n are chosen appropriately. Thus, a real-space wave packet of light typically consists of a superposition of Fock states. If the concept of particle-like photons is taken too literally, one would be forced to think that the wave packet contains a mixture of different particle numbers with weights defined by c_n. In other words, a literal particle interpretation would eventually produce a string of exceptional and cumbersome properties that can completely be avoided because the Fock-state/auxiliary-space interpretation of quantized light provides trivial explanations.

Thus, it is advisable to think of quantized light in the proper context provided by the Fock-state basis expansion of wave functions that exists in the auxiliary space of the mode-expansion coefficients. In fact, one can determine quantum-optical properties entirely within this framework, without even introducing photons. However, the photon concept makes labeling of effects easier and more intuitive. In this book, we do use the concept of the photon but always rely on the proper state-in-auxiliary-space (SIAS) frame work. For example, if one talks about single-photon sources one usually refers to the emission of light whose fluctuations are given by the Fock state $|\psi\rangle = |1\rangle$. The physically proper interpretation is assured if we implicitly apply SIAS in connection with the photon concept.

10.4.1 Quantum statistics in terms of Fock states

When we use the SIAS interpretation of quantized light, we always define the light properties through the wave function or the density matrix

$$\hat{\rho} = \sum_{n=0}^{\infty} \sum_{m=0}^{\infty} |n\rangle c_{n,m} \langle m|. \tag{10.86}$$

Since $\hat{\rho}$ determines all properties of the quantized light, it establishes one possible format to describe the quantum statistics. Alternatively, one can also use other representations such as the Wigner function, see Chapters 6 and 22.

In our computations of the $\hat{\rho}$-related properties, we often resort to the results derived in Sections 10.1–10.3. Since we use these properties extensively, we summarize here the most useful relations. The Bosonic commutators,

$$[B, B^\dagger]_- = 1 \quad [B, B]_- = 0 = [B^\dagger, B^\dagger]_-, \tag{10.87}$$

allow for efficient evaluations when they are applied in calculations together with the creation- and annihilation-operator properties

$$\left(B^\dagger\right)^k |n\rangle = \sqrt{\frac{(n+k)!}{n!}}\,|n+k\rangle, \qquad B^k |n\rangle = \sqrt{\frac{n!}{(n-k)!}}\,|n-k\rangle$$
$$n, k = 0, 1, 2, \cdots \tag{10.88}$$

Application of the annihilation operator produces zero for $n - k < 0$ because $1/(n-k)!$ vanishes, as discussed in connection with Eq. (10.73). The quadrature operators follow from

$$\left\{ \begin{array}{l} \hat{x} = \frac{B+B^\dagger}{2} \\ \hat{y} = \frac{B-B^\dagger}{2i} \end{array} \right. \Leftrightarrow \left\{ \begin{array}{l} B = \hat{x} + i\hat{y} \\ B^\dagger = \hat{x} - i\hat{y} \end{array} \right. \Leftrightarrow \left\{ \begin{array}{l} \langle x|\hat{x}|x'\rangle = \delta(x'-x)x \\ \langle x|\hat{y}|x'\rangle = -\frac{i\,\delta(x'-x)}{2}\frac{\partial}{\partial x} \end{array} \right. . \tag{10.89}$$

The x-space representation of the Fock states is

$$\begin{aligned} \psi_n(x) = \langle x|n\rangle &= \left(\frac{2}{\pi}\right)^{\frac{1}{4}} \frac{1}{\sqrt{n!2^n}} e^{-x^2} H_n\left(\sqrt{2}x\right) \\ &= \left(\frac{2}{\pi}\right)^{\frac{1}{4}} \frac{1}{\sqrt{n!2^n}} e^{-x^2} \sum\nolimits_{j=0} \frac{(-1)^j n!}{j!(n-2j)!} \left(2\sqrt{2}x\right)^{n-2j}, \end{aligned} \tag{10.90}$$

where $H_n(x)$ denotes the Hermite polynomial. As proven in connection with Eqs. (10.81)–(10.82), the Fock states form an orthogonal and complete set,

$$\langle n|n'\rangle = \delta_{n,n'}, \qquad \sum\nolimits_n |n\rangle\langle n| = \mathbb{I}. \tag{10.91}$$

The quantum statistics can be expressed via the expansion coefficients

$$\langle n|\hat{\rho}|m\rangle = c_{n,m}, \tag{10.92}$$

where the diagonal terms $c_{n,n}$ define the probability of finding the quantized light in a specific Fock state $|n\rangle$.

With the help of Eqs. (10.88) and (10.91), we find

$$\begin{aligned} \langle n'|(B^\dagger)^j (B)^k|n\rangle &= \sqrt{\frac{n!n'!}{(n-k!)(n'-j)!}}\,\langle n'-j|n-k\rangle \\ &= \delta_{n',n+j-k} \frac{\sqrt{n!(n+j-k)!}}{(n-k)!}. \end{aligned} \tag{10.93}$$

For $n = n'$, this relation reduces to

$$\langle n|(B^\dagger)^j B^k|n\rangle = \delta_{j,k}\frac{n!}{(n-k)!}. \tag{10.94}$$

The expectation value in Eq. (10.93) contains a normally ordered sequence of operators, which means that all creation operators are organized to the left of all annihilation operators. In general, operator expectation values may contain B and $B^\dagger$ operators in arbitrary order. As a typical evaluation strategy, we first convert the expressions into normal order and then complete the evaluations by using (10.93).

10.4.2 Vacuum-field fluctuations

For the single-mode case, the mode expansion of the electric field, Eq. (10.38), reduces to

$$\hat{\mathbf{E}}_1(\mathbf{r}) = i\mathcal{E}\left[\mathbf{u}(\mathbf{r})\,B - \mathbf{u}^\star(\mathbf{r})\,B^\dagger\right], \tag{10.95}$$

where $\mathcal{E}$ and $\mathbf{u}(\mathbf{r})$ refer to the vacuum field amplitude and the mode function, respectively, for the selected mode. When the system is in a Fock state, the expectation value of the electric field vanishes,

$$\langle\hat{\mathbf{E}}_1(\mathbf{r})\rangle_n = \langle n|\hat{\mathbf{E}}_1(\mathbf{r})|n\rangle = i\mathcal{E}\left[\mathbf{u}(\mathbf{r})\langle n|B|n\rangle - \mathbf{u}^\star(\mathbf{r})\langle n|B^\dagger|n\rangle\right] = 0, \tag{10.96}$$

which follows directly from (10.94). This means that all Fock states have a vanishing expectation value of the electrical field.

However, the Fock state $|n\rangle$ has an energy $E_n = \hbar\omega(n+1/2)$, which increases with the quantum number n. Thus, it is meaningful to consider the second moment of the field

$$\begin{aligned}\langle\hat{\mathbf{E}}_1(\mathbf{r})\cdot\hat{\mathbf{E}}_1(\mathbf{r})\rangle_n &= \langle n|\hat{\mathbf{E}}_1(\mathbf{r})\cdot\hat{\mathbf{E}}_1(\mathbf{r})|n\rangle \\ &= \mathcal{E}^2\Big[\mathbf{u}(\mathbf{r})\cdot\mathbf{u}(\mathbf{r})\langle n|B^2|n\rangle - \mathbf{u}(\mathbf{r})\cdot\mathbf{u}^\star(\mathbf{r})\langle n|B\,B^\dagger|n\rangle \\ &\quad - \mathbf{u}^\star(\mathbf{r})\cdot\mathbf{u}(\mathbf{r})\langle n|B^\dagger B|n\rangle \\ &\quad + \mathbf{u}^\star(\mathbf{r})\cdot\mathbf{u}^\star(\mathbf{r})\langle n|B^\dagger B^\dagger|n\rangle\Big],\end{aligned} \tag{10.97}$$

where the first and fourth contributions vanish, due to Eq. (10.94). The term $\langle n|B^\dagger B|n\rangle$ produces n. Implementing normal ordering, we see that

$$\langle n|BB^\dagger|n\rangle = \langle n|1 + B^\dagger B|n\rangle = \langle n|n\rangle + \langle n|B^\dagger B|n\rangle = 1 + n. \tag{10.98}$$

Inserting these results into (10.97), we obtain

$$\langle\hat{\mathbf{E}}_1(\mathbf{r})\cdot\hat{\mathbf{E}}_1(\mathbf{r})\rangle_n = \mathcal{E}^2|\mathbf{u}(\mathbf{r})|^2(2n+1). \tag{10.99}$$

Hence, even though the amplitude of the electric field vanishes for all Fock states, the field itself contains fluctuations that increase linearly with n. The combination of vanishing field and abundant fluctuations indicate that Fock states have strongly quantum-mechanical features without a classical counterpart. In fact, even the vacuum has nonvanishing

$$\langle\hat{\mathbf{E}}_1(\mathbf{r})\cdot\hat{\mathbf{E}}_1(\mathbf{r})\rangle_{n=0} = \mathcal{E}^2|\mathbf{u}(\mathbf{r})|^2. \tag{10.100}$$

The origin of this finite value lies in the so-called vacuum-field fluctuations that result from the quantum-mechanical Gaussian distribution of the electrical field around $E = x = 0$. This explains also why $\mathcal{E} = \sqrt{\hbar\omega/2\varepsilon_0}$ is called the vacuum-field amplitude.

The existence of the vacuum-field fluctuations can also be deduced from the shape of the probability distributions $|\psi_n(x)|^2$ presented in Fig. 10.2. For this identification, we replace B and $B^\dagger$ by their position–momentum forms, according to Eq. (10.89). This way, we find

$$\hat{\mathbf{E}}_1(\mathbf{r}) = i\mathcal{E}\left[\mathbf{u}(\mathbf{r}) - \mathbf{u}^\star(\mathbf{r})\right]\hat{x} - \mathcal{E}\left[\mathbf{u}(\mathbf{r}) + \mathbf{u}^\star(\mathbf{r})\right]\hat{y}. \tag{10.101}$$

As we can deduce from Fig. 10.2, the expectation value for the position $\langle\hat{x}\rangle_n = \int dx\, x\, |\psi_n(x)|^2$ vanishes while $\langle\hat{x}^2\rangle_n = \int dx\, x^2\, |\psi_n(x)|^2$ is clearly nonzero for any Fock state. The same conclusions follow for $\langle\hat{y}\rangle_n$ and $\langle\hat{y}^2\rangle_n$. In other words, Fock states have nonvanishing quadrature fluctuations,

$$\Delta\langle\hat{x}\rangle^2 \equiv \langle\hat{x}^2\rangle - \langle\hat{x}\rangle\langle\hat{x}\rangle, \quad \text{and} \quad \Delta\langle\hat{y}\rangle^2 \equiv \langle\hat{y}^2\rangle - \langle\hat{y}\rangle\langle\hat{y}\rangle. \tag{10.102}$$

Since $\hat{E}(\mathbf{r})$ is linearly proportional to $\hat{x}$ and $\hat{y}$, the quadrature fluctuations directly determine the fluctuations of the electric field.

For the spatial integral of the fluctuations, we find

$$\begin{aligned} \int d^3r\, n^2(\mathbf{r})\, \langle\hat{\mathbf{E}}_1(\mathbf{r})\cdot\hat{\mathbf{E}}_1(\mathbf{r})\rangle_n &= \frac{\hbar\omega_0}{2\varepsilon_0}(2n+1)\int d^3r\, n^2(\mathbf{r})\, \mathbf{u}^\star(\mathbf{r})\cdot\mathbf{u}(\mathbf{r}) \\ &= \frac{1}{\varepsilon_0}\hbar\omega\left(n+\frac{1}{2}\right) = \frac{E_n}{\varepsilon_0} \end{aligned} \tag{10.103}$$

showing that the field quadrature fluctuations are directly related to the energy of the state. In other words, $|n\rangle$ are nonclassical states whose energy is completely stored within the quadrature fluctuations while the quadrature expectation value of the field itself vanishes. It is also interesting to notice that even the vacuum displays fluctuations proportional to ε^2. The corresponding energy $E_0 = \frac{1}{2}\hbar\omega$ is nonvanishing and is often referred to as the zero-point energy.

At first sight, it seems that the zero-point energy could be eliminated simply by choosing a new zero level for the energy. However, it directly produces nontrivial effects proving that it is a true physical manifestation of light quantization. As an example, we consider a multi-mode system where each $\mathbf{q}$ mode produces the zero-point energy $\frac{1}{2}\hbar\omega_{\mathbf{q}}$. In total, we then have

$$E_{\text{zero}} = \sum_{\mathbf{q}} \frac{1}{2}\hbar\omega_{\mathbf{q}}, \tag{10.104}$$

as the collective zero-point energy. Clearly, E_{zero} depends on the geometry of the system because this determines the energetic spacing of the modes and the density of states. For example, if the system is confined between metallic walls, moving the walls by the distance ΔL changes the density of states. The corresponding change in E_{zero} then induces a force $F = \Delta E_{\text{zero}}/\Delta L$ to the metallic walls, yielding the so-called Casimir effect. This phenomenon has been experimentally measured in several systems showing that Casimir forces can become particularly strong for nanoscale systems.

The zero-point energy also has another surprising implication. Based on the current understanding, the Universe has dark energy that accounts for approximately 74 percent of all existing mass. This observation could be explained via the vacuum's capability to store large amounts of energy without the need to have real particles. Thus, the zero-point energy may be a prominent candidate to explain the origin of the dark energy.

In this book, we mostly study excitations far from vacuum such that neither the Casimir effect nor the dark energy are relevant in these investigations. Nonetheless, it is very interesting to notice that quantum optics may introduce explanations of nontrivial phenomena concerning very different aspects of nature.

Exercises

10.1 **(a)** Show that the canonical quantization (10.11) together with (10.20) yield the Bosonic commutation relations (10.21). Verify the relation (10.22). **(b)** Identify the dimensionless position and momentum $\hat{x} \equiv \frac{b+b^\dagger}{2}$ and $\hat{y} \equiv \frac{b-b^\dagger}{2i}$. How are $\Delta\langle\hat{x}\rangle^2 = \langle\hat{x}^2\rangle - \langle\hat{x}\rangle^2$ and $\Delta\langle\hat{y}\rangle^2 = \langle\hat{y}^2\rangle - \langle\hat{y}\rangle^2$ limited through the Heisenberg uncertainty relation (5.50)? Express $\Delta\langle\hat{x}\rangle$ and $\Delta\langle\hat{y}\rangle$ in terms of the operators b and $b^\dagger$.

10.2 The unitary transformations between two mode function bases are defined by Eq. (10.36). **(a)** What is $\mathbf{U}_{\mathbf{q},\mathbf{q}'}$ between $\left(u_{\sigma,s\mathbf{q}}(\mathbf{r}) = \frac{\sqrt{2}\sin\mathbf{q}\cdot\mathbf{r}}{\mathcal{L}^{3/2}}\,\mathbf{e}_q^{[\sigma]},\ u_{\sigma,c\mathbf{q}}(r) = \frac{\sqrt{2}\cos\mathbf{q}\cdot\mathbf{r}}{\mathcal{L}^{3/2}}\,\mathbf{e}_q^{[\sigma]}\right)$ and plane wave basis $u'_{\sigma,\mathbf{q}}(\mathbf{r}) = \frac{1}{\mathcal{L}^{3/2}}\,e^{i\mathbf{q}\cdot\mathbf{r}}\,\mathbf{e}_q^{[\sigma]}$? **(b)** Use this explicit form of $\mathbf{U}_{\mathbf{q},\mathbf{q}'}$ and show that the corresponding $B_\mathbf{q}$ and $B_\mathbf{q}^\dagger$ are Bosonic operators. **(c)** Show the generic nature of the mode expansion (10.38). **Hint:** *Follow the analysis steps of Eq.* (10.37).

10.3 Verify that the unitary transformation (10.36) of an orthogonal basis produces a basis that is also orthogonal. What is the corresponding completeness relation?

10.4 Verify the relations (10.50) and (10.62).

10.5 **(a)** Show that the ground-state solution (10.64) satisfies (10.63). Verify also that $\psi_0(x)$ is properly normalized. **(b)** Start from the definition (10.72) and show that Eq. (10.73) is valid for all values of k.

10.6 **(a)** Show that the Hamiltonian $\hat{H} = \hbar\omega\left(BB^\dagger + \frac{1}{2}\right)$ is the second quantized form of $\hat{H} = \frac{1}{4}\frac{\partial^2}{\partial x^2} + x^2$. **(b)** Use the ansatz $\psi_n(x) = \mathcal{N}_n\,H_n\left(\sqrt{2}x\right)e^{-x^2}$ to solve $\hat{H}\,\psi_n(x) = \hbar\omega\left(n + \frac{1}{2}\right)\psi_n(x)$. Show that $H_n(y)$ must satisfy the differential equation $H_n''(y) - 2y\,H_n'(y) + 2n\,H_n(y) = 0$, which defines the Hermite polynomials. **(c)** Show that $\int_{-\infty}^{\infty} dy\,H_n(y)\,H_{n'}(y)\,e^{-y^2} = 0$ for $n \neq n'$.

10.7 Assume that the density matrix of quantized light is described by $\hat{\rho} = \sum_{n=0}^{N} |n\rangle p_n \langle n|$, $\sum_{n=0}^{N} p_n = 1$, and $p_n \geq 0$. **(a)** Evaluate $I_K^J \equiv \langle (B^\dagger)^J B^K \rangle$ in terms of p_n. **(b)** Define the inverse transformation from I_K^J to p_n.

10.8 Assume that the multimode light field is described through $|\psi\rangle = |n_{q_1}\rangle \otimes |n_{q_2}\rangle \otimes \cdots \otimes |n_{q_N}\rangle$ where each mode occupies a Fock state $|n_{\mathbf{q}}\rangle$. **(a)** Compute $\left\langle B_q^\dagger B_{q'} \right\rangle$, $\left\langle B_q^\dagger B_{q'}^\dagger \right\rangle$, and $\langle B_q B_{q'} \rangle$. **(b)** Show that $\langle \hat{E}(\mathbf{r}) \rangle$ vanishes while $\int d^3r\, n^2(\mathbf{r}) \langle \hat{E}(\mathbf{r}) \cdot \hat{E}(\mathbf{r}) \rangle$ is nonzero. What is the interpretation of the integral?

Further reading

The quantization of light is discussed in the text books listed in Chapter 8.

Early work on the field quantization in dielectric media is discussed in:

J. M. Jauch and K. M. Watson (1948). Phenomenological quantum electrodynamics. Part II. Interaction of the field with charges, *Phys. Rev.* **74**, 950.

J. M. Jauch and K. M. Watson (1948). Phenomenological quantum electrodynamics. Part III. Dispersion, *Phys. Rev.* **75**, 1249.

The quantization of light in absorbing materials is discussed, e.g., in the following articles:

B. Huttner and S. M. Barnett (1992). Quantization of the electromagnetic field in dielectrics, *Phys. Rev. A* **46**, 4306.

R. Matloob, R. Loudon, S. M. Barnett, and J. Jeffers (1995). Electromagnetic field quantization in absorbing dielectrics, *Phys. Rev. A* **52**, 4823.

T. Gruner and D.-G. Welsch (1996). Green-function approach to the radiation-field quantization for homogeneous and inhomogeneous Kramers–Kronig dielectrics, *Phys. Rev. A* **53**, 1818.

A. Tip (1998). Linear absorptive dielectrics, *Phys. Rev. A* **57**, 4818.

N. A. R. Bhat and J. E. Sipe (2006). Hamiltonian treatment of the electromagnetic field in dispersive and absorptive structured media, *Phys. Rev. A* **73**, 063808.

11

Quasiparticles in semiconductors

Using the general framework for quantum optics presented in Chapters 9–10, we can now start from the generic system Hamiltonian (10.44) that contains the quantized matter as well as quantized light. In principle, we could directly apply it to construct the wave function or the density-matrix dynamics in the Schrödinger picture, as discussed in Section 5.3.2. Even though this approach is exact, it turns out that one can hardly solve the solid-state wave function directly due to its overwhelmingly high dimensionality. Therefore, we must develop an alternative, more flexible, formulation.

We already know that quantized light field can be expressed entirely in terms of Boson operators that appear in the mode expansion (10.43) and that define the concept of a photon. If we now apply the Heisenberg picture, we do not explicitly need to solve the cumbersome wave-function dynamics because the light properties can be expressed through operators and Bosonic commutation relations (10.21) between them. In this chapter, we review the second-quantization formalism that presents the electronic problem in terms of creation and annihilation operators. After we outline the central steps, we construct the noninteracting parts of the system Hamiltonian, which also introduces the central quasiparticles. In general, a quasiparticle is a single-particle entity that is not really an elementary particle but effectively behaves like one. For example, the electronic quasiparticle typically includes the effective interaction of the electrons with the static ionic lattice.

11.1 Second-quantization formalism

According to Eq. (10.43), the quantized electric field can always be expressed via the mode expansion

$$\hat{\mathbf{E}}(\mathbf{r}) = \sum_{\mathbf{q}} i\mathcal{E}_{\mathbf{q}} \left[\mathbf{u}_{\mathbf{q}}(\mathbf{r})\, B_{\mathbf{q}} - \mathbf{u}^{\star}_{\mathbf{q}}(\mathbf{r})\, B^{\dagger}_{\mathbf{q}}\right] \tag{11.1}$$

where $B_{\mathbf{q}}$ and $B^{\dagger}_{\mathbf{q}}$ are the Bosonic annihilation and creation operators, respectively. These operators generate a complete set of multimode Fock states

$$|n_{\mathbf{q}_1}, n_{\mathbf{q}_2}, \cdots n_{\mathbf{q}_N}\rangle \equiv \left(\prod_{j=1}^{N} \frac{1}{\sqrt{n_{\mathbf{q}_j}!}} \left[B^{\dagger}_{\mathbf{q}_j}\right]^{n_{\mathbf{q}_j}}\right) |0\rangle \tag{11.2}$$

for any set of N modes labeled by $\mathbf{q}_1, \cdots \mathbf{q}_N$. This expression follows from a direct generalization of the single-mode result (10.72) such that each identified mode $\mathbf{q}$ is described by a Fock-state $|n_{\mathbf{q}}\rangle$, and $|0\rangle$ denotes the vacuum state. The multimode basis (11.2) can be used to express any wave function

$$|\Psi_{\text{bos}}\rangle \equiv \sum_{n_{\mathbf{q}_1}=0}^{\infty} \cdots \sum_{n_{\mathbf{q}_N}=0}^{\infty} C_{n_{\mathbf{q}_1},\cdots n_{\mathbf{q}_N}} |n_{\mathbf{q}_1}, \cdots n_{\mathbf{q}_N}\rangle \tag{11.3}$$

that defines all quantum aspects of the multimode light.

One can now apply two different strategies to evaluate the quantum dynamics. One option is to start from the Schrödinger equation

$$i\hbar \frac{\partial}{\partial t} |\Psi_{\text{bos}}\rangle = \hat{H} |\Psi_{\text{bos}}\rangle, \tag{11.4}$$

where $\hat{H}$ defines the system Hamiltonian that follows from canonical quantization described in Section 3.2.3. Since $|\Psi_{\text{bos}}\rangle$ describes a large number of modes, one cannot expect to find numerically feasible solution schemes due to the overwhelming dimensionality of explicit $\langle x_{\mathbf{q}_1}, \cdots x_{\mathbf{q}_N} |\Psi_{\text{bos}}\rangle$ where the inner product presents the state in the auxiliary $(x_{\mathbf{q}_1}, \cdots x_{\mathbf{q}_N})$ space of quantum fluctuations. Obviously, any such solution algorithm involves an exponentially growing number of elements, which limits the feasibility of this approach only to few N as shown in Section 6.4.

As a second option and as an alternative to the calculations in the Schrödinger picture, one can follow the quantum dynamics in the Heisenberg picture, introduced in Section 5.3.2. In this case, we have to solve the Heisenberg equation of motion,

$$i\hbar \frac{\partial}{\partial t} \hat{O} = \left[\hat{O}, \hat{H}\right]_-, \tag{11.5}$$

for the quantities of interest $\hat{O}$.

For quantized fields, *all* operators can eventually be expressed through combinations of $B_{\mathbf{q}}$ and $B_{\mathbf{q}}^{\dagger}$ operators. Thus, one can fully describe the system dynamics in terms of creation and annihilation operators, without the explicit need to determine the wave function. In particular, we may use

$$\hat{\mathbf{E}}^{(+)}(\mathbf{r}) \equiv \sum\nolimits_{\mathbf{q}} i\mathcal{E}_{\mathbf{q}} \mathbf{u}_{\mathbf{q}}(\mathbf{r})\, B_{\mathbf{q}}, \qquad \hat{\mathbf{E}}^{(-)}(\mathbf{r}) \equiv -\sum\nolimits_{\mathbf{q}} i\mathcal{E}_{\mathbf{q}} \mathbf{u}_{\mathbf{q}}^{\star}(\mathbf{r})\, B_{\mathbf{q}}^{\dagger}\,. \tag{11.6}$$

11.1.1 Fermion many-body states

The optically active semiconductor system consists of electrons moving in a periodic potential created by the ions within the solid. As an experimentally verified fact, electrons are indistinguishable particles because one cannot distinguish one particle from another even when the many-body wave function is known exactly. All elementary particles share this property, and particles with an antisymmetric wave function Ψ are Fermions while Bosons have a symmetric Ψ. As another experimentally verified fact, electrons are Fermions such that an exchange of any two particle coordinates in Ψ will produce the same wave function with an inverted sign

$$\Psi(\mathbf{r}_1, \cdots \mathbf{r}_j, \cdots \mathbf{r}_k, \cdots \mathbf{r}_N) = -\Psi(\mathbf{r}_1, \cdots \mathbf{r}_k, \cdots \mathbf{r}_j, \cdots \mathbf{r}_N) \,. \tag{11.7}$$

As a result, two Fermions cannot occupy the exact same quantum state, which establishes the Pauli exclusion principle. In contrast to this, Bosons can occupy the same state arbitrarily many times. One can even find a situation that all Bosons occupy the same state, which produces a quantum-degenerate configuration.

The properly antisymmetrized electronic many-body states can be constructed with the help of the so-called Slater determinant

$$\begin{aligned} \Phi^{\mathrm{Sltr}}_{\lambda_1,\cdots\lambda_N}(\mathbf{r}_1, \cdots \mathbf{r}_N) &\equiv \frac{1}{\sqrt{N!}} \sum\nolimits_P (-1)^{n(P)} \prod\nolimits_{j=1}^{N} \phi_{\lambda_j}(\mathbf{r}_{P(j)}) \\ &\equiv \frac{1}{\sqrt{N!}} \det[\mathbf{M}], \qquad \mathbf{M}_{j,k} \equiv \phi_{\lambda_j}(\mathbf{r}_k) \end{aligned} \tag{11.8}$$

where $\{\phi_{\lambda_1}, \cdots \phi_{\lambda_N}\}$ is a set of orthogonal single-particle wave functions. The sum is taken over all $N!$ permutations P of the index sequence $\{1, 2, \cdots N\}$ and an individual element of the permuted sequence is generated via the mapping $P(j)$. The identified $n(P)$ gives the number of transpositions P produces with respect to the ordered sequence. Therefore, $(-1)^{n(P)}$ is +1 for even permutations and -1 for odd permutations. The constructed Φ^{Sltr} follows also from the usual determinant of a matrix $\mathbf{M}$: we simply need to identify the first index of $\mathbf{M}$ with the quantum state while the second index labels the particle coordinate, as shown in Fig. 11.1. Based on linear algebra, the Leibniz-formula for the determinant $\det[\mathbf{M}]$ contains the explicit $\sum_P$ as given in Eq. (11.8). It is also straightforward to show that the Slater determinant contains the norm $1/\sqrt{N!}$ due to $N!$ different terms appearing in the permutation, see Exercise 11.2.

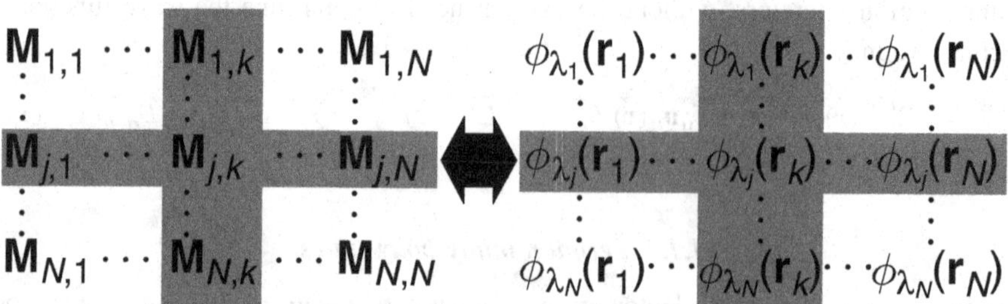

Figure 11.1 Determinant (left) vs. Slater determinant (right). The matrix elements $\mathbf{M}_{j,k}$ and wave-function components $\phi_{\lambda_j}(\mathbf{r}_k)$ have a one-to-one correspondence. The dark shaded rectangles indicate elimination of one line and one row, as defined in the Laplace expansion (11.9).

The evaluation of determinants is a standard problem in linear algebra, and we can use several alternative schemes to compute det[**M**] efficiently. In pragmatic calculations, one often applies the Laplace expansion,

$$\det[\mathbf{M}] = \sum\nolimits_{j=1}^{N} (-1)^{j+k} \det[\mathbf{M}^{j;\,k}] = \sum\nolimits_{k=1}^{N} (-1)^{j+k} \det[\mathbf{M}^{j;\,k}] \,, \tag{11.9}$$

where $\mathbf{M}^{j;\,k}$ is obtained by eliminating the j-th line and k-th row from the original $N \times N$ matrix. The elimination is highlighted in Fig. 11.1 by the dark-shaded rectangles. Since $\mathbf{M}^{j;\,k}$ is an $(N-1) \times (N-1)$ matrix, $\det[\mathbf{M}^{j;\,k}]$ is a lower-order determinant, often referred to as the subdeterminant. As a result, the Laplace expansion yields a recursive evaluation of det[**M**] in terms of subdeterminants, which can be repeated until the lowest-order contribution is the determinant of a scalar.

Also the Slater determinant can be computed with the subdeterminant expansion

$$\begin{aligned}
\Phi^{\mathrm{Sltr}}_{\lambda_1,\cdots\lambda_N}(\mathbf{r}_1,\cdots\mathbf{r}_N) &= \sum\nolimits_{J=1}^{N} \frac{(-1)^{J+K}}{\sqrt{N}} \phi_{\lambda_J}(\mathbf{r}_K) \Phi^{J;K}_{\lambda_1,\cdots\lambda_N}(\mathbf{r}_1,\cdots\mathbf{r}_N) \\
&= \sum\nolimits_{K=1}^{N} \frac{(-1)^{J+K}}{\sqrt{N}} \phi_{\lambda_J}(\mathbf{r}_K) \Phi^{J;K}_{\lambda_1,\cdots\lambda_N}(\mathbf{r}_1,\cdots\mathbf{r}_N)
\end{aligned} \tag{11.10}$$

that contains a lower-order subdeterminant

$$\Phi^{J;K}_{\lambda_1,\cdots\lambda_N}(\mathbf{r}_1,\cdots\mathbf{r}_N) \equiv \Phi^{\mathrm{Sltr}}_{\lambda_1,\cdots\not{\lambda}_J,\cdots\lambda_N}(\mathbf{r}_1,\cdots\not{\mathbf{r}}_K,\cdots\mathbf{r}_N) \tag{11.11}$$

where the state λ_J and the coordinate $\mathbf{r}_K$ is eliminated. The factor $1/\sqrt{N}$ follows from the norm of the Slater determinant. By applying the subdeterminant expansion (11.10), we quickly find

$$\begin{aligned}
\Phi^{\mathrm{Sltr}}_{\lambda_1,\lambda_2}(\mathbf{r}_1,\mathbf{r}_2) &= \sum\nolimits_{J=1}^{2} \frac{(-1)^{J+1}}{\sqrt{2}} \phi_{\lambda_J}(\mathbf{r}_1) \Phi^{J;1}_{\lambda_1,\lambda_2}(\mathbf{r}_1,\mathbf{r}_2) \\
&= \frac{1}{\sqrt{2}} \left[\phi_{\lambda_1}(\mathbf{r}_1) \Phi^{1;1}_{\lambda_1,\lambda_2}(\mathbf{r}_1,\mathbf{r}_2) - \phi_{\lambda_2}(\mathbf{r}_1) \Phi^{2;1}_{\lambda_1,\lambda_2}(\mathbf{r}_1,\mathbf{r}_2) \right] \\
&= \frac{1}{\sqrt{2}} \left[\phi_{\lambda_1}(\mathbf{r}_1)\phi_{\lambda_2}(\mathbf{r}_2) - \phi_{\lambda_2}(\mathbf{r}_1)\phi_{\lambda_1}(\mathbf{r}_2) \right]
\end{aligned} \tag{11.12}$$

for the two-particle Slater determinant. This wave function correctly describes indistinguishable particles because it is antisymmetric.

The abstract-notation Slater determinant, $|\lambda_1,\cdots\lambda_N\rangle$, satisfies antisymmetry

$$|\lambda_1,\cdots\lambda_j,\cdots\lambda_k,\cdots\lambda_N\rangle = -|\lambda_1,\cdots\lambda_k,\cdots\lambda_j,\cdots\lambda_N\rangle \tag{11.13}$$

because the determinant within Eq. (11.8) changes its sign whenever two state indices are exchanged. If we now use a complete set of single-particle wave functions $|\lambda\rangle$, we may construct a generic wave function in terms of

$$|\Psi\rangle = \sum_{N=0}^{\infty} \sum_{\lambda_1 < \lambda_2 < \cdots \lambda_N} C_{\lambda_1, \cdots \lambda_N} |\lambda_1, \cdots \lambda_N\rangle \,. \tag{11.14}$$

This expression assumes that we have formally ordered the single-particle states. After this ordering, the sum contains each state-combination only once. The corresponding quantum statistics is uniquely described by the complex-valued coefficients $C_{\lambda_1, \cdots \lambda_N}$. This formal ordering also helps in evaluating projections between two Slater determinants

$$\langle \nu_N, \cdots \nu_1 | \lambda_1, \cdots \lambda_N\rangle = \prod_{j=1}^{N} \delta_{\nu_j, \lambda_j}, \quad \text{with} \quad \lambda_j < \lambda_{j+1}, \quad \nu_j < \nu_{j+1}, \quad \forall j \,. \tag{11.15}$$

Without formal order, one must also include all permutations of otherwise identical index sequences, see Exercise 11.2.

If $|\Psi\rangle$ contains exactly N particles, we denote it by $|\Psi_N\rangle$. The infinite-dimensional space where $|\Psi_N\rangle$ exists is called the Hilbert space. It is often convenient to generalize this concept to include also wave functions with a mixed particle number, as is done in Eq. (11.14). This construction introduces the Fock space. The corresponding wave functions have the properties

$$\langle \Psi_N | \Psi_{N'}\rangle = \delta_{N,N'} \langle \Psi_N | \Psi_N\rangle, \qquad |\Psi_0\rangle = |0\rangle, \tag{11.16}$$

where $|\Psi_0\rangle$ defines the particle vacuum $|0\rangle$ that contains no particles.

The quantum dynamics of the many-body state, i.e., the time evolution of $C_{\lambda_1, \cdots \lambda_N}$, can now be solved directly from the Schrödinger equation analogous to Eq. (11.4). This first-quantization approach suffers from the same complications as the multimode analysis – the dimensionality of $C_{\lambda_1, \cdots \lambda_N}$ becomes unmanageable already when we consider few-particle systems, like it does for few modes. This follows directly from the structural similarity of Eqs. (11.3) and (11.14). Thus, the first-quantization approach outlined here is not flexible enough to describe many-body problems in a numerically feasible manner.

11.1.2 Fermion creation and annihilation operators

The many-body description of semiconductor electrons becomes significantly simpler after we apply the second-quantization formalism. As a starting point, we need to identify a systematic definition for the creation and annihilation operators, in analogy to the photon operators. Since Eqs. (11.3) and (11.14) have a similar form, we introduce Fermion creation operators that generate a Slater determinant

$$|\lambda_1, \cdots \lambda_N\rangle \equiv a^\dagger_{\lambda_1} a^\dagger_{\lambda_2} \cdots a^\dagger_{\lambda_N} |0\rangle, \quad \langle \lambda_N, \cdots \lambda_1| \equiv [|\lambda_1, \cdots \lambda_N\rangle]^\dagger \,. \tag{11.17}$$

By combining the properties (11.13) and (11.17), we find

$$|\nu, \beta, \lambda_1, \cdots \lambda_N\rangle = -|\beta, \nu, \lambda_1, \cdots \lambda_N\rangle$$
$$\Leftrightarrow \quad \left(a_\nu^\dagger a_\beta^\dagger + a_\beta^\dagger a_\nu^\dagger\right)|\lambda_1, \cdots \lambda_N\rangle = 0, \quad \forall \quad \lambda_j\,. \tag{11.18}$$

Using this in Eq. (11.14), we obtain a general relation

$$\left(a_\nu^\dagger a_\beta^\dagger + a_\beta^\dagger a_\nu^\dagger\right)|\Psi\rangle = 0\,. \tag{11.19}$$

Since $|\Psi\rangle$ is an arbitrary antisymmetric wave function in the Fock space, the term within the parentheses must vanish. This yields the Fermion anticommutation relation

$$\left[a_\nu^\dagger,\, a_\beta^\dagger\right]_+ \equiv a_\nu^\dagger a_\beta^\dagger + a_\beta^\dagger a_\nu^\dagger = 0, \tag{11.20}$$

among Fermion-creation operators.

To complete the analysis, we introduce an annihilation operator as the Hermitian adjoint to the creation operator

$$a_\lambda \equiv \left(a_\lambda^\dagger\right)^\dagger \qquad \Leftrightarrow \qquad (a_\lambda)^\dagger = a_\lambda^\dagger\,. \tag{11.21}$$

Applying these relations and using (11.17), we find

$$[\langle\lambda_N, \cdots \lambda_1|a_\nu]^\dagger = a_\nu^\dagger|\lambda_1, \cdots \lambda_N\rangle = |\nu,\, \lambda_N, \cdots \lambda_1\rangle$$
$$\Leftrightarrow \langle\lambda_N, \cdots \lambda_1|a_\nu = \langle\lambda_N, \cdots \lambda_1,\, \nu|, \tag{11.22}$$

where the last identification follows from the Hermitian conjugation.

We next assume that the single-particle states within $|\lambda_1, \cdots \lambda_N\rangle$ are formally ordered. This can be done without any loss of generality because we can choose the ordering completely freely. We then compute the projection

$$\langle\Psi|a_\nu|\lambda_1, \cdots \lambda_N\rangle = \sum_{\nu_2<\cdots<\nu_N} C^\star_{\nu_2,\cdots\nu_N}\langle\nu_N, \cdots \nu_2|a_\nu|\lambda_1, \cdots \lambda_N\rangle$$
$$= \sum_{\nu_2<\cdots<\nu_N} C^\star_{\nu_2,\cdots\nu_N}\langle\nu_N, \cdots \nu_2, \nu|\lambda_1, \cdots \lambda_N\rangle \tag{11.23}$$

where we first used the expansion (11.14) and then (11.22). Based on the orthogonality (11.16), only the $(N-1)$-particle term within $\langle\Psi|$ can contribute. If ν is different from all λ_j, this inner product vanishes because all Slater determinants with different states are orthogonal, according to Eq. (11.15). At the same time, if ν is equal to λ_1, we find

$$\langle\Psi|a_{\lambda_1}|\lambda_1, \cdots \lambda_N\rangle = \sum_{\nu_2<\cdots<\nu_N} C^\star_{\nu_2,\cdots\nu_N}\langle\nu_N, \cdots \nu_2, \lambda_1|\lambda_1, \cdots \lambda_N\rangle$$
$$= C^\star_{\lambda_2,\cdots\lambda_N} = \langle\Psi|\lambda_2, \cdots \lambda_N\rangle \tag{11.24}$$

where we used the orthogonality (11.15) and evaluated an inner product with the expansion (11.14).

Let us summarize the obtained results: Since $\langle\Psi|$ is an arbitrary wave function, the observations above imply

$$\begin{aligned}
&a_{\lambda_1}|\lambda_1,\cdots\lambda_N\rangle = |\lambda_2,\cdots\lambda_N\rangle \\
&a_{\lambda_2}|\lambda_1,\cdots\lambda_N\rangle = -a_{\lambda_2}|\lambda_2,\lambda_1,\lambda_3,\cdots\lambda_N\rangle = -|\lambda_1,\lambda_3,\cdots\lambda_N\rangle \quad \text{etc.} \\
&a_\nu|\lambda_1,\cdots\lambda_N\rangle = 0, \quad \nu \neq \lambda_j \quad \forall j\,.
\end{aligned} \tag{11.25}$$

In other words, the identified a_ν is indeed an annihilation operator. Also the annihilation operators satisfy an anticommutation relation analogous to Eq. (11.20) because we can simply take the Hermitian conjugate to produce $\left[a_\nu^\dagger,\, a_\beta^\dagger\right]_+^\dagger = \left[a_\nu,\, a_\beta\right]_+ = 0$. In the forthcoming investigations, we also need anticommutation relations for mixed creation and annihilation operators. Using the properties (11.17)–(11.18) and (11.25), we eventually find

$$\left[a_\lambda,\, a_\nu^\dagger\right]_+ = \delta_{\lambda,\nu}, \qquad \left[a_\lambda^\dagger,\, a_\nu^\dagger\right]_+ = 0 = [a_\lambda,\, a_\nu]_+, \tag{11.26}$$

where we have repeated the pure creation- and annihilation-operator anticommutators. The derivation is analogous to that producing the relations (11.20) and (11.25); this is analyzed in Exercise 11.5. It is also possible to express both creation and annihilation operators in terms of bra-ket combinations of Slater determinants; this aspect is studied in Exercise 11.6.

11.1.3 Fermions in second quantization

To present the many-body problem in second-quantized form, we need to express all operators that appear in terms of creation and annihilation operators. After that, we can evaluate the quantum dynamics using Heisenberg's equation of motion, as discussed in connection with Eq. (11.5).

It is rather typical that many of the fundamental N-particle operators – such as the system Hamiltonian – are known in their first-quantization format. To see how these are transformed into their second-quantized form, we analyze a generic N-particle operator

$$\langle\mathbf{r}_1,\cdots\mathbf{r}_N|\hat{O}|\mathbf{r}_1',\cdots\mathbf{r}_N'\rangle = \left[\prod_{j=1}^N \delta\left(\mathbf{r}_j - \mathbf{r}_j'\right)\right] O(\mathbf{r}_1,\cdots\mathbf{r}_N) \tag{11.27}$$

that is specified in its real-space representation. Typically $O(\mathbf{r}_1,\cdots\mathbf{r}_N)$ is easy to define in first quantization. For example, the system Hamiltonian (10.44) identifies a potential $O(\mathbf{r}_1,\cdots\mathbf{r}_N) = \sum_{j=1}^N QU^{\text{gen}}(\mathbf{r}_j)$ stemming from the ionic background density which acts on the identical electrons with the charge Q. The presented $O(\mathbf{r}_1,\cdots\mathbf{r}_N)$ does not have the most general functional dependency on all N coordinates because it is defined by a single-particle potential $U^{\rm gen}(\mathbf{r}_j)$. More generally, we can classify a given $\hat{O}$ as a K-particle operator when it can be expressed through

$$O_K(\mathbf{r}_1,\cdots\mathbf{r}_N) = \sum_{j_1\neq j_2\cdots\neq j_K=1}^{N} U_K(\mathbf{r}_{j_1},\mathbf{r}_{j_2},\cdots\mathbf{r}_{j_K}) \tag{11.28}$$

where U_K is a generic function with *only K different* arguments.

We will next consider how a generic single-particle operator can be expressed in the second-quantization formalism, i.e., using only creation and annihilation operators. We assume that we formally have N particles such that the identity operator in antisymmetric Hilbert space becomes

$$\mathbb{I} \equiv \sum\nolimits_{\lambda_1<\cdots<\lambda_N} |\lambda_1,\cdots\lambda_N\rangle\langle\lambda_N,\cdots\lambda_1| \tag{11.29}$$

that contains each Slater determinant only once because we have implemented a formal order in the state sum. We then sandwich a generic $\hat{O}_1$ between two identity operators, producing

$$\hat{O}_1 = \sum_{\substack{\lambda_1<\cdots<\lambda_N\\ \nu_1<\cdots<\nu_N}} |\lambda_1,\cdots\lambda_N\rangle\langle\lambda_N,\cdots\lambda_1|\hat{O}_1|\nu_1,\cdots\nu_N\rangle\langle\nu_N,\cdots\nu_1| \tag{11.30}$$

that is still expressed in its abstract format.

We continue by determining the explicit matrix element

$$\begin{aligned} I &\equiv \langle\lambda_N,\cdots\lambda_1|\hat{O}_1|\nu_1,\cdots\nu_N\rangle \\ &= \int d^{3N}\mathbf{r}\, d^{3N}\mathbf{r}'\ \langle\lambda_N,\cdots\lambda_1|\mathbf{r}_1,\cdots\mathbf{r}_N\rangle\langle\mathbf{r}_1,\cdots\mathbf{r}_N|\hat{O}_1|\mathbf{r}'_1,\cdots\mathbf{r}'_N\rangle \\ &\qquad\qquad \times\langle\mathbf{r}'_1,\cdots\mathbf{r}'_N\,|\nu_1,\cdots\nu_N\rangle \end{aligned} \tag{11.31}$$

appearing within Eq. (11.30). More explicitly, we have twice applied the projection $\mathbb{I} \equiv \int d^{3N}\mathbf{r}|\mathbf{r}_1,\cdots\mathbf{r}_N\rangle\langle\mathbf{r}_1,\cdots\mathbf{r}_N|$, which converts the expression into real-space integrals. We also have introduced $d^{3N}\mathbf{r} \equiv d^3\mathbf{r}_1\cdots d^3\mathbf{r}_N$ to shorten the notation. As we implement the definitions (11.27) and (11.28) as well as the real-space representation, i.e., $\langle\mathbf{r}_1,\cdots\mathbf{r}_N|\nu_1,\cdots\nu_N\rangle = \Phi^{\rm Sltr}_{\nu_1,\cdots\nu_N}(\mathbf{r}_1,\cdots\mathbf{r}_N)$, Eq. (11.31) becomes

$$I = \sum_{k=1}^{N}\int d^{3N}\mathbf{r}\left[\Phi^{\rm Sltr}_{\lambda_1,\cdots\lambda_N}(\mathbf{r}_1,\cdots\mathbf{r}_N)\right]^{\star} U_1(\mathbf{r}_k)\,\Phi^{\rm Sltr}_{\nu_1,\cdots\nu_N}(\mathbf{r}_1,\cdots\mathbf{r}_N). \tag{11.32}$$

We apply the subdeterminant expansion (11.10) twice to isolate the $\mathbf{r}_k$-dependent parts. This procedure eventually yields

$$I = \sum_{k,\,J,\,J'=1}^{N} \frac{(-1)^{J+k+J'+k}}{N} \int d^{3N}\mathbf{r}\, \phi^{\star}_{\lambda_J}(\mathbf{r}_k) \left[\Phi^{J;\,k}_{\lambda_1,\cdots\lambda_N}(\mathbf{r}_1,\cdots\mathbf{r}_N)\right]^{\star} U_1(\mathbf{r}_k)$$
$$\times\, \phi_{\nu_{J'}}(\mathbf{r}_k)\Phi^{J';\,k}_{\nu_1,\cdots\nu_N}(\mathbf{r}_1,\cdots\mathbf{r}_N)$$
$$= \sum_{k,\,J,\,J'=1}^{N} (-1)^{J+J'} \frac{\int d^3\mathbf{r}_k\, \phi^{\star}_{\lambda_J}(\mathbf{r}_k)U_1(\mathbf{r}_k)\phi_{\nu_{J'}}(\mathbf{r}_k)}{N}$$
$$\times \int d^{3N}_{k}\mathbf{r}\, \left[\Phi^{J;\,k}_{\lambda_1,\cdots\lambda_N}(\mathbf{r}_1,\cdots\mathbf{r}_N)\right]^{\star} \Phi^{J';\,k}_{\nu_1,\cdots\nu_N}(\mathbf{r}_1,\cdots\mathbf{r}_N) \tag{11.33}$$

where the last step follows after we reorganize the $\mathbf{r}_k$ integration to be first, use $(-1)^{J+k+J'+k} = (-1)^{J+J'+2k} = (-1)^{J+J'}$, and denote $d^{3N}_{k}\mathbf{r} \equiv d^3\mathbf{r}_1\cdots \not{d}^3\not{\mathbf{r}}_k \cdots d^3\mathbf{r}_N$.

The definition (11.11) implies that the last integral within Eq. (11.33) is in fact an inner product between two Slater determinants whose J-th or J'-th element is missing. In this context, it is convenient to introduce

$$|(\lambda_1,\cdots\lambda_N)^J\rangle \equiv |\lambda_1,\cdots\not{\lambda}_J,\cdots\lambda_N\rangle \tag{11.34}$$

where the J-th state of the original Slater determinant is eliminated. This allows us to rewrite (11.33) in the format

$$I = \sum_{k,\,J,\,J'=1}^{N} \frac{(-1)^{J+J'}\int d^3\mathbf{r}\, \phi^{\star}_{\lambda_J}(\mathbf{r})U_1(\mathbf{r})\phi_{\nu_{J'}}(\mathbf{r})}{N} \langle(\lambda_N,\cdots\lambda_1)^J|(\nu_1,\cdots\nu_N)^{J'}\rangle \tag{11.35}$$

after we change the integration variable from $\mathbf{r}_k$ into $\mathbf{r}$. We see now that neither the integration nor the inner product depends on the k index. Therefore, the k sum produces N times the same contribution, yielding

$$I = \sum_{J,\,J'=1}^{N} \int d^3\mathbf{r}\, \phi^{\star}_{\lambda_J}(\mathbf{r})U_1(\mathbf{r})\phi_{\nu_{J'}}(\mathbf{r})$$
$$\times\, \langle(\lambda_N,\cdots\lambda_1)^J|(-1)^{J+J'}|(\nu_1,\cdots\nu_N)^{J'}\rangle \tag{11.36}$$

where we have moved $(-1)^{J+J'}$ to be within the inner product.

We can now return to the basic properties of creation and annihilation operators: based on the relation (11.25), we find

$$a_{\nu_J}|\nu_1,\cdots\nu_N\rangle = (-1)^{J-1}|(\nu_1,\cdots\nu_N)^J\rangle$$
$$\Leftrightarrow \langle\nu_N,\cdots\nu_1|a^{\dagger}_{\nu_J} = (-1)^{J-1}\langle(\nu_N,\cdots\nu_1)^J| \tag{11.37}$$

that is also expressed in its Hermitian-conjugated format. As we implement this into Eq. (11.36), we obtain

$$I = \int d^3\mathbf{r}\, \langle \lambda_N, \cdots \lambda_1 | \sum_{J=1}^{N} a^\dagger_{\lambda_J} \phi^\star_{\lambda_J}(\mathbf{r})\, U_1(\mathbf{r}) \sum_{J'=1}^{N} \phi_{\nu_{J'}}(\mathbf{r}) a_{\nu_{J'}} |\nu_1, \cdots \nu_N\rangle, \tag{11.38}$$

after we organized the single-particle wave functions as well as U_1 to be inside the inner product, which can always be done because these are simple complex-valued functions and not operators.

It is now convenient to introduce the field operators for Fermions

$$\hat{\Psi}(\mathbf{r}) \equiv \sum\nolimits_\nu \phi_\nu(\mathbf{r}) a_\nu, \qquad \hat{\Psi}^\dagger(\mathbf{r}) \equiv \sum\nolimits_\nu \phi^\star_\nu(\mathbf{r}) a^\dagger_\nu\,. \tag{11.39}$$

These operators have the generic property

$$\begin{aligned} \hat{\Psi}(\mathbf{r})|\nu_1, \cdots \nu_N\rangle &= \sum\nolimits_{J=1}^{N} \phi_{\nu_J}(\mathbf{r}) a_{\nu_J} |\nu_1, \cdots \nu_N\rangle \\ \langle \lambda_N, \cdots \lambda_1 | \hat{\Psi}^\dagger(\mathbf{r}) &= \langle \lambda_N, \cdots \lambda_1 \Big| \sum\nolimits_{J=1}^{N} \phi^\star_{\lambda_J}(\mathbf{r}) a^\dagger_{\lambda_J} \end{aligned} \tag{11.40}$$

because any other ν or λ term within the field operator yields a vanishing result, based on the property (11.25). This observation converts Eq. (11.38) into

$$I = \langle \lambda_N, \cdots \lambda_1 | \int d^3\mathbf{r}\, \hat{\Psi}^\dagger(\mathbf{r}) U_1(\mathbf{r}) \hat{\Psi}(\mathbf{r}) |\nu_1, \cdots \nu_N\rangle\,. \tag{11.41}$$

As we insert this result back into (11.30), we can express the single-particle operator completely in terms of creation and annihilation operators:

$$\begin{aligned} \hat{O}_1 &= \sum\nolimits_{\lambda_1 < \cdots < \lambda_N} |\lambda_1, \cdots \lambda_N\rangle \langle \lambda_N, \cdots \lambda_1 | \int d^3\mathbf{r}\, \hat{\Psi}^\dagger(\mathbf{r}) U_1(\mathbf{r}) \hat{\Psi}(\mathbf{r}) \\ &\quad \times \sum\nolimits_{\nu_1 < \cdots < \nu_N} |\nu_1, \cdots \nu_N\rangle \langle \nu_N, \cdots \nu_1| \\ &= \int d^3\mathbf{r}\, \hat{\Psi}^\dagger(\mathbf{r}) U_1(\mathbf{r}) \hat{\Psi}(\mathbf{r}) \end{aligned} \tag{11.42}$$

where the last step follows after we realize that the state sums produce the identity operator (11.29).

The same derivation can be repeated for a generic K-particle operator. This maps a first-quantization operator (11.28) into the second-quantization operator

$$\hat{O}_K = \int d^{3K}\mathbf{r}\, \hat{\Psi}^\dagger(\mathbf{r}_K) \cdots \hat{\Psi}^\dagger(\mathbf{r}_1)\, U_K(\mathbf{r}_1, \cdots \mathbf{r}_K)\, \hat{\Psi}(\mathbf{r}_1) \cdots \hat{\Psi}(\mathbf{r}_K) \tag{11.43}$$

where the U_K function, defining the first-quantization operator in real space, is sandwiched between the field operators. We also recognize that the field operators appear in normal order, i.e., in all products the creation operators are sorted to the left while the annihilation operators are on the right. We also realize that (11.43) does not depend on the particle number N such that one can apply this formulation for any operator in the Fock space. The explicit proof of the relation (11.43) is studied further in Exercise 11.7.

11.1.4 Pragmatic formulation of second quantization

The introduction of the field operators (11.40) and the result (11.43) allow us to introduce the second quantization for Fermions in a straightforward manner. In the first step, we need to find a complete set of orthogonal single-particle wave functions $\phi_\lambda(\mathbf{r})$. We then associate the creation operator $a_\lambda^\dagger$ and the annihilation operator a_λ with each state. These operators satisfy the Fermion-anticommutation relations:

$$\left[a_\lambda,\, a_\nu^\dagger\right]_+ = \delta_{\lambda,\nu}, \qquad \left[a_\lambda^\dagger,\, a_\nu^\dagger\right]_+ = 0 = [a_\lambda,\, a_\nu]_+\,, \tag{11.44}$$

according to Eq. (11.26). We can then construct the field operators

$$\hat{\Psi}(\mathbf{r}) \equiv \sum\nolimits_\nu \phi_\nu(\mathbf{r}) a_\nu, \qquad \hat{\Psi}^\dagger(\mathbf{r}) \equiv \sum\nolimits_\nu \phi_\nu^\star(\mathbf{r}) a_\nu^\dagger\,, \tag{11.45}$$

given by Eq. (11.39). It is straightforward to show that the field operators satisfy

$$\begin{aligned} &\left[\hat{\Psi}(\mathbf{r}),\, \hat{\Psi}^\dagger(\mathbf{r}')\right]_+ = \delta(\mathbf{r}-\mathbf{r}'), \\ &\left[\hat{\Psi}^\dagger(\mathbf{r}),\, \hat{\Psi}^\dagger(\mathbf{r}')\right]_+ = 0 = \left[\hat{\Psi}(\mathbf{r}),\, \hat{\Psi}(\mathbf{r}')\right]_+ \end{aligned}, \tag{11.46}$$

based on the property (11.44) and the completeness of $\phi_\lambda(\mathbf{r})$, see Exercise 11.4. After this formulation, we can convert any K-particle operator,

$$\hat{O}_K \quad \Leftrightarrow \quad \sum_{j_1 \neq j_2 \cdots \neq j_K = 1}^{N} U_K(\mathbf{r}_{j_1}, \mathbf{r}_{j_2}, \cdots \mathbf{r}_{j_K})\,, \tag{11.47}$$

into its second-quantized form

$$\hat{O}_K = \int d^{3K}\mathbf{r}\; \hat{\Psi}^\dagger(\mathbf{r}_K)\cdots \hat{\Psi}^\dagger(\mathbf{r}_1)\, U_K(\mathbf{r}_1, \cdots \mathbf{r}_K)\, \hat{\Psi}(\mathbf{r}_1)\cdots \hat{\Psi}(\mathbf{r}_K) \tag{11.48}$$

where we used Eqs. (11.28) and (11.43). Using this second-quantization formalism, we no longer need to consider the dynamics of the wave function or the density matrix. Instead, we can compute the quantum dynamics from the Heisenberg equation of motion

$$i\hbar \frac{\partial}{\partial t} \hat{O}_K = \left[\hat{O}_K,\, \hat{H}\right]_- \tag{11.49}$$

that can be expressed completely in terms of the second-quantization operators. In particular, we can generate the dynamics of all observables $\langle \hat{O}_K \rangle = \mathrm{Tr}[\hat{O}_K \hat{\rho}(t=0)]$

straightforwardly. Due to the second-quantization formulation, this method is directly valid for any combination of particle numbers within the Fock space. These aspects make the second-quantization formalism particularly attractive for many-body and semiconductor quantum-optical investigations.

11.2 System Hamiltonian of solids

Typically, the optically active semiconductors are sandwiched between other materials that are optically inactive. Nowadays, one can routinely realize sample fabrication on a subnanometer scale such that one can systematically optimize and study nanostructure devices. Based on the analysis in Section 3.1.1 and Fig. 3.1, the nanometer length scale implies strong quantum features for electrons, which makes such structures particularly interesting for semiconductor-quantum-optics investigations. Figure 11.2 schematically presents a typical nanostructure system studied in this book. The small white-dark circles indicate the ion lattice defining where the electrons (large spheres) can move. The small dim circles indicate the charge-neutral lattice in which the nanostructure is embedded.

Both the ionic and the charge-neutral lattice yield a trivial optical response that can be described through a refractive-index profile $n(\mathbf{r})$, as indicated by the solid line in Fig. 11.2(a). Thus, the interesting quantum-optical effects stem entirely from the electronic motion (arrows in Fig. 11.2(b)) within the nanostructure. We assume, for simplicity, that $n(\mathbf{r})$ remains constant in the vicinity of the nanostructure such that the Coulomb interaction among the electrons is not modified by the nanostructure's interface, compare the discussion in Section 9.3. However, we allow for the feature that the refractive index can

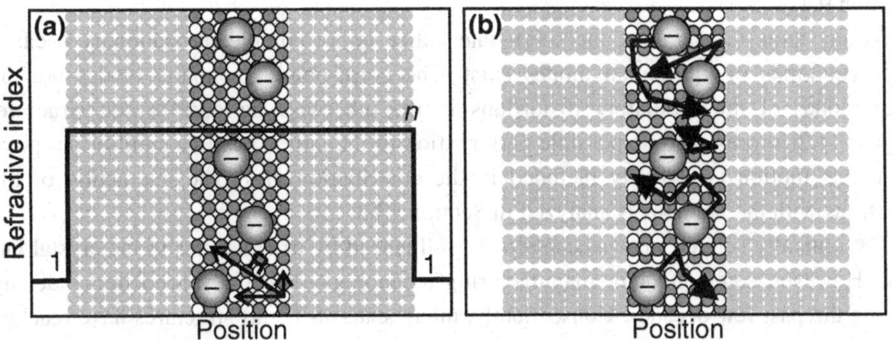

Figure 11.2 **(a)** Typical semiconductor nanostructures studied in this book. The dim and dark and white circles indicate different atomic compounds in the solid. These yield a background dielectric profile indicated by the solid line. The electrons (spheres) are quantum confined inside the nanostructure (dark and white circles). Three representative lattice vectors $\mathbf{R}$ are shown. **(b)** Schematic presentation of phonons. The phonon modes describe lattice vibrations that break the periodicity of the undisturbed system. Quantum-confined electrons scatter from the phonons but cannot escape the nanostructure (arrows indicate illustrative paths).

change at larger distances away from the active electrons. With a suitable $n(\mathbf{r})$, one can e.g. construct an optical cavity for light around the nanostructure (not shown here).

Essentially, we have already derived the system Hamiltonian in Chapter 10 for the generic structures described above. In particular, we may directly use the system Hamiltonian (10.44) as the starting point, yielding here

$$\hat{H} = \sum_{j=1}^{N} \left(\frac{\left[\hat{\mathbf{p}}_j - Q\hat{\mathbf{A}}(\mathbf{r}_j)\right]^2}{2m_0} + U(\mathbf{r}_j) \right) + \frac{1}{2} \sum_{j \neq k=1}^{N} V(\mathbf{r}_j - \mathbf{r}_k) + \sum_{\mathbf{q}} \hbar\omega_{\mathbf{q}} \left[B_{\mathbf{q}}^{\dagger} B_{\mathbf{q}} + \frac{1}{2} \right] + \hat{H}_{\text{ph}} \tag{11.50}$$

where Q is the elementary charge of the electron and m_0 is the free-electron mass. The potential $U(\mathbf{r})$ is created by *all* ions centered at the positions $\mathbf{R}$ in the solid, see arrows in Fig. 11.2(a). Since we assume that the dielectric constant n does not change at the nanostructure's interfaces, the Coulomb interaction among the electrons has the usual form

$$V(\mathbf{r}) = \frac{Q^2}{4\pi\varepsilon_0 n^2 |\mathbf{r}_k - \mathbf{r}_j|}, \tag{11.51}$$

as discussed in connection with Eq. (10.7). The vector potential that appears is given by the mode expansion

$$\hat{\mathbf{A}}(\mathbf{r}) = \sum_{\mathbf{q}} \frac{\mathcal{E}_{\mathbf{q}}}{\omega_{\mathbf{q}}} \left[\mathbf{u}_{\mathbf{q}}(\mathbf{r})\, B_{\mathbf{q}} + \mathbf{u}_{\mathbf{q}}^{\star}(\mathbf{r})\, B_{\mathbf{q}}^{\dagger} \right] \tag{11.52}$$

defined by Eq. (10.43). The mode functions are influenced by the refractive-index profile of the entire system. They can be solved using the transfer-matrix approach presented in Section 9.4.4.

At any finite temperature, the solid-state lattice is not completely stationary because different kinds of vibrational modes exist. The corresponding phonon modes are the eigenmodes of the oscillatory distortions in the otherwise periodic crystal structure. Figure 11.2(b) presents one possible perturbation of the lattice from its equilibrium position; this motion is formally included in the contribution $\hat{H}_{\text{ph}}$. The interaction of the electrons with the phonons is analyzed in Section 13.3.

The optically active electrons can also be influenced by irregularities in the crystal lattice. However, semiconductor manufacturing technologies have advanced tremendously during the past few decades. Consequently, many state-of-the-art structures have reached a quality where one can observe effects which are not or only weakly influenced by disorder. Since investigations with such high-quality samples are most attractive if one wants to observe quantum-optical effects, we focus on semiconductor systems where the main influence of disorder can be modeled by phenomenologically introduced inhomogeneous resonance broadening.

Altogether, these considerations give us clear guidelines on how to construct the system Hamiltonian. Since the general formulation is most transparent in first quantization, we start at this level when we introduce the basic concepts. Once the general properties are known, we use the formalism of second quantization to obtain a description that is more suitable for the systematic treatment of the interacting many-body problem.

11.2.1 Second quantization of system Hamiltonian

Since $n(\mathbf{r})$ is constant within the nanostructure, the condition (9.16) for the generalized Coulomb gauge reduces into the usual Coulomb gauge,

$$\nabla \cdot \hat{\mathbf{A}}(\mathbf{r}) = 0 \tag{11.53}$$

within the nanostructure. Since $\hat{\mathbf{p}} = -i\hbar\nabla$ is evaluated only within the nanostructure, we find $\hat{\mathbf{p}} \cdot \hat{\mathbf{A}}(\mathbf{r}) = \hat{\mathbf{A}}(\mathbf{r}) \cdot \hat{\mathbf{p}}$. As a result, we may rewrite Eq. (11.50) as

$$\begin{aligned} \hat{H}_N = & \sum_{j=1}^{N} \left\{ \frac{\hat{\mathbf{p}}_j^2}{2m_0} + U(\mathbf{r}_j) - \frac{Q}{m_0}\hat{\mathbf{A}}(\mathbf{r}_j) \cdot \hat{\mathbf{p}}_j + \frac{Q^2}{2m_0}\hat{\mathbf{A}}^2(\mathbf{r}_j) \right\} \\ & + \frac{1}{2}\sum_{i \neq j}^{N} V(\mathbf{r}_i - \mathbf{r}_j) + \sum_{\mathbf{q}} \hbar\omega_{\mathbf{q}} \left[B_{\mathbf{q}}^\dagger B_{\mathbf{q}} + \frac{1}{2} \right] + \hat{H}_{\text{ph}}. \end{aligned} \tag{11.54}$$

We notice that $\hat{H}_N$ contains single-particle terms like $\hat{\mathbf{p}}^2/2m_0 + U(\mathbf{r})$ and interaction terms like $\frac{Q}{m_0}\hat{\mathbf{A}}(\mathbf{r}) \cdot \hat{\mathbf{p}} + \frac{Q^2}{2m_0}\hat{\mathbf{A}}^2(\mathbf{r})$ and $V(\mathbf{r} - \mathbf{r}')$. We can convert all these terms into the second-quantization formalism using the pragmatic steps outlined in Section 11.1.4.

In principle, we can choose any complete set of single-particle wave functions $\phi_{\lambda,\mathbf{k}}$ to perform the second quantization. Nevertheless, it is convenient to choose the orthogonal basis which diagonalizes the noninteracting electron Hamiltonian,

$$\left[\frac{\hat{\mathbf{p}}^2}{2m_0} + U(\mathbf{r}) \right] \phi_{\lambda,\mathbf{k}}(\mathbf{r}) = \epsilon_{\mathbf{k}}^{\lambda} \phi_{\lambda,\mathbf{k}}(\mathbf{r}), \tag{11.55}$$

where λ denotes a discrete set of states while $\mathbf{k}$ labels the continuum of (quasi)-momentum states. This distinction is elaborated further in Section 12.1. Starting from Eq. (11.54), the second-quantization form becomes

$$\begin{aligned} \hat{H}_0 &= \int d^3r\, \hat{\Psi}^\dagger(\mathbf{r}) \left[\frac{\hat{\mathbf{p}}^2}{2m_0} + U(\mathbf{r}) \right] \hat{\Psi}(\mathbf{r}) \\ &= \sum_{\lambda,\mathbf{k},\lambda',\mathbf{k}'} a_{\lambda,\mathbf{k}}^\dagger a_{\lambda',\mathbf{k}'} \int d^3r\, \phi_{\lambda,\mathbf{k}}^\star(\mathbf{r}) \left[\frac{\mathbf{p}^2}{2m_0} + U(\mathbf{r}) \right] \phi_{\lambda',\mathbf{k}'}(\mathbf{r}) \\ &= \sum_{\lambda,\mathbf{k},\lambda',\mathbf{k}'} a_{\lambda,\mathbf{k}}^\dagger a_{\lambda',\mathbf{k}'} \int d^3r\, \phi_{\lambda,\mathbf{k}}^\star(\mathbf{r}) \epsilon_{\mathbf{k}'}^{\lambda'} \phi_{\lambda',\mathbf{k}'}(\mathbf{r}) \\ &= \sum_{\lambda,\mathbf{k}} \epsilon_{\mathbf{k}}^{\lambda}\, a_{\lambda,\mathbf{k}}^\dagger a_{\lambda,\mathbf{k}}, \end{aligned} \tag{11.56}$$

where we used Eq. (11.55) and the orthogonality of $\phi_{\lambda,\mathbf{k}}(\mathbf{r})$.

We may now construct the system Hamiltonian (11.54) via

$$\hat{H}_N = \hat{H}_0 + \hat{H}_{\rm em} + \hat{H}_{\rm ph} \\ + \sum_{j=1}^{N} \left\{ -\frac{Q}{m_0}\hat{\mathbf{A}}(\mathbf{r}_j)\cdot\hat{\mathbf{p}}_j + \frac{Q^2}{2m_0}\hat{\mathbf{A}}^2(\mathbf{r}_j) \right\} + \frac{1}{2}\sum_{i\neq j}^{N} V(\mathbf{r}_i - \mathbf{r}_j) \qquad (11.57)$$

where the noninteracting parts are collected into the first line while the interaction terms are presented in the second line. More specifically, the second-quantized Hamiltonians of the free electrons and photons are

$$\hat{H}_0 \equiv \sum_{\lambda,\mathbf{k}} \epsilon_{\mathbf{k}}^{\lambda}\, a_{\lambda,\mathbf{k}}^{\dagger} a_{\lambda,\mathbf{k}}, \qquad \hat{H}_{\rm em} \equiv \sum_{\mathbf{q}} \hbar\omega_{\mathbf{q}} \left[B_{\mathbf{q}}^{\dagger} B_{\mathbf{q}} + \frac{1}{2} \right], \qquad (11.58)$$

respectively. Besides these, we clearly need to determine also the interaction contributions as well as the phonon contributions in second quantization. The interaction terms are analyzed explicitly in Chapter 13.

11.2.2 Second quantization of lattice vibrations

Even though the periodic lattice of atoms can often be assumed to be perfect, the ions can still oscillate moderately around their equilibrium positions $\mathbf{R}_j^0$. Figure 11.3 illustrates the ions (shaded circles) in a two-dimensional square lattice. The entire solid can be divided into *unit cells* that are the smallest possible elementary volumes (shaded rectangle) around each regular lattice site whose periodic repetition generates the entire solid. In the middle line, the ions exhibit some displacement (arrows) from their equilibrium positions (dashed circles). Once an ion is dislocated, it is pulled back toward its equilibrium position via the collective Coulomb interaction with the rest of the ions. As long as the displacement is not too large, we can assume a harmonic equilibrating force (spring symbol) that leads to collective lattice vibrations in the solid. Since the optically active electrons can interact with these vibrations, we need to include them explicitly in many calculations where we want to analyze realistically experimentally relevant situations.

For this purpose, we now abandon the mean-field description of the ions for a while and consider a more microscopic treatment, first for the simplest case where we have one atom inside the unit cell. The results for more than one atom per unit cell will be mentioned at the end of this section. As the simplest model of ionic motion, we assume that ions at different lattice sites are coupled harmonically. In first quantization, the Hamiltonian of N ions has the form

$$\hat{H}_{\rm ph} = \sum_j \frac{\hat{\mathbf{P}}_j^2}{2M} + \frac{1}{2}\sum_{n\neq m} \frac{1}{2} M\Omega_{n-m}^2 (\Delta\hat{\mathbf{R}}_n - \Delta\hat{\mathbf{R}}_m)^2, \qquad (11.59)$$

where $\mathbf{P}_j$ is the momentum of the ion j with mass M. In this Hamiltonian, the ion at position $\mathbf{R}_j$ deviates from its equilibrium value $\mathbf{R}_j^0$ (dashed line) by $\Delta\mathbf{R}_j = \mathbf{R}_j - \mathbf{R}_j^0$ (arrow). The two-particle interaction introduces a harmonic force with respect to $\Delta\mathbf{R}_j$. Here, we

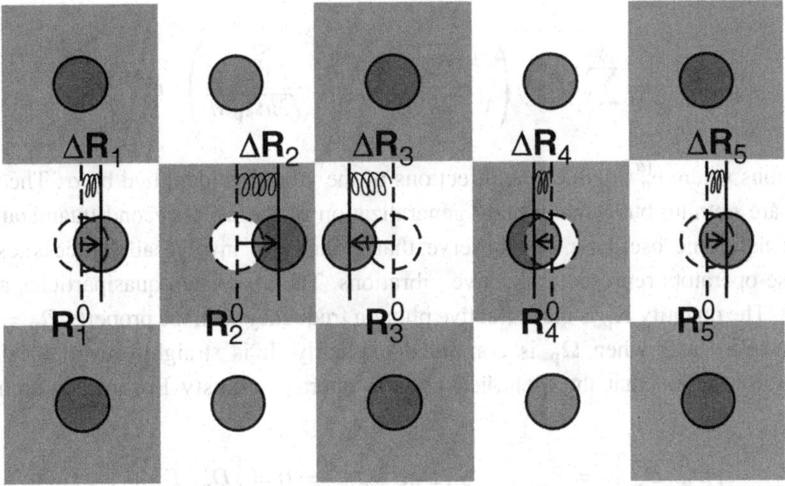

Figure 11.3 Schematic representation of the phonon model. The periodic lattice is defined by the lattice vectors $\mathbf{R}_j^0$. Phonons disturb the lattice such that the ionic position (filled circle) is displaced by $\Delta\mathbf{R}_j$ with respect to the original position (dashed circle). The restoring force is indicated by the spring symbol; the white and dark squares define the unit cells.

assume that the harmonic term depends only on the relative distance between the lattice sites such that the coupling has the form Ω^2_{n-m}. In order to check this model, we consider two situations with $\Delta\mathbf{R}_j = 0$ and $\Delta\mathbf{R}_j = \mathbf{R}$. In both cases, the interaction term vanishes such that the system does not vibrate. This feature is physically meaningful because one does not expect lattice vibrations for the equilibrium configuration ($\Delta\mathbf{R}_j = 0$) or if the entire sample is translated by $\Delta\mathbf{R}_j = \mathbf{R}$. However, if the ions are displaced differently, i.e., $\Delta\mathbf{R}_n \neq \Delta\mathbf{R}_m$, a harmonic force is present leading to vibrations in the solid.

The lattice vibrations are quantized by introducing the usual canonical commutation relations,

$$\left[\Delta\hat{\mathbf{R}}_{n,\alpha}, \hat{\mathbf{P}}_{m,\beta}\right]_- = i\hbar\delta_{n,m}\delta_{\alpha,\beta}, \tag{11.60}$$

$$\left[\Delta\hat{\mathbf{R}}_{n,\alpha}, \Delta\hat{\mathbf{R}}_{m,\beta}\right]_- = 0 = \left[\hat{\mathbf{P}}_{n,\alpha}, \hat{\mathbf{P}}_{m,\beta}\right]_-, \tag{11.61}$$

where α and β refer to the Cartesian components x, y, and z. Since Eq. (11.59) represents a genuine many-body system and the two-particle interaction is harmonic, it is – once again – convenient to adopt the formalism of second quantization. For this purpose, we introduce an annihilation operator

$$D_{\mathbf{p},\sigma} = \frac{-i}{\sqrt{N}} \sum_{j=1}^{N} e^{-i\mathbf{R}_j^0\cdot\mathbf{p}} \left(\sqrt{\frac{M\Omega_\mathbf{p}}{2\hbar}} \Delta\hat{\mathbf{R}}_j + i\frac{\hat{\mathbf{P}}_j}{\sqrt{2\hbar\Omega_\mathbf{p}M}} \right) \cdot \mathbf{e}_\mathbf{p}^{[\sigma]} \tag{11.62}$$

and a creation operator

$$D^{\dagger}_{\mathbf{p},\sigma} = \frac{i}{\sqrt{N}} \sum_{j=1}^{N} e^{i\mathbf{R}^0_j\cdot\mathbf{p}} \left(\sqrt{\frac{M\Omega_{\mathbf{p}}}{2\hbar}} \Delta\hat{\mathbf{R}}_j - i \frac{\hat{\mathbf{P}}_j}{\sqrt{2\hbar\Omega_{\mathbf{p}}M}} \right) \cdot \mathbf{e}^{[\sigma]}_{\mathbf{p}}, \tag{11.63}$$

for phonons where $\mathbf{e}^{[\sigma]}_{\mathbf{p}}$ defines the directions of the vibration identified by σ. These definitions are nothing but a many-body generalization of the usual second quantization of a single harmonic oscillator. We observe that D and $D^{\dagger}$ involve all lattice sites such that these operators represent collective vibrations. The associated quasiparticles are the *phonons*. The quantity $\Omega_{\mathbf{p}}$ is the collective phonon frequency with the property $\Omega_{\mathbf{p}} = \Omega_{-\mathbf{p}}$; this is proven later when $\Omega_{\mathbf{p}}$ is computed explicitly. It is straightforward to show – see Exercise 11.9 – that the identified phonon operators satisfy Bosonic commutation relations

$$\left[D_{\mathbf{p}}, D^{\dagger}_{\mathbf{p}'}\right]_- = \delta_{\mathbf{p},\mathbf{p}'}, \qquad \left[D_{\mathbf{p}}, D_{\mathbf{p}'}\right]_- = 0 = \left[D^{\dagger}_{\mathbf{p}}, D^{\dagger}_{\mathbf{p}'}\right]_-, \tag{11.64}$$

which is expected for the harmonic interaction potential. We have introduced here the compact notation where $\mathbf{p}$ includes the phonon-branch index σ.

Our next task is to write the Hamiltonian (11.59) in terms of phonon operators. In order to do this, we have to express the individual $\Delta\hat{\mathbf{R}}_n$ and $\hat{\mathbf{P}}_n$. For this purpose, we consider the orthogonality relation between plane waves

$$\frac{1}{N}\sum_{\mathbf{p}} e^{i(\mathbf{R}^0_n - \mathbf{R}^0_m)\cdot\mathbf{p}} = \delta_{n,m}, \tag{11.65}$$

and the completeness of the given vectors $\sum_{\sigma} \mathbf{e}^{[\sigma]}_{\mathbf{p},\beta}\, \mathbf{e}^{[\sigma]}_{\mathbf{p},\alpha} = \delta_{\alpha,\beta}$. Using these relations, we express the displacement and momentum operators

$$\begin{aligned} \Delta\hat{\mathbf{R}}_n &= \frac{i}{\sqrt{N}}\sum_{\mathbf{p}} \sqrt{\frac{\hbar}{2M\Omega_{\mathbf{p}}}} \left(D_{\mathbf{p}} e^{i\mathbf{R}^0_n\cdot\mathbf{p}} \mathbf{e}_{\mathbf{p}} - D^{\dagger}_{\mathbf{p}} e^{-i\mathbf{R}^0_n\cdot\mathbf{p}} \mathbf{e}_{\mathbf{p}} \right) \\ &= \frac{i}{\sqrt{N}}\sum_{\mathbf{p}} \sqrt{\frac{\hbar}{2M\Omega_{\mathbf{p}}}} \left(D_{\mathbf{p}} \mathbf{e}_{\mathbf{p}} - D^{\dagger}_{-\mathbf{p}} \mathbf{e}_{-\mathbf{p}} \right) e^{i\mathbf{R}^0_n\cdot\mathbf{p}} \end{aligned} \tag{11.66}$$

and

$$\begin{aligned} \hat{\mathbf{P}}_n &= \frac{1}{\sqrt{N}}\sum_{\mathbf{p}} \sqrt{\frac{\hbar M\Omega_{\mathbf{p}}}{2}} \left(D_{\mathbf{p}} e^{i\mathbf{R}^0_n\cdot\mathbf{p}} \mathbf{e}_{\mathbf{p}} + D^{\dagger}_{\mathbf{p}} e^{-i\mathbf{R}^0_n\cdot\mathbf{p}} \mathbf{e}_{\mathbf{p}} \right) \\ &= \frac{1}{\sqrt{N}}\sum_{\mathbf{p}} \sqrt{\frac{\hbar M\Omega_{\mathbf{p}}}{2}} \left(D_{\mathbf{p}} \mathbf{e}_{\mathbf{p}} + D^{\dagger}_{-\mathbf{p}} \mathbf{e}_{-\mathbf{p}} \right) e^{i\mathbf{R}^0_n\cdot\mathbf{p}}. \end{aligned} \tag{11.67}$$

In these derivations, we utilized the property that the sum over $\mathbf{p}$ includes both $+\mathbf{p}$ and $-\mathbf{p}$. These relations can now be inserted into Eq. (11.59), which leads to

$$\hat{H}_{\rm ph} = \sum_{\mathbf{p},\mathbf{p}'} \frac{1}{2M}\frac{\hbar M}{2}\sqrt{\Omega_{\mathbf{p}}\Omega_{\mathbf{p}'}}\left(D_{\mathbf{p}}\mathbf{e}_{\mathbf{p}} + D^\dagger_{-\mathbf{p}}\mathbf{e}_{-\mathbf{p}}\right)$$
$$\cdot\left(D_{-\mathbf{p}'}\mathbf{e}_{-\mathbf{p}'} + D^\dagger_{\mathbf{p}'}\mathbf{e}_{\mathbf{p}'}\right)\sum_n \frac{1}{N}e^{i\mathbf{R}^0_n\cdot(\mathbf{p}-\mathbf{p}')}$$
$$-\frac{1}{2}\sum_{\mathbf{p},\mathbf{p}'} M\frac{\hbar}{2M}\frac{1}{\sqrt{\Omega_{\mathbf{p}}\Omega_{\mathbf{p}'}}}\left(D_{\mathbf{p}}\mathbf{e}_{\mathbf{p}} - D^\dagger_{-\mathbf{p}}\mathbf{e}_{-\mathbf{p}}\right)$$
$$\cdot\left(D_{-\mathbf{p}'}\mathbf{e}_{-\mathbf{p}'} - D^\dagger_{\mathbf{p}'}\mathbf{e}_{\mathbf{p}'}\right)$$
$$\times\sum_{n,m}\frac{1}{N}e^{i\mathbf{R}^0_n\cdot(\mathbf{p}-\mathbf{p}')}\frac{\Omega^2_{n-m}}{2}\left(1-e^{i\left(\mathbf{R}^0_m-i\mathbf{R}^0_n\right)\cdot\mathbf{p}}\right)$$
$$\times\left(1-e^{-i\left(\mathbf{R}^0_m-i\mathbf{R}^0_n\right)\cdot\mathbf{p}'}\right). \tag{11.68}$$

The sums over the lattice sites can be performed analytically. The first sum produces

$$\Sigma_1 = \frac{1}{N}\sum_n e^{i\mathbf{R}^0_n\cdot(\mathbf{p}-\mathbf{p}')} = \delta_{\mathbf{p},\mathbf{p}'}, \tag{11.69}$$

and the second sum yields

$$\Sigma_2 = \sum_{n,m}\frac{1}{N}e^{i\mathbf{R}^0_n\cdot(\mathbf{p}-\mathbf{p}')}\frac{\Omega^2_{n-m}}{2}\left(1-e^{i\left(\mathbf{R}^0_m-i\mathbf{R}^0_n\right)\cdot\mathbf{p}}\right)\left(1-e^{-i\left(\mathbf{R}^0_m-i\mathbf{R}^0_n\right)\cdot\mathbf{p}'}\right)$$
$$=\sum_n\frac{1}{N}e^{i\mathbf{R}^0_n\cdot(\mathbf{p}-\mathbf{p}')}\sum_{\Delta m}\frac{\Omega^2_{\Delta m}}{2}\left(1-e^{i\left(\mathbf{R}^0_{\Delta m}\right)\cdot\mathbf{p}}\right)\left(1-e^{-i\left(\mathbf{R}^0_{\Delta m}\right)\cdot\mathbf{p}'}\right)$$
$$=\delta_{\mathbf{p},\mathbf{p}'}\sum_m\frac{\Omega^2_m}{2}\left|1-e^{i\mathbf{R}^0_m\cdot\mathbf{p}}\right|^2. \tag{11.70}$$

This result explicitly identifies the collective phonon frequency,

$$\Omega^2_{\mathbf{p}} \equiv \sum_m \frac{\Omega^2_m}{2}\left|1-e^{i\mathbf{R}^0_m\cdot\mathbf{p}}\right|^2. \tag{11.71}$$

With the help of Eqs. (11.69)–(11.71), we can now simplify the phonon Hamiltonian into the form

$$\hat{H}_{\rm ph} = \sum_{\mathbf{p}}\frac{1}{4}\hbar\Omega_{\mathbf{p}}\left(D_{\mathbf{p}}\mathbf{e}_{\mathbf{p}} + D^\dagger_{-\mathbf{p}}\mathbf{e}_{-\mathbf{p}}\right)\cdot\left(D_{-\mathbf{p}}\mathbf{e}_{-\mathbf{p}} + D^\dagger_{\mathbf{p}}\mathbf{e}_{\mathbf{p}}\right)$$
$$-\sum_{\mathbf{p}}\frac{1}{4}\hbar\Omega_{\mathbf{p}}\left(D_{\mathbf{p}}\mathbf{e}_{\mathbf{p}} - D^\dagger_{-\mathbf{p}}\mathbf{e}_{-\mathbf{p}}\right)\cdot\left(D_{-\mathbf{p}}\mathbf{e}_{-\mathbf{p}} - D^\dagger_{\mathbf{p}}\mathbf{e}_{\mathbf{p}}\right)$$
$$=\sum_{\mathbf{p}}\frac{1}{2}\hbar\Omega_{\mathbf{p}}\left(D_{\mathbf{p}}D^\dagger_{\mathbf{p}}\;\mathbf{e}_{\mathbf{p}}\cdot\mathbf{e}_{\mathbf{p}} + D^\dagger_{-\mathbf{p}}D_{-\mathbf{p}}\;\mathbf{e}_{-\mathbf{p}}\cdot\mathbf{e}_{-\mathbf{p}}\right)$$
$$=\sum_{\mathbf{p}}\hbar\Omega_{\mathbf{p}}\left(D^\dagger_{\mathbf{p}}D_{\mathbf{p}}+\frac{1}{2}\right), \tag{11.72}$$

where we have used the commutation relation (11.64). Once again, we find a Hamiltonian describing a set of harmonic oscillators.

In order to evaluate the eigenfrequencies $\Omega_{\mathbf{p}}$, we now take the long wave length limit of the dispersion relation (11.71), i.e., assume that $\mathbf{p}$ dependency of $\Omega_{\mathbf{p}}$ follows from terms linear in momentum. As a result, we find

$$\Omega_{\mathbf{p}} = c_A|\mathbf{p}|, \tag{11.73}$$

where the coefficient c_A is called the (acoustic) phonon velocity of sound. Due to this special dispersion, this class of lattice vibrations is referred to as *acoustic phonons*. Since the velocity of sound can be measured for different materials, we use these experimental values in Eqs. (11.72)–(11.73) whenever we treat acoustic-phonon effects.

Many semiconductors have unit cells consisting of more than one atom. In this case, also optical phonon excitations exist in addition to the acoustic. As it turns out, the optical phonon dispersion is nearly independent of $\mathbf{p}$ in contrast to the acoustic phonons. In GaAs based systems, optical phonons have an energy of around 36 meV. The low-dimensionality of semiconductor nanostructures often causes only minor modifications of the acoustic and optical phonons because the lattice vibrations propagate through the entire three-dimensional sample. Thus, the phonons are typically truly three dimensional. Consequently, Eq. (11.72) can be used as a starting point without dimension-dependent modifications.

In thermodynamic systems, the phonon occupation numbers closely follow the Bose–Einstein distribution

$$\left\langle D^{\dagger}_{\mathbf{p}} D_{\mathbf{p}} \right\rangle = \frac{1}{e^{\frac{\hbar\Omega_{\mathbf{p}}}{k_B T}} - 1}, \tag{11.74}$$

where k_B is the Boltzmann constant and T is the temperature of the sample. Hence, for low temperatures (below approximately $100K$ giving $k_B T = 8.617$ meV), optical phonon populations are often negligible because their occupation $\left\langle D^{\dagger}_{\mathbf{p}} D_{\mathbf{p}} \right\rangle \ll 1$. In these cases, it is usually sufficient to focus on the effects of acoustic phonons only. Since most of the quantum-optical investigations are performed under such conditions, we do not derive the optical phonon effects in detail.

Exercises

11.1 The Leibniz formula for determinants follows from

$$\det[\mathbf{M}] = \sum\nolimits_P (-1)^{n(P)} \prod\nolimits_{j=1}^{N} \mathbf{M}_{j,P(j)} \tag{11.75}$$

(a) Show that the identification in Eq. (11.8) yields the Slater determinant. **(b)** Construct all permutations for $N = 3$ and deduce $(-1)^{n(P)}$ for each P. **(c)** Construct det[**M**] for $N = 3$ using Eq. (11.75). Show that the Leibniz formula (11.75) works also when $\mathbf{M}_{j,P(j)}$ is replaced by $\mathbf{M}_{P(j),j}$.

11.2 **(a)** Show that the Slater determinant is normalized, i.e.,

$$\int d^{3N}r \left| \Phi^{\mathrm{Sltr}}_{\lambda_1,\ldots\lambda_N}(\mathbf{r}_1, \ldots \mathbf{r}_N) \right|^2 = 1$$

Hint: *Use Eq.* (11.8) *and assume that the states* $\phi_\lambda(r)$ *are orthonormalized.* **(b)** Modify the Slater determinant by implementing $\phi_{\lambda_1}(\mathbf{r}) \rightarrow \cos\theta\, \phi_{\lambda_1}(\mathbf{r}) +$

$\sin\theta\,\phi_{\lambda_2}(r)$ and construct the modified Slater determinant $\tilde{\Phi}^{\text{Sltr}}_{\lambda_1,\ldots\lambda_N}$. Normalize $\tilde{\Phi}^{\text{Sltr}}$ properly and show that the normalized $\tilde{\Phi}^{\text{Sltr}}_{\lambda_1,\ldots\lambda_N}$ is equal to $\Phi^{\text{Sltr}}_{\lambda_1,\ldots\lambda_N}$ independent of θ. **(c)** Implement the formal order (11.15) and show that $\langle\nu_N,\ldots\nu_1|\lambda_1,\ldots\lambda_N\rangle = \prod_{j=1}^{N}\delta_{\nu_j,\lambda_j}$. How is the result modified if the state sequences are not ordered?

11.3 **(a)** Show that the subdeterminant expansion (11.10) implies $\Phi^{\text{Sltr}}_{\lambda_1,\ldots\lambda_N}(\mathbf{r}_1,\ldots\mathbf{r}_N) = \sum_{J,K=1}^{N}\frac{(-1)^{J+K}}{\sqrt{N}}\phi_{\lambda_J}(\mathbf{r}_K)\,\Phi^{J;K}_{\lambda_1,\ldots\lambda_N}(\mathbf{r}_1,\ldots\mathbf{r}_N)$. **(b)** Show that (11.10) also produces

$$\Phi^{\text{Sltr}}_{\lambda_1,\ldots\lambda_N}(\mathbf{r}_1,\ldots\mathbf{r}_N) = \sum_{J_1=2}^{N}\sum_{J_2=1}^{J_1-1}\frac{M^{J_1,J_2}_{K_1,K_2}}{\sqrt{N(N-1)}}\,\Phi^{J_1,J_2;K_1,K_2}_{\lambda_1,\ldots\lambda_N}(\mathbf{r}_1,\ldots\mathbf{r}_N)$$
$$-\sum_{J_1=1}^{N-1}\sum_{J_2=J_1+1}^{N}\frac{M^{J_1,J_2}_{K_1,K_2}}{\sqrt{N(N-1)}}\,\Phi^{J_1,J_2;K_1,K_2}_{\lambda_1,\ldots\lambda_N}(\mathbf{r}_1,\ldots\mathbf{r}_N),$$

where $M^{J_1,J_2}_{K_1,K_2} \equiv (-1)^{J_1+J_2+K_1+K_2}\phi_{\lambda_{J_1}}(\mathbf{r}_{K_1})\,\phi_{\lambda_{J_2}}(\mathbf{r}_{K_2})$ and $\Phi^{J_1J_2;K_1,K_2}_{\lambda_1,\ldots\lambda_N}$ refers to a Slater determinant where the states λ_{J_1} and λ_{J_2} and the coordinates $\mathbf{r}_{K_1}$ and $\mathbf{r}_{K_2}$ are eliminated, see Eq. (11.11).

11.4 **(a)** Show that the sum $\sum_{\lambda_1<\lambda_2<\ldots\lambda_N} C_{\lambda_1\ldots\lambda_N}$ is performed over $\binom{M}{N} = \frac{M!}{N!(M-N)!}$ combinations where M denotes the total number of states λ. **(b)** Use the relation $\sum_{\lambda_1<\cdots<\lambda_N} = \binom{M}{N}$ to construct the sums $\sum_{j=1}^{S} j^P$ for $P = 1, 2,$ and 3.

11.5 **(a)** Evaluate $\left(a_\alpha a^\dagger_\beta + a^\dagger_\beta a_\alpha\right)|\lambda_1,\ldots,\lambda_N\rangle$ for different combinations of λ_j states. **(b)** Show that $\left(a_\alpha a^\dagger_\beta + a^\dagger_\beta a_\alpha\right)|\Psi\rangle = \delta_{\alpha,\beta}|\Psi\rangle$ for an arbitrary wave function. Deduce the Fermion anticommutation from this result. **(c)** Verify that the anticommutation relations (11.44) yield Eq. (11.46).

11.6 **(a)** Show that the annihilation operator can be expressed via $a_\nu = \sum_{N=0}^{\infty}\sum_{\lambda_1<\cdots<\lambda_N}|\lambda_1,\ldots\lambda_N\rangle\langle\lambda_N,\ldots\lambda_1,\nu|$. **Hint:** *Use the completeness relation* (11.29) *with all N and apply the property* (11.22). **(b)** Construct $a^\dagger_\lambda$ accordingly and compute $\left[a_\nu, a^\dagger_\lambda\right]_+$ using the explicit bra-ket format.

11.7 **(a)** Show that

$$a_{\nu_{J'}}a_{\nu_J}|\nu_1,\ldots,\nu_N\rangle = \begin{cases} (-1)^{J+J'}|(\nu_1\ldots\nu_N)^{J,J'}\rangle, & J > J' \\ 0, & J = J' \\ -(-1)^{J+J'}|(\nu_1,\ldots\nu_N)^{J,J'}, & J < J' \end{cases}.$$

(b) Assume that we know the two-particle operator in real space $\hat{O}_2(\mathbf{r}_1,\ldots,\mathbf{r}_N) = \sum_{j\neq k=1}^{N} U_2(\mathbf{r}_j,\mathbf{r}_k)$. Show that we find

$$\hat{O}_2 = \int d^3r\int d^3r'\,\hat{\Psi}^\dagger(\mathbf{r})\hat{\Psi}^\dagger(\mathbf{r}')\,U_2(\mathbf{r},\mathbf{r}')\,\hat{\Psi}(\mathbf{r}')\hat{\Psi}(\mathbf{r})$$

in the second quantization. **Hint:** *Use the recursion derived in Exercise 11.3(b) and the property derived in Exercise 11.7(a).*

11.8 Assume that the two-particle interaction reduces into the effective single particle format $U_2(\mathbf{r}, \mathbf{r}') = F(\mathbf{r}) + G(\mathbf{r}')$. **(a)** How does this simplify $O_2(\mathbf{r}_1, \ldots \mathbf{r}_N) = \sum_{j \neq k=1}^{N} U_2(\mathbf{r}, \mathbf{r}')$? **(b)** Construct the corresponding $\hat{O}_2$ in second quantization. How is this result related to the simplified first-quantization form?

11.9 **(a)** Show that the phonon operators (11.62) satisfy the Bosonic commutation relations (11.64). **(b)** Show that the number of unit cells is identical to the physically relevant p momenta, i.e., prove the relation (11.65). **Hint:** *Use the square lattice within the quantization box. Request the phonon modes to be periodic.* **(c)** Confirm that Eq. (11.68) produces (11.72).

11.10 Assume a one-dimensional lattice with the period a and the harmonic coupling $\Omega_m = \Omega_0 \, e^{-|m|\beta/2}$ where $m = 0, \pm 1, \ldots$ and $R > 0$. Compute the collective phonon frequency using Eq. (11.71) and show that

$$\Omega_p = \Omega_0 \sqrt{\coth \beta} \frac{\left|\sin \frac{ap}{2}\right|}{\left|e^{\frac{\beta}{2}} - e^{-\frac{\beta}{2}} \, e^{iap}\right|}.$$

Further reading

Determinants and their properties are discussed in many linear algebra books, cf.:

P. D. Lax (2007). *Linear Algebra and its Applications*, 2nd edition, Hoboken, NJ, Wiley & Sons.

The foundations of many-body theory are presented more thoroughly, e.g., in:

G. D. Mahan (1990). *Many-Particle Physics*, 2nd edition, New York, Plenum Press.
A. L. Fetter and J. D. Walecka (2003). *Quantum Theory of Many-Particle Systems*, New York, Dover.
J. W. Negele and H. Orland (1998). *Quantum Many-Body Systems*, Boulder, CO, Westview Press.

For further reading on central concepts in semiconductor physics, we recommend:

C. Kittel (1967). *Quantum Theory of Solids*, New York, Wiley & Sons.
J. M. Ziman (1960). *Electrons and Phonons: The Theory of Transport Phenomena in Solids*, Oxford, Oxford University Press.
C. Weisbuch and B. Vinter (1991). *Quantum Semiconductor Structures: Fundamentals and Applications*, London, Academic Press.
K. Barnham and D. Vvedensky (2001). *Low-Dimensional Semiconductor Structures: Fundamentals and Device Applications*, Cambridge, Cambridge University Press.

H. Haug and S. W. Koch (2009). *Quantum Theory of the Optical and Electronic Properties of Semiconductors*, 5th edition, Singapore, World Scientific.

P. Y. Yu and M. Cardona (2005). *Fundamentals of Semiconductors: Physics and Materials Properties*, 3rd edition, Berlin, Springer.

D. Bimberg, ed. (2008). *Semiconductor Nanostructures*, Berlin, Springer.

T. Ihn (2010). *Semiconductor Nanostructures: Quantum States and Electronic Transport*, Oxford, Oxford University Press.

T. Steiner, ed. (2004). *Semiconductor Nanostructures for Optoelectronic Applications*, Norwood, MA, Artec House.

12

Band structure of solids

In principle, the system Hamiltonian (11.57) can be used as a general starting point to investigate semiconductor quantum optics. Especially, it introduces three important classes of quasiparticles: electronic states, phonons, and photons. From these, both the phonon and the photon energy follow from a linear dispersion relation, $\hbar\omega = \hbar c|\mathbf{q}|$, connecting energy and wave vector through the speed of sound and light, respectively. However, the energies $\epsilon_{\mathbf{k}}^{\lambda}$ have a nontrivial dependency on momentum $\mathbf{k}$ in solids. Thus, we need to determine $\epsilon_{\mathbf{k}}^{\lambda}$ explicitly for a given system even to deduce its elementary single-particle properties.

In solids, the single-particle energy spectrum $\epsilon_{\mathbf{k}}^{\lambda}$ is known as the band structure. The necessary band-structure calculations are thoroughly studied in several text books mentioned at the end of this chapter. Here, we will summarize only those aspects that are relevant for a systematic theory of semiconductor quantum optics. Once the single-particle properties are known to sufficient accuracy, we can re-express also the interacting contributions within Eq. (11.57). This step is performed in Chapter 13.

12.1 Electrons in the periodic lattice potential

We start by analyzing the solutions $\phi_{\lambda,\mathbf{k}}(\mathbf{r})$ assuming a perfect ionic crystal defined by the lattice vectors $\mathbf{R}_j^0$ of the undisturbed crystal. In other words, we have to solve

$$\left[\frac{\hat{\mathbf{p}}^2}{2m_0} + U_0(\mathbf{r})\right]\phi_{\lambda,\mathbf{k}}(\mathbf{r}) = \epsilon_{\mathbf{k}}^{\lambda}\,\phi_{\lambda,\mathbf{k}}(\mathbf{r}), \tag{12.1}$$

with the perfect lattice potential $U_0(\mathbf{r})$ instead of U($\mathbf{r}$) used in Eq. (11.55). We may choose

$$\int d^3r\ \phi_{\lambda,\mathbf{k}}^{*}(\mathbf{r})\phi_{\lambda',\mathbf{k}'}(\mathbf{r}) = \delta_{\lambda,\lambda'}\,\delta_{\mathbf{k},\mathbf{k}'} \tag{12.2}$$

because these states can always be orthogonalized, cf. Section 5.2.1.

As we perform the second quantization using the basis (12.1), the system Hamiltonian (11.54) becomes

$$\hat{H}_N = \hat{H}_0 + \hat{H}_{\rm em} + \hat{H}_{\rm ph} + \sum\nolimits_j [U(\mathbf{r}_j) - U_0(\mathbf{r}_j)]$$
$$+ \sum\nolimits_{j=1}^{N} \left\{ -\frac{Q}{m_0}\hat{\mathbf{A}}(\mathbf{r}_j)\cdot\hat{\mathbf{p}}_j + \frac{Q^2}{2m_0}\hat{\mathbf{A}}^2(\mathbf{r}_j) \right\} + \frac{1}{2}\sum\nolimits_{i\neq j}^{N} V(\mathbf{r}_i - \mathbf{r}_j) \qquad (12.3)$$

that follows in the same way as Eq. (11.57). The term $U(\mathbf{r}_j) - U_0(\mathbf{r}_j)$ is relatively small because phonons typically cause only a weak perturbation of the ionic crystal. More specifically, this term accounts for the difference between the phonon-modified and the equilibrium lattice potential eventually producing the electron–phonon coupling discussed in Chapter 13.

For our purposes here, we only need to know some general features of $U_0(\mathbf{r})$, such as the symmetry and periodicity properties of the potential which reflect the structure of the crystal lattice. The periodicity of the effective lattice potential induces the translational symmetry

$$U_0(\mathbf{r}) = U_0\left(\mathbf{r} + \mathbf{R}_n^0\right), \qquad (12.4)$$

where $\mathbf{R}_n^0$ is a *lattice vector*, i.e., a vector that connects two identical sites in an infinite lattice (see Fig. 11.2(a)). Since $U_0(\mathbf{r})$ is periodic, the entire volume $\mathcal{L}^3$ can be subdivided into identical unit cells. If this is done, the positions $\mathbf{r}$ and $\mathbf{r} + \mathbf{R}_n^0$ are n unit cells apart. Thus, it is convenient to expand the lattice vectors according to

$$\mathbf{R}_n = \sum\nolimits_i n_i \mathbf{a}_i \qquad (12.5)$$

where the n_i are integers and the $\mathbf{a}_i$ are the basis vectors which span the unit cells. Note that the basis vectors are usually not unit vectors and are generally not even orthogonal. The basis vectors point in the directions of the three axes of the unit cell, which may have, e.g., a cubic, or rhombic, or more complicated shape. The basis vectors are parallel to the usual Cartesian unit vectors only in the case of orthogonal lattices such as the cubic one.

The specific symmetry of $U_0(\mathbf{r})$ implies restrictions also for $\phi_{\lambda,\mathbf{k}}(\mathbf{r})$. This symmetry requirement is expressed by the *Bloch theorem*

$$\phi_{\lambda,\mathbf{k}}\left(\mathbf{r} + \mathbf{R}_n^0\right) = e^{i\mathbf{k}\cdot\mathbf{R}_n^0}\phi_{\lambda,\mathbf{k}}(\mathbf{r}), \qquad (12.6)$$

which states that a translation by $\mathbf{R}_n^0$ can only result in a phase shift $e^{i\mathbf{k}\cdot\mathbf{R}_n^0}$. To satisfy the Bloch theorem, we make the ansatz

$$\phi_{\lambda,\mathbf{k}}(\mathbf{r}) = \frac{e^{i\mathbf{k}\cdot\mathbf{r}}}{\mathcal{L}^{3/2}} u_{\lambda,\mathbf{k}}(\mathbf{r}). \qquad (12.7)$$

Here, $u_{\lambda,\mathbf{k}}(\mathbf{r})$ is the *Bloch function* and $e^{i\mathbf{k}\cdot\mathbf{r}}$ corresponds to the effective free motion of electrons in the crystal. Therefore, the quantum number $\hbar\mathbf{k}$ is continuous. It is often referred to as the crystal momentum because it results from the special translation symmetry (12.6) in the crystal.

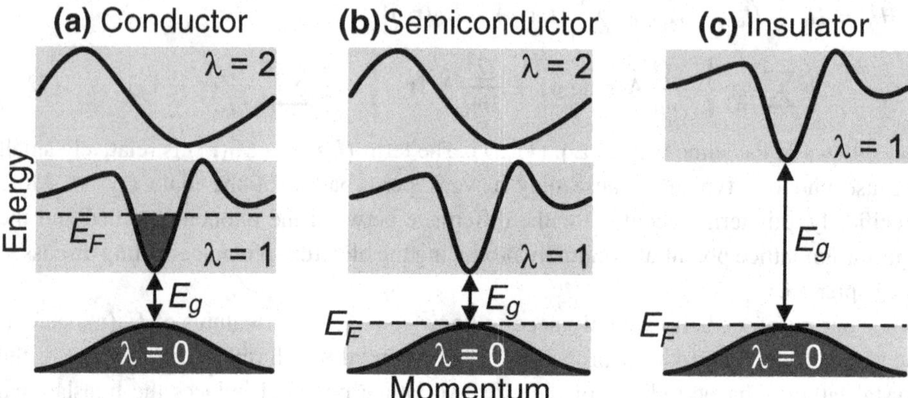

Figure 12.1 Schematic comparison of **(a)** conductors, **(b)** semiconductors, and **(c)** insulators. The solid lines present a typical band structure and the shaded areas show the occupied electronic states in the ground-state configuration. The dashed line corresponds to the Fermi energy.

The ansatz (12.7) fulfills the Bloch theorem (12.6) if u_λ is periodic in real space

$$u_{\lambda,\mathbf{k}}(\mathbf{r}) = u_{\lambda,\mathbf{k}}\left(\mathbf{r} + \mathbf{R}_n^0\right), \tag{12.8}$$

i.e., if the Bloch function has the lattice periodicity. This defines periodic boundary conditions in *all* spatial directions, which yield similar restrictions for the wave function as a confining potential does for bound-state solutions, see the discussion in Section 4.4. As a result, Eq. (12.1) can be satisfied only with a discrete set of $\epsilon_{\mathbf{k}}^{\lambda}$ values when $\mathbf{k}$ is fixed. Figure 12.1 schematically presents a typical dispersion $\epsilon_{\mathbf{k}}^{\lambda}$ (solid lines) as function of $\mathbf{k}$ in solids. As a characteristic feature, $\epsilon_{\mathbf{k}}^{\lambda}$ and $\epsilon_{\mathbf{k}}^{\nu}$ can be separated by energetically forbidden regions – the so-called band gaps E_g (arrows, and white regions). The energetically allowed regions define the bands (shaded areas) identified by the discrete band index λ.

The actual form of all $\epsilon_{\mathbf{k}}^{\lambda}$ defines the band structure of the system identifying the single-particle properties of solids. This information is sufficient to qualitatively understand many features of solids. Since electrons are Fermions, each $(\lambda, \mathbf{k})$ state can be occupied only once. Thus, the noninteracting ground state can be constructed by filling all electron states up to the so-called Fermi energy E_F (dashed line in Fig. 12.1). The last fully filled band is referred to as the valence band while the first empty or partially empty band defines the conduction band. In general, electrons cannot move in fully filled bands due to Pauli blocking. Thus, only systems with a partially filled (dark area) conduction band allow for electron transitions to neighboring states needed for electrical conductivity, see Fig. 12.1(a). When the conduction band is empty, the system is insulating (Fig. 12.1(c)) if E_g is large. Semiconductors (Fig. 12.1(b)) have an insulating ground state but one can relatively easily move electrons from the valence into the conduction band over a sufficiently small E_g.

12.1.1 k·p theory

To learn more about band-structure details, we need to know the generic form of $\epsilon_{\mathbf{k}}^{\lambda}$ for the valence and conduction band. These aspects can be efficiently investigated through the so-called $\mathbf{k}\cdot\mathbf{p}$ theory. To derive it, we apply the basic property of the momentum operator, $\hat{\mathbf{p}}\, e^{i\mathbf{k}\cdot\mathbf{r}} = \hbar\mathbf{k}\, e^{i\mathbf{k}\cdot\mathbf{r}}$, to the wave function (12.7). This leads to

$$\begin{aligned}\mathcal{L}^{3/2}\hat{\mathbf{p}}^2\phi_{\lambda,\mathbf{k}}(\mathbf{r}) &= \hat{\mathbf{p}}\cdot\hat{\mathbf{p}}\left[e^{i\mathbf{k}\cdot\mathbf{r}}u_{\lambda,\mathbf{k}}(\mathbf{r})\right] \\ &= \hat{\mathbf{p}}\cdot\left[\hbar\mathbf{k}\, e^{i\mathbf{k}\cdot\mathbf{r}}u_{\lambda,\mathbf{k}}(\mathbf{r}) + e^{i\mathbf{k}\cdot\mathbf{r}}\,\hat{\mathbf{p}}u_{\lambda,\mathbf{k}}(\mathbf{r})\right] \\ &= e^{i\mathbf{k}\cdot\mathbf{r}}\left[\hbar^2\mathbf{k}^2u_{\lambda,\mathbf{k}}(\mathbf{r}) + 2\hbar\mathbf{k}\cdot\hat{\mathbf{p}}\,u_{\lambda,\mathbf{k}}(\mathbf{r}) + \hat{\mathbf{p}}^2u_{\lambda,\mathbf{k}}(\mathbf{r})\right].\end{aligned} \tag{12.9}$$

Inserting this result into Eq. (12.1), we obtain

$$\left[\frac{\hat{\mathbf{p}}^2}{2m_0} + \frac{\hbar}{m_0}\mathbf{k}\cdot\hat{\mathbf{p}} + U_0(\mathbf{r})\right]u_{\lambda,\mathbf{k}}(\mathbf{r}) = \left[\epsilon_{\mathbf{k}}^{\lambda} - \frac{\hbar^2\mathbf{k}^2}{2m_0}\right]u_{\lambda,\mathbf{k}}(\mathbf{r}), \tag{12.10}$$

which constitutes the starting point for the $\mathbf{k}\cdot\mathbf{p}$ analysis. Once $u_{\lambda,\mathbf{k}}(\mathbf{r})$ is determined, the solution of the original Eq. (12.1) is directly obtained with the help of Eq. (12.7).

To efficiently compute the Bloch functions $u_{\lambda,\mathbf{k}}(\mathbf{r}) = \langle\mathbf{r}|\lambda,\mathbf{k}\rangle$, we apply the Dirac notation of Section 5.2.1, i.e., we present the wave functions in their abstract space formats $|\lambda,\mathbf{k}\rangle$ and $|\lambda\rangle \equiv |\lambda,\mathbf{0}\rangle$. We assume that one has solved the band-structure problem at some point $\mathbf{k}_0$ with high symmetry, which yields a discrete set of solutions $|\lambda,\mathbf{k}_0\rangle$. Here, we take this point as $\mathbf{k}_0 = 0$, which is called the Γ-point. In other words, we assume that we know all the states $|\lambda\rangle$ and the eigenvalues $\epsilon_{\mathbf{0}}^{\lambda}$.

In order to compute the Bloch functions $|\lambda,\mathbf{k}\rangle$ and the corresponding energy eigenvalues $\epsilon_{\mathbf{k}}^{\lambda}$ for $\mathbf{k}$ in the vicinity of the Γ-point, we expand the lattice-periodic function $|\lambda,\mathbf{k}\rangle$ in terms of the known functions $|\lambda\rangle$ which form a complete set. We rewrite Eq. (12.10) as

$$\left[\hat{H}_0 + \frac{\hbar}{m_0}\mathbf{k}\cdot\hat{\mathbf{p}}\right]|\lambda,\mathbf{k}\rangle = \left[\epsilon_{\mathbf{k}}^{\lambda} - \frac{\hbar^2\mathbf{k}^2}{2m_0}\right]|\lambda,\mathbf{k}\rangle, \tag{12.11}$$

with $\hat{H}_0 = \frac{\hat{\mathbf{p}}^2}{2m_0} + \hat{U}_0$. The idea now is to treat the $\mathbf{k}\cdot\mathbf{p}$ term as a perturbation to the Hamiltonian $\hat{H}_0$. Since we assume that the solutions of the eigenvalue problem $\hat{H}_0|\lambda\rangle = \epsilon_{\mathbf{0}}^{\lambda}|\lambda\rangle$ are known, we derive a perturbative solution for $|\lambda,\mathbf{k}\rangle$. Using the periodicity of $|\lambda\rangle$, see Exercise 12.2, we find that

$$\langle\lambda|\hat{\mathbf{p}}|\lambda\rangle = 0, \tag{12.12}$$

i.e., there is no first-order energy correction to $\epsilon_{\mathbf{k}}^{\lambda}$. Thus, we have to apply second-order nondegenerate perturbation theory to obtain

$$|\mathbf{k},\lambda\rangle = |\lambda\rangle + \frac{\hbar}{m_0}\sum_{\eta\neq\lambda}\frac{|\eta\rangle\mathbf{k}\cdot\langle\eta|\mathbf{p}|\lambda\rangle}{\epsilon_0^{\lambda} - \epsilon_0^{\eta}} + \mathcal{O}(\mathbf{k}^2) \tag{12.13}$$

and

$$\epsilon_{\mathbf{k}}^{\lambda} = \epsilon_0^{\lambda} + \frac{\hbar^2 \mathbf{k}^2}{2m_0} + \sum_{\eta \neq \lambda} \frac{\hbar^2}{m_0^2} \frac{(\mathbf{k} \cdot \langle \lambda | \mathbf{p} | \eta \rangle)(\mathbf{k} \cdot \langle \eta | \mathbf{p} | \lambda \rangle)}{\epsilon_0^{\lambda} - \epsilon_0^{\eta}} + \mathcal{O}(\mathbf{k}^3). \tag{12.14}$$

The accuracy of these results is best for small $\mathbf{k}$ values and gradually decreases with increasing $\mathbf{k}$.

12.1.2 Two-band approximation

In some situations, the semiconductor has a band structure where excitations from the valence to the conduction band involve only two isolated bands. This situation is illustrated in Fig. 12.2 where the rectangle identifies the region of interest (ROI) where most of the relevant excitations take place. The $\mathbf{k} \cdot \mathbf{p}$ results can now be applied directly if we associate the state $|0\rangle$ with the valence band ($0 \equiv v$) and the state $|1\rangle$ with the conduction band ($c \equiv 1$). This two-band approximation includes only the states $|c\rangle$ and $|v\rangle$ in the description of the relevant excitation dynamics.

As we apply the two-band approximation and use Cartesian coordinates with $\mathbf{p} = (p_1, p_2, p_3)$ in Eq. (12.14), we find

$$\epsilon_{\mathbf{k}}^{1} = \epsilon_0^{1} + \frac{\hbar^2 \mathbf{k}^2}{2m_0} + \sum_{i,j=1}^{3} \frac{\hbar^2 k_i k_j}{2m_0} \frac{2\mathrm{p}_i^{\star} \mathrm{p}_j}{m_0 E_g} \tag{12.15}$$

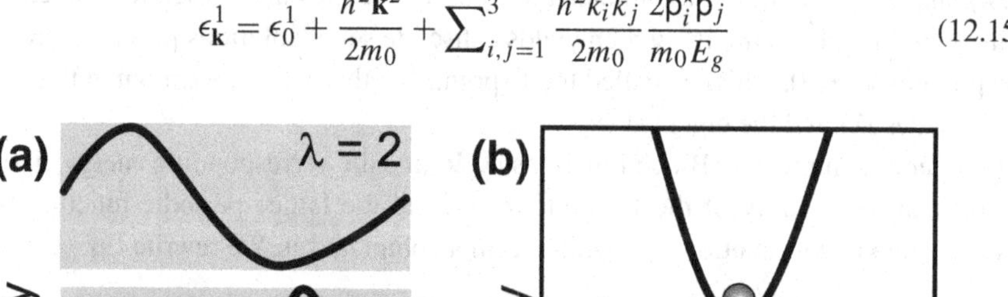

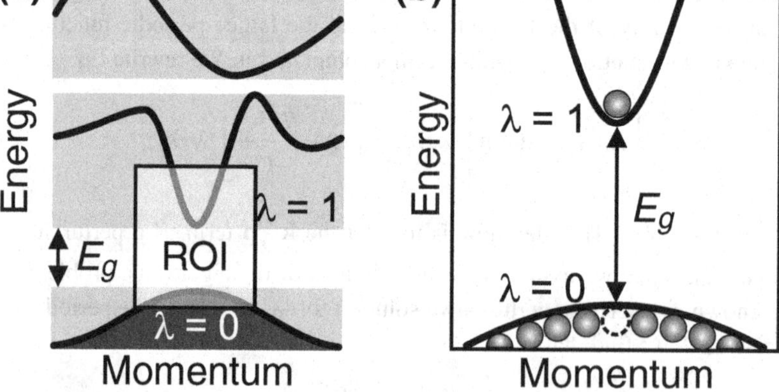

Figure 12.2 Two-band model for the semiconductor band structure. **(a)** The energetically lowest conduction band (solid line) is separated from the valence band (shaded area) by a band gap E_g (arrows). The rectangle indicates the region of interest (ROI) where all relevant excitations take place. **(b)** The shown ROI includes only two nearly parabolic bands. The fundamental excitations involve lifting electrons (spheres) from the valence to the conduction band. This process leaves holes (dashed circles) behind.

and

$$\epsilon_{\mathbf{k}}^{0} = \epsilon_{0}^{0} + \frac{\hbar^2 \mathbf{k}^2}{2m_0} - \sum\nolimits_{i,j=1}^{3} \frac{\hbar^2 k_i k_j}{2m_0} \frac{2\mathsf{p}_i^{\star} \mathsf{p}_j}{m_0 E_g}, \tag{12.16}$$

where we define the unrenormalized band gap $E_g = \epsilon_0^1 - \epsilon_0^0$ and the momentum-matrix element $\mathsf{p}_j = \langle 0|p_j|1\rangle$. Since the energy has a quadratic k-dependence, it is meaningful to introduce the effective mass tensor

$$\left(\frac{1}{m_{\text{eff}}}\right)_{ij} = \frac{1}{m_0}\left(\delta_{ij} \pm \frac{2\mathsf{p}_i^{*}\mathsf{p}_j}{m_0 E_g}\right). \tag{12.17}$$

In isotropic cases, such as in lattices with cubic symmetry, the effective masses become scalar quantities

$$m_c \equiv \frac{m_0}{1 + \frac{2\mathsf{p}^2}{m_0 E_g}} \qquad\qquad m_v = \frac{m_0}{1 - \frac{2\mathsf{p}^2}{m_0 E_g}}. \tag{12.18}$$

In this situation, the $\mathbf{k}\cdot\mathbf{p}$ energies are

$$\epsilon_{\mathbf{k}}^{c} = \epsilon_0^1 + \frac{\hbar^2 \mathbf{k}^2}{2m_e}, \qquad\qquad \epsilon_{\mathbf{k}}^{v} = \epsilon_0^0 + \frac{\hbar^2 \mathbf{k}^2}{2m_v}, \tag{12.19}$$

for the conduction and valence band, respectively.

For a sufficiently large p, Eq. (12.18) produces a negative m_v while m_c is positive. Figure 12.2(b) presents the band structure around $\mathbf{k} = 0$ for the isolated two-band system that has conduction and valence bands with positive and negative curvature (m_λ), respectively. In this situation, lifting an electron (sphere) from the valence to the conduction band leaves behind a vacancy (dashed circle) that is often referred to as a hole. Since a hole is an antiparticle of a valence-band electron, it has an opposite mass $m_h = -m_v$ that is positive definite for negative m_v. This aspect is discussed further in Section 24.1.2. Once $m_e = m_c$ and $m_h = -m_v$ are known, we may define the reduced electron–hole mass μ:

$$\frac{1}{\mu} = \frac{1}{m_e} + \frac{1}{m_h} = \frac{4\mathsf{p}^2}{m_0^2 E_g} \tag{12.20}$$

which follows directly from Eq. (12.18). This result is often used to estimate the value of p^2. In GaAs-type quantum wells, the effective masses are roughly $m_c = +0.0665\, m_0$ and $m_v = -0.235\, m_0$ where m_0 is the free-electron mass in vacuum. In other words, holes are heavier than electrons while both of them are clearly lighter than free electrons. This aspect illustrates that $\mathbf{k}$ and m_λ are related to the effective free motion of electrons and holes in a periodic crystal.

Even though the two-band approximation works very well in describing semiconductor excitations close to the bottom of the bands, it typically is not accurate enough to determine the actual values for m_c and m_v. By starting from Eq. (12.14), one obtains a more general isotropic effective mass for the band λ:

$$\frac{1}{m_\lambda} = \frac{1}{m_0} + \frac{2}{m_0^2} \sum\nolimits_{\eta\neq\lambda} \frac{\langle\lambda|p|\eta\rangle\langle\eta|p|\lambda\rangle}{\epsilon_0^\lambda - \epsilon_0^\eta}, \tag{12.21}$$

and

$$\epsilon_{\mathbf{k}}^{\lambda} = \epsilon_0^{\lambda} + \frac{\hbar^2 \mathbf{k}^2}{2m_{\lambda}}, \tag{12.22}$$

where we used the isotropic approximation

$$\mathbf{k} \cdot \langle \lambda | \mathbf{p} | \eta \rangle \mathbf{k} \cdot \langle \eta | \mathbf{p} | \lambda \rangle = \mathbf{k}^2 \langle \lambda | p | \eta \rangle \langle \eta | p | \lambda \rangle. \tag{12.23}$$

The nonisotropic generalization of Eq. (12.23) leads to an effective mass tensor in analogy to Eq. (12.17); however, we concentrate here on isotropic systems.

12.2 Systems with reduced effective dimensionality

If a planar nanostructure is grown, e.g., in the z-direction, the lattice-periodic potential in Eq. (11.55) has to be replaced by

$$U_0(\mathbf{r}) = U^i(\mathbf{r}), \quad \text{for each} \quad z_i < z < z_{i+1}, \tag{12.24}$$

where z_i indicates the positions of the different interfaces. Within each interval $z_i < z < z_{i+1}$, the lattice-periodic potential is determined by the chemical compounds in that region. This additional feature complicates the procedure to find the exact solution $\phi_{\lambda,\mathbf{k}}(\mathbf{r})$ of Eq. (12.1). However, most of the relevant results can be obtained by using an approximative approach that makes use of the fact that the multilayer structures are mesoscopic, i.e., large in comparison with the microscopic atomic scale but small in comparison with the overall sample dimensions. In other words, the active layers have a thickness $L^i = z_{i+1} - z_i$, which is much wider than the lattice unit cell while L^i is much smaller than the macroscopic sample size $\mathcal{L}$.

Figure 12.3(a) presents a typical nanostructure (white-dark circles) where one planar mesoscopic layer L is sandwiched between two identical bulk barriers (light-shaded circles). If the mesoscopic layer thickness L (arrows) exceeds the length of a few unit cells, one finds a well-defined band structure within each layer. Thus, each layer has its own $\epsilon_{\mathbf{k}}^{\lambda}$ such that the electron is effectively free to move within each seperate layer. The dashed line in Fig. 12.3(b) shows $\epsilon_0^c(z)$ as a function of position z. This shown nanostructure is known as a *quantum well* (QW) because the electrons tend to be trapped, i.e., quantum confined, within the region with the smallest $\epsilon_0^c(z)$.

For planar QW systems, it is useful to separate the three-dimensional space coordinate $\mathbf{r} = (\mathbf{r}_{\parallel}, z)$ into a two-dimensional vector $\mathbf{r}_{\parallel}$ in the QW plane and the one-dimensional coordinate z perpendicular to the QW. This system is fully periodic within the $\mathbf{r}_{\parallel}$ plane such that we may apply the Bloch theorem and the ansatz (12.7) for the planar dependency. However, the original ansatz has to be modified to include the actual z-dependence. If the chemical compounds are not considerably different, the lattice periodic Bloch function u can be assumed to be the same throughout the sample because it is determined by the microscopic unit-cell properties. However, the different layers can be assumed

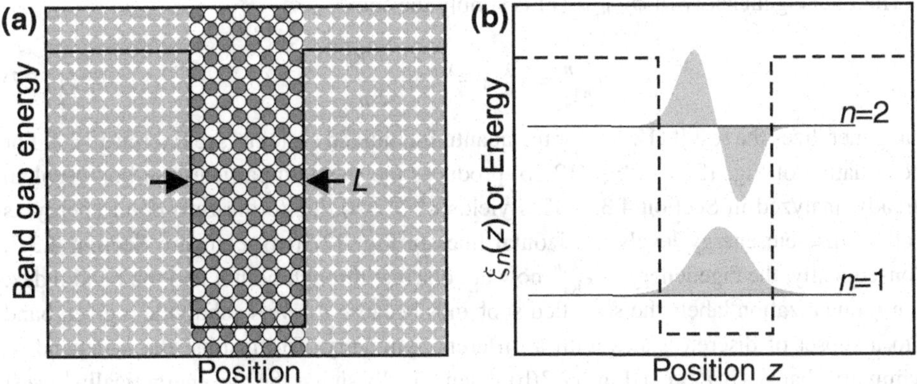

Figure 12.3 Quantum-confined systems. **(a)** The dim (material 1) and the dark-white (material 2) circles illustrate the ionic lattice in a typical semiconductor nanostructure. The solid line indicates the corresponding band offset between the materials. **(b)** The electrons are quantum confined into the low-gap material due to the particle-in-a-box-type solutions of Eq. (12.26). The confinement energies (line) and wave functions (shaded area) are shown together with the confinement potential (dashed line).

to have clearly different $\epsilon_0^c(z)$. In this situation, the QW confinement modifies only the z-dependent part of the envelope function. Thus, we introduce the *envelope-function approximation*:

$$\phi_{\lambda,n,\mathbf{k}_\parallel}(\mathbf{r}) = \xi_{\lambda,n}(z)\,\frac{1}{\sqrt{S}}\,e^{i\mathbf{k}_\parallel\cdot\mathbf{r}_\parallel}\,u_{\lambda,\mathbf{k}_\parallel}(\mathbf{r}_\parallel, z), \tag{12.25}$$

where $\xi_{\lambda,n}(z)$ is the mesoscopic confinement wave function for the level n, $\mathbf{k}_\parallel$ is the carrier momentum in the QW plane, and $u_{\lambda,\mathbf{k}_\parallel}(\mathbf{r}, z)$ is the lattice-periodic Bloch function.

Since $\xi_{\lambda,n}(z)$ describes a mesoscopic envelope, it is affected by the z-dependency of the band structure. Obviously, $\xi_{\lambda,n}(z)$ should satisfy the Schrödinger equation of a free particle with mass m_λ when all layers are identical. This limiting case follows if we determine $\xi_{\lambda,n}(z)$ approximatively from an effective Schrödinger equation

$$\left[-\frac{\hbar^2}{2m_\lambda}\frac{\partial^2}{\partial z^2} + U^\lambda_{\mathrm{conf}}(z)\right]\xi_{\lambda,n}(z) = \epsilon^{\lambda,n}\,\xi_{\lambda,n}(z), \tag{12.26}$$

with $U^\lambda_{\mathrm{conf}}(z) \equiv \epsilon_0^\lambda(z) - \epsilon_0^\lambda(0)$ where $\epsilon_0^\lambda(0)$ is defined at the position of the QW. It is straightforward to verify that this produces the correct plane-wave solutions $\xi_{\lambda,n=k_z}(z) \propto e^{ik_z z}$ in the case where all materials are identical, yielding the correct effective $\epsilon^{\lambda,n} = \hbar^2 k_z^2/2m_\lambda$, see Exercise 12.4. One can also implement other generalizations that take into account the changes in m_λ from material to material. Even though Eq. (12.26) does not yield an exact solution of Eq. (12.1) with the potential (12.24), it usually is reasonably accurate and can be used to describe most QW structures.

The total eigenenergy of $\phi_{\lambda,n,\mathbf{k}_\parallel}(\mathbf{r})$ is simply the sum

$$\epsilon_{\mathbf{k}_\parallel}^{\lambda,n} = \epsilon_0^\lambda + \epsilon^{\lambda,n} + \frac{\hbar^2 k_\parallel^2}{2m_\lambda}, \tag{12.27}$$

that generalizes the result (12.14) for the quantum-confined systems, see Exercise 12.4. For the situation of Fig. 12.3(b), Eq. (12.26) produces the standard particle-in-a-box problem already analyzed in Section 4.3.3. This yields $\epsilon^{\lambda,n} > 0$ due to the zero-point energy as well as discrete energy levels (horizontal lines in Fig. 12.3(b)), compare Section 4.3.3. Consequently, the eigenenergies $\epsilon_{\mathbf{k}_\parallel}^{\lambda,n}$ now are discrete through both n and the band index. The n quantization labels the so-called subbands because n levels subdivide a given band into a subset of discrete states with a different symmetry. To illustrate the n-dependent symmetry changes in $\xi_{c,n}(z)$, Fig. 12.3(b) schematically shows the two energetically lowest confinement wave functions (shaded areas).

Based on Eq. (12.27), the quantum confinement shifts the lowest energies of the different bands. This property can be used to a certain extent to tune the resonance energies of a structure into an experimentally desired range. In addition, if the original three-dimensional band structure has degenerate bands in the vicinity of the optical transitions, it is conceptually rather simple to remove this degeneracy in QW structures either via the confinement effects or by introducing some strain. Thus, one can design semiconductors that effectively behave like two-band systems. In this book, we mainly consider such systems as representative examples. More difficult band structures can be treated as well; however, this requires additional book keeping of the band indices and increases the numerical complexity.

12.2.1 Quasi two-, one-, and zero-dimensional systems

When the potential $U_{\rm conf}^\lambda(z)$ is deep enough, Eq. (12.26) effectively describes particles confined by infinite walls. The corresponding subband energies become then

$$\epsilon^{\lambda,n} = \frac{\hbar^2}{2|m_\lambda|}\left(\frac{\pi}{L}\right)^2 n^2, \qquad n = 1, 2, \cdots \tag{12.28}$$

This implies the energy separation

$$\Delta\epsilon^{2,1} \equiv \epsilon^{\lambda,2} - \epsilon^{\lambda,1} = \frac{\hbar^2}{2|m_\lambda|}\left(\frac{\pi}{L}\right)^2 3 = 169\left[\frac{10\,\mathrm{nm}}{L}\right]^2 \mathrm{meV} \tag{12.29}$$

between the first excited and the ground state. To obtain the numerical value for the energy splitting, we used $m_c = 0.0665 m_0$, typical for GaAs. This energy spacing is so large that one can often assume that the excitations take place only between one confinement level. We refer to this situation as the *strong confinement limit.*

In this limit, we can reduce the investigations to one strongly confined subband and write the Bloch wave function as

$$\phi_{\lambda,\mathbf{k}_\parallel}(\mathbf{r}) = \xi(z)\frac{1}{\sqrt{\mathcal{L}^2}}e^{i\mathbf{k}_\parallel\cdot\mathbf{r}_\parallel}u_{c,\mathbf{k}_\parallel}(\mathbf{r}) \qquad \lambda = c,\, v\,. \tag{12.30}$$

Since the electrons are in the lowest confinement level by definition, we omit the subband index n and use the confinement function $\xi(z)$ for both conduction and valence band. The corresponding single-particle energies are then

$$\epsilon^c_{\mathbf{k}_\parallel} = \epsilon^c_0 + \epsilon^{c,1} + \frac{\hbar^2\mathbf{k}^2_\parallel}{2m_e}, \qquad \epsilon^v_{\mathbf{k}_\parallel} = \epsilon^v_0 + \epsilon^{v,1} - \frac{\hbar^2\mathbf{k}^2_\parallel}{2m_h}, \tag{12.31}$$

where the effective-mass approximation has been applied. We then find the Hamiltonian for the noninteracting QW electrons as

$$\hat{H}_0 = \sum\nolimits_{\mathbf{k}_\parallel}\left(\epsilon^c_{\mathbf{k}_\parallel}a^\dagger_{c,\mathbf{k}_\parallel}a_{c,\mathbf{k}_\parallel} + \epsilon^v_{\mathbf{k}_\parallel}a^\dagger_{v,\mathbf{k}_\parallel}a_{v,\mathbf{k}_\parallel}\right) \tag{12.32}$$

by repeating the derivation leading to Eq. (11.56) for bulk systems. Consequently, the strongly confined QW systems show effective two-dimensional behavior. The semiconductor system can also be confined in more than one direction leading to a further reduction of the effective system dimensionality and behavior. The confinement in two directions yields one-dimensional *quantum-wire structures*, while the completely confined structures are known as effectively zero-dimensional *quantum dots*.

For quantum wires, it is useful to separate $\mathbf{r} = (\mathbf{r}_\parallel, z)$ where now $\mathbf{r}_\parallel$ denotes the two confinement directions. For this situation, the envelope-function approximation is

$$\phi_{\lambda,k_z}(\mathbf{r}) = \xi(\mathbf{r}_\parallel)\frac{1}{\sqrt{\mathcal{L}}}e^{ik_z z}u_{\lambda,k_z}(\mathbf{r}), \qquad \lambda = c,\, v \tag{12.33}$$

where we assumed confinement to the lowest level of the two-band system in analogy to the quantum-well case. For the mesoscopic quantum wire, the confinement functions can be calculated from

$$\left[-\frac{\hbar^2}{2m_\lambda}\left(\frac{\partial^2}{\partial x^2} + \frac{\partial^2}{\partial y^2}\right) + U^\lambda_{\text{conf}}(x,y)\right]\xi_{\lambda,n}(x,y) = \epsilon^{\lambda,n}\xi_{\lambda,n}(x,y), \tag{12.34}$$

with $\mathbf{r}_\parallel = (x, y)$. By choosing a harmonic confinement potential, we find for the lowest confinement level

$$\xi(\mathbf{r}_\parallel) = \frac{1}{\sqrt{\pi R^2}}e^{-\frac{1}{2R^2}\mathbf{r}^2_\parallel}, \tag{12.35}$$

where R defines the confinement scale.

In our numerical evaluations, we will often use a two-band quantum wire as a representative system to study quantum-optical effects. The corresponding noninteracting part of the Hamiltonian is

$$\hat{H}_0 = \sum_{k_z}\left(\epsilon^c_{k_z}a^\dagger_{c,k_z}a_{c,k_z} + \epsilon^v_{k_z}a^\dagger_{v,k_z}a_{v,k_z}\right), \tag{12.36}$$

with the effective-mass energies

$$\epsilon^c_{k_z} = \epsilon^c_0 + \epsilon^{c,1} + \frac{\hbar^2k_z^2}{2m_e}, \qquad \epsilon^v_{k_z} = \epsilon^v_0 + \epsilon^{c,1} - \frac{\hbar^2k_z^2}{2m_h}, \tag{12.37}$$

in analogy to the quantum-well system.

In the envelope-function approximation for quantum dots, the corresponding Bloch wave functions are

$$\phi_{\lambda,n}(\mathbf{r}) = \xi_{\lambda,n}(\mathbf{r}) u_\lambda(\mathbf{r}) , \qquad \lambda = c,\, v \tag{12.38}$$

where n refers to the quantum number of

$$\left[-\frac{\hbar^2}{2m_\lambda}\nabla^2 + U^\lambda_{\mathrm{conf}}(\mathbf{r}) \right] \xi_{\lambda,n}(\mathbf{r}) = \epsilon^{\lambda,n} \xi_{\lambda,n}(\mathbf{r}). \tag{12.39}$$

In general, these solutions consist of discrete states bound inside the quantum dot, plus energetically higher unconfined states. Since the electrons can occupy each dot level only twice (once each for spin up and down), it is natural to include many confinement levels for dots even when the confinement is strong. The corresponding noninteracting Hamiltonian is then

$$\hat{H}_0 = \sum\nolimits_n \left(\epsilon^c_n a^\dagger_{c,n} a_{c,n} + \epsilon^v_n a^\dagger_{v,n} a_{v,n} \right) , \qquad \epsilon^\lambda_n = \epsilon^\lambda_0 + \epsilon^{\lambda,n} . \tag{12.40}$$

If the energy levels of the dot are well separated, the quantum dot system allows for spectroscopy between discrete levels, similar to atomic systems.

12.2.2 Electron density of states

In the definition of the Bloch functions, we deliberately did not specify how the quantization volume $\mathcal{L}^d$ has to be chosen for the effectively d-dimensional systems under consideration. Since all real samples have a different finite size, it is useful to assume that the sample consists of many identical parts with volume $\mathcal{L}^d$. This way, we have the same quantization volume for all relevant systems if we implement periodic boundary conditions at each surface of $\mathcal{L}^d$. As a result, the plane-wave parts of the envelope functions have to fulfill the condition

$$e^{ik_j\mathcal{L}} = 1 \quad \Leftrightarrow \quad k_j\, \mathcal{L} = 2\pi n, \tag{12.41}$$

where k_j is the Cartesian component of the wave vector. Thus, each k_j is discretized according to

$$k_j = \frac{2\pi}{\mathcal{L}} n \equiv n\Delta k, \tag{12.42}$$

which defines the momentum difference $\Delta k = 2\pi/\mathcal{L}$.

For a large enough quantization volume, Δk becomes infinitesimal such that k_j becomes a continuous variable. However, by using a formally finite $\mathcal{L}$ together with the discretization (12.42), we can introduce an efficient way to handle sums over k_j, which occur quite frequently in our investigations. The typical form contains a generic function $F_\mathbf{k}$ in an expression

$$\frac{1}{\mathcal{L}^d}\sum_{\mathbf{k}} F_{\mathbf{k}} = \frac{1}{(2\pi)^d}\left(\frac{2\pi}{\mathcal{L}}\right)^d \sum_{\mathbf{k}} F_{\mathbf{k}} = \frac{1}{(2\pi)^d}\sum_{\mathbf{k}} F_{\mathbf{k}}\,(\Delta k)^d$$
$$= \frac{1}{(2\pi)^d}\int_{\mathcal{L}^d} d^d k\, F_{\mathbf{k}}, \tag{12.43}$$

where the last step follows for large $\mathcal{L}$ and infinitesimal Δk because then the second line becomes the standard definition of an integral. This property will be used several times in further derivations.

Many relevant integrals defined by Eq. (12.43) have an integrand which depends only on the magnitude $k = |\mathbf{k}|$. For these cases, it is convenient to perform the integration in either radial or spherical coordinates. The corresponding form of the integral (12.43) follows from

$$\frac{1}{\mathcal{L}^d}\sum_{\mathbf{k}} F_k = \frac{\Omega_d}{(2\pi)^d}\int_0^\infty dk\, k^{d-1} F_k, \tag{12.44}$$

where $\Omega_{d=1} = 2$, $\Omega_{d=2} = 2\pi$, and $\Omega_{d=3} = 4\pi$ contain the integral over the angles in different dimensions. Since $F_{|k|} = F(E)$ is often known as a function of energy $E = \hbar^2k^2/2m_\lambda$, one may change the integration variables to obtain

$$\frac{1}{\mathcal{L}^d}\sum_{\mathbf{k}} F_{\mathbf{k}} = \int_0^\infty dE\, g_d(E)F(E). \tag{12.45}$$

Here, the quantity

$$g_d(E) = \frac{\Omega_d}{(2\pi)^d}\frac{1}{2}\left(\frac{2m_\lambda}{\hbar^2}\right)^{d/2} E^{d/2-1} \tag{12.46}$$

is known as the *energy density of states* for the particles λ. The functional form of $g_d(E)$ depends strongly on the effective system dimension.

Exercises

12.1 **(a)** Show that the ansatz (12.7) yields (12.8) when we demand that the Bloch theorem (12.6) is satisfied. **(b)** Verify that the ansatz (12.7) converts the eigenvalue problem (12.1) into Eq. (12.9).

12.2 **(a)** Assume that $|\lambda\rangle$ is the solution to $\hat{H}_0|\lambda\rangle = E_0^\lambda|\lambda\rangle$, where $\hat{H}_0 = \hat{p}^2/2m_0 + \hat{U}_0$. Additionally, assume that $u_{\lambda,0}(r) = \langle r|\lambda\rangle$ satisfies the condition (12.8). Show that $\langle\lambda|\hat{r}|\lambda\rangle$ vanishes. Also compute $i\hbar\frac{\partial}{\partial t}\langle\lambda|\hat{r}|\lambda\rangle = \langle\lambda|[\hat{r}, \hat{H}_0]|\lambda\rangle = 0$ and show that $\langle\lambda|\hat{p}|\lambda\rangle$ must vanish. **(b)** Construct the results (12.13) and (12.14) using the standard second-order perturbation theory.

12.3 **(a)** Show that Eq. (12.14) yields the results (12.15)–(12.16) when the two-band approximation is applied. **(b)** Assume that only diagonal contributions exist in the effective-mass tensor (12.17). Verify that Eqs. (12.15)–(12.16) then produce the results (12.18)–(12.19).

12.4 (**a**) Use the ansatz (12.25) in Eq. (12.1) describing the bulk system. Show that Eq. (12.26) yields the correct plane-wave dependency, reproducing the ansatz (12.7). Verify also the relation (12.27). (**b**) Assume that $U^{\lambda}_{\text{conf}}(z) = U_0\theta(z)$ describes two semiconductors separated by an interface. Show that Eq. (12.26) prevents electrons in the low-E^c_0 material from entering the high-E^c_0 material when the kinetic energy is sufficiently low in the z direction.

12.5 Assume that $U^{\lambda}_{\text{conf}}(z)$ vanishes for $|z| \leq L/2$ and is infinite for $|z| > L/2$. (**a**) Solve Eq. (12.26) and define $\phi_{\lambda,n,k_\parallel}(r)$. (**b**) Construct the total eigenenergy of $\phi_{\lambda,n,k_\parallel}(\mathbf{r})$. Present the band structure including the three lowest subbands.

12.6 Assume that a quantum wire is confined via a harmonic potential $U^{\lambda}_{\text{conf}} = U_0\frac{r^2_\parallel}{r^2_0}$. (**a**) Solve the ground-state energy using Eq. (12.34) as an ansatz. What are R and $\epsilon^{\lambda,0}$? (**b**) For the harmonic oscillator $\epsilon^{\lambda,n} = E_0\left(n+\frac{1}{2}\right)$. Assume $U_0 = 200$ meV and $r_0 = 5$ nm. What is the energy separation of two consecutive subbands for a system with $m_c = 0.0665 m_0$?

12.7 Evaluate the integral (12.43) explicitly and verify the result (12.45).

Further reading

The band structure of solids is discussed in more depth in the following literature:

E. O. Kane (1966). The k · p method, in *Semiconductors and Semimetals*, R. K. Willardson and A. C. Beer, eds., New York, Academic Press, p. 75.

C. Kittel (1971). *Introduction to Solid State Physics*, New York, Wiley & Sons.

N. W. Ashcroft and N. D. Mermin (1976). *Solid State Physics*, Philadelphia PA, Saunders College.

The envelope function approximation and much more is studied, e.g., in:

G. Bastard (1998). *Wave Mechanics Applied to Semiconductor Heterostructures*, Paris, Les Editions de Physique.

M. G. Burt (1999). Fundamentals of envelope function theory for electronic states and photonic modes in nanostructures, *J. Phys.: Condens. Matter* **11**, 53.

B. A. Foreman (1998). Analytical envelope-function theory of interface band mixing, *Phys. Rev. Lett.* **81**, 425.

A great variety of material parameters is found, e.g., in:

Landolt-Börnstein (1982). *Numerical Data and Functional Relationships in Science and Technology*, K. H. Hellwege, ed., Vol. 17 Semiconductors, O. Madelung, M. Schulz, and H. Weiss, eds., Berlin, Springer.

13

Interactions in semiconductors

In this chapter, we complete the derivation of the basic Hamiltonian for semiconductor quantum optics. Building on the concepts introduced in the two previous chapters, we now focus on the interaction aspects. In particular, we derive the second-quantization formulation for the light–matter coupling and the interactions within the electronic system, i.e., the carrier–carrier Coulomb and the carrier–phonon interactions. The discussion is presented explicitly for quantum-well systems. However, once we have the explicit expressions, it is relatively straightforward to generalize them to obtain the Hamiltonians for systems with other effective dimensionalities.

13.1 Many-body Hamiltonian

Starting from the Hamiltonian (12.3), where the interactions are still given in the first quantization, we apply the second-quantization step (11.48) to rewrite the total system Hamiltonian:

$$\hat{H} = \hat{H}_0 + \hat{H}_{\text{em}} + \hat{H}_{\text{ph}} + \int d^3r\ \hat{\Psi}^\dagger(\mathbf{r})[U(\mathbf{r}) - U_0(\mathbf{r})]\hat{\Psi}(\mathbf{r})$$
$$+ \int d^3r\ \hat{\Psi}^\dagger(\mathbf{r})\left[-\frac{Q}{m_0}\hat{\mathbf{A}}(\mathbf{r})\cdot\hat{\mathbf{p}} + \frac{Q^2}{2m_0}\hat{\mathbf{A}}^2(\mathbf{r})\right]\hat{\Psi}(\mathbf{r})$$
$$+ \int d^3r\, d^3r'\ \hat{\Psi}^\dagger(\mathbf{r})\hat{\Psi}^\dagger(\mathbf{r}')\ V(\mathbf{r}-\mathbf{r}')\ \hat{\Psi}(\mathbf{r})\hat{\Psi}(\mathbf{r}'), \tag{13.1}$$

where $\hat{H}_0$, $\hat{H}_{\text{em}}$, and $\hat{H}_{\text{ph}}$ are given by Eqs. (11.58) and (11.72). In order to include the electron–phonon interaction, we allow $U(\mathbf{r})$ to describe lattice vibrations in the form

$$U(\mathbf{r}) = \sum\nolimits_n v(\mathbf{r} - \mathbf{R}_n), \tag{13.2}$$

where $v(\mathbf{r} - \mathbf{R}_n)$ is the effective potential of the ion at the position $\mathbf{R}_n$. In equilibrium $\mathbf{R}_n = \mathbf{R}_n^0$, the potential $U(\mathbf{r})$ is equal to the unperturbed lattice potential $U_0(\mathbf{r})$ which we used to define the Bloch basis via Eq. (12.1). The lattice vibrations induce finite deviations $\Delta\hat{\mathbf{R}}_n = \mathbf{R}_n - \mathbf{R}_n^0$ such that

$$
\begin{aligned}
U(\mathbf{r}) &= \sum_n v\left(\mathbf{r}-\mathbf{R}_n^0-\Delta\hat{\mathbf{R}}_n\right) \\
&= \sum_n v\left(\mathbf{r}-\mathbf{R}_n^0\right) - \sum_n \nabla v\left(\mathbf{r}-\mathbf{R}_n^0\right)\cdot\Delta\hat{\mathbf{R}}_n + \mathcal{O}\left(\Delta\hat{\mathbf{R}}_n^2\right) \\
&= U_0(\mathbf{r}) - \sum_n \nabla v\left(\mathbf{r}-\mathbf{R}_n^0\right)\cdot\Delta\hat{\mathbf{R}}_n + \mathcal{O}\left(\Delta\hat{\mathbf{R}}_n^2\right).
\end{aligned}
\tag{13.3}
$$

Here, we performed a Taylor expansion around the equilibrium position and, in the last step, identified the original lattice periodic potential. Since the lattice vibrations are quantized, we actually have to use the operator form $\Delta\hat{\mathbf{R}}_n$ defined by Eq. (11.66). If we neglect the higher-order corrections to the Taylor expansion, the system Hamiltonian can be written in the form

$$
\begin{aligned}
\hat{H} &= \hat{H}_0 + \hat{H}_{\mathrm{em}} + \hat{H}_{\mathrm{ph}} \\
&+ \int d^3r\, \hat{\Psi}^\dagger(\mathbf{r})\left[-\frac{Q}{m_0}\hat{\mathbf{A}}(\mathbf{r})\cdot\hat{\mathbf{p}} + \frac{Q^2}{2m_0}\hat{\mathbf{A}}^2(\mathbf{r})\right]\hat{\Psi}(\mathbf{r}) \\
&- \int d^3r\, \hat{\Psi}^\dagger(\mathbf{r})\left[\sum_n \nabla v(\mathbf{r}-\mathbf{R}_n^0)\cdot\Delta\hat{\mathbf{R}}_n\right]\hat{\Psi}(\mathbf{r}) \\
&+ \int d^3r\, d^3r'\, \hat{\Psi}^\dagger(\mathbf{r})\hat{\Psi}^\dagger(\mathbf{r}')\, V(\mathbf{r}-\mathbf{r}')\, \hat{\Psi}(\mathbf{r})\hat{\Psi}(\mathbf{r}').
\end{aligned}
\tag{13.4}
$$

With this arrangement, the first line contains the Hamiltonians of the noninteracting subsystems, the second line is the light–electron interaction, the third line describes electron–phonon interaction, while the last line contains the Coulomb interaction among the active electrons in their different bands.

13.2 Light–matter interaction

The coupling of electrons and light is described by the second line of Eq. (13.4). In order to formulate this interaction Hamiltonian more explicitly, we have to look at the contributions

$$
\hat{H}_{\mathrm{A\cdot p}} = -\int d^3r\, \hat{\Psi}^\dagger(\mathbf{r})\left[\frac{Q}{m_0}\hat{\mathbf{A}}(\mathbf{r})\cdot\hat{\mathbf{p}}\right]\hat{\Psi}(\mathbf{r})
\tag{13.5}
$$

and

$$
\hat{H}_{\mathrm{A\cdot A}} = \int d^3r\, \hat{\Psi}^\dagger(\mathbf{r})\left[\frac{Q^2}{2m_0}\hat{\mathbf{A}}^2(\mathbf{r})\right]\hat{\Psi}(\mathbf{r}).
\tag{13.6}
$$

Since we want to choose quantum wells (QW) as our representative semiconductor system, we use the explicit Bloch-electron wave function, Eq. (12.25). With this choice, we can write Eqs. (13.5) and (13.6) in the generic form

$$
\hat{H}_j = \int d^3r\, \hat{\Psi}^\dagger(\mathbf{r})O_j(\mathbf{r})\hat{\Psi}(\mathbf{r}) = \sum_{\lambda,\mathbf{k},\lambda',\mathbf{k}'} a^\dagger_{\lambda,\mathbf{k}}a_{\lambda',\mathbf{k}'}M^{\lambda,\mathbf{k}}_{\lambda',\mathbf{k}'}|_j,
\tag{13.7}
$$

where j denotes the type of interaction and

$$M_{\lambda',\mathbf{k}'}^{\lambda,\mathbf{k}}|_j \equiv \int \phi_{\lambda,\mathbf{k}}^{\star}(\mathbf{r}) O_j(\mathbf{r}) \phi_{\lambda',\mathbf{k}'}(\mathbf{r}) d^3r \tag{13.8}$$

is the matrix element between Bloch electrons. In the following, we assume strongly confined electrons such that the subband index n can be suppressed in $\xi_{\lambda,n} \rightarrow \xi_{\lambda}$. For the sake of notational simplicity, we have replaced $\mathbf{k}_{\|}$ by $\mathbf{k}$ because it is obvious from the context that $\mathbf{k}$ is the momentum within the QW plane. In the following, we will analyze this integral for $O_j(\mathbf{r}) = -\frac{Q}{m_0}\hat{\mathbf{A}}(\mathbf{r}) \cdot \hat{\mathbf{p}}$ and $O_j(\mathbf{r}) = \frac{Q^2}{2m_0}\hat{\mathbf{A}}^2(\mathbf{r})$.

To explicitly evaluate the integrals, it is convenient to consider the mode-expansion of the vector potential

$$\hat{\mathbf{A}}(\mathbf{r}) = \sum_{\mathbf{q}} \frac{\mathcal{E}_{\mathbf{q}}}{\omega_{\mathbf{q}}} \left[\mathbf{u}_{\mathbf{q}}(\mathbf{r})\, B_{\mathbf{q}} + \mathbf{u}_{\mathbf{q}}^{\star}(\mathbf{r})\, B_{\mathbf{q}}^{\dagger} \right] \tag{13.9}$$

according to Eq. (10.43). For planar systems studied here, also the eigenmodes of light can be separated into in-plane and z-dependent parts

$$\mathbf{u}_{\mathbf{q}}(\mathbf{r}_{\|}, z) = \bar{\mathbf{u}}_{\mathbf{q}}(z)\, e^{i\mathbf{q}_{\|}\cdot\mathbf{r}_{\|}}, \tag{13.10}$$

see discussion in Section 9.4.4. As a result, we may extract the plane-wave part of Eq. (13.9) to produce

$$\hat{\mathbf{A}}(\mathbf{r}, z) = \sum_{\mathbf{q}_{\|}} \hat{\mathbf{A}}_{\mathbf{q}_{\|}}(z) e^{i\mathbf{q}_{\|}\cdot\mathbf{r}_{\|}} \tag{13.11}$$

that contains

$$\hat{\mathbf{A}}_{\mathbf{q}_{\|}}(z) = \sum_{q_z} \frac{\mathcal{E}_{\mathbf{q}}}{S\omega_{\mathbf{q}}} \left[\bar{\mathbf{u}}_{\mathbf{q}_{\|},q_z}(z) B_{\mathbf{q}_{\|},q_z} + \bar{\mathbf{u}}^{\star}_{-\mathbf{q}_{\|},q_z}(z) B^{\dagger}_{-\mathbf{q}_{\|},q_z} \right] \tag{13.12}$$

where we explicitly indicated the separation $\mathbf{q} = (\mathbf{q}_{\|}, q_z)$ only when necessary.

By inserting the expansion (13.11) into Eq. (13.8) together with the explicit envelope function (12.25), we obtain

$$M_{\lambda',\mathbf{k}'}^{\lambda,\mathbf{k}}|_{A\cdot p} = -\frac{Q}{m_0} \sum_{\mathbf{q}_{\|}} \int d^3r\, e^{-i\mathbf{k}\cdot\mathbf{r}_{\|}}\, \xi_{\lambda}^{\star}(z)\, u_{\lambda,\mathbf{k}}^{\star}(\mathbf{r}) e^{i\mathbf{q}_{\|}\cdot\mathbf{r}_{\|}} \times \hat{\mathbf{A}}_{\mathbf{q}_{\|}}(z) \cdot \hat{\mathbf{p}}\, \phi_{\lambda',\mathbf{k}'}(\mathbf{r})\,. \tag{13.13}$$

Next, we have to evaluate $\hat{\mathbf{p}}$ acting on $\phi_{\lambda',\mathbf{k}'}(\mathbf{r})$ that is defined by the envelope function approximation (12.25). Using this and separating the different parts, $\hat{\mathbf{p}} = \hat{\mathbf{p}}_{\|} - i\hbar\mathbf{e}_z\frac{\partial}{\partial z}$, we find

$$\sqrt{\mathcal{S}}\hat{\mathbf{p}}\left[\phi_{\lambda',\mathbf{k}'}(\mathbf{r})\right] = \hat{\mathbf{p}}\left[e^{i\mathbf{k}'\cdot\mathbf{r}_\parallel}\xi_{\lambda'}(z)u_{\lambda',\mathbf{k}'}(\mathbf{r})\right]$$
$$= \left(\hat{\mathbf{p}}_\parallel - i\hbar\mathbf{e}_z\frac{\partial}{\partial z}\right)\left[e^{i\mathbf{k}'\cdot\mathbf{r}_\parallel}\xi_{\lambda'}(z)u_{\lambda',\mathbf{k}'}(\mathbf{r})\right]$$
$$= e^{i\mathbf{k}'\cdot\mathbf{r}_\parallel}\xi_{\lambda'}(z)\left[\hbar\mathbf{k}' + \hat{\mathbf{p}}\right]u_{\lambda',\mathbf{k}'}(\mathbf{r}) - i\hbar\mathbf{e}_z e^{i\mathbf{k}'\cdot\mathbf{r}_\parallel}u_{\lambda',\mathbf{k}'}(\mathbf{r})\frac{\partial}{\partial z}\xi_{\lambda'}(z). \tag{13.14}$$

To identify the individual contributions, we write Eq. (13.13) as

$$M^{\lambda,\mathbf{k}}_{\lambda',\mathbf{k}'}|_{\mathrm{A\cdot p}} \equiv M^{\lambda,\mathbf{k}}_{\lambda',\mathbf{k}'}|_{\mathrm{A\cdot p(1)}} + M^{\lambda,\mathbf{k}}_{\lambda',\mathbf{k}'}|_{\mathrm{A\cdot p(2)}}, \tag{13.15}$$

where

$$M^{\lambda,\mathbf{k}}_{\lambda',\mathbf{k}'}|_{\mathrm{A\cdot p(1)}} = -\frac{Q}{m_0}\frac{1}{\mathcal{S}}\sum_{\mathbf{q}_\parallel}\int d^3r\; e^{i(\mathbf{k}'+\mathbf{q}_\parallel-\mathbf{k})\cdot\mathbf{r}_\parallel}\xi^\star_\lambda(z)\hat{\mathbf{A}}_{\mathbf{q}_\parallel}(z)\xi_{\lambda'}(z)$$
$$\cdot\, u^\star_{\lambda,\mathbf{k}}(\mathbf{r})\left[\hbar\mathbf{k}' + \mathbf{p}\right]u_{\lambda',\mathbf{k}'}(\mathbf{r}) \tag{13.16}$$

and

$$M^{\lambda,\mathbf{k}}_{\lambda',\mathbf{k}'}|_{\mathrm{A\cdot p(2)}} = \frac{i\hbar Q}{m_0}\frac{1}{\mathcal{S}}\sum_{\mathbf{q}_\parallel}\int d^3r\; e^{i(\mathbf{k}'+\mathbf{q}_\parallel-\mathbf{k})\cdot\mathbf{r}_\parallel}\xi^\star_\lambda(z)$$
$$\times\, \hat{\mathbf{A}}_{\mathbf{q}_\parallel}(z)\cdot\mathbf{e}_z\frac{\partial\xi_{\lambda'}(z)}{\partial z}u^\star_{\lambda,\mathbf{k}}(\mathbf{r})u_{\lambda',\mathbf{k}'}(\mathbf{r}). \tag{13.17}$$

These two expressions correspond to the different parts of $\hat{\mathbf{p}}$ in Eq. (13.14). In order to complete the light–matter Hamiltonian, we still have to express Eq. (13.6) in the Bloch basis. This procedure introduces a matrix element

$$M^{\lambda,\mathbf{k}}_{\lambda',\mathbf{k}'}|_{\mathrm{A\cdot A}} = \frac{1}{\mathcal{S}}\sum_{\mathbf{q}_\parallel,\mathbf{q}'_\parallel}\int d^3r\; e^{i\left(\mathbf{k}'+\mathbf{q}_\parallel-\mathbf{q}'_\parallel-\mathbf{k}\right)\cdot\mathbf{r}_\parallel}\frac{Q^2}{2m_0}\hat{\mathbf{A}}_{\mathbf{q}_\parallel}(z)\cdot\hat{\mathbf{A}}_{-\mathbf{q}'_\parallel}(z)$$
$$\times\; \xi^\star_\lambda(z)u^\star_{\lambda,\mathbf{k}}(\mathbf{r})\xi_{\lambda'}(z)u_{\lambda',\mathbf{k}'}, \tag{13.18}$$

similar to $M|_{A\cdot p}$.

13.2.1 Separation of length scales

On our way to solve the remaining integrals (13.16)–(13.18), we first consider the generic form

$$M^{\lambda,\mathbf{k}}_{\lambda',\mathbf{k}'}|_G = \frac{1}{\mathcal{S}}\int d^3r\; e^{i\Delta\mathbf{Q}_\parallel\cdot\mathbf{r}_\parallel}\left[\xi^\star_\lambda(z)\hat{A}(z)\xi_{\lambda'}(z)\right]$$
$$\times\;\left[u^\star_{\lambda,\mathbf{k}}(\mathbf{r})C(\mathbf{r})u_{\lambda',\mathbf{k}'}(\mathbf{r})\right]. \tag{13.19}$$

Here, we want to make use of the fact that the different terms experience variations on quite different length scales. While the quantity $A(z)$ changes on the scale of the light field, i.e., the optical wave length, the lattice-periodic Bloch functions $u_{\lambda,\mathbf{k}}$ and $C(\mathbf{r})$ vary on the scale of the atomic unit cell. The factor $e^{i\Delta\mathbf{Q}_\parallel\cdot\mathbf{r}_\parallel}$ varies on the mesoscopic scale of the envelope function and the plane-wave part of the light field. These different characteristic length scales allow us to simplify the overall integration.

For this purpose, we express the integral (13.19) as a sum over the unit cells and an integral over the unit-cell volume $v_{\mathbf{R}}$ centered at the lattice point $\mathbf{R} = (\mathbf{R}_{\|}, Z)$:

$$M^{\lambda,\mathbf{k}}_{\lambda',\mathbf{k}'}|_G = \frac{1}{\mathcal{S}} \sum\nolimits_{\mathbf{R}} \int_{v_{\mathbf{R}}} d^3r \; e^{i\Delta\mathbf{Q}_{\|}\cdot\mathbf{r}_{\|}} \left[\xi^\star_\lambda(z)\hat{A}(z)\xi_{\lambda'}(z)\right] \\ \times \left[u^\star_{\lambda,\mathbf{k}}(\mathbf{r})C(\mathbf{r})u_{\lambda',\mathbf{k}'}(\mathbf{r})\right]. \tag{13.20}$$

Since only $u_{\lambda,\mathbf{k}}$ and $C(\mathbf{r})$ significantly vary within a unit cell, the remaining terms can be taken as constants out of the $v_{\mathbf{R}}$ integration such that

$$M^{\lambda,\mathbf{k}}_{\lambda',\mathbf{k}'}|_G = \frac{1}{\mathcal{S}} \sum\nolimits_{\mathbf{R}} e^{i\Delta\mathbf{Q}_{\|}\cdot\mathbf{R}_{\|}} \left[\xi^\star_\lambda(Z)\hat{A}(Z)\xi_{\lambda'}(Z)\right] v_{\mathbf{R}} \\ \times \frac{1}{v_{\mathbf{R}}} \int_{v_{\mathbf{R}}} d^3r \; u^\star_{\lambda,\mathbf{k}}(\mathbf{r})C(\mathbf{r})u_{\lambda',\mathbf{k}'}(\mathbf{r}). \tag{13.21}$$

Since $u_{\lambda,\mathbf{k}}$ and $C(\mathbf{r})$ are assumed to be lattice-periodic functions, the $v_{\mathbf{R}}$ integrals are equal for all unit cells. Hence, we may introduce the position-independent matrix element

$$\langle\lambda,\mathbf{k}|\hat{C}|\lambda',\mathbf{k}'\rangle \equiv \frac{1}{v_0} \int_{v_0} d^3r \; u^\star_{\lambda,\mathbf{k}}(\mathbf{r})C(\mathbf{r})u_{\lambda',\mathbf{k}'}(\mathbf{r}), \tag{13.22}$$

where v_0 denotes the common unit-cell volume.

This volume is infinitesimal compared with the scale on which the remaining terms in (13.21) vary. Hence, we can use $v_0 = d^2\mathbf{R}_{\|}dZ$ and convert the sum into an integral. With these modifications, we find

$$M^{\lambda,\mathbf{k}}_{\lambda',\mathbf{k}'}|_G = \frac{1}{\mathcal{S}} \sum_{\mathbf{R}} d^2R_{\|}dZ \; e^{i\Delta\mathbf{Q}_{\|}\cdot\mathbf{R}_{\|}} \left[\xi^\star_\lambda(Z)\hat{A}(Z)\xi_{\lambda'}(Z)\right] \langle\lambda,\mathbf{k}|\hat{C}|\lambda',\mathbf{k}'\rangle \\ = \int d^2R_{\|} \frac{e^{i\Delta\mathbf{Q}_{\|}\cdot\mathbf{R}_{\|}}}{\mathcal{S}} \int dZ \; \xi^\star_\lambda(Z)\hat{A}(Z)\xi_{\lambda'}(Z)\langle\lambda,\mathbf{k}|\hat{C}|\lambda',\mathbf{k}'\rangle. \tag{13.23}$$

This enables us to identify the envelope-function matrix element of $\hat{A}(z)$ via

$$\hat{A}^{\lambda,\lambda'} \equiv \int dz \; \xi^\star_\lambda(z)\hat{A}(z)\xi_{\lambda'}(z), \tag{13.24}$$

which only depends on the confinement structure. Furthermore, the $\mathbf{R}_{\|}$ integration can be evaluated analytically by noting that

$$\int d^2R_{\|} \frac{e^{i\Delta\mathbf{Q}_{\|}\cdot\mathbf{R}_{\|}}}{\mathcal{S}} = \delta_{\Delta\mathbf{Q}_{\|},0}. \tag{13.25}$$

With the help of these relations, the integral (13.23) becomes

$$M^{\lambda,\mathbf{k}}_{\lambda',\mathbf{k}'}|_G(\mathbf{q}) = \delta_{\Delta\mathbf{Q}_\parallel,0}\,\hat{A}^{\lambda,\lambda'}\,\langle\lambda,\mathbf{k}|\hat{C}|\lambda',\mathbf{k}'\rangle. \tag{13.26}$$

This result can be used directly to generate the explicit forms of $M^{\lambda,\mathbf{k}}_{\lambda',\mathbf{k}'}|_G$ once $\mathbf{Q}_\parallel$, $\hat{A}(z)$, and $C(\mathbf{r})$ are identified.

13.2.2 Light–matter-coupling integrals

By comparing the definition (13.19) with $M^{\lambda,\mathbf{k}}_{\lambda',\mathbf{k}'}|_{A\cdot p(1)}$ in Eq. (13.15), we find

$$e^{i\Delta\mathbf{Q}_\parallel\cdot\mathbf{r}_\parallel} \equiv \sum\nolimits_{\mathbf{q}_\parallel} e^{i(\mathbf{q}_\parallel-\mathbf{k}+\mathbf{k}')\cdot\mathbf{r}_\parallel}, \quad \hat{A}(z) = -\lambda\frac{Q}{m_0}\hat{\mathbf{A}}_{\mathbf{q}_\parallel}(z), \quad \hat{C} = \hbar\mathbf{k}' + \hat{\mathbf{p}}. \tag{13.27}$$

In the same way, $M^{\lambda,\mathbf{k}}_{\lambda',\mathbf{k}'}|_{A\cdot p(2)}$ in Eq. (13.15) consists of

$$e^{i\Delta\mathbf{Q}_\parallel\cdot\mathbf{r}_\parallel} \equiv \sum\nolimits_{\mathbf{q}_\parallel} e^{i(\mathbf{q}_\parallel-\mathbf{k}+\mathbf{k}')\cdot\mathbf{r}_\parallel}, \quad \hat{A}(z) = \frac{i\hbar Q}{m_0}\hat{\mathbf{A}}_{\mathbf{q}_\parallel}(z)\cdot\mathbf{e}_z\frac{\partial}{\partial z}, \quad \hat{C} = 1\,, \tag{13.28}$$

where $M^{\lambda,\mathbf{k}}_{\lambda',\mathbf{k}'}|_{A\cdot A}$ has $\hat{C} = 1$ and

$$e^{i\Delta\mathbf{Q}_\parallel\cdot\mathbf{r}_\parallel} \equiv \sum\nolimits_{\mathbf{q}_\parallel,\mathbf{q}'_\parallel} e^{i\left(\mathbf{q}_\parallel-\mathbf{q}'_\parallel-\mathbf{k}+\mathbf{k}'\right)\cdot\mathbf{r}_\parallel}, \quad \hat{A}(z) = \frac{Q^2}{2m_0}\hat{\mathbf{A}}_{\mathbf{q}_\parallel}(z)\cdot\hat{\mathbf{A}}_{-\mathbf{q}_\parallel}(z)\,. \tag{13.29}$$

Thus, we can directly apply the result (13.26) to produce:

$$\begin{aligned} M^{\lambda,\mathbf{k}}_{\lambda',\mathbf{k}'}|_{\mathrm{A\cdot p(1)}} &= -\sum\nolimits_{\mathbf{q}_\parallel}\delta_{\mathbf{k}',\mathbf{k}-\mathbf{q}_\parallel}\frac{Q}{m_0}\hat{\mathbf{A}}^{\lambda,\lambda'}_{\mathbf{q}_\parallel} \\ &\quad\cdot\langle\lambda,\mathbf{k}|\left[\hbar(\mathbf{k}+\mathbf{q}_\parallel)+\hat{\mathbf{p}}\right]|\lambda',\mathbf{k}-\mathbf{q}_\parallel\rangle, & (13.30)\\ M^{\lambda,\mathbf{k}}_{\lambda',\mathbf{k}'}|_{\mathrm{A\cdot p(2)}} &= \sum\nolimits_{\mathbf{q}_\parallel}\delta_{\mathbf{k}',\mathbf{k}-\mathbf{q}_\parallel}\frac{i\hbar Q}{m_0}\int dz\,\xi^\star_\lambda(z)\hat{\mathbf{A}}_{\mathbf{q}_\parallel}(z)\cdot\mathbf{e}_z\frac{\partial\xi_{\lambda'}(z)}{\partial z} \\ &\quad\times\left\langle\lambda,\mathbf{k}|\lambda',\mathbf{k}-\mathbf{q}_\parallel\right\rangle, & (13.31)\\ M^{\lambda,\mathbf{k}}_{\lambda',\mathbf{k}'}|_{\mathrm{A\cdot A}} &= \sum\nolimits_{\mathbf{q}_\parallel,\mathbf{q}'_\parallel}\delta_{\mathbf{k}'-\mathbf{q}'_\parallel,\mathbf{k}-\mathbf{q}_\parallel}\frac{Q^2}{2m_0}A^{(2),\lambda,\lambda'}_{\mathbf{q}_\parallel,-\mathbf{q}'_\parallel} \\ &\quad\times\left\langle\lambda,\mathbf{k}|\lambda',\mathbf{k}+\mathbf{q}'_\parallel-\mathbf{q}_\parallel\right\rangle. & (13.32)\end{aligned}$$

According to Eq. (13.24), the different confinement matrix elements of the vector potential are given by

$$\hat{\mathbf{A}}^{\lambda,\lambda'}_{\mathbf{q}_{\|}} = \int dz\, \xi^{\star}_{\lambda}(z)\hat{\mathbf{A}}_{\mathbf{q}_{\|}}(z)\xi_{\lambda'}(z) \equiv \hat{A}^{\lambda,\lambda'}_{\mathbf{q}_{\|}}\mathbf{e}_P, \tag{13.33}$$

$$\hat{A}^{(2),\lambda,\lambda'}_{\mathbf{q}_{\|},-\mathbf{q}'_{\|}} = \int dz\, \xi^{\star}_{\lambda}(z)\hat{\mathbf{A}}_{\mathbf{q}_{\|}}(z)\cdot\hat{\mathbf{A}}_{-\mathbf{q}'_{\|}}(z)\xi_{\lambda'}(z), \tag{13.34}$$

which defines the polarization direction $\mathbf{e}_P$ of the field.

13.2.3 Inner products within $\mathbf{k}\cdot\mathbf{p}$ theory

The final form of the matrix elements (13.30)–(13.32) can be computed once the specific forms of the Bloch functions are known. For this, we need information about the band structure which we use at the level of the $\mathbf{k}\cdot\mathbf{p}$ results, see Chapter 11. Before we evaluate Eqs. (13.30)–(13.32), we define a generic matrix element

$$\begin{aligned}\langle\lambda,\mathbf{k}|\hat{C}|\lambda',\mathbf{k}'\rangle &\equiv C^{\lambda}_{\lambda'}\left(\frac{\mathbf{k}+\mathbf{k}'}{2},\mathbf{k}-\mathbf{k}'\right)\\ &= C^{\lambda}_{\lambda'}\left(\frac{\mathbf{k}+\mathbf{k}'}{2},0\right)+\mathcal{O}\left(\mathbf{k}-\mathbf{k}'\right)\\ &= \left\langle\lambda,\frac{\mathbf{k}+\mathbf{k}'}{2}\middle|\hat{C}\middle|\lambda',\frac{\mathbf{k}+\mathbf{k}'}{2}\right\rangle+\mathcal{O}\left(\mathbf{k}-\mathbf{k}'\right),\end{aligned} \tag{13.35}$$

where we implemented the decomposition into center-of-mass and relative momenta and used a Taylor expansion in the difference momenta. Since $\mathbf{k}'-\mathbf{k}=\mathbf{q}_{\|}$ is determined by the parallel component of the light-field wave vector, $|\mathbf{k}'-\mathbf{k}|$ is roughly two orders of magnitude smaller than relevant carrier momenta $\mathbf{k}$, based on the length-scale differences of the light and carrier system. Thus, it is a very good approximation to evaluate (13.35) by omitting the linear correction in connection with Eqs. (13.30)–(13.32). We will use this approximation in our further derivations. The small corrections to this approach are discussed in Exercise 13.3.

With the help of the $\mathbf{k}\cdot\mathbf{p}$ function (12.13), we obtain

$$\begin{aligned}\langle\lambda,\mathbf{k}|\left[\hbar\mathbf{k}+\hat{\mathbf{p}}\right]|\lambda',\mathbf{k}\rangle &= \hbar\mathbf{k}\langle\lambda,\mathbf{k}|\lambda',\mathbf{k}\rangle+\langle\lambda,\mathbf{k}|\hat{\mathbf{p}}|\lambda',\mathbf{k}\rangle\\ &= \hbar\mathbf{k}\delta_{\lambda,\lambda'}+\langle\lambda|\hat{\mathbf{p}}|\lambda'\rangle\\ &\quad+\frac{\hbar\mathbf{k}}{m_0}\left(\sum_{\eta\neq\lambda}\frac{\langle\lambda|\hat{p}|\eta\rangle\langle\eta|\hat{p}|\lambda'\rangle}{\epsilon^{\lambda}_0-\epsilon^{\eta}_0}\sum_{\eta\neq\lambda'}\frac{\langle\lambda|\hat{p}|\eta\rangle\langle\eta|\hat{p}|\lambda'\rangle}{\epsilon^{\lambda}_0-\epsilon^{\eta}_0}\right)+\mathcal{O}(\mathbf{k}^2),\end{aligned} \tag{13.36}$$

when the isotropic approximation, Eq. (12.23), is made. The parity of the Bloch functions (12.12) implies that $\langle\lambda|\hat{\mathbf{p}}|\lambda'\rangle$ vanishes for equal band indices, see also Exercise 12.2. Thus, it is convenient to separate the $\lambda=\lambda'$ and $\lambda\neq\lambda'$ parts. This procedure leads to

$$\langle \lambda, \mathbf{k}| \left[\hbar\mathbf{k} + \hat{\mathbf{p}}\right] |\lambda', \mathbf{k}\rangle = \delta_{\lambda,\lambda'}\hbar\mathbf{k} \left[1 + \frac{2}{m_0} \sum\nolimits_{\eta\neq\lambda} \frac{\langle\lambda|\hat{p}|\eta\rangle\langle\eta|\hat{p}|\lambda\rangle}{\epsilon_0^\lambda - \epsilon_0^\eta} \right]$$
$$+ \left(1 - \delta_{\lambda,\lambda'}\right) \left[\langle\lambda|\hat{\mathbf{p}}|\lambda'\rangle \frac{2\hbar\mathbf{k}}{m_0} \sum\nolimits_{\eta\neq\{\lambda,\lambda'\}} \frac{\langle\lambda|\hat{p}|\eta\rangle\langle\eta|\hat{p}|\lambda'\rangle}{\epsilon_0^\lambda - \epsilon_0^\eta} \right] + \mathcal{O}(\mathbf{k}^2). \tag{13.37}$$

The second term in the first line on the RHS can be expressed using the effective-mass relation (12.21). Furthermore, we can convince ourselves that the contribution of $(1 - \delta_{\lambda,\lambda'})$ times the term in the second line vanishes for a two-band system. For multi-band systems, these contributions simply produce a **k**-dependent $p_{\lambda,\lambda'}(\mathbf{k})$ as discussed in Exercise 13.4. Since this **k** dependence is typically weak, we do not consider it explicitly in the forthcoming investigations. Hence, we are left with the simple expression

$$\langle \lambda, \mathbf{k}| \left[\hbar\mathbf{k} + \hat{\mathbf{p}}\right] |\lambda', \mathbf{k}\rangle = \delta_{\lambda,\lambda'}\hbar\mathbf{k}\frac{m_0}{m_\lambda} + \left(1 - \delta_{\lambda,\lambda'}\right) \mathbf{p}_{\lambda,\lambda'}, \tag{13.38}$$

where we used the orthogonality of the Bloch functions and defined the momentum-matrix element

$$\mathbf{p}_{\lambda,\lambda'} \equiv \langle\lambda|\hat{\mathbf{p}}|\lambda'\rangle, \tag{13.39}$$

which is nonzero only for $\lambda \neq \lambda'$ as shown in Exercise 12.2.

Using Eq. (13.35), the orthogonality (12.2), and the expression (13.38), we can now evaluate the different integrals in Eqs. (13.30)–(13.32):

$$M_{\lambda',\mathbf{k}'}^{\lambda,\mathbf{k}}|_{\mathrm{A\cdot p(1)}} = -\sum\nolimits_{\mathbf{q}_\parallel} \delta_{\mathbf{k}',\mathbf{k}-\mathbf{q}_\parallel} \left[\delta_{\lambda,\lambda'} Q \frac{\hbar\left(\mathbf{k} - \frac{1}{2}\mathbf{q}_\parallel\right)}{m_\lambda} \cdot \hat{\mathbf{A}}_{\mathbf{q}_\parallel}^{\lambda,\lambda} \right.$$
$$\left. + (1 - \delta_{\lambda,\lambda'}) \frac{Q\mathbf{p}_{\lambda,\lambda'}}{m_0} \cdot \hat{\mathbf{A}}_{\mathbf{q}_\parallel}^{\lambda,\lambda'} \right], \tag{13.40}$$

$$M_{\lambda',\mathbf{k}'}^{\lambda,\mathbf{k}}|_{\mathrm{A\cdot p(2)}} = \sum\nolimits_{\mathbf{q}_\parallel} \frac{i\hbar Q}{m_0} \delta_{\mathbf{k}',\mathbf{k}-\mathbf{q}_\parallel} \delta_{\lambda,\lambda'} \int dz\, \xi_\lambda^\star(z)$$
$$\times \hat{\mathbf{A}}_{\mathbf{q}_\parallel}(z) \cdot \mathbf{e}_z \frac{\partial \xi_\lambda(z)}{\partial z}, \tag{13.41}$$

$$M_{\lambda',\mathbf{k}'}^{\lambda,\mathbf{k}}|_{\mathrm{A\cdot A}} = \sum\nolimits_{\mathbf{q}_\parallel,\mathbf{q}'_\parallel} \delta_{\mathbf{k}'-\mathbf{q}'_\parallel,\mathbf{k}-\mathbf{q}_\parallel} \delta_{\lambda,\lambda'} \frac{Q^2}{2m_0} A_{\mathbf{q}_\parallel,-\mathbf{q}'_\parallel}^{(2),\lambda,\lambda}. \tag{13.42}$$

Since $\hat{\mathbf{A}}_{\mathbf{q}_\parallel}(z)$ varies slowly on the mesoscopic scale, the last unknown integration in Eq. (13.41) can be simplified via

$$\int \xi_\lambda^\star(z) \hat{\mathbf{A}}_{\mathbf{q}_\parallel}(z) \cdot \mathbf{e}_z \frac{\partial \xi_\lambda(z)}{\partial z} dz = \hat{\mathbf{A}}_{\mathbf{q}_\parallel}(z_{\mathrm{QW}}) \cdot \mathbf{e}_z \int \xi_\lambda^\star(z) \frac{\partial \xi_\lambda(z)}{\partial z} dz, \tag{13.43}$$

where z_{QW} denotes the position of the center of the QW. Furthermore, the confinement wave functions can be chosen as real functions, which yields the additional simplification

$$\hat{\mathbf{A}}_{\mathbf{q}_\parallel}(z_{QW}) \cdot \mathbf{e}_z \int_{-\infty}^{+\infty} \xi_\lambda^\star(z) \frac{\partial \xi_\lambda(z)}{\partial z} dz = \hat{\mathbf{A}}_{\mathbf{q}_\parallel}(z_{QW}) \cdot \mathbf{e}_z \int_{-\infty}^{+\infty} \frac{1}{2} \frac{\partial}{\partial z} |\xi_\lambda(z)|^2 dz$$
$$= \hat{\mathbf{A}}_{\mathbf{q}_\parallel}(z_{QW}) \cdot \mathbf{e}_z \Big|_{\infty}^{+\infty} |\xi_\lambda(z)|^2 = 0, \qquad (13.44)$$

i.e., this expression vanishes because the confinement wave function decays to zero for large distances. This analysis must be performed more carefully if systems with multiple confinement levels are studied. Then, ξ_λ^* becomes $\xi_{\lambda,n}^*$ and ξ_λ can have a different confinement index n', i.e., $\xi_\lambda \to \xi_{\lambda,n'}$, see Section 12.2. As a result, also $M_{A\cdot p(2)}$ is nonvanishing. However, if carriers are strongly confined to only one level $n = n' = 1$, $M_{\mathrm{A\cdot p(2)}}$ vanishes such that

$$M_{\lambda',\mathbf{k}'}^{\lambda,\mathbf{k}}|_{\mathrm{A\cdot p}} = M_{\lambda',\mathbf{k}'}^{\lambda,\mathbf{k}}|_{\mathrm{A\cdot p(1)}}$$
$$= -\sum_{\mathbf{q}_\parallel} \delta_{\mathbf{k}',\mathbf{k}-\mathbf{q}_\parallel} \left[\delta_{\lambda,\lambda'} Q \frac{\hbar\left(\mathbf{k} - \frac{1}{2}\mathbf{q}_\parallel\right)}{m_\lambda} \cdot \hat{\mathbf{A}}_{\mathbf{q}_\parallel}^{\lambda,\lambda} + (1 - \delta_{\lambda,\lambda'}) \frac{Q\mathbf{p}_{\lambda,\lambda'}}{m_0} \cdot \hat{\mathbf{A}}_{\mathbf{q}_\parallel}^{\lambda,\lambda'} \right]. \qquad (13.45)$$

The light–matter coupling is defined by Eqs. (13.42) and (13.45).

13.2.4 *Light–matter interaction in* $\mathbf{k}\cdot\mathbf{p}$ *theory*

Collecting all the results obtained for the matrix elements, we are now able to construct the final form of the light–matter interaction Hamiltonian. Inserting the matrix elements (13.42) and (13.45) into Eqs. (13.5)–(13.7) yields

$$\hat{H}_{\mathrm{em-e}} = \sum_{\lambda,\mathbf{k}} \sum_{\lambda',\mathbf{k}'} \left[M_{\lambda',\mathbf{k}'}^{\lambda,\mathbf{k}}|_{A\cdot p} + M_{\lambda',\mathbf{k}'}^{\lambda,\mathbf{k}}|_{A\cdot A} \right] a_{\lambda,\mathbf{k}}^\dagger a_{\lambda',\mathbf{k}'}$$
$$= -\sum_{\mathbf{q}_\parallel,\mathbf{k}} \sum_{\lambda} Q \frac{\hbar\mathbf{k}}{m_\lambda} \cdot \mathbf{e}_P\, \hat{A}_{\mathbf{q}_\parallel}^{\lambda,\lambda}\, a_{\lambda,\mathbf{k}+\frac{1}{2}\mathbf{q}_\parallel}^\dagger a_{\lambda,\mathbf{k}-\frac{1}{2}\mathbf{q}_\parallel}$$
$$-\sum_{\mathbf{q}_\parallel,\mathbf{k}} \sum_{\lambda\neq\lambda'} \frac{Q\mathbf{p}_{\lambda,\lambda'}}{m_0} \cdot \hat{\mathbf{A}}_{\mathbf{q}_\parallel}^{\lambda,\lambda'}\, a_{\lambda,\mathbf{k}}^\dagger a_{\lambda',\mathbf{k}-\mathbf{q}_\parallel}$$
$$+\sum_{\mathbf{q}_\parallel,\mathbf{q}_\parallel'} \sum_{\lambda,\mathbf{k}} \frac{Q^2}{2m_0} \hat{A}_{\mathbf{q}_\parallel,-\mathbf{q}_\parallel'}^{(2),\lambda,\lambda}\, a_{\lambda,\mathbf{k}-\mathbf{q}_\parallel'}^\dagger a_{\lambda,\mathbf{k}-\mathbf{q}_\parallel}. \qquad (13.46)$$

Here, we used Eq. (13.33) to identify the polarization direction $\mathbf{e}_P$ of the field.

In the first term of Eq. (13.46), the operator $A_{\mathbf{q}_\parallel}^{\lambda,\lambda}$ involves either annihilation of a photon with in-plane momentum $\mathbf{q}_\parallel$ or creation of a photon with momentum $-\mathbf{q}_\parallel$. In both cases, the overall momentum conservation is assured by the corresponding changes in carrier momenta. In other words, an electron within a single band makes a transition from a state $\mathbf{k}-\frac{1}{2}\mathbf{q}_\parallel$ to a state $\mathbf{k}+\frac{1}{2}\mathbf{q}_\parallel$ matching with the photon's in-plane momentum. This process has an *intraband* character because the band index remains unchanged, i.e., $\lambda = \lambda'$ as shown in Fig. 13.1. These intraband transitions are proportional to the current-matrix element

$$j_\lambda(\mathbf{k}) \equiv Q \frac{\hbar\mathbf{k}}{m_\lambda} \cdot \mathbf{e}_P, \qquad (13.47)$$

which contains the effective mass of the electron in the band λ.

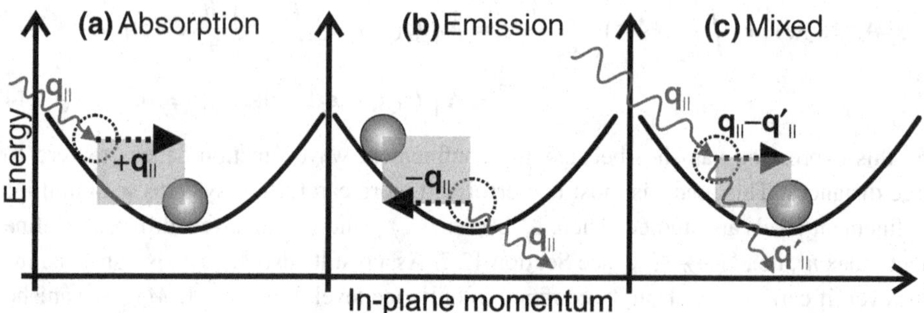

Figure 13.1 Schematic intraband processes in a parabolic band (solid line). Diagrams for **(a)** single-photon absorption, **(b)** single-photon emission, and **(c)** two-photon process. The initial (dashed circle) and final (sphere) states of the electron are shown together with the photons (wave arrow). Photon absorption (wave arrow exiting dashed circle) and emission (wave arrow entering dashed circle) add and subtract to the carrier momentum, respectively. The rectangle indicates the amount of energy (vertical direction) and momentum (horizontal direction) exchange.

Figure 13.1 schematically presents the intraband processes contained in the Hamiltonian (13.46). The dashed circle and sphere denote the initial and the final state – or the annihilation and the creation operator – of carriers within *one* band (solid line), respectively. The wave arrows entering (exiting) the dashed circle indicate photon absorption (emission). The corresponding in-plane momentum exchange is also shown (arrow). More specifically, frame (a) ((b)) describes photon absorption (emission) within the $j(\mathbf{k}) \cdot \hat{A}^{\lambda,\lambda}$ processes. Also one possible $\hat{A}^{(2),\lambda,\lambda}$ contribution is shown in Fig. 13.1(c) for a combination of photon emission and absorption accompanied with an intraband transition. In all cases, the shaded rectangle indicates the total momentum (horizontal scale) and energy transferred (vertical scale) between the photon(s) and the carrier.

The other intraband transitions follow from the last line of Eq. (13.46). In this contribution, $\hat{\mathbf{A}}^2$ contains two-photon processes where the total in-plane momentum is changed by $\mathbf{q}_{\|} - \mathbf{q}'_{\|}$ as illustrated in Fig. 13.1(c). Once again, the total in-plane momentum is conserved due to the momentum exchange of electrons in the band λ. As a distinct feature of the $\hat{\mathbf{A}}^2$ interaction, we notice that, in contrast to the $j\hat{A}$ interaction, the carrier part involves the free electron mass m_0.

The remaining parts of the light–matter Hamiltonian describe processes where photon emission or absorption is accompanied by electronic transitions between two different bands, i.e., these processes have an *interband* character. Once again, only combinations where the in-plane momentum is conserved are allowed, as shown in Fig. 13.2 for different combinations of photons and electronic interband transitions. The diagrammatic rules are the same as in Fig. 13.1.

For QW systems without disorder, the conservation of the in-plane momentum is a general property of the light–matter interaction. This conservation law reflects the fact that a QW has translational symmetry in the xy-plane. Due to this feature, we are able to

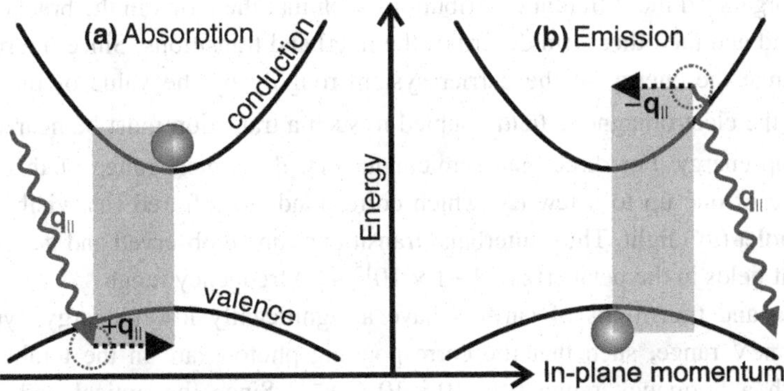

Figure 13.2 Schematic interband processes contained in Eq. (13.46). The diagrams are for **(a)** the single-photon absorption and **(b)** the single-photon emission process. The diagrammatic rules are the same as in Fig. 13.1.

express the in-plane parts of the electron-envelope function and light modes as plane waves, which introduces a well-defined in-plane momentum for both of these entities. Since the light–matter interaction does not break the translational symmetry, the total momentum of any allowed process must conserve the total in-plane momentum. However, due to the confinement, the QW does not have translational symmetry in the z-direction, such that the z-component of the momentum is not conserved.

For lower-dimensional systems, the momentum conservation becomes even more incomplete. For quantum wires, the momentum is conserved only along the z axis parallel to the wire. In quantum dots, the translational symmetry is completely lost such that the photon momentum is entirely disconnected from the carrier system. In the other extreme, i.e., in three-dimensional bulk semiconductors, one has a complete translational symmetry in all directions such that the momentum conservation requirement has to be fulfilled for the full three-dimensional momentum vector. As a common feature in all dimensions, the energy of the photon has roughly to match the energy difference of the carrier states participating in either intra- or interband transitions. This sets the basic energy scales of the different processes.

In our semiconductor quantum-optical investigations, we use the light–matter interaction in the form

$$\hat{H}_{\text{em-e}} = \sum_{\lambda,\mathbf{k}} \Big[-\sum_{\mathbf{q}_{\parallel}} j_{\lambda}(\mathbf{k})\, \hat{A}^{\lambda,\lambda}_{\mathbf{q}_{\parallel}}\, a^{\dagger}_{\lambda,\mathbf{k}+\frac{\mathbf{q}_{\parallel}}{2}} a_{\lambda,\mathbf{k}-\frac{\mathbf{q}_{\parallel}}{2}} + \sum_{\mathbf{q}_{\parallel},\mathbf{q}'_{\parallel}} \frac{Q^2}{2m_0} \hat{A}^{(2),\lambda,\lambda}_{\mathbf{q}_{\parallel},-\mathbf{q}'_{\parallel}}\, a^{\dagger}_{\lambda,\mathbf{k}+\mathbf{q}'_{\parallel}} a_{\lambda,\mathbf{k}-\mathbf{q}_{\parallel}} \Big] - \sum_{\mathbf{k},\mathbf{q}_{\parallel}} \sum_{\lambda\neq\lambda'} \frac{Q\mathbf{p}_{\lambda,\lambda'}}{m_0} \cdot \hat{\mathbf{A}}^{\lambda,\lambda'}_{\mathbf{q}_{\parallel}}\, a^{\dagger}_{\lambda,\mathbf{k}} a_{\lambda',\mathbf{k}-\mathbf{q}_{\parallel}}, \tag{13.48}$$

where we organized the different contributions such that the terms in the bracket describe the intraband and the other term describes the interband transitions. Since interband transitions change the energy of the carrier system roughly by the value of the band gap $\left|\epsilon_0^{\lambda} - \epsilon_0^{\lambda'}\right|$, the electromagnetic field coupled to such a transition must be nearly resonant with the gap energy. For direct-gap semiconductors, the typical range of this energy is in the range of one up to a few eV which corresponds to infrared (ir), visible, or even near ultraviolet (uv) light. Thus, interband transitions can be observed and generated with optical light fields in the petahertz ($0.1{-}1 \times 10^{15}\ \mathrm{s}^{-1}$) frequency range.

The intraband transitions of carriers have a significantly lower energy, typically in the 1–100 meV range, such that the corresponding photons are in the terahertz (THz) regime with a frequency range $0.1{-}10 \times 10^{12}\ \mathrm{s}^{-1}$. Since the optical and THz field are energetically well separated, they lead to very different excitation and emission dynamics.

13.3 Phonon–carrier interaction

The coupling of electrons to lattice vibrations follows from the second line of the general Hamiltonian, Eq. (13.4). If we use the quantized form of the ion displacement $\Delta\hat{\mathbf{R}}_n$ according to Eq. (11.66), the interaction Hamiltonian becomes

$$\begin{aligned}\hat{H}_{\text{ph-e}} &= -\int d^3r\,\hat{\Psi}^{\dagger}(\mathbf{r})\Big[\sum\nolimits_n \nabla v\left(\mathbf{r}-\mathbf{R}_n^0\right)\cdot\Delta\hat{\mathbf{R}}_n\Big]\hat{\Psi}(\mathbf{r}) \\ &= -\int d^3r\,\hat{\Psi}^{\dagger}(\mathbf{r})\Big[\sum\nolimits_{n,\mathbf{p}} \nabla v\left(\mathbf{r}-\mathbf{R}_n^0\right)\cdot\left(D_{\mathbf{p}}\mathbf{e}_{\mathbf{p}} - D_{-\mathbf{p}}^{\dagger}\mathbf{e}_{-\mathbf{p}}\right) \\ &\qquad \times i\sqrt{\frac{\hbar}{2NM\Omega_{\mathbf{p}}}}\,e^{i\mathbf{R}_n^0\cdot\mathbf{p}}\Big]\hat{\Psi}(\mathbf{r}) \\ &= \sum\nolimits_{\lambda,\mathbf{k},\lambda',\mathbf{k}'} M_{\lambda',\mathbf{k}'}^{\lambda,\mathbf{k}}\big|_{\text{ph-e}}\, a_{\lambda,\mathbf{k}}^{\dagger} a_{\lambda',\mathbf{k}'}.\end{aligned} \tag{13.49}$$

As in Eq. (13.7), the last form is determined by the matrix elements between the Bloch electrons. By expressing the Bloch functions via Eq. (12.25), we obtain

$$\begin{aligned}M_{\lambda',\mathbf{k}'}^{\lambda,\mathbf{k}}\big|_{\text{ph-e}} = &-\sum_{n,\mathbf{p}}\frac{1}{S}\int d^3r\, e^{i(\mathbf{k}-\mathbf{k}')\cdot\mathbf{r}_{\|}}\, e^{i\mathbf{R}_n^0\cdot\mathbf{p}}\xi_{\lambda}^{\star}(z)\xi_{\lambda'}(z) \\ &\times\ u_{\lambda,\mathbf{k}}^{\star}(\mathbf{r})\,\nabla v\left(\mathbf{r}-\mathbf{R}_n^0\right)\cdot u_{\lambda',\mathbf{k}'}(\mathbf{r}) \\ &\times\ i\sqrt{\frac{\hbar}{2NM\Omega_{\mathbf{p}}}}\left(D_{\mathbf{p}}\mathbf{e}_{\mathbf{p}} - D_{-\mathbf{p}}^{\dagger}\mathbf{e}_{-\mathbf{p}}\right),\end{aligned} \tag{13.50}$$

for strongly confined systems with $\xi_{\lambda,n} \to \xi_{\lambda}$. As for the light–matter interaction, this can be evaluated via the separation of length scales, derived in Section 13.2.1. This procedure eventually yields

$$M^{\lambda,\mathbf{k}}_{\lambda',\mathbf{k}'}|_{\text{ph-e}} = -\sum_{n,j,\mathbf{p}} \frac{1}{\mathcal{S}} e^{i(\mathbf{k}-\mathbf{k}')\cdot\mathbf{R}^0_{\parallel,j}} e^{i\mathbf{R}^0_n\cdot\mathbf{p}} \xi^\star_\lambda(Z_j)\xi_{\lambda'}(Z_j) v_0$$
$$\times \frac{1}{v_0}\int_{v_0} d^3r'\, u^\star_{\lambda,\mathbf{k}}(\mathbf{r}')\, \nabla v\left(\mathbf{r}' + \mathbf{R}^0_j - \mathbf{R}^0_n\right) \cdot u_{\lambda',\mathbf{k}'}(\mathbf{r}')$$
$$\times\, i\sqrt{\frac{\hbar}{2NM\Omega_\mathbf{p}}}\left(D_\mathbf{p}\mathbf{e}_\mathbf{p} - D^\dagger_{-\mathbf{p}}\mathbf{e}_{-\mathbf{p}}\right), \tag{13.51}$$

after a derivation analogous to the conversion of Eq. (13.19) into (13.21). Here, we again used the fact that the plane-wave parts and the confinement functions are practically constant over the unit cell. Furthermore, we performed a change of the integration variable $\mathbf{r} = \mathbf{r}' + \mathbf{R}^0_j$ and used the lattice periodicity of the Bloch functions, i.e., $u_{\lambda',\mathbf{k}'}\left(\mathbf{r}' + \mathbf{R}^0_j\right) = u_{\lambda',\mathbf{k}'}(\mathbf{r}')$.

Since $\nabla v(\mathbf{r}' + \mathbf{R})$ depends on both microscopic ($\mathbf{r}'$) and mesoscopic ($\mathbf{R}$) scales, it is convenient to introduce a Fourier expansion on the macroscopic scale

$$v_\mathbf{q}(\mathbf{r}) = \sum\nolimits_n v\left(\mathbf{r} + \mathbf{R}^0_n\right) e^{-i\mathbf{q}\cdot\mathbf{R}^0_n},$$
$$v\left(\mathbf{r} + \mathbf{R}^0_n\right) = \frac{1}{N}\sum\nolimits_\mathbf{q} v_\mathbf{q}(\mathbf{r}) e^{+i\mathbf{q}\cdot\mathbf{R}^0_n}, \tag{13.52}$$

where N is the number of unit cells. As a result, any $v\left(\mathbf{r} + \mathbf{R}^0_n\right)$ can be expressed via its microscopic part $v_\mathbf{q}(\mathbf{r})$ times the mesoscopically varying envelope $e^{i\mathbf{q}\cdot\mathbf{R}^0_n}$. With the help of this separation, we find

$$\nabla_\mathbf{r} v\left(\mathbf{r} + \mathbf{R}^0_n\right) = \nabla_\mathbf{R} v\left(\mathbf{r} + \mathbf{R}^0_n\right) = \frac{1}{N}\sum\nolimits_\mathbf{q} v_\mathbf{q}(\mathbf{r})\, \nabla_\mathbf{R} e^{i\mathbf{q}\cdot\mathbf{R}^0_n}$$
$$= \frac{1}{N}\sum\nolimits_\mathbf{q} v_\mathbf{q}(\mathbf{r})\, i\mathbf{q} e^{i\mathbf{q}\cdot\mathbf{R}^0_n}. \tag{13.53}$$

If this result is inserted into Eq. (13.51), we obtain

$$M^{\lambda,\mathbf{k}}_{\lambda',\mathbf{k}'}|_{\text{ph-e}} = \sum\nolimits_{n,j}\sum\nolimits_{\mathbf{p},\mathbf{q}} \frac{e^{i(\mathbf{k}-\mathbf{k}'+\mathbf{q}_\parallel)\cdot\mathbf{R}^0_{\parallel,j}}}{\mathcal{S}} \frac{e^{i\mathbf{R}^0_n\cdot(\mathbf{p}-\mathbf{q})}}{N} \xi^\star_\lambda(Z_j)\, e^{iq_zZ_j}\, \xi_{\lambda'}(Z_j) v_0$$
$$\times \frac{1}{v_0}\int_{v_0} d^3r'\, u^\star_{\lambda,\mathbf{k}}(\mathbf{r}')\, v_\mathbf{q}(\mathbf{r}')\, u_{\lambda',\mathbf{k}'}(\mathbf{r}')$$
$$\times \sqrt{\frac{\hbar}{2NM\Omega_\mathbf{p}}}\left(D_\mathbf{p}\mathbf{q}\cdot\mathbf{e}_\mathbf{p} - D^\dagger_{-\mathbf{p}}\mathbf{q}\cdot\mathbf{e}_{-\mathbf{p}}\right)$$
$$= \sum\nolimits_{\mathbf{p},\mathbf{q}} \frac{1}{\mathcal{S}}\int d^2R_\parallel\, e^{i(\mathbf{k}-\mathbf{k}'+\mathbf{q}_\parallel)\cdot\mathbf{R}_\parallel}$$
$$\times \frac{1}{N}\sum\nolimits_n e^{i\mathbf{R}^0_n\cdot(\mathbf{p}-\mathbf{q})}\int dZ\, \xi^\star_\lambda(Z)\, e^{iq_zZ}\, \xi_{\lambda'}(Z)$$
$$\times \frac{1}{v_0}\int_{v_0} d^3r'\, u^\star_{\lambda,\mathbf{k}}(\mathbf{r}')\, v_\mathbf{q}(\mathbf{r}')\, u_{\lambda',\mathbf{k}'}(\mathbf{r}')$$
$$\times \sqrt{\frac{\hbar}{2NM\Omega_\mathbf{p}}}\left(D_\mathbf{p}\mathbf{q}\cdot\mathbf{e}_\mathbf{p} - D^\dagger_{-\mathbf{p}}\mathbf{q}\cdot\mathbf{e}_{-\mathbf{p}}\right), \tag{13.54}$$

where the sums are converted into integrals since the unit-cell volume $v_0 = d^2\mathbf{R}_\parallel\, dZ$ can be considered to be infinitesimal on the mesoscopic scale. Now, the integral over $\mathbf{R}_\parallel$ produces $\delta_{\mathbf{k}',\mathbf{k}+\mathbf{q}_\parallel}$ while the sum over n leads to $\delta_{\mathbf{q},\mathbf{p}}$.

We define a deformation potential matrix element

$$F^{\mathbf{k},\lambda}_{\mathbf{k}',\lambda'}(\mathbf{q}) \equiv \frac{1}{v_0}\int_{v_0} d^3r'\; u^\star_{\lambda,\mathbf{k}}(\mathbf{r}')\; v_\mathbf{q}(\mathbf{r}')\; u_{\lambda',\mathbf{k}'}(\mathbf{r}'). \tag{13.55}$$

Since the phonon energy is typically orders of magnitude smaller than the interband transition energy, only intraband, i.e., $\lambda = \lambda'$ contributions are considered. Furthermore, the microscopic integral (13.55) is often approximated by a band-index-dependent deformation constant such that

$$F^{\mathbf{k},\lambda}_{\mathbf{k}',\lambda'}(\mathbf{q}) = F_\lambda\; \delta_{\lambda,\lambda'}. \tag{13.56}$$

For practical purposes, the explicit value of F_λ is often determined experimentally. If we define a confinement function

$$g^\lambda_{q_z} \equiv \int dZ\, e^{iq_z Z}\; |\xi_\lambda(Z)|^2, \tag{13.57}$$

we can eventually write

$$M^{\lambda,\mathbf{k}}_{\lambda',\mathbf{k}'}|_{\text{ph}-\text{e}} = \delta_{\lambda,\lambda'} \sum_\mathbf{p} \delta_{\mathbf{k}',\mathbf{k}+\mathbf{p}_\parallel}\; g^\lambda_{p_z} i\mathbf{p}\cdot\mathbf{e}_\mathbf{p} F^\lambda \sqrt{\frac{\hbar}{2NM\Omega_\mathbf{p}}}\left(D_\mathbf{p} + D^\dagger_{-\mathbf{p}}\right), \tag{13.58}$$

see Exercise 13.6.

For transversal phonons, $\mathbf{p}\cdot\mathbf{e}_\mathbf{p}$ vanishes because then $\mathbf{p}$ and $\mathbf{e}_\mathbf{p}$ are orthogonal. For longitudinal phonons, $\mathbf{p}$ and $\mathbf{e}_\mathbf{p} = \mathbf{p}/|\mathbf{p}|$ point in the same direction such that $\mathbf{p}\cdot\mathbf{e}_\mathbf{p} = (-\mathbf{p})\cdot\mathbf{e}_{-\mathbf{p}} = |\mathbf{p}|$. As a result, in the lowest order only longitudinal phonons contribute. By inserting this result into Eq. (13.49), we may express the phonon–electron coupling via

$$\begin{aligned}\hat{H}_{\text{ph}-\text{e}} &= \sum_{\lambda,\mathbf{k},\mathbf{p}_\parallel,p_z} G^\lambda_\mathbf{p}\left[D_{\mathbf{p}_\parallel,p_z} + D^\dagger_{-\mathbf{p}_\parallel,p_z}\right] a^\dagger_{\lambda,\mathbf{k}} a_{\lambda,\mathbf{k}-\mathbf{p}_\parallel} \\ &= \sum_{\lambda,\mathbf{k},\mathbf{p}_\parallel} \hat{\mathcal{G}}^\lambda_{\mathbf{p}_\parallel} a^\dagger_{\lambda,\mathbf{k}} a_{\lambda,\mathbf{k}-\mathbf{p}_\parallel}.\end{aligned} \tag{13.59}$$

Identifying the mass density ρ of the semiconductor material and the volume of the entire semiconductor system $\mathcal{L}^3$, we express the strength of the phonon interaction as

$$G^\lambda_\mathbf{p} = |\mathbf{p}|F^\lambda g^\lambda_{q_z}\sqrt{\frac{\hbar}{2NM\Omega_\mathbf{p}}} = F^\lambda g^\lambda_{q_z}\sqrt{\frac{\hbar|\mathbf{p}|^2}{2\rho\mathcal{L}^3 c_{\text{LA}}|\mathbf{p}|}} = F^\lambda g^\lambda_{q_z}\sqrt{\frac{\hbar|\mathbf{p}|}{2c_{\text{LA}}\rho\,\mathcal{L}^3}}, \tag{13.60}$$

where c_{LA} is the velocity of sound for the longitudinal acoustic phonons. For optical phonons, one often has both deformation-type and Fröhlich-type interactions. The Fröhlich interaction resembles the Coulomb interaction, i.e., $\left|G_{\mathbf{p}}^{\lambda}\right|^2 \propto V_{\mathbf{p}}$, because it results from the polarizability of the ionic lattice.

For both acoustic and optical phonons, we find a collective phonon field

$$\hat{\mathcal{G}}_{\mathbf{p}_{\|}}^{\lambda} \equiv \sum_{p_z} G_{\mathbf{p}}^{\lambda}\left[D_{\mathbf{p}_{\|},p_z} + D_{-\mathbf{p}_{\|},p_z}^{\dagger}\right], \tag{13.61}$$

which defines the $\mathbf{p}_{\|}$ dependent component in analogy to Eq. (13.12). We observe that the in-plane momentum is conserved in the process where a phonon is either emitted or absorbed, while the electron undergoes an intraband transition. This yields identical diagrams to those in Figs. 13.1(a–b) if the photon is simply replaced by a phonon. Thus, phonon interactions resemble intraband transitions and the momentum-conservation aspects in the systems of different dimensionality follow in the same way as for the light–matter interaction.

13.4 Coulomb interaction

The last term in the general Hamiltonian (13.1) describes the Coulomb interaction among the Bloch electrons. This is a pure two-particle contribution which can be transformed into second quantization by sandwiching it between two creation and annihilation operators, as discussed in Section 11.1.4. Implementing the basis function (12.25), the Coulomb Hamiltonian can be expressed as

$$\begin{aligned}\hat{V} &= \frac{1}{2}\int \hat{\Psi}^{\dagger}(\mathbf{r})\hat{\Psi}^{\dagger}(\mathbf{r}')\, V(\mathbf{r}'-\mathbf{r})\, \hat{\Psi}(\mathbf{r}')\hat{\Psi}(\mathbf{r}) d^3r\, d^3r' \\ &= \sum_{\mathbf{k},\lambda}\sum_{\mathbf{k}',\lambda'}\sum_{\mathbf{p}_{\|},\nu}\sum_{\mathbf{p}_{\|}',\nu'} \frac{1}{\mathcal{S}^2}\int\int e^{i\left[(\mathbf{k}'-\mathbf{k})\cdot\mathbf{R}_{\|}+(\mathbf{p}_{\|}'-\mathbf{p}_{\|})\cdot\mathbf{R}_{\|}'\right]} \\ &\quad \times\ \xi_{\lambda}^{\star}(Z)\xi_{\nu}^{\star}(Z)\, V(\mathbf{R}-\mathbf{R}')\xi_{\nu'}(Z)\xi_{\lambda'}(Z')\, d^3r\, d^3r' \\ &\quad \times\ \langle\lambda,\mathbf{k}|\lambda',\mathbf{k}'\rangle\ \langle\nu,\mathbf{p}_{\|}|\nu',\mathbf{p}_{\|}'\rangle\ a_{\lambda,\mathbf{k}}^{\dagger}a_{\nu,\mathbf{p}_{\|}}^{\dagger}a_{\nu',\mathbf{p}_{\|}'}a_{\lambda',\mathbf{k}'},\end{aligned} \tag{13.62}$$

that follows from the separation of length scales discussed in Section 13.2.1, see also Exercise 13.10.

To evaluate the mesoscopic integrals, we express the Coulomb potential via its Fourier expansion, as done before in connection with Eq. (8.43),

$$\begin{aligned}V(\mathbf{R}) &= \frac{e^2}{4\pi\epsilon\epsilon_0|\mathbf{R}|} = \sum\nolimits_{\mathbf{q}_{\|},q_z} \frac{e^2}{\epsilon\epsilon_0\mathcal{L}^3}\frac{1}{|\mathbf{q}|^2}e^{i[\mathbf{q}_{\|}\cdot\mathbf{R}_{\|}+q_zZ]} \\ &= \sum\nolimits_{\mathbf{q}_{\|}} \frac{e^2}{\epsilon\epsilon_0\mathcal{L}^2}e^{i\mathbf{q}_{\|}\cdot\mathbf{r}_{\|}}\frac{1}{\mathcal{L}}\sum\nolimits_{q_z}\frac{1}{\mathbf{q}_{\|}^2+q_z^2}e^{iq_zz} \\ &= \sum\nolimits_{\mathbf{q}_{\|}} \frac{e^2}{\epsilon\epsilon_0\mathcal{L}^2}e^{i\mathbf{q}_{\|}\cdot\mathbf{R}_{\|}}\frac{1}{2\pi}\int_{-\infty}^{\infty}dq_z\,\frac{e^{iq_zZ}}{\mathbf{q}_{\|}^2+q_z^2}.\end{aligned} \tag{13.63}$$

Here, we changed the q_z sum into an integral according to Eq. (12.43). In this context, it is customary to denote $\epsilon \equiv n^2$ as the background dielectric constant. The resulting integral can be solved analytically with the result

$$V(\mathbf{R}) = \sum_{\mathbf{q}_\parallel} \frac{e^2}{\epsilon\epsilon_0 \mathcal{S}} e^{i\mathbf{q}_\parallel \cdot \mathbf{R}_\parallel} \frac{1}{2\pi} \int_{-\infty}^{\infty} dq_z \frac{e^{iq_z Z}}{(q_z - i|\mathbf{q}_\parallel|)(q_z + i|\mathbf{q}_\parallel|)}$$
$$= \sum_{\mathbf{q}_\parallel} \frac{e^2}{2\epsilon\epsilon_0 \mathcal{S}} \frac{1}{|\mathbf{q}_\parallel|} e^{i\mathbf{q}_\parallel \cdot \mathbf{R}_\parallel} e^{-|\mathbf{q}_\parallel Z|}. \quad (13.64)$$

Here, we used Cauchy's integral theorem by extending the integration path into the complex plane such that the closed contour contains one pole at either $q_z = i|\mathbf{q}_\parallel|$ or $q_z = -i|\mathbf{q}_\parallel|$.

We use this result in Eq. (13.62) to eventually obtain

$$\hat{V} = \frac{1}{2} \sum_{\mathbf{k},\mathbf{k}',\mathbf{q}_\parallel} \sum_{\lambda,\lambda',\nu',\nu} \frac{e^2}{2\epsilon\epsilon_0 \mathcal{S}} \frac{1}{|\mathbf{q}_\parallel|}$$
$$\times \int_{-\infty}^{\infty} dZ\, dZ'\, \xi_\lambda^\star(Z) \xi_\nu^\star(Z) \frac{e^2}{2\epsilon\epsilon_0 \mathcal{S}} e^{-|\mathbf{q}_\parallel (Z-Z')|} \xi_{\nu'}(Z) \xi_{\lambda'}(Z')$$
$$\times \left\langle \lambda, \mathbf{k} | \lambda', \mathbf{k} - \mathbf{q}_\parallel \right\rangle \left\langle \nu, \mathbf{k}' | \nu', \mathbf{k}' + \mathbf{q}_\parallel \right\rangle a^\dagger_{\lambda,\mathbf{k}} a^\dagger_{\nu,\mathbf{k}'} a_{\nu',\mathbf{k}'+\mathbf{q}_\parallel} a_{\lambda',\mathbf{k}-\mathbf{q}_\parallel}, \quad (13.65)$$

see Exercise 13.8. The expression (13.65) simplifies further when we notice that

$$\left\langle \lambda, \mathbf{k} | \lambda', \mathbf{k} \pm \mathbf{q}_\parallel \right\rangle = \delta_{\lambda,\lambda'} \quad (13.66)$$

for mesoscopic $\mathbf{q}_\parallel$, see Exercise 13.3. Before we enter this result into Eq. (13.65), we define the Coulomb matrix element for a QW according to

$$V_{\mathbf{q}_\parallel}^{\lambda,\lambda'} \equiv \frac{e^2}{2\epsilon\epsilon_0 \mathcal{L}^2} \frac{1}{|\mathbf{q}_\parallel|} \int dZ\, dZ'\, |\xi_\lambda(Z)|^2 |\xi_{\lambda'}(Z')|^2\, e^{-|\mathbf{q}_\parallel (Z-Z')|}. \quad (13.67)$$

With these definitions, we cast the Coulomb interaction into its final form

$$\hat{V} = \frac{1}{2} \sum_{\mathbf{k},\mathbf{k}',\mathbf{q}_\parallel \neq 0} \sum_{\lambda,\lambda'} V_{\mathbf{q}_\parallel}^{\lambda,\lambda'}\, a^\dagger_{\lambda,\mathbf{k}} a^\dagger_{\lambda',\mathbf{k}'} a_{\lambda',\mathbf{k}'+\mathbf{q}_\parallel} a_{\lambda,\mathbf{k}-\mathbf{q}_\parallel}, \quad (13.68)$$

where the summation indices have been relabeled. This equation shows that the Coulomb interaction exchanges the momentum $\mathbf{q}_\parallel$ between two electrons. Thus, $\hat{V}$ is a genuine many-body interaction where the in-plane momentum is conserved at the elementary level. For strongly confined systems, $\hat{V}$ contains the Coulomb matrix element that approaches the strictly two-dimensional form, see Exercise 13.9. This shows that quantum confinement can indeed produce effectively two-dimensional properties also with respect to the electron–electron interactions.

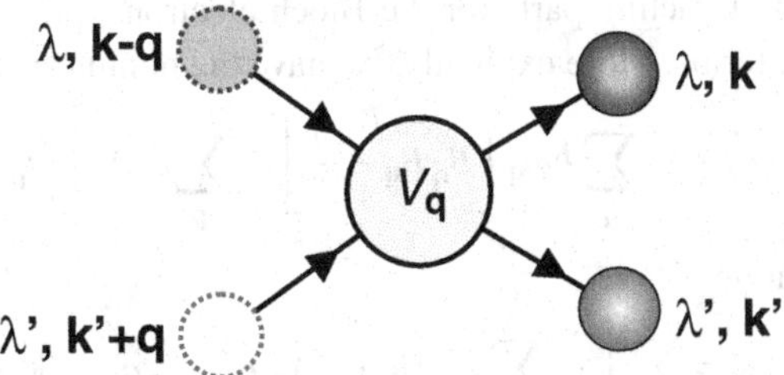

Figure 13.3 Diagrammatic representation of the Coulomb interaction (13.68). Two incoming electrons (dashed circles) exchange momentum via the Coulomb interaction. The interaction vertex contains the Coulomb-matrix element $V_{\mathbf{q}}$.

Figure 13.3 schematically presents the structure of the Coulomb interaction using the same diagrammatic rules as in Fig. 13.1. More specifically, the dashed circles correspond to particle annihilation and spheres to particle creation. We see that the Coulomb interaction scatters electrons with momentum $\mathbf{k} - \mathbf{q}_{\|}$ and $\mathbf{k}' + \mathbf{q}_{\|}$ into new momenta $\mathbf{k}$ and $\mathbf{k}'$, respectively. The momentum exchange is mediated by the Coulomb interaction defined by $V^{\lambda,\lambda'}_{\mathbf{q}_{\|}}$ indicated within the vertex (light shaded area). The arrows show the direction of the interaction.

We also notice that the $\mathbf{q}_{\|} = 0$ component would lead to a diverging energy contribution in Eq. (13.68). However, this contribution is fully compensated by the Coulomb self-energy of the charged background of ions, as shown rigorously in Exercise 13.10. This solution is often derived using the jellium model where the ions are treated as a uniform background charge. Exercise 13.10 presents a generalization to this simplified picture by describing both the electrons and ions within the second-quantization formalism. In any case, the diverging contribution $\mathbf{q}_{\|} = 0$ has to be left out of the Coulomb interactions in order to avoid unphysical features.

13.5 Complete system Hamiltonian in different dimensions

The derivation of the system Hamiltonian is pretty much independent of the dimensionality of the system. The only major changes result from the differences in the momentum conservation and the confinement-matrix elements. Since we have performed the detailed derivation for QWs, we are now able directly to construct the total Hamiltonian for a system with any given dimension.

Let us start with three-dimensional bulk structures. Since there is no confinement in this system, the full three-dimensional momentum is conserved in all microscopic processes. Once this is taken into account in Eqs. (11.58), (11.72), (13.48), (13.59), and (13.68), we find the total system Hamiltonian

$$\hat{H}^{\text{bulk}}_{\text{tot}} = \hat{H}_{\text{free}} + \hat{H}_{\text{em-e}} + \hat{H}_{\text{ph-e}} + \hat{V}, \tag{13.69}$$

which consists of noninteracting parts for the Bloch electrons, photons, and phonons, as well as the interaction terms. More explicitly, we have the noninteracting part

$$\hat{H}_{\text{free}} = \sum_{\lambda,\mathbf{k}} \epsilon^{\lambda}_{\mathbf{k}} a^{\dagger}_{\lambda,\mathbf{k}} a_{\lambda,\mathbf{k}} + \sum_{\mathbf{q}} \hbar\omega_{\mathbf{q}} \left[B^{\dagger}_{\mathbf{q}} B_{\mathbf{q}} + \frac{1}{2} \right] + \sum_{\mathbf{p}} \hbar\Omega_{\mathbf{p}} \left[D^{\dagger}_{\mathbf{p}} D_{\mathbf{p}} + \frac{1}{2} \right] \tag{13.70}$$

as well as the interactions

$$\begin{aligned} \hat{H}_{\text{em-e}} = \sum_{\lambda,\mathbf{k}} \Big[&- \sum\nolimits_{\mathbf{q}} j_{\lambda}(\mathbf{k},\mathbf{q})\, \hat{A}_{\mathbf{q}}\, a^{\dagger}_{\lambda,\mathbf{k}+\frac{\mathbf{q}}{2}} a_{\lambda,\mathbf{k}-\frac{\mathbf{q}}{2}} \\ &+ \sum\nolimits_{\mathbf{q},\mathbf{q}'} \frac{Q^2}{2m_0} \hat{A}^{(2)}_{\mathbf{q},-\mathbf{q}'}\, a^{\dagger}_{\lambda,\mathbf{k}-\mathbf{q}'} a_{\lambda,\mathbf{k}-\mathbf{q}} \Big] \\ - \sum\nolimits_{\mathbf{q},\mathbf{k}} \sum\nolimits_{\lambda\neq\lambda'} &\frac{Q\mathbf{p}_{\lambda,\lambda'}}{m_0} \cdot \hat{\mathbf{A}}_{\mathbf{q}}\; a^{\dagger}_{\lambda,\mathbf{k}} a_{\lambda',\mathbf{k}-\mathbf{q}}, \end{aligned} \tag{13.71}$$

$$\hat{H}_{\text{ph-e}} = \sum_{\lambda,\mathbf{k},\mathbf{p}} \mathcal{G}^{\lambda}_{\mathbf{p}}\, a^{\dagger}_{\lambda,\mathbf{k}} a_{\lambda,\mathbf{k}-\mathbf{p}}, \tag{13.72}$$

$$\hat{V} = \frac{1}{2} \sum_{\lambda,\mathbf{k}} \sum_{\lambda',\mathbf{k}'} \sum_{\mathbf{q}\neq 0} V_{\mathbf{q}}\, a^{\dagger}_{\lambda,\mathbf{k}} a^{\dagger}_{\lambda',\mathbf{k}'} a_{\lambda',\mathbf{k}'+\mathbf{q}} a_{\lambda,\mathbf{k}-\mathbf{q}}. \tag{13.73}$$

The light–matter interaction part contains a current-matrix element with the three-dimensional carrier momentum

$$j_{\lambda}(\mathbf{k},\mathbf{q}) \equiv \frac{Q\hbar\mathbf{k}}{m_{\lambda}} \cdot \mathbf{e}_{\mathbf{q}}, \tag{13.74}$$

where $\mathbf{e_q}$ implicitly includes the polarization direction of the light field. Since the Bloch electrons are not quantum confined in a bulk system, the envelope function is a plane wave such that the strength of the phonon–electron interaction is given by

$$G^{\lambda,3D}_{\mathbf{p}} = F^{\lambda} \sqrt{\frac{\hbar|\mathbf{p}|}{2c_{\text{LA}}\rho\mathcal{L}^3}}, \tag{13.75}$$

where the deformation potential F^{λ} contains no confinement integrals. In the same way, the Coulomb-matrix element is

$$V_{\mathbf{q}} = \frac{e^2}{\epsilon\epsilon_0\mathcal{L}^3} \frac{1}{\mathbf{q}^2}, \tag{13.76}$$

where $\mathcal{V} \equiv \mathcal{L}^3$ refers to the quantization volume.

Additionally, the different $\mathbf{q}$ components of the linear and the quadratic vector potential contributions are

$$\hat{A}_{\mathbf{q}} = \frac{\mathcal{E}_{\mathbf{q}}}{\sqrt{\mathcal{L}^3}\omega_{\mathbf{q}}} \left(B_{\mathbf{q}} + B^{\dagger}_{-\mathbf{q}} \right), \tag{13.77}$$

$$\hat{A}^{(2)}_{\mathbf{q},\mathbf{q}'} = \frac{\mathcal{E}_{\mathbf{q}}\mathcal{E}_{\mathbf{q}'}}{\mathcal{L}^3\omega_{\mathbf{q}}\omega_{\mathbf{q}'}} \left(B_{\mathbf{q}} + B^{\dagger}_{-\mathbf{q}} \right) \left(B_{\mathbf{q}'} + B^{\dagger}_{-\mathbf{q}'} \right) \mathbf{e}_{\mathbf{q}} \cdot \mathbf{e}_{\mathbf{q}'}, \tag{13.78}$$

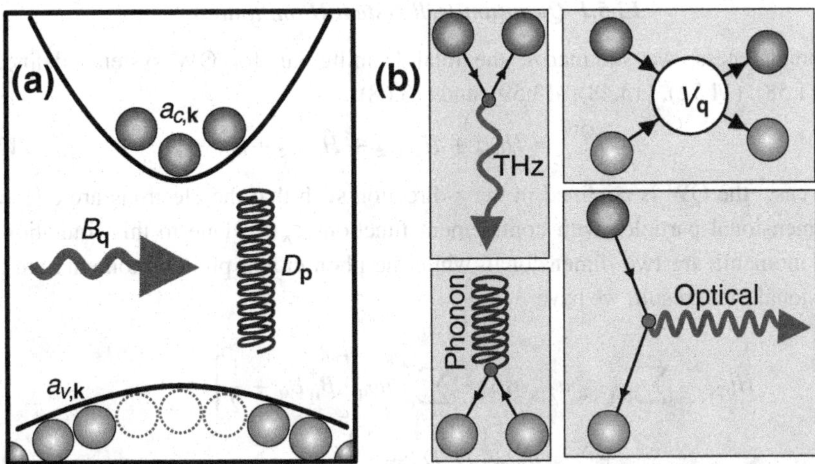

Figure 13.4 Diagrammatic form of the system Hamiltonian for a two-band semiconductor. (**a**) The spheres (dashed circles) indicate electrons (electronic vacancies, i.e. holes) in the conduction (valence) band (solid line). Photons (phonons) are presented by a wave arrow (spring). (**b**) Diagrammatic format of THz photon-carrier, carrier-phonon, Coulomb, and optical photon-carrier interactions. The diagrams are constructed as in Figs. 13.1–13.3.

respectively. In the same way, the three-dimensional phonon field is obtained from

$$\hat{\mathcal{G}}^{\lambda}_{\mathbf{p}} \equiv G^{\lambda,3D}_{\mathbf{p}} \left[D_{\mathbf{p}} + D^{\dagger}_{-\mathbf{p}} \right]. \tag{13.79}$$

The Hamiltonian (13.69) can be used as a general starting point for investigations with both visible and terahertz (THz) light fields in a bulk system.

Figure 13.4 schematically presents the system Hamiltonian containing the (a) free and (b) interaction contributions for a two-band model. A sphere represents an electron and a dashed circle defines a vacancy, i.e. a hole, in the valence band. The wave arrow corresponds to photons and the spring symbol depicts phonons. The optical intraband transitions are typically in the THz range while the interband transitions correspond to optical frequencies. The phonon interactions have an intraband character. Thus, phonon and THz interactions are structurally similar. The Coulomb part involves momentum exchange between electron pairs.

The system Hamiltonian can be generalized to include also transitions where the electron is moved between different bands by either the Coulomb or the phonon interaction. Such processes are particularly important for either spin or Auger processes where an electron scatters out of its original band or spin state. Also optical phonons may produce Auger-type scattering. Since these processes are typically relatively slow and do not occur in two-band systems, we do not include them in our quantum-optics investigations.

13.5.1 Quantum-well system Hamiltonian

For completeness, we summarize the total Hamiltonian for QW systems defined by Eqs. (11.58), (11.72), (13.48), (13.59), and (13.68):

$$\hat{H}^{\mathrm{QW}}_{\mathrm{tot}} = \hat{H}_{\mathrm{free}} + \hat{H}_{\mathrm{em-e}} + \hat{H}_{\mathrm{ph-e}} + \hat{V}. \tag{13.80}$$

In this case, the QW is confined in the z-direction such that the electrons are effectively two-dimensional particles with confinement functions $\xi_\lambda(z)$. Due to this limitation, the carrier momenta are two dimensional, while the phonon and photon momenta are three dimensional. As a result, we have

$$\hat{H}_{\mathrm{free}} = \sum\nolimits_{\lambda,\mathbf{k}} \epsilon^\lambda_{\mathbf{k}}\, a^\dagger_{\lambda,\mathbf{k}} a_{\lambda,\mathbf{k}} + \sum\nolimits_{\mathbf{q}} \hbar\omega_{\mathbf{q}} \left[B^\dagger_{\mathbf{q}} B_{\mathbf{q}} + \frac{1}{2} \right] + \sum\nolimits_{\mathbf{p}} \hbar\Omega_{\mathbf{p}} \left[D^\dagger_{\mathbf{p}} D_{\mathbf{p}} + \frac{1}{2} \right], \tag{13.81}$$

$$\begin{aligned}\hat{H}_{\mathrm{em-e}} = &\sum\nolimits_{\lambda,\mathbf{k}} \left[-\sum\nolimits_{\mathbf{q}_\parallel} j_\lambda(\mathbf{k})\, \hat{A}^{\lambda,\lambda}_{\mathbf{q}_\parallel}\, a^\dagger_{\lambda,\mathbf{k}+\frac{1}{2}\mathbf{q}_\parallel} a_{\lambda,\mathbf{k}-\frac{1}{2}\mathbf{q}_\parallel} + \sum\nolimits_{\mathbf{q}_\parallel,\mathbf{q}'_\parallel} \frac{Q^2}{2m_0} \hat{A}^{(2),\lambda}_{\mathbf{q}_\parallel,-\mathbf{q}'_\parallel}\, a^\dagger_{\lambda,\mathbf{k}-\mathbf{q}'_\parallel} a_{\lambda,\mathbf{k}-\mathbf{q}_\parallel} \right] \\ &- \sum\nolimits_{\mathbf{q}_\parallel,\mathbf{k}} \sum\nolimits_{\lambda\neq\lambda'} \frac{Q\mathbf{p}_{\lambda,\lambda'}}{m_0} \cdot \hat{\mathbf{A}}^{\lambda,\lambda'}_{\mathbf{q}_\parallel}\, a^\dagger_{\lambda,\mathbf{k}} a_{\lambda',\mathbf{k}-\mathbf{q}_\parallel},\end{aligned} \tag{13.82}$$

$$\hat{H}_{\mathrm{ph-e}} = \sum_{\lambda,\mathbf{k},\mathbf{p}_\parallel} \hat{\mathcal{G}}^\lambda_{\mathbf{p}_\parallel}\, a^\dagger_{\lambda,\mathbf{k}} a_{\lambda,\mathbf{k}-\mathbf{p}_\parallel}, \tag{13.83}$$

$$\hat{V} = \frac{1}{2} \sum\nolimits_{\mathbf{k},\mathbf{k}',\mathbf{q}_\parallel\neq 0} \sum\nolimits_{\lambda,\lambda'} V^{\lambda,\lambda'}_{\mathbf{q}_\parallel}\, a^\dagger_{\lambda,\mathbf{k}} a^\dagger_{\lambda',\mathbf{k}'} a_{\lambda',\mathbf{k}'+\mathbf{q}_\parallel} a_{\lambda,\mathbf{k}-\mathbf{q}_\parallel}. \tag{13.84}$$

Compared with the three-dimensional case, the current-matrix element is modified to

$$j_\lambda(\mathbf{k}) \equiv Q \frac{\hbar\mathbf{k}}{m_\lambda} \cdot \mathbf{e}_P, \tag{13.85}$$

where $\mathbf{e}_P$ is the polarization direction of the field. The confinement function $\xi_\lambda(z)$ alters the phonon interaction such that

$$G^\lambda_{\mathbf{p}} = F^\lambda \sqrt{\frac{\hbar|\mathbf{p}|}{2c_{\mathrm{LA}}\rho\mathcal{L}^3}} \int dZ e^{ip_z Z}\, |\xi_\lambda(Z)|^2. \tag{13.86}$$

In the same way, the quantum confinement leads to a modification of the Coulomb interaction

$$V^{\lambda,\lambda'}_{\mathbf{q}_\parallel} \equiv \frac{e^2}{2\epsilon\epsilon_0\mathcal{L}^2} \frac{1}{|\mathbf{q}_\parallel|} \int\!\!\int dZ\, dZ' |\xi_\lambda(Z)|^2 |\xi_{\lambda'}(Z')|^2\, e^{-|\mathbf{q}_\parallel(Z-Z')|}. \tag{13.87}$$

The terms containing the vector potential are

$$\hat{\mathbf{A}}^{\lambda,\lambda'}_{\mathbf{q}_\parallel} = \int dz\, \xi^\star_\lambda(z) \left[\sum_{q_z} \frac{\mathcal{E}_\mathbf{q}}{\omega_\mathbf{q}} \left(\bar{\mathbf{u}}_\mathbf{q}(z) B_{\mathbf{q}_\parallel,q_z} + \bar{\mathbf{u}}^\star_\mathbf{q}(z) B^\dagger_{-\mathbf{q}_\parallel,q_z}\right)\right] \xi_{\lambda'}(z)$$

$$= \hat{A}^{\lambda,\lambda'}_{\mathbf{q}_\parallel} \mathbf{e}_P, \tag{13.88}$$

$$\hat{A}^{(2),\lambda}_{\mathbf{q}_\parallel,\mathbf{q}'_\parallel} = \int dz\, |\xi_\lambda(z)|^2 \sum_{q_z,q'_z} \frac{\mathcal{E}_\mathbf{q}, \mathcal{E}_{\mathbf{q}'}}{\omega_\mathbf{q}\omega_{\mathbf{q}'}} \left(\bar{\mathbf{u}}_\mathbf{q}(z) B_{\mathbf{q}_\parallel,q_z} + \bar{\mathbf{u}}^\star_\mathbf{q}(z) B^\dagger_{-\mathbf{q}_\parallel,q_z}\right)$$

$$\cdot \left(\bar{\mathbf{u}}_{\mathbf{q}'}(z) B_{\mathbf{q}'_\parallel,q'_z} + \bar{\mathbf{u}}^\star_\mathbf{q}(z) B^\dagger_{-\mathbf{q}'_\parallel,q'_z}\right). \tag{13.89}$$

The collective phonon field is

$$\hat{\mathcal{G}}^\lambda_{\mathbf{p}_\parallel} \equiv \sum_{p_z} G^\lambda_\mathbf{p} \left(D_{\mathbf{p}_\parallel,p_z} + D^\dagger_{-\mathbf{p}_\parallel,p_z}\right). \tag{13.90}$$

13.5.2 Quantum-wire system Hamiltonian

From the Hamiltonian of the two-dimensional QWs, it is straightforward to generate the corresponding Hamiltonian for quantum wires. As major modifications, the confinement is now two dimensional such that $\xi_\lambda(x, y) = \xi_\lambda(\mathbf{r}_\parallel)$, while the carrier momenta are one dimensional. In this case, the system Hamiltonian is

$$\hat{H}^{\mathrm{QWI}}_{\mathrm{tot}} = \hat{H}_{\mathrm{free}} + \hat{H}_{\mathrm{em-e}} + \hat{H}_{\mathrm{ph-e}} + \hat{V}, \tag{13.91}$$

and the free-particle contributions are

$$\hat{H}_{\mathrm{free}} = \sum_{\lambda,k_z} \epsilon^\lambda_{k_z} a^\dagger_{\lambda,k_z} a_{\lambda,k_z} + \sum_\mathbf{q} \hbar\omega_\mathbf{q} \left[B^\dagger_\mathbf{q} B_\mathbf{q} + \frac{1}{2}\right] + \sum_\mathbf{p} \hbar\Omega_\mathbf{p} \left(D^\dagger_\mathbf{p} D_\mathbf{p} + \frac{1}{2}\right) \tag{13.92}$$

where both photons and phonons are three dimensional. The interaction contributions follow from

$$\hat{H}_{\mathrm{em-e}} = \sum_{\lambda,k_z} \left[-\sum_{q_z} j_\lambda(k_z)\, \hat{A}^{\lambda,\lambda}_{q_z}\, a^\dagger_{\lambda,k_z+\frac{q_z}{2}} a_{\lambda,k_z-\frac{q_z}{2}} + \sum_{q_z,q'_z} \frac{Q^2}{2m_0} \hat{A}^{(2),\lambda}_{q_z,-q'_z}\, a^\dagger_{\lambda,k_z-q'_z} a_{\lambda,k_z-q_z}\right]$$

$$- \sum_{q_z,k_z} \sum_{\lambda\neq\lambda'} \frac{Q\mathbf{p}_{\lambda,\lambda'}}{m_0} \cdot \hat{\mathbf{A}}^{\lambda,\lambda'}_{q_z}\, a^\dagger_{\lambda,k} a_{\lambda',k_z+q_z}, \tag{13.93}$$

$$\hat{H}_{\mathrm{ph-e}} = \sum_{\lambda,k_z,p_z} \hat{\mathcal{G}}^\lambda_{p_z} a^\dagger_{\lambda,k_z} a_{\lambda,k_z-p_z}, \tag{13.94}$$

$$\hat{V} = \frac{1}{2} \sum_{k_z,\lambda} \sum_{k'_z,\lambda'} \sum_{\mathbf{q}_z\neq 0} V^{\lambda,\lambda'}_{\mathbf{q}_z}\, a^\dagger_{\lambda,k_z} a^\dagger_{\lambda',k'_z} a_{\lambda',k'_z+q_z} a_{\lambda,k_z-q_z}. \tag{13.95}$$

For quantum wires, the current-matrix element

$$j_\lambda(k_z) \equiv Q\frac{\hbar k_z}{m_\lambda} \tag{13.96}$$

contains only the one-dimensional momentum k_z. The two-dimensional confinement function $\xi_\lambda(\mathbf{r}_\|)$ enters into the matrix elements for the phonon interaction

$$G^\lambda_{\mathbf{p}} = F^\lambda\sqrt{\frac{\hbar|\mathbf{p}|}{2c_{\mathrm{LA}}\rho\mathcal{L}^3}}\int d^2r_\|\; e^{i\mathbf{p}_\|\cdot\mathbf{r}_\|}\; |\xi_\lambda(\mathbf{r}_\|)|^2 \tag{13.97}$$

and for the Coulomb interaction

$$V^{\lambda,\lambda'}_{q_z} \equiv \frac{e^2}{4\pi\epsilon\epsilon_0\mathcal{L}}\int d^2r_\| d^2r'_\| dz\; \frac{|\xi_\lambda(\mathbf{r}_\|)|^2\,|\xi_{\lambda'}(\mathbf{r}'_\|)|^2}{\sqrt{\left(\mathbf{r}_\| - \mathbf{r}'_\|\right)^2 + z^2}}\; e^{-iq_z z}. \tag{13.98}$$

This yields a notoriously diverging Coulomb interaction for the strictly one-dimensional case. However, any realistic quantum wire has a finite width $|\xi_\lambda(\mathbf{r}_\|)|^2$, which always yields a converging Coulomb interaction.

The matrix elements of the vector potential are

$$\hat{A}^{\lambda,\lambda'}_{q_z} = \int d^2\mathbf{r}_\|\; \xi^\star_\lambda(\mathbf{r}_\|)\left[\sum\nolimits_{\mathbf{q}_\|}\frac{\mathcal{E}_{\mathbf{q}}}{\omega_{\mathbf{q}}}\left(\bar{\mathbf{u}}_{\mathbf{q}}(\mathbf{r}_\|)B_{\mathbf{q}_\|,q_z} + \bar{\mathbf{u}}^\star_{\mathbf{q}}(\mathbf{r}_\|)B^\dagger_{\mathbf{q}_\|,-q_z}\right)\right]\cdot\mathbf{e}_z\xi_{\lambda'}(\mathbf{r}_\|), \tag{13.99}$$

$$\hat{A}^{(2),\lambda}_{q_z,q'_z} = \int d^2\mathbf{r}_\|\; |\xi_\lambda(\mathbf{r}_\|)|^2\sum_{\mathbf{q}_\|\mathbf{q}'_\|}\frac{\mathcal{E}_{\mathbf{q}}\mathcal{E}_{\mathbf{q}'}}{\omega_{\mathbf{q}}\omega_{\mathbf{q}'}}\left(\bar{\mathbf{u}}_{\mathbf{q}}(\mathbf{r}_\|)B_{\mathbf{q}_\|,q_z} + \bar{\mathbf{u}}^\star_{\mathbf{q}}(\mathbf{r}_\|)B^\dagger_{\mathbf{q}_\|,-q_z}\right)$$
$$\cdot\left(\bar{\mathbf{u}}_{\mathbf{q}'}(\mathbf{r}_\|)B_{\mathbf{q}_\|,-q'_z} + \bar{\mathbf{u}}^\star_{\mathbf{q}}(\mathbf{r}_\|)B^\dagger_{\mathbf{q}_\|,-q'_z}\right), \tag{13.100}$$

where the full mode function is given by $\mathbf{u}_{\mathbf{q}}(\mathbf{r}) = \frac{1}{\sqrt{\mathcal{L}}}e^{iq_z z}\bar{\mathbf{u}}_{\mathbf{q}}(\mathbf{r}_\|)$. The collective phonon field is

$$\hat{\mathcal{G}}^\lambda_{p_z} \equiv \sum\nolimits_{\mathbf{p}_\|} G^\lambda_{\mathbf{p}}\left[D_{\mathbf{p}_\|,p_z} + D^\dagger_{\mathbf{p}_\|,-p_z}\right]. \tag{13.101}$$

13.5.3 Quantum-dot system Hamiltonian

For effectively zero-dimensional systems, we have three-dimensional confinement wave functions, $\xi_\lambda(x, y, z) = \xi_\lambda(\mathbf{r})$, where λ contains now both the band and subband indices. For such a quantum-dot system, the quantum properties can be determined by starting from the system Hamiltonian

$$\hat{H}^{\mathrm{QD}}_{\mathrm{tot}} = \hat{H}_{\mathrm{free}} + \hat{H}_{\mathrm{em-e}} + \hat{H}_{\mathrm{ph-e}} + \hat{V}. \tag{13.102}$$

The noninteracting parts are obtained as

$$\hat{H}_{\mathrm{free}} = \sum_\lambda \epsilon^\lambda a^\dagger_\lambda a_\lambda + \sum_{\mathbf{q}}\hbar\omega_{\mathbf{q}}\left[B^\dagger_{\mathbf{q}}B_{\mathbf{q}} + \frac{1}{2}\right] + \sum_{\mathbf{p}}\hbar\Omega_{\mathbf{p}}\left[D^\dagger_{\mathbf{p}}D_{\mathbf{p}} + \frac{1}{2}\right] \tag{13.103}$$

while the interactions follow from

$$\hat{H}_{\mathrm{em-e}} = \sum\nolimits_{\lambda}\left[-\mathbf{j}_{\lambda}\cdot\hat{\mathbf{A}}^{\lambda,\lambda}\; a_{\lambda}^{\dagger}a_{\lambda} + \frac{Q^2}{2m_0}\hat{A}^{(2),\lambda}\; a_{\lambda}^{\dagger}a_{\lambda}\right] - \sum\nolimits_{\lambda\neq\lambda'}\frac{Q\mathbf{p}_{\lambda,\lambda'}}{m_0}\cdot\hat{\mathbf{A}}^{\lambda,\lambda'}\; a_{\lambda}^{\dagger}a_{\lambda'}, \tag{13.104}$$

$$\hat{H}_{\mathrm{ph-e}} = \sum\nolimits_{\lambda}\hat{\mathcal{G}}^{\lambda}a_{\lambda}^{\dagger}a_{\lambda}, \tag{13.105}$$

$$\hat{V} = \frac{1}{2}\sum\nolimits_{\lambda,\lambda',\nu',\nu} V_{\nu',\nu}^{\lambda,\lambda'}\; a_{\lambda}^{\dagger}a_{\lambda'}^{\dagger}a_{\nu'}a_{\nu}. \tag{13.106}$$

As the major generalization of the previous cases, we have included all possible combinations of Coulomb-matrix elements because Auger-type processes are generally important for quantum dots.

For quantum dots, the current-matrix element

$$\mathbf{j}_{\lambda} \equiv Q\int d^3r\; \xi_{\lambda}^{\star}(\mathbf{r})\frac{\hat{\mathbf{p}}}{m_0}\xi_{\lambda}(\mathbf{r}) \tag{13.107}$$

vanishes for symmetric confinement functions. At the same time, the phonon matrix element becomes

$$G_{\mathbf{p}}^{\lambda} = F^{\lambda}\sqrt{\frac{\hbar|\mathbf{p}|}{2c_{\mathrm{LA}}\rho\mathcal{L}^3}}\int d^3r\; e^{i\mathbf{p}\cdot\mathbf{r}}\;|\xi_{\lambda}(\mathbf{r})|^2, \tag{13.108}$$

and the Coulomb interaction in a quantum dot has the matrix element

$$V_{\nu,\nu'}^{\lambda,\lambda'} \equiv \frac{e^2}{4\pi\epsilon\epsilon_0\mathcal{L}}\int d^3r\, d^3r'\;\frac{\xi_{\lambda}^{\star}(\mathbf{r})\xi_{\lambda'}^{\star}(\mathbf{r}')\xi_{\nu'}(\mathbf{r}')\xi_{\nu}(\mathbf{r})}{|\mathbf{r}-\mathbf{r}'|}. \tag{13.109}$$

This obviously diverges for the strict 0D limit, i.e. $\xi_{\lambda}(\mathbf{r}) \to \delta(\mathbf{r})$. However, any realistic quantum dot has a finite confinement function that completely regularizes the Coulomb matrix element. In other words, $V_{\nu,\nu'}^{\lambda,\lambda'}$ is always finite for quantum dots.

Similarly, the matrix elements of the vector potential are

$$\hat{\mathbf{A}}^{\lambda,\lambda'} = \int d^3r\; \xi_{\lambda}^{\star}(\mathbf{r})\left[\sum\nolimits_{\mathbf{q}}\frac{\mathcal{E}_{\mathbf{q}}}{\omega_{\mathbf{q}}}\left(\mathbf{u}_{\mathbf{q}}(\mathbf{r})B_{\mathbf{q}} + \mathbf{u}_{\mathbf{q}}^{\star}(\mathbf{r})B_{\mathbf{q}}^{\dagger}\right)\right]\xi_{\lambda'}(\mathbf{r}) \tag{13.110}$$

and

$$\hat{A}^{(2),\lambda} = \int d^3r\; |\xi_{\lambda}(\mathbf{r})|^2\sum\nolimits_{\mathbf{q},\mathbf{q}'}\frac{\mathcal{E}_{\mathbf{q}},\mathcal{E}_{\mathbf{q}'}}{\omega_{\mathbf{q}}\omega_{\mathbf{q}'}}\left(\mathbf{u}_{\mathbf{q}}(\mathbf{r})B_{\mathbf{q}} + \mathbf{u}_{\mathbf{q}}^{\star}(\mathbf{r})B_{\mathbf{q}}^{\dagger}\right)$$
$$\cdot\left(\mathbf{u}_{\mathbf{q}'}(\mathbf{r})B_{\mathbf{q}'} + \mathbf{u}_{\mathbf{q}'}^{\star}(\mathbf{r})B_{\mathbf{q}'}^{\dagger}\right) \tag{13.111}$$

for the linear and quadratic contributions, respectively. Both of them are constructed using the full three-dimensional mode functions $\mathbf{u}_{\mathbf{q}}(\mathbf{r})$. The phonon field is determined by

$$\hat{\mathcal{G}}^{\lambda} \equiv \sum\nolimits_{\mathbf{p}} G_{\mathbf{p}}^{\lambda}\left[D_{\mathbf{p}} + D_{\mathbf{p}}^{\dagger}\right]. \tag{13.112}$$

The explicit derivation of the quantum-dot system Hamiltonian is studied in Exercise 13.11.

Exercises

13.1 (**a**) Use a generic field operator $\hat{\Psi}(\mathbf{r}) = \sum_{\lambda,\mathbf{k}} a_{\lambda,\mathbf{k}}\,\phi_{\lambda,\mathbf{k}}(\mathbf{r})$ and show that any single-particle operator $\hat{O}_1 \equiv \int d^3r\, \hat{\Psi}^\dagger(\mathbf{r}) O_1(\mathbf{r}) \hat{\Psi}(\mathbf{r})$ is defined by Eqs. (13.7)–(13.8). (**b**) Perform $\sqrt{S}\,\hat{p}\,\phi_{\lambda',k'_\parallel}(\mathbf{r})$ explicitly for a quantum well described in the envelope-function approximation (12.25) and verify the result (13.14). (**c**) Show that the light–matter contribution $\hat{H}_{A\cdot A}$ reduces into

$$\hat{H}_{A\cdot A} \equiv \int d^3r\, \hat{\Psi}^\dagger(\mathbf{r}) \frac{Q^2}{2m_0} \hat{A}^2(\mathbf{r}) \hat{\Psi}(\mathbf{r}) = \sum_{\lambda,\mathbf{k}} \sum_{\lambda',\mathbf{k}'} M^{\lambda,\mathbf{k}}_{\lambda',\mathbf{k}'}\Big|_{A\cdot A}\, a^\dagger_{\lambda,\mathbf{k}} a_{\lambda'\mathbf{k}'}$$

defined by the matrix element (13.18).

13.2 (**a**) Verify that the definition (13.19) and Eqs. (13.16), (13.17), and (13.18) produce the identifications (13.27)–(13.29). (**b**) Show that the result (13.26) then produces Eqs. (13.30)–(13.32).

13.3 Use the $\mathbf{k}\cdot\mathbf{p}$ theory wave function (12.13) to evaluate the matrix element: (**a**) $\langle \lambda, \mathbf{k}|\lambda', \mathbf{k}+\mathbf{q}\rangle$, (**b**) $\langle \lambda, \mathbf{k}|\hat{\mathbf{p}}|\lambda', \mathbf{k}+\mathbf{q}\rangle$. Which corrections follow from $\mathbf{q}_\parallel \neq 0$?

13.4 Compute $\langle \lambda, \mathbf{k}|\hat{\mathbf{p}}|\lambda', \mathbf{k}\rangle = \mathbf{p}_{\lambda\lambda'}(\mathbf{k})$ using the $\mathbf{k}\cdot\mathbf{p}$ theory and apply the isotropic approximation (12.21) and the two-band approximation. Show that the corresponding $\mathbf{p}_{\lambda\lambda'}(\mathbf{k})$ is constant for $\lambda \neq \lambda'$.

13.5 Construct the results (13.40)–(13.42) by starting from Eqs. (13.30)–(13.32) and the relations (13.35) and (13.38).

13.6 (**a**) Construct the result (13.51) by applying the separation of length scales in Eq. (13.50). (**b**) Insert the Fourier transform (13.52) into Eq. (13.51) and show that

$$M^{\lambda\mathbf{k}_\parallel}_{\lambda'\mathbf{k}_\parallel} = \sum_{\mathbf{p}} \delta_{\mathbf{k}'_\parallel,\mathbf{k}_\parallel-\mathbf{p}_\parallel}\, g_{p_z}\, F^{\mathbf{k}_\parallel,\lambda}_{\mathbf{k}'_\parallel,\lambda'} \sqrt{\frac{\hbar}{2NM\Omega_{\mathbf{p}}}}\, \mathbf{p}\cdot\mathbf{e}_{\mathbf{p}} \left(D_{\mathbf{p}} + D^\dagger_{-\mathbf{p}}\right).$$

Apply the approximation (13.56) and verify the results (13.58)–(13.59).

13.7 (**a**) Derive the Coulomb interaction (13.62) by applying the separation of length scales. (**b**) Confirm the results (13.65) and (13.68) by implementing (13.63)–(13.64) and (13.66).

13.8 (**a**) Verify Eq. (13.64) using Cauchy's integral theorem $\oint dz \frac{f(z)}{z-z_0} = 2\pi i\, f(z_0)$. (**b**) Show that the result (13.64) converts Eq. (13.62) into (13.65).

13.9 Assume a QW system with Gaussian confinement $|\xi_\lambda(z)|^2 = \frac{1}{\sqrt{\pi\,\Delta z^2}}\, e^{-z^2/\Delta z^2}$. (**a**) Compute the Coulomb matrix element $V^{\lambda\lambda}_{\mathbf{q}_\parallel}$ by starting from Eq. (13.67). (**b**) Fourier transform the strictly two-dimensional Coulomb interaction $V^{2D}(x,y) = \frac{e^2}{4\pi\varepsilon\varepsilon_0\sqrt{x^2+y^2}}$ and show that it produces $V^{2D}_{\mathbf{q}_\parallel} = \frac{e^2}{2\varepsilon\varepsilon_0 \mathcal{L}^2}\frac{1}{|\mathbf{q}_\parallel|}$. Show that also $V^{\lambda}_{\mathbf{q}_{\|}}$ yields this result in the limit $\Delta z \to 0$. How does a nonvanishing Δz alter $V^{\lambda\lambda}_{\mathbf{q}_{\|}}$?

13.10 The total Coulomb interaction between ions and electrons follows from $\hat{V} = \frac{1}{2}\sum_{i \neq j}\left[V(\mathbf{r}_i - \mathbf{r}_j) + V(\mathbf{R}_i - \mathbf{R}_j)\right] - \sum_{i,j} V(\mathbf{r}_i - \mathbf{R}_j)$ where $\mathbf{r}_i$ denotes the electron position and $\mathbf{R}_j$ defines the ion position in first quantization. **(a)** Introduce the field operators $\hat{\Psi}(\mathbf{r}) = \sum_{\mathbf{k}} a_{\mathbf{k}} \frac{1}{\mathcal{L}^{3/2}} e^{i\mathbf{k}\cdot\mathbf{r}}$ for electrons and $\hat{\Psi}_I(\mathbf{R}) = \sum_{\mathbf{k}} A_{\mathbf{k}} \frac{1}{\mathcal{L}^{3/2}} e^{i\mathbf{k}\cdot\mathbf{R}}$ for ions. Show that

$$\hat{V} = \frac{1}{2}\sum_{\mathbf{k},\mathbf{k}',\mathbf{q}} V_{\mathbf{q}}\left[a^{\dagger}_{\mathbf{k}} a^{\dagger}_{\mathbf{k}'} a_{\mathbf{k}'+\mathbf{q}} a_{\mathbf{k}-\mathbf{q}} - 2a^{\dagger}_{\mathbf{k}} A^{\dagger}_{\mathbf{k}'} A_{\mathbf{k}'+\mathbf{q}} a_{\mathbf{k}-\mathbf{q}} + A^{\dagger}_{\mathbf{k}} A^{\dagger}_{\mathbf{k}'} A_{\mathbf{k}'+\mathbf{q}} A_{\mathbf{k}-\mathbf{q}}\right].$$

(b) Show that the Coulomb interaction becomes

$$\hat{V} = \frac{1}{2}\sum_{\mathbf{k},\mathbf{k}'}\sum_{\mathbf{q}\neq 0} V_{\mathbf{q}}\left[a^{\dagger}_{\mathbf{k}} a^{\dagger}_{\mathbf{k}'} a_{\mathbf{k}'+\mathbf{q}} a_{\mathbf{k}-\mathbf{q}} - 2a^{\dagger}_{\mathbf{k}} A^{\dagger}_{\mathbf{k}'} A_{\mathbf{k}'+\mathbf{q}} a_{\mathbf{k}-\mathbf{q}} + A^{\dagger}_{\mathbf{k}} A^{\dagger}_{\mathbf{k}'} A_{\mathbf{k}'+\mathbf{q}} A_{\mathbf{k}-\mathbf{q}}\right]$$
$$+\frac{1}{2}V_0\left[\hat{N}_a^2 - 2\hat{N}_a\hat{N}_A + \hat{N}_A^2 - \hat{N}_a - \hat{N}_A\right],$$

where $\hat{N}_a = \sum_{\mathbf{k}} a^{\dagger}_{\mathbf{k}} a_{\mathbf{k}}$ and $\hat{N}_A = \sum_{\mathbf{k}} A^{\dagger}_{\mathbf{k}} A_{\mathbf{k}}$. Show that the second term nearly vanishes for systems with charge neutrality. Argue why it can be eliminated even if V_0 diverges, showing that one should indeed omit the $q = 0$ term from Eq. (13.68). **Hint:** *Consider the structure of the diverging term and its connection to the zero-level of the single-particle energies.*

13.11 Derive the system Hamiltonian for a quantum dot using the envelope function approximation $\phi_{\lambda}(\mathbf{r}) = \xi_{\lambda}(\mathbf{r})\, u_{\lambda}(\mathbf{r})$. **Hint:** *The envelope $\xi_{\lambda}(\mathbf{r})$ can be treated as mesoscopic while the Bloch function $u_{\lambda}(\mathbf{r})$ is microscopic.*

Further reading

The properties of the interaction matrix elements are analyzed in detail, e.g., in:

C. Kittel (1974). *Quantum Theory of Solids*, New York, Wiley & Sons.
M. Lax (1974). *Symmetry Principles in Solid State and Molecular Physics*, New York, Wiley & Sons.

The Coulomb interaction potentials of quantum-well and quantum-dot systems are derived e.g. in:

Y. Z. Hu, M. Lindberg, and S. W. Koch (1990). Theory of optically excited intrinsic semiconductor quantum dots, *Phys. Rev. B* **42**, 1713.
A. Wojs, P. Hawrylak, S. Fafard, and L. Jacak (1996). Electronic structure and magneto-optics of self-assembled quantum dots, *Phys. Rev. B* **54**, 5604.

F. Jahnke, M. Kira, and S. W. Koch (1997). Linear and nonlinear optical properties of excitons in semiconductor quantum wells and microcavities, *Z. Physik B Condensed Matter* **104**, 559.
S. M. Reimann and M. Manninen (2002). Electronic structure of quantum dots, *Rev. Mod. Phys.* **74**, 1283.
T. R. Nielsen, P. Gartner, and F. Jahnke (2004). Many-body theory of carrier capture and relaxation in semiconductor quantum-dot lasers, *Phys. Rev. B* **69**, 235314.

The Coulomb interaction in quantum wires is studied, e.g., in:

R. Loudon (1959). One-dimensional hydrogen atom, *Am. J. Phys.* **27**, 649.
L. Bányai, I. Galbraith, C. Ell, and H. Haug (1987). Excitons and biexcitons in semiconductor quantum wires, *Phys. Rev. B* **36**, 6099.
W. Hoyer, M. Kira, and S. W. Koch (2003). Influence of Coulomb and phonon interaction on the exciton formation dynamics in semiconductor heterostructures, *Phys. Rev. B* **67**, 155113.

14

Generic quantum dynamics

As a broad definition, *quantum dynamics* determines how a quantum system evolves from one state to another. The overall importance of quantum dynamics becomes obvious once we realize that measurements of quantum properties rely almost exclusively on time-dependent changes in the system under investigation. This includes a broad variety of effects ranging from propagation-related phenomena like currents, diffusion, and tunneling to transition and interference effects like absorption, photoluminescence, and entanglement between different excitations. The corresponding processes typically occur on a microscopic scale, which also implies strong quantum aspects – compare the discussion in Section 3.1.1.

The Hamiltonian, derived in Chapter 13, provides a general starting point for our quantum-dynamics investigations of the optical, quantum-optical, and many-body effects in semiconductors where the elementary excitations are electrons in different bands, photons, and lattice vibrations. We will next derive the dynamics of fundamental operators for the semiconductor quantum optics. Even though our goal is to describe semiconductors, also several other systems, like solids or multiatom systems, have a structurally similar Hamiltonian. Thus, the following investigations, problems, and solutions represent rather general aspects in many-body systems.

14.1 Dynamics of elementary operators

After the second-quantization form of the Hamiltonian is known, we can directly compute the quantum dynamics of any operator using the Heisenberg equation of motion:

$$i\hbar \frac{\partial}{\partial t} \hat{O} = \left[\hat{O}, \hat{H} \right]_{-} . \tag{14.1}$$

Since we always perform the explicit derivations in this book for quantum-well (QW) structures, we start from the total Hamiltonian $\hat{H} = \hat{H}_{\text{free}} + \hat{H}_{\text{em-e}} + \hat{H}_{\text{ph-e}} + \hat{V}$ that contains the free carrier–photons–phonons in $\hat{H}_{\text{free}}$. The actual interactions are determined by the electron–photon coupling $\hat{H}_{\text{em-e}}$, the electron–phonon coupling $\hat{H}_{\text{ph-e}}$, and the Coulomb interaction $\hat{V}$ among the charge carriers. The corresponding explicit forms are given by Eqs. (13.81)–(13.90) in Section 13.5.

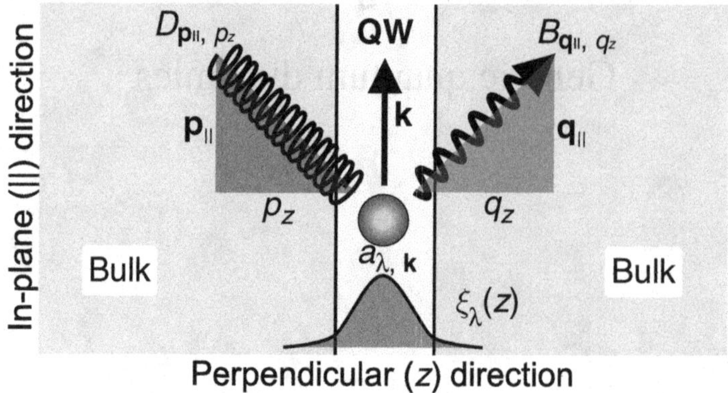

Figure 14.1 Elementary operators and geometry in planar semiconductor structures. The momentum of a carrier operator $a_{\lambda,\mathbf{k}}$ (sphere), a photon operator $B_{\mathbf{q}_\parallel,q_z}$ (wave arrow), and a phonon operator $D_{\mathbf{p}_\parallel,p_z}$ is divided into its in-plane ($\parallel$) and perpendicular (z) component; here $\mathbf{k} = \mathbf{k}_\parallel$. The shaded curve indicates one possible confinement wave function for quantum-well (QW) electrons.

Figure 14.1 schematically presents the elementary operators. The sphere denotes a Fermionic carrier operator $a_{\lambda,\mathbf{k}}$ related to an electron with momentum $\mathbf{k}$ in the band λ. Like before, λ implicitly includes the spin and confinement-level indices. The shaded curve presents one possible QW-confinement function $\xi_\lambda(z)$. Due to the confinement, the carrier momentum $\mathbf{k}$ is two dimensional. The Bosonic $B_\mathbf{q} \equiv B_{\mathbf{q}_\parallel,q_z}$ (wave arrow) and $D_{\mathbf{p}_\parallel,p_z} \equiv D_\mathbf{p}$ (spring symbol) operators define photons and phonons, respectively, that are described by three-dimensional momenta and an implicit index for the polarization direction. In connection with QWs, it is beneficial to separate the three-dimensional momenta into in-plane ($\mathbf{q}_\parallel$, $\mathbf{p}_\parallel$) and perpendicular (q_z, p_z) components.

14.1.1 Evaluation strategy

Once the operator dynamics for $\hat{O}$ is known from the evaluation of Eq. (14.1), we can deduce the dynamics for $\hat{O}^\dagger$ without any further commutations because we may use the relation

$$\begin{aligned} i\hbar\frac{\partial}{\partial t}\hat{O}^\dagger &= \left[\hat{O}^\dagger, \hat{H}\right]_- = \hat{O}^\dagger\hat{H} - \hat{H}\hat{O}^\dagger \\ &= \hat{O}^\dagger\hat{H}^\dagger - \hat{H}^\dagger\hat{O}^\dagger = \left(\hat{H}\hat{O} - \hat{O}\hat{H}\right)^\dagger = -\left[\hat{O}, \hat{H}\right]_-^\dagger, \end{aligned} \tag{14.2}$$

where the second line follows when we use $(\hat{A}\hat{B})^\dagger = \hat{B}^\dagger\hat{A}^\dagger$ for any operator pair and take into account that the Hamiltonian is always Hermitian, i.e., $\hat{H}^\dagger = \hat{H}$. This observation reduces the number of explicit commutations by a factor of two.

In general, we can represent any $\hat{O}$ in terms of $a_{\lambda,\mathbf{k}}$, $a^{\dagger}_{\lambda,\mathbf{k}}$, $B_{\mathbf{q}_{\|},q_z}$, $B^{\dagger}_{\mathbf{q}_{\|},q_z}$, $D_{\mathbf{p}_{\|},p_z}$, and $D^{\dagger}_{\mathbf{p}_{\|},p_z}$ operators. Consequently, our first task is to construct the operator dynamics for these elementary operators. From a technical point of view, we only need to derive the dynamics for the annihilation operators because we may use Eq. (14.2) to generate the corresponding creation-operator dynamics. Once the dynamics of the elementary operators is known, we may use the strategy

$$i\hbar\frac{\partial}{\partial t}\hat{O}_\alpha\,\hat{O}_\beta = \left[i\hbar\frac{\partial}{\partial t}\hat{O}_\alpha\right]\hat{O}_\beta + \hat{O}_\alpha\left[i\hbar\frac{\partial}{\partial t}\hat{O}_\beta\right] \tag{14.3}$$

multiple times, independent of which elementary operators appear in $\hat{O}$.

In the evaluations of the commutators, we frequently use the linear additivity

$$\left[\hat{A}\,,\sum_j\hat{B}_j\right]_- = \sum_j\left[\hat{A},\hat{B}_j\right]_- . \tag{14.4}$$

Another useful relation expresses the fact that products inside commutators can generally be evaluated via

$$\left[\hat{A},\hat{B}\hat{C}\right]_- = \left[\hat{A},\hat{B}\right]_-\hat{C} + \hat{B}\left[\hat{A},\hat{C}\right]_- , \tag{14.5}$$

for all operators $\hat{A}$, $\hat{B}$, and $\hat{C}$.

Since the total Hamiltonian consists of a linear combination of $\hat{H}_{\text{free}}$, $\hat{H}_{\text{em-e}}$, $\hat{H}_{\text{ph-e}}$, and $\hat{V}$, we may use the linear additivity (14.4) for a term-by-term evaluation of the commutators of the annihilation operator:

$$i\hbar\frac{\partial}{\partial t}\hat{O} = \left[\hat{O},\,\hat{H}_{\text{free}}\right]_- + \left[\hat{O},\,\hat{H}_{\text{em-e}}\right]_- + \left[\hat{O},\,\hat{H}_{\text{ph-e}}\right]_- + \left[\hat{O},\,\hat{V}\right]_- . \tag{14.6}$$

We notice immediately that the photon operators commute with $\hat{H}_{\text{ph-e}} + \hat{V}$ and that the phonon operators commute with $\hat{H}_{\text{em-e}} + \hat{V}$ because these parts of the Hamiltonian do not contain the corresponding Boson operators. Thus, we only need to evaluate the commutators

$$\begin{aligned}
C^{(1)} &\equiv \left[B_{\mathbf{q}},\,\hat{H}_{\text{free}} + \hat{H}_{\text{em-e}}\right]_- ,\\
C^{(2)} &\equiv \left[D_{\mathbf{p}},\,\hat{H}_{\text{free}} + \hat{H}_{\text{ph-e}}\right]_- ,\\
C^{(3)} &\equiv \left[a_{\mathbf{k}},\,\hat{H}_{\text{free}} + \hat{V} + \hat{H}_{\text{em-e}} + \hat{H}_{\text{ph-e}}\right]_-
\end{aligned} \tag{14.7}$$

to determine the elementary operator dynamics for semiconductors.

14.1.2 Quantum dynamics of free quasiparticles

We start the evaluation of the commutators (14.7) by including first only the noninteracting parts, $\hat{H}_{\text{free}}$, of the Hamiltonian. For the photon operator, we get

$$\begin{aligned}
\left[B_{\mathbf{q}}, \hat{H}_{\text{free}}\right]_- &= \sum_{\mathbf{q}'} \left[B_{\mathbf{q}}, \hbar\omega_{\mathbf{q}'}\left(B^\dagger_{\mathbf{q}'}B_{\mathbf{q}'} + \frac{1}{2}\right)\right]_- = \sum_{\mathbf{q}'} \hbar\omega_{\mathbf{q}'}\left[B_{\mathbf{q}}, B^\dagger_{\mathbf{q}'}B_{\mathbf{q}'}\right]_- \\
&= \sum_{\mathbf{q}'} \hbar\omega_{\mathbf{q}'}\left(\left[B_{\mathbf{q}}, B^\dagger_{\mathbf{q}'}\right]_- B_{\mathbf{q}'} + B^\dagger_{\mathbf{q}'}\left[B_{\mathbf{q}}, B_{\mathbf{q}'}\right]_-\right) \\
&= \sum_{\mathbf{q}'} \hbar\omega_{\mathbf{q}'}\left(\delta_{\mathbf{q},\mathbf{q}'}B_{\mathbf{q}'} + 0\right) = \hbar\omega_{\mathbf{q}}B_{\mathbf{q}},
\end{aligned} \tag{14.8}$$

where we have applied the linear additivity (14.4) in the first line and the relation (14.5) in the second line. The third line is then obtained by using the Boson commutation relations (10.31)–(10.32).

Following similar steps, we find for the phonon destruction operator

$$\left[D_{\mathbf{p}}, \hat{H}_{\text{free}}\right]_- = \sum_{\mathbf{p}'} \hbar\Omega_{\mathbf{p}'}\left[D_{\mathbf{p}}, D^\dagger_{\mathbf{p}'}D_{\mathbf{p}'}\right]_- = \hbar\Omega_{\mathbf{p}}D_{\mathbf{p}}, \tag{14.9}$$

and for the carrier destruction operator

$$\begin{aligned}
\left[a_{\lambda,\mathbf{k}}, \hat{H}_{\text{free}}\right]_- &= \sum_{\lambda',\mathbf{k}'} \left[a_{\lambda,\mathbf{k}}, \epsilon^{\lambda'}_{\mathbf{k}'} a^\dagger_{\lambda',\mathbf{k}'}a_{\lambda',\mathbf{k}'}\right]_- \\
&= \sum_{\lambda',\mathbf{k}'} \epsilon^{\lambda'}_{\mathbf{k}'}\left[a_{\lambda,\mathbf{k}}, a^\dagger_{\lambda',\mathbf{k}'}a_{\lambda',\mathbf{k}'}\right]_-,
\end{aligned} \tag{14.10}$$

where again the linear additivity (14.4) has been used. Since a and $a^\dagger$ obey the anticommutation relation (11.44), we do not apply the product relation (14.5), but instead follow the explicit procedure

$$\begin{aligned}
\left[a_{\lambda,\mathbf{k}}, a^\dagger_{\lambda',\mathbf{k}'}a_{\lambda',\mathbf{k}'}\right]_- &= a_{\lambda,\mathbf{k}}a^\dagger_{\lambda',\mathbf{k}'}a_{\lambda',\mathbf{k}'} - a^\dagger_{\lambda',\mathbf{k}'}a_{\lambda',\mathbf{k}'}a_{\lambda,\mathbf{k}} \\
&= \left(\delta_{\mathbf{k},\mathbf{k}'}\delta_{\lambda,\lambda'} - a^\dagger_{\lambda',\mathbf{k}'}a_{\lambda,\mathbf{k}}\right)a_{\lambda',\mathbf{k}'} - a^\dagger_{\lambda',\mathbf{k}'}a_{\lambda',\mathbf{k}'}a_{\lambda,\mathbf{k}} \\
&= \delta_{\mathbf{k},\mathbf{k}'}\delta_{\lambda,\lambda'}a_{\lambda,\mathbf{k}} - a^\dagger_{\lambda',\mathbf{k}'}\left(a_{\lambda,\mathbf{k}}a_{\lambda',\mathbf{k}'} + a_{\lambda',\mathbf{k}'}a_{\lambda,\mathbf{k}}\right) \\
&= \delta_{\mathbf{k},\mathbf{k}'}\delta_{\lambda,\lambda'}a_{\lambda,\mathbf{k}} - a^\dagger_{\lambda',\mathbf{k}'}\left[a_{\lambda,\mathbf{k}}, a_{\lambda',\mathbf{k}'}\right]_+ \\
&= \delta_{\mathbf{k},\mathbf{k}'}\delta_{\lambda,\lambda'}a_{\lambda,\mathbf{k}}.
\end{aligned} \tag{14.11}$$

Inserting this result into Eq. (14.10), we obtain

$$\left[a_{\lambda,\mathbf{k}}, \hat{H}_{\text{free}}\right]_- = \epsilon^{\lambda}_{\mathbf{k}}\, a_{\lambda,\mathbf{k}}, \tag{14.12}$$

that formally resembles the results (14.8) and (14.9).

At this point, we can now summarize the dynamics of the noninteracting carrier–photon–phonon system due to the free Hamiltonian $\hat{H}_{\text{free}} = \hat{T} + \hat{H}_{\text{em}} + \hat{H}_{\text{ph}}$. By using the definition

$$i\hbar\frac{\partial}{\partial t}\hat{O}|_{\text{free}} \equiv \left[\hat{O}, \hat{H}_{\text{free}}\right]_- , \tag{14.13}$$

together with the relation (14.2) and the results (14.8)–(14.9) and (14.12), we obtain

$$i\hbar\frac{\partial}{\partial t}a_{\lambda,\mathbf{k}}\Big|_{\text{free}} = \epsilon^{\lambda}_{\mathbf{k}}a_{\lambda,\mathbf{k}}, \qquad i\hbar\frac{\partial}{\partial t}a^{\dagger}_{\lambda,\mathbf{k}}\Big|_{\text{free}} = -\epsilon^{\lambda}_{\mathbf{k}}a^{\dagger}_{\lambda,\mathbf{k}}, \tag{14.14}$$

$$i\hbar\frac{\partial}{\partial t}B_{\mathbf{q}}\Big|_{\text{free}} = \hbar\omega_{\mathbf{q}}B_{\mathbf{q}}, \qquad i\hbar\frac{\partial}{\partial t}B^{\dagger}_{\mathbf{q}}\Big|_{\text{free}} = -\hbar\omega_{\mathbf{q}}B^{\dagger}_{\mathbf{q}}, \tag{14.15}$$

$$i\hbar\frac{\partial}{\partial t}D_{\mathbf{p}}\Big|_{\text{free}} = \hbar\Omega_{\mathbf{p}}D_{\mathbf{p}}, \qquad i\hbar\frac{\partial}{\partial t}D^{\dagger}_{\mathbf{p}}\Big|_{\text{free}} = -\hbar\Omega_{\mathbf{p}}D^{\dagger}_{\mathbf{p}}. \tag{14.16}$$

These results show that the noninteracting parts of the system Hamiltonian lead to a very simple dynamics which can be solved explicitly. In particular, the number of equations is equal to the dimension of $(\lambda, \mathbf{k})$, $\mathbf{q}$, or $\mathbf{p}$. Thus, we can evaluate the many-body Eqs. (14.14)–(14.16) with an effort equivalent to that for the single-particle problem with a matching dimensionality. This explicitly shows why the equation-of-motion approach requires much less numerical effort in comparison with any direct wave-function or density-matrix calculation when many-body systems are considered. Clearly, we have to expect considerable complications as soon as the interactions are fully included.

14.1.3 Photon-operator dynamics

In order to deal with the interactions, we need to evaluate $[\hat{O}, \hat{H}_{\text{int}}]_-$ related to the dynamics originating from $\hat{H}_{\text{int}} = \hat{H}_{\text{em-e}} + \hat{H}_{\text{ph-e}} + \hat{V}$. The derivation procedure is very similar to that in Eqs. (14.14)–(14.16) such that we show the calculations explicitly only for the photon dynamics. By using the previously derived photon–carrier interaction (13.82), we obtain

$$\begin{aligned}\left[B_{\mathbf{q}}, \hat{H}_{\text{int}}\right]_- &= \left[B_{\mathbf{q}}, \hat{H}_{\text{em-e}}\right]_- \\ &= -\sum\nolimits_{\lambda,\mathbf{k},\mathbf{q}'_{\|}} \mathbf{j}_{\lambda}(\mathbf{k}) \cdot \left[B_{\mathbf{q}}, \mathbf{A}^{\lambda,\lambda}_{\mathbf{q}'_{\|}}\right]_- a^{\dagger}_{\lambda,\mathbf{k}+\frac{1}{2}\mathbf{q}'_{\|}} a_{\lambda,\mathbf{k}-\frac{1}{2}\mathbf{q}'_{\|}} \\ &\quad + \sum\nolimits_{\lambda,\mathbf{k}} \sum\nolimits_{\mathbf{q}'_{\|},\mathbf{q}''_{\|}} \frac{Q^2}{2m_0}\left[B_{\mathbf{q}}, A^{(2),\lambda}_{\mathbf{q}'_{\|},-\mathbf{q}''_{\|}}\right]_- a^{\dagger}_{\lambda,\mathbf{k}+\mathbf{q}'_{\|}} a_{\lambda,\mathbf{k}+\mathbf{q}''_{\|}} \\ &\quad - \sum\nolimits_{\mathbf{q}'_{\|},\mathbf{k}} \sum\nolimits_{\lambda\neq\lambda'} Q\mathbf{p}_{\lambda,\lambda'} \cdot \left[B_{\mathbf{q}}, \hat{\mathbf{A}}^{\lambda,\lambda'}_{\mathbf{q}'_{\|}}\right]_- a^{\dagger}_{\lambda,\mathbf{k}} a_{\lambda',\mathbf{k}+\mathbf{q}'_{\|}}.\end{aligned} \tag{14.17}$$

Here, we again used the linear additivity (14.4), the relation for the commutation of operator products (14.5), and the fact that the photon operators always commute with carrier and phonon operators.

To simplify the notation of $\hat{\mathbf{A}}^{\lambda,\lambda'}_{\mathbf{q}_\parallel}$ and $\hat{\mathbf{A}}^{(2),\lambda}_{\mathbf{q}_\parallel,-\mathbf{q}'_\parallel}$, determined by Eqs. (13.88)–(13.89), we define a mode expansion

$$\hat{\mathbf{A}}_{\mathbf{q}_\parallel}(z) \equiv \sum_{q_z} \frac{\mathcal{E}_{\mathbf{q}}}{\omega_{\mathbf{q}}} \left(\bar{\mathbf{u}}_{\mathbf{q}}(z) B_{\mathbf{q}_\parallel,q_z} + \bar{\mathbf{u}}^\star_{\mathbf{q}}(z) B^\dagger_{-\mathbf{q}_\parallel,q_z} \right). \tag{14.18}$$

This expression contains a sum over q_z of the light modes $\bar{\mathbf{u}}_{\mathbf{q}}(z)$ obtained from Eq. (13.10) and it defines the full mode function and the vector potential via

$$\mathbf{u}_{\mathbf{q}}(\mathbf{r}) = \bar{\mathbf{u}}_{\mathbf{q}}(z)\, e^{i\mathbf{q}_\parallel \cdot \mathbf{r}_\parallel}, \qquad \hat{\mathbf{A}}(\mathbf{r}) = \sum_{\mathbf{q}_\parallel} \hat{\mathbf{A}}_{\mathbf{q}_\parallel}(z)\, e^{i\mathbf{q}_\parallel \cdot \mathbf{r}_\parallel}, \tag{14.19}$$

respectively. The z-dependent modes have the property

$$\bar{\mathbf{u}}_{\mathbf{q}_\parallel,q_z}(z) = \bar{\mathbf{u}}_{-\mathbf{q}_\parallel,q_z}(z), \tag{14.20}$$

for the planar structures studied here. With the help of Eq. (14.18), Eqs. (13.88)–(13.89) simplify to

$$\hat{\mathbf{A}}^{\lambda,\lambda'}_{\mathbf{q}_\parallel} = \int dz\, g^{\lambda,\lambda'}(z) \hat{\mathbf{A}}_{\mathbf{q}_\parallel}(z), \tag{14.21}$$

$$\hat{A}^{(2),\lambda}_{\mathbf{q}_\parallel,-\mathbf{q}'_\parallel} = \int dz\, g^{\lambda,\lambda}(z)\, \hat{\mathbf{A}}_{\mathbf{q}_\parallel}(z) \cdot \hat{\mathbf{A}}_{-\mathbf{q}'_\parallel}(z), \tag{14.22}$$

with the confinement function

$$g^{\lambda,\lambda'}(z) \equiv \xi^\star_\lambda(z) \xi_{\lambda'}(z), \tag{14.23}$$

for the electronic states within the QW, see Fig. 14.1.

For our further derivations, we need to evaluate the commutation relation

$$\begin{aligned} \left[B_{\mathbf{q}}, \hat{\mathbf{A}}_{\mathbf{q}'_\parallel}(z)\right]_- &= \sum_{q'_z} \frac{\mathcal{E}'_{\mathbf{q}}}{\omega'_{\mathbf{q}}}\, \bar{\mathbf{u}}^\star_{\mathbf{q}'}(z) \left[B_{\mathbf{q}}, B^\dagger_{-\mathbf{q}'_\parallel,q'_z}\right]_- \\ &= \sum_{q'_z} \frac{\mathcal{E}'_{\mathbf{q}}}{\omega'_{\mathbf{q}}}\, \bar{\mathbf{u}}^\star_{\mathbf{q}'}(z)\, \delta_{q'_z,q_z} \delta_{\mathbf{q}'_\parallel,-\mathbf{q}_\parallel} = \delta_{\mathbf{q}'_\parallel,-\mathbf{q}_\parallel} \frac{\mathcal{E}_{\mathbf{q}}}{\omega_{\mathbf{q}}}\, \bar{\mathbf{u}}^\star_{\mathbf{q}}(z), \end{aligned} \tag{14.24}$$

where the first line is obtained by combining the linear additivity (14.4) and the vanishing commutator for two annihilation operators. The final result follows directly from the Boson commutation relations (10.31)–(10.32) and the property (14.20). At this point, we notice that Eqs. (14.4) and (14.24) can directly be applied to evaluate the commutators in Eq. (14.17) resulting in

$$\begin{aligned}\left[B_{\mathbf{q}}, \hat{\mathbf{A}}^{\lambda,\lambda'}_{\mathbf{q}'_\parallel}\right]_- &= \int dz\, g^{\lambda,\lambda'}(z) \left[B_{\mathbf{q}}, \hat{\mathbf{A}}_{\mathbf{q}'_\parallel}(z)\right]_- \\ &= \delta_{\mathbf{q}'_\parallel,-\mathbf{q}_\parallel} \int dz\, \frac{\mathcal{E}_{\mathbf{q}}}{\omega_{\mathbf{q}}}\, \bar{\mathbf{u}}^\star_{\mathbf{q}}(z)\, g^{\lambda,\lambda'}(z).\end{aligned} \tag{14.25}$$

By additionally applying the relation (14.5) for the operator products in Eq. (14.22), we find

$$\begin{aligned}\left[B_{\mathbf{q}}, A^{(2),\lambda}_{\mathbf{q}'_\parallel,-\mathbf{q}''_\parallel}\right]_- &= \int dz\, g^{\lambda,\lambda}(z) \left[B_{\mathbf{q}}, \hat{\mathbf{A}}_{\mathbf{q}'_\parallel}(z)\cdot\hat{\mathbf{A}}_{-\mathbf{q}''_\parallel}(z)\right]_- \\ &= \int dz\, g^{\lambda,\lambda}(z) \left(\left[B_{\mathbf{q}}, \hat{\mathbf{A}}_{\mathbf{q}'_\parallel}(z)\right]_- \cdot \hat{\mathbf{A}}_{-\mathbf{q}''_\parallel}(z) + \hat{\mathbf{A}}_{\mathbf{q}'_\parallel}(z)\cdot\left[B_{\mathbf{q}}, \hat{\mathbf{A}}_{-\mathbf{q}''_\parallel}(z)\right]_-\right) \\ &= \int dz\, \frac{\mathcal{E}_{\mathbf{q}}\bar{\mathbf{u}}^\star_{\mathbf{q}}(z)}{\omega_{\mathbf{q}}}\, g^{\lambda,\lambda}(z)\cdot\left(\hat{\mathbf{A}}_{-\mathbf{q}''_\parallel}(z)\delta_{\mathbf{q}'_\parallel,-\mathbf{q}_\parallel} + \hat{\mathbf{A}}_{\mathbf{q}'_\parallel}(z)\delta_{\mathbf{q}''_\parallel,\mathbf{q}_\parallel}\right),\end{aligned} \tag{14.26}$$

where the last step follows directly from the relation (14.24).

The results (14.25)–(14.26) can now be inserted back into Eq. (14.17). The first term leads to

$$\begin{aligned}\left[B_{\mathbf{q}}, \hat{H}_{\text{int}}\right]^{(1)}_- &\equiv -\sum_{\lambda,\mathbf{k},\mathbf{q}'_\parallel} \mathbf{j}_\lambda(\mathbf{k})\cdot\left[B_{\mathbf{q}}, \mathbf{A}^{\lambda,\lambda}_{\mathbf{q}'_\parallel}\right]_- a^\dagger_{\lambda,\mathbf{k}+\frac{1}{2}\mathbf{q}'_\parallel} a_{\lambda,\mathbf{k}-\frac{1}{2}\mathbf{q}'_\parallel} \\ &= -\int dz\, \frac{\mathcal{E}_{\mathbf{q}}\bar{\mathbf{u}}^\star_{\mathbf{q}}(z)}{\omega_{\mathbf{q}}}\cdot\left(\sum_{\lambda,\mathbf{k}} g^{\lambda,\lambda}(z)\, \mathbf{j}_\lambda(\mathbf{k})\, a^\dagger_{\lambda,\mathbf{k}-\frac{1}{2}\mathbf{q}_\parallel} a_{\lambda,\mathbf{k}+\frac{1}{2}\mathbf{q}_\parallel}\right).\end{aligned} \tag{14.27}$$

The second term is given by

$$\begin{aligned}\left[B_{\mathbf{q}}, \hat{H}_{\text{int}}\right]^{(2)}_- &\equiv \sum_{\lambda,\mathbf{k}}\sum_{\mathbf{q}'_\parallel,\mathbf{q}''_\parallel} \frac{Q^2}{2m_0}\left[B_{\mathbf{q}}, A^{(2),\lambda}_{\mathbf{q}'_\parallel,-\mathbf{q}''_\parallel}\right]_- a^\dagger_{\lambda,\mathbf{k}+\mathbf{q}'_\parallel} a_{\lambda,\mathbf{k}+\mathbf{q}''_\parallel} \\ &= \int dz\, \frac{\mathcal{E}_{\mathbf{q}}\bar{\mathbf{u}}^\star_{\mathbf{q}}(z)}{\omega_{\mathbf{q}}}\cdot\sum_{\lambda,\mathbf{k}}\sum_{\mathbf{q}'_\parallel} g^{\lambda,\lambda}(z)\,\frac{Q^2}{2m_0} \\ &\quad\cdot\left(\hat{\mathbf{A}}_{-\mathbf{q}'_\parallel}(z) a^\dagger_{\lambda,\mathbf{k}-\mathbf{q}_\parallel} a_{\lambda,\mathbf{k}+\mathbf{q}'_\parallel} + \hat{\mathbf{A}}_{\mathbf{q}'_\parallel}(z) a^\dagger_{\lambda,\mathbf{k}+\mathbf{q}'_\parallel} a_{\lambda,\mathbf{k}+\mathbf{q}_\parallel}\right) \\ &= \int dz\, \frac{\mathcal{E}_{\mathbf{q}}\bar{\mathbf{u}}^\star_{\mathbf{q}}(z)}{\omega_{\mathbf{q}}}\left(\sum_{\lambda,\mathbf{k}}\sum_{\mathbf{q}'_\parallel} g^{\lambda,\lambda}(z)\frac{Q^2}{m_0}\hat{\mathbf{A}}_{\mathbf{q}'_\parallel}(z) a^\dagger_{\lambda,\mathbf{k}+\mathbf{q}'_\parallel} a_{\lambda,\mathbf{k}+\mathbf{q}_\parallel}\right),\end{aligned} \tag{14.28}$$

which follows once we implement a change of summation variables: $\mathbf{q}'_\parallel \to -\mathbf{q}'_\parallel$ and $\mathbf{k}\to\mathbf{k}+\mathbf{q}_\parallel+\mathbf{q}'_\parallel$. The remaining third term is

$$\left[B_{\mathbf{q}}, \hat{H}_{\text{int}}\right]_-^{(3)} \equiv -\sum_{\mathbf{q}'_{\|},\mathbf{k}}\sum_{\lambda\neq\lambda'} \frac{Q\mathbf{p}_{\lambda,\lambda'}}{m_0} \cdot \left[B_{\mathbf{q}}, \hat{\mathbf{A}}^{\lambda,\lambda'}_{\mathbf{q}'_{\|}}\right]_- a^{\dagger}_{\lambda,\mathbf{k}} a_{\lambda',\mathbf{k}-\mathbf{q}'_{\|}}$$

$$= -\int dz\, \frac{\mathcal{E}_{\mathbf{q}}\bar{\mathbf{u}}^{\star}_{\mathbf{q}}(z)}{\omega_{\mathbf{q}}} \cdot \left(\sum_{\lambda,\mathbf{k},\lambda'\neq\lambda} g^{\lambda,\lambda'}(z)\, \frac{Q\mathbf{p}_{\lambda,\lambda'}}{m_0}\, a^{\dagger}_{\lambda,\mathbf{k}} a_{\lambda',\mathbf{k}+\mathbf{q}_{\|}}\right). \quad (14.29)$$

14.1.4 Macroscopic matter-response operators

The photon dynamics directly couples to the matter response via the matter operators appearing in Eqs. (14.27)–(14.29). Therefore, we can now introduce the macroscopic current operator

$$\hat{\mathbf{J}}^{\lambda}_{\mathbf{q}_{\|}}(z) \equiv \frac{1}{\mathcal{S}}\sum_{\mathbf{k}}\left[\mathbf{j}_{\lambda}(\mathbf{k}) a^{\dagger}_{\lambda,\mathbf{k}-\frac{1}{2}\mathbf{q}_{\|}} a_{\lambda,\mathbf{k}+\frac{1}{2}\mathbf{q}_{\|}} - \frac{Q^2}{m_0}\sum_{\mathbf{q}'_{\|}} \hat{\mathbf{A}}_{\mathbf{q}'_{\|}}(z) a^{\dagger}_{\lambda,\mathbf{k}+\mathbf{q}'_{\|}} a_{\lambda,\mathbf{k}+\mathbf{q}_{\|}}\right], \quad (14.30)$$

and the macroscopic polarization operator

$$\frac{\partial}{\partial t}\hat{\mathbf{P}}^{\lambda,\lambda'}_{\mathbf{q}_{\|}} \equiv \frac{1}{\mathcal{S}}\sum_{\mathbf{k}} \frac{Q\mathbf{p}_{\lambda,\lambda'}}{m_0} a^{\dagger}_{\lambda,\mathbf{k}-\frac{1}{2}\mathbf{q}_{\|}} a_{\lambda',\mathbf{k}+\frac{1}{2}\mathbf{q}_{\|}}, \quad (14.31)$$

where $\mathcal{S}$ defines the normalization area in the QW system. At first sight, it may seem inconvenient to identify the polarization via its time derivative. However, the used form will get a physical justification once we study Maxwell's equations for the interacting system, Section 14.2.1. We also define the macroscopic density operator

$$\hat{\mathbf{n}}^{\lambda}_{\mathbf{p}_{\|}} \equiv \frac{1}{\mathcal{S}}\sum_{\mathbf{k}} a^{\dagger}_{\lambda,\mathbf{k}-\frac{1}{2}\mathbf{p}_{\|}} a_{\lambda,\mathbf{k}+\frac{1}{2}\mathbf{p}_{\|}}, \quad (14.32)$$

for the sake of completeness.

From a structural point of view, $\hat{\mathbf{J}}^{\lambda}_{\mathbf{q}_{\|}}(z)$ includes *intraband* processes where only the momentum of the electrons is changed. Figure 14.2 presents such a transition as a nearly horizontal arrow that connects the initial (dashed circle) and final (sphere) state within the same band. The polarization involves *interband* processes (nearly vertical arrow) where also the band index of the electron is changed. The momentum changes involve only the in-plane part of the photon, as discussed in connection with Figs. 13.1 and 13.2. The specific role of the inter- and intraband process in semiconductor quantum optics is discussed in Section 14.2.2.

The macroscopic current and the polarization response simultaneously drive the photon operators. Therefore, we introduce a generic matter response

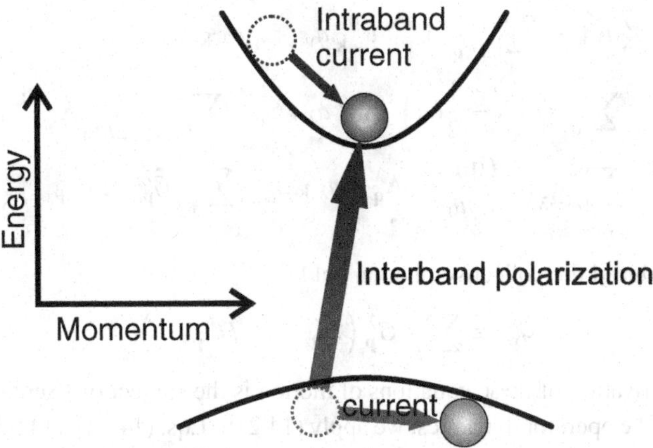

Figure 14.2 Light-induced intraband vs. interband transitions in two-band semiconductors. The dashed circle indicates the initial state of the carrier which is moved into the final state (sphere) via the transitions. The arrows define the direction of the transition. Intraband transtions yield currents while interband processes produce polarization.

$$\hat{\mathbf{R}}^{\lambda,\lambda'}_{\mathbf{q}_\parallel}(z) \equiv g_{\lambda,\lambda'}(z)\left[\delta_{\lambda,\lambda'}\hat{\mathbf{J}}^{\lambda}_{\mathbf{q}_\parallel}(z) + (1-\delta_{\lambda,\lambda'})\frac{\partial}{\partial t}\hat{\mathbf{P}}^{\lambda,\lambda'}_{\mathbf{q}_\parallel}\right] \tag{14.33}$$

that includes both of them. This allows us to collect the results (14.15) and (14.27)–(14.29) into a simple format:

$$i\hbar\frac{\partial}{\partial t}B_{\mathbf{q}_\parallel,q_z} = \hbar\omega_q\, B_{\mathbf{q}_\parallel,q_z} - \sum_{\lambda,\lambda'}\int dz\, \frac{\mathcal{E}_{\mathbf{q}}}{\omega_{\mathbf{q}}}\bar{\mathbf{u}}^{\star}_{\mathbf{q}_\parallel,q_z}(z)\cdot \mathcal{S}\,\hat{\mathbf{R}}^{\lambda,\lambda'}_{\mathbf{q}_\parallel}(z), \tag{14.34}$$

$$i\hbar\frac{\partial}{\partial t}B^{\dagger}_{\mathbf{q}_\parallel,q_z} = -\hbar\omega_q\, B^{\dagger}_{\mathbf{q}_\parallel,q_z} + \sum_{\lambda,\lambda'}\int dz\, \frac{\mathcal{E}_{\mathbf{q}}}{\omega_{\mathbf{q}}}\bar{\mathbf{u}}_{\mathbf{q}_\parallel,q_z}(z)\cdot \mathcal{S}\,\hat{\mathbf{R}}^{\lambda,\lambda'}_{-\mathbf{q}_\parallel}(z), \tag{14.35}$$

where we used the general property $\left[\hat{\mathbf{R}}^{\lambda,\lambda'}_{\mathbf{q}_\parallel}(z)\right]^{\star} = \hat{\mathbf{R}}^{\lambda',\lambda}_{-\mathbf{q}_\parallel}(z)$, see Exercise 14.2.

14.1.5 Phonon and carrier dynamics

Repeating the same derivation as in Section 14.1.3 for $D_{\mathbf{p}}$, we find

$$i\hbar\frac{\partial}{\partial t}\hat{D}_{\mathbf{p}_\parallel,p_z} = \hbar\Omega_{\mathbf{p}}\,\hat{D}_{\mathbf{p}_\parallel,p_z} + \mathcal{S}\sum\nolimits_{\lambda} G^{\lambda}_{\mathbf{p}}\,\hat{n}^{\lambda}_{\mathbf{p}_\parallel}, \tag{14.36}$$

with the phonon-matrix element $G^{\lambda}_{\mathbf{p}}$. Including the interaction part of the carrier dynamics, we get

$$
\begin{aligned}
i\hbar\frac{\partial}{\partial t}a_{\lambda,\mathbf{k}} = \epsilon^{\lambda}_{\mathbf{k}}a_{\lambda,\mathbf{k}} &+ \sum_{\nu,\mathbf{k}',\mathbf{q}_\parallel} V^{\lambda,\nu}_{\mathbf{q}_\parallel}\, a^{\dagger}_{\nu,\mathbf{k}'}a_{\nu,\mathbf{k}'+\mathbf{q}_\parallel}a_{\lambda,\mathbf{k}-\mathbf{q}_\parallel} \\
&- \sum_{\mathbf{q}_\parallel} \mathbf{j}_\lambda\left(\mathbf{k}-\frac{1}{2}\mathbf{q}_\parallel\right)\cdot\hat{\mathbf{A}}^{\lambda,\lambda}_{\mathbf{q}_\parallel}a_{\lambda,\mathbf{k}-\mathbf{q}_\parallel} + \sum_{\mathbf{q}_\parallel,\mathbf{q}'_\parallel}\frac{Q^2}{2m_0}\hat{\mathbf{A}}^{(2),\lambda}_{\mathbf{q}_\parallel,-\mathbf{q}'_\parallel}a_{\lambda,\mathbf{k}+\mathbf{q}'_\parallel-\mathbf{q}_\parallel} \\
&- \sum_{\lambda'\neq\lambda,\mathbf{q}_\parallel}\frac{Q\mathbf{p}_{\lambda,\lambda'}}{m_0}\cdot\mathbf{A}^{\lambda',\lambda}_{\mathbf{q}_\parallel}\, a_{\lambda,\mathbf{k}-\mathbf{q}_\parallel} - \sum_{\mathbf{p}_\parallel}\hat{\mathcal{G}}^{\lambda}_{\mathbf{p}_\parallel}\, a_{\lambda,\mathbf{k}-\mathbf{p}_\parallel},
\end{aligned}
\tag{14.37}
$$

where we introduced the collective phonon field

$$
\hat{\mathcal{G}}^{\lambda}_{\mathbf{p}_\parallel} \equiv \sum_{p_z} G^{\lambda}_{\mathbf{p}}\left(D_{\mathbf{p}_\parallel,p_z} + D^{\dagger}_{-\mathbf{p}_\parallel,p_z}\right).
\tag{14.38}
$$

The detailed derivation of these equations of motion is the subject of Exercise 14.4.

To complete the operator dynamics, we apply (14.2) to Eqs. (14.34), (14.36), and (14.37) yielding

$$
i\hbar\frac{\partial}{\partial t}D^{\dagger}_{\mathbf{p}_\parallel,p_z} = -\hbar\Omega_{\mathbf{p}}\,D^{\dagger}_{\mathbf{p}_\parallel,\mathbf{p}_z} - \mathcal{S}\sum_{\lambda} G^{\lambda}_{-\mathbf{p}}[\hat{n}^{\lambda}_{\mathbf{p}_\parallel}]^{\dagger},
\tag{14.39}
$$

$$
\begin{aligned}
i\hbar\frac{\partial}{\partial t}a^{\dagger}_{\lambda,\mathbf{k}} = -\epsilon^{\lambda}_{\mathbf{k}}a^{\dagger}_{\lambda,\mathbf{k}} &- \sum_{\nu,\mathbf{k}',\mathbf{q}_\parallel} V^{\lambda,\nu}_{\mathbf{q}_\parallel}\, a^{\dagger}_{\lambda,\mathbf{k}-\mathbf{q}_\parallel}a^{\dagger}_{\nu,\mathbf{k}'+\mathbf{q}_\parallel}a_{\nu,\mathbf{k}'} \\
&+ \sum_{\mathbf{q}_\parallel}\mathbf{j}_\lambda\left(\mathbf{k}-\frac{1}{2}\mathbf{q}_\parallel\right)\cdot\mathbf{A}^{\lambda,\lambda}_{-\mathbf{q}_\parallel}\, a^{\dagger}_{\lambda,\mathbf{k}-\mathbf{q}_\parallel} \\
&- \frac{Q^2}{2m_0}\sum_{\mathbf{q}_\parallel,\mathbf{q}'_\parallel}\mathbf{A}^{(2),\lambda}_{\mathbf{q}'_\parallel,-\mathbf{q}_\parallel}\, a^{\dagger}_{\lambda,\mathbf{k}+\mathbf{q}'_\parallel-\mathbf{q}_\parallel} \\
&+ \sum_{\lambda'\neq\lambda,\mathbf{q}_\parallel}\frac{Q\mathbf{p}_{\lambda,\lambda'}}{m_0}\cdot\mathbf{A}^{\lambda,\lambda'}_{-\mathbf{q}_\parallel}\, a^{\dagger}_{\lambda,\mathbf{k}-\mathbf{q}_\parallel} + \sum_{\mathbf{p}_\parallel}\hat{\mathcal{G}}^{\lambda}_{-\mathbf{p}_\parallel}\, a^{\dagger}_{\mathbf{k}-\mathbf{p}_\parallel,\lambda},
\end{aligned}
\tag{14.40}
$$

where we used the properties, $\left[\hat{\mathcal{G}}^{\lambda}_{\mathbf{p}_\parallel}\right]^{\dagger} = \hat{\mathcal{G}}^{\lambda}_{-\mathbf{p}_\parallel}$ and $\left[\mathbf{A}^{\lambda,\lambda'}_{\mathbf{q}_\parallel}\right]^{\dagger} = \mathbf{A}^{\lambda,\lambda'}_{-\mathbf{q}_\parallel}$ studied in Exercise 14.2. In contrast to the free photon–phonon–carrier system, the interaction terms in Eqs. (14.37) and (14.40) lead to an equation structure which is not closed. In other words, couplings to new terms appear, for which we don't have an explicit equation. This highly nontrivial aspect is investigated in Chapter 15.

14.2 Formal properties of light

The general structure of Eqs. (14.34) and Eq. (14.36) shows that the phonon and photon operators couple directly to pure carrier operators via the carrier–phonon and the carrier–photon interactions, respectively. For the expansion scheme (cluster expansion) that we want to develop, we need to systematically classify the complexity of general operators containing a mixture of photon, phonon, and carrier operators. This requires a formal connection between the Fermionic carrier and the Bosonic photon and phonon operators. Since there cannot be a one-to-one transformation between Fermion and Boson operators, we are looking only for a formal mapping. This will then allow us to pose strict conditions for the consistent treatment by demanding that operators and expectation values of the same complexity have to be treated equally within the approximations used.

Since the homogeneous part of Eq. (14.34) can be solved analytically, we introduce a modification of this result in the form of an ansatz

$$B_{\mathbf{q}}(t) \equiv \tilde{B}_{\mathbf{q}}(t)e^{-i\omega_{\mathbf{q}}t}, \tag{14.41}$$

where $\tilde{B}_{\mathbf{q}}(t)$ is a new operator. By inserting the ansatz (14.41) into (14.34), we obtain

$$i\hbar\frac{\partial}{\partial t}\tilde{B}_{\mathbf{q}}(t) = -\sum_{\lambda,\lambda'}\int dz\, \frac{\mathcal{E}_{\mathbf{q}}\bar{\mathbf{u}}^{\star}_{\mathbf{q}}(z)}{\omega_{\mathbf{q}}}\cdot \mathcal{S}\hat{\mathbf{R}}^{\lambda,\lambda'}_{\mathbf{q}_{\|}}(z,t)\, e^{i\omega_{\mathbf{q}}t}. \tag{14.42}$$

This can be formally evaluated by integrating both sides of the equation with respect to time, yielding

$$\tilde{B}_{\mathbf{q}}(t) = \tilde{B}_{\mathbf{q}}(0) - \frac{1}{i\hbar}\sum\nolimits_{\lambda,\lambda'}\int dz\int_0^t ds\, \frac{\mathcal{E}_{\mathbf{q}}\bar{\mathbf{u}}^{\star}_{\mathbf{q}}}{\omega_{\mathbf{q}}}(z)\cdot \mathcal{S}\hat{\mathbf{R}}^{\lambda,\lambda'}_{\mathbf{q}_{\|}}(z,s)\, e^{i\omega_{\mathbf{q}}s}. \tag{14.43}$$

This expression can now be inserted back into Eq. (14.41) yielding

$$B_{\mathbf{q}}(t) = B_{\mathbf{q}}(0)\, e^{-i\omega_{\mathbf{q}}t} - \frac{1}{i\hbar}\sum\nolimits_{\lambda,\lambda'}\int dz\int_0^t ds\, \frac{\mathcal{E}_{\mathbf{q}}\bar{\mathbf{u}}^{\star}_{\mathbf{q}}(z)}{\omega_{\mathbf{q}}}\cdot \mathcal{S}\hat{\mathbf{R}}^{\lambda,\lambda'}_{\mathbf{q}_{\|}}(z,s)\, e^{i\omega_{\mathbf{q}}(s-t)}. \tag{14.44}$$

With very similar derivation steps, we find that Eq. (14.36) produces the formal solution

$$D_{\mathbf{p}}(t) = D_{\mathbf{p}}(0)e^{-i\Omega_{\mathbf{p}}t} + \frac{1}{i\hbar}\sum\nolimits_{\lambda}\int dz\int_0^t ds\, G^{\lambda}_{\mathbf{p}}\mathcal{S}\, \hat{n}^{\lambda}_{\mathbf{p}_{\|}}(s)\, e^{i\Omega_{\mathbf{p}}(s-t)}. \tag{14.45}$$

Even though these expressions are purely formal, we use them to relate the complexities of photon, phonon, and carrier operators.

If we set both $\hat{\mathbf{R}}^{\lambda,\lambda'}_{\mathbf{q}_{\|}}$ and $\hat{\mathbf{n}}^{\lambda}_{\mathbf{p}_{\|}}$ to zero, we recover the situation where the carrier system is uncoupled from the photon–phonon system. In this case, the time-evolution (14.44)–(14.45) is determined by the simple free-propagation phase shifts. However, when $\hat{\mathbf{R}}^{\lambda,\lambda'}_{\mathbf{q}_{\|}}$ and $\hat{\mathbf{n}}^{\lambda}_{\mathbf{p}_{\|}}$ are nonzero, the photon–phonon system is coupled to the carriers leading to significant structural changes in the formal solutions. In particular, the carrier operators $\hat{\mathbf{R}}^{\lambda,\lambda'}_{\mathbf{q}_{\|}}$ and $\hat{\mathbf{n}}^{\lambda}_{\mathbf{p}_{\|}}$ couple directly to photons and phonons, respectively.

At this point, we should mention that the formal solutions cannot be used to solve the details of the coupling dynamics because $\hat{\mathbf{R}}^{\lambda,\lambda'}_{\mathbf{q}_{\|}}$ and $\hat{\mathbf{n}}^{\lambda}_{\mathbf{p}_{\|}}$ are unknown operators defined at times that are different from those of the Boson operator, according to the integrals within Eqs. (14.44)–(14.44). However, the modifications of B and D always occur via terms containing carrier-operator pairs, $a^{\dagger}_{\lambda}a_{\lambda'}$. Since a reduction below such operators is impossible, *single photon and phonon operators must correspond to a*

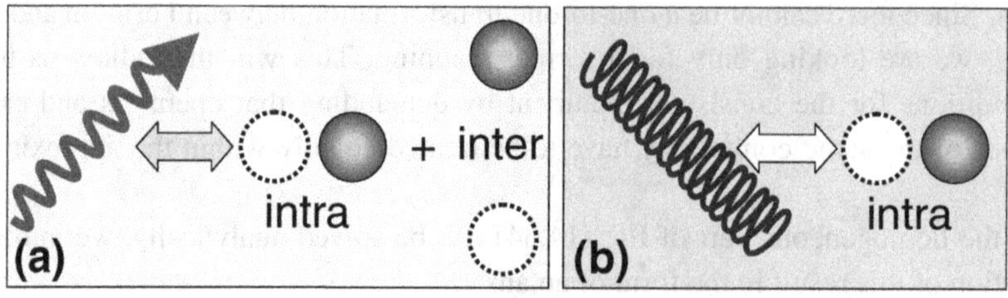

Figure 14.3 Formal analogy between **(a)** photon, **(b)** phonon and Fermionic carrier operators. The wave arrow (spring) corresponds to a single photon (phonon) operator. The carrier creation (annihilation) operator is described by a sphere (dashed circle). Intraband terms contain carrier operators within the same band while interband processes involve carrier operators in different bands.

carrier-operator pair $a^\dagger_\lambda a_{\lambda'}$ on a formal level. We will use this relation between the pure photon and phonon operators and the pure carrier operators in many of our forthcoming derivations.

Figure 14.3 symbolically represents the formal Boson-Fermion identification. A photon (wave arrow) and a phonon (spring symbol) correspond to a carrier-operator pair based on the structure of Eqs. (14.44)–(14.45). The dashed circle corresponds to particle annihilation and the sphere to particle creation. The photon is related to both intra- (horizontal pair) and interband (vertical pair) processes while phonons are involved only in intraband terms for the interactions studied here.

14.2.1 Quantized wave equation

The quantum-dynamical properties of light can fully be solved from the operator equations (14.34)–(14.35). However, some investigations become simpler when we work directly with the field operators for the vector potential or the electric field. For example, the usual light propagation problem has the form of a wave equation even at the operator level. To provide the quantum version of this approach, we now investigate the dynamics of the light fields.

Our starting point is the fully quantized vector potential (14.19) expressed via the mode expansion (14.18). As we use them together with the dynamical equations (14.34)–(14.35), we find

$$i\hbar\frac{\partial}{\partial t}\hat{\mathbf{A}}(\mathbf{r}) = \hbar\sum_{\mathbf{q}}\mathcal{E}_{\mathbf{q}}\left[\bar{\mathbf{u}}_{\mathbf{q}}(z)B_{\mathbf{q}_\parallel,q_z} - \bar{\mathbf{u}}^\star_{\mathbf{q}}(z)B^\dagger_{-\mathbf{q}_\parallel,q_z}\right]e^{i\mathbf{q}_\parallel\cdot\mathbf{r}_\parallel}$$
$$-\sum_{\lambda,\lambda',\mathbf{q}}\int dz'\,\frac{\mathcal{E}^2_{\mathbf{q}}\left[\bar{\mathbf{u}}_{\mathbf{q}}(z)\bar{\mathbf{u}}^\star_{\mathbf{q}}(z') - \bar{\mathbf{u}}^\star_{\mathbf{q}}(z)\bar{\mathbf{u}}_{\mathbf{q}}(z')\right]}{\omega^2_{\mathbf{q}}}\cdot\mathcal{S}\hat{\mathbf{R}}^{\lambda,\lambda'}_{\mathbf{q}_\parallel}(z')e^{i\mathbf{q}_\parallel\cdot\mathbf{r}_\parallel}, \tag{14.46}$$

after some reorganization of the terms. We notice that $\left[\bar{\mathbf{u}}_{\mathbf{q}}(z)\bar{\mathbf{u}}^\star_{\mathbf{q}}(z') - \bar{\mathbf{u}}^\star_{\mathbf{q}}(z)\bar{\mathbf{u}}_{\mathbf{q}}(z')\right]$ vanishes for a real-valued $\bar{\mathbf{u}}_{\mathbf{q}}(z)$. We can choose this freely because the generalized

Helmholtz equation (9.64) allows for this. Naturally, $\bar{\mathbf{u}}_{\mathbf{q}}(z)$ could also be chosen as complex valued as long as the basis is complete and orthogonal. These different basis choices are connected via the unitary transformations discussed in Section 10.2.1. Thus, the $\mathbf{q}$ sum of the last term in Eq. (14.46) must vanish for any choice of basis; this aspect is investigated more thoroughly in Exercise 14.6.

As a result, the time-dynamics of the vector potential determines the electrical field

$$\begin{aligned}\hat{\mathbf{E}}(\mathbf{r}) &\equiv -\frac{\partial}{\partial t}\hat{\mathbf{A}}(\mathbf{r})\\ &= \sum_{\mathbf{q}} i\mathcal{E}_{\mathbf{q}}\left[\bar{\mathbf{u}}_{\mathbf{q}}(z)B_{\mathbf{q}_{\parallel},q_z} - \bar{\mathbf{u}}^{\star}_{\mathbf{q}}(z)B^{\dagger}_{-\mathbf{q}_{\parallel},q_z}\right]e^{i\mathbf{q}_{\parallel}\cdot\mathbf{r}_{\parallel}}\\ &= \sum_{\mathbf{q}} i\mathcal{E}_{\mathbf{q}}\left[\mathbf{u}_{\mathbf{q}}(\mathbf{r})B_{\mathbf{q}_{\parallel},q_z} - \mathbf{u}^{\star}_{\mathbf{q}}(\mathbf{r})B^{\dagger}_{-\mathbf{q}_{\parallel},q_z}\right],\end{aligned} \tag{14.47}$$

where the relation (14.19) has been applied. This form reproduces the same definition for the electric field operator (10.43) as that obtained for noninteracting photons, which obviously verifies that the system Hamiltonian indeed reproduces the standard relations for the electromagnetic field.

In order to determine the dynamics of $\hat{\mathbf{E}}$, we use the same steps as for Eq. (14.46) to obtain

$$\begin{aligned}i\hbar\frac{\partial}{\partial t}\hat{\mathbf{E}}(\mathbf{r}) &= -i\hbar\frac{\partial^2}{\partial t^2}\hat{\mathbf{A}}(\mathbf{r})\\ &= i\hbar\sum_{\mathbf{q}}\frac{\mathcal{E}_{\mathbf{q}}}{\omega_{\mathbf{q}}}\left[\omega^2_{\mathbf{q}}\,\mathbf{u}_{\mathbf{q}}(\mathbf{r})B_{\mathbf{q}} + \omega^2_{\mathbf{q}}\,\mathbf{u}^{\star}_{\mathbf{q}}(\mathbf{r})B^{\dagger}_{\mathbf{q}}\right]\\ &\quad + \sum_{\mathbf{q},\lambda,\lambda'}\int dz'\,\frac{\bar{\mathbf{u}}_{\mathbf{q}}(z)\bar{\mathbf{u}}^{\star}_{\mathbf{q}}(z') + \bar{\mathbf{u}}^{\star}_{\mathbf{q}}(z)\bar{\mathbf{u}}_{\mathbf{q}}(z')}{2\epsilon_0}\cdot S\hat{\mathbf{R}}^{\lambda,\lambda'}_{\mathbf{q}_{\parallel}}(z')e^{i\mathbf{q}_{\parallel}\cdot\mathbf{r}_{\parallel}},\end{aligned} \tag{14.48}$$

where we have used the explicit form of the vacuum-field amplitude, i.e., Eq. (10.23), to simplify $\mathcal{E}^2_{\mathbf{q}}/\omega_{\mathbf{q}} = 1/2\epsilon_0$.

We then apply the generalized Helmholtz equation (9.64) to produce

$$\omega^2_{\mathbf{q}}\,\mathbf{u}_{\mathbf{q}}(\mathbf{r}) = \frac{c^2}{n^2(z)}\nabla\times\left[\nabla\times\mathbf{u}_{\mathbf{q}}(\mathbf{r})\right]. \tag{14.49}$$

Using similar arguments to those showing that the last term in Eq. (14.46) vanishes, we may implement $\left[\bar{\mathbf{u}}_{\mathbf{q}}(z)\bar{\mathbf{u}}^{\star}_{\mathbf{q}}(z') + \bar{\mathbf{u}}^{\star}_{\mathbf{q}}(z)\bar{\mathbf{u}}_{\mathbf{q}}(z')\right] = 2\bar{\mathbf{u}}_{\mathbf{q}}(z)\bar{\mathbf{u}}^{\star}_{\mathbf{q}}(z')$. Using these two pieces of information in Eq. (14.48), we obtain

$$\frac{\partial^2}{\partial t^2}\hat{\mathbf{A}}(\mathbf{r}) = -\frac{c^2}{n^2(z)}\nabla\times\left[\nabla\times\hat{\mathbf{A}}(\mathbf{r})\right] + \frac{1}{\epsilon_0}\sum_{\lambda,\lambda'}\sum_{\mathbf{q}}\int dz'\,\bar{\mathbf{u}}_{\mathbf{q}}(z)\bar{\mathbf{u}}^{\star}_{\mathbf{q}}(z')\cdot S\hat{\mathbf{R}}^{\lambda,\lambda'}_{\mathbf{q}_{\parallel}}(z')\,e^{i\mathbf{q}_{\parallel}\cdot\mathbf{r}_{\parallel}}, \tag{14.50}$$

where we have additionally divided both sides by $i\hbar$.

The response function that appears can be represented in real space by applying the Fourier transformation:

$$\begin{cases} \hat{\mathbf{R}}^{\lambda,\lambda'}_{\mathbf{q}_\parallel}(z) = \frac{1}{\mathcal{S}} \int d^2 r_\parallel \, \hat{\mathbf{R}}^{\lambda,\lambda'}(\mathbf{r})\, e^{-i\mathbf{q}_\parallel \cdot \mathbf{r}_\parallel} \\ \hat{\mathbf{R}}^{\lambda,\lambda'}(\mathbf{r}) = \sum_{\mathbf{q}_\parallel} \hat{\mathbf{R}}^{\lambda,\lambda'}_{\mathbf{q}_\parallel}(z)\, e^{i\mathbf{q}_\parallel \cdot \mathbf{r}_\parallel} \end{cases} \tag{14.51}$$

that is analogous to the definition (14.19). This converts the integral in Eq. (14.50) into

$$\begin{aligned} I &\equiv \int dz'\, \bar{\mathbf{u}}_{\mathbf{q}}(z) \bar{\mathbf{u}}^\star_{\mathbf{q}}(z') \cdot \mathcal{S} \hat{\mathbf{R}}^{\lambda,\lambda'}_{\mathbf{q}_\parallel}(z')\, e^{i\mathbf{q}_\parallel \cdot \mathbf{r}_\parallel} \\ &= \int d^3 r'\, \bar{\mathbf{u}}_{\mathbf{q}}(z) e^{i\mathbf{q}_\parallel \cdot \mathbf{r}_\parallel}\, \bar{\mathbf{u}}^\star_{\mathbf{q}}(z') e^{-i\mathbf{q}_\parallel \cdot \mathbf{r}'_\parallel} \cdot \hat{\mathbf{R}}^{\lambda,\lambda'}(\mathbf{r}') \\ &= \int d^3 r'\, \mathbf{u}_{\mathbf{q}}(\mathbf{r}) \mathbf{u}^\star_{\mathbf{q}}(\mathbf{r}') \cdot \hat{\mathbf{R}}^{\lambda,\lambda'}(\mathbf{r}'), \end{aligned} \tag{14.52}$$

where we reorganized the exponential functions and used (14.19). This step changes Eq. (14.50) into

$$\begin{aligned} \frac{\partial^2}{\partial t^2} \hat{\mathbf{A}}(\mathbf{r}) = &-\frac{c^2}{n^2(z)} \nabla \times \left[\nabla \times \hat{\mathbf{A}}(\mathbf{r}) \right] \\ &+ \frac{1}{\epsilon_0} \sum_{\lambda,\lambda'} \int d^3 r' \sum_{\mathbf{q}} \mathbf{u}_{\mathbf{q}}(\mathbf{r}) \mathbf{u}^\star_{\mathbf{q}}(\mathbf{r}') \cdot \hat{\mathbf{R}}^{\lambda,\lambda'}(\mathbf{r}'). \end{aligned} \tag{14.53}$$

The $\mathbf{q}$ sum that appears defines the projection in transversal space,

$$\sum\nolimits_{\mathbf{q}} \mathbf{u}^\star_{\mathbf{q}}(\mathbf{r}')\, \mathbf{u}_{\mathbf{q}}(\mathbf{r}) = \frac{1}{n^2(z)}\, \delta^T(\mathbf{r}' - \mathbf{r}), \tag{14.54}$$

according to the completeness relation (9.24).

Inserting the relation (14.54) into Eq. (14.55) and using $1/\epsilon_0 = \mu_0 c^2$, we end up with

$$\begin{aligned} \frac{\partial^2}{\partial t^2} \hat{\mathbf{A}}(\mathbf{r}) = &-\frac{c^2}{n^2(z)} \nabla \times \left[\nabla \times \hat{\mathbf{A}}(\mathbf{r}) \right] \\ &+ \mu_0 \frac{c^2}{n^2(z)} \sum_{\lambda,\lambda'} \int d^3 r'\, \delta_T(\mathbf{r}' - \mathbf{r}) \cdot \hat{\mathbf{R}}^{\lambda,\lambda'}(\mathbf{r}'). \end{aligned} \tag{14.55}$$

This integral simply projects the transversal part $\left[\hat{\mathbf{R}}^{\lambda,\lambda'}(\mathbf{r})\right]_T$ of the matter response, see the discussion in Section 9.2.2. Multiplying both sides of Eq. (14.55) by $n^2(z)/c^2$ and moving the $\nabla \times \nabla$ term to the LHS, we obtain

$$\left[\frac{n^2(z)}{c^2} \frac{\partial^2}{\partial t^2} + \nabla \times \nabla \times \right] \hat{\mathbf{A}}(\mathbf{r}) = \mu_0 \sum\nolimits_{\lambda,\lambda'} \left[\hat{\mathbf{R}}^{\lambda,\lambda'}(\mathbf{r}) \right]_T, \tag{14.56}$$

where we have additionally used the relation $c^2 = 1/\epsilon_0\mu_0$. At the same time, the application of (14.47) in Eq. (14.56) yields

$$\left[\frac{\partial^2}{\partial t^2} + \nabla\times\nabla\times\right]\hat{\mathbf{E}}(\mathbf{r}) = -\mu_0\sum\nolimits_{\lambda,\lambda'}\left[\frac{\partial}{\partial t}\hat{\mathbf{R}}^{\lambda,\lambda'}(\mathbf{r})\right]_{\mathrm{T}}. \tag{14.57}$$

In both cases, the quantized matter response follows from Eq. (14.33), defining

$$\hat{\mathbf{R}}^{\lambda,\lambda'}(\mathbf{r}) = \delta_{\lambda,\lambda'}\hat{\mathbf{J}}^{\lambda}(\mathbf{r}) + (1-\delta_{\lambda,\lambda'})\frac{\partial}{\partial t}\hat{\mathbf{P}}^{\lambda,\lambda'}(\mathbf{r}), \tag{14.58}$$

which consists of macroscopic currents, $\hat{\mathbf{J}}^{\lambda}(\mathbf{r})$, and the macroscopic polarizations, $\hat{\mathbf{P}}^{\lambda,\lambda'}(\mathbf{r})$. Even though $\hat{\mathbf{A}}$, $\hat{\mathbf{E}}$, and $\hat{\mathbf{R}}$ are now operators, the structural form of classical optics is clearly recovered. More specifically, Eqs. (14.56)–(14.57) *establish the fully quantized generalization of the wave equation* (9.19). Especially, their form is identical to that in classical optics, indicating that the identification (14.33) indeed follows from the quantum-mechanical generalizations for the current and polarization.

14.2.2 Plasmon response

We study next some practical implications of the general form of Eqs. (14.56)–(14.57). For this purpose, we assume a homogeneous system with refractive-index profile $n(z) = n$ and constant densities $\hat{n}^{\lambda}(\mathbf{r}) = \hat{n}^{\lambda}$. Furthermore, we assume that the system does not have any pure carrier-operator currents and that only polarization contributions introduce an effective mass to J. In this situation, the optical response becomes

$$\hat{\mathbf{R}}^{\lambda,\lambda'}(\mathbf{r}) \to -\delta_{\lambda,\lambda'}\frac{Q^2}{m_\lambda}\hat{n}^{\lambda}\hat{\mathbf{A}}(\mathbf{r}). \tag{14.59}$$

As the major new step, the free-electron mass is replaced by the effective mass. This simplification converts Eq. (14.56) into

$$\left[\frac{\partial^2}{\partial t^2} + \nabla\times\nabla\times\right]\hat{\mathbf{A}}(\mathbf{r},t) = -\mu_0\sum\nolimits_{\lambda}\left[\frac{Q^2}{m_\lambda}\hat{n}^{\lambda}\hat{\mathbf{A}}(\mathbf{r},t)\right]_{\mathrm{T}}. \tag{14.60}$$

In this homogeneous situation, $\hat{\mathbf{A}}(\mathbf{r},t)$ becomes transversal ($\nabla\cdot\hat{\mathbf{A}}(\mathbf{r},t) = 0$) such that $\nabla\times[\nabla\times\hat{\mathbf{A}}(\mathbf{r},t)] = -\nabla^2\hat{\mathbf{A}}(\mathbf{r},t)$ holds. These two pieces of information yield

$$\left[\nabla^2 - \frac{n^2}{c^2}\frac{\partial^2}{\partial t^2}\right]\hat{\mathbf{A}}(\mathbf{r},t) = \mu_0\epsilon_0\sum\nolimits_{\lambda}\frac{Q^2}{\epsilon_0 m_\lambda}\hat{n}^{\lambda}\hat{\mathbf{A}}(\mathbf{r},t). \tag{14.61}$$

By taking the Fourier transformation, $F(\omega) = \int dt\, F(t)\, e^{i\omega t}$ of Eq. (14.61), we obtain

$$\left[\nabla^2 - \frac{n^2}{c^2}\left(\omega^2 - \sum\nolimits_\lambda \frac{Q^2}{\epsilon_0 n^2 m_\lambda}\hat{n}^\lambda\right)\right]\hat{\mathbf{A}}(\omega, t) = 0, \tag{14.62}$$

after reorganizing the terms.

In this context, it is convenient to define a difference operator,

$$\Delta\hat{n}_\lambda \equiv \hat{n}_\lambda - \langle\hat{n}_\lambda\rangle, \tag{14.63}$$

which represents the deviation of the density operator from its expectation value. Using this in Eq. (14.62), we find

$$\left[\nabla^2 - \frac{n^2}{c^2}\left(\omega^2 - \sum\nolimits_\lambda \frac{Q^2\langle\hat{n}^\lambda\rangle}{\epsilon_0 n^2 m_\lambda}\right)\right]\hat{\mathbf{A}}(\mathbf{r}, t) = \frac{n^2}{c^2}\sum\nolimits_\lambda \frac{Q^2 \Delta\hat{n}^\lambda \hat{\mathbf{A}}(\mathbf{r}, t)}{\epsilon_0 n^2 m_\lambda}. \tag{14.64}$$

Typically, the $\langle\hat{n}^\lambda\rangle$ part includes the dominant contributions while $\Delta\hat{n}^\lambda$ identifies small quantum fluctuations around it.

When we neglect the contribution from the difference operator, we obtain the usual Helmholtz equation,

$$\left[\nabla^2 - \frac{n^2}{c^2}\left(\omega^2 - \omega_{\mathrm{PL}}^2\right)\right]\hat{\mathbf{A}}(\mathbf{r}, t) = 0, \tag{14.65}$$

where we have introduced the plasma frequency

$$\omega_{\mathrm{PL}}^2 \equiv \sum\nolimits_\lambda \frac{Q^2\langle\hat{n}^\lambda\rangle}{\epsilon_0 \epsilon_R m_\lambda}, \tag{14.66}$$

with the dielectric constant $\epsilon_R = n^2$ within the material.

In general, the three-dimensional Helmholtz equation (14.65) does not have physical solutions for $\omega < \omega_{\mathrm{PL}}$ because the corresponding mathematical solutions grow without a bound toward the edges of the system, see Exercise 14.8. This means that the frequency range $\omega < \omega_{\mathrm{PL}}$ is blocked completely in a three-dimensional system. For nanostructures, like QWs, $\omega_{\mathrm{PL}}(z)$ becomes z-dependent and $\omega_{\mathrm{PL}}(z)$ vanishes, especially in the bounding regions of the three-dimensional space. In this situation, regions with $\omega < \omega_{\mathrm{PL}}(z)$ display evanescent propagation of light as discussed in Section 9.4.3. Since evanescent propagation means exponentially decaying fields, one finds strong suppression of fields with $\omega < \omega_{\mathrm{PL}}(z)$ inside the nanostructure. Therefore, $\omega \ll \omega_{\mathrm{PL}}$ become uncoupled from the matter simply because the corresponding light is largely reflected from nanostructures. Since the observation of quantum-optical effects requires an efficient light–matter coupling, we clearly need to consider only cases with $\omega \gg \omega_{\mathrm{PL}}(z)$.

To give an estimate for the plasma frequency, we use typical GaAs values, i.e. $m_e = 0.07 m_0$, $n = 3.6$, $\langle\hat{n}^\lambda\rangle = 10^{16}\ \mathrm{cm}^{-3}$ such that $\omega_{\mathrm{PL}} = 6 \times 10^{12}\,\mathrm{s}^{-1}$, which corresponds to 4 meV in energy. This is the energy range of terahertz (THz) fields, which is more than two orders of magnitude below that of visible light. Thus, only THz fields can be strongly influenced by the plasmon in typical semiconductors. At the same time, optical fields are about a hundred times larger than this $\omega_{\rm PL}$. Thus, $\omega_{\rm PL}^2$ is 10^4 times smaller than the leading ω^2 contribution in Eq. (14.65). Hence, the plasma term can often be omitted altogether when the optical properties of semiconductors are studied. The same argumentation is valid for the intraband-current terms within $\hat{\mathbf{R}}^{\lambda,\lambda'}$ because currents correspond to THz transitions in semiconductors. As a result, we can generally apply the approximation

$$\hat{\mathbf{R}}^{\lambda,\lambda'}_{\mathbf{q}_{\|},\rm opt} = (1-\delta_{\lambda,\lambda'})\frac{\partial}{\partial t}\hat{\mathbf{P}}^{\lambda,\lambda'}_{\mathbf{q}_{\|},\rm opt} \tag{14.67}$$

to describe the response (14.58) to optical fields. Since we intend to study semiconductor quantum optics, we will apply this approximation in our forthcoming investigations. However, we should keep in mind that the current terms must be included as soon as we are interested in THz effects.

14.3 Formal properties of general operators

We can now consider a generic expectation value of an observable,

$$\langle \hat{N}\rangle \equiv \left\langle B_1^\dagger \cdots B_K^\dagger B_J \cdots B_1\; D_1^\dagger \cdots D_{K'}^\dagger D_{J'} \cdots D_1\; a_1^\dagger \cdots a_{N_a}^\dagger a_{N_a} \cdots a_1 \right\rangle,$$
$$N = N_B + N_D + N_a\,, \quad N_B = J + K\,, \quad N_D = J' + K'\,, \tag{14.68}$$

consisting of N_B photon operators, N_D phonon operators, and $2N_a$ carrier-operators. As discussed in Section 11.1.4, all carrier observables contain equally many creation and annihilation operators. It is straightforward to show that the expectation value $\langle \hat{N}\rangle$ vanishes for the systems studied, whenever Fermion creation- and annihilation-operator numbers are different, see Exercise 14.9. Therefore, we can restrict our analysis to the situation where we have an equal number N_a of carrier creation and annihilation operators.

The identification scheme determined above can be directly applied to classify $\langle \hat{N}\rangle$: Due to the formal equivalence between $a_\lambda^\dagger a_{\lambda'}$ and single photon or phonon operators – based on Eqs. (14.44)–(14.45) – all generic operators with the same value $N = N_B + N_D + N_a$ can be considered as formally equivalent. Thus, we can use the number N to classify the formal complexity of a generic operator or an expectation value. The relevant Fermion- and Boson-operator combinations are shown schematically in Fig. 14.4. The filled circles identify the formal particle number, N_B or N_D, for a term with J (K) Boson annihilation (creation) operators. The number of spheres (dashed circles) indicates the particle number N_a of a Fermionic observable with J (K) creation (annihilation) operators. For Fermions, only the cases with $J = K$ define physical observables (shaded squares), as discussed above. These simple rules allow us to construct the $N = N_B + N_D + N_a$ class for any physically relevant observable.

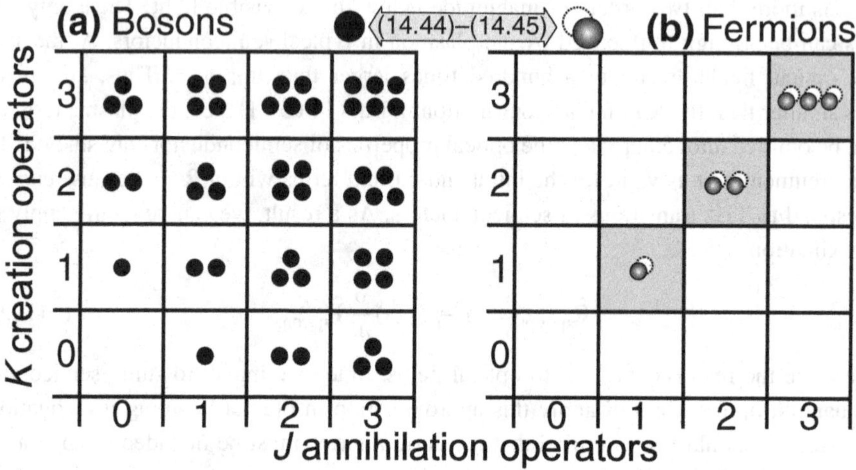

Figure 14.4 Formal identification of the number of particles involved in expectation values with **(a)** Boson and **(b)** Fermion operators. The black circles correspond to Bosonic creation or annihilation operators. The spheres (dashed circles) correspond to Fermionic creation (annihilation) operators. Physically relevant Fermion expectation values contain equally many creation and annihilation operators (shaded squares). Based on Eqs. (14.44)–(14.45), Boson operators correspond to Fermion-operator pairs. The particle number follows by counting the number of black circles and/or spheres.

We will use the terminology "N-particle operator" to denote $\hat{N}$ and "N-particle expectation value" to identify the expectation-value class $\langle \hat{N} \rangle$. For example, B is a single-particle operator while $a^\dagger a B$ is a two-particle operator. The usefulness of this classification scheme will become obvious when we derive the quantum dynamics of different operators consistently up to a given level.

14.3.1 Operator hierarchy problem

Let us first consider the possible complications in the phonon- and photon-operator dynamics with respect to the different operator classes. If we study the coupling of the phonons to the carrier system, Eqs. (14.36)–(14.39), we observe that they involve *single*-particle operators $\hat{\mathbf{n}}^{\lambda}_{\mathbf{p}_{\|}}$. The very same conclusion follows for photons if we apply the limit (14.67) in Eqs. (14.34)–(14.35). Then, a photon operator only couples to a single-particle operator $\hat{\mathbf{R}}_{\mathrm{opt}}$.

Figure 14.5(a) schematically presents the hierarchical coupling structure: a Boson operator (filled circle) is coupled (arrow) to a pair consisting of a Fermion creation (sphere) and an annihilation (dashed circle) operator, according to Eqs. (14.34)–(14.35) and (14.36)–(14.39). The "+0" within the coupling arrow indicates that the formal particle number is not increased. Thus, the photon and phonon dynamics, Eqs. (14.44) and (14.45), formally

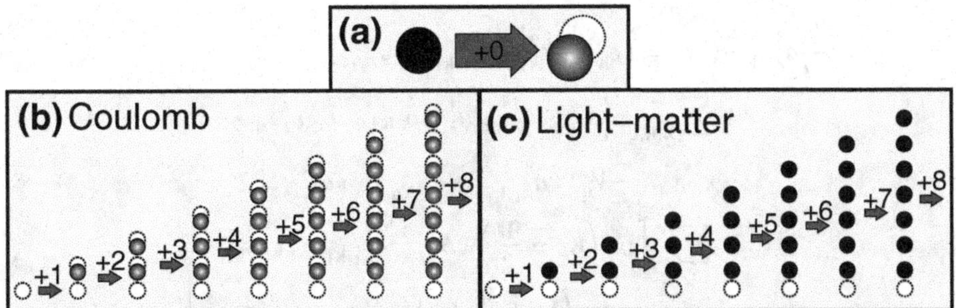

Figure 14.5 Formal structure of the operator hierarchy problem. **(a)** A Boson (black circle) operator couples to a product of Fermion creation (sphere) and annihilation (dashed circle) operators. **(b)** The Coulomb and **(c)** light–matter interaction couple a Fermion annihilation operator to one more Fermion pair and photon operator, respectively. Both interactions yield an infinite hierarchy of equations.

remain fully within the family of single-particle operators. This means that we do not have to introduce any approximations when solving the *pure* phonon dynamics. However, as soon as we look at the carrier dynamics we obtain couplings to higher-order terms, which then introduce the necessity to perform controlled approximations. Retroactively, these approximations will then affect the phonon dynamics.

A closer inspection of the Fermion-operator dynamics (14.37) and (14.40) shows that all interaction terms produce contributions where the particle number is increased by one. Thus, the Fermion-operator dynamics cannot be closed. Figures 14.5(b) and (c) present the coupling structure resulting from the Coulomb interaction and the quantum-optical coupling, respectively. The phonon coupling is formally similar to that in Fig. 14.5(c). The "$+N$" number above the arrow indicates how many additional particles are involved due to coupling compared with the original operator, here a Fermion annihilation operator (dashed circle). We see now very clearly that *all* interactions produce the same *hierarchy problem* where the operator dynamics cannot be closed. Instead, we obtain infinitely many equations where operators couple to new operator classes with one more particle. This is the distinct signature of the hierarchy problem.

14.3.2 BBGKY hierarchy problem

The discussion in Sections 14.1.2–14.3.1 summarizes to a large extent how much information about the optical and phonon properties can be obtained without the detailed knowledge of the carrier system. In order to proceed further, we have to determine the operator dynamics for pairs of carrier operators $a^\dagger a$, which follows directly by inserting Eqs. (14.37) and (14.40) into Eq. (14.3). After reorganizing the resulting expressions, we find

$$
\begin{aligned}
i\hbar\frac{\partial}{\partial t}a^{\dagger}_{\lambda_1,\mathbf{k}_1}a_{\lambda_2,\mathbf{k}_2} = &\left(\epsilon^{\lambda_2}_{\mathbf{k}_2}-\epsilon^{\lambda_1}_{\mathbf{k}_1}\right)a^{\dagger}_{\lambda_1,\mathbf{k}_1}a_{\lambda_2,\mathbf{k}_2}\\
&+\sum\nolimits_{\lambda,\mathbf{k},\mathbf{q}_{\|}}\Big[V^{\lambda_2,\lambda}_{\mathbf{q}_{\|}}a^{\dagger}_{\lambda_1,\mathbf{k}_1}a^{\dagger}_{\lambda,\mathbf{k}}a_{\lambda,\mathbf{k}+\mathbf{q}_{\|}}a_{\lambda_2,\mathbf{k}_2-\mathbf{q}_{\|}}\\
&\qquad -V^{\lambda_1,\lambda}_{\mathbf{q}_{\|}}a^{\dagger}_{\lambda_1,\mathbf{k}_1-\mathbf{q}_{\|}}a^{\dagger}_{\lambda,\mathbf{k}\mathbf{q}_{\|}}a_{\lambda,\mathbf{k}}a_{\lambda_2,\mathbf{k}_2}\Big]\\
&-\sum\nolimits_{\mathbf{q}_{\|}}\Big[\mathbf{j}_{\lambda_2}\left(\mathbf{k}_2-\frac{\mathbf{q}_{\|}}{2}\right)\cdot\mathbf{A}^{\lambda_2,\lambda_2}_{\mathbf{q}_{\|}}a^{\dagger}_{\lambda_1,\mathbf{k}_1}a_{\lambda_2,\mathbf{k}_2-\mathbf{q}_{\|}}\\
&\qquad -\mathbf{j}_{\lambda_1}\left(\mathbf{k}_1+\frac{\mathbf{q}_{\|}}{2}\right)\cdot\mathbf{A}^{\lambda_1,\lambda_1}_{\mathbf{q}_{\|}}a^{\dagger}_{\lambda_1,\mathbf{k}_1+\mathbf{q}_{\|}}a_{\lambda_2,\mathbf{k}_2}\Big]\\
&-\sum\nolimits_{\mathbf{q}_{\|},\lambda}\Big[\frac{Q\mathbf{p}_{\lambda_2,\lambda}}{m_0}\cdot\mathbf{A}^{\lambda_2,\lambda}_{\mathbf{q}_{\|}}a^{\dagger}_{\lambda_1,\mathbf{k}_1}a_{\lambda,\mathbf{k}_2-\mathbf{q}_{\|}}\\
&\qquad -\frac{Q\mathbf{p}_{\lambda,\lambda_1}}{m_0}\cdot\mathbf{A}^{\lambda,\lambda_1}_{\mathbf{q}_{\|}}a^{\dagger}_{\lambda,\mathbf{k}_1+\mathbf{q}_{\|}}a_{\lambda_2,\mathbf{k}_2}\Big]\\
&+\frac{Q^2}{2m_0}\sum\nolimits_{\mathbf{q}_{\|},\mathbf{q}'_{\|}}\Big[\mathbf{A}^{(2),\lambda_2}_{\mathbf{q}_{\|},-\mathbf{q}'_{\|}}a^{\dagger}_{\lambda_1,\mathbf{k}_1}a_{\lambda_2,\mathbf{k}_2+\mathbf{q}'_{\|}-\mathbf{q}_{\|}}\\
&\qquad -\mathbf{A}^{(2),\lambda_1}_{\mathbf{q}_{\|},-\mathbf{q}'_{\|}}a^{\dagger}_{\lambda_1,\mathbf{k}_1+\mathbf{q}_{\|}-\mathbf{q}'_{\|}}a_{\lambda,\mathbf{k}_2}\Big]\\
&-\sum\nolimits_{\mathbf{q}_{\|}}\Big[\mathcal{G}^{\lambda_2}_{\mathbf{q}_{\|}}a^{\dagger}_{\lambda_1,\mathbf{k}_1}a_{\lambda_2,\mathbf{k}_2-\mathbf{q}_{\|}}-\mathcal{G}^{\lambda_1}_{\mathbf{q}_{\|}}a^{\dagger}_{\lambda_1,\mathbf{k}_1+\mathbf{q}_{\|}}a_{\lambda_2,\mathbf{k}_2}\Big].
\end{aligned}
\tag{14.69}
$$

This dynamics features several problematic aspects because it contains:

(i) coupling to two-particle operators due to the Coulomb interaction,
(ii) mixed terms with one photon or one phonon operator combined with a carrier operator pair due to light–matter as well as phonon–carrier interaction.

As a reminder of the terminology used in this book, *an N-particle operator always corresponds to an operator class with N creation and N annihilation operators of pure carriers*. In general, already the Coulomb interaction leads to the operator dynamics where N-particle operators couple to $(N+1)$-particle operators. For the corresponding expectation values, the problem is known as the Bogolyubov–Born–Green–Kirkwood–Yvon (BBGKY) hierarchy resulting in a chain of infinitely many integrodifferential evolution equations. If the number of particles is limited, the hierarchy extends "only" to the level of the highest particle number.

Photon and phonon operators formally correspond to single-particle operators, such that the mixed operator terms in Eq. (14.69) represent couplings to two-particle operators. In addition, the $\hat{\mathbf{A}}^2$ part produces in principle also a coupling to three-particle operators, which makes the hierarchy problem even worse. However, since $\hat{\mathbf{A}}^2$ is related to the collective plasma oscillations, it does not require the full solution of the corresponding hierarchy, as discussed in Section 14.2.2.

In practice, the Coulomb interaction and the relevant parts of the quantum-mechanical photon and phonon interactions lead to similar hierarchy problems. Schematically, this can be presented as

$$i\hbar\frac{\partial}{\partial t}\langle N\rangle = T[\langle N\rangle] + \mathrm{Hi}[\langle N+1\rangle], \tag{14.70}$$

where $\langle N\rangle$ denotes a generic N-particle expectation value. Coupling to the same class of expectation values is described by a functional T while $\mathrm{Hi}[\cdots]$ symbolizes the coupling to the higher-order operator class via the Coulomb, light–matter, and carrier–phonon interactions. This schematic form clearly indicates how the infinite chain of equations emerges. The specific forms of T[. . .] and Hi[. . .] are obtained from Eqs. (14.34)–(14.36), (14.39), and (14.69).

The hierarchy problem (14.70) seems to be simple; nevertheless, it is basically responsible for all the complications in many-body and quantum-optical investigations. Current theoretical and numerical approaches provide exact solutions for such problems only in very limited cases. As a result, a significant portion of past and present research efforts are devoted to develop consistent approximation schemes for Eq. (14.70). In the course of time, these activities produced several formally very different techniques. For many-particle systems, the common approaches can roughly be classified into the following general categories:

1. The presented equation-of-motion technique belongs to the general class of density-matrix theories where the dynamics of the relevant quantities is obtained from a well-defined set of reduced density matrices or equivalently from expectation values with a lower dimension than the full density matrix. The solutions involve a single time argument corresponding to the actual physical time. The benefits of this approach follow from the direct access to physically relevant quantities. At the same time, the corresponding equation structure is relatively simple such that most of the terms that appear have a direct physical interpretation. Thus, one can often use physical arguments to develop efficient, rapidly converging approximation schemes.
2. Methods based on Feynman path integrals describe the quantum dynamics of N particles via an infinite ensemble of classical-like particle trajectories. The quantum features result from the averaging of all possible trajectories. Often, the corresponding multidimensional integrals can be evaluated numerically via Monte-Carlo sampling. This approach can, in principle, include correlations up to a very high number of particles. This approach is very efficient when thermodynamic equilibrium properties of many-body systems are studied.
3. The Green's functions approach is based on a mathematical construction where single-particle Green's functions are known analytically for the noninteracting system. The interactions can then be included via integral equations and higher-dimensional Green's functions. The method often involves multiple time indices and the hierarchy problem can be presented in terms of Feynman diagrams. This approach produces some properties analytically, e.g., the poles of the Green's functions can be related to quasiparticles. Furthermore, systematic truncation schemes for the hierarchy can be based on Feynman diagram classes.

4. The density-functional theory expresses, e.g., the system energy as a functional of the electron density. The hierarchy problem can be presented as functional integrals where systematic approximations can be introduced. This approach is in extensive use when the ground-state energy structure of many-body systems is analyzed, especially for electronic systems with a nontrivial geometry. Also, quantities directly related to densities, like currents, have straightforward descriptions.
5. When the mathematical structure of the interaction is reduced to the simplest level, one can sometimes find nearly analytical solutions to the hierarchy problem. This reduction often introduces simple models with parametrized interactions and/or spatially reduced couplings among the particles. For example, Hubbard-type models belong in this category. These approaches have been used extensively to study, e.g., metal-insulator transition, magnetism, and superconductivity.
6. Formally, the full many-body problem can be reduced to an effective single-particle problem when the limitations posed by the hierarchy problem are consistently included; this constitutes the so-called N-representability problem. Many related methods try to directly solve the dynamics with physically meaningful low-dimensional wave functions. This approach has been successfully used, e.g., in quantum chemistry, atomic, and molecular quantum optics. For example, optical transitions in molecules, time-dependent particle tunneling, collisions of cold atoms, and Bose–Einstein condensation of atoms have been studied with this kind of method.

In principle, all these methods should give us the same results if infinitely accurate solutions were possible. However, differences usually show up in practical computations because the introduced approximations are very different, both in complexity and accuracy. Thus, the specific choice of the method depends very much on the studied many-body problem and personal taste. As a good guiding principle, one should estimate which one of the approaches is the least elaborate when a predetermined level of accuracy is wanted.

Exercises

14.1 **(a)** Compute $i\hbar\frac{\partial}{\partial t}B^\dagger\,|_{\text{free}}$ and $i\hbar\frac{\partial}{\partial t}a_{\lambda k}\,|_{\text{free}}$ explicitly and verify the results (14.14)–(14.15). **(b)** Evaluate $\left[B^\dagger_{\mathbf{q}}, \hat{H}_{\text{int}}\right]_-$ using an analogous strategy to that used in the computation of (14.24)–(14.26).

14.2 Assume that $\hat{O}(\mathbf{r})$ is a Hermitian operator and that we may construct its in-plane Fourier transformation

$$\hat{O}(\mathbf{r}) = \sum\nolimits_{\mathbf{q}_\parallel} \hat{O}_{\mathbf{q}_\parallel}(z)\, e^{i\mathbf{q}_\parallel\cdot\mathbf{r}_\parallel}, \quad \hat{O}_{\mathbf{q}_\parallel}(z) = \frac{1}{S}\int d^2r_\parallel\, \hat{O}(\mathbf{r})\, e^{-i\mathbf{q}_\parallel\cdot\mathbf{r}_\parallel}.$$

(a) Show that this yields the transformation $\hat{O}^\dagger_{\mathbf{q}_\parallel}(z) = \hat{O}_{-\mathbf{q}_\parallel}(z)$. **(b)** How are $\hat{A}^{\lambda\lambda'}_{\mathbf{q}_\parallel}(z)$ and $R^{\lambda\lambda'}_{\mathbf{q}_\parallel}(z)$ transformed under Hermitian conjugation?

14.3 (a) Show that the identifications (14.30)–(14.33) convert the results (14.15) and (14.27)–(14.29) into Eqs. (14.34)–(14.35). **(b)** Derive the dynamics of the photon-annihilation operator, i.e., Eq. (14.36).

14.4 Derive the dynamics of the carrier-annihilation operator, i.e., Eq. (14.37).

14.5 A Boson operator has the dynamics

$$B_{\mathbf{q}}(t) = B_{\mathbf{q}}(0)\, e^{-i\omega_{\mathbf{q}} t} - \frac{1}{i\hbar}\int_0^t ds\, \hat{\mathcal{R}}_{\mathbf{q}}(s)\, e^{i\omega_{\mathbf{q}}(s-t)}. \tag{14.71}$$

Assume that $\hat{\mathcal{R}}_{\mathbf{q}}^{\dagger} = \hat{\mathcal{R}}_{-\mathbf{q}}(s)$ where $\hat{\mathcal{R}}_{\mathbf{q}}(t)$ is an operator describing the matter excitation, compare Eq. (14.44). **(a)** Which commutation relations are found between $B_{\mathbf{q}}(0)$, $B_{\mathbf{q}}^{\dagger}(0)$, and $\mathcal{R}_{\mathbf{q}}(s)$ based on the relation between Boson operators, satisfied at any time? **(b)** We generally have $[B_{\mathbf{q}}(t), \hat{\mathcal{R}}_{\mathbf{q}}(t)]_- = 0 = \left[B_{\mathbf{q}}^{\dagger}(t), \hat{\mathcal{R}}_{\mathbf{q}}(t)\right]_-$. Use the formal expression (14.71) to compute $B_{\mathbf{q}}(t=\varepsilon)$ and $B_{\mathbf{q}}^{\dagger}(t=\varepsilon)$ for an infinitesimal ε. Solve $\hat{\mathcal{R}}_{\mathbf{q}}(0)$ in terms of $B_{\mathbf{q}}(0)$, $B_{\mathbf{q}}^{\dagger}(0)$, $B_{\mathbf{q}}(\varepsilon)$, and $B_{\mathbf{q}}^{\dagger}(\varepsilon)$ and show that $[B_{\mathbf{q}}(0), B_{\mathbf{q}}(\varepsilon)]_- \neq 0$. What is $\left[B_{\mathbf{q}}(0), B_{\mathbf{q}}^{\dagger}(\varepsilon)\right]_-$? **(c)** Show that $\left\langle B_{\mathbf{q}}^{\dagger}(t) B_{\mathbf{q}}(t)\right\rangle = \frac{1}{\hbar^2}\int_0^t ds \int_0^t ds' \left\langle \mathcal{R}_{\mathbf{q}}^{\dagger}(s)\, \mathcal{R}_{\mathbf{q}'}(s')\right\rangle e^{i\omega_q(s'-s)}$ if the light field is initially in a vacuum state.

14.6 (a) Show that $\sum_{\mathbf{q}} \frac{\mathcal{E}_q^2}{\omega_q^2}\left[u_q(\mathbf{r})\, u_q^*(\mathbf{r}') - u_q^*(\mathbf{r})\, u_q(\mathbf{r}')\right] = 0$ based on the unitary transformations discussed in Section 10.2.1. How does this result simplify Eq. (14.46)? **Hint:** *Use the relation* (14.51). **(b)** Derive the dynamical equation for the vector potential $\hat{\mathbf{A}}(\mathbf{r}) = \sum_{\mathbf{q}} \frac{\mathcal{E}_q}{\omega_q}\left(\mathbf{u}_{\mathbf{q}}(\mathbf{r})\, B_{\mathbf{q}} + \mathbf{u}_z^*(\mathbf{r})\, B_{\mathbf{q}}^{\dagger}\right)$ by starting from Eqs. (14.34)–(14.35).

14.7 (a) Show that $\sum_{\mathbf{q}}\left[\mathbf{u}_{\mathbf{q}}(\mathbf{r})\, \mathbf{u}_{\mathbf{q}}^*(\mathbf{r}') + \mathbf{u}_{\mathbf{q}}(\mathbf{r}')\, \mathbf{u}_{\mathbf{q}}^*(\mathbf{r})\right] = \frac{2}{n^2(z)}\delta^T(\mathbf{r}-\mathbf{r}')$ based on the relation (9.24). **(b)** Derive the wave equation for the electric field $\hat{E}(\mathbf{r}) = \sum_{\mathbf{q}} i\varepsilon_{\mathbf{q}}(\mathbf{u}_{\mathbf{q}}(\mathbf{r}) B_{\mathbf{q}} - \mathbf{u}_{\mathbf{q}}^*\left(\mathbf{r}) B_{\mathbf{q}}^{\dagger}\right)$ based on Eqs. (14.34)–(14.35).

14.8 (a) Use a plane-wave ansatz $u(z) = e^{iqz}$ to solve the one-dimensional Helmholtz problem $\left[\frac{\partial^2}{0z^2} - \frac{n^2}{c^2}\left(\omega^2 - \omega_{\text{pl}}^2\right)\right] u(z) = 0$. How is q defined and which frequencies must be excluded to have a finite $u(z)$? **(b)** Consider light modes in a quantum-well system

$$\left[\frac{\partial^2}{\partial z^2} - \frac{n^2}{c^2}\left(\omega^2 - \omega_{\text{pl}}^2(z)\right)\right] u(z) = 0$$

with $\omega_{\text{pl}}(z) = \omega_{\text{pl}}$ for $|z| \leq L/2$ and $\omega_{\text{pl}}(z) = 0$ otherwise. Determine the normalized modes $u(z)$ and define $u(0)$ at the QW position as a function of ω. **Hint:** *Use the transfer-matrix method.*

14.9 Assume that the semiconductor ground state is $|\Psi_{\text{gs}}\rangle = \prod_{\mathbf{k}} a^{\dagger}_{v\mathbf{k}}|0\rangle$. **(a)** Show that the expectation values $I^J_K = \left\langle a^{\dagger}_{v_1\mathbf{k}_1} \cdots a^{\dagger}_{v_J\mathbf{k}_J}\, a_{\lambda_K\mathbf{k}_K} \cdots a_{\lambda_1\mathbf{k}_1}\right\rangle$ vanish for $J \neq K$. Conclude that the same is also true when I^J_K contains one or more normally ordered Boson operators. **(b)** Show that $I^J_{K\neq J}$ remains zero for all times if $I^J_K = 0$ for $J \neq K$ at the initial moment. **Hint:** *Use the result* **(a)** *when analyzing the equation of motion for* I^J_K.

14.10 **(a)** Derive Eq. (14.69) by starting from Eqs. (14.37) and (14.40). Deduce the functional form of Hi[...] for $\langle 1\rangle$ [see Eq. (14.70)]. **(b)** Apply the two-band approximation ($\lambda = c, v$) and assume homogeneous conditions $\left\langle a^{\dagger}_{\lambda k}\, a_{\lambda' k'}\right\rangle \equiv \delta_{k,k'}\, P^{\lambda\lambda'}_k$. Which single-particle dynamics follows from Eq. (14.69)?

Further reading

Fundamental comparisons between Fermionic and Bosonic many-body systems can be systematially formulated through the so-called Usui transformation:

T. Usui (1960). Excitations in a high density electron gas, *Prog. Theor. Phys.* **23**, 787.

Plasmons in low-dimensional systems are studied, e.g., in:

F. Stern (1967). Polarizability of a two-dimensional electron gas, *Phys. Rev. Lett.* **18**, 546.

The BBGKY hierarchy problem was originally identified in:

J. Yvon (1935). *Theorie statistique des fluides et l'equation d'etat*, Actes sientifique et industrie. No 203. Paris, Hermann.
N. N. Bogoliubov (1946). Kinetic equations, *J. Phys. USSR* **10**, 265.
M. Born and H. S. Green (1946). A general kinetic theory of liquids I. The molecular distribution functions, *Proc. Roy. Soc. A* **188**, 10.
J. G. Kirkwood (1946). The statistical mechanical theory of transport processes I. General theory, *J. Chem. Phys.* **14**, 180.

Different approaches to the BBGKY problem are investigated, e.g., in:

L. V. Keldysh (1964). Diagram technique for nonequilibrium processes, *Zh. Eksp. Teor. Fiz.* **47**, 1515 [Sov. Phys. JETP **20**, 1018].
R. Zimmerman (1988). *Theory of Highly Excited Semiconductors*, Leipzig, Taubner.
H. Haug and A.-P. Jauho (1997). *Quantum Kinetics in Transport and Optics of Semiconductors*, Heidelberg, Springer.
E. L. Economou (2006). *Green's Functions in Quantum Physics*, 3rd edition, Leipzig, Springer.
M. P. Nightingale and C. J. Umrigar, eds. (1999). *Quantum Monte Carlo Methods in Physics and Chemistry*, Dordrecht, Kluwer Academic Publishers.
I. Shavitt and R. J. Bartlett (2009). *Many-Body Methods in Chemistry and Physics: MBPT and Coupled-Cluster Theory*, Cambridge, Cambridge University Press.

F. H. L. Essler, H. Frahm, F. Göhmann, A. Klümper, and V. E. Korepin (2005). *The One-Dimensional Hubbard Model*, Cambridge, Cambridge University Press.

D. S. Sholl and J. A. Stecke (2009). *Density Functional Theory: A Practical Introduction*, Hoboken, NJ, Wiley & Sons.

H. Kleinert (2005). *Path Integrals in Quantum Mechanics, Statistics, Polymer Physics, and Financial Markets*, 3rd edition, Singapore, World Scientific.

15

Cluster-expansion representation of the quantum dynamics

In this book, we encounter the hierarchy problem when we apply the equation-of-motion technique to analyze the quantum dynamics of the coupled light–matter system. To obtain systematic approximations, we use the so-called *cluster-expansion method* where many-body quantities are systematically grouped into cluster classes based on how important they are to the overall quantum dynamics. With increasing complexity, the clusters contain

1. independent single particles (singlets),
2. interacting pairs (doublets),
3. three-particle terms (triplets),
4. coupled four-particle contributions (quadruplets),
5. as well as higher-order correlations.

In this context, the N-particle concept is somewhat formal because it refers to generic N-particle expectation values $\langle N \rangle$, which may consist of an arbitrary mixture of carrier, photon, and phonon operators, as discussed in Section 14.2. In order to truncate the hierarchy problem at a given level, $\langle N \rangle$ is approximated through a functional structure that includes all clusters up to the predetermined level while all remaining clusters with a higher rank are omitted. It is natural that the corresponding approximations can be systematically improved by increasing the number of clusters included.

To the best of our knowledge, the idea of coupled-clusters approaches was first formulated by Fritz Coester and Hermann Kümmel in the 1950s to describe nuclear many-body phenomena. The approach was then modified for the needs of quantum chemistry by Jiri Cizek 1966 to deal with many-body phenomena in atoms and molecules. Currently, it is one of the most accurate methods to compute molecular eigenstates. Explicit computations demonstrate that clusters up to doublets or triplets can accurately explain the complicated energetic structures of molecules. This is a clear indicator of the fast convergence of this expansion approach as a function of the cluster number. In solid-state physics, the coupled-clusters approach has evolved into the cluster-expansion method and we will analyze next how it can be utilized for semiconductor quantum optics.

15.1 Singlet factorization

The singlet properties of a Boson field generally follow from photon and phonon expectation values $\langle B \rangle_{\mathbf{q}} \equiv \beta_{\mathbf{q}}$ and $\langle D \rangle_{\mathbf{p}} \equiv \Delta_{\mathbf{p}}$, respectively. If the light field is defined entirely by singlets, we can apply a general substitution

$$B_{\mathbf{q}} \to \beta_{\mathbf{q}}\,, \qquad B^{\dagger}_{\mathbf{q}} \to \beta^{\star}_{\mathbf{q}}\,, \qquad D_{\mathbf{p}} \to \Delta_{\mathbf{p}}\,, \qquad D^{\dagger}_{\mathbf{p}} \to \Delta^{\star}_{\mathbf{p}}, \tag{15.1}$$

where we simply replace the operators by complex-valued amplitudes $\beta_{\mathbf{q}}$ and $\Delta_{\mathbf{p}}$. In this situation, the mode-expansion (14.47) produces the usual classical expression

$$\hat{\mathbf{E}}(\mathbf{r}) \to \sum_{\mathbf{q}} i\mathcal{E}_{\mathbf{q}} \left[\mathbf{u}_{\mathbf{q}}(\mathbf{r})\beta_{\mathbf{q}} - \mathbf{u}^{\star}_{\mathbf{q}}(\mathbf{r})\beta^{\star}_{\mathbf{q}}\right] \equiv \mathbf{E}(\mathbf{r}). \tag{15.2}$$

In other words, each light mode is described exactly by the classical field amplitude $\mathbf{E}(\mathbf{r})$. In particular, Eq. (15.1) implies that the singlet description of Boson fields yields the *classical factorization*:

$$\begin{aligned}\langle \hat{N} \rangle_{\mathrm{S}} &\equiv \left\langle B^{\dagger}_{1} \cdots B^{\dagger}_{K} D^{\dagger}_{1} \cdots D^{\dagger}_{K'}\, a^{\dagger}_{1} \cdots a^{\dagger}_{N_a} a_{N_a} \cdots a_{1}\, D_{K'} \cdots D_{1} B_{J} \cdots B_{1} \right\rangle_{\mathrm{S}} \\ &= \left\langle a^{\dagger}_{1} \cdots a^{\dagger}_{N_a} a_{N_a} \cdots a_{1} \right\rangle_{\mathrm{S}} \prod_{k=1}^{K} \beta^{\star}_{\mathbf{q}_k} \prod_{j=1}^{J} \beta_{\mathbf{q}_j} \prod_{k'=1}^{K'} \Delta^{\star}_{\mathbf{p}_{k'}} \prod_{j'=1}^{J'} \Delta_{\mathbf{p}_{j'}}\end{aligned} \tag{15.3}$$

of a *normally ordered* $N = \left(N_a + J + K + J' + K'\right)$-particle expectation value. The singlet factorization of the Fermions is discussed in Section 15.1.2.

The classical factorization implies that the quantum fluctuations (dashed line) are ignored and only the classical amplitude (solid line) of the field is included in the description. This situation is schematically depicted in Fig. 15.1, compare also Fig. 10.1.

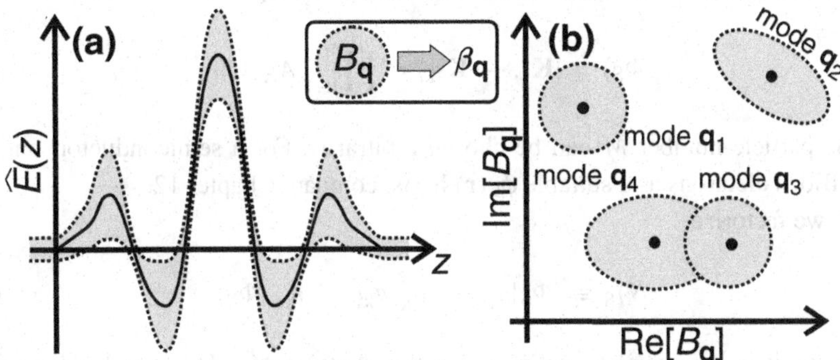

Figure 15.1 Classical factorization of light. **(a)** The quadrature fluctuations (dashed lines, shaded area) are shown in comparison with the classical light field (solid line). Mathematically, the classical limit corresponds to replacing the operator $B_{\mathbf{q}}$ by a complex-valued amplitude $\beta_{\mathbf{q}}$. **(b)** In the phase space, the quantum fluctuations (dashed, shaded circle) have a minimum extension in the quadratures $\hat{x} = \mathrm{Re}[B_{\mathbf{q}}]$ and $\hat{y} = \mathrm{Im}[B_{\mathbf{q}}]$ of the modes $\mathbf{q}_j$ (shown for j=1, 2, 3, and 4). The classical factorization is a single $(x,\ y)$ point in the phase space.

Since the singlets behave fully classically, all quantum-optical features must follow from doublets and clusters beyond. We study the relation between the different clusters and quantum-optical effects further in Chapters 21–23 and 28–30.

15.1.1 Expectation values of a Slater-determinant state

The carrier singlets, $\left\langle a^{\dagger}_{\mathbf{k}} a_{\mathbf{k}'}\right\rangle$, define those averages that contain the lowest number of Fermion operators, i.e., two as shown in Exercise 14.9. The expectation value of any single Fermion operator vanishes such that there is no single-operator factorization of the generic N-particle cluster $\langle N\rangle$. In other words, we cannot extend the classical factorization (15.3) directly to the Fermion case. Instead, the singlet factorization $\langle N\rangle_{\mathrm{S}}$ must be constructed using products of $\left\langle a^{\dagger}_{\mathbf{k}} a_{\mathbf{k}'}\right\rangle$. Therefore, we analyze $\langle N\rangle$ when the many-body wave function is a generic Slater determinant. In other words, we connect the singlet factorization of the carrier expectation value $\langle N\rangle$ with the simplest physically allowed Fermion state.

To do this systematically, we analyze the properties of the Fermion field operator

$$\hat{\Psi}(\mathbf{r}) = \sum\nolimits_{\mathbf{k}} a_{\mathbf{k}}\, \phi_{\mathbf{k}}(\mathbf{r}) = \sum\nolimits_{\mathbf{p}} A_{\mathbf{K}}\, \Phi_{\mathbf{K}}(\mathbf{r}), \tag{15.4}$$

which can be expressed in terms of any mode expansion basis. Thus, we may equivalently use either $\phi_{\mathbf{k}}(\mathbf{r})$ or $\Phi_{\mathbf{K}}(\mathbf{r})$ which both form a complete and orthogonal set of wave functions. As discussed in Section 11.1.4, both $a_{\mathbf{k}}$ and $A_{\mathbf{K}}$ then satisfy the Fermion anticommutation relation (11.26) because they are just two different second-quantization representations of one and the same Fermion system.

Utilizing this freedom, we can assume that $\langle N\rangle$ is evaluated in terms of $a_{\mathbf{k}}$ while the system's wave function is a Slater determinant defined by $A_{\mathbf{K}}$:

$$|\Phi_{\mathrm{S}}\rangle = |\mathbf{K}_1, \cdots \mathbf{K}_M\rangle = \prod\nolimits_{j=1}^{M} A_{\mathbf{K}_j}|0\rangle \tag{15.5}$$

where the particle-number M can be chosen arbitrarily. For a semiconductor, one often uses the Bloch electrons as a suitable $\phi_{\mathbf{k}}(\mathbf{r})$ basis, compare Chapter 12.

Before we factorize

$$\langle N\rangle_{\mathrm{S}} \equiv \langle \Phi_{\mathrm{S}}| a^{\dagger}_{\mathbf{k}_N} \cdots a^{\dagger}_{\mathbf{k}_1} a_{\mathbf{k}'_1} \cdots a_{\mathbf{k}'_N} |\Phi_{\mathrm{S}}\rangle \tag{15.6}$$

into singlets, it is beneficial to define $a_{\mathbf{k}}$ in the $\Phi_{\mathbf{K}}(\mathbf{r})$ basis. This can be done with a straightforward projection

$$\begin{aligned} a_{\mathbf{k}} &= \int d^3r\; \phi^{\star}_{\mathbf{k}}(\mathbf{r})\hat{\Psi}(\mathbf{r}) = \sum\nolimits_{\mathbf{K}} \int d^3r\; \phi^{\star}_{\mathbf{k}}(\mathbf{r})\Phi_{\mathbf{K}}(\mathbf{r})\, A_{\mathbf{K}} \\ &= \sum\nolimits_{\mathbf{K}} \langle \phi_{\mathbf{k}}|\Phi_{\mathbf{K}}\rangle\, A_{\mathbf{K}}, \end{aligned} \tag{15.7}$$

where we used the abstract notation of Section 5.2.1 to express the inner product between the basis states. This transformation can be notationally simplified by introducing a transformation matrix and the summation convention:

$$\mathbf{U}_{\mathbf{k},\mathbf{K}} \equiv \langle \phi_{\mathbf{k}} | \Phi_{\mathbf{K}} \rangle, \qquad \mathbf{U}_{\mathbf{K},\mathbf{k}} \equiv \mathbf{U}^{\star}_{\mathbf{k},\mathbf{K}} = \langle \Phi_{\mathbf{K}} | \phi_{\mathbf{k}} \rangle,$$
$$a_{\mathbf{k}} = \sum\nolimits_{\mathbf{K}} \mathbf{U}_{\mathbf{k},\mathbf{K}} A_{\mathbf{K}} = \mathbf{U}_{\mathbf{k},\mathbf{K}} A_{\mathbf{K}}. \tag{15.8}$$

The identified matrix $\mathbf{U}$ actually defines a unitary transformation between the operators $a_{\mathbf{k}}$ and $A_{\mathbf{K}}$, see Exercise 15.2. Especially, we do not need to specify the basis $|\Phi_{\mathbf{K}}\rangle$ explicitly such that we may define the generic singlet factorization (15.6) by only assuming that the many-body state is a Slater determinant (15.5).

As the first step, we express the singlet term

$$\left\langle a^{\dagger}_{\mathbf{k}_1} a_{\mathbf{k}'_1} \right\rangle_{\mathrm{S}} \equiv \langle \Phi_{\mathrm{S}} | a^{\dagger}_{\mathbf{k}_1} a_{\mathbf{k}'_1} | \Phi_{\mathrm{S}} \rangle = \mathbf{U}_{\mathbf{K},\mathbf{k}_1} \mathbf{U}_{\mathbf{k}'_1,\mathbf{K}'} \langle \Phi_{\mathrm{S}} | A^{\dagger}_{\mathbf{K}} A_{\mathbf{K}'} | \Phi_{\mathrm{S}} \rangle$$
$$= \mathbf{U}_{\mathbf{K},\mathbf{k}_1} \mathbf{U}_{\mathbf{k}'_1,\mathbf{K}'} \langle \mathbf{K}_M, \cdots \mathbf{K}_1 | A^{\dagger}_{\mathbf{K}} A_{\mathbf{K}'} | \mathbf{K}_1, \cdots \mathbf{K}_M \rangle. \tag{15.9}$$

We notice that $A_{\mathbf{K}'}$ can remove one state from $|\mathbf{K}_1, \cdots \mathbf{K}_M\rangle$ only if $\mathbf{K}'$ is equal to one of the $\mathbf{K}_j$ states within the Slater determinant. For other $\mathbf{K}'$ values, $A_{\mathbf{K}'}|\mathbf{K}_1, \cdots \mathbf{K}_M\rangle$ vanishes. With the same argumentation, $\langle \mathbf{K}_M, \cdots \mathbf{K}_1 | A^{\dagger}_{\mathbf{K}}$ is nonzero only if $\mathbf{K}$ is equal to one of the $\mathbf{K}_j$ states. Therefore, the $\mathbf{K}$ and $\mathbf{K}'$ sums within Eq. (15.9) contribute only at discrete momenta, such that

$$\left\langle a^{\dagger}_{\mathbf{k}_1} a_{\mathbf{k}'_1} \right\rangle_{\mathrm{S}} = \sum_{j,l=1}^{M} \mathbf{U}_{\mathbf{K}_j,\mathbf{k}_1} \mathbf{U}_{\mathbf{k}'_1,\mathbf{K}_l} \langle \mathbf{K}_M, \cdots \mathbf{K}_1 | A^{\dagger}_{\mathbf{K}_j} A_{\mathbf{K}_l} | \mathbf{K}_1, \cdots \mathbf{K}_M \rangle. \tag{15.10}$$

Here, we have expressed the summations explicitly because they consist only of the M states within the Slater determinant. It is now beneficial to apply the anticommutation relation $A^{\dagger}_{\mathbf{K}_j} A_{\mathbf{K}_l} = \delta_{j,l} - A_{\mathbf{K}_l} A^{\dagger}_{\mathbf{K}_j}$ to Eq. (15.10),

$$\left\langle a^{\dagger}_{\mathbf{k}_1} a_{\mathbf{k}'_1} \right\rangle_{\mathrm{S}} = \sum\nolimits_{j,l=1}^{M} \mathbf{U}_{\mathbf{K}_j,\mathbf{k}_1} \mathbf{U}_{\mathbf{k}'_1,\mathbf{K}_l} \, \delta_{j,l} \, \langle \mathbf{K}_M, \cdots \mathbf{K}_1 | \mathbf{K}_1, \cdots \mathbf{K}_M \rangle$$
$$- \sum\nolimits_{j,l=1}^{M} \mathbf{U}_{\mathbf{K}_j,\mathbf{k}_1} \mathbf{U}_{\mathbf{k}'_1,\mathbf{K}_l} \langle \mathbf{K}_M, \cdots \mathbf{K}_1 | A_{\mathbf{K}_l} A^{\dagger}_{\mathbf{K}_j} | \mathbf{K}_1, \cdots \mathbf{K}_M \rangle$$
$$= \sum\nolimits_{j=1}^{M} \mathbf{U}_{\mathbf{K}_j,\mathbf{k}_1} \mathbf{U}_{\mathbf{k}'_1,\mathbf{K}_j} = \sum\nolimits_{j=1}^{M} \langle \Phi_{\mathbf{K}_j} | \phi_{\mathbf{k}_1} \rangle \langle \phi_{\mathbf{k}'_1} | \Phi_{\mathbf{K}_j} \rangle. \tag{15.11}$$

Here, we have applied that $A^{\dagger}_{\mathbf{K}_j} |\mathbf{K}_1, \cdots \mathbf{K}_M\rangle$ vanishes for all j due to the Pauli exclusion principle preventing the double occupation of the state $\mathbf{K}_j$, which would result from $A^{\dagger}_{\mathbf{K}_j} |\mathbf{K}_1, \cdots \mathbf{K}_M\rangle$. In the last step of (15.11), we have used the definition (15.8) and $\langle \Phi_{\mathrm{S}} | \Phi_{\mathrm{S}} \rangle = 1$. The final result shows that the singlets consist of an average value of $|\phi_{\mathbf{k}_1}\rangle\langle\phi_{\mathbf{k}'_1}|$ with respect to all single-particle states of the Slater determinant.

15.1.2 Hartree–Fock approximation and singlet factorization

To evaluate $\langle N\rangle_{\text{S}}$, we convert (15.6) into the $\Phi_{\mathbf{K}}(\mathbf{r})$ basis defining the Slater determinant. More specifically, we apply the unitary transformation (15.8), yielding

$$\begin{aligned}\langle N\rangle_{\text{S}} &= \langle\Phi_{\text{S}}|a^{\dagger}_{\mathbf{k}_1}\cdots a^{\dagger}_{\mathbf{k}_N}a_{\mathbf{k}'_N}\cdots a_{\mathbf{k}'_1}|\Phi_{\text{S}}\rangle\\ &= \mathbf{U}_{\mathbf{K}''_1,\mathbf{k}_1}\cdots\mathbf{U}_{\mathbf{K}''_N,\mathbf{k}_N}\mathbf{U}_{\mathbf{k}'_N,\mathbf{K}'_N}\cdots\mathbf{U}_{\mathbf{k}'_1,\mathbf{K}'_1}\langle\Phi_{\text{S}}|A^{\dagger}_{\mathbf{K}''_1}\cdots A^{\dagger}_{\mathbf{K}''_N}A_{\mathbf{K}'_N}\cdots A_{\mathbf{K}'_1}|\Phi_{\text{S}}\rangle.\end{aligned} \tag{15.12}$$

As in connection with Eqs. (15.9)–(15.10), both $\mathbf{K}''_j$ and $\mathbf{K}_l$ can be limited to those $\mathbf{K}_k$ values that appear within the Slater determinant (15.5). From now on, this limitation is implicitly included with the summation convention.

This limitation also implies

$$A^{\dagger}_{\mathbf{K}''_j}|\Phi_{\text{S}}\rangle = 0\,, \qquad \text{for} \quad \mathbf{K}''_j \in \{\mathbf{K}_1,\cdots\mathbf{K}_N\} \tag{15.13}$$

such that it is a good strategy to anticommute $A^{\dagger}_{\mathbf{K}''_j}$ toward $|\Phi_{\text{S}}\rangle$ in the expression (15.12). It is straightforward to show that

$$\begin{aligned}A^{\dagger}_{\mathbf{K}''_N}A_{\mathbf{K}'_N}\cdots A_{\mathbf{K}'_1}|\Phi_{\text{S}}\rangle = \sum\nolimits_{J=1}^{M}&\delta_{\mathbf{K}''_N,\mathbf{K}'_J}(-1)^{N+J}\\ &\times A_{\mathbf{K}'_N}\cdots \not{A}_{\mathbf{K}'_J}\cdots A_{\mathbf{K}'_1}|\Phi_{\text{S}}\rangle,\end{aligned} \tag{15.14}$$

see Exercise 15.3. Implementing this in Eq. (15.12), we find

$$\begin{aligned}\langle N\rangle_{\text{S}} = \sum\nolimits_{J=1}^{M}&\delta_{\mathbf{K}''_N,\mathbf{K}'_J}(-1)^{N+J}\ \mathbf{U}_{\mathbf{K}''_N,\mathbf{k}_N}\mathbf{U}_{\mathbf{k}'_J,\mathbf{K}'_J}\\ &\times\mathbf{U}_{\mathbf{K}''_1,\mathbf{k}_1}\cdots\mathbf{U}_{\mathbf{K}''_{N-1},\mathbf{k}_{N-1}}\mathbf{U}_{\mathbf{k}'_N,\mathbf{K}'_N}\cdots\not{\mathbf{U}}_{\mathbf{k}'_J,\mathbf{K}'_J}\cdots\mathbf{U}_{\mathbf{k}'_1,\mathbf{K}'_1}\\ &\times\langle\Phi_{\text{S}}|A^{\dagger}_{\mathbf{K}''_1}\cdots A^{\dagger}_{\mathbf{K}''_{N-1}}A_{\mathbf{K}'_N}\cdots\not{A}_{\mathbf{K}'_J}\cdots A_{\mathbf{K}'_1}|\Phi_{\text{S}}\rangle.\end{aligned} \tag{15.15}$$

The first term yields

$$\begin{aligned}1\text{st} &\equiv \delta_{\mathbf{K}''_N,\mathbf{K}'_J}(-1)^{N+J}\ \mathbf{U}_{\mathbf{K}''_N,\mathbf{k}_N}\mathbf{U}_{\mathbf{k}'_J,\mathbf{K}'_J}\\ &= \sum_{\mathbf{K}''_N=\mathbf{K}_1}^{\mathbf{K}_M}\sum_{\mathbf{K}'_J=\mathbf{K}_1}^{\mathbf{K}_M}\delta_{\mathbf{K}''_N,\mathbf{K}'_J}(-1)^{N+J}\mathbf{U}_{\mathbf{K}''_N,\mathbf{k}_N}\mathbf{U}_{\mathbf{k}'_J,\mathbf{K}'_J}\\ &= (-1)^{N+J}\sum_{\mathbf{K}'=\mathbf{K}_1}^{\mathbf{K}_M}\mathbf{U}_{\mathbf{K}',\mathbf{k}_N}\mathbf{U}_{\mathbf{k}'_J,\mathbf{K}'} = (-1)^{N+J}\left\langle a^{\dagger}_{\mathbf{k}_N}a_{\mathbf{k}'_J}\right\rangle\end{aligned} \tag{15.16}$$

after we have expressed the implicit sums explicitly in terms of the $\mathbf{K}_l$ states within $|\Phi_{\text{S}}\rangle$. The last step identifies the single-particle expectation value (15.11). In this context, we have changed the summation index $\mathbf{K}''_N = \mathbf{K}'_J$ into $\mathbf{K}'$.

The second part of Eq. (15.15) becomes

$$
\begin{aligned}
2\text{nd} &\equiv \mathbf{U}_{\mathbf{K}''_1,\mathbf{k}_1} \cdots \mathbf{U}_{\mathbf{K}''_{N-1},\mathbf{k}_{N-1}} \mathbf{U}_{\mathbf{k}'_N,\mathbf{K}'_N} \cdots \cancel{\mathbf{U}_{\mathbf{k}'_J,\mathbf{K}'_J}} \cdots \mathbf{U}_{\mathbf{k}'_1,\mathbf{K}'_1} \\
&\quad \times \langle \Phi_\mathrm{S} | A^\dagger_{\mathbf{K}''_1} \cdots A^\dagger_{\mathbf{K}''_{N-1}} A_{\mathbf{K}'_N} \cdots \cancel{A_{\mathbf{K}'_J}} \cdots A_{\mathbf{K}'_1} | \Phi_\mathrm{S} \rangle \\
&= \left\langle a^\dagger_{\mathbf{k}_1} \cdots a^\dagger_{\mathbf{k}_{N-1}} a_{\mathbf{k}'_N} \cdots \cancel{a_{\mathbf{k}'_J}} \cdots a_{\mathbf{k}'_1} \right\rangle_\mathrm{S},
\end{aligned}
\tag{15.17}
$$

which is an $(N-1)$-particle expectation value where the states $\mathbf{k}_N$ and $\mathbf{k}'_J$ are removed from the original N-particle expectation value. To denote this compactly, we identify

$$
\langle N \rangle_\mathrm{S} |^{I_1,\cdots I_K}_{J_1,\cdots J_K} \equiv \left\langle a^\dagger_{\mathbf{k}_1} \cdots \cancel{a^\dagger_{\mathbf{k}_{J_1}}} \cdots \cancel{a^\dagger_{\mathbf{k}_{J_K}}} \cdots a^\dagger_{\mathbf{k}_N} a_{\mathbf{k}'_N} \cdots \cancel{a_{\mathbf{k}'_{I_K}}} \cdots \cancel{a_{\mathbf{k}'_{I_1}}} \cdots a_{\mathbf{k}'_1} \right\rangle_\mathrm{S},
\tag{15.18}
$$

where the states $\mathbf{k}'_{I_1}, \cdots \mathbf{k}'_{I_K}$ and $\mathbf{k}_{J_1}, \cdots \mathbf{k}_{J_K}$ are removed from the original $\langle N \rangle_\mathrm{S}$. This converts the second term into

$$
2\text{nd} = \langle N \rangle_\mathrm{S} |^N_J,
\tag{15.19}
$$

which is an $(N-1)$-particle expectation value evaluated with respect to the Slater determinant $|\Phi_\mathrm{S}\rangle$.

Inserting the results (15.16) and (15.19) into Eq. (15.15), we obtain

$$
\left\langle a^\dagger_{\mathbf{k}_1} \cdots a^\dagger_{\mathbf{k}_N} a_{\mathbf{k}'_N} \cdots a_{\mathbf{k}'_1} \right\rangle_\mathrm{S} = \sum_{J=1}^{N} (-1)^{N+J} \left\langle a^\dagger_{\mathbf{k}'_N} a_{\mathbf{k}_J} \right\rangle_\mathrm{S} \left\langle a^\dagger_{\mathbf{k}_1} \cdots a^\dagger_{\mathbf{k}_N} a_{\mathbf{k}'_N} \cdots a_{\mathbf{k}'_1} \right\rangle_\mathrm{S} |^N_J .
\tag{15.20}
$$

We recognize that this expression satisfies the usual subdeterminant format given by the Laplace expansion (11.9). Therefore, the expectation values of a Slater determinant can generally be expressed via

$$
\left\langle a^\dagger_{\mathbf{k}_1} \cdots a^\dagger_{\mathbf{k}_N} a_{\mathbf{k}'_N} \cdots a_{\mathbf{k}'_1} \right\rangle_\mathrm{S} = \mathrm{Det}\,[\mathbf{N}], \qquad \mathbf{N}_{j,l} \equiv \left\langle a^\dagger_{\mathbf{k}_j} a_{\mathbf{k}_l} \right\rangle.
\tag{15.21}
$$

This result defines the Hartree–Fock approximation. For example, using the recursion (15.20) for a two-particle expectation, we obtain

$$
\left\langle a^\dagger_{\mathbf{k}_1} a^\dagger_{\mathbf{k}_2} a_{\mathbf{k}'_2} a_{\mathbf{k}'_1} \right\rangle_\mathrm{S} = \left\langle a^\dagger_{\mathbf{k}_1} a^\dagger_{\mathbf{k}'_1} \right\rangle \left\langle a^\dagger_{\mathbf{k}_2} a^\dagger_{\mathbf{k}'_2} \right\rangle - \left\langle a^\dagger_{\mathbf{k}_1} a^\dagger_{\mathbf{k}'_2} \right\rangle \left\langle a^\dagger_{\mathbf{k}_2} a^\dagger_{\mathbf{k}'_1} \right\rangle.
\tag{15.22}
$$

Altogether, we see that the Hartree–Fock factorization consists of *all* antisymmetrized products of singlets and thus uniquely identifies the systematic singlet factorization. To emphasize that the Hartree–Fock approximation is just the first step of the more general cluster-expansion method, we often refer to it as the singlet factorization. In cases where the state is more complicated than a Slater determinant, the singlet factorization describes the mean-field properties of the system. In other words, the singlet factorization projects the system onto an effective single-particle state ignoring all correlation possibilities described by the higher-order clusters.

We expect that under appropriate conditions already the low-order cluster schemes yield a good description. This is the case in situations where the semiconductor electron systems are dominated by their single-particle properties. Such configurations are denoted as *uncorrelated electron states* identifying those cases where the electrons can be described at the single-particle level. This includes also situations where the electrons experience the effective potential of other particles. Only the truly coupled, not separable electron states are considered as *correlated.*

15.2 Cluster expansion

We now construct a systematic scheme to identify all clusters. For this purpose, we analyze a general expectation value $\langle N \rangle$ assuming that we formally know all the quantities from $\langle 1 \rangle$ up to $\langle N \rangle$. We start by identifying the generic singlet

$$\langle 1 \rangle = [\langle 1 \rangle]_{\mathrm{S}} . \tag{15.23}$$

Since the singlet "correlation" is equal to the single-particle expectation value, this definition is sometimes referred to as contraction. The first "real" correlated cluster is the doublet

$$\langle 2 \rangle = [\langle 2 \rangle]_{\mathrm{S}} + \Delta \langle 2 \rangle , \tag{15.24}$$

where $\Delta\langle 2 \rangle = \langle 2 \rangle - [\langle 2 \rangle]_{\mathrm{S}}$ is the correlated part. The other N-particle clusters follow recursively,

$$\langle 3 \rangle = [\langle 3 \rangle]_{\mathrm{S}} + \langle 1 \rangle \, \Delta \langle 2 \rangle + \Delta \langle 3 \rangle , \tag{15.25}$$

$$\begin{aligned} \langle N \rangle = {} & [\langle N \rangle]_{\mathrm{S}} + \left[\langle N-2 \rangle_{\mathrm{S}} \, \Delta \langle 2 \rangle + \langle N-4 \rangle_{\mathrm{S}} \, \Delta \langle 2 \rangle \, \Delta \langle 2 \rangle + \cdots \right]_{\mathrm{D}} \\ & + \left[\langle N-3 \rangle_{\mathrm{S}} \, \Delta \langle 3 \rangle + \langle N-5 \rangle_{\mathrm{S}} \, \Delta \langle 3 \rangle \, \Delta \langle 2 \rangle + \cdots \right]_{\mathrm{T}} \\ & + \sum_{J=4}^{N-1} [\langle N \rangle]_J + \Delta \langle N \rangle . \end{aligned} \tag{15.26}$$

Here the subscripts S, D, T, denote singlet, doublet, and triplet contributions, respectively. Generally, the cluster products contain *all* combinations of factorizations with the appropriate antisymmetrization and symmetrization with respect to the pertinent Fermionic and Bosonic degrees of freedom. For example, the $\langle 1 \rangle \Delta \langle 2 \rangle$ contribution in Eq. (15.25) yields

$$\langle 1\rangle \,\Delta\, \langle 2\rangle \equiv \left[\left\langle a^\dagger_{\mathbf{k}_1} a^\dagger_{\mathbf{k}_2} a^\dagger_{\mathbf{k}_3} a_{\mathbf{k}'_3} a_{\mathbf{k}'_2} a_{\mathbf{k}'_1}\right\rangle\right]_D$$
$$= \sum_{j,l=1}^{3} (-1)^{j+k} \left\langle a^\dagger_{\mathbf{k}_j} a_{\mathbf{k}'_l}\right\rangle \Delta\left\langle a^\dagger_{\mathbf{k}_1} a^\dagger_{\mathbf{k}_2} a^\dagger_{\mathbf{k}_3} a_{\mathbf{k}'_3} a_{\mathbf{k}'_2} a_{\mathbf{k}'_1}\right\rangle\Big|^j_l, \tag{15.27}$$

where $\Delta\left\langle a^\dagger_{\mathbf{k}_1} a^\dagger_{\mathbf{k}_2} a^\dagger_{\mathbf{k}_3} a_{\mathbf{k}'_3} a_{\mathbf{k}'_2} a_{\mathbf{k}'_1}\right\rangle\big|^j_l$ indicates that the operators $a^\dagger_{\mathbf{k}_j}$ and $a_{\mathbf{k}'_l}$ are removed from the original sequence, in full analogy with Eq. (15.18). The explicit evaluation of (15.27) is discussed in Exercise 15.4.

The general expression, Eq. (15.26), is organized such that the brackets $[..]_J$ denote a factorization up to terms containing at least one cluster of rank J multiplied by possible products of clusters with a lower rank than J. Consequently, the terms explicitly shown contain doublet and triplet factorizations and the last term identifies the pure N-particle correlation $\Delta\langle N\rangle$. This notation implicitly assumes that $[..]_J$ includes a sum over all possibilities to reorganize the N coordinates among singlets, doublets, triplets, and so on. This procedure identifies all truly correlated clusters $\Delta\langle j\rangle$ with j correlated particles. The corresponding classification method is presented schematically in Fig. 15.2 where the spheres indicate the particle number. The terms within the rectangle indicate the N-particle expectation value and the shaded ellipses identify the correlations used in the factorization product.

We notice from Eq. (15.26) that an expectation value can be expressed via $\langle N\rangle = \sum_{J=1}^{N} [\langle N\rangle]_J$. Consequently, Eqs. (15.23)–(15.26) serve as an exact method to identify the expectation values at any given level of complexity. At this stage, we may introduce the cluster-expansion truncation scheme where we stop the expansion at the cluster level J, i.e.,

$$\langle N\rangle_{S,D,\ldots,J} \equiv \sum\nolimits_{j=1}^{J} [\langle N\rangle]_j\,. \tag{15.28}$$

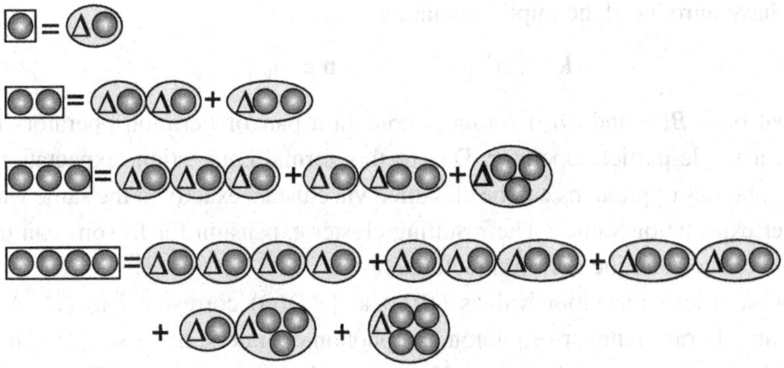

Figure 15.2 Schematic representation of the cluster-expansion-based classification. Each sphere corresponds to a particle operator, the expectation values are indicated by rectangles, and the correlated parts by ellipses. The factorization combinations are illustrated by products of clusters. The cluster expansion uniquely identifies singlets, doublets, triplets, and quadruplets from top to bottom. For each case, the product of clusters refers to *all* symmetrized or antisymmetrized permutations of the particle operators.

This procedure yields a consistent approximation if all higher-order expectation values are truncated at the same level.

Clearly, the description of the many-body properties can be systematically improved by increasing the number of clusters included in the analysis. The physical meaning of the different levels of approximation is such that:

$J = 1$ describes pure plasma properties,
$J = 2$ includes the lowest-order contribution in a correlated plasma and, e.g., the possibility of having bound pair states,
$J \geq 3$ defines higher-order correlations, like molecules and correlated droplets.

It is intuitively clear that the cluster-expansion approximation (15.28) produces rapidly converging results if clusters above a certain rank are rare.

15.2.1 Boson and Fermion factorizations

In general, the Boson and Fermion factorizations can be connected even though the classical factorization (15.3) and the Hartree–Fock approximation (15.21) have a different physical origin. To determine this connection, we start from the formal properties of Boson operators. If we omit quantum-electrodynamic effects due to plasmon oscillations, as discussed in Section 14.2.1, Eqs. (14.44)–(14.45) suggest that the photon and phonon operators have the schematic format

$$B(t) = B(0) + \int_0^t du\ \mathbf{k}^\dagger(u)\ \mathbf{p}(u), \tag{15.29}$$

$$D(t) = D(0) + \int_0^t du\ \mathbf{k}^\dagger(u)\ \mathbf{k}(u), \tag{15.30}$$

where we have introduced the implicit notation

$$\mathbf{k}^\dagger \equiv a^\dagger_{\lambda,\mathbf{k}}, \qquad \mathbf{p} \equiv a_{\nu,\mathbf{p}}. \tag{15.31}$$

We see that both $B(t)$ and $D(t)$ formally contain a pair of Fermion operators $\mathbf{k}^\dagger\ \mathbf{p}$ or $\mathbf{k}^\dagger\ \mathbf{k}$, i.e., a single-particle operator. Due to this formal connection, expectation values containing photons or phonons can be classified via clusters exactly in the same way as the pure carrier expectation values. The resulting cluster expansion for Bosons can therefore be directly connected to the Fermion factorizations.

The lowest-order expectation values $\langle B(t)\rangle$ and $\langle D(t)\rangle$ correspond to $\langle 1\rangle$. A general $\langle N\rangle$, with an arbitrary number of photon and phonon operators, has a singlet contribution which can be extracted by utilizing the Hartree–Fock approximation (15.21). Since $B(t)$ and $D(t)$ consist of an integral $\int_0^t du\ \mathbf{k}^\dagger(u)\ \mathbf{p}(u)$, we obtain the additional constraint that only those Hartree–Fock terms are included in the singlet part where $\int_0^t du\ \mathbf{k}^\dagger(u)\ \mathbf{p}(u)$ is not separated into different single-particle expectation values. The same constraint applies to identifying doublets, triplets, and higher-order clusters. Otherwise, the recursive identification procedure (15.24)–(15.26) can be applied directly. The only additional condition is that the original and factorized expectation values have to be normally ordered, i.e., all creation operators – also Bosonic ones – have to be on the LHS of the expectation values. This reorganization is relatively simple to achieve because, e.g., pure photon operators always commute with pure carrier operators.

At the lowest level, the relevant quantities are $\langle \mathbf{k}^\dagger \mathbf{p}\rangle$ for pure particles, the classical field amplitude $\langle B\rangle$, and the coherent phonon amplitude $\langle D\rangle$. Before extracting the two-particle correlations, we need to know the corresponding singlet contributions. For pure carriers, the singlet part is provided by the Hartree–Fock approximation. Using Eq. (15.22), we find

$$\langle \mathbf{k}^\dagger \alpha^\dagger \beta\ \mathbf{p}\rangle_{\mathrm{S}} = \langle \mathbf{k}^\dagger \mathbf{p}\rangle\langle \alpha^\dagger \beta\rangle - \langle \mathbf{k}^\dagger \beta\rangle\langle \alpha^\dagger \mathbf{p}\rangle. \tag{15.32}$$

This expression can also be generalized to identify the singlet contributions of expectation values with one or two Boson operators. If $\alpha^\dagger\beta \equiv \hat{O}$ corresponds to a Boson operator (photon or phonon), we have to demand that $\alpha^\dagger\beta$ cannot be separated. This procedure yields

$$\langle \mathbf{k}^\dagger \hat{O}\ \mathbf{p}\rangle_{\mathrm{S}} = \langle \mathbf{k}^\dagger \alpha^\dagger \beta\ \mathbf{p}\rangle_{\mathrm{S}} = \langle \mathbf{k}^\dagger \mathbf{p}\rangle\langle \alpha^\dagger \beta\rangle = \langle \mathbf{k}^\dagger \mathbf{p}\rangle\langle \hat{O}\rangle. \tag{15.33}$$

For two Bosons, $\alpha^\dagger\beta \equiv \hat{O}_1$ and $\mathbf{k}^\dagger\mathbf{p} \equiv \hat{O}_2$, we similarly find

$$\langle \hat{O}_1\hat{O}_2\rangle_{\mathrm{S}} = \langle \hat{O}_1\rangle\langle \hat{O}_2\rangle, \tag{15.34}$$

where we assume that $\hat{O}_1\hat{O}_2$ is normally ordered such that all Bosonic creation operators are on the LHS. We see that the resulting equations are identical to the classical factorization (15.3) showing that the Boson factorizations can indeed be derived from the Fermion factorizations when suitable constraints are enforced.

Next, we determine the explicit form of $\langle \mathbf{k}^\dagger \mathbf{p}^\dagger \alpha^\dagger \beta\ \mathbf{l}\ \mathbf{m}\rangle|_{\mathrm{S}}$ because this will be needed in the quantum-dynamics derivations for the doublets. The singlet part for the bare carriers can be obtained using the Hartree–Fock approximation (15.21),

$$\begin{aligned}\langle \mathbf{k}^\dagger \mathbf{p}^\dagger \alpha^\dagger \beta\ \mathbf{l}\ \mathbf{m}\rangle_{\mathrm{S}} = {} & \langle \mathbf{k}^\dagger \mathbf{m}\rangle\langle \mathbf{p}^\dagger \mathbf{l}\rangle\langle \alpha^\dagger \beta\rangle - \langle \mathbf{k}^\dagger \mathbf{m}\rangle\langle \mathbf{p}^\dagger \beta\rangle\langle \alpha^\dagger \mathbf{l}\rangle \\ & - \langle \mathbf{k}^\dagger \mathbf{l}\rangle\langle \mathbf{p}^\dagger \mathbf{m}\rangle\langle \alpha^\dagger \beta\rangle + \langle \mathbf{k}^\dagger \mathbf{l}\rangle\langle \mathbf{p}^\dagger \beta\rangle\langle \alpha^\dagger \mathbf{m}\rangle \\ & + \langle \mathbf{k}^\dagger \beta\rangle\langle \mathbf{p}^\dagger \mathbf{m}\rangle\langle \alpha^\dagger \mathbf{l}\rangle - \langle \mathbf{k}^\dagger \beta\rangle\langle \mathbf{p}^\dagger \mathbf{l}\rangle\langle \alpha^\dagger \mathbf{m}\rangle.\end{aligned} \tag{15.35}$$

Following the same logic as before, we find the singlet contributions with exactly one, two, or three Bosons,

$$\left\langle \mathbf{k}^\dagger \mathbf{p}^\dagger \hat{O}_1 \mathbf{l}\mathbf{m}\right\rangle_{\mathrm{S}} = \langle \mathbf{k}^\dagger \mathbf{m}\rangle\langle \mathbf{p}^\dagger \mathbf{l}\rangle\langle \hat{O}_1\rangle - \langle \mathbf{k}^\dagger \mathbf{l}\rangle\langle \mathbf{p}^\dagger \mathbf{m}\rangle\langle \hat{O}_1\rangle, \tag{15.36}$$

$$\left\langle \mathbf{k}^\dagger \hat{O}_1\hat{O}_2 \mathbf{m}\right\rangle_{\mathrm{S}} = \langle \mathbf{k}^\dagger \mathbf{m}\rangle\langle \hat{O}_2\rangle\langle \hat{O}_1\rangle, \tag{15.37}$$

$$\langle \hat{O}_1\hat{O}_2\hat{O}_3\rangle_{\mathrm{S}} = \langle \hat{O}_1\rangle\langle \hat{O}_2\rangle\langle \hat{O}_3\rangle. \tag{15.38}$$

This technique can be directly applied to also identify higher-order clusters containing a mixture of Boson and Fermion operators. Therefore, we can first construct the Fermion clusters and use these to produce the Boson factorizations as long as we do not allow for the separation of terms belonging to the same Boson operator.

15.2.2 Most relevant singlet–doublet factorizations

As background for our further derivations, we next summarize the singlet–doublet factorization up to three particles. The two-particle correlations directly follow from Eqs. (15.32)–(15.34),

$$\Delta\langle \mathbf{k}^\dagger \alpha^\dagger \beta\ \mathbf{p}\rangle = \langle \mathbf{k}^\dagger \alpha^\dagger \beta\ \mathbf{p}\rangle - \langle \mathbf{k}^\dagger \alpha^\dagger \beta\ \mathbf{p}\rangle_{\mathrm{S}}. \tag{15.39}$$

Such terms include correlations between carriers, photons, and phonons. In general, $\Delta\langle \mathbf{k}^\dagger \alpha^\dagger \beta\ \mathbf{p}\rangle$ describes the conditional dependence of two separate entities. For carriers alone, $\Delta\langle \mathbf{k}^\dagger \alpha^\dagger \beta\ \mathbf{p}\rangle$ is responsible for Coulombic correlations like bound electron–hole pairs, which we will identify as excitons in Chapter 27. If photons are involved, $\Delta\langle \mathbf{k}^\dagger \alpha^\dagger \beta\ \mathbf{p}\rangle$ leads to quantum-optical correlation effects which may result, e.g., in squeezing and entanglement, see Chapter 30.

Since a three-particle term can contain just one two-particle correlation, we only need to construct a sum over all possible and unique $\langle 1\rangle\Delta\langle 2\rangle$ contributions and multiply each term with $+1$ (-1) corresponding to the needed even (odd) permutations of the indices to construct its singlet–doublet factorization. This straightforward procedure yields

$$\begin{aligned}\langle \mathbf{k}^\dagger \mathbf{p}^\dagger \alpha^\dagger \beta\, \mathbf{l}\, \mathbf{m}\rangle_{\mathrm{SD}} = \langle \mathbf{k}^\dagger \mathbf{p}^\dagger \alpha^\dagger \beta\, \mathbf{l}\, \mathbf{m}\rangle_{\mathrm{S}} &\\ + \langle \mathbf{k}^\dagger \mathbf{m}\rangle \Delta\langle \mathbf{p}^\dagger \alpha^\dagger \beta\, \mathbf{l}\rangle - \langle \mathbf{k}^\dagger \mathbf{l}\rangle \Delta\langle \mathbf{p}^\dagger \alpha^\dagger \beta\, \mathbf{m}\rangle + \langle \mathbf{k}^\dagger \beta\rangle \Delta\langle \mathbf{p}^\dagger \alpha^\dagger \mathbf{l}\, \mathbf{m}\rangle &\\ - \langle \mathbf{p}^\dagger \mathbf{m}\rangle \Delta\langle \mathbf{k}^\dagger \alpha^\dagger \beta\, \mathbf{l}\rangle + \langle \mathbf{p}^\dagger \mathbf{l}\rangle \Delta\langle \mathbf{k}^\dagger \alpha^\dagger \beta\, \mathbf{m}\rangle - \langle \mathbf{p}^\dagger \beta\rangle \Delta\langle \mathbf{k}^\dagger \alpha^\dagger \mathbf{l}\, \mathbf{m}\rangle &\\ + \langle \alpha^\dagger \mathbf{m}\rangle \Delta\langle \mathbf{k}^\dagger \mathbf{p}^\dagger \beta\, \mathbf{l}\rangle - \langle \alpha^\dagger \mathbf{l}\rangle \Delta\langle \mathbf{k}^\dagger \mathbf{p}^\dagger \beta\, \mathbf{m}\rangle &\\ + \langle \alpha^\dagger \beta\rangle \Delta\langle \mathbf{k}^\dagger \mathbf{p}^\dagger \mathbf{l}\, \mathbf{m}\rangle &\end{aligned} \tag{15.40}$$

for pure Fermions. This result can be directly applied to generate a factorization containing mixed Fermion and Boson operators,

$$\begin{aligned}\langle \mathbf{k}^\dagger \mathbf{p}^\dagger \hat{O}_1 \mathbf{l}\, \mathbf{m}\rangle_{\mathrm{SD}} = \langle \mathbf{k}^\dagger \mathbf{p}^\dagger \hat{O}_1 \mathbf{l}\, \mathbf{m}\rangle_{\mathrm{S}} + \langle \mathbf{k}^\dagger \mathbf{m}\rangle \Delta\langle \mathbf{p}^\dagger \hat{O}_1 \mathbf{l}\rangle - \langle \mathbf{k}^\dagger \mathbf{l}\rangle \Delta\langle \mathbf{p}^\dagger \hat{O}_1 \mathbf{m}\rangle &\\ - \langle \mathbf{p}^\dagger \mathbf{m}\rangle \Delta\langle \mathbf{k}^\dagger \hat{O}_1 \mathbf{l}\rangle + \langle \mathbf{p}^\dagger \mathbf{l}\rangle \Delta\langle \mathbf{k}^\dagger \hat{O}_1 \mathbf{m}\rangle + \langle \hat{O}_1\rangle \Delta\langle \mathbf{k}^\dagger \mathbf{p}^\dagger \mathbf{l}\, \mathbf{m}\rangle, &\end{aligned} \tag{15.41}$$

$$\begin{aligned}\langle \mathbf{k}^\dagger \hat{O}_1 \hat{O}_2 \mathbf{m}\rangle_{\mathrm{SD}} = \langle \mathbf{k}^\dagger \hat{O}_1 \hat{O}_2 \mathbf{m}\rangle_{\mathrm{S}} &\\ + \langle \mathbf{k}^\dagger \mathbf{m}\rangle \Delta\langle \hat{O}_1 \hat{O}_2\rangle + \langle \hat{O}_2\rangle \Delta\langle \mathbf{k}^\dagger \hat{O}_1 \mathbf{m}\rangle + \langle \hat{O}_1\rangle \Delta\langle \mathbf{k}^\dagger \hat{O}_2 \mathbf{m}\rangle, &\end{aligned} \tag{15.42}$$

$$\begin{aligned}\langle \hat{O}_1 \hat{O}_2 \hat{O}_3\rangle_{\mathrm{SD}} = \langle \hat{O}_1 \hat{O}_2 \hat{O}_3\rangle_{\mathrm{S}} &\\ + \langle \hat{O}_3\rangle \Delta\langle \hat{O}_1 \hat{O}_2\rangle + \langle \hat{O}_2\rangle \Delta\langle \hat{O}_1 \hat{O}_3\rangle + \langle \hat{O}_1\rangle \Delta\langle \hat{O}_2 \hat{O}_3\rangle. &\end{aligned} \tag{15.43}$$

As a general feature for Bosons, the identified singlet–doublet factorizations are fully symmetric with respect to the exchange of Boson operators, as they should be. Since Fermion factorizations produce the corresponding Boson factorization after suitable constraints are introduced, also Fermion factorizations must contain a substructure that has a Bosonic character. This observation has an interesting consequence: a Fermion many-body system can also exhibit Boson features, such as particle condensates, under favorable conditions. This aspect will be investigated further in Chapter 30.

15.3 Quantum dynamics of expectation values

In semiconductor quantum optics, we encounter two-particle interactions that lead to the hierarchy problem discussed in Section 14.3.2. Based on Eq. (14.70), we generally end up with

$$i\hbar\frac{\partial}{\partial t}\langle N\rangle = L\left[\langle N\rangle\right] + \mathrm{Hp}\left[\langle N+1\rangle\right], \tag{15.44}$$

where the first term contains all those contributions involving only the same order as on the LHS and the second term denotes the interaction contributions resulting in the hierarchy problem. The explicit expressions can be generated by applying the operator equations (14.34)–(14.37) and (14.39)–(14.40) multiple times. Both Coulomb and quantum-optical interactions generate an equivalent hierarchy problem that generally does not allow for exact solutions of the N-particle dynamics. The hierarchical coupling structures of the expectation-value representation is presented in Fig. 15.3.

In general, a many-body system with $\mathcal{N}$ particles has some nonvanishing $\langle N\rangle$ elements within *all* classes of expectation values ranging from $\langle 1\rangle$ to $\langle \mathcal{N}\rangle$. Only the quantities above $N > \mathcal{N}$ vanish at all times. Thus, any unsystematic truncation of the equation hierarchy can seriously compromise the validity of the deduced quantum dynamics. Since the value of $\mathcal{N}$ is typically enormously large in solids, the hierarchy problem prevents us from finding exact solutions for the many-body problem.

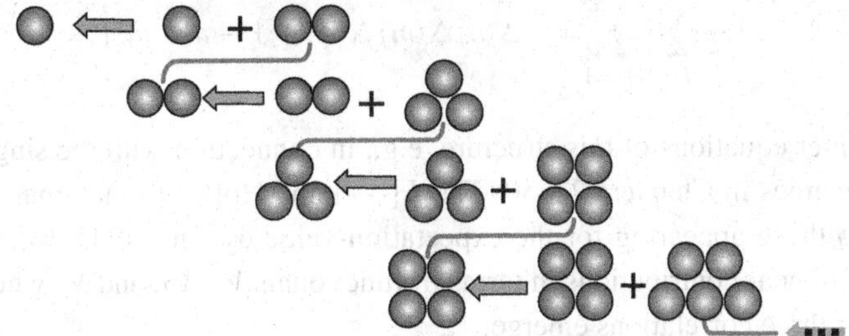

Figure 15.3 Hierarchical coupling structure in the quantum dynamics of N-particle quantities. The dynamics of one-, two-, three-, and four-particles is presented schematically. Each particle is symbolized by a circle. According to Eq. (15.44), $\langle N\rangle$ couples to both $\langle N\rangle$ and $\langle N+1\rangle$ as indicated by the arrows.

15.4 Quantum dynamics of correlations

Due to the difficulties we encounter using the expectation-value representation, we consider solving the quantum dynamics in terms of correlations. When we apply the cluster expansion (15.23)–(15.26) as a formal tool, we can exactly convert any expectation value into correlations

$$\Delta\langle N\rangle = F_N\left[\langle 1\rangle, \langle 2\rangle, \cdots \langle N\rangle\right], \tag{15.45}$$

where the functional F_N needs to be defined for each $\Delta\langle N\rangle$ separately, see Exercise 15.5. In general, F_N consists of antisymmetrized products of expectation values. Thus, the generic quantum dynamics of correlations can be solved by applying

$$i\hbar\frac{\partial}{\partial t}\Delta\langle N\rangle = \sum\nolimits_{n=1}^{N} F_N\left[\langle 1\rangle, \cdots \left(i\hbar\frac{\partial}{\partial t}\langle n\rangle\right), \cdots \langle N\rangle\right]. \tag{15.46}$$

This derivation can be performed quite easily for low $\Delta\langle N\rangle$ but the explicit evaluation becomes rather cumbersome for elevated N.

Therefore, it is desirable to develop a new formalism that yields the correlation dynamics for all N with one unified derivation. This can be done by introducing expectation-value and correlation-generating functions for a generic Fermion–Boson many-body system. Since this investigation requires a significant development of the many-body formalism beyond the scope of this book, we refer the interested reader to the discussion in the literature mentioned at the end of this chapter. Here, we just summarize the main results needed in the following chapters.

The general correlation dynamics follows from

$$\begin{aligned} i\hbar\frac{\partial}{\partial t}\Delta\langle N\rangle = {} & L\left[\Delta\langle N\rangle\right] + \text{Hp}\left[\Delta\langle N+1\rangle\right] \\ & + \sum_{n=1}^{N} V_1\left[\Delta\langle n\rangle\Delta\langle N-n\rangle\right] + \sum_{n=1}^{N} V_2\left[\Delta\langle n\rangle\Delta\langle N+1-n\rangle\right] \\ & + \sum_{n=1}^{N-1}\sum_{m=1}^{N-n} V_3\left[\Delta\langle n\rangle\Delta\langle m\rangle\Delta\langle N+1-n-m\rangle\right]. \end{aligned} \tag{15.47}$$

We will encounter equations of this structure, e.g., in connection with the singlet–doublet correlation dynamics in Chapters 27–30. The $L[\cdots]$ and $\text{Hp}[\cdots]$ functionals that appear are identical to those appearing for the expectation-value dynamics (15.44). In addition, we find new nonlinear contributions in terms of functionals V_1, V_2, and V_3 where products of either two or three correlations emerge.

At first sight, it may appear as if the correlation dynamics (15.47) is even more problematic than the expectation-value approach (15.44). After all, the correlation dynamics involves exactly the same hierarchy problem as the expectation values and additionally contains nonlinear coupling terms. However, we will see that the correlation dynamics provides a superior basis for approximations after we realize that $\Delta\langle N\rangle$ and $\langle N\rangle$ have completely different initial values.

As discussed before, all expectation-value classes, from $N = 1$ up to $N = \mathcal{N}$, are generally nonvanishing at all times, which makes the treatment of the equation hierarchy problematic. However, the correlations display an entirely different behavior. We already know that a Fermionic many-body system has several very relevant many-body configurations where only low-order clusters contribute. The investigations in Chapter 27 demonstrate that coherent excitons are described solely by singlets while incoherent exciton states can be understood in terms of the singlet–doublet factorization. At the same time, the QW system may still contain a carrier density of $10^{10}\,\mathrm{cm}^{-2}$, which implies the presence of very many particles even though only few-particle correlations describe the many-body system. Thus, the largest contributing cluster number C typically is significantly smaller than the largest particle number, i.e., $C \ll \mathcal{N}$. Thus, the complexity of solving the correlation dynamics corresponds to evolving C-particle quantities in time allowing for a much higher numerical efficiency than those of the expectation-value-based solutions. The only price to pay is the emergence of nonlinear contributions that do not produce any additional hierarchical coupling, see Exercise 15.9.

The structure of Eq. (15.47) is presented in Fig. 15.4(a) for the first three correlations. Each sphere within a shaded ellipse denotes a particle within a correlation. An ellipse with C particles refers to a C-particle correlation. The product of correlations, i.e., ellipses inside the shaded area, indicates the nonlinear contributions of Eq. (15.47). We see that the high-order correlations are exclusively driven by the nonlinear terms because the hierarchical coupling term vanishes before the corresponding correlations develop. Thus, the

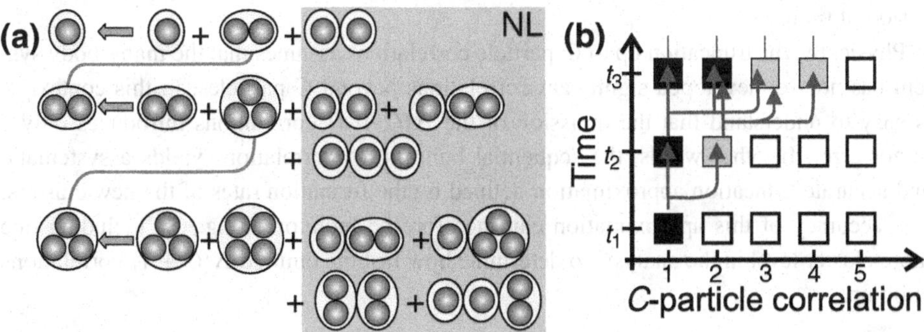

Figure 15.4 Hierarchical coupling structure of the correlations. **(a)** The dynamics of single-, two-, three-, and four-particle correlations is presented schematically. Each particle is indicated by a sphere and the correlations are enclosed by an ellipse. According to Eq. (15.47), $\Delta\langle C\rangle$ couples to both $\Delta\langle C\rangle$ and $\Delta\langle C+1\rangle$ as indicated by the arrow. The terms within the shaded area describe the coupling to the nonlinear scattering contributions. **(b)** The existing correlations (filled square) and their couplings (arrows) are shown for four different times. We assume a situation as in an optically excited system where, at the very early times, only the singlet exists. The emerging new correlations are indicated by light-shaded squares while the empty squares denote vanishing quantities. The arrows indicate which lower-order clusters appear in the nonlinear source.

high-order correlations do not contribute to the quantum dynamics if they do not exist before a given time. Consequently, *the correlations are built up sequentially* and it will take a finite time before $\Delta\langle C+1\rangle$ can become appreciable if it vanishes initially. Thus, the role of $\Delta\langle C+1\rangle$ is completely different from the $(C+1)$-particle expectation values in Eq. (15.44); $\langle C+1\rangle$ cannot be ignored because it exists from the beginning while $\Delta\langle C+1\rangle$ vanishes or it remains insignificant before it is dynamically built up.

Figure 15.4(b) schematically shows how single-particle correlations generate two-particle correlations. At the next level, combinations of singlet and two-particle correlations drive triplets and so on. The existing correlations are indicated as dark squares while the nonexisting ones are symbolized by the empty squares. The nonlinear driving terms are marked by the arrows. The light-shaded squares denote those correlations that develop to a clearly nonvanishing level during the time evolution from t_j to t_{j+1}. We see that the higher-order clusters are generated via a well-defined time sequence. Until the time t_j, only correlations up to $\Delta\langle C\rangle$ are important, until the time $t_{j+1} > t_j$ correlations $\Delta\langle C+1\rangle$ may also emerge, and so on.

Thus, we can introduce a cluster-expansion-based truncation where only clusters $\Delta\langle C+j-1\rangle$ are included until the time t_j. This approximation provides a systematic truncation scheme to the hierarchy problem and can be systematically improved by including more clusters in the analysis. It happens quite often that the formation of higher-order clusters becomes unfavorable, e.g., when lower-order clusters are energetically stable enough. For example, the formation of molecules can become very slow when atoms are already formed and experience only weakly attractive interactions. Since the molecule formation is driven by this attractive interaction, it is slowed down considerably. In these and similar situations, the dynamics of low-rank clusters defines the quantum dynamics over a long period of time.

Physically, the truncation up to C-particle correlations assumes that the many-body system has not yet developed significant correlations beyond C particles. In this context, it is easy to understand that the omission of the $\Delta\langle C+1\rangle$ correlations introduces only a minor error. In other words, the sequential build-up of correlations yields a systematic and accurate truncation approximation defined by the formation rates of the new clusters. The accuracy of this approximation can therefore be monitored, e.g., by including one more cluster level in the analysis to determine how fast the omitted $\Delta\langle C+1\rangle$ correlations build up.

15.5 Scattering in terms of correlations

The cluster expansion also allows for a solution where higher-order correlations $\Delta\langle C+1\rangle$ are described at the scattering level. To illustrate how this approach can be implemented, we consider the truncation of the hierarchy where all terms up to triplets are included. By unraveling Eq. (15.47) for singlets, doublets, and triplets, we find a closed set of equations

$$
\begin{aligned}
i\hbar\frac{\partial}{\partial t}\Delta\langle 1\rangle &= L\left[\Delta\langle 1\rangle\right] + \mathrm{Hp}\left[\Delta\langle 2\rangle\right] + V_2\left[\Delta\langle 1\rangle\Delta\langle 1\rangle\right] \\
i\hbar\frac{\partial}{\partial t}\Delta\langle 2\rangle &= L\left[\Delta\langle 2\rangle\right] + \mathrm{Hp}\left[\Delta\langle 3\rangle\right] + V_1\left[\Delta\langle 1\rangle\Delta\langle 1\rangle\right] \\
&\quad + V_2\left[\Delta\langle 1\rangle\Delta\langle 2\rangle\right] + V_3\left[\Delta\langle 1\rangle\Delta\langle 1\rangle\Delta\langle 1\rangle\right] \\
i\hbar\frac{\partial}{\partial t}\Delta\langle 3\rangle &= L\left[\Delta\langle 3\rangle\right] + \cancel{\mathrm{Hp}\left[\Delta\langle 4\rangle\right]} \\
&\quad + V_1\left[\Delta\langle 1\rangle\Delta\langle 2\rangle\right] + V_2\left[\Delta\langle 1\rangle\Delta\langle 3\rangle\right] \\
&\quad + V_2\left[\Delta\langle 2\rangle\Delta\langle 2\rangle\right] + V_3\left[\Delta\langle 1\rangle\Delta\langle 1\rangle\Delta\langle 2\rangle\right], \qquad (15.48)
\end{aligned}
$$

where the quadruplet clusters that appear are excluded based on the assumption that four-particle correlations have not been formed. We can also investigate an earlier time window where triplets remain insignificant in comparison with the singlet–doublet correlations. In this situation, we only need to solve the first and the second equation, i.e., the closed set of the singlet–doublet dynamics.

As discussed in Section 15.3, triplets are driven by the nonlinear contributions, which eventually develop them toward significance. Because singlets and doublets at early times are larger than the triplets, the driving terms $V_1\left[\Delta\langle 1\rangle\Delta\langle 2\rangle\right]$, $V_2\left[\Delta\langle 2\rangle\Delta\langle 2\rangle\right]$, and $V_3\left[\Delta\langle 1\rangle\Delta\langle 1\rangle\Delta\langle 2\rangle\right]$ dominate over $V_2\left[\Delta\langle 1\rangle\Delta\langle 3\rangle\right]$ before the triplets have grown enough. Thus, we may simplify the triplet terms within Eq. (15.48) to describe the early stages of the triplet correlations. In particular, we can always introduce the approximation

$$
L\left[\Delta\langle 3\rangle\right] + V_2\left[\Delta\langle 1\rangle\Delta\langle 3\rangle\right] \rightarrow \Delta E\ \Delta\langle 3\rangle, \qquad (15.49)
$$

where only those contributions are included that can be associated with a mean-field energy ΔE. An explicit form of ΔE is investigated in Exercise 15.10. This approximation produces a new scheme to include a part of the triplets before strong three-particle correlations emerge. In particular, this approximation yields results that are between the fully consistent singlet–doublet and the singlet–doublet–triplet approximation.

Making the approximation (15.49) in Eq. (15.48), we find

$$
\begin{aligned}
i\hbar\frac{\partial}{\partial t}\Delta\langle 3\rangle &= \left(\Delta E - i\gamma\right)\Delta\langle 3\rangle + V_1\left[\Delta\langle 1\rangle\Delta\langle 2\rangle\right] \\
&\quad + V_2\left[\Delta\langle 2\rangle\Delta\langle 2\rangle\right] + V_3\left[\Delta\langle 1\rangle\Delta\langle 1\rangle\Delta\langle 2\rangle\right], \qquad (15.50)
\end{aligned}
$$

where we also included a phenomenological dephasing constant γ to account for the formation of quadruplets from triplets. Equation (15.50) shows clearly that triplets are generated via terms involving products of singlets and doublets describing microscopic collision processes.

Since Eq. (15.50) has a relatively simple homogeneous part, we can solve it,

$$
\begin{aligned}
\Delta\langle 3\rangle &= \int_{-\infty}^{t} du\ \frac{V_1\left[\Delta\langle 1\rangle\Delta\langle 2\rangle\right](u) + V_2\left[\Delta\langle 2\rangle\Delta\langle 2\rangle\right](u) + V_3\left[\Delta\langle 1\rangle\Delta\langle 1\rangle\Delta\langle 2\rangle\right](u)}{i\hbar} \\
&\quad \times e^{i[\Delta E - i\gamma](u-t)/\hbar}. \qquad (15.51)
\end{aligned}
$$

In cases where the nonlinear source changes slowly enough, we can implement the Markov approximation, producing

$$\Delta\langle 3\rangle = -\frac{V_1\left[\Delta\langle 1\rangle\Delta\langle 2\rangle\right] + V_2\left[\Delta\langle 2\rangle\Delta\langle 2\rangle\right] + V_3\left[\Delta\langle 1\rangle\Delta\langle 1\rangle\Delta\langle 2\rangle\right]}{\Delta E - i\gamma}. \tag{15.52}$$

This approximation can be analyzed from the cluster-expansion point of view. The three-particle correlations are not allowed to lead to the formation of bound triplets because the triplets are proportional to products of singlets and doublets, i.e., scattering within the singlet–doublet manifold. Thus, the triplets are included at the scattering level when Eq. (15.52) is implemented to the full singlet–doublet dynamics within Eq. (15.48).

An equivalent mean-field approximation (15.51) can also be introduced for the doublet correlations. In this limit, we find

$$\Delta\langle 2\rangle_{\text{scatt}} = -\frac{V_1\left[\Delta\langle 1\rangle\Delta\langle 1\rangle\Delta\langle 1\rangle\right] + V_2\left[\Delta\langle 1\rangle\Delta\langle 1\rangle\Delta\langle 1\rangle\right]}{\Delta E - i\gamma}, \tag{15.53}$$

which describes the doublets at the scattering level. This approximation produces the so-called second Born approximation, applied to describe scattering among densities and polarization in solids. Thus, the cluster expansion includes the second-Born approximation as a limiting case. Obviously, the inclusion of the full doublet structure and/or triplets at the scattering level is a major improvement over the second Born approximation.

Since the full cluster-expansion solution includes phenomena beyond the second Born approach, we may anticipate that it also generalizes other approximation schemes. To verify this, we consider an initially unexcited semiconductor where the real many-body dynamics starts after the system is exposed to light. Excitation with an optical pulse generates polarization, carrier densities, and higher-order correlations within the system. The dynamics-controlled truncation (DCT) scheme assumes that we use a classical excitation pulse such that each excitation contribution can be classified as a power series of the incoming field E. For weak enough excitations, contributions only up to certain power E^M have to be included. The limiting M depends on the physically relevant quantities investigated. The corresponding truncation is often referred to as $\chi^{(M)}$ result, which refers to higher-order optical susceptibilities. The DCT method has been proven to be converging if the excitation is weak enough and the incoherent dynamics, which exists independently also for $E = 0$, can be neglected. The approach has successfully been used to analyze, e.g., Coulomb correlation effects in the four-wave mixing signal.

In the DCT scheme, the lowest-order contributions to the single-particle expectation values are $P_{\mathbf{k}} \equiv \left\langle a^\dagger_{v,\mathbf{k}} a_{c,\mathbf{k}}\right\rangle \propto E$, $f^e_{\mathbf{k}} \equiv \left\langle a^\dagger_{c,\mathbf{k}} a_{c,\mathbf{k}}\right\rangle \propto E^2$, and $f^h_{\mathbf{k}} \equiv \left\langle a_{v,\mathbf{k}} a^\dagger_{v,\mathbf{k}}\right\rangle \propto E^2$. When just the classical light–matter interaction is included, the $\chi^{(2)}$ limit of the higher-order correlations gives, e.g.,

$$\left\langle \mathbf{k}^\dagger_c \mathbf{k}'^\dagger_v \mathbf{k}'_c \mathbf{k}_v \right\rangle_{\chi^{(2)}} = P^\star_{\mathbf{k}} P_{\mathbf{k}'} - \delta_{\mathbf{k},\mathbf{k}'} f^e_{\mathbf{k}}, \tag{15.54}$$

where we have assumed that singlets are diagonal in **k**. With the help of the factorization (15.22), the singlet approximation of the same term yields

$$\begin{aligned}\left\langle \mathbf{k}_c^\dagger \mathbf{k}_v'^\dagger \mathbf{k}_c' \mathbf{k}_v \right\rangle_S &= \left\langle \mathbf{k}_c^\dagger \mathbf{k}_v \right\rangle \left\langle \mathbf{k}_v'^\dagger \mathbf{k}_c' \right\rangle - \left\langle \mathbf{k}_c^\dagger \mathbf{k}_c' \right\rangle \left\langle \mathbf{k}_v'^\dagger \mathbf{k}_v \right\rangle \\ &= P_{\mathbf{k}}^\star P_{\mathbf{k}'} - \delta_{\mathbf{k},\mathbf{k}'} \left(f_{\mathbf{k}}^e - f_{\mathbf{k}}^e f_{\mathbf{k}}^h \right). \end{aligned} \tag{15.55}$$

This approximation fully includes the $\chi^{(2)}$ limit but has an extra term $f_{\mathbf{k}}^e f_{\mathbf{k}}^h$, which is at least a contribution of the order of E^4. Hence, we see that the $\chi^{(2)}$ limit is a subset of the singlet factorization in the fully coherent case. For more general conditions, including phonons and the possibility of having incoherent populations, the singlet truncation fully includes only the $\chi^{(1)}$ limit. If we include an additional doublet cluster to the analysis, the $\langle N \rangle_{\text{SD}}$ truncation fully includes the $\chi^{(3)}$ limit. Similar relations hold as the number of clusters is increased, i.e., $\chi^{(2C-1)}$ is a subset of the C-th order cluster-expansion truncation. As a result, also the cluster expansion becomes more accurate as the number of clusters is increased. This verifies indirectly the convergence of the cluster expansion when the conditions for the $\chi^{(2C-1)}$ truncation are satisfied. However, the validity of the cluster expansion is not limited to low E or to the coherent regime because each cluster can actually contain E in infinite power.

Exercises

15.1 We assume a single-mode Boson field B and investigate normally ordered expectation values $I_K^J = \langle (B^\dagger)^J B^K \rangle$ and antinormally ordered ones $A_K^J = \langle B^K (B^\dagger)^J \rangle$. **(a)** What is the general connection between I_2^2 and A_2^2? **(b)** The classical factorization $B \to \beta$, $B^\dagger \to \beta^\star$ can directly be applied for I_K^J. What is the classical factorization of A_2^2? **(c)** Find the singlet–doublet factorizations of I_2^2 and A_2^2. **(d)** We define $\hat{x} \equiv \frac{1}{2}(B + B^\dagger)$ and $\hat{y} = \frac{1}{2i}(B - B^\dagger)$ and their fluctuations $\Delta x^2 \equiv \langle \hat{x}^2 \rangle - \langle \hat{x} \rangle^2$ and $\Delta y^2 \equiv \langle \hat{y}^2 \rangle - \langle \hat{y} \rangle^2$. Show that the classical factorization yields $\Delta x \Delta y = 1/4$, which satisfies the Heisenberg uncertainty relation.

15.2 A unitary transformation $\mathbf{U}$ of the known Fermion operator $a_{\mathbf{k}}$ produces a new operator $A_{\mathbf{K}} = \mathbf{U}_{\mathbf{K},\mathbf{k}}\, a_{\mathbf{k}}$. The condition of unitarity implies

$$\mathbf{U}_{\mathbf{K},\mathbf{k}}^\star\, U_{\mathbf{K},\mathbf{k}'} = \delta_{\mathbf{k},\mathbf{k}'} \quad \text{and} \quad \mathbf{U}_{\mathbf{K},\mathbf{k}}^\star\, \mathbf{U}_{\mathbf{K}',\mathbf{k}} = \delta_{\mathbf{K},\mathbf{K}'}\,.$$

(a) Show that $\left[A_{\mathbf{K}}, A_{\mathbf{K}'}^\dagger\right]_+ = \delta_{\mathbf{K},\mathbf{K}'}$ and $[A_{\mathbf{K}}, A_{\mathbf{K}'}]_+ = 0 = \left[A_{\mathbf{K}}^\dagger, A_{\mathbf{K}'}^\dagger\right]_+$. **(b)** Construct the field operator $\hat{\Psi}(\mathbf{r}) = \sum_{\mathbf{k}} a_{\mathbf{k}} \phi_{\mathbf{k}}(\mathbf{r})$ in terms of A_K. What is the single-particle wave function connected with the modes? **(c)** Assume that we know $\hat{\Psi}(\mathbf{r}) = \sum_{\mathbf{K}} A_{\mathbf{K}}\, \Phi_{\mathbf{K}}(\mathbf{r})$. Show that $U_{\mathbf{K},\mathbf{k}} = \langle \Phi_{\mathbf{K}} | \phi_{\mathbf{k}} \rangle$.

15.3 Verify the result (15.14). **Hint:** *Anticommute* $A_{\mathbf{K}_N''}^\dagger$ *toward* $|\Phi_S\rangle$ *and apply the property* (15.13). Show that $\langle N \rangle_J I_J^N$ produces Eq. (15.17).

15.4 Construct a $\langle 1\rangle\Delta\langle 2\rangle$ factorization explicitly using Eq. (15.27). Show that the exchange of any pair of creation or annihilation operators yields a sign change.

15.5 **(a)** Construct an explicit $\Delta\langle 2\rangle\Delta\langle 2\rangle$ factorization for Bosons. **(b)** Construct an explicit $\Delta\langle 2\rangle\Delta\langle 2\rangle$ factorization for Fermions. Verify that the Fermion factorization reduces to the Boson $\Delta\langle 2\rangle\Delta\langle 2\rangle$ if we add the constraint that the products $a_k^\dagger a_k$ corresponding to Boson operators are inseparable.

15.6 **(a)** Define $\Delta\langle 1\rangle$, $\Delta\langle 2\rangle$, $\Delta\langle 3\rangle$, and $\Delta\langle 4\rangle$ in terms of expectation values. **(b)** Express the triplet correlation in terms of Fermion or Boson operators.

15.7 Construct a factorization $\langle 1\rangle\Delta\langle 2\rangle$ for a term with one, two, or three Boson operators. Use the symmetrization rule introduced in the context of Eqs. (15.23)–(15.26). Verify the validity of the results (15.41)–(15.43).

15.8 Assume an incoherent single-mode Boson field, i.e., $\langle B\rangle = \langle B^\dagger\rangle = 0$, for which only the doublet correlation $\Delta\langle B^\dagger B\rangle$ is nonzero. Show that the quantum statistics then follows from $I_K^J = \langle [B^\dagger]^J B^K\rangle = \delta_{J,K}\, J![\Delta\langle B^\dagger B\rangle]^J$. **Hint:** *Denote* $B^K = B_1 B_2 \ldots B_K$ *and define all doublet factorizations recursively with the constraint* $\Delta\left\langle B_j^\dagger B_k^\dagger\right\rangle = \Delta\langle B_j\, B_k\rangle = 0$ *and* $\Delta\left\langle B_j^\dagger\, B_k\right\rangle = \Delta\langle B^\dagger B\rangle$.

15.9 **(a)** Show that the nonlinear terms of Eq. (15.47) do not produce a hierarchy problem. **(b)** Assume that the initial value of the correlations below $\Delta\langle C\rangle$ vanishes. Show that all correlations above $\Delta\langle 2C+1\rangle$ remain zero for some time, based on Eq. (15.47).

15.10 **(a)** Start from the operator equations (14.37) and (14.40) and derive $i\hbar\frac{\partial}{\partial t}\left\langle a_{\mathbf{k}_1}^\dagger\, a_{\mathbf{k}_2}^\dagger\, a_{\mathbf{k}_3}\, a_{\mathbf{k}_4}\right\rangle$ and $i\hbar\frac{\partial}{\partial t}\left\langle a_{\mathbf{k}_1}^\dagger\, a_{\mathbf{k}_2}^\dagger\, a_{\mathbf{k}_3}\, a_{\mathbf{k}_4}\right\rangle_s$.
(b) Construct $i\hbar\frac{\partial}{\partial t}\Delta\left\langle a_{\mathbf{k}_1}^\dagger\, a_{\mathbf{k}_2}^\dagger\, a_{\mathbf{k}_3}\, a_{\mathbf{k}_4}\right\rangle$ and deduce the $L[\Delta\langle 2\rangle]$ contributions. Evaluate also the singlet source and construct the Markov limit, in analogy to Eq. (15.52).

Further reading

The cluster expansion is a variant of the cumulant expansion and the coupled-cluster method. Thorvald. N. Thiele first formulated the "half-invariants" – later known as the cumulants – in 1889. A historical account as well as a translation of the original work can be found in:

S. L. Lauritzen (2002). *Thiele: a Pioneer in Statistics*, Oxford, Oxford University Press.

The work on coupled clusters is based on:

F. Coester (1958). Bound states of a many-particle system, *Nucl. Phys.* **7**, 421.

F. Coester and H. Kümmel (1960). Short-range correlations in nuclear wave functions, *Nucl. Phys.* **17**, 477.

J. Čiček (1966). On the correlation problem in atomic and molecular systems. Calculation of wavefunction components in Ursell-type expansion using quantum-field theoretical methods, *J. Chem. Phys.* **45**, 4256.

The application of the cluster expansion in quantum chemistry, solid-state systems, and quantum-optics problems is described, e.g., in:

H. W. Eyld and B. D. Fried (1963). Quantum mechanical kinetic equations, *Ann. Phys.* **23**, 374.

J. Fricke (1996). Transport equations including many-particle correlations for an arbitrary quantum system: A general formalism, *Ann. Phys.* **252**, 479.

R. J. Bartlett, ed. (1997). *Recent Advances in Coupled-Cluster Methods*, Singapore, World Scientific.

M. Kira and S. W. Koch (2008). Cluster-expansion representation in quantum optics, *Phys. Rev. A* **78**, 022102.

P. Carsky, J. Paldus, and J. Pittner, eds. (2010). *Recent Progress in Coupled Cluster Methods: Theory and Applications*, Dordrecht, Springer.

An advanced discussion of the cluster-expansion approach, details of the method involving correlation-generating functions, and much more can be found in:

M. Kira and S. W. Koch (2006). Many-body correlations and excitonic effects in semiconductor spectroscopy, *Prog. Quantum Electr.* **30**, 155.

The dynamics-controlled truncation method was introduced in:

V. M. Axt and A. Stahl (1994). A dynamics-controlled truncation scheme for the hierarchy of density matrices in semiconductor optics, *Z. Physik B Condensed Matter* **93**, 195.

See also:

M. Lindberg, Y. Z. Hu, R. Binder, and S. W. Koch (1994). $\chi^{(3)}$ formalism in optically excited semiconductors and its applications in four-wave-mixing spectroscopy, *Phys. Rev. B* **50**, 18060.

16

Simple many-body systems

In the previous chapters, we have developed the formalisms of quantum optics and electronic many-body theory to such a level that it can be used to analyze a broad range of phenomena. Based on the discussion in Chapter 13, a very general class of system Hamiltonians can be derived. With the help of the cluster-expansion approach, Chapter 15, we can treat the resulting many-body and quantum-optical dynamics at a consistent level to study interesting problems in semiconductor quantum optics.

For many realistic situations, the quantum-statistical hierarchy problem is rather complex, posing significant analytical and numerical challenges even though its structural format is relatively simple, compare Section 14.3.2. In order to familiarize ourselves with these problems and their solutions, we start by discussing light–matter interaction effects for idealized but physically relevant model systems. In this chapter, we begin these investigations by focusing on the conceptually simplest many-body state consisting of two oppositely charged particles.

16.1 Single pair state

In many ways, the hydrogen atom constitutes the simplest "many-body" system. It consists of a single electron moving in the attractive potential of the nucleus. To analyze this case, we start from the generic many-body Hamiltonian (10.44) but have to account for the special aspects of the hydrogen problem. Since the system contains only two particles, it is not meaningful to treat the positively charged ion as an effective background. Thus, the longitudinal fields are entirely included via the usual Coulomb interaction between the electron and the ion with the charges Q_e and Q_i, respectively. Clearly, in such an atomic system there is no effective lattice potential $U^{\text{gen}}(\mathbf{r}_j)$ and $V^{\text{gen}}_{j,k}(\mathbf{r}_e, \mathbf{r}_i) = \frac{Q_e Q_i}{4\pi\varepsilon\epsilon_0|\mathbf{r}_e - \mathbf{r}_i|}$ has its usual free-space form. For later generalizations, we assume that the electron and the ion move in a dielectric environment with relative permittivity ε, where $\varepsilon = 1$ for vacuum. With these considerations, Eq. (10.44) simplifies to

$$\hat{H} = \frac{\left[\hat{\mathbf{p}}_e - Q_e\,\hat{\mathbf{A}}(\mathbf{r}_e)\right]^2}{2m_e} + \frac{\left[\hat{\mathbf{p}}_i - Q_i\,\hat{\mathbf{A}}(\mathbf{r}_i)\right]^2}{2m_i} + \frac{Q_e Q_i}{4\pi\varepsilon\,\epsilon_0|\mathbf{r}_e - \mathbf{r}_i|} + \sum_{\mathbf{q}} \hbar\omega_{\mathbf{q}}\left[B^\dagger_{\mathbf{q}}\,B_{\mathbf{q}} + \frac{1}{2}\right],$$

$$\hat{\mathbf{A}}(\mathbf{r}) = \sum_{\mathbf{q}} \frac{\mathcal{E}_{\mathbf{q}}}{\omega_{\mathbf{q}}}\left[\mathbf{u}_{\mathbf{q}}(\mathbf{r})\,B_{\mathbf{q}} + \mathbf{u}^\star_{\mathbf{q}}(\mathbf{r})\,B^\dagger_{\mathbf{q}}\right], \tag{16.1}$$

where $(\mathbf{r}_e, \hat{\mathbf{p}}_e = -i\hbar\nabla_{\mathbf{r}_e})$ and $(\mathbf{r}_i, \hat{\mathbf{p}}_i = -i\hbar\nabla_{\mathbf{r}_i})$ refer to the quantized canonical coordinate pair of the electron and the ion, respectively. The vector potential that appears is expressed with the help of the mode-expansion relation (10.43).

Since the Hamiltonian (16.1) is defined in the Coulomb gauge, we have

$$\hat{\mathbf{p}}_\lambda \cdot \hat{\mathbf{A}}(\mathbf{r}_\lambda) = \left[-i\hbar\nabla_\lambda \cdot \hat{\mathbf{A}}(\mathbf{r}_\lambda)\right] + \hat{\mathbf{A}}(\mathbf{r}_\lambda)\cdot\hat{\mathbf{p}}_\lambda = \hat{\mathbf{A}}(\mathbf{r}_\lambda)\cdot\hat{\mathbf{p}}_\lambda. \tag{16.2}$$

Using this relation in Eq. (16.1) yields

$$\hat{H} = \hat{H}_0 - \frac{Q_e}{m_e}\hat{\mathbf{A}}(\mathbf{r}_e)\cdot\hat{\mathbf{p}}_e - \frac{Q_i}{m_i}\hat{\mathbf{A}}(\mathbf{r}_i)\cdot\hat{\mathbf{p}}_i + \frac{\left[Q_e\hat{\mathbf{A}}(\mathbf{r}_e)\right]^2}{2m_e} + \frac{\left[Q_i^2\,\hat{\mathbf{A}}(\mathbf{r}_i)\right]^2}{2m_i} + \hat{H}_F$$

$$\hat{H}_0 = \frac{\hat{\mathbf{p}}_e^2}{2m_e} + \frac{\hat{\mathbf{p}}_i^2}{2m_i} + \frac{Q_e Q_i}{4\pi\varepsilon\,\epsilon_0|\mathbf{r}_e - \mathbf{r}_i|}, \qquad \hat{H}_F \equiv \sum_{\mathbf{q}} \hbar\omega_{\mathbf{q}}\left[B^\dagger_{\mathbf{q}}\,B_{\mathbf{q}} + \frac{1}{2}\right], \tag{16.3}$$

where $\hat{H}_0$ and $\hat{H}_F$ identify the noninteracting parts of the matter and light Hamiltonians, respectively. The remaining contributions lead to the so-called $\hat{\mathbf{A}}\cdot\mathbf{p}$ interaction between charged particles and light, including the nonlinear $\hat{\mathbf{A}}^2$ contributions.

Before we analyze the full interacting light–matter problem, we first discuss the quantum dynamics of the isolated atom. The stationary Schrödinger equation of the atom is

$$\hat{H}_0|\psi_E\rangle = E|\psi_E\rangle. \tag{16.4}$$

We transform the eigenvalue problem into relative and center-of-mass coordinates,

$$\begin{cases} \mathbf{r} \equiv \mathbf{r}_e - \mathbf{r}_i \\ \mathbf{R} \equiv \frac{m_e}{M}\mathbf{r}_e + \frac{m_i}{M}\mathbf{r}_i \end{cases} \Leftrightarrow \begin{cases} \mathbf{r}_e = \mathbf{R} + \frac{m_i}{M}\mathbf{r} \\ \mathbf{r}_i = \mathbf{R} - \frac{m_e}{M}\mathbf{r} \end{cases} \tag{16.5}$$

and introduce the total and reduced mass

$$M \equiv m_e + m_i, \qquad \frac{1}{\mu} \equiv \frac{1}{m_e} + \frac{1}{m_i}, \tag{16.6}$$

respectively. Using the chain rule of differentiation, the momentum operators can be expressed as

$$\hat{\mathbf{p}}_e = -i\hbar\nabla_{\mathbf{r}} - i\hbar\frac{m_e}{M}\nabla_{\mathbf{R}} \equiv \hat{\mathbf{p}} + \frac{m_e}{M}\hat{\mathbf{P}},$$
$$\hat{\mathbf{p}}_i = +i\hbar\nabla_{\mathbf{r}} - i\hbar\frac{m_i}{M}\nabla_{\mathbf{R}} \equiv -\hat{\mathbf{p}} + \frac{m_i}{M}\hat{\mathbf{P}}. \tag{16.7}$$

This simplifies the electron–ion Hamiltonian into

$$\hat{H}_0 = \hat{H}_{\text{rel}} + \hat{T}_{\text{COM}}, \quad \hat{H}_{\text{rel}} \equiv \frac{\hat{\mathbf{p}}^2}{2\mu} - V(\mathbf{r}), \quad V(\mathbf{r}) \equiv \frac{Q_e^2}{4\pi\varepsilon\,\epsilon_0|\mathbf{r}|},$$
$$\hat{T}_{\text{COM}} \equiv \frac{\hat{\mathbf{P}}^2}{2M}. \tag{16.8}$$

As a consequence of their opposite charges, i.e., $Q_i = -Q_e$, the Coulombic electron–ion interaction is attractive.

16.1.1 Electron–hole system

The same form (16.8) of the Hamiltonian can also be derived when we start from the quantum-confined many-body Hamiltonian of solids, discussed in Section 13.5. Figure 16.1(b) illustrates the semiconductor ground state (slab) that consists of filled valence and empty conduction bands. If exactly one electron is excited into the conduction band, one vacancy is left in the valence band. To account for such valence-band vacancies, it is convenient to introduce hole operators

$$h_{\mathbf{k}} \equiv a^{\dagger}_{v,-\mathbf{k}}, \qquad h^{\dagger}_{\mathbf{k}} \equiv a_{v,-\mathbf{k}} \tag{16.9}$$

that have a fully Fermionic statistics because they merely describe the missing electrons. Since the creation of a hole corresponds to the annihilation of a valence-band electron, a hole can be considered as the antiparticle of a valence-band electron.

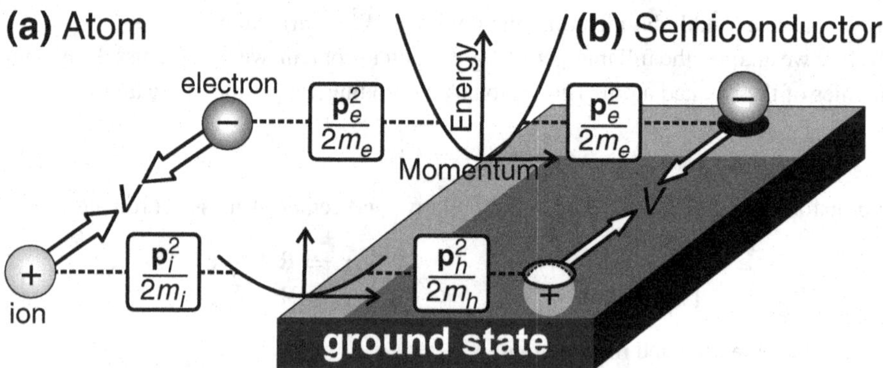

Figure 16.1 Comparison of simple many-body systems consisting of a single atom or an electron–hole pair in a solid. **(a)** The atomic system consists of an oppositely charged electron and ion (spheres). The attractive Couloumb force is indicated by the arrows. **(b)** A single electron (sphere) is lifted above the ground state (slab) leaving behind a valence-band vacancy (white, transparent sphere) that defines the hole. The electron and hole experience Coulomb attraction in analogy to the atom. The quadratic energy dispersion of the electron and ion/hole is indicated at the border of the slab. Whereas in the atom case m_e and m_i are the true, i.e., free-space masses of the electron and ion, respectively, in the semiconductor m_e and m_h denote the effective masses defined by parabolic approximations to the band structure.

The Coulomb interaction among conduction- and valence-band electrons in Eqs. (13.87), (13.98), and (13.109) can be expressed via

$$\begin{aligned}\hat{V}_{\text{eh}} &\equiv \sum\nolimits_{\mathbf{k},\mathbf{k}',\mathbf{q}\neq 0} V_{\mathbf{q}}^{c,v} a^{\dagger}_{c,\mathbf{k}} a^{\dagger}_{v,\mathbf{k}'} a_{v,\mathbf{k}'+\mathbf{q}} a_{c,\mathbf{k}-\mathbf{q}} \\ &= -\sum\nolimits_{\mathbf{k},\mathbf{k}',\mathbf{q}\neq 0} V_{\mathbf{q}}^{c,v} a^{\dagger}_{c,\mathbf{k}} a_{v,\mathbf{k}'+\mathbf{q}} a^{\dagger}_{v,\mathbf{k}'} a_{c,\mathbf{k}-\mathbf{q}} \\ &= -\sum\nolimits_{\mathbf{k},\mathbf{k}',\mathbf{q}\neq 0} V_{\mathbf{q}}^{c,v} a^{\dagger}_{c,\mathbf{k}} h^{\dagger}_{-\mathbf{k}'-\mathbf{q}} h_{-\mathbf{k}'} a_{c,\mathbf{k}-\mathbf{q}} \\ &= -\sum\nolimits_{\mathbf{k},\mathbf{k}',\mathbf{q}\neq 0} V_{\mathbf{q}}^{c,v} a^{\dagger}_{c,\mathbf{k}} h^{\dagger}_{\mathbf{k}'} h_{\mathbf{k}'+\mathbf{q}} a_{c,\mathbf{k}-\mathbf{q}} \end{aligned} \tag{16.10}$$

after we have exchanged the order of two valence-band operators using the Fermionic anticommutation relation. Furthermore, we have introduced hole operators and changed the summation variable $\mathbf{k}' \to -\mathbf{k}' - \mathbf{q}$. Equation (16.10) shows that the repulsive Coulomb interaction between conduction- and valence-band electrons can equally well be described as an attractive interaction between electrons and holes. Since the interaction changes sign, the hole formally has an inverted charge $Q_h = -Q_e$ in comparison with the valence-band electron, which is in accordance with the antiparticle interpretation.

For parabolic bands, the kinetic energy of the holes is reversed relative to that of the valence-band electrons, producing

$$E^c_{\mathbf{k}} = \frac{\hbar^2\mathbf{k}^2}{2m_c} + E_{\text{gap}}, \qquad E^h_{\mathbf{k}} = -E^v_{\mathbf{k}} = \frac{\hbar^2\mathbf{k}^2}{2m_h}, \tag{16.11}$$

where $m_h = -m_v$ is positive and E_{gap} defines the energy of the semiconductor band gap. When the system consists of a single conduction-band electron and a single hole, we find a complete analogy with the hydrogen atom after having associated the conduction-band electron with the atomic electron and the hole with the ion, compare Fig. 16.1.

16.1.2 Separation of relative and center-of-mass motion

Since the Hamiltonian $\hat{H}_0$ in Eq. (16.3) can be divided into relative- and center-of-mass-dependent parts, we may use a separation ansatz,

$$|\psi_E\rangle \equiv |\phi_\nu\rangle \otimes |\mathbf{K}\rangle \equiv |\nu, \mathbf{K}\rangle. \tag{16.12}$$

This allows us to split the stationary Schrödinger equation into two simpler ones,

$$\begin{aligned} &\hat{H}_0\,|\nu,\mathbf{K}\rangle = E\,|\nu,\mathbf{K}\rangle, \quad E = E_\nu + \frac{\hbar^2\mathbf{K}^2}{2M} \\ \Rightarrow\quad &\hat{H}_{\text{rel}}\,|\phi_\nu\rangle = E_\nu\,|\phi_\nu\rangle, \quad \hat{T}_{\text{COM}}\,|\mathbf{K}\rangle = \frac{\hbar^2\mathbf{K}^2}{2M}\,|\mathbf{K}\rangle. \end{aligned} \tag{16.13}$$

The total eigenenergy is the sum of the relative E_ν and the center-of-mass $E_{\mathbf{K}} \equiv \hbar^2\mathbf{K}^2/2M$ contributions.

Based on the discussion in Section 2.3.1, the center-of-mass eigenstates are plane-wave solutions,

$$\langle \mathbf{R}|\phi_\nu\rangle = \frac{1}{\sqrt{2\pi}}\, e^{i\mathbf{K}\cdot\mathbf{R}}, \tag{16.14}$$

where we used the quantization in infinite space, compare Eq. (5.29). Hence, the eigenstate of the hydrogen problem is completely delocalized with respect to its center-of-mass coordinate. Clearly, a real atom can always be in a plane-wave superposition state describing a localized but dynamically moving configuration. Such an atomic wave packet both moves and spreads in width as discussed in Section 3.1.1.

The relative-coordinate equation $\hat{H}_{\rm rel}\,|\phi_\nu\rangle = E_\nu\,|\phi_\nu\rangle$ defines the standard stationary hydrogen problem that can be solved analytically in real space,

$$\begin{aligned} &\langle \mathbf{r}|\hat{H}_{\rm rel}\,|\phi_\nu\rangle = E_\nu\,\langle \mathbf{r}|\phi_\nu\rangle, \qquad \langle \mathbf{r}|\phi_\nu\rangle \equiv \phi_\nu(\mathbf{r}) \\ \Rightarrow\quad &\left[-\frac{\hbar^2\nabla^2}{2\mu} - V(\mathbf{r})\right]\phi_\nu(\mathbf{r}) = E_\nu\,\phi_\nu(\mathbf{r}). \end{aligned} \tag{16.15}$$

As a direct generalization to the usual hydrogen problem, we can consider this equation to define a generic D-dimensional configuration. The corresponding $V(\mathbf{r})$ is discussed in Section 13.5 where the Hamiltonian for quantum-confined semiconductors is derived. For all cases, the potential $V(\mathbf{r})$ yields a Helmholtz equation with discrete bound-state solutions for $E_\nu < 0$. For $E_\nu > 0$, we have a continuum of freely propagating solutions as classified in Section 4.4.

16.2 Hydrogen-like eigenstates

Since the strict three- and two-dimensional attractive two-particle systems have well-known analytic solutions, we merely outline the results here. The omitted details can be found in many quantum-mechanics textbooks, see the references at the end of this chapter. In three dimensions, it is convenient to define the state index $\nu = (n,\, l,\, m)$ through a set of three discrete quantum numbers. Historically, n is referred to as the principal quantum number while l and m define the orbital and the magnetic quantum numbers, respectively. The corresponding bound states can be expressed in spherical coordinates,

$$\begin{aligned} &\phi_{n,l,m}(r,\varphi,\theta) = \mathcal{N}_{n,l}\,\rho^l\, e^{-\frac{\rho}{2}}\, L^{2l+1}_{n-l-1}(\rho)\, Y_{l,m}(\theta,\varphi) \\ &\rho = \frac{2r}{n\,a_0}, \qquad \mathcal{N}_{n,l} = \left(\frac{2r}{n\,a_0}\right)^{\frac{3}{2}}\sqrt{\frac{(n-l-1)!}{2n(n+l)!)}} \\ &n = 1, 2, \cdots, \quad l = 0, 1, \cdots, n-1, \quad m = 0, \pm 1, \cdots, \pm l. \end{aligned} \tag{16.16}$$

The corresponding discrete eigenenergies are given by

$$E_\nu = E_{n,l,m} = -\frac{E_B}{n^2}, \tag{16.17}$$

which is defined by the Bohr radius and the binding energy

$$a_0 = \frac{4\pi\,\varepsilon\epsilon_0\,\hbar^2}{Q_e^2\,\mu} \quad \text{and} \quad E_B = \frac{\hbar^2}{2\mu\,a_0^2}, \tag{16.18}$$

respectively. The special functions that appear $L_l^m(x)$ and $Y_l^m(\theta,\varphi)$ are the associated Laguerre polynomial and the spherical harmonics, respectively. The associated Laguerre polynomial is given by

$$L_n^j(x) \equiv \sum_{k=0}^{n} \frac{(-1)^k\,(n+j)!\,x^k}{k!(k+j)!(n-k)!}, \tag{16.19}$$

and the spherical harmonics,

$$Y_l^m(\theta,\varphi) \equiv \sqrt{\frac{(2l+1)\,(l-m)!}{4\pi\,(l+m)!}}\;P_l^m(\cos\theta)\,e^{im\varphi}$$

$$P_l^m(x) \equiv \left(\frac{1+x}{1-x}\right)^{\frac{m}{2}} \sum\nolimits_{k=m}^{n} \frac{(n+k)!}{k!(k-m)!(n+k)!}\left(\frac{x-1}{2}\right)^k, \tag{16.20}$$

are defined by the associated Legendre polynomial $P_l^m(x)$. The actual forms of these special functions can be found in many mathematical table books. For later reference, we explicitly list a few of the lowest-order expressions,

$$P_0^0(x) = 1,\; P_1^{-1}(x) = \frac{1}{2}\sqrt{1-x^2},\; P_1^0(x) = x,\; P_1^1(x) = \sqrt{1-x^2},$$

$$L_0^0(x) = 1,\; L_1^{-1}(x) = -x,\; L_1^0(x) = 1-x,\; L_1^1(x) = 2-x, \tag{16.21}$$

which follow directly from Eqs. (16.19)–(16.20).

To get an idea of the typical range of the Bohr radius and binding energy defined in Eq. (16.18), we have tabulated a_0 and E_B for a hydrogen atom without neutrons and for a bulk GaAs system. The model hydrogen atom then consists of the negatively charged electron and the positively charged ion, giving $Q_e = -e$. The explicit values for m_e, m_{ion}, and ε are given in Table 16.1. For solids, the charged particles move in a lattice producing a background dielectric constant ε as well as effective masses for charge carriers, see Chapter 11. The explicit values for the electron and hole masses as well as ε are given in

Table 16.1 *Bohr radius and binding energy for hydrogen and bulk GaAs.*

System	m_e $[m_0]$	m_i or m_h $[m_0]$	ε	a_0	E_B
Hydrogen	1	1836	1	0.5295 Å	13.6 eV
Bulk GaAs	0.0665	0.457	13.74	125 Å	4.18 meV

Table 16.1. The masses are defined in terms of the free-electron mass $m_0 = 9.109\,382\,15 \times 10^{-31}$ kg. The elementary electron charge is $e = 1.602\,176\,487 \times 10^{-19}$ C.

Using these values as input to Eq. (16.18), we find that a_0 is in the range of 1Å and E_B is of the order of ten electron volts (eV) for the atomic system. The semiconductor's a_0 is larger by roughly two orders of magnitude due to the lower effective mass and the higher dielectric constant. As a result, the binding energy is in the range of some meV. The bound electron–hole pairs are commonly referred to as excitons. The physical relevance of excitonic resonances is discussed in Chapters 26–27 where we analyze the optical response of semiconductors.

16.2.1 Low-dimensional systems

The excitonic bound-state solutions are also interesting for systems with reduced dimensionality because the electron–hole excitations in semiconductors can be confined to two, one, or zero dimensions, as discussed in Chapter 11. The purely one-dimensional system yields an infinite binding energy and is thus unphysical. However, quasi-one-dimensional systems, i.e., quantum-wire (QWI) structures with a small but finite radius can be realized experimentally. In those systems, the excitonic eigenstates depend on the wire radius and other system parameters.

In reality, most two-dimensional systems also have a finite extension in the third dimension, making them quasi-two-dimensional in a strict sense. However, the idealized perfectly two-dimensional system provides a good reference since the exciton problem can be solved analytically. The corresponding eigenstates can be labeled with two discrete quantum numbers n and m,

$$\phi_{n,m}(r,\varphi) = \mathcal{N}_{n,m}\,\rho^{|m|}\,e^{-\frac{\rho}{2}}\,L^{2|m|}_{n-|m|}(\rho)\,e^{im\varphi}$$

$$\rho = \frac{4r/a_0}{1+2n}, \qquad \mathcal{N}_{n,m} = (-1)^n\sqrt{\frac{1}{\pi a_0^2}\left(\frac{2}{1+2n}\right)^3\frac{(n-|m|)!}{(n+|m|)!}}$$

$$n = 0,\,1,\,\cdots, \qquad m = 0,\,\pm 1,\,\cdots,\,\pm n. \tag{16.22}$$

These expressions have a clear structural similarity to the three-dimensional solution (16.16). Thus, one often labels the two-dimensional states in the same tradition as the three-dimensional ones. This naming follows a historic tradition and the specific letters stem from the words **s**harp, **p**rincipal, **d**iffuse, and **f**undamental while the $m \geq 4$ orbitals are denoted simply in alphabetical order starting from g and omitting s and p from the sequence.

Among the solutions (16.22), the lowest radially symmetric state $\phi_{1s}(\mathbf{r}) \equiv \phi_{0,0}(r)$ defines the ground state. The first excited state is given by the radially symmetric $\phi_{2s}(\mathbf{r}) \equiv \phi_{1,0}(r)$ and by the orthogonal $2p$ states

$$\phi_{2p_x}(\mathbf{r}) \equiv \frac{1}{\sqrt{2}}\left[\phi_{1,+1}(r,\varphi) + \phi_{1,-1}(r,\varphi)\right],$$
$$\phi_{2p_y}(\mathbf{r}) \equiv \frac{1}{\sqrt{2}i}\left[\phi_{1,+1}(r,\varphi) - \phi_{1,-1}(r,\varphi)\right] \tag{16.23}$$

aligned along the x and the y axis, respectively. The two-dimensional exciton has the energy eigenvalues

$$E_\nu = E_{n,m} = -\frac{4E_B}{(1+2n)^2}, \qquad n = 0, 1, \cdots \tag{16.24}$$

having a $(2n+1)$-fold degeneracy due to the m-independent E_ν. From the ground state, we see that the two-dimensional system has a four times larger binding energy than the three-dimensional one.

Plots of the two-dimensional 1s (solid line), 2s (shaded area), and 2p (dashed line) wave functions are presented in Fig. 16.2(a) as slices along the x axis. Since all s-like states are rotationally symmetric in the xy-plane, a single x slice fully determines their properties. Functionally, $\psi_{1s}(x,0)$ peaks at the origin and decays monotonously as $|x|$ is increased. The 2s slice displays a more complicated structure because $\psi_{2s}(x,0)$ has a minimum at the origin and two peaks at $|x| = 9a_0/4$ resulting in two nodes at $|x| = 3a_0/4$. In contrast to these, the $\psi_{2p_x}(x,0)$ slice (dashed line in Fig. 16.2(a)) has a node at the origin, a peak at $x = -3a_0/2$, and a minimum at $x = +3a_0/2$. This state has an asymmetric functional form because it is not rotationally symmetric in the xy-plane. To present the full xy dependence, Fig. 16.2(b) displays the contour plot of $\psi_{2p_x}(x,y)$, where white (black) indicates maximally positive (negative)-valued regions. We see that $\psi_{2p_x}(x,y)$ is indeed aligned along the x axis producing a sign reversal,

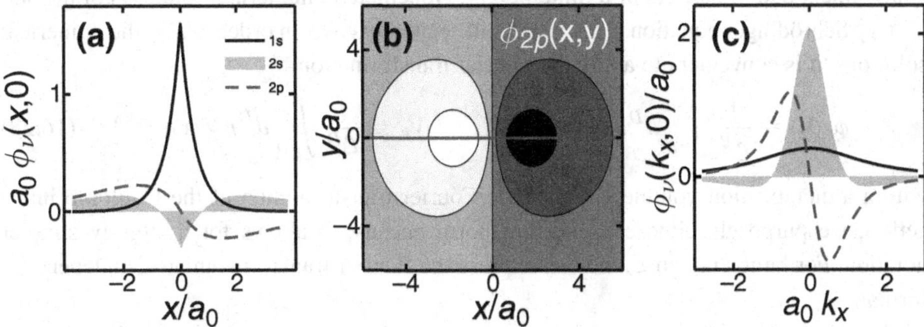

Figure 16.2 Exciton eigenstates for a strictly two-dimensional system. **(a)** A cross section of $\phi_\nu(\mathbf{r})$ is shown along the x axis for $\nu = 1s$ (solid line), $\nu = 2p_x$ (dashed line), and $\nu = 2s$ (shaded area) state. **(b)** Contour plot for $\phi_{2p_x}(x,y)$ where the contours are identified at 0, ±0.25, and ±0.75 of the peak value. The white region indicate positive and the black region negative values. The horizontal line indicates the x slice presented in frame (a). **(c)** The Fourier transformed $\phi_\nu(\mathbf{k})$ is presented as a function of k_x for $\nu = 1s$ (solid line), $\nu = 2p_x$ (dashed line), and $\nu = 2s$ (shaded area).

$$\psi_{np_x}(-\mathbf{r}) = -\psi_{np_x}(\mathbf{r}), \qquad n = 2, 3, \cdots, \tag{16.25}$$

under the transformation $\mathbf{r} \to -\mathbf{r}$.

The same transformation can also be expressed in terms of a π rotation $R_\pi[\mathbf{r}] = -\mathbf{r}$ in the xy-plane. In general, the ν states are classified as p-like if they display the symmetry

$$\psi_\nu^x\left(R_{\frac{2\pi}{m}}[\mathbf{r}]\right) = +\psi_\nu^x(\mathbf{r}), \qquad \psi_\nu^x\left(R_{\frac{\pi}{m}}[\mathbf{r}]\right) = -\psi_\nu^x(\mathbf{r}), \quad m = 1, 2, \cdots \tag{16.26}$$

only for $m = 1$. If these relations are satisfied only for $m = 2, 3 \cdots$, ν refers to a d-like, f-like etc. state. The s-like states are invariant under any rotation $R_\varphi[\mathbf{r}]$ yielding a fully symmetric function,

$$\psi_{ns}(R_\varphi[\mathbf{r}]) = +\psi_{ns}(\mathbf{r}), \quad \forall\varphi \quad \Rightarrow \quad \psi_{ns}(-\mathbf{r}) = +\psi_{ns}(\mathbf{r}). \tag{16.27}$$

In one-dimensional (1D) systems, one can only apply the rotations by $\varphi = 0$ and $\varphi = \pi$ ($m = 0$) to remain within the same 1D structure. Consequently, 1D systems have either symmetric s-like or antisymmetric p-like eigenstates.

Figure 16.2(c) presents the Fourier transform of the $1s$ (solid line), $2s$ (shaded area), and $2p_x$ (dashed line) states as a function of k_x. We see that the symmetry properties of the Fourier transformed $\phi_\nu(\mathbf{k})$ are identical to those of $\phi_\nu(\mathbf{r})$. In general, the $\phi_\nu(\mathbf{k})$ representation allows for simpler numerical solutions than the $\mathbf{r}$-space representation, as discussed next.

16.2.2 Numerical solutions of bound and unbound states

The exciton eigenstates are known analytically only for the idealized three- and two-dimensional cases. The Coulomb interaction does not have a simple format in realistic low-dimensional systems where the dimensional reduction results from the quantum confinement due to barriers of a finite height. Nonetheless, numerical solutions of the stationary Schrödinger equation (16.15) are still relatively easy. In order to find the numerical solutions, it is convenient to apply the Fourier transformation

$$\phi_\nu(\mathbf{k}) \equiv \frac{1}{\mathcal{L}^{\frac{D}{2}}}\int_{\mathcal{L}^D} d^D r\, \phi_\nu(\mathbf{r})e^{-i\mathbf{k}\cdot\mathbf{r}}, \qquad V_\mathbf{k} \equiv \frac{1}{\mathcal{L}^D}\int_{\mathcal{L}^D} d^D r\, V(\mathbf{r})e^{-i\mathbf{k}\cdot\mathbf{r}} \tag{16.28}$$

within a quantization volume $\mathcal{L}^D$. For the Fourier transformation of the Coulomb interaction, we purposely choose a different norm because it allows for a slightly simpler notation. For large enough $\mathcal{L}$, we can express the Fourier transforms in an $\mathcal{L}$-independent format,

$$\begin{aligned}
\bar{\phi}_\nu(\mathbf{k}) &\equiv \lim_{\mathcal{L}\to\infty}\frac{\phi_\nu(\mathbf{k})}{[\Delta k]^{\frac{D}{2}}} = \frac{1}{[2\pi]^{\frac{D}{2}}}\int d^D r\, \phi_\nu(\mathbf{r})e^{-i\mathbf{k}\cdot\mathbf{r}}, \qquad \Delta k \equiv \frac{2\pi}{\mathcal{L}} \\
\bar{V}_\mathbf{k} &\equiv \lim_{\mathcal{L}\to\infty}\frac{V_\mathbf{k}}{[\Delta k]^D} = \frac{1}{[2\pi]^D}\int d^D r\, V(\mathbf{r})e^{-i\mathbf{k}\cdot\mathbf{r}},
\end{aligned} \tag{16.29}$$

where Δk defines the discretization via the quantization volume, compare Section 8.4.1.

With the help of the Fourier transformation (16.28), Eq. (16.15) becomes

$$\frac{\hbar^2\mathbf{k}^2}{2\mu}\phi_\nu(\mathbf{k}) - \sum_{\mathbf{k}'} V_{\mathbf{k}'-\mathbf{k}}\,\phi_\nu(\mathbf{k}') = E_\nu\,\phi_\nu(\mathbf{k}). \tag{16.30}$$

Here, we have used the convolution theorem for the Fourier transformation of the product $V(\mathbf{r})\,\phi_\nu(\mathbf{r})$. In the $\mathcal{L}\to\infty$ limit, Eq. (16.30) becomes an integral equation. For finite $\mathcal{L}$, this can be treated as a typical matrix eigenvalue problem. Since this is a standard problem in linear algebra, one can easily find several algorithms and/or library routines to solve Eq. (16.30) computationally. Thus, one can efficiently determine the numerical solutions of Eq. (16.30).

Among the quantum confined semiconductor systems, the effectively zero-dimensional quantum dots (QD) present a special case where the kinetic energy $\hbar^2\mathbf{k}^2/2\mu$ is replaced by a discrete confinement energy and $V_{\mathbf{k}'-\mathbf{k}}$ becomes a matrix element between the confinement levels, see Section 13.5. For strongly confined QDs, one can use a model with a single electron and hole state. The resulting Eq. (16.30) then reduces to a particularly simple diagonal format yielding a trivial solution. In fact, even the generic QD system is so simple that one can look for direct diagonalization of the full many-body problem with N electrons and holes. Since QDs are distinctly different from two- and one-dimensional systems, we mainly concentrate here on the exciton states in QWs and QWIs.

Before we solve Eq. (16.30), we summarize some important results following from Eqs. (16.22), (16.23), and (16.29) for the strictly two-dimensional eigenstates,

$$\begin{aligned}
&\phi_{1s}(\mathbf{r}) = \sqrt{\frac{8}{\pi a_0^2}}\,e^{-2\frac{r}{a_0}}, && \bar{\phi}_{1s}(\mathbf{k}) = \sqrt{\frac{a_0^2}{2\pi}}\,\frac{1}{[1+(ka_0/2)^2]^{\frac{3}{2}}}\\
&\phi_{2s}(\mathbf{r}) = \sqrt{\frac{8}{27\pi a_0^2}}\left(\frac{4r}{3a_0}-1\right)e^{-\frac{2}{3}\frac{r}{a_0}}, && \bar{\phi}_{2s}(\mathbf{k}) = \sqrt{\frac{27a_0^2}{2\pi}}\,\frac{1-(3ka_0/2)^2}{[1+(3ka_0/2)^2]^{\frac{5}{2}}}\\
&\phi_{2p_x}(\mathbf{r}) = -\sqrt{\frac{128}{243\pi a_0^2}}\,\frac{x}{a_0}\,e^{-\frac{2}{3}\frac{r}{a_0}}, && \bar{\phi}_{2p_x}(\mathbf{k}) = \sqrt{\frac{27a_0^2}{2\pi}}\,\frac{3k_y a_0}{[1+(3ka_0/2)^2]^{\frac{5}{2}}}\\
&\phi_{2p_y}(\mathbf{r}) = -\sqrt{\frac{128}{243\pi a_0^2}}\,\frac{y}{a_0}\,e^{-\frac{2}{3}\frac{r}{a_0}}, && \bar{\phi}_{2p_y}(\mathbf{k}) = -\sqrt{\frac{27a_0^2}{2\pi}}\,\frac{3k_x a_0}{[1+(3ka_0/2)^2]^{\frac{5}{2}}}\\
&k_x = k\cos\varphi, \qquad k_y = k\sin\varphi.
\end{aligned} \tag{16.31}$$

The real-space dependence of these functions is presented in Figs. 16.2(a)–(b). The corresponding slices of $\bar{\phi}_\nu(k_x, 0)$ are shown in Fig. 16.2(c) for the $1s$ (solid line), $2s$ (shaded area), and $2p$ (dashed line) states. The $\mathbf{r}$- and $\mathbf{k}$-space formats of the two-dimensional solutions are qualitatively similar. As a major difference, a broad feature in **r** space becomes a narrow feature in the **k** space and vice versa.

To illustrate the properties of excitonic states for realistic quantum-confined systems, we solve Eq. (16.30) numerically for: (i) an 8 nm thick QW and (ii) a 20 nm thick QWI. The material parameters are chosen to match a typical quantum-confined GaAs system with $\varepsilon = 13.74$, $m_e = 0.0665\, m_0$, and $m_h = 0.235\, m_0$. Note that the confinement typically induces a reduction of the effective hole mass in comparison with the bulk defined in Table 16.1.

The Coulomb interaction follows from Eqs. (13.87) and (13.98), producing

$$V_{\mathbf{q}} = \frac{e^2}{2\epsilon\epsilon_0 \mathcal{L}^2} \frac{1}{|\mathbf{q}|} \int\int dZ\, dZ' g(Z)\, g(Z')\; e^{-|\mathbf{q}(Z-Z')|} \tag{16.32}$$

for the QW and

$$V_{q_z} = \frac{e^2}{4\pi\epsilon\epsilon_0 \mathcal{L}} \int d^2r\, d^2r'\, dz\, \frac{g(\mathbf{r})\, g(\mathbf{r}')}{\sqrt{(\mathbf{r}-\mathbf{r}')^2 + z^2}} e^{-iq_z z} \tag{16.33}$$

for the QWI. The confinement functions that appear are

$$g(Z) = \begin{cases} \frac{2}{L}\cos^2 \frac{\pi}{L} Z, & |Z| \le \frac{L}{2} \\ 0, & |Z| > \frac{L}{2} \end{cases}, \qquad g(\mathbf{r}) = \frac{2}{\pi \Delta R^2}\, e^{-\frac{2\mathbf{r}^2}{\Delta R^2}}, \tag{16.34}$$

where $L = 8$ nm and $\Delta R = 10$ nm define the width of the QW and the effective radius of the QWI, respectively. The chosen $g(Z)$ follows from the ground-state solution (4.58) which describes the envelope of the strongly confined carriers, as discussed in Section 12.2. In the same way, $g(\mathbf{r}) = |\xi(\mathbf{r})|^2$ results from the confinement function $\xi(\mathbf{r}) = \sqrt{\frac{2}{\pi\Delta R^2}}\, e^{-\frac{\mathbf{r}^2}{\Delta R^2}}$ which gives a good approximation of the ground-state confinement in the QWI if ΔR is adjusted appropriately, see Section 12.2.

Figure 16.3 shows the computed wave functions corresponding to the three lowest energy eigenvalues for a QWI ((a)–(b)) and a QW ((c)–(d)). The **r**-space wave functions are shown on the LHS while the **k**-representations are displayed on the RHS. The QW wave functions are presented as cross sections along the r_x and k_x axes in analogy to the strictly two-dimensional case analyzed in Fig. 16.2.

For both QWs and QWIs the ground state is symmetric with respect to the transformation $\mathbf{r} \to -\mathbf{r}$ for the QW and $x \to -x$ for the QWI. In other words, we find the typical behavior of the symmetric s-like state according to Eq. (16.27). Thus, we denote the ground state as the $1s$ state (solid line), in generalization of the strictly two-dimensional case. The first excited state is antisymmetric, identifying it as the $2p$ state (dashed line), according to Eq. (16.25). The third excited state (shaded area) is the symmetric $2s$ state.

The discussed ordering structure holds more generally. If we organize all the s- and p-like states in ascending order in terms of their eigenenergy E_ν, we find a sequence where s- and p-like states alternate. We also see that the **r** and **k** representations of these states qualitatively resemble the ideal two-dimensional wave functions, presented in Fig. 16.2.

However, there are also details that differ from the idealized case. For example, the analyzed QW has a $1s$ binding energy of $2.45\, E_B$ while the wire has $3\, E_B$. Thus, the

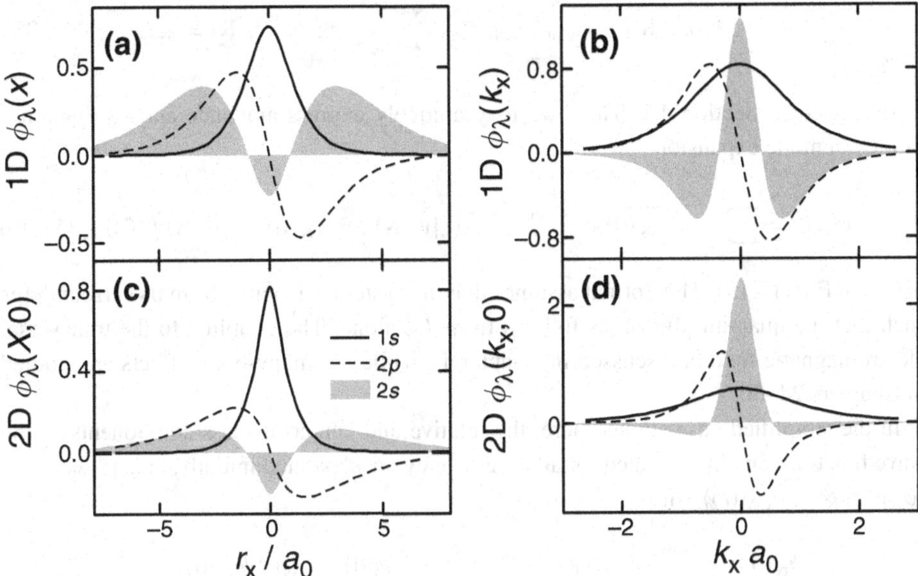

Figure 16.3 Wave functions of the three lowest exciton states for a GaAs-type quantum well (QW) and a wire (QWI). The wave functions are presented **(a)** in real-space and **(b)** in momentum space for the QWI system with an effective $\Delta R = 10$ nm radius. The corresponding wave functions for the 8 nm QW are shown in frames **(c)** and **(d)** as cross sections along the r_x and k_x axes. Reprinted from *Progress in Quantum Electronics*, **30/5**, M. Kira, S. W. Koch, "Many-body correlations and excitonic effects in semiconductor spectroscopy," pp. 155–296, 2006, with permission from Elsevier.

effective confinement reduces relative to the ideal value of $4\,E_B$. This is a very typical trend for QWs. For QWIs, however, the binding energy can even exceed $4\,E_B$ if the radius ΔR becomes very small. As mentioned earlier, the ground-state binding energy diverges in the mathematical limit of $\Delta R \to 0$. Clearly, realistic semiconductor QWIs always have a finite diameter and thus a finite binding energy.

For the example QW, the first excited state is found at $0.51\,E_B$ and the second excited state is at $0.38\,E_B$ below the gap. Hence, even though the $2s$ and $2p$ states are energetically very close, it is interesting to notice that the realistic QW confinement, i.e., the quasi-two-dimensional configuration, lifts the degeneracy between the $2s$ and $2p$ states which exists for the strictly two-dimensional system. For the QWI system, the corresponding $2p$ and $2s$ states are found at $1.49\,E_B$ and $0.99\,E_B$ below the gap, respectively. In comparison with the QW case, the QWI system has similar but less degenerate $2p$ and $2s$ states.

16.3 Optical dipole

In principle, the continuum of unbound states with $E_\nu > 0$ also exists for the generalized hydrogen problem (16.30). As for the bound states, one finds analytic expressions only for the strictly two- and three-dimensional cases. All the states together form a complete and orthogonal set,

$$\langle \nu, \mathbf{K}|\nu', \mathbf{K}'\rangle = \delta_{\nu,\nu'}\,\delta_{\mathbf{K},\mathbf{K}'} \qquad \sum_{\nu,\mathbf{K}} |\nu, \mathbf{K}\rangle\langle \nu, \mathbf{K}| = \mathbb{I}, \tag{16.35}$$

as discussed in Section 4.1. Thus, we may uniquely express any state and its dynamics through a mode expansion,

$$|\psi(t)\rangle = \sum_{\nu,\mathbf{K}} c_{\nu,\mathbf{K}}(0)\, e^{-\frac{i}{\hbar}\left[E_\nu + \frac{\hbar^2\mathbf{K}^2}{2M}\right]t}\, |\nu, \mathbf{K}\rangle\,, \quad c_{\nu,\mathbf{K}}(0) = \langle \nu, \mathbf{K}|\psi(0)\rangle\,, \tag{16.36}$$

based on Eq. (16.13). This format assumes that the system is isolated from its surroundings such that the quantum dynamics follows from $\hat{H}_0$ alone. The coupling to the transversal electromagnetic field is discussed in Chapter 17 while the many-body effects are studied in Chapters 24–30.

In the simplified case studied here, the relative and center-of-mass components of the wave function remain separated for all times if they are separated initially, i.e., $|\psi_{\text{sep}}(t)\rangle = |\psi_{\text{rel}}(t)\rangle \otimes |\psi_{\text{COM}}(t)\rangle$ with

$$\begin{aligned} |\psi_{\text{rel}}(t)\rangle &= \sum_{\nu} c_\nu(0)\, |\phi_\nu\rangle\, e^{-\frac{i}{\hbar}E_\nu t}, & c_\nu(0) &= \langle \phi_\nu|\psi_{\text{rel}}(0)\rangle \\ |\psi_{\text{COM}}(t)\rangle &= \sum_{\mathbf{K}} c_{\mathbf{K}}(0)\, e^{-\frac{i\hbar\mathbf{K}^2 t}{2m}}\, |\mathbf{K}\rangle, & c_{\mathbf{K}}(0) &= \langle \mathbf{K}|\psi_{\text{COM}}(0)\rangle. \end{aligned} \tag{16.37}$$

The center-of-mass contribution displays the typical wave-packet dynamics, investigated in Sections 2.3.2 and 3.1.1.

To determine the dynamics as a function of the relative-motion coordinate, we consider a strictly two-dimensional system such that we can express $|\psi_{\text{rel}}(t)\rangle$ analytically. As an example, we discuss the dynamics of an equally weighted superposition of the $\nu = 1s$ and $\nu = 2p_x$ states,

$$|\psi_{1s,\,2p}(t)\rangle \equiv \frac{1}{\sqrt{2}}\left(|1s\rangle + |2p_x\rangle\, e^{-i\omega_{2p,\,1s}t}\right) e^{-i\omega_{1s}t}, \qquad \omega_\nu \equiv \frac{E_\nu}{\hbar}, \tag{16.38}$$

where $\omega_{2p,\,1s} = \omega_{2p} - \omega_{1s}$ defines the difference between the eigenfrequencies of the two lowest bound states. If we use $\hbar\omega_{2p,\,1s} = 1\,\text{eV}$, we find $\omega_{2p,\,1s} = 1.52 \times 10^{15}\,\text{s}^{-1}$, which gives the ultrafast oscillation period $T_{2p,\,1s} = \frac{2\pi}{\omega_{2p,\,1s}} = 4.14$ fs.

The knowledge of the explicit state $|\psi_{1s,\,2p}(t)\rangle$ yields the probability distribution

$$P_{1s,\,2p}(\mathbf{r},\, t) \equiv \left|\langle \mathbf{r}|\psi_{1s,\,2p}(t)\rangle\right|^2, \tag{16.39}$$

which defines the probability of finding the electron at the distance $\mathbf{r}$ from the ion. Figure 16.4 presents a contour plot of $P_{1s,\,2p}(\mathbf{r},\, t)$ for (a) the initial time $t = 0$, (b) $t = 0.25\,T_{2p,1s}$, and (c) $t = 0.5\,T_{2p,1s}$. We observe that the electron–ion separation is moving along the x axis. To quantify this motion, we compute the average electron–ion displacement $\langle x(t)\rangle = \int dx\,dy\,x\,P_{1s,\,2p}(x,\, y,\, t)$. The result is depicted as the solid line in Fig. 16.4(d). Here, the filled circles correspond to times when the snapshots of the entire

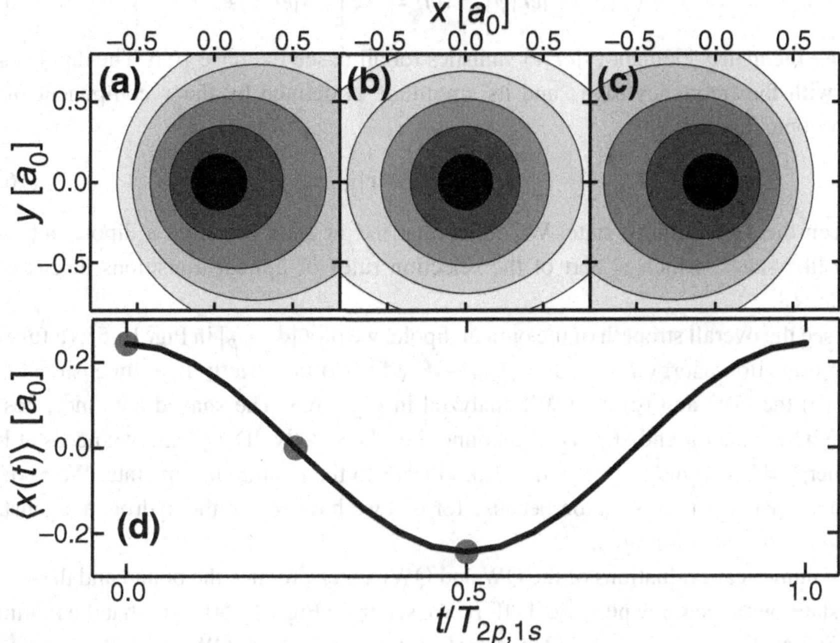

Figure 16.4 Oscillating dipole state constructed of the $1s$-$2p$ superposition, according to Eq. (16.38). The contour plot of $P_{1s,\,2p}(\mathbf{r},\,t)$ is shown for the times **(a)** $t=0$, **(b)** $t=\frac{T_{2p,\,1s}}{4}$, and **(c)** $t=\frac{T_{2p,\,1s}}{2}$, where $T_{2p,\,1s}$ defines the period of oscillation. The contours are shown at $1/32$, $1/16$, $1/8$, $1/4$, and $1/2$ of the peak value. **(d)** The computed evolution of the displacement $\langle\hat{x}\rangle$ as a function of time. The circles indicate the probability-distribution snapshots shown in frames **a–c**.

$P_{1s,\,2p}(x,\,y,\,t)$ are plotted in Figs. 16.4(a)–(c). We see that the electron–ion displacement oscillates with the amplitude $0.244\,a_0$.

Generally, if the system consists of bound states, a_0 defines the range of the typical electron–ion separation. Clearly, this is the expected behavior because, in a bound state, the electron should always be found in the vicinity of the ion. However, we also see that the oscillations of the interparticle separation can be appreciable, almost as large as the typical binding distance a_0. Since electron and ion are oppositely charged, the $P_{1s,\,2p}(\mathbf{r},\,t)$ dynamics also implies an oscillating dipole, which should radiate and couple efficiently with light having a frequency comparable with $\omega_{2p,\,1s}$. This aspect is studied further in Chapters 17–18.

As a generalization of Eq. (16.38), we take a look at

$$|\psi_{1s,\,\nu}(t)\rangle \equiv \frac{1}{\sqrt{2}}\left(|1s\rangle + |\nu\rangle\, e^{-i\omega_{\nu,\,1s}t}\right) e^{-i\omega_{1s}t}, \quad \omega_{\nu,\lambda} \equiv \frac{E_\nu - E_\lambda}{\hbar}, \tag{16.40}$$

describing a superposition of the $1s$ state with an arbitrary ν state. The corresponding dipole in the x direction follows from the product of charge and displacement

$$\langle e\hat{r}\rangle \equiv \langle\psi_{1s,\,\nu}(t)|e\hat{r}|\psi_{1s,\,\nu}(t)\rangle = \mathrm{Re}\left[\langle 1s|e\hat{r}|\nu\rangle\, e^{-i\omega_{\nu,\,1s}t}\right] \tag{16.41}$$

because the matrix element $\langle\nu|e\hat{\mathbf{r}}|\nu\rangle$ vanishes for all ν, see Exercise 16.6. The dipole oscillates with the frequency $\omega_{\nu,\,1s}$ and its amplitude is defined by the x component of the microscopic dipole

$$\mathbf{d}_{1s,\,\nu} \equiv \langle 1s|e\hat{\mathbf{r}}|\nu\rangle \tag{16.42}$$

between the $1s$ and the ν state. We notice that the $1s$ state produces a dipole only with the p-like states, which is part of the selection rules of optical transitions discussed in Chapter 18.

To see the overall strength of the optical dipole, we plot $|\mathbf{d}_{1s,\,np}|$ in Fig. 16.5 as a function of the transition energy $\hbar\omega_{np,\,1s} = E_{np} - E_{1s}$ for (a) the strictly two-dimensional (2D) case, (b) the QW, and (c) the QWI analyzed in Fig. 16.3. The shaded area indicates the region where the np end-states are unbound. For the strictly 2D system, one needs at least the energy $4\,E_B$ to make a transition from the $1s$ to the continuum np state. We have not presented $|\mathbf{d}_{1s,\,np}|$ in this region because for this we have to use the hydrogen eigenstates for the ionization continuum.

The numerical evaluations of the QW and QWI states produce the bound and the continuum states with the same numerical effort. We see from Figs. 16.5(b)–(c) that the minimum energy $2.45\,E_B$ and $3\,E_B$ is needed to make a transition to the QW and QWI continuum, respectively. As a general trend for all cases studied, the dipole strength has its maximum at the $1s$-to-$2p$ energy and decreases monotonously as the border to the continuum (shaded area) is approached. At the same time, the energetic distance between the neighboring np states becomes smaller.

The numerically treated QW and QWI problems have only a finite number of states because we have used a discrete $\mathbf{k}$ grid to determine the eigenstates of Eq. (16.30). As

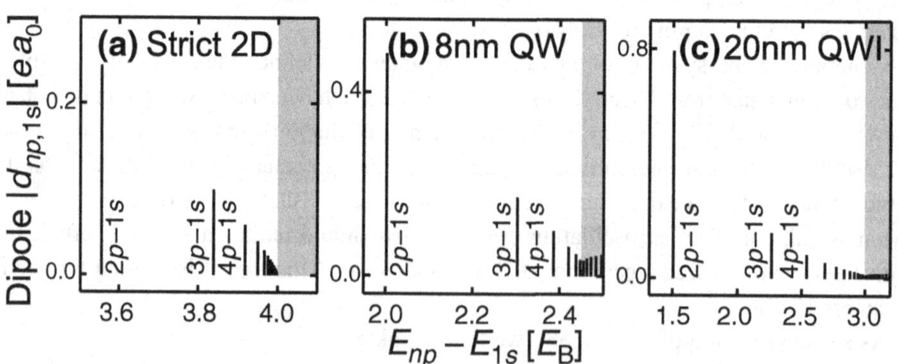

Figure 16.5 Spectral strength of dipole oscillations involving the $1s$ and different np states. Each vertical line corresponds to $|d_{np,\,1s}(\omega)|$ defined by Eq. (16.42) and is centered at the corresponding transition energy $E_{np} - E_{1s}$. The shaded region indicates the position of the ionization continuum. **(a)** The strictly two-dimensional case is compared with the results for **(b)** the 8 nm QW and **(c)** the 20 nm QWI analyzed in Fig. 16.3.

a result, we only find a finite number of bound and unbound states. Thus, the numerically obtained unbound states actually form a quasicontinuum where each discrete energy represents an energy interval. This feature can easily be tested by increasing the number of **k** values used in the numerical computation. One finds that the numerical solutions rapidly approach the correct continuum limit. We also see that $|\mathbf{d}_{1s,\,np}|$ between bound and continuum states increases after the continuum border is crossed.

The Fourier spectrum of $\langle e\hat{x}\rangle$ produces discrete lines at positions matching $\hbar\omega = E_{np} - E_{1s}$. Thus, we may anticipate that the dipole emits light exactly at the discrete energies indicated in Fig. 16.5. Since these energies depend on the specific system, the spectral lines can be used as a spectroscopic tool, e.g., in atomic systems to characterize the presence and abundance of a given atom occupying a given state. One can use this method to identify different atoms or molecules via their emission or absorption spectra. Interestingly, the direct generalization of this principle to semiconductor emission is more complicated due to the nontrivial many-body effects involved. A thorough discussion of these aspects is found in Chapter 28.

16.3.1 Momentum-matrix elements

Since the average oscillating dipole involves a movement of the electron relative to the ion, it is relevant also to consider the average relative momentum. Before we compute $\langle \mathbf{p}\rangle$ for an oscillating dipole state (16.40), we first evaluate the commutation relation

$$\left[\hat{\mathbf{r}},\, \hat{H}_{\rm rel}\right]_- = \sum_{j=1}^{3} \mathbf{e}_j \left[\hat{r}_j,\, \hat{H}_{\rm rel}\right]_- = \sum_{j=1}^{3} \mathbf{e}_j \left[\hat{r}_j,\, \frac{\hat{\mathbf{p}}^2}{2\mu} + \hat{V}\right]_- = \sum_{j=1}^{3} \mathbf{e}_j \left[\hat{r}_j,\, \frac{\hat{\mathbf{p}}^2}{2\mu}\right]_-$$
$$= \sum_{j,k=1}^{3} \frac{\mathbf{e}_j}{2\mu}\left[\hat{r}_j,\, \hat{p}_k\right]_- \hat{p}_k + \sum_{j,k=1}^{3} \frac{\mathbf{e}_j}{2\mu} \hat{p}_k \left[\hat{r}_j,\, \hat{p}_k\right]_-, \tag{16.43}$$

where we expressed the relative position in terms of its Cartesian components. Since $\hat{V}$ refers to an ordinary **r**-dependent function, it always commutes with $\hat{r}_j$. The last step in Eq. (16.43) is obtained if we use the generic operator relation (3.36) after we express $\hat{\mathbf{p}}^2$ through its Cartesian components $\sum_{k=1}^{3} \hat{p}_k \hat{p}_k$. The remaining commutations are evaluated using the canonical commutation relation (3.35).

We can then write Eq. (16.43) as

$$\left[\hat{\mathbf{r}},\, \hat{H}_{\rm rel}\right]_- = \frac{i\hbar}{2\mu}\hat{\mathbf{p}} \quad \Leftrightarrow \quad \hat{\mathbf{p}} = \frac{\mu}{i\hbar}\left[\hat{\mathbf{r}},\, \hat{H}_{\rm rel}\right]_- \tag{16.44}$$

allowing us to express the momentum operator in terms of $\hat{\mathbf{r}}$ and $\hat{H}_{\rm rel}$. Thus, the momentum-matrix element between two states follows from

$$\mathbf{p}_{\nu,\lambda} \equiv \langle \nu|\hat{\mathbf{p}}|\lambda\rangle = \frac{\mu}{i\hbar}\langle\nu| \left(\hat{\mathbf{r}}\,\hat{H}_{\rm rel} - \hat{H}_{\rm rel}\,\hat{\mathbf{r}}\right)|\lambda\rangle$$
$$= \frac{\mu}{i\hbar}\langle\nu|\left(\hat{\mathbf{r}}\,E_\lambda - E_\nu\hat{\mathbf{r}}\right)|\lambda\rangle = \frac{\mu}{i\hbar}\left(E_\lambda - E_\nu\right)\langle\nu|\hat{\mathbf{r}}\,|\lambda\rangle. \tag{16.45}$$

We can replace $\hat{H}_{\rm rel}$ by its energy eigenvalue if the states are the eigenstates $\langle\nu|$ or $|\lambda\rangle$, see Eq. (16.13). Consequently, $\mathbf{p}_{\nu,\lambda}$ can be expressed in terms of the dipole-matrix element (16.42) and the frequency-difference $\omega_{\nu,\lambda} = \frac{E_\nu - E_\lambda}{\hbar}$, yielding

$$\mathbf{p}_{\nu,\lambda} = \langle\nu|\hat{\mathbf{p}}|\lambda\rangle = \frac{\mu\,\omega_{\lambda,\nu}}{ie}\mathbf{d}_{\nu,\lambda} \Leftrightarrow \mathbf{d}_{\nu,\lambda} \equiv \langle\nu|e\hat{\mathbf{r}}|\lambda\rangle = \frac{ie}{\mu\,\omega_{\lambda,\nu}}\mathbf{p}_{\nu,\lambda}. \quad (16.46)$$

These results show that the momentum and dipole-matrix elements are directly related via a simple transformation. Especially, the oscillating dipole can also be understood as an oscillating current, defined by $e\mathbf{p}_{\nu,\lambda}/\mu$. This aspect is elaborated more in Chapter 17.

A very similar derivation eventually yields the following sum rules:

$$\sum\nolimits_\nu \frac{[\mathbf{p}_{\nu,\lambda}]_j[\mathbf{p}_{\lambda,\nu}]_k}{\omega_{\nu,\lambda}} = \frac{\mu\hbar}{2}\delta_{j,k}, \quad \sum\nolimits_\nu \omega_{\nu,\lambda}[\mathbf{d}_{\lambda,\nu}]_j[\mathbf{d}_{\nu,\lambda}]_k = \frac{\hbar e^2}{2\mu}\delta_{j,k}. \quad (16.47)$$

The explicit derivation of these expressions is the topic of Exercise 16.16.7. If we now set λ to $1s$ and multiply both sides of Eq. (16.47) by $\hbar/e^2a_0^2$, we find

$$\sum_\nu (E_\nu - E_{1s})\left|\frac{d^x_{1s,\nu}}{e\,a_0}\right|^2 = \frac{\hbar^2}{2\mu a_0} = E_B \Leftrightarrow \sum_\nu \frac{E_\nu - E_{1s}}{E_B}\left|\frac{d^x_{1s,\nu}}{e\,a_0}\right|^2 = 1, \quad (16.48)$$

where we identified the three-dimensional binding energy with the help of the relation (16.18). Since only the $\nu = np$ states contribute to the sum and E_ν is greater than E_{1s}, we can convert the sum rule (16.48) into the form

$$\sum_{n=2}^{\infty} \frac{E_{np} - E_{1s}}{E_B}\left|\frac{d^x_{1s,np}}{e\,a_0}\right|^2 \le 1 \Rightarrow \sum_{n=2}^{\infty} \frac{E_{np} - E_{1s}}{E_B}\left|\frac{\mathbf{d}_{1s,np}}{e\,a_0}\right|^2 \le 1, \quad (16.49)$$

where the inequality follows because we omitted the unbound states. We may replace the x component of the dipole-matrix element by $|\mathbf{d}_{1s,np}| = \left|d^x_{1s,np_x}\right|$ because of our freedom to choose the direction of the dipole.

Since each sum term in (16.49) is positive definite, it is convenient to introduce a normalized dipole-matrix element

$$\bar{d}_{1s,np} = \sqrt{\frac{E_{np} - E_{1s}}{E_B}}\left|\frac{\mathbf{d}_{1s,np}}{e\,a_0}\right|, \quad (16.50)$$

which has the properties

$$\sum\nolimits_{n=2}^{\infty} |\bar{d}_{1s,np}|^2 = 1, \qquad |\bar{d}_{1s,np}| \le 1. \quad (16.51)$$

For the strictly two-dimensional case, the normalized $\bar{d}_{1s,\,2p}$ has the value 0.46. For the QW and QWI of Fig. 16.5, we find $\bar{d}_{1s,\,2p} = 0.66$ and $\bar{d}_{1s,\,2p} = 0.93$, respectively, showing that the lowest $1s$-to-$2p$ transition can be truly dominant.

The relation (16.51) sets strict limits for the magnitude of the dipole- and thus the momentum-matrix elements for the dominant $1s$-to-$2p$ transition. By combining Eq. (16.46) with Eqs. (16.50)–(16.51), we find

$$\begin{aligned} |\mathbf{d}_{1s,\,2p}| &\leq \sqrt{\bar{\omega}_{2p,1s}}\; e\, a_0 \qquad\qquad \bar{\omega}_{2p,1s} \equiv \frac{E_{np} - E_{1s}}{E_B} \\ |\mathbf{p}_{1s,\,2p}| &\leq \sqrt{\bar{\omega}_{2p,1s}}\; \mu\, \omega_{2p,\,1s}\, a_0. \end{aligned} \tag{16.52}$$

Since $v_{1s,\,2p} \equiv |\mathbf{p}_{1s,\,2p}|/\mu$ can be associated with the relative peak velocity between the electron and ion/hole, we find the limit

$$\begin{aligned} |v_{1s,\,2p}| &\leq \sqrt{\bar{\omega}_{2p,1s}}\; \omega_{2p,\,1s}\, a_0 \\ &= 1.5193 \times 10^5\, \frac{\mathrm{m}}{\mathrm{s}} \sqrt{\bar{\omega}_{2p,1s}} \left[\frac{\hbar\omega_{2p,\,1s}}{\mathrm{eV}}\right] \left[\frac{a_0}{\text{Å}}\right]. \end{aligned} \tag{16.53}$$

Using the values in Table 16.1, we obtain $|\mathbf{v}_{1s,\,2p}| \leq 10^5 \frac{\mathrm{m}}{\mathrm{s}}$ for GaAs. Even though this velocity is appreciable, it is considerably smaller than the speed of light.

16.3.2 Long-wave-length limit for the $\hat{\mathbf{A}} \cdot \hat{\mathbf{p}}$ interaction

We can now use our knowledge of the typical interparticle distances to introduce a simplifying approximation in the generic Hamiltonian. The bound states produce probability distributions that decay exponentially fast when the electron–ion separation becomes much larger than a_0. As a rough estimate, a_0 ranges on the scale of less than 1 Å up to some 100 Å. At the same time, we anticipate that the oscillating dipole couples strongly to resonant fields whose energy satisfies $\hbar\omega \approx E_{2p} - E_{1s}$. The corresponding optical wave length is

$$\lambda = \frac{2\pi}{q} = \frac{2\pi c}{\omega} = \frac{2\pi\hbar c}{E} = 1240\,\mathrm{nm}\left[\frac{\mathrm{eV}}{E}\right]. \tag{16.54}$$

Since atomic transitions are typically in the range of one to many eV, the light field changes on the length scale of 10^{-6} m. This is two to four orders of magnitude larger than typical ion–electron separations. Thus, the vector potential varies only negligibly over the length scales relevant for atomic transitions. Therefore, we can usually employ the so-called long-wave-length approximation

$$\hat{A}(\mathbf{r}_e) = \hat{A}\left(\mathbf{R} + \frac{m_e}{M}\mathbf{r}\right) \to \hat{A}(\mathbf{R}), \;\; \hat{A}(\mathbf{r}_h) = \hat{A}\left(\mathbf{R} - \frac{m_h}{M}\mathbf{r}\right) \to \hat{A}(\mathbf{R}), \tag{16.55}$$

where we neglect the $\mathbf{r}$ dependence, as motivated above. With this approximation, the Hamiltonian (16.3) becomes

$$\begin{aligned}\hat{H} &= \frac{\hat{\mathbf{P}}^2}{2M} + \frac{\hat{\mathbf{p}}^2}{2\mu} - \frac{Q_e}{\mu}\hat{\mathbf{A}}(\mathbf{R}) + \frac{Q_e^2}{2\mu}\hat{\mathbf{A}}^2(\mathbf{R}) - \frac{Q_e^2}{4\pi\varepsilon\,\epsilon_0|\mathbf{r}|} + \hat{H}_F \\ &= \frac{\hat{\mathbf{P}}^2}{2M} + \frac{[\hat{\mathbf{p}} - Q_e\hat{\mathbf{A}}(\mathbf{R})]^2}{2\mu} - \frac{Q_e^2}{4\pi\varepsilon\,\epsilon_0|\mathbf{r}|} + \sum_{\mathbf{q}} \omega_{\mathbf{q}}\left[B_{\mathbf{q}}^\dagger B_{\mathbf{q}} + \frac{1}{2}\right]. \end{aligned} \quad (16.56)$$

Since the vector potential no longer depends on the inner structure of the electron–ion dipole, this simplification is sometimes referred to as the *dipole approximation*. It is very accurate in those cases where the elementary dipoles have much smaller sizes than the typical optical wave length.

When we discuss extended states in a solid, we use the separation of length scales for the Bloch functions with respect to their unit-cell and crystal-momentum dependent parts. Whenever $\hat{\mathbf{A}}$ changes slowly on the length scale of the unit cell, we can simplify the system Hamiltonian as in Eq. (16.56). The remaining crystal-momentum dependence then appears in the argument of $\hat{\mathbf{A}}(\mathbf{R})$ together with the center-of-mass dependence.

Exercises

16.1 (**a**) Apply the transformation to relative and center-of-mass coordinates, i.e., Eqs. (16.5) and (16.7), to (16.1). Show that this step yields Eq. (16.8). (**b**) Derive the corresponding second-quantized Hamiltonian from Eq. (16.3) by introducing a field operator $\hat{\Psi}_E(\mathbf{r}) = \sum_{\mathbf{k}} e_{\mathbf{k}} \frac{1}{\sqrt{\mathcal{V}}} e^{i\mathbf{k}\cdot\mathbf{r}}$ for electrons and $\hat{\Psi}_i(\mathbf{r}) = \sum_{\mathbf{k}} h_{\mathbf{k}} \frac{1}{\sqrt{\mathcal{V}}} e^{i\mathbf{k}\cdot\mathbf{r}}$ for ions. Show that the result reduces to the semiconductor Hamiltonian (13.69)–(13.73) when it is limited to the subspace of one electron–hole pair.

16.2 (**a**) The kinetic energy of a free particle is $E_{\mathbf{k}} = \hbar^2\mathbf{k}^2/2m$ and the Coulomb potential is $V(\mathbf{r}) = e^2/4\pi\varepsilon\varepsilon_0(\mathbf{r})$. Show that Eq. (16.18) defines them via $E_{\mathbf{k}} = E_B \frac{\mu}{m}|\mathbf{k}a_0|^2$ and $V(\mathbf{r}) = 2E_B\frac{a_0}{|\mathbf{r}|}$. (**b**) Fourier transform $\left[-\frac{\hbar^2}{2\mu}\mathbf{p}^2 - V|\mathbf{r}|\right]\phi_\lambda(\mathbf{r}) = E_\lambda\,\phi_\lambda(\mathbf{r})$ and show that we find $\frac{\hbar^2k^2}{2\mu}\phi_\lambda(\mathbf{k}) - \sum_{\mathbf{k}'} V_{\mathbf{k}-\mathbf{k}'}\,\phi_\lambda(\mathbf{k}') = E_\lambda\,\phi_\lambda(\mathbf{k})$ in momentum space, where $V_{\mathbf{q}} = \frac{1}{\mathcal{L}^3}\int d^3r\, V(\mathbf{r})\, e^{-i\mathbf{q}\cdot\mathbf{r}} = e^2/\varepsilon\varepsilon_0\mathbf{q}^2$.

16.3 An effective one-dimensional potential follows from

$$V^{\text{eff}}(z) = \frac{e^2}{4\pi\varepsilon_0\varepsilon}\int d^2r \int d^2r' \frac{g(\mathbf{r})\,g(\mathbf{r}')}{\sqrt{(\mathbf{r}-\mathbf{r}')^2 + z^2}}.$$

(**a**) Assume a Gaussian confinement $g(r) = \frac{2}{\pi\Delta R^2}\, e^{-\frac{2r^2}{\Delta R^2}}$ and show that:

$V^{\text{eff}}(z) = \frac{e^2}{4\pi\varepsilon_0\varepsilon}\frac{\sqrt{\pi}}{\Delta R}\, e^{\frac{z^2}{\Delta R^2}}\left[1 - \text{erf}\left(\left|\frac{z}{\Delta R}\right|\right)\right]$ where $\text{erf}(z) = \frac{2}{\sqrt{\pi}}\int_0^z dx\, e^{-x^2}$ is the error function. (**b**) The error function can be accurately approximated via the relation $\sqrt{\pi}\,e^{z^2}(1 - \text{erf}(z)) \cong 1/\sqrt{z^2+\gamma^2}$ with the constant $\gamma \approx 0.6$. Show that $V_{q_z} = \frac{1}{\mathcal{L}}\int_{-\mathcal{L}/2}^{\mathcal{L}/2} dz\, V^{\text{eff}}(z)\, e^{-iq_z z} \cong \frac{e^2}{4\pi\varepsilon_0\varepsilon}\frac{2}{\mathcal{L}} K_0(|q_z\Delta R\gamma|)$ where $K_0(x) \equiv \frac{1}{\sqrt{\pi}}\int_0^\infty ds\, \frac{\cos s}{\sqrt{s^2+x^2}}$ is a modified Bessel function of the second kind. (**c**) Show that

V_{q_z} is alternatively $V_{q_z} = \frac{e^2}{4\pi\varepsilon_0\varepsilon}\frac{4\sqrt{\pi}}{\mathcal{L}}\int_0^\infty dy\, y\, K_0(q_z\,\Delta R y)\, e^{-y^2}$. Show numerically that $V_{q_z} \cong \frac{e^2}{4\pi\varepsilon_0\varepsilon}\frac{2}{\mathcal{L}}\sqrt{\frac{\pi}{2|q_z\Delta R\gamma|}}\, e^{-|q_z\Delta R\gamma|}$ yields an accurate approximation for V_{q_z}, especially for large q_z.

16.4 (**a**) Derive the solutions (16.31) from Eq. (16.22). (**b**) Based on Eq. (16.32), actual quantum wells have a Coulomb-matrix element

$$V_{\mathbf{q}} = V_{\mathbf{q}}^{2\mathrm{D}} \int dz\, dz'\, g(z)\, g(z')\, e^{-|\mathbf{q}(z-z')|},$$

where $V_{\mathbf{q}}^{2\mathrm{D}}$ is strictly two-dimensional. Calculate $V_{\mathbf{q}}$ for a Gaussian confinement $g(z) = \frac{1}{\sqrt{\pi\Delta z^2}} e^{-z^2/\Delta z^2}$. (**c**) Use first-order perturbation theory to show that E_{1s} is lower than $E_{1s}^{2\mathrm{D}}$. Show also that the $2s$ and $2p$ states become nondegenerate in this case.

16.5 Assume that the atom is initially described by $\psi(\mathbf{r},\mathbf{R},0) = \frac{1}{\sqrt{2}}[\varphi_{1s}(\mathbf{r}) + \varphi_{2p}(\mathbf{r})]\left[\frac{1}{\sqrt{\pi}\Delta R}\right]^3 e^{-\mathbf{R}^2/2\Delta R^2}$. (**a**) Calculate $\psi(\mathbf{r},\mathbf{R},t)$ when the time evolution follows from $\hat{H}_0$. **Hint:** *Make use of the investigations in Section 3.1.1 and use strictly two-dimensional* $\varphi_\lambda(r)$. (**b**) Compute $\mathbf{d}_{1s,2p}(t) = \int d^3r \int d^3R\, \psi^\star(\mathbf{r},\mathbf{R},t)\, e\,\mathbf{r}\, \psi(\mathbf{r},\mathbf{R},t)$.

16.6 Assume that the relative part of the atomic state follows from Eq. (16.40). (**a**) Show that $\langle\nu|\hat{\mathbf{r}}|\nu\rangle$ vanishes for all atomic states. (**b**) Construct $\langle\hat{\mathbf{r}}\rangle = \langle\psi(t)|\hat{\mathbf{r}}|\psi(t)\rangle$ for state (16.40). Which ν states produce the dynamics $\langle\hat{\mathbf{r}}\rangle$ for a strictly two-dimensional system? (**c**) Fourier transform $\langle\hat{\mathbf{r}}(t)\rangle$. Which frequencies are detectable in $\langle\hat{\mathbf{r}}(\omega)\rangle$?

16.7 (**a**) Derive the sum rules given by Eq. (16.47). (**b**) Verify the validity of the inequalities (16.52).

16.8 The correction term to the dipole approximation follows from $\hat{\mathbf{A}}(\mathbf{r}_E) = \hat{\mathbf{A}}(\mathbf{R}) + \frac{m_E}{M}\mathbf{r}\cdot\nabla_{\mathbf{R}}\hat{\mathbf{A}}(\mathbf{R})$ and $\hat{\mathbf{A}}(\mathbf{r}_H) = \hat{\mathbf{A}}(\mathbf{R}) - \frac{m_H}{M}\mathbf{r}\cdot\nabla_R\hat{\mathbf{A}}(\mathbf{R})$. Implement this into the Hamiltonian (16.3). What are the contributions beyond the dipole approximation? Assume $\nabla_{\mathbf{R}}\hat{\mathbf{A}}(\mathbf{R}) \cong i\mathbf{q}\,\hat{\mathbf{A}}(\mathbf{R})$ to show that the dipole approximation is well justified for typical optical fields and atomic dipoles.

Further reading

The hydrogen problem is discussed in most quantum-mechanics books, cf. the references in Chapter 3. Special functions can be found, e.g., in:

G. B. Arfken and H. J. Weber (2005). *Mathematical Methods for Physicists*, 6th edition, New York, Elsevier.

Excitons in systems with different dimensionality are thoroughly discussed in:

H. Haug and S. W. Koch (2009). *Quantum Theory of the Optical and Electronic Properties of Semiconductors*, 5th edition, Singapore, World Scientific.

The hydrogen material parameters are retrieved from:

P. J. Mohr, B. N. Taylor, and D. B. Newell (2008). CODATA recommended values of the fundamental physical constants: 2006, *Rev. Mod. Phys.* **80**, 633.

The same constants can be found at:

http://physics.nist.gov/cuu/Constants/index.html

17

Hierarchy problem for dipole systems

In the previous chapter, we have shown that an isolated atomic system produces an oscillating dipole if it is in an appropriate superposition of hydrogen-like eigenstates. According to classical optics, such an oscillating dipole emits light. To investigate this aspect in a quantum-optical setting, we derive the equations of motion for the interacting light–matter system in this chapter. We show that the quantum-mechanical treatment of light produces a hierarchy problem similar to that arising from the many-body interactions in complex systems. This problem is unsolvable even though some rather general simplifications can be introduced by analyzing how the Lorentz force acts upon charged particles. We also investigate the possible simplifications gained by converting the original $\hat{\mathbf{A}} \cdot \hat{\mathbf{p}}$ picture into the $\hat{\mathbf{E}} \cdot \hat{\mathbf{x}}$ picture.

In our studies of the dipole emission, we have to introduce a set of systematic approximations in order to obtain a solution and to explore the quantum dynamics. As a first step, we truncate the hierarchy at the level equivalent to a single-particle description in the many-body theory. We also assume that the dipole is unaltered by the field it generates. This analysis produces the traditional classical emission for a point-like dipole and a planar arrangement of dipoles. The analytic relations derived here are used and extended in Chapters 18–26 where we derive the fully self-consistent solution including the back action of the emitted light on dipole.

17.1 Quantum dynamics in the $\hat{\mathbf{A}} \cdot \hat{\mathbf{p}}$ picture

Once the system Hamiltonian is known, we may directly evaluate the full quantum dynamics by solving the Heisenberg equation of motion (5.57). Here, we apply this technique to discuss the quantum dynamics of a single atomic dipole system coupled to a generic quantized light field. Since the dipole approximation is typically very accurate for both atoms and excitons we may start from the Hamiltonian (16.56):

$$\hat{H} = \frac{\hat{\mathbf{P}}^2}{2M} + \frac{\hat{\mathbf{p}}^2}{2\mu} - \frac{Q_e}{\mu}\hat{\mathbf{p}}\cdot\hat{\mathbf{A}}(\mathbf{R}) + \frac{Q_e^2}{2\mu}\hat{\mathbf{A}}^2(\mathbf{R}) - \frac{Q_e^2}{4\pi\varepsilon\,\epsilon_0|\mathbf{r}|} + \hat{H}_F$$
$$= \frac{\hat{\mathbf{P}}^2}{2M} + \frac{[\hat{\mathbf{p}} - Q_e\hat{\mathbf{A}}(\mathbf{R})]^2}{2\mu} - \frac{Q_e^2}{4\pi\varepsilon\,\epsilon_0|\mathbf{r}|} + \sum_{\mathbf{q}}\omega_{\mathbf{q}}\left[B_{\mathbf{q}}^\dagger B_{\mathbf{q}} + \frac{1}{2}\right],$$
$$\hat{\mathbf{A}}(\mathbf{r}) = \sum_{\mathbf{q}}\frac{\mathcal{E}_{\mathbf{q}}}{\omega_{\mathbf{q}}}\left[\mathbf{u}_{\mathbf{q}}(\mathbf{r})\,B_{\mathbf{q}} + \mathbf{u}_{\mathbf{q}}^\star(\mathbf{r})\,B_{\mathbf{q}}^\dagger\right], \quad \mathcal{E}_{\mathbf{q}} = \sqrt{\frac{\hbar\omega_{\mathbf{q}}}{2\varepsilon_0}}. \tag{17.1}$$

Here, we have also listed the mode-expansion relation (10.43) for the vector potential and expressed the vacuum-field amplitude $\mathcal{E}_{\mathbf{q}}$ explicitly, based on Eq. (10.23). The pairs $(\hat{\mathbf{r}}, \hat{\mathbf{p}})$ and $(\hat{\mathbf{R}}, \hat{\mathbf{P}})$ refer to the relative and center-of-mass coordinates, respectively. More specifically, $\hat{\mathbf{r}}$ $(\hat{\mathbf{R}})$ determines the electron–ion separation (center-of-mass position) while $\hat{\mathbf{p}}$ $(\hat{\mathbf{P}})$ is the related momentum.

In the following, we directly apply the Bosonic and canonic commutators

$$\left[B_{\mathbf{q}}, B_{\mathbf{q}'}^\dagger\right]_- = \delta_{\mathbf{q},\mathbf{q}'}, \quad \left[B_{\mathbf{q}}, B_{\mathbf{q}'}\right]_- = 0 = \left[B_{\mathbf{q}}^\dagger, B_{\mathbf{q}'}^\dagger\right]_-,$$
$$\left[\hat{r}_j, \hat{p}_k\right]_- = i\hbar\delta_{j,k}, \quad \left[\hat{R}_j, \hat{P}_k\right]_- = i\hbar\delta_{j,k},$$
$$\left[\hat{r}_j, \hat{r}_k\right]_- = \left[\hat{R}_j, \hat{R}_k\right]_- = \left[\hat{r}_j, \hat{R}_k\right]_- = 0 = \left[\hat{r}_j, \hat{P}_k\right]_-,$$
$$\left[\hat{p}_j, \hat{p}_k\right]_- = \left[\hat{P}_j, \hat{P}_k\right]_- = \left[\hat{p}_j, \hat{P}_k\right]_- = 0 = \left[\hat{R}_j, \hat{p}_k\right]_-, \tag{17.2}$$

which are obtained from Eqs. (10.31)–(10.32) and (3.35). With the help of these, we construct the Heisenberg equations of motion (5.57) for the Boson operators

$$i\hbar\frac{\partial}{\partial t}B_{\mathbf{q}} = \hbar\omega_{\mathbf{q}}\,B_{\mathbf{q}} - \frac{\mathcal{E}_{\mathbf{q}}}{\omega_{\mathbf{q}}}\,\mathbf{u}^\star(\mathbf{R})\cdot\hat{\mathbf{J}}_A$$
$$i\hbar\frac{\partial}{\partial t}B_{\mathbf{q}}^\dagger = -\hbar\omega_{\mathbf{q}}\,B_{\mathbf{q}}^\dagger + \frac{\mathcal{E}_{\mathbf{q}}}{\omega_{\mathbf{q}}}\,\mathbf{u}(\mathbf{R})\cdot\hat{\mathbf{J}}_A, \tag{17.3}$$

where we identified the kinetic current and velocity

$$\hat{\mathbf{J}}_A \equiv Q_e\hat{\mathbf{v}}, \qquad \hat{\mathbf{v}} \equiv \frac{\hat{\mathbf{p}} - Q_e\hat{\mathbf{A}}(\mathbf{R})}{\mu}, \tag{17.4}$$

respectively. In the derivation of Eq. (17.3), we follow steps identical to those leading to Eq. (14.34). The result shows that the photon-operator dynamics is coupled to both the relative and center-of-mass motion of the system. To construct the Heisenberg equations for these, we evaluate the commutators that appear to obtain

$$\frac{\partial}{\partial t}\hat{\mathbf{r}} = \frac{1}{i\hbar}\left[\hat{\mathbf{r}}, \hat{H}\right]_{-} = \hat{\mathbf{v}}, \qquad \frac{\partial}{\partial t}\hat{\mathbf{R}} = \frac{\hat{\mathbf{P}}}{M},$$

$$\frac{\partial}{\partial t}\hat{\mathbf{v}} = -\frac{Q_e}{2\mu}\sum_{j=1}^{3}\left[\left(\nabla_{\mathbf{R}_j}\hat{\mathbf{A}}(\mathbf{R})\right)\frac{\hat{P}_j}{M} + \frac{\hat{P}_j}{M}\left(\nabla_{\mathbf{R}_j}\hat{\mathbf{A}}(\mathbf{R})\right)\right] + \frac{Q_e\,\hat{\mathbf{E}}(\mathbf{R})}{\mu} + \frac{\mathbf{F}(\mathbf{r})}{\mu},$$

$$\frac{\partial}{\partial t}\hat{\mathbf{P}} = \frac{Q_e}{2}\sum_{j=1}^{3}\left[\left(\nabla_{\mathbf{R}}\hat{\mathbf{A}}_j(\mathbf{R})\right)\hat{v}_j + \hat{v}_j\left(\nabla_{\mathbf{R}}\hat{\mathbf{A}}_j(\mathbf{R})\right)\right]. \tag{17.5}$$

Here, we identified the electric field

$$\hat{\mathbf{E}}(\mathbf{r}) \equiv -\frac{\partial}{\partial t}\hat{\mathbf{A}}(\mathbf{r}) = \sum_{\mathbf{q}} i\mathcal{E}_{\mathbf{q}}\left[\mathbf{u}_{\mathbf{q}}(\mathbf{r})\,B_{\mathbf{q}} - \mathbf{u}_{\mathbf{q}}^{\star}(\mathbf{r})\,B_{\mathbf{q}}^{\dagger}\right] \tag{17.6}$$

and the attractive Coulomb force

$$\mathbf{F}(\mathbf{r}) = -\frac{Q_e\,Q_e}{4\pi\varepsilon\varepsilon_0}\frac{\mathbf{r}}{|\mathbf{r}|^3} \tag{17.7}$$

between the electron and the ion. The differentiation $\nabla_{\mathbf{R}}$ that appears acts exclusively on the vector potential. To emphasize this, we put parentheses around this contribution. As usual, $\nabla_{R_j} = \partial/\partial R_j$ defines the differentiation with respect to the x $(j=1)$, y $(j=2)$, and z $(j=3)$ components of the $\mathbf{R}$ vector. Even though the quantum dynamics in Eqs. (17.3) and (17.5) are derived by assuming a real-space representation of the canonical pair, we can directly adopt an abstract form by replacing $\mathbf{r} \to \hat{\mathbf{r}}$ and $\mathbf{R} \to \hat{\mathbf{R}}$ within these equations. Thus, Eqs. (17.3) and (17.5) represent the quantum dynamics of the elementary operators.

Based on the analysis in Section 14.3, we may associate $\hat{\mathbf{r}}$, $\hat{\mathbf{v}}$, $\hat{\mathbf{R}}$, and $\hat{\mathbf{P}}$ with single photon operators. Hence, in the formal cluster-expansion sense, these quantities as well as $B_{\mathbf{q}}$ and $B_{\mathbf{q}}^{\dagger}$ are all single-particle operators. We see from Eq. (17.5) that $\hat{\mathbf{v}}$ couples to $\mathbf{F}(\hat{\mathbf{r}})$ and $\hat{\mathbf{A}}(\hat{\mathbf{R}})$. These terms contain $\hat{\mathbf{r}}^J$ and $\hat{\mathbf{R}}^J$ in all orders of J. Thus, even for the single atom case, we are confronted with an infinite hierarchy of equations, as illustrated in Fig. 17.1. In practice, this means that the equations of motion of $\mathbf{F}(\hat{\mathbf{r}})$ and $\hat{\mathbf{A}}(\mathbf{R})$ couple to an infinite sequence of new operator equations containing different mixtures of $\hat{\mathbf{r}}$, $\hat{\mathbf{p}}$, $\hat{\mathbf{R}}$, $\hat{\mathbf{P}}$, $B_{\mathbf{q}}$, and $B_{\mathbf{q}}^{\dagger}$.

Even though Eqs. (17.3) and (17.5) yield a seemingly simple structure for the operator dynamics, they cannot be solved analytically due to the equation hierarchy. Thus, the

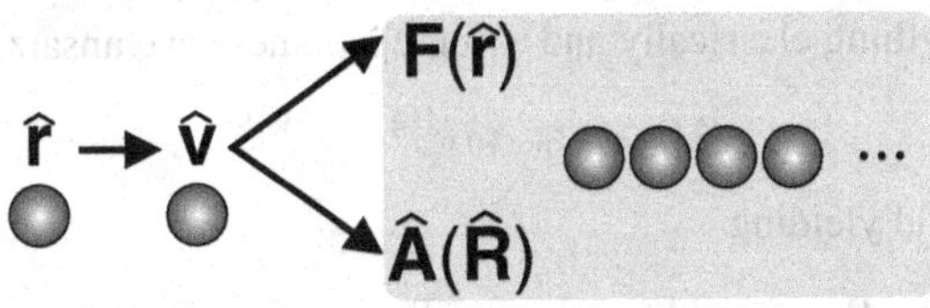

Figure 17.1 Hierarchical coupling structure of the coupling between quantized light and atom. The relative position operator is coupled to the kinetic velocity. Both represent single-particle quantities ($\hat{\mathbf{r}}$, $\hat{\mathbf{v}}$, $\hat{\mathbf{R}}$ $\leftrightarrow$ single sphere). Equation (17.5) relates them to the contributions $\mathbf{F}(\hat{\mathbf{r}})$ and $\hat{\mathbf{A}}(\hat{\mathbf{R}})$ inducing a coupling to clusters of all orders (many spheres) because the attractive Coulomb force and vector potential are nontrivial functions.

quantum-optical problem is fundamentally as complicated as the one resulting from the Coulomb interaction in nontrivial many-body systems. To study how the quantum dynamics of atomic dipoles proceeds under the influence of quantized light, we must apply a set of systematic approximations.

17.1.1 Lorentz force

To gain some insight, we first consider the dynamics of the kinetic velocity based on Eq. (17.5). By using the properties of the differentiation operator ∇ and the cross product, we eventually find

$$\begin{aligned}\frac{\partial}{\partial t}\hat{\mathbf{v}} &= -\frac{Q_e}{2\mu}\sum_{j=1}^{3}\left[\left(\nabla_{\mathbf{R}_j}\hat{\mathbf{A}}(\mathbf{R})\right)\frac{\hat{P}_j}{M}+\frac{\hat{P}_j}{M}\left(\nabla_{\mathbf{R}_j}\hat{\mathbf{A}}(\mathbf{R})\right)\right]+\frac{Q_e\,\hat{\mathbf{E}}(\mathbf{R})}{\mu}+\frac{\mathbf{F}(\mathbf{r})}{\mu}\\ &= \frac{Q_e}{\mu}\left[\frac{1}{2}\left(\hat{\mathbf{V}}\times\hat{\mathbf{B}}(\mathbf{R})+\hat{\mathbf{B}}(\mathbf{R})\times\hat{\mathbf{V}}-\nabla\times\left(\hat{\mathbf{A}}(\mathbf{R})\times\hat{\mathbf{V}}\right)\right)\right]\\ &\quad+\frac{Q_e\,\hat{\mathbf{E}}(\mathbf{R})}{\mu}+\frac{\mathbf{F}(\mathbf{r})}{\mu}, \qquad \hat{\mathbf{B}}(\mathbf{x})=\nabla\times\hat{\mathbf{A}}(\mathbf{x}), \quad \hat{\mathbf{V}}\equiv\frac{\hat{\mathbf{P}}}{M}.\end{aligned} \tag{17.8}$$

Here, we have identified the kinetic center-of-mass velocity $\hat{\mathbf{V}}$ and the transversal magnetic field based on the definition (7.12). The explicit derivation of the final result in (17.8) is discussed in connection with Exercise 17.3. The sum of $Q_e\,\hat{\mathbf{E}}(\mathbf{R})$ and the term within the brackets very much resemble the classical Lorentz force (7.30). Similarly, the $\mathbf{F}(\mathbf{r})$ contribution introduces the pairwise attraction between the electron and the ion. Since $\hat{\mathbf{V}}$ and $\hat{\mathbf{B}}(\mathbf{R})$ do not commute, the simple classical expression is replaced by a properly symmetrized version of the classical expression $\mathbf{V}\times\hat{\mathbf{B}}(\mathbf{R})$. A similar derivation can be used for the center-of-mass momentum dynamics leading to

$$\begin{aligned}\frac{\partial}{\partial t}\hat{\mathbf{P}} &= \frac{Q_e}{2}\sum_{j=1}^{3}\left[\left(\nabla_{\mathbf{R}}\hat{\mathbf{A}}_j(\mathbf{R})\right)\hat{v}_j+\hat{v}_j\left(\nabla_{\mathbf{R}}\hat{\mathbf{A}}_j(\mathbf{R})\right)\right]\\ &= \nabla_{\mathbf{R}}\left[Q_e\hat{\mathbf{r}}\cdot\hat{\mathbf{E}}(\mathbf{R})\right]+\frac{\partial}{\partial t}Q_e\nabla_{\mathbf{R}}\left(\hat{\mathbf{r}}\cdot\hat{\mathbf{A}}(\mathbf{R})\right),\end{aligned} \tag{17.9}$$

see Exercise 17.4.

To estimate the strength of the magnetic vs. electric components of the Lorentz force (16.8), we treat everything classically and assume a plane-wave ansatz,

$$\mathbf{A}(\mathbf{x},\,t)=\mathbf{A}_0\,e^{i(\mathbf{q}\cdot\mathbf{x}-c|\mathbf{q}|t)} \tag{17.10}$$

for the vector potential yielding

$$\mathbf{E}(\mathbf{x},\,t)=ic|\mathbf{q}|\mathbf{A}(\mathbf{x},\,t), \qquad \mathbf{B}(\mathbf{x},\,t)=i\mathbf{q}\times\mathbf{A}(\mathbf{x},\,t). \tag{17.11}$$

For the components of the Lorentz force $\mathbf{F}_L\equiv\mathbf{F}_B+\mathbf{F}_E$, we thus obtain

$$\begin{aligned}\mathbf{F}_B &\equiv Q_e\,\mathbf{B}(\mathbf{R},\,t)\times\mathbf{V}=i\,Q_e\,[\mathbf{q}\times\mathbf{A}(\mathbf{R},\,t)]\times\mathbf{V},\\ \mathbf{F}_E &\equiv Q_e\,\mathbf{E}(\mathbf{R},\,t)=i\,Q_e\,c\,|\mathbf{q}|\mathbf{A}(\mathbf{R},\,t).\end{aligned} \tag{17.12}$$

We see that the magnetic component of the Lorentz force has the magnitude

$$|\mathbf{F}_B| \equiv |Q_e\,[\mathbf{q} \times \mathbf{A}(\mathbf{R},\,t)] \times \mathbf{V}| \leq |Q_e\,|\mathbf{q}|\,\mathbf{A}(\mathbf{R},\,t)|\;|\mathbf{V}|$$
$$\leq |i\,Q_e\,c\,|\mathbf{q}|\,\mathbf{A}(\mathbf{R},\,t)|\;\frac{|\mathbf{V}|}{c} = |Q_e\,\mathbf{E}(\mathbf{R},\,t)|\;\frac{|\mathbf{V}|}{c} \ll |\mathbf{F}_E|\,, \tag{17.13}$$

which is much smaller than the electric component because we are analyzing nonrelativistic atoms, moving much slower than the speed of light. Even the faster electron–ion motion satisfies the condition (17.13) extremely well, as shown in Section 16.3.1. Therefore, the Lorentz force is typically dominated by its electrical component for nonrelativistically moving atoms. Thus, we introduce the approximation where we include only the dominant Lorentz-force contribution in Eq. (17.8). In other words, we omit the terms within the brackets while we keep the $Q_e\hat{\mathbf{E}}(\mathbf{r})$ part.

For a more general $\hat{\mathbf{A}}(\mathbf{R})$, this approximation can be applied whenever $|\mathbf{v}_{\text{ple}} \times \mathbf{B}| \ll |\mathbf{E}|$ is satisfied. Both sides of this inequality can be represented via the vector potential using

$$\hat{\mathbf{E}}(\mathbf{r}) = -\frac{\partial}{\partial t}\hat{\mathbf{A}}(\mathbf{r}) \quad \text{and} \quad \hat{\mathbf{B}}(\mathbf{r}) = \nabla \times \hat{\mathbf{A}}(\mathbf{r}), \tag{17.14}$$

based on the definitions (17.6) and (17.8). By introducing a typical frequency ω for the field and using the particle velocity v_{ple}, the inequality becomes

$$\left|\nabla_{\mathbf{R}} \times \hat{\mathbf{A}}(\mathbf{R})\right| \ll \frac{\omega}{v_{\text{ple}}}\left|\hat{\mathbf{A}}(\mathbf{R})\right| = \frac{1}{l_{\text{ple}}}\left|\hat{\mathbf{A}}(\mathbf{R})\right|, \qquad l_{\text{ple}} \equiv \frac{v_{\text{ple}}}{\omega}, \tag{17.15}$$

where $2\pi\, l_{\text{ple}}$ defines the propagation distance of a particle during an optical cycle. Hence, as long as the vector potential does not change much on the length scale of l_{ple}, we may omit the magnetic component of the Lorentz force from Eq. (17.8). To estimate l_{ple}, we may use the interparticle velocity (16.53) as the upper limit of v_{ple} and set ω to match the $1s$-to-$2p$ frequency, giving

$$l_{\text{ple}} \leq \frac{|v_{1s,\,2p}|}{\omega_{1p,\,1s}} = \sqrt{\frac{E_{2p} - E_{1s}}{E_B}}\; a_0 \approx a_0. \tag{17.16}$$

Here, the simplification in the last step follows because the $1s$-to-$2p$ transition energy is close to the binding energy. The result shows that within the optical cycle the particles typically move only over distances that are smaller than the atomic Bohr radius. In particular, $l_{\text{ple}} \leq a_0$ is significantly smaller than the optical wave-length, based on the long-wave-length limit discussed in Section 16.3.2. In fact, the system Hamiltonian (17.1) is valid only in this limit such that it is fully consistent to drop the magnetic part of the Lorentz force from Eq. (17.8), see Exercise 17.5.

17.1.2 Time scale for the center-of-mass motion

The validity of the condition (17.15) could be understood to suggest that one might also omit the second $\hat{\mathbf{A}}$-dependent contribution from Eq. (17.9). As it turns out, most of this term can indeed be dropped but the argumentation must be modified in order to keep the leading-order contribution. For this purpose, we Fourier transform the operators

$$\hat{\mathbf{A}}(\mathbf{R},\omega) \equiv \frac{1}{\sqrt{2\pi}}\int dt\,\hat{\mathbf{A}}(\mathbf{R},t)\,e^{+i\omega t}, \quad \hat{\mathbf{E}}(\mathbf{R},\omega) = i\omega\,\hat{\mathbf{A}}(\mathbf{R},\omega)$$
$$\hat{\mathbf{P}}(\Omega) \equiv \frac{1}{\sqrt{2\pi}}\int dt\,\hat{\mathbf{P}}(t)\,e^{-i\Omega t}, \quad \hat{\mathbf{r}}(\omega) \equiv \frac{1}{\sqrt{2\pi}}\int dt\,\hat{\mathbf{r}}(t)\,e^{i\omega t} \tag{17.17}$$

and Eq. (17.9) to obtain

$$\begin{aligned}
-i\Omega\,\hat{\mathbf{P}}(\Omega) &= Q_e\nabla_{\mathbf{R}}\int d\omega\Big[\hat{\mathbf{r}}(\Omega-\omega)\cdot\hat{\mathbf{E}}(\mathbf{R},\omega) - i\Omega\,\hat{\mathbf{r}}(\Omega-\omega)\cdot\hat{\mathbf{A}}(\mathbf{R},\omega)\Big] \\
&= -iQ_e\nabla_{\mathbf{R}}\int d\omega\,[-\omega+\Omega]\,\hat{\mathbf{r}}(\Omega-\omega)\cdot\hat{\mathbf{A}}(\mathbf{R},\omega) \\
\Rightarrow\ \hat{\mathbf{P}}(\Omega) &= Q_e\nabla_{\mathbf{R}}\int d\omega\Big[-\frac{\omega}{\Omega}+1\Big]\,\hat{\mathbf{r}}(\Omega-\omega)\cdot\hat{\mathbf{A}}(\mathbf{R},\omega).
\end{aligned} \tag{17.18}$$

Since both $\hat{\mathbf{r}}(\omega)$ and $\hat{\mathbf{A}}(\mathbf{R},\omega)$ typically exist only for frequencies $\omega = \pm\omega_{2p,\,1s}$, the convolution has its dominant contribution for $\Omega \approx 0$. In this situation, $(-\omega/\Omega+1)$ is dominated by $-\omega/\Omega$ and the constant can be ignored. This is equivalent to omitting the second term in Eq. (17.9). More physically, the relevant frequencies ω and Ω identify the time scales $T_{\mathrm{opt}} = 2\pi/\omega$ and $T_{\mathrm{COM}} = 2\pi/\Omega$ for the optical oscillations and the center-of-mass motion, respectively. Whenever the center-of-mass motion proceeds slowly on the time scale of an optical cycle, i.e., $T_{\mathrm{COM}} \gg T_{\mathrm{opt}} \equiv \frac{\omega}{\Omega} \gg 1$, we can ignore $\frac{\partial}{\partial t}\hat{\mathbf{r}}\cdot\hat{\mathbf{A}}(\mathbf{R})$ relative to the $\hat{\mathbf{r}}\cdot\hat{\mathbf{E}}(\mathbf{R})$ contribution.

We have found two simplifications related to the center-of-mass motion: (i) Whenever the vector potential does not change much over the length l_{ple} which the atom propagates during an optical cycle, the magnetic part of the Lorentz force can be ignored. (ii) Whenever the center-of-mass motion is negligible during the optical cycle, we can omit the second term in Eq. (17.9), or more mathematically,

$$\begin{aligned}
&\hat{\mathbf{P}}(t) \to \hat{\mathbf{P}}(\Omega)\,e^{-i\Omega t}\ \ \&\ \ T_{\mathrm{COM}} = \frac{2\pi}{\Omega}, \quad \hat{\mathbf{A}}(t) \to \hat{\mathbf{A}}(\omega)\,e^{-i\omega t}\ \ \&\ \ T_{\mathrm{opt}} = \frac{2\pi}{\omega}, \\
&T_{\mathrm{COM}} \gg T_{\mathrm{opt}} \quad \Leftrightarrow \quad \frac{\omega}{\Omega} \gg 1 \\
&\Rightarrow \hat{\mathbf{r}}\cdot\hat{\mathbf{E}}(\mathbf{R}) \text{ dominates over } \frac{\partial}{\partial t}\hat{\mathbf{r}}\cdot\hat{\mathbf{A}}(\mathbf{R}) \text{ in Eq. (17.9).}
\end{aligned} \tag{17.19}$$

Typically, the conditions (17.15) and (17.19) are excellently satisfied in solids such that we can simplify Eq. (17.5) into

$$\begin{cases} \dfrac{\partial}{\partial t}\hat{\mathbf{r}} = \hat{\mathbf{v}} \\ \dfrac{\partial}{\partial t}\hat{\mathbf{v}} = \dfrac{Q_e\,\hat{\mathbf{E}}(\mathbf{R})}{\mu} + \dfrac{\mathbf{F}(\mathbf{r})}{\mu} \end{cases}, \qquad \begin{cases} \dfrac{\partial}{\partial t}\hat{\mathbf{R}} = \dfrac{\hat{\mathbf{P}}}{M} \\ \dfrac{\partial}{\partial t}\hat{\mathbf{P}} = \nabla_{\mathbf{R}}\Big[Q_e\hat{\mathbf{r}}\cdot\hat{\mathbf{E}}(\mathbf{R})\Big] \end{cases}, \tag{17.20}$$

where we have grouped the relative-motion and the center-of-mass operators together. Even though the application of the conditions (17.15) and (17.19) simplify the investigations, Eq. (17.20) still contains the hierarchy problem sketched in Fig. 17.1. It is also interesting to notice that we cannot produce Eq. (17.20) with a simple approximation to the system Hamiltonian (17.1) because the interaction $\hat{\mathbf{A}}\cdot\hat{\mathbf{p}}$ yields both dominant and negligible terms. In Section 17.4, we introduce an alternative approach that allows for an implementation of the conditions (17.15) and (17.19) directly in the system Hamiltonian.

17.2 Light–matter coupling

The quantum dynamics of the light operators can also be computed using the vector potential or the electric field. Following the steps of Section 14.2.1 and using Eq. (17.3), we eventually obtain

$$\left[\nabla^2 - \frac{n^2(\mathbf{x})}{c^2}\frac{\partial^2}{\partial t^2}\right]\hat{\mathbf{A}}(\mathbf{x}) = -\mu_0\left[\hat{\mathbf{J}}_A\,\delta(\mathbf{x}-\mathbf{R})\right]_T, \tag{17.21}$$

in full analogy to Eq. (14.56). Here, the transversal part of the current is evaluated at the atomic center-of-mass position. For the sake of generality, we assume that the dielectric refractive index $n(\mathbf{x})$ can have an arbitrary spatial dependence.

With the help of Eq. (17.20), the current can be expressed in terms of a dipole,

$$\hat{\mathbf{J}}_A = Q_e\hat{\mathbf{v}} = Q_e\frac{\partial\,\hat{\mathbf{r}}}{\partial t} = \frac{\partial\,\hat{\mathbf{d}}}{\partial t}, \qquad \hat{\mathbf{d}} \equiv Q_e\,\hat{\mathbf{r}}. \tag{17.22}$$

This converts the wave equation (17.21) into

$$\left[\nabla^2 - \frac{n^2(\mathbf{x})}{c^2}\frac{\partial^2}{\partial t^2}\right]\hat{\mathbf{A}}(\mathbf{x}) = -\mu_0\left[\frac{\partial\hat{\mathbf{d}}}{\partial t}\,\delta(\mathbf{x}-\mathbf{R})\right]_T. \tag{17.23}$$

Taking the time derivative of both sides of Eq. (17.23) and using (17.14), we find the wave equation for the transverse electric field

$$\left[\nabla^2 - \frac{n^2(\mathbf{x})}{c^2}\frac{\partial^2}{\partial t^2}\right]\hat{\mathbf{E}}(\mathbf{x}) = \mu_0\left[\frac{\partial^2\hat{\mathbf{d}}}{\partial t^2}\,\delta(\mathbf{x}-\mathbf{R})\right]_T. \tag{17.24}$$

This equation looks almost classical however, it has no closed-form solution due to the infinite hierarchy of equations emerging from the coupling to Eq. (17.20).

17.3 Dipole emission

To investigate some elementary implications of the light–matter coupling, we look at the expectation value of Eq. (17.24),

$$\left[\nabla^2 - \frac{1}{c^2}\frac{\partial^2}{\partial t^2}\right]\langle\hat{\mathbf{E}}(\mathbf{x})\rangle = \mu_0\left[\frac{\partial^2\langle\hat{\mathbf{d}}\rangle}{\partial t^2}\,\delta(\mathbf{x}-\mathbf{R})\right]_T. \tag{17.25}$$

Here, we assumed a constant refractive index, i.e., $n = 1$, for simplicity. The treatment of Eq. (17.25) is a standard problem in classical electrodynamics discussed in many textbooks. Thus, we skip the explicit derivations and proceed directly to the formal inhomogeneous solution

$$\langle \hat{\mathbf{E}}_{\text{dip}}(\mathbf{x}) \rangle = -\frac{\mu_0}{4\pi} \int d^3x' \, \frac{\mathbf{S}(\mathbf{x}', t')}{|\mathbf{x}' - \mathbf{x}|}, \qquad t' = t - \frac{|\mathbf{x}' - \mathbf{x}|}{c},$$
$$\mathbf{S}(\mathbf{x}, t) \equiv \left[\frac{\partial^2 \langle \hat{\mathbf{d}} \rangle}{\partial t^2} \delta(\mathbf{x} - \mathbf{R}) \right]_T . \tag{17.26}$$

Here, t' defines the retarded time for light traveling between $\mathbf{x}$ and $\mathbf{x}'$. The expression for $\mathbf{S}(\mathbf{x}, t)$ describes the dipole source of the wave equation. The full solution of Eq. (17.25) contains a sum of the incoming homogeneous field and the field $\langle \hat{\mathbf{E}}_{\text{dip}}(\mathbf{x}) \rangle$ emitted by the dipole.

To compute $\mathbf{S}(\mathbf{x}, t)$ entering $\langle \hat{\mathbf{E}}_{\text{dip}}(\mathbf{x}) \rangle$, we need to evaluate $\hat{\mathbf{d}}$. However, $\hat{\mathbf{d}}$ couples to $\hat{\mathbf{J}}_A$ and to all orders of the photon, position, and kinetic velocity operators due to the hierarchy problem discussed in Section 17.1. To gain some preliminary insight into how light–matter coupling works, we evaluate how a *known* oscillating dipole emits light. Here, we only have to deal with the singlets $\langle \hat{\mathbf{E}} \rangle$ and $\langle \hat{\mathbf{d}} \rangle$ allowing us to express $\langle \hat{\mathbf{E}} \rangle$ in terms of $\langle \hat{\mathbf{d}} \rangle$. We do not need to solve the full hierarchy problem because the dipole dynamics is assumed to be known and we disregard its back reaction to the field it creates.

Based on our discussion in Section 16.3, Eq. (16.41) directly produces

$$\langle \hat{\mathbf{d}} \rangle = \text{Re}\left[d_{12} \, e^{-i\omega_{21} t} \right] \mathbf{e}_z, \tag{17.27}$$

where we choose the dipole orientation as $\mathbf{e}_z$. The strength is defined by the dipole-matrix element $d_{12} = d_{1s,\,2p}$ between the atomic $1s$ and $2p$ levels and $\omega_{21} = \omega_{2p,\,1s} = \frac{E_{2p}-E_{1s}}{\hbar}$ gives the frequency of the $1s$-to-$2p$ transition. Altogether, this defines the dipole source

$$\begin{aligned} \mathbf{S}(\mathbf{x}, t) &= -\text{Re}\left[d_{12} \, \omega_{2,1}^2 \, e^{-i\omega_{21} t} \right] \theta(t) \left[\mathbf{e}_z \, \delta(\mathbf{x} - \mathbf{R}) \right]_T \\ &= S(t) \left[\mathbf{e}_z \, \delta(\mathbf{x} - \mathbf{R}) \right]_T , \\ S(t) &\equiv -\text{Re}\left[d_{12} \, \omega_{21}^2 \, e^{-i\omega_{21} t} \right] \theta(t), \end{aligned} \tag{17.28}$$

according to Eq. (17.26). Here, we multiplied the source by $\theta(t)$ to describe a scenario where the dipole is switched on at the time $t = 0$.

The integration of Eq. (8.34) yields the transversal part of the vector field

$$\mathbf{S}(\mathbf{x}, t) = S(t) \left(\delta(\mathbf{x} - \mathbf{R}) \, \mathbf{e}_z - (\mathbf{e}_z \cdot \nabla_{\mathbf{R}}) \nabla_{\mathbf{R}} \frac{1}{4\pi \, |\mathbf{x} - \mathbf{R}|} \right), \tag{17.29}$$

where the differential operators act on the center-of-mass position of the dipole. Inserting Eq. (17.29) into Eq. (17.26), evaluating the integration and differentiation, we eventually obtain

$$
\begin{aligned}
\mathbf{E}_{\mathrm{dip}}(\mathbf{x},t) = & \left(\frac{(x_z-R_z)(\mathbf{x}-\mathbf{R})}{|\mathbf{x}-\mathbf{R}|^2}-\mathbf{e}_z\right)\mathrm{Re}\left[E_{\mathrm{near}}(|\mathbf{x}-\mathbf{R}|,t)\right] \\
& + \left(3\frac{(x_z-R_z)(\mathbf{x}-\mathbf{R})}{|\mathbf{x}-\mathbf{R}|^2}-\mathbf{e}_z\right) \\
& \times \mathrm{Re}\left[E_{\mathrm{far}}(|\mathbf{x}-\mathbf{R}|,t)+E_{\mathrm{asym}}(|\mathbf{x}-\mathbf{R}|,t)\right]. \qquad (17.30)
\end{aligned}
$$

Here, we identified

$$
\begin{aligned}
E_{\mathrm{far}}(r,t) &\equiv \frac{d_{12}}{4\pi\varepsilon_0}\frac{q_{21}^2}{r}\,e^{iq_{21}(r-ct)}\,\theta(ct-r), \qquad q_{21}\equiv\frac{\omega_{21}}{c}, \\
E_{\mathrm{near}}(r,t) &\equiv \frac{d_{12}}{4\pi\varepsilon_0}\frac{e^{iq_{21}(r-ct)}}{r^3}\left(e^{-iq_{21}r}-1+iq_{21}r\right)\theta(ct-r), \\
E_{\mathrm{asym}}(r,t) &\equiv \frac{d_{12}}{4\pi\varepsilon_0}\frac{1}{r^3}\left(e^{-iq_{21}ct}-1+iq_{21}ct\right)\theta(r-ct)\theta(t). \qquad (17.31)
\end{aligned}
$$

Since $E_{\mathrm{far}}(r,t)$ scales like one over distance, it describes a spherical wave. Due to the θ-function that appears, the front edge of the wave propagates exactly with the speed of light. The leading order of $E_{\mathrm{near}}(r,t)$ and $E_{\mathrm{asym}}(r,t)$ scale like $1/r^2$ and $1/r^3$, respectively. Thus, both fields have appreciable magnitudes only in the near-field regime. However, the $\theta(r/c-t)\theta(t)$ factor shows that $E_{\mathrm{asym}}(r,t)$ is limited to the far field $r > ct$, where it usually describes vanishingly small contributions.

17.3.1 Dipole-emission dynamics

The time sequence in Fig. 17.2 shows how the intensity

$$
I_{\mathrm{dip}}(\mathbf{x},t)\equiv|\mathbf{x}|^2\left|\mathbf{E}_{\mathrm{dip}}(\mathbf{x}+\mathbf{R},t)\right|^2 \qquad (17.32)
$$

of the dipole field evolves after the dipole starts to oscillate. To obtain these results, we have used the wave length $\lambda_{21}=30a_0$ defining the wave vector $q_{21}=2\pi/\lambda_{21}$ and placed the dipole at the origin orienting it along the z axis. Since the resulting $I_{\mathrm{dip}}(\mathbf{x},t)$ is invariant for any rotation around the z axis, we present only contour plots of its xz cross section.

In Fig. 17.2, we see a typical dipole field which has negligible magnitude along the z axis and propagates as two lobes of a spherical wave with the speed of light. Frames 17.2(b)–(c) show that we have almost unchanged contour patterns for $I_{\mathrm{dip}}(\mathbf{x},t)$ as long as the snapshot times are separated by an oscillation period $T=2\pi/q_{21}c=30a_0/c$. For time separations shorter than T, the emission pattern naturally shows all phases of the oscillation. For the largest t, the nodes and the peaks of the dipole field are particularly clear in the far field where oscillating rings propagate away from the origin. At short distances, the $I_{\mathrm{dip}}(\mathbf{x},t)$ pattern exhibits a nontrivial form due to the near-field contributions.

This analysis shows that the coupling of a dipole to the electromagnetic field leads to the emission of light. Consequently, the dipole gradually loses energy, which leads to a damping of its oscillations. This damping due to the light–matter coupling yields radiative

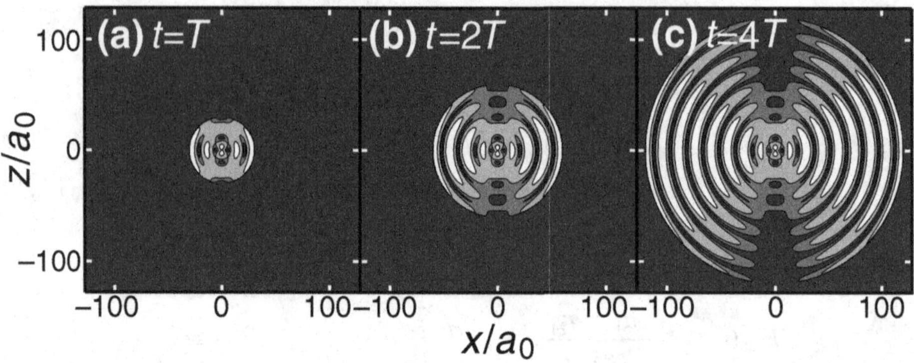

Figure 17.2 Dipole emission pattern. The dipole is positioned at the origin and oriented along $\mathbf{e}_z$. Contour plots of $I_{\text{dip}}(\mathbf{x}, t)$ are shown at (a) $t = T$, (b) $t = 2T$, and (c) $t = 4T$ after the dipole is switched on. The oscillation period of the dipole is T and the dipole emission has the wave length $\lambda = 30\, a_0$ where a_0 is the Bohr radius. The contours are plotted at 1/32, 1/8, and 1/2 of the maximum of $I_{\text{dip}}(\mathbf{x}, t)$. The emission pattern is invariant with respect to rotations around the z axis.

broadening of the emission spectra, as discussed in connection with Fig. 16.5. The broadening is often referred to as the natural line width or radiative width. Our current analysis is not yet capable of describing the radiative broadening because we have not included the dipole-field interaction self-consistently. This analysis is presented in Chapter 26 where the radiative decay process is evaluated.

17.3.2 Emission of planar dipoles

For semiconductors, the atomic dipole-emission results are relevant only in very few cases, e.g., when the radiation of a single quantum dot is analyzed. More commonly, semiconductor nanostructures are planar systems such that the dipoles are confined into a plane. To study such a scenario, we consider a fixed arrangement of dipoles in the xy-plane where the dipoles are oriented along the $\mathbf{e}_x$ direction. We can still use the expression (17.26) for the dipole source with the following modifications that (i) the δ-function becomes a one-dimensional $\delta(z)$ describing the two-dimensional confinement and (ii) the individual dipole $\langle \hat{\mathbf{d}} \rangle$ is replaced by the dipole density, i.e., the polarization $P(t)$ multiplied by $\mathbf{e}_x$. Thus, we find a planar dipole source,

$$\mathbf{S}_{2\text{D}}(\mathbf{x}, t) \equiv \left[\frac{\partial^2 P(t)}{\partial t^2} \mathbf{e}_x\, \delta(z) \right]_T = \frac{\partial^2 P(t)}{\partial t^2} \mathbf{e}_x\, \delta(z) \tag{17.33}$$

that is transversal ($\nabla \cdot \mathbf{S}_{2\text{D}}(\mathbf{x}, t) = 0$) by definition. The identified $P(t)$ can have any time dependence such that it, in principle, can also describe the fully self-consistent solution. As a generalization, $\mathbf{S}_{2\text{D}}(\mathbf{x}, t)$ can additionally contain a plane-wave factor, i.e., we can replace $\delta(z)$ by $\delta(z)\, e^{i\mathbf{q}\cdot\mathbf{x}}$ where $\mathbf{q}$ defines the direction of emission. This case is analyzed in Exercise 17.10.

The source (17.33) is so simple that we can evaluate the integral in Eq. (17.26) analytically for all situations. When the polarization vanishes at $t = -\infty$, we eventually find that the planar dipoles emit the field

$$\langle \hat{\mathbf{E}}_{\text{dip}}(\mathbf{x}, t) \rangle = -\frac{\mu_0 c}{2} \mathbf{e}_x \frac{\partial}{\partial t} P \left(t - \frac{|z|}{c} \right). \tag{17.34}$$

These are plane waves propagating with the speed of light. It is interesting to see that the emission does not produce near-field contributions because the transversal polarization, Eq. (17.33), is completely localized unlike its atomic counterpart defined by Eq. (17.29). However, as discussed in Exercise 17.10, we also obtain near-field contributions in cases where the polarization is directional.

To illustrate the propagation of the planar dipole emission, we consider a fixed oscillator

$$P(t) = P_0 \cos \omega t \, \theta(t), \tag{17.35}$$

which produces the field

$$\langle \hat{\mathbf{E}}_{\text{fix}}(\mathbf{x}, t) \rangle = \frac{\mu_0 c \, \omega}{2} P_0 \, \mathbf{e}_x \sin \omega \left(t - \frac{|z|}{c} \right) \theta(ct - |z|). \tag{17.36}$$

We see that the electrical field oscillates as $\langle \hat{E}_{\text{fix}}(0, t) \rangle \propto \sin \omega t$ in the plane of the semiconductor material. Hence, the emitted light and the polarization are out of phase by $\pi/2$. As discussed in Chapter 26, this phase difference leads to the radiative decay in planar semiconductor structures.

Figure 17.3(a) presents the time evolution of the intensity distribution $I_{\text{2D}}(z, t) = |\langle \hat{E}_{\text{fix}}(\mathbf{x}, t) \rangle|^2$ emitted by a planar polarization. The result is compared with $I_{\text{dip}}(x, 0, 0, t)$ for a simple atomic emitter in Fig. 17.3(b). Here, we follow the cross sections of the dipole-emission pattern along the x axis of Fig. 17.2. Both cases have $\lambda = 30 a_0$ and the intensities are normalized to one.

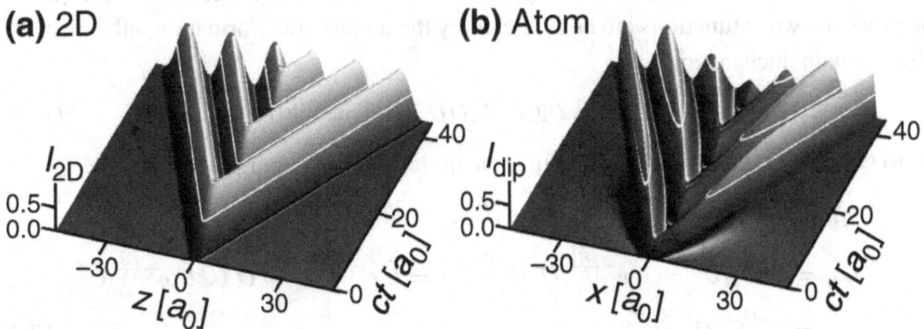

Figure 17.3 Time evolution of the dipole emission along the main axis. (**a**) The emission $I_{\text{2D}}(z, t)$ induced by a planar polarization in the direction $\mathbf{e}_x$ is compared with (**b**) the emission $I_{\text{dip}}(x, 0, 0, t)$ from a point-like dipole aligned along $\mathbf{e}_z$. The dipole is switched on at time $t = 0$ having the emission wave length $\lambda = 30\,a_0$. This corresponds to an oscillation period of $cT = 30 a_0$. Both intensities are normalized to one and the contour is defined at the value 1/2.

We see that the planar polarization yields light emission in a completely sinusoidal form that propagates with the speed of light away from the center of the emitter located at $z = 0$. The wave front does not alter its form after it has been created at $z = 0$. In contrast to this, the single dipole (Fig. 17.3b) displays some modification during the propagation, which results from the decaying near-field contributions. Otherwise, the far-field dynamics of $I_{2D}(z, t)$ and $I_{dip}(x, 0, 0, t)$ are qualitatively very similar.

17.4 Quantum dynamics in the $\hat{\mathbf{E}} \cdot \hat{\mathbf{x}}$ picture

Dipoles play a central role in studies of the light–matter coupling because their presence induces the emission of electromagnetic waves $\langle \hat{E}_{dip} \rangle$. However, the dipole does not appear directly in the Hamiltonian (17.1) and emerges only in the equations of motion (17.20). In other words, even though the concept of the dipole is fully included in the $\hat{\mathbf{A}} \cdot \hat{\mathbf{p}}$ picture, it becomes explicitly visible only if the Heisenberg equations of motion are evaluated, e.g., to produce the wave equation (17.24). Here, the dipole contribution $\hat{\mathbf{E}}(\mathbf{r}) \cdot \hat{\mathbf{d}}$ dominates over terms such as $\frac{\partial}{\partial t} \hat{\mathbf{r}} \cdot \hat{A}(\mathbf{r})$ if the condition (17.19) is satisfied.

To further elucidate the dipole-emission properties and to gain further insights, we proceed to an alternative formulation of the light–matter interaction where the dipole directly appears in the Hamiltonian. Then, we can implement the condition (17.19) directly at the level of the system Hamiltonian to produce the correct dynamics (17.20) without further approximations.

17.4.1 Göppert-Mayer transformation

We start by reminding ourselves of the properties of a generic unitary transformation $\hat{U}$,

$$\hat{O}' = \hat{U}\,\hat{O}\,\hat{U}^\dagger, \quad |\Psi'(t)\rangle = \hat{U}|\Psi(t)\rangle, \quad \text{with} \quad \hat{U}\hat{U}^\dagger = \hat{U}^\dagger\hat{U} = \mathbb{I}, \tag{17.37}$$

where the last relation imposes the unitarity for $\hat{U}$. Here, we have assumed that we work in the Schrödinger picture with time-independent operators and dynamic states. Even though operators and wave functions are both altered by the unitary transformation, all expectation values remain unchanged,

$$\langle \hat{O}' \rangle = \langle \Psi' | \hat{O}' | \Psi' \rangle = \langle \Psi | \hat{U}^\dagger \hat{U} \hat{O} \hat{U}^\dagger \hat{U} | \Psi \rangle = \langle \Psi | \hat{O} | \Psi \rangle = \langle \hat{O} \rangle, \tag{17.38}$$

due to (17.37). In the Heisenberg picture, the unitary transformation yields

$$\begin{aligned}
\hat{H}' &= \hat{U}^\dagger\,\hat{H}\,\hat{U}, \qquad \hat{O}_H = e^{-\frac{i\hat{H}t}{\hbar}}\,\hat{O}\,e^{+\frac{i\hat{H}t}{\hbar}} \quad \Rightarrow \\
\hat{O}'_H &= \hat{U}\,\hat{O}_H\hat{U}^\dagger = \hat{U}e^{-\frac{i\hat{H}t}{\hbar}}\,\hat{O}\,e^{+\frac{i\hat{H}t}{\hbar}}\hat{U}^\dagger = \hat{U}e^{-\frac{i\hat{H}t}{\hbar}}\,\hat{U}^\dagger\hat{U}\,\hat{O}\,\hat{U}^\dagger\hat{U}\,e^{+\frac{i\hat{H}t}{\hbar}}\hat{U}^\dagger \\
&= e^{-\frac{i\hat{H}'t}{\hbar}}\,\hat{O}'\,e^{+\frac{i\hat{H}'t}{\hbar}}.
\end{aligned} \tag{17.39}$$

As a consequence, the Heisenberg equation of motion becomes

$$i\hbar\frac{\partial}{\partial}\langle \hat{O}'_H \rangle = \left[\hat{O}'_H, \hat{H}'\right]_-. \tag{17.40}$$

Thus, we can study the quantum dynamics entirely in the transformed picture.

To simplify the Hamiltonian (16.56), we compute $\hat{H}'$ using the so-called Göppert-Mayer transformation,

$$\hat{U} = e^{\hat{S}}, \qquad \hat{U}^\dagger = e^{-\hat{S}}, \qquad \hat{S} = -\frac{i}{\hbar} Q_e \mathbf{r} \cdot \hat{\mathbf{A}}(\mathbf{R}). \tag{17.41}$$

To implement this transformation, we evaluate Eq. (17.37) with the help of the general operator relation (see Exercise 3.6),

$$\hat{O}' = e^{\hat{S}}\, \hat{O}\, e^{-\hat{S}} = \hat{S} + \left[\hat{S}, \hat{O}\right]_- + \frac{1}{2!}\left[\hat{S}, \left[\hat{S}, \hat{O}\right]_-\right]_- + \cdots \tag{17.42}$$

We obtain

$$\begin{aligned} \hat{\mathbf{r}}' &= \hat{\mathbf{r}}, & \hat{\mathbf{p}}' &= \hat{\mathbf{p}} + Q_e \hat{\mathbf{A}}(\mathbf{R}), \\ \hat{\mathbf{R}}' &= \hat{\mathbf{R}}, & \hat{\mathbf{P}}' &= \hat{\mathbf{P}} + Q_e \left[\nabla_{\mathbf{R}} \left(\hat{\mathbf{r}} \cdot \hat{\mathbf{A}}(\mathbf{R})\right)\right], \\ B'_{\mathbf{q}} &= B_{\mathbf{q}} + \frac{i\mathcal{E}_{\mathbf{q}}}{\hbar\omega_{\mathbf{q}}} Q_e \mathbf{r} \cdot \mathbf{u}^\star_{\mathbf{q}}(\mathbf{R}), \end{aligned} \tag{17.43}$$

where the explicitly written $\nabla_{\mathbf{R}}$ acts exclusively on $\hat{\mathbf{A}}(\mathbf{R})$. This contribution describes the dependence of the electromagnetic field on the center-of-mass coordinate of the atom. Interestingly, $\hat{\mathbf{P}}'$ contains a contribution $\hat{\mathbf{r}} \cdot \hat{\mathbf{A}}(\mathbf{R})$ that is expected to be small in comparison to $\hat{\mathbf{r}} \cdot \hat{\mathbf{E}}(\mathbf{R})$ if the condition (17.19) is satisfied.

If we now insert $\hat{\mathbf{r}}'$ and $\hat{\mathbf{p}}'$ into the transformed Hamiltonian (16.1), we find

$$\hat{H}' = \frac{\hat{\mathbf{P}}' \cdot \hat{\mathbf{P}}'}{2M} + \frac{\hat{\mathbf{p}}^2}{2\mu} - \frac{Q_e^2}{4\pi\varepsilon\,\epsilon_0 |\mathbf{r}|} + \sum_{\mathbf{q}} \omega_{\mathbf{q}} \left[\left(B'_{\mathbf{q}}\right)^\dagger B'_{\mathbf{q}} + \frac{1}{2}\right], \tag{17.44}$$

where the typical minimal-substitution form is completely removed. At the same time, the free-photon part becomes

$$\begin{aligned} \hat{H}'_F &= \sum_{\mathbf{q}} \hbar\omega_{\mathbf{q}} \left[\left(B^\dagger_{\mathbf{q}} - \frac{i\mathcal{E}_{\mathbf{q}}}{\hbar\omega_{\mathbf{q}}} Q_e \mathbf{r} \cdot \mathbf{u}_{\mathbf{q}}(\mathbf{R})\right)\left(B_{\mathbf{q}} + \frac{i\mathcal{E}_{\mathbf{q}}}{\hbar\omega_{\mathbf{q}}} Q_e \mathbf{r} \cdot \mathbf{u}^\star_{\mathbf{q}}(\mathbf{R})\right) + \frac{1}{2}\right] \\ &= \sum_{\mathbf{q}} \hbar\omega_{\mathbf{q}} \left[B^\dagger_{\mathbf{q}} B_{\mathbf{q}} + \frac{1}{2}\right] - Q_e \mathbf{r} \cdot \sum_{\mathbf{q}} i\mathcal{E}_{\mathbf{q}} \left[B_{\mathbf{q}} \mathbf{u}_{\mathbf{q}}(\mathbf{R}) - B^\dagger_{\mathbf{q}} \mathbf{u}^\star_{\mathbf{q}}(\mathbf{R})\right] \\ &\quad + \sum_{\mathbf{q}} \frac{\mathcal{E}^2_{\mathbf{q}}}{\hbar\omega_{\mathbf{q}}} \left[Q_e \mathbf{r} \cdot \mathbf{u}_{\mathbf{q}}(\mathbf{R})\right]\left[Q_e \mathbf{r} \cdot \mathbf{u}^\star_{\mathbf{q}}(\mathbf{R})\right], \end{aligned} \tag{17.45}$$

which follows directly by sorting out the terms containing the product of two, one, or zero mode functions.

We recognize that the first term on the RHS of Eq. (17.45) reproduces the original free-photon Hamiltonian. To find the physical interpretation of the second term, we evaluate the transformation of the electric field,

$$\begin{aligned}\hat{\mathbf{E}}'(\mathbf{x}) &= \sum_{\mathbf{q}} i\mathcal{E}_{\mathbf{q}}\left[B'_{\mathbf{q}}\mathbf{u}_{\mathbf{q}}(\mathbf{x}) - \left(B'_{\mathbf{q}}\right)^{\dagger}\mathbf{u}^{\star}_{\mathbf{q}}(\mathbf{x})\right] \\ &= \sum_{\mathbf{q}} i\mathcal{E}_{\mathbf{q}}\left[B_{\mathbf{q}}\mathbf{u}_{\mathbf{q}}(\mathbf{x}) - (B_{\mathbf{q}})^{\dagger}\mathbf{u}^{\star}_{\mathbf{q}}(\mathbf{x})\right] \\ &\quad - \frac{Q_e\mathbf{r}}{2\varepsilon_0}\cdot\sum_{\mathbf{q}}\left[\mathbf{u}^{\star}_{\mathbf{q}}(\mathbf{R})\mathbf{u}_{\mathbf{q}}(\mathbf{x}) + \mathbf{u}^{\star}_{\mathbf{q}}(\mathbf{x})\mathbf{u}_{\mathbf{q}}(\mathbf{R})\right].\end{aligned} \tag{17.46}$$

Here, we used (17.43) together with $\mathcal{E}^2_{\mathbf{q}}/\hbar\omega_{\mathbf{q}} = 1/2\varepsilon_0$, based on Eq. (17.1). The first contribution in (17.46) is the electric field in the original $\hat{\mathbf{A}}\cdot\hat{\mathbf{p}}$ picture, as given by the mode-expansion (17.6). The additional contribution produces the transversal δ-function $\sum_{\mathbf{q}}\mathbf{u}^{\star}_{\mathbf{q}}(\mathbf{R})\mathbf{u}_{\mathbf{q}}(\mathbf{x}) = \boldsymbol{\delta}^T(\mathbf{x}-\mathbf{R})$ due to the completeness relation (8.66) of the transversal fields. As a result, the transformed electric field casts into the form

$$\begin{aligned}\hat{\mathbf{E}}'(\mathbf{x}) &= \hat{\mathbf{E}}(\mathbf{x}) - \frac{Q_e\mathbf{r}}{\varepsilon_0}\cdot\boldsymbol{\delta}^T(\mathbf{x}-\mathbf{R}) = \hat{\mathbf{E}}(\mathbf{x}) - \frac{1}{\varepsilon_0}\hat{\mathbf{P}}(\mathbf{x}) \\ \hat{\mathbf{P}}(\mathbf{x}) &\equiv Q_e\mathbf{r}\cdot\boldsymbol{\delta}^T(\mathbf{x}-\mathbf{R}) = \left[\hat{\mathbf{d}}\,\delta(\mathbf{x}-\mathbf{R})\right]_T,\end{aligned} \tag{17.47}$$

where $\hat{\mathbf{P}}(\mathbf{x})$ identifies the transversal polarization of the dipole $\hat{\mathbf{d}} \equiv Q_e\hat{\mathbf{r}}$, where $\hat{\mathbf{r}}$ defines the electron–ion distance.

Interestingly, the Hamiltonian (17.45) contains the transformed dielectric displacement operator

$$\hat{\mathbf{D}}'(\mathbf{x}) \equiv \hat{\mathbf{E}}'(\mathbf{x}) - \frac{1}{\epsilon_0}\hat{\mathbf{P}}(\mathbf{x}) = \sum_{\mathbf{q}} i\mathcal{E}_{\mathbf{q}}\left[B_{\mathbf{q}}\mathbf{u}_{\mathbf{q}}(\mathbf{x}) - (B_{\mathbf{q}})^{\dagger}\mathbf{u}^{\star}_{\mathbf{q}}(\mathbf{x})\right], \tag{17.48}$$

yielding

$$\begin{aligned}\hat{H}'_F &= \sum_{\mathbf{q}}\hbar\omega_{\mathbf{q}}\left[B^{\dagger}_{\mathbf{q}}B_{\mathbf{q}} + \frac{1}{2}\right] - Q_e\hat{\mathbf{r}}\cdot\hat{\mathbf{D}}(\mathbf{R}) \\ &\quad + \sum_{\mathbf{q}}\frac{\mathcal{E}^2_{\mathbf{q}}}{\hbar\omega_{\mathbf{q}}}\left[Q_e\hat{\mathbf{r}}\cdot\mathbf{u}_{\mathbf{q}}(\mathbf{R})\right]\left[Q_e\hat{\mathbf{r}}\cdot\mathbf{u}^{\star}_{\mathbf{q}}(\mathbf{R})\right].\end{aligned} \tag{17.49}$$

We see that the light–matter interaction is now described by the usual dipole interaction $-Q_e\mathbf{r}\cdot\hat{\mathbf{D}}(\mathbf{R})$. In classical electromagnetic theory, a dipole indeed has the energy $\mathbf{d}\cdot\mathbf{E}$ in an electric field. In the quantum description, $\hat{\mathbf{E}}$ is divided into the displacement field and the dipole self-energy described by the last term of Eq. (17.49). Since it is clear from the context, we omit the "prime" on $\hat{\mathbf{D}}$ from now on.

17.4.2 Dipole self-energy

The last contribution in Eq. (17.49) can be simplified by expressing the vacuum-field amplitude $\mathcal{E}_{\mathbf{q}} = \sqrt{\frac{\hbar\omega_{\mathbf{q}}}{2\varepsilon_0}}$ explicitly. This converts the dipole self-energy into

$$\hat{H}_{\text{dip}} \equiv \frac{1}{2\varepsilon_0}\sum_{\mathbf{q}}\left[\hat{\mathbf{d}}\cdot\mathbf{u}_{\mathbf{q}}(\mathbf{R})\right]\cdot\left[\hat{\mathbf{d}}\cdot\mathbf{u}^{\star}_{\mathbf{q}}(\mathbf{R})\right] \tag{17.50}$$

after we have identified the atomic dipole operator $Q_e\hat{\mathbf{r}}$. To get a feeling for the magnitude of the self-energy, we consider the average value

$$\langle \hat{H}_{\rm dip}\rangle = \frac{1}{2\varepsilon_0}\sum_{\mathbf{q}} \langle \left|Q_e\mathbf{r}\cdot\mathbf{u}_{\mathbf{q}}(\mathbf{R})\right|^2\rangle \leq \frac{1}{2\varepsilon_0}\sum_{\mathbf{q}} \langle |Q_e\mathbf{r}|^2 \left|\mathbf{u}_{\mathbf{q}}(\mathbf{R})\right|^2\rangle, \qquad (17.51)$$

which follows from the Cauchy–Schwartz inequality discussed in Exercise 5.6. Since the atomic system is in free space, we may choose plane-wave modes $\mathbf{u}_{\mathbf{q}}(\mathbf{R}) = \frac{1}{\mathcal{L}^{3/2}} e^{i\mathbf{q}\cdot\mathbf{r}}$, yielding $|\mathbf{u}_{\mathbf{q}}(\mathbf{R})|^2 = 1/\mathcal{L}^3$ that is proportional to the inverse of the quantization volume.

With this information, we find an upper bound for the dipole self-energy

$$\langle \hat{H}_{\rm dip}\rangle \leq \frac{\langle |Q_e\mathbf{r}|^2\rangle}{2\varepsilon_0}\frac{1}{\mathcal{L}^3}\sum_{\mathbf{q}} = \frac{\langle |Q_e\mathbf{r}|^2\rangle}{2\varepsilon_0}\frac{1}{(2\pi)^3}\int d^3q, \qquad (17.52)$$

which is converted into the mode integral using the large quantization-volume limit (8.61). This integral seems to diverge. However, we have started from the long-wave-length-limit Hamiltonian (17.1) based on the approximation (16.55). Thus, all the involved electromagnetic fields must have a wave length λ that is large relative to the size of the dipole. In other words, fields with short λ, such as gamma radiation, are not to be described with the transformed Hamiltonian (17.45). Thus, it is reasonable to impose upper limits to the wave vector $|\mathbf{q}| \leq 2\pi/\lambda_{\rm min} \equiv q_{\rm max}$ and to the photon energy $\hbar\omega \leq \hbar\omega_{\rm max} = \hbar c q_{\rm max}$, yielding $\int d^3q \to \frac{4\pi}{3}q_{\rm max}^3$. Using these limits in Eq. (17.52), we find

$$\langle \hat{H}_{\rm dip}\rangle \leq \frac{\langle |Q_e\mathbf{r}|^2\rangle\, q_{\rm max}^3}{12\pi^2\varepsilon_0} = \frac{e^2\langle |\mathbf{r}|^2\rangle\,[\hbar\omega_{\rm max}]^3}{12\pi^2\varepsilon_0\hbar^3c^3}$$

$$\Rightarrow \quad \langle \hat{H}_{\rm dip}\rangle \leq 1.99\times 10^{-10}\,\text{eV}\left[\frac{\langle |\mathbf{r}|^2\rangle}{\text{Å}^2}\right]\left[\frac{\hbar\omega_{\rm max}}{eV}\right]^3. \qquad (17.53)$$

Hence, the dipole self-energy is negligible for typical Å-sized dipoles and light fields on the eV energy scale.

17.4.3 System Hamiltonian

Even though we could assure a finite dipole self-energy by introducing a cutoff for the photon energy, it is beneficial not to do this and to keep the Hamiltonian of Eq. (17.50). As it turns out, the formally diverging term conveniently provides the polarization part of the electric field in the relevant equations of motion, see Exercise 17.11. In other words, we can always apply the relation (17.48) to identify the electric field $\hat{\mathbf{E}} = \hat{\mathbf{D}} + \frac{1}{\epsilon_0}\hat{\mathbf{P}}$ in the dynamical operator equations such that the inclusion of the formally diverging $\hat{H}_{\rm dip}$ is not at all problematic because all physical equations remain finite.

Combining Eqs. (17.44), (17.49), and (17.52), we obtain the transformed system Hamiltonian. As the only approximation, we neglect the $\nabla_{\mathbf{R}}\left(\hat{\mathbf{r}}\cdot\hat{\mathbf{A}}(\mathbf{R})\right)$ contribution from

$\hat{\mathbf{P}}'$ with the intention to eliminate all the terms that are vanishingly small relative to the $\hat{\mathbf{r}} \cdot \hat{\mathbf{E}}(\mathbf{R})$ contributions. As suggested by the condition (17.19), this smallness assumes that the center-of-mass motion proceeds on a time scale that is much longer than the oscillation period of the field. With these considerations, the long-wave-length-limit Hamiltonian casts into the form

$$\hat{H}' = \frac{\hat{\mathbf{P}}^2}{2M} + \frac{\hat{\mathbf{p}}^2}{2\mu} - \frac{Q_e^2}{4\pi \varepsilon \, \epsilon_0 |\mathbf{r}|} + \sum_{\mathbf{q}} \hbar\omega_{\mathbf{q}} \left[(B_{\mathbf{q}})^\dagger B_{\mathbf{q}} + \frac{1}{2} \right]$$
$$- Q_e \mathbf{r} \cdot \hat{\mathbf{D}}(\mathbf{R}) + \sum_{\mathbf{q}} \frac{\left[\hat{\mathbf{d}} \cdot \mathbf{u}_{\mathbf{q}}(\mathbf{R})\right] \cdot \left[\hat{\mathbf{d}} \cdot \mathbf{u}^\star_{\mathbf{q}}(\mathbf{R})\right]}{2\varepsilon_0},$$
$$\hat{\mathbf{D}}(\mathbf{x}) = \sum_{\mathbf{q}} i\mathcal{E}_{\mathbf{q}} \left[B_{\mathbf{q}} \mathbf{u}_{\mathbf{q}}(\mathbf{x}) - (B_{\mathbf{q}})^\dagger \mathbf{u}^\star_{\mathbf{q}}(\mathbf{x}) \right], \tag{17.54}$$

where we have repeated the definition (17.48) for the sake of completeness. We use this Hamiltonian as the starting point for our investigations of the quantum dynamics in the $\hat{\mathbf{E}} \cdot \hat{\mathbf{x}}$ picture. Since all operator commutation relations are unaltered, we may determine the operator dynamics without denoting the "prime" explicitly. The original Hamiltonian (16.56) is clearly nonlinear in the $\hat{\mathbf{A}} \cdot \hat{\mathbf{p}}$ picture whereas the Hamiltonian in the $\hat{\mathbf{E}} \cdot \mathbf{r}$ picture contains an entirely linear light–matter coupling. In addition, the often negligible Lorentz-force terms are already eliminated by the $\hat{\mathbf{P}}' \to \hat{\mathbf{P}}$ approximation. This is a clear benefit and can yield simpler derivations of optical properties.

17.4.4 Quantum dynamics

As in the $\hat{\mathbf{A}} \cdot \hat{\mathbf{p}}$ picture, we can determine the quantum dynamics of the system using the Heisenberg equation of motion (17.40) for the elementary operators in the $\hat{\mathbf{E}} \cdot \hat{\mathbf{x}}$ picture. Evaluating the commutators with the Hamiltonian (17.54), we obtain

$$i\hbar \frac{\partial}{\partial t} B_{\mathbf{q}} = \hbar\omega_{\mathbf{q}} \, B_{\mathbf{q}} + i\mathcal{E}_{\mathbf{q}} \, \mathbf{u}^\star(\mathbf{R}) \cdot Q_e \mathbf{r}$$
$$i\hbar \frac{\partial}{\partial t} B^\dagger_{\mathbf{q}} = -\hbar\omega_{\mathbf{q}} \, B^\dagger_{\mathbf{q}} - i\mathcal{E}_{\mathbf{q}} \, \mathbf{u}(\mathbf{R}) \cdot Q_e \mathbf{r}. \tag{17.55}$$

We see that the photon operators are directly coupled to the dipole and not to the current as in Eq. (17.3). In analogy to the derivations in Section 14.2.1, we can transform Eq. (17.55) into

$$\left[\nabla^2 - \frac{n^2(\mathbf{x})}{c^2} \frac{\partial^2}{\partial t^2} \right] \hat{\mathbf{A}}(\mathbf{r}) = -\mu_0 \left[\frac{\partial \, Q_e \hat{\mathbf{r}}}{\partial t} \, \delta(\mathbf{x} - \mathbf{R}) \right]_T, \tag{17.56}$$

which directly contains the coupling of the vector potential to the atomic dipole $\hat{\mathbf{d}} = Q_e \hat{\mathbf{r}}$. Performing the differentiation $-\partial/\partial t$ on both sides of Eq. (17.56) and using the identification (17.14), we end up with

$$\left[\nabla^2 - \frac{n^2(\mathbf{x})}{c^2} \frac{\partial^2}{\partial t^2} \right] \hat{\mathbf{E}}(\mathbf{r}) = \mu_0 \left[\frac{\partial^2 \, Q_e \hat{\mathbf{r}}}{\partial t^2} \, \delta(\mathbf{x} - \mathbf{R}) \right]_T, \tag{17.57}$$

which is identical to the wave equation (17.24). Hence, even though the photon-operator dynamics depends on the picture used, the wave equations (17.56)–(17.57) are independent. This property is related to the gauge invariance of physical observables such as the electric field and the dipole.

Evaluating the respective commutators, we obtain the dynamics of the relative and center-of-mass coordinates as

$$\frac{\partial}{\partial t}\hat{\mathbf{r}} = \frac{1}{i\hbar}\left[\hat{\mathbf{r}}, \hat{H}'\right]_- = \hat{\mathbf{v}}_E, \qquad \hat{\mathbf{v}}_E \equiv \frac{\hat{\mathbf{p}}}{\mu}, \qquad \frac{\partial}{\partial t}\hat{\mathbf{R}} = \frac{\hat{\mathbf{P}}}{M}. \tag{17.58}$$

Here, the velocity $\hat{\mathbf{v}}_E$ is identical to the canonical velocity $\hat{\mathbf{p}}/\mu$. In other words, the kinetic and canonical velocity are equal in the $\hat{\mathbf{E}}\cdot\hat{\mathbf{x}}$ picture, unlike Eq. (17.4) in the $\hat{\mathbf{A}}\cdot\hat{\mathbf{p}}$ picture. The momentum dynamics follows from similar commutations, producing

$$\frac{\partial}{\partial t}\hat{\mathbf{v}}_E = \frac{Q_e\,\hat{\mathbf{E}}(\mathbf{R})}{\mu} + \frac{\mathbf{F}(\mathbf{r})}{\mu}, \qquad \frac{\partial}{\partial t}\hat{\mathbf{P}} = \nabla_{\mathbf{R}}\left(Q_e\hat{\mathbf{r}}\cdot\hat{\mathbf{E}}(\mathbf{r})\right). \tag{17.59}$$

In this context, the electric field operator $\hat{\mathbf{E}}(\mathbf{R}) = \hat{\mathbf{D}}(\mathbf{R}) - \frac{1}{\epsilon_0}\hat{\mathbf{P}}(\mathbf{R})$ can be identified via Eq. (17.48) only if the dipole self-energy is properly included in the system Hamiltonian.

Collecting the results (17.57)–(17.59), we find the dynamical equations for the elementary operators,

$$\left[\nabla^2 - \frac{n^2(\mathbf{x})}{c^2}\frac{\partial^2}{\partial t^2}\right]\hat{\mathbf{E}}(\mathbf{r}) = \mu_0\left[\frac{\partial^2 Q_e\hat{\mathbf{r}}}{\partial t^2}\,\delta(\mathbf{x}-\mathbf{R})\right]_T$$
$$\begin{cases} \frac{\partial}{\partial t}\hat{\mathbf{r}} = \hat{\mathbf{v}} \\ \frac{\partial}{\partial t}\hat{\mathbf{v}} = \frac{Q_e\,\hat{\mathbf{E}}(\mathbf{R})}{\mu} + \frac{\mathbf{F}(\mathbf{r})}{\mu} \end{cases}, \qquad \begin{cases} \frac{\partial}{\partial t}\hat{\mathbf{R}} = \frac{\hat{\mathbf{P}}}{M} \\ \frac{\partial}{\partial t}\hat{\mathbf{P}} = \nabla_{\mathbf{R}}\left[Q_e\hat{\mathbf{r}}\cdot\hat{\mathbf{E}}(\mathbf{R})\right] \end{cases}. \tag{17.60}$$

These equations are identical to Eqs. (17.20) and (17.24), defining the dynamics in the $\hat{\mathbf{A}}\cdot\hat{\mathbf{p}}$ picture. Hence, we obtain the same quantum dynamics in both pictures.

We also observe that the approximated Hamiltonian (17.54) correctly produces the dominant contributions to the operator dynamics. Thus, the $\hat{\mathbf{E}}\cdot\hat{\mathbf{x}}$ picture separates the large and small terms already in the system Hamiltonian. The negligible contributions can be eliminated by applying the simple approximation $\hat{\mathbf{P}}' \to \hat{\mathbf{P}}$. In the $\hat{\mathbf{A}}\cdot\hat{\mathbf{p}}$ picture, the light–matter interaction contains a mixture of small and large contributions. The small terms have to be eliminated by implementing the condition (17.19) in the equations of motion, not in the Hamiltonian itself. Thus, one typically needs to perform a more careful analysis in the $\hat{\mathbf{A}}\cdot\hat{\mathbf{p}}$ than in the $\hat{\mathbf{E}}\cdot\hat{\mathbf{x}}$ picture if one wants to systematically approximate the quantum dynamics.

Exercises

17.1 **(a)** Start from the system Hamiltonian (17.1) and derive Eq. (17.3). **(b)** Show that $[\hat{\mathbf{v}}, \hat{\mathbf{P}}^2] = i\hbar\frac{Q}{\mu}\sum_{j=1}^{3}[(\nabla_{R_j}\hat{\mathbf{A}}(\mathbf{R}))\hat{P}_j + \hat{P}_j(\nabla_{R_j}\hat{\mathbf{A}}(\mathbf{R}))]$, where $\hat{\mathbf{v}} \equiv \frac{1}{\mu}[\hat{\mathbf{p}} - Q\hat{\mathbf{A}}(\mathbf{R})]$. **(c)** Derive Eq. (17.5). **Hint:** *Use result (b).*

17.2 Assume that the vector potential $\hat{\mathbf{A}}(\mathbf{R})$ is transversal and use the relations $\hat{\mathbf{B}}(\mathbf{R}) = \nabla \times \hat{\mathbf{A}}(\mathbf{R})$ and $\hat{\mathbf{P}} = -i\hbar\nabla_{\mathbf{R}}$. **(a)** Show that $\hat{\mathbf{P}} \times \hat{\mathbf{A}}(\mathbf{R}) = -i\hbar\,\hat{\mathbf{B}}(\mathbf{R}) - \hat{\mathbf{A}}(\mathbf{R}) \times \hat{\mathbf{P}}$ and $\hat{\mathbf{P}} \cdot \hat{\mathbf{A}}(\mathbf{R}) = \hat{\mathbf{A}}(\mathbf{R}) \cdot \hat{\mathbf{P}}$. **(b)** Verify the following relations:

$$\hat{\mathbf{P}} \times (\hat{\mathbf{A}}(\mathbf{R}) \times \hat{\mathbf{P}}) + (\hat{\mathbf{A}}(\mathbf{R}) \times \hat{\mathbf{P}}) \times \hat{\mathbf{P}} = -i\hbar\; \nabla \times (\hat{\mathbf{A}}(\mathbf{R}) \times \hat{\mathbf{P}})$$

$$(\hat{\mathbf{P}} \times \hat{\mathbf{A}}(\mathbf{R})) \times \hat{\mathbf{P}} + (\hat{\mathbf{A}}(\mathbf{R}) \times \hat{\mathbf{P}}) \times \hat{\mathbf{P}} = -i\hbar\; \hat{\mathbf{B}}(\mathbf{R}) \times \hat{\mathbf{P}}$$

$$\hat{\mathbf{P}} \times (\hat{\mathbf{P}} \times \hat{\mathbf{A}}(\mathbf{R})) + \hat{\mathbf{P}} \times (\hat{\mathbf{A}}(\mathbf{R}) \times \hat{\mathbf{P}}) = -i\hbar\; \hat{\mathbf{P}} \times \hat{\mathbf{B}}(\mathbf{R})$$

$$\hat{\mathbf{P}} \times (\hat{\mathbf{P}} \times \hat{\mathbf{A}}(\hat{\mathbf{R}})) + (\hat{\mathbf{P}} \times \hat{\mathbf{A}}(\mathbf{R})) \times \hat{\mathbf{P}} = i\hbar \sum_{j=1}^{3} \hat{P}_j(\nabla_{R_j} \hat{\mathbf{A}}(\mathbf{R}))$$

$$\hat{\mathbf{P}} \times (\hat{\mathbf{A}}(\mathbf{R}) \times \hat{\mathbf{P}}) + (\hat{\mathbf{A}}(\mathbf{R}) \times \hat{\mathbf{P}}) \times \hat{\mathbf{P}} = i\hbar \sum_{j=1}^{3} (\nabla_{R_j} \hat{\mathbf{A}}(\mathbf{R}))\hat{\mathbf{P}}_j.$$

17.3 The first contribution in Eq. (17.8) can be interpreted as quantum-mechanically symmetrized Lorentz force

$$\hat{\mathbf{F}}_B \equiv -\frac{Q}{2M} \sum_{j=1}^{3} \left[(\nabla_{R_j} \hat{\mathbf{A}}(\mathbf{R}))\hat{\mathbf{P}}_j + \frac{\hat{\mathbf{P}}_j}{M}(\nabla_{R_j} \hat{\mathbf{A}}(\mathbf{R})) \right].$$

By using the relations in Exercise 17.2(b), show that **(a)**

$$\hat{\mathbf{F}}_B = \frac{iQ}{2\hbar} \Big[\hat{\mathbf{P}} \times (\hat{\mathbf{P}} \times \hat{\mathbf{A}}(\hat{\mathbf{R}})) + (\hat{\mathbf{P}} \times \hat{\mathbf{A}}(\mathbf{R})) \times \hat{\mathbf{V}}$$
$$-\hat{\mathbf{P}} \times (\hat{\mathbf{A}}(\mathbf{R}) \times \hat{\mathbf{P}}) - (\hat{\mathbf{A}}(\mathbf{R}) \times \hat{\mathbf{P}}) \times \hat{\mathbf{P}} \Big],$$

(b) $\hat{\mathbf{F}}_B = Q\left[\frac{1}{2}(\hat{\mathbf{V}} \times \hat{\mathbf{B}}(\mathbf{R}) + \hat{\mathbf{B}}(\mathbf{R}) \times \hat{\mathbf{V}}) - \nabla \times (\hat{\mathbf{A}}(\mathbf{R}) \times \hat{\mathbf{V}})\right]$ with $\hat{\mathbf{V}} \equiv \frac{\hat{\mathbf{P}}}{M}$.

17.4 The transversal electric field follows from $\hat{\mathbf{E}}(\mathbf{R}) \equiv -\frac{\partial}{\partial t}\hat{\mathbf{A}}(\mathbf{R})$ and the relative-motion velocity is $\hat{\mathbf{v}} \equiv \frac{\partial}{\partial t}\hat{\mathbf{r}}$. Start from Eq. (17.5) and show that it can be converted into $\frac{\partial \hat{\mathbf{P}}}{\partial t} = \nabla_R[Q\hat{\mathbf{r}} \cdot \hat{\mathbf{E}}(\mathbf{R})] + \frac{\partial}{\partial t} Q_E\, \nabla_R(\hat{\mathbf{r}} \cdot \hat{\mathbf{A}}(\mathbf{R}))$. **Hint:** *Use the property* $\frac{\partial}{\partial t}\hat{\mathbf{r}} \cdot \hat{\mathbf{A}}(\mathbf{R}) = \frac{\partial \hat{\mathbf{r}}}{\partial t} \cdot \hat{\mathbf{A}}(\mathbf{R}) + \hat{\mathbf{r}} \cdot \frac{\partial \hat{\mathbf{A}}(\mathbf{R})}{\partial t}$.

17.5 A generic Lorentz force is given by $\mathbf{F}_L \equiv Q(\mathbf{v} \times \mathbf{B}(\mathbf{R}) + \mathbf{E}(\mathbf{R}))$. **(a)** Use Eq. (17.14) and show that the magnetic component of $\mathbf{F}_L$ is negligibly small whenever $|\nabla_R \times \mathbf{A}(\mathbf{R})| \ll \frac{1}{l_{\text{ple}}}|\mathbf{A}(\mathbf{R})|$ with $l_{\text{ple}} \equiv v_{\text{ple}}/\omega$ where v_{ple} is the typical particle velocity and ω is the central frequency of light. **(b)** Verify that this condition implies (17.16). **(c)** Rewrite Eq. (17.18) using the limit (17.15).

17.6 **(a)** Derive Eq. (17.18) from Eq. (17.9). **(b)** Apply the conditions (17.15) and (17.19) to Eq. (17.5) and use this to derive Eq. (17.20). **(c)** Derive Eq. (17.21) by starting from the Hamiltonian (17.1).

17.7 **(a)** Use the property $\nabla^2 \frac{1}{|\mathbf{x}-\mathbf{x}'|} = -4\pi\delta(\mathbf{x}-\mathbf{x}')$ to prove $\left(\nabla^2 - \frac{1}{c^2}\frac{\partial^2}{\partial t^2}\right)\langle \hat{\mathbf{E}}_{\text{dip}}(\mathbf{x}, t)\rangle = \mu_0 S(\mathbf{x}, t)$ if $\langle \hat{\mathbf{E}}_{\text{dip}}(\mathbf{x}, t)\rangle = -\frac{\mu_0}{4\pi}\int d^3x' \frac{S(\mathbf{x}',t')}{|\mathbf{x}'-\mathbf{x}|}$ with retarded time $t' = t - \frac{|\mathbf{x}'-\mathbf{x}|}{c}$. **(b)** Show that the source (17.28) yields Eq. (17.29) after the projection (8.34) is applied.

17.8 Assume that the dipole source (17.29) is centered at $\mathbf{R}$ and that it is defined by $\mathbf{S}(\mathbf{x},t) = R_E[\mathbf{S}_1(\mathbf{x},t)+\mathbf{S}_2(\mathbf{x},t)]$, where $\mathbf{S}_1(\mathbf{x},t) = \mathbf{e}_z\, S(t)\, \delta(\mathbf{x}-\mathbf{R})$ and $\mathbf{S}_2(\mathbf{x},t) = \hat{D}_{\mathbf{R}} \frac{S(t)}{|\mathbf{x}-\mathbf{R}|}$ with $S(t) \equiv -\omega_{21}^2\, d_{21}\, e^{-i\omega_{21}t}\, \theta(t)$ and $\hat{D}_{\mathbf{R}} \equiv \frac{1}{4\pi} \frac{\partial}{\partial R_z} \nabla_{\mathbf{R}}$. Compute the dipole field $E_{\rm dip}^{(j)}(\mathbf{x},t) = -\frac{\mu_0}{4\pi} \int d^3x'\, \frac{\mathbf{S}_j(\mathbf{x}',t')}{|\mathbf{x}'-\mathbf{x}|}$ according to Eq. (17.26). Show that $E_{\rm dip}^{(1)}(\mathbf{x},t) = -\frac{\mu_0}{4\pi}\, e_z \frac{S_{21}\left(t-\frac{|\mathbf{x}-\mathbf{R}|}{c}\right)}{|\mathbf{x}-\mathbf{R}|}$ and

$$E_{\rm dip}^{(2)}(\mathbf{x},t) = -\frac{4\pi c^2 d_{21}}{\Delta x} \begin{cases} e^{iq_{21}(\Delta x - ct)} - iq_{21}\Delta x - e^{-i\omega_{21}t}, & \Delta x < ct \\ 1 - i\omega_{21}\, t - e^{-i\omega_{21}t}, & \Delta x > ct \end{cases},$$

where $q_{21} = \omega_{21}/c$ and $\Delta x = |\mathbf{R}-\mathbf{x}|$.

17.9 Use the result of Exercise 17.8 to generate Eqs. (17.30)–(17.31).

17.10 **(a)** Show that the source (17.33) is transversal. **(b)** Prove that $\mathbf{E}_{\rm dip}(z,t) = -\frac{\mu_0 c}{2}\, \mathbf{e}_x \frac{\partial}{\partial t} P\left(t - \frac{|z|}{c}\right)$ satisfies $\left[\frac{\partial^2}{\partial z^2} - \frac{1}{c^2}\frac{\partial^2}{\partial t^2}\right] \mathbf{E}_{\rm dip}(z,t) = \mu_0\, \mathbf{S}_{\rm 2D}(z,t)$. **(c)** Compute the explicit form of $\mathbf{S}_{\rm 2D}(\mathbf{x},t) \equiv \left[\frac{\partial^2 P(t)}{\partial t^2}\, \mathbf{e}_x\, \delta(x)\, e^{iqz}\right]_T$ with the help of the transformation (8.34). **(d)** Compute $E_q(\mathbf{x},t) = -\frac{\mu_0}{4\pi} \int d^3\mathbf{x}'\, \frac{S_q(\mathbf{x},t)}{|x'-x|}$.

17.11 **(a)** Produce the elementary transformations using the Göppert-Mayer transformation. **(b)** Transform the Hamiltonian (17.1) into (17.44)–(17.45). **(c)** Derive Eqs. (17.58)–(17.59) starting from Eq. (17.54). Show that the dipole self-energy is needed to identify the $\hat{\mathbf{E}}$ that appears.

Further reading

Lorentz force, dipole, dipole emission, and gauges are standard concepts in classical electromagnetism; see, e.g., the books mentioned in Chapter 2. The role of the gauge transformations in the light–matter interaction is discussed further, e.g., in:

C. Cohen-Tannoudji, J. Dupont-Roc, and G. Grynberg (1997). *Photons & Atoms: Introduction to Quantum Electrodynamics*, New York, Wiley & Sons.

C. Aversa and J. E. Sipe (1995). Nonlinear optical susceptibilities of semiconductors: Results with a length-gauge analysis, *Phys. Rev. B* **52**, 14636.

M. Kira, F. Jahnke, W. Hoyer, and S. W. Koch (1999). Quantum theory of spontaneous emission and coherent effects in semiconductor microstructures, *Prog. Quantum Electron.* **23**, 189.

The Göppert-Mayer transformation was introduced in:

M. Göppert-Mayer (1930). Über Elementarakte mit zwei Quantensprüngen (Elementary processes with two quantum jumps), *Ann. Phys.* **9**, 273.

The relation between the dipole and momentum matrix elements is studied further, e.g., in:

E. Blount (1962). *Solid State Physics: Advances in Research and Applications Vol. 13*, New York, Academic Press.

M. Lax (1974). *Symmetry Principles in Solid State and Molecular Physics*, New York, Wiley & Sons.

H. Haug and S. W. Koch (2009). *Quantum Theory of the Optical and Electronic Properties of Semiconductors*, 5th edition, Singapore, World Scientific.

18

Two-level approximation for optical transitions

In a classical picture, an atom constitutes a natural dipole if its negatively charged electrons oscillate relative to the positively charged ionic nucleus. Microscopically, the dipole results from a superposition of energetically bound states, allowing only specific discrete energies, compare Section 16.3. Therefore, the corresponding dipole oscillations give rise to light emission only with certain characteristic frequencies, see Chapter 17.

The light–matter coupling also works in the reversed direction. Properly chosen light sources are able to excite dipole oscillations in matter if they are resonant with the characteristic dipole-oscillation frequencies. In other words, optical excitation can induce transitions between the atomic eigenstates. We investigate this and related phenomena in this chapter with the focus on exploring basic properties of light–matter coupling. As shown in Chapter 17, the full quantum description of light and matter yields an infinite hierarchy problem even for the simplest atomic systems. Thus, we first study how and when a few-level approximation can be applied to explain optical transitions caused by classical light.

18.1 Classical optics in atomic systems

In Section 17.3, we derive a closed-form expression for the emission of a known atomic or planar dipole source. Describing the dipole via the polarization as in Eq. (17.33), the wave equation can be written as

$$\left[\nabla^2 - \frac{n^2(\mathbf{x})}{c^2}\frac{\partial^2}{\partial t^2}\right]\langle\hat{\mathbf{E}}(\mathbf{x})\rangle = \mu_0 \frac{\partial^2\,[\mathbf{P}(\mathbf{x})]_T}{\partial t^2},$$

$$\mathbf{P}(\mathbf{x}) = \begin{cases} \langle\hat{\mathbf{d}}\rangle\delta(\mathbf{x}-\mathbf{R}), & \text{atom} \\ \mathbf{P}(t)\,\mathbf{e}_{\parallel}\,\delta(z), & \text{planar solid} \end{cases}, \tag{18.1}$$

where $\mathbf{d} = Q_e\mathbf{r}$ is the atomic dipole, $\mathbf{r}$ is the electron–ion separation, and $\mathbf{R}$ defines the center-of-mass coordinate of the electron–ion system. According to Eq. (17.33), the polarization of a planar solid is restricted to the xy-plane pointing along $\mathbf{e}_{\parallel}$ perpendicular to the confinement direction z. Only the transversal part of $\mathbf{P}(\mathbf{x})$ enters as a source in the wave equation.

Since the wave equation of classical electrodynamics agrees with Eq. (18.1), $\langle \hat{\mathbf{E}}(\mathbf{x}) \rangle$ represents the classical aspects of light. As an extension of our discussion in Chapter 17, where we investigate the classical light emission by a known dipole source, we now want to treat the inverse problem by computing the matter excitations induced by classical light. For this purpose, we need, in addition to Eq. (18.1), a model for the inhomogeneous polarization source generated by light. Thus, we must directly investigate the matter dynamics resulting from the Hamiltonian (17.44) or (17.54). The hierarchy problem does not appear as long as we treat light classically by replacing the operator of the electric field by the complex-valued field. In other words, we implement the *classical limit* for the light,

$$\hat{\mathbf{E}}(\mathbf{x}) \to \mathbf{E}(\mathbf{x}) \quad \Leftrightarrow \quad B_{\mathbf{q}} \to \beta_{\mathbf{q}} \quad \text{and} \quad B^{\dagger}_{\mathbf{q}} \to \beta^{\star}_{\mathbf{q}}. \tag{18.2}$$

Introducing this classical limit in the system Hamiltonian (17.54), the free-field term becomes a constant number, i.e., $B^{\dagger}_{\mathbf{q}} B_{\mathbf{q}} \to |\beta_{\mathbf{q}}|^2$, that can be omitted from all further investigations. We can also replace $-Q_e \mathbf{r} \cdot \hat{\mathbf{D}}(\mathbf{R})$ by its classical counterpart $-Q_e \mathbf{r} \cdot \mathbf{E}(\mathbf{R})$. In this approximation, we furthermore omit a term proportional to the dipole self-energy which is negligible as shown in Section 17.4.2 and Exercise 18.1. With these approximations, Eq. (17.54) can be converted into the semiclassical system Hamiltonian

$$\hat{H}_{\text{sc}} = \hat{H}_0 - Q_e \mathbf{r} \cdot \mathbf{E}(\mathbf{R}), \qquad \hat{H}_0 \equiv \frac{\hat{\mathbf{P}}^2}{2M} + \frac{\hat{\mathbf{p}}^2}{2\mu} - \frac{Q_e^2}{4\pi\varepsilon\,\epsilon_0 |\mathbf{r}|}, \tag{18.3}$$

where only the atomic part is described quantum mechanically. The light–matter interaction now follows from the semiclassical dipole interaction term $-Q_e \mathbf{r} \cdot \mathbf{E}(\mathbf{R})$.

In Section 16.1.2, we discuss the properties of $\hat{H}_0$, separate the relative and center-of-mass contributions using Eq. (16.12), and obtain the complete set of eigenstates,

$$\begin{cases} \hat{H}_0 \,|\nu, \mathbf{K}\rangle = E_{\nu,\mathbf{K}} \,|\nu, \mathbf{K}\rangle \\ |\nu, \mathbf{K}\rangle \equiv |\phi_\nu\rangle \otimes |\mathbf{K}\rangle \\ E_{\nu,\mathbf{K}} = E_\nu + \frac{\hbar^2 \mathbf{K}^2}{2M} \end{cases}, \qquad \begin{cases} \langle \nu', \mathbf{K}' | \nu, \mathbf{K}\rangle = \delta_{\nu',\nu} \delta_{\mathbf{K}',\mathbf{K}} \\ \sum_{\nu,\mathbf{K}} |\nu, \mathbf{K}\rangle\langle \nu, \mathbf{K}| = \mathbb{I} \end{cases}. \tag{18.4}$$

The center-of-mass motion is described through the plane-wave solution (16.14),

$$\langle \mathbf{R} | \mathbf{K} \rangle = \frac{1}{\mathcal{L}^{\frac{3}{2}}} e^{i\mathbf{K}\cdot\mathbf{R}}, \tag{18.5}$$

where we assumed a finite quantization volume $\mathcal{L}^3$. Thus, we know all quantum-mechanical eigenstates of the matter, which considerably simplifies our further investigations.

At any moment in time, the atomic system is described by the superposition state

$$|\psi(t)\rangle = \sum_{\nu,\mathbf{K}} c_{\nu,\mathbf{K}}(t) \,|\nu, \mathbf{K}\rangle, \tag{18.6}$$

where $|c_{\nu,\mathbf{K}}(t)|^2$ defines the probability of finding the system in the state $|\nu, \mathbf{K}\rangle$, as discussed in Section 5.3. If the light field vanishes, i.e., $\mathbf{E}(\mathbf{x}) = 0$, $|c_{\nu,\mathbf{K}}(t)|^2$ remains constant in time according to Eq. (16.37). Consequently, no transitions occur despite the atom's dipolar properties. In order to convert one configuration $c_{\nu,\mathbf{K}}$ into another, i.e., in order to induce optical transitions, the dipole interaction term is needed.

To analyze this scenario, we directly evaluate the time-dependent Schrödinger equation,

$$i\hbar \frac{\partial}{\partial t}|\psi(t)\rangle = \hat{H}_{\rm sc}|\psi(t)\rangle. \tag{18.7}$$

Inserting the expansion (18.6) allows us to determine the dynamics of any coefficient $c_{\nu,\mathbf{K}}$ by projecting both sides of Eq. (18.7) onto $\langle \nu, \mathbf{K}| = \langle \nu| \otimes \langle \mathbf{K}|$. A straightforward calculation produces

$$i\hbar \frac{\partial}{\partial t} c_{\nu,\mathbf{K}} = E_{\nu,\mathbf{K}}\, c_{\nu,\mathbf{K}} - \sum\nolimits_{\nu',\mathbf{K}'} \langle \nu|Q_e \mathbf{r}|\nu'\rangle \cdot \langle \mathbf{K}|\mathbf{E}(\hat{\mathbf{R}})|\mathbf{K}'\rangle\, c_{\nu',\mathbf{K}'}. \tag{18.8}$$

Here, we used the fact that the states $|\nu\rangle$ involve only the relative-coordinate space and $|\mathbf{K}\rangle$ involves only the center-of-mass coordinate space, respectively. The matrix element $\langle \nu|Q_e \mathbf{r}|\nu'\rangle$ identifies a microscopic dipole

$$\mathbf{d}_{\nu,\nu'} \equiv \langle \nu|Q_e \mathbf{r}|\nu'\rangle, \tag{18.9}$$

based on the definition (16.42). Using (18.5), the center-of-mass matrix element can be evaluated in real space,

$$\langle \mathbf{K}|\mathbf{E}(\hat{\mathbf{R}})|\mathbf{K}'\rangle = \frac{1}{\mathcal{L}^3}\int d^3R\;\mathbf{E}(\mathbf{R}, t)\, e^{-i(\mathbf{K}-\mathbf{K}')\cdot\mathbf{R}} = \mathbf{E}(\mathbf{K}-\mathbf{K}', t), \tag{18.10}$$

simply yielding the Fourier transformation of the electric field. Altogether, the quantum dynamics (18.8) can now be expressed through

$$i\hbar \frac{\partial}{\partial t} c_{\nu,\mathbf{K}} = E_{\nu,\mathbf{K}}\, c_{\nu,\mathbf{K}} - \sum\nolimits_{\nu',\mathbf{K}'} \mathbf{d}_{\nu,\nu'} \cdot \mathbf{E}(\mathbf{K}-\mathbf{K}', t)\, c_{\nu',\mathbf{K}'}. \tag{18.11}$$

This equation shows that light can induce transitions from the state $|\nu', \mathbf{K}'\rangle$ to $|\nu, \mathbf{K}\rangle$ as long as both the corresponding dipole-matrix element and $\mathbf{E}(\mathbf{K}-\mathbf{K}', t)$ are nonvanishing. However, based on the analysis in Section 16.3, we know that the dipole-matrix element is nonvanishing only between a well-defined subset of states. Thus, $\mathbf{d}_{\nu,\nu'}$ defines the standard dipole selection rules for the optical transitions. For example, if the atom initially occupies a radially symmetric s state, there is no direct optical coupling to other s states, however, the light can convert the s state into a state with p-like symmetry.

If the electromagnetic field has a large spatial extension relative to the atomic center-of-mass distribution, we can approximate it by the transversal plane-wave solution

$$\mathbf{E}(\mathbf{x}, t) = E_0 e^{i(\mathbf{q}\cdot\mathbf{x}-\omega_\mathbf{q} t)}\mathbf{e}_T, \quad \omega_\mathbf{q} = c|\mathbf{q}| \tag{18.12}$$

of the homogeneous part of the wave equation (18.1). As we insert this into Eq. (18.10), we find

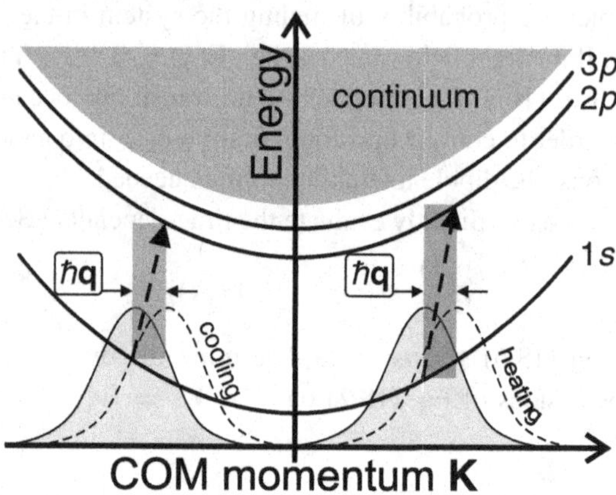

Figure 18.1 Optical transitions for atomic systems. The parabolas display the energy dispersion of atoms occupying the $1s$ or np states as well as the ionization continuum (shaded area). Two initial (Gaussian areas) and final (dashed line) center-of-mass distributions are also shown. An optical transition (dashed arrow) changes the center-of-mass momentum by $\hbar\mathbf{q}$.

$$\mathbf{E}(\mathbf{K}-\mathbf{K}',\, t) = \delta_{\mathbf{K}',\mathbf{K}-\mathbf{q}}\,\mathbf{E}_0(t), \quad \mathbf{E}_0(t) \equiv E_0\,\mathbf{e}_T\, e^{-i\omega_{\mathbf{q}}t}. \tag{18.13}$$

The corresponding atomic quantum dynamics then follows from

$$i\hbar\frac{\partial}{\partial t}c_{\nu,\mathbf{K}} = E_{\nu,\mathbf{K}}\,c_{\nu,\mathbf{K}} - \sum\nolimits_{\nu'} \mathbf{d}_{\nu,\nu'}\cdot\mathbf{E}_0(t)\,c_{\nu',\mathbf{K}-\mathbf{q}}. \tag{18.14}$$

We see that light changes the center-of-mass momentum of the atom by $\hbar\mathbf{q}$ in the transition between the states ν and ν'.

Figure 18.1 presents the relevant energy structure $E_{\nu,\mathbf{K}}$ for the situation where the atom is initially in the $1s$ state with a given center-of-mass distribution. Due to the dipole selection rules, this state couples only to np states. The corresponding $E_{1s,\mathbf{K}}$ and $E_{2p,\mathbf{K}}$ have a parabolic dispersion as a function of the center-of-mass momentum $\hbar\mathbf{K}$. When excited by a plane wave with wave vector $\mathbf{q}$, the atom changes its momentum by $\hbar\mathbf{q}$ indicated by the shaded rectangle and dashed arrows in Fig. 18.1. Schematically, we also show two possible center-of-mass distributions and the resulting end state after the optical transition. In cases where the distribution center is shifted into a different (in the same) direction as $\mathbf{q}$, it becomes cooled (heated) due to the momentum exchange in the optical interaction. This is a key element in laser cooling of atoms.

Mathematically, Eq. (18.14) constitutes an infinite-dimensional matrix equation. In practice, however, the coefficients $c_{\nu,\mathbf{K}}$ are nonvanishing only for a finite momentum range allowing us to limit the number of $(\nu, \mathbf{K})$ states included in the analysis. Under these conditions, the optical excitation dynamics can be determined from the straightforward numerical solution of Eq. (18.14) by starting from a given $c_{\nu,\mathbf{K}}$.

One can easily generalize Eq. (18.14) to incorporate a center-of-mass dependent potential that could, e.g., lead to spatial localization of the atom. In principle, one can even use ν and ν'-dependent pseudopotentials to describe optical transitions in molecules. In these investigations, one can apply combinations of attractive, repulsive, or mixed potentials, such as the Lennard-Jones potential, to describe binding and dissociation of molecular bonds.

18.1.1 Separation of relative and center-of-mass motion

In many relevant situations, we can introduce simplifying assumptions with the goal to analytically investigate the principal features of optical transitions. For instance, the photon momentum $\hbar\mathbf{q}$ can often be considered as insignificantly small relative to typical values of the center-of-mass momentum $\hbar\mathbf{K}$ of an atomic wave packet. To quantify this limit, we can estimate that an optical transition produces the momentum uncertainty of $\Delta P_{\mathrm{opt}} = |\hbar\mathbf{q}| = 2\pi\frac{\hbar}{\lambda}$ that corresponds to the optically induced momentum change. According to the Heisenberg relation (5.51), this momentum uncertainty corresponds to the position uncertainty

$$\Delta R_{\mathrm{opt}} \geq \frac{\hbar}{2\Delta P_{\mathrm{opt}}} = \frac{\hbar}{2\hbar|\mathbf{q}|} = \frac{\lambda}{4\pi}. \tag{18.15}$$

The optical wave length λ is of the order of 10^{-6}m. Hence, it significantly exceeds the typical width of the atom's center-of-mass distribution for which we take its dipole length scale, i.e., a few Ångströms. Consequently, the position uncertainty of an atom, ΔR_{atom}, is easily two to three orders of magnitude smaller than ΔR_{opt}.

Consequently, the atomic momentum uncertainty must significantly exceed $\hbar|\mathbf{q}|$, i.e., $\Delta P_{\mathrm{atom}} \geq \frac{\hbar}{2\Delta R_{\mathrm{atom}}} \gg \frac{\hbar}{2\Delta R_{\mathrm{opt}}} = \Delta P_{\mathrm{opt}} = \hbar|\mathbf{q}|$ greatly exceeds ΔP_{opt}. Thus, we can conclude that the optically induced momentum change is very small relative to the typical width of the atomic momentum distribution. Hence, we can often neglect the momentum transfer in Eq. (18.14), simplifying it to

$$i\hbar\frac{\partial}{\partial t}c_{\nu,\mathbf{K}} = E_{\nu,\mathbf{K}}\,c_{\nu,\mathbf{K}} - \sum\nolimits_{\nu'}\mathbf{d}_{\nu,\nu'}\cdot\mathbf{E}_0(t)\,c_{\nu',\mathbf{K}}. \tag{18.16}$$

This approximation allows us to decouple the relative and center-of-mass motion using the separation ansatz

$$c_{\nu,\mathbf{K}} \equiv a_\nu\,C_{\mathbf{K}} \tag{18.17}$$

in the product form. Since the eigenstate energy is the sum of the E_ν and $\hbar^2\mathbf{K}^2/2M$ contributions, Eq. (18.16) can be separated into two uncoupled equations,

$$\begin{cases} i\hbar\dfrac{\partial}{\partial t}a_\nu = E_\nu\,a_\nu - \sum_{\nu'}\mathbf{d}_{\nu,\nu'}\cdot\mathbf{E}_0(t)\,a_{\nu'} \\ i\hbar\dfrac{\partial}{\partial t}C_{\mathbf{K}} = \dfrac{\hbar^2\mathbf{K}^2}{2M}C_{\mathbf{K}} \end{cases}. \tag{18.18}$$

We see that the optical field only influences the relative-motion part a_ν of the system whereas the center-of-mass part describes free propagation. For laser-cooling investigations, one obviously must start from the original Eq. (18.14) because multiple cycles of ν' to ν transitions can eventually change the atomic momentum appreciably.

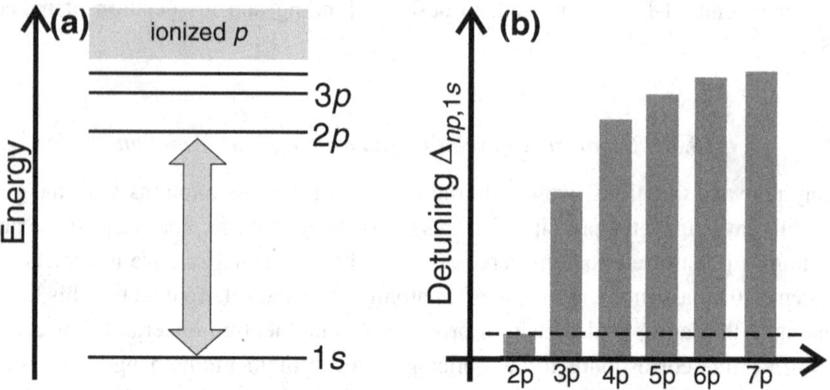

Figure 18.2 Near-resonance transitions in atomic systems. **(a)** The energetic structure (horizontal lines) of a hydrogen-like atomic system is shown together with the photon energy (arrow). **(b)** The detuning between the exciting field and the dipole-allowed transition energies is plotted for the different p states. Only the $1s$-to-$2p$ transition is near resonant. The horizontal dashed line indicates the level of near-resonance detuning.

18.1.2 Formal aspects of the optical excitation

As a representative example, we schematically plot the internal energy level structure of the hydrogen atom in Fig. 18.2(a). In such a system, optical transitions are particularly strong between the $1s$ and $2p$ states. Thus, it is interesting to study a situation where the central frequency ω_L of the exciting light is nearly resonant with the $1s$-to-$2p$ transition such that ω_L and the frequency of the oscillating dipole in Eq. (16.41) are matched sufficiently well. To quantify such a near-resonance condition, we introduce the detuning

$$\Delta_{\nu,\,\nu'} \equiv \omega_L - (\omega_\nu - \omega_{\nu'}), \tag{18.19}$$

where $\omega_\nu = E_\nu/\hbar$. The detuning defines the frequency difference between the light and the ν-to-ν' transition.

Since hydrogen-like bound states have a nonequidistant spacing of their E_ν levels, we can choose ω_L such that the near-resonance condition applies only for the $1s$-to-$2p$ dipole transition. The arrow in Fig. 18.2(a) indicates the energy $\hbar\omega_L$ of the exciting light. Since $\hbar\omega_L$ is chosen close to the energetic difference of the $2p$ and $1s$ states, the detuning $|\Delta_{2p,\,1s}|$ is much smaller than that for any other transition. This condition is illustrated in Fig. 18.2(b). This near-resonance condition can be expressed through

$$\begin{aligned} &|\Delta_{2p,\,1s}| \ll |\omega_{2p} - \omega_{1s}|, \\ &|\Delta_{2p,\,1s}| \ll |\Delta_{\nu,\lambda}|, \quad \text{with } \nu \neq 2p,\ \lambda \neq 1s. \end{aligned} \tag{18.20}$$

In the following, we ignore the transitions to the continuum of ionized states (shaded rectangle in Fig. 18.2(a)) because the corresponding dipole matrix elements are small and the transitions are strongly nonresonant. In principle, the analysis of semiclassical excitations involves the a_ν dynamics of Eq. (18.18) coupled to the wave equation (18.1). However, to gain some first insights into how the optically induced transitions between different matter states proceed, we start by investigating how a known, i.e., unchanged, $E(t)$ couples to the a_ν states. The fully self-consistent solution is presented in Section 20.4.1.

In the following, we assume a pulsed excitation that is switched on at time $t = 0$ and oscillates with the central frequency ω_L. Even though the physical field is a real-valued quantity, it is often mathematically convenient to use a representation with complex amplitudes. In particular, we introduce the separation

$$\mathbf{E}(t) \equiv \mathbf{E}^{(+)}(t) + \mathbf{E}^{(-)}(t), \quad \mathbf{E}^{(\pm)}(t) = \mathbf{E}_0\, \theta(t)\, e^{\mp i\omega_L t}\, e^{-\gamma t}, \tag{18.21}$$

which follows the sign convention often used in quantum optics, i.e., the oscillation of $\mathbf{E}^{(+)}(t)$ is given by $e^{-i\omega_L t}$. The separation (18.21) has the following properties,

$$\mathbf{E}^{(\pm)}(t) = \left[\mathbf{E}^{(\mp)}(t)\right]^\star \quad \Rightarrow \quad \mathbf{E}(t) = 2\mathrm{Re}\left[\mathbf{E}^{(\pm)}(t)\right]. \tag{18.22}$$

In Eq. (18.21), we model the situation where $\mathbf{E}^{(\pm)}(t)$ is switched off exponentially on a time scale γ^{-1}. The factor $\theta(t)$ describes an abrupt switch-on of the excitation while the exponential decay factor γ determines the duration of the light pulse. Even though the temporal shape of this pulse is somewhat artificial, it is well suited to illustrate the principal properties of optical transitions because it allows for analytic expressions.

To solve Eq. (18.18), we separate the free evolution of a_ν using the ansatz

$$a_\nu = \tilde{a}_\nu e^{-i\omega_\nu t}, \qquad \text{with } \hbar\omega_\nu = E_\nu. \tag{18.23}$$

Inserting Eqs. (18.21) and (18.23) into Eq. (18.18), we obtain

$$i\hbar\frac{\partial}{\partial t}\tilde{a}_\nu = -\sum\nolimits_\lambda \mathbf{d}_{\nu,\lambda}\cdot\mathbf{E}_0\,\tilde{a}_\lambda\,\theta(t)\left[e^{-i\Delta_{\nu,\lambda}t-\gamma t} + e^{+i\Delta_{\lambda,\nu}t-\gamma t}\right] \tag{18.24}$$

after having identified the detuning using (18.19). We abbreviate the terms on the RHS of (18.24) as

$$\begin{cases} \dfrac{\partial}{\partial t}K^{(+)}_{\nu,\lambda}(t) = -\frac{\mathbf{d}_{\nu,\lambda}\cdot\mathbf{E}_0}{i\hbar}\,\tilde{a}_\lambda\,\theta(t)\,e^{-i\Delta_{\nu,\lambda}t-\gamma t} \\ \dfrac{\partial}{\partial t}K^{(-)}_{\nu,\lambda}(t) = -\frac{\mathbf{d}_{\nu,\lambda}\cdot\mathbf{E}_0}{i\hbar}\,\tilde{a}_\lambda\,\theta(t)\,e^{i\Delta_{\lambda,\nu}t-\gamma t} \end{cases}, \tag{18.25}$$

where the sign convention is used as in Eq. (18.21). Note that besides the different sign in front of the detuning, also the order of the detuning's state indices is exchanged between $K^{(+)}_{\nu,\lambda}(t)$ and $K^{(-)}_{\nu,\lambda}(t)$.

With the help of (18.25), we can rewrite Eq. (18.24) as

$$i\hbar\frac{\partial}{\partial t}\tilde{a}_\nu = \sum\nolimits_\lambda i\hbar\frac{\partial}{\partial t}\left[K^{(+)}_{\nu,\lambda}(t) + K^{(-)}_{\nu,\lambda}(t)\right]. \tag{18.26}$$

The formal solution of this equation is

$$\tilde{a}_\nu(t) = \tilde{a}_\nu(0) + \sum\nolimits_\lambda \left(K^{(+)}_{\nu,\lambda}(t) + K^{(-)}_{\nu,\lambda}(t)\right), \tag{18.27}$$

where $K^{(\pm)}_{\nu,\lambda}(t)$ vanishes for $t < 0$. The explicit expression for $K^{(\pm)}_{\nu,\lambda}(t)$ is obtained through the integration of (18.25),

$$\begin{cases} K^{(+)}_{\nu,\lambda}(t) = -\frac{\mathbf{d}_{\nu,\lambda}\cdot\mathbf{E}_0}{i\hbar} \int_0^t du\, \tilde{a}_\lambda(u)\, e^{(-i\Delta_{\nu,\lambda}-\gamma)u} \\ K^{(-)}_{\nu,\lambda}(t) = -\frac{\mathbf{d}_{\nu,\lambda}\cdot\mathbf{E}_0}{i\hbar} \int_0^t du\, \tilde{a}_\lambda(u)\, e^{(+i\Delta_{\lambda,\nu}-\gamma)u} \end{cases} . \tag{18.28}$$

The remaining integrand contains the slowly varying $\tilde{a}_\lambda(u)$ and the detuning term. According to the condition (18.21), this term is rapidly oscillating for large detunings such that both $K^{(-)}_{\nu,\lambda}(t)$ and $K^{(+)}_{\nu,\lambda}(t)$ have appreciable values only for very few (ν, λ) combinations.

18.1.3 Two-level approximation

To determine which (ν, λ) combinations are relevant, we evaluate the RHS of (18.28) using the mean-value theorem of integration. In the present case, this means that we take $\tilde{a}_\lambda(u)$ outside the integral with the time argument $u \to t^*$ where $0 \le t^* \le t$. Thus, we may estimate the relative strength of the different $K^{(\pm)}_{\nu,\lambda}(t)$ using

$$\begin{cases} K^{(+)}_{\nu,\lambda}(t) = \mathbf{d}_{\nu,\lambda} \cdot \mathbf{E}_0\, \tilde{a}_\lambda(t^*)\, g(-\Delta_{\nu,\lambda}, t) \\ K^{(-)}_{\nu,\lambda}(t) = \mathbf{d}_{\nu,\lambda} \cdot \mathbf{E}_0\, \tilde{a}_\lambda(t^*)\, g(+\Delta_{\lambda,\nu}, t) \end{cases} , \tag{18.29}$$

where

$$g(\omega, t) \equiv \frac{i}{\hbar} \int_0^t du\, e^{(+i\omega-\gamma)u} = \frac{e^{+i\omega t-\gamma t} - 1}{\hbar\omega + i\hbar\gamma}. \tag{18.30}$$

Applying the near-resonance condition (18.21), we find

$$\begin{aligned} &|g(-\Delta_{2p,\,1s}, t)| \approx |g(+\Delta_{2p,\,1s}, t)|, \\ &|g(\pm\Delta_{\nu,\lambda}, t)| \ll |g(\pm\Delta_{2p,\,1s}, t)|, \quad \text{for } \nu \neq 2p,\ \lambda \neq 1s, \end{aligned} \tag{18.31}$$

which is valid for sufficiently small values of γ. The inequalities (18.31) allow us to neglect the nonresonant $K^{(\pm)}_{\nu,\lambda}(t)$ contributions in Eq. (18.26). This way, we introduce the *two-level approximation* (TLA)

$$K^{(+)}_{\nu,\lambda}(t) = \delta_{\nu,2p}\delta_{\lambda,1s} K^{(+)}_{2p,1s}(t), \qquad K^{(-)}_{\nu,\lambda}(t) = \delta_{\nu,1s}\delta_{\lambda,2p} K^{(-)}_{1s,2p}(t), \tag{18.32}$$

which only includes the dominant, near-resonant contributions. Implementing the TLA into Eq. (18.26), we find

$$\begin{aligned} i\hbar\frac{\partial}{\partial t}\tilde{a}_{np} &= \sum\nolimits_\lambda i\hbar\frac{\partial}{\partial t}\left[K^{(+)}_{np,\lambda}(t) + K^{(-)}_{np,\lambda}(t)\right]_{2\text{LS}} \\ &= \delta_{n,2}\, i\hbar\frac{\partial}{\partial t}K^{(+)}_{2p,1s}(t) = -\mathbf{d}_{2p,1s} \cdot \mathbf{E}^{(+)}\, a_{1s}\, e^{i\omega_{2p}t}\, \delta_{n,2} \\ i\hbar\frac{\partial}{\partial t}\tilde{a}_{ns} &= \sum\nolimits_\lambda i\hbar\frac{\partial}{\partial t}\left[K^{(+)}_{1s,\lambda}(t) + K^{(-)}_{1s,\lambda}(t)\right]_{2\text{LS}} \\ &= \delta_{n,1}\, i\hbar\frac{\partial}{\partial t}K^{(-)}_{1s,2p}(t) = -\mathbf{d}_{1s,2p} \cdot \mathbf{E}^{(-)}\, a_{2p}\, e^{i\omega_{1s}t}\, \delta_{n,1}. \end{aligned} \tag{18.33}$$

We see that only two a_ν states are altered by the light whereas all the other states evolve without perturbation.

As it turns out, one can frequently find experimental conditions in atomic systems where the optical excitation can be described by the two-level model,

$$\begin{cases} i\hbar \frac{\partial}{\partial t} a_{2p} = E_{2p}\, a_{2p} - \mathbf{d}_{2p,1s} \cdot \mathbf{E}^{(+)}(t)\, a_{1s} \\ i\hbar \frac{\partial}{\partial t} a_{1s} = E_{1s}\, a_{1s} - \mathbf{d}_{1s,2p} \cdot \mathbf{E}^{(-)}(t)\, a_{2p} \end{cases}, \tag{18.34}$$

which is expressed here in the original frame using the relation (18.23) in the reversed direction. At this level of approximation, the full quantum problem (18.18) is reduced into two linearly coupled differential equations. These equations are fully applicable for arbitrary shapes of $\mathbf{E}^{(\pm)}(t)$ as long as the field is nearly resonant only with the $1s$-to-$2p$ transition. The same analysis can also be repeated to describe dipole-allowed transition between any two states ν_1 and ν_2. If the corresponding dipole-matrix element is nonvanishing and the field is nearly resonant only with the analyzed transition, we can directly apply (18.34) after we implement the following notational changes: $2 \leftrightarrow 2p$ denotes the excited state and $1 \leftrightarrow 1s$ defines the ground state of the optical excitation.

Using this notation, we can express Eq. (18.34) as a 2×2 matrix problem,

$$i\hbar \frac{\partial}{\partial t} \begin{pmatrix} a_2 \\ a_1 \end{pmatrix} = \begin{pmatrix} E_2 & -\mathbf{d}_{2,1} \cdot \mathbf{E}^{(-)}(t) \\ -\mathbf{d}_{1,2} \cdot \mathbf{E}^{(+)}(t) & E_1 \end{pmatrix} \begin{pmatrix} a_2 \\ a_1 \end{pmatrix}. \tag{18.35}$$

This has exactly the form of the Schrödinger equation if we identify a state vector $\mathbf{a}$ and an interaction matrix $\mathbf{H}_{2\mathrm{LS}}$ via

$$\mathbf{a} \equiv \begin{pmatrix} a_2 \\ a_1 \end{pmatrix}, \qquad \mathbf{H}_{2\mathrm{LS}}(t) \equiv \begin{pmatrix} E_2 & -\mathbf{d}_{2,1} \cdot \mathbf{E}^{(+)}(t) \\ -\mathbf{d}_{1,2} \cdot \mathbf{E}^{(-)}(t) & E_1 \end{pmatrix}$$

$$i\hbar \frac{\partial}{\partial t} \mathbf{a} = \mathbf{H}_{2\mathrm{LS}}(t)\, \mathbf{a}. \tag{18.36}$$

Clearly, this simple equation allows for a straightforward numerical solution even if the driving field has an arbitrary, but nearly resonant form. The actual solutions of Eq. (18.36) are discussed in Section 18.2. A typical two-level system (TLS) analysis is schematically shown in Fig. 18.3. The lower (upper) state is labeled by $|1\rangle$ ($|2\rangle$) and the energy of the optical excitation is indicated by the thick arrow. Since we did not quantize the light field in the present discussion, the current notation should not be confused with the Fock states.

In many relevant situations, one can utilize two or more light pulses to excite the system. Each pulse can have its individual frequency $\omega_{L,j}$ such that a suitable $(\omega_{L,1}, \omega_{L,2}, \cdots)$ combination increases the number of states with which the optical excitation can interact under near-resonant conditions. Among other things, a sequence of excitations can move the system between states that are dipole-forbidden for a single-pulse excitation. For example, one can connect the $1s$ to the $2p$ state with one pulse and the $2p$ to the $3d$ state with another one, resulting in an effective coupling between the $1s$ and the $3d$ states.

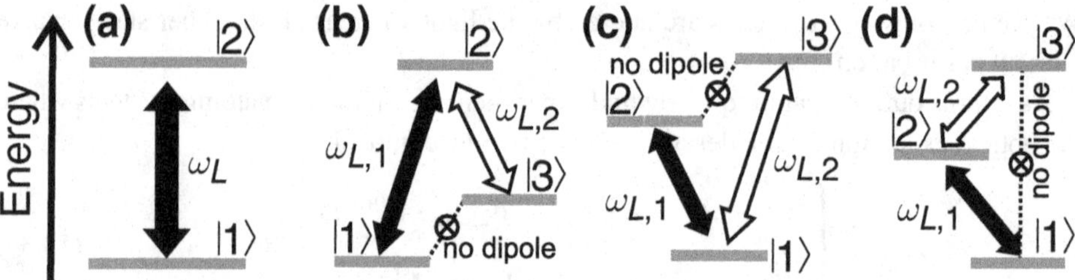

Figure 18.3 Few-level models representing near-resonant excitation conditions. **(a)** The two-level model assumes excitation by a single source with central frequency ω_L. Two-pulse excitation is characterized by two frequencies $\omega_{L,1}$ and $\omega_{L,2}$. The near-resonance conditions can yield **(b)** Λ-, **(c)** V-, and **(d)** ladder-like excitations. The arrows indicate the dipole-allowed transitions. They are labeled by the corresponding $\omega_{L,j}$. The dashed lines indicate dipole-forbidden transitions.

Figure 18.3(b)–(d) presents a selection of typical two-pulse excitation schemes yielding coupling among three states. Due to their schematic geometric form, these excitations are often referred to as (b) "Λ", (c) "V", and (d) ladder excitations.

18.1.4 Rotating-wave approximation (RWA)

The TLA (18.32) not only reduces the full system dynamics (18.18) to just two levels but it also includes only a part of the full electrical field in (18.36). To isolate the physical meaning of this scheme, we apply the TLA to Eq. (18.26) and analyze the $\tilde{a}_2$ evolution,

$$i\hbar\frac{\partial}{\partial t}\tilde{a}_2^{2\text{LS}} = i\hbar\frac{\partial}{\partial t}\left[K_{2,1}^{(+)}(t) + K_{2,1}^{(-)}(t)\right]_{2\text{LS}} = i\hbar\,\frac{\partial}{\partial t}\left[K_{2,1}^{(+)}(t) + \cancel{K_{2,1}^{(-)}(t)}\right]$$
$$= -\mathbf{d}_{2,1}\cdot\mathbf{E}^{(+)}(t)\,e^{i(\omega_2-\omega_1)t}\,\tilde{a}_1. \tag{18.37}$$

Even though the $K_{2,1}^{(+)}(t)$ contribution does not play any role within the TLA, we can keep it to identify the full electric field in terms of

$$i\hbar\,\frac{\partial}{\partial t}\left[K_{2,1}^{(+)}(t) + K_{2,1}^{(-)}(t)\right] = -\mathbf{d}_{2,1}\cdot\tilde{\mathbf{E}}(t)\,\tilde{a}_1. \tag{18.38}$$

This follows after we apply Eqs. (18.19), (18.21), and (18.25) to identify

$$\tilde{\mathbf{E}}(t) \equiv \mathbf{E}(t)\,e^{i\omega_{21}t} \quad \Leftrightarrow \quad \mathbf{E}(t) \equiv \tilde{\mathbf{E}}(t)\,e^{-i\omega_{21}t},$$
$$\omega_{21} \equiv \omega_2 - \omega_1. \tag{18.39}$$

Thus, the full electrical field appears in the a_2 dynamics only if we go beyond the TLA.

We see that $\tilde{\mathbf{E}}(t)$ is the total electric field represented in the rotating frame of the $\tilde{a}_2$-to-$\tilde{a}_1$ transition, described by its frequency ω_{21}. Physically, this corresponds to a representation of the dynamics in a rotating frame where all matter coefficients $\tilde{a}_\nu$ are stationary without the light–matter interaction. Since this yields the separation of the pure matter dynamics from the interaction-induced dynamics, the ansatz (18.23) actually implements a unitary transformation into the interaction picture. Within this representation, the total electric field becomes

$$\begin{aligned}\tilde{\mathbf{E}}(t) &= \mathbf{E}_0\theta(t)\, e^{+i\omega_{21}t} e^{-\gamma t}\left[e^{-i\omega_L t} + e^{+i\omega_L t}\right] \\ &= \mathbf{E}_0\theta(t)\, e^{-i\Delta_{2,1}t} e^{-\gamma t}\left[1 + e^{+2i\,\omega_L t}\right],\end{aligned} \tag{18.40}$$

where the detuning is identified via (18.19) and (18.39). We see that the rotating-frame $\tilde{\mathbf{E}}(t)$ consists of a nearly constant part, $\propto e^{-i\Delta_{2,1}t}$ and a fast-oscillating part $\propto e^{-i\Delta_{2,1}t}\, e^{+2i\,\omega_L t}$, which is eliminated by the TLA where $|\Delta_{2,1}|$ is assumed to be much smaller than ω_L.

To investigate the dynamics of $\tilde{a}_2$, we have to solve

$$i\hbar\frac{\partial}{\partial t}\tilde{a}_2^{2\text{LS}} = -\left[\mathbf{d}_{2,1}\cdot\tilde{\mathbf{E}}(t)\,\tilde{a}_1\right]_{2\text{LS}} = -\left[\mathbf{d}_{2,1}\cdot\tilde{\mathbf{E}}(t)\,\tilde{a}_1\right]_{\text{RWA}}, \tag{18.41}$$

where the fast-oscillating contribution $e^{+2i\,\omega_L t}$ to $\tilde{\mathbf{E}}(t)$ is eliminated via the TLA. Thus, the TLA can also be viewed as the rotating wave approximation (RWA) that eliminates all rapidly oscillating contributions from the sources in the interaction picture.

To see that the RWA and the TLA produce the same result, we formally integrate (18.41) to get

$$\begin{aligned}\tilde{a}_2^{\text{RWA}}(t) &\equiv \tilde{a}_2(0) - \frac{1}{i\hbar}\left[\int_0^t du\,\mathbf{d}_{2,1}\cdot\tilde{\mathbf{E}}(u)\,\tilde{a}_1(u)\right]_{\text{RWA}} \\ &= \tilde{a}_2(0) - \frac{\tilde{a}_1(t^\star)}{i\hbar}\,\mathbf{d}_{2,1}\cdot\left[\int_0^t du\,\tilde{\mathbf{E}}(u)\right]_{\text{RWA}}.\end{aligned} \tag{18.42}$$

Here, only the slowly varying part of $\tilde{\mathbf{E}}(u)$ produces an appreciable contribution to the integral because the fast-oscillating part $\propto e^{+2i\omega_L t}$ averages out to a large degree. This justifies the RWA, i.e., the inclusion of only the slowly varying part $\mathbf{E}_0 e^{-i\Delta_{2,1}t}e^{-\gamma t}$ of the electric field. Thus, the RWA produces

$$\begin{aligned}\tilde{a}_2^{\text{RWA}}(t) &= \tilde{a}_2(0) - \frac{\mathbf{d}_{2,1}\cdot\mathbf{E}_0}{i\hbar}\int_0^t du\; e^{-i\Delta_{2,1}u}e^{-\gamma u}\,\tilde{a}_1(u) \\ &= \tilde{a}_2(0) + K_{2,1}^{(+)}(t) = \tilde{a}_2^{2\text{LS}}(t),\end{aligned} \tag{18.43}$$

based on Eq. (18.40) and the definition (18.25). The final step identifies $a_2^{2\text{LS}}$, which is obtained by inserting the TLA (18.32) into Eq. (18.27). Equation (18.43) shows that the TLA indeed implements the RWA.

To get a feeling for the accuracy of the RWA under different conditions, we compare it to the full electric field,

$$\tilde{E}_{\text{RWA}}(t) \equiv \theta(t)\,E_0\,e^{-i\Delta_{2,1}t}e^{-\gamma t}, \quad \tilde{E}_{\text{tot}}(t) \equiv \tilde{E}_{\text{RWA}}(t)(1 + e^{2i\omega_L t}). \tag{18.44}$$

Here, all fields are given in the rotating frame identified in Eq. (18.40). To define the effect of RWA and non-RWA contributions on atomic excitations, we compute the corresponding integrals

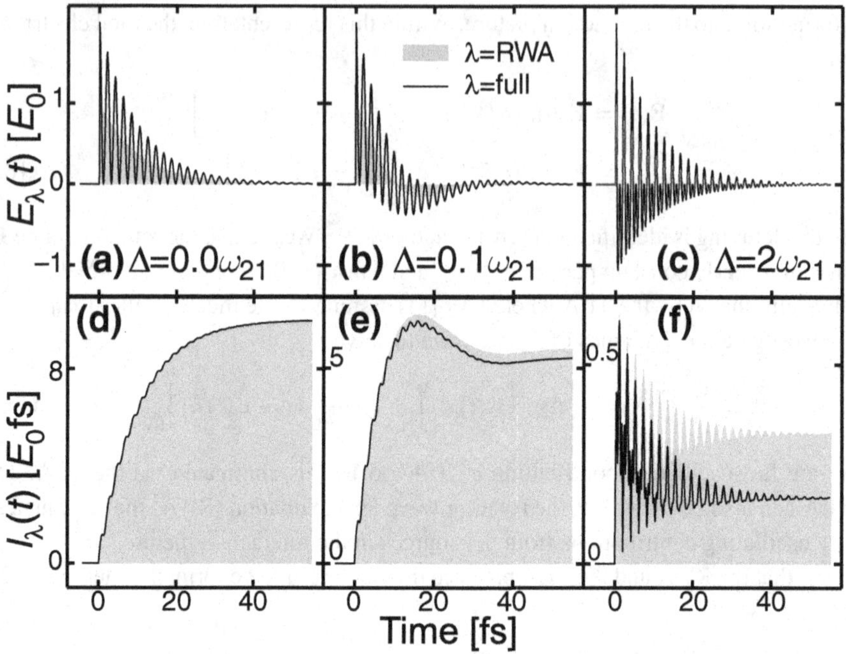

Figure 18.4 Effect of contributions beyond the rotating wave approximation (RWA). The full electric field is shown as solid line and the RWA is presented as the shaded area for the detuning: **(a)** $\Delta_{21} = 0$, **(b)** $\Delta_{21} = 0.1\omega_{21}$, and **(c)** $\Delta_{21} = 2\omega_{21}$ where ω_{21} is the transition frequency. The corresponding excitation integrals (18.45) are shown in frames **(d)–(f)** for the full (solid line) and the RWA (shaded area) solution.

$$I_{\text{RWA}}(t) \equiv \int_0^t du\, \tilde{\mathbf{E}}_{\text{RWA}}(u), \qquad I_{\text{tot}}(t) \equiv \int_0^t du\, \tilde{\mathbf{E}}_{\text{tot}}(u), \tag{18.45}$$

defining the actual $\bar{a}_2$ dynamics via Eq. (18.42). Figure 18.4 presents the RWA (shaded area) and the full electric field (solid line) in the upper row and the corresponding integrals in the lower row. Here, we assumed the transition energy $\hbar\omega_{21}$=1 eV for the $\tilde{a}_1$-to-$\tilde{a}_2$ processes. The corresponding field frequency can be expressed in terms of the detuning relation (18.19), yielding $\omega_L = \omega_{21} + \Delta_{2,1}$.

To determine the effect of the non-RWA contributions, Fig. 18.4 presents results for: (a) and (d) resonant excitation $\Delta_{2,1} = 0$, (b) and (e) nearly resonant excitation $\Delta_{2,1} = 0.1\omega_{21}$, as well as (c) and (f) strongly nonresonant excitation $\Delta_{2,1} = 0.5\omega_{21}$. As a general trend, $\tilde{E}_{\text{RWA}}(t)$ is a slowly varying quantity whereas the non-RWA contributions produce fast oscillations. For the chosen detunings, the non-RWA contributions oscillate with the period of (a) 2.07 fs, (b) 1.88 fs, and (c) 0.69 fs and have $\gamma = 100\frac{1}{\text{ps}}$. If we integrate the electric fields, only the RWA part yields a significant contribution whereas the non-RWA parts nearly average out due to their fast oscillations. Thus, the I_{tot} is dominated by I_{RWA} even if the detuning is very large. In the investigated case, one needs a detuning of at least $\Delta_{2,1} = 2\omega_{21}$ in order to produce deviations larger than 50 percent.

As a typical trend, the non-RWA contributions introduce rapid oscillations around the RWA result. The amplitude of these oscillations as well as the deviation from the RWA result decreases drastically as the detuning becomes smaller, which makes the RWA very accurate for near-resonant excitations. If we look for numerical solutions of Eq. (18.18), the RWA considerably reduces the numerical efforts because it limits the dynamics to few levels and eliminates the negligible but fast oscillations. This means that near-resonant excitations can be treated by including only few states and relatively long time steps. If we want to or need to include the non-RWA contributions, we often have to incorporate many non-resonant levels and apply extremely short time steps to resolve the rapidly oscillating terms. It is clearly beneficial to avoid these numerical issues when near-resonant excitations are studied by implementing the RWA and/or the TLA whenever they are physically justified.

18.2 Two-level system solutions

Mathematically, the two-level equation (18.36) is an ordinary differential equation for a 2×1 vector. Thus, we can always solve it numerically with relative ease. Nevertheless, Eq. (18.36) is not yet in the best possible form as far as the numerics is concerned. In particular, $a_j(t)$ rapidly oscillates $\propto e^{-i\omega_j t}$ and one must, in principle, resolve this dynamics using a very fine time discretization. To avoid this numerical complication, it is much better to evaluate Eq. (18.36) in the rotating frame. The most symmetric TLS solution is obtained by introducing a new rotating frame

$$\begin{cases} a_1(t) = \bar{a}_1(t)\, e^{-i\omega_1 t}\, e^{+\frac{i}{2}\Delta_{21}t} \\ a_2(t) = \bar{a}_2(t)\, e^{-i\omega_2 t}\, e^{-\frac{i}{2}\Delta_{21}t} \end{cases}, \qquad \Delta_{21} \equiv \Delta_{2,1}, \tag{18.46}$$

which is only a slight modification to the interaction picture introduced by Eq. (18.23). This transformation converts Eq. (18.34) into

$$\begin{cases} i\hbar\dfrac{\partial}{\partial t}\bar{a}_2 = -\dfrac{\hbar\Delta_{21}}{2}\bar{a}_2 - \mathbf{d}_{2,1}\cdot\mathbf{E}^{(+)}(t)\, e^{i\omega_L t}\,\bar{a}_1 \\ i\hbar\dfrac{\partial}{\partial t}\bar{a}_1 = +\dfrac{\hbar\Delta_{21}}{2}\bar{a}_1 - \mathbf{d}_{1,2}\cdot\mathbf{E}^{(-)}(t)\, e^{-i\omega_L t}\,\bar{a}_2 \end{cases}. \tag{18.47}$$

To simplify the notation further, we introduce the envelope of the Rabi frequency

$$\begin{aligned} \hbar\bar{\Omega}^{(+)}(t) &\equiv 2\mathbf{d}_{2,1}\cdot\mathbf{E}^{(+)}(t)\, e^{i\omega_L t}, \\ \hbar\bar{\Omega}^{(-)}(t) &\equiv 2\mathbf{d}_{1,2}\cdot\mathbf{E}^{(-)}(t)\, e^{-i\omega_L t}. \end{aligned} \tag{18.48}$$

The expressions are connected through complex conjugation

$$\bar{\Omega}^{(+)}(t) \equiv \left[\bar{\Omega}^{(-)}(t)\right]^{\star}. \tag{18.49}$$

The factor 2 in the definition of the Rabi frequency produces a more symmetric matrix form for the $\bar{a}_j$ dynamics. Altogether, we obtain the actual Rabi frequency as

$$\Omega(t) = 2\text{Re}\left[\bar{\Omega}^{(\pm)}(t)e^{-i\omega_L t}\right] = \bar{\Omega}^{(\pm)}(t)e^{-i\omega_L t} + \bar{\Omega}^{(-)}(t)e^{+i\omega_L t}$$
$$\equiv \Omega^{(+)}(t) + \Omega^{(-)}(t), \tag{18.50}$$

where the sign convention is chosen as in Eq. (18.21).

With the help of the identification (18.49), Eq. (18.47) can be expressed compactly via

$$i\frac{\partial}{\partial t}\bar{a} = \mathbf{M}(t)\,\bar{a},$$
$$\bar{a} \equiv \begin{pmatrix} \bar{a}_2 \\ \bar{a}_1 \end{pmatrix}, \quad \mathbf{M}(t) \equiv \frac{1}{2}\begin{pmatrix} -\Delta_{21} & -\bar{\Omega}^{(+)}(t) \\ -\bar{\Omega}^{(-)}(t) & +\Delta_{21} \end{pmatrix}. \tag{18.51}$$

The $\mathbf{M}(t) = \mathbf{M}^H(t)$ matrix that appears is Hermitian. Since we investigate near-resonance excitation conditions, the detuning Δ_{21} is small. The resulting equation structure is clearly more symmetric than Eq. (18.34) and the fast oscillations of $\mathbf{E}^{\pm}(t)$ are eliminated through the multiplications with $e^{\pm i\omega_L t}$ in the definition (18.49). As a result, Eq. (18.51) provides a numerically feasible approach where all quantities vary slowly for the near-resonant excitations studied here.

18.2.1 Analytic solution of the two-level system

To determine the principal features of the TLS excitations, we first analyze the formal aspects of Eq. (18.51). Since Eq. (18.51) is a first-order differential equation for a 1×2 vector, we can always write its formal solution as

$$\bar{\mathbf{a}}(t) = \hat{T}\left[e^{-i\int_0^t du\,\mathbf{M}(u)}\right]\bar{\mathbf{a}}(0), \tag{18.52}$$

where $\hat{T}$ introduces a time order when we evaluate the exponential function. For a general time dependence of $\mathbf{M}(t)$, Eq. (18.52) can be solved only numerically because $\mathbf{M}(t_1)$ and $\mathbf{M}(t_2)$ do not necessarily commute for $t_1 \neq t_2$.

Thus, we consider a simplified excitation where $\bar{\Omega}^{(\pm)}(t)$ is constant over the time interval $0 \leq t \leq T$ and vanishes for all other times. In this situation, the time ordering in Eq. (18.52) can be ignored because $[\mathbf{M}(t_1), \mathbf{M}(t_2)]_- = 0$ for all $0 \leq t_j \leq T$. Under these conditions, we find

$$\bar{a}(t) = e^{-i\mathbf{M}(0)\,t}\,\bar{a}(0), \qquad 0 \leq t \leq T. \tag{18.53}$$

The exponent can be evaluated after expressing the Hermitian matrix $\mathbf{M}(t)$ via the general decomposition

$$\mathbf{M}(t) \equiv \mathbf{T}\,\Lambda\,\mathbf{T}^{-1}$$
$$\mathbf{T} \equiv \begin{pmatrix} \cos\theta & -e^{i\varphi}\sin\theta \\ e^{-i\varphi}\sin\theta & \cos\theta \end{pmatrix}, \quad \Lambda \equiv \frac{1}{2}\begin{pmatrix} +\Upsilon & 0 \\ 0 & -\Upsilon \end{pmatrix}. \tag{18.54}$$

The similarity transformation $\mathbf{T}$ that appears has been chosen to satisfy $\mathbf{T}^{-1} = \mathbf{T}^H$, which is possible because $\mathbf{M}$ is Hermitian. A straightforward matrix-eigenvalue problem identifies the constants that appear as

$$\Upsilon = \sqrt{\Delta^2 + |\bar{\Omega}^{(+)}|^2}, \qquad \Omega^{(+)} = \Upsilon e^{i\varphi} \sin 2\theta, \qquad \Delta_{21} = \Upsilon \cos 2\theta, \tag{18.55}$$

where Υ defines the generalized Rabi frequency. Since $\Omega^{(\pm)}$ and Δ are constant under the investigated conditions, Υ, θ, and φ remain constant too.

With the help of the transformation (18.54), the solution (18.53) casts into the form

$$\bar{a}(t) = e^{-i\mathbf{M}t}\,\bar{a}(0) = \mathbf{T}\,e^{-i\Lambda t}\,\mathbf{T}^H\,\bar{a}(0) = \mathbf{T}\begin{pmatrix} e^{\frac{i}{2}\Upsilon t} & 0 \\ 0 & e^{-\frac{i}{2}\Upsilon t} \end{pmatrix}\mathbf{T}^H\,\bar{a}(0)$$
$$= \begin{pmatrix} \cos\frac{\Upsilon t}{2} + i\frac{\Delta_{21}}{\Upsilon}\sin\frac{\Upsilon t}{2} & i\frac{\Omega^{(+)}}{\Upsilon}\sin\frac{\Upsilon t}{2} \\ i\frac{\Omega^{(-)}}{\Upsilon}\sin\frac{\Upsilon t}{2} & \cos\frac{\Upsilon t}{2} - i\frac{\Delta_{21}}{\Upsilon}\sin\frac{\Upsilon t}{2} \end{pmatrix}\bar{a}(0) \tag{18.56}$$

for $0 \le t \le T$. If the TLS is initially unexcited, we simply set $\bar{a}_2(0)$ to zero while $\bar{a}_1(0) = 1$. This initial condition produces the excitation dynamics

$$\bar{a}_2(t) = \begin{cases} 0, & t < 0 \\ i\frac{\Omega^{(+)}}{\Upsilon}\sin\frac{\Upsilon t}{2}, & 0 \le t \le T \\ \bar{a}_2(T)\,e^{+i\frac{\Delta_{21}(t-T)}{2}}, & t > T \end{cases}$$
$$\bar{a}_1(t) = \begin{cases} 1, & t < 0 \\ \cos\frac{\Upsilon t}{2} - i\frac{\Delta_{21}}{\Upsilon}\sin\frac{\Upsilon t}{2}, & 0 \le t \le T \\ \bar{a}_1(T)\,e^{-i\frac{\Delta_{21}(t-T)}{2}}, & t > T \end{cases}. \tag{18.57}$$

For times $t > T$, the amplitude of $\bar{a}_j(t)$ has the constant value $\bar{a}_j(T)$ while the phases of the coefficients follow from $e^{\pm i\frac{\Delta_{21}}{2}(t-T)}$ describing the free evolution of the atomic states after the time T.

18.2.2 Bloch-vector representation

The result (18.57) presents the system dynamics through the coefficients of the two states involved. However, one often wants to evaluate relevant physical quantities such as the occupation of the excited state or the magnitude of the dipole created by the field. For this purpose, we first construct the wave function starting from its generic form Eq. (18.6). Based on the factorization (18.17) and the TLA (18.32), the optical excitations follow from

$$|\psi(t)\rangle = (a_1\,|1\rangle + a_2\,|2\rangle) \otimes \sum\nolimits_{\mathbf{K}} C_{\mathbf{K}}|\mathbf{K}\rangle$$
$$= \left(\bar{a}_1\,e^{-i\left(\omega_1 - \frac{\Delta_{21}}{2}\right)t}\,|1\rangle + \bar{a}_2\,e^{-i\left(\omega_2 + \frac{\Delta_{21}}{2}\right)t}\,|2\rangle\right) \otimes \sum_{\mathbf{K}} C_{\mathbf{K}}|\mathbf{K}\rangle. \tag{18.58}$$

This directly defines the probabilities

$$f_2(t) \equiv |\bar{a}_2(t)|^2, \qquad f_1(t) \equiv |\bar{a}_1(t)|^2 \tag{18.59}$$

of finding the system in the excited and the ground state, respectively. In many cases, it is also interesting to know the inversion factor

$$P_z(t) \equiv \frac{1}{2}\left(|\bar{a}_2(t)|^2 - |\bar{a}_1(t)|^2\right), \tag{18.60}$$

which can assume values between $-1/2$ and $+1/2$. In the ground state, $P_z = -1/2$ while the other extreme, $P_z = 1/2$, corresponds to a completely inverted, i.e., maximally excited system.

The knowledge of $\bar{a}_2(t)$ and $\bar{a}_1(t)$ allows us to compute the dipole of the system, as discussed in connection with Eq. (16.41). We obtain

$$\begin{aligned} \langle Q_e\hat{\mathbf{r}}\rangle &= 2\,\mathrm{Re}\left[\langle 1|Q_e\hat{\mathbf{r}}|2\rangle\,\bar{a}_1^\star\bar{a}_2\,e^{-i(\omega_{21}+\Delta)t}\right] \equiv 2\,\mathrm{Re}\left[\mathbf{d}_{1,2}\,\bar{P}\,e^{-i\omega_L t}\right] \\ \bar{P} &\equiv \bar{a}_1^\star\bar{a}_2, \end{aligned} \tag{18.61}$$

where $\bar{P}$ identifies the polarization between the ground state and the excited state. The dipole-matrix element is identified through Eq. (18.9). Since $\bar{P}$ is defined in the rotating frame, the actual $P = \bar{P}\,e^{-i\omega_L t}$ oscillates with the frequency comparable to that of the near-resonant optical excitation.

When the system is described by the wave function (18.58), the inversion and polarization are connected through

$$\begin{aligned} P_z^2 + |P|^2 &= \frac{1}{4}\left(|\bar{a}_2(t)|^2 - |\bar{a}_1(t)|^2\right)^2 + |\bar{a}_1|^2|\bar{a}_2|^2 \\ &= \frac{1}{4}\left(|\bar{a}_2(t)|^2 + |\bar{a}_1(t)|^2\right)^2 = \frac{1}{4} \end{aligned} \tag{18.62}$$

because the wave-function coefficients are normalized to unity. The relation (18.62) implies that the amplitude of the inversion and the polarization rotate on a circle whose radius is $1/2$. This is true also for a general excitation as long as the relative and center-of-mass motion remain separated and the relative motion is a pure state, given by Eq. (18.58).

Since we may express $|P|^2$ through orthogonal real and imaginary parts, we may also identify these orthogonal contributions as vectorial x and y components of the polarization,

$$\left\{\begin{array}{l} \bar{P}_x \equiv \mathrm{Re}\left[\bar{P}\right] \\ \bar{P}_y \equiv \mathrm{Im}\left[\bar{P}\right] \end{array}\right. \quad \Rightarrow \quad \left\{\begin{array}{l} \bar{P} \equiv \bar{P}_x + i\bar{P}_y \\ \bar{\mathbf{P}} \equiv (\bar{P}_x,\ \bar{P}_y,\ \bar{P}_z) \end{array}\right., \tag{18.63}$$

while the z component is identified through the inversion (18.60). Together, these components identify the so-called Bloch vector $\bar{\mathbf{P}}$. In particular, the conservation law (18.62) can be expressed through

$$P_x^2 + P_y^2 + P_z^2 = \frac{1}{4}. \tag{18.64}$$

Hence, the Bloch vector stays within a sphere of radius $1/2$, which is often referred to as the *Bloch sphere*.

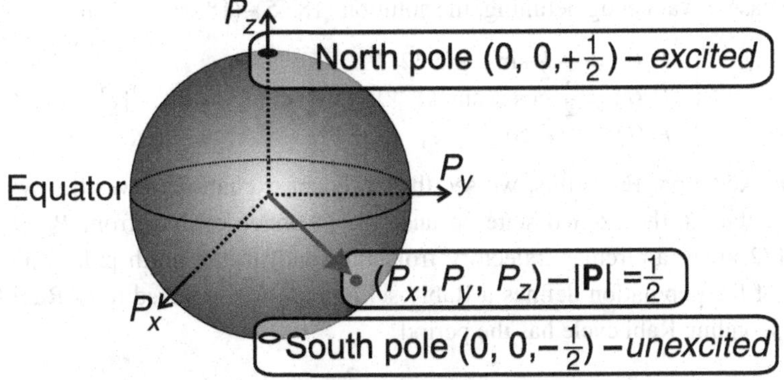

Figure 18.5 Special points of the Bloch sphere. The filled circles indicate the north and the south pole. The equator is given by the horizontal circle. A possible Bloch vector is indicated via the arrow.

We can identify many qualitative properties of the atomic two-level excitations based on the motion of $\bar{\mathbf{P}}$ on the Bloch sphere. Since the Bloch sphere resembles the planet Earth, it is customary to identify special points with a geographical terminology. For example, when the Bloch vector is at the position $\bar{\mathbf{P}} = (0,\ 0,\ -1/2)$, we say that it is at the "south pole," as shown in Fig. 18.5. In this situation, the polarization components of $\bar{\mathbf{P}}$ vanish and $\bar{P}_z = -1/2$ refers to a system in its ground state. The opposite limit, $\bar{\mathbf{P}} = (0,\ 0,\ +1/2)$, characterizes the "north pole" describing a completely excited system. The equator $\bar{\mathbf{P}} = (\bar{P}_x,\ \bar{P}_y,\ 0)$ is a horizontal circle with the radius 1/2. This situation produces the maximal magnitude for the polarization while the inversion vanishes.

18.2.3 Rabi oscillations

To construct the trajectory of the Bloch vector under different conditions, we use the solution (18.57) to determine $\bar{\mathbf{P}}(t)$ for constant $\bar{\Omega}^{(+)}$. A straightforward substitution into (18.60) and (18.61) produces

$$\bar{P}_z(t) = -\frac{1}{2\Upsilon^2}\left(\Delta^2 + |\bar{\Omega}^{(+)}|^2 \sin \Upsilon t\right), \qquad \Upsilon = \sqrt{|\bar{\Omega}^{(+)}|^2 + \Delta_{21}^2},$$
$$\bar{P}(t) = \frac{\bar{\Omega}^{(+)}}{2\Upsilon}\left[\frac{\Delta_{21}}{\Upsilon}(\cos \Upsilon t - 1) + i \sin \Upsilon t\right], \tag{18.65}$$

during the time of the pulse. The full evolution is then given by

$$\left[\bar{P}(t),\ \bar{P}_z(t)\right] = \begin{cases} \left[0,\ -\frac{1}{2}\right], & t < 0 \\ \left[\bar{P}(t),\ \bar{P}_z(t)\right], & 0 \le t \le T \\ \left[\bar{P}(T)e^{i\Delta_{21}(t-T)},\ \bar{P}_z(T)\right], & t > T \end{cases}. \tag{18.66}$$

The x and y components of $\bar{\mathbf{P}}$ directly follow from the real and imaginary parts of $\bar{P}(t)$, as given by Eq. (18.63).

For the case of vanishing detuning, the solution (18.65)–(18.66) leads to

$$\begin{cases} \bar{P}_x(t) = -\frac{1}{2}\sin\varphi\,\sin|\bar{\Omega}^{(+)}|t \\ \bar{P}_y(t) = \frac{1}{2}\cos\varphi\,\sin|\bar{\Omega}^{(+)}|t \\ \bar{P}_z(t) = -\frac{1}{2}\cos|\bar{\Omega}^{(+)}|t \end{cases}, \quad \bar{\Omega}^{(+)} = |\bar{\Omega}^{(+)}|\,e^{i\varphi} \tag{18.67}$$

during the excitation. From this, we see that resonant excitation moves the system from the ground state to the excited state because the inversion evolves from $\bar{P}_z = -1/2$ to $\bar{P}_z = +1/2$ along a circular trajectory from the south to the north pole. This periodic evolution of the population defines a Rabi oscillation, often referred to as Rabi flopping. The corresponding Rabi cycle has the period

$$T_{\text{Rabi},0} \equiv \frac{2\pi}{|\bar{\Omega}^{(+)}|} \tag{18.68}$$

for the resonant square-pulse excitation. It is often useful to use $T_{\text{Rabi},0}$ to define the time scale of light-induced processes.

Equation (18.65) shows that Rabi oscillations are also observed for the detuned case. However, we have some clear modifications with respect to the resonant Rabi-flopping behavior. According to Eq. (18.65), detuned Rabi oscillations proceed with the frequency Υ such that one Rabi cycle has the period

$$T_{\text{Rabi},\,\Delta_{21}} \equiv \frac{2\pi}{\Upsilon} = \frac{T_{\text{Rabi},\,0}}{\sqrt{1+\left|\frac{\Delta_{21}}{\bar{\Omega}^{(+)}}\right|^2}}. \tag{18.69}$$

Hence, the Rabi cycle becomes shorter for larger detunings. Since the corresponding oscillation frequency, Υ, grows without bounds, the time scale of the optical excitations can approach the one defined by the transition frequency $\omega_2 - \omega_1$. In this situation, the near-resonance condition (18.21) is clearly no longer satisfied and the TLA breaks down. In these situations, one must analyze the optical excitation dynamics using the more general expansion (18.18).

The effect of the detuning is also seen directly as a decrease in the excitation level. According to Eq. (18.65), the detuned atomic system reaches the maximum occupation of

$$f_2^{\max} \equiv \left(R_z + \frac{1}{2}\right)_{\max} = \frac{|\bar{\Omega}^{(+)}|^2}{|\bar{\Omega}^{(+)}|^2 + \Delta_{21}^2} = \frac{1}{1+\left|\frac{\Delta_{21}}{\bar{\Omega}^{(+)}}\right|^2} \tag{18.70}$$

during the Rabi cycle. Thus, a complete excitation, i.e., $f_2^{\max} = 1$, can only be reached under fully resonant conditions whereas the maximum excitation level is strongly reduced for larger detunings.

18.2.4 Pulse area and Rabi flopping

The Rabi-flopping scenario can be described completely analytically only for very few time-dependent fields due to the time-order complications arising from Eq. (18.52).

However, the resonant case $\Delta_{21} = 0$ can always be solved analytically if $\bar{\Omega}^{(+)}(t) = |\bar{\Omega}^{(+)}(t)|\, e^{i\varphi(t)}$ has a temporally constant phase $\varphi(t) = \varphi$. In this situation, the decomposition (18.54)–(18.55) must have $\theta = \pi/4$, such that

$$\mathbf{M}(t) = \mathbf{T}\Lambda(t)\mathbf{T}^H$$

$$\Lambda(t) = \begin{pmatrix} -\frac{|\Omega^{(+)}(t)|}{2} & 0 \\ 0 & \frac{|\Omega^{(-)}(t)|}{2} \end{pmatrix}, \quad \mathbf{T} = \begin{pmatrix} \frac{1}{\sqrt{2}} & -\frac{e^{i\varphi}}{\sqrt{2}} \\ \frac{e^{-i\varphi}}{\sqrt{2}} & \frac{1}{\sqrt{2}} \end{pmatrix}, \tag{18.71}$$

and the dynamics is fully contained in the diagonal matrix. Consequently, the formal solution (18.52) can be evaluated analytically for any $|\Omega^{(+)}(t)|$ leading to

$$\begin{aligned} \bar{\mathbf{a}}(t) &= \left[e^{-i\int_{-\infty}^{t} du\, \mathbf{M}(u)}\right]_T \bar{\mathbf{a}}(-\infty) = \mathbf{T}e^{-i\int_{-\infty}^{t} du\, \Lambda(u)}\mathbf{T}^H\, \bar{\mathbf{a}}(-\infty) \\ &= \mathbf{T}\begin{pmatrix} e^{\frac{i}{2}S(t)} & 0 \\ 0 & e^{-\frac{i}{2}S(t)} \end{pmatrix}\mathbf{T}^H\, \bar{\mathbf{a}}(-\infty) \\ &= \begin{pmatrix} \cos\frac{S(t)}{2} & ie^{i\varphi}\sin\frac{S(t)}{2} \\ ie^{-i\varphi}\sin\frac{S(t)}{2} & \cos\frac{S(t)}{2} \end{pmatrix}\begin{pmatrix} \bar{a}_2(-\infty) \\ \bar{a}_1(-\infty) \end{pmatrix}, \end{aligned} \tag{18.72}$$

where

$$S(t) \equiv \int_{-\infty}^{t} du\, |\Omega^{(+)}(u)| \tag{18.73}$$

is the temporal area of the excitation pulse. For the special case of a square pulse, this area is given by

$$S_{\text{squ}}(t) \equiv \begin{cases} 0, & t < 0 \\ |\bar{\Omega}^{(+)}|t, & 0 \le t \le T \\ |\bar{\Omega}^{(+)}|T, & t > T \end{cases}. \tag{18.74}$$

As a general trend, $S(t)$ grows monotonously and increases linearly with the magnitude of the exciting electric field.

For an initially unexcited system, Eq. (18.72) yields

$$\bar{a}_2(t) = ie^{i\varphi}\sin\frac{S(t)}{2}, \qquad \bar{a}_1(t) = \cos\frac{S(t)}{2} \tag{18.75}$$

with the polarization and inversion factor

$$\bar{P}(t) = \frac{i}{2}e^{i\varphi}\sin S(t), \qquad \bar{P}_z = -\frac{1}{2}\cos S(t), \tag{18.76}$$

respectively. This result correctly reduces to Eq. (18.67) when the pulse area is given by the square-pulse relation (18.74).

We see that the pulse area defines the final level of excitation left in the system after the pulse is gone. Maximum excitation, i.e., $\bar{P}_z = 1/2$, is reached whenever the total pulse area satisfies

$$S(+\infty) = \pi(2n+1), \qquad n = 0, 1, 2, \cdots \tag{18.77}$$

In other words, one needs a π, 3π, $\cdots$ pulse to produce maximal excitation of the atomic system. In the case of $2\pi n$ pulses, where the total temporal area amounts to

$$S(+\infty) = 2\pi n, \qquad n = 1, 2, 3, \cdots, \tag{18.78}$$

the system is returned to the ground state after the pulse, as seen from Eq. (18.76).

18.2.5 Square-pulse excitation

To illustrate the Rabi-flopping aspects, we analyze the trajectory of the Bloch vector for the special case of a square pulse. In particular, we follow the solution (18.65)–(18.66) as a function of time for different detunings. We assume a 3π pulse which moves the system from the ground to the excited state under resonant excitation conditions. In other words, the square pulse has the duration $T = \frac{3}{2}\,T_{\mathrm{Rabi},0}$. We also assume that the excitation pulse is real valued, i.e., φ vanishes. Figure 18.6 presents the resulting evolution of the Bloch vector for near-resonant excitations. The corresponding $|\bar{\Omega}^{(+)}(t)|$ (scaled) is shown as a thin dashed line in frames (a)–(b). To determine the effect of detuning we compare (a) resonant $\Delta_{21} = 0$ with (b) detuned $\Delta_{21} = |\bar{\Omega}^{(+)}(0)|$ excitation. In both cases, we follow the evolution of the Bloch vector's R_x, R_y, and R_z components.

We see that the resonant excitation (Fig. 18.6(a)) produces a $\bar{P}_z$ (solid line) that oscillates with the frequency $|\bar{\Omega}^{(+)}|$, as predicted by Eq. (18.67). It starts from the value $-1/2$ and peaks at $+1/2$ such that the system is oscillating between the unexcited and the excited state. Since the square pulse has the duration $T = \frac{3}{2}\,T_{\mathrm{Rabi},0}$, the system performs 1.5 Rabi flops, leaving it in the excited state after the interaction. Thus, the final Rabi vector remains

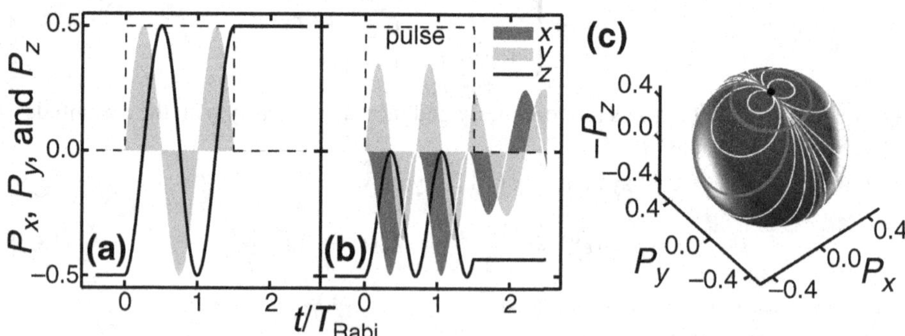

Figure 18.6 Near-resonant excitation of a two-level system. The time evolution of the excitation pulse (dashed line) is shown together with $\bar{P}_x$ (dark area), $\bar{P}_y$ (light area), and $\bar{P}_z$ (solid line) for **(a)** resonant and **(b)** detuned $\Delta_{21} = |\Omega^{(+)}(t)|$ excitation. The time scale is expressed through the typical interaction time $T_{\mathrm{Rabi}} = \frac{2\pi}{|\Omega^{(+)}(t)|}$ such that the results are independent of the actual values of $|\Omega^{(+)}(t)|$ and the transition energy. **(c)** The trajectories (gray thick lines) of the Bloch vector are shown on the Bloch sphere. For clarity, the south pole of the Bloch sphere is shown on top of the figure. The white lines indicate the resulting trajectories if the detuning is given by Eq. (18.79). The angles used are $\alpha = \pm 10°, \pm 30°, \pm 50°, \pm 75°, \pm 85°$.

stationary in the north-pole position $\bar{\mathbf{P}} = (0,\ 0,\ 1/2)$ on the Rabi sphere. At the same time, $\bar{P}_x$ vanishes for all times while $\bar{P}_y$ (bright area) performs Rabi oscillations that are $\pi/2$ out of phase relative to $\bar{P}_z$. Both Rabi-flopping aspects are verified by Eq. (18.67) where we see that $\bar{P}_x$ (dark area) evolves only if the Rabi frequency becomes genuinely complex valued, i.e., $\varphi \neq 0$.

For detuned excitation, we find a somewhat more complicated but generally similar dynamics as in the resonant-excitation case. The major difference is that $\bar{P}_z$ (solid line in Fig. 18.6(b)) now oscillates with the generalized Rabi frequency Υ instead of $|\bar{\Omega}^{(+)}|$. We also see that both polarization and some excitation remain in the system, even though we are treating the situation of a 3π pulse. Obviously, the difference between the generalized Rabi frequency and the Rabi frequency accounts for this discrepancy. In addition, both $\bar{P}_x$ (bright area) and $\bar{P}_y$ (dark area) oscillate. These oscillations even continue after the excitation pulse. During the pulse, the frequency is Υ while it is $|\bar{\Omega}^{(+)}|$ after the pulse. For the particular detuning chosen here, we have $\Upsilon = \sqrt{2}|\bar{\Omega}^{(+)}|$, which yields visibly faster oscillations than those in frame 18.6(a) for $\Delta_{21} = 0$. At the same time, the amplitude of the $\bar{P}_z$ oscillations is 50 percent smaller than for the resonant excitation, as predicted by Eq. (18.70). Thus, under detuned conditions, the excitation of the two-level system is never perfect, i.e., the Rabi vector cannot climb all the way up to the north pole.

We may also view the excitation in terms of the motion of the Bloch vector. Figure 18.6(c) presents the actual $\bar{\mathbf{P}} = (\bar{P}_x,\ \bar{P}_y,\ \bar{P}_z)$ trajectories (gray thick lines) corresponding to the cases $\Delta_{21} = 0$ and $\Delta_{21} = |\bar{\Omega}^{(+)}|$ studied in Figs. 18.6(a) and 18.6(b), respectively. The resonant excitation produces a single circular trajectory oscillating directly between the south pole (black circle) and the north pole (not shown). In order to see the trajectories more clearly, we have positioned the south pole of the Bloch sphere at the top of Fig. 18.6(c). The detuned case produces a two-fold trajectory. First it performs a circular motion with an angle that is tilted relative to the z axis. This motion takes place during the excitation. After the excitation, $\bar{\mathbf{P}}_z$ freezes and only the $\bar{\mathbf{P}}_x$ and $\bar{\mathbf{P}}_y$ components oscillate. This is seen as emergence of an additional circular motion. This time, it proceeds horizontally, i.e., perpendicular to the z axis.

To see the effect of detuning, we also present the trajectory of the Rabi vector during the excitation for the sequence of detunings given by

$$\Delta_{21} \equiv \tan\alpha\, |\bar{\Omega}^{(+)}| \tag{18.79}$$

with $\alpha = \pm 10°, \pm 30°, \pm 50°, \pm 75°, \pm 85°$. These are plotted as white lines in Fig. 18.6(c). All of them are circles on the Bloch sphere. As a notable feature, these circles become more and more tilted with respect to the z axis as the magnitude of the detuning is increased. For the largest detuning, we have $|\Delta_{21}| = 35.96|\bar{\Omega}^{(+)}|$. In this case, the Bloch vector moves only negligibly away from the ground state, i.e., the south pole (black filled circle).

These examples illustrate that the Bloch vector can be used nicely to present the quantum dynamics of optically excited atomic states. We also see very clearly the qualitative changes in trajectories caused by the detuning. The spherical symmetry of the solutions suggests that the optical excitations can be described alternatively as geometric rotations of the Bloch vector. This aspect is investigated in the next chapter. In this context, we also compute the self-consistent coupling between the light and the atomic dipole and generalize the two-level description to the case of quantized light.

Exercises

18.1 Start with the system Hamiltonian (17.54), given in the $\hat{\mathbf{E}} \cdot \hat{\mathbf{x}}$ picture. **(a)** Express $\hat{H}$ using $\hat{\mathbf{E}}(\mathbf{R})$ instead of $\hat{\mathbf{D}}(\mathbf{R})$ with the help of Eq. (17.47). Show that the dipole self-energy term cannot be eliminated. **(b)** Apply the classical factorization (18.2) and show that the Hamiltonian (18.3) follows if we omit the remaining term resembling the dipole self-energy.

18.2 **(a)** Insert the expansion (18.6) into Eq. (18.7) to reproduce Eq. (18.11). **(b)** When $\mathbf{E}(\mathbf{R})$ is broad with respect to the atomic wave packet, we may introduce the approximation $\mathbf{E}(\mathbf{R}) \rightarrow \mathbf{E}_0(t)$ in Eq. (18.8). Show that this assumption leads to Eq. (18.16). Use the ansatz (18.17) to produce (18.18). **(c)** Show that the identification (18.23) converts Eq. (18.18) into Eq. (18.24).

18.3 The momentum exchange in atomic transitions is described by Eq. (18.14). **(a)** Assume that an atom is initially in the state $C_{1s,\mathbf{K}}(0) = 1$. Use Eq. (18.14) to make a prediction on which $C_{np,\mathbf{K}'}$ states are generated by a weak field. **Hint:** *Assume a weak field such that* $C_{1s,\mathbf{K}} = e^{-iE_{1s,\mathbf{K}}t/\hbar}$ *propagates freely.* **(b)** Based on the result in (a), show that a light field with the energy $\hbar\omega_{\mathbf{q}} = E_\nu - E_{1s} + \frac{\hbar^2\mathbf{q}^2}{2M} + \left(\frac{\mathbf{v}}{c} \cdot \mathbf{e_q}\right) \hbar\omega_{\mathbf{q}}$ drives the state $C_{\nu,\mathbf{K}+\mathbf{q}}$ most efficiently. Here, $\mathbf{v} = \frac{\hbar\mathbf{K}}{M}$ is the velocity of the particle and $\mathbf{q} = |\mathbf{q}|\, \mathbf{e_q}$. Verify that a Doppler shift is observed depending on the propagation direction of the light, i.e., $\hbar\omega_{\mathbf{q}}$ is increased when the light and particle propagation directions are aligned.

18.4 **(a)** Show that the insertion of Eq. (18.28) into Eq. (18.26) produces Eq. (18.24). **(b)** Show that the condition (18.31) converts Eq. (18.29) into the two-level approximation (18.32). **(c)** Apply Eq. (18.32) to Eq. (18.26) and produce the two-level model (18.34).

18.5 Assume a field $E_1(t)$ which is resonant only with the $\lambda_1 \leftrightarrow \lambda_2$ transition and a field $E_2(t)$ which is resonant with the $\lambda_2 \leftrightarrow \lambda_3$ transition. Formulate the equations for the corresponding three-level model. Verify the level structure in Fig. 18.3(b)–(d). **Hint:** *Use the steps identical to those applied in the two-level analysis.*

18.6 **(a)** Apply the transformation (18.46) to Eq. (18.34) to produce Eq. (18.51). **(b)** Show that Eq. (18.52) is a formal solution of Eq. (18.51). **(c)** Diagonalize the matrix $\mathbf{M}(t) = \mathbf{T}(t)\mathbf{\Lambda}(t)\,\mathbf{T}^H(t)$ in Eq. (18.51). Introduce a new basis $\mathbf{b}(t) \equiv \mathbf{T}^H(t)\,\mathbf{a}(t)$. What is the dynamical equation for $\mathbf{b}(t)$ and when is the solution for $\mathbf{b}(t)$ expressed as a diagonal problem?

18.7 **(a)** Show that Eq. (18.54) diagonalizes $M(t)$ through the parametrization (18.55). **(b)** Construct the explicit solutions (18.56) and (18.57). **(c)** Construct the inversion $\mathbf{P}_z(t)$ and the polarization $\mathbf{P}(t)$ of Eq. (18.65).

18.8 **(a)** Show that Eq. (18.72) produces the general solution for resonant excitations. **(b)** Verify that the square-pulse result (18.57), with $\Delta = 0$, is also included in Eq. (18.72).

Further reading

Laser cooling is a standard approach in atomic optics. It is discussed in the following reviews where also many references to the original literature can be found:

S. Stenholm (1986). The semiclassical theory of laser cooling, *Rev. Mod. Phys.* **58**, 699.

C. S. Adams and E. Riis (1997). Laser cooling and trapping of neutral atoms, *Prog. Quantum Electron.* **21**, 1.

S. Chu (1998). Nobel Lecture: The manipulation of neutral particles, *Rev. Mod. Phys.* **70**, 685.

C. N. Cohen-Tannoudji (1998). Nobel Lecture: Manipulating atoms with photons, *Rev. Mod. Phys.* **70**, 707.

W. D. Phillips (1998). Nobel Lecture: Laser cooling and trapping of neutral atoms, *Rev. Mod. Phys.* **70**, 721.

Laser-induced transitions between atomic states are reviewed in:

A. H. Zewail (2000). Femtochemistry: Atomic-scale dynamics of the chemical bond using ultrafast lasers (Nobel Lecture), *Angew. Chem.* **39**, 2586.

Nonresonant excitations, such as multiphoton transitions, require a theoretical description beyond the rotating-wave approximation. For investigations in solid-state systems, cf.:

P. Guyot-Sionnest, W. Chen, and Y. R. Shen (1986). General considerations on optical second-harmonic generation from surfaces and interfaces, *Phys. Rev. B* **33**, 8254.

J. E. Sipe, D. J. Moss, and H. M. van Driel (1987). Phenomenological theory of optical second- and third-harmonic generation from cubic centrosymmetric crystals, *Phys. Rev. B* **35**, 1129.

J. T. Steiner, M. Kira, and S. W. Koch (2008). Optical nonlinearities and Rabi flopping of an exciton population in a semiconductor interacting with strong terahertz fields, *Phys. Rev. B* **77**, 165308.

Optical excitations in atomic two-level systems are discussed in:

L. Allen and J. H. Eberly (1975). *Optical Resonance and Two-Level Atoms*, New York, Wiley & Sons.

19

Self-consistent extension of the two-level approach

In this chapter, we extend the discussion of the two-level model for the optically induced atomic dipole by including the field modifications caused by the reradiation of the atomic dipole system. For this purpose, we have to treat the light–matter coupling self-consistently, including the decay of the matter's dipole polarization through its self-induced radiation field. Besides the radiative decay, we also obtain an energy shift and show that both effects can lead to nonlinear modifications of the matter excitation dynamics. Generally, these self-consistent light–matter coupling effects are critically important when we want to understand how light is transmitted, reflected, and absorbed by matter.

19.1 Spatial coupling between light and two-level system

As the starting point of our analysis, we use the semiclassical system Hamiltonian (18.3),

$$\hat{H}_{\mathrm{sc}} = \hat{H}_0^{\mathrm{CM}} + \hat{H}_0^{\mathrm{rel}} - Q_e \mathbf{r} \cdot \mathbf{E}(\mathbf{R}),$$
$$\hat{H}_0^{\mathrm{CM}} \equiv \frac{\hat{\mathbf{P}}^2}{2M}, \qquad \hat{H}_0^{\mathrm{rel}} \equiv \frac{\hat{\mathbf{p}}^2}{2\mu} - \frac{Q_e^2}{4\pi\varepsilon\,\epsilon_0 |\mathbf{r}|}. \tag{19.1}$$

Here, we have separated the contributions depending only on the center-of-mass $\mathbf{R}$ and relative $\mathbf{r}$ coordinates of the electron–ion system. The term $-Q_e \mathbf{r} \cdot \mathbf{E}(\mathbf{R})$ represents the light–matter interaction.

Even though this light–matter interaction correlates the relative and center-of-mass motion, the optical transitions predominantly change only the relative-motion part of the system while the center-of-mass dynamics remains relatively inert. As discussed in Section 18.1, it is therefore usually a good approximation to separate the relative and center-of-mass motion parts, unless effects of multiple photon absorption and/or emission are analyzed, e.g., with the goal to explore laser cooling.

The separation ansatz (18.17) corresponds to a factorizable division of the relative and the center-of-mass contributions,

$$|\psi(t)\rangle = |\psi_{\mathrm{rel}}(t)\rangle \otimes |\psi_{\mathrm{CM}}(t)\rangle, \tag{19.2}$$

for *all* times. The center-of-mass dynamics corresponds to the wave packet,

$$|\psi_{CM}(t)\rangle = e^{-\frac{i}{\hbar}\hat{H}_0^{CM}t}\,|\psi_{CM}(0)\rangle, \tag{19.3}$$

that propagates freely as given by Eq. (18.18).

19.1.1 Center-of-mass distribution in optical coupling

If we use the separation ansatz (19.2), we can write the Schrödinger equation in the form

$$\begin{aligned} i\hbar\frac{\partial}{\partial t}|\psi(t)\rangle &= \hat{H}_{sc}|\psi_{rel}(t)\rangle \otimes |\psi_{CM}(t)\rangle \\ &= \left[\hat{H}_0^{rel}\,|\psi_{rel}(t)\rangle\right] \otimes |\psi_{CM}(t)\rangle + |\psi_{rel}(t)\rangle \otimes \hat{H}_0^{CM}\,|\psi_{CM}(t)\rangle \\ &\quad - \left[Q_e\hat{\mathbf{r}}\cdot|\psi_{rel}(t)\rangle\right] \otimes \mathbf{E}(\hat{\mathbf{R}})|\psi_{CM}(t)\rangle. \end{aligned} \tag{19.4}$$

Using Eq. (19.3), the LHS can also be expressed as

$$\begin{aligned} i\hbar\frac{\partial}{\partial t}|\psi(t)\rangle &= i\hbar\frac{\partial|\psi_{rel}(t)\rangle}{\partial t} \otimes |\psi_{CM}(t)\rangle + |\psi_{rel}(t)\rangle \otimes i\hbar\frac{\partial|\psi_{CM}(t)\rangle}{\partial t} \\ &= i\hbar\frac{\partial|\psi_{rel}(t)\rangle}{\partial t} \otimes |\psi_{CM}(t)\rangle + |\psi_{rel}(t)\rangle \otimes \hat{H}_{CM}|\psi_{CM}(t)\rangle. \end{aligned} \tag{19.5}$$

Combining (19.4) and (19.5) yields

$$i\hbar\frac{\partial|\psi_{rel}\rangle}{\partial t} \otimes |\psi_{CM}\rangle = \hat{H}_0^{rel}\,|\psi_{rel}\rangle \otimes |\psi_{CM}\rangle - Q_e\hat{\mathbf{r}}\cdot|\psi_{rel}\rangle \otimes \mathbf{E}(\hat{\mathbf{R}})|\psi_{CM}\rangle\,. \tag{19.6}$$

Projecting both sides of this equation onto $\langle\psi_{CM}(t)|$, we obtain

$$\begin{aligned} i\hbar\frac{\partial}{\partial t}|\psi_{rel}(t)\rangle &= \hat{H}_0^{rel}\,|\psi_{rel}(t)\rangle \\ &\quad - Q_e\hat{\mathbf{r}}|\psi_{rel}(t)\rangle \cdot \langle\psi_{CM}(t)|\mathbf{E}(\hat{\mathbf{R}})|\psi_{CM}(t)\rangle. \end{aligned} \tag{19.7}$$

At this stage, we have a Schrödinger equation for the relative-motion part with a parametric center-of-mass dependence. The comparison with Eq. (18.10) reveals that the use of the separation ansatz (19.2) neglects the momentum exchange between photon and atom.

From Eq. (19.7), we note that we can introduce an effective electric field via

$$\begin{aligned} \mathbf{E}(t) \equiv \langle\psi_{CM}(t)|\mathbf{E}(\hat{\mathbf{R}})|\psi_{CM}(t)\rangle &= \int d^3x\,|\psi_{CM}(\mathbf{x},\,t)|^2\,\mathbf{E}(\mathbf{x},\,t) \\ &\equiv \int d^3x\,g(\mathbf{x}-\mathbf{R}_0)\,\mathbf{E}(\mathbf{x},\,t). \end{aligned} \tag{19.8}$$

Here, the function $g(\mathbf{x}) \equiv |\psi_{CM}(\mathbf{x}+\mathbf{R}_0,\,t)|^2$ describes the probability distribution of the atom's center-of-mass position $\mathbf{x}$ around $\mathbf{R}_0$. Therefore, the relative-motion dynamics is driven by the field averaged over the atomic center-of-mass distribution,

$$i\hbar\frac{\partial}{\partial t}|\psi_{rel}(t)\rangle = \hat{H}_0^{rel}\,|\psi_{rel}(t)\rangle - Q_e\hat{\mathbf{r}}|\psi_{rel}(t)\rangle \cdot \mathbf{E}(t). \tag{19.9}$$

In the spirit of the dipole approximation, $g(\mathbf{x})$ is assumed to be much smaller than the wave length of the exciting field, i.e., $g(\mathbf{x})$ is a relatively narrow distribution. Since we investigate the dynamics in a situation where laser-cooling effects can be neglected, the central position of the atom, $\mathbf{R}_0$, can be considered as stationary on the time scale of the optical transitions.

Making the two-level approximation (TLA) and assuming that the system is initially in its ground state, the atomic wave function (19.2) can be written as

$$
\begin{aligned}
|\psi(t)\rangle &= (a_1\,|1\rangle + a_2\,|2\rangle) \otimes |\psi_{\mathrm{CM}}(t)\rangle \\
&= \left(\bar{a}_1\, e^{-i\left(\omega_1 - \frac{\Delta}{2}\right)t}\,|1\rangle + \bar{a}_2\, e^{-i\left(\omega_2 + \frac{\Delta}{2}\right)t}\,|2\rangle\right) \otimes |\psi_{\mathrm{CM}}(t)\rangle, \qquad (19.10)
\end{aligned}
$$

based on Eq. (18.58). As in Fig. 18.3(a), we adopt here the notation where $|1\rangle$ refers to the ground state and $|2\rangle$ denotes the excited state. Furthermore, we express the state coefficients $\bar{a}_\lambda$ in the rotating frame (18.46),

$$
\begin{cases} a_1(t) = \bar{a}_1(t)\, e^{-i\omega_1 t}\, e^{+\frac{i}{2}\Delta_{2,1}t} \\ a_2(t) = \bar{a}_2(t)\, e^{-i\omega_2 t}\, e^{-\frac{i}{2}\Delta_{2,1}t} \end{cases}, \qquad \begin{cases} \omega_\lambda = \dfrac{E_\lambda}{\hbar}, \qquad \lambda = 1,\, 2 \\ \Delta_{2,1} \equiv \omega_L - (\omega_2 - \omega_1) \end{cases}. \qquad (19.11)
$$

The ω_λ that appears defines the eigenfrequency associated with the eigenstate energy E_λ while $\Delta_{2,1}$ is the detuning between the central light frequency ω_L and the transition frequency.

Within this frame, the excitation dynamics follows from Eq. (18.51),

$$
\begin{cases} i\dfrac{\partial}{\partial t}\bar{a}_2 = -\dfrac{\Delta}{2}\,\bar{a}_2 - \dfrac{\bar{\Omega}^{(+)}(t)}{2}\,\bar{a}_1 \\ i\dfrac{\partial}{\partial t}\bar{a}_1 = +\dfrac{\Delta}{2}\,\bar{a}_1 - \dfrac{\bar{\Omega}^{(-)}(t)}{2}\,\bar{a}_2 \end{cases}, \qquad (19.12)
$$

which also includes the rotating-wave approximation (RWA), as defined in Section 18.1.4. Based on (19.8), we generalize the Rabi frequency as

$$
\begin{aligned}
&\hbar\bar{\Omega}^{(+)}(t) \equiv 2\mathbf{d}_{2,1}\cdot\mathbf{E}^{(+)}(t)\, e^{i\omega_L t} = 2\mathbf{d}_{2,1}\cdot\int d^3x\, g(\mathbf{x})\,\mathbf{E}^{(+)}(\mathbf{x},t)e^{i\omega_L t}, \\
&\bar{\Omega}^{(-)}(t) \equiv \left[\bar{\Omega}^{(+)}(t)\right]^\star, \qquad \mathbf{E}(\mathbf{x},t) \equiv 2\mathrm{Re}\left[\mathbf{E}^{(+)}(\mathbf{x},t)\right]. \qquad (19.13)
\end{aligned}
$$

We see that the RWA picks only one of the $\mathbf{E}^{(\pm)}(t)$ components to appear within the two-level system dynamics (19.12).

19.1.2 Optical Bloch equations

For our further analysis, it is instructive to consider an alternative formulation of the excitation dynamics (19.12). As shown in Section 18.2.2, the experimental observables consist of the polarization and the inversion factor

$$\bar{P}_- \equiv \bar{a}_1^\star \bar{a}_2, \qquad \bar{P}_z \equiv \frac{|\bar{a}_2|^2 - |\bar{a}_1|^2}{2}, \tag{19.14}$$

respectively. They uniquely identify the Bloch vector in the rotating frame

$$\bar{\mathbf{P}} \equiv \begin{pmatrix} \bar{P}_x \\ \bar{P}_y \\ \bar{P}_z \end{pmatrix}, \qquad \begin{cases} \bar{P}_x = \mathrm{Re}\left[\bar{\mathbf{P}}_-\right] \\ \bar{P}_y = \mathrm{Im}\left[\bar{\mathbf{P}}_-\right] \end{cases} \Leftrightarrow \quad \bar{P} = \bar{P}_x + i\bar{P}_y, \tag{19.15}$$

which satisfies the condition $|\mathbf{P}| = 1/2$ whenever the approximation (19.10) is valid. As discussed in Section 20.3, the Bloch vector $\bar{\mathbf{P}}$ varies on the surface of the Bloch sphere as long as neither dephasing nor quantum-optical effects are significant.

Since both the polarization and the inversion factor can – in principle – be measured, it is interesting to compute their dynamics directly. With the help of (19.12) we find for the polarization dynamics,

$$i\frac{\partial}{\partial t}\bar{P} = \left(i\frac{\partial}{\partial t}\bar{a}_1^\star\right)\bar{a}_2^\star + \bar{a}_1^\star\left(i\frac{\partial}{\partial t}\bar{a}_2\right) = -\Delta_{2,1}\,\bar{P} + \bar{P}_z\,\bar{\Omega}^{(+)}. \tag{19.16}$$

Extracting the real and imaginary parts, we identify the dynamics of the x and y components of the Bloch vector,

$$\begin{cases} \frac{\partial}{\partial t}\bar{P}_x = -\Delta_{2,1}\,\bar{P}_y + \bar{P}_z\,\bar{\Omega}_y \\ \frac{\partial}{\partial t}\bar{P}_y = +\Delta_{2,1}\,\bar{P}_x - \bar{P}_z\,\bar{\Omega}_x \end{cases}, \qquad \begin{cases} \bar{\Omega}_x \equiv \mathrm{Re}\left[\bar{\Omega}^{(+)}\right] \\ \bar{\Omega}_y \equiv \mathrm{Im}\left[\bar{\Omega}^{(+)}\right] \end{cases}. \tag{19.17}$$

In the same way, Eq. (19.12) directly generates the dynamics of the inversion factor

$$\frac{\partial}{\partial t}\bar{P}_z = \mathrm{Im}\left[\left(\bar{\Omega}^{(+)}\right)^\star \bar{P}\right] = \bar{\Omega}_x\,\bar{P}_y - \bar{\Omega}_y\,\bar{P}_x, \tag{19.18}$$

where the last form follows from the definitions (19.15) and (19.17). Introducing

$$\bar{\Omega} \times \bar{\mathbf{P}} = \begin{pmatrix} \bar{\Omega}_2 P_z - \bar{\Omega}_3 P_y \\ -\bar{\Omega}_1 P_z + \bar{\Omega}_3 P_x \\ \bar{\Omega}_1 P_y - \bar{\Omega}_2 P_x \end{pmatrix}, \qquad \bar{\Omega} \equiv \begin{pmatrix} \bar{\Omega}_x \\ \bar{\Omega}_y \\ \Delta_{2,1} \end{pmatrix} \tag{19.19}$$

allows us to reduce Eqs. (19.17)–(19.18) to

$$\frac{\partial}{\partial t}\bar{\mathbf{P}} = \bar{\Omega} \times \bar{\mathbf{P}}. \tag{19.20}$$

Hence, we find exactly the form of the Bloch equations describing the spin precession in a magnetic field. The analogy between optical and magnetic Bloch equations follows if $\bar{\Omega}$ is assumed to correspond to the magnetic field and $\bar{\mathbf{P}}$ corresponds to the magnetization. Since we are studying optical excitations, Eqs. (19.17)–(19.18) or Eq. (19.20) are commonly referred to as the optical Bloch equations.

We can easily test that the optical Bloch equations conserve $|\bar{\mathbf{P}}|$ because

$$\frac{\partial}{\partial t}\bar{\mathbf{P}} \cdot \bar{\mathbf{P}} = 2\bar{\mathbf{P}} \cdot \frac{\partial}{\partial t}\bar{\mathbf{P}} = 2\bar{\mathbf{P}} \cdot \left[\bar{\Omega} \times \bar{\mathbf{P}}\right] = 0 \tag{19.21}$$

based on Eq. (19.20). In other words, in the case without dephasing, the Bloch vector strictly remains within the Bloch sphere $|\bar{\mathbf{P}}| = 1/2$.

19.1.3 Angle parametrization of Bloch vector

The optical Bloch equations form a closed set of equations describing the same excitation dynamics as the original equations (19.12) for the coefficients $\bar{a}_\lambda$ of the two-level system. Structurally, Eqs. (19.17)–(19.18) contain three real-valued quantities while Eq. (19.12) describes the time evolution of two complex-valued quantities. Hence, it seems that the number of independent variables does not match. However, the condition $|\bar{a}_1|^2 + |\bar{a}_2|^2 = 1$ reduces the effective degrees of freedom by one such that Eq. (18.51) and Eqs. (19.17)–(19.18) both have three independent variables.

In fact, the motion of the two-level system can even be parametrized with only two independent variables as long as the approximation (19.10) is valid. To see this, we consider the relative-motion part of the wave function (19.10) that can be written as

$$\begin{aligned}
|\psi_{2\mathrm{LS}}\rangle &= e^{-i\left(\omega_1 - \frac{i}{2}\Delta_{2,1}\right)t}\left[\bar{a}_1\,|1\rangle + \bar{a}_2\,e^{-i\omega_L t}\,|2\rangle\right] \\
&= e^{-i\left(\omega_1 + \frac{\Delta_{2,1}}{2}\right)t}\left[|\bar{a}_1(t)|\,e^{i\varphi_1(t)}\,|1\rangle + |\bar{a}_2(t)|\,e^{i(\varphi_2(t) - \omega_L t)}\,|2p\rangle\right] \\
&= e^{-i\left[\varphi_1 + \left(\omega_1 + \frac{\Delta_{2,1}}{2}\right)t\right]}\left[|\bar{a}_1|\,|1\rangle + |\bar{a}_2|\,e^{i(\varphi_2 - \varphi_1 - \omega_L t)}\,|2\rangle\right],
\end{aligned} \tag{19.22}$$

after having expressed the complex-valued $\bar{a}_j = |\bar{a}_j(t)|\,e^{i\varphi_j(t)}$ in terms of its magnitude and phase. We occasionally write $|\bar{a}_j(t)|$ and $\varphi_j(t)$ without the time argument for the sake of brevity. From Eq. (19.22), we deduce that only the phase difference

$$\Delta\phi(t) \equiv \varphi_2(t) - \varphi_1(t) \tag{19.23}$$

between the states $|1\rangle$ and $|2\rangle$ is physically relevant because one can always eliminate the arbitrary common phase from the wave function when constructing observables.

Since $|\psi_{2\mathrm{LS}}\rangle$ is normalized, the amplitudes can be parametrized via

$$|\bar{a}_1| \equiv \cos\frac{\vartheta(t)}{2}, \qquad |\bar{a}_2| \equiv \sin\frac{\vartheta(t)}{2}. \tag{19.24}$$

Inserting (19.23) and (19.24) into Eq. (19.14), we find

$$\begin{aligned}
\bar{P} &= \bar{a}_1^\star \bar{a}_2 = |\bar{a}_1|\,|\bar{a}_2|\,e^{i\Delta\phi(t)} = \cos\frac{\vartheta(t)}{2}\sin\frac{\vartheta(t)}{2}\,e^{i\Delta\phi(t)} = \frac{1}{2}\sin\vartheta(t)\,e^{i\Delta\phi(t)} \\
\bar{P}_z &= \frac{|\bar{a}_2|^2 - |\bar{a}_1|^2}{2} = \frac{1}{2}\left(\sin^2\frac{\vartheta(t)}{2} - \cos^2\frac{\vartheta(t)}{2}\right) = -\frac{1}{2}\cos\vartheta(t).
\end{aligned} \tag{19.25}$$

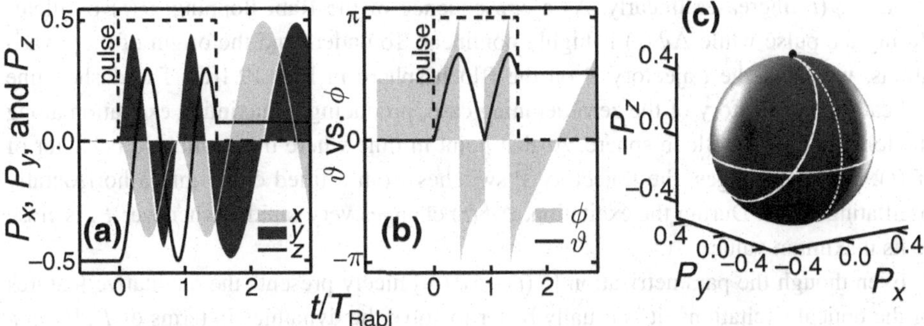

Figure 19.1 Temporal evolution of the Bloch-vector components and the angle-parametrized representation of the optical excitations. The results are shown for a square pulse excitation with duration $T = 1.5\, T_{\text{Rabi}}$ and detuning $\Delta_{21} = 0.5\, |\Omega^{(+)}|$. **(a)** The resulting dynamics of the Bloch vector's x (light area), y (dark area), and z (solid line) components are plotted as a function of time. **(b)** The corresponding angle representation consists of the azimuthal angle ϑ (solid line) and the polar angle (shaded area). The dashed line in **(a)** and **(b)** indicates the scaled pump pulse. The time is given in units of $T_{\text{Rabi}} = 2\pi/|\Omega^{(+)}|$ such that the representation does not depend on $|\Omega^{(+)}|$. **(c)** The full Bloch-vector (white solid line) evolution. The direct trajectory from pole-to-pole is shown as the dashed line. The solution follows from Eq. (19.20).

This constructs the three-dimensional Bloch vector

$$\bar{\mathbf{P}} = \frac{1}{2} \begin{pmatrix} \sin\vartheta(t)\cos\Delta\phi(t) \\ \sin\vartheta(t)\sin\Delta\phi(t) \\ -\cos\vartheta(t) \end{pmatrix}, \tag{19.26}$$

based on the definition (19.15). Equation (19.26) represents the Bloch vector in spherical coordinates showing that the physics of the two-level problem can be parametrized via the two independent variables, $\vartheta(t)$ and $\Delta\phi(t)$, whenever the Bloch vector remains at the surface of the Bloch sphere.

Figure 19.1 presents the motion of the Bloch vector and its angle parameters during a square-pulse excitation. The 3π pulse used has duration $T = 1.5\, T_{\text{Rabi}} = \frac{3\pi}{|\Omega^{(+)}|}$ and detuning $\Delta_{2,1} = 0.5|\Omega^{(+)}|$ relative to the transition energy. The resulting $\bar{P}_x$ (bright area), $\bar{P}_y$ (dark area), and $\bar{P}_z$ (solid line) are shown in Fig. 19.1(a) together with the square pulse $|\Omega^{(+)}|$ (dashed line, scaled). As a consequence of the detuning, all $\bar{P}_j$ components oscillate during the pulse, performing slightly more than 1.5 Rabi cycles.

The corresponding $\vartheta(t)$ (solid line) and $\Delta\phi(t)$ (shaded area) are shown in Fig. 19.1(b) together with the square pulse (dashed line). During the pulse, ϑ starts to grow from zero but does not reach the value π corresponding to a fully excited system. As discussed in connection with Fig. 18.6, the detuning prevents the system from becoming completely excited. Instead, the detuned case produces fast oscillations also for $\Delta\phi(t)$, describing the motion of the Bloch vector within the xy-plane.

The parametrization $[\vartheta(t),\ \Delta\phi(t)]$ nicely reveals the qualitative changes of the Bloch-vector motion during and after the excitation. After the excitation, $\vartheta(t)$ becomes constant while $\Delta\phi(t)$ increases linearly. As a consequence of the Rabi flopping, $\vartheta(t)$ oscillates during the pulse while $\Delta\phi(t)$ is highly nonlinear. To understand the origin of these variations, we show the trajectory $\bar{\mathbf{P}}$ on the Bloch sphere in Fig. 19.1(c). The dashed line indicates the trajectory of the zero-detuning case, producing a maximum excitation along the longitude of the Bloch sphere. At that point in time where the qualitative behavior of $[\vartheta(t),\ \Delta\phi(t)]$ changes, the trajectory $\bar{\mathbf{P}}$ switches from a tilted circle into a horizontally oscillating circle. During the excitation, $\Delta\phi(t)$ changes very rapidly whenever $\bar{P}_z$ is close to its maximum value.

Even though the parametrization $[\vartheta(t),\ \Delta\phi(t)]$ nicely presents the qualitative features of the optical excitations, it is usually better to solve the dynamics in terms of $\bar{P}$. From a completely numerical point of view, $[\vartheta(t),\ \Delta\phi(t)]$ can change much faster than the vectorial components of $\bar{P}$ do, which makes $\bar{P}$-based computations more feasible even though one more degree of freedom is needed. In addition, the $\bar{P}$ description is the most suitable one for the discussion of dephasing effects and the quantum-optical generalizations of the model.

19.2 Maxwell-optical Bloch equations

Many phenomena in nature depend self-consistently on both the external source causing the effect and the effect's self-induced modifications to the total source. Therefore, one must describe the effect self-consistently, i.e., one must fully account for both the creation and the back action of the effect. So far, we have analyzed the optical Bloch equations only for the situation where the atomic dipole is driven by a predetermined external field $\mathbf{E}(t)$. However, as discussed in Section 17.3, the induced atomic polarization also radiates a dipole field $\mathbf{E}_{\text{dip}}(t)$, which leads to modifications of the actual field $\mathbf{E}(t)$. Hence, the connection between the exciting field $\mathbf{E}_0(t)$ and polarization must be computed self-consistently to determine the total field $\mathbf{E}(t)$, i.e. $\mathbf{E}(t) = \mathbf{E}_0(t) + \mathbf{E}_{\text{dip}}(t)$ driving the atomic dipole. This can be done after we solve the Maxwell's equations together with the optical Bloch equations.

Generally, the light propagation follows from the wave equation (18.1) containing the atomic dipole at its center-of-mass position $\mathbf{R}_0$,

$$\left[\nabla^2 - \frac{1}{c^2}\frac{\partial^2}{\partial t^2}\right]\langle\hat{\mathbf{E}}(\mathbf{x})\rangle = \mu_0\left[\frac{\partial^2\langle Q_e\hat{\mathbf{r}}\rangle}{\partial t^2}\,\delta(\mathbf{x}-\mathbf{R}_0)\right]_T . \tag{19.27}$$

As shown in connection with Eq. (16.41), the wave function (19.10) produces an atomic dipole

$$\langle Q_e\hat{\mathbf{r}}\rangle = \mathbf{d}_{1,2}\,\bar{P}\,e^{-i\omega_L t} + \mathbf{d}_{2,1s}\,\bar{P}^\star\,e^{+i\omega_L t} . \tag{19.28}$$

Since this expression presents the linear source to the electric field in Eq. (19.27), we may decompose the total field into parts oscillating with $e^{-i\omega_L t}$ or $e^{+i\omega_L t}$. This division produces

$$\left[\nabla^2 - \frac{1}{c^2}\frac{\partial^2}{\partial t^2}\right]\langle\hat{\mathbf{E}}^{(+)}(\mathbf{x})\rangle = \mu_0\left[d_{1,2}\frac{\partial^2}{\partial t^2}\bar{P}\,e^{-i\omega_L t}\,\delta(\mathbf{x}-\mathbf{R}_0)\right]_T, \tag{19.29}$$

where the corresponding equation for $\langle\hat{\mathbf{E}}^{(-)}(\mathbf{x})\rangle = \left[\langle\hat{\mathbf{E}}^{(+)}(\mathbf{x})\rangle\right]^\star$ follows from complex conjugation.

The self-consistent solution is found if Eq. (19.29) is solved together with Eqs. (19.16) and (19.18), resulting in the Maxwell-optical Bloch equations,

$$\begin{cases} i\frac{\partial}{\partial t}\bar{P} = -\Delta_{2,1}\,\bar{P} + \bar{P}_z\,\bar{\Omega}^{(+)}, \qquad \frac{\partial}{\partial t}\bar{P}_z = \mathrm{Im}\left[\left(\bar{\Omega}^{(+)}\right)^\star\,\bar{P}\right] \\ \left[\nabla^2 - \frac{1}{c^2}\frac{\partial^2}{\partial t^2}\right]\langle\hat{\mathbf{E}}^{(+)}(\mathbf{x})\rangle = \mu_0\left[\mathbf{d}_{1,2}\frac{\partial^2}{\partial t^2}\,\bar{P}\,e^{-i\omega_L t}\,\delta(\mathbf{x}-\mathbf{R}_0)\right]_T \end{cases}. \tag{19.30}$$

The atomic dipole interacts with an effective electric field given by the convolution (19.8). In this case, the Rabi frequency follows from Eq. (19.13)

$$\hbar\bar{\Omega}^{(+)}(t) = 2\mathbf{d}_{2,1}\cdot\int d^3x\,g(\mathbf{x}-\mathbf{R}_0)\,\langle\hat{\mathbf{E}}^{(+)}(\mathbf{x},\,t)\rangle e^{i\omega_L t}, \tag{19.31}$$

which is a convolution between the center-of-mass distribution of the atom and the full electrical field. At this level, the back reaction between the dipole and the electric field is fully included.

19.2.1 Radiative decay of the atomic dipole

Since $\bar{P}$ varies in the rotating frame, it is nearly constant on the time scale defined by the optical frequency. Thus, we may introduce the approximation

$$\frac{\partial^2}{\partial t^2}\bar{P}\,e^{-i\omega_L t} \to -\omega_L^2\bar{P}\,e^{-i\omega_L t}, \tag{19.32}$$

which converts the RHS of Eq. (19.29) into

$$\left[\nabla^2 - \frac{1}{c^2}\frac{\partial^2}{\partial t^2}\right]\langle\hat{\mathbf{E}}^{(+)}(\mathbf{x})\rangle = \mu_0\left[-d_{1,2}\,\bar{P}\mathbf{e}_z\,\omega_L^2\,e^{-i\omega_L t}\,\delta(\mathbf{x}-\mathbf{R}_0)\right]_T, \tag{19.33}$$

where $\bar{P}$ is the slowly varying envelope of the polarization. Here, we assumed that the dipole is excited in the direction $\mathbf{e}_z$. Equation (19.33) is identical to Eq. (17.25) with the source (17.28). Thus, we may directly use (17.30)–(17.31) to identify the inhomogeneous solution of Eq. (19.33) as

$$\begin{aligned}\mathbf{E}_{\text{dip}}(\mathbf{x},\,t) = \frac{d_{1,2}\,\bar{P}}{4\pi\varepsilon_0}&\left[\left(\mathbf{e}_z - \frac{(x_z-R_z)(\mathbf{x}-\mathbf{R}_0)}{|\mathbf{x}-\mathbf{R}_0|^2}\right)\frac{q_L^2}{|\mathbf{x}-\mathbf{R}_0|}\,e^{iq_L(|\mathbf{x}-\mathbf{R}_0|-ct)}\right.\\ &\left.+\left(\mathbf{e}_z - 3\frac{(x_z-R_z)(\mathbf{x}-\mathbf{R}_0)}{|\mathbf{x}-\mathbf{R}_0|^2}\right)\frac{e^{-iq_L|\mathbf{x}-\mathbf{R}_0|}-1+iq_L|\mathbf{x}-\mathbf{R}_0|}{|\mathbf{x}-\mathbf{R}_0|^3}\,e^{iq_L(|\mathbf{x}-\mathbf{R}_0|-ct)}\right],\\ \text{if } |\mathbf{x}-\mathbf{R}_0| < ct, &\qquad q_L \equiv \frac{\omega_L}{c}. \end{aligned} \tag{19.34}$$

Since we are mostly interested in the field close to the dipole, we omit here the asymptotic part of Eq. (17.30).

The total field is now obtained as the sum of the homogeneous and inhomogeneous solutions,

$$\langle \hat{\mathbf{E}}^{(+)}(\mathbf{x},\, t)\rangle = \mathbf{E}_0^{(+)}(\mathbf{x},\, t) + \mathbf{E}_{\rm dip}(\mathbf{x},\, t). \tag{19.35}$$

While we have taken $\mathbf{E}_0^{(+)}(\mathbf{R}_0,\, t) = \mathbf{E}^{(+)}(t)$ as the field driving the two-level system in Chapter 18, we may now improve this approximation using the fully self-consistent field provided by Eq. (19.35). In other words, we could use the exact $\langle \hat{\mathbf{E}}^{(+)}(\mathbf{R}_0,\, t)\rangle = \mathbf{E}^{(+)}(t)$ as the source driving the optical Bloch equation in Eq. (19.30). However, this attempt yields a severe problem because $\mathbf{E}_{\rm dip}(\mathbf{x},\, t)$ diverges as $\mathbf{x}$ approaches $\mathbf{R}_0$. Clearly, this divergence is unphysical. It appears as a consequence of our approximation to treat the center-of-mass distribution of the atomic dipole as a δ-function instead of a proper distribution. The physically correct form presented in Eq. (19.31) yields an effective driving field

$$\mathbf{E}(t) = \int d^3x\, g(\mathbf{x}-\mathbf{R}_0)\, \mathbf{E}_0^{(+)}(\mathbf{x},\, t) + \int d^3x\, g(\mathbf{x}-\mathbf{R}_0)\, \mathbf{E}_{\rm dip}(\mathbf{x},\, t), \tag{19.36}$$

after we have expressed $\langle \hat{\mathbf{E}}^{(+)}(\mathbf{x},\, t)\rangle$ in terms of the incoming and the dipole-radiation fields, identified by Eq. (19.35). As before, $g(\mathbf{x})$ defines the probability distribution for the center-of-mass extension of the atom. For a sufficiently narrow distribution, the incoming field is nearly constant on the scale of $g(\mathbf{x})$. Thus, we only need to compute the integral containing the dipole-generated field,

$$\mathbf{E}(t) = \mathbf{E}_0^{(+)}(\mathbf{R}_0,\, t) + \int d^3r\, g(\mathbf{r})\, \mathbf{E}_{\rm dip}(\mathbf{r}+\mathbf{R}_0,\, t), \tag{19.37}$$

where we exchanged the integration variable by $\mathbf{x} \to \mathbf{r}+\mathbf{R}_0$.

Since the atom has no preferred direction, it is reasonable to assume a radially symmetric $g(\mathbf{r})$ distribution. Consequently, we use

$$\mathbf{E}_{\rm dip}(t) \equiv \int d^3r\, g(r)\, \mathbf{E}_{\rm dip}(\mathbf{r}+\mathbf{R}_0,\, t) \tag{19.38}$$

in the optical Bloch equations. For a radially symmetric $g(\mathbf{r}) = g(r)$, only the $\mathbf{e}_z$ component of $\mathbf{E}_{\rm dip}(\mathbf{r}+\mathbf{R}_0,\, t)$ contributes, yielding

$$\begin{aligned}\mathbf{E}_{\rm dip}(t) &= \frac{d_{1,2}\, \mathbf{e}_z\, \bar{P}e^{-i\omega_L t}}{12\pi\varepsilon_0}\, 4\pi \int_0^\infty dr\, r^2\, g(r)\, q_L^2 \frac{e^{iq_L r}}{r} \\ &= \frac{\mathbf{d}_{1,2}}{12\pi\varepsilon_0}\, 4\pi \int_0^\infty dr\, r^2\, g(r)\, q_L^2 \frac{e^{iq_L r}}{r}\, \bar{P}e^{-i\omega_L t},\end{aligned} \tag{19.39}$$

where we expressed the dipole matrix element in its vectorial form. Since $g(r)$ extends only to distances much shorter than the optical wave length $1/q_L = \lambda_L/2\pi$, we may evaluate the integral by Taylor expanding the plane-wave factor, with the result

$$\mathbf{E}_{\text{dip}}(t) = \frac{\mathbf{d}_{1,2}}{12\pi\varepsilon_0} 4\pi \int_0^\infty dr\, r^2\, g(r)\, q_L^2 \frac{1 + iq_L r}{r}\, \bar{P} e^{-i\omega_L t}$$
$$= \frac{\mathbf{d}_{1,2}\, q_L^2}{12\pi\varepsilon_0} 4\pi \int_0^\infty dr \left(r + iq_L r^2\right) g(r)\, \bar{P} e^{-i\omega_L t}. \tag{19.40}$$

Due to the normalization of the probability distribution $g(r)$, we have

$$\int d^3r\, g(r) = 4\pi \int_0^\infty dr\, r^2 g(r) = 1 \tag{19.41}$$

allowing us to simplify Eq. (19.40) into

$$\mathbf{d}_{2,1} \cdot \mathbf{E}_{\text{dip}}(t) = \frac{\mathbf{d}_{2,1} \cdot \mathbf{d}_{1,2}}{12\pi\varepsilon_0} \left(q_L^2 \frac{\int_0^\infty dr\, r\, g(r)}{\int_0^\infty dr\, r^2 g(r)} + iq_L^3 \right) \bar{P} e^{-i\omega_L t}$$
$$\equiv \hbar \frac{\Delta_{\text{dip}} + i\Gamma_{2,1}}{2} \bar{P} e^{-i\omega_L t}. \tag{19.42}$$

For the self-consistent Rabi frequency (19.31) we find

$$\hbar\bar{\Omega}^{(+)}(t) = 2\mathbf{d}_{2,1} \cdot \mathbf{E}(t) = 2\mathbf{d}_{2,1} \cdot \mathbf{E}_0^{(+)}(\mathbf{R}_0,\, t) + 2\mathbf{d}_{2,1} \cdot \mathbf{E}_{\text{dip}}(t)$$
$$\equiv \hbar\bar{\Omega}_0^{(+)}(t) + \hbar \left(\Delta_{\text{dip}} + i\Gamma_{2,1}\right) \bar{P} e^{-i\omega_L t}. \tag{19.43}$$

This expression shows that the Rabi frequency contains both the known driving term $\bar{\Omega}_0^{(+)}(t)$ as well as the back reaction from the induced atomic polarization.

The quantity $\Gamma_{2,1}$ describes the radiative decay of the dipole resulting from the reradiation-induced decay of the polarization. Its explicit form follows from Eq. (19.42), producing

$$\hbar\Gamma_{21} \equiv \frac{|\mathbf{d}_{2,1}|^2}{6\pi\varepsilon_0\hbar} q_L^3 = 1.249 \times 10^{-9}\,\text{eV} \left[\frac{|\mathbf{d}_{21}|}{\text{e}\text{Å}}\right]^2 \left[\frac{\hbar\omega_L}{\text{eV}}\right]^3. \tag{19.44}$$

The corresponding radiative decay time is

$$\tau_{21} \equiv \frac{1}{\Gamma_{21}} = 5.268 \times 10^{-7}\,\text{s} \left[\frac{\text{e}\text{Å}}{|\mathbf{d}_{21}|}\right]^2 \left[\frac{\text{eV}}{\hbar\omega_L}\right]^3. \tag{19.45}$$

This time scale is very long in the quantum-mechanical sense, see Fig. 3.1. Thus, radiative dephasing is typically much slower than other processes. Nonetheless, it determines the natural line width for atomic transitions, which sets the lower bound for the atomic resonance line width in free space.

The quantity $\hbar\Delta_{\text{dip}}$ represents the energetic shift created by the dipolar back reaction to the exciting field. From Eq. (19.42), we see that

$$\hbar\Delta_{\text{dip}} \equiv \frac{|\mathbf{d}_{2,1}|^2}{6\pi\varepsilon_0\hbar} q_L^2 \frac{\int_0^\infty dr\, r\, g(r)}{\int_0^\infty dr\, r^2 g(r)} = \frac{2\hbar\Gamma_{21}}{\sqrt{\pi} q_L \Delta R}, \tag{19.46}$$

where we assumed a Gaussian distribution $g(r) \propto e^{-\frac{r^2}{\Delta R^2}}$. Since the atomic wave-packet size, ΔR, ranges from a few nanometers to a micron, Δ_{dip} varies roughly from Γ_{21} to $100\Gamma_{21}$. In all cases, we only find very small energetic shifts. Interestingly, Γ_{21} is completely independent of the size of the atomic wave packet while Δ_{dip} is influenced by it.

19.2.2 Radiative decay of planar dipoles

The self-consistent coupling between the optical Bloch equations and Maxwell's wave equation depends very much on the overall geometry of the polarization and the light propagation. For example, when the two-level dipoles have a planar arrangement, the dipole emission follows from Eq. (17.34) such that

$$\mathbf{E}_{\text{dip}}^{(+)}(\mathbf{r}, t) = -\frac{1}{2\varepsilon_0 c}\mathbf{e}_x \frac{\partial}{\partial t} P_{\text{macr}}\left(t - \frac{|z|}{c}\right). \tag{19.47}$$

Here, we assumed that the dipoles lie in the xy-plane at $z = 0$ pointing in the $\mathbf{e}_x$ direction. The density of the macroscopic polarization, $P_{\text{macr}}(t)$, is then simply the dipole polarization of the single two-level system multiplied by the density of such systems, $n_{2\text{D}}$ in the xy-plane. Using the rotating frame for the two-level polarization, $d_{1,2}\bar{P}\, e^{-i\omega_L t}$, we obtain

$$\frac{\partial}{\partial t} P_{\text{macr}}(t) = \frac{\partial}{\partial t} d_{1,2}\, n_{2\text{D}}\, \bar{P}\, e^{-i\omega_L t} \to -i\omega_L\, d_{1,2}\, n_{2\text{D}}\, \bar{P}\, e^{-i\omega_L t}, \tag{19.48}$$

where the last step follows because $\bar{P}$ in the rotating-frame is slowly varying in comparison with the optical frequency. Thus, the dipole-generated field becomes

$$\mathbf{E}_{\text{dip}}^{(+)}(t) \equiv \mathbf{E}_{\text{dip}}^{(+)}(z = 0, t) \to i\frac{\mathbf{d}_{1,2}\, n_{2\text{D}}\, q_L}{2\varepsilon_0}\, \bar{P}\, e^{-i\omega_L t} \tag{19.49}$$

at the plane of the dipoles. Here, we identified the vectorial dipole $\mathbf{d}_{1,2} = d_{1,2}\,\mathbf{e}_x$ and used the relation $q_L = \omega_L/c$.

Since the planar dipole emission is not singular in the dipole plane, we may directly apply $\mathbf{E}_{\text{dip}}^{(+)}(z = 0, t)$ to construct the self-consistent Rabi frequency whenever the planar emitters are much thinner than the optical wave length. Inserting (19.49) into Eq. (19.43), we find

$$\begin{aligned}\hbar\bar{\Omega}_{2\text{D}}^{(+)}(t) &= 2\mathbf{d}_{2,1}\cdot\mathbf{E}_0^{(+)}(z = 0, t)\, e^{+i\omega_L t} + 2\mathbf{d}_{2,1}\cdot\mathbf{E}_{\text{dip}}(z = 0, t)\, e^{+i\omega_L t}\\ &\equiv \hbar\bar{\Omega}_0^{(+)}(t) + i\hbar\Gamma_{2\text{D}}\bar{P}.\end{aligned} \tag{19.50}$$

Here, we have identified the radiative decay

$$\hbar\Gamma_{2D} \equiv \frac{|\mathbf{d}_{2,1}|^2}{2\varepsilon_0\hbar} n_{2D}\, q_L = \hbar\Gamma_{2,1}\frac{3\lambda_L^2 n_{2D}}{4\pi}, \tag{19.51}$$

which follows from Eq. (19.44) after we have expressed $q_L = 2\pi/\lambda_L$ through the wave length. We see that the planar dipoles produce a back reaction to the driving field which is analogous to that of the single atom. As a major difference, the planar dipoles only produce radiative decay while all radiative shifts are missing. In addition, the planar system can show a significantly enhanced Γ_{2D} in comparison with the atomic $\Gamma_{2,1}$ if the system has many dipoles within the area defined by the wave length.

19.3 Optical Bloch equations with radiative coupling

The presented investigations of the atomic and planar dipoles show that we can incorporate the self-consistent driving field into the optical Bloch equations simply by replacing the external Rabi frequency with either Eq. (19.43) or (19.50). This way, we obtain the *self-consistent optical Bloch equations*,

$$\begin{cases} i\frac{\partial}{\partial t}\bar{P} = -\Delta_{2,1}\,\bar{P} + \bar{P}_z\,\bar{\Omega}_0^{(+)} + \Delta_{\text{dip}}\,\bar{P}_z\,\bar{P} + i\Gamma_{2,1}\,\bar{P}_z\,\bar{P} \\ \frac{\partial}{\partial t}\bar{P}_z = \text{Im}\left[\left(\bar{\Omega}_0^{(+)}\right)^\star \bar{P}\right] - \Gamma_{2,1}\,|\bar{P}|^2 \end{cases}. \tag{19.52}$$

For planar dipoles, we simply replace $\Gamma_{2,1}$ by Γ_{2D} and remove the Δ_{dip} contribution.

In the form of Eq. (19.52), the radiative back coupling produces additional nonlinear sources. In general, Eq. (19.52) can therefore only be solved numerically. However, for the case of weak excitations $\bar{\Omega}_0^{(+)}$ close to the south pole of the Bloch sphere, we may approximate $\bar{P}_z$ by $-1/2$, yielding

$$i\frac{\partial}{\partial t}\bar{P} = -\left(\Delta_{2,1} + \frac{\Delta_{\text{dip}}}{2} + i\frac{\Gamma_{2,1}}{2}\right)\bar{P} - 2\frac{\mathbf{d}_{2,1}}{\hbar}\cdot\bar{\mathbf{E}}_0^{(+)}(t), \tag{19.53}$$

where we have expressed the known external Rabi frequency explicitly. We see now that $\Gamma_{2,1}$ introduces an exponential decay of the polarization toward the ground state, i.e., toward the south pole of the Bloch sphere. Since this decay mechanism exists for both planar and point-like dipole, we identify this decay of the light-generated excitations as the main effect of the self-consistent solution.

To deduce the actual motion in the Bloch sphere, we express Eq. (19.52) in its vectorial form, i.e., we apply the identifications (19.15) and (19.19) to obtain

$$\frac{\partial}{\partial t}\bar{\mathbf{P}} = \Omega_0 \times \bar{\mathbf{P}} - \Delta_{\text{dip}}\,\bar{P}_z\left(\mathbf{e}_z \times \bar{\mathbf{P}}\right) + \Gamma_{2,1}\,\bar{\mathbf{P}} \times \left(\mathbf{e}_z \times \bar{\mathbf{P}}\right). \tag{19.54}$$

Here, Ω_0 contains the real and imaginary parts of $\bar{\Omega}_0^{(+)}$ as its x and y components, respectively. For the magnitude of the Bloch vector, we find

$$\frac{\partial}{\partial t}\bar{\mathbf{P}}\cdot\bar{\mathbf{P}} = 2\left[\Omega_0\times\bar{\mathbf{P}}\right]\cdot\bar{\mathbf{P}} - 2\Delta_{\text{dip}}\,\bar{P}_z\,\left(\mathbf{e}_z\times\bar{\mathbf{P}}\right)\cdot\bar{\mathbf{P}}$$
$$+\,2\Gamma_{2,1}\left[\bar{\mathbf{P}}\times\left(\mathbf{e}_z\times\bar{\mathbf{P}}\right)\right]\cdot\bar{\mathbf{P}} = 0, \tag{19.55}$$

which vanishes because of the properties of the cross product. Hence, $\bar{\mathbf{P}}$ remains within the Bloch sphere for all times even if the nonlinear contributions are included.

The full effect of the radiative decay and the associated energy shift can be evaluated only numerically due to the nonlinearity of Eq. (19.54). To study effects common to both an atomic dipole and a planar system, we first set the radiative shift to zero and investigate an 11π square pulse with detuning $\Delta_{2,1} = 0.5|\Omega^{(+)}|$. We start from an initially unexcited system, i.e., $\bar{P}_x = \bar{P}_y = 0$ and $\bar{P}_z = -1/2$. Figure 19.2 compares the dynamics of the excitation with and without radiative decay. Figure 19.2(a) presents the time evolution of $\bar{P}_z$ when the radiative decay is $\Gamma_{2,1} = 0$ (shaded area) and $\Gamma_{2,1} = 0.5|\Omega^{(+)}|$ (solid line). Since the time is expressed in terms of the zero-detuning Rabi period, $T_{\text{Rabi}} = 2\pi/|\Omega^{(+)}|$, the results are invariant with respect to the specific choice of $|\Omega^{(+)}|$.

For $\Gamma_{2,1} = 0$, the state inversion (shaded area) performs completely sinusoidal oscillations. In the presence of radiative decay, $\bar{R}_z$ (solid line) decays exponentially already during the excitation pulse. As a result, the state inversion evolves – in this case rather slowly – toward a steady state. Once the excitation pulse is switched off at $t = 11\,T_{\text{Rabi}}$, $\bar{R}_z$ starts to decay fast toward the unexcited system, i.e., the south pole of the Rabi sphere.

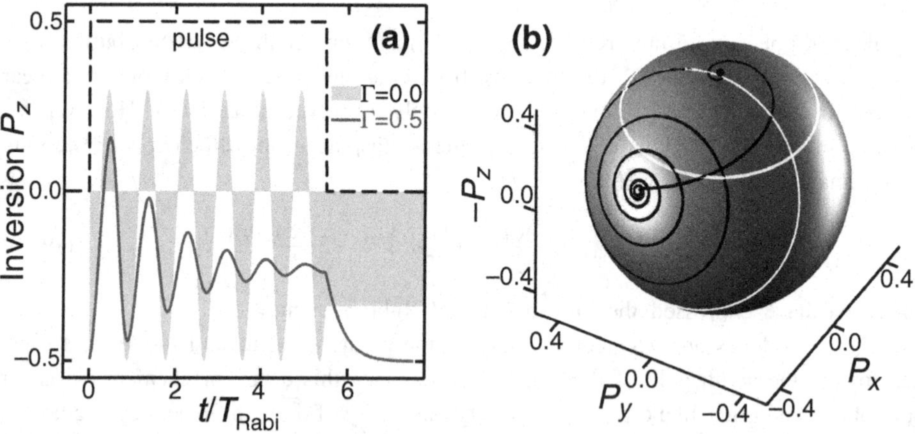

Figure 19.2 Effect of radiative decay on the optical excitations. We used an 11π square pulse (dashed line) with the detuning $\Delta = 0.5|\Omega^{(+)}|$. **(a)** The time evolution of the state inversion is shown for the case without (shaded area) and with radiative decay $\Gamma_{\text{rad}} = 0.5|\Omega^{(+)}|$ (solid line). The time is given in units of $T_{\text{Rabi}} = 2\pi/|\Omega^{(+)}|$ such that the representation becomes independent of the specific choice of $|\Omega^{(+)}|$. **(b)** The computed trajectory of the Bloch vector is shown with (black line) and without (white line) radiative decay.

For the sake of visibility, we plot the Bloch sphere (shaded sphere) upside down to track the Bloch vector's motion close to the ground state (south pole).

It is notable that the decay during the pulse can be significantly slower than after the pulse. This behavioral difference is particularly pronounced when the detuning is reduced while $\Gamma_{2,1}$ is kept constant. Only if $\Gamma_{2,1}$ becomes very large, do we see equally strong decay during and after the pulse. In fact, the behavior changes rather abruptly – typically around $\Gamma_{2,1} > 2|\Omega^{(+)}|$. This effect clearly results from the nonlinearity of the radiative decay phenomena.

The full temporal trajectory of the Bloch vector is presented in Fig. 19.2(b) for both cases analyzed in Fig. 19.2(a). The black circle indicates the south pole of the Bloch sphere, corresponding to the initially unexcited system. Without radiative decay, $\bar{\mathbf{P}}$ performs a circular motion on the Bloch sphere. As before, this motion is divided into two distinctly different parts: during the excitation pulse, the circle is slightly tilted with respect to the z axis, and after the pulse, the direction of the circular path abruptly switches to a horizontal motion where $\bar{P}_z$ remains constant. We see that the radiative decay introduces several changes. The Bloch vector $\bar{\mathbf{P}}$ (black line) no longer follows a circular path but spirals toward an equilibrium configuration and reaches a constant value. After the pulse, the direction of the spiraling motion changes such that $\bar{\mathbf{P}}$ decays very quickly toward the ground state, i.e., the south pole of the Bloch sphere.

The effect of the radiative shift is analyzed in Fig. 19.3 where we assume $\Delta_{\text{dip}} = 2|\Omega^{(+)}(t)|$ and resonant excitation, $\Delta_{2,1} = 0$. Otherwise, the excitation conditions are

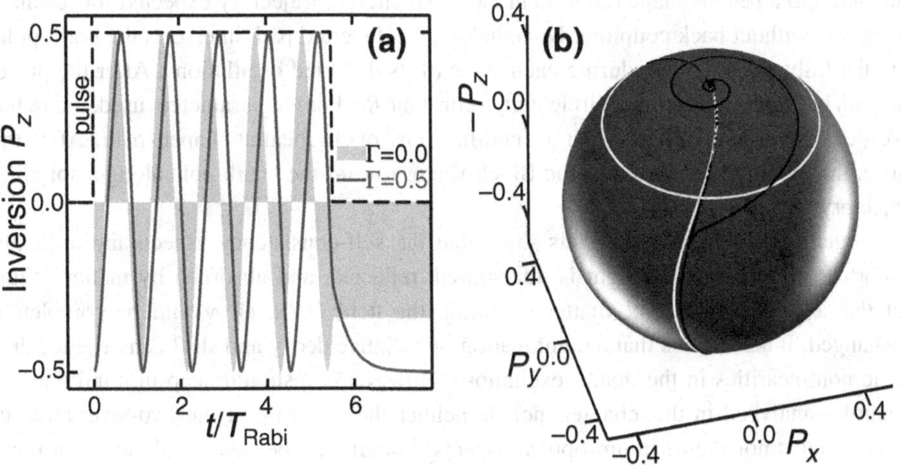

Figure 19.3 Influence of the radiative shift on the optical excitations. A resonant, i.e., $\Delta_{2,1} = 0$, 11π square pulse (dashed line) is applied and the radiative shift is set to $\Delta_{\text{dip}} = 2|\Omega^{(+)}|$. **(a)** The time evolution of the state inversion is shown for the case without (shaded area) and with radiative decay $\Gamma_{\text{rad}} = 0.5|\Omega^{(+)}|$ (solid line). The time is scaled by $T_{\text{Rabi}} = 2\pi/|\Omega^{(+)}|$. **(b)** The corresponding trajectory of the Bloch vector is shown for the case with (black line) and without (white line) radiative decay.

the same as in Fig. 19.2. Figure 19.3 presents the inversion factor as a function of time for $\Gamma_{\text{rad}} = 0$ (shaded area) and $\Gamma_{\text{rad}} = 0.5|\Omega^{(+)}(t)|$ (solid line).

Under the fully resonant conditions chosen and without radiative coupling effects, we see in Fig. 19.3(a) the expected sinusoidal Rabi flopping of $\bar{P}_z$ between its ground-state value $-1/2$ and the excited-state value $1/2$. For a finite Δ_{dip}, $\bar{P}_z$ still reaches its extreme values $\pm 1/2$. However, a closer inspection of the shaded area in Fig. 19.3(a) shows that $\bar{P}_z$ no longer follows a sinusoidal path. Especially, $\bar{P}_z$ remains in the vicinity of its extreme values for shorter times than in the case of a sinusoidal oscillation. This behavior persists even in the presence of radiative decay because during the excitation, the corresponding $\bar{P}_z$ (solid line) is only slightly distorted away from the $\Gamma_{\text{rad}} = 0$ case.

We also see very clearly that continuous resonant pumping can almost completely compensate for the radiative decay when Γ_{rad} is small enough. For the resonant excitation, i.e., $\Delta_{2,1} = 0$, major modifications are observed only for larger dephasing levels $\Gamma_{\text{rad}} > 2|\Omega^{(+)}(t)|$. Clearly, dephasing has a large effect after the pumping ends. Whereas $\bar{P}_z$ simply remains constant without Γ_{rad}, it decays rapidly toward its ground state in the presence of dephasing. This means that only continuous pumping of the system can maintain the excited state. After the pump, we have $\Gamma_{\text{rad}} \gg 2|\Omega^{(+)}(t)|$ showing the dominance of dephasing. This switch-on behavior of the dephasing action follows from the same nonlinear nature of the radiative coupling effects already encountered in Fig. 19.2.

Figure 19.3(b) shows the full trajectory of the Bloch vector for a situation with (black line) and without (white line) radiative decay. We see that the radiative shift alone distorts the path into a peanut shape rather than the usual circular trajectory expected for resonant excitation without back coupling. Nonetheless, the distorted path intersects the north pole, i.e., the fully excited state, during each cycle of its distorted oscillations. After the pulse, the path becomes a horizontal circle with a constant $\bar{R}_z$. For the parameters used, the radiative decay does not introduce large modifications of the peanut-shaped path. After the pulse, the radiative decay forces the Bloch vector toward the south pole along a spiraling trajectory.

Altogether, the discussed results show that the self-consistency aspects are critically important in defining how light is transmitted, reflected, and absorbed by matter. Without the self-consistent light–matter coupling, the field $\langle \hat{E}(\mathbf{x}, t) \rangle$ would be completely unchanged. It also seems that a combination of radiative decay and shift can produce dramatic nonlinearities in the matter excitations. However, we should keep in mind that the examples analyzed in this chapter include neither the coupling of the two-level atom to other systems nor the quantum-optical aspects. Looking at the numerical values, it turns out that the values of $\Gamma_{2,1}$ and Δ_{dip} are incredibly small in comparison with dephasings and shifts created by other sources or quantum-optical effects. Thus, a direct observation of the pure radiative decay and shift-related effects discussed in the present chapter is quite challenging for realistic systems. Nonetheless, they can often mix with more dominant effects to produce nontrivial overall features such that one should not eliminate them lightheartedly from a systematic description.

Exercises

19.1 **(a)** Apply the separation ansatz (19.2)–(19.3) to reduce the Schrödinger equation (19.4) into Eq. (19.9). **(b)** Apply the two-level approximation to construct Eq. (19.12).

19.2 **(a)** Start from Eq. (19.12) and derive the dynamics of the Bloch vector, i.e., Eqs. (19.16)–(19.18). **(b)** Show that the identification (19.19) yields the optical Bloch equation (19.20). **(c)** For resonant conditions, the coordinate system can be chosen such that $\Omega_y = 0$ and $\Omega_x \neq 0$. Use Eq. (19.20) and compute the motion of the Bloch vector to verify the usual Rabi-flopping scenario.

19.3 What is the connection between the parameters φ, Δ, and Υ in Eqs. (18.65)–(18.67) and the angle parametrization (19.26)? Note that we are looking at general ($\Delta \neq 0$) excitations with a square pulse.

19.4 **(a)** Show that the real-valued Eq. (19.27) can generally be solved through the complex-valued Eq. (19.29). **(b)** Show that the solution (19.35) satisfies Eq. (19.33) for any incoming field.

19.5 Show that a radially symmetric $g(\mathbf{x})$ produces an $E_{\text{dip}}(t)$ only in the direction $\mathbf{e}_z$ as Eq. (19.34) is inserted into Eq. (19.38). Verify the validity of Eq. (19.39). Show also that the near-field part of $E_{\text{dip}}(x,t)$ does not contribute to $E_{\text{dip}}(t)$.

19.6 **(a)** Start from Eq. (19.39) and show that Eq. (19.42) yields the identifications (19.44) and (19.46). **(b)** Use a Gaussian $g(\mathbf{R})$ and show that $\hbar\Delta_{\text{dip}} = \frac{2\hbar\Gamma_{21}}{\sqrt{\pi}\, q_L\, \Delta R}$ where ΔR defines the width of the center-of-mass distribution. Evaluate $\Delta_{\text{dip}}/\Gamma_{\text{dip}}$ in terms of realistic ΔR. **(c)** Construct the radiative decay of planar structures by starting from Eq. (19.49).

19.7 **(a)** Show that the self-consistent field, i.e., Eq. (19.43), converts Eq. (19.30) into Eq. (19.52). **(b)** Solve the weak-excitation equation (19.53) and show that the polarization decays after the excitation. **(c)** Show that Eq. (19.54) reproduces Eq. (19.52).

19.8 Linearize Eq. (19.52) around $\mathbf{P}_0$. Show that the polarization decays at the southern hemisphere and grows at the northern hemisphere. Why does the radiative decay tend to move the Bloch vector toward the south pole? How does Δ_{dip} alter the detuning at the hemisphere of the Bloch sphere?

Further reading

Radiative decay is quite often discussed in electromagnetism text books through the Poynting vector, cf.:

J. D. Jackson (1999). *Classical Electrodynamics*, 3rd edition, New York, Wiley & Sons.

J. R. Reitz, F. J. Milford, and R. W. Christy (2008). *Foundations of Electromagnetic Theory*, 4th edition, New York, Addison-Wesley.

A more thorough discussion of radiative decay is found, e.g., in:

P. W. Milonni (1976). Semiclassical and quantum-electrodynamical approaches in nonrelativistic radiation theory, *Phys. Rep.* **25**, 1.

L. C. Andreani, F. Tassone, and F. Bassani (1991). Radiative lifetime of free excitons in quantum wells, *Solid State Commun.* **77**, 641.

M. Kira and S. W. Koch (2006). Many-body correlations and excitonic effects in semiconductor spectroscopy, *Prog. Quantum Electr.* **30**, 155.

20

Dissipative extension of the two-level approach

In most realistic situations, a two-level atom has to be treated as an open system since it cannot be completely isolated from its environment. As a result, the two-level system (TLS) is influenced not only by the optical interactions but also by its coupling to the outside world. Consequently, the TLS experiences some disturbances that expose the light-generated excitations to various decay channels. At the simplest level, the outside world can be modeled as a dissipative reservoir that damps the excitation dynamics.

In this chapter, we analyze how the two-level approach can be extended to systematically include dephasing phenomena. We first introduce a phenomenological formulation of the combined effects of dephasing and radiative decay. Then, we develop the so-called master-equation approach that allows us to describe relaxation phenomena of a TLS at the operator level.

20.1 Spin representation of optical excitations

To extend our two-level analysis, we want to express the optical Bloch equations through operators. For this purpose, we return to the original two-level description provided by Eq. (19.12) and express it in a matrix form,

$$i\frac{\partial}{\partial t}\begin{pmatrix} \bar{a}_2 \\ \bar{a}_1 \end{pmatrix} = \frac{1}{2}\begin{pmatrix} -\Delta & -\bar{\Omega}^{(+)}(t) \\ -\bar{\Omega}^{(-)}(t) & \Delta \end{pmatrix}\begin{pmatrix} \bar{a}_2 \\ \bar{a}_1 \end{pmatrix},$$

$$\hbar\bar{\Omega}^{(+)}(t) = 2\mathbf{d}_{2,1}\cdot\int d^3x\, g(\mathbf{x}-\mathbf{R}_0)\,\langle\hat{\mathbf{E}}^{(+)}(\mathbf{x},t)\rangle e^{i\omega_L t}. \tag{20.1}$$

For completeness, we have repeated here the definition of the self-consistent Rabi frequency (19.31). The coefficients $\bar{a}_2$ and $\bar{a}_1$ that appear can be compactly written as a state vector

$$\bar{\mathbf{a}} \equiv \begin{pmatrix} \bar{a}_2 \\ \bar{a}_1 \end{pmatrix} \quad \Leftrightarrow \quad |\psi_{2\mathrm{LS}}\rangle = e^{-i\left(\omega_1-\frac{i}{2}\Delta_{2,1}\right)t}\left[\bar{a}_1\,|1\rangle + \bar{a}_2\,e^{-i\omega_L t}\,|2\rangle\right], \tag{20.2}$$

which defines the wave function of the TLS, based on Eq. (19.22). Due to the one-to-one mapping between $|\psi_{2\mathrm{LS}}\rangle$ and $\bar{\mathbf{a}}$, the state vector provides one possible quantum-statistical TLS representation.

The 2×2 matrix that appears can be expressed in terms of the identity matrix and the Pauli spin matrices,

$$\sigma_z = \frac{1}{2}\begin{pmatrix} 1 & 0 \\ 0 & -1 \end{pmatrix}, \quad \sigma_+ = \begin{pmatrix} 0 & 1 \\ 0 & 0 \end{pmatrix}, \quad \sigma_- = \begin{pmatrix} 0 & 0 \\ 1 & 0 \end{pmatrix}. \tag{20.3}$$

Combining this and (20.2) allows us to convert Eq. (20.1) into

$$i\hbar \frac{\partial \bar{\mathbf{a}}}{\partial t} = -\hbar \Delta_{2,1}\, \sigma_z\, \bar{\mathbf{a}} - \int d^3x\, g(\mathbf{x} - \mathbf{R}_0) \Big[\mathbf{d}_{2,1} \cdot \langle \hat{\mathbf{E}}^{(-)}(\mathbf{x}, t) \rangle\, e^{-i\omega_L t}\, \sigma_- + \mathbf{d}_{1,2} \cdot \langle \hat{\mathbf{E}}^{(+)}(\mathbf{x}, t) \rangle\, e^{+i\omega_L t}\, \sigma_+ \Big] \bar{\mathbf{a}}. \tag{20.4}$$

Hence, we obtain exactly the form of the usual Schrödinger equation

$$i\hbar \frac{\partial}{\partial t} \bar{\mathbf{a}} = \hat{H}_{2\mathrm{LS}}\, \bar{\mathbf{a}}, \tag{20.5}$$

after having identified the semiclassical system Hamiltonian

$$\hat{H}_{2\mathrm{LS}} = -\hbar \Delta_{2,1}\, \sigma_z - \int d^3x\, g(\mathbf{x} - \mathbf{R}_0) \Big[\mathbf{d}_{2,1} \cdot \langle \hat{\mathbf{E}}^{(-)}(\mathbf{x}, t) \rangle\, e^{-i\omega_L t}\, \sigma_- + \mathbf{d}_{1,2} \cdot \langle \hat{\mathbf{E}}^{(+)}(\mathbf{x}, t) \rangle\, e^{+i\omega_L t}\, \sigma_+ \Big] \tag{20.6}$$

for the TLS. We notice that this Hamiltonian is explicitly time dependent. The dipole interaction is described via the $\langle \hat{\mathbf{E}}^{(\pm)}(\mathbf{x}, t) \rangle\, \sigma_\pm$ contributions instead of the full electrical field. This reduction is a consequence of the rotating-wave approximation included in the two-level approximation, as discussed in Section 18.1.4.

20.2 Dynamics of Pauli spin matrices

The two-level basis (20.2) can directly be used to identify the TLS density matrix,

$$\begin{aligned} \hat{\rho}_{2\mathrm{LS}} &\equiv |\psi_{2\mathrm{LS}}\rangle \langle \psi_{2\mathrm{LS}}| \\ &= \Big[\bar{a}_1 |1\rangle + \bar{a}_2\, e^{-i\omega_L t} |2\rangle \Big] \Big[\bar{a}_1^\star \langle 1| + \bar{a}_2^\star\, e^{+i\omega_L t} \langle 2| \Big] \\ &= |\bar{a}_1|^2 |1\rangle\langle 1| + \bar{a}_1^\star \bar{a}_2\, e^{-i\omega_L t} |2\rangle\langle 1| + \bar{a}_1 \bar{a}_2^\star\, e^{+i\omega_L t} |1\rangle\langle 2| + |\bar{a}_2|^2 |2\rangle\langle 2| \\ &= p_2 |2\rangle\langle 2| + \bar{P}_-\, e^{-i\omega_L t} |2\rangle\langle 1| + \bar{P}_-^\star\, e^{+i\omega_L t} |1\rangle\langle 2| + p_1 |1\rangle\langle 1|, \end{aligned} \tag{20.7}$$

where we used Eq. (19.14) to identify the polarization and $p_j \equiv |\bar{a}_j|^2$ defines the probability of finding the system in the state $|j\rangle$. We see that $|j\rangle\langle k|$ identifies an orthogonal 2×2 basis for the density matrix which can be written in the form

$$\hat{\rho}_{2\mathrm{LS}} \equiv \begin{pmatrix} p_2 & \bar{P}_-\, e^{-i\omega_L t} \\ \bar{P}_-^\star\, e^{+i\omega_L t} & p_1 \end{pmatrix}. \tag{20.8}$$

This density matrix is Hermitian and properly normalized since $\mathrm{Tr}\left[\hat{\rho}_{2\mathrm{LS}}\right] = p_2 + p_1 = 1$ due to the two-level approximation. Using the basis $|j\rangle\langle k|$, we can represent the Pauli spin matrices (20.3),

$$\sigma_z = \frac{1}{2}\left(|2\rangle\langle 2| - |1\rangle\langle 1|\right), \qquad \sigma_+ = |2\rangle\langle 1|, \qquad \sigma_- = |1\rangle\langle 2|, \tag{20.9}$$

using the abstract form. Even though we constructed $\hat{\rho}_{2\mathrm{LS}}$ assuming a pure state, Eq. (20.8) is generally valid also for any mixed state. We can show that Eq. (20.8) has an unphysical negative eigenvalue if the Bloch vector is outside the Bloch sphere (see Exercise 20.1). When the system is in a pure state (20.2), the Bloch vector ends on the surface of the Bloch sphere.

Utilizing the general properties of density matrices presented in Section 6.3, we may evaluate the expectation values of the Pauli matrices

$$\begin{aligned} \langle\sigma_z\rangle &= \mathrm{Tr}\left[\sigma_z\,\hat{\rho}_{2\mathrm{LS}}\right] = \frac{1}{2}\left[p_2 - p_1\right] = \bar{P}_z, \\ \langle\sigma_-\rangle &= \mathrm{Tr}\left[\sigma_-\,\hat{\rho}_{2\mathrm{LS}}\right] = \bar{P}_-\,e^{-i\omega_L t} = P = \left(\bar{P}_x + i\bar{P}_y\right)e^{-i\omega_L t}, \\ \langle\sigma_+\rangle &= \langle\sigma_-\rangle^\star = P_-^\star = \left(\bar{P}_x - i\bar{P}_y\right)e^{+i\omega_L t}. \end{aligned} \tag{20.10}$$

Here, we used the relation (19.15) to express the expectation values in terms of the Bloch-vector components. We see that σ_z defines the operator for the inversion of the system while $\sigma_\pm$ is the polarization operator.

The dynamics of these operators can be determined using Heisenberg's equation of motion. Since $e^{-i\omega_L t}$ is a rapidly oscillating trivial phase, it is usually preferable to discuss the dynamics in the rotating frame (19.11). In practice, we just have to use the system Hamiltonian (20.6) that is already presented in the rotating frame. For the evaluation of the Heisenberg equations, we need the commutation relations for the Pauli spin matrices,

$$\left[\sigma_-, \sigma_+\right]_- = -2\sigma_z, \quad \left[\sigma_\pm, \sigma_z\right]_- = \mp\sigma_\pm \quad \left[\sigma_j, \sigma_j\right]_- = 0\,,\, j = \pm,\, z. \tag{20.11}$$

With the help of these relations, we find

$$\begin{aligned} i\hbar\frac{\partial}{\partial t}\sigma_- &= \left[\sigma_-,\, \hat{H}_{2\mathrm{LS}}\right]_- = -\hbar\Delta_{2,1}\,\sigma_- + \hbar\bar{\Omega}^{(+)}(t)\,\sigma_z \\ i\hbar\frac{\partial}{\partial t}\sigma_z &= \left[\sigma_z,\, \hat{H}_{2\mathrm{LS}}\right]_- = +\hbar\bar{\Omega}^{(-)}(t)\,\sigma_- - \hbar\bar{\Omega}^{(+)}(t)\,\sigma_+, \end{aligned} \tag{20.12}$$

where we identified the Rabi frequency via Eq. (19.13).

The corresponding expectation values in the rotating frame follow from

$$\bar{P}_j(t) = \mathrm{Tr}\left[\sigma_j(t)\,\hat{\rho}_{2\mathrm{LS}}(0)\right], \qquad j = \pm,\, z, \tag{20.13}$$

where $\rho_{2\mathrm{LS}}(0)$ is the density matrix at the initial time. Multiplying both sides of Eq. (20.12) with a static $\hat{\rho}_{2\mathrm{LS}}(0)$ and taking the trace, we obtain

$$i\hbar\frac{\partial}{\partial t}\bar{P}_- = -\hbar\Delta_{2,1}\,\bar{P}_- + \hbar\bar{\Omega}^{(+)}(t)\,\bar{P}_z$$
$$i\hbar\frac{\partial}{\partial t}\bar{P}_z = +\hbar\,\mathrm{Im}\left[\bar{\Omega}^{(-)}(t)\,\bar{P}_-\right]. \tag{20.14}$$

These equations for the polarization and inversion dynamics are identical to Eqs. (19.16) and (19.18), respectively. Thus, we have verified that the description via the Pauli matrices gives the same results as our original wave-function approach. As a major advantage, however, the σ_j dynamics allows for the direct generalization to statistically mixed states described by the density matrix.

20.3 Phenomenological dephasing

In many relevant situations, the atomic TLS is not completely isolated from the outside world but weakly coupled to a reservoir of other states. Examples of such reservoirs can be the radiation field and/or collisions with other atoms. The resulting interactions are essentially random and lead to the dissipation of the excitations and relaxation of the TLS to its ground state. These dissipative processes can be modeled by introducing phenomenological decay terms to the optical Bloch equations. These terms are chosen such that (i) the population inversion $\bar{P}_z$ relaxes exponentially toward its unexcited value $-1/2$ with the rate γ_1 and (ii) the polarization $\bar{P}_-$ decays to zero with the rate γ_2. Implementing these decay processes, we modify Eq. (20.12) to

$$i\frac{\partial}{\partial t}\bar{P}_- = (-\Delta_{2,1} - i\gamma_2)\,\bar{P}_- + \bar{\Omega}^{(+)}(t)\,\bar{P}_z$$
$$\frac{\partial}{\partial t}\bar{P}_z = \mathrm{Im}\left[\bar{\Omega}^{(-)}(t)\,\bar{P}_-\right] - \gamma_1\left(P_z + \frac{1}{2}\right). \tag{20.15}$$

For historical reasons, the decay rates are often represented through T_1 and T_2 times using the relation $\gamma_j = 1/T_j$.

As an illustration, we consider a situation where $\bar{\Omega}^{(+)} \equiv \bar{\Omega}$ is constant and we have zero detuning. In this case, the Bloch vector approaches a steady state that can be determined by setting the RHS of Eq. (20.15) to zero,

$$\begin{cases} -i\gamma_2\,\bar{P}_-^{\mathrm{ste}} + \bar{\Omega}\,\bar{P}_z^{\mathrm{ste}} = 0 \\ \mathrm{Im}\left[\bar{\Omega}^\star\,\bar{P}_-^{\mathrm{ste}}\right] - \gamma_1\left(P_z^{\mathrm{ste}} + \frac{1}{2}\right) = 0 \end{cases} \Rightarrow \begin{cases} \bar{P}_-^{\mathrm{ste}} = \dfrac{i}{2}\dfrac{\gamma_1\,\bar{\Omega}}{|\bar{\Omega}|^2 + \gamma_1\gamma_2} \\ \bar{P}_z^{\mathrm{ste}} = -\dfrac{1}{2}\dfrac{\gamma_1\gamma_2}{|\bar{\Omega}|^2 + \gamma_1\gamma_2} \end{cases}. \tag{20.16}$$

The resulting Bloch vector has the length

$$|\bar{\mathbf{P}}^{\mathrm{ste}}|^2 = \left|\bar{P}_-^{\mathrm{ste}}\right|^2 + \left|\bar{P}_z^{\mathrm{ste}}\right|^2 = \frac{1}{4}\frac{\gamma_1^2|\bar{\Omega}|^2 + \gamma_1^2\gamma_2^2}{\left(|\bar{\Omega}|^2 + \gamma_1\gamma_2\right)^2}. \tag{20.17}$$

Clearly, $\bar{\mathbf{P}}^{\mathrm{ste}}$ must reside within the Bloch sphere, i.e., $|\bar{\mathbf{P}}^{\mathrm{ste}}|^2 \leq 1/4$, for all $|\bar{\Omega}|$, see Exercise 20.1. The solution (20.17) can satisfy this condition only if

$$\gamma_1 \leq 2\gamma_2 \quad \Leftrightarrow \quad 2T_1 \geq T_2. \tag{20.18}$$

In other words, $2\gamma_2$ must always be greater than or equal to γ_1.

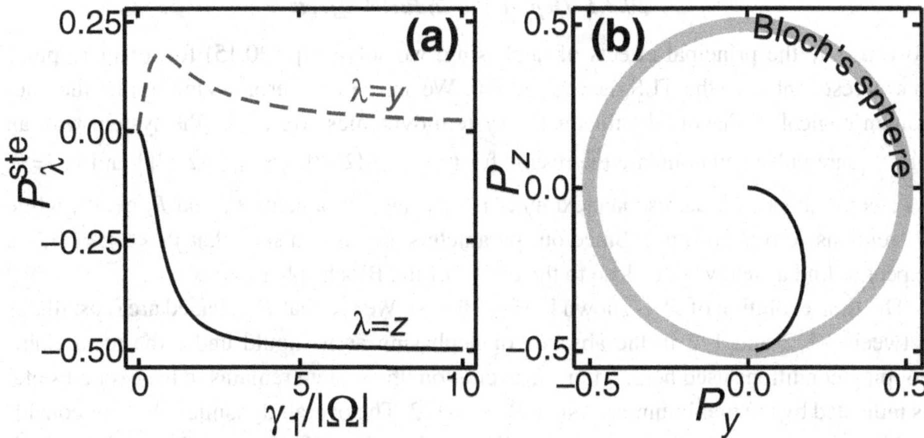

Figure 20.1 Steady-state excitation configurations for different decay rates. We assume resonant optical pumping with the constant amplitude $|\Omega^{(+)}|$. **(a)** The steady-state polarization (dashed line) and the inversion (solid line) are plotted as functions of γ_1; γ_2 is taken at its minimum value satisfying $2\gamma_2 = \gamma_1$. **(b)** The corresponding steady-state positions of the Bloch vector (solid line) are shown within the Bloch sphere (thick solid line). For vanishing decay rates, the Bloch vector starts at the center of the sphere and then curves via the eastern hemisphere toward the unexcited state $(0,\ 0,\ -1/2)$ for elevated γ_1. The steady-state values are evaluated using Eq. (20.16).

Figure 20.1 presents the steady-state configuration $\mathbf{P}^{\rm ste} = \left(0,\ P_y^{\rm ste},\ P_z^{\rm ste}\right)$ for the case where $\gamma_1 = 2\gamma_2$. In Fig. 20.1(a), we plot $P_y^{\rm ste}$ (dashed line) and $P_z^{\rm ste}$ (solid line) as functions of γ_1, based on Eq. (20.17). As a general trend, we see that $P_z^{\rm ste}$ decays monotonously from zero toward its unexcited value $-1/2$. The y-component, $P_y^{\rm ste}$, first increases from zero for low γ_1 before it decays back to zero for larger decay rates.

It is interesting to see that $\gamma_1 \to 0$ produces $\mathbf{P}^{\rm ste} \to (0,\ 0,\ 0)$, which is exactly at the center of the Bloch sphere. Hence, dephasing causes a reduction of the Bloch vector's length, moving it from the surface of the Bloch sphere to its inside. Clearly, the steady-state solution at the center of the Bloch sphere is intuitively acceptable for any finite γ_1. However, for $\gamma_1 = 0$ one would expect sustained Rabi oscillations corresponding to the Bloch vector's motion on the surface of the sphere. This apparent contradiction shows that it is not really meaningful to identify a steady state with the ongoing Rabi flopping. Instead one should always relate the temporal average of $\bar{\mathbf{P}}$ to the steady state. In fact, even an infinitesimal γ_1 is sufficient to bring the steady state to the center of the Bloch sphere.

In Fig. 20.1(b), we illustrate the steady-state configurations reached for different values of $\gamma_1 = 2\gamma_2$. We see that $\bar{\mathbf{P}}^{\rm ste} = \left(0,\ \bar{P}_y^{\rm ste},\ \bar{P}_z^{\rm ste}\right)$ starts from the center of the Bloch sphere for low γ_1 and then curves via the eastern hemisphere toward the unexcited state $(0,\ 0,\ -1/2)$ for elevated γ_1. Thus, a large dephasing brings the excitation of the TLS back into its ground state. Generally, the magnitude of the dephasing should be compared with $|\Omega^{(-)}|$. Only situations with $\gamma_1 \ll |\Omega^{(-)}|$ can be considered to have weak dephasing.

20.3.1 Dephasing-induced effects

To illustrate the principal effects of dephasing, we solve Eq. (20.15) for a square pulse that is resonant with the TLS, i.e., $\Delta_{2,1} = 0$. We set $\Gamma_{2,1}$ to zero, assuming that the phenomenological dephasing dominates the system dynamics. We excite the system with an 11π square pulse and compare the results for $\left(\gamma_1 = \frac{1}{4}|\bar{\Omega}^{(+)}|,\ \gamma_2 = \frac{1}{8}|\bar{\Omega}^{(+)}|\right)$ and $\gamma_j = 0$. As a consequence of the resonant excitation, only the components $\bar{P}_y$ and $\bar{P}_z$ evolve while $\bar{P}_x$ remains zero at all times. Since our parameters are chosen such that $\gamma_1 \ll |\bar{\Omega}^{(+)}|$, we expect to find a steady state close to the center of the Bloch sphere.

The time evolution of $\bar{P}_z$ is shown in Fig. 20.2(a). We see that $\bar{P}_z$ (shaded area) oscillates between $-1/2$ and $1/2$ in the absence of dephasing as it should under the ideal Rabi-flopping conditions used here. After the excitation, the system remains in the excited state, as indicated by the maximum inversion $\bar{P}_z = +1/2$. The overall dynamics changes considerably as soon as dephasing is included. We see from Fig. 20.2(a) that $\bar{P}_z$ (solid line) still oscillates, however, the oscillation amplitude decays toward the steady state already during the excitation. Furthermore, $\bar{P}_z$ no longer reaches its maximum value $+1/2$. The inversion decays monotonously after the excitation.

The full Bloch vector is presented in Fig. 20.2(b). Without dephasing, the Bloch vector (shaded line) remains on a circular trajectory as expected. In the presence of dephasing (solid line), the Bloch vector follows a spiraling trajectory to approach a steady state during

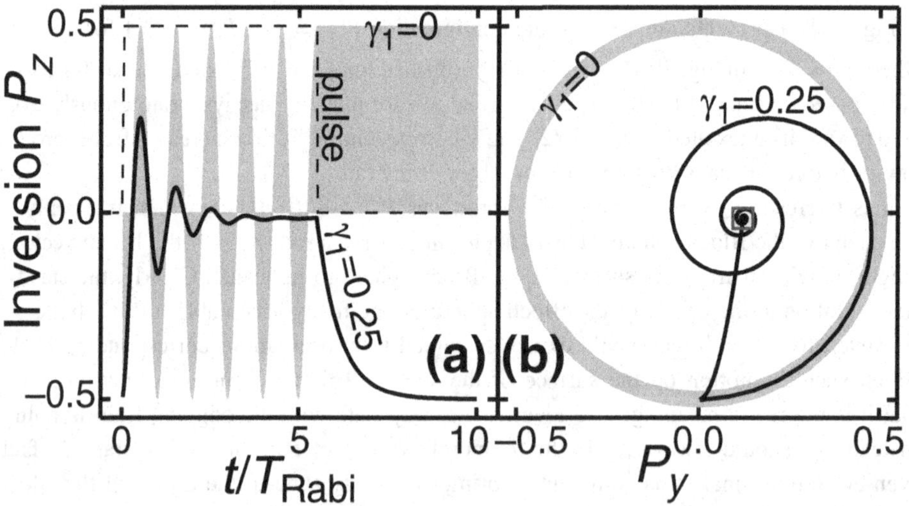

Figure 20.2 Effect of phenomenological dephasing on optical excitations. A resonant, i.e., $\Delta_{2,1} = 0$, 11π square pulse is applied. **(a)** The time evolution of the state inversion is shown for the case without (shaded area) and with decay ($\gamma_1 = |\Omega^{(+)}|/4$, $\gamma_2 = |\Omega^{(+)}|/8$, solid line). The dashed line indicates the excitation pulse. The time is scaled by $T_{\text{Rabi}} = 2\pi/|\Omega^{(+)}|$. **(b)** The corresponding trajectory of the Bloch vector is shown with (black line) and without (shaded line) decay. The square indicates the steady state according to Eq. (20.16). Due to the resonant excitation, $\bar{P}_x$ remains zero.

the pulse. The square indicates the actual steady state configuration $\bar{\mathbf{P}}^{\text{ste}} = \left(0,\ \bar{P}_y^{\text{ste}},\ \bar{P}_z^{\text{ste}}\right)$, based on Eq. (20.16). After the excitation, the Bloch vector abruptly moves away from the steady-state configuration and eventually reaches the south-pole value $\bar{\mathbf{P}}^{\text{ste}} = (0,\ 0,\ -1/2)$ corresponding to a completely unexcited system.

20.3.2 Dephasing and radiative decay

As discussed in Chapter 19, the radiative decay follows from the self-consistent coupling between the light and the TLS. Omitting the radiative shift, we can simply study the combined influence of phenomenological dephasing and radiative decay by adding

$$i\frac{\partial}{\partial t}\bar{P}_-\bigg|_{\text{rad}} = +i\Gamma_{2,1}\,\bar{P}_z\,\bar{P}_-, \qquad \frac{\partial}{\partial t}\bar{P}_z\bigg|_{\text{rad}} = -\Gamma_{2,1}\,|\bar{P}_-|^2 \tag{20.19}$$

to the excitation dynamics (20.15), compare (19.52). As a general feature, the radiative decay makes the excitation dynamics nonlinear such that we must resort to numerical solutions.

The combined influence of phenomenological and radiative decay is analyzed in Fig. 20.3. We use the same resonant square-pulse excitation as in Fig. 20.2 and set the phenomenological dephasing to $\gamma_1 = \frac{1}{4}|\Omega^{(+)}|$ and $\gamma_2 = \frac{1}{8}|\Omega^{(+)}|$. Figure 20.3(a) presents

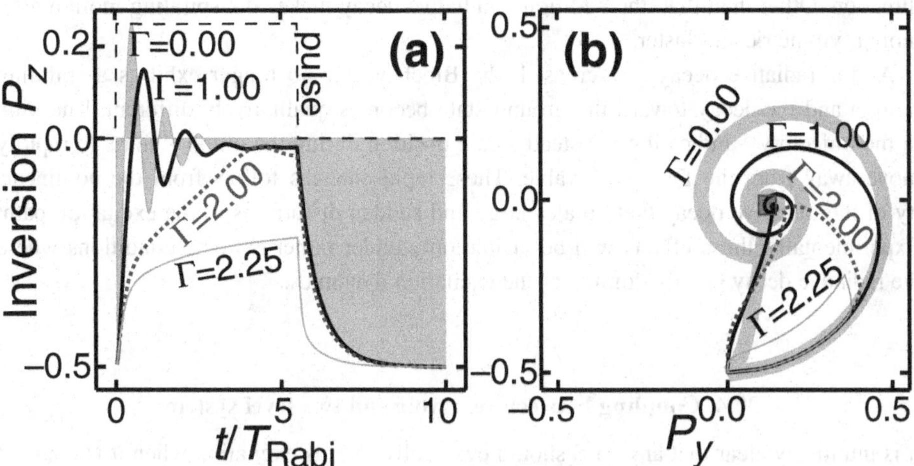

Figure 20.3 Combined influence of dephasing and radiative decay on optical excitations. Results are shown for a resonant, i.e., $\Delta_{2,1} = 0$, 11π square pulse using the dephasing rates $\gamma_1 = |\Omega^{(+)}|/4$ and $\gamma_2 = |\Omega^{(+)}|/8$. **(a)** The time evolution of the state inversion is plotted for different values of the radiative decay: $\Gamma_{2,1} = 0$ (shaded area), $\Gamma_{2,1} = |\Omega^{(+)}|$ (solid line), $\Gamma_{2,1} = 2|\Omega^{(+)}|$ (dotted line), and $\Gamma_{2,1} = 2.25|\Omega^{(+)}|$ (thin solid line). The time is scaled by $T_{\text{Rabi}} = 2\pi/|\Omega^{(+)}|$. **(b)** The corresponding trajectory of the Bloch vector is depicted for $\Gamma_{2,1} = 0$ (thick shaded line), $\Gamma_{2,1} = |\Omega^{(+)}|$ (solid line), $\Gamma_{2,1} = 2|\Omega^{(+)}|$ (dotted line), and $\Gamma_{2,1} = 2.25|\Omega^{(+)}|$ (thin solid line). The square indicates the steady state according to Eq. (20.16). Due to the resonant excitation, $\bar{P}_x$ remains zero.

the inversion factor $\bar{P}_z$ as function of time for different values of $\Gamma_{2,1}$ and the dashed line identifies the excitation pulse. We see that even if the radiative dephasing becomes as large as $\Gamma_{2,1} = |\Omega^{(+)}|$, it only leads to a slight increase of the overall dephasing in $\bar{P}_z$ and stretching of the oscillation period. These changes are relatively modest considering that $\Gamma_{2,1} = |\Omega^{(+)}|$ is eight times larger than γ_2, which suggests that γ_1 and γ_2 dominate the decay for a broad range of physically relevant parameters.

To observe major modifications, $\Gamma_{2,1}$ must become significantly larger than the other decay mechanisms. We find a major change in the qualitative behavior when the radiative decay reaches $2|\Omega^{(+)}|$. For this situation, $\bar{P}_z$ converges directly toward the steady state without significant oscillations. We also see that even the relatively small change from $\Gamma_{2,1} = 2|\Omega^{(+)}|$ to $\Gamma_{2,1} = 2.25|\Omega^{(+)}|$ alters the $\bar{P}_z$ evolution considerably. As discussed in connection with Figs. 19.2–19.3, this sudden change follows from the nonlinearity of the radiative decay, which becomes clearly visible as soon as the radiative decay is the dominating decay channel.

The nonlinear transition in the excitation dynamics is observed particularly well when we follow the trajectories of the Bloch vector in the yz-plane, see Fig. 20.3(b). Due to the resonant excitation, $\bar{P}_x$ vanishes in all cases. For low $\Gamma_{2,1}$, we see that the Bloch vector first exhibits a spiraling motion toward the steady state followed by a decay toward the unexcited state after the pulse. Comparing the cases with $\Gamma_{2,1} = 0$ and $\Gamma_{2,1} = |\Omega^{(+)}|$, we see that the additional radiative decay moves the steady state slightly to the positive $\bar{P}_y$ direction. Other than that, the additional radiative decay makes the spiraling motion a bit more asymmetric and faster.

As the radiative decay is increased, the Bloch vector no longer exhibits a spiraling motion and the decay toward the ground state becomes qualitatively different. The kink in the trajectory – indicating the steady-state position during the pump – starts to rapidly move away from the $\Gamma_{2,1} = 0$ value. These rapid changes follow from the nonlinearity of the radiative decay that creates large and sudden distortions to the excitation path. Experimentally, these effects will be visible only under rather extreme conditions where the radiative decay greatly dominates the excitation dynamics.

20.4 Coupling between reservoir and two-level system

It is intuitively clear that any TLS should eventually lose its excitation when it is coupled to the external world. So far, we have simply accounted for these facts by introducing phenomenological decay rates. To systematically improve our model, we now analyze the interaction of a TLS with a reservoir. Such a coupling is shown schematically in Fig. 20.4 where the reservoir consists of many simple systems (dashed circles) that can be, e.g., harmonic oscillators or other TLS. The excitation of the TLS is transferred to some of the reservoir states from where it then spreads throughout the entire reservoir due to the subsequent interactions. As soon as a few such interactions have happened, it is very unlikely that the excitation released by the TLS ever returns. Instead, the excitation is dissipated

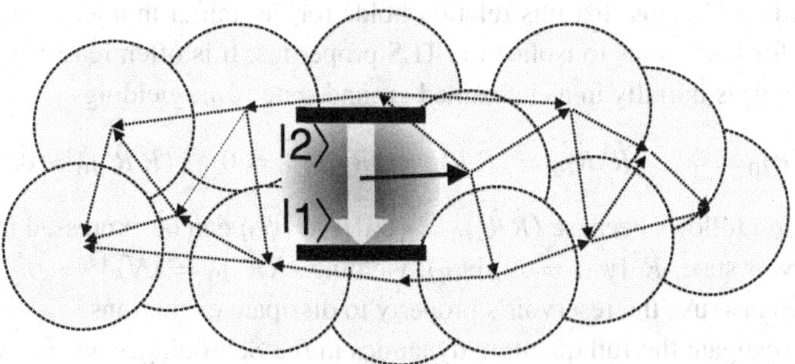

Figure 20.4 Schematic illustration of the coupling of a two-level system to a reservoir. The ground state and the excited state of the two-level system are indicated by $|1\rangle$ and $|2\rangle$, respectively. The reservoir coupling causes de-excitation (shaded arrow) that moves the excitation (solid arrow) to some of the reservoir states (dashed circles). The excitation is then rapidly dissipated (dotted arrows) throughout the reservoir.

throughout the entire reservoir via a network of interactions leading to an effectively irreversible absorption of the TLS excitations. For any true reservoir, we can assume that the dissipation of excitations is so efficient that any trace of it is rapidly lost.

To present a simple model for a reservoir–TLS interaction, we introduce a total system Hamiltonian

$$\hat{H}_{\text{tot}} = \hat{H}_{\text{2LS}} + \hat{W} + \hat{H}_{\text{R}}. \tag{20.20}$$

The Hamiltonian for the isolated TLS alone is given by Eq. (20.6). The interaction between the TLS and the reservoir is modeled as

$$\hat{W} = r\,(\hat{U}^\dagger + \hat{U}), \qquad \hat{U}^\dagger \equiv \hat{R}^\dagger\,\hat{L}, \tag{20.21}$$

where the operator $\hat{U}$ transfers excitation from the TLS into the reservoir. More specifically, $\hat{R}^\dagger$ creates excitation in the reservoir while $\hat{L}$ re-excites the two-level system. One possible choice could be $\hat{L} = \sigma_-$, but we do not need such an explicit form at this point. The constant r defines the coupling strength between the TLS and the reservoir. The last contribution, $\hat{H}_{\text{R}}$, defines interactions or transitions within the reservoir. We do not need its explicit form as long as it has the inherent tendency to irreversibly destroy all excitations.

20.4.1 Master-equation description of dephasing

To investigate the quantum dynamics induced by the Hamiltonian (20.20), we assume a system that is initially defined by the density matrix

$$\hat{\rho}(0) = \hat{\rho}_{\text{2LS}} \otimes \hat{\rho}_R. \tag{20.22}$$

Here, $\hat{\rho}_{\text{2LS}}$ and $\hat{\rho}_R$ describe the TLS and the reservoir, respectively. The density matrix of the TLS alone is obtained by tracing over the reservoir degrees of freedom

$$\hat{\rho}_{\text{2LS}} \equiv \text{Tr}_{\text{R}}\left[\hat{\rho}\right]. \tag{20.23}$$

Even though it is obvious that this relation holds for the initial moment, it can generally also be used for later times to isolate the TLS properties. It is often reasonable to assume that the reservoir is initially in its unexcited ground state $|\psi_0\rangle$, yielding

$$\langle \hat{R} \rangle_0 = 0, \quad \langle \hat{R}^\dagger \hat{R} \rangle_0 = \langle \hat{R}\,\hat{R} \rangle_0 = \langle \hat{R}^\dagger \hat{R}^\dagger \rangle_0 = 0, \quad \langle \hat{R}\,\hat{R}^\dagger \rangle_0 > 0. \tag{20.24}$$

The last relation follows because $\langle \hat{R}\hat{R}^\dagger \rangle_0 = \langle \psi_0 | \hat{R}\,\hat{R}^\dagger | \psi_0 \rangle$ can be expressed in terms of an excited reservoir state, $\hat{R}^\dagger |\psi_0\rangle = \mathcal{N}_R |\psi_R\rangle$, yielding $\langle \hat{R}\hat{R}^\dagger \rangle_0 = |\mathcal{N}_R|^2 > 0$. As discussed above, we will also use the reservoir's property to dissipate excitations.

We first investigate the full quantum dynamics in the Schrödinger picture where $\hat{\rho}(t)$ is the only time-dependent operator. Using the system Hamiltonian (20.20) and the additivity of commutators, we can write the Liouville equation as

$$i\hbar \frac{\partial}{\partial t} \hat{\rho}(t) = \left[\hat{H}_{\text{tot}}, \hat{\rho} \right]_- = \left[\hat{H}_{\text{2LS}}, \hat{\rho} \right]_- + \left[\hat{W}, \hat{\rho} \right]_- + \left[\hat{H}_{\text{R}}, \hat{\rho} \right]_- . \tag{20.25}$$

Since we are eventually interested only in the TLS dynamics, we can trace both sides of this equation over the reservoir degrees of freedom,

$$\begin{aligned} i\hbar \frac{\partial}{\partial t} \hat{\rho}_{\text{2LS}}(t) &= \text{Tr}_{\text{R}} \left[\left[\hat{H}_{\text{2LS}}, \hat{\rho} \right]_- + \left[\hat{W}, \hat{\rho} \right]_- + \left[\hat{H}_{\text{R}}, \hat{\rho} \right]_- \right] \\ &= \left[\hat{H}_{\text{2LS}}, \text{Tr}_{\text{R}} \left[\hat{\rho} \right] \right]_- + \text{Tr}_{\text{R}} \left[\left[\hat{W}, \hat{\rho} \right]_- + \left[\hat{H}_{\text{R}}, \hat{\rho} \right]_- \right] \\ &= \left[\hat{H}_{\text{2LS}}, \hat{\rho}_{\text{2LS}} \right]_- + \text{Tr}_{\text{R}} \left[\left[\hat{W}, \hat{\rho} \right]_- + \left[\hat{H}_{\text{R}}, \hat{\rho} \right]_- \right]. \end{aligned} \tag{20.26}$$

Here, we utilized the additivity of traces, the independence of $\hat{H}_{\text{2LS}}$ on the reservoir degrees of freedom, and the definition (20.23).

Since we do not want to resolve the specific reservoir dynamics, we can simplify the $\hat{W}$ and $\hat{H}_{\text{R}}$ contributions in Eq. (20.26). One idea could be to try to implement a simple dephasing rate by replacing these terms with $-i\gamma \hat{\rho}_{\text{2LS}}$. However, this approximation is too crude because it does not conserve the norm $\text{Tr}\left[\hat{\rho}_{\text{2LS}}\right] = 1$. Instead, we use a perturbation approach assuming that reservoir coupling is weak such that we can expand Eq. (20.26) in terms of r.

Since $\hat{W}$ is static in the Schrödinger picture, we can analyze the dynamics of $\text{Tr}_{\text{R}}\left[\hat{W}\,\hat{\rho}\right]$ by inserting Eq. (20.25),

$$\begin{aligned} i\hbar \frac{\partial}{\partial t} \text{Tr}_{\text{R}} \left[\hat{W}\,\hat{\rho}(t) \right] &= \text{Tr}_{\text{R}} \left[\hat{W} \left[\hat{H}_{\text{2LS}}, \hat{\rho} \right]_- + \hat{W} \left[\hat{H}_{\text{R}}, \hat{\rho} \right]_- \right] \\ &\quad + \text{Tr}_{\text{R}} \left[\hat{W} \left[\hat{W}, \hat{\rho} \right]_- \right]. \end{aligned} \tag{20.27}$$

Here, the last term is quadratic in r because $\hat{W}$ is proportional to r. We also know that any evolution in the reservoir degrees of freedom for $\hat{\rho}(t)$ must be at least of order r because they are initiated by the $\hat{W}$ contributions. Consequently, we may implement an expansion

$$\mathrm{Tr_R}\left[\hat{W}\left[\hat{W},\,\hat{\rho}\right]_-\right]=\mathrm{Tr_R}\left[\hat{W}\left[\hat{W},\,\hat{\rho}_{\mathrm{2LS}}(t)\otimes\hat{\rho}_R(0)\right]_-\right]+\mathcal{O}\left(r^3\right), \tag{20.28}$$

where the lowest-order term evolves only the TLS. We see that as a consequence of the initial condition (20.22), the TLS and the reservoir are decoupled in the dominant contribution. A straightforward commutation produces

$$\begin{aligned}\hat{W}\left[\hat{W},\,\hat{\rho}\right]_- &= r^2\left(\hat{L}^\dagger\hat{R}\hat{R}^\dagger\hat{L}\hat{\rho}-\hat{R}^\dagger\hat{L}\hat{\rho}\hat{L}^\dagger\hat{R}\right)\\ &\quad+r^2\left(\hat{R}^\dagger\hat{L}\hat{L}^\dagger\hat{R}+\hat{R}^\dagger\hat{L}\hat{R}^\dagger\hat{L}+\hat{L}^\dagger\hat{R}\hat{L}^\dagger\hat{R}\right)\hat{\rho}\\ &\quad-r^2\left(\hat{L}^\dagger\hat{R}\hat{\rho}\hat{R}^\dagger\hat{L}+\hat{R}^\dagger\hat{L}\hat{\rho}\hat{R}^\dagger\hat{L}+\hat{L}^\dagger\hat{R}\hat{\rho}\hat{L}^\dagger\hat{R}\right),\end{aligned} \tag{20.29}$$

where we used the definition (20.21) to express $\hat{W}$ explicitly. Combining this result with the expansion (20.28), we obtain

$$\begin{aligned}\mathrm{Tr_R}\left[\hat{W}\left[\hat{W},\,\hat{\rho}\right]_-\right] &= r^2\mathrm{Tr_R}\left[\hat{L}^\dagger\hat{R}\hat{R}^\dagger\hat{L}\hat{\rho}_{\mathrm{2LS}}\otimes\hat{\rho}_R-\hat{R}^\dagger\hat{L}\hat{\rho}_{\mathrm{2LS}}\otimes\hat{\rho}_R\hat{L}^\dagger\hat{R}\right]\\ &\quad+r^2\mathrm{Tr_R}\left[\left(\hat{R}^\dagger\hat{L}\hat{L}^\dagger\hat{R}+\hat{R}^\dagger\hat{L}\hat{R}^\dagger\hat{L}+\hat{L}^\dagger\hat{R}\hat{L}^\dagger\hat{R}\right)\hat{\rho}\right]\\ &\quad-r^2\mathrm{Tr_R}\left[\hat{L}^\dagger\hat{R}\hat{\rho}\hat{R}^\dagger\hat{L}+\hat{R}^\dagger\hat{L}\hat{\rho}\hat{R}^\dagger\hat{L}+\hat{L}^\dagger\hat{R}\hat{\rho}\hat{L}^\dagger\hat{R}\right]\\ &\quad+\mathcal{O}\left(r^3\right),\end{aligned} \tag{20.30}$$

where we substituted $\hat{\rho}\to\hat{\rho}_{\mathrm{2LS}}(t)\otimes\hat{\rho}_R(0)$ only in the first two terms and omitted the time arguments, for the sake of notational brevity. The first term produces

$$\begin{aligned}\mathrm{Tr_R}\left[\hat{L}^\dagger\hat{R}\hat{R}^\dagger\hat{L}\hat{\rho}_{\mathrm{2LS}}\otimes\hat{\rho}_R\right] &= \hat{L}^\dagger\hat{L}\hat{\rho}_{\mathrm{2LS}}\mathrm{Tr_R}\left[\hat{R}\hat{R}^\dagger\hat{\rho}_R(0)\right]\\ &= \hat{L}^\dagger\hat{L}\hat{\rho}_{\mathrm{2LS}}\langle\hat{R}\hat{R}^\dagger\rangle_0.\end{aligned} \tag{20.31}$$

When the partial trace over the reservoir is taken, it does not influence the two-level operators which can be moved outside the trace. The remaining trace then defines the expectation value $\langle\hat{R}\hat{R}^\dagger\rangle_0$ at the initial time, which has a finite value due to Eq. (20.24).

This way, we can also evaluate the other terms in Eq. (20.30) using the fact that the reservoir operators can be cyclically permuted within the trace. As a result, we find that only the first line of Eq. (20.30) produces a nonvanishing contribution. The other terms are proportional to $\langle\hat{R}\hat{R}\rangle_0$, $\langle\hat{R}^\dagger\hat{R}\rangle_0$, or $\langle\hat{R}^\dagger\hat{R}^\dagger\rangle_0$, all of which are vanishing according to Eq. (20.24). Thus, we obtain

$$\mathrm{Tr_R}\left[\hat{W}\left[\hat{W},\,\hat{\rho}\right]_-\right]=r^2\langle\hat{R}\hat{R}^\dagger\rangle_0\left(\hat{L}^\dagger\hat{L}\hat{\rho}_{\mathrm{2LS}}-\hat{L}\hat{\rho}_{\mathrm{2LS}}\hat{L}^\dagger\right), \tag{20.32}$$

as the leading-order contribution. Substituting this into Eq. (20.33), we get

$$\begin{aligned}i\hbar\frac{\partial}{\partial t}\mathrm{Tr_R}\left[\hat{W}\hat{\rho}(t)\right] &= \mathrm{Tr_R}\left[\hat{W}\left[\hat{H}_{\mathrm{2LS}},\,\hat{\rho}\right]_-+\hat{W}\left[\hat{H}_{\mathrm{R}},\,\hat{\rho}\right]_-\right]\\ &\quad+r^2\langle\hat{R}\hat{R}^\dagger\rangle_0\left(\hat{L}^\dagger\hat{L}\hat{\rho}_{\mathrm{2LS}}-\hat{L}\hat{\rho}_{\mathrm{2LS}}\hat{L}^\dagger\right).\end{aligned} \tag{20.33}$$

Hence, $\hat{W}$ indeed generates transitions between the TLS and the reservoir whenever $\hat{L}^\dagger \hat{L} \hat{\rho}_{2LS} - \hat{L} \hat{\rho}_{2LS} \hat{L}^\dagger$ is nonvanishing.

Due to the dissipative nature of the reservoir, we can always assume that $\mathrm{Tr}_R \left[\hat{W}\, \hat{\rho}(t) \right]$ decays. As the simplest approximation, we replace the first two terms in Eq. (20.33) by a phenomenological decay model,

$$i\hbar \frac{\partial}{\partial t} \mathrm{Tr}_R \left[\hat{W}\, \hat{\rho}(t) \right] = -i\hbar\, \kappa\, \mathrm{Tr}_R \left[\hat{W}\, \hat{\rho}(t) \right] + r^2 \langle \hat{R} \hat{R}^\dagger \rangle_0 \left(\hat{L}^\dagger \hat{L} \hat{\rho}_{2LS} - \hat{L} \hat{\rho}_{2LS} \hat{L}^\dagger \right), \tag{20.34}$$

where κ is the relaxation rate of the excitation in the reservoir. If this relaxation is fast enough, the system rapidly reaches its steady state

$$\mathrm{Tr}_R \left[\hat{W}\, \hat{\rho}(t) \right] = i\hbar\gamma \left(\hat{L} \hat{\rho}_{2LS} \hat{L}^\dagger - \hat{L}^\dagger \hat{L} \hat{\rho}_{2LS} \right), \quad \gamma \equiv \frac{r^2 \langle \hat{R} \hat{R}^\dagger \rangle_0}{\hbar^2 \kappa}. \tag{20.35}$$

Here, we introduced the effective decay rate γ. The result (20.35) immediately yields

$$\mathrm{Tr}_R \left[\hat{\rho}(t) \hat{W} \right] = \mathrm{Tr}_R \left[\hat{W}\, \hat{\rho}(t) \right]^\dagger = -i\hbar\gamma \left(\hat{L} \hat{\rho}_{2LS} \hat{L}^\dagger - \hat{\rho}_{2LS} \hat{L}^\dagger \hat{L} \right) \tag{20.36}$$

because both $\hat{W}$ and $\hat{\rho}(t)$ are Hermitian. Subtracting (20.36) from (20.35) yields the commutation relation

$$\mathrm{Tr}_R \left[\left[\hat{W}, \hat{\rho}(t) \right]_- \right] = i\hbar\gamma \left(2\hat{L} \hat{\rho}_{2LS} \hat{L}^\dagger - \hat{L}^\dagger \hat{L} \hat{\rho}_{2LS} - \hat{\rho}_{2LS} \hat{L}^\dagger \hat{L} \right) \tag{20.37}$$

in steady state.

Inserting this result (20.37) into Eq. (20.26), we obtain the *master equation*

$$i\hbar \frac{\partial}{\partial t} \hat{\rho}_{2LS}(t) = \left[\hat{H}_{2LS}, \hat{\rho}_{2LS} \right]_- + i\hbar\gamma \left(2\hat{L} \hat{\rho}_{2LS} \hat{L}^\dagger - \hat{L}^\dagger \hat{L} \hat{\rho}_{2LS} - \hat{\rho}_{2LS} \hat{L}^\dagger \hat{L} \right). \tag{20.38}$$

Here, we set $\mathrm{Tr}_R \left[\left[\hat{H}_R, \hat{\rho} \right]_- \right]$ to zero because the steady-state approximation (20.35) already assumes that quasiequilibrium is reached between the TLS and the reservoir. In other words, the reservoir is assumed to dissipate excitations fast enough on the time scale of the interesting physical processes. The master equation shows that the reservoir coupling produces contributions that cannot be expressed via commutators with an ordinary Hamiltonian. Nevertheless, these terms fully conserve the norm, i.e., $\mathrm{Tr}\left[\hat{\rho}_{2LS} \right] = 1$, such that the master equation is a formally suitable extension of the theory, allowing us to systematically include the effects of dissipation.

The additional reservoir-generated contributions have the so-called Lindblad form that provides a general mathematical framework to treat dissipation in an open system. At the level discussed so far, we ignore all memory effects between the system and the reservoir because Eq. (20.33) has been solved using a steady-state approximation. To explain dynamical efects in semiconductors, one often needs to go beyond the Lindblad form because the relaxation processes may display significant quantum-kinetic aspects.

20.4.2 Master-equation for two-level system

In many investigations, it is preferable to treat the system dynamics in the Heisenberg picture rather than analyzing the temporal evolution of the density matrix in the Schrödinger picture. To see how the master equation changes the operator dynamics, we multiply both sides of Eq. (20.38) by $\hat{O}$ and take a trace,

$$i\hbar\frac{\partial}{\partial t}\mathrm{Tr}\left[\hat{O}\,\hat{\rho}_{2LS}(t)\right] = \mathrm{Tr}\left[\hat{O}\left[\hat{H}_{2LS},\,\hat{\rho}_{2LS}(t)\right]_-\right]$$
$$+\,i\hbar\gamma\,\mathrm{Tr}\left[\hat{O}\left(2\hat{L}\hat{\rho}_{2LS}\hat{L}^\dagger - \hat{L}^\dagger\hat{L}\hat{\rho}_{2LS}(t) - \hat{\rho}_{2LS}(t)\hat{L}^\dagger\hat{L}\right)\right]. \tag{20.39}$$

We now perform cyclic permutations to move $\hat{\rho}_{2LS}$ to the RHS of all other operators,

$$i\hbar\frac{\partial}{\partial t}\mathrm{Tr}\left[\hat{O}\,\hat{\rho}_{2LS}(t)\right] = \mathrm{Tr}\left[\left[\hat{O},\,\hat{H}_{2LS}\right]_-\,\hat{\rho}_{2LS}(t)\right]$$
$$+\,i\hbar\gamma\,\mathrm{Tr}\left[\left(\hat{L}^\dagger\left[\hat{O},\,\hat{L}\right]_- + \left[\hat{L}^\dagger,\,\hat{O}\right]_-\hat{L}\right)\hat{\rho}_{2LS}(t)\right]. \tag{20.40}$$

Since the traces define expectation values, we can write this equation in the form

$$i\hbar\frac{\partial}{\partial t}\langle\hat{O}\rangle = \langle\left[\hat{O},\,\hat{H}_{2LS}\right]_-\rangle + i\hbar\gamma\langle\hat{L}^\dagger\left[\hat{O},\,\hat{L}\right]_- + \left[\hat{L}^\dagger,\,\hat{O}\right]_-\hat{L}\rangle. \tag{20.41}$$

Thus, the usual expression in the Heisenberg picture, i.e., $\langle\left[\hat{O},\,\hat{H}_{2LS}\right]_-\rangle$, is supplemented by additional dephasing contributions resulting from the Lindblad form (20.38).

To obtain explicit expressions, we investigate how the polarization operator dynamics is modified by the coupling of the TLS to the reservoir. Choosing $\hat{L} = \sigma_-$ and evaluating the commutators, we obtain

$$\hat{L}^\dagger\left[\hat{O},\,\hat{L}\right]_- + \left[\hat{L}^\dagger,\,\hat{O}\right]_-\hat{L} = \begin{cases} -\sigma_-\,, & \hat{O} = \sigma_- \\ -\sigma_+\,, & \hat{O} = \sigma_+ \\ -2(\frac{1}{2}+\sigma_z)\,, & \hat{O} = \sigma_z \end{cases}. \tag{20.42}$$

Using these results together with the operator dynamics described by (20.12) in Eq. (20.41), we eventually find

$$i\hbar\frac{\partial}{\partial t}\bar{P}_- = -\hbar\Delta_{2,1}\,\bar{P}_- + \hbar\bar{\Omega}^{(+)}(t)\,\bar{P}_z - i\hbar\gamma\,\bar{P}_-$$
$$i\hbar\frac{\partial}{\partial t}\bar{P}_z = +\hbar\,\mathrm{Im}\left[\bar{\Omega}^{(-)}(t)\,\bar{P}_-\right] - 2i\hbar\gamma\left(\frac{1}{2}+\bar{P}_z\right). \tag{20.43}$$

We see that the resulting equations are identical to the phenomenological Eq. (20.15) if we identify $\gamma_1 = \gamma$ and $\gamma_2 = 2\gamma_1$. Hence, we can conclude that the master equation indeed produces dephasing of the two-level dynamics. It is interesting to see that the $\hat{L} = \sigma_-$ used leads to the largest possible inversion decay according to (20.18). A situation with $\gamma_1 < 2\gamma_2$ can be realized, e.g., by choosing $\hat{L} = \sigma_- + \varepsilon\sigma_z$ with $\varepsilon > 0$, see Exercise 20.7.

Exercises

20.1 (**a**) Show that the Hamiltonian (20.6) yields the usual form, i.e., (20.1). (**b**) Solve the eigenvalues of $\hat{\rho}_{2SL}$ defined by Eq. (20.8). Show that $\hat{\rho}_{2ls}$ is positive definite only if $|\mathbf{P}| \leq 1/2$.

20.2 (**a**) Verify the commutation relations (20.11). (**b**) Compute the Heisenberg equations of motion for σ_j and reproduce Eq. (20.12). (**c**) Show that Eq. (20.12) yields the optical Bloch equations (20.14).

20.3 (**a**) Use the steady-state result (20.16) in the density matrix (20.8). Diagonalize $\hat{\rho}_{2LS}$ and show that it is not a pure state, i.e., $\hat{\rho}_{2LS} \neq |\psi\rangle\langle\psi|$. (**b**) Which condition is found for a pure state?

20.4 Solve the optical Bloch equation (20.15) for a resonant square-pulse excitation. Assume finite dephasing and show that the steady state (20.16) is reached even for infinitesimal γ_j.

20.5 (**a**) Show that the replacement of the reservoir terms in Eq. (20.22) by $-i\gamma\,\hat{\rho}_{2LS}$ yields a decaying $T_R[\hat{\rho}_{2LS}]$. (**b**) Verify the result (20.28) and evaluate Eq. (20.29) explicitly.

20.6 (**a**) Show that the leading order reservoir effects follow from (20.33). (**b**) Show that the relaxation approximation (20.34) produces the results (20.35)–(20.37). Confirm that this yields the master equation (20.38). (**c**) Show that the master equation (20.38) conserves $T_R[\hat{\rho}_{2LS}]$.

20.7 (**a**) Choose $\hat{L} = \sigma_-$ and verify Eq. (20.42). (**b**) Choose $\hat{L} = \sigma_- + \varepsilon\,\sigma_z$ and compute $\hat{L}^\dagger[\hat{O}, \hat{L}]_- + [\hat{L}^\dagger, \hat{O}]_-\,\hat{L}$ for the relevant Pauli matrices. How is the dephasing in the optical Bloch equations (20.43) modified?

Further reading

More details of damping and dephasing effects in two-level systems can be found in:

L. Allen and J. H. Eberly (1975). *Optical Resonance and Two-Level Atoms*, New York Wiley & Sons.

The master equation is a central concept in quantum optics, cf.:

C. W. Gardiner and M. J. Collett (1985). Input and output in damped quantum-systems – quantum stochastic differential-equations and the master equation, *Phys. Rev. A* **31**, 3761.

L. Mandel and E. Wolf (1995). *Optical Coherence and Quantum Optics*, Cambridge, Cambridge University Press.

M. O. Scully and M. S. Zubairy (2002). *Quantum Optics*, Cambridge, Cambridge University Press.

P. Meystre and M. Sargent III (2007). *Elements of Quantum Optics*, 4th edition, Berlin, Springer.

D. F. Walls and G. J. Milburn (2008). *Quantum Optics*, 2nd edition, Berlin, Springer.

H. J. Carmichael (2008). *Statistical Methods in Quantum Optics: Non-Classical Fields*, Berlin, Springer.

The mathematical formulation of the Lindblad form was first introduced in:

G. Lindblad (1976). On the generators of quantum dynamical semigroups, *Commun. Math. Phys.* **48**, 119.

21

Quantum-optical extension of the two-level approach

The two-level-system (TLS) investigations in Chapters 18–20 are based on a completely classical treatment of light. We now extend the analysis into the quantum-optical regime following the canonical quantization scheme of Chapter 10. In this chapter, we formulate the fully self-consistent operator equations that are the starting point for most quantum-optical studies. We then focus on the case where the TLS is placed inside a single-mode cavity. This situation yields a reduced quantum-optical problem, the so-called Jaynes–Cummings model that nicely demonstrates a new class of quantum-optical effects. We also investigate the role and origin of quantum-optical correlations between light and matter.

21.1 Quantum-optical system Hamiltonian

We use the semiclassical system Hamiltonian (19.1),

$$\hat{H}_{sc} = \hat{H}_0^{CM} + \hat{H}_0^{rel} - Q_e \mathbf{r} \cdot \langle \hat{\mathbf{E}}(\mathbf{R}) \rangle, \tag{21.1}$$

as a convenient starting point for the quantum-optical generalization of the TLS treatment presented in the previous chapters. For this purpose, we put

$$\begin{cases} \langle \hat{\mathbf{E}}(\mathbf{R}) \rangle \rightarrow \hat{\mathbf{E}}(\mathbf{R}) = \sum_{\mathbf{q}} i\mathcal{E}_{\mathbf{q}} \left[\mathbf{u}_{\mathbf{q}}(\mathbf{r})\, B_{\mathbf{q}} - \mathbf{u}^{\star}_{\mathbf{q}}(\mathbf{r})\, B^{\dagger}_{\mathbf{q}} \right] \\ \text{include } \hat{H}_F = \sum_{\mathbf{q}} \hbar\omega_{\mathbf{q}} \left[B^{\dagger}_{\mathbf{q}}\, B_{\mathbf{q}} + \frac{1}{2} \right] \end{cases}, \tag{21.2}$$

i.e., we simply replace $\langle \hat{\mathbf{E}}(\mathbf{R}, t) \rangle$ by its operator form $\hat{\mathbf{E}}(\mathbf{R}, t)$ and include the free-field Hamiltonian as in Eq. (10.44). The light quantization converts (21.1) into

$$\hat{H} = \hat{H}_0^{CM} + \hat{H}_0^{rel} - Q_e \mathbf{r} \cdot \hat{\mathbf{E}}(\mathbf{R}) + \sum_{\mathbf{q}} \hbar\omega_{\mathbf{q}} \left[B^{\dagger}_{\mathbf{q}}\, B_{\mathbf{q}} + \frac{1}{2} \right]. \tag{21.3}$$

In principle, this form is equivalent to the original Eq. (10.44) assuming that we use the interaction in the $\mathbf{r} \cdot \hat{\mathbf{E}}$ picture and apply the dipole approximation, as discussed in Chapter 17. Compared with Eq. (17.54), we have omitted the dipole self-energy and replaced $\mathbf{r} \cdot \hat{\mathbf{D}}(\mathbf{R})$ by $\mathbf{r} \cdot \hat{\mathbf{E}}(\mathbf{R})$. As discussed in Section 18.1, these minor approximations are well justified for atomic systems.

21.1.1 Reduction to two-level system

To gain some analytic understanding of the basic quantum-optical effects, it is instructive to start with the two-level approximation. For this purpose, we extend the semiclassical derivation of Section 20.1 to see how the quantization step (21.2) can be applied to the Hamiltonian (20.6). In order to obtain this Hamiltonian, we have made the approximation that the center-of-mass motion is not changed by the light and that the relative and center-of-mass motion can be separated via the ansatz (19.2). To remove the center-of-mass degrees of freedom, we take the corresponding average of the Hamiltonian

$$\begin{aligned}\hat{H}_{\rm sc}^{(2)} &\equiv \langle\psi_{\rm CM}(t)|\hat{H}_{\rm sc}|\psi_{\rm CM}(t)\rangle - \langle\psi_{\rm CM}(t)|\hat{H}_0^{\rm CM}|\psi_{\rm CM}(t)\rangle \\ &= \hat{H}_0^{\rm rel} - Q_e\mathbf{r}\cdot\int d^3x\, g(\mathbf{x}-\mathbf{R}_0)\,\langle\hat{\mathbf{E}}(\mathbf{x},t)\rangle \end{aligned} \tag{21.4}$$

and omit the constant center-of-mass energy. The matrix element between the atom's center-of-mass degree of freedom and the electric field produces the distribution function $g(\mathbf{x}-\mathbf{R}_0) \equiv |\langle\mathbf{x}|\psi_{\rm CM}(t)\rangle|^2$ around the atom's central position $\mathbf{R}_0$, compare Eq. (19.8).

The two-level approximation (TLA) (18.32) restricts the relative-motion problem to the subspace of a ground state and a single excited state. Formally, we simply project $\hat{H}_{\rm sc}^{(2)}$ using the identity operator $\mathbb{I} = |1\rangle\langle 1| + |2\rangle\langle 2|$,

$$\begin{aligned}\hat{H}_{\rm sc}^{(2)} &= (|1\rangle\langle 1| + |2\rangle\langle 2|)\,\hat{H}_{\rm sc}^{(2)}\,(|1\rangle\langle 1| + |2\rangle\langle 2|) \\ &= (|1\rangle\langle 1| + |2\rangle\langle 2|)\left(\hat{H}_0^{\rm rel}|1\rangle\langle 1| + \hat{H}_0^{\rm rel}|2\rangle\langle 2|\right) \\ &\quad - \Big[|1\rangle\langle 1|Q_e\mathbf{r}|1\rangle\langle 1| + |1\rangle\langle 1|Q_e\mathbf{r}|2\rangle\langle 2| \\ &\quad + |2\rangle\langle 2|Q_e\mathbf{r}|1\rangle\langle 1| + |2\rangle\langle 2|Q_e\mathbf{r}|2\rangle\langle 2|\Big] \\ &\quad \cdot\int d^3x\, g(\mathbf{x}-\mathbf{R}_0)\,\langle\hat{\mathbf{E}}(\mathbf{x},t)\rangle. \end{aligned} \tag{21.5}$$

Since $|j\rangle$ is an eigenstate of $\hat{H}_0^{\rm rel}$, we have

$$\hat{H}_0^{\rm rel}|j\rangle = E_j|j\rangle\,, \quad \langle j|k\rangle = \delta_{j,k}\,, \quad \langle j|Q_e\mathbf{r}|k\rangle = \mathbf{d}_{j,k}, \tag{21.6}$$

where $\mathbf{d}_{j,k}$ denotes the dipole matrix element between the two atomic states. Due to the dipole selection rules, the diagonal parts $\mathbf{d}_{1,1}$ and $\mathbf{d}_{2,2}$ vanish. Thus, Eq. (21.5) casts into the form

$$\begin{aligned}\hat{H}_{\rm sc}^{(2)} &= E_1\,|1\rangle\langle 1| + E_2\,|2\rangle\langle 2| - |1\rangle\langle 2|\,\mathbf{d}_{1,2}\cdot\int d^3x\, g(\mathbf{x}-\mathbf{R}_0)\,\langle\hat{\mathbf{E}}(\mathbf{x},t)\rangle \\ &\quad - |2\rangle\langle 1|\,\mathbf{d}_{2,1}\cdot\int d^3x\, g(\mathbf{x}-\mathbf{R}_0)\,\langle\hat{\mathbf{E}}(\mathbf{x},t)\rangle. \end{aligned} \tag{21.7}$$

The operators $|j\rangle\langle k|$ that appear represent processes where the state $|k\rangle$ is converted into $|j\rangle$. As discussed in connection with Eq. (20.9), we may introduce a matrix representation in terms of the Pauli spin matrices

$$\sigma_- \equiv |1\rangle\langle 2|, \qquad \sigma_+ \equiv |2\rangle\langle 1|,$$
$$\begin{cases} \mathbb{I} \equiv |1\rangle\langle 1| + |2\rangle\langle 2| \\ \sigma_z \equiv \frac{1}{2}\left(|2\rangle\langle 2| - |1\rangle\langle 1|\right) \end{cases} \Leftrightarrow \begin{cases} |1\rangle\langle 1| = \frac{\mathbb{I}}{2} - \sigma_z \\ |2\rangle\langle 2| = \frac{\mathbb{I}}{2} + \sigma_z \end{cases}. \tag{21.8}$$

This way, the semiclassical Hamiltonian assumes the form

$$\begin{aligned} \hat{H}^{(2)}_{\mathrm{sc}} &= \frac{\hbar(\omega_1+\omega_2)}{2}\mathbb{I} + \hbar\omega_L\,\sigma_z \\ &\quad - \hbar\Delta_{2,1}\,\sigma_z - \mathbf{d}_{1,2}\cdot\int d^3x\, g(\mathbf{x}-\mathbf{R}_0)\,\sigma_-\,\langle\hat{\mathbf{E}}(\mathbf{x},t)\rangle \\ &\quad - \mathbf{d}_{2,1}\cdot\int d^3x\, g(\mathbf{x}-\mathbf{R}_0)\,\sigma_+\,\langle\hat{\mathbf{E}}(\mathbf{x},t)\rangle, \end{aligned} \tag{21.9}$$

where we also use $\hbar\omega_j \equiv E_j$ and identify the detuning $\Delta_{2,1} = \omega_L - [\omega_2 - \omega_1]$. Here, ω_L defines a reference frequency that can be chosen, e.g., as the central frequency of the exciting light pulse. The first line in (21.9) is the noninteracting part

$$\hat{H}_0 = \frac{\hbar(\omega_1+\omega_2)}{2}\mathbb{I} + \hbar\omega_L\,\sigma_z \tag{21.10}$$

and the remaining two lines represent the semiclassical interaction contributions.

21.1.2 Rotating-wave approximation

The TLA (18.32) also introduces the rotating-wave approximation (RWA) that includes only $E^{(\pm)}$ part of the field that is resonant with the operator $\sigma_\pm$. To see exactly how the RWA must be implemented, we investigate the dynamics in the rotating frame. More precisely, we study the wave function in the interaction picture

$$|\psi_I(t)\rangle = e^{+i\hat{H}_0 t/\hbar}|\psi(t)\rangle. \tag{21.11}$$

Its dynamic evolution follows from

$$\begin{aligned} i\hbar\frac{\partial}{\partial t}|\psi_I(t)\rangle &= i\hbar\frac{\partial}{\partial t}\left[e^{+i\hat{H}_0 t/\hbar}\right]|\psi(t)\rangle + e^{+i\hat{H}_0 t/\hbar}\, i\hbar\frac{\partial}{\partial t}|\psi(t)\rangle \\ &= -\hat{H}_0|\psi_I(t)\rangle + e^{+i\hat{H}_0 t/\hbar}\,\hat{H}^{(2)}_{\mathrm{sc}}\,|\psi(t)\rangle \\ &= \left(-\hat{H}_0 + e^{+i\hat{H}_0 t/\hbar}\,\hat{H}^{(2)}_{\mathrm{sc}}\,e^{-i\hat{H}_0 t/\hbar}\right)|\psi_I(t)\rangle. \end{aligned} \tag{21.12}$$

This introduces the interaction-picture Hamiltonian

$$\begin{aligned} \hat{H}^{(2)}_{\mathrm{sc},I} &\equiv -\hat{H}_0 + e^{+i\hat{H}_0 t/\hbar}\,\hat{H}^{(2)}_{\mathrm{sc}}\,e^{-i\hat{H}_0 t/\hbar} \\ &\equiv -e^{+i\hat{H}_0 t/\hbar}\,\hat{H}_0\,e^{-i\hat{H}_0 t/\hbar} + e^{+i\hat{H}_0 t/\hbar}\,\hat{H}^{(2)}_{\mathrm{sc}}\,e^{-i\hat{H}_0 t/\hbar} \\ &\equiv e^{+i\hat{H}_0 t/\hbar}\left(-\hat{H}_0 + \hat{H}^{(2)}_{\mathrm{sc}}\right)e^{-i\hat{H}_0 t/\hbar}, \end{aligned} \tag{21.13}$$

which removes the trivial $\hat{H}_0$ contribution from the original Hamiltonian.

The Pauli spin matrices are transformed using

$$e^{+i\hat{H}_0 t/\hbar}\,\sigma_z\,e^{-i\hat{H}_0 t/\hbar} = \sigma_z\,, \qquad e^{+i\hat{H}_0 t/\hbar}\,\sigma_\pm\,e^{-i\hat{H}_0 t/\hbar} = \sigma_\pm\,e^{\pm i\omega_L t}. \tag{21.14}$$

Inserting these expressions into Eq. (21.13), we obtain

$$\hat{H}^{(2)}_{\mathrm{sc},I} = -\hbar\Delta_{2,1}\,\sigma_z - \mathbf{d}_{1,2}\cdot\int d^3x\, g(\mathbf{x}-\mathbf{R}_0)\,\sigma_-\,\langle\hat{\mathbf{E}}(\mathbf{x},t)\rangle\, e^{+i\omega_L t}$$
$$- \mathbf{d}_{2,1}\cdot\int d^3x\, g(\mathbf{x}-\mathbf{R}_0)\,\sigma_+\,\langle\hat{\mathbf{E}}(\mathbf{x},t)\rangle\, e^{-i\omega_L t} \tag{21.15}$$

in the original frame. We see that this Hamiltonian is identical to Eq. (20.6) if we implement the rotating-wave approximation

$$\sigma_\pm\,\langle\hat{\mathbf{E}}(\mathbf{x},t)\rangle\, e^{\pm i\omega_L t}\Big|_{\mathrm{RWA}} = \sigma_\pm\,\langle\hat{\mathbf{E}}^{(\pm)}(\mathbf{x},t)\rangle\, e^{\pm i\omega_L t}. \tag{21.16}$$

The noninteracting field evolves as $B_{\mathbf{q}}(t) = B_{\mathbf{q}}(0)\,e^{-i\omega_{\mathbf{q}}t}$, allowing us to identify

$$\hat{\mathbf{E}}^{(+)}(\mathbf{x}) = \sum_{\mathbf{q}} i\mathcal{E}_{\mathbf{q}}\mathbf{u}_{\mathbf{q}}(\mathbf{x})\,B_{\mathbf{q}}, \qquad \hat{\mathbf{E}}^{(-)}(\mathbf{x}) = \left[\hat{\mathbf{E}}^{(+)}(\mathbf{x})\right]^\dagger \tag{21.17}$$

based on Eq. (21.2).

We may generalize the semiclassical two-level approximation (21.15)–(21.16) by applying the quantization step (21.2). This yields the quantized TLS Hamiltonian

$$\hat{H}_{\mathrm{QED}} = -\hbar\Delta_{2,1}\,\sigma_z + \sum_{\mathbf{q}}\hbar\omega_{\mathbf{q}}\left[B^\dagger_{\mathbf{q}}B_{\mathbf{q}} + \frac{1}{2}\right]$$
$$-\int d^3x\, g(\mathbf{x}-\mathbf{R}_0)\left[\mathbf{d}_{1,2}\,\sigma_-\cdot\hat{\mathbf{E}}^{(-)}(\mathbf{x})\,e^{-i\omega_L t}\right.$$
$$\left. + \mathbf{d}_{2,1}\,\sigma_+\cdot\hat{\mathbf{E}}^{(+)}(\mathbf{x})\,e^{+i\omega_L t}\right]. \tag{21.18}$$

Formally, this result could have been guessed directly from the semiclassical form (20.6). However, our derivation confirms that the TLA is not altered by the light quantization and we see that $\hat{H}_{\mathrm{QED}}$ is given in the interaction picture.

The structure of the system Hamiltonian can be simplified after we have introduced the quantized form of the Rabi frequency

$$\hbar\hat{\Omega}^{(+)} \equiv 2\mathbf{d}_{2,1}\cdot\int d^3x\, g(\mathbf{x}-\mathbf{R}_0)\,\hat{\mathbf{E}}^{(+)}(\mathbf{x})\,e^{+i\omega_L t} \equiv 2\sum_{\mathbf{q}} i\mathcal{F}_{\mathbf{q}}\,B_{\mathbf{q}}\,e^{+i\omega_L t}$$
$$\mathcal{F}_{\mathbf{q}} \equiv \mathbf{d}_{2,1}\mathcal{E}_{\mathbf{q}}\cdot\tilde{\mathbf{u}}_{\mathbf{q}}, \qquad \tilde{\mathbf{u}}_{\mathbf{q}} \equiv \int d^3x\, g(\mathbf{x}-\mathbf{R}_0)\,\mathbf{u}_{\mathbf{q}}(\mathbf{x}), \tag{21.19}$$

which generalizes the classical Eq. (19.13). In its quantum form, $\hat{\Omega}^{(+)}$ contains the effective mode strength $\tilde{\mathbf{u}}_{\mathbf{q}}$ multiplied by the dipole-matrix element and the vacuum-field amplitude $\mathcal{E}_{\mathbf{q}}$. Altogether, these quantities define the light–matter coupling strength $\mathcal{F}_{\mathbf{q}}$. This identification converts the two-level Hamiltonian (21.18) into

$$\hat{H}_{\mathrm{QED}} = -\hbar\Delta_{2,1}\,\sigma_z + \sum_{\mathbf{q}}\hbar\omega_{\mathbf{q}}\left[B^\dagger_{\mathbf{q}}B_{\mathbf{q}} + \frac{1}{2}\right] - \frac{\hbar}{2}\left[\sigma_-\hat{\Omega}^{(-)} + \hat{\Omega}^{(+)}\sigma_+\right], \tag{21.20}$$

which will be used as the starting point of our further investigations.

21.1.3 Operator dynamics

To study the quantum dynamics of the coupled light–matter system, we set up the Heisenberg equations of motion for the relevant operators. As in connection with Eq. (20.12), we start by analyzing the dynamics of the Pauli operators,

$$\begin{aligned} i\hbar\frac{\partial}{\partial t}\sigma_- &= \left[\sigma_-,\, \hat{H}_{\text{QED}}\right]_- = -\hbar\Delta_{2,1}\,\sigma_- + \hbar\hat{\Omega}^{(+)}\,\sigma_z \\ i\hbar\frac{\partial}{\partial t}\sigma_z &= \left[\sigma_z,\, \hat{H}_{\text{QED}}\right]_- = +\frac{\hbar}{2}\hat{\Omega}^{(-)}\,\sigma_- - \frac{\hbar}{2}\hat{\Omega}^{(+)}\,\sigma_+. \end{aligned} \tag{21.21}$$

We see that the Pauli operators are directly coupled to photon operators through $\hat{\Omega}^{(\pm)}$ that contains a mode sum over all single photon operators, see Eq. (21.19). Thus, we also need to know the dynamics of the photon operators,

$$i\hbar\frac{\partial}{\partial t}B_{\mathbf{q}} = \left[B_{\mathbf{q}},\, \hat{H}_{\text{QED}}\right]_- = -\hbar\omega_{\mathbf{q}}\,B_{\mathbf{q}} + i\hbar\mathcal{F}^{\star}_{\mathbf{q}}\,\sigma_-\,e^{-i\omega_L t}, \tag{21.22}$$

which follows after we have evaluated the commutator with the Hamiltonian. We see that the photon operator directly couples to the σ_- dynamics. Using Eqs. (21.21)–(21.22) and their Hermitian adjoints, we can construct the dynamics of all possible light–matter operator combinations. Formally, Eqs. (21.21)–(21.22) generalize the Maxwell-optical Bloch equations (19.30) into the quantum-optical regime.

However, we now encounter a major complication. While the photon-operator dynamics contains a single operator $\sigma_\pm$, the light–matter interaction induces a coupling to mixed operators consisting of products of photon and σ_λ operators. Thus, we must additionally solve the dynamics of quantities such as $\hat{B}_{\mathbf{q}}\sigma_\lambda$ which again couple to even higher-order terms of the type $\hat{B}_{\mathbf{q}'}\hat{B}_{\mathbf{q}}\sigma_{\lambda'}$ and $\hat{B}^\dagger_{\mathbf{q}'}\hat{B}_{\mathbf{q}}\sigma_{\lambda'}$. This coupling structure is shown schematically in Fig. 21.1. We see that the σ dynamics is the sole cause of the quantum-optical hierarchy

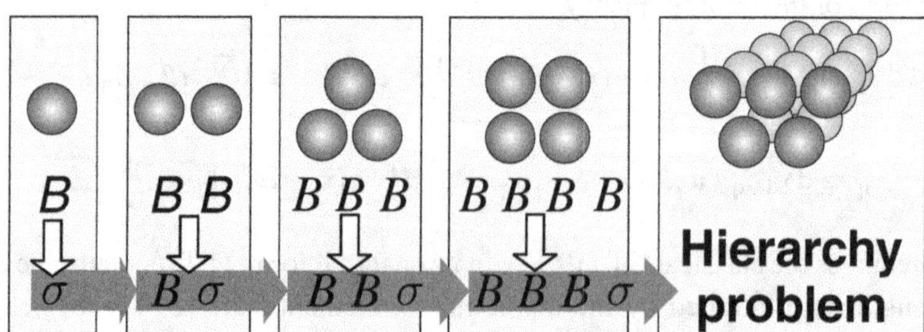

Figure 21.1 Quantum-optical hierarchy problem caused by the dipole interaction. The symbols for B and σ refer to photon and Pauli operators, respectively. The white arrow indicates the light–matter coupling created by the B dynamics (21.22) while the shaded arrow indicates the coupling in the σ dynamics (21.21). The spheres indicate the formal number of particles involved.

problem, compare Chapter 16. To obtain an analytic solution, we need to introduce additional approximations.

Before we develop the needed approximations, we discuss the primary implications of the light quantization. Taking the expectation value of Eq. (21.23) yields

$$i\frac{\partial}{\partial t}\langle\sigma_-\rangle = -\Delta_{2,1}\langle\sigma_-\rangle + \left\langle\hat{\Omega}^{(+)}\sigma_z\right\rangle$$
$$i\frac{\partial}{\partial t}\langle\sigma_z\rangle = +\frac{1}{2}\left\langle\hat{\Omega}^{(-)}\sigma_-\right\rangle - \frac{1}{2}\left\langle\hat{\Omega}^{(+)}\sigma_+\right\rangle. \tag{21.23}$$

These equations are structurally similar to the semiclassical Eq. (20.14) when $\hat{\Omega}^{(\pm)}$ is replaced by a complex-valued amplitude, i.e., $\langle\hat{\Omega}^{(\pm)}\rangle$. More precisely, the quantum-optical description reduces to the semiclassical description if we apply the classical factorization

$$\langle B_{\mathbf{q}}\,\sigma_\lambda\rangle_{\text{cl}} \equiv \langle B_{\mathbf{q}}\rangle\,\langle\sigma_\lambda\rangle \qquad \lambda = \pm,\, z, \tag{21.24}$$

compare also Eq. (15.3). At this level, only the classical part of the light field, i.e., $\langle B_{\mathbf{q}}\rangle$, defines the properties. We can easily convince ourselves that this approximation converts Eq. (21.23) into the optical Bloch equations (19.20), studied in Section 19.1.2. We can also show that Eqs. (21.21)–(21.22) generalize the Maxwell-optical Bloch equations, see Exercise 21.2.

21.1.4 Quantum-optical hierarchy problem

To proceed beyond the semiclassical limit, we now investigate how the cluster-expansion method, introduced in Chapter 15, can be applied to systematically unravel the quantum-optical hierarchy problem resulting from Eq. (21.21). In the cluster-expansion scheme, each σ_λ and each photon operator corresponds to a single-particle operator and an N-fold product of them produces an N-particle expectation value. This identification refers to the formal many-body structure of operators and does not imply that photon and/or Pauli operators represent real particles. A detailed discussion of photons and their interpretation is given in Section 10.4.

By introducing the effective counting of the particle numbers and the resulting ordering of clusters, we can classify the contributions and their relevance in the equations of motion. For example, the hierarchy problem – resulting from Eq. (21.21) – couples N-particle operators directly to $(N+1)$-particle operators. The cluster expansion (15.24)–(15.26) allows for a systematic scheme to identify the relevant correlations that define the quantum fluctuations.

The classical factorization (21.24) identifies the simplest contributions to the light–matter interaction. To discuss genuine quantum-optical effects, the description must include at least the two-particle correlations,

$$\Delta\langle B_{\mathbf{q}}\,\sigma_\lambda\rangle \equiv \langle B_{\mathbf{q}}\,\sigma_\lambda\rangle - \langle B_{\mathbf{q}}\rangle\,\langle\sigma_\lambda\rangle,$$
$$\Delta\left\langle B^\dagger_{\mathbf{q}}\,B_{\mathbf{q}'}\right\rangle \equiv \left\langle B^\dagger_{\mathbf{q}}\,B_{\mathbf{q}'}\right\rangle - \left\langle B^\dagger_{\mathbf{q}}\right\rangle\langle B_{\mathbf{q}'}\rangle,$$
$$\Delta\langle B_{\mathbf{q}}\,B_{\mathbf{q}'}\rangle \equiv \langle B_{\mathbf{q}}\,B_{\mathbf{q}'}\rangle - \langle B_{\mathbf{q}}\rangle\,\langle B_{\mathbf{q}'}\rangle, \tag{21.25}$$

resulting from the difference between the full expectation values and their classical factorizations, as discussed in Section 15.2.2.

These correlations generalize the optical Bloch equations (19.20) into

$$\frac{\partial}{\partial t}\langle\hat{\mathbf{P}}\rangle = \langle\hat{\Omega}\rangle \times \langle\hat{\mathbf{P}}\rangle + \Delta\langle\hat{\Omega} \times \hat{\mathbf{P}}\rangle$$
$$\hat{\mathbf{P}} \equiv \left(\hat{P}_x, \hat{P}_y, \hat{P}_z\right), \quad \hat{\mathbf{P}}_x \equiv \frac{\sigma_- + \sigma_+}{2}, \quad \hat{\mathbf{P}}_y \equiv \frac{\sigma_- - \sigma_+}{2i}, \quad \hat{\mathbf{P}}_z \equiv \sigma_z,$$
$$\hat{\Omega} \equiv \left(\hat{\Omega}_x, \hat{\Omega}_y, \Delta_{2,1}\right), \quad \hat{\Omega}_x \equiv \frac{\hat{\Omega}^{(-)} + \hat{\Omega}^{(+)}}{2}, \quad \hat{\Omega}_y \equiv \frac{\hat{\Omega}^{(-)} - \hat{\Omega}^{(+)}}{2i}. \tag{21.26}$$

We see that the correlations directly produce the quantum-optical contributions. Hence, in quantum-optical investigations, we have to treat the dynamics of the two-particle correlations – and beyond – to determine the system behavior.

However, before we enter into such quantum-dynamics investigations, we want to reanalyze the physical origin of the quantum-optical hierarchy problem. The hierarchical operator coupling in Eq. (21.21) clearly follows from the light–matter interaction part of Eq. (21.20),

$$\hat{H}_{\mathrm{lm}} = -\frac{\hbar}{2}\left[\sigma_-\,\hat{\Omega}^{(-)} + \hat{\Omega}^{(+)}\sigma_+\right] \equiv -\frac{\hbar}{2}\left(\hat{P}_-\,\hat{\Omega}^{(-)} + \hat{\Omega}^{(+)}\,\hat{P}_-^\dagger\right), \tag{21.27}$$

which is expressed in terms of the general polarization operator $\hat{P}_-$. For photon operators, we always find that the interaction described by $\hat{H}_{\mathrm{lm}}$ induces the coupling

$$i\hbar\frac{\partial}{\partial t}B_{\mathbf{q}} = \left[B_{\mathbf{q}}, \hat{H}_{\mathrm{lm}}\right]_- = -\frac{\hbar}{2}\left(\hat{P}_-\left[B_{\mathbf{q}}, \hat{\Omega}^{(-)}\right]_- + \left[B_{\mathbf{q}}, \hat{\Omega}^{(+)}\right]_-\hat{P}_-^\dagger\right) \tag{21.28}$$

because $B_{\mathbf{q}}$ commutes with the matter polarization. Using the definition (21.19), we can evaluate the commutation relations with the result

$$\left[B_{\mathbf{q}}, \hbar\hat{\Omega}^{(-)}\right]_- = -2\sum\nolimits_{\mathbf{q}'} i\mathcal{F}^\star_{\mathbf{q}'}\left[B_{\mathbf{q}}, B^\dagger_{\mathbf{q}'}\right]_- e^{-i\omega_L t} = -2i\mathcal{F}^\star_{\mathbf{q}}\,e^{-i\omega_L t}$$
$$\left[B_{\mathbf{q}}, \hbar\hat{\Omega}^{(+)}\right]_- = 2\sum\nolimits_{\mathbf{q}'} i\mathcal{F}_{\mathbf{q}'}\left[B_{\mathbf{q}}, B_{\mathbf{q}'}\right]_- e^{+i\omega_L t} = 0. \tag{21.29}$$

Since the photon operators satisfy Bosonic commutation relations, the commutators produce a complex number or zero – not an operator. Consequently, the light–matter coupling does not induce a hierarchy problem through the photon operators.

Analyzing the matter polarization dynamics, we obtain

$$i\hbar\frac{\partial}{\partial t}\hat{P}_- = \left[\hat{P}_-, \hat{H}_{\mathrm{lm}}\right]_- = -\frac{\hbar}{2}\left(\left[\hat{P}_-, \hat{P}_-\right]_-\hat{\Omega}^{(-)} + \hat{\Omega}^{(+)}\left[\hat{P}_-, \hat{P}_-^\dagger\right]_-\right)$$
$$= -\frac{\hbar}{2}\hat{\Omega}^{(+)}\left[\hat{P}_-, \hat{P}_-^\dagger\right]_-. \tag{21.30}$$

If the polarization operator could obey Bosonic commutation statistics, just as the photons, the remaining commutator would be a constant and no hierarchical coupling would emerge. However, for the atomic polarization, the commutator is

$$\left[\hat{P}_-, \hat{P}_-^\dagger\right]_- = [\sigma_-, \sigma_+]_- = -2\sigma_z, \tag{21.31}$$

which is the nontrivial Fermionic inversion operator. These Fermionic properties of the atomic excitation are the direct cause of the quantum-optical hierarchy problem. This means that even if light initially has a classical form, the light–matter coupling inevitably produces quantum-optical correlations through the Fermionic aspects of the atomic polarization. If the matter excitations were purely Bosonic, an interaction of the type (21.27) could not alter the light quantum statistics.

To define when to expect strong quantum-optical effects, we take the expectation value of the commutation relation (21.31),

$$\left\langle\left[\hat{P}_-, \hat{P}_-^\dagger\right]_-\right\rangle = -2\langle\sigma_z\rangle = 1 - 2\left(P_z + \frac{1}{2}\right), \tag{21.32}$$

where we identified the state-inversion component of the Bloch vector, based on Eq. (20.10). We see that the polarization exhibits a Bosonic commutation relation, i.e., $\langle[\hat{P}_-, \hat{P}_-^\dagger]_-\rangle = 1$, only when the Bloch vector is at the south pole of the Bloch sphere, i.e., $P_z = -1/2$. Thus, only negligibly weak excitations close to the south pole do not produce appreciable quantum-optical correlations. The maximal deviation from this behavior is found close to the north pole, producing $\langle[\hat{P}_-, \hat{P}_-^\dagger]_-\rangle = 1$. Thus, the quantum-optical effects become most prominent for an inverted TLS where its true Fermionic nature is most visible. Thus, we may expect that strong excitations yielding, e.g., a Rabi-flopping dynamics, must generate pronounced quantum-optical effects.

21.2 Jaynes–Cummings model

To gain some analytic insight into the hierarchy problem, one often searches for a simplified version of the system Hamiltonian (21.20) that still includes the essential features of the quantum-optical coupling between the Bosonic light and the Fermionic excitations of matter. As a physical motivation, we can consider the situation shown in Fig. 21.2(a) where the TLS is positioned inside a cavity. Figure 21.2(b) presents a typical form of the mode strength $|\tilde{\mathbf{u}}_\mathbf{q}|^2$ as a function of $\omega_\mathbf{q}$. We see that this mode strength is strongly concentrated around a single resonance frequency, ω_c, denoted as the cavity mode.

Since $\mathcal{F}_\mathbf{q} \propto \tilde{\mathbf{u}}_\mathbf{q}$ defines the strength of the light–matter interaction, it is intuitively clear that the atom predominantly interacts with the cavity mode. Thus, it is reasonable to simplify the system Hamiltonian (21.20) by making a single-mode approximation. As a first step, we introduce an effective photon operator for the cavity mode

$$B_c \equiv \frac{1}{\hbar g_c} \sum\nolimits_\mathbf{q} i\mathcal{F}_\mathbf{q} B_\mathbf{q}, \qquad \hbar g_c \equiv \sqrt{\sum\nolimits_\mathbf{q} |\mathcal{F}_\mathbf{q}|^2}. \tag{21.33}$$

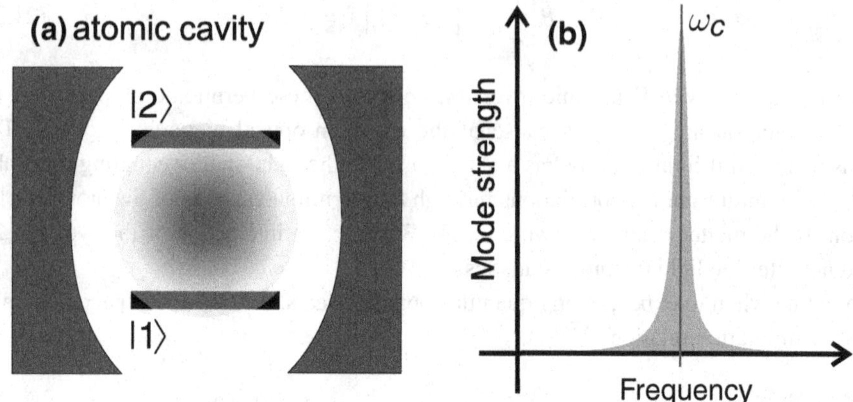

Figure 21.2 Two-level system inside a high-quality cavity. **(a)** The horizontal lines indicate the excited $|2\rangle$ and the ground state $|1\rangle$. The convex structures symbolize mirrors producing a standing wave (shaded area) inside the cavity. The blurred area is the center-of-mass distribution of an atom. **(b)** A typical mode strength $|\mathbf{u_q}|^2$ for an atom inside a cavity.

The constant $\hbar g_c$ is chosen such that the Bosonic commutation relations,

$$\left[B_c,\, B_c^\dagger\right]_- = 1\,, \qquad \left[B_c^\dagger,\, B_c^\dagger\right]_- = [B_c,\, B_c]_- = 0, \tag{21.34}$$

are satisfied.

Using (21.33) in the quantized system Hamiltonian (21.20), we obtain

$$\begin{aligned} \hat{H}_{\rm QED} = & -\hbar\Delta_{2,1}\,\sigma_z + \sum\nolimits_{\mathbf{q}} \hbar\omega_{\mathbf{q}}\left[B_{\mathbf{q}}^\dagger\, B_{\mathbf{q}} + \frac{1}{2}\right] \\ & - \hbar g_c\left[\sigma_-\, e^{-i\omega_L t}\, B_c^\dagger + \sigma_+\, e^{+i\omega_L t}\, B_c\right]. \end{aligned} \tag{21.35}$$

At this level, the TLS exclusively interacts with a single quantized mode B_c. Even though the single-mode concept is simplest for a single-mode cavity, the reduction into an effective single-mode B_c is possible even without a cavity because Eqs. (21.33)–(21.35) are valid for any $\mathcal{F}_{\mathbf{q}}$. Thus, the light–matter interaction can always be reduced into the coupling between the TLS and an effective single mode. The resulting single-mode light field may have unusual properties because $\mathcal{F}_{\mathbf{q}}$ can have an arbitrary spectral shape. The identified g_c then defines the real-valued coupling strength between the TLS and the single mode of quantized light, whichever shape it has.

All possible multimode complications arise from the free-field part of the Hamiltonian because it inevitably includes all light modes. To approximate these effects, we consider a strongly peaked $\mathcal{F}_{\mathbf{q}}$ as in Fig. 21.2(b). In this case, only $\omega_{\mathbf{q}}$ modes in the vicinity of ω_c can be influenced by the TLS. Thus, we may approximate the $\mathbf{q}$ sum to include only the near resonant modes. To search for the simplest possible form, we analyze

$$\hbar\omega_c B_c^\dagger B_c = \sum\nolimits_{\mathbf{q},\mathbf{q}'} \hbar\omega_c \frac{\mathcal{F}_\mathbf{q}^\star \mathcal{F}_{\mathbf{q}'}}{\hbar^2 g_c^2} B_\mathbf{q}^\dagger B_{\mathbf{q}'}$$

$$= \sum_\mathbf{q} \hbar\omega_\mathbf{q} B_\mathbf{q}^\dagger B_\mathbf{q} + \sum_{\mathbf{q},\mathbf{q}'} \frac{\mathcal{F}_{\mathbf{q}'} B_\mathbf{q}^\dagger}{\hbar g_c^2} \left(\omega_c \mathcal{F}_\mathbf{q}^* B_{\mathbf{q}'} - \omega_\mathbf{q} \mathcal{F}_{\mathbf{q}'}^* B_\mathbf{q}\right). \qquad (21.36)$$

We see that the $\mathbf{q}' = \mathbf{q}$ contribution, corresponding to ω_c, cancels in the last term. At the same time, this contribution should be dominant in a situation where $\mathcal{F}_\mathbf{q}$ is strongly peaked, see Exercise 21.5. Thus, we introduce the single-mode approximation by omitting the sum-term in (21.36) altogether and replace the free-photon Hamiltonian by $\hbar\omega_c B_c^\dagger B_c$. This approximation allows for a reasonably good description of quantum-optical effects in cavities. Formally, this approximation directly converts Eq. (21.35) into the so-called Jaynes–Cummings Hamiltonian

$$\hat{H}_{\mathrm{JC}} = \hbar\omega_{2,1}\,\sigma_z + \hbar\omega_c\left(B_c^\dagger B_c + \frac{1}{2}\right) - \hbar g_c\left[\sigma_-\, B_c^\dagger + \sigma_+\, B_c\right]. \qquad (21.37)$$

In our derivation of this Hamiltonian, ω_L defines the relevant reference frame. If we chose $\omega_L = \omega_c$, we work in the rotating frame defined by the cavity mode. Here, we describe the physics in a stationary frame by setting $\omega_L = 0$, leading to a time-independent Hamiltonian with a transition frequency $-\Delta_{2,1} = \omega_2 - \omega_1 = \omega_{2,1}$. This particular frame therefore corresponds to the Schrödinger picture. Additionally, we have only included the zero-point energy of the remaining single mode, neglecting all the other modes and their zero-point energies.

The Jaynes–Cummings Hamiltonian constitutes one of the standard models in quantum optics. Despite its relative simplicity, it fully includes the quantum-optical hierarchy problem because the light–matter coupling is exact, as shown by Eq. (21.35). However, the multimode aspects are simplified to the level where Eq. (21.37) involves only two degrees of freedom. Therefore, we can analytically solve the eigenvalue problem.

As an extension of the rigorous single-mode approximation, the coupling to the other light modes can be viewed as reservoir coupling leading to the decay of light out of the cavity and/or decay of the two-level excitations. These aspects can still be incorporated in the single-mode theory as long as the reservoir coupling is treated at the master-equation level introduced in Section 20.4.1. Hence, the Jaynes–Cummings model provides a rather flexible platform to gain an understanding of the principal quantum-optical effects.

21.2.1 Eigenstates

The quantum properties of a microscopic entity are known once all of its eigenstates are determined. For this purpose, we next analyze the eigenstates of the Jaynes–Cummings model. We start from the noninteracting problem

$$\hat{H}^0_{\text{JC}}\,|\phi_E\rangle = E|\phi_E\rangle, \qquad \hat{H}^0_{\text{JC}} \equiv \hbar\omega_{2,1}\,\sigma_z + \hbar\omega_c\left(B_c^\dagger B_c + \frac{1}{2}\right), \tag{21.38}$$

where $\omega_{2,1} = \omega_2 - \omega_1$ defines the frequency difference between the excited and the ground state. We know from Section 10.4 that the Fock states are the proper eigenstates of a quantized single-mode light field. These states satisfy the stationary Schrödinger equation,

$$\hbar\omega_c\left(B_c^\dagger B_c + \frac{1}{2}\right)|n\rangle = \hbar\omega_c\left(n + \frac{1}{2}\right)|n\rangle, \qquad n = 0,\,1,\,2,\cdots, \tag{21.39}$$

for all values of n.

For the noninteracting TLS, we have the excited state $|+\rangle \equiv |2\rangle$ and the ground state $|-\rangle \equiv |1\rangle$. To avoid notational confusion with the Fock states $|1\rangle$ and $|2\rangle$, we now label the TLS states as $|\pm\rangle$. Equation (20.9) allows us to use the representation $\sigma_z \equiv \frac{1}{2}\left(|+\rangle\langle+| - |-\rangle\langle-|\right)$, which directly produces

$$\hbar\omega_{2,1}\,\sigma_z\,|\pm\rangle = \pm\frac{\hbar\omega_{2,1}}{2}\,|\pm\rangle, \tag{21.40}$$

solving the atomic eigenvalue problem.

The product states $|\pm,\,n\rangle \equiv |\pm\rangle \otimes |n\rangle$ define a complete set because they are the eigenstates of the noninteracting system,

$$\hat{H}^0_{\text{JC}}\,|\pm,\,n\rangle = \hbar\left(\omega_c\left(n + \frac{1}{2}\right) \pm \frac{\omega_{2,1}}{2}\right)|\pm,\,n\rangle \equiv E^0_{\pm,n}\,|\pm,\,n\rangle. \tag{21.41}$$

To rewrite the eigenenergy of the noninteracting system, we identify the detuning between the cavity and the TLS

$$\Delta_{2,1} \equiv \omega_c - \omega_{2,1}. \tag{21.42}$$

Hence, we can write the eigenenergies as

$$\begin{aligned} E^0_{-,n} &= \hbar\left(\omega_c\,n + \frac{\Delta_{2,1}}{2}\right), \qquad n = 0,\,1,\,2,\cdots, \\ E^0_{+,n-1} &= \hbar\left(\omega_c\,n - \frac{\Delta_{2,1}}{2}\right), \qquad n = 1,\,2,\,3,\cdots \end{aligned} \tag{21.43}$$

In other words, the states $|-,\,n\rangle$ and $|-,\,n+1\rangle$ are separated by the energy $\hbar\Delta_{2,1}$. Clearly, these states are degenerate for vanishing detuning. Only the ground state $|-,\,0\rangle$ is nondegenerate. Altogether, the states $|-,\,n\rangle$ (or $|+,\,n\rangle$) form a discrete set with a separation of $\hbar\omega_c$, just as in any harmonic-oscillator problem.

21.2.2 Interacting Jaynes–Cummings states

To solve the full Jaynes–Cummings model, we use

$$\begin{aligned} B|n\rangle &= \sqrt{n}|n-1\rangle, \qquad & B^\dagger|n\rangle &= \sqrt{n+1}|n+1\rangle, \\ \sigma_\pm|\mp\rangle &= |\pm\rangle, \qquad & \sigma_\pm|\pm\rangle &= 0, \end{aligned} \tag{21.44}$$

which follows directly from Eqs. (10.88) and (20.9). The noninteracting ground state produces

$$\begin{aligned}\hat{H}_{\rm JC}\,|-,\,0\rangle &= \left(\hat{H}^0_{\rm JC} - \hbar g_c\left[B_c^\dagger\,\sigma_- + \sigma_+\,B_c\right]\right)|-,\,0\rangle\\ &= E^0_{-,0}\,|-,\,0\rangle - \hbar g_c\,B_c^\dagger\,\sigma_-|-,\,0\rangle - \hbar g_c\,\sigma_+\,B_c|-,\,0\rangle\\ &= E^0_{-,0}\,|-,\,0\rangle \end{aligned}\tag{21.45}$$

because the interaction gives a vanishing contribution. Thus, the noninteracting ground state is an eigenstate also for the interacting Jaynes–Cummings model.

To define the excited eigenstates, we make an ansatz for the wave function,

$$|\phi_n\rangle \equiv c_n^-\,|-,\,n\rangle + c_n^+\,|+,\,n-1\rangle, \qquad n = 1,\,2,\,\cdots \tag{21.46}$$

using the nearly degenerate eigenstates of the noninteracting system. The coefficients $c_{\pm,\,n}$ have to be determined through the eigenvalue problem,

$$\begin{aligned}&\hat{H}_{\rm JC}\,|\phi_n\rangle = E_n\,|\phi_n\rangle\\ \Leftrightarrow\quad &\hat{H}^0_{\rm JC}\,|\phi_n\rangle - \hbar g_c\,B_c^\dagger\,\sigma_-|\phi_n\rangle - \hbar g_c\sigma_+\,B_c\,|\phi_n\rangle = E_n\,|\phi_n\rangle.\end{aligned}\tag{21.47}$$

Using (21.41) in a term-by-term evaluation, we find

$$\begin{aligned}\hat{H}^0_{\rm JC}\,|\phi_n\rangle &= c_n^-\,\hat{H}^0_{\rm JC}\,|-,\,n\rangle + c_n^+\,\hat{H}^0_{\rm JC}\,|+,\,n-1\rangle\\ &= E^0_{-,n}\,c_n^-\,|-,\,n\rangle + E^0_{+,n-1}\,c_n^+\,|+,\,n-1\rangle\end{aligned}\tag{21.48}$$

for the first contribution. The second contribution follows from relation (21.44),

$$\begin{aligned}g_c\,B_c^\dagger\,\sigma_-|\phi_n\rangle &= c_n^-\,g_c\,B_c^\dagger\,\cancel{\sigma_-\,|-,\,n\rangle} + c_n^+\,g_c\,B_c^\dagger\,\sigma_-\,|+,\,n-1\rangle\\ &= g_c\sqrt{n}\,c_n^+\,|-,\,n\rangle,\end{aligned}\tag{21.49}$$

where we used the fact that $\sigma_-\,|-\rangle$ vanishes. A similar derivation produces the third contribution

$$\begin{aligned}g_c\,\sigma_+\,B_c|\phi_n\rangle &= c_n^-\,g_c\,B_c\,\sigma_+\,|-,\,n\rangle + c_n^+\,g_c\,B_c\,\cancel{\sigma_+\,|+,\,n-1\rangle}\\ &= g_c\sqrt{n}\,c_n^-\,|+,\,n-1\rangle\,,\end{aligned}\tag{21.50}$$

and so on. We see that the effective light–matter coupling

$$g_n \equiv \sqrt{n}\,g_c, \qquad n = 1,\,2,\,\cdots \tag{21.51}$$

explicitly depends on the Fock-state number n.

Combining the results (21.48)–(21.50), the eigenvalue problem (21.47) can be summarized as

$$\begin{aligned}&\left(E^0_{-,n}\,c_n^- - \hbar g_n\,c_n^+\right)|-,\,n\rangle + \left(E^0_{+,n-1}\,c_n^+ - \hbar g_n\,c_n^-\right)|+,\,n-1\rangle\\ &\qquad = E_n\left[c_n^-\,|-,\,n\rangle + c_n^+\,|+,\,n-1\rangle\right].\end{aligned}\tag{21.52}$$

Since $|-, n\rangle$ and $|+, n-1\rangle$ are orthogonal states, the coefficients of the corresponding terms on the LHS and RHS must be equal. This observation can be represented in the matrix form

$$\begin{pmatrix} E^0_{+,n-1} & -\hbar g_n \\ -\hbar g_n & E^0_{-,n} \end{pmatrix} \begin{pmatrix} c^+_n \\ c^-_n \end{pmatrix} = E_n \begin{pmatrix} c^+_n \\ c^-_n \end{pmatrix}. \tag{21.53}$$

The resulting 2×2 eigenvalue problem has the two solutions

$$E^\pm_n = \frac{E^0_{+,n-1} + E^0_{-,n}}{2} \pm \sqrt{\frac{1}{4}\left(E^0_{+,n-1} - E^0_{-,n}\right)^2 + \hbar^2 g^2_n}$$
$$= \hbar\omega_c n \pm \hbar\sqrt{\frac{1}{4}\Delta^2_{2,1} + g^2_n}. \tag{21.54}$$

This energy structure is illustrated in Fig. 21.3 and discussed further in Section 21.2.3.

The evaluation of the eigenstates produces (see Exercise 21.6)

$$|\phi^-_n\rangle \equiv \cos\theta_n\,|-, n\rangle - \sin\theta_n|+, n-1\rangle, \qquad E_n = E^-_n,$$
$$|\phi^+_n\rangle \equiv \cos\theta_n|+, n-1\rangle + \sin\theta_n\,|-, n\rangle, \qquad E_n = E^+_n, \tag{21.55}$$

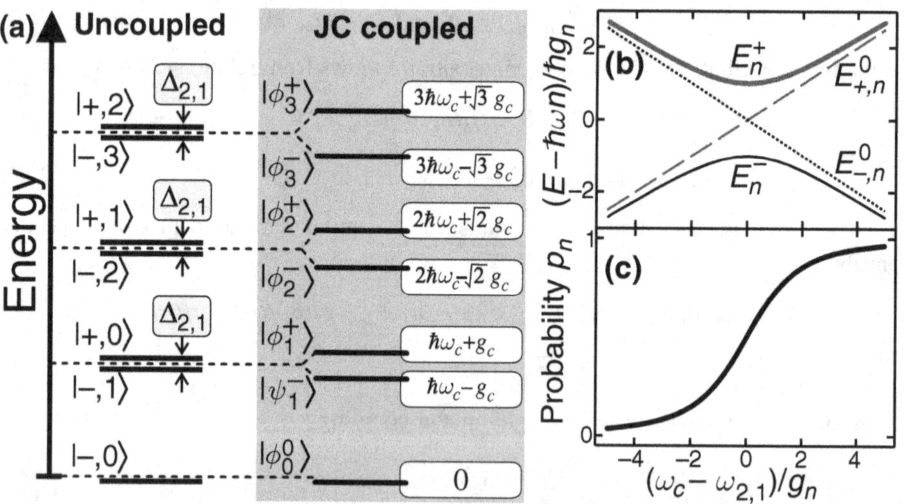

Figure 21.3 Jaynes–Cummings ladder structure. **(a)** Left: energetic levels (horizontal lines) of the noninteracting states $|\pm, n\rangle$ for slightly positive detuning $\Delta_{2,1} > 0$. Right: the corresponding structure for the coupled light–matter system under resonant conditions ($\Delta_{2,1} = 0$). The dashed lines indicate the position of the pure cavity-mode energies $\hbar\omega_c\, n$. These lines bifurcate into two nondegenerate quantum rungs. **(b)** The actual E^+_n (thick solid line) and E^-_n (thin solid line) are shown as functions of detuning, based on Eq. (21.54). The corresponding noninteracting $E^0_{+,n}$ (dashed line) and $E^0_{-,n}$ (dotted line) are computed from Eq. (21.43). **(c)** The probability p_n of finding $|\phi_{+,n}\rangle$ in the state $|+, n\rangle$ as a function of detuning, obtained from Eqs. (21.57)–(21.58).

which clearly are orthogonal for all $n \geq 1$. In this notation, the ground-state wave function is given by

$$\left|\phi_n^0\right\rangle \equiv |-, 0\rangle, \qquad E_0 = -\frac{\hbar\Delta_{2,1}}{2} \tag{21.56}$$

resulting from (21.45). The angle parameter θ_n in Eq. (21.55) is defined by

$$\cos 2\theta_n = \frac{\Delta_{2,1}}{\sqrt{\Delta_{2,1}^2 + 4g_n^2}}, \qquad \sin 2\theta_n = \frac{2g_n}{\sqrt{\Delta_{2,1}^2 + 4g_n^2}} \tag{21.57}$$

if we limit it to the interval $\theta_n \in [0, \pi[$.

21.2.3 Jaynes–Cummings ladder

Using the properties of the trigonometric functions, we find for the probability

$$p_n \equiv \left|\langle +, n-1 \,|\phi_n^+\rangle\right|^2 \equiv \left|\langle -, n \,|\phi_n^-\rangle\right|^2 = \cos^2\theta_n = \frac{1}{2}\left(1 + \cos 2\theta_n\right). \tag{21.58}$$

For example, the resonant case ($\Delta_{2,1} = 0$) produces $2\theta_n = \pi/2$ and the probability $p_n = 1/2$ of finding the system in either one of the originally degenerate states. More specifically, for resonant conditions Eq. (21.55) yields

$$\left|\phi_n^-\right\rangle \equiv \frac{1}{\sqrt{2}}\left(|-, n\rangle - |+, n-1\rangle\right), \qquad E_n^- = \hbar\omega_c\, n - \hbar g_n,$$

$$\left|\phi_n^+\right\rangle \equiv \frac{1}{\sqrt{2}}\left(|-, n\rangle + |+, n-1\rangle\right), \qquad E_n^+ = \hbar\omega_c\, n + \hbar g_n. \tag{21.59}$$

These states are separated by the energy $2\hbar g_n$ showing that the interaction lifts the degeneracy between the eigenstates $|-, n\rangle$ and $|+, n-1\rangle$ of the noninteracting system. This removal of the degeneracy is an example of the general quantum-mechanical feature that any interaction among initially degenerate states lifts the degeneracy if it couples the original states. Here, the light–matter interaction, i.e., the finite value of $g_n = \sqrt{n}\, g_c$, produces the coupling. Only for vanishing integration, i.e., $g_c = 0$, do we return to the original degenerate basis states (21.41).

As another intriguing feature, the solution (21.59) shows that the new eigenstates $\left|\phi_n^-\right\rangle$ cannot be factorized into products of atomic and Fock states. Therefore, the light–matter wave function is entangled, i.e.,

$$|\psi\rangle \neq |\psi_{\text{mat}}\rangle \otimes |\psi_{\text{light}}\rangle\,. \tag{21.60}$$

One generally speaks of light–matter entanglement whenever the combined wave function is not separable into a product of separate light and matter contributions. An entangled wave function corresponds to true quantum phenomena because light and matter are correlated in a nonclassical manner. Since the eigenstates are inseparable, one often introduces a new quasiparticle concept – the cavity polariton – to describe the physics of the coupled

TLS and cavity photon. The state $|\phi_{+,n}\rangle$ defines the upper polariton branch because its energy E_n^+ is greater than E_n^- of the lower polariton branch related to $|\phi_{-,n}\rangle$. The polariton concept is often used also in a more general context to describe mixed states of light and matter.

Besides the entanglement, we observe as an additional quantum effect that the consecutive eigenstates are no longer equidistantly spaced. The energy difference,

$$\Delta E_n^{\pm} \equiv E_{n+1}^{\pm} - E_n^{\pm} = \hbar\omega_c \pm \hbar g_c \left(\sqrt{n+1} - \sqrt{n}\right), \tag{21.61}$$

depends on the Fock-state number n. This anharmonicity of the Jaynes–Cummings eigenenergies results from their n-dependence, which is a direct consequence of the light quantization. Therefore, the anharmonicity is a true quantum phenomenon. The resulting energy structure $E_n^{\pm}$ is often referred to as the Jaynes–Cummings ladder.

A typical form of this ladder is plotted on the RHS of Fig. 21.3(a) for $\Delta_{2,1} = 0$. Here, each solid vertical line identifies a nondegenerate eigenstate, i.e., a rung of the Jaynes–Cummings ladder. The splitting of the states depends on the photon number such that the energetic distance to the nearest, next nearest, and all higher rungs depends on n. For comparison, the LHS of Fig. 21.3(a) shows the energy eigenstates (solid lines) without the light–matter coupling. Based on Eq. (21.43), $E^0_{\pm,n+1} - E^0_{\pm,n} = \Delta_{2,1}$ are equidistant for all n. This harmonic behavior is natural because its n-dependence follows exclusively from the harmonic-oscillator problem defined by the noninteracting part of the quantized light.

The splitting of the Jaynes–Cummings eigenstates is

$$E_n^+ - E_n^- = \hbar\sqrt{\Delta_{2,1}^2 + 4g_c^2 n} \geq 2\hbar g_c \sqrt{n} \tag{21.62}$$

for arbitrary detuning. The corresponding transition from degenerate into split states is depicted in Fig. 21.3(b). For both interacting and uncoupled systems, we find the same average energy

$$\frac{1}{2}\left(E_n^+ + E_n^-\right) = \frac{1}{2}\left(E^0_{+,n} + E^0_{-,n}\right) = \hbar\omega_c n. \tag{21.63}$$

The figure shows how the interacting $E_n^+ - \hbar n\omega_c$ (thick solid line) and $E_n^- - \hbar n\omega_c$ (thin solid line) depend on the detuning $\Delta_{2,1} = \omega_c - \omega_{2,1}$. For comparison, also the noninteracting limit is plotted (dashed lines). In the uncoupled case, $E^0_{+,n} - \hbar n\omega_c$ increases linearly with detuning while $E^0_{-,n} - \hbar n\omega_c$ decreases linearly. As a result, we find a level crossing for zero detuning, i.e., the states are degenerate at this point. The interacting system produces an anticrossing such that $E_n^+ - \hbar n\omega_c$ and $E_n^- - \hbar n\omega_c$ never cross, i.e., the degeneracy is completely removed. Instead, a gap of the size $2g_n = 2g_c\sqrt{n}$ opens up in the energetic spectrum. The minimum splitting is observed for resonant conditions with ($\Delta_{2,1} = 0$).

Due to the anticrossing, $E_n^{\pm} - \hbar n\omega_c$ is a nonlinear function of the detuning. For example, $E_n^{\pm} - \hbar n\omega_c$ asymptotically approaches $E^0_{-,n}$ for large negative detunings while it converges to the $E^0_{+,n}$ branch for large positive detunings. To determine the physical origin of this crossover, we analyze the $|+, n-1\rangle$ content of the eigenstate $|\phi_{+,n}\rangle$ as function of detuning. Figure 21.3(c) depicts the probability p_n showing that $|\phi_{+,n}\rangle$ consists mostly of the $|-, n\rangle$ component for large negative detunings while $|+, n-1\rangle$ dominates at large detunings. The value $p_n = 1/2$ is observed exactly on resonance. Thus, the nonlinearity of the anticrossing follows from a combination of anharmonic coupling and modified state mixing as function of detuning.

To appreciate the unusual properties of the Jaynes–Cummings ladder, we contrast it with the coupling of the two degenerate harmonic oscillators with characteristic frequency. In other words, we replace the Fermionic two-level system in Eq. (21.37) by a Boson field B_2. The corresponding Hamiltonian is

$$\hat{H}_{\text{BOS}} = \hbar\omega_c \left[B_c^\dagger B_c + B_2^\dagger B_2\right] - \hbar g_c \left[B_c^\dagger B_2 + B_2^\dagger B_c\right]. \tag{21.64}$$

This can also be viewed as a system where a Bosonic mode B_c couples to Bosonic polarization B_2, compare Eq. (21.27). Thus, we may expect that $\hat{H}_{\text{BOS}}$ coupling produces neither the quantum-optical hierarchy problem nor quantum features for B_c. Indeed, $\hat{H}_{\text{BOS}}$ can be expressed through two independent Boson modes

$$\hat{H}_{\text{BOS}} = \hbar\omega_+ \, B_+^\dagger B_+ + \hbar\omega_- \, B_-^\dagger B_-, \tag{21.65}$$

see Exercise 21.7. The coupling simply removes the degeneracy, producing $\omega_\pm = \omega_c \pm g$ such that the energetic spacing is always equidistant, i.e., harmonic,

$$\left(E^\pm_{n+1,\text{BOS}} - E^\pm_{n,\text{BOS}}\right)_{\text{class}} = \hbar\omega_c \pm \hbar g_c, \tag{21.66}$$

for all "ladder" states corresponding to Eq. (21.61). Hence, we see again that the anharmonicity in Eq. (21.61) is a genuine quantum effect. We also notice that the quantum anharmonicity tends to make the eigenstate spacing smaller (larger) than classically allowed, for the $-(+)$ branch, see Exercise 21.7.

As shown in Exercise 21.8, the energetic properties of $\hat{H}_{\text{BOS}}$ can be understood completely classically due to its harmonicity. However, our comparison of the classical and quantum cases also reveals that the mere emergence of split resonances $E^\pm_n$ is not sufficient to conclude that the respective system possesses true quantum features. As a direct verification of quantum anharmonicity, one has to additionally detect at least the second rung of the Jaynes–Cummings ladder; otherwise, one cannot distinguish between the quantum relation (21.61) and the classical Eq. (21.66). If the second rung is observed, this not only demonstrates the true quantum nature of light but also shows that the analyzed system can be used to study other quantum-optical features, such as light–matter entanglement. Thus, one can consider the detection of the second rung of the Jaynes–Cummings ladder as the first important step toward intriguing quantum-optical phenomena. The corresponding experimental observations have been reported for atomic cavity quantum-electrodynamics (Brune *et al.* 1996).

Exercises

21.1 (**a**) Verify that the projection (21.5) produces the Hamiltonian (21.9). (**b**) Confirm that the implementation of the interaction picture, i.e., Eq. (21.11), yields the Hamiltonian (21.15). (**c**) Show that the quantization step eventually yields Eq. (21.20) once we implement the rotating-wave approximation.

21.2 (**a**) Derive Eqs. (21.21)–(21.22) using $\hat{H}_{\rm QED}$ in the Heisenberg equations of motion. (**b**) Show that Eqs. (21.22)–(21.23) yield the Maxwell-optical Bloch equation as soon as the classical factorization (21.24) is used.

21.3 Apply the singlet–doublet factorization to Eq. (21.23) and construct Eq. (21.26). What is the explicit form of the singlet–doublet contribution $\Delta\langle \mathbf{\Omega} \times \mathbf{P}\rangle$?

21.4 Assume that the polarization is described by $\hat{P}_- = v^\dagger c$ where v and c are two types of Fermion operators. (**a**) Compute the quantum-optical dynamics for $\hat{P}_-$, $c^\dagger c$, and $v^\dagger v$ due to the light–matter part (21.27). Show that a hierarchy problem arises in each of the cases. (**b**) Compute $\left\langle \left[\hat{P}_-, \hat{P}_-^\dagger\right]_- \right\rangle$. Under which conditions is this expectation value nearly Bosonic? Express the $\langle \hat{P}_- \rangle$ dynamics in terms of Bosonic and Fermion contributions.

21.5 (**a**) Show that the cavity-mode operator B_c, defined by Eq. (21.33) is Bosonic. (**b**) Verify the relation (21.36) using (21.33). (**c**) Assume that a TLS excites one photon into the mode B_c, i.e., $|1\rangle = B_c^\dagger|0\rangle$. Show that $\hat{\Delta}_{q,q'} \equiv \omega_c\, \mathcal{F}_q^*\, B_{q'} - \omega_q\, \mathcal{F}_{q'}^*\, B_q$ yields a negligible $\hat{\Delta}_{q,q'}|1\rangle$ for a high-quality cavity. Why does this imply that the $\hat{\Delta}_{q,q'}$ part within (21.36) can be neglected?

21.6 (**a**) The general TLS wave function follows from

$$|\psi\rangle = C_0^-|-,0\rangle + \sum\nolimits_{n=1}^{\infty}\left[C_n^-|-,n\rangle + C_n^+|+,n-1\rangle\right].$$

Identify a vector $\mathbf{C} = \left(C_N^+, C_N^-, C_{N-1}^+, C_{N-1}^-, \cdots, C_0^-\right)$. Construct the matrix form of $\hat{H}_{\rm JC}|\psi\rangle = \lambda|\psi\rangle$ and show that this eigenvalue problem can be reduced to the Bloch-matrix relation (21.53). (**b**) Show that Eq. (21.55) is an eigensolution to (21.53) when the parametrization (21.57) is used.

21.7 Assume that the coupling between degenerate harmonic oscillators is described through the Hamiltonian

$$\hat{H}_2 = \hbar\omega_c\left(B_1^\dagger B_1 + B_2^\dagger B_2\right) + \hbar g\left(B_1^\dagger B_2 + B_2^\dagger B_1\right).$$

(**a**) Apply the unitary transformation $B_1 = \cos\theta\, B + \sin\theta\, \tilde{B}$ and $B_2 = -\sin\theta\, B + \cos\theta\, \tilde{B}$ where B and $\tilde{B}$ are independent Bosonic operators. Show that $[B_i, B_j]_- = \delta_{i,j}$, as it should. (**b**) Show that the Hamiltonian becomes $\hat{H}_2 = \hbar\omega_-\, B^\dagger B + \hbar\omega_+ \tilde{B}^\dagger \tilde{B}$, with $\omega_\pm = \omega \pm g$. Which θ value yields this diagonalization? (**c**) Conclude that the eigenenergies of this system follow from

$E_{n,n'} = \hbar\omega_- n + \hbar\omega_+ n'$, $n = 0, 1, \cdots$ **(d)** Show that $\hat{H}_2$ becomes the Jaynes–Cummings model if the substitution $B_2 \to \sigma_-$ is made. Which differences emerge as B_1 is coupled with a Boson vs. a Fermion?

21.8 Assume that the coupling between classical harmonic oscillators follows from

$$H = \frac{p_1^2}{2m} + \frac{1}{2}m\omega^2 x_1^2 + \frac{p_2^2}{2m} + \frac{1}{2}m\omega^2 x_2^2 - g\left[m\,\omega\, x_1 x_2 + \frac{p_1 p_2}{\omega\, m}\right].$$

(a) Derive the corresponding Hamilton equations and show that the problem has only two characteristic frequencies $\omega^{\pm} = \omega \pm g$. **(b)** Express the solutions via $x(t) = \mathrm{Re}[C_- e^{-i\omega_- t} + C_+ e^{-i\omega_+ t}]$. Show that $[x(t)]^N$ has a harmonic spectrum, i.e., integer multiples of the elementary frequencies. What is their value?

Further reading

The Jaynes–Cummings model was originally proposed in:

E. T. Jaynes and F. W. Cummings (1963). Comparison of quantum and semiclassical radiation theories with application to the beam maser, *Proc. IEEE* **51**, 89.

The Jaynes–Cummings model plays a central role for the explanation of a number of quantum-optical effects, cf.:

B. W. Shore and P. L. Knight (1993). The Jaynes–Cummings model, *J. Mod. Opt.* **40**, 1195.

P. Meystre and M. S. Zubairy (1982). Squeezed states in the Jaynes–Cummings model, *Phys. Lett. A* **89**, 390.

C. C. Gerry and P. L. Knight (2005). *Introductory Quantum Optics*, Cambridge, Cambridge University Press.

See also the books mentioned at the end of Chapter 20.

Quantum Rabi oscillations were directly measured for the first time in:

M. Brune, F. Schmidt-Kaler, A. Maali, *et al.* (1996). Quantum Rabi oscillation: A direct test of field quantization in a cavity, *Phys. Rev. Lett.* **76**, 1800.

22

Quantum dynamics of two-level system

The quantum anharmonicity of the Jaynes–Cummings ladder is a source of a multitude of quantum-optical effects. Therefore, we study next the dynamic evolution of optically generated excitations using the Jaynes–Cummings model. In particular, we formulate a general theory for the quantum-state dynamics and consider light fields represented by either Fock or coherent states. This allows us to study phenomena such as quantum Rabi flopping, collapses and revivals. We emphasize the similarities and characteristic differences to a classical excitation scenario.

22.1 Formal quantum dynamics

In general, the joint light–matter wave function evolves according to the time-dependent Schrödinger equation

$$i\hbar\frac{\partial}{\partial t}|\phi(t)\rangle = \hat{H}_{\mathrm{JC}}\,|\phi(t)\rangle \qquad \Leftrightarrow \qquad |\phi(t)\rangle = e^{-\frac{i}{\hbar}\hat{H}_{\mathrm{JC}}t}\,|\phi(0)\rangle. \tag{22.1}$$

Here, we use the Jaynes–Cummings Hamiltonian to describe the full quantum interaction between a two-level system (TLS) and a single quantized light mode,

$$\hat{H}_{\mathrm{JC}} = \hbar\omega_{2,1}\,\sigma_z + \hbar\omega\left(B^\dagger\,B + \frac{1}{2}\right) - \hbar g\left[\sigma_-\,B^\dagger + \sigma_+\,B\right]. \tag{22.2}$$

In comparison with Eq. (21.37), we simplify the notation by omitting the index c from the photon operator B, the photon frequency ω, and the coupling constant g. The TLS is described by the Pauli spin operators, the transition energy $\hbar\omega_{2,1}$, and the eigenvalue problem

$$\hat{H}_{\mathrm{JC}}\left|\phi_0^0\right\rangle = E_0^0\left|\phi_0^0\right\rangle, \qquad \hat{H}_{\mathrm{JC}}\left|\phi_n^\pm\right\rangle = E_n^\pm\left|\phi_n^\pm\right\rangle, \quad n = 1, 2, \cdots. \tag{22.3}$$

According to Section 21.2.2, the wave functions are

$$\left|\phi_0^0\right\rangle = |-, 0\rangle, \qquad \begin{cases} \left|\phi_n^-\right\rangle = \cos\theta_n\,|-, n\rangle - \sin\theta_n|+, n-1\rangle \\ \left|\phi_n^+\right\rangle = \cos\theta_n|+, n-1\rangle + \sin\theta_n\,|-, n\rangle \end{cases}, \tag{22.4}$$

where $|\pm, n\rangle \equiv |\pm\rangle \otimes |n\rangle$. Here, n identifies the Fock-state $|n\rangle$ of the light field and $|+\rangle$ ($|-\rangle$) defines the TLS in its excited (ground) state. Furthermore, we have

$$\Delta \equiv \omega - \omega_{2,1}, \qquad g_n \equiv g\sqrt{n}, \qquad \Upsilon_n \equiv \sqrt{\Delta^2 + 4g_n^2},$$
$$\begin{cases} E_0^0 = \frac{1}{2}\hbar\Delta \\ E_n^{\pm} = \hbar\omega n \pm \frac{1}{2}\hbar\Upsilon_n \end{cases}, \qquad \begin{cases} \sin 2\theta_n = 2g_n/\Upsilon_n \\ \cos 2\theta_n = \Delta/\Upsilon_n \end{cases} \tag{22.5}$$

compare Eqs. (21.43), (21.45), (21.51), and (21.54)–(21.57). The inverse relation to Eq. (22.4) is

$$|-, 0\rangle = \left|\phi_0^0\right\rangle, \qquad \begin{cases} |+, n-1\rangle = \cos\theta_n \left|\phi_n^+\right\rangle - \sin\theta_n \left|\phi_n^-\right\rangle \\ |-, n\rangle = \sin\theta_n \left|\phi_n^+\right\rangle + \cos\theta_n \left|\phi_n^-\right\rangle \end{cases}, \tag{22.6}$$

as can be verified by direct substitution.

22.1.1 Wave-function dynamics

In cases where we only know the initial wave function in terms of the noninteracting eigenstates $|\pm, n\rangle$ rather than $\left|\phi_n^\lambda\right\rangle$, we use the expansion

$$|\phi(0)\rangle = \sum_{n=0}^{\infty}\left[c_n^-|-, n\rangle + c_n^+|+, n\rangle\right], \qquad \sum_{n=0}^{\infty}\left[\left|c_n^-\right|^2 + \left|c_n^+\right|^2\right] = 1. \tag{22.7}$$

Since $|\pm, n\rangle$ forms a complete orthogonal set of states, the $c_n^\pm$ coefficients define the probability $\left|c_n^\pm\right|^2$ of finding the system in the respective state $|\pm, n\rangle$. The corresponding quantum dynamics follows as we insert the expansion (22.7) into Eq. (22.1)

$$\begin{aligned} |\phi(t)\rangle &= \sum_{n=0}^{\infty}\left[c_n^- e^{-\frac{i}{\hbar}\hat{H}_{\rm JC}t}\,|-, n\rangle + c_n^+ e^{-\frac{i}{\hbar}\hat{H}_{\rm JC}t}\,|+, n\rangle\right] \\ &\equiv \sum_{n=0}^{\infty}\left[c_n^-\,|-, n, t\rangle + c_n^+\,|+, n, t\rangle\right]. \end{aligned} \tag{22.8}$$

Applying the transformation (22.6) and using (22.3), we find

$$\begin{aligned} |-, n, t\rangle &\equiv e^{-\frac{i}{\hbar}\hat{H}_{\rm JC}t}\,|-, n\rangle = e^{-\frac{i}{\hbar}\hat{H}_{\rm JC}t}\left(\sin\theta_n\left|\phi_n^+\right\rangle + \cos\theta_n\left|\phi_n^-\right\rangle\right) \\ &= \sin\theta_n\, e^{-i\frac{E_n^+}{\hbar}t}\left|\phi_n^+\right\rangle + \cos\theta_n\, e^{-i\frac{E_n^-}{\hbar}t}\left|\phi_n^-\right\rangle \\ &= e^{-i\omega n t}\left(e^{-i\Upsilon_n t/2}\sin\theta_n\left|\phi_n^+\right\rangle + e^{i\Upsilon_n t/2}\cos\theta_n\left|\phi_n^-\right\rangle\right), \end{aligned} \tag{22.9}$$

where we expressed the eigenergies using Eq. (22.5).

Converting $\left|\phi_n^\pm\right\rangle$ with the help of (22.4) allows us to express $|-, n, t\rangle$ in terms of the noninteracting states,

$$\begin{aligned} |-, n, t\rangle &= e^{-i\omega n t}\, e^{-i\Upsilon_n t/2}\sin\theta_n\,(\cos\theta_n\,|+, n-1\rangle + \sin\theta_n|-, n\rangle) \\ &\quad + e^{-i\omega n t}\, e^{i\Upsilon_n t/2}\cos\theta_n\,(\cos\theta_n\,|-, n\rangle - \sin\theta_n|+, n-1\rangle) \\ &= e^{-i\omega n t}\left[-i\sin 2\theta_n\,\sin\Upsilon_n t/2\,|+, n-1\rangle\right. \\ &\quad \left. + (\cos\Upsilon_n t/2 + i\cos 2\theta_n\,\sin\Upsilon_n t/2)\,|-, n\rangle\right]. \end{aligned} \tag{22.10}$$

To simplify the notation, we use (22.5) and

$$\begin{cases} Z_n(t) \equiv \cos \Upsilon_n t/2 + i \cos 2\theta_n \, \sin \Upsilon_n t/2 = \cos \Upsilon_n t/2 + i \frac{\Delta \sin \Upsilon_n t/2}{\Upsilon_n} \\ S_n(t) \equiv \sin 2\theta_n \, \sin \Upsilon_n t/2 = \frac{2g_n \sin \Upsilon_n t/2}{\Upsilon_n}, \qquad n = 1, 2, \cdots \end{cases}$$
$$S_0(t) \equiv 0, \quad Z_0(t) \equiv e^{i\Delta t/2}, \qquad \text{for } n = 0. \tag{22.11}$$

Here, we extended the n range to include the case $n = 0$. These definitions allow us to condense (22.10) and the corresponding expression for $|+, n-1, t\rangle$ into

$$\begin{aligned} |-, n, t\rangle &= e^{-\frac{i}{\hbar}\hat{H}_{\text{JC}}t} |-, n\rangle = e^{-i\omega n t} \left[-i S_n(t) \, |+, n-1\rangle + Z_n(t) \, |-, n\rangle\right] \\ |+, n, t\rangle &= e^{-\frac{i}{\hbar}\hat{H}_{\text{JC}}t} |+, n\rangle = \\ &= e^{-i\omega(n+1)t} \left[Z^\star_{n+1}(t)|+, n\rangle - i S_{n+1}(t) \, |-, n+1\rangle\right]. \end{aligned} \tag{22.12}$$

We can easily verify that these relations are valid for all $n = 0, 1, 2, \cdots$ provided that we apply the $n = 0$ extension of Eq. (22.11).

22.1.2 Dynamics of density matrices

Whenever the system is characterized by a statistical mixture instead of a pure state, we must replace the wave function $|\psi(t)\rangle$ by the density matrix $\hat{\rho}(t)$. To analyze this situation, we look at a TLS in its ground state with the initial

$$\hat{\rho}(0) = |-\rangle\langle-| \otimes \sum_{n,n'=0}^{\infty} |n\rangle c_{n,n'} \langle n'| = \sum_{n,n'=0}^{\infty} |-, n\rangle c_{n,n'} \langle -, n'|. \tag{22.13}$$

Here, $c_{n,n'}$ defines the general quantum statistics of the light before the light–matter interaction starts. Since the TLS is in its ground state, the light and matter parts are initially uncorrelated.

The quantum dynamics follows from

$$\begin{aligned} \hat{\rho}(t) &= e^{-\frac{i}{\hbar}\hat{H}_{\text{JC}}t} \, \hat{\rho}(0) \, e^{+\frac{i}{\hbar}\hat{H}_{\text{JC}}t} \\ &= \sum_{n,n'=0}^{\infty} e^{-\frac{i}{\hbar}\hat{H}_{\text{JC}}t} \, |-, n\rangle c_{n,n'} \langle -, n'| \, e^{+\frac{i}{\hbar}\hat{H}_{\text{JC}}t} \\ &= \sum_{n,n'=0}^{\infty} |-, n, t\rangle c_{n,n'} \langle -, n', t|, \end{aligned} \tag{22.14}$$

where we applied Eq. (22.12) to identify the dynamics of the basis states. Inserting the definition (22.11) into Eq. (22.14) yields

$$\begin{aligned} \hat{\rho}(t) = \sum_{n,n'=0}^{\infty} e^{-i\omega(n-n')t} \big[&|+, n-1\rangle S_n(t) c_{n,n'} S_{n'}(t) \langle +, n'-1| \\ &- i|+, n-1\rangle S_n(t) c_{n,n'} Z^\star_{n'}(t) \langle -, n'| \\ &+ i|-, n\rangle Z_n(t) c_{n,n'} S_{n'}(t) \langle +, n'-1| \\ &+ |-, n\rangle Z_n(t) c_{n,n'} Z^\star_{n'}(t) \langle -, n'| \big]. \end{aligned} \tag{22.15}$$

Quite often, we are only interested in the pure matter or in the pure light properties. These are described by the reduced density matrices

$$\hat{\rho}_{2\mathrm{LS}}(t) \equiv \mathrm{Tr}_B\left[\hat{\rho}(t)\right] = \sum\nolimits_{n=0}^{\infty} \langle n|\hat{\rho}(t)|n\rangle$$
$$\hat{\rho}_B(t) \equiv \mathrm{Tr}_{2\mathrm{LS}}\left[\hat{\rho}(t)\right] = \sum\nolimits_{\lambda=\pm} \langle \lambda|\hat{\rho}(t)|\lambda\rangle. \tag{22.16}$$

Using (22.15), we find for the reduced TLS density matrix

$$\hat{\rho}_{2\mathrm{LS}}(t) = c_{+,+}|+\rangle\langle+| + c_{+,-}|+\rangle\langle-| + c_{-,+}|-\rangle\langle+| + c_{-,-}|-\rangle\langle-|$$
$$c_{+,+} \equiv \sum\nolimits_{n=0}^{\infty} c_{n,n}|S_n(t)|^2, \quad c_{+,-} \equiv -i\, e^{-i\omega t}\sum\nolimits_{n=0}^{\infty} S_{n+1}(t)c_{n+1,n}Z_n^{\star}(t),$$
$$c_{-,+} \equiv c_{+,-}^{\star}, \qquad c_{-,-} \equiv \sum_{n=0}^{\infty} c_{n,n}|Z_n(t)|^2, \tag{22.17}$$

which follows from the orthogonality of $|n\rangle$ states. The resulting four orthogonal elements are defined by the coefficients $c_{\pm,\pm}$ and $c_{\pm,\mp}$.

For the reduced density matrix of the quantized light field, we find

$$\hat{\rho}_B(t) = \sum\nolimits_{n,n'=0}^{\infty} e^{-i\omega(n-n')t}\, |n\rangle\, p_{n,n'}\langle n'|$$
$$p_{n,n'} \equiv \left[S_{n+1}(t)c_{n+1,n'+1}S_{n'+1}(t) + Z_n(t)c_{n,n'}Z_{n'}^{\star}(t)\right]. \tag{22.18}$$

Here, the diagonal element $p_{n,n}$ defines the probability of detecting light with exactly n photons, i.e., to find the system in the state $|n\rangle$. Thus, $p_n \equiv p_{n,n}$ is often referred to as the photon statistics or photon number distribution. Clearly, since p_n only describes the diagonal part of the density matrix, it does not constitute a generic representation of the full quantum statistics, unlike $p_{n,n'}$.

22.2 Quantum Rabi flopping

As an interesting example, we consider an initial configuration where the TLS is in its excited state and the light is in Fock state $|n\rangle$. This initial state is described by the factorizable wave function $|\psi(0)\rangle = |+\rangle \otimes |n\rangle = |+, n\rangle$. We use Eq. (22.12) for $|+, n, t\rangle$ to describe the quantum dynamics at any later time t. Since

$$\sigma_-|+, n, t\rangle = e^{-i\omega(n+1)t}\left[Z_{n+1}^{\star}(t)\sigma_-|+, n\rangle - i\,S_{n+1}(t)\,\cancel{\sigma_-|-, n+1\rangle}\right]$$
$$= e^{-i\omega(n+1)t}\, Z_{n+1}^{\star}(t)|-, n\rangle, \tag{22.19}$$

is orthogonal to the state $\langle+, n, t|$, the polarization $P_- = \langle+, n, t|\sigma_-|+, n, t\rangle$ vanishes at all times. However, the state inversion has the dynamics

$$
\begin{aligned}
P_z &\equiv \langle +,\, n,\, t|\sigma_z|+,\, n,\, t\rangle \\
&= \langle +,\, n,\, t|e^{-i\omega(n+1)t}\left[Z^\star_{n+1}(t)\sigma_z|+,\, n\rangle - iS_{n+1}(t)\,\sigma_z|-,\, n+1\rangle\right] \\
&= \left[Z_{n+1}(t)\langle +,\, n| + iS_{n+1}(t)\,\langle -,\, n+1|\right] \\
&\quad\times\left[Z^\star_{n+1}(t)\frac{1}{2}|+,\, n\rangle - iS_{n+1}(t)\left(-\frac{1}{2}\right)|-,\, n+1\rangle\right] \\
&= \frac{1}{2}|Z_{n+1}(t)|^2 - \frac{1}{2}|S_{n+1}(t)|^2.
\end{aligned}
\tag{22.20}
$$

To obtain the final expression, we expanded $|+,\, n,\, t\rangle$ in the first step using Eq. (22.12). Then, we have used $\sigma_z|\pm,\, n\rangle = \pm\frac{1}{2}|\pm,\, n\rangle$ and the orthogonality of the basis states. Inserting the definitions (22.11) into Eq. (22.20) and using (22.5), we obtain

$$
P_z = \frac{1}{2} - \frac{2g^2(n+1)}{\Delta^2 + 4g^2(n+1)}\left(1 - \cos\Upsilon_{n+1}t\right), \quad n = 0,\, 1,\, 2, \ldots \tag{22.21}
$$

after some algebra. The identified P_z changes in time.

22.2.1 Observation of quantum rungs

Under resonant conditions $\Delta = 0$, Eq. (22.21) yields the simpler form $P_z = -\frac{1}{2}\cos\Upsilon_{n+1}t$ with $\Upsilon_{n+1} = 2g\sqrt{n+1}$. The time evolution of this P_z is plotted in Fig. 22.1 for $n = 0$ (shaded area) and $n = 1$ (solid line). We see that P_z oscillates sinusoidally between its extreme values $\pm 1/2$. This inversion dynamics resembles very much the Rabi flopping encountered in Section 18.2.3. However, we now have two very distinct quantum signatures: (i) no polarization is created during the Rabi cycle because P_- remains zero throughout the evolution, and (ii) the Rabi frequency is discretized. Since these features cannot be explained classically, this phenomenon is called *quantum Rabi flopping*. It has the period

$$
T_n \equiv \frac{2\pi}{\Upsilon_n} = \frac{2\pi}{\sqrt{\Delta^2 + 4g(n+1)}} \tag{22.22}
$$

depending on the rung of the Jaynes–Cummings ladder. Here, $n = 0$ corresponds to the first rung while $n = 1$ corresponds to the second rung etc. The difference between the $n = 0$ and $n = 1$ cases is clearly visible in Fig. 22.1(a) where the second rung ($n = 1$) is $\sqrt{2} \approx 1.41$ times faster than the first rung ($n = 0$).

In Fig. 22.1(b), we plot the trajectory $\mathbf{P} = (0,\, P_y,\, P_z)$ (solid line) showing that the Bloch vector oscillates directly between the north pole (excited TLS) and the south pole (ground state). This motion is very different from a classical trajectory which always stays on the surface of the Bloch sphere (dotted circle) unless the system is exposed to external dephasing.

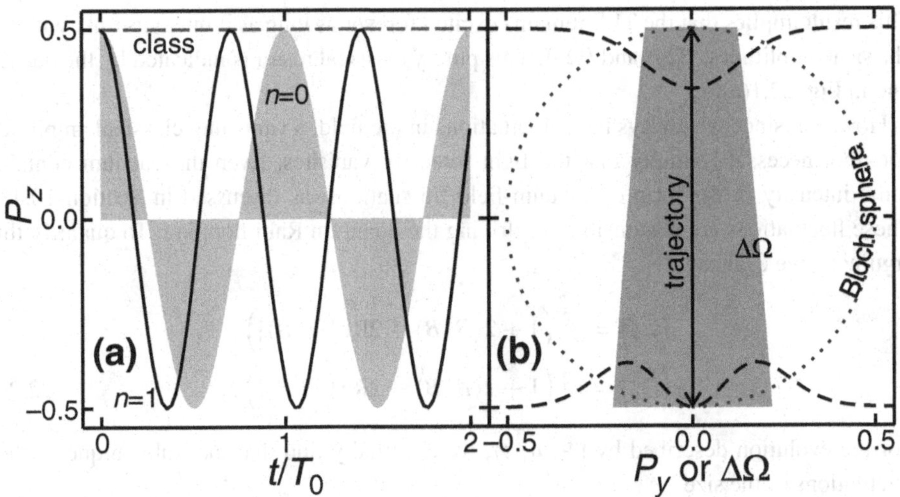

Figure 22.1 Quantum Rabi flopping resulting from Fock-state excitations. The two-level system is initially ($t = 0$) in its excited state $|+\rangle$ and resonant ($\Delta = 0$) with the light. (**a**) The dynamics of the population inversion P_z [computed from Eq. (22.21)] is shown for the excitation with a vacuum $|0\rangle$ (shaded area) or a single-photon state $|1\rangle$ (solid line); the time T_0 is given by Eq. (22.22). The dashed line shows the result of a completely classical analysis with vanishing field, i.e., $\mathbf{E} = \Omega = 0$ at $t = 0$. For this case, the Maxwell–Bloch equations (19.30) produce a constant P_z. (**b**) The trajectory (solid line, with arrows) of the Bloch vector $\bar{\mathbf{P}} = (0,\ 0,\ P_z)$ is shown together with the extent of the quantum fluctuations $\Delta\Omega$ (shaded area) during the P_z evolution for the vacuum-state excitation. The dashed lines indicate the marginal distribution for $|0\rangle$ (upside down, across the north pole) and $|1\rangle$ (across the south pole). The dotted line denotes the Bloch sphere.

22.2.2 Semiclassical interpretation

To understand the physical origin of the quantum Bloch-vector motion, we consider the quantized form of the Rabi frequency

$$\hat{\Omega}^{(+)} \equiv 2g\, B, \qquad \hat{\Omega}^{(-)} \equiv 2g\, B^\dagger, \tag{22.23}$$

compare Eq. (21.19). The combinations

$$\hat{\Omega}_x \equiv \frac{\hat{\Omega}^{(+)} + \hat{\Omega}^{(-)}}{2} = 2g\frac{B + B^\dagger}{2} \equiv 2g\,\hat{x},$$
$$\hat{\Omega}_y \equiv \frac{\hat{\Omega}^{(+)} - \hat{\Omega}^{(-)}}{2i} = 2g\frac{B - B^\dagger}{2i} \equiv 2g\,\hat{y} \tag{22.24}$$

are directly proportional to the $\hat{x}$ and $\hat{y}$ quadratures of the quantized single-mode light, see Eq. (10.52). Following similar steps as in the derivation of Eq. (22.21), we obtain the expectation values of the Rabi frequency components as

$$\langle\hat{\Omega}_x\rangle = 2g\,\mathrm{Re}\,[\langle B\rangle] = 2g\,\mathrm{Re}\,[\langle +,\, n,\, t|B|+,\, n,\, t\rangle] = 0$$
$$\langle\hat{\Omega}_y\rangle = 2g\,\mathrm{Im}\,[\langle B\rangle] = 2g\,\mathrm{Im}\,[\langle +,\, n,\, t|B|+,\, n,\, t\rangle] = 0\,. \tag{22.25}$$

This result implies that the TLS remains excited forever as long as it interacts only with the classical amplitudes, $\langle\hat{\Omega}_x\rangle$ and $\langle\hat{\Omega}_y\rangle$. This purely classical result is indicated by the dashed line in Fig. 22.1(a).

However, since we always have fluctuations in the field, a vanishing classical amplitude does not necessarily imply that the light intensity vanishes. Even the vacuum contains some intensity in the form of vacuum-field fluctuations, as discussed in Section 10.4.2. These fluctuations are responsible for driving the quantum Rabi flopping. To quantify this argument, we evaluate

$$\begin{aligned}\left\langle\hat{\Omega}_x^2\right\rangle &= g^2\left(1+2\langle B^\dagger B\rangle + 2\mathrm{Re}\left[\langle B\,B\rangle\right]\right)\\ \left\langle\hat{\Omega}_y^2\right\rangle &= g^2\left(1+2\langle B^\dagger B\rangle - 2\mathrm{Re}\left[\langle B\,B\rangle\right]\right).\end{aligned} \tag{22.26}$$

For the evolution described by $|+,\,n,\,t\rangle$, we eventually find that the Rabi frequency has fluctuations of the size

$$\Delta\Omega^2 \equiv \left\langle\hat{\Omega}_x^2\right\rangle_n = \left\langle\hat{\Omega}_y^2\right\rangle_n = 2g^2\left[1+n-P_z(t)\right]. \tag{22.27}$$

This result shows that the fluctuations depend on the initial Fock-state number and the state inversion at any given time. Since $|P_z|$ cannot exceed 1/2, Eq. (22.27) shows that $\Delta\Omega^2$ never vanishes. The extent of these fluctuations during the quantum Rabi oscillations for the initial vacuum state $|+,\,n=0,\,t\rangle$ is illustrated by the shaded area in Fig. 22.1(b). We clearly see that $\hat{\Omega}_x$ has nonvanishing fluctuations.

In principle, the notion of purely classical excitations can be extended by applying the concept of classical averaging, introduced in Section 1.2.2. There, a single excitation is replaced by a probabilistic average over an ensemble of initial conditions. We can thus realize a situation where an ensemble of Ω_x values is randomly picked in the range between $-\Delta\Omega$ and $+\Delta\Omega$ while the average of all Ω_x vanishes.

Since any single realization j most likely has a nonzero $\Omega_{x,\,j}$, it will generate "the usual" classical motion of $\mathbf{P}_j(t)$ along the Bloch sphere. If we then average over all trajectories $\mathbf{P}_j(t)$, both P_x and P_y will average to zero because Ω_x and Ω_y also vanish on average. However, the P_z component will evolve in time since the individual realizations will nudge the two-level system out of its excited state. This happens due to the $\Delta\Omega$ fluctuations in the driving field.

Even though it is highly simplified, this line of thinking shows that a semiclassical ensemble of states does explain the qualitative features of the Bloch-vector dynamics in Fig. 22.1(b). However, the discrete nature of the quantum Rabi flopping cannot be understood semiclassically because each individual $\Omega_{x,\,j}$ would produce a different, continuously spread Rabi frequency. Thus, the quantum Rabi flopping is a genuine quantum effect. Physically, it results from the discrete nature of quantized light leading to discrete n-dependent $\Delta\Omega$ fluctuations, as shown in (22.27).

Due to the $\Delta\Omega$ fluctuations, the excited matter ($R_z = +1/2$) starts to relax toward its ground state ($R_z = -1/2$). This process appears to be spontaneous, especially when the

TLS interacts with the vacuum state. In a semiclassical interpretation, the random fluctuations of the light trigger the decay of the TLS, which results in the conversion of matter excitation into light. Thus, this process is often referred to as *spontaneous emission*. For the Jaynes–Cummings model, the spontaneous emission is fully reversible, yielding the quantum Rabi oscillations. The spontaneous emission also emerges in multimode systems. However, one often loses the reversibility because the emitted light can propagate away from the system without having a chance of being reabsorbed.

To get an idea of how the light field is distributed during $|+, n = 0, t\rangle$, we show the quadrature distribution

$$\begin{aligned} p_n(x, t) &\equiv \langle x | \hat{\rho}_B | x \rangle \\ &= |Z_{n+1}(t)|^2 |\langle x|n\rangle|^2 + |S_{n+1}(t)|^2 |\langle x|n+1\rangle|^2 \end{aligned} \tag{22.28}$$

for the initial time and the time $t = T_0/2$ in Fig. 22.1(b). Here, the factor $|\langle x|n\rangle|^2$ identifies the probability distribution of the Fock-state $|n\rangle$. For the $|+, n = 0, t\rangle$ state, $\hat{\Omega}_x \propto \hat{x}$ initially only has the vacuum-state fluctuations. However, this develops into the fluctuations of the Fock-state $|1\rangle$ when the TLS reaches the south pole. Thus, reversible spontaneous emission corresponds to situations with cyclical single-photon emission and reabsorption.

22.3 Coherent states

To uniquely identify the quantum properties of light, we consider normally ordered expectation values where all creation operators are organized to the left. In this formulation, the classical factorization becomes

$$\left\langle [B^\dagger]^J B^K \right\rangle_{\text{cl}} = \langle B^\dagger \rangle^J \langle B \rangle^K \equiv [\alpha^\star]^J \alpha^K, \qquad \text{with } \alpha \equiv \langle B \rangle, \tag{22.29}$$

in analogy to the definition (21.24) and the cluster-expansion factorization (15.3). Introducing $|\psi_\alpha\rangle$ as the eigenstate of the annihilation operator,

$$B|\psi_\alpha\rangle = \alpha|\psi_\alpha\rangle, \tag{22.30}$$

we can express even the most classical field completely quantum mechanically. Since B is not a Hermitian operator, the eigenvalue α is generally complex valued. Any single-mode field is expressible through an expansion in terms of Fock states, i.e.,

$$|\psi_{\alpha_0}\rangle = \sum_{n=0}^{\infty} c_n(\alpha_0)\,|n\rangle. \tag{22.31}$$

Hence, despite the fact that the Fock states $|n\rangle$ are genuine quantum states, a superposition of them may behave more or less classically, provided that it satisfies the eigenvalue problem (22.30).

Inserting the expansion (22.31), we can evaluate the LHS of Eq. (22.30),

$$\begin{aligned} B|\psi_\alpha\rangle &= \sum_{n=0}^{\infty} c_n(\alpha) B|n\rangle = \sum_{n=1}^{\infty} c_n(\alpha)\sqrt{n}\,|n-1\rangle \\ &= \sum_{n=0}^{\infty} \sqrt{n+1}\, c_{n+1}(\alpha)|n\rangle, \end{aligned} \tag{22.32}$$

where we used Eq. (10.73) for $B|n\rangle$. Since the Fock states are complete and orthogonal, $B|\psi_\alpha\rangle$ can be equal to $\alpha|\psi_\alpha\rangle = \sum_{n=0}^{\infty} \alpha\, c_n(\alpha)|n\rangle$ only if

$$\sqrt{n+1}\, c_{n+1}(\alpha) = \alpha\, c_n(\alpha) \quad \Leftrightarrow \quad c_{n+1}(\alpha) = \frac{\alpha}{\sqrt{n+1}}\, c_n(\alpha). \tag{22.33}$$

This simple recursion problem has the unique solution

$$c_n(\alpha) = \frac{\alpha^n}{\sqrt{n!}}\, c_0(\alpha). \tag{22.34}$$

For $|\psi_\alpha\rangle$ to be normalized, we demand that

$$\langle\psi_\alpha|\psi_\alpha\rangle = \sum_{n=0}^{\infty} |c_n(\alpha)|^2 = \sum_{n=0}^{\infty} \frac{|\alpha|^{2n}}{n!} |c_0(\alpha)|^2 = e^{|\alpha|^2} |c_0(\alpha)|^2 \tag{22.35}$$

is equal to one. The emerging series is the Taylor expansion of $e^{|\alpha|^2}$ such that $c_0(\alpha)$ is given by

$$c_0(\alpha) = e^{-\frac{|\alpha|^2}{2}} \tag{22.36}$$

if we set the allowed arbitrary phase factor to one.

As we combine the results (22.34) and (22.36) with the expansion (22.31), we find the *coherent state*

$$|\alpha\rangle \equiv |\psi_\alpha\rangle = \sum_{n=0}^{\infty} \frac{\alpha^n}{\sqrt{n!}} |n\rangle\, e^{-\frac{1}{2}|\alpha|^2}. \tag{22.37}$$

This state has the properties

$$B^J|\alpha\rangle = \alpha^J|\alpha\rangle, \qquad \langle\alpha|(B^\dagger)^J = \langle\alpha|(\alpha^\star)^J. \tag{22.38}$$

For the creation operator $B^\dagger$, we get a similarly simple relation only if it operates on $\langle\alpha|$.

Combining the obtained results, we see that the coherent state $|\alpha\rangle$ produces normally ordered expectation values according to

$$\left(I_K^J\right)_\alpha = \langle (B^\dagger)^J B^K \rangle_\alpha = \langle\alpha|(B^\dagger)^J B^K|\alpha\rangle = (\alpha^\star)^J \alpha^K. \tag{22.39}$$

In other words, the normally ordered coherent-state expectation values behave as if the Boson creation and annihilation operators were simply substituted by a complex-valued amplitude, i.e., $B \to \alpha$ and $B^\dagger \to \alpha^\star$. The very same factorization follows when light is treated completely classically, i.e., each operator is replaced by a complex-valued amplitude, compare Eq. (15.3). Thus, the result (22.39) is the same as the classical factorization (22.29). Hence, the classical factorization is exact for the coherent states such that, in many ways, they describe the field with the most classical behavior.

22.3.1 Displacement operator

To get a better feeling for the coherent states, we evaluate their real-space representation using Eq. (22.37),

$$\langle x|\alpha\rangle = \sum_{n=0}^{\infty} \frac{\alpha^n}{\sqrt{n!}} \langle x|n\rangle\, e^{-\frac{1}{2}|\alpha|^2} = \left(\frac{2}{\pi}\right)^{\frac{1}{4}} e^{-x^2} \sum_{n=0}^{\infty} \frac{\left[\frac{\alpha}{\sqrt{2}}\right]^n}{n!} H_n\left(\sqrt{2}x\right). \tag{22.40}$$

Here, we implemented Eqs. (10.50)–(10.81) for the explicit form of the Fock state. The sum appearing in (22.40) can be expressed via the Hermite polynomial or it can be converted into known functions using either mathematical tables or by evaluating the sums with the help of the expansion (10.80), see Exercise 22.7. This procedure eventually yields

$$\langle x|\alpha\rangle = \left(\frac{2}{\pi}\right)^{\frac{1}{4}} e^{-(x-\alpha_1)^2}\, e^{i\alpha_2(2x-\alpha_1)}, \qquad \alpha \equiv \alpha_1 + i\alpha_2, \tag{22.41}$$

where we expressed α in terms of its real and imaginary parts. By comparing this result with Eq. (10.90), we observe that $\langle x|\alpha\rangle$ for $\alpha_2 = 0$ is identical to the wave function $\psi_0(x - \alpha_1)$ of the vacuum state displaced by α_1. Similarly, if $\alpha_1 = 0$, the momentum representation of the coherent state corresponds to a vacuum state displaced by α_2. For a generic α, the vacuum state is displaced by α_1 in position and by α_2 in momentum. This connects the coherent state with the displaced vacuum state such that α is often referred to as the displacement parameter.

To elaborate the displacement properties further, we consider a general displacement operator

$$e^{-\alpha_1 \frac{\partial}{\partial x}} f(x) = f(x - \alpha_1). \tag{22.42}$$

This relation can be verified by Taylor expanding both the exponential operator and the generic analytic function $f(x)$. Using the definition (10.89) of the quadrature operator $\hat{y}$, we can express the displacement operator as

$$\langle x|e^{-i2\alpha_1\hat{y}}|x'\rangle = \langle x|x'\rangle\, e^{-i2\alpha_1 \frac{-i}{2}\frac{\partial}{\partial x'}} = \delta(x - x')\, e^{-\alpha_1 \frac{\partial}{\partial x'}}. \tag{22.43}$$

Expressing the quadrature operator in terms of photon operators $\hat{y} = \frac{B - B^\dagger}{2i}$, we identify the displacement operator as

$$e^{-i2\alpha_1\hat{y}} = e^{\alpha_1 B^\dagger - \alpha_1 B} \equiv D(\alpha_1). \tag{22.44}$$

By combining the properties (22.42)–(22.44), we can evaluate all displacement operations via

$$\begin{aligned} \langle x|D(\alpha_1)|f\rangle &= \int dx'\, \langle x|D(\alpha_1)|x'\rangle\langle x'|f\rangle = \int dx'\, \delta(x - x')\, e^{-\alpha_1 \frac{\partial}{\partial x'}} f(x') \\ &= \int dx'\, \delta(x - x')\, f(x' - \alpha_1) = f(x - \alpha_1), \end{aligned} \tag{22.45}$$

where the completeness relation (5.18), i.e., $\int dx'|x'\rangle\langle x'| = \hat{\mathbb{I}}$, is also used.

In its most general format, the displacement operator is given by

$$D(\alpha) = e^{\alpha B^\dagger - \alpha^\star B}. \tag{22.46}$$

It generates the coherent state

$$|\alpha\rangle \equiv D(\alpha)|0\rangle \tag{22.47}$$

as the displaced vacuum state. This form is commonly used as a starting point in many quantum-optics investigations.

22.3.2 Shot-noise limit

To define the Fock-state occupation, we determine the photon statistics of the coherent state

$$p_\alpha(n) \equiv |\langle n|\alpha\rangle|^2 = \frac{|\alpha|^{2n}}{n!} e^{-|\alpha|^2}, \tag{22.48}$$

which follows directly from Eq. (22.37). This results shows that the coherent state contains exactly n photons with $p_\alpha(n)$ given by a Poisson distribution.

The photon statistics can also be viewed as the probability distribution of the photon-number operator $\hat{n} \equiv B^\dagger B$. In particular, all $\hat{n}$-related expectation values follow from

$$\bar{n}^J \equiv \langle \hat{n}^J \rangle = \sum\nolimits_{n=0}^{\infty} n^j \, p_n. \tag{22.49}$$

We can easily verify that coherent states produce $\bar{n} = |\alpha|^2$ and $\langle \hat{n}^2 \rangle = |\alpha|^4 + |\alpha|^2 = \bar{n}^2 + \bar{n}$ due to their Poisson-distribution form. This result implies that the coherent state has the photon-number fluctuations $[\Delta\bar{n}]^2 = \langle \hat{n}^2 \rangle - \langle \hat{n} \rangle^2 = \bar{n} = |\alpha|^2$.

Since the Poisson distribution approaches a Gaussian shape for sufficiently large $\bar{n}$, we can approximate $p_\alpha(n)$ in this limit through a Gaussian centered at $\bar{n}$ having a width proportional to $\Delta\bar{n}$. A very accurate match with the Poisson distribution is reached through the approximation

$$p_\alpha(n) \to \frac{1}{\sqrt{\pi 2[\Delta\bar{n}]^2}} e^{-\frac{(n-\bar{n})^2}{2[\Delta\bar{n}]^2}} = \frac{1}{\sqrt{2\pi|\alpha|^2}} e^{-\frac{(n-|\alpha|^2)^2}{2|\alpha|^2}}, \text{ for } |\alpha| \gg 1. \tag{22.50}$$

The verification of this result is the topic of Exercise 22.8.

The relation (22.50) shows that $p_\alpha(n)$ converges toward a distribution that is narrow relative to its displacement. In other words, the relative fluctuations of a coherent state scale like

$$\frac{\Delta\bar{n}}{\bar{n}} = \frac{1}{\sqrt{\bar{n}}}, \tag{22.51}$$

which introduces the so-called shot-noise limit for a light source. An ideal single-mode laser operates in this regime. Since the relative fluctuations approach zero for very high intensities, i.e., $\bar{n} \gg 1$, these lasers approach the classical limit where the quantum fluctuations become negligible relative to the average intensity. This behavior shows that coherent states are very close to a situation in which light can be described classically.

22.4 Quantum-optical response to superposition states

Since the Jaynes–Cummings model produces different quantum Rabi oscillations for each Fock state, it is not a priori clear what kind of dynamics is generated for light that initially is in a superposition of Fock states. Clearly, this scenario is not describable by the classical averaging of Section 1.2.2 because the superposition averaging of waves produces nonclassical interferences, as shown in Section 2.1.2. We may thus anticipate that these interferences introduce interesting new quantum effects beyond quantum Rabi flopping.

We know already that a coherent state $|\alpha\rangle$ is a superposition state with features close to classical light. It also is interesting to study to what degree the quantum-optical hierarchy problem generates quantum features in connection with the interference effects. To study these phenomena, we start from an initial density matrix (22.13) in the form

$$\hat{\rho}(0) = |-\rangle\langle -| \otimes |\alpha\rangle\langle\alpha| = |-\rangle\langle -| \otimes \sum\nolimits_{n,n'=0}^{\infty} |n\rangle c_{n,\,n'}^{|\alpha\rangle} \langle n'|,$$

$$\Rightarrow \quad c_{n,\,n'}^{|\alpha\rangle} = a_n (a_{n'})^{\star}, \qquad a_n = \frac{\alpha^n}{\sqrt{n!}} e^{-\frac{1}{2}|\alpha|^2}. \tag{22.52}$$

These expressions define the elements of the density matrix based on the superposition expansion (22.37).

We can directly apply the result (22.17) to determine the inversion created by coherent light,

$$\begin{aligned} P_z(t) = \langle\sigma_z\rangle &= \mathrm{Tr}\left[\sigma_z\, \hat{\rho}_{2\mathrm{LS}}(t)\right] \\ &= \frac{1}{2} \sum\nolimits_{n=0}^{\infty} \left[|S_n(t)|^2 - |Z_n(t)|^2\right] c_{n,\,n}^{|\alpha\rangle}. \end{aligned} \tag{22.53}$$

For resonant conditions, i.e., $\Delta = 0$, Eqs. (22.11) and (22.52)–(22.53) produce

$$P_z^{|\alpha\rangle}(t) = -\frac{1}{2} \sum_{n=0}^{\infty} \frac{|\alpha|^{2n}}{n!} e^{-|\alpha|^2} \cos 2g\sqrt{n}\, t. \tag{22.54}$$

The quantum-statistical changes of light can be monitored by analyzing the temporal evolution of the marginal distribution

$$\begin{aligned} p(x,\,t) &\equiv \langle x|\hat{\rho}_{2\mathrm{LS}}(t)|x\rangle = \sum\nolimits_{n,n'=0}^{\infty} e^{-i\omega(n-n')t}\; p_{n,\,n'}\, \langle x|n\rangle\langle n'|x\rangle \\ p_{n,n'} &\equiv \left[S_{n+1}(t) c_{n+1,\,n'+1}^{|\alpha\rangle} S_{n'+1}(t) + Z_n(t) c_{n,\,n'}^{|\alpha\rangle} Z_{n'}^{\star}(t)\right], \end{aligned} \tag{22.55}$$

which follows directly from Eq. (22.18). The wave functions of the Fock states $\phi_n(x) = \langle x|n\rangle$ are given in Eq. (10.90). Following an analysis similar to that leading to (22.55), we can also deduce the dynamics of the polarization components P_x and P_y as well as the photon statistics.

22.4.1 Collapses and revivals of excitations

For illustration purposes, we now focus on a case where the TLS is resonantly excited by a coherent state with $\alpha = \sqrt{6}$. This field contains $\langle \hat{n} \rangle = \langle B^\dagger B \rangle = |\alpha|^2 = 6$ photons on average. A purely classical analysis yields the Rabi frequency $|\langle \Omega^{(+)} \rangle| = 2g|\alpha|$ corresponding to a time-dynamics in the form of sinusoidal oscillations with the classical period

$$T_{\text{cl}} \equiv \frac{2\pi}{|\langle \Omega^{(+)} \rangle|} = \frac{\pi}{g|\alpha|}, \tag{22.56}$$

see Section 18.2.3. Interestingly, also the resulting quantum Rabi flopping for the state $|n = 6\rangle$ produces $P_z(t) = -\frac{1}{2}\cos 2\pi \frac{t}{T_{\text{cl}}}$ based on Eq. (22.21). However, this agreement is purely formal because the states $|n\rangle$ only allow for discrete $\langle \Omega^{(+)} \rangle$ whereas the classical results have continuous frequencies. Furthermore, unlike the classical result, the Fock-state excitation induces no polarization.

The shaded area in Fig. 22.2(a) represents $P_z^{\text{cl}} = -\frac{1}{2}\cos 2\pi \frac{t}{T_{\text{cl}}}$ as a function of time. This result is compared with $P_z^{|\alpha\rangle}$ (solid line) corresponding to an excitation with a coherent state $|\alpha = \sqrt{6}\rangle$. We see that during the first Rabi cycle, $P_z^{|\alpha\rangle}$ follows the classical excitation rather closely. However, the fully quantum-mechanical result shows drastic deviations from the classical behavior after a few Rabi cycles.

As the first distinct difference, we observe that $P_z^{|\alpha\rangle}$ collapses toward a steady state roughly within two Rabi cycles, i.e., $t \approx 2T_{\text{cl}}$. This collapse to a nearly vanishing inversion is a pure quantum effect because the classical description predicts persistent oscillations at all times. The collapse is then followed by an interval $t \in [2, 6]T_{\text{cl}}$ where the inversion remains close to zero. However, the excitation is revived during the interval $t \in [6, 12]T_{\text{cl}}$ and exhibits a largely oscillating behavior. This revival is centered around $T_{\text{rev}} = 12T_{\text{cl}}$. The signal then collapses once more, just to be revived again after another $12T_{\text{cl}}$-long evolution. Thus, a coherent state $|\alpha\rangle$ generates a sequence of collapses and revivals which have no classical analogy. As a general trend, the collapse–revival sequence presents increasingly irregular oscillations as the order of the revival is increased.

22.4.2 Origin of collapses and revivals

To explain why the collapse of P_z emerges, we use the approximation (22.50) for $\bar{n} = |\alpha|^2$ to solve Eq. (22.54),

$$\begin{aligned} P_z^{|\alpha\rangle}(t) &= -\frac{1}{2}\sum_{n=0}^{\infty} \frac{|\alpha|^{2n}}{n!} e^{-|\alpha|^2} \cos 2g_n t = -\frac{1}{2}\text{Re}\left[\sum\nolimits_{n=0}^{\infty} \frac{|\alpha|^{2n}}{n!} e^{-|\alpha|^2} e^{i2g\sqrt{n}t}\right] \\ &\to -\frac{1}{2}\text{Re}\left[\sum_{n=0}^{\infty} \frac{1}{\sqrt{2\pi\bar{n}}} e^{-\frac{(n-\bar{n})^2}{2\bar{n}}} e^{i2g\sqrt{n}\,t}\right], \quad \text{for } \bar{n} \gg 1. \end{aligned} \tag{22.57}$$

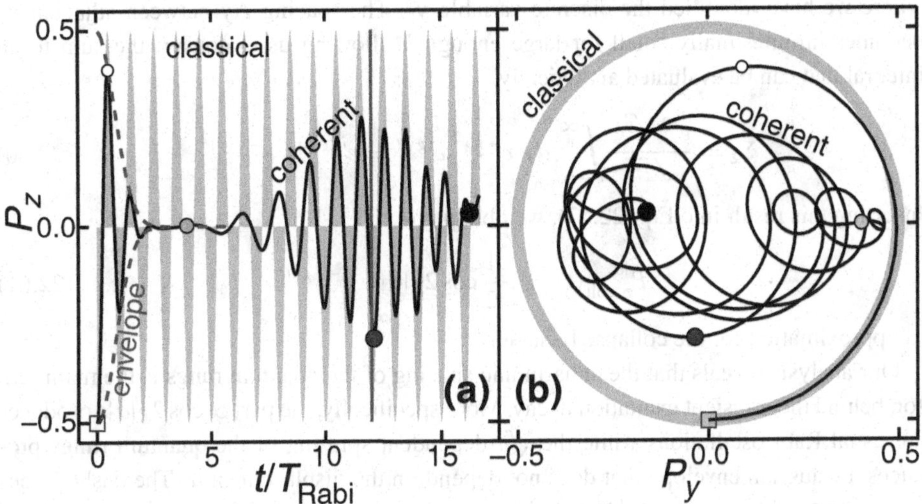

Figure 22.2 Excitation of the Jaynes–Cummings system with coherent light. The two-level system is initially unexcited ($|-\rangle$) and the light is in a coherent state $|\alpha\rangle = \sqrt{6}\rangle$ corresponding to $\bar{n} = |\alpha|^2 = 6$ photons on average. **(a)** The solid line shows the resulting inversion dynamics, Eq. (22.54), as a function of time. The purely classical result is shown by the shaded area. The dashed lines indicate the expected envelope of collapse, based on Eq. (22.61). The vertical line denotes the expected temporal position of the revival, based on Eq. (22.63). **(b)** The corresponding trajectory of the Bloch vector $\bar{\mathbf{P}} = (0,\ \bar{P}_y,\ \bar{P}_z)$ (solid line) is shown together with the Bloch sphere (thick, shaded line). To synchronize this motion with $P_z(t)$, we have added circles that grow in size and darken for increasing time. The initial moment is indicated by a square in both frames.

For later reference, we denote the sum within the real part as Σ_z. For large enough $\bar{n}$, the terms in Σ_z are strongly concentrated around $n \approx \bar{n}$ due to the shot-noise limit discussed in Section 22.3.2. In other words, the terms under the sum decay extremely fast as n deviates significantly from $\bar{n}$ such that we can extend the summation limits from $-\infty$ to ∞ without introducing major errors. Replacing the summation variable by $n = j + \bar{n}$, the sum becomes

$$\Sigma_z = \sum_{j=-\infty}^{\infty} \frac{1}{\sqrt{2\pi\bar{n}}} e^{-\frac{j^2}{2\bar{n}}} e^{i2g\sqrt{\bar{n}+j}\,t}. \tag{22.58}$$

Here, the dominant arguments are close to $j = 0$ whenever $\bar{n}$ is large. In this limit, we can express $2\sqrt{\bar{n}+j} = 2\sqrt{\bar{n}} + j/\sqrt{\bar{n}}$ through its first two Taylor-expansion terms. This approximation yields

$$\begin{aligned}
\Sigma_z &= \frac{1}{\sqrt{2\pi}} e^{i2g\sqrt{\bar{n}}\,t} \sum_j \frac{1}{\sqrt{\bar{n}}} e^{-\frac{1}{2}\left(\frac{j}{\sqrt{\bar{n}}}\right)^2} e^{ig\frac{j}{\sqrt{\bar{n}}}t} \\
&= \frac{e^{i2g\sqrt{\bar{n}}\,t}}{\sqrt{2\pi}} \sum_j \Delta y\, e^{-\frac{1}{2}y_j^2} e^{igy_j t}, \quad \Delta y \equiv \frac{1}{\sqrt{\bar{n}}} \quad y_j \equiv j\Delta y,
\end{aligned} \tag{22.59}$$

where we have identified the discrete variable y_j. The spacing Δy between adjacent y_j becomes infinitesimally small for large enough $\bar{n}$ allowing us to convert the sum to an integral that can be evaluated analytically,

$$\Sigma_z = \frac{e^{i2g\sqrt{\bar{n}}\,t}}{\sqrt{2\pi}} \int_{-\infty}^{\infty} dy\; e^{-\frac{1}{2}y^2}\, e^{igyt} = e^{i2g\sqrt{\bar{n}}\,t}\, e^{-\frac{1}{2}g^2t^2}. \tag{22.60}$$

Inserting this result into Eq. (22.57), we obtain

$$P_{z,\,\mathrm{coll}}^{|\alpha\rangle}(t) = -\frac{1}{2}\cos 2g|\alpha|t\;\; e^{-\frac{1}{2}g^2t^2} \tag{22.61}$$

as approximation for the collapse behavior.

Our analysis reveals that the anharmonic spacing of the quantum rungs is the main reason behind the transient excitation decay. More specifically, the part $\propto \cos 2g|\alpha|t$ produces the usual Rabi oscillations while the $\sqrt{n}$-dependent splitting of the quantum rungs produces a Gaussian envelope that does not depend on the displacement α. The dashed lines in Fig. 22.2(a) show the $\pm e^{-\frac{1}{2}g^2t^2}$ envelope demonstrating that the approximative analysis nicely reproduces the collapse behavior.

To explain the revivals, we consider the quasiperiodicity in (22.57) caused by the quantum rungs. Initially, at $t = 0$ two neighboring n states are in phase. However, for any later time t, their phases differ by $\Delta\phi_n \equiv 2(g_{n+1} - g_n)\,t = 2g(\sqrt{n+1} - \sqrt{n})\,t$. Due to the shot-noise limit, the dominant terms under the sum are very close to $n = \bar{n}$ for a large $\bar{n}$. Consequently, we apply the Taylor expansion

$$\Delta\phi_n = 2g(\sqrt{n+1} - \sqrt{n})\,t = \frac{g\,t}{\sqrt{n}} + \mathcal{O}(n^{-\frac{3}{2}}) \tag{22.62}$$

to deduce the relevant phase shifts. For large enough $\bar{n}$, all the dominant terms have the same phase difference $\Delta\phi_n = \Delta\phi_{\bar{n}} = g\,t/\sqrt{\bar{n}} = g\,t/|\alpha|$ between two consecutive quantum rungs. We therefore find rephasing of the individual contributions at the time when $\Delta\phi_{\bar{n}}$ becomes 2π. This revival time is

$$\Delta\phi_{\bar{n}} = \frac{g\,T_{\mathrm{rev}}}{|\alpha|} = 2\pi \qquad \Leftrightarrow \qquad T_{\mathrm{rev}} = \frac{2\pi\,|\alpha|}{g} = 2|\alpha|^2 T_{\mathrm{cl}}, \tag{22.63}$$

where the last form follows directly from the definition (22.56). Hence, a revival should be observed at $T_{\mathrm{rev}} = 12T_{\mathrm{cl}}$ for the parameters used to generate Fig. 22.2(a). The horizontal line indicates the predicted revival time showing that a very good agreement with the full results is observed already for $\bar{n} = 6$.

Typically, the revival behavior becomes more and more pronounced with increasing $|\alpha|$. For large enough $|\alpha|$, one can even detect multiple collapses and revivals. The revivals are then spaced by integer multiples of T_{rev}. However, as a consequence of the higher-order corrections to Eq. (22.62), the revivals are only partial. The rephasing of each n term in Eq. (22.54) is influenced by terms beyond the linear approximation (22.62). These corrections prevent a perfect revival, i.e., rephasing, especially for small $\bar{n}$ or large times. Thus, the higher-order revivals are typically weaker than the lower-order ones.

It is interesting to compare the physical origin of the collapses and the revivals. Both of them stem from the $\sqrt{n}$ anharmonicity of the Jaynes–Cummings ladder structure, see Section 21.2.3. Thus, they are both genuine quantum effects. However, the collapses do not rely on the discrete nature of the quantum Rabi splitting. In fact, we took the continuum limit when evaluating Eq. (22.60). Thus, the collapse stems only from the quantum anharmonicity of the Jaynes–Cummings system. The revivals, however, are possible only due to the discreteness of the quantum rungs. Thus, the revivals are even a stronger indication of quantum-optical effects than the collapses.

22.4.3 Quantum-statistical modifications

To gain further insights into the collapse–revival dynamics, we show the trajectories of the Bloch vector $\bar{\mathbf{P}} = (0, \bar{P}_y, \bar{P}_z = P_z)$ (solid line) in Fig. 22.2(b). For this plot, we have used a rotating frame of $e^{-i\omega t}$. Due to the resonance conditions, $\bar{P}_x$ remains zero for all times, just as in the purely classical analysis. However, the trajectory of the Bloch vector becomes extraordinarily complicated. In fact, it seems to perform a chaotic motion that is very far from regular oscillations or from the spiraling dynamics observed for a fully classical investigation, see e.g., Fig. 20.2.

To understand the complicated Bloch-vector behavior, we analyze the marginal distribution (Fig. 22.3(a)) and the photon statistics (Fig. 22.3(b)) at three different times: $t = 0$ (shaded area), $t = 4\,T_{\text{cl}}$ (solid line) just after the collapse, and $t = 12\,T_{\text{cl}}$ (solid line) at the peak of the revival. The parameters are the same as in Fig. 22.2. We see that the initial $p(x, t = 0)$ (shaded area) is a displaced Gaussian centered at $\alpha = \sqrt{6} \approx 2.45$.

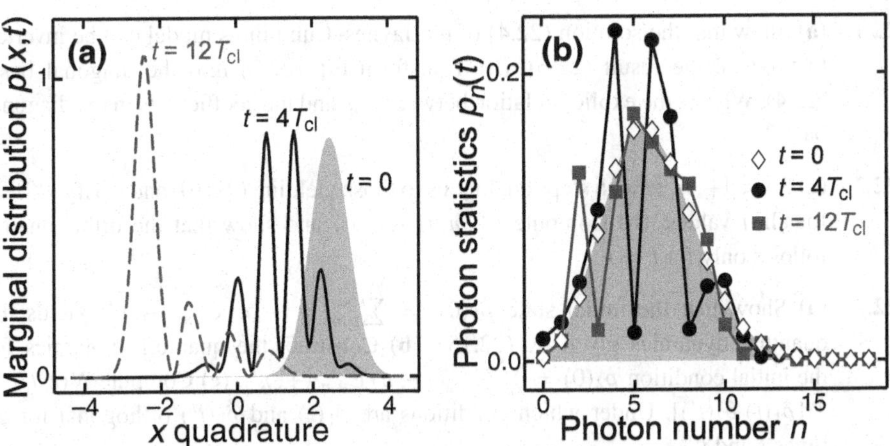

Figure 22.3 The light quantum statistics during the collapse–revival dynamics. **(a)** The marginal distribution and **(b)** the photon statistics are shown at the initial time $t = 0$ (shaded area/+ diamonds), at $t = 4T_{\text{cl}}$ (solid line/+ circles), and at $t = 12T_{\text{cl}}$ (dashed line/+ squares). The chosen times are the same as those indicated by the three last circles in Fig. 22.2. We have applied Eq. (22.55) to produce $p(x, t)$ and $p_n(t) = p_{n,n}$ under the same conditions as in Fig. 22.2.

As discussed in Section 22.3.1, this has exactly the shape of the vacuum's quadrature distribution. With increasing time, $p(x,t)$ spreads and develops an interference pattern (solid line) after the collapse of the excitation has occurred. This strong deviation from a Gaussian shape indicates that the light has become strongly nonclassical. The Gaussian is partially recovered at the time of the revival. Even though this recovery is not complete, the revival indicates that the system is back to a more classical behavior.

As we can see in Fig. 22.3(b), the photon statistics $p_n(t)$ shows a similar development of quantum features. In the beginning, $p_n(t = 0)$ (shaded area, diamonds) is Poisson distributed and very close to a Gaussian as suggested by the approximation (22.50). For the collapsed excitation, $p_n(t)$ (solid line, circles) displays a strongly oscillating photon statistics indicating pronounced quantum features. At the revival time, the photon statistics (dashed line, squares) regains some of its original Poisson-like form.

Altogether, this discussion shows that the Jaynes–Cummings model can produce profound deviations from the classical behavior. We have identified the interferences between different quantum rungs as the main reason to produce highly nonclassical states during the excitation dynamics. As a result, the classical contributions to the optical Bloch equations are accompanied by strong quantum-optical correlations, see Eq. (21.26). In other words, the quantum-hierarchy problem starts to involve higher-order clusters corresponding to correlated light fields and light–matter entanglement. As a result, we find a very complicated Bloch-vector dynamics whose trajectory is significantly different from a classical one.

Exercises

22.1 **(a)** Show that the solution (22.4) of the Jaynes–Cummings model can be inverted to produce the result (22.6). **(b)** Transform Eq. (22.7) into the diagonal basis (22.4). What is the explicit relation between $C_n^{\pm}$ and the coefficients in the diagonal basis?

22.2 **(a)** Solve $|+, n, t\rangle$ with steps analogous to those yielding (22.10) and verify (22.12) for all n values. **(b)** Compute $\langle +, n, t|-, n, t'\rangle$ and show that the orthogonality follows only for $t = t'$.

22.3 **(a)** Show that the initial state $\hat{\rho}_1(0) = \sum_{n,n'=0}^{\infty} |-, n\rangle\, C_{n,n'} \langle -, n'|$ yields the quantum dynamics given by (22.15). **(b)** Construct the quantum dynamics for the initial condition $\hat{\rho}_2(0) = \sum_{n,n'=0}^{\infty} |+, n\rangle\, C_{n,n'} \langle +, n'|$. **(c)** Compute $N(t,t') \equiv \mathrm{Tr}[\hat{\rho}_1(t)\, \hat{\rho}_2(t')]$. Under which conditions are $\hat{\rho}_1(t)$ and $\hat{\rho}_2(t')$ orthogonal for all times t and t'?

22.4 **(a)** Show that the light properties of a TLS follow from the reduced density matrix, i.e., $\langle [\hat{B}^\dagger]^J\, B^K \rangle = \mathrm{Tr}[[B^\dagger]^J\, B^K\, \hat{\rho}_B(t)]$. **(b)** Construct $\hat{\rho}_{2\mathrm{LS}}(t)$ and $\hat{\rho}_B(t)$ from Eq. (22.15). **(c)** Diagonalize $\hat{\rho}_{2\mathrm{LS}}(t)$ and show that it is generally not a pure state.

22.5 Construct $P_z(t)$ for the initial state **(a)** $|\psi_1(0)\rangle = |+\rangle \otimes |n\rangle$ and **(b)** $|\psi_2(0)\rangle = |-\rangle \otimes |n+1\rangle$. **(c)** For which time difference Δt is $\langle\psi_1(t)|\psi_2(t+\Delta t)\rangle$ maximized/minimized?

22.6 **(a)** Show that the initial state $|\psi(0)\rangle = |+\rangle \otimes |0\rangle$ yields $\Delta\Omega^2(t)$ fluctuations described by Eq. (22.27). **(b)** According to Eq. (18.67), classical Rabi oscillations follow from $P_x(\Omega, t) = 0$, $P_y(\Omega, t) = \frac{1}{2}\sin\Omega t$, and $P_z(\Omega, t) = \frac{1}{2}\cos\Omega t$. Develop a semiclassical model of vacuum-field fluctuations, i.e., assume that the Rabi frequency is distributed according to $p(\Omega) = \frac{1}{\sqrt{2\pi\Delta\Omega^2}} e^{-\Omega^2/2\Delta\Omega^2}$. Show that classical averaging $\langle\langle P_j(t)\rangle\rangle = \int d\Omega\; p(\Omega)\; P_j(\Omega, t)$ yields $\langle\langle P_{x,y}(\Omega, t)\rangle\rangle = 0$ and $\langle\langle P_{z_1}(\Omega, t)\rangle\rangle = \frac{1}{2} e^{-t^2\Delta\Omega^2/2}$. **(c)** How has one to choose $\Delta\Omega$ in the semiclassical approach to reproduce quantum Rabi flopping at early times? Which evident differences appear between the semiclassical and the full quantum model? **(d)** Refine the semiclassical model by assuming $p(\Omega) = \frac{1}{2}(\delta(\Omega - \Delta\Omega) + \delta(\Omega + \Delta\Omega))$. Compute $\langle\langle P_j(t)\rangle\rangle$ and match $\Delta\Omega$ to produce quantum Rabi flopping. Show that only discrete $\Delta\Omega$ values are allowed. Why is this $p(\Omega)$ model inconsistent with the quantum-optical one? **Hint:** *Inspect Eq.* (22.28).

22.7 **(a)** Start from the definition (22.37) and show that $B^J|\alpha\rangle = \alpha^J|\alpha\rangle$ and $\langle\alpha|(B^\dagger)^J = \langle\alpha|(\alpha^*)^J$ but $B^\dagger|\alpha\rangle \neq \beta^*|\alpha\rangle$. **(b)** Express the Hermite polynomial via the relation (10.80) and convert Eq. (22.40) into the result (22.41). **(c)** Verify the relation (22.42) and show that the coherent state is generated by the displacement operator, as given in Eqs. (22.46)–(22.47).

22.8 **(a)** Show that the coherent state yields the Poisson-distributed photon statistics $p_\alpha(n) = |\langle n|\alpha\rangle|^2 = \frac{|\alpha|^{2n}}{n!} e^{-|\alpha|^2}$. **(b)** Show that $\mathbf{n} \equiv \langle\hat{n}\rangle_{|\alpha\rangle} = |\alpha|^2$ and $\langle\hat{n}^2\rangle = \langle\hat{n}^2\rangle_{|\alpha\rangle} = |\alpha|^4 + |\alpha|^2$. **(c)** Convert the Poisson distribution (22.48) into a Gaussian distribution by demanding that $\sum_n n^J\, p(n) = \int_{-\infty}^{\infty} dn\; p_{\rm gauss}(n)$ yields the same result for $J = 0$, 1, and 2. Verify numerically that the approximation (22.50) is indeed accurate for large $|\alpha|$.

22.9 Classical Rabi flopping follows from $P_z(\Omega, t) = -\frac{1}{2}\cos\Omega t$ where Ω is the Rabi frequency of a classical field. **(a)** Assume semiclassical averaging where Ω is distributed according to $p(\Omega) = \frac{1}{\sqrt{2\pi\Delta\Omega^2}} e^{-(\Omega-\Omega_0)^2/2\Delta\Omega^2}$. How must Ω_0 and $\Delta\Omega$ be chosen in order to reproduce the collapse result (22.61) semiclassically? **Hint:** *See Exercise 22.6.* **(b)** Define $\hat{x} = \frac{B+B^\dagger}{2}$ and use $p(x) = |\langle x|\alpha\rangle|^2$ to describe $p(\Omega)$. Show that the collapse can be understood through vacuum-field fluctuations. **(c)** A classical field is described by a coherent state $|\alpha\rangle$ with $(B^J)|\alpha\rangle = \alpha^J|\alpha\rangle$. Show that *normally ordered* expectation values are generated by a sharp $P(\alpha') = \delta^{(2)}(\alpha' - \alpha)$ distribution. This is actually the Glauber–Sudarshan function of the coherent state. Why does the $P(\alpha')$ contradict $p(x)$ of part **(b)**? **Hint:** *Analyze the symmetry of quadrature operators.*

22.10 Assume that a laser is initially defined by a coherent state$|\alpha\rangle$. As long as the light propagates freely, its dynamics follows from $\hat{H}_0 = \hbar\omega\left(B^\dagger B + \frac{1}{2}\right)$. Show that the laser light yields $|\alpha(t)\rangle = e^{-i\hat{H}_0 t/\hbar}\,|\alpha\rangle$ with $\alpha(t) = \alpha\, e^{-i\omega t}$. Calculate the dynamics of the marginal distribution $p_\alpha(x) \equiv |\langle x|\alpha(t)\rangle|^2$ and the photon statistics $p_n(t) = |\langle n|\alpha(t)\rangle|^2$.

Further reading

Coherent states were first introduced in:

R. J. Glauber (1963). Coherent and incoherent states of the radiation field, *Phys. Rev.* **131**, 2766.

See also:

R. J. Glauber (2006). Nobel lecture: One hundred years of light quanta, *Rev. Mod. Phys.* **78**, 1267.

Collapses and revivals were first predicted in:

J. H. Eberly, N. B. Narozhny, and J. J. Sanchez–Mondragon (1980). Periodic spontaneous collapse and revival in a simple quantum model, *Phys. Rev. Lett.* **44**, 1323.

The experimental confirmation of collapses and revivals is reported in:

G. Rempe, H. Walther, and N. Klein (1987). Observation of quantum collapse and revival in a one-atom maser, *Phys. Rev. Lett.* **58**, 353.

23

Spectroscopy and quantum-optical correlations

In the previous chapters, we have seen that quantum-optical correlations can produce effects that have no classical explanation. In particular, the matter excitations can depend strongly on the specific form of the quantum fluctuations, i.e., the quantum statistics of the light source. For example, Fock-state sources can produce quantum Rabi flopping with discrete frequencies while coherent-state sources generate a sequence of collapses and revivals in atomic excitations. Hence, not only the intensity or the classical amplitude of the field is relevant but also the quantum statistics of the exciting light influences the matter response. Even if we take sources with identical intensities, the resulting atomic excitations are fully periodic for a Fock-state excitation while a coherent-state excitation produces a chaotic Bloch-vector trajectory with multiple collapses and revivals.

In this chapter, we use this fundamental observation as the basis to develop the concept of quantum-optical spectroscopy. We show in Chapter 30 that this method yields a particularly intriguing scheme to characterize and control the quantum dynamics in solids. Since one cannot imagine how to exactly compute the many-body wave function or the density matrix, we also study how principal quantum-optical effects can be described with the help of the cluster-expansion scheme.

23.1 Quantum-optical spectroscopy

Historically, the continued refinement of optical spectroscopy and its use to manipulate the states of matter has followed a very distinct path where one simultaneously tries to control and characterize light with increased accuracy. At the beginning, spectrally broad weak-intensity sources were used, provided, e.g., by incandescent light bulbs that essentially emit white light. In this case, one can, e.g., identify and quantify the constituents of the sample by analyzing the material-specific absorption lines in the spectrum of the transmitted light. At the next level, one reduces the spectral width and increases the intensity of the light source with the goal to induce desired transitions in matter by resonant excitation of specific transitions. In the simplest case, the light–matter interaction can then be reduced to a two-level scheme as discussed in Chapter 18. Most

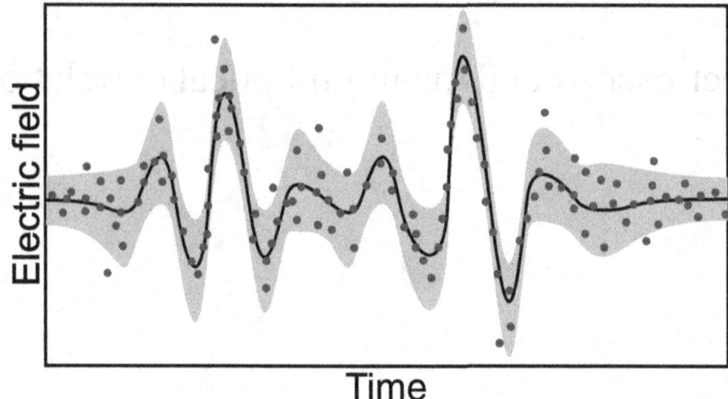

Figure 23.1 Schematic representation of a tailored light pulse. The circles indicate an ensemble of measurement outcomes for the time-resolved electric field $\hat{E}(t)$. Its average value (solid line) is shown together with the extent of the fluctuations (shaded area) defined by the quantum statistics.

of the important spectroscopic developments have become possible through the invention of lasers, which are nearly ideal sources that deliver spectrally narrow and intense light.

During the last few decades, the control and characterization of laser light could be perfected to a level that one can stabilize its phase and temporal profile to a very high degree. The solid line in Fig. 23.1 schematically illustrates the temporal evolution of the electric field in such a designed laser pulse. This kind of pulse can be applied to control and monitor a wide variety of phenomena including electron–hole excitations in solids, chemical reactions, molecular motion etc. with unprecedented accuracy. State-of-the-art experiments can nowadays manipulate and characterize these systems on the femto- down to zepto-second time scales, which is sufficient to capture all nonrelativistic aspects of the interesting quantum-mechanical processes, compare Fig. 3.1.

We already know that light is not purely classical but also has important quantum-mechanical features. For example, if we perform electric-field measurements, we can directly determine the value $\langle \hat{E}(t) \rangle$ (solid line, in Fig. 23.1). However, each individual measurement produces only one of many possible values, all of which are scattered around the average field $\langle \hat{E}(t) \rangle$. The filled circles in Fig. 23.1 schematically present the results of a measurement ensemble. According to Born's theorem, the outcome of each measurement is arbitrary but on average follows the quantum-mechanical distribution, compare Section 5.3. The shaded area in Fig. 23.1 indicates the half width of the quantum-mechanical distribution.

The detailed distribution of the random intrinsic fluctuations is determined by the laws of quantum mechanics and is referred to as the quantum statistics of the light field. In practice, the measurement of the electromagnetic field properties can be performed, e.g., with the help of the homodyne detection scheme that essentially measures $\langle \hat{E} \rangle$ via an ensemble of detections. For a single-mode field, $\hat{E}(t)$ corresponds to a position quadrature operator $\hat{x}$ such that one can construct the marginal distribution $p(x)$ from an ensemble of measured $E(t)$ values, as described in Section 5.3.

We have already seen that the optical response of a two-level system (TLS) can become extremely sensitive to the quantum statistics of the light source. Thus, the quantum statistics produces a new level of characterization and control over light–matter interactions, in addition to the classical intensity, spectral, temporal, and phase extensions of the laser light. To understand which material aspects are affected by quantum-optical fluctuations, we return to the main conclusion of Section 21.1.4. There, we have shown that it is the nonlinear, Fermionic aspects of matter that are influenced by the quantum statistics of the exciting field. In other words, the quantum-optical fluctuations of light characterize and control to a certain degree the nonlinear matter response. This is the principal idea behind the concept of quantum-optical spectroscopy. These effects are quite extreme in a TLS because it represents a perfect Fermion.

Since the electronic excitations in solids constitute a Fermionic many-body state, it is obvious that their optical response must eventually become nonlinear under sufficiently strong laser excitation. We explain in Chapter 30 how the quantum statistics of the exciting light can indeed be used to control the many-body quantum dynamics of semiconductors. The full utilization of the resulting possibilities provides a new challenge and intriguing prospects for semiconductor optical spectroscopy. For this purpose, one needs detailed knowledge of how the many-body state and the quantum fluctuations of the generating light source are connected. As an experimental challenge, one needs to develop strong-intensity quantum optical sources. Research in this direction is still at an early stage, however, we discuss some of the possible implications for semiconductor quantum optics in Chapter 30.

23.2 Quantum-statistical representations

To understand the degree to which the quantum statistics of the exciting light influences the nonlinear response of matter, we first trace the differences between the Fock state $|n_0 = 6\rangle$ and the coherent-state $|\alpha = \sqrt{6}\rangle$ excitation. For example, we see in Fig. 22.2 that the inversion factor $\bar{P}_z$ evolves completely differently for the Fock state (shaded area) and for the coherent state (solid line) even though both sources have the same average intensity. Thus, the quantum statistics indeed controls many aspects of the matter response according to the principle of quantum-optical spectroscopy.

To quantify the light sources in more detail, we can follow, e.g., the photon statistics $p_n \equiv \langle n|\hat{\rho}|n\rangle$ or the marginal distribution $p(x) \equiv \langle x|\hat{\rho}|x\rangle$. For the coherent state, we find

$$p^{|\alpha\rangle}(x) = \sqrt{\frac{2}{\pi}} e^{-2(x-\alpha_1)^2}, \qquad \alpha \equiv \alpha_1 + i\alpha_2$$
$$p_n^{|\alpha\rangle} = |\langle n|\alpha\rangle|^2 = \frac{|\alpha|^{2n}}{n!} e^{-|\alpha|^2}, \tag{23.1}$$

based on Eqs. (22.41) and (22.48). In the same way, the Fock state yields

$$p^{|n_0\rangle}(x) = |\langle x|n_0\rangle|^2 = \sqrt{\frac{2}{\pi}} \frac{H_{n_0}^2(\sqrt{2}x)}{n_0!\, 2^{n_0}} e^{-2x^2},$$
$$p_n^{|n_0\rangle} = |\langle n|n_0\rangle|^2 = \delta_{n,n_0}, \tag{23.2}$$

based on Eq. (10.81) and the orthogonality of the Fock states. Figure 23.2(a) presents the marginal distribution for the Fock state $|n_0 = 6\rangle$ (shaded area) and the coherent state $|\alpha = \sqrt{6}\rangle$ (solid line), both having $\langle B^\dagger B\rangle = 6$ photons on average. The corresponding photon statistics is shown in Fig. 23.2(b).

We see that the probability distributions for the Fock and the coherent state are very different. The distribution $p^{|\alpha\rangle}(x)$ is a Gaussian centered at $\alpha = \sqrt{6} \approx 2.449$ while the Fock state has a broad $p^{|n_0\rangle}(x)$ with a pronounced interference pattern. Thus, the $E(t)$ counts for the coherent state are evenly distributed as in Fig. 23.1 whereas the Fock state produces counts in stripes that are separated by regions without any counts.

The quantum-statistical difference in the photon statistics of these sources is also very significant. The Fock-state p_n has only one nonvanishing value while the coherent state yields a broad distribution with a peak close to $n = |\alpha|^2 = 6$. These results represent the fact that the Fock state has no photon-number fluctuations while the fluctuations of the coherent state are shot-noise limited, as discussed in Section 22.3.2. Hence, these two fields are quantum statistically very different even though they have exactly the same average

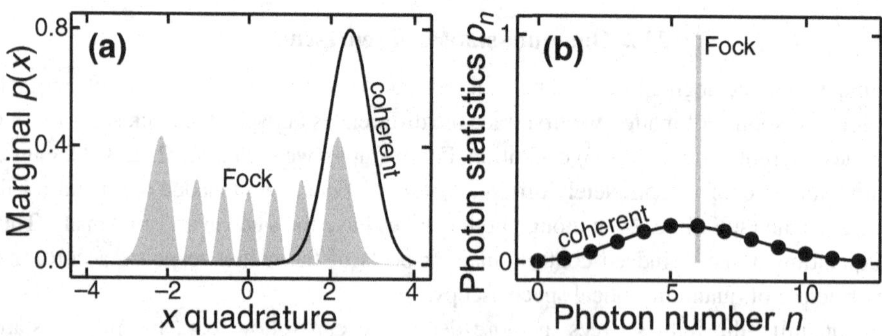

Figure 23.2 **(a)** Marginal distribution for the Fock state $|n_0 = 6\rangle$ (shaded area) and the coherent state $|\alpha = \sqrt{6}\rangle$ (solid line), both having $\langle B^\dagger B\rangle = 6$ photons on average. **(b)** The corresponding photon statistics.

intensity. Thus, we can conclude that the significantly different matter response – analyzed in Fig. 22.2 – for the two exciting fields is a direct consequence of their quantum-statistical properties.

23.2.1 Wigner function

We already know that a coherent state has the most classical quantum-statistical form among all states. The interference pattern in the x distribution of the Fock state results from wave properties in the x quadrature, which cannot be explained classically. As discussed in Chapter 6, this interference feature can be analyzed nicely with the help of the Wigner function. In particular, we expect the Fock state to produce negative-valued regions of $W(x, y)$ in the phase space. To inspect this more closely, we generalize the definition of the Wigner function

$$W(x, y) \equiv \frac{1}{\pi} \int dr \left\langle x - \frac{r}{2} \middle| \hat{\rho} \middle| x + \frac{r}{2} \right\rangle e^{2iy\,r} \tag{23.3}$$

for the case of quantized light. This expression is a logical extension of the original definition (6.9) if we use the scaled quadrature position x and momentum y as coordinates. The prefactor $1/\pi$ results if we demand that the Wigner function is properly normalized, i.e., $\int dx\, dy\, W(x, y) = 1$.

For the coherent state, we can use $\langle x|\hat{\rho}^{|\alpha\rangle}|x'\rangle = \langle x|\alpha\rangle\langle\alpha|x'\rangle$ together with (22.41) to obtain

$$\left\langle x - \frac{r}{2} \middle| \hat{\rho}^{|\alpha\rangle} \middle| x + \frac{r}{2} \right\rangle = \sqrt{\frac{2}{\pi}}\, e^{-2(x-\alpha_1)^2}\, e^{-\frac{r^2}{2} - 2i\alpha_2 r}. \tag{23.4}$$

Inserting this into Eq. (23.3) produces a Gaussian integration kernel, eventually yielding

$$W^{|\alpha\rangle}(x, y) = \frac{2}{\pi}\, e^{-2(x-\alpha_1)^2}\, e^{-2(y-\alpha_2)^2}. \tag{23.5}$$

Thus, the Wigner function of a coherent state is a Gaussian centered around $(x, y) = (\alpha_1, \alpha_2)$. This distribution is positive definite and has the quadrature fluctuations,

$$\begin{aligned} \Delta X^2 &= \int dx\, dy\, x^2\, W^{|\alpha\rangle}(x, y) - \left[\int dx\, dy\, x\, W^{|\alpha\rangle}(x, y)\right]^2 = \frac{1}{4}, \\ \Delta Y^2 &= \int dx\, dy\, y^2 W^{|\alpha\rangle}(x, y) - \left[\int dx\, dy\, y W^{|\alpha\rangle}(x, y)\right]^2 = \frac{1}{4}. \end{aligned} \tag{23.6}$$

Looking at the Heisenberg uncertainty relation (5.50), the canonical quadrature pair with $[\hat{x}, \hat{y}]_- = i/2$ yields

$$\Delta X \Delta Y \geq \frac{1}{4}. \tag{23.7}$$

Thus, the coherent state has the minimally allowed quantum fluctuations, very much in accordance with what one would expect for such a maximally classical state.

The Wigner function of the Fock states can be derived using steps similar to those yielding $W^{|\alpha\rangle}(x, y)$. Technically, we simply insert the definition (10.81) into (23.3). A lengthy but straightforward calculation eventually produces

$$\begin{aligned} W^{|n\rangle}(x, y) &= \frac{2}{\pi} e^{-2(x^2+y^2)} \sum_{K=0}^{n} \frac{n!(-1)^{n-K}}{K!K!(n-K)!} (4(x^2+y^2))^K \\ &= \frac{2}{\pi} e^{-2(x^2+y^2)} (-1)^n L_n(4(x^2+y^2)), \end{aligned} \tag{23.8}$$

for the Fock state $|n\rangle$. The special function that appears is the well-known Laguerre polynomial defined by

$$\begin{aligned} &L_n(x) \equiv \sum_{K=0}^{n} \frac{n!\,(-x)^K}{K!K!(n-K)!}, \qquad L_0(x) = 1, \qquad L_1(x) = 1 - x \\ &L_2(x) = \frac{1}{2}(x^2 - 4x + 2) \qquad L_3(x) = \frac{1}{6}(-x^3 + 9x^2 - 18x + 6). \end{aligned} \tag{23.9}$$

Here, we have also presented the explicit forms of the four lowest-order Laguerre polynomials.

The Wigner function of the coherent state $|\alpha = \sqrt{6}\rangle$ and $W^{|n=6\rangle}(x, y)$ are presented in Fig. 23.3(a) and (b), respectively. We again note the pronounced quantum-statistical differences between both these states. Most importantly, we see that $W^{|n=6\rangle}(x, y)$ has negatively valued phase-space regions, unlike the coherent-state Wigner function. To show the negativity of $W^{|n=6\rangle}(x, y)$ more clearly, we have also plotted the cross sections $W(x, 0)$ (solid lines) along the x axis in Fig. 23.3. As already discussed in Section 6.1.3, this negativity is a genuine quantum feature and there is no classical counterpart for the state producing such a quasiprobability distribution. All Fock states $|n \geq 1\rangle$ share this prominent quantum feature.

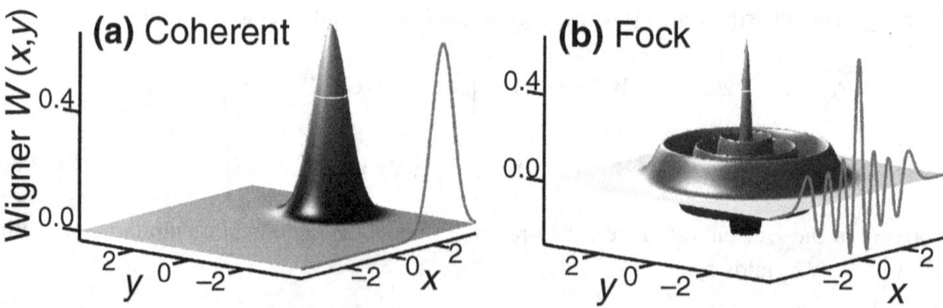

Figure 23.3 Wigner function of **(a)** the coherent state $|\alpha = \sqrt{6}\rangle$ and **(b)** the Fock state $|n = 6\rangle$. The solid lines show the respective cross section $W(x, 0)$.

23.2.2 Quantum-statistical pluralism

It is often very useful to describe the full quantum statistics of an object in several equivalent ways. For example, the density matrix $\hat{\rho}$ defines the full quantum statistics of an object, while the marginal distribution $p(x) = \langle x|\hat{\rho}|x\rangle$ or the photon statistics $p_n = \langle n|\hat{\rho}|n\rangle$ provide only partial projections of the full result. In order to find alternative representations, we can, e.g., use the mapping between the Wigner function and the density matrix by making the Fourier transformation,

$$\int dy\, W(x, y)\, e^{-2iyr} = \left\langle x - \frac{r}{2}\right|\hat{\rho}\left|x + \frac{r}{2}\right\rangle. \tag{23.10}$$

Together with Eq. (23.3), this establishes a one-to-one mapping between $\hat{\rho}$ and W. In quantum optics, one often studies other phase-space distributions such as the Husimi Q function,

$$Q(x, y) = \frac{2}{\pi}\int dx'\, dy'\, W(x', y')\, e^{-2(x'-x)^2-2(y'-y)^2}, \tag{23.11}$$

and the Glauber–Sudarshan P function

$$W(x, y) = \frac{2}{\pi}\int dx'\, dy'\, P(x', y')\, e^{-2(x'-x)^2-2(y'-y)^2}. \tag{23.12}$$

All of these can be formally connected through one-to-one mappings such that also the Q and P functions uniquely define the systems.

In principle, there are infinitely many different ways to represent one and the same quantum statistics. Thus, the quantum statistics can be analyzed from many different points of view, which introduces the concept of quantum-statistical pluralism. One of the great benefits of this pluralism is that one can highlight specific quantum aspects of a given source by changing the respective quantum-statistical representation. Thus, one typically needs to simultaneously consider many representations to reveal all quantum aspects of the analyzed phenomena.

For example, it is not very easy to conclude from $\langle x|\hat{\rho}|x'\rangle$ that a Fock state is a genuine quantum state. However, the Wigner function reveals the quantum nature directly as partial negativity of the quasiprobability distribution. Moreover, one can also find genuine quantum fields that have a positive definite $W(x, y)$ even though they have no classical counterpart. Their quantum nature can unambiguously be identified via their Glauber–Sudarshan function. Any genuine quantum field not only has a partially negative-valued $P(x, y)$ but this function also becomes nonanalytic for fields without classical counterpart.

It is often inconvenient or even impossible to directly compute the high-dimensional quasiprobability distributions when one wants to analyze the case of a multimode light field interacting with many-body states, which is the typical scenario in semiconductor quantum optics. Also here, the quantum-statistical pluralism helps because both the expectation values as well the correlated clusters identify possible representations of the full quantum statistics. To illustrate this feature, we investigate in Section 23.4 how the cluster-expansion approach can be applied to analyze the principal effects of quantum-optical spectroscopy in TLS.

23.3 Thermal state

Before we investigate the correlation dynamics, we want to introduce a state that has maximal fluctuations both in its marginal distribution as well as in its photon statistics. We construct this state using a diagonal density matrix of Fock states

$$\hat{\rho} = \sum\nolimits_n |n\rangle \, p_n \, \langle n|, \qquad p_n \geq 0. \tag{23.13}$$

This is clearly not the most general $\hat{\rho}$ because only the probabilities, here photon statistics, p_n appear. At the same time, this is not a superposition state but rather resembles the classical averaging algorithm introduced in Section 1.2.2.

The light state is defined to be a thermal state $\hat{\rho}^{\text{th}}$ when it contains the probabilities

$$p_n^{\text{th}} = \frac{1}{1+\bar{n}} \left(\frac{\bar{n}}{1+\bar{n}} \right)^n . \tag{23.14}$$

Unlike the coherent state, the p_n^{th} decays monotonously as a function of n. The corresponding photon statistics spreads across a wide range of states when $\bar{n}$ becomes large, which implies strong intensity fluctuations. The circles in Fig. 23.4 show the photon statistics of a thermal state with $\bar{n} = 6$ photons, on average. The solid line is a guide to the eye. For comparison, the shaded area shows the coherent (frame (a)) and the Fock state (frame (b)), both having six photons, on average. We see that the thermal state has a distinctly different photon statistics compared with both other cases.

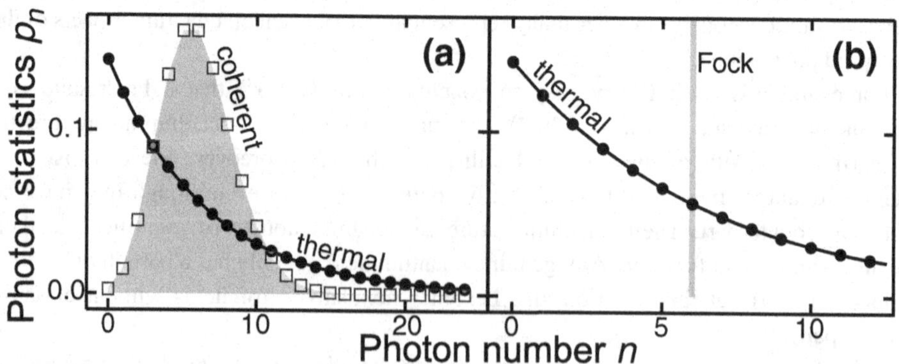

Figure 23.4 Photon statistics of the thermal, the coherent, and the Fock state, each containing six photons on average. **(a)** Thermal (circles) vs. coherent (squares) state. **(b)** Thermal (circles) vs. Fock state (thick, shaded line). The solid line and the shaded area are guides to the eye.

23.3.1 Thermal fluctuations

As a special property, the normally ordered expectation values of thermal light can be evaluated to yield

$$\langle [B^\dagger]^J B^K \rangle_{\text{th}} = \delta_{J,K}\, J!\, \bar{n}^J. \tag{23.15}$$

This relation can be proved using the properties of the geometric series. We see that the thermal state does not satisfy the classical factorization (22.29). Instead, it is entirely defined by the intensity fluctuations

$$\Delta\langle B^\dagger B\rangle \equiv \langle B^\dagger B\rangle - \langle B^\dagger\rangle\langle B\rangle = \bar{n} \tag{23.16}$$

relative to the classical factorization. The photon-number fluctuations are

$$\begin{aligned} \Delta n^2 &\equiv \langle \hat{n}^2\rangle - \langle \hat{n}\rangle^2 = \langle B^\dagger B\, B^\dagger B\rangle - \langle B^\dagger B\rangle^2 \\ &= \langle B^\dagger B^\dagger B\, B\rangle + \langle B^\dagger B\rangle - \langle B^\dagger B\rangle^2 = \bar{n}^2 + \bar{n}, \end{aligned} \tag{23.17}$$

which follow by normally ordering the expressions and using the relation (23.15). We see that the relative photon-number fluctuations of the thermal state yield $\Delta n^2/\bar{n}^2 = 1 + 1/\bar{n}$ showing that thermal light greatly exceeds the shot-noise limit (22.51). Because of these large fluctuations, $\hat{\rho}^{\text{th}}$ is also referred to as chaotic light.

To estimate the character of the fluctuations, we study the entropy of the thermal light

$$S_{\bar{n}} = -k_B \sum\nolimits_n p_n \ln p_n = -k_B \ln \frac{1}{1+\bar{n}} \left(\frac{\bar{n}}{1+\bar{n}}\right)^{\bar{n}}, \tag{23.18}$$

where we have introduced the Boltzmann constant, as is usual for the entropy in thermodynamics. The explicit derivation of (23.18) is the topic of Exercise 23.4.

We know that the average energy of the single-mode system is given by

$$E \equiv \langle \hat{H}_1\rangle = \hbar\omega\left(\langle B^\dagger B\rangle + \frac{1}{2}\right) = \hbar\omega\left(\bar{n} + \frac{1}{2}\right). \tag{23.19}$$

By using the standard thermodynamic relation,

$$\frac{1}{T} = \frac{\partial S}{\partial E} = \frac{1}{\hbar\omega}\frac{\partial S}{\partial \bar{n}} = -\frac{k_B}{\hbar\omega}\ln\frac{\bar{n}}{1+\bar{n}} \quad \Leftrightarrow \quad \bar{n} = \frac{1}{e^{\frac{\hbar\omega}{k_B T}} - 1}, \tag{23.20}$$

we can identify the temperature as a parameter in the expression of the average photon number. This is one of the reasons why $\hat{\rho}^{\text{th}}$ is called the *thermal state*. In particular, the functional form of $\bar{n}$ is given by the typical Bose–Einstein distribution.

23.3.2 Quantum-optical spectroscopy with thermal source

The real-space representation of the thermal state's Wigner function can be computed directly by combining (23.13)–(23.14) with the explicit Fock-state function (10.81).

A more elegant derivation is outlined in Exercise 23.5. From both of these calculations, we obtain

$$\left\langle x - \frac{r}{2} \middle| \hat{\rho}^{\text{th}} \middle| x + \frac{r}{2} \right\rangle = \sqrt{\frac{2}{\pi(1+2\bar{n})}}\, e^{-2\frac{x^2}{1+2\bar{n}}}\, e^{-\frac{1+2\bar{n}}{2} r^2}. \tag{23.21}$$

This leads to a Gaussian integration kernel if the Wigner function is constructed via Eq. (23.3). This integral eventually yields

$$W^{\text{th}}(x, y) = \frac{2}{\pi(1+2\bar{n})}\, e^{-2\frac{x^2+y^2}{1+2\bar{n}}}, \tag{23.22}$$

which is a Gaussian distribution around the center of the phase space, similar to a vacuum field. However, thermal light has $\sqrt{1+2\bar{n}}$ times larger fluctuations than the vacuum.

One can easily verify that a Gaussian Glauber–Sudarshan function,

$$P^{\text{th}}(x, y) = \frac{1}{\pi\bar{n}}\, e^{-\frac{x^2+y^2}{\bar{n}}}, \tag{23.23}$$

is the correct Gaussian kernel producing the Wigner function via the transformation (23.12). Since $P^{\text{th}}(x, y)$ is positive definite, thermal light is essentially a semiclassical field whose fluctuations are enhanced via classical averaging. We can also evaluate the marginal distribution of the thermal state,

$$p^{\text{th}}(x) = \int dy\, W^{\text{th}}(x, y) = \langle x|\hat{\rho}^{\text{th}}|x\rangle = \sqrt{\frac{2}{\pi(1+2\bar{n})}}\, e^{-2\frac{x^2}{1+2\bar{n}}}. \tag{23.24}$$

Again, this shows enhanced fluctuations around the origin. The marginal distributions of the coherent state (shaded area) and the thermal state (solid line) are compared in Fig. 23.5(a) for the case with $\bar{n} = 6$ photons. We see that the thermal state indeed has elevated fluctuations also in its quadrature distribution.

To characterize the excitation of a TLS by a thermal state, we compute the state-inversion $\bar{P}_z$ using the expression (22.53) where $c_{n,n}^{|\alpha\rangle}$ is replaced by p_n^{th}. To have a clear-cut comparison between thermal light and coherent-state excitation, we use $\bar{n} = 6$ photons for both cases. Figure 23.5(b) presents the time evolution of $\bar{P}_z(t)$ for the thermal excitation (solid line) and compares it with the coherent-state excitation (shaded area). In both cases, we observe strong deviations from the classical Rabi flopping. In particular, the thermal state does not produce a collapse–revival dynamics for $\bar{P}_z(t)$.

Whenever the system's optical response is modified by the quantum statistics of the light source used, we are in the regime of quantum-optical spectroscopy. Also in solids, the quantum-optical effects result from Fermion nonlinearities, as discussed in Section 21.1.4. In complicated systems, these nonlinearities typically appear only if high-intensity sources

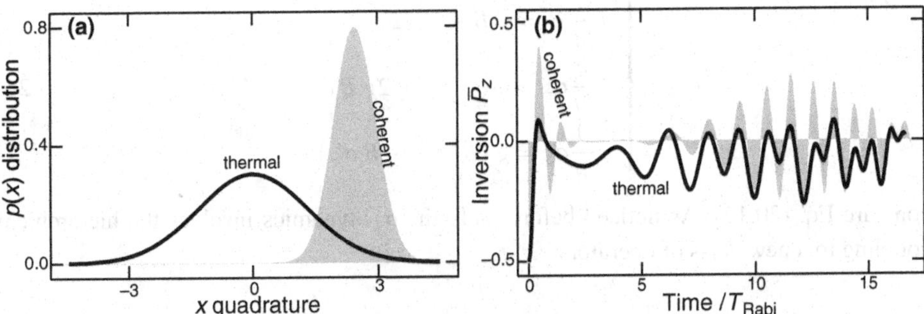

Figure 23.5 Comparison of the TLS excitation by a thermal or a coherent state. The thermal (solid line) and the coherent (shaded area) source have six photons, on average, and the TLS is initially unexcited. **(a)** The initial marginal distributions $p(x)$ are shown for both light sources. **(b)** The resulting excitation dynamics of the TLS is presented via the inversion factor $\bar{P}_z$.

are applied. Therefore, a comparison using thermal and coherent fields seems to offer a promising way to study such effects in such systems. One of the reasons is that it is technically easier to generate high-intensity sources with thermal or coherent statistics than Fock states with high n.

23.4 Cluster-expansion dynamics

In all the Jaynes–Cummings-model examples studied so far, light and matter were initially uncorrelated, i.e., the total density matrix initially was expressible as a product $\hat{\rho}(0) = \hat{\rho}_{2LS} \otimes \hat{\rho}_B$. In such cases, all the effects related to quantum-optical spectroscopy exclusively arise from the build-up of light–matter correlations. The increasing order of the correlations is directly related to the hierarchy problem, as shown in Section 21.1.3. Due to the simplicity of the Jaynes–Cummings model, we can solve the dynamics of $\hat{\rho}(t)$ either analytically or numerically. This luxury immediately vanishes once we tackle more complicated systems because their quantum statistics usually involves a $\hat{\rho}(t)$ with an overwhelmingly high dimension.

However, we already know that the cluster-expansion scheme allows for an excellent description of the build-up of correlations for any many-body and/or quantum-optical systems. To see how the development of quantum-optical correlations can be captured by the clusters, we next analyze the Jaynes–Cummings model using the cluster expansion. We start from the system Hamiltonian

$$\hat{H}_{JC} = \hbar\omega_{2,1}\,\sigma_z + \hbar\omega\left(B^\dagger B + \frac{1}{2}\right) - \hbar g\left[\sigma_-\,B^\dagger + \sigma_+\,B\right] \tag{23.25}$$

and compute the Heisenberg equations of motion. From this, we obtain the operator equations,

$$\begin{cases} i\frac{\partial}{\partial t}B = \omega\, B - g\,\sigma_- \\ i\frac{\partial}{\partial t}\sigma_- = \omega_{2,1}\,\sigma_- + 2g\;B\,\sigma_z, \\ i\frac{\partial}{\partial t}\sigma_z = g\left(B^\dagger\,\sigma_- - B\,\sigma_+\right) \end{cases} \tag{23.26}$$

compare Eq. (20.12). As noticed before, only the σ_j dynamics involves the hierarchical coupling to a new class of operators.

23.4.1 Identification of correlated clusters

Before we implement the cluster expansion, we summarize the pragmatic steps based on the discussion in Chapter 15. As a general classification, the normally ordered expectation values,

$$\langle J+K+L\rangle = \left\langle [B^\dagger]^J B^K \sigma_\lambda^L \right\rangle, \quad J,K=0,1,\cdots, \quad L=0,1, \tag{23.27}$$

identify all $(J+K+L)$-particle quantities. The Pauli matrices have the indices $\lambda = \pm,\ z$, as usual. As discussed in Section 22.3, the classical factorization,

$$\langle J+K+L\rangle_{\rm S} = \left\langle [B^\dagger]^J B^K \sigma_\lambda^L \right\rangle_{\rm cl} = \langle B^\dagger\rangle^J \langle B\rangle^K \langle\sigma_\lambda\rangle^L, \tag{23.28}$$

treats photon operators as classical variables, implementing the substitution $B \to \langle B\rangle = \beta$ and $B^\dagger \to \langle B^\dagger\rangle = \beta^\star$. In the terminology of the cluster expansion, the classical factorization identifies the singlet contribution of the $(J+K+L)$-particle quantities. The remaining two-particle contributions

$$\begin{aligned} &\langle B^\dagger B\rangle = \langle B^\dagger\rangle\langle B\rangle + \Delta\langle B^\dagger B\rangle, \quad \langle B\,B\rangle = \langle B\rangle^2 + \Delta\langle B\,B\rangle \\ &\langle B^\dagger\sigma_\lambda\rangle = \langle B^\dagger\rangle\langle\sigma_\lambda\rangle + \Delta\langle B^\dagger\sigma_\lambda\rangle, \end{aligned} \tag{23.29}$$

identify the correlated doublets $\Delta\langle 2\rangle$. The definition of $\Delta\langle B^\dagger B^\dagger\rangle$ and $\Delta\langle B\,\sigma_\lambda\rangle$ follows from straightforward complex conjugation.

The cluster-expansion factorization

$$\langle N\rangle_C = \sum\nolimits_{\text{all } C_j \text{ combinations}} \prod\nolimits_j \Delta\langle C_j\rangle\Big|_{C_j\le C} \tag{23.30}$$

generates a sum over all possibilities to construct a factorization in terms of cluster products describing up to C-particle correlations. For example, the factorization up to triplets, i.e., three-particle correlations, yields

$$\langle B^\dagger B\,\sigma_\lambda\rangle_1 = \langle B^\dagger\rangle\langle B\rangle\langle\sigma_\lambda\rangle$$
$$\langle B^\dagger B\,\sigma_\lambda\rangle_2 = \langle B^\dagger\rangle\langle B\rangle\langle\sigma_\lambda\rangle + \Delta\langle B^\dagger B\rangle\langle\sigma_\lambda\rangle + \Delta\langle B^\dagger\sigma_\lambda\rangle\langle B\rangle + \Delta\langle B\sigma_\lambda\rangle\langle B^\dagger\rangle$$
$$\langle B^\dagger B\,\sigma_\lambda\rangle_3 = \langle B^\dagger B\,\sigma_\lambda\rangle_2 + \Delta\langle B^\dagger B\,\sigma_\lambda\rangle. \tag{23.31}$$

For all cases, $\langle N\rangle_{C=N}$ identifies an exact factorization while $\langle N\rangle_{C<N}$ represents a systematic approximation of the N-particle quantity in terms of the lower-order clusters.

23.4.2 Beyond Maxwell-optical Bloch equations

Even without the light–matter interaction ($g = 0$), Eqs. (23.26) produce rapid oscillations for the operator dynamics of B and $\sigma_\pm$. It is convenient to handle this trivial feature by working in the interaction picture, as discussed in Section 21.1.2. In the current study, we can simply express the operators in the rotating frame,

$$\begin{cases} B \equiv \bar{B}\, e^{-i\omega t} \\ B^\dagger \equiv \bar{B}^\dagger\, e^{+i\omega t} \end{cases}, \quad \begin{cases} \sigma_- \equiv \bar{\sigma}_-\, e^{-i\omega t} \\ \sigma_+ \equiv \bar{\sigma}_+\, e^{+i\omega t} \end{cases}, \quad \begin{cases} \sigma_z \equiv \bar{\sigma}_z \\ \Delta_{2,1} \equiv \omega - \omega_{2,1} \end{cases}, \tag{23.32}$$

where we identified the detuning between the cavity mode and the TLS. These definitions convert the dynamics (23.26) of the elementary operators into

$$\begin{cases} i\frac{\partial}{\partial t}\bar{B} = -g\,\bar{\sigma}_- \\ i\frac{\partial}{\partial t}\bar{\sigma}_- = -\Delta_{2,1}\,\bar{\sigma}_- + 2g\,\bar{B}\,\bar{\sigma}_z \\ i\frac{\partial}{\partial t}\bar{\sigma}_z = g\left(\bar{B}^\dagger\,\bar{\sigma}_- - \bar{B}\,\bar{\sigma}_+\right) \end{cases}. \tag{23.33}$$

The rotating frame does not influence the cluster-expansion-based factorizations because the expressions

$$\langle N\rangle = \left\langle [\bar{B}^\dagger]^J\,\bar{B}^K\bar{\sigma}_\lambda^L\right\rangle,\quad \Delta\langle N\rangle = \Delta\left\langle [\bar{B}^\dagger]^J\,\bar{B}^K\bar{\sigma}_\lambda^L\right\rangle,\quad J+K+L=N, \tag{23.34}$$

remain invariant. We construct the dynamics of singlets

$$\begin{cases} i\frac{\partial}{\partial t}\langle\bar{B}\rangle = -g\,\langle\bar{\sigma}_-\rangle \\ i\frac{\partial}{\partial t}\langle\bar{\sigma}_-\rangle = -\Delta_{2,1}\,\langle\bar{\sigma}_-\rangle + 2g\,\langle\bar{B}\,\rangle\langle\bar{\sigma}_z\rangle + 2g\,\Delta\langle\bar{B}\,\bar{\sigma}_z\rangle \\ i\frac{\partial}{\partial t}\langle\bar{\sigma}_z\rangle = i\mathrm{Im}\left[2g\langle\bar{B}^\dagger\rangle\,\langle\bar{\sigma}_-\rangle + 2g\Delta\langle\bar{B}^\dagger\,\bar{\sigma}_-\rangle\right] \end{cases} \tag{23.35}$$

by taking the average of both sides of Eq. (23.33) and applying the singlet–doublet factorization (23.29) to the two-particle contributions that appear. This dynamics is exact, provided that the two-particle correlations are known.

The singlets that appear identify the polarization, the inversion factor, and the Rabi frequency,

$$\bar{P}_- \equiv \langle\bar{\sigma}_-\rangle,\quad \bar{P}_z \equiv \langle\bar{\sigma}_z\rangle,\quad \bar{\Omega}^{(+)} \equiv 2g\langle\bar{B}\rangle \tag{23.36}$$

respectively, compare Eqs. (19.13) and (20.13). We can therefore convert Eq. (23.35) into the *Maxwell–Bloch equations with quantum-optical corrections*

$$\left\{\begin{array}{l} i\frac{\partial}{\partial t}\bar{P}_- = -\Delta_{2,1}\,\bar{P}_- + \bar{\Omega}^{(+)}\bar{P}_z + 2g\,\Delta\langle\bar{B}\,\bar{\sigma}_z\rangle \\ \frac{\partial}{\partial t}\bar{P}_z = \mathrm{Im}\left[(\bar{\Omega}^{(+)})^\star\,\bar{P}_- + 2g\Delta\langle\bar{B}^\dagger\,\bar{\sigma}_-\rangle\right] \\ i\frac{\partial}{\partial t}\langle\bar{B}\rangle = -g\,\bar{P}_- \end{array}\right. . \tag{23.37}$$

It is interesting to notice that Maxwell's wave equation is replaced by a first-order partial differential equation due to the cavity assumption. It produces the self-consistency of the classical light–matter coupling, yielding reversible decay of the polarization within the system. The doublets that appear produce the quantum-optical hierarchy problem with new contributions beyond Eq. (19.30). Since the classical description neither explains quantum Rabi flopping nor the collapse–revival dynamics, we obviously need to solve at least the doublet equations to understand the principal quantum-optical effects.

23.4.3 General singlet–doublet dynamics

We start to derive the doublet dynamics by evaluating

$$\begin{aligned} i\frac{\partial}{\partial t}\Delta\langle\bar{B}^\dagger\bar{\sigma}_-\rangle &= i\frac{\partial}{\partial t}\left[\langle\bar{B}^\dagger\bar{\sigma}_-\rangle - \langle\bar{B}^\dagger\rangle\langle\bar{\sigma}_-\rangle\right] \\ &= -\Delta_{2,1}\langle\bar{B}^\dagger\bar{\sigma}_-\rangle + 2g\langle\bar{B}^\dagger\bar{B}\bar{\sigma}_z\rangle + g\langle\bar{\sigma}_+\bar{\sigma}_-\rangle \\ &\quad - \left(-\Delta_{2,1}\langle\bar{B}^\dagger\rangle\langle\bar{\sigma}_-\rangle + \langle\bar{B}^\dagger\rangle 2g\langle\bar{B}\bar{\sigma}_z\rangle + g\langle\bar{\sigma}_+\rangle\langle\bar{\sigma}_-\rangle\right), \end{aligned} \tag{23.38}$$

which follows directly from the operator equations (23.33). The $\bar{\sigma}_+\bar{\sigma}_-$ contribution appears to be a two-particle contribution. However, the operator space of the TLS is limited such that one can express any two-level operator in terms of $\bar{\sigma}_\pm$, $\bar{\sigma}_z$, and the identity operator. In the present case, we find

$$\bar{\sigma}_+\bar{\sigma}_- = \begin{pmatrix} 0 & 1 \\ 0 & 0 \end{pmatrix}\begin{pmatrix} 0 & 0 \\ 1 & 0 \end{pmatrix} = \begin{pmatrix} 1 & 0 \\ 0 & 0 \end{pmatrix} = \frac{1}{2} + \bar{\sigma}_z, \tag{23.39}$$

where we used the matrix-representation (20.3) of the Pauli spin matrices.

Inserting (23.39) and applying the factorizations (23.29) and (23.31) in Eq. (23.38), we eventually find

$$\begin{aligned} i\frac{\partial}{\partial t}\Delta\langle\bar{B}^\dagger\bar{\sigma}_-\rangle &= -\Delta_{2,1}\,\Delta\langle\bar{B}^\dagger\bar{\sigma}_-\rangle + 2g\left[\Delta\langle\bar{B}^\dagger\bar{B}\rangle\bar{P}_z + \langle\bar{B}\rangle\langle\bar{B}^\dagger\bar{\sigma}_z\rangle\right] \\ &\quad + g\left[\frac{1}{2} + \bar{P}_z - |\bar{\mathbf{P}}_-|^2\right] + 2g\Delta\langle\bar{B}^\dagger\bar{B}\bar{\sigma}_z\rangle. \end{aligned} \tag{23.40}$$

Here, we identified the Bloch vector corresponding to the singlets $\langle \bar{\sigma}_\lambda \rangle \equiv \bar{P}_\lambda$. Equation (23.40) shows that the doublet dynamics couples to triplet correlations such that the equation is formally exact if the triplets are known.

Even if we do not have any light–matter correlations initially, they are eventually generated via the equation hierarchy. This correlation generation proceeds sequentially in the sense that the lower-order correlations $\Delta\langle C < N \rangle$ are the sources for the higher-order ones. Thus, there must always be a time window where the clusters up to the level of $\Delta\langle C \rangle$-particle correlations accurately describe the quantum dynamics. This allows for a simple truncation where the correlations $\Delta\langle C+1 \rangle$ are simply omitted from the dynamics. More mathematically, one can introduce a cluster-expansion truncation, $\langle N \rangle \to \langle N \rangle_C$, that is accurate at the times before clusters above $\Delta\langle C \rangle$ are effectively generated, compare Section 15.4.

In the following, we are interested in determining the onset of the simplest quantum-optical correlations. For this purpose, we need to follow the dynamics up to doublets. To simplify the notation, we abbreviate the light–matter correlations as

$$\Pi_\lambda \equiv \Delta\langle \bar{B}^\dagger \bar{\sigma}_\lambda \rangle, \qquad \lambda = \pm, z. \tag{23.41}$$

Following an analogous derivation as in Eq. (23.40), we can eventually construct the dynamics for a closed set of doublets (see Exercise 23.8),

$$\begin{aligned}
i\frac{\partial}{\partial t}\Pi_- &= -\Delta_{2,1}\,\Pi_- + 2g\,\Delta\langle \bar{B}^\dagger \bar{B}\rangle\,\bar{P}_z + \Omega^{(+)}\,\Pi_z \\
&\quad + g\left[\frac{1}{2} + \bar{P}_z - |\bar{P}_-|^2\right] + 2g\Delta\langle \bar{B}^\dagger \bar{B}\bar{\sigma}_z\rangle \\
i\frac{\partial}{\partial t}\Pi_+ &= \Delta_{2,1}\,\Pi_+ - 2g\,\Delta\langle \bar{B}^\dagger \bar{B}^\dagger\rangle\,\bar{P}_z - (\Omega^{(+)})^\star\,\Pi_z \\
&\quad - g\bar{P}_-^\star \bar{P}_-^\star - 2g\Delta\langle \bar{B}^\dagger \bar{B}^\dagger\bar{\sigma}_z\rangle \\
i\frac{\partial}{\partial t}\Pi_z &= \frac{(\Omega^{(+)})^\star\Pi_- - \Omega^{(+)}\,\Pi_+}{2} + g\left(\Delta\langle \bar{B}^\dagger \bar{B}^\dagger\rangle\,\bar{P}_- - \Delta\langle \bar{B}^\dagger \bar{B}\rangle\,\bar{P}_-^\star\right) \\
&\quad - g\bar{P}_-^\star\left[\frac{1}{2} + \bar{P}_z\right] + g\left[\Delta\langle \bar{B}^\dagger \bar{B}^\dagger\bar{\sigma}_-\rangle - \Delta\langle \bar{B}^\dagger \bar{B}\bar{\sigma}_+\rangle\right] \\
i\frac{\partial}{\partial t}\Delta\langle \bar{B}^\dagger \bar{B}\rangle &= -2i\,\mathrm{Im}\left[g\,\Pi_-\right], \qquad i\frac{\partial}{\partial t}\Delta\langle \bar{B}^\dagger \bar{B}^\dagger\rangle = 2g\,\Pi_+.
\end{aligned} \tag{23.42}$$

The structure of the Π_λ dynamics very much resembles the form of Eq. (23.37) for $\bar{P}_\lambda$. We note the appearance of pure singlet sources in the second line of each Π_λ, such as $g\left[1/2 + \bar{P}_z - |\bar{P}_-|^2\right]$. These terms represent the singlet sources for the light–matter correlations. Physically, they can be traced back to the Fermionic character of the TLS, as discussed in Section 21.1.4. In cases where no significant triplet correlations have been generated, we can simply neglect all triplets. In this situation, the singlet–doublet equations

(23.37) and (23.42) are closed and we can use them to describe the onset of several quantum-optical effects.

As a typical trend, the formation of higher-order clusters is hindered by dephasing. Thus, the singlet–doublet approximation is even more valid for open quantum-optical systems. The dephasing can be implemented systematically via the master-equation approach introduced in Section 20.4.1. Using Eq. (20.43) with the reservoir operator $\hat{L} = \bar{\sigma}_z$, we find the following additional dephasing contributions to the singlet–doublet dynamics,

$$i\frac{\partial}{\partial t}\bar{P}_{\pm}\bigg|_{\text{deph}} = -i\gamma\bar{P}_{\pm}, \qquad i\frac{\partial}{\partial t}\Pi_{\pm}\bigg|_{\text{deph}} = -i\gamma\Pi_{\pm}, \tag{23.43}$$

which underlines the formal similarity of the polarization and the $\Pi_{\pm}$ correlations. The other contributions remain unchanged.

23.4.4 Luminescence equations for two-level system

The existence of a polarization implies that the TLS is in a superposition state between its excited and ground state. At the same time, a nonvanishing $\bar{P}_{-}$ induces transitions between these states via the dipole interaction, yielding emission or absorption of light. Thus, a nonvanishing polarization is typically a relatively short-lived transient phenomenon in contrast to the state occupation or $1/2+\hat{P}_z$, which can remain finite long after the transition process is over. The polarization also directly generates classical light, i.e., a coherent state, as can be seen from Eq. (23.37). The same conclusion follows from the dipole-radiation relation (17.34). Thus, the emission from two polarization sources can interfere like ordinary waves. Obviously, the occupation sources do not have this wave property.

Due to its connection to classical waves and coherent states, the polarization is a coherent quantity defining the coherences within the TLS. Once all coherences vanish, only the incoherent quantities remain. This division into incoherent and coherent quantities allows us to identify long-living and transient quantities also in solids. Furthermore, the division is helpful to analyze the coupling structure in terms of coherent and incoherent correlations, which often simplifies the analysis. As a general definition, $\Delta\langle C\rangle$ is a coherent correlation if it is finite only in the presence of the field or the polarization source. All other correlations are incoherent. We see from Eq. (23.42) that in the absence of the polarization only Π_{-} and $\Delta\langle\bar{B}^{\dagger}\bar{B}\rangle$ can be spontaneously generated through the singlet source $g\left[1/2+\bar{P}_z\right]$. Thus, Π_{-} and $\Delta\langle\bar{B}^{\dagger}\bar{B}\rangle$ are incoherent doublets while Π_{+}, Π_z, and $\Delta\langle\bar{B}^{\dagger}\bar{B}^{\dagger}\rangle$ are coherent doublets.

The quantum-optical system can display nontrivial transitions also in the incoherent regime. For example, we have shown in Sections 22.2.1–22.2.2 that also the vacuum-field fluctuations produce quantum Rabi flopping even though $\bar{P}_{-}$ vanishes. To see this even clearer, we now consider the incoherent singlet–doublet dynamics based on Eqs. (23.37) and (23.42). By setting P_{-} and the coherent field $\bar{B}$ to zero, we find that only $\bar{P}_z$, Π_{-}, and $\Delta\langle\bar{B}^{\dagger}\bar{B}\rangle$ are incoherent. The corresponding dynamics follows from the *luminescence equations*,

$$\frac{\partial}{\partial t}\bar{P}_z = 2\mathrm{Im}\left[g\,\Pi_-\right]$$
$$i\frac{\partial}{\partial t}\Pi_- = -\left(\Delta_{2,1} + i\gamma\right)\Pi_- + 2g\,\Delta\langle\bar{B}^\dagger\bar{B}\rangle\;\bar{P}_z + g\left[\frac{1}{2} + \bar{P}_z\right] + 2g\Delta\langle\bar{B}^\dagger\bar{B}\bar{\sigma}_z\rangle$$
$$\frac{\partial}{\partial t}\Delta\langle\bar{B}^\dagger\bar{B}\rangle = -2\mathrm{Im}\left[g\,\Pi_-\right], \tag{23.44}$$

for the TLS. This structure is formally exact under incoherent conditions because the triplet that appears is also incoherent. We have also added dephasing to describe the coupling to a reservoir.

The luminescence equations have a relatively simple interpretation. The $\Pi_- = \Delta\langle\bar{B}^\dagger\bar{P}_-\rangle$ that appears identifies a photon-assisted polarization. In other words, Π_- defines a correlated process where the TLS decays from the excited to the ground state by emitting a photon. According to Eq. (23.44), this process is spontaneously driven by the incoherent singlet source

$$S_-^{\mathrm{inc}} \equiv g\left[\frac{1}{2} + \bar{P}_z\right] = g\begin{pmatrix} 1 & 0 \\ 0 & 0 \end{pmatrix} = g\,f_+, \tag{23.45}$$

which is proportional to the occupation of the excited state f_+. Thus, only an excited TLS can produce spontaneous recombination/emission. In particular, S_-^{inc} can be viewed as the spontaneous-emission source that is directly generated by the vacuum-field fluctuations, as discussed in Section 22.2.2. Once Π_- is generated, it starts to reduce the excitation and increase the photon number, as shown by Eq. (23.44). In particular, Eq. (23.44) yields a general conservation law

$$\frac{\partial}{\partial t}\bar{P}_z = \frac{\partial}{\partial t}f_+ = -\frac{\partial}{\partial t}\Delta\langle\bar{B}^\dagger\bar{B}\rangle \tag{23.46}$$

showing that lowering the TLS from the upper to the ground state increases the photon number exactly by one.

Besides the spontaneous-emission source, the Π_- dynamics contains a contribution $2g\,\Delta\langle\bar{B}^\dagger\bar{B}\rangle\;\bar{P}_z$. This part describes the self-consistent back action of the light field to the generated incoherent emission. It is called the stimulated contribution because it can considerably enhance the rate of photon emission. At the same time, it produces reversibility via the coupling between the $\Delta\langle\bar{B}^\dagger\bar{B}\rangle$ and Π_- dynamics. As a result, this contribution can explain the quantum Rabi features encountered in Section 22.2.1. Since the luminescence equations are nonlinear, they can generally be solved only numerically.

23.5 Quantum optics at the singlet–doublet level

To see how well the singlet–doublet approximation describes quantum Rabi flopping, we solve Eq. (23.44) without dephasing and triplet correlations. The system is assumed to be initially fully excited while the light field is in the Fock state $|n = 5\rangle$, i.e., the light–matter wave function is $|\psi(0)\rangle = |+\rangle \otimes |5\rangle$. The corresponding initial values are then $\bar{P}_z(t=0) = 1/2$, $\Pi_- = 0$, and $\Delta\langle\bar{B}^\dagger\bar{B}\rangle = 5$. Figure 23.6(a) shows the time evolution

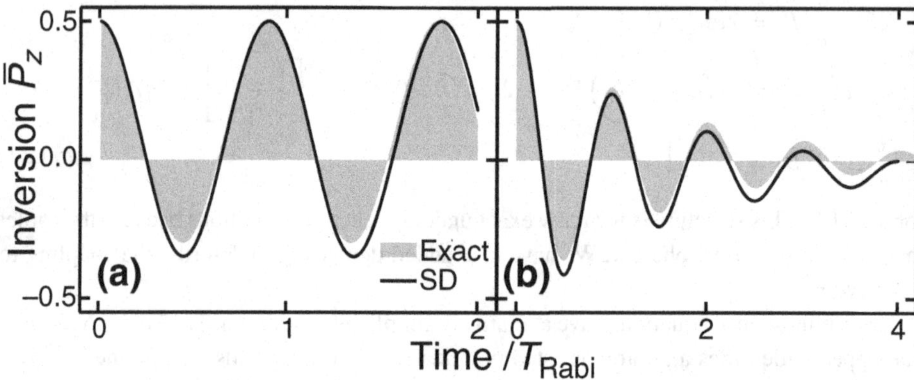

Figure 23.6 Dynamics of the atomic inversion for a Fock-state $|n = 5\rangle$ source. The shaded area indicates the full solution while the solid line shows the results of the singlet–doublet computation. We have assumed either **(a)** no dephasing and detuning $\Delta_{2,1} = \sqrt{6}\,g$ or **(b)** $\gamma = g$ and $\Delta_{2,1} = 0$.

of $\bar{P}_z$ for the singlet–doublet computation (solid line) and the exact solution (shaded area), evaluated from Eq. (22.21). In both calculations we used the detuning $\Delta_{2,1} = \sqrt{6}$, producing the generalized Rabi frequency $\Upsilon_6 = \sqrt{30}\,g$, based on the relation (22.5). This detuned case increases the quantum Rabi flopping frequency by $\sqrt{30/24} \approx 1.118$ compared with the fully resonant case having the frequency $2\Upsilon_6 = \sqrt{24}\,g$. Thus, we can expect only minor modifications to the fully resonant analysis examined in Section 22.2.1. The time scale is expressed in terms of the quantum Rabi cycle $T_{\rm Rabi} = 2\pi/\Upsilon_6$.

Figure 23.6 clearly shows that the singlet–doublet analysis (solid line) captures the main features of the full solution (shaded area). Both cases produce reversible quantum Rabi flopping with virtually identical frequencies. The singlet–doublet analysis dips slightly lower than the exact solution. As a general tendency, this difference becomes smaller as the photon number is increased. Clearly, these differences stem from the omission of the higher-order correlations. It is also interesting to note that fully classical analysis, i.e., the pure singlet computation, produces a strictly constant $\bar{R}_z = 1/2$. Thus, one must include at least doublets to explain the principal features of quantum Rabi flopping.

In the presence of dephasing, the system gradually loses energy. In this case, it is not a priori clear how the quantum Rabi frequency is altered. To study this scenario, we assume a dephasing of $\gamma = g$ and resonant $\Delta_{2,1} = 0$ conditions. Figure 23.6(b) shows the corresponding singlet–doublet result (solid line) together with the full solution (shaded area) computed by solving the density-matrix dynamics with the corresponding master equation (20.38). We clearly see that both cases produce a very similar decay of $\bar{R}_z$ while the quantum Rabi frequency remains nearly unaltered. As in the case without dephasing, the singlet–doublet result dips to slightly lower values of $\bar{R}_z$ than the exact result.

To see whether the singlet–doublet approach can also explain more complicated phenomena, we next analyze collapses and revivals resulting from a coherent-state excitation of the TLS. More specifically, we assume the initial state to be a coherent state with displacement $\alpha = \sqrt{6}$ while the TLS is originally unexcited, i.e., the light–matter wave function is $|\psi(0)\rangle = |-\rangle \otimes |\alpha = \sqrt{6}\rangle$ as in the examples shown in Figs. 22.2–22.3. We also include a dephasing of $\gamma = 2g$, which is quite a bit lower than the expected Rabi frequency of $\Upsilon_6 = \sqrt{24}\, g$ corresponding to a field with $|\alpha|^2 = 6$ photons, on average.

Altogether, the initial condition $|\psi(0)\rangle = |-\rangle \otimes |\alpha = \sqrt{5}\rangle$ yields

$$\begin{cases} \bar{P}_- = 0\,, & \bar{P}_z = -\frac{1}{2}, \quad \langle \bar{B}\rangle = \sqrt{6} \\ \Delta\langle \bar{B}^\dagger \bar{B}\rangle = \Delta\langle \bar{B}^\dagger \bar{B}^\dagger\rangle = 0, & \Pi_\lambda = 0, \qquad \lambda = \pm, z \end{cases} . \tag{23.47}$$

With these, we numerically solve the nonlinear singlet–doublet equations (23.37) and (23.42). The same problem is evaluated exactly by evolving the density matrix directly via the master equation (20.38) implementing the $\hat{L} = \sigma_z$ dephasing model. For comparison, we have also performed computations with only the singlet terms, i.e., solving only the Maxwell–Bloch equations (23.37).

Figure 23.7(a) presents the time evolution of the inversion factor $\bar{P}_z$ obtained from the pure singlet (dashed), the singlet–doublet (solid line), and the full (shaded area) computations. We see that all computations produce $\bar{P}_z$ that eventually decays. The singlet computation produces an exponential decay with a relatively large oscillation amplitude. The full and singlet–doublet results decay much faster. The additional decay follows from the collapse of Rabi oscillations, discussed in Section 22.4.1. We can also conclude that collapse is a correlation effect, not present in classical, i.e., singlet analysis.

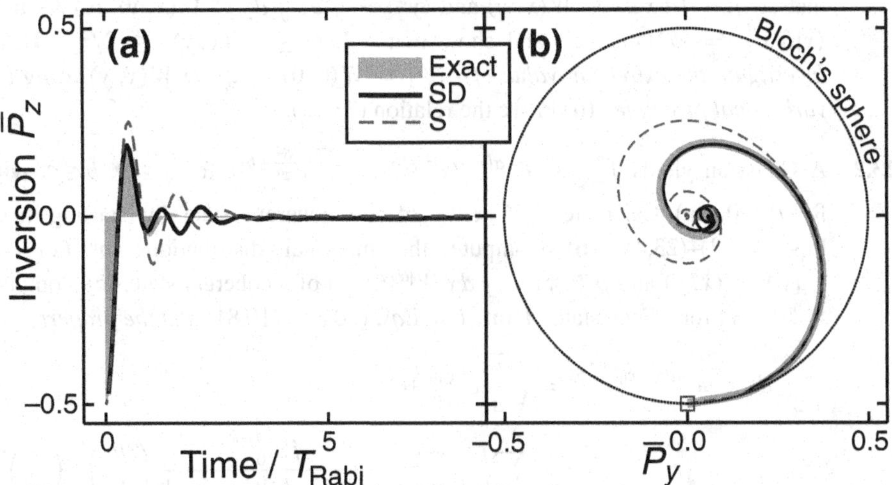

Figure 23.7 Coherent-state $|\alpha = \sqrt{6}\rangle$ excitation of the TLS. The computations are performed under resonant conditions with $\gamma = 2g$. **(a)** The time evolution of the atomic inversion $\bar{P}_z$ is plotted and **(b)** shows the corresponding Bloch-vector trajectory $\mathbf{P} = (0,\ \text{Im}[\bar{\mathbf{P}}_-],\ \bar{\mathbf{P}}_z)$. The full solution (shaded area/thick shaded line) is compared with the results of the singlet–doublet (solid line) and the pure singlet (dashed line) computations.

Due to the dephasing used in the calculations, even the full analysis does not show a revival at the expected $T_{rev} = 12T_{\rm Rabi}$ (not shown). In general, the revival appears only if γ is low enough because dephasing destroys the coherence needed to rephase the system toward its original state. Generally, the collapse–rephasing–revival cycle generates very strong quantum-optical correlations which are beyond the singlet–doublet analysis. Thus, one typically needs to include higher-order correlations to reproduce the revival behavior for low-dephasing investigations.

To get an alternative view of the analysis, we have also plotted the trajectories of the Bloch vector in Fig. 23.7(b). The thick solid line corresponds to the full computation while the singlet and the singlet–doublet analysis is denoted by the solid and the dashed line, respectively. The dotted line indicates the extent of the Bloch sphere. In all cases, the Bloch vector lies in the yz-plane, i.e., $\bar{\mathbf{P}} = (0, \text{Im}[\bar{P}_-], \bar{P}_z)$. We see very clearly that the singlet–doublet analysis reproduces the spiraling collapse of the full calculation while the simple singlet analysis yields a topologically very different result.

The presented analysis shows that the singlet–doublet approach is very successful in describing the creation of correlations. Thus, it is well suited to explain a number of important effects in semiconductor quantum optics.

Exercises

23.1 **(a)** Show that the Wigner function is normalized $\int dx\,dy\, W(x, y) = 1$. **(b)** Show that $\langle \hat{x}^J \rangle = \int dx\,dy\, x^J\, W(x, y)$ and $\langle \hat{y}^J \rangle = \int dx \int dy\, y^J\, W(x, y)$. **(c)** Assume $\langle x|\hat{\rho}|x'\rangle = \psi^*(x)\,\psi(x')$ and show that $-2/\pi \leq W(x, y) \leq 2/\pi$. **Hint:** *Investigate the extremum value of* $L[\psi] = W(0, 0) - \lambda \int dx\, W(x, y)$ *using the variational principle.* **(d)** Prove the relation (23.10).

23.2 A Gaussian yields $\int_{-\infty}^{\infty} dx\, e^{-a(x-x_0)^2+bx} = \sqrt{\frac{\pi}{a}}\, e^{\frac{b^2}{4a}+bx_0}$ for $a \in \not{C}$, $b \in \not{C}$ with $\text{Re}[a] > 0$. **(a)** Calculate $W^{|\alpha\rangle}(x, y)$ of the coherent state by starting from Eqs. (23.3)–(23.4). **(b)** Compute the marginal distributions $p^{|\alpha\rangle}(x) = \int dy\, W^{|\alpha\rangle}(x, y)$ and $p^{|\alpha\rangle}(y) = \int dx\, W^{|\alpha\rangle}(x, y)$ of a coherent state. **(c)** Compute $W^{|n\rangle}(x, y)$ for a Fock state. **Hint:** *Use Eqs.* (10.80)–(10.81) *and the property*

$$\int_{-\infty}^{\infty} dx\, x^j\, e^{-a(x-x_0)^2+bx} = \sqrt{\frac{\pi}{a}}\, e^{bx_0+\frac{b^2}{4a}}$$
$$\times \sum_{k=0}^{j} \sum_{n=0}^{k/2} \frac{j!\, x_0^{j-k}}{n!(j-k)!(k-2n)!} \left(\frac{b}{2a}\right)^k \left(\frac{a}{b^2}\right)^n.$$

23.3 **(a)** Calculate the Husimi function for the coherent state $|\alpha\rangle$ and the Fock state $|n\rangle$ by applying the relation (23.11). **(b)** The Glauber–Sudarshan function of the coherent state follows from $P^{|\alpha\rangle}(\alpha') = \delta^{(2)}(\alpha' - \alpha)$. Verify that the correct

$W^{|\alpha\rangle}(x, y)$ is obtained through Eq. (23.12). **(c)** The function $P^{n.a.}(x, y) = \left[1 - \frac{\partial^2}{\partial x^2} - \frac{\partial^2}{\partial y^2}\right] \delta(x)\, \delta(y)$ is clearly nonanalytic. Compute $W^{n.a.}(x, y)$ by using the relation (23.12) and show that the resulting Wigner function corresponds to the Fock state $|1\rangle$.

23.4 Assume that we have a thermal light source, defined by Eqs. (23.13)–(23.14). **(a)** Verify the relations (23.15)–(23.17). **(b)** Compute the entropy of the thermal light and verify the results (23.18) and (23.20).

23.5 Wigner function of thermal state. **(a)** The diffusion problem $\frac{\partial}{\partial t} u(x, t) = \frac{\partial^2}{\partial x^2} u(x, t)$ can be formally solved via

$$u(x, t) = e^{t\frac{\partial^2}{\partial x^2}} u(x, 0) = \int dx'\, g(x', t)\, u(x', t) \text{ with } g(x, t) = \frac{1}{\sqrt{4\pi t}} e^{-\frac{x^2}{4\Delta t^2}}.$$

Verify these formal relations. **(b)** Show that

$$\langle x|\hat{\rho}^{\text{th}}|x'\rangle = \sqrt{\frac{2}{\pi}} \frac{1}{1+\bar{n}} e^{-(x^2+x'^2)} e^{-\frac{1}{8}\left[\frac{\partial^2}{\partial x^2}+\frac{\partial^2}{\partial x'^2}\right]} e^{4pxx'}$$

where $p = \frac{\bar{n}}{1+\bar{n}}$. **Hint:** *Use the property* $H_n(\alpha x) = e^{-\frac{1}{4\alpha^2}\frac{\partial^2}{\partial x^2}} [2\alpha x]^n$. **(c)** Show that $\langle R - \frac{r}{2}|\hat{\rho}^{\text{th}}|R + \frac{r}{2}\rangle = \sqrt{\frac{2}{\pi(1+2\bar{n})}} e^{-\frac{2R^2}{1+2\bar{n}}} e^{-\frac{1+2\bar{n}}{2} r^2}$ and construct the corresponding Wigner function. **Hint:** *Use the results of part* **(a)**.

23.6 **(a)** Compute the marginal distribution (23.24) by starting from Eq. (23.22). **(b)** Verify that Eq. (23.23) defines the Glauber–Sudarshan function of the thermal state. **(c)** Construct $P_z(t)$ of a TLS resulting from excitation by thermal light. **Hint:** *Apply the relation* (22.55).

23.7 The cluster-expansion factorizations of Boson operators can be conveniently derived by labeling each photon operator by j, constructing all unique permutations of B_j, and then applying indistinguishability $B_j \to B$ in the last step. Use this technique to show that:
(a) $\langle B^\dagger B^\dagger \sigma_\lambda\rangle_{SD} = \langle B^\dagger\rangle\langle B^\dagger\rangle\langle\sigma_\lambda\rangle + \Delta\langle B^\dagger B^\dagger\rangle\langle\sigma_\lambda\rangle + 2\Delta\langle B^\dagger\sigma_\lambda\rangle\langle B^\dagger\rangle$.
(b) $\langle B^\dagger B^\dagger B^\dagger\rangle_{SD} = \langle B^\dagger\rangle\langle B^\dagger\rangle\langle B^\dagger\rangle + 3\langle B^\dagger\rangle\, \Delta\langle B^\dagger B^\dagger\rangle$.
(c) $\langle B^\dagger B^\dagger B B\rangle_{SD} = \langle B^\dagger\rangle\langle B^\dagger\rangle\langle B\rangle\langle B\rangle + \Delta\langle B^\dagger B^\dagger\rangle\Delta\langle BB\rangle$
$+ 2\Delta\langle B^\dagger B\rangle\Delta\langle B^\dagger B\rangle$.
(d) $\langle (B^\dagger)^J B^K\rangle_{SD} = \delta_{J,K}\; J!\,(\Delta\langle B^\dagger B\rangle)^J$ for incoherent fields. What is the connection of this to thermal light?

23.8 **(a)** Derive the operator dynamics (23.26) of the Jaynes–Cummings model (23.25) and reproduce (23.33). Apply the singlet–doublet factorization and construct Eq. (23.35). **(b)** Derive the quantum dynamics of the doublets $\Delta\langle\bar{B}^\dagger\bar{B}\rangle$, $\Delta\langle\bar{B}^\dagger\bar{B}^\dagger\rangle$, and $\Pi_\lambda \equiv \Delta\langle\bar{B}^\dagger\sigma_\lambda\rangle$, i.e., verify Eq. (23.42). **(c)** Use the master equation (20.41)–(20.42) to describe dephasing within the singlet–doublet dynamics. Use $\hat{L} = \boldsymbol{\sigma}_z$ and verify Eq. (23.43).

23.9 **(a)** Construct the incoherent singlet–doublet equations from Eqs. (23.35) and (23.42) for purely incoherent conditions. **(b)** Assume that the contributions $\Delta\langle \bar{B}^\dagger \bar{B}\rangle$ and $\Delta\langle \bar{B}^\dagger \bar{B}\bar{\sigma}_z\rangle$ only lead to weak modifications of the Π_- dynamics. Solve Π_- in steady state and show that photon emission follows from $\frac{\partial}{\partial t}\langle \bar{B}^\dagger \bar{B}\rangle = \frac{g^2 \gamma f_+}{\Delta_{21}^2+\gamma^2}$ where $f_+ = \frac{1}{2} + \mathbf{P}_z$ defines the TLS excitation level. What is the physical interpretation of this result?

23.10 Use the luminescence equation (23.44) without triplets and make an ansatz $\Pi_- = b^+(t)\, p_-(t)$ and $\Delta\langle B^\dagger B\rangle = p_-^*(t)\, p_-(t)$ with $i\frac{\partial}{\partial t} p_- = -\Delta_{21} p_- + 2g\, b\, \mathbf{P}_z$ and $i\frac{\partial}{\partial t} b = -g\, p_-$. **(a)** Show that the luminescence equations are converted into a format where the spontaneous emission source is given by $g\left[1/2 + \mathbf{P}_z\right] - g\,(1/2 + \mathbf{P}_z)^2$. **(b)** Show that the luminescence equations produce reversible Rabi flopping if the quadratic correction $(1/2 + \mathbf{P}_z)^2$ is included in the spontaneous emission source. Under which conditions is this a reasonable assumption? Which terms could be responsible for this contribution?

Further reading

Many important aspects of laser spectroscopy are discussed, e.g., in:

S. Stenholm (1984). *Foundations of Laser Spectroscopy*, New York, Wiley & Sons.
W. Demtöder (2003). *Laser Spectroscopy Basic Concepts and Instrumentation*, 3rd edition, Berlin, Springer.

Quantum-optical spectroscopy for semiconductors is conceptualized in:

M. Kira and S. W. Koch (2006). Quantum-optical spectroscopy in semiconductors, *Phys. Rev. A* **73**, 013813.
S. W. Koch, M. Kira, G. Khitrova, and H. M. Gibbs (2006). Semiconductor excitons in new light, *Nature Materials* **5**, 523.

Homodyne detection of optical fields was experimentally demonstrated in:

H. P. Yuen and J. H. Shapiro (1980). Optical communication with two-photon coherent states–Part III: Quantum measurements realizable with photoemissive detectors, *IEEE Trans. Inf. Theory* **26**, 78.
H. P. Yuen and V. W. S. Chan (1983). Noise in homodyne and heterodyne detection, *Opt. Lett.* **8**, 177.

The principles of quantum-state tomography for light are outlined in:

K. Vogel and H. Risken (1989). Determination of quasiprobability distributions in terms of probability distributions for the rotated quadrature phase, *Phys. Rev. A* **40**, 2847.

The first experimental demonstration of quantum-state tomography is reported in:

D. T. Smithey, M. Beck, M. G. Raymer and A. Faridani (1993). Measurement of the Wigner distribution and the density matrix of a light mode using optical homodyne tomography: Application to squeezed states and the vacuum, *Phys. Rev. Lett.* **70**, 1244.

Husimi Q and Glauber–Sudarshan P functions are introduced in:

K. Husimi (1940). Some formal properties of the density matrix, *Proc. Phys. Math. Soc. Jpn.* **22**, 264.
R. J. Glauber (1963). Coherent and incoherent states of the radiation field, *Phys. Rev.* **131**, 2766.
E. C. G. Sudarshan (1963). Equivalence of semiclassical and quantum mechanical descriptions of statistical light beams, *Phys. Rev. Lett.* **10**, 277.

Different quantum-statistical distributions of light are reviewed, e.g., in:

G. S. Agarwal and E. Wolf (1970). Calculus for functions of noncommuting operators and general phase-space methods in quantum mechanics. 1. Mapping theorems and ordering of functions of noncommuting operators, *Phys. Rev. D* **2**, 2161.
P. D. Drummond and C. W. Gardiner (1980). Generalized P-representation in quantum optics, *J. Phys. A-Math. and Gen.* **13**, 2353.
M. Hillery, R. F. O'Connell, M. O. Scully, and E. P. Wigner (1984). Distribution-functions in physics – fundamentals, *Phys. Rep.-Rev. Sec. of Phys. Lett.* **106**, 121.
U. Leonhardt (1997). *Measuring the Quantum State of Light*, Cambridge, Cambridge University Press.
W. Vogel and D.-G. Welsch (2006). *Quantum Optics*, 3rd edition, Darmstadt, Wiley.
D. F. Walls and G. J. Milburn (2008). *Quantum Optics*, 2nd edition, Berlin, Springer.

The cluster-expansion transformation (CET) is introduced in:

M. Kira and S. W. Koch (2008). Cluster-expansion representation in quantum optics, *Phys. Rev. A* **78**, 022102.

24

General aspects of semiconductor optics

The optically generated excitations in semiconductors constitute a genuine many-body system. To describe its quantum-optical features, we have to expand significantly the theoretical models used so far. However, the important insights of Chapters 16–23 are already presented in a form in which most of them can directly be used and generalized to analyze central properties of the optical excitations in solids. As for atoms, the optical transitions in semiconductors are induced via dipole interaction between photons and electrons. We can thus efficiently construct a systematic quantum-optical theory for semiconductors by following the cluster-expansion approach.

One of the main differences from atoms is that the electronic excitations in semiconductors form a strongly interacting many-body system. Thus, we must systematically treat the arising Coulomb-induced hierarchy problem together with the quantum-optical one. Moreover, the coupling of electrons to lattice vibrations, i.e., the phonons, produces yet another hierarchy problem. In addition, in solid-state spectroscopy one often uses multimode light fields such that one cannot rely on the single-mode simplifications to study semiconductor quantum optics.

As shown in Chapter 15, the Coulomb-, phonon-, and photon-induced hierarchy problems have formally an identical structure. Thus, we start the analysis by investigating how semiconductor quantum optics emerges from the dynamics of correlated clusters. We first focus on the basic properties of the optical transitions in the classical regime. This means investigating the fundamental optical phenomena resulting from the singlets. The full singlet–doublet approach is presented in Chapters 28–30.

24.1 Semiconductor nanostructures

In most investigations, we consider light excitation, propagation, and the emission characteristics for a system of planar semiconductor nanostructures. As shown in Fig. 24.1(a), such systems can consist of multiple layers of optically passive material surrounding the optically active semiconductor structure. Usually, we treat nanostructures that consist of either quantum wells (QW) or a planar arrangement of quantum wires (QWI). An individual QW and QWI layer is shown schematically in Fig. 24.1(b) and 24.1(c), respectively.

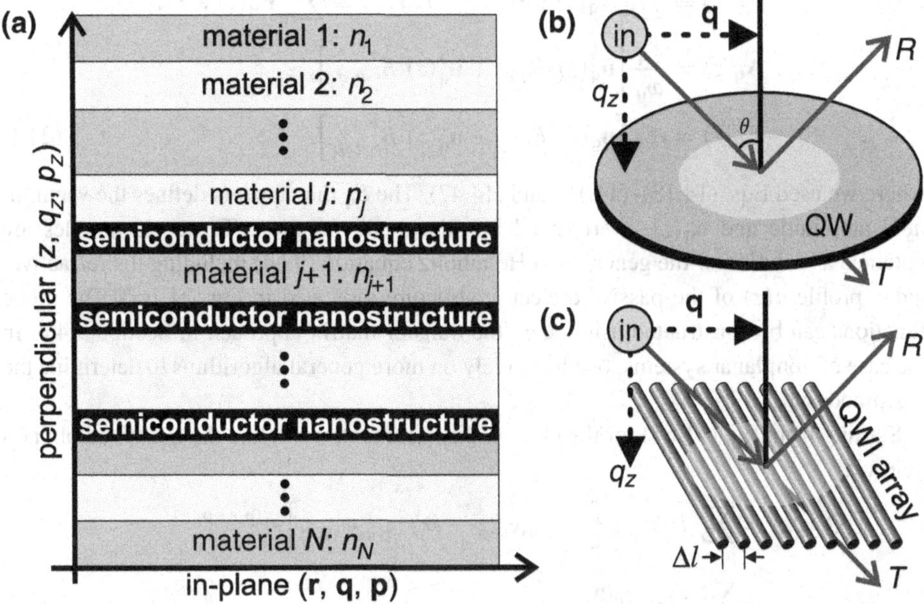

Figure 24.1 Schematic representation of planar semiconductor heterostructures. **(a)** Typically, semiconductor nanostructures are stacked between multiple layers of optically inactive materials. The optical properties of the passive materials are characterized by a refractive index n_j. **(b)** The prototype of a planar semiconductor nanostructure is a quantum well (QW). The plane-wave components of light (arrows) are separated into in-plane **q** and perpendicular q_z components. The propagation direction is defined by the angle θ. The transmitted (T) and reflected light (R) are indicated by arrows. **(c)** Example of a planar semiconductor nanostructure consisting of quantum wires (QWI).

For the QWI, we assume that the distance between the individual wires is much less than the optical wave-length but still large enough such that there is no electronic coupling between them. In this case, light is not diffracted appreciably and its propagation can be described on a plane-wave basis. In all our numerical evaluations, we deliberately choose the system parameters for the QW and QWI configurations such that their optical properties are as close as possible to each other. Thus, a direct comparison of the results and an investigation of the influence of the electronic degrees of freedom, i.e., the quasi-two-dimensional mobility in the QW and the quasi-one-dimensional behavior in the QWI, becomes possible.

When planar structures are considered, it is useful to separate the three-dimensional wave vectors of photons $\mathbf{q}_{3D} \equiv (\mathbf{q}, q_z)$ and phonons $\mathbf{p}_{3D} \equiv (\mathbf{p}, p_z)$ into their in-plane and perpendicular components, denoted by $\mathbf{q}$, $\mathbf{p}$ and q_z, p_z, respectively. Based on Eqs. (14.56) and (14.57), the dynamics of the transversal light field follows from the quantized vector potential and the electric field. We separate their in-plane and z dependency in the mode expansion

$$\hat{\mathbf{A}}(\mathbf{r},z) \equiv \sum\nolimits_{\mathbf{q}} \hat{\mathbf{A}}_{\mathbf{q}}(z)\, e^{i\mathbf{q}\cdot\mathbf{r}}, \qquad \hat{\mathbf{E}}(\mathbf{r},z) \equiv \sum\nolimits_{\mathbf{q}} \hat{\mathbf{E}}_{\mathbf{q}}(z)\, e^{i\mathbf{q}\cdot\mathbf{r}},$$
$$\hat{\mathbf{A}}_{\mathbf{q}}(z) = \frac{\mathcal{E}_{\mathbf{q}}}{\omega_{\mathbf{q}}}\left[\bar{\mathbf{u}}_{\mathbf{q}}(z)\; B_{\mathbf{q},q_z} + \bar{\mathbf{u}}^{\star}_{\mathbf{q}}(z)\; B^{\dagger}_{-\mathbf{q},q_z}\right],$$
$$\hat{\mathbf{E}}_{\mathbf{q}}(z) = i\mathcal{E}_{\mathbf{q}}\left[\bar{\mathbf{u}}_{\mathbf{q}}(z)\; B_{\mathbf{q},q_z} - \bar{\mathbf{u}}^{\star}_{\mathbf{q}}(z)\; B^{\dagger}_{-\mathbf{q},q_z}\right], \tag{24.1}$$

where we used Eqs. (14.18)–(14.19) and (14.47). The $\mathcal{E}_{\mathbf{q}}$ that appears defines the vacuum-field amplitude and $\mathbf{u}_{\mathbf{q}}(z) = \bar{\mathbf{u}}_{\mathbf{q}}(\mathbf{r})\, e^{i\mathbf{q}\cdot\mathbf{r}}$ is the mode function. These eigenmodes are obtained as solution of the generalized Helmholtz equation (9.64) including the refractive-index profile $n(z)$ of the passive dielectric structure indicated in Fig. 24.1(a). The mode functions can be constructed using, e.g., the transfer-matrix approach in Section 9.4.4. In the case of nonplanar systems, one has to rely on more general algorithms to determine the eigenmodes.

Starting from Eq. (11.67) for the phonon displacement, we define the quantized phonon field via

$$\hat{\mathbf{Q}}(\mathbf{r},z) = \sum_{\mathbf{p},p_z} i\mathcal{Q}_{\alpha,\mathbf{p}_{3D}}\left[D_{\alpha,\mathbf{p}_{3D}}\mathbf{e}_{\mathbf{p}_{3D}} - D^{\dagger}_{\alpha,-\mathbf{p}_{3D}}\mathbf{e}_{-\mathbf{p}_{3D}}\right] e^{i(\mathbf{p}\cdot\mathbf{r}+p_z z)},$$
$$= \sum\nolimits_{\mathbf{p}} \hat{\mathbf{Q}}_{\mathbf{p}}(z) e^{i\mathbf{p}\cdot\mathbf{r}},$$
$$\hat{\mathbf{Q}}_{\mathbf{p}}(z) = \sum_{p_z} i\mathcal{Q}_{\alpha,\mathbf{p}_{3D}}\left[D_{\alpha,\mathbf{p},p_z} e^{ip_z z}\mathbf{e}_{\mathbf{p}_{3D}} - D^{\dagger}_{\alpha,-\mathbf{p},p_z} e^{-ip_z z}\mathbf{e}_{-\mathbf{p}_{3D}}\right], \tag{24.2}$$

where $\mathcal{Q}_{\alpha,\mathbf{p}}$ denotes the vacuum-displacement amplitude for type-α phonons and $\mathbf{e}_{\mathbf{p}_{3D}} = \frac{\mathbf{p}_{3D}}{|\mathbf{p}_{3D}|}$ is the unit vector pointing in the direction of the three-dimensional momentum $\mathbf{p}_{3D} \equiv (\mathbf{p}, p_z)$. The identified in-plane components have the general property

$$\hat{\mathbf{A}}^{\dagger}_{\mathbf{q}}(z) = \hat{\mathbf{A}}_{-\mathbf{q}}(z), \quad \hat{\mathbf{E}}^{\dagger}_{\mathbf{q}}(z) = \hat{\mathbf{E}}_{-\mathbf{q}}(z), \quad \hat{\mathbf{Q}}^{\dagger}_{\mathbf{p}}(z) = \hat{\mathbf{Q}}_{-\mathbf{p}}(z). \tag{24.3}$$

As we will see, the in-plane components appear in a similar form in the equations for all Boson fields. Based on this observation, we can identify a general coupling structure between Bosonic fields and Fermionic matter excitations within the singlet–doublet dynamics.

24.1.1 Homogeneous many-body states

To classify the singlets describing the semiconductor many-body systems, we only need to consider quantities containing pure Bloch-electron operators. These quantities can always be generated from $G_{\lambda,\lambda'}(\mathbf{x},\mathbf{y}) \equiv \left\langle \hat{\Psi}^{\dagger}_{\lambda}(\mathbf{x})\hat{\Psi}_{\lambda'}(\mathbf{y})\right\rangle$. Inserting the field operator (11.45) with a Bloch state (12.7) yields

$$G_{\lambda,\lambda'}(\mathbf{x},\mathbf{y}) = \sum\nolimits_{\mathbf{k},\mathbf{k}'} \left\langle a^{\dagger}_{\lambda,\mathbf{k}} a_{\lambda',\mathbf{k}'}\right\rangle \phi^{\star}_{\lambda}(\mathbf{x})\phi_{\lambda'}(\mathbf{y})$$
$$= \frac{1}{\mathcal{L}^d}\sum\nolimits_{\mathbf{k},\mathbf{k}'} \left\langle a^{\dagger}_{\lambda,\mathbf{k}} a_{\lambda',\mathbf{k}'}\right\rangle e^{i(\mathbf{k}'\cdot\mathbf{y}-\mathbf{k}\cdot\mathbf{x})}, \tag{24.4}$$

where the last form is obtained when only the mesoscopic plane-wave dependency is considered. In practice, one simply uses the field operators in the form

$$\hat{\Psi}_\lambda(\mathbf{x}) \equiv \frac{1}{\mathcal{L}^{\frac{d}{2}}} \sum_{\mathbf{k}} e^{i\mathbf{k}\cdot\mathbf{x}}\, a_{\lambda,\mathbf{k}} \tag{24.5}$$

for type λ Fermions, where d refers to the dimension of the quantum-confined system. Formally, this identification follows after we integrate the microscopic $G_{\lambda,\lambda'}$ over the unit cell as well as the confinement function, see Exercise 24.1.

Microscopically, $\langle a^\dagger_{\lambda,\mathbf{k}} a_{\lambda',\mathbf{k}'}\rangle$ describes a transition from the electronic state $(\lambda', \mathbf{k}')$ to $(\lambda, \mathbf{k})$. In Section 18.2.2, we have shown that these transitions and the coherent polarization are directly related, such that $\langle a^\dagger_{\lambda,\mathbf{k}} a_{\lambda',\mathbf{k}'}\rangle$ identifies the microscopic polarization within the solid. At the same time, the diagonal $\langle a^\dagger_{\lambda,\mathbf{k}} a_{\lambda,\mathbf{k}}\rangle$ counts how many electrons occupy the state $(\lambda, \mathbf{k})$ defining the microscopic density distribution.

When these quantities are summed over all possible states, we find the quantities $G_{\lambda,\lambda'}(\mathbf{x},\mathbf{x})$ and $G_{\lambda,\lambda}(\mathbf{x},\mathbf{x})$ defining the spatial distribution of the macroscopic polarization and the density, respectively. If the many-body system can be assumed to be homogeneous,

$$G_{\lambda,\lambda'}(\mathbf{x},\mathbf{x}) = \frac{1}{\mathcal{L}^d} \sum_{\mathbf{k},\mathbf{k}'} \left\langle a^\dagger_{\lambda,\mathbf{k}} a_{\lambda',\mathbf{k}'} \right\rangle e^{i(\mathbf{k}'-\mathbf{k})\cdot\mathbf{x}} \tag{24.6}$$

must be independent of the position $\mathbf{x}$ where the system is probed. This position independence requires that

$$\left\langle a^\dagger_{\lambda,\mathbf{k}} a_{\lambda',\mathbf{k}'} \right\rangle = \delta_{\mathbf{k}',\mathbf{k}} \left\langle a^\dagger_{\lambda,\mathbf{k}} a_{\lambda',\mathbf{k}} \right\rangle \tag{24.7}$$

because Eq. (24.6) is formally a Fourier transformation. This relation is often also applied as an approximation for inhomogeneous systems. This simplification is then called the random-phase approximation (RPA).

To analyze the effect of homogeneity on two-particle quantities, we introduce a generic pair-correlation function

$$G^{(2)}(\mathbf{x},\mathbf{y}) \equiv \left\langle \hat{\Psi}^\dagger_{\lambda_1}(\mathbf{x}) \hat{\Psi}^\dagger_{\lambda_2}(\mathbf{y}) \hat{\Psi}_{\lambda_3}(\mathbf{y}) \hat{\Psi}_{\lambda_4}(\mathbf{x}) \right\rangle. \tag{24.8}$$

For the combination $\lambda_1 = \lambda_4$ and $\lambda_2 = \lambda_3$, it *defines the conditional probability of finding a particle of type* λ_1 *at position* $\mathbf{x}$ *when another particle of type* λ_2 *is positioned at* $\mathbf{y}$. Since we are interested in $G^{(2)}(\mathbf{x},\mathbf{y})$ only on a mesoscopic scale, we include neither the lattice periodic nor the confinement functions to the field operators. Hence, we insert the definition (24.5) into Eq. (24.8) to produce

$$G^{(2)}(\mathbf{x},\mathbf{y}) = \sum_{\mathbf{k}_1,\mathbf{k}_2} \sum_{\mathbf{k}_3,\mathbf{k}_4} \left\langle a^\dagger_{\lambda_1,\mathbf{k}_1} a^\dagger_{\lambda_2,\mathbf{k}_2} a_{\lambda_3,\mathbf{k}_3}\, a_{\lambda_4,\mathbf{k}_4} \right\rangle \frac{e^{i[(\mathbf{k}_4-\mathbf{k}_1)\cdot\mathbf{x}+(\mathbf{k}_3-\mathbf{k}_2)\cdot\mathbf{y}]}}{\mathcal{L}^{2d}}. \tag{24.9}$$

This expression shows that the microscopic $\left\langle a^\dagger_{\lambda_1,\mathbf{k}_1} a^\dagger_{\lambda_2,\mathbf{k}_2} a_{\lambda_3,\mathbf{k}_3}\, a_{\lambda_4,\mathbf{k}_4} \right\rangle$ is connected with the pair-correlation function via a multidimensional Fourier transformation.

To take full advantage of the system's homogeneity, we introduce relative and center-of-mass coordinates

$$\begin{cases} \mathbf{r} = \mathbf{y} - \mathbf{x} \\ \mathbf{R} = \frac{\mathbf{x}+\mathbf{y}}{2} \end{cases} \quad \Leftrightarrow \quad \begin{cases} \mathbf{x} = \mathbf{R} - \frac{\mathbf{r}}{2} \\ \mathbf{y} = \mathbf{R} + \frac{\mathbf{r}}{2} \end{cases} . \tag{24.10}$$

As before, the relative coordinate defines the interparticle distance. For homogeneous systems, also the two-particle contributions should depend only on $\mathbf{r}$ while they should be independent of $\mathbf{R}$. The transformation (24.10) converts Eq. (24.9) into

$$\begin{aligned} G^{(2)}(\mathbf{x}, \mathbf{y}) &\equiv \sum_{\mathbf{k}_1,\mathbf{k}_2} \sum_{\mathbf{k}_3,\mathbf{k}_4} \left\langle a^\dagger_{\lambda_1,\mathbf{k}_1} a^\dagger_{\lambda_2,\mathbf{k}_2} a_{\lambda_3,\mathbf{k}_3}\, a_{\lambda_4,\mathbf{k}_4} \right\rangle \\ &\quad \times \frac{1}{\mathcal{L}^{2d}}\, e^{i\left[(\mathbf{k}_4+\mathbf{k}_3-\mathbf{k}_2-\mathbf{k}_1)\cdot\mathbf{R} + \frac{\mathbf{k}_1+\mathbf{k}_3-\mathbf{k}_2-\mathbf{k}_4}{2}\cdot\mathbf{r}\right]} \\ &= \sum_{\mathbf{k}_1,\mathbf{k}_2} \sum_{\mathbf{q}_1,\mathbf{q}_2} \left\langle a^\dagger_{\lambda_1,\mathbf{k}_1} a^\dagger_{\lambda_2,\mathbf{k}_2} a_{\lambda_3,\mathbf{k}_2+\mathbf{q}_2}\, a_{\lambda_4,\mathbf{k}_1+\mathbf{q}_1} \right\rangle \\ &\quad \times \frac{1}{\mathcal{L}^{2d}}\, e^{i\left[(\mathbf{q}_2-\mathbf{q}_1)\cdot\mathbf{R} + \frac{\mathbf{q}_1+\mathbf{q}_2}{2}\cdot\mathbf{r}\right]}, \end{aligned} \tag{24.11}$$

after we exchanged the summation variables $\mathbf{k}_3 \to \mathbf{k}_2 + \mathbf{q}_2$ and $\mathbf{k}_4 \to \mathbf{k}_1 - \mathbf{q}_1$. We see that $G_2(\mathbf{x}, \mathbf{y})$ does not depend on the center-of-mass coordinate if the plane-wave component with the $\mathbf{R}$ dependence yields a factor unity. This happens if $\left\langle a^\dagger_{\lambda_1,\mathbf{k}_1} a^\dagger_{\lambda_2,\mathbf{k}_2} a_{\lambda_3,\mathbf{k}_2+\mathbf{q}_2}\, a_{\lambda_4,\mathbf{k}_1-\mathbf{q}_1} \right\rangle$ exists only for $\mathbf{q}_1 = \mathbf{q}_2$ and vanishes for any other case. Thus, the two-particle expectation values of a homogeneous system must have the general property

$$\begin{aligned} &\left\langle a^\dagger_{\lambda_1,\mathbf{k}_1} a^\dagger_{\lambda_2,\mathbf{k}_2} a_{\lambda_3,\mathbf{k}_2+\mathbf{q}_2}\, a_{\lambda_4,\mathbf{k}_1-\mathbf{q}_1} \right\rangle \\ &\qquad = \delta_{\mathbf{q}_1,\mathbf{q}_2} \left\langle a^\dagger_{\lambda_1,\mathbf{k}_1} a^\dagger_{\lambda_2,\mathbf{k}_2} a_{\lambda_3,\mathbf{k}_2+\mathbf{q}_1}\, a_{\lambda_4,\mathbf{k}_1-\mathbf{q}_1} \right\rangle . \end{aligned} \tag{24.12}$$

This analysis can be directly generalized for an N-particle expectation value, producing

$$\begin{aligned} \langle N \rangle &= \left\langle a^\dagger_{\lambda_1,\mathbf{k}_1} \cdots a^\dagger_{\lambda_N,\mathbf{k}_N} a_{\nu_N,\mathbf{p}_N} \cdots a_{\nu_1,\mathbf{p}_1} \right\rangle \\ &= \delta_{\sum_{j=1}^N \mathbf{k}_j,\, \sum_{j=1}^N \mathbf{p}_j} \left\langle a^\dagger_{\lambda_1,\mathbf{k}_1} \cdots a^\dagger_{\lambda_N,\mathbf{k}_N} a_{\nu_N,\mathbf{p}_N} \cdots a_{\nu_1,\mathbf{p}_1} \right\rangle \end{aligned} \tag{24.13}$$

for homogeneous excitations. In other words, the total momentum of all the creation operators must equal that of all the annihilation operators.

24.1.2 Two-band model and electron–hole picture

In intrinsic, i.e., undoped semiconductors, the ground-state configuration does not contain partially filled bands. The energetically highest, fully filled electronic band is called the valence band. The next highest is then the conduction band that remains unoccupied in the semiconductor ground state. In fact, we can even write down the exact form of the uncorrelated semiconductor ground state

$$|\Psi_G\rangle \equiv a^\dagger_{v,\mathbf{k}_1} a^\dagger_{v,\mathbf{k}_2} \cdots a^\dagger_{v,\mathbf{k}_N} |0\rangle, \tag{24.14}$$

where all Bloch states $(v,\ \mathbf{k}_j)$ of the valence band(s) are occupied. The $|0\rangle$ that appears is a particle vacuum with no electrons in the valence or conduction band(s). The wave function (24.14) is a Slater determinant because it describes an antisymmetrized product of single-particle Bloch functions $|\phi_{v,\mathbf{k}_j}\rangle$.

Even though general semiconductor excitations typically comprise a large number of Bloch electrons, optical excitations often involve only a small subset of them. As for the atomic systems, this allows us to describe the excitation dynamics with a reduced model. In this context, the system Hamiltonian developed in Chapter 13 shows that an optical field can transfer sufficient energy to the valence-band electron to lift it into the conduction band. The likelihood of this transition is determined by the dipole- or momentum-matrix element (13.38) and by the energetic match between the photon energy and the energy difference between conduction and valence band. These two limitations impose strict selection rules for the possible optical excitations of semiconductors. In the following, we consider situations where only one conduction and one valence band participate in the optical excitations. This allows us to reduce the analysis into an effective two-band model shown in Fig. 24.2(a). This is a direct generalization of the two-level approximation introduced for atomic excitations in Chapter 18.

In many relevant semiconductor nanostructures, the two optically coupled bands are nearly parabolic, as discussed in Section 12.1.1. In addition, the optical transitions are direct and therefore involve only a negligible wave-vector change of the Bloch electron. An example of such a direct transition is indicated in Fig. 24.2(b) as a shaded arrow. Generally,

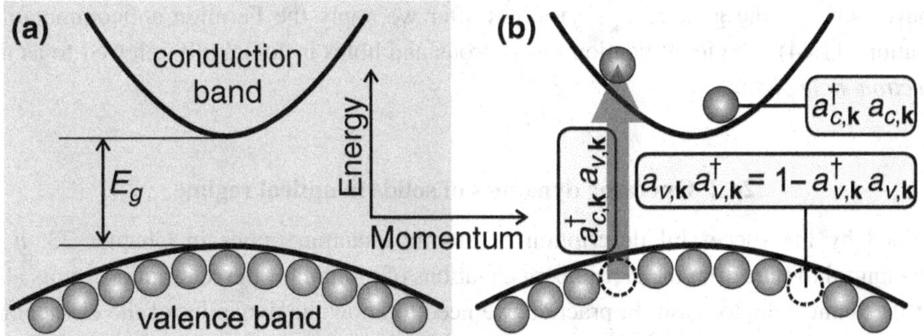

Figure 24.2 Schematic representation of the possible excitations involving two bands. **(a)** Before the excitation, all electrons (spheres) are in the valence band and the conduction band is empty. **(b)** Optical excitations move electrons directly from the valence to the conduction band (arrow). As an electron is lifted from the valence to the conduction band, it leaves behind a hole (dashed circle), i.e., a vacancy in the valence band.

it is straightforward to generalize the approach to multiband systems. In GaAs-type semiconductors, the multiband aspects can arise from the near-degeneracy of the valence bands, the so-called heavy- and light-hole bands, as well as multiple confinement levels in cases where the nanostructure confinement is weak.

The direct optical transitions are described using the diagonal microscopic polarization,

$$P_{\mathbf{k}}^{\lambda,\lambda'} \equiv \left\langle a_{\lambda,\mathbf{k}}^{\dagger} a_{\lambda',\mathbf{k}} \right\rangle, \tag{24.15}$$

between two different bands ($\lambda \neq \lambda'$). The shaded arrow in Fig. 24.2(b) symbolically illustrates how $a_{c,\mathbf{k}}^{\dagger} a_{v,\mathbf{k}}$ moves an electron from the valence ($\lambda' = v$) to the conduction ($\lambda = c$) band. Such a transition is accompanied by a nonvanishing polarization $P_{\mathbf{k}}^{c,v}$, as discussed in Section 18.2.2. When electrons are created in the conduction band, the state occupation is determined by the Fermion-occupation operator $a_{c,\mathbf{k}}^{\dagger} a_{c,\mathbf{k}}$. Its expectation value,

$$f_{\mathbf{k}}^{e,\lambda} \equiv P_{\mathbf{k}}^{\lambda,\lambda} = \left\langle a_{\lambda,\mathbf{k}}^{\dagger} a_{\lambda,\mathbf{k}} \right\rangle, \tag{24.16}$$

determines the probability distribution for the electrons in the band λ. Due to the Pauli exclusion principle, the microscopic occupation cannot exceed unity, i.e., $0 \leq f_{\mathbf{k}}^{e,\lambda} \leq 1$. The quantity $f_{\mathbf{k}}^{e,\lambda}$ is often referred to as the electron distribution function.

Besides conduction-band electrons, the optical excitation simultaneously creates vacancies in the initially fully occupied valence bands. These valence-band vacancies are often referred to as *holes*. As we will see later on, the holes may experience quantum kinetic interactions that are independent of the conduction-band electrons. Thus, it is necessary to monitor the number and distribution of the holes together with those of the electrons. The presence of a hole is detected using the operator $a_{v,\mathbf{k}} a_{v,\mathbf{k}}^{\dagger}$ that produces one if the valence-band state is empty, i.e., the hole state is occupied. The corresponding observable,

$$f_{\mathbf{k}}^{h,\lambda'} = \left\langle a_{\lambda',\mathbf{k}} a_{\lambda',\mathbf{k}}^{\dagger} \right\rangle = 1 - \left\langle a_{\lambda',\mathbf{k}}^{\dagger} a_{\lambda',\mathbf{k}} \right\rangle = 1 - P_{\mathbf{k}}^{\lambda',\lambda'}, \tag{24.17}$$

determines the occupation probability of holes, i.e., the hole distribution function. The conversion into the generic $P_{\mathbf{k}}^{\lambda,\lambda'}$ follows after we apply the Fermion anticommutation relation (11.44). The identification of electrons and holes is commonly referred to as the *electron–hole picture*.

24.2 Operator dynamics of solids in optical regime

Guided by the successful description of atomic quantum optics in Chapter 23, it is meaningful to start from the Heisenberg equations of motion for the relevant operators also for the semiconductor case. In practice, we need to know the dynamics of the elementary Fermion operators $a_{\lambda,\mathbf{k}}$ and $a_{\lambda,\mathbf{k}}^{\dagger}$ as well as the Bosonic photon and phonon operators $B_{\mathbf{q},q_z}$, $B_{\mathbf{q},q_z}^{\dagger}$, $D_{\mathbf{p},p_z}$, and $D_{\mathbf{p},p_z}^{\dagger}$, respectively. In the following investigations, we concentrate exclusively on a two-band model where the carriers are either in the conduction band $\lambda = c$ or in the valence band $\lambda = v$, compare Section 24.1.2. Furthermore, we only treat processes related to light-induced interband transitions. Thus, we can eliminate the current and ponderomotive contributions from the dynamics because they typically involve fields in the THz range, i.e., electromagnetic radiation whose energy is two to three orders of magnitude lower than that of the optical fields discussed here.

24.2.1 Dynamics of elementary operators

To separate the optical interband and the THz-range intraband contributions, we first investigate the dynamics of the photon operators. The generic form is derived in Section 14.1.2 with the end result (14.34),

$$i\hbar\frac{\partial}{\partial t}B_{\mathbf{q},q_z} = \hbar\omega_q\, B_{q_z,\mathbf{q}} - \mathcal{S}\frac{\mathcal{E}_{\mathbf{q}}}{\omega_{\mathbf{q}}}\int dz\,\mathbf{\bar{u}}^{\star}_{q_z,\mathbf{q}}(z)$$
$$\cdot\sum_{\lambda}\left[g_{\lambda,\lambda}(z)\hat{\mathbf{J}}^{\lambda}_{\mathbf{q}}(z) + \sum_{\lambda'\neq\lambda} g_{\lambda,\lambda'}(z)\frac{\partial}{\partial t}\hat{\mathbf{P}}^{\lambda,\lambda'}_{\mathbf{q}}\right]. \tag{24.18}$$

Since the generation of the current $\hat{\mathbf{J}}^{\lambda}$ requires a THz light source, we can ignore this contribution for the optical frequency range studied here. Thus, we only have to include the optical interband polarization defined in Eq. (14.31),

$$\frac{\partial}{\partial t}\hat{\mathbf{P}}^{\lambda,\lambda'}_{\mathbf{q}} = \frac{1}{\mathcal{S}}\sum\nolimits_{\mathbf{k}}\frac{Q\,\mathbf{p}_{\lambda,\lambda'}}{m_0}\,a^{\dagger}_{\lambda,\mathbf{k}}a_{\lambda',\mathbf{k}+\mathbf{q}}$$
$$= \mathbf{e}_P\frac{1}{\mathcal{S}}\sum\nolimits_{\mathbf{k}}\frac{Q\,p_{\lambda,\lambda'}}{m_0}\,a^{\dagger}_{\lambda,\mathbf{k}}a_{\lambda',\mathbf{k}+\mathbf{q}} \equiv \mathbf{e}_P\frac{\partial}{\partial t}\hat{P}^{\lambda,\lambda'}_{\mathbf{q}}. \tag{24.19}$$

Here, Q is the electron charge and m_0 refers to the free-electron mass. As a simplification, we ignore the $\mathbf{k}$ dependency of the momentum-matrix element $\mathbf{p}_{\lambda,\lambda'} = p_{\lambda,\lambda'}\,\mathbf{e}_P$, where $\mathbf{e}_P$ denotes the direction of the light polarization within the QW or QWI plane. This approximation is valid close to the extrema of the parabolic bands. The quantization area for the QWI is $\mathcal{S} = \mathcal{L}\Delta l$, i.e., the product of the quantization length $\mathcal{L}$ and the separation Δl between two neighboring QWIs, see Fig. 24.1.

Since the two-band model is valid only if the confinement is strong enough, we can assume that the confinement wave functions of the carriers are band independent,

$$g_{\lambda,\lambda'}(z) = \xi^{\star}_{\lambda}(z)\xi_{\lambda'}(z) = |\xi(z)|^2 \equiv g(z), \tag{24.20}$$

based on the definition (14.23). For the two-band model, it is convenient to introduce the operation $\bar{\lambda}$ with the properties

$$\bar{\lambda} = (\bar{\sigma},\,\bar{\lambda}) = (\sigma,\,\bar{\lambda}), \qquad \bar{c} = v, \qquad \bar{v} = c. \tag{24.21}$$

In other words, the bar above a band index changes that band index into the complementary one but leaves the spin index σ untouched.

Implementing the two-band model and (24.20)–(24.21), the polarization Eq. (24.19) can be simplified into

$$\sum_{\lambda}\sum_{\lambda'\neq\lambda} g_{\lambda,\lambda'}(z)\frac{\partial}{\partial t}\hat{\mathbf{P}}^{\lambda,\lambda'}_{\mathbf{q}} = \mathbf{e}_P\, g(z)\sum\nolimits_{\lambda}\frac{\partial}{\partial t}\hat{P}^{\lambda,\bar{\lambda}}_{\mathbf{q}}. \tag{24.22}$$

Omitting the THz-generated currents and using (24.22), Eq. (24.18) becomes

$$\begin{aligned} i\hbar\frac{\partial}{\partial t}B_{\mathbf{q},q_z} &= \hbar\omega_q\, B_{\mathbf{q},q_z} - \frac{\mathcal{E}_{\mathbf{q}}}{\omega_{\mathbf{q}}}\int dz\,\bar{\mathbf{u}}^{\star}_{q_z,\mathbf{q}}(z)\cdot\mathbf{e}_P g(z)\sum\nolimits_{\lambda}\frac{\partial}{\partial t}\hat{P}^{\lambda,\bar{\lambda}}_{\mathbf{q}} \\ &= \hbar\omega_q\, B_{\mathbf{q},q_z} - \mathcal{E}_{\mathbf{q}}\,\tilde{u}^{\star}_{\mathbf{q},q_z}\frac{\partial}{\partial t}\frac{\hat{P}^{v,c}_{\mathbf{q}}+\hat{P}^{c,v}_{\mathbf{q}}}{\omega_{\mathbf{q}}}, \end{aligned} \tag{24.23}$$

where we identified the effective mode function

$$\tilde{u}_{\mathbf{q},q_z} = \int dz\, g(z)\,\bar{\mathbf{u}}_{\mathbf{q},q_z}(z)\cdot\mathbf{e}_P. \tag{24.24}$$

We are only interested in planar structures that are much thinner than the optical wavelength. Therefore, we can approximate $\tilde{u}_{\mathbf{q},q_z} \approx u_{\mathbf{q},q_z}(z_j)$ where z_j is the central position of the QW or QWI layer j. Taking the Hermitian conjugate on both sides of Eq. (24.23) and using $\left[\hat{P}^{\lambda,\bar{\lambda}}_{\mathbf{q}}\right]^{\dagger} = \hat{P}^{\bar{\lambda},\lambda}_{-\mathbf{q}}$, we find

$$i\hbar\frac{\partial}{\partial t}B^{\dagger}_{\mathbf{q},q_z} = -\hbar\omega_q\, B^{\dagger}_{\mathbf{q},q_z} + \mathcal{E}_{\mathbf{q}}\,\tilde{u}_{\mathbf{q},q_z}\frac{\partial}{\partial t}\frac{\hat{P}^{v,c}_{-\mathbf{q}}+\hat{P}^{c,v}_{-\mathbf{q}}}{\omega_{\mathbf{q}}} \tag{24.25}$$

via straightforward Hermitian conjugation.

The photon–operator dynamics can equally well be obtained from the general wave equation (14.57). Here, we use this equation for the chosen light propagation direction perpendicular to the planar structure, i.e., $\mathbf{q} = 0$. When the THz contributions are eliminated, Eq. (14.56) reduces to

$$\left[\nabla^2 - \frac{n^2(z)}{c^2}\frac{\partial^2}{\partial t^2}\right]\hat{A}_{\mathbf{0}}(z,t) = -\mu_0 g(z)\frac{\partial}{\partial t}\left(\hat{P}^{v,c}_{\mathbf{0}} + \hat{P}^{c,v}_{\mathbf{0}}\right). \tag{24.26}$$

Here, we use the scalar form by identifying $\hat{\mathbf{A}} \equiv \hat{A}\,\mathbf{e}_P$ because the perpendicular propagation condition aligns the direction of the transversal field and its polarization. The $n(z)$ that appears defines the refractive-index profile of the passive planar structure omitting the optical response of the QWs or QWIs.

24.2.2 Explicit operator dynamics

Using the phonon- and carrier-operator dynamics derived in Section 14.1.2, we can summarize the complete set of equations for the elementary operators,

$$i\hbar\frac{\partial}{\partial t}B_{\mathbf{q},q_z} = \hbar\omega_q\, B_{\mathbf{q},q_z} - \mathcal{E}_{\mathbf{q},q_z}\,\tilde{u}^{\star}_{\mathbf{q},q_z}\,\frac{\partial}{\partial t}\frac{\hat{P}^{v,c}_{\mathbf{q}} + \hat{P}^{c,v}_{\mathbf{q}}}{\omega_{\mathbf{q}}}, \tag{24.27}$$

$$i\hbar\frac{\partial}{\partial t}D_{\mathbf{p},p_z} = \hbar\Omega_{\mathbf{p}}\, D_{\mathbf{p},p_z} + G^{c}_{\mathbf{p},p_z}\,\hat{n}^{c}_{\mathbf{p}} + G^{v}_{\mathbf{p},p_z}\,\hat{n}^{v}_{\mathbf{p}}, \tag{24.28}$$

$$\begin{aligned} i\hbar\frac{\partial}{\partial t}a_{\lambda,\mathbf{k}} &= \epsilon^{\lambda}_{\mathbf{k}}a_{\lambda,\mathbf{k}} + \sum_{\nu}\sum_{\mathbf{k}',\mathbf{q}} V_{\mathbf{q}}\, a^{\dagger}_{\nu,\mathbf{k}'}a_{\nu,\mathbf{k}'+\mathbf{q}}a_{\lambda,\mathbf{k}-\mathbf{q}} \\ &\quad - \sum_{\mathbf{q}}\frac{Qp_{\bar{\lambda},\lambda}}{m_0}\hat{\mathcal{A}}^{\lambda}_{\mathbf{q}}\, a_{\bar{\lambda},\mathbf{k}-\mathbf{q}} - \sum_{\mathbf{q}}\hat{\mathcal{G}}^{\lambda}_{\mathbf{q}}\, a_{\mathbf{k}-\mathbf{q},\lambda}, \end{aligned} \tag{24.29}$$

$$\begin{aligned} i\hbar\frac{\partial}{\partial t}a^{\dagger}_{\lambda,\mathbf{k}} &= -\epsilon^{\lambda}_{\mathbf{k}}a^{\dagger}_{\lambda,\mathbf{k}} - \sum_{\nu}\sum_{\mathbf{k}',\mathbf{q}} V_{\mathbf{q}}\, a^{\dagger}_{\lambda,\mathbf{k}-\mathbf{q}}a^{\dagger}_{\nu,\mathbf{k}'+\mathbf{q}}a_{\nu,\mathbf{k}'} \\ &\quad + \sum_{\lambda'\neq\lambda}\sum_{\mathbf{q}}\hat{\mathcal{A}}^{\bar{\lambda}}_{-\mathbf{q}}\, a^{\dagger}_{\bar{\lambda},\mathbf{k}-\mathbf{q}} + \sum_{\mathbf{p}}\mathcal{G}^{\lambda}_{-\mathbf{p}}\, a^{\dagger}_{\mathbf{k}-\mathbf{p},\lambda}, \end{aligned} \tag{24.30}$$

based on Eqs. (14.36), (14.39), (14.37), and (14.40). Here, we introduced the collective photon field,

$$\begin{aligned} \hat{\mathcal{A}}^{\lambda}_{\mathbf{q}} &\equiv \frac{Q\mathbf{p}_{\bar{\lambda},\lambda}}{m_0}\cdot\int dz\; g(z)\hat{\mathbf{A}}_{\mathbf{q}}(z) \\ &= \sum_{q_z}\left(\mathcal{F}^{\lambda}_{\mathbf{q},q_z}\, B_{\mathbf{q},q_z} + \mathcal{F}^{\bar{\lambda},\star}_{-\mathbf{q},q_z}\, B^{\dagger}_{-\mathbf{q},q_z}\right), \end{aligned} \tag{24.31}$$

and defined the interaction matrix element

$$\mathcal{F}^{\lambda}_{\mathbf{q}} \equiv \frac{Qp_{\bar{\lambda},\lambda}}{m_0\,\omega_{\mathbf{q}}}\mathcal{E}_{\mathbf{q}}\mathbf{e}_P\cdot\int dz\; g(z)\;\bar{\mathbf{u}}_{\mathbf{q},q_z}(z) = \frac{Qp_{\bar{\lambda},\lambda}}{m_0\,\omega_{\mathbf{q}}}\mathcal{E}_{\mathbf{q}}\;\tilde{u}_{\mathbf{q},q_z}, \tag{24.32}$$

by implementing the definitions (14.19) and (14.21). In the same way, we identify a collective phonon field

$$\begin{aligned} \hat{\mathcal{G}}^{\lambda}_{\mathbf{p}} &\equiv -iF_{\lambda}\,\mathbf{p}\cdot\int dz\; g(z)\mathbf{Q}_{\mathbf{p}}(z) \\ &= \sum_{p_z}\left(G^{\lambda}_{\mathbf{p},p_z}D_{\mathbf{p},p_z} + G^{\lambda,\star}_{\mathbf{p},p_z}D^{\dagger}_{-\mathbf{p},p_z}\right), \end{aligned} \tag{24.33}$$

where the interaction matrix element is given by

$$G^{\lambda}_{\mathbf{p},p_z} \equiv |\mathbf{p}|F_{\lambda}\sqrt{\frac{\hbar|\mathbf{p}|}{2c_{\mathrm{LA}}\rho\mathcal{L}^3}}\int dz g(z)e^{iq_z z}, \tag{24.34}$$

according to Eq. (13.86).

The collective quantities obey the following symmetry relations,

$$\left(\hat{\mathcal{A}}^{\lambda}_{\mathbf{q}}\right)^{\dagger} = \hat{\mathcal{A}}^{\bar{\lambda}}_{-\mathbf{q}}, \qquad \left(\hat{\mathcal{G}}^{\lambda}_{\mathbf{p}}\right)^{\dagger} = \hat{\mathcal{G}}^{\lambda}_{-\mathbf{p}}, \tag{24.35}$$

which are used frequently in the subsequent derivations. In analogy to the macroscopic polarization, we also use the macroscopic density operator

$$\hat{\mathbf{n}}^{\lambda}_{\mathbf{p}} \equiv \frac{1}{S}\sum\nolimits_{\mathbf{k}} a^{\dagger}_{\lambda,\mathbf{k}} a_{\lambda,\mathbf{k}+\mathbf{p}}, \tag{24.36}$$

compare Eq. (14.32). The Coulomb matrix element $V_{\mathbf{q}}$ that appears is independent of the band index since we are working in the ideal limit of a strongly confined two-band system.

24.3 Cluster-expansion dynamics

The analysis in Chapters 16–23 shows that classical light excites the system by inducing optical polarization. This process can be described using the Maxwell–Bloch equations (19.30) that follow from a singlet analysis, as shown in Section 23.4.2. Part of the generated singlets are then converted into doublets which then may lead to even higher-order correlations. This sequential build-up is a generic feature embedded within the many-body quantum dynamics, see Chapter 15. Consequently, we can directly apply the cluster-expansion approach to systematically describe both the Coulomb and the quantum optically induced effects at the equivalent level.

If we know that correlations beyond pairs are not formed under the conditions of interest, a singlet–doublet analysis is appropriate and produces a closed set of equations. In particular, this approximation is sufficient when it is justified to ignore the merging of bound pairs into molecules, i.e., biexciton populations, or into electron–hole droplets consisting of many electrons and holes. Since these processes are often very slow and/or occur only in certain externally controllable parameter ranges, we can treat a wide variety of semiconductor properties within a singlet–doublet theory. At the same time, the singlet–doublet approach already describes many intriguing quantum-optical effects, as shown in Chapter 23. Thus, we formulate the generic singlet–doublet description of semiconductor excitations in this and the following chapter.

Starting from Eq. (15.48), the closed set of equations for the singlet–doublet correlation dynamics follows from

$$\begin{aligned} i\hbar\frac{\partial}{\partial t}\Delta\langle 1\rangle &= L\left[\Delta\langle 1\rangle\right] + \mathrm{Hp}\left[\Delta\langle 2\rangle\right] + V_2\left[\Delta\langle 1\rangle\Delta\langle 1\rangle\right] \\ i\hbar\frac{\partial}{\partial t}\Delta\langle 2\rangle &= L\left[\Delta\langle 2\rangle\right] + \cancel{\mathrm{Hp}\left[\Delta\langle 3\rangle\right]} + V_1\left[\Delta\langle 1\rangle\Delta\langle 1\rangle\right] \\ &\quad + V_2\left[\Delta\langle 1\rangle\Delta\langle 2\rangle\right] + V_3\left[\Delta\langle 1\rangle\Delta\langle 1\rangle\Delta\langle 1\rangle\right], \end{aligned} \tag{24.37}$$

where the triplet correlations that appear are omitted and we still have to determine the detailed form of the functionals $L[\cdots]$, $\mathrm{Hp}[\cdots]$, and $V_j[\cdots]$. In practice, we explicitly evaluate the terms appearing in the singlet–doublet dynamics (24.37) using the operator Eqs. (14.34)–(14.37) and (14.35)–(14.40). In this chapter, we unravel the singlet dynamics and discuss its physical content. The explicit doublet dynamics is derived in Chapters 28–30.

24.4 Relevant singlets and doublets

To have a well-defined situation, we assume that the system is excited by a classical laser emitting a pulsed plane wave that propagates perpendicular to the planar structure, i.e., the angle $\theta = 0$ in Fig. 24.1(a). Since the excitation is homogeneous in the plane, it can only induce homogeneous excitations in the semiconductor nanostructures. This feature considerably simplifies our quantum-dynamics investigations but is still general enough to allow for the analysis of the most important effects in semiconductor quantum optics.

Beams that are incident under an angle or configurations where several beams propagate in multiple directions can be implemented with analogous derivations. Under such conditions, the semiconductor excitations may not remain spatially homogeneous. As a major consequence, e.g., the polarization and the carrier densities may display a position-dependent grating that scatters light into new directions. This effect is the basis of four-wave mixing (FWM) investigations. Since most of the current FWM experiments are not yet focused on quantum-optical aspects, we concentrate in this book on the simpler scenario of homogeneous excitations.

Before we derive the full singlet–doublet dynamics, we categorize the relevant observables. The singlets of the carrier–photon–phonon system are determined by

$$\langle 1 \rangle = \left\{ \left\langle a^{\dagger}_{\lambda,\mathbf{k}} a_{\lambda',\mathbf{k}'} \right\},\ \langle B_{\mathbf{q},q_z} \rangle,\ \text{or}\ \langle D_{\mathbf{p},p_z} \rangle \right\}. \tag{24.38}$$

Since we study excitation conditions where the laser excites the system perpendicular to the layered structure, the photon singlets exist only for vanishing in-plane momenta,

$$\langle B_{\mathbf{q},q_z} \rangle = \delta_{\mathbf{q},\mathbf{0}} \langle B_{\mathbf{0},q_z} \rangle. \tag{24.39}$$

Similarly, the phonons satisfy the conditions

$$\langle D_{\mathbf{p},p_z} \rangle = \delta_{\mathbf{p},\mathbf{0}} \langle D_{\mathbf{0},p_z} \rangle, \tag{24.40}$$

as long as no inhomogeneities are generated. For the carriers, the resulting completely homogeneous excitation implies

$$\left\langle a^{\dagger}_{\lambda,\mathbf{k}} a_{\lambda',\mathbf{k}'} \right\rangle = \delta_{\mathbf{k},\mathbf{k}'} \left\langle a^{\dagger}_{\lambda,\mathbf{k}} a_{\lambda',\mathbf{k}} \right\rangle. \tag{24.41}$$

As introduced in Section 24.1.2, we use the compact notations

$$P^{\lambda,\lambda'}_{\mathbf{k}} = \left\langle a^{\dagger}_{\lambda,\mathbf{k}} a_{\lambda',\mathbf{k}} \right\rangle, \quad n^{\lambda}_{\mathbf{k}} \equiv P^{\lambda,\lambda}_{\mathbf{k}} = \left\langle a^{\dagger}_{\lambda,\mathbf{k}} a_{\lambda,\mathbf{k}} \right\rangle, \tag{24.42}$$

where $P^{\lambda,\lambda'\neq\lambda}_{\mathbf{k}}$ determines the microscopic polarization and $n^{\lambda}_{\mathbf{k}}$ is the microscopic carrier occupation probability for the band λ.

Based on the discussion in Section 24.1.1, the homogeneous excitation configuration implies for the carrier doublets,

$$\Delta\left\langle a^{\dagger}_{\lambda,\mathbf{k}} a^{\dagger}_{\lambda',\mathbf{k}'} a_{\nu',\mathbf{k}'+\mathbf{q}'}\, a_{\nu,\mathbf{k}-\mathbf{q}}\right\rangle = \delta_{\mathbf{q},\mathbf{q}'}\ \Delta\left\langle a^{\dagger}_{\lambda,\mathbf{k}} a^{\dagger}_{\lambda',\mathbf{k}'} a_{\nu',\mathbf{k}'+\mathbf{q}}\, a_{\nu,\mathbf{k}-\mathbf{q}}\right\rangle$$
$$\equiv \delta_{\mathbf{q},\mathbf{q}'}\ c^{\mathbf{q},\mathbf{k}',\mathbf{k}}_{\lambda,\lambda',\nu',\nu}\,. \tag{24.43}$$

Sometimes, $c^{\mathbf{q},\mathbf{k}',\mathbf{k}}_{\lambda,\lambda',\nu',\nu}$ is called four-point correlation instead of two-particle correlation (doublets) because it follows from the correlated part of expectation values with four Fermion operators. In our subsequent calculations, we often identify $\mathbf{q}$ as the center-of-mass momentum of the correlated two-particle entities.

Since the doublet correlations with mixed combinations of carrier, photon, and phonon operators obey the same conservation law for the in-plane momentum as the pure carrier correlations, we find that only the combinations

$$\Delta\langle 2\rangle_{\text{mix}} = \left\{\Delta\left\langle B^{\dagger}_{\mathbf{q},q_z} a^{\dagger}_{\nu,\mathbf{k}-\mathbf{q}} a_{\nu',\mathbf{k}}\right\rangle,\ \Delta\left\langle D^{\dagger}_{\mathbf{p},p_z} a^{\dagger}_{\nu,\mathbf{k}-\mathbf{p}} a_{\nu',\mathbf{k}}\right\rangle,\right.$$
$$\left.\Delta\left\langle B^{\dagger}_{\mathbf{q},q_z} D_{\mathbf{q},p_z}\right\rangle,\ \Delta\left\langle B_{\mathbf{q},q_z} D_{-\mathbf{q},p_z}\right\rangle,\ \text{c.c.}\right\} \tag{24.44}$$

are allowed. Here, "c.c." denotes complex-conjugated versions of those already expressed explicitly. The pure photon and phonon correlations have the generic form

$$\Delta\langle 2\rangle_{\text{bos}} = \left\{\Delta\left\langle B^{\dagger}_{\mathbf{q},q_z} B_{\mathbf{q},q'_z}\right\rangle,\ \Delta\left\langle B_{\mathbf{q},q_z} B_{-\mathbf{q},q'_z}\right\rangle,\right.$$
$$\left.\Delta\left\langle D^{\dagger}_{\mathbf{p},p_z} D_{\mathbf{p},p'_z}\right\rangle,\ \Delta\left\langle D_{\mathbf{p},p_z} D_{-\mathbf{p},p'_z}\right\rangle,\ \text{c.c.}\right\}. \tag{24.45}$$

Again, the in-plane momentum must be conserved between the creation and annihilation operators to satisfy the homogeneity requirements. Equations (24.43)–(24.45) define the generic two-particle correlations of homogeneous systems.

24.5 Dynamics of singlets

The photon-singlet dynamics is obtained by taking the expectation value of both sides of Eq. (24.27),

$$i\hbar\frac{\partial}{\partial t}\langle B_{\mathbf{q},q_z}\rangle = \hbar\omega_q\,\langle B_{\mathbf{q},q_z}\rangle - \mathcal{E}_{\mathbf{q},q_z}\,\tilde{u}^{\star}_{\mathbf{q},q_z}\,\frac{\partial}{\partial t}\frac{\left\langle \hat{P}^{v,c}_{\mathbf{q}}\right\rangle + \left\langle \hat{P}^{c,v}_{\mathbf{q}}\right\rangle}{\omega_{\mathbf{q}}}. \tag{24.46}$$

As a consequence of the homogeneous excitation, the macroscopic polarization vanishes for $\mathbf{q} \neq \mathbf{0}$. Hence, only the $\mathbf{q} = \mathbf{0}$ part of the photon-singlet dynamics,

$$i\hbar\frac{\partial}{\partial t}\langle B_{\mathbf{q},q_z}\rangle = \hbar\omega_q\,\langle B_{\mathbf{q},q_z}\rangle - \delta_{\mathbf{q},\mathbf{0}}\ \mathcal{E}_{\mathbf{0},q_z}\,\tilde{u}^{\star}_{\mathbf{0},q_z}\,\frac{\partial}{\partial t}\frac{\left\langle \hat{P}^{v,c}_{\mathbf{0}}\right\rangle + \left\langle \hat{P}^{c,v}_{\mathbf{0}}\right\rangle}{\omega_{\mathbf{q}}}, \tag{24.47}$$

is coupled to the matter excitations. Since we initially have a vanishing $\langle B_{q_z,\mathbf{q}}\rangle = 0$ for any $\mathbf{q} \neq \mathbf{0}$, the singlet field only exists in the propagation direction perpendicular to the planar structure. Sometimes, it is advantageous to use the wave-equation formulation (24.26),

$$\left[\frac{\partial^2}{\partial z^2} - \frac{n^2(z)}{c^2}\frac{\partial^2}{\partial t^2}\right]\langle \hat{\mathbf{A}}(z,t)\rangle = -\mu_0 g(z)\frac{\partial}{\partial t}\left(\left\langle \hat{P}_{\mathbf{0}}^{v,c}\right\rangle + \left\langle \hat{P}_{\mathbf{0}}^{c,v}\right\rangle\right), \tag{24.48}$$

instead of the mode-expansion format (24.46).

Similar to the photons, the phonon singlets are defined by $\langle D_{\mathbf{p},\mathbf{p_z}}\rangle$. Due to the homogeneous excitation conditions, we have to treat only incoherent phonons with

$$\langle D_{\mathbf{p},p_z}\rangle = \left\langle D^{\dagger}_{\mathbf{p},p_z}\right\rangle = 0, \tag{24.49}$$

for all values of $\mathbf{p}$. Thus, we do not need to solve any phonon-singlet equations.

The photon-singlet dynamics (24.48) involves the optical interband polarization in the semiconductor. To complete the singlet investigation, we therefore have to derive the dynamics of $P_{\mathbf{k}}^{\lambda,\lambda'} = \left\langle a^{\dagger}_{\lambda,\mathbf{k}} a_{\lambda',\mathbf{k}}\right\rangle$:

$$i\hbar\frac{\partial}{\partial t}P_{\mathbf{k}}^{\lambda,\lambda'} = \left\langle i\hbar\frac{\partial}{\partial t}\left(a^{\dagger}_{\lambda,\mathbf{k}}\right)a_{\lambda',\mathbf{k}}\right\rangle + \left\langle a^{\dagger}_{\lambda,\mathbf{k}}\, i\hbar\frac{\partial}{\partial t}\left(a_{\lambda',\mathbf{k}}\right)\right\rangle. \tag{24.50}$$

Inserting Eqs. (24.29) and (24.30) into Eq. (24.50) produces

$$\begin{aligned}
i\hbar\frac{\partial}{\partial t}P_{\mathbf{k}}^{\lambda,\lambda'} &= \left(\varepsilon_{\mathbf{k}}^{\lambda'} - \varepsilon_{\mathbf{k}}^{\lambda}\right)P_{\mathbf{k}}^{\lambda,\lambda'} \\
&+ \sum_{\nu,\mathbf{k}',\mathbf{q}\neq 0} V_{\mathbf{q}}\left[\left\langle a^{\dagger}_{\lambda,\mathbf{k}} a^{\dagger}_{\nu,\mathbf{k}'} a_{\nu,\mathbf{k}'+\mathbf{q}} a_{\lambda',\mathbf{k}-\mathbf{q}}\right\rangle - \left\langle a^{\dagger}_{\lambda,\mathbf{k}-\mathbf{q}} a^{\dagger}_{\nu,\mathbf{k}'+\mathbf{q}} a_{\nu,\mathbf{k}'} a_{\lambda',\mathbf{k}}\right\rangle\right] \\
&- \sum_{\mathbf{q}}\left[\left\langle \hat{\mathcal{A}}_{\mathbf{q}}^{\lambda'} a^{\dagger}_{\lambda,\mathbf{k}} a_{\bar{\lambda}',\mathbf{k}-\mathbf{q}}\right\rangle - \left\langle \hat{\mathcal{A}}_{-\mathbf{q}}^{\bar{\lambda}} a^{\dagger}_{\bar{\lambda},\mathbf{k}-\mathbf{q}} a_{\lambda',\mathbf{k}}\right\rangle\right] \\
&+ \sum_{\mathbf{p}}\left[\left\langle \hat{\mathcal{G}}_{\mathbf{p}}^{\lambda'} a^{\dagger}_{\lambda,\mathbf{k}} a_{\lambda',\mathbf{k}-\mathbf{p}}\right\rangle - \left\langle \hat{\mathcal{G}}_{-\mathbf{p}}^{\lambda} a^{\dagger}_{\lambda,\mathbf{k}-\mathbf{p}} a_{\lambda',\mathbf{k}}\right\rangle\right].
\end{aligned} \tag{24.51}$$

We see that the Coulomb, the light–matter, and the phonon interactions all produce couplings of the single-particle contributions to two-particle quantities resulting in the hierarchy problem, compare Section 14.3.2.

24.5.1 Separation of singlets and doublets

The emerging hierarchy problem can be systematically truncated by identifying the relevant clusters, as discussed in Section 24.3. For this purpose, we separate the two-particle expectation values $\langle 2\rangle$ appearing in Eq. (24.51) into their singlet and doublet contributions. In practice, we apply the singlet factorizations (15.32)–(15.34) and use Eq. (15.39) for the identification of the doublet correlations. For example, Eq. (15.39) separates the singlet and doublet contributions from the first Coulomb term in Eq. (24.51) via

$$\begin{aligned}\sum \langle 2\rangle^{1\text{st}} &\equiv \sum_{\nu,\mathbf{k}',\mathbf{q}\neq 0} V_{\mathbf{q}} \left\langle a^{\dagger}_{\lambda,\mathbf{k}} a^{\dagger}_{\nu,\mathbf{k}'} a_{\nu,\mathbf{k}'+\mathbf{q}} a_{\lambda',\mathbf{k}-\mathbf{q}} \right\rangle \\ &= \sum_{\nu,\mathbf{k}',\mathbf{q}\neq 0} V_{\mathbf{q}} \left[\left\langle a^{\dagger}_{\lambda,\mathbf{k}} a^{\dagger}_{\nu,\mathbf{k}'} a_{\nu,\mathbf{k}'+\mathbf{q}} a_{\lambda',\mathbf{k}-\mathbf{q}} \right\rangle_S \right. \\ &\qquad \left. + \Delta \left\langle a^{\dagger}_{\lambda,\mathbf{k}} a^{\dagger}_{\nu,\mathbf{k}'} a_{\nu,\mathbf{k}'+\mathbf{q}} a_{\lambda',\mathbf{k}-\mathbf{q}} \right\rangle \right] \\ &= \sum_{\nu,\mathbf{k}',\mathbf{q}\neq 0} V_{\mathbf{q}} \left[\left\langle a^{\dagger}_{\lambda,\mathbf{k}} a^{\dagger}_{\nu,\mathbf{k}'} a_{\nu,\mathbf{k}'+\mathbf{q}} a_{\lambda',\mathbf{k}-\mathbf{q}} \right\rangle_S + c^{\mathbf{q},\mathbf{k}',\mathbf{k}}_{\lambda,\nu,\nu,\lambda'} \right],\end{aligned} \tag{24.52}$$

where we expressed the Fermionic doublet using Eq. (24.43). To construct the singlet part, we implement the Hartree–Fock factorization (15.32),

$$\begin{aligned}\sum \langle 2\rangle^{1\text{st}}_S &\equiv \sum_{\nu,\mathbf{k}',\mathbf{q}\neq 0} V_{\mathbf{q}} \left\langle a^{\dagger}_{\lambda,\mathbf{k}} a^{\dagger}_{\nu,\mathbf{k}'} a_{\nu,\mathbf{k}'+\mathbf{q}} a_{\lambda',\mathbf{k}-\mathbf{q}} \right\rangle_S \\ &= \sum_{\nu,\mathbf{k}',\mathbf{q}\neq 0} V_{\mathbf{q}} \left[\left\langle a^{\dagger}_{\lambda,\mathbf{k}} a_{\lambda',\mathbf{k}-\mathbf{q}} \right\rangle \left\langle a^{\dagger}_{\nu,\mathbf{k}'} a_{\nu,\mathbf{k}'+\mathbf{q}} \right\rangle \delta_{\mathbf{q},\mathbf{0}} \right. \\ &\qquad \left. - \left\langle a^{\dagger}_{\lambda,\mathbf{k}} a_{\nu,\mathbf{k}'+\mathbf{q}} \right\rangle \left\langle a^{\dagger}_{\nu,\mathbf{k}'} a_{\lambda',\mathbf{k}-\mathbf{q}} \right\rangle \delta_{\mathbf{k}',\mathbf{k}-\mathbf{q}} \right],\end{aligned} \tag{24.53}$$

where the δ-functions appear due to the condition (24.41). Since the $\mathbf{q}$ sum excludes any contribution with $\mathbf{q} = \mathbf{0}$, the first part – i.e., the Hartree contribution – vanishes such that the singlets are entirely determined via the second, i.e., the Fock part. As a result, the singlet factorization reduces into

$$\begin{aligned}\sum \langle 2\rangle^{1\text{st}}_S &= -\left\langle a^{\dagger}_{\lambda,\mathbf{k}} a_{\nu,\mathbf{k}} \right\rangle \sum_{\nu,\mathbf{q}\neq 0} V_{\mathbf{q}} \left\langle a^{\dagger}_{\nu,\mathbf{k}-\mathbf{q}} a_{\lambda',\mathbf{k}-\mathbf{q}} \right\rangle \\ &= -\sum_{\nu,\mathbf{k}'} V_{\mathbf{k}-\mathbf{k}'} P^{\lambda,\nu}_{\mathbf{k}} P^{\nu,\lambda'}_{\mathbf{k}'},\end{aligned} \tag{24.54}$$

where the last step follows from (24.42) and the exchange of summation variables, $\mathbf{q} \rightarrow \mathbf{k} - \mathbf{k}'$. In order to somewhat lighten our notation, we no longer explicitly indicate that the terms with $\mathbf{k}' \neq \mathbf{k}$ have to be omitted from the sum. From now on, we have to keep in mind that any contribution that yields a vanishing momentum in the Coulomb-matrix element has to be excluded. In practice, $P^{\nu,\lambda'}_{\mathbf{k}'}$ is an analytic, smooth function such that the omission of a single $\mathbf{k}' = \mathbf{k}$ does not change the sum that approaches an integral for sufficiently densely spaced $\mathbf{k}'$.

In the expression for the singlet factorization, we now use the fact that the ν sum yields only the two contributions $\nu = \lambda'$ and $\nu = \bar{\lambda}'$ allowed for the two-band model studied here. Thus, we find

$$\sum \langle 2 \rangle_S^{1\text{st}} = -P_{\mathbf{k}}^{\lambda,\lambda'} \sum\nolimits_{\mathbf{k}'} V_{\mathbf{k}-\mathbf{k}'} P_{\mathbf{k}'}^{\lambda',\lambda'} - P_{\mathbf{k}}^{\lambda,\bar{\lambda}'} \sum\nolimits_{\mathbf{k}'} V_{\mathbf{k}-\mathbf{k}'} P_{\mathbf{k}'}^{\bar{\lambda}',\lambda'}$$
$$= -P_{\mathbf{k}}^{\lambda,\lambda'} \sum\nolimits_{\mathbf{k}'} V_{\mathbf{k}-\mathbf{k}'} n_{\mathbf{k}'}^{\lambda'} - P_{\mathbf{k}}^{\lambda,\bar{\lambda}'} \sum\nolimits_{\mathbf{k}'} V_{\mathbf{k}-\mathbf{k}'} P_{\mathbf{k}'}^{\bar{\lambda}',\lambda'}, \tag{24.55}$$

where we identified the carrier distribution $n_{\mathbf{k}'}^{\lambda'}$ using Eq. (24.42). Combining the results (24.52) and (24.55), we obtain the singlet–doublet factorization

$$\sum \langle 2 \rangle^{1\text{st}} = -P_{\mathbf{k}}^{\lambda,\lambda'} \sum_{\mathbf{k}'} V_{\mathbf{k}-\mathbf{k}'} n_{\mathbf{k}'}^{\lambda'} - P_{\mathbf{k}}^{\lambda,\bar{\lambda}'} \sum_{\mathbf{k}'} V_{\mathbf{k}-\mathbf{k}'} P_{\mathbf{k}'}^{\bar{\lambda}',\lambda'}$$
$$+ \sum_{\nu,\mathbf{k}',\mathbf{q}\neq 0} V_{\mathbf{q}}\, c_{\lambda,\nu,\nu,\lambda'}^{\mathbf{q},\mathbf{k}',\mathbf{k}}. \tag{24.56}$$

A very similar derivation of the singlet–doublet contributions can be performed for all two-particle expectation values appearing in Eq. (24.51). For example, the first photon-carrier contribution casts into the form

$$\sum \langle 2 \rangle^{1\text{st}} \equiv -\sum\nolimits_{\mathbf{q}} \left\langle \hat{\mathcal{A}}_{\mathbf{q}}^{\lambda'} a_{\lambda,\mathbf{k}}^{\dagger} a_{\bar{\lambda}',\mathbf{k}-\mathbf{q}} \right\rangle$$
$$= -\sum\nolimits_{\mathbf{q}} \left[\left\langle \hat{\mathcal{A}}_{\mathbf{q}}^{\lambda'} \right\rangle \left\langle a_{\lambda,\mathbf{k}}^{\dagger} a_{\bar{\lambda}',\mathbf{k}-\mathbf{q}} \right\rangle \delta_{\mathbf{q},\mathbf{0}} + \Delta \left\langle \hat{\mathcal{A}}_{\mathbf{q}}^{\lambda'} a_{\lambda,\mathbf{k}}^{\dagger} a_{\bar{\lambda}',\mathbf{k}-\mathbf{q}} \right\rangle \right]$$
$$= -\left\langle \hat{\mathcal{A}}_{\mathbf{0}}^{\lambda'} \right\rangle P_{\mathbf{k}}^{\lambda,\bar{\lambda}'} - \sum\nolimits_{\mathbf{q}} \Delta \left\langle \hat{\mathcal{A}}_{\mathbf{q}}^{\lambda'} a_{\lambda,\mathbf{k}}^{\dagger} a_{\bar{\lambda}',\mathbf{k}-\mathbf{q}} \right\rangle, \tag{24.57}$$

after we implement the singlet factorization (15.33) and identify the photon-carrier doublet via the relation (15.39). The last form follows after we implement the fact that the carrier and the photon singlets exist only for vanishing in-plane momentum $\mathbf{q}$. Furthermore, the carrier singlet is expressed via its generic distribution form (24.41).

24.5.2 Closed set of singlets

Evaluating all singlet–doublet contributions in Eq. (24.51), we eventually find

$$i\hbar \frac{\partial}{\partial t} P_{\mathbf{k}}^{\lambda,\lambda'} = \left(\tilde{\epsilon}_{\mathbf{k}}^{\lambda'} - \tilde{\epsilon}_{\mathbf{k}}^{\lambda} \right) P_{\mathbf{k}}^{\lambda,\lambda'} + \Omega_{\mathbf{k}}^{\lambda} P_{\mathbf{k}}^{\bar{\lambda},\lambda'} - P_{\mathbf{k}}^{\lambda,\bar{\lambda}'} \Omega_{\mathbf{k}}^{\lambda'}$$
$$+ \sum_{\nu,\mathbf{k}',\mathbf{q}\neq 0} V_{\mathbf{q}} \left[c_{\lambda,\nu;\nu,\lambda'}^{\mathbf{q},\mathbf{k}',\mathbf{k}} - \left(c_{\lambda',\nu;\nu,\lambda}^{\mathbf{q},\mathbf{k}',\mathbf{k}} \right)^{\star} \right]$$
$$- \sum_{\mathbf{q}} \left[\Delta \left\langle \hat{\mathcal{A}}_{\mathbf{q}}^{\lambda'} a_{\lambda,\mathbf{k}}^{\dagger} a_{\bar{\lambda}',\mathbf{k}-\mathbf{q}} \right\rangle - \Delta \left\langle \left(\hat{\mathcal{A}}_{\mathbf{q}}^{\lambda} \right)^{\dagger} a_{\bar{\lambda},\mathbf{k}-\mathbf{q}}^{\dagger} a_{\lambda',\mathbf{k}} \right\rangle \right]$$
$$+ \sum_{\mathbf{q}} \left[\Delta \left\langle \hat{\mathcal{G}}_{\mathbf{q}}^{\lambda'} a_{\lambda,\mathbf{k}}^{\dagger} a_{\lambda',\mathbf{k}-\mathbf{q}} \right\rangle - \Delta \left\langle \left(\hat{\mathcal{G}}_{\mathbf{q}}^{\lambda} \right)^{\dagger} a_{\lambda,\mathbf{k}-\mathbf{q}}^{\dagger} a_{\lambda',\mathbf{k}} \right\rangle \right]. \tag{24.58}$$

Here, we introduced the renormalized kinetic energy

$$\tilde{\epsilon}^{\lambda}_{\mathbf{k}} \equiv \epsilon^{\lambda}_{\mathbf{k}} - \sum_{\mathbf{k}'} V_{\mathbf{k}'-\mathbf{k}} n^{\lambda}_{\mathbf{k}'}, \tag{24.59}$$

and the renormalized Rabi frequency

$$\begin{aligned} \Omega^{\lambda}_{\mathbf{k}} &\equiv \left\langle \hat{A}^{\lambda}_{\mathbf{0}} \right\rangle + \sum_{\mathbf{k}'} V_{\mathbf{k}'-\mathbf{k}} P^{\lambda,\bar{\lambda}}_{\mathbf{k}'} \\ &= \frac{Q p_{\lambda,\bar{\lambda}}}{m_0} \int dz\, g(z)\, \langle \hat{A}(z,t) \rangle + \sum_{\mathbf{k}'} V_{\mathbf{k}'-\mathbf{k}} P^{\lambda,\bar{\lambda}}_{\mathbf{k}'}, \end{aligned} \tag{24.60}$$

where the last step implements the definition (24.31). Typically, the classical light problem is simplest when we solve the wave equation (24.48).

Equations (24.58) and (24.48) are formally exact for homogeneous systems. We also notice that Eq. (24.58) satisfies the structural form of the singlet dynamics in Eq. (24.37). Especially, the linear contributions result from the hierarchical coupling to doublets while the nonlinearities appear as products of two singlets. When the doublet contributions are not significant, they can be ignored with the consequence that Eqs. (24.48) and (24.58) form a closed set of equations.

Exercises

24.1 **(a)** Confirm that (24.1) and (24.2) define generic Boson fields in solids, based on the discussion in Chapter 13. Verify also the relation (24.3). **(b)** Express the field operators in terms of Bloch electrons, i.e., $\hat{\Psi}(r) = \sum_{\mathbf{k},\lambda} a_{\lambda,k} \frac{1}{\mathcal{L}^{d/2}} e^{i\mathbf{k}\cdot\mathbf{r}}\, u_{\lambda,\mathbf{k}}(\mathbf{r})$ where $u_{\lambda,\mathbf{k}}(\mathbf{r})$ is the lattice periodic Bloch function. Show that

$$G_{\lambda,\lambda'}(\mathbf{x},\mathbf{y}) = \frac{1}{\Omega_0^2} \int_{\Omega_0} d^3r \int_{\Omega_0} d^3r'\; u_{\lambda,0}(\mathbf{r})\, u^{\star}_{\lambda',0}(\mathbf{r}') \langle \hat{\Psi}^{\dagger}(\mathbf{x}+\mathbf{r}) \hat{\Psi}(\mathbf{y}+\mathbf{r}') \rangle$$

yields Eq. (24.4) when the separation of length scales is applied to solve the integral over the unit cell. Assume that $\mathbf{x}$ and $\mathbf{y}$ correspond to the lattice vectors and that the $\mathbf{k}$ dependence of $u_{\lambda,\mathbf{k}}(\mathbf{r})$ can be neglected. Why can the confinement functions be eliminated with a similar procedure?

24.2 A general N-particle expectation value follows from

$$G^{(N)}(x_1 \cdots x_N; y_N \cdots y_1) = \langle \hat{\Psi}^{\dagger}(\mathbf{x}_1) \cdots \hat{\Psi}^{\dagger}(\mathbf{x}_N)\; \hat{\Psi}(\mathbf{y}_N) \cdots \hat{\Psi}(\mathbf{y}_1) \rangle \,.$$

(a) In homogeneous systems $G^{(N)}$ remains invariant under the translation $\mathbf{x}_j \to \mathbf{x}_j + \mathbf{R}$ and $\mathbf{y}_j \to \mathbf{y}_j + \mathbf{R}$. Show that this implies the condition (24.13). **(b)** Verify that the RPA as well as Eq. (24.12) follow from Eq. (24.13).

24.3 Introduce hole operators $h_{\mathbf{k}} \equiv a^{\dagger}_{v_1-\mathbf{k}}$ and $h^{\dagger}_{\mathbf{k}} \equiv a_{v_1-\mathbf{k}}$ that annihilate and create, respectively, an antiparticle of the valence-band electron. **(a)** Show that the hole operators are Fermionic. **(b)** Show that the polarization $\hat{P}_{\mathbf{k}} = a^{\dagger}_{v,\mathbf{k}}\, a_{c,\mathbf{k}}$ destroys an electron–hole pair. **(c)** Compute $\left[\hat{P}_{\mathbf{k}}, \hat{P}^{\dagger}_{\mathbf{k}'}\right]_{-}$. In which limit does $\hat{P}_{\mathbf{k}}$ become nearly Bosonic? How is the condition (24.13) modified in the electron–hole picture?

24.4 **(a)** Derive the dynamics of $P^{\lambda,\lambda'}_{\mathbf{k}}$ by starting from Eqs. (24.29)–(24.30) and verify Eq. (24.51). **(b)** Identify the singlet and doublet contributions from the $P^{\lambda,\lambda'}_{\mathbf{k}}$ dynamics by applying the explicit factorizations discussed in Sections 15.2.1–15.2.2. Verify that Eq. (24.58) follows.

24.5 Start from Eq. (24.58) and construct the different functions appearing in Eq. (15.47). Which contributions are missing and why?

24.6 **(a)** Verify that Eqs. (24.38) and (24.43)–(24.45) define all relevant singlets and doublets. **(b)** Compute the dynamic equation of $\left\langle B^{\dagger}_{\mathbf{q},q_z}\, B_{\mathbf{q}',q'_z}\right\rangle$. Apply the singlet–doublet factorization and show that in homogeneous systems only the contributions with $\mathbf{q} = \mathbf{q}'$ are driven by the singlets.

24.7 The renormalized kinetic energy is given by $\tilde{\varepsilon}^{\lambda}_{\mathbf{k}} = \varepsilon^{\lambda}_{\mathbf{k}} - \sum_{\mathbf{k}'} V_{\mathbf{k}-\mathbf{k}'}\, n^{\lambda}_{\mathbf{k}'}$. **(a)** Apply the two-band model and use the electron–hole picture to show that $\tilde{\varepsilon}^{\mathrm{c}}_{\mathbf{k}} = \varepsilon^{\mathrm{c}}_{\mathbf{k}} - \sum_{\mathbf{k}'} V_{\mathbf{k}-\mathbf{k}'}\, f^{\mathrm{e}}_{\mathbf{k}'}$ and $\tilde{\varepsilon}^{\mathrm{h}}_{\mathbf{k}} \equiv -\tilde{\varepsilon}^{\mathrm{v}}_{\mathbf{k}} = \varepsilon^{\mathrm{h}}_{\mathbf{k}} - \sum_{\mathbf{k}'} V_{\mathbf{k}-\mathbf{k}'}\, f^{\mathrm{h}}_{\mathbf{k}'}$ where $\varepsilon^{\mathrm{h}}_{\mathbf{k}} \equiv -\varepsilon^{\mathrm{v}}_{\mathbf{k}} + \sum_{\mathbf{k}'} V_{\mathbf{k}'}$. **(b)** The identified $\varepsilon^{\mathrm{h}}_{\mathbf{k}}$ contains the Coulomb renormalization of the completely filled valence band. Show that $\varepsilon^{\mathrm{h}}_{\mathbf{k}} \equiv -\varepsilon^{\mathrm{v}}_{\mathbf{k}} + V(\mathbf{r}=0)$ where $V(0)$ is diverging. **(c)** Show that this contribution yields the average kinetic energy

$$\langle \hat{T}_v \rangle = \sum_{\mathbf{k}} \varepsilon^{\mathrm{v}}_{\mathbf{k}} \left\langle a^{\dagger}_{v,\mathbf{k}}\, a_{v,\mathbf{k}} \right\rangle = V(\mathbf{r}=0) \sum_{\mathbf{k}} \left\langle a^{\dagger}_{v,\mathbf{k}}\, a_{v,\mathbf{k}} \right\rangle - \sum_{\mathbf{k}} \varepsilon^{\mathrm{h}}_{\mathbf{k}} \left(1 - f^{\mathrm{h}}_{\mathbf{k}}\right)$$

among the valence-band electrons. Show that their Coulomb energy is $\langle \hat{V}_v \rangle_S = -\frac{1}{2} \sum_{\mathbf{k},\mathbf{q}} V_q \left\langle a^{\dagger}_{v,\mathbf{k}}\, a_{v,\mathbf{k}} \right\rangle \left\langle a^{\dagger}_{v,\mathbf{k}-\mathbf{q}}\, a_{v,\mathbf{k}-\mathbf{q}} \right\rangle$ in the singlet approximation. **(d)** Show that an analogous exchange term follows for the ion–ion interactions such that a charge-neutral system has $\langle \hat{V} \rangle_{\mathrm{exc}} = -V(0) \sum_{\mathbf{k}} \left\langle a^{\dagger}_{v,\mathbf{k}} a_{v,\mathbf{k}} \right\rangle$. Since the energy per particle, i.e., $\langle \hat{H} \rangle_S / \sum_k \left\langle a^{\dagger}_{v,\mathbf{k}} a_{v,\mathbf{k}} \right\rangle$ must be finite, conclude that $\varepsilon^{\mathrm{h}}_{\mathbf{k}}$ is finite while $\varepsilon^{\mathrm{v}}_{\mathbf{k}}$ diverges as $V(\mathbf{r}=0)$.

Further reading

Many central concepts of semiconductor optics are discussed in:

J. Shah (1999). *Ultrafast Spectroscopy of Semiconductors and Semiconductor Nanostructures*, 2nd edition, Berlin, Springer.

W. Schäfer and M. Wegener (2002). *Semiconductor Optics and Transport Phenomena*, Heidelberg, Springer.

P. Y. Yu and M. Cardona (2005). *Fundamentals of Semiconductors: Physics and Materials Properties*, 3rd edition, Berlin, Springer.

T. Meier, P. Thomas, and S. W. Koch (2007). *Coherent Semiconductor Optics: From Basic Concepts to Nanostructure Applications*, Heidelberg, Springer.
C. F. Klingshirn (2007). *Semiconductor Optics*, 3rd edition, Berlin, Springer.
H. Haug and S. W. Koch (2009). *Quantum Theory of the Optical and Electronic Properties of Semiconductors*, 5th edition, Singapore, World Scientific.

25
Introductory semiconductor optics

In the previous chapter, we discuss how the excitation with a laser pulse generates an excited state of semiconductor electrons and holes via the light–matter dipole interaction. In this chapter, we introduce a useful division into coherent and incoherent quantities by generalizing the discussion of Section 23.4.4. Applying the rotating-wave approximation, we find a separation of time scales between fast and slowly varying contributions, which we can use to speed-up significantly our numerical evaluations. This improvement is needed to make the full singlet–doublet analysis of Coulomb and quantum-optical effects feasible.

25.1 Optical Bloch equations

Under homogeneous conditions, the singlet dynamics of the semiconductor excitations is described by Eq. (24.58). The homogeneous part of this equation is

$$i\hbar \frac{\partial}{\partial t} P_{\mathbf{k}}^{\lambda,\lambda'} \bigg|_{\text{hom}} = \left(\epsilon_{\mathbf{k}}^{\lambda'} - \epsilon_{\mathbf{k}}^{\lambda}\right) P_{\mathbf{k}}^{\lambda,\lambda'}, \tag{25.1}$$

while all the remaining contributions are either nonlinear or involve a hierarchical coupling. If these contributions are left out, the singlet dynamics describes pure oscillations $P_{\mathbf{k}}^{\lambda,\lambda'} = \bar{P}\, e^{-\frac{i}{\hbar}\left(\epsilon_{\mathbf{k}}^{\lambda'} - \epsilon_{\mathbf{k}}^{\lambda}\right)t}$ where $\bar{P}$ is constant. The λ and λ' that appear refer to the different electronic bands of the semiconductor. As before, we make the two-band approximation where only one valence and one conduction band participate in the optical transitions.

To see the principal effect of the nonlinear coupling, we neglect both the doublets as well as the Coulomb contributions from Eq. (24.58) to find

$$\begin{aligned} i\hbar \frac{\partial}{\partial t} P_{\mathbf{k}}^{v,c} &= \left(\epsilon_{\mathbf{k}}^{c} - \epsilon_{\mathbf{k}}^{v}\right) P_{\mathbf{k}}^{v,c} + \Omega_{\mathbf{k}}^{v} P_{\mathbf{k}}^{c,c} - P_{\mathbf{k}}^{v,v} \Omega_{\mathbf{k}}^{v} \\ &= \left(\epsilon_{\mathbf{k}}^{c} - \epsilon_{\mathbf{k}}^{v}\right) P_{\mathbf{k}}^{v,c} - \left(1 - f_{\mathbf{k}}^{e} - f_{\mathbf{k}}^{h}\right) \frac{Q p_{c,v}}{m_0} \langle \hat{A}(0,t) \rangle \end{aligned} \tag{25.2}$$

for the band-index combination, $\lambda = v,\ \lambda' = c$ of the two-band model, see Fig. 24.2. We identified the electron $f_{\mathbf{k}}^{e} \equiv n_{\mathbf{k}}^{c} = P_{\mathbf{k}}^{c,c}$ and hole $f_{\mathbf{k}}^{h} = 1 - n_{\mathbf{k}}^{v} = 1 - P_{\mathbf{k}}^{v,v}$ distributions using Eqs. (24.16)–(24.17). Additionally, we applied the strong-confinement-limit

$g(z) \to \delta(z)$ to evaluate the Rabi frequency (24.60). This approximation is justified whenever $\langle \hat{A}(z,t)\rangle$ changes on a much larger length scale than the confinement function. This condition is usually satisfied because the typical quantum wells (QWs) or quantum wires (QWIs) are much thinner than the wave length of light. This set of approximations reduces Eq. (24.58) into

$$i\hbar\frac{\partial}{\partial t}f^e_{\mathbf{k}} = i\hbar\frac{\partial}{\partial t}f^h_{\mathbf{k}} = 2i\mathrm{Im}\left[\frac{Qp_{v,c}}{m_0}\langle\hat{A}(0,t)\rangle P^{v,c}_{\mathbf{k}}\right], \tag{25.3}$$

where we expressed $P^{c,c}_{\mathbf{k}}$ and $P^{v,v}_{\mathbf{k}}$ in the electron–hole picture discussed in Section 24.1.2. The resulting Eqs. (25.2)–(25.3) form a closed set.

25.1.1 Two-level aspects of semiconductors

The simplified singlet analysis of Eqs. (25.2)–(25.3) produces a dynamics that is formally analogous to the singlet dynamics of two-level systems, investigated in Section 23.4.2. To show this connection more clearly, we introduce the notation

$$\begin{aligned} S^-_{\mathbf{k}} &\equiv P^{v,c}_{\mathbf{k}}, & S^z_{\mathbf{k}} &\equiv \frac{1}{2}\left(f^e_{\mathbf{k}} + f^h_{\mathbf{k}} - 1\right) \\ \hbar\Omega(t) &\equiv \frac{Qp_{c,v}}{m_0}\langle\hat{A}(0,t)\rangle, & \hbar\omega_{g,\mathbf{k}} &\equiv \epsilon^c_{\mathbf{k}} - \epsilon^v_{\mathbf{k}}, \end{aligned} \tag{25.4}$$

where $\Omega(t)$ stands for the Rabi frequency and $\hbar\omega_{g,\mathbf{k}}$ denotes the momentum-dependent transition energy. These definitions convert Eqs. (25.2)–(25.3) into the familiar format of the optical Bloch equations,

$$\begin{cases} i\frac{\partial}{\partial t}S^-_{\mathbf{k}} = \omega_{g,\mathbf{k}}\,S^-_{\mathbf{k}} + 2S^z_{\mathbf{k}}\,\Omega(t) \\ \frac{\partial}{\partial t}S^z_{\mathbf{k}} = +2\mathrm{Im}\left[\Omega^\star(t)\,S^-_{\mathbf{k}}\right] \end{cases}, \tag{25.5}$$

investigated in Chapter 18.

Equations (25.5) show very clearly that, under the highly idealized conditions discussed here, each electronic $\mathbf{k}$ state forms an independent two-level system described by a Bloch vector,

$$\begin{aligned} \mathbf{S}_{\mathbf{k}} &\equiv \left(\mathrm{Re}\left[S^-_{\mathbf{k}}\right], \mathrm{Im}\left[S^-_{\mathbf{k}}\right], S^z_{\mathbf{k}}\right) \\ &= \left(\mathrm{Re}\left[P^{v,c}_{\mathbf{k}}\right], \mathrm{Im}\left[P^{v,c}_{\mathbf{k}}\right], \frac{1}{2}\left(f^e_{\mathbf{k}} + f^h_{\mathbf{k}} - 1\right)\right), \end{aligned} \tag{25.6}$$

in analogy to Eq. (19.15). In particular, we can describe the dynamics (25.5) as the Bloch-vector rotation following from

$$\frac{\partial}{\partial t}\mathbf{S}_{\mathbf{k}} = \mathbf{\Omega}_{\mathbf{k}} \times \mathbf{S}_{\mathbf{k}}, \qquad \mathbf{\Omega}_{\mathbf{k}} \equiv \left(2\mathrm{Re}\left[\Omega(t)\right], 2\mathrm{Im}\left[\Omega(t)\right], -\omega_{g,\mathbf{k}}\right), \tag{25.7}$$

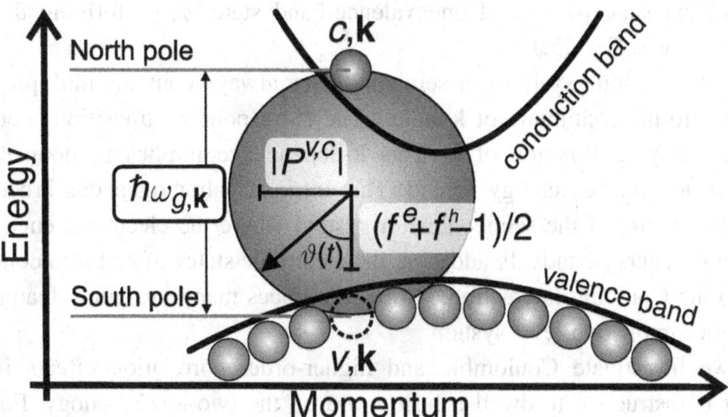

Figure 25.1 Schematic representation of optical semiconductor excitations in a two-band model. Without doublets and Coulomb effects, the excitation can be described via a Bloch vector for each electronic momentum state. The Bloch vector (arrow) has perpendicular components representing the polarization $|P_{\mathbf{k}}|$ and the excitation density defining the inversion factor $\frac{1}{2}\left(f^e_{\mathbf{k}} + f^h_{\mathbf{k}} - 1\right)$.

based on Eq. (19.20) and the identification (19.19). With this notation, the unexcited $\left(f^e_{\mathbf{k}} = f^h_{\mathbf{k}} = 0\right)$ and the fully excited $\left(f^e_{\mathbf{k}} = f^h_{\mathbf{k}} = 1\right)$ electronic states correspond to the Bloch vectors

$$\mathbf{S}_{\mathbf{k}} = \left(0,\, 0,\, -\frac{1}{2}\right) \quad \text{and} \quad \mathbf{S}_{\mathbf{k}} = \left(0,\, 0,\, +\frac{1}{2}\right), \tag{25.8}$$

respectively. When the semiconductor is initially in its ground state, i.e., $f^e_{\mathbf{k}} = f^h_{\mathbf{k}} = 0$, the excitation via $\Omega(t)$ starts to move $\mathbf{S}_{\mathbf{k}}$ from the south pole towards the north pole of the Bloch sphere, as illustrated in Fig. 25.1. This process always involves the generation of polarization, i.e., $P^{v,c}_{\mathbf{k}} \neq 0$, which correctly implies a nonzero transition amplitude to move the electron from the valence to the conduction band. For sufficiently strong excitations, $\mathbf{S}_{\mathbf{k}}$ oscillates on the Bloch sphere defined by the surface $|\mathbf{S}_{\mathbf{k}}| = 1/2$ (large sphere in Fig. 25.1), which eventually produces Rabi flopping as discussed in Section 18.2.3.

25.1.2 Electronic wave function in the singlet analysis

The simple two-band noninteracting electron picture describes the optical semiconductor excitations as a single-electron motion labeled by the momentum $\mathbf{k}$. We may even identify an associated two-level wave function,

$$|\psi_{2\text{LS},\mathbf{k}}(t)\rangle \equiv \cos\frac{1}{2}\vartheta_{\mathbf{k}}\,|\phi_{v,\mathbf{k}}\rangle + e^{-i\Delta\phi_{\mathbf{k}}}\sin\frac{1}{2}\vartheta_{\mathbf{k}}|\phi_{c,\mathbf{k}}\rangle, \tag{25.9}$$

for each electron, based on Eqs. (19.22) and (19.25). This identification implies that each individual electron is described through a superposition consisting of only one conduction-band state $|\phi_{c,\mathbf{k}}\rangle$ and one valence-band state $|\phi_{v,\mathbf{k}}\rangle$ forming the two-level system, as shown in Fig. 25.1.

In contrast to atomic systems, a semiconductor always contains multiple "two-level systems" due to the continuum of **k** states. The corresponding transition energies $\hbar\omega_{g,\mathbf{k}}$ vary continuously as function of **k**. This **k**-dependence implies pronounced inhomogeneous broadening, i.e., energy spread. This intrinsic inhomogeneous broadening is a characteristic feature of the semiconductor system where the electronic energies constitute continuous energy bands. In addition, the electronic states of real semiconductors are coupled via the Coulomb interaction, which introduces many intriguing features that are missing in the simple two-level system.

Before we investigate Coulombic and higher-order correlation effects in the next chapters, it is instructive to dwell a little more on the two-level analogy. For example, we can extend the current analysis relatively easily to multiband systems as long as the Coulomb effects among the Bloch electrons are excluded. We then see that the multiband generalization corresponds to extending the two-level approximation (25.9) to the multilevel model, such as Eq. (18.6). We can therefore follow a derivation very similar to that in Sections 18.1.2–18.1.3 to show which levels really couple to the light. In cases where the laser light is near-resonant with only one dipole-allowed transition energy, i.e., $p_{v,c} \neq 0$ and $\omega \approx \omega_{g,\mathbf{k}}$, the optical excitation indeed involves only two bands. In some situations, laser pulses are resonant with more than two bands. The pertinent generalizations lead to, e.g., a three-band model with either two valence bands or two conduction bands, corresponding to "Λ" and "V" transitions, respectively, see the discussion in connection with Fig. 18.3. Similarly, the N-band generalizations of the two-band model follow in a straightforward manner.

25.2 Linear response

As we can see from Eq. (25.2), the polarization in lowest order is linear in terms of the exciting field $\langle \hat{A}(0, t)\rangle$. Due to the driving term $\Omega^{\star}(t)P_{\mathbf{k}}^{v,c}$ that appears, the excited carrier occupations vary quadratically with the exciting field. If the semiconductor is initially unexcited, we have $f_{\mathbf{k}} \equiv f_{\mathbf{k}}^{e} = f_{\mathbf{k}}^{h} \propto \left|P_{\mathbf{k}}^{v,c}\right|^2 \propto |\langle \hat{A}\rangle|^2$. As it turns out, this scaling law is generally valid for weak excitations even when Coulomb effects and doublet contributions are fully included. Thus, we can deduce the relevant coherence properties of semiconductors by investigating the leading order response to weak excitations.

25.2.1 Linear polarization

To determine how the optical excitation proceeds in two-band systems, we analyze Eq. (25.2) in the limit of weak excitation assuming an originally unexcited semiconductor. For this purpose, we only need to include the lowest-order contributions in terms of the field. Since $P^{v,c}_{\mathbf{k}}$ is linear in the field while $f_{\mathbf{k}}$ is at least quadratic, the density contributions in Eq. (25.2) can be ignored. As a result, the linear polarization response follows from

$$i\frac{\partial}{\partial t}P^{v,c}_{\mathbf{k}} = \left(\omega_{g,\mathbf{k}} - i\gamma\right) P^{v,c}_{\mathbf{k}} - \Omega(t), \tag{25.10}$$

where we included a phenomenological polarization dephasing γ and identified the Rabi frequency using Eq. (25.4). Physically, γ results from the doublet terms within the full Eq. (24.58).

Equation (25.10) is mathematically equivalent to the problem of an externally driven, damped oscillator with the analytic solution

$$P^{v,c}_{\mathbf{k}}(t) = i\int_{-\infty}^{t} du\ \Omega(u)\ e^{(i\omega_{g,\mathbf{k}}+\gamma)(u-t)}\ . \tag{25.11}$$

For later reference, we introduce the Rabi frequency in the rotating frame

$$\tilde{\Omega}(t) \equiv \Omega(t)\ e^{+i\omega_{g,\mathbf{k}}t}\ , \tag{25.12}$$

where the free polarization oscillation is removed. This identification of $\tilde{\Omega}(t)$ converts Eq. (25.11) into

$$P^{v,c}_{\mathbf{k}}(t) = i\ e^{-(i\omega_{g,\mathbf{k}}+\gamma)t}\int_{-\infty}^{t} du\ \tilde{\Omega}(u)\ e^{+\gamma u}\ . \tag{25.13}$$

If a pulsed optical excitation is applied, $\tilde{\Omega}$ is nonvanishing only for times shorter than the duration of the pulse, T. Alternatively, we could also consider a situation where the pulse decays faster than $e^{-\gamma t}$ after its peak such that the integral of Eq. (25.11) does not change for $t > T$. Both of these scenarios yield

$$P^{v,c}_{\mathbf{k}}(t) = i\bar{P}\ e^{-i\omega_{g,\mathbf{k}}t}\ e^{-\gamma t}, \quad \bar{P} \equiv \int_{-\infty}^{T} du\ \tilde{\Omega}(u)\ e^{+\gamma t}, \quad t > T \tag{25.14}$$

after the pulse, showing that the created polarization decays with the rate defined by γ. Thus, the polarization vanishes rather rapidly after the end of the excitation.

25.2.2 Weak excitation of densities

To determine the density dynamics for the conditions used in Section 25.2.1, we need to solve Eq. (25.3). Since $f^e_{\mathbf{k}}$ and $f^h_{\mathbf{k}}$ are generated by the identical source, it is sufficient to analyze

$$\frac{\partial}{\partial t}f_{\mathbf{k}} = \frac{\partial}{\partial t}f^e_{\mathbf{k}} = \frac{\partial}{\partial t}f^h_{\mathbf{k}} = 2\mathrm{Im}\left[\tilde{\Omega}^{\star}(t)\ e^{+i\omega_{g,\mathbf{k}}t}\ P^{v,c}_{\mathbf{k}}\right], \tag{25.15}$$

where we used (25.12). For linear excitation, the densities do not couple back to the polarization dynamics. In this situation, the polarization only acts as a source for $f_{\mathbf{k}}$ such that we may substitute the result (25.13) into Eq. (25.15) leading to

$$\begin{aligned}\frac{\partial}{\partial t} f_{\mathbf{k}} &= 2\mathrm{Im}\left[\tilde{\Omega}^\star(t) i\, e^{-\gamma t} \int_{-\infty}^{t} du\, \tilde{\Omega}(u)\, e^{+\gamma u}\right], \\ &= 2\mathrm{Re}\left[\int_{-\infty}^{t} du\, \tilde{\Omega}^\star(t)\, \tilde{\Omega}(u)\, e^{-\gamma(u-t)}\right], \\ &= \int_{-\infty}^{t} du \left[\tilde{\Omega}^\star(t)\, \tilde{\Omega}(u) + \tilde{\Omega}(t)\, \tilde{\Omega}^\star(u)\right] e^{-\gamma(t-u)}. \end{aligned} \tag{25.16}$$

Since the RHS of this differential equation is just a time-dependent source, we perform a straightforward integration

$$f_{\mathbf{k}}(t) = \int_{-\infty}^{t} dt' \int_{-\infty}^{t'} du \left[\tilde{\Omega}^\star(t')\, \tilde{\Omega}(u) + \tilde{\Omega}(t')\, \tilde{\Omega}^\star(u)\right] e^{-\gamma(t'-u)}, \tag{25.17}$$

which can always be evaluated numerically if the temporal shape of the excitation pulse is known.

To understand the implications of (25.17), we insert the generic relation

$$\tilde{\Omega}^\star(t')\, \tilde{\Omega}(u) + \tilde{\Omega}(t')\, \tilde{\Omega}^\star(u) = \frac{|\tilde{\Omega}(t') + \tilde{\Omega}(u)|^2}{2} - \frac{|\tilde{\Omega}(t') - \tilde{\Omega}(u)|^2}{2} \tag{25.18}$$

to obtain

$$\begin{aligned} f_{\mathbf{k}}(t) = &\int_{-\infty}^{t} dt' \int_{-\infty}^{t'} du\, \frac{|\tilde{\Omega}(t') + \tilde{\Omega}(u)|^2}{2}\, e^{-\gamma(t'-u)} \\ &- \int_{-\infty}^{t} dt' \int_{-\infty}^{t'} du\, \frac{|\tilde{\Omega}(t') - \tilde{\Omega}(u)|^2}{2}\, e^{-\gamma(t'-u)}. \end{aligned} \tag{25.19}$$

Both integrands are positive for all t' and u such that the integrals always yield positive values, i.e., $f_{\mathbf{k}}(t) \equiv I_{1\mathrm{st}}(t) - I_{2\mathrm{nd}}(t)$ with positive definite integrals $I_j(t)$. Since the integrands that appear are not the same, the corresponding $I_{1\mathrm{st}}$ and $I_{2\mathrm{nd}}$ generally yield different results. We see that the second integrand vanishes for $u = t'$ while the first integrand simply equals $|\tilde{\Omega}(t')|^2$. Thus, $I_{1\mathrm{st}}(t)$ is typically larger than $I_{2\mathrm{nd}}(t)$. Consequently, we find a positive nonvanishing $f_{\mathbf{k}}(+\infty) \equiv I_{1\mathrm{st}}(+\infty) - I_{2\mathrm{nd}}(+\infty) > 0$ after the $\tilde{\Omega}$ pulse. Thus, we can conclude that some carrier distributions remain in the system even if the polarization has decayed away.

25.2.3 Weak square-pulse excitation

To have a simple analytically solvable situation, we consider an excitation that is described by a square pulse

$$\tilde{\Omega}(t) = \Omega_0 \left[\theta\left(t + \frac{T}{2}\right) - \theta\left(t - \frac{T}{2}\right)\right], \tag{25.20}$$

where the Heaviside functions produce a box-like envelope for the excitation that starts at $t = -T/2$ and ends at $t = +T/2$. The straightforward solution of Eqs. (25.13) and (25.17) produces

$$P_{\mathbf{k}}(t) = i\Omega_0 T e^{-i\omega_{g,\mathbf{k}}t} \begin{cases} 0, & t < -\frac{T}{2} \\ \frac{1-e^{-\gamma\left(t+\frac{T}{2}\right)}}{\gamma T}, & -\frac{T}{2} \leq t \leq \frac{T}{2} \\ \frac{e^{\gamma\frac{T}{2}}-e^{-\gamma\frac{T}{2}}}{\gamma T} e^{-\gamma t}, & t > \frac{T}{2} \end{cases} \tag{25.21}$$

for the polarization and

$$f_{\mathbf{k}}(t) = 2|\Omega_0 T|^2 \begin{cases} 0, & t < -\frac{T}{2} \\ \frac{\gamma\left(t+\frac{\gamma T}{2}\right)+e^{-\gamma\left(t+\frac{T}{2}\right)}-1}{\gamma^2 T^2}, & -\frac{T}{2} \leq t \leq \frac{T}{2}, \\ \frac{\gamma T+e^{-\gamma T}-1}{\gamma^2 T^2}, & t > \frac{T}{2} \end{cases} \tag{25.22}$$

for the densities, see Exercise 25.3. Rigorously, the results (25.21)–(25.22) are valid only for that electronic state $\mathbf{k}$ which is resonant with the field because only this allows us to use a time-independent Ω_0.

For resonant excitation, $\Omega_0 T$ defines the area under the Rabi frequency defined by

$$A_{\text{Rabi}} \equiv \int_{-\infty}^{\infty} dt\, \tilde{\Omega}(t). \tag{25.23}$$

This is the prefactor multiplying the expressions for both the polarization and the density. The condition of weak excitation, i.e., $f_{\mathbf{k}} \ll 1$, can be satisfied only if we have $|A_{\text{Rabi}}|^2 \ll 1$. Otherwise, $f_{\mathbf{k}}$ can grow above unity, which is a physical impossibility for Fermions. For large A_{Rabi}, one must include the nonlinear contributions to all orders because, e.g., $f_{\mathbf{k}}$ eventually starts to oscillate in time, which cannot be described in terms of an A_{Rabi} expansion. In other words, the Rabi-flopping behavior cannot be analyzed perturbatively, see Exercise 25.4.

In Fig. 25.2, we show $|\tilde{\Omega}(t)|$ (normalized, shaded area), $|P_{\mathbf{k}}(t)|^2$ (dashed line), and $f_{\mathbf{k}}$ (solid line) as functions of time using the weak-excitation limit results (25.21)–(25.22). For the evaluations, we assumed $T = 1$ ps and $|A_{\text{Rabi}}|^2 = 0.1$ and three different dephasing values. For the low dephasing, we see that $f_{\mathbf{k}}$ and $|P_{\mathbf{k}}|^2$ exhibit a nearly identical quadratic increase as long as the square pulse is present. To explain this behavior, we consider the $\gamma = 0$ analysis that forces the Bloch vector (25.6) to remain on the Bloch sphere, i.e., $|\mathbf{S}_{\mathbf{k}}| = 1/2$. It is straightforward to show that this establishes the strict condition

$$f_{\mathbf{k}} = \frac{1}{2} - \sqrt{\frac{1}{4} - \left|P_{\mathbf{k}}^{v,c}\right|^2} \to \left|P_{\mathbf{k}}^{v,c}\right|^2, \quad \text{for } \gamma = 0 \text{ and } \left|P_{\mathbf{k}}^{v,c}\right| \ll 1 \tag{25.24}$$

between carrier density and polarization. This result is often referred to as the coherent limit. It is also valid when the Coulomb interaction is included before the onset of scattering, as shown in Section 27.1.1.

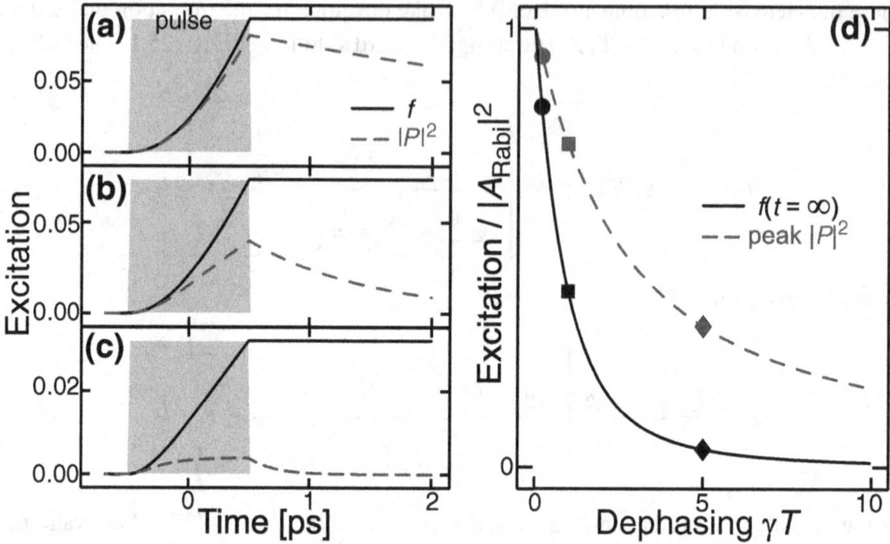

Figure 25.2 Weak excitation of a single electronic **k** state when the Coulomb interaction and the doublet contributions are omitted. The shaded area indicates the box-like excitation pulse. The temporal evolution of the polarization $|P_{\mathbf{k}}|^2$ (dashed line) and the density $f_{\mathbf{k}}$ (solid line) are shown for the dephasings: **(a)** $\gamma = 0.2\,\text{ps}^{-1}$, **(b)** $\gamma = 1\,\text{ps}^{-1}$, and **(c)** $\gamma = 5\,\text{ps}^{-1}$. **(d)** The peak polarization squared (dashed line) is compared with the residual density (solid line) for different dephasings. The circles, squares, and diamonds indicate the cases analyzed in frames **a–c**. All computations are performed for a low-intensity field with $A_{\text{Rabi}} = 0.1$.

For the conditions assumed in Fig. 25.2(a), the polarization and the density almost ideally satisfy the coherent-limit condition during the excitation because the assumed dephasing is relatively small. As the dephasing is increased, the dynamics of $f_{\mathbf{k}}$ and $|P_{\mathbf{k}}|^2$ coincide only at early times, allowing the density to become significantly larger than $|P_{\mathbf{k}}|^2$. For the largest $\gamma = 5\,\text{ps}^{-1}$, the polarization starts to saturate as seen clearly in Fig. 25.2(c). For all cases, the polarization decays exponentially after the pulse while a constant density level remains in the system, just as predicted by Eq. (25.19). Thus, the Bloch vector eventually approaches the z axis, i.e., $\mathbf{S}_{\mathbf{k}}(+\infty) = \left(0,\, 0,\, \frac{1-2f_{\mathbf{k}}(+\infty)}{2}\right)$, which is not on the Bloch sphere.

25.2.4 Transient polarization vs. stationary density

To quantify how the increasing dephasing distorts the polarization dynamics away from the Bloch sphere, we compare the maximum value of the polarization

$$P_{\max} = P_{\mathbf{k}}\left(\frac{T}{2}\right) = i\,A_{\text{Rabi}}\frac{1 - e^{-\gamma T}}{\gamma T} \tag{25.25}$$

with the constant density level $f_{\text{const}} \equiv f_{\mathbf{k}}(T/2) = f_{\mathbf{k}}(+\infty)$ remaining after the excitation. If we divide both $P_{\mathbf{k}}$ and $f_{\mathbf{k}}$ by $|A_{\text{Rabi}}|^2$ the excitation curves depend only on γT. The corresponding $|P_{\max}|^2/|A_{\text{Rabi}}|^2$ (dashed line) and $f_{\text{const}}/|A_{\text{Rabi}}|^2$ (solid line) are shown in Fig. 25.2(d) as functions of γT. We observe that the final level of the density and of the peak-polarization squared are equal only for vanishing dephasing. In other words, the coherent-limit result (25.24) is realized only if the system has no dephasing. As γT becomes appreciable, $|P_{\max}|^2$ decays rapidly but the densities remain significant even though they are slowly diminished for increasing γ. In other words, we find a broad range of dephasing conditions where densities remain in the system long after the excitation and the polarization have vanished. Generally, a polarization exists only transiently in semiconductors while a quasi-stationary excitation remains in the form of a density.

In real systems, the density also experiences some decay which eventually brings it back to the unexcited state. In isolated atoms, the density and polarization decay rates obey the relation $\gamma_f = 2\gamma_P$, compare (20.18) and the analysis in Section 20.4.2. Physically, the atomic γ_f and γ_P result from the same spontaneous recombination processes such that the overall excitation, i.e., $|P|^2 \propto e^{-2\gamma_P t}$ and $f \propto e^{-\gamma_f t}$, are destroyed at the same rate. In contrast to this idealized scenario, semiconductors are typically in a very different regime because the Coulomb and the phonon interactions produce strong dephasing of the polarization. Since these decay processes conserve the particle number, they do not change the total densities such that nonradiative impurity related effects or the very slow radiative recombination removes the densities from the system. As a result, typical semiconductors are characterized by a γ_P for the polarization that is orders of magnitude larger than the γ_f for the densities.

In general, the γ_P and γ_f values of semiconductors can be computed only at the level of a fully microscopic analysis. In particular, the master-equation approach of Section 20.4.1 is usually not sufficient to describe decay in solids. Instead, one needs to solve the full quantum kinetics of the dissipative processes to fully include the nonlinearity and the quasiequilibrium aspects of the relaxation and dephasing processes.

25.3 Coherent vs. incoherent quantities

Without going into the details of the systematic many-body analysis, the transient nature of the polarization and the quasistationary character of the densities can be understood qualitatively. As a relevant example, Fig. 25.3 presents the results of a fully microscopic investigation where the Coulomb effects and the doublet contributions are fully included. In the figure, we show the total polarization $|P|^2 \equiv \left|\frac{1}{\mathcal{L}}\sum_k P_k\right|^2$ (shaded area) and the total density $n^e \equiv \frac{1}{\mathcal{L}}\sum_k f_k^e$ (solid line) resulting from a Gaussian excitation pulse (dashed line) of a quantum wire (QWI).

We observe that $|P|^2$ and the carrier density indeed display a behavior qualitatively similar to that predicted by the simple analysis in Fig. 25.2. The density (solid line)

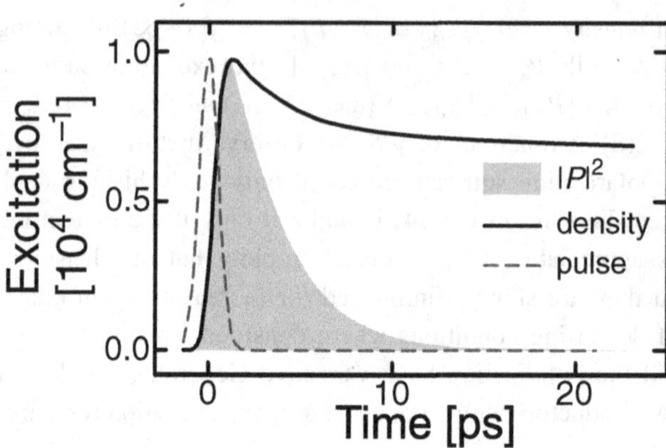

Figure 25.3 Weak-intensity excitation dynamics following from the full singlet–doublet analysis including both the Coulomb and the phonon scattering. The shaded area indicates the Gaussian pulse used. The solid line corresponds to the generated carrier density and the dashed line represents the total polarization squared. Reprinted from *Progress in Quantum Electronics*, **30/5**, M. Kira, S. W. Koch, "Many-body correlations and excitonic effects in semiconductor spectroscopy," pp. 155–296, 2006, with permission from Elsevier.

and $|P|^2$ (shaded area) more or less agree with one another as long as the resonant pulse (dashed line) is present verifying that the excitation is sufficiently low to apply the coherent-limit relation (25.24). However, due to the microscopic scattering processes, the polarization starts to decay rapidly, which is especially visible after the pulse. Consequently, a significant amount of electron and hole densities remain in the system after the pulse. In other words, the results of the simplified analysis are qualitatively verified by the fully microscopic computation. This generalizes our analytic observation that the polarization requires the presence of the pump pulse to exist whereas the densities remain in the system long after the pulse fades away.

As before, we can use our observations to define coherent and incoherent quantities. In particular, it is meaningful to classify the semiconductor excitations based on whether or not they are still present at times exceeding the polarization dephasing time $\tau \equiv 1/\gamma$ after the excitation pulse. As a general definition, *coherent quantities* exist only if external pumping is present while *incoherent quantities* remain in the system for long times $t \gg \tau$. Our discussions above clearly show that the polarization is a coherent quantity whereas the densities are incoherent.

As we analyze the structure of $P^{v,c}_{\mathbf{k}} = \left\langle a^{\dagger}_{v,\mathbf{k}} a_{c,\mathbf{k}} \right\rangle$, we realize that it describes a transition between conduction and valence band. In other words, the polarization measures the strength of the simplest *interband* transition. In the same way, the dynamics of the carrier occupations, e.g., $f^e_{\mathbf{k}} = \left\langle a^{\dagger}_{c,\mathbf{k}} a_{c,\mathbf{k}} \right\rangle$, can be viewed as *intraband* transitions because they involve electrons within the same band. Obviously, interband transitions need an external driving field with an energy on the scale of the band gap, typically in the range of 1 to 4 eV in solids. In contrast, the intraband transitions involve energies that are orders of magnitude smaller. Thus, it is intuitively clear that interband transitions must be externally driven, making them transient. At the same time, intraband transitions can exist without external driving because they involve only minor energy exchanges within the many-body system. This conclusion makes a simple qualitative argument to identify any C-particle correlation as either coherent or incoherent based on whether it describes inter- or intraband transitions, respectively.

25.3.1 Coherent vs. incoherent correlations

The coherence or incoherence of a given C-particle correlation can be judged by analyzing its free propagation,

$$i\hbar\frac{\partial}{\partial t}\Delta\langle C\rangle\Big|_{\text{free}} = E_C\,\Delta\langle C\rangle, \tag{25.26}$$

which can be deduced from the dynamics generated by the noninteracting Hamiltonian $\hat{H}_0$. If the energy $|E_C|$ is roughly $n = 1, 2, \cdots$ times the band-gap energy, $\Delta\langle C\rangle$ describes correlations involving one or more interband transitions. Just like the polarization, such transitions experience strong Coulomb and phonon scattering leading to rapid decay. Hence, these contributions are classified as coherent quantities that need the presence of external fields to be sustained. In contrast to this, $E_C \approx 0$ refers to intraband contributions that can exist for a long time within the system because scattering may equilibrate $\Delta\langle C\rangle$ but does not destroy it.

To show how the classification by Eq. (25.26) works in practice, we consider the free propagation of $\Delta\left\langle B^\dagger_{\mathbf{q},q_z} a^\dagger_{v,\mathbf{k}-\mathbf{q}} a_{c,\mathbf{k}}\right\rangle$ that produces

$$\begin{aligned} i\hbar\frac{\partial}{\partial t}\Delta\left\langle B^\dagger_{\mathbf{q},q_z} a^\dagger_{v,\mathbf{k}-\mathbf{q}} a_{c,\mathbf{k}}\right\rangle\Big|_{\text{free}} &= E_C\,\Delta\left\langle B^\dagger_{\mathbf{q},q_z} a^\dagger_{v,\mathbf{k}-\mathbf{q}} a_{c,\mathbf{k}}\right\rangle \\ E_C = E^c_{\mathbf{k}} - E^v_{\mathbf{k}-\mathbf{q}} - \hbar\omega_{\mathbf{q},q_z} &= \hbar\left(\omega_{g,\mathbf{k}} - \omega_{\mathbf{q},q_z}\right) + \left(E^v_{\mathbf{k}} - E^v_{\mathbf{k}-\mathbf{q}}\right), \end{aligned} \tag{25.27}$$

compare Eqs. (24.27)–(24.30). By regrouping the energies, we find that E_C consists of the energy difference between the band-gap energy and the photon and intraband-transition energy. Thus, $\Delta\left\langle B^\dagger_{\mathbf{q},q_z} a^\dagger_{v,\mathbf{k}-\mathbf{q}} a_{c,\mathbf{k}}\right\rangle$ is an incoherent correlation whenever the photon energy is near resonant with the band gap. This classification is also intuitively clear because $\Delta\left\langle B^\dagger_{\mathbf{q},q_z} a^\dagger_{v,\mathbf{k}-\mathbf{q}} a_{c,\mathbf{k}}\right\rangle$ describes a microscopic process where an electron is moved from the conduction to the valence band via the correlated emission of a photon. Obviously, this process must be nearly energy conserving to be efficient.

In general, the classification into incoherent and coherent contributions is very valuable because it allows us to distinguish the transient from the long-living correlations. This information can be utilized if we want to develop efficient numerical evaluation schemes. The coherent quantities must be resolved on a fast time scale while they can be ignored at long times due to the rapid decay of coherences. The incoherent quantities must be followed for all times but the time scales involved are slower than for the coherent processes. In addition, we can identify a quantum-dynamical coupling structure that yields a transfer of energy between coherent and incoherent quantities, which helps us to understand the generation of the characteristic many-body and quantum-optical correlations.

The classification of coherent vs. incoherent quantities requires some degree of knowledge of the band structure and of the relevant photon and phonon energies. For the examples in this book, we will mostly analyze direct-gap semiconductors that can be reduced to a two-band model, see Section 24.1.2. We also assume that the photon energy matches well with the band-gap energy while the phonon energy is on the scale of the intraband transitions. A process involving a single phonon operator is classified as coherent because it is subject to strong reservoir damping in the solid-state lattice.

The arguments above identify the coherent singlet–doublet contributions of Section 24.4 as

$$\begin{aligned}
\Delta\langle 1\rangle_{\text{coh}} &= \left\{\left\langle a^\dagger_{v,\mathbf{k}}a_{c,\mathbf{k}}\right\rangle, \quad \langle B_{\mathbf{0},q_z}\rangle, \quad \langle D_{\mathbf{0},p_z}\rangle, \quad \& \quad \text{c.c.}\right\},\\
\Delta\langle 2\rangle_{\text{coh}} &= \Big\{c^{\mathbf{q},\mathbf{k}',\mathbf{k}}_{v,v,c,c}, \quad c^{\mathbf{q},\mathbf{k}',\mathbf{k}}_{v,c,c,c}, \quad c^{\mathbf{q},\mathbf{k}',\mathbf{k}}_{v,v,v,c}, \quad \Delta\langle B_{\mathbf{q},q_z}B_{-\mathbf{q},q'_z}\rangle,\\
&\quad \Delta\langle D_{\mathbf{p},p_z}D_{-\mathbf{p},p'_z}\rangle, \quad \Delta\left\langle B^\dagger_{\mathbf{q},q_z}a^\dagger_{c,\mathbf{k}-\mathbf{q}}a_{v,\mathbf{k}}\right\rangle, \quad \Delta\left\langle B^\dagger_{\mathbf{q},q_z}a^\dagger_{\lambda,\mathbf{k}-\mathbf{q}}a_{\lambda,\mathbf{k}}\right\rangle,\\
&\quad \Delta\left\langle D^\dagger_{\mathbf{p},p_z}a^\dagger_{\lambda,\mathbf{k}-\mathbf{q}}a_{\lambda'\neq\lambda,\mathbf{k}}\right\rangle, \Delta\left\langle B^\dagger_{\mathbf{q},q_z}D_{\mathbf{q},p_z}\right\rangle, \quad \Delta\langle B_{\mathbf{q},q_z}D_{-\mathbf{q},p_z}\rangle, \quad \& \quad \text{c.c.}\Big\}.
\end{aligned} \tag{25.28}$$

The remaining contributions,

$$\begin{aligned}
\Delta\langle 1\rangle_{\text{inc}} &= \left\{\left\langle a^\dagger_{c,\mathbf{k}}a_{c,\mathbf{k}}\right\rangle \equiv f^e_{\mathbf{k}}, \quad \left\langle a^\dagger_{v,\mathbf{k}}a_{v,\mathbf{k}}\right\rangle \equiv 1 - f^h_{\mathbf{k}}\right\},\\
\Delta\langle 2\rangle_{\text{inc}} &= \Big\{c^{\mathbf{q},\mathbf{k}',\mathbf{k}}_{v,c,v,c}, \quad c^{\mathbf{q},\mathbf{k}',\mathbf{k}}_{c,c,c,c}, \quad c^{\mathbf{q},\mathbf{k}',\mathbf{k}}_{v,v,v,v}, \quad \Delta\left\langle B^\dagger_{\mathbf{q},q_z}B_{\mathbf{q},q'_z}\right\rangle, \quad \Delta\left\langle D^\dagger_{\mathbf{p},p_z}D_{\mathbf{p},p'_z}\right\rangle,\\
&\quad \Delta\left\langle B^\dagger_{\mathbf{q},q_z}a^\dagger_{v,\mathbf{k}-\mathbf{q}}a_{c,\mathbf{k}}\right\rangle, \Delta\left\langle D^\dagger_{\mathbf{p},p_z}a^\dagger_{\lambda,\mathbf{k}-\mathbf{q}}a_{\lambda,\mathbf{k}}\right\rangle, \&\ \text{sym}\Big\},
\end{aligned} \tag{25.29}$$

are incoherent.

The coherent and incoherent classes of doublets also include correlations that are obtained from those explicitly written via generic symmetry transformations, indicated by "sym":

$$\begin{aligned}
c^{\mathbf{q},\mathbf{k}',\mathbf{k}}_{\lambda,\nu;\nu',\lambda'} &= c^{-\mathbf{q},\mathbf{k},\mathbf{k}'}_{\nu,\lambda;\lambda',\nu'} = -c^{\mathbf{k}-\mathbf{k}'-\mathbf{q},\mathbf{k}',\mathbf{k}}_{\lambda,\nu;\lambda',\nu'}\\
&= -c^{\mathbf{k}'+\mathbf{q}-\mathbf{k},\mathbf{k},\mathbf{k}'}_{\nu,\lambda;\lambda',\nu'} = \left[c^{-\mathbf{q},\mathbf{k}'+\mathbf{q},\mathbf{k}-\mathbf{q}}_{\lambda',\nu';\nu,\lambda}\right]^\star,
\end{aligned} \tag{25.30}$$

which follow from the Fermionic commutation relations of $a_{\lambda_j,\mathbf{k}_j}$ and the complex conjugation of the correlations. The division into coherent and incoherent quantities is not needed to produce and to solve the singlet–doublet dynamics as such. However, this distinction is very useful if we want to distinguish transient from long-lasting quantum dynamics, which helps us in developing efficient numerical simulations.

25.3.2 Coherence in quantum optics

The concept of coherence somewhat depends on the physics investigated. In semiconductor quantum optics, it is good to adopt the pragmatic approach separating the transient from long-living contributions. In pure quantum optics, one often applies a more sophisticated distinction where one studies n-th order coherences that are deduced from quantum-statistical correlations rather than from the lifetimes of the excitations in matter.

Nonetheless, the excitation- and quantum-optics-based coherences have some clear connections. For example, the analysis above shows that $\langle B_{q_z,\mathbf{q}} \rangle$ is a coherent quantity. If the laser field is defined entirely by $\langle B_{q_z,\mathbf{q}} \rangle$, it must have the quantum statistics of a coherent state, introduced in Section 22.3.1. In the quantum-optical sense, a field displays n-th-order coherence only if

$$I_n^n \equiv \langle [B^\dagger]^n B^n \rangle = \langle B^\dagger \rangle^n \langle B \rangle^n \tag{25.31}$$

is satisfied. A coherent state fulfills this to all orders because the condition (25.31) is a subset of the classical factorization (22.29) that uniquely defines the coherent state, as shown in Section 22.3. Thus, the identification of $\langle B_{q_z,\mathbf{q}} \rangle$ as coherent and the definition of n-th-order coherence are related. Since the coherent state is the best possible representation of classical fields, it naturally shows the usual interference effects when combined with another beam of coherent light, which connects coherence to interference effects.

Generally, the excitation- and quantum-optics-based definitions of coherence refer to different classes of phenomena. For example, $\Delta\langle B^\dagger B^\dagger B\, B \rangle$ is always an incoherent quantity based on the excitation-based definition. Yet, one can study the second-order coherence via the $g^{(2)}$ function,

$$g^{(2)} \equiv \frac{\langle B^\dagger B^\dagger B\, B \rangle}{\langle B^\dagger B \rangle \langle B^\dagger B \rangle}. \tag{25.32}$$

Using this definition, a coherent state produces $g^{(2)} = 1$ while quantum light can display the effect of antibunching with $g^{(2)} < 1$. These phenomena relate to the quantum-statistical concepts of coherence in light sources and should not be confused with the excitation-based definition that is applied to classify all the correlations generated in semiconductor quantum optics.

In general, the $g^{(n)}$ of the light source can be used to deduce the largest number of clusters needed to describe the desired phenomenon. However, this quantum-statistical coherence cannot be used to find simplifying additional structures within the dynamics of the C-particle clusters. This can be done only by using the excitation-based definition of coherent and incoherent correlations. Thus, we always start from the excitation-based coherence definition in order to find subsets within the correlated clusters. As we separate the coherent from the incoherent quantities, we know which contributions must be computed with higher accuracy in the transient regime. We then apply concepts like $g^{(n)}$ to characterize what kind of light emission the semiconductor produces, not to classify the equation structure. This analysis will eventually lead to the concept of quantum-optical spectroscopy, introduced in Chapter 23 where the quantum-statistical coherence clearly plays a central role.

25.4 Temporal aspects in semiconductor excitations

Our model calculations summarized in Fig. 25.2(d) show that also the duration T of the optical pulse influences the excited state because, e.g., γT defines the relative strength of polarization and densities based on the pulse duration in units of the characteristic coherence time $\tau_{\rm coh} \equiv 1/\gamma$. Besides its duration, the light pulse has the additional relevant time scale related to its characteristic oscillations. In free space, the light propagation results from

$$\langle B_{q_z,\mathbf{0}}\rangle \equiv \langle \bar{B}_{q_z,\mathbf{0}}\rangle\, e^{-i\omega_q t}, \qquad \left\langle B^{\dagger}_{q_z,\mathbf{0}}\right\rangle \equiv \left\langle \bar{B}^{\dagger}_{q_z,\mathbf{0}}\right\rangle\, e^{+i\omega_q t}. \tag{25.33}$$

Hence, $\tau_{\rm opt} \equiv 1/\omega_q$ defines a time scale for optical oscillations while $\bar{B}$ includes the time scale of the envelope part related to the pulse duration $T_{\rm env}$.

To estimate how fast the optical oscillations are relative to $T_{\rm env}$, we analyze $e^{-i\omega_q\,t}$ by assuming that the typical photon energy, $\hbar\omega_{\rm opt}$, matches the band gap of the semiconductor, i.e., $\hbar\omega_{\rm opt} = E_g = \hbar\omega_{g,0}$. For $E_g = 1$ eV, we find $\omega_{\rm opt} \equiv \frac{1\,{\rm eV}}{\hbar} = 2\pi \times 2.4180 \times 10^{14}\,\frac{1}{\rm s}$ resulting in $\tau_{\rm opt} = 0.6582$ fs. Thus, we obtain the order-of-magnitude relations

$$\omega_{\rm opt} \equiv 2\pi \times 2.4180 \times 10^{14}\,\frac{1}{\rm s} \times \left[\frac{E_g}{1{\rm eV}}\right], \quad \tau_{\rm opt} \equiv 0.6582\,{\rm fs} \times \left[\frac{1{\rm eV}}{E_g}\right]. \tag{25.34}$$

These estimates show that the typical time scale of the optical oscillations, $\tau_{\rm opt}$, is much faster than the characteristic times of the optical transitions observed, e.g., in Fig. 25.3.

In many relevant cases, the optical pulses are much longer than $\tau_{\rm opt}$. Examples of typical temporal pulses are plotted in Fig. 25.4(a)–(b) assuming a Gaussian envelope $\bar{A}(t) = A_0\, e^{-\frac{t^2}{T_{\rm env}^2}}$ (shaded area) together with the actual $A(t) = \bar{A}(t)\cos\omega_{\rm opt}t$ field (solid line) for: (a) $T_{\rm env} = 10$ fs and (b) $T_{\rm env} = 2.5$ fs pulses. The central frequency corresponds to $\hbar\omega_{\rm opt} = 1$ eV. The dashed line indicates the relative correction $\delta E(t)$ of the envelope contributions with respect to $e^{-i\omega_{\rm opt}t}$ oscillations when the relation $E = -\partial A/\partial t$ is used, compare Section 25.4.3.

In Fig. 25.4(a)–(b), we notice multiple field oscillations under the pulse envelope. Only when $T_{\rm env}$ approaches the sub-fs regime does the envelope part become faster than the $\tau_{\rm opt}$ oscillations, giving rise to single-cycle or even shorter pulses. The spectrum of the Gaussian $A(t) = \bar{A}(t)\cos\omega_{\rm opt}t$ is

$$A(\omega) \equiv \frac{1}{\sqrt{2\pi}}\int dt\, A(t)\, e^{+i\omega t} = \frac{T_{\rm env}A_0}{\sqrt{2}}\, \frac{e^{-\frac{T_{\rm env}^2}{4}(\omega-\omega_{\rm opt})^2} + e^{-\frac{T_{\rm env}^2}{4}(\omega+\omega_{\rm opt})^2}}{2} \tag{25.35}$$

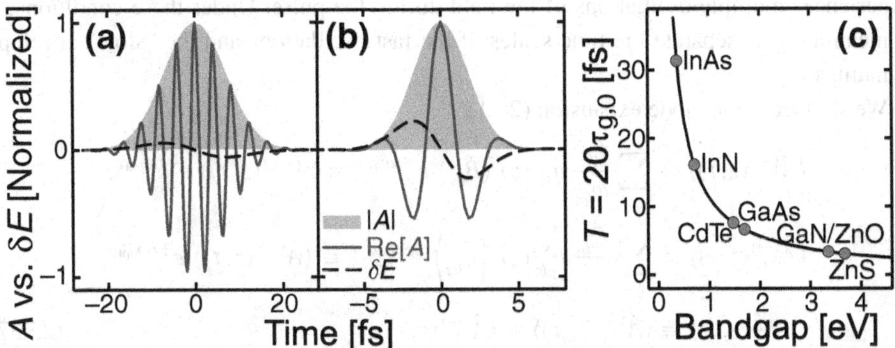

Figure 25.4 Time scale of optical oscillations vs. envelope dynamics for a field with $\hbar\omega_{\rm opt} = 1\,{\rm eV}$. The shaded area shows the envelope $|A(t)|$ and the solid line presents the actual $A(t)$ for: **(a)** $T = 10\,{\rm fs}$- and **(b)** $T = 2.5\,{\rm fs}$-long Gaussian pulses. The dashed line indicates the relative magnitude of $\delta E(t) = -\frac{\partial \bar{A}}{\partial t}/[\omega_{\rm opt}\bar{A}]$ correction. **(c)** The long-pulse limit $T_{\rm env} = 20\,\tau_{\rm opt}$ (solid line) is plotted as a function of the band gap. The circles indicate typical band gap values for InAs (0.36 eV), InN (0.7 eV), GaAs (1.48 eV), CdTe (1.7 eV), GaN (3.36 eV), ZnO (3.36 eV), and ZbS (3.68 eV).

showing peaks around the frequencies $\omega = \pm\omega_{\rm opt}$ with a width of $\Delta\omega = 2/T_{\rm env}$. The energy spectrum of the pulse is very broad if $\Delta E = \hbar\Delta\omega = 2\hbar/T_{\rm env}$ is larger than the central energy. This condition sets a limit between the envelope and oscillation time scales,

$$\Delta E > \hbar\omega_{\rm opt} \quad \Leftrightarrow \quad T_{\rm env} < 2\tau_{\rm osc}. \tag{25.36}$$

Thus, subsingle-cycle pulses are always spectrally very broad. For semiconductors, the corresponding ΔE can range over several band gap energies. Consequently, subsingle-cycle pulses are simultaneously resonant with multiple bands. To study only excitations near the band edge with only two bands, we must restrict the analysis to multicycle pulses with $T_{\rm env} \gg \tau_{\rm osc}$.

We choose $T_{20} = 20\,\tau_{\rm opt}$ to define a rough limit for multicycle pulses allowing for a two-band analysis. Figure 25.4(c) presents the T_{20} limit as a function of the band gap. The circles indicate typical band gaps for the binary semiconductor compounds InAs, InN, GaAs, CdTe, GaN, ZnO, and ZbS. For our GaAs-based investigations, multicycle pulses have to be longer than 10 fs in order to justify the approximation via a two-band model. Such pulse durations ranging from 10 fs to 1 ps are still fast enough to induce nontrivial and ultrafast quantum kinetic effects in the system, see also Fig. 3.1. Thus, we may study ultrafast two-band excitation effects in GaAs-type nanostructures with pulses that have roughly 20 to 2000 oscillations within the pulse.

25.4.1 Rotating-wave approximation

In our investigations, $\tau_{\rm opt}$ is one to three orders of magnitude faster than both the coherence time $\tau_{\rm coh}$ and the typical pulse duration $T_{\rm env}$. As a result, each microscopic process experiences multiple oscillations of the field during the pulse. Under these conditions, it is meaningful to separate the time scales of the fast oscillations and the "slow" envelope dynamics.

We start from the mode expansion (24.1),

$$\langle \hat{A}^{(+)}(z,t)\rangle \equiv \sum_{q_z} \frac{\mathcal{E}_{q_z}}{\omega_{q_z}} u_{q_z}(z)\, \langle \bar{B}_{q_z}\rangle\, e^{-i\omega_{q_z} t} \equiv \langle \bar{A}^{(+)}(z,t)\rangle\, e^{-i\omega_{\rm opt} t},$$

$$\langle \hat{A}^{(-)}(z,t)\rangle \equiv \sum_{q_z} \frac{\mathcal{E}_{q_z}}{\omega_{q_z}} u^{\star}_{q_z}(z)\, \left\langle \bar{B}^{\dagger}_{q_z}\right\rangle\, e^{+i\omega_{q_z} t} \equiv \langle \bar{A}^{(-)}(z,t)\rangle\, e^{+i\omega_{\rm opt} t},$$

$$\langle \hat{A}(z,t)\rangle \equiv \langle \hat{A}^{(+)}(z,t)\rangle + \langle \hat{A}^{(-)}(z,t)\rangle, \tag{25.37}$$

where $\langle \hat{A}^{(+)}(z)\rangle$ contains the contributions oscillating with $e^{-i\omega_{\rm opt} t}$ while $\langle \hat{A}^{(-)}(z)\rangle$ has only the $e^{+i\omega_{\rm opt} t}$-like terms, where $\omega_{\rm opt}$ refers to the central frequency of the light pulse. We have chosen the signs $\hat{A}^{(\pm)}$ according to the convention where "+" refers to the clockwise oscillations in the complex plane while "–" is associated with the counterclockwise direction. Based on Eq. (25.4), an analogous identification can be made for the Rabi frequency

$$\Omega(t) \equiv \Omega^{(+)}(t) + \Omega^{(-)}(t), \qquad \Omega^{(\pm)}(t) \equiv \bar{\Omega}^{(\pm)}(t)\, e^{\mp i\omega_{\rm opt} t}, \tag{25.38}$$

where $\bar{\Omega}^{(\pm)}(t)$ denotes the slowly varying envelope. The usefulness of this separation follows after we consider how it simplifies the equations for the semiconductor excitation dynamics.

For this purpose, we analyze the linear polarization response (25.11) producing the integral

$$P^{v,c}_{\mathbf{k}}(t) = i e^{-(i\omega_{g,\mathbf{k}}+\gamma)t} \int_{-\infty}^{t} du\, \Omega(u)\, e^{(i\omega_{g,\mathbf{k}}+\gamma)u}\,, \tag{25.39}$$

which can be decomposed into

$$\begin{aligned} P^{v,c}_{\mathbf{k}}(t) &= i e^{-(i\omega_{g,\mathbf{k}}+\gamma)t} \int_{-\infty}^{t} du\, \left[\Omega^{(+)}(u) + \Omega^{(-)}(u)\right]\, e^{(i\omega_{g,\mathbf{k}}+\gamma)u} \\ &= i e^{-(i\omega_{g,\mathbf{k}}+\gamma)t} \int_{-\infty}^{t} du\, \left[\bar{\Omega}^{(+)}(u)\, e^{(i[\omega_{g,\mathbf{k}}-\omega_{\rm opt}]+\gamma)u} \right. \\ &\quad \left. +\bar{\Omega}^{(-)}(u)\, e^{(i[\omega_{g,\mathbf{k}}+\omega_{\rm opt}]+\gamma)u}\right] \end{aligned} \tag{25.40}$$

with the help of Eq. (25.38). We see that $e^{(i[\omega_{g,\mathbf{k}}-\omega_{\rm opt}]+\gamma)u}$ is a slowly varying integrand for near-resonant conditions with $\omega_{\rm opt} \approx \omega_{g,\mathbf{k}}$. The second term is clearly nonresonant because the $e^{(i[\omega_{g,\mathbf{k}}+\omega_{\rm opt}]+\gamma)u}$ part oscillates rapidly. Thus, only the resonant term contributes to the significant build-up of polarization while the nonresonant term almost averages to zero in the integration.

We encounter formally similar integrals in Section 18.1.2. In particular, the conversion of the integral (18.28) into (18.29) shows that only the $\Omega^{(+)}$ parts generate transitions. Thus, we can accurately approximate the integral by including only the resonant contribution

$$P^{v,c}_{\mathbf{k}}(t)|_{\text{RWA}} = ie^{-(i\omega_{g,\mathbf{k}}+\gamma)t}\int_{-\infty}^{t} du\, \bar{\Omega}^{(+)}(u)\, e^{(i[\omega_{g,\mathbf{k}}-\omega_{\text{opt}}]+\gamma)u}. \tag{25.41}$$

This corresponds to the rotating-wave approximation (RWA) discussed in Section 18.1.4. Its overall validity is investigated in Fig. 18.4 showing that the RWA can be applied for near-resonance conditions. It is also clear that if $\bar{\Omega}^{(+)}(t)$ becomes too short, we cannot leave out the contribution $\bar{\Omega}^{(-)}(t)$. However, the RWA is extremely accurate for optical multicycle excitations.

25.4.2 Separation of time scales

In the following, we concentrate on cases where the excitation pulses are resonant and sufficiently long. In this situation, the singlet-dynamics (25.2)–(25.3) can be solved accurately by applying the RWA,

$$\begin{cases} i\frac{\partial}{\partial t}P^{v,c}_{\mathbf{k}} = \omega_{g,\mathbf{k}}\, P^{v,c}_{\mathbf{k}} - \left(1 - f^e_{\mathbf{k}} - f^h_{\mathbf{k}}\right)\bar{\Omega}^{(+)}e^{-i\omega_{\text{opt}}t} \\ \frac{\partial}{\partial t}f^e_{\mathbf{k}} = \frac{\partial}{\partial t}f^h_{\mathbf{k}} = 2\text{Im}\left[[\bar{\Omega}^{(+)}(t)]^{\star}P^{v,c}_{\mathbf{k}}\, e^{+i\omega_{\text{opt}}t}\right] \end{cases}, \tag{25.42}$$

which simply eliminates the $\bar{\Omega}^{(-)}(t)$ contributions. It is now beneficial to express the polarization in the rotating frame,

$$P^{v,c}_{\mathbf{k}} \equiv \bar{P}^{v,c}_{\mathbf{k}}\, e^{-i\omega_{\text{opt}}t}, \tag{25.43}$$

defined by the central light frequency. This converts the optical Bloch equations into

$$\begin{cases} i\frac{\partial}{\partial t}\bar{P}^{v,c}_{\mathbf{k}} = (\omega_{g,\mathbf{k}} - \omega_{\text{opt}})\, \bar{P}^{v,c}_{\mathbf{k}} - \left(1 - f^e_{\mathbf{k}} - f^h_{\mathbf{k}}\right)\bar{\Omega}^{(+)} \\ \frac{\partial}{\partial t}f^e_{\mathbf{k}} = \frac{\partial}{\partial t}f^h_{\mathbf{k}} = 2\text{Im}\left[[\bar{\Omega}^{(+)}(t)]^{\star}\bar{P}^{v,c}_{\mathbf{k}}\right] \end{cases}, \tag{25.44}$$

which contains only the envelope and incoherent quantities. Thus, as a consequence of the RWA and transforming into the rotating frame, we obtain a separation of time scales allowing us to follow the excitation dynamics on the relatively slow scale of T_{env}. As a consequence, we gain considerable speed in computations of the temporal development of the semiconductor excitations. The increase in computation speed scales roughly like $T_{\text{env}}/\tau_{\text{opt}}$, which can easily be a factor of a hundred in realistic computations. The validity of the approximations is limited by the non-RWA contributions that are often negligible, as shown in Fig. 18.4.

A fully self-consistent singlet analysis requires that we also solve the wave equation (24.48),

$$\left[\frac{\partial^2}{\partial z^2} - \frac{n^2(z)}{c^2}\frac{\partial^2}{\partial t^2}\right]\left(\langle\bar{A}^{(+)}(z)\rangle\, e^{-i\omega_{\text{opt}}t} + \langle\bar{A}^{(-)}(z)\rangle\, e^{+i\omega_{\text{opt}}t}\right)$$
$$= -\mu_0\delta(z)\frac{\partial}{\partial t}\left(\left\langle\bar{P}^{v,c}_{\mathbf{0}}\right\rangle e^{-i\omega_g t} + \left\langle\bar{P}^{c,v}_{\mathbf{0}}\right\rangle e^{+i\omega_g t}\right), \tag{25.45}$$

where we have separated the envelope variations and the optical oscillations. Due to the linearity of the field dynamics, it can be treated equally well using the complex-valued wave equation

$$\left[\frac{\partial^2}{\partial z^2} - \frac{n^2(z)}{c^2}\frac{\partial^2}{\partial t^2}\right]\langle \hat{A}^{(+)}(z)\rangle = -\mu_0 \delta(z)\frac{\partial}{\partial t}\left(\left\langle \bar{P}_{\mathbf{0}}^{v,c}\right\rangle e^{-i\omega_{\mathrm{opt}}t}\right), \tag{25.46}$$

which includes only clockwise optical oscillations. The physical real-valued solutions are obtained by taking the real part $\langle \hat{A}\rangle = 2\mathrm{Re}\left[\langle \hat{A}^{(+)}\rangle\right]$.

As discussed in connection with Eq. (17.25), the RHS of Eq. (25.46) produces a planar dipole emission while the LHS describes the propagation of light within the dielectric environment. For constant $n(z) = n$, we may directly apply the results (17.33)–(17.34) to obtain an exact solution

$$\begin{aligned} \langle \hat{A}^{(+)}(z,t)\rangle &= \langle \hat{A}^{(+)}(z,t)\rangle_0 + \frac{\mu_0 c}{2n}\left\langle \bar{P}_{\mathbf{0}}^{v,c}\left(t - \frac{n|z|}{c}\right)\right\rangle e^{-i\omega_{\mathrm{opt}}t} \\ \Rightarrow \quad \langle \bar{A}^{(+)}(0,t)\rangle &= \langle \bar{A}^{(+)}(0,t)\rangle_0 + \frac{\mu_0 c}{2n}\left\langle \bar{P}_{\mathbf{0}}^{v,c}\right\rangle, \end{aligned} \tag{25.47}$$

where $\langle \bar{A}^{(+)}(0,t)\rangle_0$ defines the incoming field. We see now that even the self-consistent field can be described through its envelope. A similar solution can also be constructed for a generic $n(z)$, as discussed in Exercise 25.10.

25.4.3 Electrical field and dipole interaction

In the singlet analysis, it is customary to express the optical field through its electrical field rather than the vector potential. For this purpose, one needs to evaluate

$$\begin{aligned} \langle \hat{E}^{(+)}(0,t)\rangle &\equiv -\frac{\partial}{\partial t}\langle \hat{A}^{(+)}(0,t)\rangle \\ &= -\frac{\partial}{\partial t}\left[\langle \bar{A}^{(+)}(0,t)\rangle_0 e^{-i\omega_{\mathrm{opt}}t} + \frac{\mu_0 c}{2n}\left\langle \bar{P}_{\mathbf{0}}^{v,c}\right\rangle e^{-i\omega_{\mathrm{opt}}t}\right], \end{aligned} \tag{25.48}$$

based on $\hat{\mathbf{E}} = -\frac{\partial}{\partial t}\hat{\mathbf{A}}$ and (25.47). The envelope of the electric field is then

$$\begin{aligned} \langle \bar{E}^{(+)}(0,t)\rangle &\equiv e^{+i\omega_{\mathrm{opt}}t}\,\langle \hat{E}^{(+)}(0,t)\rangle \\ &= i\omega_{\mathrm{opt}}\langle \bar{A}^{(+)}(0,t)\rangle_0 - \frac{\mu_0 c}{2n}\frac{1}{S}\sum_{\mathbf{k}}\frac{Qp_{v,c}}{m_0}\bar{P}_{\mathbf{k}}^{v,c} \\ &\quad - \frac{\partial\langle \bar{A}^{(+)}(0,t)\rangle_0}{\partial t}, \end{aligned} \tag{25.49}$$

where the derivative of the polarization is evaluated using (24.19). The last term in (25.49) is typically very small because the envelope of the field varies much more slowly than the optical cycle, as long as the condition $T_{\mathrm{env}} \gg \tau_{\mathrm{osc}}$ is satisfied. To estimate the size of this contribution, we evaluate the ratio

$$\delta E(t) \equiv \frac{1}{\omega_{\mathrm{opt}}\langle \bar{A}^{(+)}(0,t)\rangle_0}\frac{\partial\langle \bar{A}^{(+)}(0,t)\rangle_0}{\partial t}. \tag{25.50}$$

The size of $\delta E(t)$ is plotted in Fig. 25.4(a)–(b) as a dashed line. There, we see that these corrections indeed become negligible for longer pulses. As a result, the incoming electrical field can be accurately approximated through

$$\langle \bar{E}^{(+)}(0,\,t)\rangle_0 \to i\omega_{\rm opt}\langle \bar{A}^{(+)}(0,\,t)\rangle_0 \tag{25.51}$$

for the sufficiently long excitation pulses studied here.

As shown in connection with Eq. (16.46), the momentum matrix element can be converted into

$$\frac{Q\mathbf{p}_{v,c}}{m_0} = \frac{E_c - E_v}{i\hbar}\langle v|Q\hat{\mathbf{x}}|c\rangle \equiv -i\omega_{g,0}\,\mathbf{d}_{v,c}. \tag{25.52}$$

The last step identifies the dipole matrix element (16.42). As we combine this result with Eq. (25.49), we find

$$\langle \bar{E}^{(+)}(0,\,t)\rangle = \langle \bar{E}^{(+)}(0,\,t)\rangle_0 + i\omega_{g,0}\frac{\mu_0 c}{2n}\frac{1}{S}\sum\nolimits_{\mathbf{k}} d_{v,c}\bar{P}^{v,c}_{\mathbf{k}}. \tag{25.53}$$

It is straightforward to show that the wave equation for the electric field,

$$\left[\frac{\partial^2}{\partial z^2} - \frac{n^2(z)}{c^2}\frac{\partial^2}{\partial t^2}\right]\langle \hat{E}^{(+)}(z)\rangle = \mu_0\delta(z)\frac{\partial^2}{\partial t^2}\left(\frac{1}{S}\sum\nolimits_{\mathbf{k}} d_{v,c}\,P^{v,c}_{\mathbf{k}}\right), \tag{25.54}$$

produces the same result as long as $\omega_{g,0}$ and $\omega_{\rm opt}$ are nearly identical. Obviously, the conditions needed to satisfy the two-band approximation, i.e., near-resonance and $T_{\rm env} \gg \tau_{\rm osc}$, allow us to replace $\hat{A}$ by $\hat{E}$ in a straightforward manner.

More generally, whenever we are justified to use a two-band model, we can always utilize the operator-level conversions

$$\frac{Qp_{\lambda,\bar{\lambda}}}{m_0}\hat{A} \to d_{\lambda,\bar{\lambda}}\hat{E}, \qquad \hat{P}^{\lambda,\bar{\lambda}}_{\mathbf{q}} \to \frac{1}{S}\sum\nolimits_{\mathbf{k}} d_{\lambda,\bar{\lambda}}\,a^\dagger_{\lambda,\mathbf{k}}a_{\bar{\lambda},\mathbf{k}+\mathbf{q}}, \tag{25.55}$$

without introducing additional approximations. This, e.g., converts the Rabi frequency (25.4) into

$$\begin{aligned}\hbar\bar{\Omega}^{(+)}(t) &= \frac{Qp_{c,v}}{m_0}\langle \bar{A}^{(+)}(0,t)\rangle = \frac{\omega_{g,0}}{\omega_{\rm opt}}\left[\frac{Qp_{c,v}}{i\omega_{g,0}\,m_0}\right]\left[i\omega_{\rm opt}\langle \bar{A}^{(+)}(0,t)\rangle\right]\\ &\to d_{c,v}\langle \bar{E}^{(+)}(0,t)\rangle, \qquad \text{for } \omega_{\rm opt} \to \omega_{g,0}.\end{aligned} \tag{25.56}$$

These considerations essentially show the equivalence of the $\hat{\mathbf{A}}\cdot\hat{\mathbf{p}}$- and the $\hat{\mathbf{E}}\cdot\hat{\mathbf{r}}$-pictures for optical excitations within a two-band model. For more general investigations, one should apply the Göppert-Mayer transformation introduced in Section 17.4.1. This transformation yields the appropriate modifications to the system Hamiltonian. This extension is necessary when the optical pulse is either resonant with multiple bands or nonresonant with all bands. If the excitation pulse has an excessively large intensity, it can also induce multiphoton transitions between nonresonant and dipole-forbidden transitions. Clearly, such investigations require an extension beyond the two-band model.

Exercises

25.1 **(a)** Convert the generic Eq. (24.58) into the two-band model Eqs. (25.2)–(25.3) by omitting the doublets and the Coulomb interactions. **(b)** Make the identifications (25.4) and (25.6) and reproduce Eq. (25.7) using the results derived in part **(a)**.

25.2 **(a)** Show that the solution (25.11) fulfills Eq. (25.10) and verify the result (25.14). **(b)** Verify the relation (25.18) and show that Eq. (25.19) satisfies (25.15).

25.3 Use the square pulse (25.20) in Eqs. (25.13) and (25.17). Show that the integration produces Eqs. (25.21)–(25.22).

25.4 **(a)** Convert Eq. (25.5) into the rotating frame via the transformation $S^-_{\mathbf{k}} = \bar{S}^-_{\mathbf{k}}\, e^{-i\omega_{g,\mathbf{k}}t}$ and $\Omega^{(+)}(t) = \bar{\Omega}^{(+)}(t)\, e^{-i\omega_{g,\mathbf{k}}t}$. **(b)** Assume the resonant pumping $\bar{\Omega}^{(+)}(t) = i\Omega_0\, \theta(t)$ and show that $\bar{S}^-_{\mathbf{k}} = -\frac{1}{2}\sin 2\Omega_0 t$ and $S^z_{\mathbf{k}} = -\frac{1}{2}\cos 2\Omega_0 t$ for $t > 0$ when $S^-_{\mathbf{k}}(0) = 0$ and $S^z_{\mathbf{k}}(0) = -1/2$. Conclude that this scenario corresponds to Rabi flopping. **(c)** Taylor expand $\bar{S}^-_{\mathbf{k}}(t)$ within $P^{v,c}_{\mathbf{k}} = \bar{S}_{\mathbf{k}}(t)\, e^{-i\omega_{g,\mathbf{k}}t}$ and show that the first-order contribution reproduces the result (25.14) if we set $\gamma = 0$. Show that one needs terms up to all orders in Ω_0 to reproduce the Rabi flopping. **Hint:** *Take the limit $T \to \infty$ for the Taylor expanded result. Study numerically which is the largest t for which the Taylor expansion up to first, second, and third order can be used.*

25.5 **(a)** Show that the coherent-limit relation yields $f^{\mathrm{coh}}_{\mathbf{k}} = \left|P^{v,c}_{\mathbf{k}}\right|$ for low polarization levels. **(b)** Define the difference $\Delta f_{\mathbf{k}} = f_{\mathbf{k}} - f^{\mathrm{coh}}_{\mathbf{k}}$ based on the linearized results (25.21)–(25.22). Show that $\Delta f_{\mathbf{k}} \to 0$ for $\gamma \to 0$. **(c)** Show that $f^{\mathrm{coh}}_{\mathbf{k}}(T/2)$ scales like $1/(\gamma T)^2$ while $f_{\mathbf{k}}(T/2) \propto 1/\gamma T$ for large γT.

25.6 Construct the interaction-free evolution of **(a)** $\Delta\left\langle B^\dagger_{\mathbf{q},q_z} B_{\mathbf{q},q'_z}\right\rangle$, **(b)** $\Delta\left\langle B_{\mathbf{q},q_z}, B_{-\mathbf{q},q_z}\right\rangle$, and **(c)** $c^{\mathbf{q},\mathbf{k}',\mathbf{k}}_{\lambda,v;v'\lambda'} \equiv \Delta\left\langle a^\dagger_{\lambda,\mathbf{k}}\, a^\dagger_{v,\mathbf{k}'}\, a_{v',\mathbf{k}'+\mathbf{q}}\, a_{\lambda',\mathbf{k}-\mathbf{q}}\right\rangle$. Identify the coherent and incoherent doublet correlations within the two-band approximation. **(d)** Verify symmetry relations (25.30). How many structurally different correlations are found in a two-band analysis?

25.7 In quantum optics, second-order coherence is studied via the $g^{(2)}$ function identified by Eq. (25.32). **(a)** Apply the cluster expansion to express $\langle B^\dagger B\rangle$ as well as $\langle B^\dagger B^\dagger B B\rangle$ in terms of singlet–doublet–triplet–quadruplet correlations. **(b)** Which of the emerging contributions are coherent as defined in semiconductor optics? **(c)** Show that the coherent state produces $g^{(2)} = 1$ and that even incoherent correlations can produce a second-order coherence, i.e., $g^{(2)} \neq 1$.

25.8 **(a)** Show that $g^{(2)} = 1 - 1/n$ for the Fock state $|n\rangle$ and $g^{(2)} = 2$ for the thermal state with n photons on average. Why do both states have second-order coherences? **(b)** Compute $\Delta\langle (B^\dagger)^J\, B^K\rangle$ for the combinations $J, K = 0, 1, 2$ for the

Fock state $|n\rangle$ and the thermal state. Conclude that both states yield only incoherent correlations. Why is there an apparent contradiction between the classification via clusters and n-th-order coherence?

25.9 **(a)** Assume that $\mathbf{\Omega}^{(\pm)}(t)$ is a box function with magnitude Ω_0 within $-T/2 \leq t \leq T/2$; $\mathbf{\Omega}^{(\pm)}(t)$ vanishes for other times. Compute $P_k^{v,c}(T/2)$ from Eq. (25.40). Study the ratio of RWA and non-RWA contributions for the resonant case. **(b)** Show that Eq. (25.42) is compatible with the RWA (25.41). **(c)** Construct the solution (25.53) from (25.54) by applying (17.33)–(17.34) as well as the separation of time scales.

25.10 Consider the wave equation $\left[\frac{\partial^2}{\partial z^2} - \frac{n^2(z)}{c^2}\frac{\partial^2}{\partial t^2}\right]\langle E^{(+)}(z,t)\rangle = 0$ when the index step appears at $z = z_j$, i.e.,

$$n(z) = n_j \quad \text{for} \quad z_j < z \leq z_{j+1} \text{ and}$$
$$n(z) = n_{j+1} \quad \text{for} \quad z_{j+1} < z \leq z_{j+2}.$$

(a) Show that within each interval $z_j \leq z \leq z_{j+1}$, the propagation problem has a general solution $\langle E^{(+)}(z,t)\rangle = E_j^{\rightarrow}\left(z - z_j - \frac{c}{n_j}t\right) + E_j^{\leftarrow}\left(z - z_{j+1} + \frac{ct}{n_j}\right)$. **(b)** Demand that $\langle E^{(+)}(z,t)\rangle$ and $\frac{\partial}{\partial z}\langle E^{(+)}(z,t)\rangle$ are continuous and show that

$$E_{j+1}^{\rightarrow}\left(\frac{ct}{n_{j+1}}\right) = T_j^{\rightarrow} E_j^{\rightarrow}\left(\frac{c(t-\Delta t_j)}{n_j}\right) + R_j^{\leftarrow} E_{j+1}^{\leftarrow}\left(\frac{c(t-\Delta t_{j+1}}{n_{j+1}}\right)$$
$$E_j^{\leftarrow}\left(\frac{ct}{n_j}\right) = R_j^{\rightarrow} E_j^{\rightarrow}\left(\frac{c(t-\Delta t_j)}{n_j}\right) + T_j^{\leftarrow} E_{j+1}^{\leftarrow}\left(\frac{c(t-\Delta t_{j+1})}{n_{j+1}}\right)$$

where $\Delta t_j = \frac{z_{j+1}-z_j}{c}n_j$, $T_j^{\rightarrow} = \frac{2n_j}{n_j+n_{j+1}}$, $R_j^{\rightarrow} = \frac{n_j-n_{j+1}}{n_j+n_{j+1}}$, $T_j^{\leftarrow} = \frac{2n_{j+1}}{n_j+n_{j+1}}$, and $R_j^{\leftarrow} = \frac{n_{j+1}-n_j}{n_j+n_{j+1}}$. **(c)** How can one apply the result of part **(b)** to generate the transfer-matrix solution in the time domain? How can the QW or QWI polarization be added exactly? **Hint:** *Discretize the space into Δz_j intervals with equal optical length. Apply the result* (25.53) *to describe the reradiation of the polarization.*

Further reading

The dynamics controlled truncation (DCT) is based on the scaling between the exciting light field E and carrier quantities it generates. See the discussion in Section 15.4 and the references in Chapter 15.

The fully quantum kinetic analysis of Coulomb and phonon scattering rates is reviewed, e.g., in:

M. Kira and S. W. Koch (2006). Many-body correlations and excitonic effects in semiconductor spectroscopy, *Prog. Quantum Electr.* **30**, 155.

The concept of quantum-optical coherence was first formulated in:

R. J. Glauber (1963). The quantum theory of optical coherence, *Phys. Rev.* **130**, 2529.

For textbook discussions of quantum-optical coherence, cf. the books listed at the end of Chapter 6 and:

L. Mandel and E. Wolf (1995). *Optical Coherence and Quantum Optics*, Cambridge, Cambridge University Press.

The $g^{(2)}$ of a light field was first measured in the so-called Hanbury Brown–Twiss setup:

R. Hanbury Brown and R. Q. Twiss (1956). A test of a new type of stellar interferometer on Sirius, *Nature* **178**, 1046.

Photon antibunching was first reported in:

H. J. Kimble, M. Dagenais, and L. Mandel (1977). Photon antibunching in resonance fluorescence, *Phys. Rev. Lett.* **39**, 691.

The explicit Göppert-Mayer transformation in two-band semiconductors is worked out, e.g., in:

M. Kira, F. Jahnke, W. Hoyer, and S. W. Koch (1999). Quantum theory of spontaneous emission and coherent effects in semiconductor microstructures, *Prog. Quantum Electron.* **23**, 189.

26

Maxwell-semiconductor Bloch equations

For a more quantitative description of semiconductor optics, we need to extend the simple model of the previous chapter by additionally including the many-body Coulomb effects. Fortunately, the qualitative arguments detailing which bands are coupled by light, which correlations are coherent or incoherent, and when to apply the rotating-wave approximation remain valid. However, the Coulomb interaction defines the exact strength and spectral shape of the optical transitions and it determines the scattering effects and the related coherence times. Beyond this, the simple model has to be further expanded whenever quantum-optical aspects become significant.

In this chapter, we treat the case where a coherent-state laser pulse excites a planar semiconductor nanostructure. For this analysis, we have to include the Coulomb, photon, and phonon effects into the optical Bloch equations (25.2)–(25.3). All this will be done directly at the level of the singlet dynamics derived in Section 24.5.

26.1 Semiconductor Bloch equations

We start from Eqs. (24.48), (24.49), and (24.51) which have been derived for homogeneous excitation perpendicular to the quantum well (QW) or the quantum-wire (QWI) array. In this situation, the relevant singlets are the microscopic polarization

$$P_{\mathbf{k}} \equiv P_{\mathbf{k}}^{v,c} = \left\langle a_{v,\mathbf{k}}^{\dagger} a_{c,\mathbf{k}} \right\rangle, \tag{26.1}$$

as well as electron and hole distributions, $f_{\mathbf{k}}^{e} = P_{\mathbf{k}}^{c,c}$ and $f_{\mathbf{k}}^{h} = 1 - P_{\mathbf{k}}^{v,v}$, respectively. These were originally defined by Eqs. (24.15)–(24.17) implementing the electron–hole picture, compare Fig. 24.2. The polarization is a coherent quantity while $f_{\mathbf{k}}^{e}$ and $f_{\mathbf{k}}^{h}$ are incoherent, as discussed in Section (25.3). For simplicity, we assume that the laser light is near resonant only with one dipole-allowed transition such that we can apply the two-band model. In this situation, the polarization can be identified through $P_{\mathbf{k}}$ without the band indices.

The full dynamics of the relevant singlets is derived in Section 24.4. The inclusion of the Coulomb, photon, and phonon interactions generalizes the optical Bloch equations (25.42) to the semiconductor Bloch equations (SBE),

$$i\hbar\frac{\partial}{\partial t}P_{\mathbf{k}} = \tilde{\epsilon}_{\mathbf{k}}P_{\mathbf{k}} - \left[1 - f^e_{\mathbf{k}} - f^h_{\mathbf{k}}\right]\Omega^{(+)}_{\mathbf{k}} + \Gamma^{v,c}_{\mathbf{k}} + \Gamma^{\mathrm{QED}}_{v,c;\mathbf{k}}, \tag{26.2}$$

$$\hbar\frac{\partial}{\partial t}f^e_{\mathbf{k}} = 2\mathrm{Im}\left[P_{\mathbf{k}}\left[\Omega^{(+)}_{\mathbf{k}}\right]^\star + \Gamma^{c,c}_{\mathbf{k}} + \Gamma^{\mathrm{QED}}_{c,c;\mathbf{k}}\right], \tag{26.3}$$

$$\hbar\frac{\partial}{\partial t}f^h_{\mathbf{k}} = 2\mathrm{Im}\left[P_{\mathbf{k}}\left[\Omega^{(+)}_{\mathbf{k}}\right]^\star - \Gamma^{v,v}_{\mathbf{k}} - \Gamma^{\mathrm{QED}}_{v,v;\mathbf{k}}\right], \tag{26.4}$$

based on Eq. (24.58). The doublets $\Gamma^{\lambda,\lambda'}_{\mathbf{k}}$ and $\Gamma^{\mathrm{QED}}_{\mathbf{k}}$ that appear are discussed in Section 26.1.2.

The singlet factorization produces the renormalized electron–hole-pair transition energy

$$\tilde{\epsilon}_{\mathbf{k}} \equiv \epsilon^c_{\mathbf{k}} - \epsilon^v_{\mathbf{k}} + \sum_{\mathbf{k}'} V_{\mathbf{k}-\mathbf{k}'} - \sum_{\mathbf{k}'} V_{\mathbf{k}-\mathbf{k}'}\left(f^e_{\mathbf{k}'} + f^h_{\mathbf{k}'}\right). \tag{26.5}$$

The sum $\sum_{\mathbf{k}'} V_{\mathbf{k}-\mathbf{k}'}$ that appears represents the formally diverging repulsive Coulomb energy of the electrons in the fully filled valence band. However, this contribution is already included in the band-structure computations such that it has to be dropped from the equations when we use experimental values for the band gap, see also Exercise 24.7. If the bands can be approximated as parabolic, compare Section 12.1.1, we find

$$\tilde{\epsilon}_{\mathbf{k}} = \frac{\hbar^2\mathbf{k}^2}{2\mu} + E_{g,0} - \sum_{\mathbf{k}'} V_{\mathbf{k}-\mathbf{k}'}\left(f^e_{\mathbf{k}'} + f^h_{\mathbf{k}'}\right), \quad \frac{1}{\mu} \equiv \frac{1}{m_e} + \frac{1}{m_h}, \tag{26.6}$$

where $E_{g,0}$ is the band gap of the unexcited system and μ defines the effective mass.

As long as we ignore the Coulomb interaction, the electron and hole states are separated by $E_{g,\mathbf{k}} = \hbar^2\mathbf{k}^2/2\mu + E_{g,0}$. From the simplified model of Section 25.1.1, we know that an optical field lifts electrons from the valence into the conduction band. Each momentum state $\mathbf{k}$ represents an isolated two-level system, as shown in Fig. 26.1. The arrows indicate the optically coupled states. Equation (26.6) identifies one of the simplest Coulomb-induced effects. The presence of electron and hole occupations decreases the band gap via $-\sum_{\mathbf{k}'} V_{\mathbf{k}-\mathbf{k}'}\left(f^e_{\mathbf{k}'} + f^h_{\mathbf{k}'}\right)$. This renormalization is both finite and $\mathbf{k}$ dependent, showing that the electronic excitations produce direct modifications of the band structure, see Fig. 26.1(b).

Additionally, the Rabi frequency is also changed due to the Coulomb interaction,

$$\Omega^{(+)}_{\mathbf{k}} \equiv \frac{Qp_{c,v}}{m_0}\int dz\, g(z)\langle\hat{A}^{(+)}(z,t)\rangle + \sum_{\mathbf{k}'} V_{\mathbf{k}-\mathbf{k}'}P_{\mathbf{k}'}$$
$$= d_{c,v}\int dz\, g(z)\,\langle\hat{E}^{(+)}(z,t)\rangle + \sum_{\mathbf{k}'} V_{\mathbf{k}-\mathbf{k}'}P_{\mathbf{k}'}. \tag{26.7}$$

In these expressions, we used the rotating-wave approximation (RWA) and the conversion

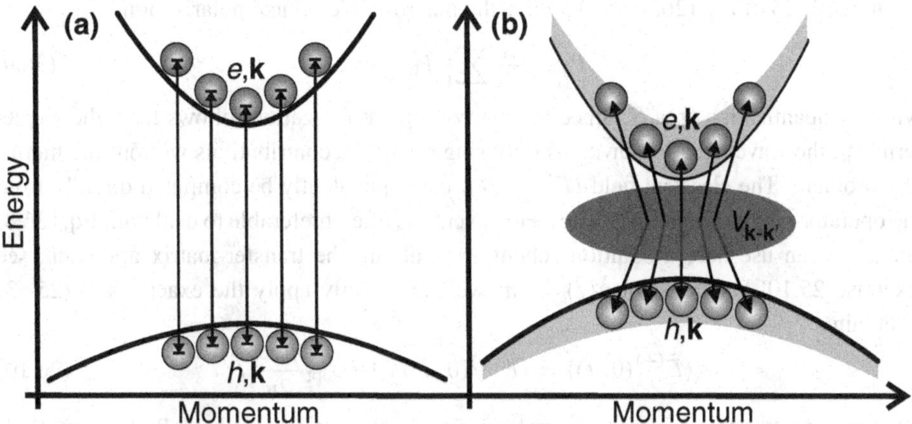

Figure 26.1 Schematic representation of optical semiconductor excitations **(a)** without and **(b)** with Coulomb coupling. Without Coulomb interaction, the light couples (arrows) an individual electron and hole (spheres) via a direct transition over the band gap. The Coulomb interaction couples *all* microscopic polarizations (arrows); the dark shaded oval indicates the coupling. Additionally, the Coulomb interaction shifts the electron–hole bands as symbolized by the light-shaded areas.

(25.55), both of which are extremely well justified for near-resonant interband excitations. We typically investigate situations where the QW or QWI confinement function $g(z)$ is much narrower than the optical wave length. Therefore, also the approximation $g(z) \to \delta(z)$ is very accurate.

The Coulomb renormalization in (26.7), i.e., $\sum_{\mathbf{k}'} V_{\mathbf{k}-\mathbf{k}'} P_{\mathbf{k}'}$, directly couples the polarization $P_{\mathbf{k}}$ to all other polarizations $P_{\mathbf{k}' \neq \mathbf{k}}$ created in the system. As a result, the semiconductor response is no longer given by the sum over the responses of independent two-level systems. The Coulomb-induced coupling structure among the electron–hole states is schematically indicated in Fig. 26.1(b) by the shaded dark oval. To compute the optical response, we have to determine how the Coulomb coupling among the infinitely many states can be evaluated systematically.

26.1.1 Maxwell-semiconductor Bloch equations

In a fully self-consistent approach, we have to solve the SBE (26.2)–(26.4) simultaneously together with the wave equation

$$\left[\frac{\partial^2}{\partial z^2} - \frac{n^2(z)}{c^2}\frac{\partial^2}{\partial t^2}\right]\langle \hat{E}^{(+)}(z)\rangle = -\mu_0 \delta(z)\frac{\partial^2}{\partial t^2}\left(\frac{1}{\mathcal{S}}\sum_{\mathbf{k}} d_{v,c}\, P^{v,c}_{\mathbf{k}}\right). \quad (26.8)$$

This coupled set of equations constitutes the Maxwell-semiconductor Bloch equations (MSBE) which can be used as a general starting point to investigate excitations induced by classical light fields in direct-gap semiconductors. The quality of the results critically depends on the accuracy with which the doublet contributions $\Gamma^{\lambda,\lambda'}$ and Γ^{QED} can be evaluated.

On the RHS of Eq. (26.8), we identify the macroscopic optical polarization

$$P \equiv \frac{d_{vc}}{\mathcal{S}} \sum\nolimits_{\mathbf{k}} P_{\mathbf{k}} + \text{c.c.}, \tag{26.9}$$

with the quantization area $\mathcal{S}$. Since the macroscopic polarization follows from the singlet term $P_{\mathbf{k}}$, the wave equation involves only single-particle contributions without the hierarchy problem. The classical field $\langle E^{(+)}(z,t)\rangle$ can equivalently be computed directly from the operator-singlet equation (24.46). However, it is often preferable to deal with Eq. (26.8) since we can use here a solution scheme resembling the transfer-matrix approach, see Exercise 25.10. For constant $n(z) = n$, we can directly apply the exact result (25.53) to obtain

$$\langle \bar{E}^{(+)}(0,t)\rangle = \langle \bar{E}^{(+)}(0,t)\rangle_0 + i\omega_{\text{opt}} \frac{\mu_0 c}{2n} \bar{P} \tag{26.10}$$

after we identify the macroscopic polarization in the rotating frame, $P = \bar{P}\, e^{-i\omega_{\text{opt}} t}$, where ω_{opt} denotes the central frequency of the incoming field $\langle E^{(+)}(0,t)\rangle_0 \equiv \langle \bar{E}^{(+)}(0,t)\rangle_0\, e^{-i\omega_{\text{opt}} t}$. We can then efficiently treat the MSBE by also converting $P_{\mathbf{k}} \equiv \bar{P}_{\mathbf{k}}\, e^{-i\omega_{\text{opt}} t}$ into the rotating frame, as discussed in Section 25.4.2.

26.1.2 Coupling to doublet correlations

In Eqs. (26.2)–(26.4), the doublet contributions show up as quantum-optical correlations Γ^{QED} and as microscopic scattering terms

$$\begin{aligned}\Gamma^{\lambda,\lambda'}_{\mathbf{k}} \equiv & \sum_{\nu,\mathbf{k}',\mathbf{q}} V_{\mathbf{q}} \left[c^{\mathbf{q},\mathbf{k}',\mathbf{k}}_{\lambda,\nu;\nu,\lambda'} - \left(c^{\mathbf{q},\mathbf{k}',\mathbf{k}}_{\lambda',\nu;\nu,\lambda} \right)^{\star} \right] \\ & + \sum_{\mathbf{q}} \left[\Delta\left\langle Q^{\lambda'}_{\mathbf{q}} a^{\dagger}_{\lambda,\mathbf{k}} a_{\lambda',\mathbf{k}-\mathbf{q}} \right\rangle - \Delta\left\langle \left(Q^{\lambda}_{\mathbf{q}}\right)^{\dagger} a^{\dagger}_{\lambda,\mathbf{k}-\mathbf{q}} a_{\lambda',\mathbf{k}} \right\rangle \right]\end{aligned} \tag{26.11}$$

resulting from the Coulomb (first line) and the phonon–carrier (second line) interactions. The explicit format of the pure carrier doublets is given by Eq. (24.43). In general, the doublet terms $\Gamma^{\lambda,\lambda'}$ introduce microscopic coupling of the singlets to the two-particle Coulomb and phonon correlations, which describe dephasing, energy renormalizations, screening of the Coulomb interaction, as well as relaxation of the carrier densities toward steady-state distributions. It is not necessary to introduce any additional coupling to an additional reservoir because the Coulomb and phonon interactions produce intrinsic scattering within the many-body system itself. In fact, the resulting quantum kinetics is so complicated that its quantitative features cannot be mapped into a simple master-equation approach described in Chapter 20.

The quantum-optical two-particle correlations have the explicit form

$$\Gamma^{\text{QED}}_{\lambda,\lambda';\mathbf{k}} \equiv - \sum_{\mathbf{q},\sigma=\pm} \left[\Delta\left\langle \hat{\mathcal{A}}^{\bar{\lambda}'(\sigma)}_{\mathbf{q}} a^{\dagger}_{\lambda,\mathbf{k}} a_{\bar{\lambda}',\mathbf{k}-\mathbf{q}} \right\rangle - \Delta\left\langle \hat{\mathcal{A}}^{\bar{\lambda}(\sigma)}_{-\mathbf{q}} a^{\dagger}_{\bar{\lambda},\mathbf{k}-\mathbf{q}} a_{\lambda',\mathbf{k}} \right\rangle \right], \tag{26.12}$$

where we divided the collective photon-field operator $\hat{A}^{\lambda}_{\mathbf{q}} \equiv \hat{A}^{\lambda(+)}_{\mathbf{q}} + \hat{A}^{\lambda(-)}_{\mathbf{q}}$ into separate parts containing the photon annihilation (+) and creation (−) operators. Based on the definitions (24.31)–(24.32), we find the explicit expressions

$$\hat{A}^{\lambda(\sigma)}_{\mathbf{q}} \equiv \sum_{q_z} \mathcal{E}_{\mathbf{q},q_z} \frac{Q\mathbf{p}_{\bar{\lambda},\lambda}}{m_0\omega_{\mathbf{q},q_z}} \cdot \begin{cases} \tilde{\mathbf{u}}_{\mathbf{q},q_z} B_{\mathbf{q},q_z}, & \text{for } \sigma = + \\ \tilde{\mathbf{u}}^{\star}_{\mathbf{q},q_z} B^{\dagger}_{-\mathbf{q},q_z}, & \text{for } \sigma = - \end{cases}, \tag{26.13}$$

which contain the effective mode function

$$\tilde{\mathbf{u}}_{\mathbf{q},q_z} \equiv \int dz\, g(z)\mathbf{u}_{\mathbf{q},q_z}(z). \tag{26.14}$$

For sufficiently strong QW or QWI confinement, we may use $\tilde{\mathbf{u}}_{\mathbf{q},q_z} = \mathbf{u}_{\mathbf{q},q_z}(z=0)$ defined at the position of the nanostructure.

The momentum matrix element that appears can be simplified with the help of Eq. (25.52),

$$\frac{Q\mathbf{p}_{c,v}}{m_0\omega_{\mathbf{q},q_z}} = i\frac{\omega_{g,0}}{\omega_{\mathbf{q},q_z}} d_{c,v} \to i d_{c,v}, \qquad \frac{Q\mathbf{p}_{v,c}}{m_0\omega_{\mathbf{q},q_z}} = -i\frac{\omega_{g,0}}{\omega_{\mathbf{q},q_z}} d_{v,c} \to -i d_{v,c}, \tag{26.15}$$

where the last step follows because the photon energy essentially matches the band-gap energy $\hbar\omega_{g,0}$. This simplifies the collective photon operators into

$$\begin{cases} \hat{A}^{v(+)}_{\mathbf{q}} = \mathbf{d}_{c,v} \cdot \tilde{\mathbf{E}}^{(+)}_{\mathbf{q}} \\ \hat{A}^{c(-)}_{\mathbf{q}} = \mathbf{d}_{v,c} \cdot \tilde{\mathbf{E}}^{(-)}_{\mathbf{q}} \end{cases}, \qquad \begin{cases} \tilde{\mathbf{E}}^{(+)}_{\mathbf{q}} \equiv \sum_{q_z} i\mathcal{E}_{\mathbf{q},q_z} \tilde{\mathbf{u}}_{\mathbf{q},q_z} B_{\mathbf{q},q_z} \\ \tilde{\mathbf{E}}^{(-)}_{\mathbf{q}} \equiv -\sum_{q_z} i\mathcal{E}_{\mathbf{q},q_z} \tilde{\mathbf{u}}^{\star}_{\mathbf{q},q_z} B^{\dagger}_{-\mathbf{q},q_z} \end{cases}. \tag{26.16}$$

The analysis in Chapters 28–30 shows that the correlations $\Delta\langle \tilde{\mathbf{E}}^{(\pm)}_{\mathbf{q}} a^{\dagger} a\rangle$ build up spontaneously without directional preference. Thus, Γ^{QED} contributes to the SBE influencing all light-propagation directions even if the classical light component exists only in the normal direction $\mathbf{q} = 0$.

The quantum-optical effects influence the polarization through $\Gamma^{\text{QED}}_{v,c;\mathbf{k}}$ containing photon-density correlations of the type $\Delta\langle B a^{\dagger}_{\lambda} a_{\lambda}\rangle$ and $\Delta\langle B^{\dagger} a^{\dagger}_{\lambda} a_{\lambda}\rangle$. According to our classification (25.28), both of them are coherent quantities. However, we see that only $\Delta\langle B a^{\dagger}_{\lambda} a_{\lambda}\rangle$ oscillates with $e^{-i\omega_{\text{opt}}t}$ in the same way as $P_{\mathbf{k}}$ does. Since $\Delta\langle B^{\dagger} a^{\dagger}_{\lambda} a_{\lambda}\rangle$ oscillates with the opposite phase $e^{+i\omega_{\text{opt}}t}$, its contributions to $P_{\mathbf{k}}$ are strongly nonresonant. Based on the RWA arguments of Section 25.4.1, only the $\Delta\langle B a^{\dagger}_{\lambda} a_{\lambda}\rangle$ parts are relevant for the near-resonant excitations studied here, i.e., only the $\hat{A}^{v(+)}_{\mathbf{q}}$ part of $\Gamma^{\text{QED}}_{v,c;\mathbf{k}}$ can efficiently contribute to the polarization dynamics.

An analogous analysis can be made for the correlations $\Gamma^{\text{QED}}_{c,c;\mathbf{k}}$ and $\Gamma^{\text{QED}}_{v,v;\mathbf{k}}$ driving the densities. At the RWA level, we eventually obtain

$$\Gamma^{\text{QED}}_{v,c;\mathbf{k}} = \sum_{\mathbf{q}} \left[\Delta\left\langle \hat{\mathcal{A}}^{v(+)}_{-\mathbf{q}} a^{\dagger}_{c,\mathbf{k}-\mathbf{q}} a_{c,\mathbf{k}} \right\rangle - \Delta\left\langle \hat{\mathcal{A}}^{v(+)}_{\mathbf{q}} a^{\dagger}_{v,\mathbf{k}} a_{v,\mathbf{k}-\mathbf{q}} \right\rangle \right]$$

$$\Gamma^{\text{QED}}_{c,c;\mathbf{k}} = \sum_{\mathbf{q}} \left[\Delta\left\langle \hat{\mathcal{A}}^{c(-)}_{-\mathbf{q}} a^{\dagger}_{v,\mathbf{k}-\mathbf{q}} a_{c,\mathbf{k}} \right\rangle - \Delta\left\langle \hat{\mathcal{A}}^{v(+)}_{\mathbf{q}} a^{\dagger}_{c,\mathbf{k}} a_{v,\mathbf{k}-\mathbf{q}} \right\rangle \right]$$

$$\Gamma^{\text{QED}}_{v,v;\mathbf{k}} = \sum_{\mathbf{q}} \left[\Delta\left\langle \hat{\mathcal{A}}^{v(+)}_{-\mathbf{q}} a^{\dagger}_{c,\mathbf{k}-\mathbf{q}} a_{v,\mathbf{k}} \right\rangle - \Delta\left\langle \hat{\mathcal{A}}^{c(-)}_{\mathbf{q}} a^{\dagger}_{v,\mathbf{k}} a_{c,\mathbf{k}-\mathbf{q}} \right\rangle \right]. \tag{26.17}$$

These expressions show that the polarization is driven by processes where a photon is absorbed while the carriers make an intraband transition. At the same time, the carrier densities are modified through $\hat{\mathcal{A}}^{c(-)}_{\mathbf{q}}$ containing $B^{\dagger}_{\mathbf{q},q_z}$. This term describes correlated processes where the electron makes a transition from the conduction to the valence band by emitting a photon. The reverse process is also possible.

The role of all these correlations is studied further in Chapters 28–30. In the remainder of this chapter, we concentrate on situations where the effect induced by the Coulomb interaction and the classical light field dominate and the photon-assisted correlation terms can be omitted. In particular, we are interested in seeing which physical effects can be described by the MSBE including $\Gamma^{\lambda,\lambda'}$ at different levels of approximation.

26.2 Excitonic states

We start by analyzing the principal structure of the SBE, i.e., the singlet part of the induced polarization dynamics in Eq. (26.2),

$$\begin{aligned} i\hbar\frac{\partial}{\partial t}P_{\mathbf{k}} = &\left(E_{g,0} + \frac{\hbar^2\mathbf{k}^2}{2\mu} - \sum_{\mathbf{k}'} V_{\mathbf{k}-\mathbf{k}'}\left(f^e_{\mathbf{k}'} + f^h_{\mathbf{k}'}\right)\right) P_{\mathbf{k}} \\ &- \left[1 - f^e_{\mathbf{k}} - f^h_{\mathbf{k}}\right]\left(\sum_{\mathbf{k}'} V_{\mathbf{k}-\mathbf{k}'} P_{\mathbf{k}'} + d_{c,v}\langle \hat{E}^{(+)}(0,t)\rangle\right), \end{aligned} \tag{26.18}$$

where we inserted the explicit forms of the renormalized kinetic energy and the Rabi frequency. As described in connection with Fig. 26.1, we see that the polarization $P_{\mathbf{k}}$ couples to all other $P_{\mathbf{k}'}$ as a consequence of the Coulomb interaction. In the limit of continuous $\mathbf{k}$ values, we therefore have to solve an integral equation to deduce the actual polarization.

Under typical semiconductor excitation conditions, the coherent polarization vanishes without the driving field while the electron–hole densities may exist for longer times, compare Section 25.3. Thus, it is meaningful to study semiconductor optics for a case where $P_{\mathbf{k}}$ is only generated by a weak probe pulse whereas $f^e_{\mathbf{k}}$ and $f^h_{\mathbf{k}}$ can have arbitrary finite values. In the linear regime, the probe pulse induces a linear polarization $P_{\mathbf{k}}$ but it does not change the densities, see Section 25.2. In this situation, only $d_{c,v}\langle \hat{E}^{(+)}(0,t)\rangle$ drives the Coulomb-coupled polarization. For this situation, we can find an analytical solution if we first construct the eigenstates for the homogeneous part of Eq. (26.18),

$$\left[\frac{\hbar^2\mathbf{k}^2}{2\mu} - \sum_{\mathbf{k}'} V_{\mathbf{k}-\mathbf{k}'}\left(f^e_{\mathbf{k}'} + f^h_{\mathbf{k}'}\right)\right]\phi^R_\lambda(\mathbf{k})$$
$$- \left(1 - f^e_{\mathbf{k}} - f^h_{\mathbf{k}}\right)\sum_{\mathbf{k}'} V_{\mathbf{k}-\mathbf{k}'}\phi^R_\lambda(\mathbf{k}') = E_\lambda \phi^R_\lambda(\mathbf{k}). \quad (26.19)$$

For vanishing densities, this eigenvalue problem has a one-to-one correspondence to the Schrödinger equation (16.30). In other words, the Coulomb interaction couples the different $P_{\mathbf{k}}$ as if these were the wave functions of the hydrogen problem. The eigenvalue problem (26.19) defines the *Wannier equation* and the solutions are often referred to as Wannier excitons.

In the hydrogen problem, the Coulomb attraction between the oppositely charged constituents leads to the existence of bound states, compare Section 16.2. Since the electron and hole are also oppositely charged, bound-state solutions of Eq. (26.19) exist. These define the exciton states, which have a formal analogy to the hydrogen problem. However, one has to be careful not to overstress this analogy. In particular, we show in Chapter 27 that the appearance of excitonic resonances in the polarization does not automatically imply that electrons and holes have formed truly bound pairs. To have a precise terminology, we refer to "*excitons*" only if incoherent populations of bound pairs can be identified in the many-body state while we call a given property *excitonic* if it only relies on the formal analogy between the eigenstates of the Wannier and the hydrogen Schrödinger equation.

In many ways, excitonic states are of central importance to understand the semiconductor response in the weak-excitation limit. For a strictly two-dimensional system at zero density, the bound excitonic states have the energies $E_n = -4E_B/(1+2n)^2$ for integer n, compare Eq. (16.24). For GaAs, the three-dimensional exciton binding energy is $E_B = 4.18\,\text{meV}$ based on Table 16.1. Thus, the QW polarization can have bound resonances about 10 meV below the band gap.

26.2.1 Left- and right-handed excitonic states

As soon as carrier populations are present, $f^e_{\mathbf{k}}$ and $f^h_{\mathbf{k}}$ assume finite values with the consequence that Eq. (26.19) deviates from the original hydrogen problem and becomes a non-Hermitian equation. Consequently, Eq. (26.19) has both left-handed, $\phi^L_\lambda(\mathbf{k})$, and right-handed, $\phi^R_\lambda(\mathbf{k})$, solutions. The left-handed eigenvalue problem is defined by

$$(\tilde{\varepsilon}_{\mathbf{k}} - E_{g,0})\phi^L_\lambda(\mathbf{k}) - \sum_{\mathbf{k}'}\phi^L_\lambda(\mathbf{k}')\left(1 - f^e_{\mathbf{k}'} - f^h_{\mathbf{k}'}\right)V_{\mathbf{k}-\mathbf{k}'} = E_\lambda\phi^L_\lambda(\mathbf{k}). \quad (26.20)$$

Looking at the asymmetry between Eqs. (26.19) and (26.20), we notice that the left- and right-handed solutions are connected via a simple transformation

$$\phi^L_\lambda(\mathbf{k}) = \frac{\phi^R_\lambda(\mathbf{k})}{1 - f^e_{\mathbf{k}} - f^h_{\mathbf{k}}}, \quad (26.21)$$

see Exercise 26.3. The left- and right-handed solutions can be normalized such that

$$\sum_{\mathbf{k}} \phi_\lambda^L(\mathbf{k})\phi_\nu^R(\mathbf{k}) = \delta_{\lambda,\nu}. \tag{26.22}$$

Since Eqs. (26.19) and (26.20) define a real-valued eigenvalue problem, we may choose the eigenstates to be real-valued in momentum space.

The asymmetry between the left- and right-handed solutions is a consequence of the inversion factor, often also called phase-space filling factor, $(1 - f_{\mathbf{k}}^e - f_{\mathbf{k}}^h)$ multiplying the Coulomb matrix element. One can combine this factor with $V_{\mathbf{k}-\mathbf{k}'}$ to produce an effective Coulomb matrix element

$$\left(1 - f_{\mathbf{k}}^e - f_{\mathbf{k}}^h\right) V_{\mathbf{k}-\mathbf{k}'} \equiv V_{\mathbf{k},\mathbf{k}'}^{\text{eff}}. \tag{26.23}$$

Using this definition allows us to write the left- and right-handed eigenvalue problems as

$$\begin{aligned} &(\tilde{\varepsilon}_{\mathbf{k}} - E_{g,0})\phi_\lambda^R(\mathbf{k}) - \sum_{\mathbf{k}'} V_{\mathbf{k},\mathbf{k}'}^{\text{eff}} \phi_\lambda^R(\mathbf{k}') = E_\lambda\, \phi_\lambda^R(\mathbf{k}) \\ &(\tilde{\varepsilon}_{\mathbf{k}} - E_{g,0})\phi_\lambda^L(\mathbf{k}) - \sum_{\mathbf{k}'} \phi_\lambda^L(\mathbf{k}') V_{\mathbf{k}',\mathbf{k}}^{\text{eff}} = E_\lambda\, \phi_\lambda^L(\mathbf{k}). \end{aligned} \tag{26.24}$$

These two equations are different because the effective Coulomb interaction is not symmetric, i.e., $V_{\mathbf{k},\mathbf{k}'}^{\text{eff}} \neq V_{\mathbf{k}',\mathbf{k}}^{\text{eff}}$. At this stage, we see that one of the principal consequences of elevated carrier densities is that $(1 - f^e - f^h)$ can become negative for some range of momentum values. As a consequence, V^{eff} changes its sign, switching from attractive for low densities to repulsive for sufficiently large densities. The k-value at which $1 - f_{\mathbf{k}}^e - f_{\mathbf{k}}^h = 0$ defines the inversion point. Since the phase-space filling effect is a direct consequence of the Fermionic Pauli blocking, one often says that the resulting net repulsive interaction manifests itself as a Fermi pressure. The switch from an attractive to a repulsive effective interaction leads to the vanishing of the bound-state solutions. In other words, the excitonic states are ionized with increasing carrier density. This effect is often referred to as the excitonic Mott transition.

26.2.2 Density-dependent aspects of the 1*s* resonance

To get some impression of how the densities influence the excitonic wave function, we solve Eq. (26.19) using the coherent-limit relation (25.24),

$$f_{\mathbf{k}} = f_{\mathbf{k}}^e = f_{\mathbf{k}}^h \equiv \frac{1}{2} - \sqrt{\frac{1}{4} - \left|P_{\mathbf{k}}^{1s}\right|^2}, \qquad P_{\mathbf{k}}^{1s} = p_{1s}\,\frac{\phi_{1s}(\mathbf{k})}{\phi_{1s}(0)}. \tag{26.25}$$

Here, p_{1s} determines the density level. Since the densities are described by real-valued quantities, we have to restrict the analysis to $p_{1s} \leq 1/2$.

Figure 26.2(a) presents the ground-state wave function $\phi^R(\mathbf{k})$ for vanishing densities (shaded area, $p_{1s} = 0$) and for $p_{1s} = 0.497$ (solid line) using typical GaAs material parameters. The corresponding real-space representation is shown in Fig. 26.2(b). We see that the

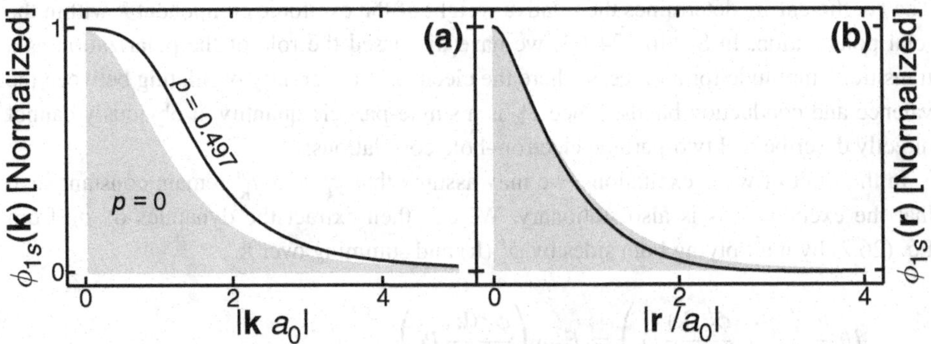

Figure 26.2 Exciton wave function for a GaAs-type system. The scaled wave functions are shown **(a)** in momentum and **(b)** in real space. The solid line indicates the low-density limit ($p_{1s} = 0$) while the shaded area indicates the result for $p_{1s} = 0.497$. The exciton Bohr radius is $a_0 = 12.5$ nm.

$1s$-wave function broadens and flattens in momentum space while it becomes narrower in real space as the density is increased. The value of $p_{1s} = 0.497$ corresponds to the total density of $4 \times 10^{10}\,\mathrm{cm}^{-2}$ computed using

$$n_{e(h)} \equiv \frac{1}{S} \sum_{\mathbf{k}} f_{\mathbf{k}}^{e(h)}. \tag{26.26}$$

It is interesting to see that the excitonic wave function remains very much centered close to $\mathbf{k} = 0$ even though $p_{1s} = 0.497$ reduces the Coulomb attraction for the small momentum states almost to zero. It is also typical that the resonance energy E_{1s} of the $1s$ state is not much affected by the density. This conclusion can be understood via the structure of Eq. (26.19), where the energy renormalization, $\Sigma_{\mathbf{k}}^{\mathrm{ren}} \equiv -\sum_{\mathbf{k}'} V_{\mathbf{k}'-\mathbf{k}} \left(f_{\mathbf{k}'}^{e} + f_{\mathbf{k}'}^{h}\right)$, tends to reduce the effective band gap as the density is decreased, see Fig. 26.1(b). This effect alone produces a red shift of the detected $1s$ resonance, i.e., E_{1s} becomes more negative. At the same time, however, the phase-space filling reduces the excitonic binding energy because the effective Coulomb attraction, $V_{\mathbf{k},\mathbf{k}'}^{\mathrm{eff}}$, becomes smaller. This alone produces a blue shift to the $1s$ resonance, i.e., E_{1s} becomes more positive. Since the modifications of $\Sigma_{\mathbf{k}}^{\mathrm{ren}}$ and $V_{\mathbf{k},\mathbf{k}'}^{\mathrm{eff}}$ occur simultaneously and nearly compensate each other, one typically finds no or at most a small shift in the total E_{1s}.

26.3 Semiconductor Bloch equations in the exciton basis

Since excitonic states diagonalize the homogeneous part of the SBE, it can be beneficial to transform the problem into the exciton basis using the expansion

$$P_{\mathbf{k}} = \sum_{\lambda} \sqrt{S}\, p_{\lambda}\, \phi_{\lambda}^{R}(\mathbf{k}), \qquad p_{\lambda} = \sum_{\mathbf{k}} \frac{1}{\sqrt{S}} \phi_{\lambda}^{L}(\mathbf{k})\, P_{\mathbf{k}}. \tag{26.27}$$

The coefficient p_λ determines the relative weight of the excitonic component λ within the total polarization. In Section 24.1.1, we have discussed the role of the polarization as a transition amplitude for a process where the electron is coherently oscillating between the valence and conduction bands. Since $P_{\mathbf{k}}$ is a single-particle quantity, it obviously cannot directly describe real two-particle electron–hole correlations.

In the limit of weak excitations, we may assume that $f^e_{\mathbf{k}}$ and $f^h_{\mathbf{k}}$ remain constant such that the exciton basis is also stationary. We can then extract the dynamics of p_λ from Eq. (26.2) by multiplying both sides by $\phi^L_\lambda(\mathbf{k})$ and summing over $\mathbf{k}$,

$$
\begin{aligned}
i\hbar\frac{\partial}{\partial t}\left(\sum_{\mathbf{k}}\frac{\phi^L_\lambda(\mathbf{k})}{\sqrt{S}}P_{\mathbf{k}}\right) &= E_{g,0}\left(\frac{\phi^L_\lambda(\mathbf{k})}{\sqrt{S}}P_{\mathbf{k}}\right)\\
&+\sum_{\mathbf{k}}\frac{\phi^L_\lambda(\mathbf{k})}{\sqrt{S}}\left[(\tilde{\epsilon}_{\mathbf{k}}-E_{g,0})P_{\mathbf{k}}-\left(1-f^e_{\mathbf{k}}-f^h_{\mathbf{k}}\right)\sum_{\mathbf{k}'}V_{\mathbf{k}'-\mathbf{k}}P_{\mathbf{k}'}\right]\\
&-\sum_{\mathbf{k}}\frac{\phi^L_\lambda(\mathbf{k})}{\sqrt{S}}\left(1-f^e_{\mathbf{k}}-f^h_{\mathbf{k}}\right)\mathbf{d}_{c,v}\cdot\left\langle\tilde{\mathbf{E}}^{(+)}_0\right\rangle\\
&+\sum_{\mathbf{k}}\frac{\phi^L_\lambda(\mathbf{k})}{\sqrt{S}}\Gamma^{v,c}_{\mathbf{k}}+\sum_{\mathbf{k}}\frac{\phi^L_\lambda(\mathbf{k})}{\sqrt{S}}\Gamma^{\mathrm{QED}}_{v,c;\mathbf{k}}.
\end{aligned}
\tag{26.28}
$$

Here, we expressed the driving light source using $d_{c,v}\int dz\, g(z)\,\langle\hat{E}^{(+)}(z,t)\rangle=\mathbf{d}_{c,v}\cdot\langle\tilde{\mathbf{E}}^{(+)}_0\rangle$ that follows from the definition (26.16).

Using Eq. (26.27), we see that the projections in the first line of Eq. (26.28) simply yield p_λ. We transform the second term in Eq. (26.28) into

$$
\begin{aligned}
2\text{nd} &\equiv \sum_{\mathbf{k},\nu}\phi^L_\lambda(\mathbf{k})\,p_\nu\left[(\tilde{\epsilon}_{\mathbf{k}}-E_{g,0})\phi^R_\nu(\mathbf{k})-\left(1-f^e_{\mathbf{k}}-f^h_{\mathbf{k}}\right)\sum_{\mathbf{k}'}V_{\mathbf{k}'-\mathbf{k}}\phi^R_\nu(\mathbf{k}')\right]\\
&=\sum_\nu p_\nu\sum_{\mathbf{k}}\phi^L_\lambda(\mathbf{k})\,E_\nu\,\phi^R_\nu(\mathbf{k})=\sum_\nu E_\nu\,p_\nu\,\delta_{\nu,\lambda}=E_\lambda\,p_\nu,
\end{aligned}
\tag{26.29}
$$

where we used the eigenvalue problem (26.19) and the orthogonality (26.22) of the excitonic states. This way, Eq. (26.28) becomes

$$
\begin{aligned}
i\hbar\frac{\partial}{\partial t}p_\lambda &= (E_{g,0}+E_\lambda)p_\lambda-\sum_{\mathbf{k}}\frac{\phi^L_\lambda(\mathbf{k})}{\sqrt{S}}\left(1-f^e_{\mathbf{k}}-f^h_{\mathbf{k}}\right)\mathbf{d}_{c,v}\cdot\left\langle\tilde{\mathbf{E}}^{(+)}_0\right\rangle\\
&+\sum_{\mathbf{k}}\phi^L_\lambda(\mathbf{k})\Gamma^{v,c}_{\mathbf{k}}+\sum_{\mathbf{k}}\phi^L_\lambda(\mathbf{k})\Gamma^{\mathrm{QED}}_{v,c;\mathbf{k}}.
\end{aligned}
\tag{26.30}
$$

The optical driving term can be simplified by converting the left-handed eigenstate into a right-handed one via the transformation (26.21),

$$
\sum_{\mathbf{k}}\frac{\phi^L_\lambda(\mathbf{k})}{\sqrt{S}}\left(1-f^e_{\mathbf{k}}-f^h_{\mathbf{k}}\right)=\sum_{\mathbf{k}}\frac{\phi^R_\lambda(\mathbf{k})}{\sqrt{S}}=\phi^R_\lambda(\mathbf{r}=0).
\tag{26.31}
$$

In the final step, we identified the Fourier-transformed exciton wave function in real space,

$$\phi_\lambda^R(\mathbf{r}) \equiv \frac{1}{\sqrt{S}} \sum_{\mathbf{k}} \phi_\lambda^R(\mathbf{k})\, e^{i\mathbf{k}\cdot\mathbf{r}}. \tag{26.32}$$

Combining all the steps, we can express the main parts of the polarization equation as

$$i\hbar \frac{\partial}{\partial t} p_\lambda = \left(E_{g,0} + E_\lambda\right) p_\lambda - d_{cv}\, \phi_\lambda^R(\mathbf{r}=0) \langle E(t) \rangle - i\Gamma_\lambda, \tag{26.33}$$

where Γ_λ includes the dephasing induced by scattering and quantum-optical doublets. In general, Γ_λ is a complicated functional which cannot be simplified further in a fully systematic computation. However, to gain some elementary insights, we use in the following an approximative form of Γ_λ.

Equation (26.33) not only shows that the different λ states are uncoupled but also that the singlet response is completely linear as long as E_λ and $\phi_\lambda^R(\mathbf{r} = 0)$ remain constant. Formally, the structure of this equation resembles that of an optically driven Bosonic system instead of a two-level system with a purely Fermionic response, compare Section 21.1.4. Hence, we see that even though we are dealing with a Fermionic system of charge carriers, we can obtain some Bosonic features due to the collective Coulomb coupling.

As an additional important difference from the optics of atoms, we note that the optical coupling in the semiconductor systems involves $d_{c,v}\phi_\lambda^R(\mathbf{r} = 0)\langle E(t)\rangle$ where the Rabi frequency is enhanced by $\phi_\lambda^R(\mathbf{r} = 0)$. This enhancement is a pure Coulomb effect. Furthermore, one can easily show that $\phi_\lambda^R(\mathbf{r} = 0)$ is nonzero only for the s states because other classes of excitonic states are not symmetric. Thus, the optical semiconductor response directly involves only the s-like excitonic states.

Figure 26.3(a) shows the so-called oscillator strength $\left|\phi_\lambda^R(\mathbf{r} = 0)\right|^2$ as a function of the excitonic energies for an 8-nm-thick GaAs-type QW. We see the prominent contributions of the bound s-like states and note that the continuum states (shaded area) have a very much reduced oscillator strength. In particular, the attractive Coulomb interaction leads to a strong enhancement of the $1s$ component at the expense of all other states. Thus, for nearly resonant and not too strong excitation conditions it is often justified to concentrate only on the $1s$ resonance.

Based on Eq. (26.33), each λ component of the polarization has a characteristic energy $E_{g,0} + E_\lambda$ that defines the frequency of the related dipole emission. The spectral positions of $E_{g,0} + E_\lambda$ are shown in Fig. 26.3(b) as solid lines. The dashed lines illustrate the conduction and valence bands whose energetic difference defines the characteristic transition energies of the noninteracting system. We see that the bound excitonic resonances can be detected as a Rydberg-like series below the band gap. For example, the $1s$-polarization component is roughly 10 meV below the band gap for GaAs-type QWs. Thus, the Coulomb

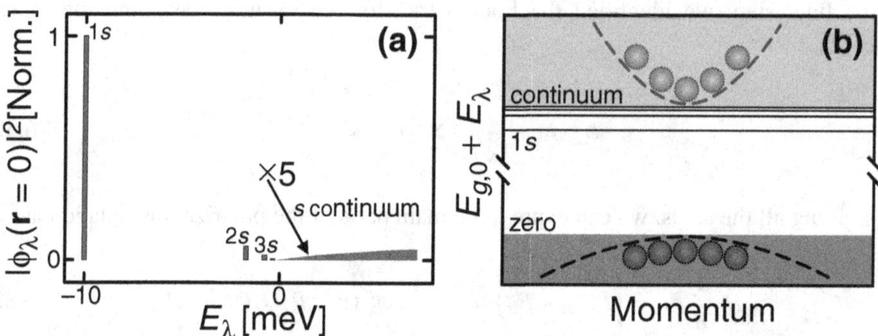

Figure 26.3 Excitonic resonances in the optical response. (**a**) The excitonic oscillator strength is shown for the energetically lowest bound states (bars) and the continuum states (shaded area, multiplied by five). (**b**) The characteristic polarization energies $E_{g,0} + E_\lambda$ are shown for the bound (horizontal lines) and the continuum states (light area). The zero-energy level is denoted by the dark area. The dashed lines (with spheres) indicate the energy range for the free-electron bands.

coupling produces optical resonances in the energetic range that is forbidden without this interaction.

Altogether, these discussions at the singlet level already show that the optical properties of semiconductors are qualitatively very different from those of a simple Fermionic two-level system. This conclusion remains valid also when the doublets are included. As we will see, the doublets produce significant quantitative changes that are essential to understand, in particular, the nonlinear optical semiconductor response. Even though Eq. (26.33) has a seemingly simple form, the consistent solution with the two-particle correlation terms Γ_λ is highly nontrivial. In addition, the exciton basis becomes genuinely time dependent in the nonlinear regime due to its dependence on $f^e_{\mathbf{k}}$ and $f^h_{\mathbf{k}}$. As a consequence of these complications, it is in most cases advantageous to numerically evaluate the full problem in the original electron–hole picture where all the microscopic processes are treated in the **k** basis.

26.4 Linear optical response

The conceptually simplest experiment probes the linear response of the semiconductor with a weak classical pulse spectrally overlapping the dominant transitions in the vicinity of the band-gap energy, see Fig. 26.3(b). The basic measurable quantities in this setup follow from the linear susceptibility,

$$\chi(\omega) \equiv \frac{P(\omega)}{\epsilon_0 E(\omega)}, \tag{26.34}$$

which is obtained as the probe-induced macroscopic polarization, $P(\omega)$, divided by the probe field, $E(\omega)$.

26.4.1 Elliott formula

Before we discuss the full microscopic correlation contributions to the SBE, we use a phenomenological expression for Γ because this already allows us to identify the principal effects of dephasing. To simplify the analysis, we start from an incoherent semiconductor system, i.e., all polarizations vanish before the system is excited. In the linear limit used, $f^e_{\mathbf{k}}$ and $f^h_{\mathbf{k}}$ remain zero while only a small – linear – polarization $P_{\mathbf{k}}$ is generated. Replacing the microscopic Γ by a phenomenological value $-i\gamma p_\lambda$ and Fourier-transforming into frequency space modifies Eq. (26.33) into

$$\hbar\omega p_\lambda(\omega) = \left(E_{g,0} + E_\lambda - i\gamma\right) p_\lambda(\omega) - d_{vc}\,\phi^R_\lambda(r=0)\langle E(\omega)\rangle. \tag{26.35}$$

Inverting Eq. (26.35) yields

$$p_\lambda(\omega) \equiv \frac{1}{2\pi}\int dt\; p_\lambda(t)\,e^{+i\omega t} = \frac{d_{vc}\,\phi^R_\lambda(r=0)\langle E(\omega)\rangle}{E_{g,0} + E_\lambda - \hbar\omega - i\gamma}, \tag{26.36}$$

which can be directly used in the macroscopic polarization relation (26.9),

$$\begin{aligned} P(\omega) &= \frac{d_{c,v}}{S}\sum\nolimits_{\mathbf{k}} P_{\mathbf{k}}(\omega) = \frac{d_{c,v}}{\sqrt{S}}\sum\nolimits_{\mathbf{k},\lambda} \phi^R_\lambda(\mathbf{k})\,p_\lambda(\omega) = \\ &= \sum_\lambda d_{c,v}\phi^R_\lambda(\mathbf{r}=0)\,p_\lambda(\omega) = \sum_\lambda \frac{\left|d_{cv}\,\phi^R_\lambda(\mathbf{r}=0)\right|^2 \langle \hat{E}(\omega)\rangle}{E_{g,0} + E_\lambda - \hbar\omega - i\gamma}. \end{aligned} \tag{26.37}$$

Inserting this result into Eq. (26.34), we find the famous *Elliott formula* for the linear semiconductor susceptibility

$$\chi(\omega) = \frac{|d_{cv}|^2}{\epsilon_0}\sum_\lambda \frac{\left|\phi^R_\lambda(r=0)\right|^2}{E_{g,0} + E_\lambda - \hbar\omega - i\gamma}. \tag{26.38}$$

Since the linear absorption is proportional to the imaginary part of the susceptibility, see Section 26.4.2, the semiconductor absorption shows characteristic resonances at the frequencies $\omega = E_{g,0} + E_\lambda/\hbar$ corresponding to the excitonic energies. These absorption peaks appear below the band gap of the semiconductor as shown in Fig. 26.3.

The presence of excitonic resonances in $\chi(\omega)$ should not be taken as evidence for the existence of exciton populations in the probed system. In fact, the largest and best-defined resonances are observed for an originally unexcited low-temperature semiconductor. In this case, the system cannot contain any exciton populations because the ground-state system has no electron–hole pairs. Therefore, the weak probe field merely tests the transition possibilities of the interacting system. In other words, the linear response is exclusively determined by the linear polarization that defines the strengths of the different allowed optical transitions.

To illustrate the basic features of the linear optical properties, Fig. 26.4 shows the computed $\mathrm{Im}\,[\chi(\omega)]$ for an unexcited GaAs-type QWI (frame (a)) and QW (frame (b))

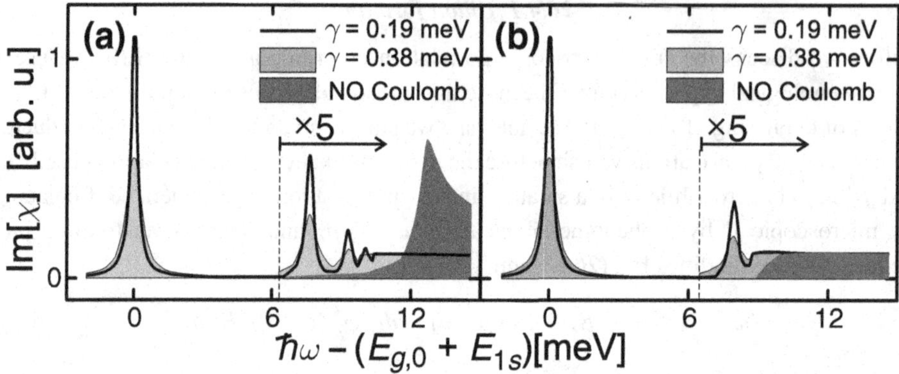

Figure 26.4 Imaginary part of the susceptibility, $\mathrm{Im}[\chi(\omega)]$, obtained by evaluating the Elliott formula with a constant dephasing γ. Results for **(a)** a quantum-wire (QWI) and **(b)** a quantum well (QW) system are shown. To enhance the visibility of the higher excitonic and the continuum resonances, the corresponding part of the spectrum is multiplied by a factor 5. The calculated spectra for $\gamma = 0.19$ meV are plotted as a solid line and the light-shaded area presents the results for $\gamma = 0.38$ meV. For comparison, we plot as a dark-shaded area the spectra obtained from a calculation without Coulomb interaction ($\gamma = 0.38$ meV). The frequency detuning is chosen with respect to the $1s$ resonance. Adapted from *Progress in Quantum Electronics*, **30/5**, M. Kira, S. W. Koch, "Many-body correlations and excitonic effects in semiconductor spectroscopy," pp. 155–296, 2006, with permission from Elsevier.

system. Here, we used the phenomenological dephasing constants $\gamma = 0.19$ meV (solid line) and $\gamma = 0.38$ meV (light-shaded area). Comparing the QW and the QWI results, we immediately notice that the spectra look very similar. As predicted by the oscillator-strength analysis of Fig. 26.3, the absorption is dominated by the bound excitonic resonances energetically below the band gap. By comparing the results for $\gamma = 0.19$ meV (solid line) with those for $\gamma = 0.38$ meV (light-shaded area), we see that the spectral width of the $1s$ resonance is determined by the phenomenological dephasing constant. From the energetically higher excitons, only the $2s$-state is well resolved. The other resonances merge with the onset of the continuum absorption. Due to the negligibly small energetic separation between the individual states, the spectral shape of the continuum absorption is not much affected by the γ used. At the same time, we see that the continuum absorption has a level that is approximately one order-of-magnitude smaller than that of the $1s$-state. Thus, the linear optical response of semiconductors is often dominated by the $1s$-exciton contribution.

The difference between the QW and QWI systems is more pronounced when we switch off the Coulomb interaction (dark-shaded area in Fig. 26.4). In particular, we obtain a peak just above the band gap energy (12 meV above the $1s$ resonance) for the QWI case as a consequence of the broadened $1/\sqrt{\hbar\omega}$ singularity of the 1D density of states, see Eq. (12.46). For the QW system, the density of states is a step function, resulting in a spectrally flat response. It is interesting to see that the Coulomb interaction tends to flatten the true continuum response (solid line) compared with the noninteracting cases. This effect happens also for the response of 3D bulk semiconductors. Thus, the Coulomb coupling not only creates pronounced bound-state resonances below the band gap but it also modifies the response for energies above the band gap, see Exercise 26.6.

26.4.2 Self-consistent optical response

In the derivation of the Elliott formula, we have assumed an unmodified external field by solving the SBE without the coupling to the wave equation. If we want to fully account for the self-consistent coupling between the polarization and the light field, we additionally have to include the wave equation (26.8).

To gain some analytic insight, we assume that the QW and the barrier materials have a negligible difference in their background refractive index such that we can take the same value for n everywhere. With these simplifying assumptions, Eq. (26.8) has the solution

$$\langle \hat{E}(z,t)\rangle = \langle \hat{E}(z,t)\rangle_0 - \mu_0 \frac{c}{2n}\frac{\partial}{\partial u}P(u)\bigg|_{u=t-|nz|/c}, \tag{26.39}$$

based on (25.47). In other words, the incoming light pulse, $\langle \hat{E}(z,t)\rangle_0$, is refracted into the transmitted and the reflected part

$$\langle \hat{E}_R(z,t)\rangle = -\mu_0 \frac{c}{2n}\frac{\partial}{\partial u}P(u)\bigg|_{u=t-|nz|/c}, \qquad z \le 0, \tag{26.40}$$

$$\langle \hat{E}_T(z,t)\rangle = \langle E(z,t)\rangle_0 - \mu_0 \frac{c}{2n}\frac{\partial}{\partial u}P(u)\bigg|_{u=t-|nz|/c}, \qquad z \ge 0, \tag{26.41}$$

respectively, at the QW or QWI position $z = 0$. Taking the Fourier transform of both expressions at $z = 0$, we find the spectrum of the reflected and transmitted fields

$$\langle \hat{E}_R(\omega)\rangle = \frac{1}{2\pi}\int dt\, \langle \hat{E}_R(0,t)\rangle\, e^{+i\omega t} = i\mu_0 \frac{c}{2n}\omega P(\omega), \tag{26.42}$$

$$\langle \hat{E}_T(\omega)\rangle = \frac{1}{2\pi}\int dt\, \langle \hat{E}_T(0,t)\rangle\, e^{+i\omega t} = \langle \hat{E}(\omega)\rangle_0 + i\mu_0 \frac{c}{2n}\omega P(\omega), \tag{26.43}$$

which establishes exact relations for both linear and nonlinear excitation conditions.

In the linear case, we have $P(\omega) = \epsilon_0 \chi(\omega)\langle \hat{E}_T(\omega)\rangle$ because the macroscopic polarization is directly proportional to the field $\langle \hat{E}_T(\omega)\rangle$ at the QW position, see Eq. (26.34). Inserting these relations into Eqs. (26.42)–(26.43), we can uniquely obtain the linear reflection and transmission spectra,

$$\langle \hat{E}_R(\omega)\rangle = \frac{i\xi(\omega)}{1 - i\xi(\omega)}\langle \hat{E}(\omega)\rangle_0, \qquad \langle \hat{E}_T(\omega)\rangle = \frac{1}{1 - i\xi(\omega)}\langle \hat{E}(\omega)\rangle_0, \tag{26.44}$$

respectively. This simple form follows after we identify the scaled susceptibility

$$\xi(\omega) \equiv \frac{1}{2}\frac{\omega}{n\,c}\chi(\omega) = \sum_\lambda \frac{\Gamma^{\mathrm{rad}}_{\lambda,\lambda}}{E_{g,0} + E_\lambda - \hbar\omega - i\gamma}, \tag{26.45}$$

where $c = 1/\sqrt{\mu_0\epsilon_0}$ is the speed of light in vacuum. We have also defined the radiative coupling constant

$$\Gamma^{\rm rad}_{\lambda,\nu} \equiv \frac{1}{2}\frac{|d_{v,c}|^2}{\epsilon_0}\frac{\omega_{g,0}}{n\,c}\phi^R_\lambda(\mathbf{r}=0)\phi^R_\nu(\mathbf{r}=0). \tag{26.46}$$

In principle, $\Gamma^{\rm rad}_{\lambda,\nu}$ should contain the frequency ω instead of $\omega_{g,0} \equiv E_g/\hbar$ corresponding to the band-gap energy. However, since all interesting interband transitions are within a few percent of $\omega_{g,0}$, we may ignore the full frequency dependency in Eq. (26.46). For a QWI system, also the density of wires, $n_{\rm wire}$, has to be included to provide

$$\Gamma^{\rm rad}_{\lambda,\nu} \equiv \frac{1}{2}\frac{|d_{v,c}|^2}{\epsilon_0}\frac{\omega_{g,0}}{n\,c}n_{\rm wire}\phi^R_\lambda(\mathbf{r}=0)\phi^R_\nu(\mathbf{r}=0). \tag{26.47}$$

In our numerical evaluations, we adjust $n_{\rm wire}$ such that the QWI array produces the same $\Gamma^{\rm rad}_{1s,1s}$ as the corresponding QW system. The presented $\Gamma^{\rm rad}_{\lambda,\nu}$ analysis has full analogy with the radiative-decay investigation of planar atoms, discussed in Section 12.2.2.

Along the lines of these arguments, we can now identify the reflection and transmission coefficients in the linear regime as

$$R(\omega) \equiv \frac{\langle \hat{E}_R(\omega)\rangle}{\langle \hat{E}(\omega)\rangle_0} = \frac{i\xi(\omega)}{1-i\xi(\omega)}, \quad T(\omega) \equiv \frac{\langle \hat{E}_T(\omega)\rangle}{\langle \hat{E}(\omega)\rangle_0} = \frac{1}{1-i\xi(\omega)}, \tag{26.48}$$

respectively. From these, we obtain the linear absorption as

$$\alpha(\omega) \equiv 1 - |R(\omega)|^2 - |T(\omega)|^2 = \frac{2{\rm Im}\left[\xi(\omega)\right]}{1+|\xi(\omega)|^2 + 2{\rm Im}\left[\xi(\omega)\right]}, \tag{26.49}$$

which is given as the scaled intensity difference between the incoming and the sum of the reflected and the transmitted fields.

For sufficiently low densities, $\xi(\omega)$ is strongly dominated by the $1s$ resonance as shown in Fig. 26.4. Consequently, in the spectral range of the $1s$-exciton resonance, $\xi(\omega)$ can often be approximated by the first term in the sum (26.45). This approximation produces

$$R(\omega) = \frac{i\Gamma^{\rm rad}_{1s,1s}}{E_{1s} - \hbar\omega - i\left(\gamma + \Gamma^{\rm rad}_{1s,1s}\right)}, \quad T(\omega) = \frac{E_{1s} - \hbar\omega - i\gamma}{E_{1s} - \hbar\omega - i\left(\gamma + \Gamma^{\rm rad}_{1s,1s}\right)}, \tag{26.50}$$

$$\alpha(\omega) \equiv \frac{2\gamma\Gamma^{\rm rad}_{1s,1s}}{(E_{1s}-\hbar\omega)^2 + \left(\gamma + \Gamma^{\rm rad}_{1s,1s}\right)^2}, \tag{26.51}$$

showing that the self-consistent light–matter coupling introduces an additional broadening $\Gamma^{\rm rad}_{1s,1s}$ of the absorption resonance, known as the radiative broadening, see also the discussion in Chapter 19.

If we assume that Im $[\xi(\omega)]$ and $|\xi(\omega)|^2$ are small in magnitude ($\ll 1$), we find that the absorption spectrum reduces to the form

$$\alpha_{\text{Elliott}}(\omega) = 2\text{Im}\,[\xi(\omega)] = \sum_{\lambda} \frac{2\gamma\Gamma^{\text{rad}}_{\lambda,\lambda}}{(E_\lambda + E_{q,0} - \hbar\omega)^2 + \gamma^2}, \qquad (26.52)$$

which connects the Elliott formula (26.38) and the QW absorption. In the Elliott result, the light–matter coupling is not included fully self-consistently because the excitonic resonances are not additionally broadened by $\Gamma^{\text{rad}}_{\lambda,\lambda}$. Nevertheless, $\alpha_{\text{Elliott}}(\omega)$ should describe the QW absorption accurately as long as the nonradiative homogeneous dephasing strongly exceeds the radiative dephasing, i.e., $\gamma \gg \Gamma^{\text{rad}}_{\lambda,\lambda}$, which often is the case in real semiconductor structures.

Equations (26.48)–(26.49) present general relations for the linear reflection, transmission, and absorption of a thin QW even if the response function $\xi(\omega)$ does not have a Lorentzian form. In particular, these relations tell us how the self-consistent light–matter coupling transforms the linear susceptibility $\chi(\omega)$ into measurable quantities. For example, the true single QW absorption, Eq. (26.49), can deviate significantly from its inconsistent counterpart, $\alpha_{\text{Elliott}}(\omega) = 2\text{Im}\,[\xi(\omega)] = \frac{\omega}{nc}\text{Im}\,[\chi(\omega)]$, for situations where the radiative coupling is dominant.

To investigate how the self-consistent light–matter coupling influences the absorption, we compute the actual absorption for a QW and a QWI system using Eq. (26.49). The results (solid line) are presented in Fig. 26.5 and compared with those obtained from Eq. (26.52) (shaded area) for different phenomenological dephasing values γ. More specifically, we use $\gamma = \Gamma^{\text{rad}}_{1s,1s}$ (frames (a) and (c)) and $\gamma = 10\,\Gamma^{\text{rad}}_{1s,1s}$ (frames (b) and (d)) for both QWI (upper frames) and QW (lower frames). We see that for γ close to the radiative damping, the light–matter coupling strongly modifies the absorption at the $1s$ resonance. The true $1s$ absorption becomes considerably smaller in magnitude and broader in width, as predicted by Eq. (26.51). Overall, the radiative coupling reduces α mostly at the $1s$ peak. For the higher excitonic states, the coupling effects are much weaker. These features are still clearly observable for the case of larger dephasing, presented in Figs. 26.5(b) and (d). As a general tendency, the differences between the full solution (26.49) and the approximative Eq. (26.52) become smaller as γ is increased.

26.4.3 Quantitative measurements and dephasing

It is interesting to note that Eqs. (26.50)–(26.51) suggest an experiment-based scheme to extract both γ and $\Gamma_{1s,1s}$ using the measured width of the $1s$ resonance and the measured value of the peak absorption, transmission, or reflection. As an illustration, we consider a case where the strongest $1s$ resonance has a dominantly Lorentzian line shape. In this situation, we may directly apply the relation (26.51) to obtain the peak absorption at $\hbar\omega = E_{1s}$,

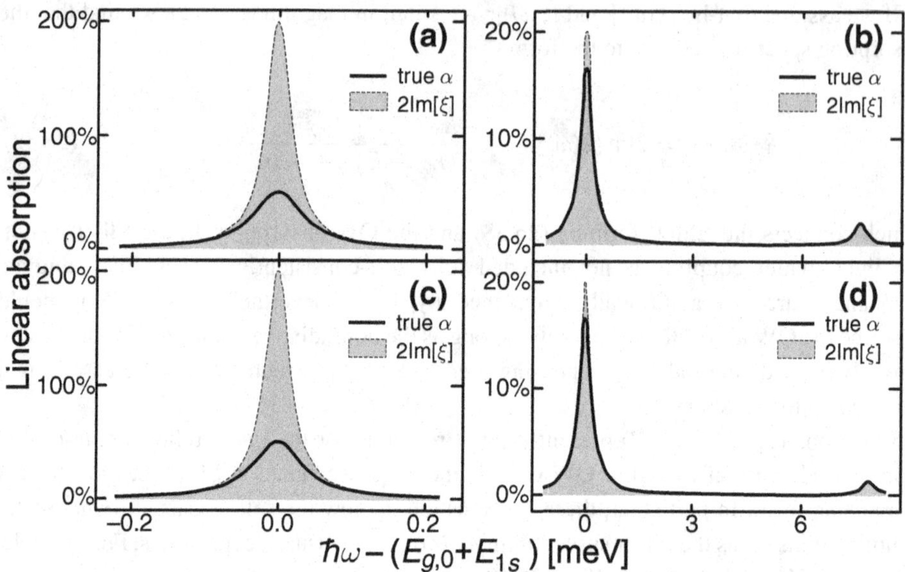

Figure 26.5 Influence of the self-consistent light–matter coupling on the linear absorption. The true absorption (solid line) is compared with α_{Elliott} (dashed, shaded area) which does not include radiative coupling. Frames **(a)** and **(b)** [**(c)** and **(d)**] present the results for a QWI [QW] system using $\Gamma^{\text{rad}}_{1s,1s} = 20\,\mu\text{eV}$. The phenomenological dephasing is taken as $\gamma = \Gamma^{\text{rad}}_{1s,1s}$ in **(a)** and **(c)**, and as $\gamma = 10\Gamma^{\text{rad}}_{1s,1s}$ in **(b)** and **(d)**. Adapted from *Progress in Quantum Electronics*, **30/5**, M. Kira, S. W. Koch, "Many-body correlations and excitonic effects in semiconductor spectroscopy," pp. 155–296, 2006, with permission from Elsevier.

$$\alpha_{1s} \equiv \frac{2\gamma\Gamma^{\text{rad}}_{1s,1s}}{\left(\gamma + \Gamma^{\text{rad}}_{1s,1s}\right)^2} = 2\left[\frac{\Gamma^{\text{rad}}_{1s,1s}}{\gamma}\right] \Bigg/ \left(1 + \left[\frac{\Gamma^{\text{rad}}_{1s,1s}}{\gamma}\right]^2\right), \tag{26.53}$$

which depends only on the ratio $\Gamma^{\text{rad}}_{1s,1s}/\gamma$. Furthermore, the measured half width at half maximum of the $1s$ resonance is given by

$$\gamma_{\text{meas}} = \gamma + \Gamma^{\text{rad}}_{1s,1s}, \tag{26.54}$$

based on Eq. (26.51).

The same analysis can be made for QWs or QWIs embedded within different dielectric layers, described by the refractive-index profile $n(z)$. As long as the linear response is nearly Lorentzian, $1s$- the peak absorption depends only on the ratio $\gamma/\Gamma^{\text{rad}}_{1s,1s}$ through a function α_{uni} that is "universal" for the given dielectric structure. Similarly, the measured half width is still defined by Eq. (26.54). These two pieces of information are enough to retrieve the actual γ and $\Gamma^{\text{rad}}_{1s,1s}$ of the systems via the inversion,

$$\left\{ \begin{array}{l} \alpha_{\text{meas}} = \alpha_{\text{uni}}\left(\frac{\gamma}{\Gamma^{\text{rad}}_{1s,1s}}\right) \\ \gamma_{\text{meas}} = \gamma + \Gamma^{\text{rad}}_{1s,1s} \end{array} \right. \Rightarrow \left\{ \begin{array}{l} \gamma = \gamma(\alpha_{\text{meas}}, \gamma_{\text{meas}}) \\ \Gamma^{\text{rad}}_{1s,1s} = \Gamma^{\text{rad}}_{1s,1s}(\alpha_{\text{meas}}, \gamma_{\text{meas}}) \end{array} \right. . \tag{26.55}$$

For $n(z) = n$, the measured values of the $1s$-peak absorption, α_{meas}, and its half width, γ_{meas}, Eq. (26.55) can be inverted analytically,

$$\gamma = \frac{\gamma_{\text{meas}}}{2}\left(1 \mp \sqrt{1-2\alpha_{\text{meas}}}\right), \quad \Gamma^{\text{rad}}_{1s,1s} = \frac{\gamma_{\text{meas}}}{2}\left(1 \pm \sqrt{1-2\alpha_{\text{meas}}}\right). \tag{26.56}$$

The measurement of the single-QW absorption in absolute units can also be used for the generic α_{uni} to extract the actual values of the radiative and nonradiative dephasing constants.

Figure 26.6 presents the universal peak absorption α_{uni} (solid line) computed from $\xi(\omega)$ with all excitonic states. This result is compared with the analytical calculation of the $1s$-peak absorption (shaded area) including only the $1s$ resonance and the $2\text{Im}[\xi]$ approximation (dashed line) given by Eq. (26.52). Even though the analytic formula includes only the $1s$ state, it reproduces the full result with great accuracy even for elevated values of γ. The Elliott result $\text{Im}[\xi]$ approaches the full α_{1s} only for large dephasing values. The level of deviation between $\text{Im}[\xi]$ and α_{1s} gives an estimate of how accurately the absorption peak height must be measured in order to reliably extract the actual γ and $\Gamma_{1s,1s}$ using Eq. (26.56).

We also notice that α vanishes for $\gamma = 0$. In this situation, the excited semiconductor reradiates all excitation, which yields 100 percent reflection at the $1s$ resonance, as predicted by Eq. (26.50). In the opposite limit, the linear absorption is maximized if γ equals the radiative dephasing. For this case, 50 percent of the light is absorbed because only the cosine part (antinode at the QW) of the plane wave can be absorbed. This maximum absorption is changed when we include, e.g., the air–QW–air index steps in the analysis since the structure then acts as a low-quality optical cavity where the light is multiply reflected resulting in standing-wave contributions. This is studied in Exercise 26.9.

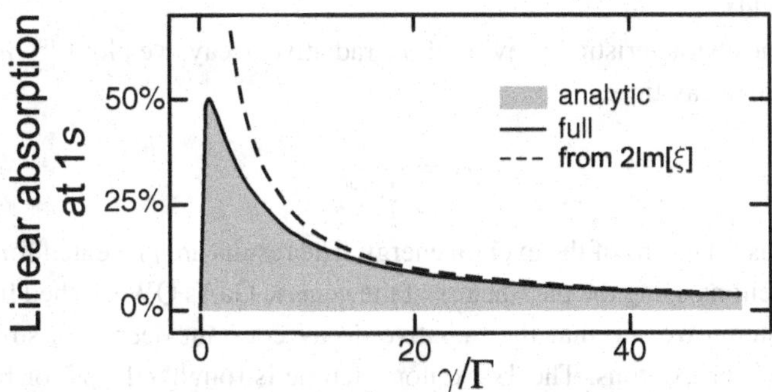

Figure 26.6 The peak absorption at the $1s$ resonance is presented as a function of dephasing for the full calculation (solid line), the analytic result from Eq. (26.53) (shaded area), and the Elliott result, $\text{Im}[\xi]_{1s} = \Gamma^{\text{rad}}_{1s,1s}/\gamma$ (dashed line). Adapted from *Progress in Quantum Electronics*, **30/5**, M. Kira, S. W. Koch, "Many-body correlations and excitonic effects in semiconductor spectroscopy," pp. 155–296, 2006, with permission from Elsevier.

26.4.4 Radiative polarization decay

The exact solution (26.39) for the classical field allows us to fully include the self-consistent light–matter coupling. In particular, applying Eq. (26.39) at the QW position $z = 0$, we obtain

$$\begin{aligned}\langle E(0,t)\rangle &= \langle E(0,t)\rangle_0 - \mu_0 \frac{c}{2n}\frac{\partial}{\partial t}P(t)\\ &= \langle E(0,t)\rangle_0 - \frac{d_{v,c}}{2n\epsilon_0 c}\sum_\lambda \phi^R_\lambda(\mathbf{r}=0)\frac{\partial}{\partial t}p_\lambda,\end{aligned} \tag{26.57}$$

where $\langle E(0,t)\rangle_0$ is the incoming laser field and the macroscopic polarization is expressed in the exciton basis (26.27). We notice from Eq. (26.33) that each different excitonic component has its eigenfrequency $\omega_\lambda = E_\lambda/\hbar$. However, since these frequencies and the excitation frequency are all within a few percent of the band gap frequency $\omega_{g,0} \equiv E_{g,0}/\hbar$, we may use $\frac{\partial}{\partial t}p_\lambda = -i\omega_{g,0}\, p_\lambda$ to simplify (26.57),

$$\langle E(0,t)\rangle = \langle E(0,t)\rangle_0 + i\,\frac{d_{v,c}\omega_{g,0}}{2\epsilon_0 n\, c}\sum_\lambda \phi^R_\lambda(\mathbf{r}=0)p_\lambda(t), \tag{26.58}$$

just as in Eq. (26.46). Inserting this solution back into Eq. (26.33) yields

$$i\hbar\frac{\partial}{\partial t}p_\lambda = E_\lambda p_\lambda - d_{vc}\phi^R_\lambda(\mathbf{r}=0)\langle E(t)\rangle_0 - i\sum_\beta \Gamma^{\rm rad}_{\lambda,\beta}\, p_\beta - i\Gamma_\lambda, \tag{26.59}$$

where the radiative decay constant $\Gamma^{\rm rad}$ is identified according to Eq. (26.46).

If the excitation is resonant with a particular exciton energy E_λ, we may use the approximation that only $p_\nu = \delta_{\nu,\lambda}p_\lambda$ is excited. In this case, Eq. (26.59) predicts that this polarization component decays exponentially with the rate $\Gamma^{\rm rad}_{\lambda,\lambda}/\hbar$ if the additional scattering contribution Γ_λ can be omitted. This radiative decay of the QW polarization is a consequence of the momentum-conservation violation in the z direction of the quantum-confined system. For an infinite bulk material, this radiative decay is replaced by reversible polariton oscillations resulting from the self-consistent coupling between the polarization and the light field.

To analyze the characteristic behavior of the radiative decay, we plot $\Gamma^{\rm rad}_{\lambda,\lambda}$ in Fig. 26.7(a) and the radiative decay time,

$$\tau_\lambda = \frac{\hbar}{2\Gamma^{\rm rad}_{\lambda,\lambda}}, \tag{26.60}$$

in Fig. 26.7(b) as functions of the exciton energy. The results are presented for the optically active s-like excitons using the parameters of the generic GaAs QW (dashed line) and QWI (solid line) system. We see that the radiative decay constant decreases strongly for the energetically higher excitons. The $1s$-exciton lifetime is roughly 15 ps for both the QWI and the QW configuration. For the other states, the lifetime approaches the nanosecond range, which clearly indicates that these states are only weakly coupled to the light field. In Fig. 26.7(b), we note that the transition to the continuum states is indicated by a sharp increase of τ_λ.

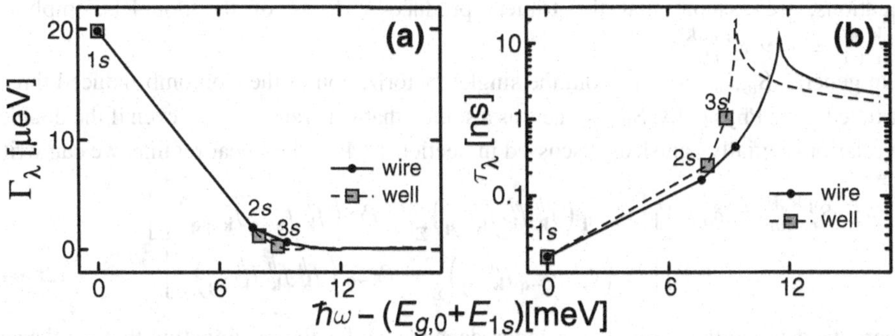

Figure 26.7 Radiative decay rate for the *s*-like exciton states. **(a)** The computed $\Gamma^{\mathrm{rad}}_{\lambda,\lambda}$ and **(b)** the radiative decay time are presented as functions of the exciton energy. These quantities are evaluated using Eqs. (26.47) and (26.60), respectively. The QWI results (solid line) are compared with those for a QW (dashed line). The filled circles (QWI) and squares (QW) indicate the three first bound states. The line between these points is only a guide to the eye since there are no real states in between these points. Adapted from *Progress in Quantum Electronics*, **30/5**, M. Kira, S. W. Koch, "Many-body correlations and excitonic effects in semiconductor spectroscopy," pp. 155–296, 2006, with permission from Elsevier.

26.5 Excitation-induced dephasing

So far, our analysis shows that the coherent optical polarization vanishes from the system either via radiative decay or nonradiatively via the scattering-induced Γ_λ contributions. We next investigate the microscopic origin of Γ_λ by returning to Eq. (26.2). More specifically, we focus on the Coulomb correlations,

$$i\hbar\frac{\partial}{\partial t}P_{\mathbf{k}} = \tilde{\epsilon}_{\mathbf{k}}P_{\mathbf{k}} - \left[1 - f^e_{\mathbf{k}} - f^h_{\mathbf{k}}\right]\Omega_{\mathbf{k}} + \Gamma^{v,c}_{\mathbf{k}}, \tag{26.61}$$

$$\Gamma^{v,c}_{\mathbf{k}} \equiv \sum_{\nu,\mathbf{k}',\mathbf{q}\neq 0} V_{\mathbf{q}}\left[c^{\mathbf{q},\mathbf{k}',\mathbf{k}}_{v,v;v,c} + c^{\mathbf{q},\mathbf{k}',\mathbf{k}}_{v,c;c,c} - \left(c^{\mathbf{q},\mathbf{k}',\mathbf{k}}_{c,v;v,v} + c^{\mathbf{q},\mathbf{k}',\mathbf{k}}_{c,c;c,v}\right)^{\star}\right]. \tag{26.62}$$

In this form, the self-consistent light–matter coupling follows as we use the result (26.57) to define the renormalized Rabi frequency $\Omega_{\mathbf{k}}$. To evaluate Eqs. (26.61)–(26.62), we need the dynamics of the coherent carrier correlations $c_{v,v;v,c}$, $c_{v,c;c,c}$, $c_{c,v;v,v}$, and $c_{c,c;c,v}$. The corresponding doublet dynamics, $i\hbar\frac{\partial}{\partial t}c_{\lambda,\lambda',\nu',\nu}$, can be derived with steps similar to those of Sections 24.5–24.5.2 to produce a complete set of dynamic singlet equations.

For the details of the derivation, see (Kira and Koch, 2006). Here, we directly proceed to the resulting equation for the doublet dynamics,

$$\begin{aligned} i\hbar\frac{\partial}{\partial t}c^{\mathbf{q},\mathbf{k}',\mathbf{k}}_{v,v;v,c} = {} & \left(\tilde{\epsilon}^e_{\mathbf{k}-\mathbf{q}} + \tilde{\epsilon}^h_{\mathbf{k}} - \tilde{\epsilon}^h_{\mathbf{k}'+\mathbf{q}} + \tilde{\epsilon}^h_{\mathbf{k}'} - i\gamma\right)c^{\mathbf{q},\mathbf{k}',\mathbf{k}}_{v,v;v,c} \\ & + \mathrm{S}^{\mathbf{q},\mathbf{k}',\mathbf{k}}_{v,v;v,c} + \mathrm{D}^{\mathbf{q},\mathbf{k}',\mathbf{k}}_{v,v;v,c} + \mathrm{T}^{\mathbf{q},\mathbf{k}',\mathbf{k}}_{v,v;v,c}, \end{aligned} \tag{26.63}$$

where S, D, and T contain the singlet, doublet, and triplet contributions, respectively. As for the singlet dynamics (26.2), the highest-order clusters – here the triplets – contain the residual hierarchical coupling. As the simplest approximation to close the equations, we assume that the triplets produce a decay of the doublets, implying $\mathrm{T}^{\mathbf{q},\mathbf{k}',\mathbf{k}}_{v,v;v,c} = -i\gamma\, c^{\mathbf{q},\mathbf{k}',\mathbf{k}}_{v,v;v,c}$.

In general, $\mathrm{S}_{v,v;v,c}$ results from the singlet factorization of the Coulomb-induced three-particle terms. Physically, $\mathrm{S}_{v,v;v,c}$ acts as a source that generates $c_{v,v;v,c}$ even if the doublet correlations initially vanish as discussed in Section 15.4. In the linear regime, we can write

$$\mathrm{S}^{\mathbf{q},\mathbf{k}',\mathbf{k}}_{v,v;v,c} \equiv \delta_{\sigma,\sigma'} V_{\mathbf{j}} \left[P_{\mathbf{k}-\mathbf{q}} \left(f^h_{\mathbf{k}} f^h_{\mathbf{k}'} \bar{f}^h_{\mathbf{k}'+\mathbf{q}} \right)_\Sigma - P_{\mathbf{k}'} \left(f^h_{\mathbf{k}} f^e_{\mathbf{k}-\mathbf{q}} \bar{f}^h_{\mathbf{k}'+\mathbf{q}} \right)_\Sigma \right] + V_{\mathbf{q}} \left[P_{\mathbf{k}} \left(f^h_{\mathbf{k}'} f^e_{\mathbf{k}-\mathbf{q}} \bar{f}^h_{\mathbf{k}'+\mathbf{q}} \right)_\Sigma - P_{\mathbf{k}-\mathbf{q}} \left(f^h_{\mathbf{k}} f^h_{\mathbf{k}'} \bar{f}^h_{\mathbf{k}'+\mathbf{q}} \right)_\Sigma \right], \tag{26.64}$$

where we denoted the explicit spin-index dependency for the combination that is relevant for optical excitations,

$$c_{v,v;v,c} \equiv c_{(v,\sigma),(v,\sigma');(v,\sigma'),(c,\sigma)}, \qquad S_{v,v;v,c} \equiv S_{(v,\sigma),(v,\sigma');(v,\sigma'),(c,\sigma)}. \tag{26.65}$$

We also introduced the abbreviations

$$\mathbf{j} \equiv \mathbf{k}' + \mathbf{q} - \mathbf{k}, \qquad \bar{f}^\lambda_{\mathbf{k}} = 1 - f^\lambda_{\mathbf{k}}, \tag{26.66}$$

$$\left(f^\lambda_{\mathbf{k}} f^{\lambda'}_{\mathbf{k}'} \bar{f}^{\lambda''}_{\mathbf{k}''} \right)_\Sigma \equiv f^\lambda_{\mathbf{k}} f^{\lambda'}_{\mathbf{k}'} \left(1 - f^{\lambda''}_{\mathbf{k}''}\right) + \left(1 - f^\lambda_{\mathbf{k}}\right) \left(1 - f^{\lambda'}_{\mathbf{k}'}\right) \bar{f}^{\lambda''}_{\mathbf{k}''}. \tag{26.67}$$

We note that all terms in $\mathrm{S}_{v,v;v,c}$ contain P. Hence, $c_{v,v;v,c}$ is generated only if a polarization is present in the system, identifying it as a coherent correlation according to our classification in Section 25.3.

The term $(f^\lambda_{\mathbf{k}} f^{\lambda'}_{\mathbf{k}'} \bar{f}^{\lambda''}_{\mathbf{k}''})_\Sigma$ in Eq. (26.67) has a rather simple scattering interpretation: the $f^\lambda_{\mathbf{k}} f^{\lambda'}_{\mathbf{k}'} (1 - f^{\lambda''}_{\mathbf{k}''})$ part requires occupation of the states $(\lambda, \mathbf{k})$ and $(\lambda', \mathbf{k}')$ and vacancy of the state $(\lambda'', \mathbf{k}'')$. Thus, it describes a scattering process where one particle is removed from the system, which is possible only if polarization is present, since the polarization describes the transition amplitude to remove electrons and holes. This polarization-density scattering $P_{\mathbf{k}''}(f^\lambda_{\mathbf{k}} f^{\lambda'}_{\mathbf{k}'} \bar{f}^{\lambda''}_{\mathbf{k}''})_\Sigma$ requires the Coulomb interaction to provide the momentum exchange among the participating singlets. Physically, these microscopic processes yield the dephasing of the polarization. Since this dephasing results from the electron–hole excitation in the system, it is often referred to as the *excitation-induced dephasing*. In addition, the singlet dynamics (26.61) experiences Coulomb-induced energy renormalizations as well as the Lindhard screening of the Coulomb interaction due to doublet coupling.

Once $c_{v,v;v,c}$ is generated by the singlet source $S_{v,v;v,c}$, it is modified by more complicated terms that consist of products of the type $\Sigma V\langle 1\rangle \Delta\langle 2\rangle$. These contributions may describe the formation of new quasiparticle correlations such as bound doublet correlations among polarization and densities. Nevertheless, the principal effect of the polarization scattering can be understood already without including the $D_{v,v;v,c}$ contribution. This approach corresponds to the so-called second Born approximation when we neglect $D_{v,v;v,c}$ and simplify $\mathrm{T}^{\mathbf{q},\mathbf{k}',\mathbf{k}}_{v,v;v,c} = -i\gamma c^{\mathbf{q},\mathbf{k}',\mathbf{k}}_{v,v;v,c}$ in Eq. (26.63), see discussion in Section 15.5.

The solution of Eqs. (26.61)–(26.62) yields a fully microscopic description for the linear response to a classical probe. Here, we have presented only the dynamics of $c_{v,v;v,c}$. However, the other correlations can be obtained from Eqs. (26.63)–(26.64) using the simple substitution rules

$$v \leftrightarrow c, \quad f^e \to 1 - f^h, \quad f^h \to 1 - f^e \tag{26.68}$$

and/or complex conjugation. In the following, we investigate the excitation-induced dephasing resulting when a linear probe is used to measure the semiconductor absorption for various carrier concentrations using the full microscopic theory. We will next apply the full singlet–doublet approach to explain the details of the excitation-induced dephasing. The explicit forms of $D_{v,v;v,c}$ are relatively lengthy and are elaborated in (Kira and Koch, 2006).

26.5.1 Density-dependent absorption

The essence of the excitation-induced dephasing can be understood as we investigate the linear response of a semiconductor in the presence of a quasistationary plasma density, i.e., we assume that all doublets vanish initially while the distributions $f^e_{\mathbf{k}}$ and $f^h_{\mathbf{k}}$ exist. In practice, this situation can be realized, e.g., for elevated lattice temperatures and/or elevated carrier densities. As discussed above, one can then omit $D_{v,v;v,c}$ from the full $c_{v,v;v,c}$ dynamics (26.63), i.e., apply the second Born approximation. At this level, one obtains a computationally feasible scheme for semiconductor systems of any dimensionality. Since the second Born linear-response results have already shown excellent agreement between theory and experiments for a wide range of parameters, it can be concluded that the underlying assumptions represent a good approximation to the conditions realized in the respective experiments.

For our numerical evaluations, we assume that we probe a system in which we have an incoherent electron–hole plasma with Fermi–Dirac quasiequilibrium distributions of electrons and holes at the lattice temperature, $T = 40$ K. We solve the full singlet–doublet dynamics within (26.63). Figures 26.8(a) and (c) present the computed absorption spectra for a QWI and a QW, respectively, for three representative carrier densities. Figures 26.8(b) and (d) show $(f^e_{\mathbf{k}} + f^h_{\mathbf{k}})$ to quantify the level of excitation.

For the lowest density, the population factor $(f^e_{\mathbf{k}} + f^h_{\mathbf{k}})$ is way below unity. Consequently, we observe clear absorption resonances at the $1s$ and the $2s$ energies in the corresponding spectra. A closer look reveals that the $2s$ resonance is spectrally broader than the $1s$ peak already for the lowest density. This shows one of the basic features of Coulomb-induced dephasing, i.e., the higher excitonic states experience more dephasing than the lower ones. This trend is clearly opposite to that of pure radiative dephasing, compare Fig. 26.7. We can estimate from Fig. 26.8 that the excitation-induced dephasing produces a broadening in the range of $\gamma = 1$ meV for the highest density used. Thus, even moderate densities already lead to dephasing rates which largely exceed the radiative decay $\Gamma^{\text{rad}}_{1s,1s} = 20\,\mu$eV. Hence, for the conditions chosen here, the self-consistent

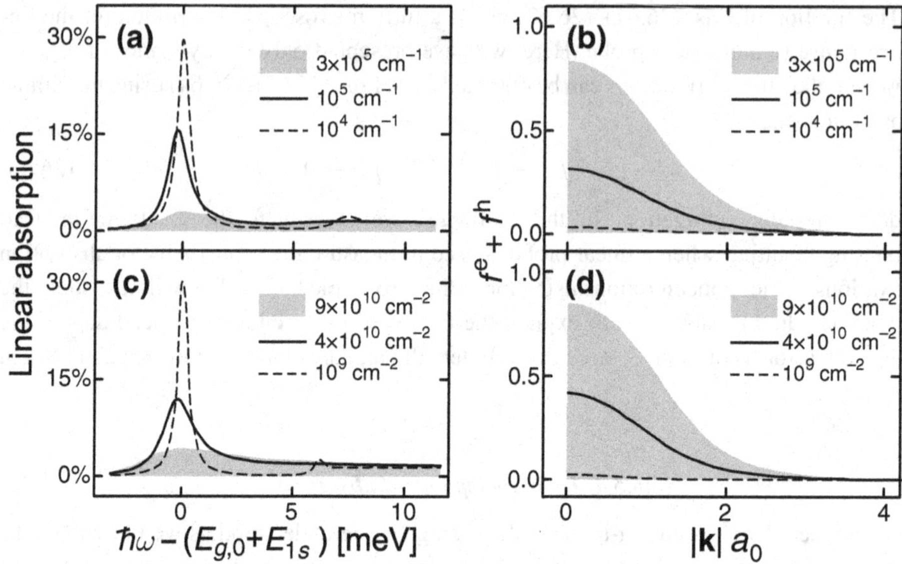

Figure 26.8 Fully microscopically computed self-consistent absorption spectra. The QWI and QW spectra for three different carrier densities are shown in **(a)** and **(c)**, respectively. Frames **(b)** and **(d)** show the assumed Fermi–Dirac quasiequilibrium distributions $(f^e + f^h)$ at 40 K. Adapted from *Progress in Quantum Electronics*, **30/5**, M. Kira, S. W. Koch, "Many-body correlations and excitonic effects in semiconductor spectroscopy," pp. 155–296, 2006, with permission from Elsevier.

light–matter coupling effects become less prominent. As a general trend, we observe that the QW system experiences a bit larger excitation-induced dephasing than the QWI since the phase-space for Coulomb scattering events is larger in two dimensions than in one.

For elevated densities, also the 1*s* resonance is broadened and the absorption dip between the bound and continuum states is gradually filled. This absorption increase is not a consequence of the band-gap shift but is caused by the frequency-dependent scattering. Even for situations where $\left(f^e_{\mathbf{k}} + f^h_{\mathbf{k}}\right)$ is still relatively low, the 2*s* and higher excitons are already bleached. We also see that the spectral position of the 1*s* resonance remains basically unchanged for different carrier densities, indicating that microscopic scattering leads to energy renormalizations that largely compensate each other. This compensation is highly dependent on the excitation configuration and the actual shift of the 1*s* resonance can be sensitively controlled by changing the configuration of the many-body state.

As the density is increased, we see that the 1*s* resonance is nearly completely bleached. The corresponding $\left(f^e_{\mathbf{k}} + f^h_{\mathbf{k}}\right)$ is close to unity, indicating strong phase-space filling effects which eventually eliminate the bound exciton states. Only ionized excitons exist beyond this Mott transition. As the density is increased further, the system enters the regime of negative absorption, i.e., optical gain. The second Born scattering provides an extremely accurate description of semiconductor lasers, even to such a level that one can apply microscopic computations to provide accurate design predictions for quantum-well lasers.

Our numerical evaluations show that the full QWI computation and the second Born results are very similar for the investigated conditions. Hence, we conclude that the contributions of $D_{v,v;v,c}$ are not significant for the plasma conditions analyzed here. However, in cases where quasiparticle correlations are present, these doublets must be included in the analysis to reproduce corresponding experiments. This aspect is further discussed in Chapter 30.

26.5.2 Diffusive model

The principal aspect of excitation-induced dephasing can be understood using an analytic model that includes the important symmetries of the Coulomb-induced scattering. To identify these symmetries, we start from Eqs. (26.61)–(26.62) and notice that Coulomb scattering must satisfy the general conservation law

$$\sum_{\mathbf{k}} \Gamma^{v,c}_{\mathbf{k}} = 0. \tag{26.69}$$

This result suggests that the total polarization is conserved in the presence of pure excitation-induced dephasing. As a result, the Coulomb scattering has a diffusive character in the sense that it redistributes the microscopic polarization without changing the total macroscopic polarization.

For a simple model where $\Gamma^{v,c}_{\mathbf{k}}$ is approximated by the constant dephasing $-i\gamma P_{\mathbf{k}}$, the condition (26.69) cannot be satisfied because the diffusive scattering is replaced by a genuine decay. To improve this approximation, we introduce a diffusive dephasing model

$$\Gamma^{\text{diff}}_{\mathbf{k}} \equiv -i\gamma \left[P_{\mathbf{k}} - \frac{1}{2\pi} \int d\theta \, P_{\mathbf{k}+\mathbf{K}_\theta} \right], \tag{26.70}$$

where $\mathbf{K}_\theta$ is a typical momentum exchange in a Coulomb-scattering process. More specifically, $\mathbf{K}_\theta$ points in all θ directions, as shown in Fig. 26.9. The original $\mathbf{k}$ (black arrow) is compared with a set of scattering $\mathbf{K}_\theta$ (gray arrows) on a circle with the radius K. Since the diffusive model includes all θ, the scattering has no directional preference. For a

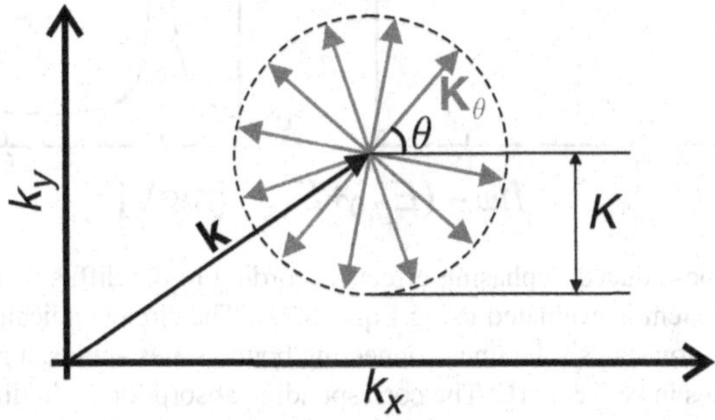

Figure 26.9 Schematic representation of the diffusive scattering model (26.70). The polarization with momentum **k** (black arrow) is scattered on a ring (dashed line). The corresponding scattering momenta are indicated by gray arrows.

QWI system, we can include only angles $\theta = 0$ and $\theta = \pi$ on a line such that the $1/2\pi$ prefactor is replaced by $1/2$ and the integral reduces to a sum. For both QW and QWI, the diffusive model (26.70) clearly satisfies the fundamental relation (26.69). If we use the ansatz (26.70) in the exciton basis Eq. (26.59), we obtain

$$i\hbar\frac{\partial}{\partial t}p_\lambda = \left(E_\lambda - i\Gamma^{\text{rad}}_{\lambda,\lambda} - i\gamma_\lambda\right)p_\lambda - d_{vc}\,\phi^R_\lambda(\mathbf{r}=0)\langle E(t)\rangle_0, \tag{26.71}$$

$$\gamma_\lambda \equiv \gamma\left[1 - \frac{1}{2\pi}\int d\theta \sum_{\mathbf{k}} \phi^L_\lambda(\mathbf{k})\phi^R_\lambda(\mathbf{k}+\mathbf{K}_\theta)\right], \tag{26.72}$$

where we assumed that only the exciton state p_λ is excited.

As an illustration, we present in Figs. 26.10(a) and (c) results for γ_λ obtained by evaluating Eq. (26.72) for a QWI and a QW system, respectively. We clearly see the strong variation of the dephasing for the different exciton states. The dephasing constant is several times larger for the $2s$ than for the $1s$ state, which suffers the weakest excitation-induced dephasing effects. According to our simplified model, the precise value of the dephasing for the energetically higher states shows some nonmonotonic variations, which can be

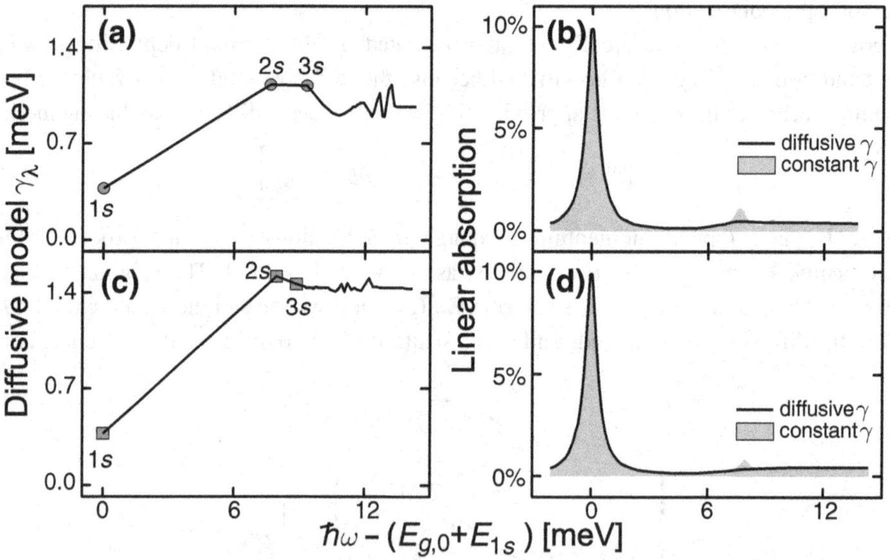

Figure 26.10 Excitation-induced dephasing effects according to the diffusive model. **(a)** The computed γ_λ for a QWI system is evaluated using Eq. (26.72). The circles indicate the positions of the three lowest bound exciton states. The lines connecting bound states are just a guide to the eye since there are no actual states in between. **(b)** The corresponding absorption (solid line) is compared with a constant γ computation (shaded area). The same analysis for a QW system is presented in frames **(c)** and **(d)**. The scattering parameters, $Ka_0 = 1.25$ and γ, are chosen to produce $\gamma_{1s} = 380\,\mu\text{eV}$ for both systems. Reprinted from *Progress in Quantum Electronics*, **30/5**, M. Kira, S. W. Koch, "Many-body correlations and excitonic effects in semiconductor spectroscopy," pp. 155–296, 2006, with permission from Elsevier.

partly attributed to the discrete nature of the diffusive model where the magnitude of the scattering momentum is fixed. Inserting the computed dephasing into the Elliott formula, we obtain the results shown in Figs. 26.10(b) and (d) which qualitatively reproduce the main features of the full calculations, Fig. 26.8. In particular, we immediately understand why the $1s$ state shows a clear resonance while higher excitonic resonances merge into one common resonance close to the $2s$ state.

Exercises

26.1 **(a)** Verify that Eq. (24.58) generates the semiconductor Bloch equations (26.2)–(26.4) when the two-band mode is applied. **(b)** Show that $\tilde{E}_{\mathbf{k}}$ contains a formally diverging contribution $\sum_{\mathbf{k}} V_{\mathbf{k}-\mathbf{k}'}$. This part is included in the energy gap, as shown in Exercise 24.7. **(c)** Verify the different forms of the renormalized Rabi frequency, Eq. (26.7).

26.2 **(a)** Verify the transformations (26.16) assuming that the transformation (26.15) is valid. Start from the mode expansion (24.31). **(b)** Confirm that the RWA converts Eq. (26.12) into (26.17).

26.3 **(a)** Show that the singlet part of Eq. (26.2) produces (26.18). **(b)** The homogeneous part of Eq. (26.18) yields the Wannier equation (26.19). Show that the left- and the right-handed solutions are connected via the relation (26.21). **(c)** Show that $\sum_{\mathbf{k}} \phi^L_\lambda(\mathbf{k})\, \phi^R_\nu(\mathbf{k}) = 0$ for $E_\lambda \neq E_\nu$. Why can we eventually use the relation (26.22)?

26.4 **(a)** Show that $\sum_{\mathbf{k}} V_{\mathbf{k}-\mathbf{k}'}\, \phi^R_\lambda(\mathbf{k}') = [E_\lambda + E_{g,0} - \tilde{\varepsilon}_{\mathbf{k}}]\, \phi^L_\lambda(\mathbf{k})$. **(b)** Show that $\sum_{\mathbf{k},\mathbf{k}'} \phi^R_\nu(\mathbf{k})\, V_{\mathbf{k}-\mathbf{k}'}\, \phi^R_\lambda(\mathbf{k}') = (E_\lambda + E_{g,0})\, \delta_{\lambda,\nu} - \sum_{\mathbf{k}} \phi^R_\nu(\mathbf{k})\, \tilde{\varepsilon}_{\mathbf{k}}\, \phi^L_\lambda(\mathbf{k})$.

26.5 Show that $\sum_{\mathbf{k}} \phi^L_\lambda(\mathbf{k})[i\hbar\nabla_{\mathbf{k}}]\, \phi^R_\nu(\mathbf{k}) = \frac{i\hbar}{E_\nu - E_\lambda} \sum_{\mathbf{k}} \phi^L_\lambda(\mathbf{k})\, \mathbf{k}\, \phi^R_\nu(\mathbf{k})$. **Hint:** *Note that* $i\hbar\nabla_{\mathbf{k}} \hat{=} \hat{\mathbf{x}}$ *corresponds to the position and* $\hbar\mathbf{k} \hat{=} \hat{p}$ *to the momentum.* Show that the Wannier equation can be written as $\hat{H}\,|\phi^R_\lambda\rangle = E_\lambda\,|\phi^R_\lambda$ $\langle\phi^L_\lambda|\,\hat{H} = E_\lambda\,\langle\phi^L_\lambda|$ and compute $\langle\phi^L_\lambda|[\hat{\mathbf{x}}, \hat{H}]_-|\phi^R_\lambda\rangle$ in different ways, see also the discussion in Section 16.3.1.

26.6 **(a)** Implement the approximation $\Gamma_\lambda = -i\,\gamma\, p_\lambda(t)$ into Eq. (26.33) and Fourier transform it via $p_\lambda(\omega) = \frac{1}{2\pi}\int_{-\infty}^{\infty} dt\, p_\lambda(t)\, e^{+i\omega t}$. Show that Eq. (26.35) follows. **(b)** Solve the resulting $p_\lambda(\omega)$ and construct the susceptibility $\chi(\omega) = \frac{1}{S\varepsilon_0}\sum_{\mathbf{k}} d_{cv}\, p_\lambda(\omega)/\langle\hat{E}(\omega)\rangle$. **(c)** Construct $\chi(\omega)$ without Coulomb interaction and show that absorption is possible only for $\hbar\omega > E_{g,0}$. Show also that the continuum absorption depends on the density of states, unlike for the case with Coulomb interaction.

26.7 **(a)** When light interacts with a QW at $z = 0$, it generally follows from

$$\langle\hat{E}(z,t)\rangle = \begin{cases} \langle\hat{E}(z,t)\rangle_0 + \langle\hat{E}_R(z,t)\rangle, & z < 0 \\ \langle\hat{E}_T(z,t)\rangle, & z > 0 \end{cases}.$$

Show that the general solution (26.39) produces the identifications (26.40)–(26.41). **(b)** Fourier transform $\langle E(z,t)\rangle$ and derive the relations (26.44)–(26.45) for $P(\omega) = \varepsilon_0 \chi(\omega)\langle E_T(\omega)\rangle$.

26.8 **(a)** Construct the true QW absorption (26.49) from (26.48). **(b)** Include only the $1s$ resonance to $\xi(\omega)$ and show that Eqs. (26.50)–(26.51) follow. **(c)** Define $\Gamma^{\text{rad}}_{1s,1s}$ for an unexcited QW that is strictly two dimensional. Use $a_0 = 125$ Å, $d_{cv} = e\,5$ Å, and $n = 3.6$.

26.9 Assume that the dielectric profile of the semiconductor system is described by $n(z) = \theta(z - L) + n\ \theta(L - z)$ where $L > 0$ defines the position of the semiconductor–vacuum interface with respect to the QW position. Assume that the QW has the linear response $\chi(\omega) = \frac{d_{cv}|\phi_{1s}(0)|^2/\varepsilon_0}{E_{g,0}+E_{1s}-\hbar\omega-i\gamma}$. Compute the true QW absorption for this radiative environment. **Hint:** *Use the transfer-matrix method outlined in Exercise 24.7.*

26.10 **(a)** Show that the Coulomb contributions within Eq. (26.62) yield a conservation law $\sum_{\mathbf{k}} \Gamma^{v,c}_{\mathbf{k}} = 0$. **(b)** Show that the diffuse model, Eq. (26.70), satisfies this condition. **(c)** Project $\Gamma^{\text{diff}}_{\mathbf{k}}$ into the exciton basis, i.e., compute $\Gamma^{\text{diff}}_{\lambda} = \sum_{\mathbf{k}} \phi^L_\lambda(\mathbf{k})\,\Gamma^{\text{diff}}_{\mathbf{k}}$ and verify Eq. (26.72).

Further reading

For the derivation of the semiconductor Bloch equations see:

M. Lindberg and S. W. Koch (1988). Effective Bloch equations for semiconductors, *Phys. Rev. B* **38**, 3342.

The Wannier exciton was introduced in:

G. Wannier (1937). The structure of electronic excitation levels in insulating crystals, *Phys. Rev.* **52**, 191.

The Elliott formula was developed in:

R. J. Elliott (1963). In *Polarons and Excitons*, C. G. Kuper and D. G. Whitefield, eds., New York, Plenum Press, p. 269.

The connection of radiative decay and unconserved momentum in QWs was pointed out in:

L. C. Andreani, F. Tassone, and F. Bassani (1991). Radiative lifetime of free excitons in quantum wells, *Solid State Commun.* **77**, 641.

The polariton concept was introduced in:

J. J. Hopfield (1958). Theory of the contribution of excitons to the complex dielectric constant of crystals, *Phys. Rev.* **112**, 1555.

For a text-book discussion of Lindhard screening, cf.:

H. Haug and S. W. Koch (2009). *Quantum Theory of the Optical and Electronic Properties of Semiconductors*, 5th edition, Singapore, World Scientific.

For early work on excitation-induced dephasing, cf.:

H. M. Gibbs, A. C. Gossard, S. L. McCall, *et al.* (1979). Saturation of the free exciton resonance in GaAs, *Solid State Commun.* **30**, 271.

S. Schmitt–Rink, D. Chemla, and D. A. B. Miller (1989). Linear and nonlinear optical properties of semiconductor quantum wells, *Adv. Phys.* **38**, 89.

H. Wang, K. Ferrio, D. G. Steel, *et al.* (1993). Transient nonlinear optical response from excitation induced dephasing in GaAs, *Phys. Rev. Lett.* **71**, 1261.

The theory of excitation-induced dephasing and other many-body effects is reviewed in:

M. Kira and S. W. Koch (2006). Many-body correlations and excitonic effects in semiconductor spectroscopy, *Prog. Quantum Electr.* **30**, 155.

The sensitivity of excitation-induced dephasing on the specific form of the many-body state can be applied to construct many-body distributions from quantitative absorption measurements, see:

R. P. Smith, J. K. Wahlstrand, A. C. Funk, *et al.* (2010). Extraction of many-body configurations from nonlinear absorption in semiconductor quantum wells, *Phys. Rev. Lett.* **104**, 247401.

Examples of microscopic laser calculations are shown, e.g., in:

W. W. Chow and S. W. Koch (1999). *Semiconductor-Laser Fundamentals: Physics of the Gain Materials*, Berlin, Springer.

J. V. Moloney, J. Hader, and S. W. Koch (2007). Quantum design of semiconductor active materials: Laser and amplifier applications, *Laser & Photonics Rev.* **1**, 24.

27

Coherent vs. incoherent excitons

In the previous chapter, we discussed excitonic states by solving the homogeneous part of the semiconductor Bloch equations (SBE). In some cases, we use the exciton basis to transform the SBE into an analytically simpler format. The resulting description is particularly straightforward if the densities are constant and the polarization varies linearly with the applied electric field. As we expand the polarization into the exciton basis, we eventually find the Elliott formula (26.38) that describes the linear semiconductor absorption in terms of strongly peaked resonances at the bound exciton states. The excitonic wave functions themselves are obtained by solving the Wannier equation (26.19) that can be related to the standard hydrogen problem. However, despite these similarities, we always have to keep in mind that the polarization itself cannot describe an atom-like electron–hole binding because it is fundamentally a single-particle quantity and not a two-particle correlation function.

To improve our understanding of excitonic states, we start from the general singlet excitations where the density and polarization dynamics are included in all orders in the field. Once we know the exact many-body wave function corresponding to the singlet excitations, we can define the concept of coherent excitons. Next, we show that the coherent excitons can transform into incoherent exciton populations consisting of truly bound electron–hole pairs. The appearance of these excitonic populations requires the build-up of pair-wise correlations among the electrons and holes.

27.1 General singlet excitations

The classical light field generates singlet excitations in the form of the microscopic polarization $P_{\mathbf{k}}$ and the carrier densities $f^e_{\mathbf{k}}$ and $f^h_{\mathbf{k}}$, as discussed in Chapters 25–26. Their quantum dynamics follows from the Maxwell-semiconductor Bloch equations (MSBE) (26.2)–(26.4) and (26.8),

$$\left[\frac{\partial^2}{\partial z^2} - \frac{n^2(z)}{c^2}\frac{\partial^2}{\partial t^2}\right]\langle \hat{E}^{(+)}(z,t)\rangle = -\mu_0 \delta(z)\frac{\partial^2}{\partial t^2}P, \tag{27.1}$$

$$\begin{cases} i\hbar\frac{\partial}{\partial t}P_{\mathbf{k}} = \tilde{\epsilon}_{\mathbf{k}}P_{\mathbf{k}} - \left[1 - f^e_{\mathbf{k}} - f^h_{\mathbf{k}}\right]\Omega^{(+)}_{\mathbf{k}} + \Gamma^{v,c}_{\mathbf{k}} + \Gamma^{\mathrm{QED}}_{v,c;\mathbf{k}} \\ \hbar\frac{\partial}{\partial t}f^e_{\mathbf{k}} = \mathrm{Im}\left[2P_{\mathbf{k}}\left[\Omega^{(+)}_{\mathbf{k}}\right]^\star + \Gamma^{c,c}_{\mathbf{k}} + \Gamma^{\mathrm{QED}}_{c,c;\mathbf{k}}\right] \\ \hbar\frac{\partial}{\partial t}f^h_{\mathbf{k}} = \mathrm{Im}\left[2P_{\mathbf{k}}\left[\Omega^{(+)}_{\mathbf{k}}\right]^\star - \Gamma^{v,v}_{\mathbf{k}} - \Gamma^{\mathrm{QED}}_{v,v;\mathbf{k}}\right] \end{cases} . \tag{27.2}$$

The macroscopic polarization is $P \equiv \frac{1}{S}\sum_{\mathbf{k}} d_{v,c}\;P_{\mathbf{k}}$, the renormalized band gap is $\tilde{\epsilon}_{\mathbf{k}} = E_{g,\mathbf{k}} - \sum_{\mathbf{k}'} V_{\mathbf{k}-\mathbf{k}'}\left(f^e_{\mathbf{k}'} + f^h_{\mathbf{k}'}\right)$, and the renormalized Rabi frequency is given by

$$\Omega^{(+)}_{\mathbf{k}} = d_{c,v}\langle \hat{E}^{(+)}(0,t)\rangle + \sum\nolimits_{\mathbf{k}'} V_{\mathbf{k}-\mathbf{k}'}P_{\mathbf{k}'}, \tag{27.3}$$

which assumes an optically thin QW positioned at $z = 0$. The coupling to doublets follows from the Coulomb and the phonon correlations within $\Gamma^{\lambda,\nu}_{\mathbf{k}}$ and from the quantum-optical correlations $\Gamma^{\mathrm{QED}}_{\lambda,\nu;\mathbf{k}}$. Their explicit form is given by Eqs. (26.11)–(26.12).

The MSBE describe the general excitations of semiconductors to all orders in the electric field providing the basis for the analysis of linear and nonlinear excitation situations. Due to the dynamic nature of the densities, it is not useful to expand the equation into the exciton basis (26.27) to simplify the problem. However, since the equation structure is relatively simple, we can determine the exact many-body function corresponding to the excited singlets regardless of how nonlinear the excitation is.

27.1.1 Coherent limit

In our earlier analysis of the singlet dynamics without the Coulomb terms, it was convenient to express the singlets in terms of the Bloch vector

$$\mathbf{S}_{\mathbf{k}} \equiv \left(\mathrm{Re}\left[P_{\mathbf{k}}\right],\, \mathrm{Im}\left[P_{\mathbf{k}}\right],\, \frac{1}{2}\left(f^e_{\mathbf{k}} + f^h_{\mathbf{k}} - 1\right)\right), \tag{27.4}$$

based on Eq. (25.6). Without dephasing and Coulomb interaction, the length of the Bloch vector is conserved regardless of the exciting field strength. More specifically, we have seen in Section 25.1.1 that

$$|\mathbf{S}_{\mathbf{k}}|^2 = |P_{\mathbf{k}}|^2 + \frac{1}{4}\left(f^e_{\mathbf{k}} + f^h_{\mathbf{k}} - 1\right)^2 \tag{27.5}$$

has a fixed value of 1/4 in the so-called coherent limit (25.24) where dephasing mechanisms are not present.

We next analyze how $|\mathbf{S}_{\mathbf{k}}|$ evolves when the Coulomb interaction is included but the doublet correlations are still ineffective such that we can set $\Gamma^{\lambda,\nu}$ and Γ^{QED} to zero in Eq. (27.2). This produces the dynamics

$$\begin{aligned} i\hbar\frac{\partial}{\partial t}|\mathbf{S}_{\mathbf{k}}|^2 &= P^\star_{\mathbf{k}} i\hbar\frac{\partial}{\partial t}P_{\mathbf{k}} + P_{\mathbf{k}} i\hbar\frac{\partial}{\partial t}P^\star_{\mathbf{k}} + \frac{1}{2}\left(f^e_{\mathbf{k}} + f^h_{\mathbf{k}} - 1\right)\left(i\hbar\frac{\partial}{\partial t}f^e_{\mathbf{k}} + i\hbar\frac{\partial}{\partial t}f^h_{\mathbf{k}}\right) \\ &= 2i\,\mathrm{Im}\left[P^\star_{\mathbf{k}} i\hbar\frac{\partial}{\partial t}P_{\mathbf{k}}\right] + 2i\left(f^e_{\mathbf{k}} + f^h_{\mathbf{k}} - 1\right)\mathrm{Im}\left[P_{\mathbf{k}}\,[\Omega^{(+)}_{\mathbf{k}}]^\star\right]. \end{aligned} \tag{27.6}$$

Here, the second step follows by connecting the first and the second derivatives of the polarization via complex conjugation. Furthermore, we have used (27.2) for the density dynamics to produce the last term. Inserting Eq. (27.2) into Eq. (27.6) yields

$$\begin{aligned} i\hbar\frac{\partial}{\partial t}|\mathbf{S}_\mathbf{k}|^2 &= 2i\,\mathrm{Im}\left[\tilde{\epsilon}_\mathbf{k}P^\star_\mathbf{k}P_\mathbf{k} - \left[1-f^e_\mathbf{k}-f^h_\mathbf{k}\right]P^\star_\mathbf{k}\Omega^{(+)}_\mathbf{k}\right] \\ &\quad + 2i\left(f^e_\mathbf{k}+f^h_\mathbf{k}-1\right)\mathrm{Im}\left[P_\mathbf{k}\,[\Omega^{(+)}_\mathbf{k}]^\star\right] \\ &= 2i\left(f^e_\mathbf{k}+f^h_\mathbf{k}-1\right)\mathrm{Im}\left[P^\star_\mathbf{k}\Omega^{(+)}_\mathbf{k} + P_\mathbf{k}\,[\Omega^{(+)}_\mathbf{k}]^\star\right] = 0. \end{aligned} \tag{27.7}$$

Consequently, as long as the doublets vanish, $|\mathbf{S}_\mathbf{k}|^2$ is a constant of the motion even when the Coulomb interaction couples the different polarization components.

If the doublets can be neglected, the equations for the electron and hole distributions have identical source terms, according to Eq. (27.2). Thus, we can express them through a common electron–hole distribution $f_\mathbf{k} \equiv f^e_\mathbf{k} = f^h_\mathbf{k}$. Furthermore, $|\mathbf{S}_\mathbf{k}|^2 = 1/4$, as discussed in Section 25.2.3 such that we identify the strict conservation law

$$\left(f_\mathbf{k}-\frac{1}{2}\right)^2 + |P_\mathbf{k}|^2 = \frac{1}{4}. \tag{27.8}$$

This equation is valid in the coherent limit before the formation of doublet correlations.

27.1.2 Many-body state of singlet excitations

Since the coherent-limit system does not have any correlations, the Hartree–Fock factorization is exact and we should be able to find the many-body wave function in the form of a Slater determinant. Thus, we make the ansatz,

$$|\Psi_{\mathrm{coh}}\rangle = \prod_\mathbf{k} L^\dagger_\mathbf{k}|\Psi_0\rangle, \tag{27.9}$$

where $|\Psi_0\rangle$ is the state of the completely empty semiconductor and $L^\dagger_\mathbf{k}$ is a Fermionic creation operator. The abstract notation $|\Psi_{\mathrm{coh}}\rangle$ corresponds to a real-space Slater determinant, which can be identified uniquely if we can find the $L^\dagger_\mathbf{k}$ that produces $f_\mathbf{k}$ and $P_\mathbf{k}$ satisfying the condition (27.8).

By choosing

$$L^\dagger_\mathbf{k} = \cos\frac{\vartheta_\mathbf{k}}{2}\,a^\dagger_{v,\mathbf{k}} + \mathrm{e}^{-i\Delta\phi_\mathbf{k}}\sin\frac{\vartheta_\mathbf{k}}{2}\,a^\dagger_{c,\mathbf{k}}, \tag{27.10}$$

each electron is created in a state described by the two-level wave function identified in connection with Eq. (25.9). The operators $L_\mathbf{k}$ and $L^\dagger_\mathbf{k}$ satisfy the usual Fermionic anticommutation relations,

$$[L_\mathbf{k}, L_{\mathbf{k}'}]_+ = \left[L^\dagger_\mathbf{k}, L^\dagger_{\mathbf{k}'}\right]_+ = 0, \qquad \left[L_\mathbf{k}, L^\dagger_{\mathbf{k}'}\right]_+ = \delta_{\mathbf{k},\mathbf{k}'}. \tag{27.11}$$

Evaluating the expectation values with the coherent-state wave function (27.9), we can determine all singlets

$$\left\langle a^\dagger_{c,\mathbf{k}} a_{c,\mathbf{k}'} \right\rangle_{\text{coh}} = \left\langle a_{v,\mathbf{k}'} a^\dagger_{v,\mathbf{k}} \right\rangle_{\text{coh}} = \delta_{\mathbf{k},\mathbf{k}'} \sin^2 \frac{\vartheta_\mathbf{k}}{2} \equiv \delta_{\mathbf{k},\mathbf{k}'} f_\mathbf{k}, \tag{27.12}$$

$$\left\langle a^\dagger_{v,\mathbf{k}} a_{c,\mathbf{k}'} \right\rangle_{\text{coh}} = \delta_{\mathbf{k},\mathbf{k}'} \sin \frac{\vartheta_\mathbf{k}}{2} \cos \frac{\vartheta_\mathbf{k}}{2} \, e^{-i\Delta\phi_\mathbf{k}} \equiv \delta_{\mathbf{k},\mathbf{k}'} P_\mathbf{k}. \tag{27.13}$$

We see that $|\Psi_{\text{coh}}\rangle$ defines a homogeneous excitation because only diagonal contributions exist, compare the discussion in Section 24.1.1. The polarization, the density, and the angle parameters are

$$\begin{cases} P_\mathbf{k} = \frac{1}{2} e^{-i\Delta\phi_\mathbf{k}} \sin \vartheta_\mathbf{k} \\ f_\mathbf{k} = \frac{1}{2}(1 - \cos \vartheta_\mathbf{k}) \end{cases} \Leftrightarrow \begin{cases} e^{-i\Delta\phi_\mathbf{k}} = \frac{P_\mathbf{k}}{|P_\mathbf{k}|} \\ \vartheta_\mathbf{k} = \arccos(1 - 2 f_\mathbf{k}) \end{cases}, \tag{27.14}$$

which uniquely specifies the many-body wave function (27.9). We recognize a clear connection of $\Delta\phi_\mathbf{k}$ and $\vartheta_\mathbf{k}/2$ to the angle-parametrized representation of Bloch vectors, discussed in Section 19.1.3.

We can easily show that the parameters $(\Delta\phi_\mathbf{k}, \vartheta_\mathbf{k})$ indeed yield a singlet state $|\Psi_{\text{coh}}\rangle$ that satisfies the coherent limit condition (27.8),

$$(f_\mathbf{k} - \frac{1}{2})^2 + |P_\mathbf{k}|^2 = \frac{1}{4}\left(\cos^2 \vartheta_\mathbf{k} + \sin^2 \vartheta_\mathbf{k}\right) = \frac{1}{4}. \tag{27.15}$$

Hence, any excited singlet state can always be uniquely expressed via the many-body wave function (27.9). In a situation where the polarization exists only in the $1s$ state, we find

$$\begin{cases} P_\mathbf{k} = |p_{1s}| \, e^{-i\omega_{1s} t} \phi^R_{1s}(\mathbf{k}) \\ f_\mathbf{k} = \frac{1}{2}(1 - \cos \vartheta_\mathbf{k}) \end{cases} \Rightarrow \begin{cases} \Delta\phi_\mathbf{k} = \omega_{1s} t \\ \vartheta_\mathbf{k} = \arcsin 2|p_{1s}|\phi^R_{1s}(\mathbf{k}) \end{cases},$$

which can be constructed from the exciton-basis expansion (26.27) containing only the $1s$ component. Here, we find the characteristic oscillation frequency $\omega_{1s} \equiv (E_{g,0} + E_{1s})/\hbar$ that follows from the homogeneous part of Eq. (26.30).

The coherent-limit wave function shows that for each $\mathbf{k}$, the electron is in a superposition between its conduction- and valence-band state. The mixing ratio is defined by $\sin^2 \frac{\vartheta_\mathbf{k}}{2}$ and $\cos^2 \frac{\vartheta_\mathbf{k}}{2}$ identifying the probabilities of finding the electron in the conduction and valence band, respectively. At the same time, this superposition yields the polarization $P_\mathbf{k}$, which can have excitonic features. Figure 27.1(a) shows a schematic example of how an excitonic polarization can be represented in terms of $\vartheta_\mathbf{k}$. More specifically, the relative height of the dark portion of each bar presents $\vartheta_\mathbf{k}$ as a function of $\mathbf{k}$ for a representative $|\Psi_{\text{coh}}\rangle$ state. The radius of the spheres at the $\vartheta_\mathbf{k} = 0$ ($\vartheta_\mathbf{k} = \pi$) curve is proportional to the valence-band content $\cos^2 \frac{\vartheta_\mathbf{k}}{2}$ (conduction-band content $\sin^2 \frac{\vartheta_\mathbf{k}}{2}$) of the individual electron states.

We see that each charge carrier state $\mathbf{k}$ is described by a superposition of the electron–hole single-particle wave function. Clearly, the excitonic polarization does not describe true excitons because those require the binding of an electron to *another* hole. The description of this bound state requires pair-wise, i.e., doublet, correlations. In cases where the coherent system shows a dominantly $1s$ polarization, the exciton wave function $\phi^R_{1s}(\mathbf{k})$ is an order

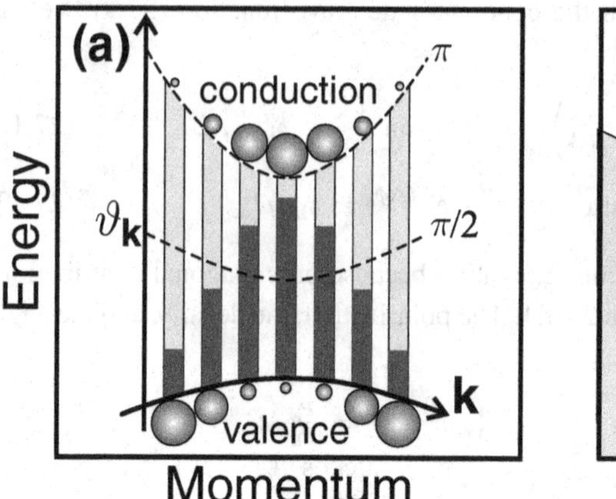

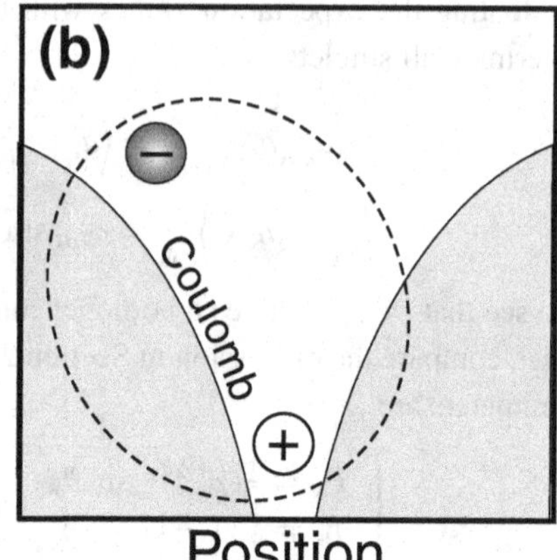

Figure 27.1 Schematic presentation of single-particle states using the coherent-limit Slater determinant. **(a)** The $\vartheta_{\mathbf{k}}$ parameter (dark bars) is plotted as a function of carrier momenta. Each bar is scaled with respect to the corresponding full bar. The respective size of the circles at the $\vartheta_{\mathbf{k}} = 0$ and $\vartheta_{\mathbf{k}} = \pi$ curves indicates the conduction- and valence-band content of each electron. The curves $\vartheta_{\mathbf{k}} = 0$ and $\vartheta_{\mathbf{k}} = \pi$ depict the usual conduction and valence-band energies. **(b)** Illustrative picture of an exciton where an electron (sphere) is bound to a hole (open circle) through the attractive Coulomb potential (shaded area).

parameter that determines for each momentum state which electron fraction is found in the conduction band. To distinguish an excitonic polarization from true excitons, we refer to this polarization as *coherent excitons*.

27.1.3 Coherent excitonic polarization

Since the coherent limit can also be discussed in the exciton basis, Eq. (26.27), it is worthwhile to study how excitonic features enter the exact wave function, Eq. (27.9). For this purpose, it is convenient to introduce an exciton operator,

$$X_{\lambda,\mathbf{q}} \equiv \sum_{\mathbf{k}} \phi_\lambda^L(\mathbf{k}) a^\dagger_{v,\mathbf{k}-\mathbf{q}_h} a_{c,\mathbf{k}+\mathbf{q}_e}, \tag{27.16}$$

containing the center-of-mass momentum $\mathbf{q}$ and

$$\mathbf{q}_e = \frac{m_e}{m_e + m_h}\mathbf{q}, \qquad \mathbf{q}_h = \frac{m_h}{m_e + m_h}\mathbf{q}. \tag{27.17}$$

The inverse transformation from the exciton to the electron–hole picture follows from

$$a^\dagger_{v,\mathbf{k}-\mathbf{q}_h} a_{c,\mathbf{k}+\mathbf{q}_e} = \sum_\lambda \phi_\lambda^R(\mathbf{k}) X_{\lambda,\mathbf{q}}, \tag{27.18}$$

where we used the completeness relation,

$$\sum_{\lambda} \phi_{\lambda}^{R}(\mathbf{k}') \phi_{\lambda}^{L}(\mathbf{k}) = \delta_{\mathbf{k},\mathbf{k}'}, \tag{27.19}$$

to simplify the λ sum.

Sometimes, the optical properties of semiconductors are modeled by treating $X_{\lambda,\mathbf{q}}$ as a Bosonic operator. However, this approximation can seriously compromise the validity of the description because exciton operators have a Fermionic substructure, which cannot be ignored in most of the relevant semiconductor investigations, see Exercise 27.2. Thus, we always perform our explicit calculations in the electron–hole picture. This way, we not only include the correct Fermionic properties but we can also evaluate the excitonic features by using the transformations (27.16) and (27.18).

Next, we introduce the operator

$$S = \sum_{\lambda} \left[c_{\lambda} X_{\lambda,0}^{\dagger} - c_{\lambda}^{\star} X_{\lambda,0} \right], \tag{27.20}$$

which has the interesting property

$$\begin{aligned} e^{S} a_{v,\mathbf{k}}^{\dagger} e^{-S} &= a_{v,\mathbf{k}}^{\dagger} + \left[S, a_{v,\mathbf{k}}^{\dagger} \right]_{-} + \frac{1}{2!} \left[S, \left[S, a_{v,\mathbf{k}}^{\dagger} \right]_{-} \right]_{-} + \cdots \\ &= \mathrm{e}^{-i\Delta\phi_{\mathbf{k}}} \sin\frac{\vartheta_{\mathbf{k}}}{2}\, a_{c,\mathbf{k}}^{\dagger} + \cos\frac{\vartheta_{\mathbf{k}}}{2}\, a_{v,\mathbf{k}}^{\dagger} = L_{\mathbf{k}}^{\dagger}. \end{aligned} \tag{27.21}$$

This transformation produces the Fermionic $L_{\mathbf{k}}^{\dagger}$ operator when we make the identification

$$\frac{1}{2} \vartheta_{\mathbf{k}} \, \mathrm{e}^{-i\Delta\phi_{\mathbf{k}}} \equiv \sum_{\lambda} c_{\lambda} \phi_{\lambda}^{L}(\mathbf{k}). \tag{27.22}$$

The inverse relation is

$$c_{\lambda} = \sum_{\mathbf{k}} \phi_{\lambda}^{R}(\mathbf{k}) \, \frac{1}{2} \, \vartheta_{\mathbf{k}} \, \mathrm{e}^{-i\Delta\phi_{\mathbf{k}}}. \tag{27.23}$$

Consequently, we can write $L_{\mathbf{k}}^{\dagger} = e^{S} a_{v,\mathbf{k}}^{\dagger} e^{-S}$. Using this general connection, the coherent-limit wave function can be expressed as

$$\begin{aligned} |\Psi_{\mathrm{coh}}\rangle &= \prod_{\mathbf{k}} \left(e^{S} a_{v,\mathbf{k}}^{\dagger} e^{-S} \right) |\Psi_0\rangle \\ &= e^{S} \left(\prod_{\mathbf{k}} a_{v,\mathbf{k}}^{\dagger} \right) e^{-S} |\Psi_0\rangle = e^{S} \prod_{\mathbf{k}} a_{v,\mathbf{k}}^{\dagger} |\Psi_0\rangle \end{aligned} \tag{27.24}$$

since $e^{S} e^{-S} = 1$ and $e^{-S}|\Psi_0\rangle = |\Psi_0\rangle$. With the help of this form, we see that $\prod_{\mathbf{k}} a_{v,\mathbf{k}}^{\dagger} |\Psi_0\rangle \equiv |G\rangle$ is the ground state of a semiconductor defined by Eq. (24.14). Thus, the coherent-limit state can equivalently be presented as

$$|\Psi_{\mathrm{coh}}\rangle = e^{S} |G\rangle, \tag{27.25}$$

showing that the operator e^S generates coherent-limit excitations from the semiconductor ground state.

To study the coherent limit from an alternative point of view, we define the functional displacement operator

$$D[c] \equiv e^{S[c]} = e^{\sum_\lambda \left(c_\lambda X^\dagger_{\lambda,0} - c^\star_\lambda X_{\lambda,0}\right)}, \tag{27.26}$$

which creates the coherent-limit wave function, $|\Psi_{\rm coh}\rangle = D[c]|G\rangle$, according to Eq. (27.25). The functional form of D resembles the displacement operator (22.47) generating coherent states for fully Bosonic fields, as presented in Section 22.3.1. Based on this formal analogy, we can argue that $|\Psi_{\rm coh}\rangle$ describes *coherent exciton states*. This observation connects an excitonic polarization to coherent excitons, as suggested in Section 27.1.2. However, we should always keep in mind that the operator $X_{\lambda,0}$ is fundamentally non-Bosonic such that $|\Psi_{\rm coh}\rangle$ should not be interpreted as a Bosonic exciton. Moreover, $|\Psi_{\rm coh}\rangle$ is still a Slater determinant of single-electron functions in a conduction–valence-band superposition state, as discussed in Section 27.1.2. Hence, such a coherent exciton state is fundamentally different from an incoherent population of truly bound excitons, which is a strongly correlated two-particle electron–hole-pair state.

27.2 Incoherent excitons

As a very broad definition, an exciton is a quasiparticle that consists of a Coulomb-bound electron–hole pair. Even though this statement seems simple and straightforward, one cannot even define a rigorous exciton number operator once one seriously considers the Fermionic nature of the constituent electron and hole, see also Exercise 27.2. To identify excitons in practice, we refer to exciton populations in a system only if we have correlations between at least one electron and hole. Furthermore, true excitons exhibit Coulombic *correlation in the relative coordinate between the electron and hole* and the dependency of the pair wave function is governed by *the generalized Wannier equation*. The *bound (ionized) state solutions* define *bound (ionized) excitons*.

These requirements are illustrated in Fig. 27.2. If the pair-wise correlation between electron and hole resembles, e.g., a $1s$ distribution (shaded area), the electron is clearly bound to a hole. This many-body configuration contains at least one exciton that is presented schematically on the left (shaded arrow). A flat pair-wise correlation (dashed line) indicates that electrons and holes move freely with respect to each other, implying that the system contains an electron–hole plasma. The corresponding "free" motion is illustrated on the RHS of the figure.

We find a natural way to identify excitons once we evaluate the electron–hole pair-correlation function

$$g_{\rm eh}(\mathbf{r}) \equiv \left\langle \Psi^\dagger_e(\mathbf{r})\Psi^\dagger_h(0)\Psi_h(0)\Psi_e(\mathbf{r})\right\rangle, \tag{27.27}$$

$$\Psi_e(\mathbf{r}) = \frac{1}{\sqrt{S}}\sum\nolimits_{\mathbf{k}} a_{c,\mathbf{k}} e^{i\mathbf{k}\cdot\mathbf{r}}, \qquad \Psi^\dagger_h(\mathbf{r}) = \frac{1}{\sqrt{S}}\sum\nolimits_{\mathbf{k}} a_{v,\mathbf{k}} e^{i\mathbf{k}\cdot\mathbf{r}}. \tag{27.28}$$

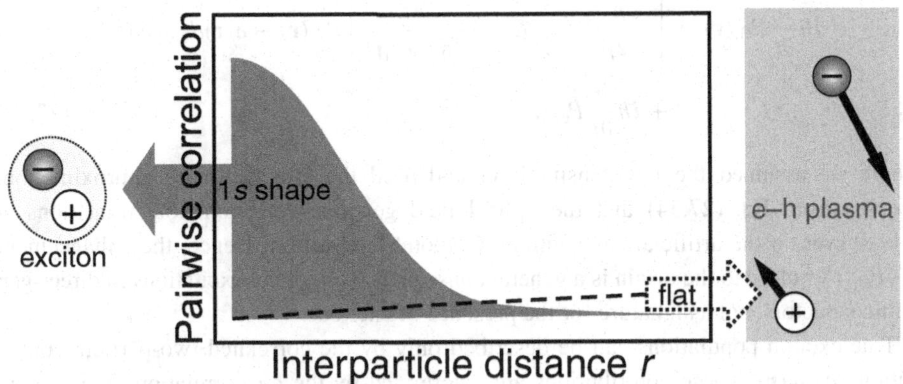

Figure 27.2 Pairwise correlations among electrons and holes. The presented 1*s* shape (shaded area) indicates the existence of true excitons (left) while a flat shape implies that the system is in an electron–hole-plasma state (right).

Here, only the envelope part of the field operator $\hat{\Psi}_\lambda(\mathbf{r})$ is included because the characteristic dimensions of Wannier excitons involve length scales much longer than the lattice periodic part of the Bloch functions. In general, $g_{\text{eh}}(\mathbf{r})$ defines the conditional probability of finding an electron at the position $\mathbf{r}$ if the hole is positioned at the origin. Thus, $g_{\text{eh}}(\mathbf{r})$ allows us to identify the relative-motion correlations of the electron–hole pairs in the many-body system.

It is useful to divide the pair-correlation function into singlet and doublet contributions. Assuming homogeneous excitation conditions, i.e., Eqs. (24.41) and (24.43), we can write

$$g_{\text{eh}}(\mathbf{r}) \equiv g^S_{\text{eh}}(\mathbf{r}) + \Delta g_{\text{eh}}(\mathbf{r}), \tag{27.29}$$

$$g^S_{\text{eh}}(\mathbf{r}) \equiv n_e n_h + |P_{\text{eh}}(\mathbf{r})|^2, \tag{27.30}$$

$$n_{e(h)} \equiv \frac{1}{\mathcal{S}} \sum_{\mathbf{k}} f^{e(h)}_{\mathbf{k}}, \qquad P_{\text{eh}}(\mathbf{r}) \equiv \frac{1}{\mathcal{S}} \sum_{\mathbf{k}} P_{\mathbf{k}} e^{i\mathbf{k}\cdot\mathbf{r}}, \tag{27.31}$$

$$\Delta g_{\text{eh}}(\mathbf{r}) \equiv \frac{1}{\mathcal{S}^2} \sum_{\mathbf{k},\mathbf{k}',\mathbf{q}} c_X^{\mathbf{q},\mathbf{k}',\mathbf{k}} e^{i(\mathbf{k}'+\mathbf{q}-\mathbf{k})\cdot\mathbf{r}}, \tag{27.32}$$

which contains the exciton correlation

$$c_X^{\mathbf{q},\mathbf{k}',\mathbf{k}} = \Delta\langle a^\dagger_{c,\mathbf{k}} a^\dagger_{v,\mathbf{k}'} a_{c,\mathbf{k}'+\mathbf{q}} a_{v,\mathbf{k}-\mathbf{q}} \rangle, \tag{27.33}$$

as defined by Eq. (24.43). We observe that the singlet part, $g^S_{\text{eh}}(\mathbf{r})$, consists of a background contribution, $n_e n_h$, showing that the probability of simultaneously finding an electron and a hole has a simple contribution, which is nothing but the product of their densities. This density product is flat and $\mathbf{r}$ independent.

We also find a coherent singlet contribution $|P_{\text{eh}}(\mathbf{r})|^2$, which expresses the fact that optical transitions connect those electrons and holes that are spatially close to each other. This can be seen by transforming the semiconductor Bloch equation (27.2) into $\mathbf{r}$-space,

$$i\hbar\frac{\partial}{\partial t}P_{\text{eh}}(\mathbf{r}) = \left[-\frac{\hbar^2}{2\mu}\nabla^2 + E_g - \frac{e^2}{4\pi\epsilon_0\epsilon_r|\mathbf{r}|}\right]P_{\text{eh}}(\mathbf{r}) - \delta(\mathbf{r})d_{v,c}\langle E(t)\rangle$$
$$+ i\hbar\frac{\partial}{\partial t}P_{\text{eh}}(\mathbf{r})\bigg|_{\text{scatt}}, \tag{27.34}$$

where we assumed the low-density limit and used the effective-mass approximation. We see from Eq. (27.34) that the optical field generates electron–hole transitions at $\mathbf{r} = 0$ even in the artificial case without Coulomb interaction. Hence, the enhancement of $|P_{\text{eh}}(\mathbf{r})|^2$ close to the origin is a general consequence of optical excitations in direct-gap semiconductors, not a signature for the presence of excitons.

True exciton populations can be described only by the correlated two-particle contributions $\Delta g_{\text{eh}}(\mathbf{r})$. These contributions are determined by the c_X correlation, which is an incoherent quantity based on the classification made in Section 25.3. Thus, true excitons are incoherent by nature such that they may exist long after the coherences have decayed.

27.2.1 Dynamics of exciton correlations

We now take a closer look at the c_X dynamics that has a singlet–doublet–triplet structure analogous to that of Eq. (26.63). Since the explicit derivation of the c_X dynamics is lengthy but straightforward, we concentrate only on the main features and refer the interested reader to the details derived in (Kira and Koch, 2006). The starting point is

$$i\hbar\frac{\partial}{\partial t}c_X^{\mathbf{q},\mathbf{k}',\mathbf{k}} = \left(\tilde{\epsilon}^e_{\mathbf{k}'+\mathbf{q}} + \tilde{\epsilon}^h_{\mathbf{k}'} - \tilde{\epsilon}^e_{\mathbf{k}} - \tilde{\epsilon}^h_{\mathbf{k}-\mathbf{q}}\right)c_X^{\mathbf{q},\mathbf{k}',\mathbf{k}} + S_X^{\mathbf{q},\mathbf{k}',\mathbf{k}}$$
$$+ \left(1 - f^e_{\mathbf{k}} - f^h_{\mathbf{k}-\mathbf{q}}\right)\sum_{\mathbf{l}} V_{\mathbf{l}-\mathbf{k}}c_X^{\mathbf{q},\mathbf{k}',\mathbf{l}} - \left(1 - f^e_{\mathbf{k}'+\mathbf{q}} - f^h_{\mathbf{k}'}\right)\sum_{\mathbf{l}} V_{\mathbf{l}-\mathbf{k}'}c_X^{\mathbf{q},\mathbf{l},\mathbf{k}}$$
$$+ G_{X,\text{Coul}}^{\mathbf{q},\mathbf{k}',\mathbf{k}} + G_{X,\text{phon}}^{\mathbf{q},\mathbf{k}',\mathbf{k}} + D_{X,\text{rest}}^{\mathbf{q},\mathbf{k}',\mathbf{k}} + T_X^{\mathbf{q},\mathbf{k}',\mathbf{k}}. \tag{27.35}$$

Here, the first line is the sum of the renormalized kinetic energy of the particles plus the singlet source

$$S_X^{\mathbf{q},\mathbf{k}',\mathbf{k}} \equiv \delta_{\sigma,\sigma'}V_{\mathbf{j}}\Big[\left(f^e_{\mathbf{k}'+\mathbf{q}}f^h_{\mathbf{k}'}\bar{f}^e_{\mathbf{k}}\bar{f}^h_{\mathbf{k}-\mathbf{q}}\right)_\Sigma$$
$$+ P^\star_{\mathbf{k}}P_{\mathbf{k}'+\mathbf{q}}\left(f^h_{\mathbf{k}-\mathbf{q}} - f^h_{\mathbf{k}'}\right) + P^\star_{\mathbf{k}-\mathbf{q}}P_{\mathbf{k}'}\left(f^e_{\mathbf{k}} - f^e_{\mathbf{k}'+\mathbf{q}}\right)\Big]$$
$$+ V_{\mathbf{q}}\Big[P^\star_{\mathbf{k}}P_{\mathbf{k}'}\left(f^h_{\mathbf{k}-\mathbf{q}} - f^e_{\mathbf{k}'+\mathbf{q}}\right) - P^\star_{\mathbf{k}-\mathbf{q}}P_{\mathbf{k}'+\mathbf{q}}\left(f^h_{\mathbf{k}'} - f^e_{\mathbf{k}}\right)$$
$$- P^\star_{\mathbf{k}-\mathbf{q}}P_{\mathbf{k}'}\left(f^e_{\mathbf{k}} - f^e_{\mathbf{k}'+\mathbf{q}}\right) + P^\star_{\mathbf{k}}P_{\mathbf{k}'+\mathbf{q}}\left(f^h_{\mathbf{k}'} - f^h_{\mathbf{k}-\mathbf{q}}\right)\Big], \tag{27.36}$$

where $\mathbf{j} \equiv \mathbf{k}' + \mathbf{q} - \mathbf{k}$. The expression (27.36) contains the singlet factorization of the Coulomb-induced two- and three-particle terms. For clarity, we explicitly write here the spin dependence following from the sequence $c_X \equiv c_{(c,\sigma),(v,\sigma');(c,\sigma'),(v,\sigma)}$. Additionally, we introduce the abbreviation

$$\left(f_{\mathbf{k}}^{\lambda} f_{\mathbf{k}'}^{\lambda'} \bar{f}_{\mathbf{k}''}^{\lambda''} \bar{f}_{\mathbf{k}'''}^{\lambda'''}\right)_{\Sigma} \equiv f_{\mathbf{k}}^{\lambda} f_{\mathbf{k}'}^{\lambda'} \left(1 - f_{\mathbf{k}''}^{\lambda''}\right)\left(1 - f_{\mathbf{k}'''}^{\lambda'''}\right) - \left(1 - f_{\mathbf{k}}^{\lambda}\right)\left(1 - f_{\mathbf{k}'}^{\lambda'}\right) f_{\mathbf{k}''}^{\lambda''} f_{\mathbf{k}'''}^{\lambda'''}, \tag{27.37}$$

which identifies Boltzmann-type in- and out-scattering terms among carrier distributions. These terms also act as a source of the c_X dynamics in the purely incoherent regime.

The second line in Eq. (27.35) contains the two most important contributions of the incoherent Coulomb-induced doublet correlations. These are the so-called main-sum terms describing the attractive interaction between electrons and holes, which allows them to become truly bound excitons. The third line of Eq. (27.35) contains $G_{X,\mathrm{Coul}}$ and $G_{X,\mathrm{phon}}$, which include couplings to coherent quantities in the form

$$G_{X,\mathrm{Coul}}^{\mathbf{q},\mathbf{k}',\mathbf{k}} \equiv V_{\mathbf{q}}\left(P_{\mathbf{k}}^{\star} - P_{\mathbf{k}-\mathbf{q}}^{\star}\right)\sum_{\mathbf{l}}\left(c_{c,v;c,c}^{\mathbf{q},\mathbf{k}',\mathbf{l}} + c_{v,v;c,v}^{\mathbf{q},\mathbf{k}',\mathbf{l}}\right) - V_{\mathbf{q}}(P_{\mathbf{k}'+\mathbf{q}} - P_{\mathbf{k}'})\sum_{\mathbf{l}}\left(c_{c,c;c,v}^{\mathbf{q},\mathbf{l},\mathbf{k}} + c_{c,v;v,v}^{\mathbf{q},\mathbf{l},\mathbf{k}}\right), \tag{27.38}$$

$$G_{X,\mathrm{phon}}^{\mathbf{q},\mathbf{k}',\mathbf{k}} \equiv P_{\mathbf{k}}^{\star}\Delta\left\langle\left[Q_{\mathbf{q}}^{v}\right]^{\dagger} a_{v,\mathbf{k}'}^{\dagger} a_{c,\mathbf{k}'+\mathbf{q}}\right\rangle - P_{\mathbf{k}-\mathbf{q}}^{\star}\Delta\left\langle\left[Q_{\mathbf{q}}^{c}\right]^{\dagger} a_{v,\mathbf{k}'}^{\dagger} a_{c,\mathbf{k}'+\mathbf{q}}\right\rangle + P_{\mathbf{k}'}\Delta\left\langle Q_{\mathbf{q}}^{c} a_{c,\mathbf{k}}^{\dagger} a_{v,\mathbf{k}-\mathbf{q}}\right\rangle - P_{\mathbf{k}'+\mathbf{q}}\Delta\left\langle Q_{\mathbf{q}}^{v} a_{c,\mathbf{k}}^{\dagger} a_{v,\mathbf{k}-\mathbf{q}}\right\rangle. \tag{27.39}$$

We notice that $G_{X,\mathrm{Coul}}$ and $G_{X,\mathrm{phon}}$ contain exactly the same coherent correlations as those appearing in the polarization scattering terms $\Gamma_{\mathbf{k},\mathrm{Coul}}^{v,c}$ and $\Gamma_{\mathbf{k},\mathrm{phon}}^{v,c}$ in Eq. (26.11). While $\Gamma_{\mathbf{k},\mathrm{Coul}}^{v,c}$ and $\Gamma_{\mathbf{k},\mathrm{phon}}^{v,c}$ dissipate polarization via excitation- and phonon-induced dephasing, the corresponding $G_{X,\mathrm{Coul}}$ and $G_{X,\mathrm{phon}}$ act as sources generating exciton correlations out of polarization. Thus, the decay of polarization yields a simultaneous creation of exciton correlations. This process is known as the polarization-to-population conversion.

The remaining two-particle contributions denoted as D_{rest} describe the terms beyond the main-sum, $G_{X,\mathrm{Coul}}$, and $G_{X,\mathrm{phon}}$ contributions. These can be interpreted as microscopic processes where a correlated two-particle quantity scatters from a singlet. They result in quantum-optical effects and describe the equilibration within the system of the created correlations. The c_X dynamics also contains the hierarchical coupling to the triplets T_X. As presented here, the c_X dynamics (27.35) is formally exact and the accuracy of the numerical solutions depends only on the accuracy with which the three-particle correlation terms can be included in the analysis.

27.2.2 Polarization-to-population transfer

In order to study the quasiparticle populations generated during the nonradiative decay of the optically induced polarization, we follow the relevant processes by mapping how energy is transferred between coherent and incoherent states. For this purpose, we start from the pure carrier part of the system Hamiltonian (13.81)–(13.84),

$$H_{\rm carr} = \sum_{\lambda,\mathbf{k}} \varepsilon^{\lambda}_{\mathbf{k}} a^{\dagger}_{\lambda,\mathbf{k}} a_{\lambda,\mathbf{k}} + \frac{1}{2}\sum_{\lambda,\lambda'} \sum_{\mathbf{k},\mathbf{k}',\mathbf{q}\neq 0} V_{\mathbf{q}}\, a^{\dagger}_{\lambda,\mathbf{k}} a^{\dagger}_{\lambda',\mathbf{k}'} a_{\lambda',\mathbf{k}'+\mathbf{q}} a_{\lambda,\mathbf{k}-\mathbf{q}}. \tag{27.40}$$

Each electron–hole-pair excitation process increases the system energy roughly by E_g because an electron is moved across the band gap from the valence into the conduction band. Thus, using $E_{g,0}$ as the zero energy level, we define the average carrier energy as

$$E_{\rm carr} \equiv \langle H_{\rm carr}\rangle - E_{g,0}\sum_{\mathbf{k}} f^h_{\mathbf{k}} - E_{GS} \equiv E^S_{\rm carr} + E^D_{\rm carr}, \tag{27.41}$$

$$E^S_{\rm carr} = \sum_{\mathbf{k}} \left[\bar{\varepsilon}^e_{\mathbf{k}} f^e_{\mathbf{k}} + \bar{\varepsilon}^h_{\mathbf{k}} f^h_{\mathbf{k}}\right] - \frac{1}{2}\sum_{\mathbf{k},\mathbf{k}'} V_{\mathbf{k}-\mathbf{k}'}\left[f^e_{\mathbf{k}} f^e_{\mathbf{k}'} + f^h_{\mathbf{k}} f^h_{\mathbf{k}'}\right] - \sum_{\mathbf{k},\mathbf{k}'} V_{\mathbf{k}-\mathbf{k}'} P^{\star}_{\mathbf{k}} P_{\mathbf{k}'}, \tag{27.42}$$

$$E^D_{\rm carr} = \sum_{\mathbf{k},\mathbf{k}',\mathbf{q}} \left[\frac{1}{2} V_{\mathbf{q}}\left(c^{\mathbf{q},\mathbf{k}',\mathbf{k}}_{c,c;c,c} + c^{\mathbf{q},\mathbf{k}',\mathbf{k}}_{v,v;v,v}\right) - V_{\mathbf{k}'+\mathbf{q}-\mathbf{k}}\, c^{\mathbf{q},\mathbf{k}',\mathbf{k}}_{X}\right], \tag{27.43}$$

relative to the unexcited ground-state energy $E_{GS} \equiv \langle G|\hat{H}_{\rm carr}|G\rangle$. This choice fully compensates the gap within $\varepsilon^{\lambda}_{\mathbf{k}}$, see Exercise 27.7, such that

$$\bar{\varepsilon}^{\beta}_{\mathbf{k}} = \frac{\hbar^2\mathbf{k}^2}{2m_{\beta}}, \qquad \beta = e,\, h \tag{27.44}$$

denotes the carrier energy without the gap. At the level of the singlet–doublet decomposition, the sum of the singlet $E^S_{\rm carr}$ and the doublet $E^D_{\rm carr}$ energy contributions determines the total energy of the carrier system. While both coherent and incoherent singlets add to the energy of the system, it is interesting to see that only the incoherent two-particle correlations appear in Eq. (27.43). Thus, the energy transfer from the decaying polarization only occurs in incoherent quantities.

We now apply the microscopic equations at the singlet–doublet level to analyze the energy redistribution due to the different scattering mechanisms. We furthermore investigate which quasiparticle states are created or destroyed during the quantum-dynamical processes. To monitor the energy transfer, we introduce the scaled energies

$$\bar{E} \equiv \bar{E}^S + \bar{E}^D, \qquad \bar{E}^S \equiv \frac{E^S_{\rm carr}}{\sum_{\mathbf{k}} f^e_{\mathbf{k}}}, \qquad \bar{E}^D \equiv \frac{E^D_{\rm carr}}{\sum_{\mathbf{k}} f^e_{\mathbf{k}}}, \tag{27.45}$$

where $\sum_{\mathbf{k}} f^e_{\mathbf{k}} = \sum_{\mathbf{k}} f^h_{\mathbf{k}}$ defines the total number of optically excited carriers and $\bar{E}$ defines the total energy per particle.

To analyze the states that are generated via classical excitation of the originally unexcited semiconductor, we compare the results of two calculations at different levels of approximation for our quantum-wire (QWI) system. In one case, we perform the analysis only at the singlet level. Here, we just solve the MSBE without the scattering terms. In the other case,

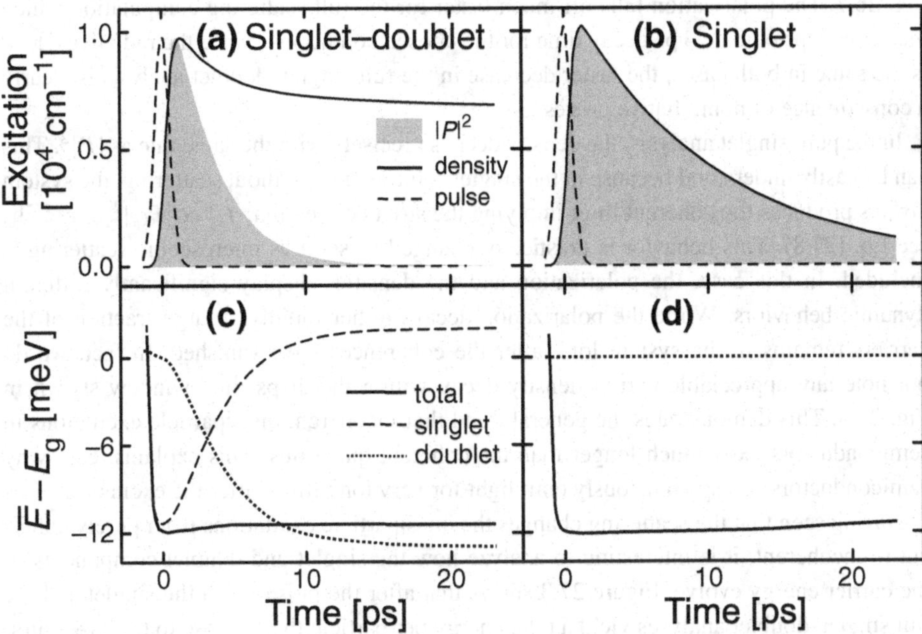

Figure 27.3 Comparison of the consistent singlet–doublet and singlet computations for a resonantly excited quantum wire (QWI). The optical polarization (shaded area), the carrier density (solid line), and the excitation pulse (dashed line) are shown for **(a)** the full singlet–doublet computation and **(b)** the singlet computation. Also the corresponding energy per particle is shown; **(c)** the total energy (solid line) of the full singlet–doublet is also split into its singlet (dashed line) and doublet (dotted line) contributions according to Eq. (27.45). **(d)** The energy for the singlet computation consists only of the singlet part. Adapted from *Progress in Quantum Electronics*, **30/5**, M. Kira, S. W. Koch, "Many-body correlations and excitonic effects in semiconductor spectroscopy," pp. 155–296, 2006, with permission from Elsevier.

we evaluate the full singlet–doublet dynamics with the scattering terms (26.11)–(26.12) fully included. We always assume a relatively weak excitation with a 500 fs pulse that is resonant with the excitonic $1s$-absorption peak. This way, we have strong absorption and at the same time we can keep the excitation level sufficiently low such that most of the interpretations used for the linear absorption case are still relevant. The phonons are treated as a bath with a lattice temperature of 10 K. In this situation, only the acoustical phonons are relevant.

The upper panels of Fig. 27.3 present the laser pulse $I(t) = |\langle \hat{E}(0,t)\rangle|^2$ (dashed line), the generated macroscopic polarization $|P|^2$ (shaded area), and the total carrier density $n_{\text{eh}} \equiv n_e = n_h$ (solid line). The lower panels of Fig. 27.3 show the total average energy per particle $\bar{E}$ (solid line) and its subdivision into singlet $\bar{E}^S$ (dashed line) and doublet $\bar{E}^D$ (dotted line) energies. The result of the full singlet–doublet analysis is presented in the left column while the singlet computation is presented in the right column of Fig. 27.3.

We notice that the optically generated polarization in the singlet analysis vanishes purely radiatively at the scale of the radiative decay time $\tau_{1s} = 15\,\text{ps}$ determined earlier, see

Fig. 26.7. The polarization falls off much faster for the full scattering computation which yields an approximate 4 ps decay time for the chosen conditions. Since the radiative decay is the same in both cases, the faster decrease in the full singlet–doublet analysis is clearly a consequence of nonradiative processes.

In the pure singlet analysis, the density decays precisely with the same rate as $|P|^2$. This can be easily understood because in the singlet solution, i.e., without scattering, the system always produces the coherent limit implying the strict connection, $f_{k_\parallel} \propto |P_{k_\parallel}|^2 \propto e^{-t/\tau_{1s}}$, see Eq. (27.8). This behavior is drastically changed as soon as microscopic scattering is included. In this case, the polarization and the densities display significantly different dynamic behaviors. While the polarization decays rather rapidly, a large fraction of the carriers remains in the system long after the coherences have vanished. In fact, we do not note any appreciable carrier-density decay within the 20 ps time window shown in Fig. 27.3. This demonstrates the general trend that incoherent quasiparticle excitations in semiconductors exist much longer than the coherent quantities. This explains, e.g., why semiconductors can spontaneously emit light for very long times after the excitation.

Having seen that the scattering changes the quasiparticle excitations from purely coherent to incoherent, it is interesting to analyze how the singlet and doublet components of the carrier energy evolve. Figure 27.3 shows that after the pulse. Both the singlet and the full singlet–doublet analyses yield a total energy per particle that is close to the $1s$-exciton energy of -12 meV. Since this is also the average energy of the excitation, both computations conserve the total energy. However, in the pure singlet analysis the doublet energy (dotted line) is always zero by definition. In other words, the coherent-limit quasiparticles exhibit energy conservation exclusively via the singlet contributions. If the scattering is systematically included, we see in Fig. 27.3(c) that the light pulse first pumps energy into the singlets which then transfer their energy almost completely to the doublets during the decay of the polarization. As we see from Eq. (27.43), the doublets consist of incoherent quasiparticle correlations besides the long-living densities.

We now come back to the question of which incoherent quasiparticles are generated by the nonradiative scattering of the polarization. As the polarization decays, we enter into the incoherent regime where the singlet part of the pair-correlation function reduces to $g^S_{\text{eh}}(\mathbf{r}) = n_e n_h$ such that it does not exhibit any $\mathbf{r}$ dependence. However, since the coherent polarization may also be converted into incoherent c_X correlations, the corresponding pair-correlation function, $\Delta g_{\text{eh}}(\mathbf{r})$, may also show a genuine $\mathbf{r}$ dependence. Thus, we may anticipate that the polarization can be directly converted into incoherent excitons under suitable conditions.

To confirm this conclusion, we analyze again Fig. 27.3 and plot the normalized

$$\Delta \bar{g}_{\text{eh}}(\mathbf{r}) \equiv \Delta g_{\text{eh}}(\mathbf{r})/(n_e n_h) \tag{27.46}$$

in Fig. 27.4(a) as a solid line at the final time $t = 21.8$ ps. At this moment, the polarization-to-population conversion is completed, as can be seen in Fig. 27.3. We also plot $|\phi^R_{1s}(\mathbf{r})|^2$ (shaded area) defined by the Wannier equation (26.19) in order to show the spatial dependency of the $1s$ exciton polarization. We observe a perfect match between $\Delta \bar{g}_{\text{eh}}(\mathbf{r})$ and

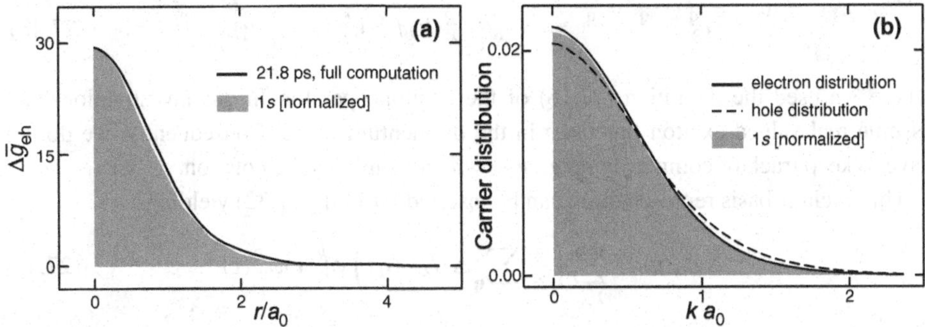

Figure 27.4 **(a)** The pair-correlation function (solid line) after the polarization-to-population transfer. The shaded area shows the extension of $\left|\phi_{1s}^{R}(\mathbf{r})\right|^2$ for a $1s$-exciton state. The pair-correlation function is determined at 21.8 ps, i.e., the final time in Fig. 27.3. **(b)** The corresponding electron (solid line) and hole (dashed line) distributions. The shaded area shows the normalized $\left|\phi_{1s}^{R}(\mathbf{k})\right|^2$. Adapted from *Progress in Quantum Electronics*, **30/5**, M. Kira, S. W. Koch, "Many-body correlations and excitonic effects in semiconductor spectroscopy," pp. 155–296, 2006, with permission from Elsevier.

the $1s$-exciton state demonstrating that electrons and holes are correlated to the degree described by the $1s$-exciton wave function. Since all criteria for a true exciton are satisfied, we have thus shown that nonradiative scattering can convert a coherent polarization, i.e., coherent excitons, into true incoherent exciton populations. As further evidence, we observe in Fig. 27.3 that incoherent excitons can have an energy per particle that corresponds to the $1s$-exciton binding energy. The corresponding electron and hole distributions in Fig. 27.4(b) also follow very accurately the momentum distribution of the $1s$ wave function. This is to be expected if electrons and holes are truly bound to each other.

27.3 Electron–hole correlations in the exciton basis

In the previous section, we have shown that the c_X correlation dynamics can be used to describe microscopically the generation of incoherent excitons. However, since c_X can also include a more general form of electron–hole correlations, it is interesting to identify the most general class of incoherent quasiparticle excitations described by c_X. With this knowledge, we can then precisely specify the incoherent many-body states that can be observed in semiconductors as long as no clusters of higher order than doublets occur. In other words, we are then able to define the full phase-space of singlet–doublet quasiparticle excitations and their relevance under different physical conditions.

We start this investigation by taking another look at $\Delta g_{eh}(\mathbf{r})$. For this purpose, we transform it into an exciton basis,

$$\Delta\left\langle X_{\lambda,\mathbf{q}}^{\dagger} X_{\nu,\mathbf{q}}\right\rangle = \sum_{\mathbf{k},\mathbf{k}'} \phi_{\lambda}^{L}(\mathbf{k})\phi_{\nu}^{L}(\mathbf{k}')c_{X}^{\mathbf{q},\mathbf{k}'-\mathbf{q}_h,\mathbf{k}+\mathbf{q}_e} \equiv \Delta N_{\lambda,\nu}(\mathbf{q}), \tag{27.47}$$

$$c_{\mathrm{X}}^{\mathbf{q},\mathbf{k}'-\mathbf{q}_h,\mathbf{k}+\mathbf{q}_e} = \sum_{\lambda,\nu} \phi_\lambda^R(\mathbf{k})\phi_\nu^R(\mathbf{k}')\Delta N_{\lambda,\nu}(\mathbf{q}), \tag{27.48}$$

where we used the definition (27.16) of the exciton operator. In our investigations, we assume real-valued exciton functions in the momentum space. Consequently, we do not have to keep track of complex conjugation, which simplifies the notation.

This exciton-basis representation can be inserted into Eq. (27.32) yielding

$$\Delta g_{\mathrm{eh}}(\mathbf{r}) = \sum_{\lambda,\nu} \left(\frac{1}{S} \sum_{\mathbf{q}} \Delta N_{\lambda,\nu}(\mathbf{q}) \right) \phi_\lambda^R(\mathbf{r})\phi_\nu^R(\mathbf{r}). \tag{27.49}$$

In situations where $\Delta g_{\mathrm{eh}}(\mathbf{r})$ exclusively displays a $\left|\phi_\beta^R(\mathbf{r})\right|^2$ dependence, it is obvious that the $\sum_{\lambda,\nu}$ sum is dominated by the element $\lambda = \nu = \beta$. Since this corresponds to the case where only β excitons exist, we conclude that

$$\Delta n_\beta = \Delta n_{\beta,\beta} \equiv \frac{1}{S} \sum_{\mathbf{q}} \Delta N_{\beta,\beta}(\mathbf{q}) \tag{27.50}$$

defines the density of β excitons in the system and

$$\Delta N_\beta(\mathbf{q}) \equiv \Delta N_{\beta,\beta}(\mathbf{q}) \tag{27.51}$$

is their momentum distribution.

Since only the diagonal part of $\Delta\left\langle X_\lambda^\dagger X_\nu \right\rangle$ corresponds to excitons, the off-diagonal contributions must correspond to some other correlated quasiparticle states. To resolve the character and relevance of these many-body states, we apply Eq. (27.48) to Eq. (27.35) and eventually obtain

$$\begin{aligned} i\hbar\frac{\partial}{\partial t}\Delta N_{\lambda,\nu}(\mathbf{q}) &= (E_\nu - E_\lambda)\,\Delta N_{\lambda,\nu}(\mathbf{q}) \\ &\quad + (E_\nu - E_\lambda)\,N_{\lambda,\nu}(\mathbf{q})_{\mathrm{S}} + S_{\mathrm{coh}}^{\lambda,\nu}(\mathbf{q}) \\ &\quad + iG^{\lambda,\nu}(\mathbf{q}) + D_{\mathrm{rest}}^{\lambda,\nu}(\mathbf{q}) + T^{\lambda,\nu}(\mathbf{q}), \end{aligned} \tag{27.52}$$

where the incoherent part of the singlet scattering, S_X, in Eq. (27.35) produces a source

$$N_{\lambda,\nu}(\mathbf{q})_{\mathrm{S}} \equiv \left\langle X_{\lambda,\mathbf{q}}^\dagger X_{\nu,\mathbf{q}} \right\rangle_{\mathrm{S}} = \sum_{\mathbf{k}} \phi_\lambda^L(\mathbf{k}) f_{\mathbf{k}+\mathbf{q}_e}^e f_{\mathbf{k}-\mathbf{q}_h}^h \phi_\nu^L(\mathbf{k}). \tag{27.53}$$

This contribution has a finite value in the incoherent regime whenever we have any quasiparticle excitation in the system. Particularly, it drives exclusively the nondiagonal $\Delta\left\langle X_\lambda^\dagger X_\nu \right\rangle$ because it exists in Eq. (27.52) only if $\lambda \neq \nu$. Thus, we can already conclude at this point that the incoherent part of the singlet scattering generates new correlated quasiparticles that do not have the character of excitons.

The other sources of c_X are expressed symbolically as $S_{\mathrm{coh}}^{\lambda,\nu}(\mathbf{q})$ containing the coherent contributions of the singlet scattering. The polarization-to-population scattering is determined by $G^{\lambda,\nu}(\mathbf{q}) \equiv G^{\lambda,\nu}_{X,\mathrm{Coul}}(\mathbf{q}) + G^{\lambda,\nu}_{X,\mathrm{phon}}(\mathbf{q})$. The contributions $D^{\lambda,\nu}_{\mathrm{rest}}(\mathbf{q})$ result from the remaining doublet correlations and the triplet-scattering term is denoted as $T^{\lambda,\nu}(\mathbf{q})$. Since these remaining terms are cumbersome to write down in the exciton basis, it is beneficial to perform the fully systematic calculations in the original electron–hole picture as given by Eq. (27.35). The exciton basis is useful mostly for the interpretation of the results.

Our discussion in Section 27.2.2 shows that polarization-to-population conversion can lead to a semiconductor state consisting mainly of incoherent exciton populations. Thus, it is obvious that $G^{\lambda,\nu}(\mathbf{q})$ can be dominantly diagonal. We also see from the structure of Eq. (27.52) that the generated $\Delta\left\langle X^{\dagger}_{\beta} X_{\beta}\right\rangle$ changes only due to the $D^{\beta,\beta}_{\mathrm{rest}}$- and $T^{\beta,\beta}$-scattering terms as the incoherent regime is reached. It can be shown that this scattering becomes weak for densities and lattice temperatures that are so low that we can identify a physically relevant phase space where true excitons exist in the system for very long times.

27.3.1 Correlated electron–hole plasma

Besides the identified excitons, the nature of the other class of incoherent quasiparticles is still unknown. To determine their properties, we assume completely incoherent conditions such that $S^{\lambda,\nu}_{\mathrm{coh}}(\mathbf{q})$ and $G^{\lambda,\nu}(\mathbf{q})$ vanish for all times. In addition, we introduce an approximation where $D^{\lambda,\nu}_{\mathrm{rest}}(\mathbf{q})$ is neglected while the triplet scattering is described phenomenologically via $T^{\lambda,\nu}(\mathbf{q}) = -i\gamma\Delta\left\langle X^{\dagger}_{\lambda,\mathbf{q}} X_{\nu,\mathbf{q}}\right\rangle$ in Eq. (27.52). This way, we obtain a simple steady-state solution

$$\Delta N_{\lambda,\nu}(\mathbf{q}) = -\frac{E_\nu - E_\lambda}{E_\nu - E_\lambda - i\gamma}\left\langle X^{\dagger}_{\lambda,\mathbf{q}} X_{\nu,\mathbf{q}}\right\rangle_{\mathrm{S}} \tag{27.54}$$

whenever the carrier densities remain constant. In this form, we see that the new class of incoherent excitations excludes excitons and depends only on carrier densities. Thus, this part of the many-body state can be referred to as *correlated electron–hole plasma* which is a state where electrons and holes attract each other without forming excitons.

The principal characteristics of the correlated electron–hole plasma is visualized in Fig. 27.5 for QW and QWI systems. These results are obtained by inserting Eq. (27.54) into Eq. (27.49) for the dilute carrier densities of $n_{\mathrm{eh}} = 10^4\ \mathrm{cm}^{-1}$ for the QWI and $n_{\mathrm{eh}} = 2\times 10^9\mathrm{cm}^{-2}$ for the QW, respectively. In the numerical evaluations we used 40 K Fermi–Dirac distributions for the electrons and holes in both cases. We notice in Fig. 27.5 that the pair-correlation functions of the QW and the QWI systems are qualitatively similar. In particular, $\Delta\bar{g}_{\mathrm{eh}}(r)$ shows an elevated probability of finding the electron and the hole close together. This probability is only weakly dependent on the precise value of the dephasing constant γ.

As a distinctive difference between the correlated plasma and the results for exciton populations (see Fig. 27.4), we note that for the correlated plasma $\Delta\bar{g}_{\mathrm{eh}}(r)$ becomes negative

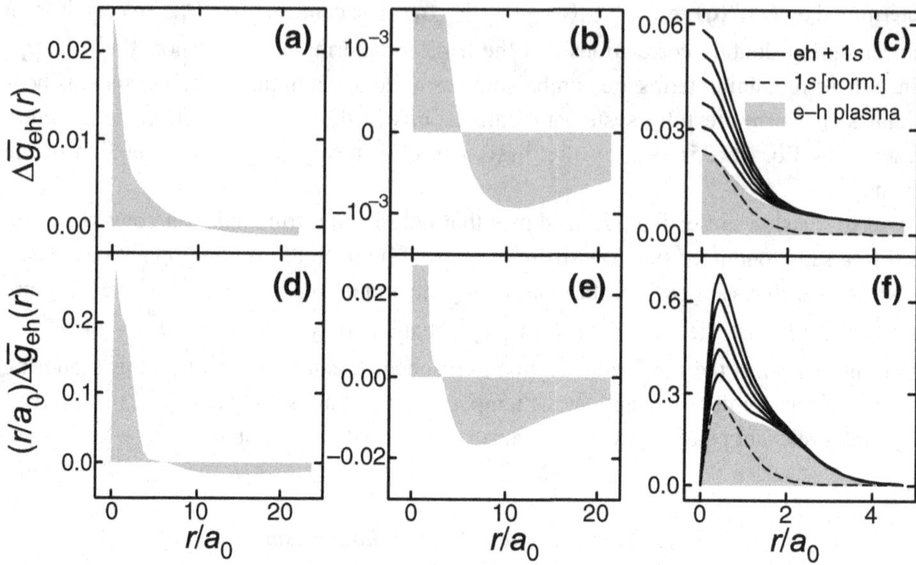

Figure 27.5 Pair-correlation function of a correlated electron–hole plasma obtained by evaluating Eqs. (27.49) with (27.54). **(a)** The normalized $\Delta\bar{g}_{\text{eh}}$ is shown for the QWI system with the carrier density $10^4\ \text{cm}^{-1}$ at 40 K using $\gamma = 0.42$ meV. **(b)** Magnification of the large-distance part of frame **(a)**. **(c)** The influence of real exciton populations is analyzed by evaluating $\Delta\bar{g}_{\text{eh}}$. Here, a fraction of 0.16%–0.8% (from bottom to top) of $1s$-excitons is added to the correlated electron–hole plasma; $|\phi_{1s}(r)|^2$ (dashed line) and the correlated plasma result (shaded area) are also shown. **(d)–(f)** The same analysis as in **(a)–(c)** is presented for a QW system with $2 \times 10^9\ \text{cm}^{-2}$ carrier density at 40 K. Here, $\Delta\bar{g}_{\text{eh}}$ has been multiplied by r to obtain the radial probability distribution. Reprinted from *Progress in Quantum Electronics*, **30/5**, M. Kira, S. W. Koch, "Many-body correlations and excitonic effects in semiconductor spectroscopy," pp. 155–296, 2006, with permission from Elsevier.

for larger distances. This indicates that the electron–hole pairs reduce their average separation by reorganizing the long- and short-range parts of the joint probability distribution. In the integrals $\int dr \Delta\bar{g}_{\text{eh}}(r)$ in 1D and $\int dr r \Delta\bar{g}_{\text{eh}}(r)$ in 2D the negative and the positive valued parts compensate each other to yield a vanishing result, see Exercise 27.10. We see in Figs. 27.5(c) and (f) that already the addition of a small amount of exciton populations to the correlated plasma modifies $\Delta\bar{g}_{\text{eh}}(r)$ toward a pronounced $1s$-like shape, which produces an overall positive integration area. This indicates that $\Delta\bar{g}_{\text{eh}}(r)$ provides a sensitive measure for the presence of truly bound excitons.

27.3.2 Energy considerations

In order to understand the energetic differences between the correlated electron–hole plasma and the exciton populations for a general incoherent many-body state, we analyze the total average energy by expressing Eqs. (27.41)–(26.43) in the exciton basis

$$E_{\text{carr}} = \sum_{\mathbf{k}} \left[\bar{\varepsilon}^e_{\mathbf{k}} f^e_{\mathbf{k}} + \bar{\varepsilon}^h_{\mathbf{k}} f^h_{\mathbf{k}} \right] - \sum_{\mathbf{k},\mathbf{k}'} V_{\mathbf{k}-\mathbf{k}'} \frac{f^e_{\mathbf{k}} f^e_{\mathbf{k}'} + f^h_{\mathbf{k}} f^h_{\mathbf{k}'}}{2} + \sum_{\mathbf{k},\mathbf{k}',\mathbf{q},\lambda} V_{\mathbf{q}} \frac{c^{\mathbf{q},\mathbf{k}',\mathbf{k}}_{\lambda,\lambda;\lambda,\lambda}}{2}$$
$$+ \sum_{\lambda,\mathbf{q}} \left[E_\lambda(\mathbf{q}) - \sum\nolimits_{\mathbf{k}} \left(\bar{\varepsilon}^e_{\mathbf{k}} + \bar{\varepsilon}^h_{\mathbf{k}} \right) \phi^L_\lambda(\mathbf{k}) \phi^R_\lambda(\mathbf{k}) \right] \Delta N_\lambda(\mathbf{q})$$
$$- \sum_{\lambda \neq \nu,\mathbf{q}} \left[\sum\nolimits_{\mathbf{k}} \left(\bar{\varepsilon}^e_{\mathbf{k}} + \bar{\varepsilon}^h_{\mathbf{k}} \right) \phi^L_\lambda(\mathbf{k}) \phi^R_\nu(\mathbf{k}) \right] \Delta N_{\lambda,\nu}(\mathbf{q})\rangle, \tag{27.55}$$

$$E_\lambda(\mathbf{q}) \equiv E_\lambda + \frac{\hbar^2 \mathbf{q}^2}{2(m_e + m_h)}. \tag{27.56}$$

Here, $E_\lambda(\mathbf{q})$ defines the energy of an exciton with a center-of-mass momentum $\mathbf{q}$.

If the incoherent carrier system exists dominantly in the form of an exciton population $\Delta N_\lambda(\mathbf{q})$, we may omit the nondiagonal $\Delta \left\langle X^\dagger_\lambda X_\nu \right\rangle$ contributions to obtain

$$E_{\text{carr}}|_{\lambda-\text{pop}} = \sum_{\mathbf{k}} \bar{\varepsilon}^e_{\mathbf{k}} \left[f^e_{\mathbf{k}} - \sum_{\mathbf{q}} \phi^L_\lambda(\mathbf{k}) \phi^R_\lambda(\mathbf{k}) \Delta N_\lambda(\mathbf{q}) \right]$$
$$+ \sum_{\mathbf{k}} \bar{\varepsilon}^h_{\mathbf{k}} \left[f^h_{\mathbf{k}} - \sum_{\mathbf{q}} \phi^L_\lambda(\mathbf{k}) \phi^R_\lambda(\mathbf{k}) \Delta N_\lambda(\mathbf{q}) \right] + \sum_{\mathbf{q}} E_\lambda(\mathbf{q}) \Delta N_\lambda(\mathbf{q})$$
$$- \sum_{\mathbf{k},\mathbf{k}'} V_{\mathbf{k}-\mathbf{k}'} \frac{f^e_{\mathbf{k}} f^e_{\mathbf{k}'} + f^h_{\mathbf{k}} f^h_{\mathbf{k}'}}{2} + \sum_{\mathbf{k},\mathbf{k}',\mathbf{q}} \frac{1}{2} V_{\mathbf{q}} \left(c^{\mathbf{q},\mathbf{k}',\mathbf{k}}_{c,c;c,c} + c^{\mathbf{q},\mathbf{k}',\mathbf{k}}_{v,v;v,v} \right). \tag{27.57}$$

In addition, the carrier distributions follow from

$$f^e_{\mathbf{k}} = f^h_{\mathbf{k}} = \sum_{\mathbf{q}} \phi^L_\lambda(\mathbf{k}) \phi^R_\lambda(\mathbf{k}) \Delta N_\lambda(\mathbf{q}) \tag{27.58}$$

when the system contains nearly 100 percent excitons. In this situation, each electron is bound to a hole and the resulting state is described by an excitonic wave function. Indeed, the results shown in Fig. 27.4 verify this statement.

Consequently, the first two kinetic energy terms practically compensate each other when the exciton fraction is close to 100 percent. Since the remaining terms also vanish for sufficiently dilute densities, we find

$$\lim_{n_{e,h} \to 0} E_{\text{carr}}|_{\lambda-\text{pop}} = \sum\nolimits_{\mathbf{q}} E_\lambda(\mathbf{q}) \Delta N_\lambda(\mathbf{q}). \tag{27.59}$$

Hence, the average energy per particle is close to E_λ, which can be significantly below the band-gap energy, see Fig. 27.3. Due to the energy difference between the uncorrelated electron–hole plasma ($E_{\text{carr}} > 0$) and the exciton populations ($E_{\text{carr}} < 0$), the formation of excitons in an electron–hole plasma requires significant dynamical rearrangements between these two many-body states. Hence, we conclude from Eq. (27.52) that the phenomenological scattering model for $T^{\lambda,\nu}(\mathbf{q})$ is not good enough to describe the formation of excitons in the incoherent regime because such an approach only produces a correlated electron–hole plasma.

If the system only contains the correlated electron–hole plasma, we have

$$E_{\text{carr}}|_{\text{cp}} = \sum_{\mathbf{k}} \left[\bar{\varepsilon}^e_{\mathbf{k}} f^e_{\mathbf{k}} + \bar{\varepsilon}^h_{\mathbf{k}} f^h_{\mathbf{k}}\right] - \sum_{\mathbf{k},\mathbf{k}'} V_{\mathbf{k}-\mathbf{k}'} \frac{f^e_{\mathbf{k}} f^e_{\mathbf{k}'} + f^h_{\mathbf{k}} f^h_{\mathbf{k}'}}{2}$$
$$- \sum_{\lambda \neq \nu, \mathbf{q}} \left[\sum_{\mathbf{k}} \left(\bar{\varepsilon}^e_{\mathbf{k}} + \bar{\varepsilon}^h_{\mathbf{k}}\right) \phi^L_\lambda(\mathbf{k}) \phi^R_\nu(\mathbf{k})\right] \Delta N_{\lambda,\nu}(\mathbf{q})$$
$$+ \sum_{\mathbf{k},\mathbf{k}',\mathbf{q}} \frac{1}{2} V_{\mathbf{q}} \frac{c^{\mathbf{q},\mathbf{k}',\mathbf{k}}_{c,c;c,c} + c^{\mathbf{q},\mathbf{k}',\mathbf{k}}_{v,v;v,v}}{2}. \tag{27.60}$$

Since here all correlation terms are nonlinear with respect to the carrier density, we can understand that the formation of such a correlated electron–hole plasma yields only a very small energy reduction compared with the uncorrelated plasma state. Since the correlated and the uncorrelated electron–hole plasma have a very similar positively valued energy, the uncorrelated electron–hole plasma can be converted rather easily into a correlated state where electrons and holes attract each other without true exciton formation. This small-scale rearrangement can be well modeled by implementing a phenomenological dephasing $T^{\lambda,\nu}(\mathbf{q})$.

Exercises

27.1 Introduce two classes of Fermion operators $L_{1,\mathbf{k}} \equiv L_{\mathbf{k}}$, defined by Eq. (27.10), and $L^\dagger_{2,\mathbf{k}} = -e^{-i\Delta\phi_{\mathbf{k}}} \sin\frac{\vartheta_{\mathbf{k}}}{2} a^\dagger_{v,\mathbf{k}} + \cos\frac{\vartheta_{\mathbf{k}}}{2} a^\dagger_{c,\mathbf{k}}$. **(a)** Construct $a^\dagger_{v,\mathbf{k}}$ and $a^\dagger_{c,\mathbf{k}}$ in terms of $L^\dagger_{\lambda,\mathbf{k}}$. **(b)** Show that $\left[L_{\lambda,\mathbf{k}}, L^\dagger_{\lambda',\mathbf{k}'}\right]_+ = \delta_{\mathbf{k},\mathbf{k}'}$ and $[L_{\lambda,\mathbf{k}}, L_{\lambda',\mathbf{k}'}]_+ = 0 = \left[L^\dagger_{\lambda,\mathbf{k}}, L^\dagger_{\lambda',\mathbf{k}'}\right]_+$. **(c)** Show that $\langle L^\dagger_{\lambda,\mathbf{k}} L_{\lambda',\mathbf{k}'}\rangle = \delta_{\mathbf{k},\mathbf{k}'}\,\delta_{\lambda,1},\,\delta_{\lambda',1}$. **(d)** Verify relations (27.12). **Hint:** *Use the results derived in parts (a) and (c).*

27.2 Compute the explicit form of the Fermionic corrections $\hat{F}^{\nu,\mathbf{q}}_{\lambda,\mathbf{q}'}$ in $\left[X_{\nu,\mathbf{q}}, X^\dagger_{\lambda,\mathbf{q}'}\right]_- = \delta_{\nu,\lambda}\,\delta_{\mathbf{q},\mathbf{q}'} + \hat{F}^{\nu,\mathbf{q}}_{\lambda,\mathbf{q}'}$. **(b)** Compute $\langle \hat{F}^{\nu,\mathbf{q}}_{\lambda,\mathbf{q}'}\rangle$ for a homogeneous system. When can one expect to see large Fermion corrections? **(c)** Show that $\sum_{\nu,\mathbf{q}} X^\dagger_{\nu\mathbf{q}} X_{\nu\mathbf{q}} = \sum_{\mathbf{k},\mathbf{k}'} \frac{a^\dagger_{c\mathbf{k}} a_{v\mathbf{k}'} a^\dagger_{v\mathbf{k}'} a_{c\mathbf{k}}}{1-f^e_{\mathbf{k}}-f^h_{\mathbf{k}'}}$. Show that this defines $\hat{N}_e\,\hat{N}_h$ for low densities where $\hat{N}_e$ ($\hat{N}_h$) is the number operator for electrons (holes). Why can $\hat{N}_e\hat{N}_h$ not describe the exciton number?

27.3 Calculate $e^S\,a_{v,\mathbf{k}}\,e^{-S}$ explicitly, i.e., verify Eq. (27.21) with Eq. (27.20) as an input. Verify that the transformation (27.22)–(27.23) for c_λ into the angle parameters $(\vartheta_{\mathbf{k}}, \Delta\phi_{\mathbf{k}})$ yields a unique connection between coherent state (27.25) and coherent limit excitons.

27.4 **(a)** Show that the coherent exciton state satisfies the completeness relation

$$\left[\prod_{\mathbf{k}} \int_{-\pi}^{\pi} d\vartheta_{\mathbf{k}}\right] |\Psi_{\text{coh}}\rangle \langle\Psi_{\text{coh}}| = \mathbb{I}_N \text{ and}$$

$$\left[\prod_{\mathbf{k}} \int_{0}^{2\pi} d\,\Delta\phi_{\mathbf{k}} \int_{-\pi}^{\pi} d\vartheta_{\mathbf{k}}\right] |\Psi_{\text{coh}}\rangle \langle\Psi_{\text{coh}}| = 2\pi\mathbb{I}_N$$

where $\mathbb{I}_N$ is the identity operator with N electrons either in the conduction or the valence band. **(b)** Show that $\langle\Psi_{\text{coh}}|\Psi'_{\text{coh}}\rangle$ does not necessarily vanish if $|\Psi_{\text{coh}}\rangle \neq |\Psi'_{\text{coh}}\rangle$. This, together with part (a), shows that coherent excitons form an overcomplete set of states.

27.5 **(a)** Construct the pair-correlation function $g_{\text{eh}}(\mathbf{r}) \equiv \langle\hat{\Psi}_e^\dagger(r)\,\hat{\Psi}_h^\dagger(0)\,\hat{\Psi}_h(0)\,\hat{\Psi}_e(\mathbf{r})\rangle$ in terms of singlet and doublet contributions. Use the condition of homogeneous excitation as well as (27.31) and (27.33).
(b) Apply $P_{eh}(r) \equiv \frac{1}{S}\sum_{\mathbf{k}} P_{\mathbf{k}} e^{i\mathbf{k}\cdot\mathbf{r}}$ to the SBE (27.2) and show that Eq. (27.34) follows in the low-density limit.

27.6 Analyze Eq. (27.35) and show that it conforms to the structure of the generic Eq. (15.47). Show that Eq. (27.37) can also be written in the $\langle 1\rangle\langle 1\rangle\langle 1\rangle$ format. Show that the main-sum terms are only one possible reorganization of $\langle 1\rangle\Delta\langle 2\rangle$ terms; hypothesize other possibilities.

27.7 Identify the singlet and doublet contributions for a pure carrier Hamiltonian $\langle\hat{H}_{carr}\rangle$ defined by Eq. (27.40). Use the two-band model and assume homogeneous conditions to verify Eqs. (27.41)–(27.43).

27.8 **(a)** Use the exciton basis (27.48) to prove Eq. (27.49). **(b)** Show that $N_{\nu,\lambda}(\mathbf{q})_S \equiv \sum_{\mathbf{k},\mathbf{k}'} \phi_\lambda^L(\mathbf{k})\phi_\nu^L(\mathbf{k}')\, S_x^{\mathbf{q},\mathbf{k}'-\mathbf{q}_h,\mathbf{k}+\mathbf{q}_e}$ becomes Eq. (27.53) under incoherent conditions. **(c)** Project Eq. (27.35) into the exciton basis using Eqs. (27.47)–(27.48) and prove that Eq. (27.52) follows. **Hint:** *Use the properties of the exciton states derived in Sections 27.2–27.2.1.* In fact, one needs to extend the exciton basis by using $f_{\mathbf{k}}^e \to f_{\mathbf{k}+\mathbf{q}_e}^e$ and $f_{\mathbf{k}}^h \to f_{\mathbf{k}-\mathbf{q}_h}^h$ in Eqs. (27.19)–(27.20). Explain why this modification is justified.

27.9 **(a)** Show that the contribution $c_{c,v;v,c}$ within $\Gamma_{\mathbf{k}}^{c,c}$ (see Eq. (26.11)) can be written as $\Gamma_{\mathbf{k},X}^{c,c} = -2i\,\text{Im}\left[\sum_{\mathbf{k}',\mathbf{q}} V_{\mathbf{k}-\mathbf{k}'-\mathbf{q}}\, c_X^{\mathbf{q},\mathbf{k}',\mathbf{k}}\right]$. **(b)** Implement the exciton basis (27.48) and show that

$$\Gamma_{\mathbf{k},X}^{c,c} = -2i\,\text{Im}\left[\sum_{\mathbf{q},\lambda} \sum_{\nu\neq\lambda} T_{\lambda,\nu}\,\Delta N_{\lambda,\nu}(\mathbf{q})\right] \text{ with}$$

$$T_{\lambda,\nu} \equiv \sum_{\mathbf{k}} \phi_\lambda^L(\mathbf{k}) \left[\varepsilon_{\mathbf{k}}^e + \varepsilon_{\mathbf{k}}^h\right] \phi_\lambda^R(\mathbf{k})$$

using the properties of the exciton states, derived in Sections 27.2–27.2.1. This result implies that only exciton transitions influence the carrier dynamics.

27.10 **(a)** Show that Eq. (27.52) yields the steady-state result (27.54) if triplets are approximated as constant dephasing contributions and the main-sum approximation is applied. **(b)** Use Eq. (27.52) in Eq. (27.49) and verify that the correlated electron–hole plasma yields $\int d^2r\ \Delta g_{\rm eh}(\mathbf{r}) = 0$ for low densities.

27.11 **(a)** An exciton basis with center-of-mass momentum follows from $\left(\varepsilon^e_{\mathbf{k}+\mathbf{q}_e} + \varepsilon^h_{\mathbf{k}-\mathbf{q}_h}\right) \phi^R_\lambda(\mathbf{k}) - \left(1 - f^e_{\mathbf{k}+\mathbf{q}_e} - f^h_{\mathbf{k}-\mathbf{q}_h}\right) \sum_{\mathbf{k}'} V_{\mathbf{k}'-\mathbf{k}}\, \phi^R_\lambda(\mathbf{k}') = E_\lambda(\mathbf{q})\, \phi^R_\lambda(\mathbf{k})$. Verify that parabolic bands yield $E_\lambda(\mathbf{q}) = E_\lambda + \frac{\hbar^2\mathbf{q}^2}{2(m_e+m_h)}$ where E_λ is the solution for $\mathbf{q} = 0$ at dilute densities. Why can one use this to define the exciton basis via the transformations (27.47)–(27.48)? **(b)** Convert the carrier energy, i.e., Eqs. (27.41)–(27.43), into the exciton basis form to verify the result (27.55).

Further reading

Exciton operators cannot be treated as Bosonic operators due to the fundamental incompatibility between Fermionic pairs and Bosons, cf.:

T. Usui (1960). Excitations in a high density electron gas, *Prog. Theor. Phys.* **23**, 787.
M. D. Girardeau (1970). Pair occupation, Fermi condensation, and phase transitions in many-Fermion systems, *J. Math. Phys.* **11**, 684.

For earlier work on excitons, cf.:

S. A. Moskalenko (1962). Reversible optico-hydrodynamic phenomena in a nonideal exciton gas, *Sov. Phys. Solid State* **4**, 199.
C. G. Kuper and D. G. Whitefield, eds. (1963). *Polarons and Excitons*, New York, Plenum Press.
R. S. Knox (1963). *Theory of Excitons*, New York, Academic Press.
L. V. Keldysh and A. N. Kozlov (1968). Collective properties of excitons in semiconductors, *Sov. Phys. JETP-USSR* **27**, 521.
C. Klingshirn and H. Haug (1981). Optical properties of highly excited direct gap semiconductors, *Phys. Rep.* **70**, 315.
E. L. Rashba and M. D. Sturge (1982). *Excitons*, Amsterdam, North-Holland.
B. Hönerlage, R. Levy, J. B. Grun, C. Klingshirn, and K. Bohnert (1985). The dispersion of excitons, polaritons, and biexcitons in direct-gap semiconductors, *Phys. Rep.* **124**, 161.
C. I. Ivanov, H. Barentzen, and M. D. Girardeau (1987). On the theory of dense exciton systems, *Physica A* **140**, 612.

The main-sum approximation as well as the exciton-dynamics equations were introduced in:

M. Kira, W. Hoyer, T. Stroucken, and S. W. Koch (2001). Exciton formation in semiconductors and influence of a photonic environment, *Phys. Rev. Lett.* **87**, 176401.
W. Hoyer, M. Kira, and S. W. Koch (2003). Influence of Coulomb and phonon interaction on the exciton-formation dynamics in semiconductor heterostructures, *Phys. Rev. B* **67**, 155113.

Polarization-to-population conversion is studied in:

M. Kira and S. W. Koch (2004). Exciton-population inversion and terahertz gain in semiconductors excited to resonance, *Phys. Rev. Lett.* **93**, 076402.

The full singlet–doublet quantum kinetics is reviewed in:

M. Kira and S. W. Koch (2006). Many-body correlations and excitonic effects in semiconductor spectroscopy, *Prog. Quantum Electr.* **30**, 155.

Alternative formulations of the exciton problem are found, e.g., in:

M. Gulia, F. Rossi, E. Molinari, P. E. Selbmann, and P. H. Lugli (1997). Phonon-assisted exciton formation and relaxation in GaAs/Al_xGa_{1-x}As quantum wells, *Phys. Rev. B* **55**, R16049.

C. Ciuti, V. Savona, C. Piermarocchi, A. Quattropani, and P. Schwendimann (1998). Role of the exchange of carriers in elastic exciton-exciton scattering in quantum wells, *Phys. Rev. B* **58**, 7926.

K. Siantidis, V. M. Axt, and T. Kuhn (2001). Dynamics of exciton formation for near band-gap excitations, *Phys. Rev. B* **65**, 035303.

E. Runge (2002). Excitons in semiconductor nanostructures, *Solid State Phys.* **57**, 149.

M. Combescot and O. Betbeder-Matibet (2010), General many-body formalism for composite quantum particles, *Phys. Rev. Lett.* **104**, 206404. Here, the interested reader can also find many references to the earlier work of the authors on the exciton problem.

28
Semiconductor luminescence equations

Our discussion in Chapter 27 shows that incoherent carrier excitations can remain in the system after the coherences have vanished. This incoherent regime may last for an extended time interval such that several intriguing many-body and quantum-optical phenomena can develop before the semiconductor eventually returns to its ground state via radiative and/or nonradiative recombination processes.

To learn how the quantum aspects of light affect the quantum dynamics of incoherent semiconductors, we investigate the spontaneous emission of light and the related electron–hole-pair recombination processes. We derive the *semiconductor luminescence equations* and discuss their solutions under different conditions. In particular, we analyze the contributions of bound and unbound electron–hole pairs, i.e., excitons and electron–hole plasma to the spontaneous emission.

28.1 Incoherent photon emission

We start from the dynamics of the photon operators,

$$\begin{aligned}
i\hbar\frac{\partial B_{\mathbf{q},q_z}}{\partial t} &= \hbar\omega_q B_{\mathbf{q},q_z} - \sum_{\mathbf{k}} \frac{\mathcal{E}_{\mathbf{q},q_z}\tilde{u}^{\star}_{\mathbf{q},q_z}}{\omega_{\mathbf{q}}}\left(\frac{Q\,p_{v,c}}{m_0}\,a^{\dagger}_{v,\mathbf{k}-\mathbf{q}}a_{c,\mathbf{k}} + \frac{Q\,p_{c,v}}{m_0}\,a^{\dagger}_{c,\mathbf{k}-\mathbf{q}}a_{v,\mathbf{k}}\right) \\
&= \hbar\omega_q\, B_{\mathbf{q},q_z} + i\mathcal{E}_{\mathbf{q},q_z}\,\tilde{u}^{\star}_{\mathbf{q},q_z}\sum_{\mathbf{k}}\left(d_{v,c}\,a^{\dagger}_{v,\mathbf{k}-\mathbf{q}}a_{c,\mathbf{k}} - d_{c,v}\,a^{\dagger}_{c,\mathbf{k}-\mathbf{q}}a_{v,\mathbf{k}}\right) \\
&= \hbar\omega_q\, B_{\mathbf{q},q_z} + i\sum\nolimits_{\mathbf{k}} d_{v,c}\,\mathcal{E}_{\mathbf{q},q_z}\,\tilde{u}^{\star}_{\mathbf{q},q_z}\,a^{\dagger}_{v,\mathbf{k}-\mathbf{q}}a_{c,\mathbf{k}},
\end{aligned} \tag{28.1}$$

where we combined Eq. (24.27) with the expression (24.19) for the interband polarization. Furthermore, we applied the transformation (26.15) to convert the momentum-matrix element into the dipole-matrix element. The operator product $a^{\dagger}_{c,\mathbf{k}-\mathbf{q}}a_{v,\mathbf{k}}$ behaves like a polarization $P^{\star}_{\mathbf{k}}$ that rotates opposite, i.e., nonresonantly, relative to $P_{\mathbf{k}}$ and $B_{\mathbf{q},q_z}$. Based on the rotating-wave argumentation of Section 25.4.1, such nonresonant contributions have a negligible influence on the near-resonant optical transitions. Thus, it is justified to drop the nonresonant part already at the operator level, as is done in the last step in (28.1). A more thorough investigation of this approximation is discussed in Exercise 28.1.

In Eq. (28.1), we see that the strength of the light–matter coupling is determined by

$$\mathcal{F}_{\mathbf{q}_{3D}} \equiv \begin{cases} \mathbf{d}_{c,v}\, \mathcal{E}_{\mathbf{q},q_z} \cdot \tilde{\mathbf{u}}_{\mathbf{q},q_z} = \frac{\mathbf{d}_{c,v}\, \mathcal{E}_{\mathbf{q}_{3D}}}{\sqrt{S}} \cdot \int dz\, g(z)\, \mathbf{u}_{\mathbf{q}_{3D}}(z)\,, & |\mathbf{q}| \leq q_{\text{rad}} \\ 0\,, & |\mathbf{q}| > q_{\text{rad}} \end{cases}, \tag{28.2}$$

which contains an overlap integral between the mode function $\mathbf{u}_{\mathbf{q}_{3D}}(z)$ and the confinement function $g(z)$, based on Eq. (26.14). Furthermore, we introduced the notation $\mathbf{q}_{3D} \equiv (\mathbf{q},\, q_z)$ to identify the full three-dimensional (3D) wave vector of the photons. For sufficiently strong confinement, $\tilde{\mathbf{u}}_{\mathbf{q}_{3D}}$ becomes $\mathbf{u}_{\mathbf{q}_{3D}}(0)$, evaluated at the QW position. To generalize $\mathcal{F}_{\mathbf{q}_{3D}}$ for all in-plane momenta, we introduced the cutoff at $q_{\text{rad}} = \omega_{\mathbf{q}_{3D}}/c = 2\pi/\lambda$. In other words, we simply set the coupling to zero when $\mathbf{q}$ exceeds q_{rad} which is the largest possible in-plane momentum in free space. In the most general case, the dipole-matrix element $\mathbf{d}_{c,v}$ and therefore $\mathcal{F}_{\mathbf{q}_{3D}}$ are $\mathbf{k}$ dependent. For simplicity, we do not explicitly denote this dependence because it can be easily included after the incoherent emission equations are derived, see Exercise 28.3.

With the help of Eq. (28.2), the dynamical equations for the photon operator can be written compactly as

$$\begin{cases} i\hbar\frac{\partial}{\partial t} B_{\mathbf{q},q_z} = \hbar\omega_{\mathbf{q}_{3D}}\, B_{\mathbf{q},q_z} + i \sum_{\mathbf{k}} \mathcal{F}^{\star}_{\mathbf{q},q_z} a^{\dagger}_{v,\mathbf{k}-\mathbf{q}} a_{c,\mathbf{k}} \\ i\hbar\frac{\partial}{\partial t} B^{\dagger}_{\mathbf{q},q_z} = -\hbar\omega_{\mathbf{q}_{3D}}\, B^{\dagger}_{\mathbf{q},q_z} + i \sum_{\mathbf{k}} \mathcal{F}_{\mathbf{q},q_z} a^{\dagger}_{c,\mathbf{k}} a_{v,\mathbf{k}-\mathbf{q}} \end{cases}. \tag{28.3}$$

These equations can be used as the starting point to investigate quantum-optical modifications of the semiconductor interband transitions. The photon–singlet dynamics is obtained simply by taking the average,

$$\begin{aligned} i\hbar\frac{\partial}{\partial t}\langle B_{\mathbf{q},q_z}\rangle &= \hbar\omega_{\mathbf{q}_{3D}}\, \langle B_{\mathbf{q},q_z}\rangle + i\sum_{\mathbf{k}} \mathcal{F}^{\star}_{\mathbf{q},q_z} \left\langle a^{\dagger}_{v,\mathbf{k}-\mathbf{q}} a_{c,\mathbf{k}}\right\rangle \\ &= \hbar\omega_{\mathbf{q}_{3D}}\, \langle B_{\mathbf{q},q_z}\rangle + i\,\delta_{\mathbf{q},0}\sum_{\mathbf{k}} \mathcal{F}^{\star}_{\mathbf{q},q_z} P_{\mathbf{k}}, \end{aligned} \tag{28.4}$$

where we assumed homogeneous excitation conditions. Even though both $\langle B_{\mathbf{q},q_z}\rangle$ and $P_{\mathbf{k}}$ are coherent quantities, we see that only the component $\langle B_{\mathbf{q}=0,q_z}\rangle$ can be driven by the coherent polarization. In the fully incoherent regime, both the polarization and $\langle B_{\mathbf{q},q_z}\rangle$ vanish for all argument values.

Even though the purely classical extension of light $\langle B_{\mathbf{q}_{3D}}\rangle$ vanishes in the incoherent regime, that does not necessarily imply that the light intensity is zero. To show this, we look at the photon number dynamics,

$$i\hbar\frac{\partial}{\partial t}\left\langle B^{\dagger}_{\mathbf{q}_{3D}}\, B_{\mathbf{q}_{3D}}\right\rangle = 2i\,\text{Re}\left[\sum_{\mathbf{k}} \mathcal{F}^{\star}_{\mathbf{q}_{3D}} \left\langle B^{\dagger}_{\mathbf{q},q_z}\, a^{\dagger}_{v,\mathbf{k}-\mathbf{q}} a_{c,\mathbf{k}}\right\rangle\right], \tag{28.5}$$

which directly follows from Eq. (28.3). We see that the photon number is driven by $\langle B^{\dagger}_{\mathbf{q},q_z}\, a^{\dagger}_{v,\mathbf{k}-\mathbf{q}} a_{c,\mathbf{k}}\rangle$, which is an incoherent quantity according to our classification (25.29). Consequently, both $\langle B^{\dagger}_{\mathbf{q}_{3D}}\, B_{\mathbf{q}_{3D}}\rangle$ and $\langle B^{\dagger}_{\mathbf{q},q_z}\, a^{\dagger}_{v,\mathbf{k}-\mathbf{q}} a_{c,\mathbf{k}}\rangle$ can exist even when all coherences vanish.

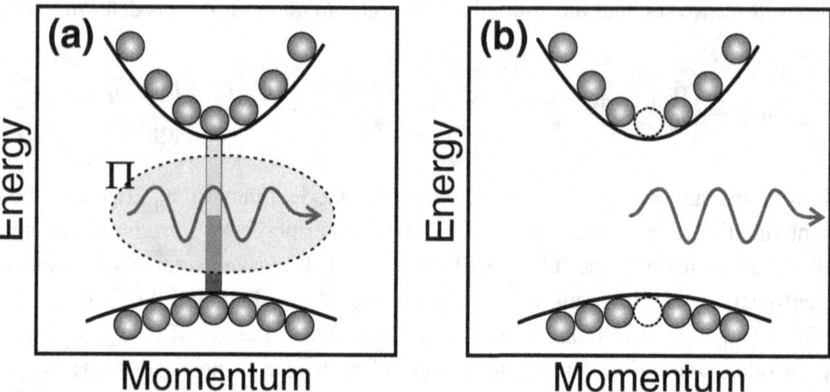

Figure 28.1 Schematic representation of a spontaneous-emission process in a direct-gap semiconductor. **(a)** The vacuum-field fluctuations of the light (wave arrow) induce a virtual polarization (bar), which generates a photon-assisted polarization correlation Π (dashed ellipse). **(b)** When Π is built up, electron–hole pairs start to recombine radiatively via photon emission (wave arrow). The spheres indicate electrons (holes) in the conduction (valence) band.

For general excitation conditions, we follow the incoherent doublet correlations

$$\Pi_{\mathbf{k};\,\mathbf{q}_{3D}} = \Pi_{\mathbf{k};\,\mathbf{q},q_z} \equiv \Delta\left\langle B^\dagger_{\mathbf{q}_{3D}} a^\dagger_{v,\mathbf{k}-\mathbf{q}} a_{c,\mathbf{k}}\right\rangle$$
$$= \left\langle B^\dagger_{\mathbf{q}_{3D}} a^\dagger_{v,\mathbf{k}-\mathbf{q}} a_{c,\mathbf{k}}\right\rangle - \left\langle B^\dagger_{\mathbf{q}_{3D}}\right\rangle\left\langle a^\dagger_{v,\mathbf{k}-\mathbf{q}} a_{c,\mathbf{k}}\right\rangle,$$
$$\Delta\left\langle B^\dagger_{\mathbf{q},q_z} B_{\mathbf{q},q'_z}\right\rangle \equiv \left\langle B^\dagger_{\mathbf{q},q_z} B_{\mathbf{q},q'_z}\right\rangle - \left\langle B^\dagger_{\mathbf{q},q_z}\right\rangle \langle B_{\mathbf{q},q'_z}\rangle, \tag{28.6}$$

where the correlated part is the difference between the two-particle expectation value and its classical factorization. In the purely incoherent regime, these two-particle expectation values are defined entirely by the correlated doublets. The expression $\Delta\left\langle B^\dagger a^\dagger_v a_c\right\rangle$ can also be interpreted as *photon-assisted polarization* because it contains the product of the photon and the polarization-type operator $a^\dagger_v a_c$. Figure 28.1(a) schematically depicts a Π correlation (dashed ellipse) between a polarization state (dark-light bar) and a photon (wave arrow). As net effect, the photon-assisted polarization initiates the simultaneous photon emission and electron–hole-pair recombination, as illustrated in Fig. 28.1(b).

28.1.1 Photon emission vs. electron–hole recombination

To determine the carrier quantum dynamics, it is useful to analyze the equations of motion for $f^e_\mathbf{k}$ and $f^h_\mathbf{k}$. Under incoherent conditions, we obtain from (27.2)

$$\hbar\frac{\partial}{\partial t} f^e_\mathbf{k} = \mathrm{Im}\left[\Gamma^{c,c}_\mathbf{k} + \Gamma^{\mathrm{QED}}_{c,c;\,\mathbf{k}}\right], \qquad \hbar\frac{\partial}{\partial t} f^h_\mathbf{k} = -\mathrm{Im}\left[\Gamma^{v,v}_\mathbf{k} + \Gamma^{\mathrm{QED}}_{v,v;\,\mathbf{k}}\right]. \tag{28.7}$$

Since the Coulomb-induced doublets do not alter the particle number, we can concentrate on the quantum-optical contributions $\Gamma^{\rm QED}_{\lambda,\lambda;\,\mathbf{k}}$. As we combine the definitions (26.16), (26.17), and (28.6), we find

$$\hbar\frac{\partial}{\partial t} f^e_{\mathbf{k}}\bigg|_{\rm QED} = -2\,{\rm Re}\left[\sum\nolimits_{\mathbf{q}_{3D}} \mathcal{F}^\star_{\mathbf{q}_{3D}}\,\Pi_{\mathbf{k};\,\mathbf{q}_{3D}}\right] \tag{28.8}$$

$$\hbar\frac{\partial}{\partial t} f^h_{\mathbf{k}}\bigg|_{\rm QED} = -2\,{\rm Re}\left[\sum\nolimits_{\mathbf{q}_{3D}} \mathcal{F}^\star_{\mathbf{q}_{3D}}\,\Pi_{\mathbf{k}+\mathbf{q};\,\mathbf{q}_{3D}}\right]. \tag{28.9}$$

These equations show that not only the photon number but also the electron- and hole-density dynamics couples to a correlated photon-assisted polarization, $\Pi \equiv \Delta\langle B^\dagger a^\dagger_v a_c\rangle$. Thus, Π also describes the correlated *photon-assisted electron–hole recombination*, as illustrated in Fig. 28.1. In this process, the momentum $\mathbf{q}$ is conserved because the recombining electron–hole pair transfers its in-plane momentum to the generated photon. However, there is no momentum conservation in the q_z direction perpendicular to the planar structure because the confinement breaks the translational symmetry.

Using the definition of Π, we can write the equation for the photon-number correlations as

$$\hbar\frac{\partial}{\partial t}\Delta\left\langle B^\dagger_{\mathbf{q}_{3D}}\,B_{\mathbf{q}_{3D}}\right\rangle = 2\,{\rm Re}\left[\sum\nolimits_{\mathbf{k}} \mathcal{F}^\star_{\mathbf{q}_{3D}}\,\Pi_{\mathbf{k};\,\mathbf{q}_{3D}}\right], \tag{28.10}$$

where we focus on the correlated dynamics by subtracting the pure singlet dynamics (28.4) from Eq. (28.5). Combining Eqs. (28.8)–(28.10) allows us to construct a general conservation law

$$\frac{\partial}{\partial t}\sum_{\mathbf{q},q_z}\Delta\left\langle B^\dagger_{\mathbf{q},q_z}\,B_{\mathbf{q},q_z}\right\rangle = -\frac{\partial}{\partial t}\sum_{\mathbf{k}} f^e_{\mathbf{k}}\bigg|_{\rm QED} = -\frac{\partial}{\partial t}\sum_{\mathbf{k}} f^h_{\mathbf{k}}\bigg|_{\rm QED}, \tag{28.11}$$

which is valid for an arbitrary form of the photon-assisted polarization. This result shows that the total number of emitted photons must equal the number of the removed electrons and holes. This observation agrees with the intuitive picture of spontaneous emission generated by spontaneous electron–hole recombination, illustrated schematically in Fig. 28.1(b).

28.1.2 Exciton-correlation dynamics

We discuss in Chapter 27 that the incoherent many-body state may also contain a significant amount of incoherent excitons besides the unbound electrons and holes. The excitons are treated using the correlation function $c_X^{\mathbf{q},\mathbf{k}',\mathbf{k}} = \Delta\langle a^\dagger_{c,\mathbf{k}} a^\dagger_{v,\mathbf{k}'} a_{c,\mathbf{k}'+\mathbf{q}}\, a_{v,\mathbf{k}-\mathbf{q}}\rangle$ defined earlier in connection with Eq. (27.33). For the incoherent regime, the doublet contributions to the light–matter interactions produce the quantum-optical term

$$\hbar\frac{\partial}{\partial t}c_X^{\mathbf{q},\mathbf{k}',\mathbf{k}}\bigg|_{\mathrm{QED}} = -\left(1-f^e_{\mathbf{k}}-f^h_{\mathbf{k}-\mathbf{q}}\right)\sum_{q_z}\mathcal{F}^{\star}_{\mathbf{q}_{3\mathrm{D}}}\Pi_{\mathbf{k}'+\mathbf{q};\,\mathbf{q}_{3\mathrm{D}}} - \left(1-f^e_{\mathbf{k}'+\mathbf{q}}-f^h_{\mathbf{k}'}\right)\sum_{q_z}\mathcal{F}_{\mathbf{q}_{3\mathrm{D}}}\Pi^{\star}_{\mathbf{k};\,\mathbf{q}_{3\mathrm{D}}}, \tag{28.12}$$

that occurs additively in the exciton-correlation dynamics (27.35). Exercise 28.2 derives this result directly from the three-particle contributions when the doublets are fully included.

Equation (28.12) shows that also the exciton correlations couple to the photon-assisted polarization $\Pi = \Delta\langle B^{\dagger}a^{\dagger}_{v}a_c\rangle$. However, unlike the carrier dynamics (28.8)–(28.9), Π appears in the c_X dynamics *without* the sum over the in-plane momentum $\mathbf{q}$ of the photon. More precisely, the center-of-mass momentum of $c_X^{\mathbf{q},\mathbf{k}',\mathbf{k}}$ matches exactly with the in-plane momentum of the photons. To show this even more clearly, we expand the photon-assisted polarization into the exciton basis

$$\begin{cases} \Pi_{\mathbf{k}+\mathbf{q}_e;\,\mathbf{q}_{3\mathrm{D}}} = \sum_{\lambda}\sqrt{\mathcal{S}}\,\Pi_{\lambda;\,\mathbf{q}_{3\mathrm{D}}}\phi^R_{\lambda}(\mathbf{k}) \\ \Pi_{\lambda;\,\mathbf{q}_{3\mathrm{D}}} = \frac{1}{\sqrt{\mathcal{S}}}\sum_{\mathbf{k}}\phi^L_{\lambda}(\mathbf{k})\,\Pi_{\mathbf{k}+\mathbf{q}_e;\,\mathbf{q}_{3\mathrm{D}}} \end{cases}. \tag{28.13}$$

As it turns out, it is beneficial to shift the momentum by $\mathbf{q}_e = \frac{m_e}{m_e+m_h}\mathbf{q}$ to simplify the center-of-mass momentum dependency of the excitons. Using this and the transformation (27.48) in Eq. (28.12), we can express the quantum-optical contributions to the exciton-correlation dynamics via

$$\hbar\frac{\partial}{\partial t}\Delta\left\langle X^{\dagger}_{\lambda,\mathbf{q}}X_{\nu,\mathbf{q}}\right\rangle\bigg|_{\mathrm{QED}} = -\sum_{q_z}\mathcal{S}\left[\phi^R_{\lambda}(\mathbf{r}=0)\mathcal{F}^{\star}_{\mathbf{q}_{3\mathrm{D}}}\Pi_{\nu;\,\mathbf{q},q_z} + \phi^R_{\nu}(\mathbf{r}=0)\mathcal{F}_{\mathbf{q}_{3\mathrm{D}}}\Pi^{\star}_{\lambda;\,\mathbf{q},q_z}\right], \tag{28.14}$$

where the real-space representation $\phi^R_{\nu}(\mathbf{r})$ appears according to (26.32). The diagonal contributions $\lambda = \nu$ of $\Delta\langle X^{\dagger}_{\lambda,\mathbf{q}}X_{\nu,\mathbf{q}}\rangle$ define the incoherent exciton populations while the nondiagonal contributions $\lambda \neq \nu$ are related to the correlated electron–hole plasma, as discussed in Section 27.3.1. The influence of the quantum-optical fluctuations on the exciton population is given by

$$\hbar\frac{\partial\Delta\left\langle X^{\dagger}_{\lambda,\mathbf{q}}X_{\lambda,\mathbf{q}}\right\rangle}{\partial t}\Bigg|_{\mathrm{QED}} = -2\mathcal{S}\,\mathrm{Re}\left[\phi^R_{\lambda}(\mathbf{r}=0)\sum_{q_z}\mathcal{F}^{\star}_{\mathbf{q}_{3\mathrm{D}}}\Pi_{\lambda;\,\mathbf{q},q_z}\right] = -2\,\mathrm{Re}\left[\sum_{q_z,\mathbf{k},\mathbf{k}'}\phi^R_{\lambda}(\mathbf{k}')\phi^L_{\lambda}(\mathbf{k})\mathcal{F}^{\star}_{\mathbf{q}_{3\mathrm{D}}}\Pi_{\mathbf{k}+\mathbf{q}_e;\,\mathbf{q},q_z}\right]. \tag{28.15}$$

Taking the sum over all exciton states produces

$$\sum_\lambda \hbar \frac{\partial \Delta\langle X^\dagger_{\lambda,\mathbf{q}} X_{\lambda,\mathbf{q}}\rangle}{\partial t}\bigg|_{\text{QED}} = -2\,\text{Re}\left[\sum_{q_z,\mathbf{k},\mathbf{k}'}\sum_\lambda \phi^R_\lambda(\mathbf{k}')\phi^L_\lambda(\mathbf{k})\mathcal{F}^\star_{\mathbf{q}3\text{D}}\Pi_{\mathbf{k}+\mathbf{q}_e;\,\mathbf{q},q_z}\right]$$
$$= -2\,\text{Re}\left[\sum_{q_z,\mathbf{k}} \mathcal{F}^\star_{\mathbf{q}3\text{D}}\Pi_{\mathbf{k};\,\mathbf{q},q_z}\right], \tag{28.16}$$

after we apply the completeness relation $\sum_\lambda \phi^R_\lambda(\mathbf{k}')\phi^L_\lambda(\mathbf{k}) = \delta_{\mathbf{k},\mathbf{k}'}$ among the exciton states.

Comparing Eqs. (28.10) and (28.16), we note that the summation of the photon equation over the momentum q_z yields the additional conservation law

$$\frac{\partial}{\partial t}\sum_\lambda \Delta\langle X^\dagger_{\lambda,\mathbf{q}} X_{\lambda,\mathbf{q}}\rangle\bigg|_{\text{QED}} = -\left[\frac{\partial}{\partial t}\sum_{q_z}\Delta\langle B^\dagger_{\mathbf{q},q_z} B_{\mathbf{q},q_z}\rangle\right]. \tag{28.17}$$

This relation holds separately for each in-plane momentum $\mathbf{q}$, again verifying that the center-of-mass momentum is conserved in the emission process. Interestingly, the conservation law involves *all* exciton states including the continuum of ionized states. Hence, the emission of a particular photon does not necessarily imply the recombination of a bound electron–hole pair. As shown in Section 28.4, even the photon emission at the excitonic resonances can stem from an unbound plasma.

28.2 Dynamics of photon-assisted correlations

The processes of photon emission and electron–hole-pair recombination are both initiated by the photon-assisted polarization, $\Pi = \Delta\langle B^\dagger a^\dagger_v a_c\rangle$. To determine when and how Π-related effects occur, we need to analyze the corresponding quantum dynamics. For this purpose, we start from

$$i\hbar\frac{\partial}{\partial t}\hat{\Pi}_{\mathbf{k};\,\mathbf{q},q_z} = \left[i\hbar\frac{\partial B^\dagger_{\mathbf{q},q_z}}{\partial t}\right]a^\dagger_{v,\mathbf{k}-\mathbf{q}}a_{c,\mathbf{k}} + B^\dagger_{\mathbf{q},q_z}\left[i\hbar\frac{\partial a^\dagger_{v,\mathbf{k}-\mathbf{q}}a_{c,\mathbf{k}}}{\partial t}\right]. \tag{28.18}$$

Here, the first part contains the time derivative of the Bosonic photon operator. Using Eq. (28.3), we obtain

$$i\hbar\frac{\partial}{\partial t}\hat{\Pi}^{\text{bos}}_{\mathbf{k};\,\mathbf{q}3\text{D}} \equiv -\hbar\omega_{\mathbf{q}3\text{D}}\,B^\dagger_{\mathbf{q},q_z}a^\dagger_{v,\mathbf{k}-\mathbf{q}}a_{c,\mathbf{k}} + i\sum_{\mathbf{k}'}\mathcal{F}_{\mathbf{q},q_z}a^\dagger_{c,\mathbf{k}'}a_{v,\mathbf{k}'-\mathbf{q}}a^\dagger_{v,\mathbf{k}-\mathbf{q}}a_{c,\mathbf{k}}$$
$$= -\hbar\omega_{\mathbf{q}3\text{D}}\,\hat{\Pi}_{\mathbf{k};\,\mathbf{q},q_z}$$
$$+ i\mathcal{F}_{\mathbf{q},q_z}\left(a^\dagger_{c,\mathbf{k}}a_{c,\mathbf{k}} + \sum_{\mathbf{k}'}a^\dagger_{c,\mathbf{k}'}a^\dagger_{v,\mathbf{k}-\mathbf{q}}a_{c,\mathbf{k}}a_{v,\mathbf{k}'-\mathbf{q}}\right), \tag{28.19}$$

where we normally ordered the term containing the Fermi operators. Equation (28.19) contains all contributions resulting from the photon-operator dynamics. The remaining Fermion-operator part is discussed in Section 28.2.2.

We can now construct the correlated dynamics resulting from the Bosonic part alone. For this purpose, we take the expectation value of Eq. (28.19) and subtract the pure singlet dynamics,

$$\begin{aligned}
i\hbar\frac{\partial}{\partial t}\Pi^{\text{bos}}_{\mathbf{k};\,\mathbf{q}_{3D}} &= i\hbar\frac{\partial}{\partial t}\left\langle\hat{\Pi}^{\text{bos}}_{\mathbf{k};\,\mathbf{q}_{3D}}\right\rangle - \left[i\hbar\frac{\partial\left\langle B^{\dagger}_{\mathbf{q}_{3D}}\right\rangle}{\partial t}\right]\left\langle a^{\dagger}_{v,\mathbf{k}-\mathbf{q}}a_{c,\mathbf{k}}\right\rangle \\
&= -\hbar\omega_{\mathbf{q}_{3D}}\,\Pi_{\mathbf{k};\,\mathbf{q}_{3D}} + i\mathcal{F}_{\mathbf{q}_{3D}}\left[\left\langle a^{\dagger}_{c,\mathbf{k}}a_{c,\mathbf{k}}\right\rangle + \sum\nolimits_{\mathbf{k}'}\left\langle a^{\dagger}_{c,\mathbf{k}'}a^{\dagger}_{v,\mathbf{k}-\mathbf{q}}a_{c,\mathbf{k}}a_{v,\mathbf{k}'-\mathbf{q}}\right\rangle\right. \\
&\quad \left. - \sum_{\mathbf{k}'}\left\langle a^{\dagger}_{c,\mathbf{k}'}a_{v,\mathbf{k}'-\mathbf{q}}\right\rangle\left\langle a^{\dagger}_{v,\mathbf{k}-\mathbf{q}}a_{c,\mathbf{k}}\right\rangle\right] \\
&= -\hbar\omega_{\mathbf{q}_{3D}}\,\Pi_{\mathbf{k};\,\mathbf{q}_{3D}} + i\mathcal{F}_{\mathbf{q}_{3D}}\mathrm{S}^{\text{spnt}}_{\mathbf{k};\,\mathbf{q}}.
\end{aligned} \tag{28.20}$$

The last term can be simplified by applying the homogeneity condition (24.7) and by separating the singlet and the doublet contributions using (15.32),

$$\begin{aligned}
\mathrm{S}^{\text{spnt}}_{\mathbf{k};\,\mathbf{q}} &= f^e_{\mathbf{k}} + \sum_{\mathbf{k}'}\left[\left\langle a^{\dagger}_{c,\mathbf{k}'}a^{\dagger}_{v,\mathbf{k}-\mathbf{q}}a_{c,\mathbf{k}}a_{v,\mathbf{k}'-\mathbf{q}}\right\rangle - \left\langle a^{\dagger}_{c,\mathbf{k}'}a_{v,\mathbf{k}'-\mathbf{q}}\right\rangle\left\langle a^{\dagger}_{v,\mathbf{k}-\mathbf{q}}a_{c,\mathbf{k}}\right\rangle\right] \\
&= f^e_{\mathbf{k}}\,f^h_{\mathbf{k}-\mathbf{q}} + \sum\nolimits_{\mathbf{k}'} c_X^{\mathbf{q},\mathbf{k}-\mathbf{q},\mathbf{k}'},
\end{aligned} \tag{28.21}$$

after we identify the exciton correlations.

28.2.1 Spontaneous-emission source

We notice that $\mathrm{S}^{\text{spnt}}_{\mathbf{k};\,\mathbf{q}}$ does not contain any polarization because the singlet contributions appear only through the product of electron and hole occupations. As we insert the result (28.21) into Eq. (28.20), we find

$$\begin{aligned}
i\hbar\frac{\partial}{\partial t}\Pi^{\text{bos}}_{\mathbf{k};\,\mathbf{q}_{3D}} &= -\hbar\omega_{\mathbf{q}_{3D}}\,\Pi_{\mathbf{k};\,\mathbf{q}_{3D}} + i\mathcal{F}_{\mathbf{q}_{3D}}\mathrm{S}^{\text{spnt}}_{\mathbf{k};\,\mathbf{q}} \\
&= -\hbar\omega_{\mathbf{q}_{3D}}\,\Pi_{\mathbf{k};\,\mathbf{q}_{3D}} + i\mathcal{F}_{\mathbf{q}_{3D}}\left[f^e_{\mathbf{k}}\,f^h_{\mathbf{k}-\mathbf{q}} + \sum\nolimits_{\mathbf{k}'} c_X^{\mathbf{q},\mathbf{k}-\mathbf{q},\mathbf{k}'}\right].
\end{aligned} \tag{28.22}$$

Since $\mathrm{S}^{\text{spnt}}_{\mathbf{k};\,\mathbf{q}}$ can also have a finite value when all coherences vanish, the photon-assisted polarization is also driven in the incoherent regime. We see from Eq. (28.2) that the coupling factor $\mathcal{F}_{\mathbf{q}_{3D}}$ is defined by the product of the dipole matrix element, the vacuum-field amplitude, and the mode strength at the position of the QW or QWI.

Since $\mathcal{F}_{\mathbf{q}_{3D}}\mathrm{S}^{\text{spnt}}_{\mathbf{k};\,\mathbf{q}}$ is nonvanishing even without the presence of coherent light, one can argue that it must stem from vacuum-field fluctuations. To support this interpretation, we investigate which carrier states contribute to the spontaneous source. Clearly, an electron with the momentum $\hbar\mathbf{k}$ can recombine with a hole having the momentum $\hbar(\mathbf{q}-\mathbf{k})$ if their center-of-mass momentum, i.e., $\hbar\mathbf{k}+\hbar(\mathbf{q}-\mathbf{k}) = \hbar\mathbf{q}$, is transferred to the spontaneously emitted photon. When electrons and holes move freely, the likelihood of this process is defined by the product of the electron and hole occupations, i.e., $f^e_{\mathbf{k}}\, f^h_{\mathbf{k}-\mathbf{q}}$. Note that the state $f^h_{\mathbf{k}-\mathbf{q}} \equiv 1 - \langle a^\dagger_{v,\mathbf{k}-\mathbf{q}} a_{v,\mathbf{k}-\mathbf{q}}\rangle$ refers to a hole with an inverted momentum $-\hbar(\mathbf{k}-\mathbf{q})$ because the hole is the antiparticle of the valence-band electron.

An emission source, purely consisting of the carrier distribution product, is realized in an uncorrelated electron–hole plasma, which is described fully by the mean-field contribution, i.e., the singlet part of the many-body system. When electrons and holes are bound together to form exciton populations, the relative electron and hole momenta are automatically matched to produce the conditions that are favorable for radiative recombination. Then only the center-of-mass momentum of the pair, i.e., $\mathbf{q}$, must match the in-plane momentum of the photon and the correlated pairs can recombine with a probability that is directly proportional to the pairwise exciton correlation with a matching $\mathbf{q}$. Thus, the part $\sum_{\mathbf{k}'} c_X^{\mathbf{q},\mathbf{k}-\mathbf{q},\mathbf{k}'}$ of the spontaneous-emission source must result from the presence of excitons or a correlated electron–hole plasma. Since neither the uncorrelated plasma nor the correlated sources depend on the photon energy in any way, all of them can initiate photon emission – that is observed as photoluminescence – in certain frequency ranges. This aspect is studied further in Section 28.4.

In the regime of not too high excitations, i.e., as long as the electron distribution is not degenerate, the $f^e f^h$ source of the spontaneous emission scales roughly quadratically with the total density while the c_X source can yield a linear density dependence. Furthermore, the densities as well as the exciton correlations may have nonzero values also for wave vectors in the range of $|\mathbf{k}| \propto 1/a_0$ where the Bohr radius a_0 is roughly 10 nm in GaAs-type systems. However, the prefactor $\mathcal{F}_{\mathbf{q},q_z}$ vanishes for all in-plane wave vectors larger than $q_{\text{rad}} = 2\pi/\lambda$, compare Eq. (28.2). For GaAs systems, the relevant wave length is roughly $\lambda = 800$ nm, which gives $q_{\text{rad}} = 2\pi/\lambda a_0 \frac{1}{a_0} \approx 0.1/a_0$, which is much smaller than typical carrier wave vectors. Even though we may find finite values for $c_X^{\mathbf{q},\mathbf{k}',\mathbf{k}}$ for a wide range of center-of-mass momenta, the condition $|\mathbf{q}| < q_{\text{rad}} \ll 1/a_0$ limits the possibility of the corresponding excitons to directly contribute to the luminescence. In contrast, the plasma source $f^e_{\mathbf{k}}\, f^h_{\mathbf{k}-\mathbf{q}}$ contributes to the spontaneous emission for *all* electron and hole states. Thus, the possibility of spontaneous recombination is very momentum selective for excitons, whereas it is nearly momentum independent for an uncorrelated electron–hole plasma.

The momentum dependence of the spontaneous emission source is shown schematically in Fig. 28.2 for the direction $\mathbf{q} = 0$ perpendicular to the QW. The area of the spheres (circles) in Fig. 28.2(a) indicates the level of electron (hole) occupation and the bar chart indicates the strength of $\mathrm{S}^{\text{spnt}}_{\mathbf{k};\,\mathbf{q}=0}$ as a function of carrier momentum. For $f^e_{\mathbf{k}} f^h_{\mathbf{k}}$, all momentum states of the uncorrelated plasma contribute. If the system contains excitons, we can describe it through the corresponding $\Delta\langle X^\dagger_{\lambda,\mathbf{q}} X^\dagger_{\lambda,\mathbf{q}}\rangle$ distribution defining how strongly a given center-of-mass momentum state is occupied. Figure 28.2(b) schematically shows a

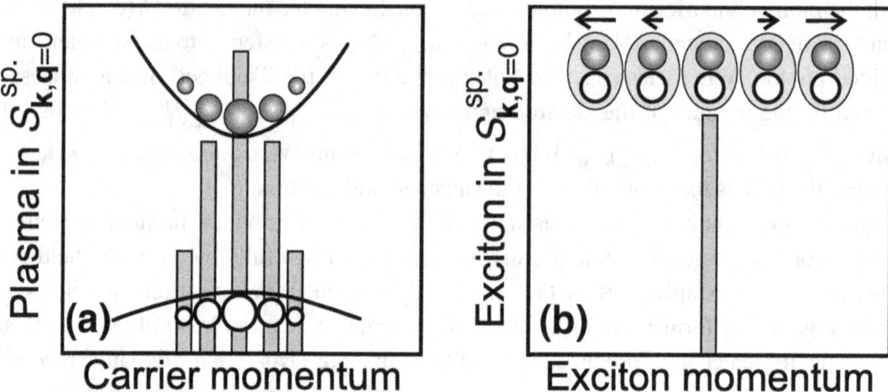

Figure 28.2 Dependence of the spontaneous-emission source on the many-body state. **(a)** The electron (hole) densities are symbolized by the size of the spheres (open circles). The bars indicate the corresponding plasma source $f^e_{\mathbf{k}} f^h_{\mathbf{k}}$. **(b)** Exciton correlations with different momenta (arrows) are shown. The bar chart indicates their contribution to the spontaneous-emission source in the same $\mathbf{q} = 0$ direction as the plasma part. The division of spontaneous emission into plasma and correlated part follows from Eq. (28.21).

set of bound exciton states with the momenta $\hbar\mathbf{q}$ indicated by the arrows above the pair states (sphere and open circle within shaded ellipse). As a consequence of the momentum conservation between the recombining excitons and the emitted photons, only the $\mathbf{q} = 0$ excitons contribute to the spontaneous emission source (bar).

28.2.2 Semiconductor luminescence equations

The Fermion part of Eq. (28.18) can be determined after we evaluate the dynamics for the polarization operator

$$\begin{aligned} i\hbar\frac{\partial}{\partial t} a^\dagger_{v,\mathbf{k}-\mathbf{q}} a_{c,\mathbf{k}} &= \left(\epsilon^c_{\mathbf{k}} - \epsilon^v_{\mathbf{k}-\mathbf{q}}\right) a^\dagger_{v,\mathbf{k}-\mathbf{q}} a_{c,\mathbf{k}} \\ &+ \sum_{\lambda,\mathbf{k}',\mathbf{q}'} V_{\mathbf{q}'} \left(a^\dagger_{v,\mathbf{k}-\mathbf{q}} a^\dagger_{\lambda,\mathbf{k}'+\mathbf{q}'} a_{\lambda,\mathbf{k}'} a_{c,\mathbf{k}+\mathbf{q}'} - a^\dagger_{v,\mathbf{k}-\mathbf{q}-\mathbf{q}'} a^\dagger_{\lambda,\mathbf{k}'+\mathbf{q}'} a_{\lambda,\mathbf{k}'} a_{c,\mathbf{k}}\right) \\ &- \sum_{\mathbf{q}'} \left(\hat{\mathcal{A}}^v_{\mathbf{q}'}\, a^\dagger_{v,\mathbf{k}-\mathbf{q}} a_{v,\mathbf{k}-\mathbf{q}'} + \hat{\mathcal{A}}^v_{-\mathbf{q}'}\, a^\dagger_{c,\mathbf{k}-\mathbf{q}-\mathbf{q}'} a_{c,\mathbf{k}}\right) \\ &- \sum_{\mathbf{p}} \left(\hat{\mathcal{G}}^c_{\mathbf{p}}\, a^\dagger_{v,\mathbf{k}} a_{c,\mathbf{k}-\mathbf{p}} + \mathcal{G}^v_{-\mathbf{p}}\, a^\dagger_{v,\mathbf{k}-\mathbf{q}-\mathbf{p}} a_{c,\mathbf{k}}\right), \end{aligned} \tag{28.23}$$

where we used Eqs. (24.29)–(24.30). By combining this and (28.22) with Eq. (28.18), we eventually find

$$i\hbar\frac{\partial \Pi_{\mathbf{k},\mathbf{q}_{3D}}}{\partial t} = \left(\tilde{\epsilon}^c_{\mathbf{k}} - \tilde{\epsilon}^v_{\mathbf{k}-\mathbf{q}} - \hbar\omega_{\mathbf{q}_{3D}}\right)\Pi_{\mathbf{k},\mathbf{q}_{3D}} - \left(1 - f^e_{\mathbf{k}} - f^h_{\mathbf{k}-\mathbf{q}}\right)\sum_{\mathbf{l}} V_{\mathbf{k}-\mathbf{l}}\Pi_{\mathbf{l};\,\mathbf{q}_{3D}}$$
$$+ i\mathcal{F}_{\mathbf{q}_{3D}} S^{\mathrm{spnt}}_{\mathbf{k},\mathbf{q}} - i\left[1 - f^e_{\mathbf{k}} - f^h_{\mathbf{k}-\mathbf{q}}\right]\Delta\left\langle B^\dagger_{\mathbf{q}_{3D}} B_{\mathbf{q},\Sigma}\right\rangle$$
$$+ D^{\mathrm{coh}}_{\mathbf{k};\,\mathbf{q}_{3D}} + T^{\Pi}_{\mathbf{k};\,\mathbf{q}_{3D}}, \tag{28.24}$$

where we identified the effective photon operator,

$$B_{\mathbf{q}\Sigma} \equiv \sum_{q_z} \mathcal{F}_{\mathbf{q}_{3D}} B_{\mathbf{q},q_z}. \tag{28.25}$$

The term $D^{\mathrm{coh}}_{\mathbf{k};\,\mathbf{q}_{3D}}$ contains the coherent singlet–doublet contributions, which vanish in the incoherent regime. These are studied in Exercise 28.4. The remaining triplet contributions are given by

$$T^{\Pi}_{\mathbf{k};\,\mathbf{q}_{3D}} \equiv -i\sum_{\mathbf{l}}\left(\Delta\left\langle B^\dagger_{\mathbf{q}_{3D}} B_{\mathbf{l},\Sigma} a^\dagger_{v,\mathbf{k}-\mathbf{q}} a_{v,\mathbf{k}-\mathbf{l}}\right\rangle - \Delta\left\langle B^\dagger_{\mathbf{q}_{3D}} B_{\mathbf{q}-\mathbf{l},\Sigma} a^\dagger_{c,\mathbf{k}-\mathbf{l}} a_{c,\mathbf{k}}\right\rangle\right)$$
$$+\sum_{\mathbf{l}}\left(\Delta\left\langle B^\dagger_{\mathbf{q}_{3D}} Q^c_{\mathbf{l}} a^\dagger_{v,\mathbf{k}-\mathbf{q}} a_{c,\mathbf{k}-\mathbf{l}}\right\rangle - \Delta\left\langle B^\dagger_{\mathbf{q}_{3D}} Q^v_{\mathbf{q}-\mathbf{l}} a^\dagger_{v,\mathbf{k}-\mathbf{l}} a_{c,\mathbf{k}}\right\rangle\right)$$
$$+\sum_{\nu,\mathbf{k}',\mathbf{l}}\left(V_{\mathbf{l}}\Delta\left\langle B^\dagger_{\mathbf{q}_{3D}} a^\dagger_{v,\mathbf{k}-\mathbf{q}} a^\dagger_{\nu,\mathbf{k}'+\mathbf{q}} a_{\nu,\mathbf{k}'+\mathbf{l}} a_{c,\mathbf{k}-\mathbf{l}}\right\rangle\right.$$
$$\left.- V_{\mathbf{l}-\mathbf{q}}\Delta\left\langle B^\dagger_{\mathbf{q}_{3D}} a^\dagger_{v,\mathbf{k}-\mathbf{l}} a^\dagger_{\nu,\mathbf{k}'+\mathbf{l}} a_{\nu,\mathbf{k}'} a_{c,\mathbf{k}}\right\rangle\right). \tag{28.26}$$

Here, the first two sums contain higher-order correlations due to the coupling between photons and phonons while the last sum describes the influence of the Coulomb-induced scattering on the Π dynamics. The treatment of the Coulomb-induced triplets is explicitly discussed in (Kira and Koch, 2006) and (Hoyer *et al.*, 2007).

The second line of Eq. (28.24) contains photon-number-like correlations that are particularly large if the semiconductor material is either inside an optical cavity or optically pumped with incoherent light fields. This contribution provides either stimulated coupling or direct excitation effects due to external incoherent fields. To solve the corresponding dynamics, we apply Eq. (28.3) resulting in

$$i\hbar\frac{\partial}{\partial t}\Delta\left\langle B^\dagger_{\mathbf{q},q_z} B_{\mathbf{q},q'_z}\right\rangle = \hbar\left(\omega_{q'_{3D}} - \omega_{\mathbf{q}_{3D}}\right)\Delta\left\langle B^\dagger_{\mathbf{q},q_z} B_{\mathbf{q},q'_z}\right\rangle$$
$$+ i\hbar\sum_{\mathbf{k}}\left[\mathcal{F}_{\mathbf{q}_{3D}}\Pi^\star_{\mathbf{k};\,\mathbf{q},q'_z} + \mathcal{F}^\star_{q'_{3D}}\Pi_{\mathbf{k};\,\mathbf{q},q_z}\right], \tag{28.27}$$

that contains again only the incoherent correlations.

In the incoherent regime, the singlet–doublet equations additionally also include the carrier dynamics (28.7)–(28.9), the exciton-correlation dynamics (27.35) and (28.12), as well as the dynamics of the pure electron–electron and hole–hole correlations, i.e., $c^{\mathbf{k},\mathbf{k}',\mathbf{k}}_{\lambda,\lambda,\lambda,\lambda}$, which are, however, not directly coupled with the quantum-optical Π correlations, see Exercise 28.5. Together with Eqs. (28.24) and (28.27), we have thus identified the complete set of incoherent singlet–doublet equations that constitute the *semiconductor luminescence equations* (SLE). The SLE are formally exact and the accuracy of their solution depends only on how precisely the triplet contributions can be treated.

If the carrier system is close to a quasiequilibrium situation, the pure carrier quantities are often nearly constant. Especially, the contributions described by $c^{\mathbf{k},\mathbf{k}',\mathbf{k}}_{\lambda,\lambda,\lambda,\lambda}$ are not directly influenced, e.g., by the spontaneous recombination within the system. If $f^e_{\mathbf{k}}$, $f^h_{\mathbf{k}}$, and $c^{\mathbf{k},\mathbf{k}',\mathbf{k}}_{\lambda,\nu,\lambda,\nu}$ are quasistationary, we can concentrate on the principal structure of the SLE given by Eqs. (28.24) and (28.27). In this context, the semiconductor emits a constant flux of photons that produces the steady-state luminescence spectrum according to

$$I_{\rm PL}(\omega_{\mathbf{q}_{3D}}) = \frac{\partial}{\partial t}\Delta\left\langle B^\dagger_{\mathbf{q},q_z} B_{\mathbf{q},q_z}\right\rangle = 2{\rm Re}\Big[\sum\nolimits_{\mathbf{k}} \mathcal{F}^\star_{\mathbf{q},q_z}\Pi_{\mathbf{k};\,\mathbf{q},q_z}\Big]. \tag{28.28}$$

One can additionally introduce a dynamical detection model to compute the time-dependent luminescence spectra resulting from a dynamically changing many-body state. References to this extension are given at the end of this chapter.

28.3 Analytic investigation of the semiconductor luminescence

We next develop an analytic model that captures the essential features of the spontaneous-emission processes initiated by the photon-assisted polarization. For this purpose, we investigate a situation where the pure carrier many-body state changes sufficiently slowly to model it as quasistationary. Furthermore, we assume that the semiconductor system is not within a cavity such that each emitted photon escapes fast enough and the stimulated contributions can be omitted. In this situation, Eq. (28.24) reduces to

$$\begin{aligned} i\hbar\frac{\partial}{\partial t}\Pi_{\mathbf{k},\mathbf{q},q_z} &= \left(\tilde\epsilon^c_{\mathbf{k}} - \tilde\epsilon^v_{\mathbf{k}-\mathbf{q}} - \hbar\omega_{\mathbf{q},q_z} - i\gamma\right)\Pi_{\mathbf{k},\mathbf{q},q_z} \\ &\quad - \left(1 - f^e_{\mathbf{k}} - f^h_{\mathbf{k}-\mathbf{q}}\right)\sum\nolimits_{\mathbf{l}} V_{\mathbf{k}-\mathbf{l}}\,\Pi_{\mathbf{l};\,\mathbf{q},q_z} + i\mathcal{F}_{\mathbf{q},q_z} S^{\rm spnt}_{\mathbf{k},\mathbf{q}}. \end{aligned} \tag{28.29}$$

Since the triplet contributions mainly describe scattering events in the many-body system, we approximate them through a phenomenological dephasing constant γ, compare the discussion in Section 26.4.1. Under these conditions, the Π dynamics is closed and we can look for analytic solutions similar to those of the semiconductor Bloch equations (SBE). For this purpose, we first determine the analytic steady-state solution of Π. We then use these results to evaluate the strength of the photon-assisted recombination that initiates a constant emission flux of photons, according to Eq. (28.28), as well as the spontaneous recombination of electron–hole pairs and excitons, based on Eqs. (28.8)–(28.9) and (28.15).

The photon-assisted polarization is driven by a constant spontaneous emission source $S^{\rm spnt}_{\mathbf{k},\mathbf{q}}$. The contributing range of momentum states can be rather broad because the plasma

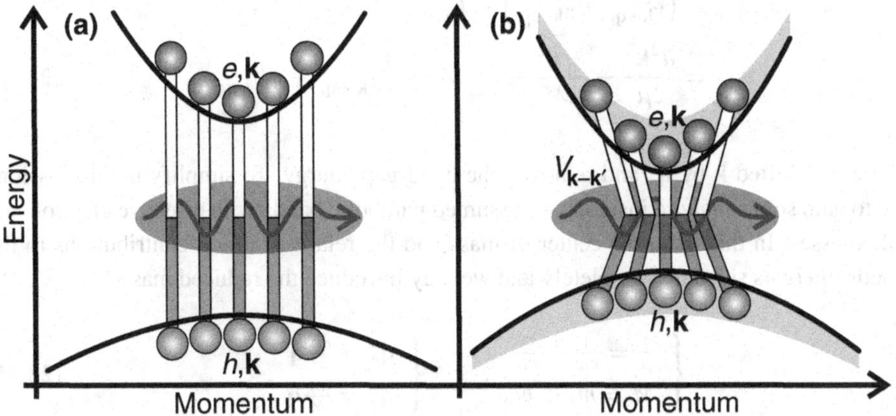

Figure 28.3 Schematic representation of the build-up and the Coulomb coupling of the photon-assisted polarization. **(a)** The spontaneous-emission source (28.21) initiates Π (bars and wave arrow) independently in all momentum states. **(b)** The Coulomb interaction in Eq. (28.29) bundles all Π together yielding a collective photon-assisted polarization.

source, i.e., $f^e_{\mathbf{k}} f^h_{\mathbf{k}-\mathbf{q}}$, may cover a wide $\mathbf{k}$ range. Figure 28.3(a) illustrates how $\Pi_{\mathbf{k},\mathbf{q},q_z}$ is generated in all those momentum states that simultaneously contain an electron and a hole (spheres in the parabolic bands). Without the Coulomb coupling, the induced $\Pi_{\mathbf{k};\,\mathbf{q},q_z}$ are mutually uncoupled as if each $\mathbf{k}$ state formed an individual two-level system (bar chart). However, in the Coulomb interacting semiconductor system, the different $\Pi_{\mathbf{k},\mathbf{q},q_z}$ are coupled such that one must solve $\Pi_{\mathbf{k},\mathbf{q},q_z}$ collectively rather than analyze a set of individual $\mathbf{k}$-parametrized two-level systems. Figure 28.3(b) schematically depicts how the Coulomb interaction bundles, i.e., couples the different $\Pi_{\mathbf{k},\mathbf{q},q_z}$ together to act as a collective and Coulomb-correlated photon-assisted process. In fact, the Coulomb-coupling structure is identical to that of the SBE, see Fig. 26.1. Thus, we may anticipate that the interactions among the different $\Pi_{\mathbf{k},\mathbf{q},q_z}$ can yield excitonic features in the collective photon-assisted polarizations. As for the coherent polarization, these excitonic features do not necessarily imply the presence of true excitons. This conclusion is supported by the fact that the uncorrelated and the correlated electron–hole plasma, as well as incoherent excitons appear additively in the source of $\Pi_{\mathbf{k},\mathbf{q},q_z}$. This aspect is studied further in Section 28.4.

28.3.1 Wannier excitons with finite center-of-mass momentum

In our analysis of the SBE, we used the expansion into the complete set of solutions of the homogeneous Wannier equation (26.19). Due to the obvious structural analogy between the SBE and the SLE, we therefore need a suitable exciton basis for the SLE. For this purpose, we first analyze the single-particle carrier energies appearing in Eq. (28.29),

$$E_{\mathbf{k},\mathbf{q}} \equiv \left(\tilde{\epsilon}^{c}_{\mathbf{k}+\mathbf{q}_e} - \tilde{\epsilon}^{v}_{\mathbf{k}-\mathbf{q}_h}\right) - E_{g,0}$$
$$= \frac{\hbar^2\mathbf{k}^2}{2\mu} + \frac{\hbar^2\mathbf{q}^2}{2M} - \sum_{\mathbf{k}'} V_{\mathbf{k}'-\mathbf{k}}\left(f^e_{\mathbf{k}'+\mathbf{q}_e} + f^h_{\mathbf{k}'-\mathbf{q}_h}\right), \tag{28.30}$$

where we shifted $\mathbf{k}$ by $\mathbf{q}_e$ and removed the band-gap energy. To simplify the discussions and to gain some analytic insights, we assumed parabolic bands with effective electron and hole masses. In this case, the center-of-mass and the relative-motion contributions to the kinetic energies separate completely and we may introduce the reduced mass

$$\begin{cases} \frac{1}{\mu} = \frac{1}{m_e} + \frac{1}{m_h} \\ M = m_e + m_h \end{cases}, \qquad \begin{cases} \mathbf{q}_e = \frac{m_e}{M}\mathbf{q} \\ \mathbf{q}_h = \frac{m_h}{M}\mathbf{q} \end{cases}, \tag{28.31}$$

where $\mathbf{q}_e$ and $\mathbf{q}_h$ denote the fractions of the center-of-mass momentum $\mathbf{q} = \mathbf{q}_e + \mathbf{q}_h$ assigned to the electron and the hole, respectively.

Substituting $\mathbf{k} \to \mathbf{k} + \mathbf{q}_e$ in Eq. (28.29) allows us to diagonalize its homogeneous part via the eigenvalue problem,

$$E_{\mathbf{k},\mathbf{q}}\,\phi^R_{\nu,\mathbf{q}}(\mathbf{k}) - \left(1 - f^e_{\mathbf{k}+\mathbf{q}_e} - f^h_{\mathbf{k}-\mathbf{q}_h}\right)\sum_{\mathbf{l}} V_{\mathbf{k}-\mathbf{l}}\,\phi^R_{\nu,\mathbf{q}}(\mathbf{k}') = E_{\nu,\mathbf{q}}\,\phi^R_{\nu,\mathbf{q}}(\mathbf{k}). \tag{28.32}$$

For vanishing $\mathbf{q}$, this is identical to the Wannier equation (26.19). Consequently, the SLE diagonalization can also be performed using excitonic states. For a nonzero $\mathbf{q}$, the exciton wave functions $\phi_{\nu,\mathbf{q}}(\mathbf{k}')$ depend on the center-of-mass momentum. This dependence is parametric, i.e., each $\mathbf{q}$ produces a $\mathbf{q}$-dependent Wannier equation for $\phi_{\nu,\mathbf{q}}(\mathbf{k})$ without coupling between the different $\mathbf{q}$ states. For sufficiently low densities, the $f^e_{\mathbf{k}+\mathbf{q}_e}$ and $f^h_{\mathbf{k}-\mathbf{q}_h}$ dependencies are ineffective such that $\phi_{\nu,\mathbf{q}}(\mathbf{k})$ varies only weakly with $\mathbf{q}$. In this situation, the $\mathbf{q}$ dependence of $E_{\nu,\mathbf{q}}$ is essentially given by the center-of-mass energy $\hbar^2\mathbf{q}^2/2M$, see Exercise 27.11.

As in the case of the usual Wannier equation, the left- and right-handed solutions form a complete set of orthogonal states and can be chosen as real valued. In the following derivations, we use the properties

$$\begin{cases} \sum_{\mathbf{k}} \phi^L_{\lambda,\mathbf{q}}(\mathbf{k})\phi^R_{\nu,\mathbf{q}}(\mathbf{k}) = \delta_{\lambda,\nu} \\ \sum_{\lambda} \phi^L_{\lambda,\mathbf{q}}(\mathbf{k})\phi^R_{\lambda,\mathbf{q}}(\mathbf{k}') = \delta_{\mathbf{k},\mathbf{k}'} \end{cases}, \qquad \phi^L_{\lambda,\mathbf{q}}(\mathbf{k}) = \frac{\phi^R_{\lambda,\mathbf{q}}(\mathbf{k})}{1 - f^e_{\mathbf{k}+\mathbf{q}_e} - f^h_{\mathbf{k}-\mathbf{q}_h}}, \tag{28.33}$$

which follow as a straightforward generalization of the analysis in Section 26.2.1. Since the $\mathbf{q}$ dependence of $\phi^R_{\lambda,\mathbf{q}}(\mathbf{k})$ is parametric, it is often convenient to drop the $\mathbf{q}$ index when the center of mass of the exciton wave function is obvious from the context. This notational simplification has already been implemented in Eq. (27.48) where we represent the electron–hole correlation c_X in the exciton basis.

28.3.2 Elliott formula for luminescence

To solve the SLE analytically, we transform the photon-assisted polarization into the exciton basis using Eq. (28.13). We also need the transformation

$$\begin{cases} \Delta N_{\lambda,\nu}(\mathbf{q}) = \sum_{\mathbf{k},\mathbf{k}'} \phi^{L}_{\lambda}(\mathbf{k})\phi^{L}_{\nu}(\mathbf{k}') c_{\mathrm{X}}^{\mathbf{q},\mathbf{k}'-\mathbf{q}_h,\mathbf{k}+\mathbf{q}_e} \\ c_{\mathrm{X}}^{\mathbf{q},\mathbf{k}'-\mathbf{q}_h,\mathbf{k}+\mathbf{q}_e} = \sum_{\lambda,\nu} \phi^{R}_{\lambda}(\mathbf{k})\phi^{R}_{\nu}(\mathbf{k}')\,\Delta N_{\lambda,\nu}(\mathbf{q}) \end{cases}, \tag{28.34}$$

provided by Eq. (27.48). In both of these transformations, we suppress the $\mathbf{q}$ index of the exciton wave function because it is clear from the context. As before, the diagonal part $\Delta N_{\lambda,\lambda}(\mathbf{q})$ defines the momentum distribution of the incoherent excitons while the nondiagonal part identifies the correlated electron–hole plasma, as discussed in Section 27.3.

We now apply the transformation (28.13) to extract the λ components of Eq. (28.29). Using (28.32) and (28.33), we eventually find

$$\begin{aligned} i\hbar\frac{\partial}{\partial t}\Pi_{\lambda,\mathbf{q}_{3\mathrm{D}}} &= \left(E_{\lambda,\mathbf{q}} + E_{g,0} - \hbar\omega_{\mathbf{q}_{3\mathrm{D}}} - i\gamma_{\lambda}\right)\Pi_{\lambda,\mathbf{q}_{3\mathrm{D}}} \\ &\quad + i\mathcal{F}_{\mathbf{q}_{3\mathrm{D}}}\frac{1}{\sqrt{\mathcal{S}}}\sum_{\mathbf{k}}\phi^{L}_{\lambda}(\mathbf{k})\, S^{\mathrm{spnt}}_{\mathbf{k}+\mathbf{q}_e,\mathbf{q}}, \end{aligned} \tag{28.35}$$

which is now in a diagonalized form. In order to generalize the model, we introduced an exciton-state-dependent γ_{λ}, which allows for the inclusion of the diffusive character of scattering, in analogy to our investigations in Section 26.5.2 (see also Exercise 28.6). Furthermore, we converted the spontaneous-emission source into the exciton basis,

$$\begin{aligned} S^{\mathrm{spnt}}_{\lambda,\mathbf{q}} &\equiv \frac{1}{\sqrt{\mathcal{S}}}\sum_{\mathbf{k}}\phi^{L}_{\lambda}(\mathbf{k})S^{\mathrm{spnt}}_{\mathbf{k}+\mathbf{q}_e,\mathbf{q}} \\ &= \frac{1}{\sqrt{\mathcal{S}}}\sum_{\mathbf{k}}\phi^{L}_{\lambda}(\mathbf{k})\left[f^{e}_{\mathbf{k}-\mathbf{q}_e} f^{h}_{\mathbf{k}+\mathbf{q}_e} + \sum_{\mathbf{k}'} c_{X}^{\mathbf{q},\mathbf{k}-\mathbf{q}_h,\mathbf{k}'+\mathbf{q}_e}\right], \end{aligned} \tag{28.36}$$

which follows directly from Eq. (28.21).

When $S^{\mathrm{spnt}}_{\lambda,\mathbf{q}}$ is a quasistationary source, Eq. (28.35) has the steady-state solution

$$\Pi^{\mathrm{std}}_{\lambda,\mathbf{q}_{3\mathrm{D}}} = \frac{i\mathcal{F}_{\mathbf{q}_{3\mathrm{D}}}S^{\mathrm{spnt}}_{\lambda,\mathbf{q}}}{\hbar\omega_{\mathbf{q}_{3\mathrm{D}}} - E_{g,0} - E_{\lambda,\mathbf{q}} + i\gamma_{\lambda}}. \tag{28.37}$$

A direct substitution of $\Pi^{\mathrm{std}}_{\lambda,\mathbf{q}_{3\mathrm{D}}}$ in Eq. (28.28) produces the *Elliott formula for luminescence*,

$$\begin{aligned} I_{\mathrm{PL}}(\omega_{\mathbf{q}}) &= 2\mathrm{Re}\left[\sum_{\lambda}\mathcal{F}^{\star}_{\mathbf{q}_{3\mathrm{D}}}\mathcal{S}\,\phi^{R}_{\lambda}(\mathbf{r}=0)\,\Pi^{\mathrm{std}}_{\lambda;\,\mathbf{q}_{3\mathrm{D}}}\right] \\ &= 2\mathrm{Re}\left[\sum_{\lambda}\frac{i\mathcal{S}\,|\mathcal{F}_{\mathbf{q}_{3\mathrm{D}}}|^{2}\phi^{R}_{\lambda}(\mathbf{r}=0)\,S^{\mathrm{spnt}}_{\lambda,\mathbf{q}}}{\hbar\omega_{\mathbf{q}_{3\mathrm{D}}} - E_{g,0} - E_{\lambda,\mathbf{q}} + i\gamma_{\lambda}}\right]. \end{aligned} \tag{28.38}$$

This result strongly resembles the Elliott formula (26.38) for the semiconductor absorption. As we can see from its resonance denominator, Eq. (28.38) predicts emission peaks at the photon energies $\hbar\omega = E_{g,0} + E_{\lambda,\mathbf{q}}$ if the corresponding $S^{\mathrm{spnt}}_{\lambda,\mathbf{q}}$ is nonvanishing. Hence, one can detect photoluminescence below the band gap at the excitonic resonance energies. These signatures clearly result from the Coulomb-coupling structure among the photon-assisted polarizations. The source term $S^{\mathrm{spnt}}_{1s,\mathbf{q}}$ contains contributions from both the electron–hole plasma and from true excitons but it has no frequency selectivity. Thus, excitonic signatures in the semiconductor luminescence do not necessarily imply that exciton populations are present in the system.

28.3.3 Plasma vs. population source in photoluminescence

To explain how the plasma and the excitons contribute to the excitonic luminescence, i.e., the spontaneous emission at the exciton resonance energy, we need to analyze the explicit form of $S^{\mathrm{spnt}}_{1s,\mathbf{q}}$ in more detail. For this purpose, we express the correlated part of the spontaneous-emission source (28.36) in the exciton basis,

$$\begin{aligned}S^{\mathrm{corr}}_{\lambda,\mathbf{q}} &\equiv \sum_{\mathbf{k},\mathbf{k}'} \frac{\phi^L_\lambda(\mathbf{k})}{\sqrt{S}} c_X^{\mathbf{q},\mathbf{k}-\mathbf{q}_h,\mathbf{k}'+\mathbf{q}_e} = \sum_{\mathbf{k}',\alpha,\beta} \phi^R_\beta(\mathbf{k}') \frac{\sum_{\mathbf{k}} \phi^L_\lambda(\mathbf{k})\,\phi^R_\alpha(\mathbf{k})}{\sqrt{S}} \Delta N_{\beta,\alpha}(\mathbf{q}) \\ &= \sum_{\mathbf{k}',\beta} \phi^R_\beta(\mathbf{k}') \frac{\delta_{\alpha,\lambda}}{\sqrt{S}} \Delta N_{\beta,\alpha}(\mathbf{q}) = \sum_{\mathbf{k}',\beta} \frac{\phi^R_\beta(\mathbf{k}')}{\sqrt{S}} \Delta N_{\beta,\lambda}(\mathbf{q}).\end{aligned} \tag{28.39}$$

We know from the analysis in Section 27.3 that only the diagonal part $\Delta N_{\lambda,\lambda}(\mathbf{q})$ identifies incoherent excitons while the nondiagonal contributions $\Delta N_{\beta\neq\lambda,\lambda}(\mathbf{q}) \equiv \Delta N^{\mathrm{eh}}_{\beta\neq\lambda,\lambda}(\mathbf{q})$ describe the correlated electron–hole plasma. Using Eqs. (27.53)–(27.54), we can bring the correlated plasma part into the form

$$\begin{aligned}\Delta N^{\mathrm{eh}}_{\lambda,\nu}(\mathbf{q}) &= \left(\delta_{\lambda,\nu} - 1\right) \left\langle X^\dagger_{\lambda,\mathbf{q}} X_{\nu,\mathbf{q}} \right\rangle_{\mathrm{S}} \\ &= \left(\delta_{\lambda,\nu} - 1\right) \sum_{\mathbf{k}} \phi^L_\lambda(\mathbf{k})\, f^e_{\mathbf{k}+\mathbf{q}_e} f^h_{\mathbf{k}-\mathbf{q}_h} \phi^L_\nu(\mathbf{k})\end{aligned} \tag{28.40}$$

if γ in Eq. (27.54) is sufficiently small. This approximation is also valid when scattering is included microscopically, compare (Hoyer *et al.*, 2003).

Together, the exciton and correlated plasma contributions are

$$\begin{aligned}\Delta N_{\lambda,\nu}(\mathbf{q}) &= \delta_{\nu,\lambda} \Delta N_{\lambda,\lambda}(\mathbf{q}) + \Delta N^{\mathrm{eh}}_{\lambda,\nu}(\mathbf{q}) \\ &= \delta_{\nu,\lambda} \Delta\left\langle X^\dagger_{\lambda,\mathbf{q}} X_{\lambda,\mathbf{q}} \right\rangle + \Delta N^{\mathrm{eh}}_{\lambda,\nu}(\mathbf{q}),\end{aligned} \tag{28.41}$$

where the exciton distributions are defined by the correlated part of the exciton operators, as shown in Section 27.3. Similarly, the correlated plasma is defined by $N^{\mathrm{S}}_{\lambda,\nu}(\mathbf{q}) = \langle X^\dagger_{\lambda,\mathbf{q}} X_{\nu,\mathbf{q}} \rangle_{\mathrm{S}}$, which contains the singlet part of the exciton operators (27.16). It stems from the Fermionic substructure of the excitons.

Equations (28.39)–(28.40) allow us to analytically evaluate the contribution of the correlated electron–hole plasma to the spontaneous emission source

$$S^{\rm eh}_{\lambda,\mathbf{q}} \equiv \sum_{\mathbf{k}',\beta} \frac{\phi^R_\beta(\mathbf{k}')}{\sqrt{S}} \Delta N^{\rm eh}_{\beta,\lambda}(\mathbf{q}) = \sum_{\mathbf{k}',\beta} \frac{\phi^R_\beta(\mathbf{k}')}{\sqrt{S}} \left(\delta_{\beta,\lambda} - 1\right) N^{\rm S}_{\beta,\lambda}(\mathbf{q})$$
$$= \phi^R_\lambda(\mathbf{r}=0) N^{\rm S}_{\lambda,\lambda}(\mathbf{q}) - \sum\nolimits_{\mathbf{k}',\beta} \frac{\phi^R_\beta(\mathbf{k}')}{\sqrt{S}} N^{\rm S}_{\beta,\lambda}(\mathbf{q}) \tag{28.42}$$

after we convert $\sum_{\mathbf{k}',\beta} \frac{\phi^R_\beta(\mathbf{k}')}{\sqrt{S}}$ into $\phi^R_\beta(\mathbf{r}=0)$ using the transformation (26.32) into real space. The remaining last contribution can also be simplified,

$$S^{\rm last}_{\lambda,\mathbf{q}} \equiv -\sum_{\mathbf{k}',\beta} \frac{\phi^R_\beta(\mathbf{k}')}{\sqrt{S}} N^{\rm S}_{\beta,\lambda}(\mathbf{q}) = -\sum_{\mathbf{k}',\beta} \frac{\phi^R_\beta(\mathbf{k}')}{\sqrt{S}} \sum_{\mathbf{k}} \phi^L_\beta(\mathbf{k}) f^e_{\mathbf{k}+\mathbf{q}_e} f^h_{\mathbf{k}-\mathbf{q}_h} \phi^L_\lambda(\mathbf{k})$$
$$= -\sum_{\mathbf{k}',\mathbf{k}} \left[\sum\nolimits_\beta \phi^R_\beta(\mathbf{k}')\phi^L_\beta(\mathbf{k})\right] f^e_{\mathbf{k}+\mathbf{q}_e} f^h_{\mathbf{k}-\mathbf{q}_h} \frac{\phi^L_\lambda(\mathbf{k})}{\sqrt{S}}$$
$$= -\sum_{\mathbf{k}',\mathbf{k}} \delta_{\mathbf{k},\mathbf{k}'} f^e_{\mathbf{k}+\mathbf{q}_e} f^h_{\mathbf{k}-\mathbf{q}_h} \frac{\phi^L_\lambda(\mathbf{k})}{\sqrt{S}} = -\sum_{\mathbf{k}} \frac{\phi^L_\lambda(\mathbf{k})}{\sqrt{S}} f^e_{\mathbf{k}+\mathbf{q}_e} f^h_{\mathbf{k}-\mathbf{q}_h}, \tag{28.43}$$

after we implement the definition (28.40) and the completeness relation (28.33). Inserting (28.43) into Eq. (28.42) shows that the correlated electron–hole plasma contributes to the spontaneous-emission source through

$$S^{\rm eh}_{\lambda,\mathbf{q}} = \phi^R_\lambda(\mathbf{r}=0)\, N^{\rm S}_{\lambda,\lambda}(\mathbf{q}) - \sum\nolimits_{\mathbf{k}} \frac{\phi^L_\lambda(\mathbf{k})}{\sqrt{S}} f^e_{\mathbf{k}+\mathbf{q}_e} f^h_{\mathbf{k}-\mathbf{q}_h}. \tag{28.44}$$

In total, the correlated spontaneous-emission source (28.39) is

$$S^{\rm corr}_{\lambda,\mathbf{q}} \equiv \phi^R_\lambda(\mathbf{r}=0) \left[\Delta N_{\lambda,\lambda}(\mathbf{q}) + N^{\rm S}_{\lambda,\lambda}(\mathbf{q})\right] - \sum_{\mathbf{k}} \frac{\phi^L_\lambda(\mathbf{k})}{\sqrt{S}} f^e_{\mathbf{k}+\mathbf{q}_e} f^h_{\mathbf{k}-\mathbf{q}_h}. \tag{28.45}$$

Here, the last term of $S^{\rm corr}_{\lambda,\mathbf{q}}$ cancels exactly with the singlet part of the spontaneous-emission source (28.36), which leads to

$$S^{\rm spnt}_{\lambda,\mathbf{q}} = \phi^R_\lambda(\mathbf{r}=0) \left[\Delta N_{\lambda,\lambda}(\mathbf{q}) + N^{\rm S}_{\lambda,\lambda}(\mathbf{q})\right]. \tag{28.46}$$

Consequently, the spontaneous-emission source contains exciton populations as well as a diagonal projection of the plasma contribution $f^e f^h$. Using these results, we can simplify the Elliott luminescence formula (28.38) into the form

$$
\begin{aligned}
I_{\mathrm{PL}}(\omega_{\mathbf{q}_{3\mathrm{D}}}) &= \sum_{\lambda} \frac{2\gamma_\lambda \left|\mathcal{F}^{\lambda}_{\mathbf{q}_{3\mathrm{D}}}\right|^2 \left[\Delta N_\lambda(\mathbf{q}) + N^{\mathrm{S}}_\lambda(\mathbf{q})\right]}{(\hbar\omega_{\mathbf{q}_{3\mathrm{D}}} - E_{g,0} - E_{\lambda,\mathbf{q}})^2 + \gamma_\lambda^2}, \\
\mathcal{F}^{\lambda}_{\mathbf{q}_{3\mathrm{D}}} &\equiv \sqrt{\mathcal{S}}\,\phi^R_\lambda(\mathbf{r}=0)\mathcal{F}_{\mathbf{q}_{3\mathrm{D}}}, \\
\Delta N_\lambda(\mathbf{q}) &\equiv \Delta N_{\lambda,\lambda}(\mathbf{q}) = \Delta\left\langle X^\dagger_{\lambda,\mathbf{q}} X_{\lambda,\mathbf{q}}\right\rangle, \\
N^{\mathrm{S}}_\lambda(\mathbf{q}) &\equiv \sum_{\mathbf{k}} \left|\phi^L_\lambda(\mathbf{k})\right|^2 f^e_{\mathbf{k}+\mathbf{q}_e} f^h_{\mathbf{k}-\mathbf{q}_h},
\end{aligned}
\tag{28.47}
$$

where we identified an effective oscillator strength $\mathcal{F}^{\lambda}_{\mathbf{q}_{3\mathrm{D}}}$ and shortened the definitions for the exciton distribution $\Delta N_\lambda(\mathbf{q})$ and the plasma part of $N^{\mathrm{S}}_\lambda(\mathbf{q})$.

28.4 Excitonic signatures in the semiconductor luminescence

As long as we have carriers in the semiconductor, the corresponding distributions $f^e_{\mathbf{k}}$ and $f^h_{\mathbf{k}}$ produce a nonvanishing source $N^{\mathrm{S}}_\lambda(\mathbf{q})$ for the luminescence. This excitation configuration does not necessarily contain incoherent excitons because exciton formation requires a very special many-body arrangement that includes the formation of the doublets $\Delta N_\lambda(\mathbf{q})$. Thus, it is clear that $N^{\mathrm{S}}_\lambda(\mathbf{q})$ exists in a much broader phase-space of excited states than exciton distributions do.

In order to determine the explicit role of plasma vs. excitons in the QW luminescence, we compare the spectra $I_{\mathrm{PL}}(\omega_{\mathbf{q}})$ resulting from a quasistationary plasma with those of a system that additionally contains a distribution of $1s$ excitons. More specifically, we assume here that 10 percent of electrons and holes form bound excitons. In both cases, we analyze the photoluminescence emitted into the normal direction $\mathbf{q} = 0$ by evaluating the Elliott formula (28.47). For these results, the scattering γ_λ is computed fully microscopically. Figure 28.4 presents the luminescence spectrum for the carrier densities $n_{\mathrm{eh}} = 2 \times 10^9\,\mathrm{cm}^{-2}$ (solid line) and $n_{\mathrm{eh}} = 10^{10}\,\mathrm{cm}^{-2}$ (shaded area). The upper line presents the spectra normalized with respect to their peak maximum whereas the lower frames display the actual spectra on a logarithmic scale.

In the calculations, we assumed a temperature of 20 K for the quasistationary many-body system, i.e., we take for $f^e_{\mathbf{k}}$ and $f^e_{\mathbf{k}}$ the corresponding Fermi–Dirac distributions and for $\Delta N_\lambda(\mathbf{q}) = \delta_{1s,\lambda}\Delta N_{1s}(\mathbf{q})$ a Bose–Einstein distribution at this temperature. We use the parameters of a typical GaAs system, yielding the $1s$-binding energy of $E_{1s} = -10\,\mathrm{meV}$. The scaled densities correspond to $na_0^2 = 1.56 \times 10^{-2}$ for $n_{\mathrm{eh}} = 10^{10}\,\mathrm{cm}^{-2}$. This implies that we are in the relatively low-density regime where the charge carriers are separated by at least an average distance of $r_{\mathrm{eh}} = a_0/\sqrt{n_{\mathrm{eh}}a_0^2} = 8.0\,a_0$ that is far greater than the exciton Bohr radius. Under these conditions, we may expect to observe clear excitonic resonances in the luminescence spectrum.

The left column of Fig. 28.4 presents the computed photoluminescence spectra resulting from the plasma contribution alone while the right column shows results for a calculation

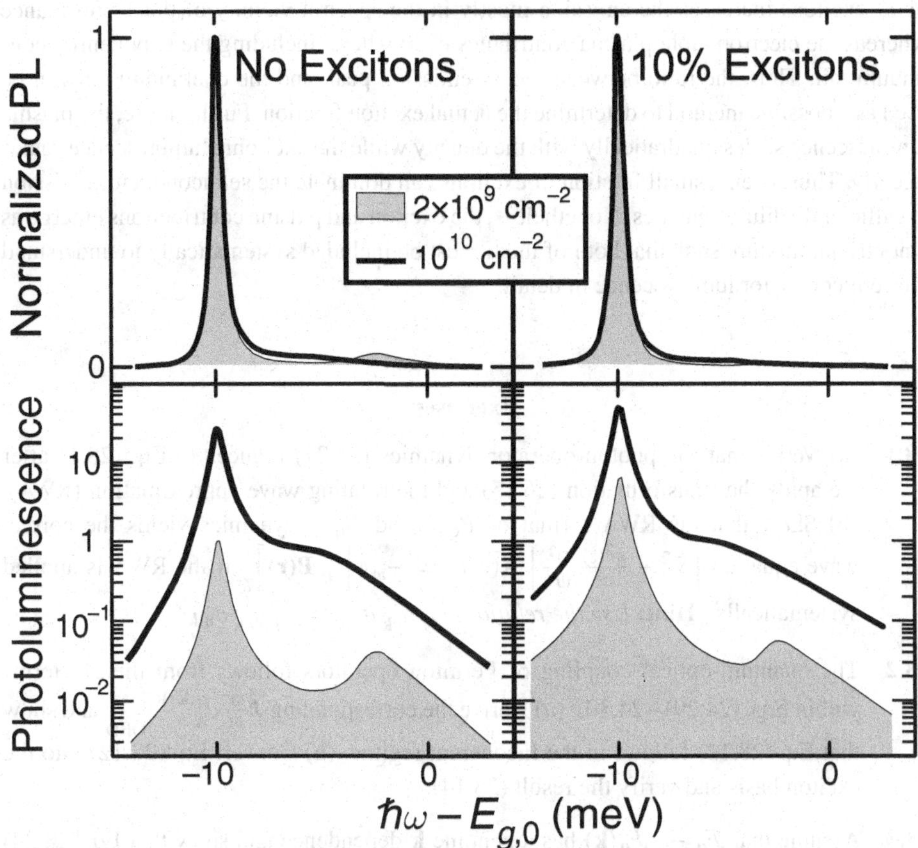

Figure 28.4 The influence of exciton populations on the excitonic photoluminescence spectra of semiconductor quantum wells. The left column shows the computed luminescence spectra for an 8 nm InGaAs/GaAs structure on a linear (top) and a logarithmic (bottom) scale assuming a quasiequilibrium electron–hole population at a temperature of 20 K without any excitonic population. The corresponding figures in the right column present the results for the same situation where 10 percent excitons in a thermal distribution have been added. The shaded (black) curves have been computed for carrier densities of $2 \times 10^9\ \mathrm{cm}^{-2}$ ($10^{10}\ \mathrm{cm}^{-2}$). The linear spectra are normalized, whereas the spectra on the logarithmic scale are in absolute units relative to the peak value of the $1s$ peak at $2 \times 10^9\ \mathrm{cm}^{-2}$ without excitons (black curve lower left frame) whose value was set to unity. From (Koch *et al.*, 2006).

where we additionally include 10 percent of incoherent excitons. We see that both cases produce distinct emission resonance -10 meV below the band gap $E_{g,0}$ and an exponentially decaying emission tail at higher energies. In particular, there are no pronounced *qualitative* differences between the excitonic plasma luminescence and the incoherent exciton emission.

However, a *quantitative* analysis reveals subtle differences in the influence of the finite carrier density and the excitons on the luminescence spectra. In particular, the presence of $1s$ excitons increases the emission mostly in the spectral vicinity of the $1s$ resonance whereas the electron–hole plasma contributes everywhere, including the exponential continuum tail. Thus, the ratio between the $1s$ emission peak and the continuum tail can be used as a possible method to determine the actual exciton fraction. Furthermore, the plasma luminescence scales quadratically with the density while the excitonic luminescence scales linearly. Thus, even a small fraction of excitons can dominate the semiconductor emission at sufficiently dilute densities. Nonetheless, the exciton and plasma contributions emerge as a nontrivial mixture such that both of them must be included systematically to understand the semiconductor luminescence in detail.

Exercises

28.1 **(a)** Verify that the photon-operator dynamics (24.27) reduces to Eq. (28.1) after we apply the transformation (26.15) and the rotating wave approximation (RWA). **(b)** Show that the RWA format of $B_{\mathbf{q},q_z}$ and $B^{\dagger}_{\mathbf{q},q_z}$ dynamics yields the correct wave equation $\left[\nabla^2 - \frac{n^2(z)}{c^2}\frac{\partial^2}{\partial t^2}\right]\hat{\mathbf{A}}(\mathbf{r},t) = -\mu_0\left[\frac{\partial}{\partial t}\hat{\mathbf{P}}(\mathbf{r})\right]_T$ if the RWA is applied systematically. **Hint:** *Use the relation* $i\omega_q\, a^{\dagger}_{c\mathbf{k}}\, a_{v,\mathbf{k}'} = \frac{\partial}{\partial t} a^{\dagger}_{c\mathbf{k}}\, a_{v,\mathbf{k}'}$.

28.2 The quantum-optical coupling of Fermion operators follows from the $\hat{A}^{\lambda}$ terms within Eqs. (24.29)–(24.30). **(a)** Derive the corresponding $\hbar\frac{\partial}{\partial t} c_X^{\mathbf{q},\mathbf{k}',\mathbf{k}}\Big|_{\mathrm{QED}}$ and show that Eq. (28.12) follows in the incoherent regime. **(b)** Convert Eq. (28.12) into the exciton basis and verify the result (28.14).

28.3 Assume that $\mathcal{F}_{\mathbf{q}} \to \mathcal{F}_{\mathbf{q}}(\mathbf{k})$ has a genuine $\mathbf{k}$-dependence and show that Eq. (28.22) becomes

$$i\hbar\frac{\partial}{\partial t}\Pi^{\mathrm{bos}}_{\mathbf{k};\mathbf{q}_{3\mathrm{D}}} = -\hbar\omega_q\,\Pi_{\mathbf{k};\mathbf{q}_{3\mathrm{D}}} + i\mathcal{F}_{\mathbf{q}_{3\mathrm{D}}}(\mathbf{q})\, f^e_{\mathbf{k}}\, f^h_{\mathbf{k}-\mathbf{q}} + i\sum_{\mathbf{k}'}\mathcal{F}_{\mathbf{q}_{3\mathrm{D}}}(\mathbf{k}')\, c_X^{\mathbf{q},\mathbf{k}-\mathbf{q},\mathbf{k}'}.$$

28.4 **(a)** Derive Eq. (28.23) by starting from Eqs. (24.29)–(24.30). Implement the transformation (25.55) to develop the expression further. **(b)** Apply this result to derive Eq. (28.24) via the identification of singlet, doublet, and triplet contributions. Separate the coherent from the incoherent contributions. What is the explicit form of $D^{\mathrm{coh}}_{\mathbf{k};\mathbf{q}_{3\mathrm{D}}}$?

28.5 Start from the quantum-optical A_q contributions in Eq. (24.29) and show that the dynamics of $i\hbar\frac{\partial}{\partial t} c^{\mathbf{q},\mathbf{k}',\mathbf{k}}_{\lambda,\lambda,\lambda,\lambda}\Big|_{\mathrm{QED}}$ has no coupling to quantum-optical correlations in the incoherent regime.

28.6 **(a)** Convert Eq. (28.29) into Eq. (28.35) by using the Wannier equation (28.32) and the exciton basis (28.34). **(b)** Repeat the calculation with a diffusive scattering model where $i\gamma\,\Pi_{\mathbf{k},\mathbf{q}_{3D}}$ is replaced by $i\gamma\left[\Pi_{\mathbf{k},\mathbf{q}_{3D}} - \frac{1}{2\pi}\int d\theta\,\Pi_{\mathbf{k}+\mathbf{K}_\theta,\mathbf{q}_{3D}}\right]$, in analogy to Eq. (26.70).

28.7 When the dipole-matrix element has a **k** dependence, the spontaneous-emission source becomes $S^{\text{spnt}}_{\mathbf{k},\mathbf{q}} = d(\mathbf{k})\, f^e_{\mathbf{k}}\, f^h_{\mathbf{k}-\mathbf{q}} + \sum_{\mathbf{k}'} d(\mathbf{k}'+\mathbf{q}_h)\, c_X^{\mathbf{q},\mathbf{k}-\mathbf{q},\mathbf{k}'}$. Show that $S^{\text{spnt}}_{\lambda,\mathbf{q}} = \frac{1}{\sqrt{S}}\sum_{\mathbf{k}} \phi^L_\lambda(\mathbf{k})\, S^{\text{spnt}}_{\mathbf{k}+\mathbf{q}_e,\mathbf{q}}$ becomes $S^{\text{spnt}}_{\lambda,\mathbf{q}} = \sum_{\mathbf{k}'} \frac{d(\mathbf{k}')\,\phi^R_\lambda(\mathbf{k}')}{\sqrt{S}}\left[N^S_\lambda(\mathbf{q}) + \Delta N_\lambda(\mathbf{q})\right]$. **Hint:** *Follow and generalize the steps producing Eq.* (28.46).

28.8 **(a)** Assume that $f^e_{\mathbf{k}+\mathbf{q}_e}$ and $f^h_{\mathbf{k}-\mathbf{q}_h}$ are Fermi–Dirac distributions. Show that $\left(1 - f^e_{\mathbf{k}+\mathbf{q}_e} - f^h_{\mathbf{k}-\mathbf{q}_h}\right) = \frac{f^e_{\mathbf{k}+\mathbf{q}_e}\, f^h_{\mathbf{k}-\mathbf{q}_h}}{f^{\text{BE}}_{\mathbf{k},\mathbf{q}}}$ with the Bose–Einstein distribution $f^{\text{BE}}_{\mathbf{k},\mathbf{q}} = \frac{1}{e^{\beta(E_{\mathbf{k},\mathbf{q}}-\mu_e-\mu_h)}-1}$ with $\beta = \frac{1}{k_0 T}$ and the chemical potentials μ_e and μ_h. **(b)** Show that the singlet source (28.47) can also be expressed via $N^S_\lambda(\mathbf{q}) = \sum_{\mathbf{k}} \phi^L_\lambda(\mathbf{k})\, f^{\text{BE}}_{\mathbf{k},\mathbf{q}}\, \phi^R_\lambda(\mathbf{k})$. **(c)** Show that $f^\lambda_{\mathbf{k}} \ll 1/2$ implies $f^{\text{BE}}_{\mathbf{k}} \ll 1$. Use this information to show that $N^S_\lambda(\mathbf{q}) \ll 1$ for low excitations.

Further reading

The semiconductor luminescence equations were introduced in:

M. Kira, F. Jahnke, S. W. Koch, *et al.* (1997). Quantum theory of nonlinear semiconductor microcavity luminescence explaining "Boser" experiments, *Phys. Rev. Lett.* **79**, 5170.

Many details of the semiconductor luminescence equations are derived in:

M. Kira and S. W. Koch (2006). Many-body correlations and excitonic effects in semiconductor spectroscopy, *Prog. Quantum Electr.* **30**, 155.

W. Hoyer, M. Kira, S. W. Koch, J. Hader, and J. V. Moloney (2007). Coulomb effects on quantum-well luminescence spectra and radiative recombination times, *J. Opt. Soc. Am. B* **24**, 1344.

Excitons, their role in the semiconductor luminescence and more features are reviewed in:

S.W. Koch, M. Kira, G. Khitrova, and H. M. Gibbs (2006). Semiconductor excitons in new light, *Nature Materials* **5**, 523.

The Elliott formula for the luminescence was first applied in:

S. Chatterjee, C. Ell, S. Mosor, *et al.* (2004). Excitonic photoluminescence in semiconductor quantum wells: Plasma versus excitons, *Phys. Rev. Lett.* **92**, 067402.

The dynamic detection model has been developed, e.g., in:

H. J. Carmichael and D. F. Walls (1976). Proposal for the measurement of the resonant Stark effect by photon correlation techniques, *J. Phys. B* **9**, L43.

J. H. Eberly and K. Wódkiewicz (1977). The time-dependent physical spectrum of light, *J. Opt. Soc. Am. A* **67**, 1252.

M. Kira, F. Jahnke, W. Hoyer, and S. W. Koch (1999). Quantum theory of spontaneous emission and coherent effects in semiconductor microstructures, *Prog. Quantum Electron.* **23**, 189.

The nature of exciton correlations is studied, e.g., in:

W. Hoyer, M. Kira, and S. W. Koch (2003). Influence of Coulomb and phonon interaction on the exciton formation dynamics in semiconductor heterostructures, *Phys. Rev. B* **67**, 155113.

29

Many-body aspects of excitonic luminescence

In the previous chapter, we have already discussed that the microscopic origin of spontaneous light emission in semiconductors is significantly more complicated than for the atomistic situation where a single entity, i.e., an isolated electron–ion pair, emits a photon. In the artificial situation of a purely excitonic population without any Fermionic substructure, all the plasma contributions vanish and the semiconductor luminescence equations (SLE) predict that, e.g., the emission at the $1s$ energy stems only from the $1s$-exciton population. The very same conclusion follows from the simplified atomic picture analyzed in Chapters 16–23 because the isolated atomic entities are uniquely defined by their eigenstates $|\phi_\lambda\rangle$ and the eigenenergies E_λ. For example, the two-level luminescence Eq. (23.44) yields an emission that is proportional to the population of the excited state.

In reality, the semiconductor excitations are hardly ever dilute enough for the electron–hole pairs to be treated as isolated entities. Instead, the excited quasiparticles interact collectively with the emitted photons. Consequently, it is not justified to omit the Fermionic aspects from Eq. (28.47). In addition, semiconductors with a continuous band structure always have a much greater number of available plasma than exciton states and both of them usually contribute to the excitonic luminescence.

29.1 Origin of excitonic plasma luminescence

Besides an exciton population, also an electron–hole plasma can be the source for light emission at the exciton resonance. To investigate how such a plasma can emit a photon with an energy below the band gap, we consider a single photon-emission event in a many-body system. In this situation, the carriers have the average energy per particle E_{ini} before and E_{fin} after the emission. We assume that the semiconductor has $N_{\text{eh}} \gg 1$ electrons and holes before the spontaneous recombination and that the emitted photon has the energy $\hbar\omega_{1s}$ corresponding to the excitonic $1s$ resonance. Simple energy conservation requires that

$$E_{\text{ini}} N_{\text{eh}} = E_{\text{fin}} \left(N_{\text{eh}} - 1\right) + \hbar\omega_{1s} \tag{29.1}$$

must be satisfied. From this relation, we obtain the average final energy of the carrier system,

$$E_{\text{fin}} = E_{\text{ini}} + \frac{E_{\text{ini}} - \hbar\omega_{1s}}{N_{\text{eh}} - 1} \tag{29.2}$$

indicating that there is a negligible difference between initial and final energy for large particle numbers. This tiny energy difference can be absorbed by the many-body system, leading, e.g., to a small amount of carrier heating whenever the spontaneous recombination of an electron–hole pair and the emission of a below band-gap photon takes place. It is intuitively clear that one can find multiple many-body configurations whose average energy infinitesimally differs from E_{ini}. Thus, neither the availability of many-body states nor the conservation of energy disallow excitonic plasma luminescence from an interacting electron–hole plasma.

Taking the few-body limit of (29.2), we notice a divergency as $N_{\text{eh}} \to 1$. Hence, to treat the emission of an isolated electron–hole pair, we must use Eq. (29.1) requiring that $E_{\text{ini}} = \hbar\omega_{1s}$. Since only a $1s$ exciton has this average energy, this special case allows for excitonic luminescence only if the electron–hole pair is bound as an exciton. However, this simple argument does not hold in cases where many-body effects are important.

29.1.1 Energy redistribution dynamics

To gain further insight into the many-body emission process, we investigate the energy redistribution dynamics within the interacting light–matter system in the incoherent regime. It is particularly instructive to follow how the spontaneous emission influences the average system energy

$$E_{\text{inc}} \equiv \langle \hat{H} \rangle \equiv \langle \hat{H}_{\text{carr}} \rangle + \langle \hat{H}_{\text{dip}} \rangle + \langle \hat{H}_{\text{phot}} \rangle. \tag{29.3}$$

Here, the different terms describe the energy of the incoherent carriers, of the dipole interaction, and of the photons,

$$\begin{aligned}
\langle \hat{H}_{\text{carr}} \rangle &\equiv \sum\nolimits_{\mathbf{k}} \left[\left(E_{g,0} + \bar{\varepsilon}^{e}_{\mathbf{k}}\right) f^{e}_{\mathbf{k}} + \bar{\varepsilon}^{h}_{\mathbf{k}} f^{h}_{\mathbf{k}} \right] - \sum\nolimits_{\mathbf{k},\mathbf{k}'} V_{\mathbf{k}-\mathbf{k}'} \frac{f^{e}_{\mathbf{k}} f^{e}_{\mathbf{k}'} + f^{h}_{\mathbf{k}} f^{h}_{\mathbf{k}'}}{2} \\
&\quad + \sum\nolimits_{\mathbf{k},\mathbf{k}',\mathbf{q}} \left[\frac{1}{2} V_{\mathbf{q}} \left(c^{\mathbf{q},\mathbf{k}',\mathbf{k}}_{c,c;c,c} + c^{\mathbf{q},\mathbf{k}',\mathbf{k}}_{v,v;v,v} \right) - V_{\mathbf{k}'+\mathbf{q}-\mathbf{k}}\, c^{\mathbf{q},\mathbf{k}',\mathbf{k}}_{X} \right], \\
\langle \hat{H}_{\text{dip}} \rangle &\equiv +\text{Re}\left[\sum\nolimits_{\mathbf{k},\mathbf{q}_{3\text{D}}} i\mathcal{F}^{\star}_{\mathbf{q}_{3\text{D}}} \Pi_{\mathbf{k};\mathbf{q}_{3\text{D}}} \right], \\
\langle \hat{H}_{\text{phot}} \rangle &\equiv \sum\nolimits_{\mathbf{q}_{3\text{D}}} \hbar\omega_{\mathbf{q}_{3\text{D}}}\, \Delta\left\langle B^{\dagger}_{\mathbf{q}_{3\text{D}}} B_{\mathbf{q}_{3\text{D}}} \right\rangle,
\end{aligned} \tag{29.4}$$

respectively. As defined in connection with Eq. (28.2), we again use the notation $\mathbf{q}_{3\text{D}} = (\mathbf{q},\, q_z)$ for the three-dimensional (3D) wave vector of the photon.

The different energy contributions follow directly from the system Hamiltonian (13.81)–(13.84) in the incoherent regime. The division into singlets and doublets is exact such that the accuracy of E_{inc} depends only on the precision of the singlet–doublet solutions. As we already know the quantum-optical dynamics of each singlet and doublet contribution in Eq. (29.4), we can use Eqs. (28.8)–(28.10), (28.12), and (28.24) to study how the quantum-optical processes alter the system energy in the incoherent regime.

Since the dynamical equations always involve a photon-assisted polarization, we can classify the contributions according to their Π correlations. The corresponding energy changes of the carrier singlets, the carrier doublets, the dipole-interaction, and the $\Delta\langle B^\dagger B\rangle$ contributions are

$$\begin{aligned}
\frac{\partial\left\langle \hat{H}^S_{\rm carr}\right\rangle}{\partial t}\Bigg|^{\Pi} &= \frac{2}{\hbar}\sum_{\lambda,\mathbf{q}_{\rm 3D}} {\rm Re}\left[-E_{g,0}\,(\mathcal{F}^{\lambda}_{\mathbf{q}_{\rm 3D}})^{\star}\Pi_{\lambda;\mathbf{q}_{\rm 3D}} - \sum\nolimits_{\mathbf{k}} \tilde{E}_{\mathbf{k}}\,(\mathcal{F}^{\lambda}_{\mathbf{q}_{\rm 3D}})^{\star}\Pi_{\lambda;\mathbf{q}_{\rm 3D}}\phi^{R}_{\lambda}(\mathbf{k})\right],\\
\frac{\partial\left\langle \hat{H}^D_{\rm carr}\right\rangle}{\partial t}\Bigg|^{\Pi} &= \frac{2}{\hbar}\sum_{\lambda,\mathbf{q}_{\rm 3D}} {\rm Re}\left[-E_{\lambda,\mathbf{q}}\,(\mathcal{F}^{\lambda}_{\mathbf{q}_{\rm 3D}})^{\star}\Pi_{\lambda;\mathbf{q}_{\rm 3D}} + \sum\nolimits_{\mathbf{k}} \tilde{E}_{\mathbf{k}}\,(\mathcal{F}^{\lambda}_{\mathbf{q}_{\rm 3D}})^{\star}\Pi_{\lambda;\mathbf{q}_{\rm 3D}}\phi^{R}_{\lambda}(\mathbf{k})\right],\\
\frac{\partial\langle \hat{H}_{\rm dip}\rangle}{\partial t}\Bigg|^{\Pi} &= \frac{2}{\hbar}\sum_{\lambda,\mathbf{q}_{\rm 3D}} {\rm Re}\left[\left(E_{\lambda,\mathbf{q}} + E_{g,0} - \hbar\omega_{\mathbf{q}_{\rm 3D}} - i\gamma_{\lambda}\right)(\mathcal{F}^{\lambda}_{\mathbf{q}_{\rm 3D}})^{\star}\Pi_{\lambda;\mathbf{q}_{\rm 3D}}\right],\\
\frac{\partial\langle \hat{H}_{\rm phot}\rangle}{\partial t}\Bigg|^{\Pi} &= \frac{2}{\hbar}\sum_{\lambda,\mathbf{q}_{\rm 3D}} {\rm Re}\left[\hbar\omega_{\mathbf{q}_{\rm 3D}}(\mathcal{F}^{\lambda}_{\mathbf{q}_{\rm 3D}})^{\star}\Pi_{\lambda;\mathbf{q}_{\rm 3D}}\right],
\end{aligned} \tag{29.5}$$

respectively. Here, the $\langle\hat{H}^D_{\rm carr}\rangle$ and the $\langle\hat{H}_{\rm dip}\rangle$ dynamics contain triplet contributions that actually cancel each other in the sum. Nevertheless, it is convenient to approximate the triplets by an exciton-state-dependent dephasing constant γ_λ because this allows us to check whether such a dephasing approximation introduces any additional energy loss.

Combining all parts of Eq. (29.5), we find that the quantum-optical contributions alter the system energy via

$$\hbar\frac{\partial E_{\rm inc}}{\partial t}\Bigg|^{\Pi} = -2\sum_{\lambda,\mathbf{q}_{\rm 3D}} {\rm Re}\left[i\gamma_{\lambda}(\mathcal{F}^{\lambda}_{\mathbf{q}_{\rm 3D}})^{\star}\Pi_{\lambda;\mathbf{q}_{\rm 3D}}\right], \tag{29.6}$$

which contains only the dephasing contribution. In other words, the quantum-optical correlations fully conserve the system energy if the triplets are either omitted or fully included. A simple dephasing approximation can, in principle, result in energy loss. To identify the consequence of the dephasing approximation, we use the steady-state expression (28.37) for Π to convert Eq. (29.6) into

$$\hbar\frac{\partial E_{\rm inc}}{\partial t}\Bigg|^{\Pi} = 2\sum_{\lambda}\gamma_{\lambda}\phi^{R}_{\lambda}(\mathbf{r}=0)\,{\rm Re}\left[\sum_{\mathbf{q}_{\rm 3D}}\frac{|\mathcal{F}^{\lambda}_{\mathbf{q}_{\rm 3D}}|^2 S^{\rm spt}_{\lambda,\mathbf{q}}}{\hbar\omega_{\mathbf{q}_{\rm 3D}} - E_{g,0} - E_{\lambda,\mathbf{q}} + i\gamma_{\lambda}}\right]. \tag{29.7}$$

In free space, $|\mathcal{F}_{\mathbf{q}_{\rm 3D}}|^2$ is only weakly q_z dependent. Without this factor, the q_z sum produces a completely imaginary result. Thus, the RHS of Eq. (29.7) vanishes even in the presence of phenomenological dephasing showing that the dephasing approximation does not alter the energy flow under steady-state emission conditions.

Besides the energy conservation, the SLE predict conservation relations between the number of carriers, excitons, and photons, as shown in Sections 28.1.1–28.1.2. Furthermore, the quantum-optical contributions fully conserve the total momentum and angular momentum within the carrier system. The proof of this statement is the topic of Exercise 29.2. In summary, the presented description of the many-body and quantum-optical emission dynamics fulfills all the fundamental conservation laws.

29.1.2 Energy flow between many-body states and photons

A closer examination of Eq. (29.5) reveals which contributions cancel each other to establish the overall conservation of energy. The carrier singlet $\langle \hat{H}^{S}_{\rm carr} \rangle$ contains parts which depend on the band gap or the kinetic energy. From these, the $E_{g,0}$ contribution cancels with the second term in the dipole-energy dynamics in Eq. (29.5) while the $\tilde{E}_{\mathbf{k}}$ part is compensated by the second term in the carrier doublet $\langle \hat{H}^{D}_{\rm carr} \rangle$. These cancellations can be interpreted to show that the Π processes simultaneously move energy between the carrier singlets and doublets and between the singlets and the dipole energy. The first $E_{\lambda,\mathbf{q}}$ contribution of the carrier doublets is compensated by the energy increase in the dipole energy. The remaining third term of $\langle \hat{H}_{\rm dip} \rangle$ is compensated by the energy increase in the photon-energy term showing how the energy of the many-body system is eventually transferred to the photons.

Thus, we can construct an illustrative chart for the energy flow in the many-body system. Figure 29.1 presents how the energy of the carrier singlets is converted into the energy of the carrier doublets via the contributions $\tilde{E}_{\mathbf{k}}$ or into the dipole energy via the contributions $E_{g,0}$ in Eq. (29.5). The possible energy transfer directions are indicated by arrows. It is interesting to notice that carriers do not transfer their energy directly to photons. Instead, the quantum-optical processes induce energy transfer either between carriers

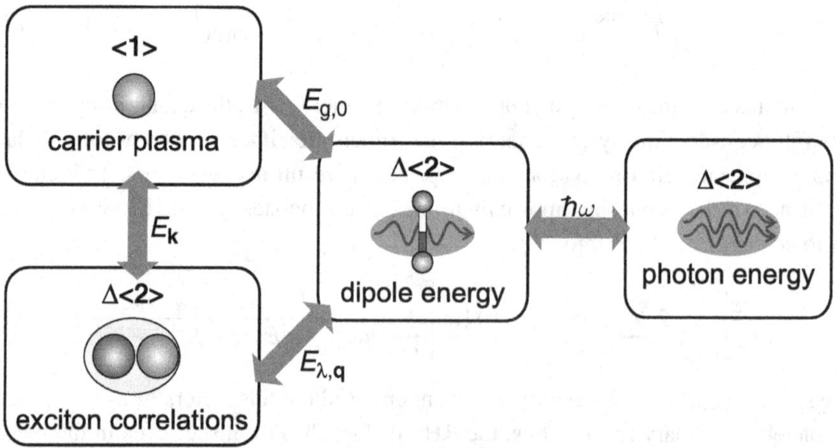

Figure 29.1 Schematic representation of the energy transfer in the semiconductor luminescence dynamics. The photon-assisted contribution converts energy between carrier singlets and doublets through $E_{\mathbf{k}}$. They are coupled to the dipole energy through the $E_{g,0}$ (singlets) and $E_{\lambda,\mathbf{q}}$ (doublets) contributions. The dipole energy is eventually converted into photons ($\hbar\omega$). Each term is identified based on Eq. (29.5) and the coupling is indicated by the arrows.

and the dipoles or between the carrier plasma (singlets) and the correlated electron–hole pairs (doublets). This internal energy redistribution is an inherent part of the spontaneous emission in semiconductors. In the final step, only the dipole energy is directly transferred to the photon field.

We see that the photon-assisted polarization has multiple roles in the semiconductor-luminescence process. It is responsible for the energy rearrangements within the many-body system, it stores carrier energy into the dipole energy, and eventually converts it to photons. Thus, $\Pi_{\mathbf{k};\mathbf{q}_{3D}}$ mediates not only the photon emission processes but also the transition to the appropriate final many-body states. Due to the Coulomb coupling among the different $\Pi_{\mathbf{k};\mathbf{q}_{3D}}$, see Fig. 28.3, the emission at the excitonic resonance energies becomes favorable even for a plasma configuration via the simultaneous increase in the kinetic energy, i.e., plasma heating.

29.2 Excitonic plasma luminescence

To demonstrate that excitonic plasma luminescence can be detected experimentally, we perform a computation where an initially unexcited system is nonresonantly laser excited in the interband continuum. In this situation, the average energy per particle is clearly above the band gap, i.e., $E_{\text{ini}} > E_{g,0}$ such that the excitation predominantly creates an electron–hole plasma. For demonstration purposes, we switch off the c_X contributions and solve the SLE together with the semiconductor Bloch equations (SBE). Without the c_X contributions, there is no possibility to have excitons in the system. The excitation-to-emission cycle is illustrated in Fig. 29.2(a). The excitation pulse mainly generates carriers that are initially ($t = 0\,\text{ps}$) distributed into relatively high momentum states. Coulomb and phonon scattering events equilibrate the carrier distributions toward the bottom of the bands.

In the numerical evaluations, we assumed a 500 fs excitation pulse 16.7 meV above the unrenormalized band edge and an initial lattice temperature of 4 K. Furthermore, we used the parameters corresponding to a single 8 nm-wide GaAs quantum well having $E_{1s} = -2.45\,E_B = -10\,\text{meV}$ and a Bohr radius $a_0 = 12.5\,\text{nm}$. Additionally, we performed a calculation where the Coulomb-interaction effects in the photon-assisted processes are switched off, i.e., the Coulomb terms within Eq. (28.24) are eliminated. The comparison allows us to determine how the photoluminescence is altered when the emission-mediating Π processes are not Coulomb coupled via the main part of the SLE. Figure 29.2(b) presents time-resolved luminescence spectra with (solid line) and without (shaded area) the Coulomb interaction for three representative times after the excitation. The corresponding electron (solid line) and hole (shaded area) distributions are shown in Fig. 29.2. Implementing a time-dependent detector model allows us to convert the computed $\frac{\partial}{\partial t}\langle B^{\dagger}_{\mathbf{q}} B_{\mathbf{q}}\rangle$ into the corresponding luminescence spectra, cf., the references in Chapter 28 for further details. In the present calculations we assumed a 1 ps detector time resolution, which corresponds to a 0.6 meV energy resolution.

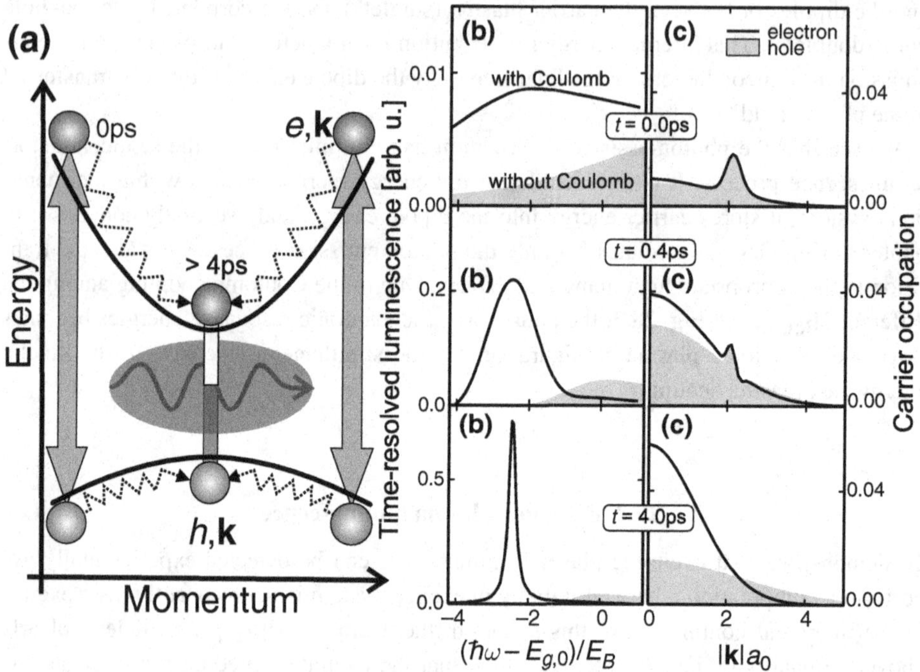

Figure 29.2 (**a**) Schematic picture showing the occurrence of photoluminescence after optical excitation into the interband absorption continuum. (**b**) The time evolution of the bare quantum-well luminescence is computed with (solid line) and without (shaded area) the Coulomb correlations. The terms describing formation of true exciton populations have been omitted from the equations. (**c**) The carrier distribution functions are shown at different times. The final carrier density is $n = 1.8 \times 10^{10}\text{cm}^{-2}$. In the computations, the band-gap energy is taken as E_G =1.485 eV and the dephasing constant is 0.4 meV. Adapted from (Kira *et al.*, 1998).

Originally, the optically generated electron and hole distributions are peaked at the pump energy 16.7 meV and strongly deviate from thermal Fermi–Dirac distributions. However, *the luminescence spectrum is already centered around the exciton resonance energy* even though it is very broad. At the time $t = 0.4$ ps, the steady-state carrier density $n = 1.8 \times 10^{10}$ cm^{-2} is reached and the carrier occupation functions approach Fermi–Dirac distributions where the low-energy **k**-states are increasingly populated. At the same time, the luminescence spectrum becomes considerably narrower and its peak value increases. After roughly $t = 4$ ps, the spectral shape becomes time-independent (quasistationary), peaking all the time at the 1*s*-exciton energy even though exciton populations have been omitted from the calculations.

As can be seen from the test calculations, the shape of the emission spectrum (shaded area in Fig. 29.2(b)) changes radically when we completely neglect the Coulomb interaction among the different $\Pi_{\mathbf{k};\mathbf{q}_{3D}}$. In comparison with the full calculations, the strength of the emission drops roughly by an order of magnitude (late times) and it is peaked at the band edge describing band-to-band luminescence. This switch-off analysis verifies that the Coulomb coupling among Π creates the excitonic resonance and that the excitonic luminescence can very well originate from an electron–hole plasma.

29.2.1 Radiative recombination of carriers vs. excitons

To determine how fast radiative recombination alters the many-body state, we first evaluate the recombination rates following from the steady-state result. Combining Eqs. (28.37) and (28.46) yields

$$\Pi^{\text{std}}_{\lambda,\mathbf{q}_{3\text{D}}} = \frac{i\mathcal{F}_{\lambda,\mathbf{q}_{3\text{D}}}\left[\Delta N_\lambda(\mathbf{q}) + N^{\text{S}}_\lambda(\mathbf{q})\right]}{\hbar\omega_{\mathbf{q}_{3\text{D}}} - E_{g,0} - E_{\lambda,\mathbf{q}} + i\gamma_\lambda} \tag{29.8}$$

and the exciton-recombination relation (28.15) becomes

$$\hbar\frac{\partial N_\lambda(\mathbf{q})}{\partial t}\bigg|_{\text{QED}} = -2\mathcal{S}\,\text{Re}\left[\sum_{q_z}\frac{i|\mathcal{F}_{\lambda,\mathbf{q}_{3\text{D}}}|^2\left[\Delta N_\lambda(\mathbf{q}) + N^{\text{S}}_\lambda(\mathbf{q})\right]}{\hbar\omega_{\mathbf{q}_{3\text{D}}} - E_{g,0} - E_{\lambda,\mathbf{q}} + i\gamma_\lambda}\right]. \tag{29.9}$$

To evaluate this expression, we need to know the explicit form of the prefactor $|\mathcal{F}_{\lambda,\mathbf{q}_{3\text{D}}}|^2$. When the QW is in free space, it is convenient to use plane waves $\mathbf{u}_{\mathbf{q}_{3\text{D}}}(z) = \frac{1}{\sqrt{n\mathcal{S}\mathcal{L}}}e^{iq_z z}\,\mathbf{e}^{[\sigma]}_{\mathbf{q}_{3\text{D}}}$ as the basis functions, compare Section 9.4. Due to the normalization (9.23), the mode functions contain the refractive index n and the quantization length $\mathcal{L}$. For each $\mathbf{q}_{3\text{D}}$, we have two independent transversal modes identified by their polarization direction $\mathbf{e}^{[\sigma]}_{\mathbf{q}_{3\text{D}}}$.

For a fixed photon energy $\hbar\omega_{\mathbf{q}_{3\text{D}}}$, the in-plane momentum of the photon is limited to $|\mathbf{q}| \leq \omega_{\mathbf{q}_{3\text{D}}}/c = q_{\text{rad}}$, which defines the largest possible photon momentum, i.e., the largest value of $\mathbf{q}$ for which $|\mathcal{F}_{\lambda,\mathbf{q}_{3\text{D}}}|^2$ exists based on Eq. (28.2). As calculated in connection with Fig. 28.2, GaAs-type systems have $q_{\text{rad}}a_0 \approx 0.1/a_0 \ll 1/a_0$ such that only a small portion of the excitons can couple to the light field. Based on this information, we can separate the full exciton distribution into optically bright, i.e., radiatively coupled excitons ($|\mathbf{q}| \leq q_{\text{rad}}$) and dark, i.e., nonradiative excitons ($|\mathbf{q}| > q_{\text{rad}}$). Combining Eqs. (28.2) and (28.47), we can write

$$|\mathcal{F}_{\lambda,\mathbf{q}_{3\text{D}}}|^2 = \begin{cases} \frac{|d_{vc}\phi^R_\lambda(\mathbf{r}=0)|^2}{2\epsilon_0 n}\frac{\hbar\omega_{g,0}}{\mathcal{S}\mathcal{L}}, & |\mathbf{q}| \leq q_{\text{rad}} \\ 0, & \text{otherwise} \end{cases}. \tag{29.10}$$

Here, we used $\mathcal{E}^2_{\mathbf{q}_{3\text{D}}} = \hbar\omega_{g,0}/2\varepsilon_0$ because the photon energies of interest are close to the band gap $\hbar\omega_{g,0}$. So far, our results apply for s-polarized light. For p-polarized modes, an additional angle-dependent factor appears. This aspect is the topic of Exercise 29.4.

Inserting the result (29.10) into Eq. (29.9) shows that bright excitons recombine with the rate

$$\hbar \frac{\partial N_\lambda(\mathbf{q})}{\partial t}\bigg|_{\mathrm{QED}} = -2\frac{\left|d_{vc}\phi_\lambda^R(0)\right|^2}{2\epsilon_0}\frac{\omega_{g,0}}{nc}\mathrm{Re}\left[\frac{\hbar c}{\mathcal{L}}\sum_{q_z}\frac{i\left[\Delta N_\lambda(\mathbf{q})+N_\lambda^{\mathrm{S}}(\mathbf{q})\right]}{\hbar\omega_{\mathbf{q}_{3\mathrm{D}}}-E_{g,0}-E_{\lambda,\mathbf{q}}+i\gamma_\lambda}\right]$$
$$= -2\Gamma_{\lambda,\lambda}^{\mathrm{rad}}\left[\Delta N_\lambda(\mathbf{q})+N_\lambda^{\mathrm{S}}(\mathbf{q})\right]\Sigma_{\mathbf{q}}^{\mathrm{rad}}, \qquad \text{for } |\mathbf{q}|\le q_{\mathrm{rad}}$$
$$\Sigma_{\mathbf{q}}^{\mathrm{rad}} \equiv \mathrm{Re}\left[\frac{\hbar c}{\mathcal{L}}\sum_{q_z}\frac{i}{\hbar\omega_{\mathbf{q}_{3\mathrm{D}}}-E_{g,0}-E_{\lambda,\mathbf{q}}+i\gamma_\lambda}\right]. \tag{29.11}$$

Interestingly, the prefactor agrees with the radiative coupling constant (26.46) describing the radiative decay of the polarization. The expression for $\Sigma_{\mathbf{q}}^{\mathrm{rad}}$ yields one for low $\mathbf{q}$, see Exercise 29.5. For $\Delta N_\lambda(\mathbf{q}) \gg N_\lambda^{\mathrm{S}}(\mathbf{q})$, i.e., in situations where excitons dominate the spontaneous emission, they decay exponentially as

$$\hbar \frac{\partial N_\lambda(\mathbf{q})}{\partial t}\bigg|_{\mathrm{QED}} = \begin{cases} -2\Gamma_{\lambda,\lambda}^{\mathrm{rad}}\,\Delta N_\lambda(\mathbf{q}), & |\mathbf{q}|\le q_{\mathrm{rad}} \\ 0, & |\mathbf{q}|> q_{\mathrm{rad}} \end{cases}. \tag{29.12}$$

For GaAs-type systems, the radiative decay time is in the range of 10 ps, which is short in comparison with the typical nanosecond lifetime of incoherent semiconductor excitations. The plasma-related recombination rate is obtained by replacing $\Delta N_\lambda(\mathbf{q})$ by the singlet part $N_\lambda^S(\mathbf{q})$. In thermodynamic equilibrium, we always have $N_\lambda^S(\mathbf{q}) \ll 1$ and $N_\lambda^S(\mathbf{q}) \ll \Delta N_\lambda(\mathbf{q})$, see Exercise 28.8. Therefore, the exciton decay due to $N_\lambda^S(\mathbf{q})$ is extremely slow.

We repeat the analysis to determine the recombination dynamics of the total electron and hole density $n_{\mathrm{eh}} \equiv \frac{1}{S}\sum_{\mathbf{k}} f_{\mathbf{k}}^e = \frac{1}{S}\sum_{\mathbf{k}} f_{\mathbf{k}}^e$. We substitute the steady-state $\Pi_{\lambda,\mathbf{q}_{3\mathrm{D}}}^{\mathrm{std}}$ into Eqs. (28.8)–(28.9) to obtain

$$\hbar \frac{\partial n_{\mathrm{eh}}}{\partial t}\bigg|_{\mathrm{QED}} \equiv -\gamma_{\mathrm{rad}}(n_{\mathrm{eh}})\,n_{\mathrm{eh}} = -\sum_\lambda \frac{5[q_0a_0]^2\Gamma_\lambda^{\mathrm{rad}}}{6\pi}\frac{N_\lambda^S(\mathbf{q}\cong 0)}{a_0^2} \tag{29.13}$$

when the plasma dominates the photoluminescence, see Exercise 29.6. The identified decay rate γ_{rad} depends on the plasma density n_{eh}. Due to this almost linear dependency, the largest decay rates are found for elevated densities. Nevertheless, they are typically orders of magnitude slower than the exciton-decay rate $\Gamma_\lambda^{\mathrm{rad}}$ because $[q_0a_0]^2 \approx 0.01 \ll 1$ and $N_\lambda^S(\mathbf{q}=0)/a_0^2 < n_{\mathrm{eh}}$. This increase of the radiative lifetime is a consequence of the two-step energy transfer process needed to generate emission at the exciton resonance frequency through the additional heating of the remaining plasma. The increased carrier lifetime also explains why, even in direct-gap semiconductors, the incoherent excitations exist for nanoseconds rather than for some tens of picoseconds as the bright excitons do. As a consequence of these large differences in their radiative decay times, we usually have a situation where the bright exciton populations are nearly depleted. Under these conditions, the photoluminescence intensity is relatively weak and has a significant plasma contribution.

29.2.2 Hole burning in exciton distributions

Due to fast radiative decay of bright excitons, we expect that the $1s$ distribution always displays a strongly nonthermal shape where bright excitons are nearly depleted while a significant amount of dark excitons can remain in the system. To analyze this nonequilibrium effect, Fig. 29.3(a) shows the $1s$-exciton distribution at different time moments after the assumed coherent resonant $1s$-excitation. The corresponding electron distributions are plotted in Fig. 29.3(b). The distribution function of the holes is similar to that of the electrons and thus not explicitly shown. In the numerical calculations leading to the results in Fig. 29.3, we have solved the full singlet–doublet equations for the same planar arrangement of quantum wires as in the previous chapters. We have assumed conditions similar to those in Fig. 27.3(a) and used a 1 ps excitation pulse with a sufficiently low intensity.

The snapshot times in Fig. 29.3 are chosen such that the polarization-to-population conversion is already complete. For these times, we see that both the carrier and the $1s$-exciton distributions have a wide momentum spread due to the lack of momentum sensitivity in the Coulomb-scattering-induced polarization-to-population conversion, compare the discussion in Section 27.2.2. After their generation, the excitons in the very low momentum states, i.e., those with $|\mathbf{q}|a_0 < 0.1 = q_{\rm rad}$, show a fast decay due to their photoluminescence related recombination.

Figure 29.3 illustrates how spontaneous emission predominantly recombines the bright excitons, producing a significant hole burning in the exciton distributions. This hole burning is quite prominent because the typical times for excitons to scatter from the dark into the bright states are rather long relative to the 10 ps radiative decay time. Hence, radiative recombination rapidly depletes the bright exciton population leaving the majority of the excitons in optically inactive dark states. As a consequence, *spontaneous emission is never a weak perturbation for exciton distributions in direct-gap semiconductors.* The pronounced hole burning leads to significant deviations from a quasiequilibrium thermodynamic distribution function.

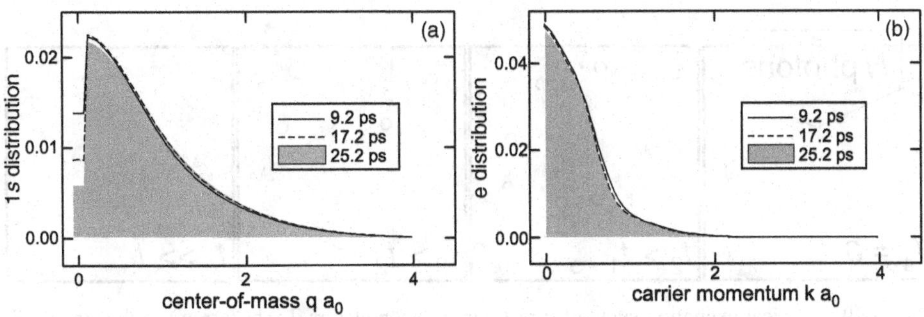

Figure 29.3 **(a)** Computed $1s$-exciton and **(b)** carrier distributions for the three indicated times after the resonant $1s$ excitation with classical light. Reprinted from *Progress in Quantum Electronics*, **30/5**, M. Kira, S. W. Koch, "Many-body correlations and excitonic effects in semiconductor spectroscopy," pp. 155–296, 2006, with permission from Elsevier.

Once the bright excitons have decayed, only the plasma part remains as the source of the luminescence. However, no hole burning occurs here because an electron with an arbitrary momentum **k** can recombine with a hole in the matching momentum state, $\mathbf{k}-\mathbf{q}$. Therefore, all electron and hole states contribute to the emission and the radiative recombination only leads to slow changes of the overall distributions on a nanosecond time scale. Since the Coulomb- and phonon-induced scattering processes are much faster, they rapidly bring the electron and hole distributions into the Fermi–Dirac form and a quasiequilibrium description is justified.

29.3 Direct detection of excitons

Our discussions of the absorption and emission processes clearly reveal many fundamental differences between atomic and semiconductor systems. Clearly, in atomic vapors, the optical fields do not alter the number of electrons and nuclei. However, each absorbed or emitted optical photon alters the electron–hole number in a semiconductor by one such that the interaction with optical fields always involves transitions between many-body states with different numbers of particles. In contrast, the particle number is strictly constant in atom optics.

A typical excitation cycle of a semiconductor is illustrated in Fig. 29.4. In the beginning ($t_1 = 0$), the semiconductor is in its ground state without excited electrons or holes. When the light arrives at the semiconductor, it creates the electron–hole pairs ($t_2 > t_1$) which, under suitable conditions, may develop into bound excitons. Due to vacuum-field fluctuations, electrons (spheres) and holes (open circles) start to recombine ($t_3 > t_2$) via the spontaneous emission of photons (wavy arrows). Also nonradiative recombination is possible. After a sufficiently long time ($t_4 \gg t_3$), the system returns back to its ground state.

Since one expects neither creation nor destruction of atoms by optical fields, it is obvious that semiconductor and atomic optics are profoundly different such that results from one system may not necessarily be used to explain the properties of the other. One interesting

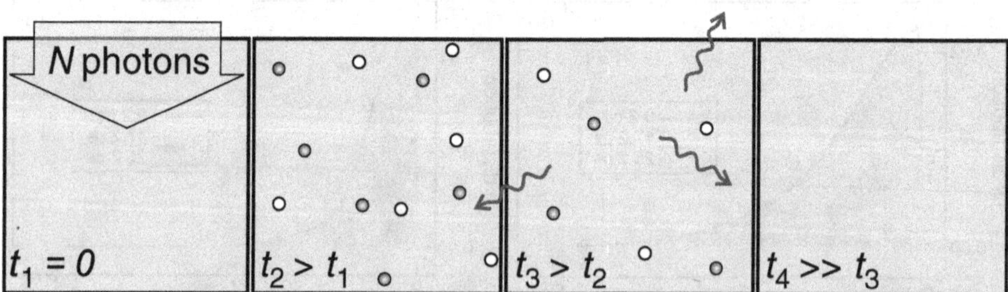

Figure 29.4 Typical excitation cycle of direct-gap semiconductors. Light (arrow) excites the semiconductor from its ground state ($t_1 = 0$). The absorbed photons create electrons (spheres) and holes (open circles) that remain in the system ($t_2 > t_1$). Radiative recombination destroys electron–hole pairs and creates photons (wave arrows) for subsequent times $t_3 > t_2$. The semiconductor eventually returns back to its ground state (void, $t_4 \gg t_3$).

difference is found in the origin of resonances detected within absorption and emission spectra. Whereas an atomic transition can happen only in the presence of the atom, the mere appearance of excitonic resonances does not imply the presence of excitons. Instead, one needs a detailed many-body analysis to identify true exciton populations.

However, there is a relatively direct alternative method for exciton identification where one can use concepts of atomic or dilute-gas spectroscopy. Here, one often detects small concentrations of a particular species of atoms or molecules by using an optical probe that is sensitive to transitions between the eigenstates of the respective species. If the characteristic absorption resonances are observed in the probe spectrum, the atoms or molecules must be present, and one can deduce their relative concentration through proper normalization of the respective transition strengths.

In semiconductors, this means that one should look for transitions between different excitonic states, e.g., from the $1s$ to the $2p$ state. For this purpose, we have to use photon energies in the terahertz (THz) regime, i.e., in the range of one to tens of meV, depending on the system. Since such THz photons have an energy much smaller than $E_{g,0}$, they cannot lift electrons from the valence to the conduction band as long as the intensity is low enough to prevent multiphoton processes. Hence, the use of THz fields allows us to perform spectroscopy with a fixed number of particles, just as in the case of atomic optics.

The detailed many-body theory for THz absorption, presented in (Kira and Koch, 2006) and (Kira *et al.*, 2004), includes all singlet and doublet contributions. Here, we only use the result

$$\alpha_{\mathrm{THz}}(\omega) \propto \frac{\omega}{c} \mathrm{Im}\left[\sum_{\lambda} \left(S_\lambda(\omega) - [S_\lambda(-\omega)]^\star\right) \Delta n_\lambda\right] , \tag{29.14}$$

$$S_\lambda(\omega) = \sum_\beta \frac{|D_{\lambda,\beta}|^2}{E_\beta - E_\lambda - \hbar\omega - i\kappa_\lambda}, \tag{29.15}$$

for the linear THz absorption, where $D_{\lambda,\beta} = \langle \phi^L_\lambda | Q\mathbf{r} \cdot \mathbf{e}_P \left| \phi^R_\beta \right\rangle$ defines the dipole-matrix element between the exciton states λ and β when $\mathbf{e}_P$ is the THz polarization direction. The dephasing between intraband levels is defined by κ_λ. Equation (29.15) has the typical form of an atomic absorption spectrum when the different atomic levels are populated according to $\Delta n_\lambda = \frac{1}{S}\sum_{\mathbf{q}} \Delta N_\lambda(\mathbf{q})$. Interestingly, as discussed in Section 27.3, Δn_λ contains the *total number of excitons via the doublet correlations*. Thus, the concept of incoherent excitons has a one-to-one correspondence to the bound states of atoms, which verifies the physical meaning of the $\Delta N_\lambda(q)$ doublet. Furthermore, we see from Eq. (29.15) that THz spectroscopy can directly detect excitons unlike interband spectroscopy.

To illustrate these features, we show in Fig. 29.5(a) the computed THz absorption (solid line) from a postulated $1s$ distribution. This shows a pronounced peak at the energy corresponding to the difference of $2p$- and $1s$-exciton resonance. Since semiconductors have a substantial broadening κ_λ, the individual $1s$-to-np transitions (vertical lines) typically merge into an asymmetric tail that extends to the high-energy side. In other words, one can typically resolve only the $1s$-to-$2p$ resonance in GaAs-based systems. Nevertheless, the

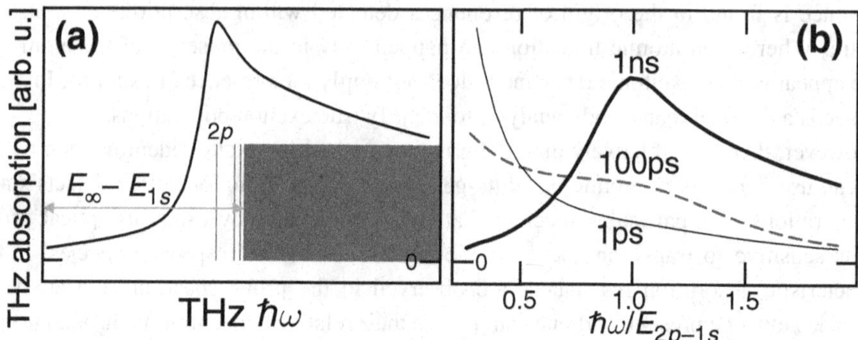

Figure 29.5 Exciton THz-absorption spectra. **(a)** Generic structure of the THz absorption (solid line) in the presence of a 1s-exciton population. The dipole-allowed transitions between the excitonic eigenstates are indicated by vertical lines and the shaded area indicates transitions from the $1s$ state into the excitonic ionization continuum. **(b)** THz spectra for different times (thin solid = 1 ps, dashed = 100 ps, thick solid = 1 ns) after the short-pulse interband excitation of a semiconductor quantum-wire system. At the early times, one sees a Drude-like response whereas the characteristic excitonic resonance due to the $1s$ to $2p$ transition becomes increasingly pronounced as the percentage of incoherent excitons increases. Reprinted from *Solid State Communications*, **129/11**, M. Kira, W. Hoyer, S. W. Koch, "Terahertz signatures of the exciton formation dynamics in non-resonantly excited semiconductors," pp. 733–739, 2004, with permission from Elsevier.

emergence of this resonance is a unique signature for the presence of $1s$ excitons in the system.

To study this further, we solve the equations resulting from the full singlet–doublet analysis assuming nonresonant excitation 10 meV above the band gap. Figure 29.5(b) shows the time evolution of the resulting THz spectrum. At early times, we observe the characteristic Drude-like spectrum (thin solid line) of an electron–hole plasma, which exhibits an increase toward low frequencies. As we see in Fig. 29.5(b), the plasma signatures disappear at later times (dashed and thick-solid line) and the $1s$-to-$2p$ transition resonance gradually develops. The strength of this resonance increases with increasing exciton concentration.

Exercises

29.1 **(a)** Show that the Hamiltonian (13.81)–(13.84) yields the system energy (29.4). Choose the zero level as in Eq. (28.41). **(b)** Include all contributions due to phonon-assisted polarizations, Π, to derive Eq. (29.5). **Hint:** *Use the semiconductor luminescence equations.* **(c)** Study the photon-assisted triplet contributions due to $\langle \hat{H}^{\mathrm{D}}_{\mathrm{carr}} \rangle$ and $\langle \hat{H}_{\mathrm{dip}} \rangle$ and verify their mutual compensation.

29.2 Assume that the system is initially homogeneous and incoherent and use a plane-wave basis for the light modes. **(a)** Show that the SLE conserve $f^{\lambda}_{R[\mathbf{k}]} = f^{\lambda}_{\mathbf{k}}$, $c^{\mathbf{q},\mathbf{k}',\mathbf{k}} = c^{R[\mathbf{q}],R[\mathbf{k}'],R[\mathbf{k}]}$, $\mathcal{F}^{\star}_{R[\mathbf{q}]}\,\Pi_{R[\mathbf{q}];R[\mathbf{k}]} = \mathcal{F}^{\star}_{\mathbf{q}}\,\Pi_{\mathbf{q},\mathbf{k}}$, and $\Delta\langle B^{\dagger}_{R[\mathbf{q}]}\,B_{R[\mathbf{q}']}\rangle$ for any rotation $R[\ldots]$ in three-dimensional space. **(b)** Show that the total momentum

$\sum_{\mathbf{k}} \hbar\mathbf{k}\, f_{\mathbf{k}}^{E}$ and the total angular momentum $\sum_{\mathbf{k}}[-i\,\nabla_{\mathbf{k}} \times \hbar\mathbf{k}]\, f_{\mathbf{k}}^{\lambda}$ are conserved due to the rotational symmetry.

29.3 Time-resolved emission measurements can be modeled via a projection $B_{T,\tau}(\Omega) \equiv \int_{-\infty}^{\infty} dt\, G_{\tau}(t-T)\, \hat{E}(z_0, t)\, e^{+i\Omega(t-T)}$ where $G_{\tau}(t) = \frac{1}{\sqrt{\pi\tau^2}}\, e^{-t^2/\tau^2}$ defines the gate function of the frequency Ω detection. Assume that $\hat{E}(z_0, t)$ propagates freely and express the time-resolved intensity $I_{T,\tau}(\Omega) = \Delta\langle B_{T,\tau}^{\dagger}(\Omega)\, B_{T,\tau}(\Omega)\rangle$ in terms of the known $\Delta\langle B_q^{\dagger}\, B_{q'}\rangle$ values and the detector function filtering the emission in q space.

29.4 **(a)** Show that p-polarized modes yield a reduction of $|\mathcal{F}_{\lambda,\mathbf{q}3\mathrm{D}}|^2$ by $q_z^2/|\mathbf{q}|^2$ in comparison to Eq. (29.10). **Hint:** *Use Eq.* (8.33) *to determine the polarization vectors.* **(b)** How is Eq. (29.11) modified for p-polarized modes?

29.5 **(a)** Derive Eq. (29.11) for the QW emission in free space. **(b)** Implement the continuum limit $\mathcal{L} \to \omega$ in Eq. (29.11) and show that $\sum_{\mathbf{q}}^{\mathrm{rad}} = 1$ for small enough $\mathbf{q}$. **(c)** Evaluate the corresponding integral for p-polarized modes.

29.6 Assume that a semiconductor has no exciton populations and that the emission is governed by the steady state formula (29.8). The electron–hole density is $n_{eh} = \frac{1}{S}\sum_{\mathbf{k}} f_{\mathbf{k}}^{e}$. **(a)** Show that the s-polarized modes yield $\hbar\, \frac{\partial n_{eh}}{\partial t}\Big|_{\mathrm{QED}}^{S} = -\sum_{\lambda} \Gamma_{\lambda}^{\mathrm{rad}}\, \frac{q_0^2}{2\pi^2}\, N_{\lambda}^{S}(\mathbf{q})$. **(b)** Show that the p-polarized modes yield $\hbar\, \frac{\partial n_{eh}}{\partial t}\Big|_{\mathrm{QED}}^{P} = -\sum_{\lambda} \Gamma_{\lambda}^{\mathrm{rad}}\, \frac{q_0^2}{3\pi^2}\, N_{\lambda}^{S}(\mathbf{q})$ and construct Eq. (29.13) via the sum of s- and p-polarized decay rates.

29.7 Assume a low-density approximation for $f_{\mathbf{k}}^{\lambda} = N_{\lambda}\, e^{-\frac{\hbar^2\mathbf{k}^2}{2m_{\lambda}}\beta}$. Approximate also $\phi_{1s}^{L,R}(\mathbf{k}) = N\, e^{-[F\,\mathbf{k}a0]^2}$ where F is an adjustable parameter close to $F \approx 1$. **(a)** Fix the N_e, N_h, and N values to produce the correct norms. **(b)** Compute an estimate for $N_{1s}^{S}(\mathbf{q})$ based on (28.47). Show that $N_{1s}^{S}(\mathbf{q})$ is only weakly $\mathbf{q}$ dependent. **(c)** Compute an estimate of $\gamma_{\mathrm{rad}}(n_{eh})$ for the density $n_{eh} = 10^{10}\ \mathrm{cm}^{-2}$. How does $\gamma_{\mathrm{rad}}(n_{eh})$ scale as a function of n_{eh}? What is the maximum n_{eh} for which this approximative $\gamma_{\mathrm{rad}}(n_{eh})$ can be used?

29.8 Assume that the system has initially either distributed $\Delta N_{1s}(\mathbf{q}) = A\, e^{-[a_0\mathbf{q}]^2}$ or quantum-degenerate $\Delta N_{1s}(\mathbf{q}) = \delta_{\mathbf{q},0}\, B$ $1s$-exciton distribution. Describe the exciton recombination via Eq. (29.12) without singlets. **(a)** Define constants A and B such that both situations have initially the density N_{1s}. **(b)** Compute both $\Delta N_{1s}(\mathbf{q}, t)$ and $\Delta n_{1s}(t) = \frac{1}{S}\sum_{\mathbf{q}} \Delta N_{1s}(q, t)$ resulting from radiative decay. Why is it difficult to maintain exciton condensates in direct-gap semiconductors? **Hint:** *Define* $q_{\mathrm{rad}}\, a_0 = 0.1$ *to separate bright from dark excitons.*

29.9 Assume that the system has no bright excitons such that the carrier densities decay radiatively via $\dot{n}_{\mathrm{EH}} = -\beta\, n_{\mathrm{EH}}\, n_{\mathrm{EH}}$. Solve this numerically for different values of β. **(a)** Show that n_{EH} decays nonexponentially. **(b)** Estimate $\beta = \Gamma^{\mathrm{rad}}_{1s} \frac{N^S_{1s}(q=0)}{n_E n_H}$ for the density $n_{\mathrm{EH}} = 10^{10}\ \mathrm{cm}^{-2}$ using typical GaAs parameters. What is the typical half lifetime of the excitation?

29.10 **(a)** Evaluate the THz absorption from the $1s$-exciton population Δn_{1s}. **(b)** Compute $D_{1s,\beta}$ analytically for a strictly two-dimensional quantum well and construct $S_{1s}(\beta)$. **Hint:** *Use Eq.* (16.22). **(c)** Compute $\alpha_{THz}(\omega)$ from $1s$ populations using different values for $\kappa_\lambda \equiv \kappa$.

Further reading

The excitonic luminescence without the need of exciton populations was predicted in:

M. Kira, F. Jahnke, and S. W. Koch (1998). Microscopic theory of excitonic signatures in semiconductor photoluminescence, *Phys. Rev. Lett.* **81**, 3263.

Experimental verifications of such excitonic plasma luminescence can, e.g., be found in:

M. Oestreich, D. Hägele, J. Hübner, and W. W. Rühle (2000). Excitons, or no excitons, that is the question, *Phys. Status Solidi A* **178**, 27.

S. Chatterjee, C. Ell, S. Mosor, *et al.* (2004). Excitonic photoluminescence in semiconductor quantum wells: Plasma versus excitons, *Phys. Rev. Lett.* **92**, 067402.

I. Galbraith, R. Chari, S. Pellegrini, *et al.* (2005). Excitonic signatures in the photoluminescence and terahertz absorption of a GaAs/$\mathrm{Al_xGa_{1-x}As}$ multiple quantum well, *Phys. Rev. B* **71**, 073302.

W. Hoyer, C. Ell, M. Kira, *et al.* (2005). Many-body dynamics and exciton formation studied by time-resolved photoluminescence, *Phys. Rev. B* **72**, 075324.

S. W. Koch, M. Kira, G. Khitrova, and H. M. Gibbs (2006). Semiconductor excitons in new light, *Nature Mat.* **5**, 523.

Nonexponential decay due to excitonic plasma luminescence has also been observed in quantum-dot systems, cf.:

M. Schwab, H. Kurtze, T. Auer, *et al.* (2006). Radiative emission dynamics of quantum dots in a single cavity micropillar, *Phys. Rev. B* **74**, 045323.

The extremely nonthermal shapes of the semiconductor photoluminescence were first reported in:

R. F. Schnabel, R. Zimmermann, D. Bimberg, *et al.* (1992). Influence of exciton localization on recombination line shapes: $\mathrm{In_xGa_{1-x}As}$/GaAs quantum wells as a model, *Phys. Rev. B* **46**, 9873.

During the past few years, THz technology has developed tremendously as a versatile tool to study different aspects of many-body systems, cf.:

R. Huber, F. Tauser, A. Brodschelm, *et al.* (2001). How many-particle interactions develop after ultrafast excitation of an electron–hole plasma, *Nature* **414**, 286.

R. A. Kaindl, M. A. Carnahan, D. Hägele, R. Lövenich, and D. S. Chemla (2003). Ultrafast terahertz probes of transient conducting and insulating phases in an electron–hole gas, *Nature* **423**, 734.

M. Kira, W. Hoyer, and S. W. Koch (2004). Terahertz signatures of the exciton formation dynamics in non-resonantly excited semiconductors, *Solid State Comm.* **129**, 733.

J. R. Danielson, Y.-S. Lee, J. P. Prineas, *et al.* (2007). Interaction of strong single-cycle terahertz pulses with semiconductor quantum wells, *Phys. Rev. Lett.* **99**, 237401.

30
Advanced semiconductor quantum optics

The discussions in the previous chapters show that already the spontaneous emission, i.e., the simplest manifestation of the quantum-optical fluctuations, exhibits highly nontrivial features as soon as the quantum light is coupled to interacting many-body systems. We have seen, for example, that pronounced resonances in the semiconductor luminescence originate from a nontrivial mixture of exciton and plasma contributions in contrast to the simple transitions between the eigenstates of isolated atomic systems. As a consequence of the conservation laws inherent to the light–matter coupling, the photon emission induces rearrangements in the entire many-body system leading, e.g., to pronounced hole burning in the exciton distribution. As discussed in Chapter 29, this depletion of the optically active excitons leads to a reduction of the total radiative recombination and the appearance of non-thermal luminescence even when the electron–hole system is in quasiequilibrium. Already these observations show that the *coupled* quantum-optical and many-body interactions induce new intriguing phenomena that are not explainable by the concepts of traditional quantum optics or classical semiconductor physics alone.

The foundations of quantum optics are based on systematic investigations of simple systems interacting with few quantized light modes. In this context, one can evaluate and even measure the exact eigenstates or the density matrix with respect to both the photonic and the atomic degrees of freedom. In semiconductor systems, currently the investigations using one or a few quantum dots (QD) are closest to the atomic studies because, due to their discrete eigenstates, one can treat strongly confined quantum dots to some extent like artificial atoms. However, semiconductors with at least one continuous motional degree of freedom, such as the effectively one-dimensional quantum wires (QWI), the effectively two-dimensional quantum wells (QW), or the three-dimensional bulk systems, have an astronomically large number of excitable many-body states. In such systems, only the reconstruction of single- and two-particle distributions in QWs has been established (Smith *et al.*, 2010) so far.

Typically, semiconductor quantum optics explores phenomena depending on the coupled many-body and quantum-optical multi-photon correlations rather than measuring the extremely high-dimensional density matrix directly. Since these investigations immediately invoke an immense number of available states, it is usually not sufficient to rely on single-particle concepts, simplifying Bosonic approximations, or phenomenological predictions to explain the emerging quantum-optical correlations. Instead, one must develop the means to analyze systematically and understand the quantum-optical processes in many-body systems, which makes the topic of semiconductor quantum optics both intriguing and challenging.

It should not be surprising that semiconductors can exhibit a larger variety of quantum-optical features than atoms, simply because solids have so many available states that can interact with the quantized light. Clearly, one can also view this abundance as a curse because, in comparison with simple systems, all quantitative measurements in semiconductor quantum optics are much more tedious and the interpretation of the results is much more involved. Thus, the field of semiconductor quantum optics poses many exciting challenges for theory and experiments. Currently, this research field is vigorously emerging with several first-principle demonstrations of exciting phenomena.

This book establishes the foundations of a bridge between atomic and semiconductor quantum optics. For this purpose, we rely on the systematic cluster-expansion approach which allows us to identify and describe the relevant correlations in the coupled quantized light–semiconductor system. As simple applications, we evaluate the quantum-optical features considering the doublet contributions in the incoherent regime. These investigations result in the mutually consistent treatment of the many-body state and the light-field quantum statistics.

As shown in Section 22.3.1, the classical theory follows from the singlet part of the light field that is described by the coherent state. To understand the semiconductor luminescence, we also have to include the intensity fluctuations, defined by the doublet $\Delta\langle B^\dagger B\rangle$ correlation. The emitted light can be described via the thermal state, which has a semiclassical quantum statistics, see Section 23.3.2. Including also higher-order clusters in our description, we can analyze true quantum fields that have no classical counterparts. To appreciate the most prominent prospects in semiconductor quantum optics, we briefly summarize some recent developments in this final chapter.

30.1 General singlet–doublet dynamics

So far, we have discussed quantum-optical features only in the incoherent region. However, as shown in Section 23.5, coherent interactions can produce a new class of quantum-optical effects in atomic systems already at the singlet–doublet level. Thus, we also expect additional contributions in the semiconductor luminescence equations (SLE) for coherently driven systems. For this purpose, we take another look at the main structure of the SLE (28.24)–(28.26) and express also the triplets and the coherent correlations explicitly,

$$
\begin{aligned}
i\hbar\frac{\partial\,\Pi_{\mathbf{k},\mathbf{q}_{3D}}}{\partial t} &= \left(\tilde{\epsilon}^{c}_{\mathbf{k}} - \tilde{\epsilon}^{v}_{\mathbf{k}-\mathbf{q}} - \hbar\omega_{\mathbf{q}_{3D}}\right)\Pi_{\mathbf{k},\mathbf{q}_{3D}} \\
&\quad - \left(1 - f^{e}_{\mathbf{k}} - f^{h}_{\mathbf{k}-\mathbf{q}}\right)\sum\nolimits_{\mathbf{l}} V_{\mathbf{k}-\mathbf{l}}\Pi_{\mathbf{l};\,\mathbf{q}_{3D}} + i\mathcal{F}_{\mathbf{q}_{3D}} S^{\mathrm{spnt}}_{\mathbf{k},\mathbf{q}} \\
&\quad - i\left(1 - f^{e}_{\mathbf{k}} - f^{h}_{\mathbf{k}-\mathbf{q}}\right)\Delta\left\langle B^{\dagger}_{\mathbf{q}_{3D}} B_{\mathbf{q},\Sigma}\right\rangle
\end{aligned}
$$

$$
\begin{aligned}
&+\sum_{\mathbf{l}}\left(\Delta\left\langle B^{\dagger}_{\mathbf{q}_{3\mathrm{D}}}Q^{c}_{\mathbf{l}}a^{\dagger}_{v,\mathbf{k}-\mathbf{q}}a_{c,\mathbf{k}-\mathbf{l}}\right\rangle-\Delta\left\langle B^{\dagger}_{\mathbf{q}_{3\mathrm{D}}}Q^{v}_{\mathbf{q}-\mathbf{l}}a^{\dagger}_{v,\mathbf{k}-\mathbf{l}}a_{c,\mathbf{k}}\right\rangle\right)\\
&-i\sum_{\mathbf{l}}\left(\Delta\left\langle B^{\dagger}_{\mathbf{q}_{3\mathrm{D}}}B_{\mathbf{l},\Sigma}a^{\dagger}_{v,\mathbf{k}-\mathbf{q}}a_{v,\mathbf{k}-\mathbf{l}}\right\rangle-\Delta\left\langle B^{\dagger}_{\mathbf{q}_{3\mathrm{D}}}B_{\mathbf{q}-\mathbf{l},\Sigma}a^{\dagger}_{c,\mathbf{k}-\mathbf{l}}a_{c,\mathbf{k}}\right\rangle\right)\\
&+\sum_{v,\mathbf{k}',\mathbf{l}}\Big[V_{\mathbf{l}}\Delta\left\langle B^{\dagger}_{\mathbf{q}_{3\mathrm{D}}}a^{\dagger}_{v,\mathbf{k}-\mathbf{q}}a^{\dagger}_{v,\mathbf{k}'+\mathbf{q}}a_{v,\mathbf{k}'+\mathbf{l}}a_{c,\mathbf{k}-\mathbf{l}}\right\rangle\\
&-V_{\mathbf{l}-\mathbf{q}}\Delta\left\langle B^{\dagger}_{\mathbf{q}_{3\mathrm{D}}}a^{\dagger}_{v,\mathbf{k}-\mathbf{l}}a^{\dagger}_{v,\mathbf{k}'+\mathbf{l}}a_{v,\mathbf{k}'}a_{c,\mathbf{k}}\right\rangle\Big]+D^{\mathrm{coh}}_{\mathbf{k};\,\mathbf{q}_{3\mathrm{D}}}.
\end{aligned}
\tag{30.1}
$$

Here, the two first lines contain the parts already analyzed in Chapters 28–29. The physical consequences of the remaining terms will be discussed in the following sections.

The phonon contributions that appear can be expressed via an effective field

$$
Q^{\lambda}_{\mathbf{p}}\equiv\left[D^{\lambda}_{\mathbf{p}\Sigma}+\left(D^{\lambda}_{-\mathbf{p}\Sigma}\right)^{\dagger}\right],\qquad D^{\lambda}_{\mathbf{p}\Sigma}\equiv\sum_{p_z}\mathsf{G}^{\lambda}_{\mathbf{p},p_z}D_{\mathbf{p},p_z},
\tag{30.2}
$$

where λ is the band index and $\mathsf{G}^{\lambda}_{\mathbf{p},p_z}$ defines the generic λ-dependent phonon matrix element. To simplify the notation, we use the effective light mode

$$
B_{\mathbf{q},\Sigma}\equiv\sum_{q_z}\mathcal{F}_{\mathbf{q}_{3\mathrm{D}}}B_{\mathbf{q},q_z},
\tag{30.3}
$$

where the explicit form of the light–matter coupling $\mathcal{F}_{\mathbf{q}_{3\mathrm{D}}}$ is given by Eq. (28.2).

30.1.1 Coherent coupling in the $\Pi^{v,c}$ dynamics

Generally, coherent contributions are an additional source for the photon-assisted polarization as symbolized by the term $D^{\mathrm{coh}}_{\mathbf{k};\,\mathbf{q}_{3\mathrm{D}}}$ in Eq. (30.1). Starting from Eqs. (28.3) and (28.23) and systematically identifying the singlet–doublet terms, we obtain

$$
\begin{aligned}
D^{\mathrm{coh}}_{\mathbf{k};\,\mathbf{q}_{3\mathrm{D}}}=&\;P_{\mathbf{k}}\sum_{\mathbf{l}}V_{\mathbf{l}-\mathbf{k}}\left[\Pi^{v,v}_{\mathbf{l};\,\mathbf{q}_{3\mathrm{D}}}-\Pi^{c,c}_{\mathbf{l};\,\mathbf{q}_{3\mathrm{D}}}\right]+\Omega_{\mathbf{k}-\mathbf{q}}\,\Pi^{c,c}_{\mathbf{k};\,\mathbf{q}_{3\mathrm{D}}}-\Omega_{\mathbf{k}}\,\Pi^{v,v}_{\mathbf{k};\,\mathbf{q}_{3\mathrm{D}}}\\
&+V_{\mathbf{q}}\left(P_{\mathbf{k}-\mathbf{q}}-P_{\mathbf{k}}\right)\sum_{\beta,\mathbf{l}}\Pi^{\beta,\beta}_{\mathbf{l};\,\mathbf{q}_{3\mathrm{D}}}\\
&+P_{\mathbf{k}-\mathbf{q}}\Delta\left\langle B^{\dagger}_{\mathbf{q}_{3\mathrm{D}}}Q^{c}_{\mathbf{q}}\right\rangle-P_{\mathbf{k}}\Delta\left\langle B^{\dagger}_{\mathbf{q}_{3\mathrm{D}}}Q^{v}_{\mathbf{q}}\right\rangle.
\end{aligned}
\tag{30.4}
$$

The photon-assisted carrier correlations

$$
\Pi^{\lambda',\lambda}_{\mathbf{k};\mathbf{q}_{3\mathrm{D}}}\equiv\Delta\left\langle B^{\dagger}_{\mathbf{q},q_z}a^{\dagger}_{\lambda',\mathbf{k}-\mathbf{q}}a_{\lambda,\mathbf{k}}\right\rangle
\tag{30.5}
$$

describe the correlated photon emission via an electronic transition $(\lambda,\,\mathbf{k})\rightarrow(\lambda',\,\mathbf{k}-\mathbf{q})$ in the solid. Here, $\Pi^{v,c}\equiv\Pi$ is the photon-assisted polarization discussed in the previous chapters. For coherent excitations, we additionally need $\Pi^{c,c}$ and $\Pi^{v,v}$, which describe photon-assisted density correlations. They oscillate with the driving field as $e^{i\omega t}$. Hence, both contributions are coherent according to our classification in Section 25.3.

Equation (30.4) also contains singlets in the form of the polarization $P_{\mathbf{k}}$ and the renormalized Rabi frequency

$$
\Omega_{\mathbf{k}}\equiv\langle\hat{E}^{c}\rangle+\sum_{\mathbf{k}'}V_{\mathbf{k}-\mathbf{k}'}P_{\mathbf{k}'}=i\langle B_{\mathbf{q};\,\Sigma}\rangle+\sum_{\mathbf{k}'}V_{\mathbf{k}-\mathbf{k}'}P_{\mathbf{k}'},
\tag{30.6}
$$

compare Eqs. (26.7), (26.16), and (30.3). For later reference, we introduce the band-index dependent notation,

$$\Omega_{\mathbf{k}}^{\lambda} \equiv \begin{cases} \Omega_{\mathbf{k}}, & \text{for } \lambda = c \\ \Omega_{\mathbf{k}}^{\star}, & \text{for } \lambda = v \end{cases}, \qquad \hat{E}_{\mathbf{q}}^{\lambda} \equiv \begin{cases} iB_{\mathbf{q};\,\Sigma}, & \text{for } \lambda = c \\ -iB_{-\mathbf{q};\,\Sigma}^{\dagger}, & \text{for } \lambda = v \end{cases}. \tag{30.7}$$

Here, $\hat{E}_{\mathbf{q}}^{c}$ identifies the rotating-wave part of the electric field oscillating as $e^{-i\omega t}$ and $\hat{E}_{\mathbf{q}}^{v}$ oscillates with $e^{i\omega t}$.

30.1.2 General $\Pi^{\lambda,\lambda'}$ dynamics

The generic dynamics of the photon-assisted carrier correlations can be derived starting from Eqs. (28.3) and (28.23). A calculation analogous to that resulting in Eq. (30.1) eventually yields

$$\begin{aligned} i\hbar\frac{\partial}{\partial t}\Pi_{\mathbf{k};\mathbf{q}_{3D}}^{\lambda',\lambda} = {} & \left(\tilde{\epsilon}_{\mathbf{k}}^{\lambda} - \tilde{\epsilon}_{\mathbf{k}-\mathbf{q}}^{\lambda'} - \hbar\omega_{\mathbf{q}}\right)\Pi_{\mathbf{k};\mathbf{q}_{3D}}^{\lambda',\lambda} + \left(n_{\mathbf{k}}^{\lambda} - n_{\mathbf{k}-\mathbf{q}}^{\lambda'}\right)\sum_{\mathbf{l}} V_{\mathbf{l}-\mathbf{k}}\Pi_{\mathbf{l};\mathbf{q}_{3D}}^{\lambda',\lambda} \\ & + i\mathcal{F}_{\mathbf{q}}\left[P_{\mathbf{k}}^{c,\lambda}\left(\delta_{\lambda',v} - P_{\mathbf{k}-\mathbf{q}}^{\lambda',v}\right) + \sum_{v,\mathbf{k}'} c_{c,\lambda';\lambda,v}^{\mathbf{q},\mathbf{k}-\mathbf{q},\mathbf{k}'}\right] \\ & + P_{\mathbf{k}}^{\bar{\lambda}',\lambda}\Delta\left\langle B_{\mathbf{q}_{3D}}^{\dagger}\hat{E}_{\mathbf{q}}^{\bar{\lambda}'}\right\rangle - P_{\mathbf{k}-\mathbf{q}}^{\lambda',\bar{\lambda}}\Delta\left\langle B_{\mathbf{q}_{3D}}^{\dagger}\hat{E}_{\mathbf{q}}^{\lambda}\right\rangle \\ & + P_{\mathbf{k}}^{\bar{\lambda},\lambda}\sum_{\mathbf{l}} V_{\mathbf{l}-\mathbf{k}}\Pi_{\mathbf{l};\mathbf{q}_{3D}}^{\lambda',\bar{\lambda}} - P_{\mathbf{k}}^{\lambda',\bar{\lambda}'}\sum_{\mathbf{l}} V_{\mathbf{l}-\mathbf{k}}\Pi_{\mathbf{l};\mathbf{q}_{3D}}^{\bar{\lambda}',\lambda} \\ & + \Pi_{\mathbf{k};\mathbf{q}_{3D}}^{\bar{\lambda}',\lambda}\Omega_{\mathbf{k}-\mathbf{q}}^{\bar{\lambda}'} - \Pi_{\mathbf{k};\mathbf{q}_{3D}}^{\lambda',\bar{\lambda}}\Omega_{\mathbf{k}}^{\lambda} + V_{\mathbf{q}}\left(P_{\mathbf{k}-\mathbf{q}}^{\lambda',\lambda} - P_{\mathbf{k}}^{\lambda',\lambda}\right)\sum_{v,\mathbf{l}}\Pi_{\mathbf{l};\mathbf{q}_{3D}}^{v,v} \\ & + P_{\mathbf{k}-\mathbf{q}}^{\lambda',\lambda}\Delta\left\langle B_{\mathbf{q}_{3D}}^{\dagger}Q_{\mathbf{q}}^{\lambda}\right\rangle - P_{\mathbf{k}}^{\lambda',\lambda}\Delta\langle B_{\mathbf{q}_{3D}}^{\dagger}Q_{\mathbf{q}}^{\lambda'}\rangle + T_{\mathbf{k};\mathbf{q}_{3D}}^{\lambda,\lambda'}. \end{aligned} \tag{30.8}$$

This expression shows that the quantum dynamics of all photon-assisted doublets has the same structural form, regardless of whether an interband or an intraband carrier transition is involved. Similarly, the Coulomb-induced triplet contribution $T^{\lambda,\lambda'}$ resembles the one appearing in Eq. (30.1). They produce a hierarchical coupling due to many-body and quantum-optical interactions. We do not need the explicit form of the triplets in this book, for details the interested reader may consult (Kira and Koch, 2006).

As in Eq. (30.1), also the general dynamics of $\Pi^{\lambda,\lambda'}$ contains spontaneous and stimulated contributions in the second and third line of Eq. (30.8), respectively. The contributions in lines four and five also couple $\Pi^{\lambda,\lambda'}$ to other types of $\Pi^{\alpha,\beta}$ correlations, in the same way as D^{coh} in Eq. (30.1). This coupling yields an additional mixing between coherent and incoherent correlations, yielding a new class of quantum-optical effects. As a consequence, we also have to extend the analysis of Chapters 28–29.

Generally, Eq. (30.8) shows that we have to compute $\Pi^{\lambda,\lambda'}$ for all band-index combinations when coherences are present. Furthermore, the third line of Eq. (30.8) contains pure photon-operator terms of the type $\Delta\langle B^{\dagger}B\rangle$ and $\Delta\langle B^{\dagger}B^{\dagger}\rangle$ that follow directly from Eqs. (28.25) and (30.7). From these, only $\Delta\langle B^{\dagger}B\rangle$ is incoherent and its dynamics is already included in the SLE via Eq. (28.27). The two-photon emission correlations $\Delta\langle B^{\dagger}B^{\dagger}\rangle$ are coherent and their dynamics is determined by

$$i\hbar\frac{\partial}{\partial t}\Delta\left\langle B^{\dagger}_{\mathbf{q},q_z} B^{\dagger}_{-\mathbf{q},q'_z}\right\rangle = -\hbar\left(\omega_{\mathbf{q}'_{3D}} + \omega_{\mathbf{q}_{3D}}\right)\Delta\left\langle B^{\dagger}_{\mathbf{q},q_z} B^{\dagger}_{-\mathbf{q},q'_z}\right\rangle + i\sum_{\mathbf{k}}\left[\mathcal{F}^{\star}_{\mathbf{q}_{3D}}\Pi^{c,v}_{\mathbf{q}'_{3D};\mathbf{k}-\mathbf{q}} + \mathcal{F}^{\star}_{\mathbf{q}'_{3D}}\Pi^{c,v}_{\mathbf{q}_{3D};\mathbf{k}}\right]. \tag{30.9}$$

We see that these correlations can be generated only in the coherent regime because they are exclusively driven by the coherent $\Pi^{c,v}$ correlation. Since they are directly connected via complex conjugation, the emergence of $\Delta\langle B^{\dagger} B^{\dagger}\rangle$ also implies a nonvanishing $\Delta\langle B\, B\rangle = \left[\Delta\langle B^{\dagger} B^{\dagger}\rangle\right]^{\star}$.

30.1.3 General dynamics of carrier doublets

The structural form of the dynamic equations for the pure carrier doublets $c^{\mathbf{q},\mathbf{k}',\mathbf{k}}_{\lambda,\nu;\nu'\lambda'}$ does not depend on the indices λ, ν, ν', and λ'. Therefore, we do not elaborate the dynamics for all the 16 index combinations in $c^{\mathbf{q},\mathbf{k}',\mathbf{k}}_{\lambda,\nu;\nu'\lambda'}$ but focus on the generic coupling aspects of the coherent and the incoherent correlations. For this purpose, we can directly use Eq. (27.35) that defines the quantum dynamics of the exciton correlation, $c_X \equiv c_{c,v;c,v}$, without the quantum-optical field. Equation (27.38) shows how c_X is additionally driven by the coherent correlations $c^{\mathbf{q},\mathbf{k}',\mathbf{k}}_{v,c;c,c}$ and $c^{\mathbf{q},\mathbf{k}',\mathbf{k}}_{c,v;v,v}$. Thus, the presence of coherences induces a coupling between the incoherent and the coherent correlations due to the Coulomb interaction.

Additionally, $c^{\mathbf{q},\mathbf{k}',\mathbf{k}}_{X}$ is coupled to coherent quantum-optical correlations via,

$$\hbar\frac{\partial}{\partial t}c^{\mathbf{q},\mathbf{k}',\mathbf{k}}_{\mathrm{X}}\bigg|^{\mathrm{coh}}_{\mathrm{QED}} = P_{\mathbf{k}'}\Delta\left\langle \hat{E}^{v}_{-\mathbf{j}}\, a^{\dagger}_{c,\mathbf{k}} a_{c,\mathbf{k}+\mathbf{j}}\right\rangle - P_{\mathbf{k}'+\mathbf{q}}\Delta\left\langle \hat{E}^{v}_{\mathbf{j}}\, a^{\dagger}_{v,\mathbf{k}'} a_{v,\mathbf{k}'-\mathbf{j}}\right\rangle + P^{\star}_{\mathbf{k}}\Delta\left\langle \hat{E}^{c}_{\mathbf{j}}\, a^{\dagger}_{v,\mathbf{k}'} a_{v,\mathbf{k}'-\mathbf{j}}\right\rangle - P^{\star}_{\mathbf{k}-\mathbf{q}}\Delta\left\langle \hat{E}^{c}_{-\mathbf{j}}\, a^{\dagger}_{c,\mathbf{k}} a_{c,\mathbf{k}+\mathbf{j}}\right\rangle. \tag{30.10}$$

Here, we used the abbreviation $\mathbf{j} \equiv \mathbf{k}' + \mathbf{q} - \mathbf{k}$. The explicit derivation of Eq. (30.10) can be found in (Kira and Koch, 2006) and the incoherent contributions follow from Eq. (28.12). The RHS of Eq. (30.10) contains products of the polarization and different coherent photon-assisted correlations $\Pi^{\lambda,\lambda}$ showing that also the quantum-optical coupling mixes coherent and incoherent quantities.

30.1.4 Singlet–doublet correlations and beyond

In Section 26.1, we discuss the singlet dynamics including quantum-optical correlations via Eq. (26.12). Here, the singlet structure consists of the Maxwell-semiconductor Bloch equations (MSBE) (26.2)–(26.4) and (26.8). The quantum-optical hierarchy problem couples the densities $f^{e(h)}_{\mathbf{k}}$ to incoherent $\Pi^{v,c}$ correlations while the polarization $P_{\mathbf{k}}$ is altered by a coherent $\Pi^{\lambda,\lambda}$ correlation, based on Eq. (26.12). At the same time, the photon operators have no quantum-optical coupling at the singlet level. In other words, the quantum-optical hierarchy problem follows completely from the Fermionic character of matter, compare Section 21.1.4.

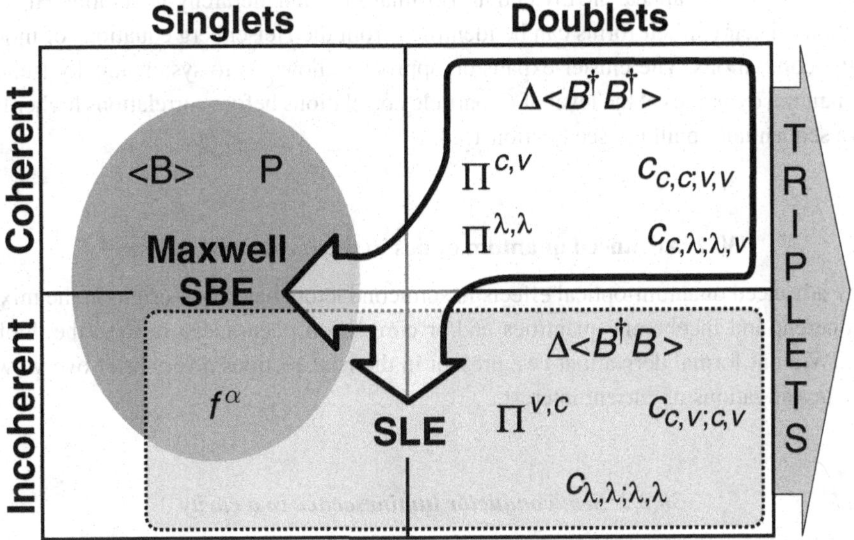

Figure 30.1 Principal structure of the singlet–doublet equations. The quantities are classified as coherent and incoherent and identified as singlets, doublets, and triplets. The core structure of the semiconductor Bloch (luminescence) equations is indicated by the shaded ellipse (dashed box). The coupling between the semiconductor Bloch and the luminescence equations follows through the coherent correlations indicated inside the box. The arrows indicate the coupling.

To provide an overview, we schematically present the relevant contributions to the MSBE and SLE in Fig. 30.1. The four-field graph identifies the singlet and doublet, as well as the coherent and incoherent terms. As we can see from the shaded ellipse, we have the electric field, the polarization, and the densities as the relevant singlet quantities in the MSBE (26.2)–(26.4) and (26.8), while only the densities appear as singlet terms in the SLE (dashed box). Whenever coherences are present, they lead to couplings between all singlets and doublets. The pure Coulombic terms introduce excitation-induced dephasing and relaxation for the singlet dynamics, as discussed in Section 26.5. The $\Pi^{\lambda,\lambda'}$ correlations produce quantum-optical corrections to the motion of the Bloch vector, i.e., $P_{\mathbf{k}}$ and $f^{\alpha}_{\mathbf{k}}$, see Eq. (26.17) as well as the discussion in Chapter 23.

Analyzing the doublet dynamics and noting that all the doublet terms containing carrier operators are additionally coupled to triplets due to both Coulomb and quantum-optical correlations, we eventually identify the equation structure

$$\begin{cases} i\hbar\frac{\partial}{\partial t}\langle 1\rangle = O_1\left[\langle 1\rangle\right] + \mathrm{Hi}_1\left[\langle 1\rangle, \Delta\langle 2\rangle\right] \\ i\hbar\frac{\partial}{\partial t}\Delta\langle 2\rangle = O_2\left[\langle 1\rangle, \Delta\langle 2\rangle\right] + \mathrm{Hi}_2\left[\Delta\langle 3\rangle\right] \\ i\hbar\frac{\partial}{\partial t}\Delta\langle 3\rangle = O_3\left[\langle 1\rangle, \langle 2\rangle, \Delta\langle 3\rangle\right] + \mathrm{Hi}_3\left[\Delta\langle 4\rangle\right] \\ \qquad\cdots \end{cases} . \tag{30.11}$$

The functionals that appear are divided into ordinary O_j and hierarchy-generating Hi_j contributions. Their explicit forms can be identified from the Heisenberg equations of motion for the correlations. The cluster-expansion approach allows us to systematically truncate the quantum dynamics at the level of C-particle correlations before correlations higher than C are sequentially built up, see Section 15.4.

30.2 Advanced quantum optics in the incoherent regime

Many advanced quantum-optical effects in semiconductors have their origin in the mixing of coherent and incoherent quantities and/or correlation phenomena beyond the doublet level. Without formal derivations, we present in the next sections a very brief overview of some investigations of current interest.

30.2.1 Semiconductor luminescence in a cavity

It is well known from traditional quantum optics that the coupling of atoms to a high-quality cavity makes it possible to observe a number of interesting phenomena. To study similar effects in semiconductors, we have to extend the treatment of Chapters 28–29 to include also the stimulated contribution in Eq. (30.1). The SLE then describe, e.g., a planar material that is placed inside a cavity such that the emitted photons return back to the semiconductor. This cavity feedback gives rise to a significant stimulated contribution $\Delta\langle B^{\dagger}_{\mathbf{q}_{3D}} B_{\mathbf{q},\Sigma}\rangle$ making it necessary to evaluate both the semiconductor correlations and the evolution of the collective mode $B_{\mathbf{q},\Sigma}$. Generally, such studies can be done only numerically (Kira *et al.*, 1999) unless some approximations are implemented.

From a pragmatic point of view, the cavity simply modifies $\mathcal{F}_{\mathbf{q}_{3D}}$ into a form where it is strongly peaked around the wave vector $q_z = q_c$ corresponding to the eigenmode. Under suitable conditions, the coupling between the cavity mode and the excitonic resonance can lead to a double-peaked spectrum in the photoluminescence, the reflection, and transmission. More generally, this scenario is often referred to as *normal-mode coupling* (NMC) or vacuum Rabi splitting when only one two-level system is coupled with the cavity. As an important step in quantum optics, vacuum Rabi splitting was first measured in (Thompson *et al.*, 1992). At the same time, NMC was detected (Weisbuch *et al.*, 1992) in a semiconductor microcavity having quantum wells. Semiconductor vacuum Rabi splitting was later demonstrated in (Reithmaier *et al.*, 2004; Yoshie *et al.*, 2004) where a single quantum dot was placed inside a cavity.

Even when the system is not in the NMC regime, the cavity considerably alters the radiative environment of the QW. This not only leads to characteristic spectral changes but also modifies the radiative decay rates. As it turns out, the radiative recombination rates can be enhanced or suppressed by appropriate positioning of the planar semiconductor structure. More efficient radiative decay is observed if the emitting system is placed at the maximum of the cavity mode. Alternatively, the recombination is suppressed by positioning the system at the node. These modifications of the spontaneous emission via the radiative environment are often referred to as the Purcell effect (Purcell, 1946).

30.2.2 Interference effects in incoherent luminescence

One of the most curious features of quantum mechanics is the occurrence of entangled states and the possibility of a quantum particle being at different positions at the same time. Both phenomena are important in so-called which-way experiments where, e.g., a light beam is first split into two partial beams that show interference fringes according to the relative path difference when recombined into a common detector. The most counter-intuitive properties are that: (a) these interference fringes persist even when the intensity of the light beam is lowered to the single-photon level and (b) the interference fringes go away when two entangled particles are used and a measurement is performed on one of them, which in principle allows us to say which path the other particle took.

The lack of translational invariance perpendicular to the plane of a single QW causes equal probability for spontaneous emission to the left or right of the QW allowing for a new form of which-way experiment. Here, the mesoscopic semiconductor QW system is entangled with its spontaneously emitted photons. Experiments of this kind are reported in (Hoyer *et al.*, 2004) where the normal QW emission to both sides is directed onto a common detector. These two emission paths yield either a clear or no interference pattern depending on whether the emission paths are indistinguishable or distinguishable, respectively. Hence, this experiment with incoherent QW emission demonstrates the analogy to the traditional which-way experiments. As an added bonus, the use of quantum wells makes it possible to study multiple-quantum-well structures. Interestingly, the presence or absence of interference fringes is strongly sensitive on the QW spacing. The SLE also explain a more general class of interference effects in incoherent QW emission. For example, the multi-QW luminescence can also exhibit sub- and superradiant effects, i.e., suppression or enhancement of radiative decay due to mutual coupling of QWs, as shown in (Kira *et al.*, 1999).

30.2.3 Phonon sidebands

Besides electron–hole pairs and/or excitons, the quantum-optical emission processes may also involve phonons. In particular, optical phonons, which typically have a nearly constant energy $\hbar\Omega$ in the range of tens of meV, can participate in triplet correlations of the type $\Delta\langle D^\dagger B^\dagger a_v^\dagger a_c\rangle$. As we can see in the fourth line of Eq. (30.1), the phonon-assisted processes produce a triplet source whose energy is reduced by $\hbar\Omega$ relative to that of $\Delta\langle B^\dagger a_v^\dagger a_c\rangle$. Thus, this triplet term describes phonon-assisted processes where the emitted photons have an energy $\hbar\omega = E_{1s} - \hbar\Omega$. Since $\hbar\Omega$ is discrete, one can detect the so-called phonon sideband resonance in polar semiconductor materials energetically well below the usual $1s$-exciton line.

The cluster-expansion analysis of the phonon sideband emission is developed in (Feldtmann *et al.*, 2010). As for the ordinary, i.e., zero-phonon luminescence, the spontaneous source for the phonon-sideband emission is identified as $\Delta N_\lambda(\mathbf{q}) + N_\lambda^S(\mathbf{q})$ showing that both plasma and excitons can contribute. The overall energy conservation is always satisfied via the many-body interactions in the emitting system, compare Chapter 29.

30.2.4 Quantum-dot emission

Quantum dots are conceptually somewhat simpler than extended semiconductors because the electronic states have fewer configuration possibilities. One can, e.g., compute the exact many-body wave functions of isolated dots containing N_e electrons and N_h holes using the method of direct diagonalization. However, real QDs are often grown onto a quantum well, the so-called "wetting layer." As a result, such QDs are electronically open systems because of the possible carrier transfer into and out of the wetting layer and the coupling to the phonons of the surrounding host material. Also in this situation, the cluster-expansion approach allows for a systematic treatment of all the different correlation levels. For example, the semiconductor luminescence equations (Feldtmann *et al.*, 2006) have successfully been adapted and applied to explain nontrivial features in the quantum-dot luminescence. As for extended solids, also the "excitonic luminescence" of open QDs may originate from exciton and/or plasma populations. This observation explains, e.g., why the radiative recombination displays strong nonlinearities for elevated excitation levels, compare the discussion in Section 29.2.1.

In the current literature, an increasing number of investigations is reported where advanced quantum-optical effects are studied in QD systems. Examples are the generation of photons-on-demand and photon antibunching effects (Michler *et al.*, 2000), entanglement in the exciton-biexciton cascade (Akopian *et al.*, 2006; Stevenson *et al.*, 2006), the photon statistics of QD microcavity lasers (Ulrich *et al.*, 2007), as well as strong-coupling investigations (Reithmaier *et al.*, 2004; Yoshie *et al.*, 2004). So far, the higher rungs of the Jaynes–Cummings ladder could not be observed, mostly because of the prohibitive influence of the dephasing effects. However, in (Schneebeli *et al.*, 2008), a scheme has been published that should allow the experimentalists to overcome these obstacles.

30.3 Advanced quantum optics in the coherent regime

A mixing of coherent and incoherent correlations occurs in situations where the semiconductor is excited with a coherent light source, i.e., a source where $\langle B \rangle$ is finite. From quantum-optical studies of atoms, it is known that a resonantly driven system can display remarkably strong effects in the resonance fluorescence. Important examples are the observations of squeezing and entanglement in atomic systems (Carmichael and Walls, 1970; Kimble *et al.*, 1977). In semiconductor quantum optics, the coherent contributions

$\Delta\langle B^\dagger B^\dagger\rangle$, $\Pi^{c,v}$, and $\Pi^{\lambda,\lambda}$ provide significant modifications of the emission properties. Thus, the experimental investigation of resonantly driven systems, especially of their resonance fluorescence properties, should provide access to advanced quantum-optical effects also in solids.

30.3.1 Squeezing in the resonance fluorescence

Quantization effects of the light fields can be investigated by monitoring the fluctuations directly. Here, one uses the single-mode quadratures

$$\hat{X} \equiv \frac{B + B^\dagger}{2}, \qquad \hat{Y} \equiv \frac{B - B^\dagger}{2i} \tag{30.12}$$

as the relevant observables. These can be measured, e.g., by using the method of homodyne detection, as discussed in Section 23.1. It is straightforward to show that $\hat{X}$ and $\hat{Y}$ satisfy the canonical commutation relation

$$\left[\hat{X}, \hat{Y}\right]_- = \frac{i}{2}. \tag{30.13}$$

As a consequence of the generic limitation (5.50), the quadrature fluctuations must satisfy the Heisenberg uncertainty relation

$$\Delta X \, \Delta Y \geq \frac{1}{4}. \tag{30.14}$$

Using Eq. (30.12), we can express the quadrature fluctuations as

$$\begin{cases} \Delta X^2 = \langle [\hat{X} - \langle X\rangle]^2\rangle = \frac{1}{4} + \frac{1}{2}\Delta\langle B^\dagger B\rangle + \frac{1}{2}\mathrm{Re}\left[\Delta\langle B^\dagger B^\dagger\rangle\right] \\ \Delta Y^2 = \langle [\hat{Y} - \langle Y\rangle]^2\rangle = \frac{1}{4} + \frac{1}{2}\Delta\langle B^\dagger B\rangle - \frac{1}{2}\mathrm{Re}\left[\Delta\langle B^\dagger B^\dagger\rangle\right] \end{cases}. \tag{30.15}$$

These expressions contain only the two-particle correlations of the light field. In addition to $\Delta\langle B^\dagger B\rangle$, which is determined by the semiconductor luminescence equations (30.8)–(30.9), we see from Eq. (30.15) that the existence of a two-photon-emission correlation $\mathrm{Re}\left[\Delta\langle B^\dagger B^\dagger\rangle\right]$ always implies $\Delta X \neq \Delta Y$. Such a case is known as light-field squeezing. In cases where the fluctuations in one of the quadratures are squeezed below the minimum uncertainty limit, i.e., $\Delta X < 1/2$ or $\Delta Y < 1/2$, this effect has no classical counterpart.

To clearly identify squeezing, it is beneficial to analyze the quadrature fluctuations in such emission directions where no classical field exists. This situation can be realized in planar quantum-well structures which are nearly free of disorder. In such systems, a light pulse propagating perpendicular to the structure leads to transmission and reflection of nearly classical light only in directions parallel to the excitation axis. If the detection is performed under an angle, this so-called secondary emission contains no coherent components and originates exclusively from clusters higher than singlets. Thus, the secondary emission can display strong quantum-optical effects under suitable conditions.

From atomic optics it is well known that the resonance fluorescence shows pronounced quantum signatures when the regime of Rabi flopping is reached (Kimble *et al.*, 1977).

Thus, it is meaningful to consider this situation also for semiconductors. For this purpose, one may use a relatively strong, resonant excitation pulse to produce Rabi flopping of the carrier density. This feature stems from the Fermi nature, i.e., the Pauli blocking effect of the electron–hole excitations. Whenever the states become almost fully excited, the probability for stimulated recombination exceeds the absorption probability. For a de-excited system, only photon absorption is possible. As a result, we observe periodic excitation and de-excitation analogous to the Rabi flopping of a strongly excited two-level system. The fully microscopic analysis (Kira *et al.*, 1999a) shows that the two-photon emission correlation $|\Delta\langle B^\dagger B^\dagger\rangle|$ can then become larger than $\Delta\langle B^\dagger B\rangle$, implying distinct quantum-mechanical squeezing of the semiconductor emission.

30.3.2 Coherent quantum-optical correlations

When a semiconductor is excited with an external light pulse, entanglement-related correlations couple the classical and the quantum emission dynamics. The resulting interplay between the semiconductor Bloch and luminescence equations is mainly mediated by the correlations between photons and electron/hole densities. These effects can be studied, e.g., in a microcavity configuration where a strong pump pulse generates large quantum correlations whose effect is then measured by analyzing the response to a weak probe beam.

The investigations in (Lee *et al.*, 1999) show that the probe reflection displays long-living oscillations as a function of the pump–probe time delay, i.e., the phase difference between the pump and the probe. The microscopic analysis reveals that the oscillatory probe reflection can be explained only by properly including the quantum-optical correlations of the type $\Pi^{\lambda,\lambda}$ in the theory. If one omits the $\Pi^{\lambda,\lambda}$ contributions, the computed probe reflection becomes independent of the pump–probe phase difference. Thus, the observations in (Lee *et al.*, 1999) constitute a direct observation of quantum-optical correlation effects, which are specific for semiconductor systems.

The contributions described by $\Pi^{\lambda,\lambda}$ can also alter the NMC characteristics. The experiment–theory comparison in (Ell *et al.*, 2000) analyzes a situation where the pump pulse is energetically located between the two NMC resonances inducing an additional third peak. Its energetic position and magnitude are determined by the pump pulse. Furthermore, a decrease of the probe intensity increases the relative effect of the correlations such that the third peak grows in the probe transmission. Properly including the $\Pi^{\lambda,\lambda}$ correlations generated by the pump pulse, the theory nicely reproduces all the observed experimental features.

30.3.3 Quantum-optical spectroscopy

Chapter 23 introduces the concept of quantum-optical spectroscopy, which is based on a simple two-way principle. In a direct way, the quantum statistics of the light source defines the quantum state of the generated excitations, which may then be modified by the many-body interactions. In the opposite way, the existing excitation state determines the quantum statistics of the emitted light. The quantum-optical investigations summarized above clearly satisfy some of these generic aspects. However, quantum-optical spectroscopy defines the general framework to explore new quantum-optical effects in many-body systems.

Quantum-optical spectroscopy is particularly interesting in extended semiconductor systems because they have an infinite variety of many-body configurations that can couple to the light field. As suggested in Section 23.1, one may look for nonlinear Fermionic effects that depend on the quantum statistics of the light source. In other words, one wants to characterize and control the nonlinear quantum dynamics and response of solids using the connection between the quantum-optical and the many-body correlations.

Theoretical investigations of this scheme are presented in (Kira and Koch, 2006a) modeling a planar quantum-wire structure resonantly excited with pulsed single-mode light. It is shown that the quantum statistics of the exciting single-mode field is directly mapped into the many-body correlations of the extended system,

$$\Delta\langle\left[B^{\dagger}_{\mathbf{q}\Sigma}\right]^{J}\left[B_{\mathbf{q}\Sigma}\right]^{K}\rangle \rightarrow \Delta\langle\left[X^{\dagger}_{1s,\mathbf{q}}\right]^{J}\left[X_{1s,\mathbf{q}}\right]^{K}\rangle. \tag{30.16}$$

This mapping follows directly from the microscopic light–matter interaction that converts photons with energy $\hbar\omega_{1s}$ into states with matching energy and momentum. It is satisfied before significant Coulomb and/or phonon scattering occurs.

Figures 30.2(a)–(b) illustrate the light–matter–light conversion process for a single-mode excitation source where only one $\mathbf{q}$ state is occupied. While this aspect is completely trivial for any single-mode laser source, it is unusual for the matter excitations to be in such a quantum-degenerate single-momentum state. Since the light field excites the

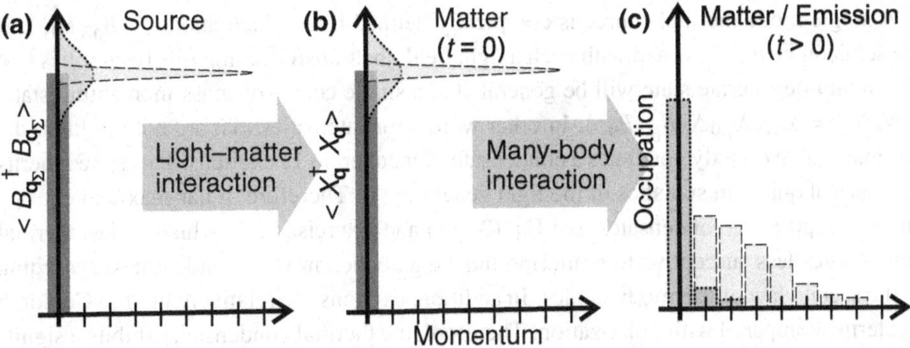

Figure 30.2 Concept of quantum-optical spectroscopy of an extended state. **(a)** The quantum statistics of the exciting light field is schematically presented as a narrow or a broad distribution (dotted and dashed area). **(b)** The induced quantum statistics of the matter excitations before the onset of scattering is defined by the exciting light field. **(c)** The resulting matter state evolving from the many-body dynamics. The matter quantum statistics is mapped into the spontaneously emitted light.

matter excitations directly into this quantum-degenerate state, it is clear that the transformation (30.16) has no connection with the thermodynamic Bose–Einstein condensation. As long as the excitation densities are not too high, the Fermionic substructure of the excitons yields only minor corrections in the generation process. However, the Fermionic aspects are critically important in the many-body dynamics following the excitation of the quantum-degenerate state.

The transformation (30.16) also predicts that the light-field quantum fluctuations are directly converted into equivalent many-body correlations, which has interesting consequences much beyond the "trivial" generation of the quantum-degenerate state. Figures 30.2(a)–(b) display two representative distributions (dashed and dotted area) of the light and the created many-body state. The generated many-body correlations then determine how the semiconductor excitations evolve in time. Especially, the nonlinear Coulomb scattering strongly depends on the quantum statistics even when the electron–hole number is kept constant. Figure 30.2(c) illustrates the many-body state at later times (dashed and dotted bars) after two quantum-statistically different excitations. The resulting excitation state is then directly mapped into the quantum statistics of the emission and is thus directly characterizable through a quantum-statistical measurement of the emitted light.

This concept is illustrated in (Kira and Koch, 2006a) by evaluating the many-body dynamics after the excitation with a single-mode coherent-light and a thermal-light source. Since the coherent excitation is completely defined by the singlets $\langle B_{\mathbf{0}_\Sigma}\rangle$, it induces only $\langle X_{1s,\mathbf{0}}\rangle$, according to Eq. (30.16). Since $\langle X_{1s,\mathbf{0}}\rangle$ corresponds to a coherent polarization, we see again that coherent light first generates a coherent polarization before the onset of scattering. This is in full agreement with the coherent limit studied in Section 27.1.1. The many-body scattering converts the polarization into incoherent excitons in a broad distribution of momentum states (dashed bars in Fig. 30.2(c)), as discussed in Section 27.2.2.

A single-mode thermal source is completely defined by its fluctuations $\Delta\langle B^\dagger_{\mathbf{0}_\Sigma} B_{\mathbf{0}_\Sigma}\rangle$. If the semiconductor is excited with such a light field, the transformation (30.16) implies that a quantum-degenerate state will be generated at a single center-of-mass-momentum state, $\Delta N_\lambda(\mathbf{q}) = \delta_{\lambda,1s}\,\delta_{\mathbf{q},0}\Delta\langle B^\dagger_{\mathbf{0}_\Sigma} B_{\mathbf{0}_\Sigma}\rangle$. In other words, for intensities that are not too high, the thermal source directly generates a condensate of incoherent $1s$ excitons. This state inherits the thermal quantum statistics of the light generating it. Therefore, it has maximal entropy among all possible condensates, see Eq. (23.18) and Exercise 30.7, which makes thermal condensates less susceptive to scattering than, e.g., coherent state condensates, according to the second law of thermodynamics. In addition, excitons are relatively inert to Coulomb scattering compared with polarization. Therefore, the thermal condensate exhibits a significantly reduced nonlinear Coulomb scattering in comparison with the strong polarization scattering observed for the coherent-excitation case. In other words, thermal excitation yields a distribution $\Delta N_{1s}(\mathbf{q})$ that stays close to a quantum-degenerate state (dotted bars in Fig. 30.2(c)), unlike the many-body state evolution following the excitation by a coherent light source.

The example above nicely demonstrates the principles of quantum-optical spectroscopy: (i) Quantum light can be applied to generate semiconductor quasiparticle excitations in a well-defined quantum state. (ii) The quantum statistics of the generated state influences its many-body interactions while (iii) the properties of the state and its subsequent evolution show up in the quantum emission. In extended systems, the mapping from the source to the many-body quantum statistics follows from Eq. (30.16) while the onset of many-body interactions, especially the nonlinearities, depend strongly on the quantum-statistical state generated.

30.3.4 Quantum optics in simple vs. complicated systems

It is instructive to apply the principles of quantum-optical spectroscopy also to the more traditional quantum-optical investigations in atomic or ionic systems. However, we must then keep in mind that in contrast to semiconductors, where the optical excitation generates the many-body state, optical interactions do not generate or destroy atoms, but merely modify their electronic configuration. In other words, the excitation scheme does not follow from Eq. (30.16). Moreover, the quantum statistics of the Bosonic light cannot be mapped into the few available degrees of freedom. For example, any two-level system constitutes an ideal Fermion, which cannot store Bosonic features. Instead, the light–matter interaction in the two-level case involves a Boson–Fermion mapping that inherently distorts the quantum statistics. As a result, already the light–matter coupling directly introduces an abundance of quantum-optical effects, both in the matter excitations and in the resulting light emission. Unlike the many-body system in semiconductors, there is no need for interactions among the excited entities in atomic systems to generate nontrivial quantum-optical effects. In fact, such matter interactions are often harmful because they induce, e.g., dephasing that masks the quantum-optical effects.

The Fermion–Boson coupling of simple systems is the most obvious source for their nonlinear and quantum-optical properties. Due to the existence of the Jaynes–Cummings ladder, e.g., a single- and a two-photon source will produce very different responses, compare Section 21.2.3. In complicated systems, the Fermionic aspects also produce nonlinear features via the Pauli blocking, many-body interaction, and the different light–matter correlations. Therefore, the concept of quantum-optical spectroscopy actually establishes the general framework where these nonlinearities are utilized, controlled, and characterized regardless of the system's complexity.

Especially, both simple and complicated systems display the most spectacular quantum-optical effects when the nonlinear Fermion effects dominate. One can already reach this regime in simple systems because few-photon quantum sources are sufficient to generate Fermionic nonlinearities in few-level systems. In complicated systems, one needs strong-intensity quantum sources to access and to control nonlinearities, which makes these investigations challenging. We show in (Kira and Koch, 2006) how the principles of quantum-optical spectroscopy can be realized to follow and to control nonlinear quantum dynamics of many-body states with the highest possible precision.

Exercises

30.1 Derive the general Π dynamics (30.1) under coherent and homogeneous excitation conditions. **Hint:** *Start from Eqs.* (28.3) *and* (28.23) *and identify all relevant clusters.*

30.2 **(a)** Start from Eqs. (24.27) and (24.30) and derive the generic $\Pi^{\lambda',\lambda}$ dynamics. **(b)** Derive dynamics (30.9).

30.3 Derive the general dynamics of $c_X^{\mathbf{q},\mathbf{k}',\mathbf{k}}$ due to light–matter interaction. Especially, verify Eq. (30.10).

30.4 **(a)** Confirm that the generic singlet–doublet dynamics has the structure illustrated in Fig. 30.1. **(b)** What does functional $\mathrm{O}_2[\ldots]$ and $\mathrm{Hi}_2[\ldots]$ look like for $\Pi^{v,c}$ based on Eq. (30.1)?

30.5 Assume that two QWs contain incoherent excitations. Show that the QWs are coupled via the correction $\Delta\langle \hat{P}^{\dagger}_{\mathbf{k},1}\, \hat{P}_{\mathbf{k},2}\rangle$ where $\hat{P}_{\mathbf{k},n}$ is the polarization operator for QW_j. Show that this coupling depends on the QW spacing.

30.6 Verify the result (30.15). Which condition is posed between $\Delta\langle B^{\dagger}\, B\rangle$ and $\Delta\langle B\, B\rangle$ due to the Heisenberg uncertainty relation?

30.7 Compute the entropy of coherent vs. thermal exciton condensates. Conclude that thermal condensates maximize the entropy among all possible condensates. **Hint:** *Maximize* $S = -k_B\langle \ln\hat{\rho}\rangle$ *with the constraint of a fixed average particle number for a given quantum-degenerate state.* Show that the variational principle yields Eq. (23.18).

Further reading

It is impossible to present a fair and comprehensive list of references concerning all the relevant work in quantum optics within the constraints of this book. Therefore, we only briefly highlight some research that illustrates the long evolutionary arch of "traditional" quantum optics as a perpetual generator of new concepts and demonstrations. The quantum nature of light was first encountered in:

M. Planck (1901). Über das Gesetz der Energieverteilung im Normalspectrum (On the law of distribution of energy in the normal spectrum), *Ann. Phys.* **309**, 553.

The development of the laser was the single most important invention in bringing quantum optics to life. This development was possible due to pioneering work dating already from the early 1900s. For instance, Albert Einstein, Willis E. Lamb, Theodore H. Mainman, Charles H. Townes, Nikolay Basov, Aleksandr Prokhorov, Arthur L. Schawlow, Gordon Gould, Ali Javan, William R. Bennett, Donald Herriot, Robert N. Hall, Nick Holonyak,

Jr., and Zhores Alferov had a major impact in the early development of the laser until the 1960s. Based on this early work, quantum optics started to develop into multiple directions:

E. M. Purcell (1946). Spontaneous emission probabilities at radio frequencies, *Phys. Rev.* **69**, 681.

R. Hanbury Brown and R. Q. Twiss (1956). A test of a new type of stellar interferometer on Sirius, *Nature* **178**, 1046.

R. J. Glauber (1963). Coherent and incoherent states of the radiation field, *Phys. Rev.* **131**, 2766.

E. T. Jaynes and F. W. Cummings (1963). Comparison of quantum and semiclassical radiation theories with application to the beam maser, *Proc. IEEE* **51**, 89.

B. R. Mollow (1969). Power spectrum of light scattered by two-level systems, *Phys. Rev.* **188**, 1969.

S. J. Freedman and J. F. Clauser (1972). Experimental test of local hidden-variable theories, *Phys. Rev. Lett.* **28**, 938.

H. J. Carmichael and D. F. Walls (1976). A quantum-mechanical master equation treatment of the dynamical Stark effect, *J. Phys. B* **9**, 1199.

H. J. Kimble, M. Dagenais, and L. Mandel (1977). Photon anti-bunching in resonance fluorescence, *Phys. Rev. Lett.* **39**, 691.

C. M. Caves (1981). Quantum-mechanical noise in an interferometer, *Phys. Rev. D* **23**, 1693.

A. Aspect, P. Grangier, and R. Gérard (1981). Experimental tests of realistic local theories via Bell's Theorem, *Phys. Rev. Lett.* **47**, 460.

D. F. Walls and P. Zoller (1981). Reduced quantum fluctuations in resonance fluorescence, *Phys. Rev. Lett.* **47**, 709.

M. O. Scully and K. Drühl (1982). Quantum eraser: A proposed photon correlation experiment concerning observation and "delayed choice" in quantum mechanics, *Phys. Rev. A* **25**, 2208.

D. F. Walls (1983). Squeezed states of light, *Nature* **306**, 141.

R. E. Slusher, L. W. Hollberg, B. Yurke and J. C. Mertz (1985). Observation of squeezed states generated by four-wave mixing in an optical cavity, *Phys. Rev. Lett.* **55**, 2409.

D. Meschede, H. Walther, and G. Müller (1985). One-atom maser, *Phys. Rev. Lett.* **54**, 551.

P. Filipowicz, J. Javanainen, and P. Meystre (1986). Theory of a microscopic maser, *Phys. Rev. A* **34**, 3077.

L.-A. Wu, H. J. Kimble, J. L. Hall, and H. Wu (1986). Generation of squeezed states by parametric down conversion, *Phys. Rev. Lett.* **57**, 2520.

C. K. Hong, Z. Y. Ou, and L. Mandel (1987). Measurement of subpicosecond time intervals between two photons by interference, *Phys. Rev. Lett.* **59**, 2044.

R. Loudon and P. L. Knight (1987). Squeezed light, *J. Mod. Opt.* **34**, 709.

G. Rempe, H. Walther, and N. Klein (1987). Observation of quantum collapse and revival in a one-atom maser, *Phys. Rev. Lett.* **58**, 353.

W. Schleich and J. A. Wheeler (1987). Oscillations in photon distribution of squeezed states and interference in phase-space, *Nature* **326**, 574.

K. Vogel and H. Risken (1989). Determination of quasiprobability distributions in terms of probability distributions for the rotated quadrature phase, *Phys. Rev. A* **40**, 2847.

K. J. Boller, A. Imamoglu, and S. E. Harris (1991). Observation of electromagnetically induced transparency, *Phys. Rev. Lett.* **66**, 2593.

R. J. Thompson, G. Rempe, and H. J. Kimble (1992). Observation of normal-mode splitting for an atom in an optical cavity, *Phys. Rev. Lett.* **68**, 1132.

D. T. Smithey, M. Beck, M. G. Raymer and A. Faridani (1993). Measurement of the Wigner distribution and the density matrix of a light mode using optical homodyne tomography: Application to squeezed states and the vacuum, *Phys. Rev. Lett.* **70**, 1244.

P. W. Shor (1994). Algorithms for quantum computation: Discrete logarithms and factoring, in *Proceedings of the 35th Annual Symposium on the Foundations of Computer Science*, S. Goldwasser, ed., Los Alamitos, CA, IEEE Computer Society.

M. H. Anderson, J. R. Ensher, M. R. Matthews, C. E. Wieman, and E. A. Cornell (1995). Observation of Bose–Einstein condensation in a dilute atomic vapor, *Science* **269**, 198.

K. B. Davis, M.-O. Mewes, M. R. Andrews, *et al.* (1995). Bose–Einstein condensation in a gas of sodium atoms, *Phys. Rev. Lett.* **75**, 3969.

J. I. Cirac and P. Zoller (1995). Quantum computations with cold trapped ions, *Phys. Rev. Lett.* **74**, 4091.

P. G. Kwiat, K. Mattle, H. Weinfurter, *et al.* (1995). New high-intensity source of polarization-entangled photon pairs, *Phys. Rev. Lett.* **75**, 4337.

C. Monroe, D. M. Meekhof, B. E. King, W. M. Itano, and D. J. Wineland (1995). Demonstration of a fundamental quantum logic gate, *Phys. Rev. Lett.* **75**, 4714.

M. Brune, F. Schmidt-Kaler, A. Maali, *et al.* (1996). Quantum Rabi oscillation: A direct test of field quantization in a cavity, *Phys. Rev. Lett.* **76**, 1800.

D. Deutsch, A. Ekert, R. Josza, *et al.* (1996). Quantum privacy amplification and the security of quantum cryptography over noisy channels, *Phys. Rev. Lett.* **77**, 2818.

V. Vedral, M. B. Plenio, M. A. Rippin, and P. L. Knight (1997). Quantifying entanglement, *Phys. Rev. Lett.* **78**, 2275.

G. J. Milburn, J. Corney, E. M. Wright, and D. F. Walls (1997). Quantum dynamics of an atomic Bose–Einstein condensate in a double-well potential, *Phys. Rev. A* **55**, 4318.

D. Bouwmeester, J.-W. Pan, K. Mattle, *et al.* (1997). Experimental quantum teleportation, *Nature* **390**, 575.

A. Furusawa, J. L. Sorensen, S. L. Braunstein, *et al.* (1998). Unconditional quantum teleportation, *Science* **282**, 706.

L. Vestergaard Hau, S. E. Harris, Z. Dutton, and C. H. Behrooz (1999). Light speed reduction to 17 metres per second in an ultracold atomic gas, *Nature* **397**, 594.

M. Greiner, O. Mandel, T. Esslinger, T. W. Hänsch, and I. Bloch (2002). Quantum phase transition from a superfluid to a Mott insulator in a gas of ultracold atoms, *Nature* **415**, 39.

M. S. Bigelow, N. N. Lepeshkin, and R. W. Boyd (2003). Superluminal and slow light propagation in a room-temperature solid, *Science* **301**, 200.

A. Ourjoumtsev, R. Tualle-Brouri, J. Laurat, and P. Grangier (2006). Generating optical Schrödinger kittens for quantum information processing, *Science* **312**, 83.

A. Kubanek, M. Koch, C. Sames, *et al.* (2009). Photon-by-photon feedback control of a single-atom trajectory, *Nature* **462**, 898.

K. Hammerer, A. S. Sorensen, and E. S. Polzik (2010). Quantum interface between light and atomic ensembles, *Rev. Mod. Phys.* **82**, 1041.

Semiconductor quantum optics is a strongly emerging field as illustrated by the following examples:

S. Machida, Y. Yamamoto, and Y. Itaya (1987). Observation of amplitude squeezing in a constant-current-driven semiconductor laser, *Phys. Rev. Lett.* **58**, 1000.

V. Langer, H. Stolz, and W. von der Osten (1990). Observation of quantum beats in the resonance fluorescence of free excitons, *Phys. Rev. Lett.* **64**, 854.

C. Weisbuch, M. Nishioka, A. Ishikawa, and Y. Arakawa (1992). Observation of the coupled exciton–photon mode splitting in a semiconductor quantum microcavity, *Phys. Rev. Lett.* **69**, 3314.

M. Kira, F. Jahnke, S. W. Koch, *et al.* (1997). Quantum theory of nonlinear semiconductor microcavity luminescence explaining "Boser" experiments, *Phys. Rev. Lett.* **79**, 5170.

N. H. Bonadeo, J. Erland, D. Gammon, *et al.* (1998). Coherent optical control of the quantum state of a single quantum dot, *Science* **282**, 1473.

M. Kira, F. Jahnke, and S. W. Koch (1999a). Quantum theory of secondary emission in optically excited semiconductor quantum wells, *Phys.Rev. Lett.* **82**, 3544.

A. Imamoglu, D. D. Awschalom, G. Burkard, *et al.* (1999). Quantum information processing using quantum dot spins and cavity QED, *Phys. Rev. Lett.* **83**, 4204.

M. Kira, F. Jahnke, W. Hoyer, and S. W. Koch (1999). Quantum theory of spontaneous emission and coherent effects in semiconductor microstructures, *Prog. Quantum Electron.* **23**, 189.

Y.-S. Lee, T. B. Norris, M. Kira, *et al.* (1999). Quantum correlations and intraband coherences in semiconductor cavity QED, *Phys. Rev. Lett.* **83**, 5338.

P. G. Savvidis, J. J. Baumberg, R. M. Stevenson, *et al.* (2000). Angle-resonant stimulated polariton simplifier, *Phys. Rev. Lett.* **84**, 1547.

E. Biolatti, R. C. Iotti, P. Zanardi, and F. Rossi (2000). Quantum information processing with semiconductor macroatoms, *Phys. Rev. Lett.* **85**, 5647.

C. Ell, P. Brick, M. Hübner, E. S. Lee, *et al.* (2000). Quantum correlations in the nonperturbative regime of semiconductor microcavities, *Phys. Rev. Lett.* **85**, 5392.

P. Michler, A. Kiraz, C. Becher, *et al.* (2000). A quantum dot single-photon turnstile device, *Science* **290**, 2282.

W. Hoyer, M. Kira, S. W. Koch, *et al.* (2004). Control of entanglement between photon and quantum well, *Phys. Rev. Lett.* **93**, 067401.

J. P. Reithmaier, G. Sek, A. Löffler, *et al.* (2004). Strong coupling in a single quantum dot-semiconductor microcavity system, *Nature* **432**, 197.

T. Yoshie, A. Scherer, J. Hendrickson, *et al.* (2004). Vacuum Rabi splitting with a single quantum dot in a photonic crystal nanocavity, *Nature* **432**, 200.

P.-C. Ku, F. Sedgwick, C. J. Chang-Hasnain, *et al.* (2004). Slow light in semiconductor quantum wells, *Optics Lett.* **29**, 229.

M. Kira and S. W. Koch (2006a). Quantum-optical spectroscopy of semiconductors, *Phys. Rev. A* **73**, 013813.

S. W. Koch, M. Kira, G. Khitrova, and H. M. Gibbs (2006). Semiconductor excitons in new light, *Nature Mat.* **5**, 523.

J. Kasprzak, M. Richard, S. Kundermann, *et al.* (2006). Bose–Einstein condensation of exciton polaritons, *Nature* **443**, 409.

M. Kira and S. W. Koch (2006). Many-body correlations and excitonic effects in semiconductor spectroscopy, *Prog. Quantum Electr.* **30**, 155.

R. M. Stevenson, R. J. Young, P. Atkinson, *et al.* (2006). A semiconductor source of triggered entangled photon pairs, *Nature* **439**, 179.

N. Akopian, N. H. Lindner, E. Poem, *et al.* (2006). Entangled photon pairs from semiconductor quantum dots, *Phys. Rev. Lett.* **96**, 130501.

H. M. Gibbs, G. Khitrova, M. Kira, S. W. Koch, and A. Scherer (2006). Vacuum Rabi splitting in semiconductors, *Nature Phys.* **2**, 81.

T. Feldtmann, L. Schneebeli, M. Kira, and S. W. Koch (2006). Quantum theory of light emission from a semiconductor quantum-dot, *Phys. Rev. B* **73**, 155319.

S. M. Ulrich, C. Gies, S. Ates, *et al.* (2007). Photon statistics of semiconductor microcavity lasers, *Phys. Rev. Lett.* **98**, 043906.

L. Schneebeli, M. Kira, and S. W. Koch (2008). Characterization of strong light–matter coupling in semiconductor quantum-dot microcavities via photon-statistics spectroscopy, *Phys. Rev. Lett.* **101**, 097401.

J. Wiersig, C. Gies, F. Jahnke, *et al.* (2009). Direct observation of correlations between individual photon emission events of a microcavity laser, *Nature* **460**, 245.

G. Günter, A. A. Anappara, J. Hees, *et al.* (2009). Sub-cycle switch-on of ultrastrong light–matter interaction, *Nature* **458**, 178.

T. Feldtmann, M. Kira, and S. W. Koch (2010). Theoretical analysis of higher-order phonon sidebands in semiconductor luminescence spectra, *J. Luminescence* **130**, 107.

A. Carmele, M. Richter, W. W. Chow, and A. Knorr (2010). Antibunching of thermal radiation by a room-temperature phonon bath: A numerically solvable model for a strongly interacting light–matter-reservoir system, *Phys. Rev. Lett.* **104**, 156801.

R. P. Smith, J. K. Wahlstrand, A. C. Funk, *et al.* (2010). Extraction of many-body configurations from nonlinear absorption in semiconductor quantum wells, *Phys. Rev. Lett.* **104**, 247401.

Appendix

Conservation laws for the transfer matrix

In this appendix, we analyze the transfer matrix elements $\mathcal{T}^{\pm}$ and $\mathcal{R}^{\pm}$ and construct conservation laws. Our starting point is the Helmholtz equation (4.8)

$$\frac{\partial^2}{\partial z^2}\phi_E(z) + k^2(z)\,\phi_E(z) = 0, \qquad k(z) \equiv \sqrt{\frac{2m}{\hbar^2}\left[E - U(z)\right]} \tag{A.1}$$

and its discretized version

$$\frac{\partial^2}{\partial z^2}\phi_{k_0}(z) + k_j^2\,\phi_{k_0}(z) = 0, \quad \text{for } z_j \le z < z_{j+1}, \tag{A.2}$$

defined by Eq. (4.11). The k_j that appears is the wave vector representing the interval $z_j \le z < z_{j+1}$ as shown in Fig. 4.1. The wave-function solution is

$$\phi_{k_0}(z) \equiv \phi_j(z) = \mathbf{u}_j^T(z)\,\mathbf{s}_j, \qquad \text{for } z_j \le z < z_{j+1}$$
$$\mathbf{u}_j(z) = \begin{pmatrix} e^{ik_j(z-z_j)} \\ e^{-ik_j(z-z_j)} \end{pmatrix}, \quad \mathbf{s}_j = \begin{pmatrix} A_j \\ B_j \end{pmatrix}, \tag{A.3}$$

where we have represented Eq. (4.12) in terms of two vectors. The transfer matrix then connects the different $\mathbf{s}_j$ elements through the relation

$$\mathbf{s}_j = \mathcal{M}_{j,l}\,\mathbf{s}_l, \qquad \mathcal{M}_{j,l} \equiv \frac{1}{\mathcal{T}^{\leftarrow}_{j,l}}\begin{pmatrix} \mathcal{D}_{j,l} & \mathcal{R}^{\leftarrow}_{j,l} \\ -\mathcal{R}^{\rightarrow}_{j,l} & 1 \end{pmatrix}, \tag{A.4}$$

where $\mathcal{T}$ and $\mathcal{R}$ identify the transmission and reflection coefficients, respectively, and $\mathcal{D}_{j,l} = \mathcal{T}^{\leftarrow}_{j,l}\mathcal{T}^{\rightarrow}_{j,l} - \mathcal{R}^{\leftarrow}_{j,l}\mathcal{R}^{\rightarrow}_{j,l}$. The transfer matrix can be constructed either directly via matrix multiplication (4.24) or even more efficiently using the recursive approach introduced in Exercise 4.4.

A.1 Wronskian-induced constraints

The transfer-matrix approach above is explicitly formulated using the plane-wave basis. We also can solve the Helmholtz equation (A.1) using any linearly independent basis. In order to do this, we consider the more general form

$$\phi_E(z) = \begin{cases} A_l\, e^{ik_l(z-z_l)} + B_l\, e^{-ik_l(z-z_l)} & z_l \le z < z_{l+1} \\ C_1\, F(z) + C_2\, G(z) & z_{l+1} \le z < z_j \\ A_j\, e^{ik_j(z-z_j)} + B_j\, e^{-ik_j(z-z_j)} & z_j \le z < z_{j+1} \end{cases} \tag{A.5}$$

where we expressed the solution within the interval $z_{l+1} \le z < z_j$ in terms of two linearly independent solutions $F(z)$ and $G(z)$ instead of the plane-wave form used in Eq. (A.3). To make the notation simpler in the forthcoming calculations, we take the limit $z_{l+1} \to z_l$. This can be done without any loss of generality because we may choose the discretization arbitrarily.

We can now express the transfer matrix entirely in terms of the $F(z)$ and $G(z)$ functions. More specifically, we demand that $\phi_E(z)$ and $\frac{\partial}{\partial z}\phi_E(z)$ are continuous at the $z = z_{l+1} = z_l$ and z_j boundaries. A straightforward calculation eventually produces

$$\mathcal{M}_{j,l} = \frac{1}{2ik_j}\begin{pmatrix} ik_j & 1 \\ ik_j & -1 \end{pmatrix}\begin{pmatrix} f(z_j, z_{l+1}) & g(z_j, z_l) \\ \frac{\partial f(z_j,z_l)}{\partial z_j} & \frac{\partial g(z_j,z_l)}{\partial z_j} \end{pmatrix}\begin{pmatrix} 1 & 1 \\ ik_l & -ik_l \end{pmatrix} \tag{A.6}$$

which contains the functions

$$f(z_j, z_l) \equiv \frac{F(z_j)G'(z_l) - F'(z_l)G(z_j)}{W_l}, \qquad W_l \equiv F(z_l)G'(z_l) - F'(z_l)G(z_l)$$

$$g(z_j, z_l) \equiv \frac{F(z_j)G(z_l) - F(z_l)G(z_j)}{W_l}. \tag{A.7}$$

Mathematically, W_l is the Wronskian that is always nonzero for linearly independent F and G.

We may now solve the Wronskian part of Eq. (A.6); a straightforward matrix inversion and multiplication produces

$$\mathcal{W} \equiv \begin{pmatrix} f(z_j, z_{l+1}) & g(z_j, z_l) \\ \frac{\partial f(z_j,z_l)}{\partial z_j} & \frac{\partial g(z_j,z_l)}{\partial z_j} \end{pmatrix}$$
$$= \begin{pmatrix} \frac{\mathcal{D}_{j,l}+1+\mathcal{R}^{\leftarrow}_{j,l}-\mathcal{R}^{\rightarrow}_{j,l}}{2\mathcal{T}^{\leftarrow}_{j,l}} & \frac{\mathcal{D}_{j,l}-1-\mathcal{R}^{\leftarrow}_{j,l}-\mathcal{R}^{\rightarrow}_{j,l}}{2ik_l\,\mathcal{T}^{\leftarrow}_{j,l}} \\ ik_j\frac{\mathcal{D}_{j,l}-1+\mathcal{R}^{\leftarrow}_{j,l}+\mathcal{R}^{\rightarrow}_{j,l}}{2\mathcal{T}^{\leftarrow}_{j,l}} & \frac{k_j}{k_l}\frac{\mathcal{D}_{j,l}+1-\mathcal{R}^{\leftarrow}_{j,l}+\mathcal{R}^{\rightarrow}_{j,l}}{2\mathcal{T}^{\leftarrow}_{j,l}} \end{pmatrix}. \tag{A.8}$$

Since the transfer matrix produces a unique solution of Eq. (A.2), $\mathcal{W}$ cannot depend on the choice of the functions F and G. At the same time, Eq. (A.1) establishes a real-valued eigenvalue problem such that we may choose $F(z)$ and $G(z)$ to be real-valued functions. Therefore, $\mathcal{W}$ must also be real valued under all circumstances. This restriction establishes the first connection between the elements of the transfer matrix.

A.2 Current-induced constraints

We can also use the solutions of the Helmholtz equation (A.1) to express the time-dependent wave function in the form

$$\psi(z, t) = \phi_E(z)e^{-i\frac{Et}{\hbar}}. \tag{A.9}$$

It is straightforward to show that $\psi(z,\,t)$ satisfies the Schrödinger equation (3.1). Therefore, the introduced $\psi(z,\,t)$ must satisfy the continuity equation, discussed in Exercise 3.2. More explicitly, we can apply

$$i\hbar\frac{\partial}{\partial t}\left[\psi(z,\,t)\,\psi^{\star}(z,\,t)\right]=-\frac{\hbar^2}{2m}\frac{\partial}{\partial z}\left(\psi^{\star}(z,\,t)\frac{\partial\psi(z,\,t)}{\partial z}-\psi(z,\,t)\frac{\partial\psi^{\star}(z,\,t)}{\partial z}\right). \quad (A.10)$$

For a stationary eigenstate (A.9), $\psi^{\star}(z,\,t)\psi(z,\,t)$ is independent of time such that Eq. (A.10) reduces into

$$\frac{\partial}{\partial z}\left(\psi(z,\,t)^{\star}\frac{\partial\psi(z,\,t)}{\partial z}-\psi\frac{\partial\psi^{\star}(z,\,t)}{\partial z}\right)=0. \quad (A.11)$$

This also means that the solution to the Helmholtz equation (A.1) defines a constant current

$$J\equiv\psi^{\star}(z,\,t)\frac{\partial\psi(z,\,t)}{\partial z}-\psi(z,\,t)\frac{\partial\psi^{\star}(z,\,t)}{\partial z}=\phi_E^{\star}(z)\frac{\partial\phi_E(z)}{\partial z}-\phi_E(z)\frac{\partial\phi_E^{\star}(z)}{\partial z} \quad (A.12)$$

for all investigated intervals. This yields additional restrictions between the elements of the $\mathcal{M}$-matrix.

Inserting solution (A.3) into (A.12) produces

$$\begin{aligned}J&=\mathbf{s}_l^H\mathcal{M}_{j,l}^H\left(\mathbf{u}_j^{\star}(z)\frac{\partial\mathbf{u}_j^T(z)}{\partial z}-\frac{\partial\mathbf{u}_j^{\star}(z)}{\partial z}\mathbf{u}_j^T(z)\right)\mathcal{M}_{j,0}\,\mathbf{s}_l\\&\equiv\mathbf{s}_l^H\,\mathcal{M}_{j,0}^H\mathbf{U}_j(z)\mathcal{M}_{j,l}\,\mathbf{s}_l,\end{aligned} \quad (A.13)$$

where we defined a new 2-by-2 matrix $\mathbf{U}_j(z)$. Since all intervals produce the same J, we can rewrite the condition (A.13) in the generic form

$$\mathbf{s}_l^H\mathcal{M}_{j,l}^H\mathbf{U}_j(z)\mathcal{M}_{j,l}\mathbf{s}_l=\mathbf{s}_l^H\mathbf{U}_l(z_l)\mathbf{s}_l. \quad (A.14)$$

Since $\mathbf{s}_l$ is an arbitrary complex vector, the transfer matrix must also satisfy the condition

$$\mathcal{M}_{j,l}^H\mathbf{U}_j(z)\mathcal{M}_{j,l}=\mathbf{U}_l(z_l), \quad (A.15)$$

for all j and l combinations. This establishes the second restriction among the elements of $\mathcal{M}$.

To specify condition (A.15) completely explicitly, we insert $\mathbf{u}_j(z)$ from Eq. (A.3) into Eq. (A.13) and obtain

$$\mathbf{U}_j(z)=i\begin{pmatrix}\left(k_j+k_j^{\star}\right)e^{i\left(k_j-k_j^{\star}\right)\Delta z} & -\left(k_j-k_j^{\star}\right)e^{-i\left(k_j+k_j^{\star}\right)\Delta z}\\ \left(k_j-k_j^{\star}\right)e^{i\left(k_j+k_j^{\star}\right)\Delta z} & \left(k_j+k_j^{\star}\right)e^{-i\left(k_j-k_j^{\star}\right)\Delta z}\end{pmatrix}, \quad (A.16)$$

where $\Delta z = z - z_j$. Generally, Eq. (A.1) produces k_j that is either real or purely imaginary. A real k_j implies $k_n^\star = k_n$ while a purely imaginary k_n yields $k_n^\star = -k_n$. These two possibilities simplify $\mathbf{U}_j(z)$ into

$$\mathbf{U}_j(z) = \begin{cases} 2ik_j \begin{pmatrix} 1 & 0 \\ 0 & -1 \end{pmatrix}, & k_j \in \mathbb{R} \\ 2ik_j \begin{pmatrix} 0 & -1 \\ 1 & 0 \end{pmatrix}, & k_j = i\kappa_j, \quad \kappa_j \in \mathbb{R} \end{cases}, \tag{A.17}$$

which can be applied to evaluate condition (A.15) explicitly.

A.3 Explicit conservation laws

Condition (A.15) yields four different cases depending on whether k_l and k_j are real or purely imaginary numbers. This means that we have four different cases in total. We will perform the derivations in detail for the case with real-valued k_j and k_l; the other cases are obtained analogously.

Case 1: The condition $k_l \in \mathbb{R}$ and $k_j \in \mathbb{R}$ produces the $\mathbf{U}$ matrices

$$U_j = 2ik_j \begin{pmatrix} 1 & 0 \\ 0 & -1 \end{pmatrix}, \qquad U_l = 2ik_l \begin{pmatrix} 1 & 0 \\ 0 & -1 \end{pmatrix}, \tag{A.18}$$

based on the result (A.17). As we insert this together with the explicit transfer matrix (A.4) into (A.15), we eventually obtain the three conditions

$$\begin{cases} \mathcal{D}_{j,l}\mathcal{D}^\star_{j,l} - \mathcal{R}^{\rightarrow}_{j,l}\left(\mathcal{R}^{\rightarrow}_{j,l}\right)^\star = \frac{k_l}{k_j}\mathcal{T}^{\leftarrow}_{j,l}\left(\mathcal{T}^{\leftarrow}_{j,l}\right)^\star \\ \mathcal{R}^{\leftarrow}_{j,l}\left(\mathcal{R}^{\leftarrow}_{j,l}\right)^\star - 1 = -\frac{k_l}{k_j}\mathcal{T}^{\leftarrow}_{j,l}\left(\mathcal{T}^{\leftarrow}_{j,l}\right)^\star \\ \mathcal{D}^\star_{j,l}\mathcal{R}^{\leftarrow}_{j,l} + \left(\mathcal{R}^{\rightarrow}_{j,l}\right)^\star = 0 \end{cases}. \tag{A.19}$$

We can now write $\mathcal{D}_{j,l} = \mathcal{T}^{\rightarrow}_{j,l}\mathcal{T}^{\leftarrow}_{j,l} - \mathcal{R}^{\rightarrow}_{j,l}\mathcal{R}^{\leftarrow}_{j,l}$, based on Eq. (4.24). If we additionally assume that $\mathcal{T}^{\pm}_{j,l}$ does not vanish, Eq. (A.19) casts into the form

$$\begin{cases} \frac{k_l}{k_j}\left|\mathcal{T}^{\leftarrow}_{j,l}\right|^2 + \left|\mathcal{R}^{\leftarrow}_{j,l}\right|^2 = 1 \\ \frac{k_j}{k_l}\left|\mathcal{T}^{\rightarrow}_{j,l}\right|^2 + \left|\mathcal{R}^{\rightarrow}_{j,l}\right|^2 = 1 \\ \left(\mathcal{T}^{\rightarrow}_{j,l}\right)^\star \mathcal{R}^{\leftarrow}_{j,l} + \frac{k_0}{k_n}\mathcal{T}^{\leftarrow}_{j,l}\left(\mathcal{R}^{\rightarrow}_{j,l}\right)^\star = 0 \end{cases}. \tag{A.20}$$

The first two relations imply the conservation of the probability current. Originally, the $\mathcal{M}$-matrix has eight unknown parameters because it is a complex-valued 2×2 matrix. The relations (A.20) determine four different conditions: the first and the second relation produce one constraint each and the third relation yields two constraints because it is a complex-valued condition.

We find one more condition by applying Eq. (A.8) in the studied situation. More explicitly, we demand that $\mathcal{W}$ is real valued to find

$$\begin{cases} \left(\mathcal{T}_{\overleftarrow{j,l}}\right)^{\star}\mathcal{D}_n - \mathcal{T}_{\overleftarrow{j,l}} = 0 \\ \left(\mathcal{T}_{\overleftarrow{j,l}}\right)^{\star}\mathcal{R}_{\overrightarrow{j,l}} + \mathcal{T}_{\overleftarrow{j,l}}\left(\mathcal{R}_{\overleftarrow{j,l}}\right)^{\star} = 0 \end{cases}. \tag{A.21}$$

As we combine this with Eq. (A.20), Eq. (A.21) reduces into

$$\mathcal{T}_{\overleftarrow{j,l}} = \frac{k_n}{k_0}\mathcal{T}_{\overrightarrow{j,l}}. \tag{A.22}$$

Therefore, the transfer matrix has the following conservation laws:

$$\begin{cases} \mathcal{T}_{\overleftarrow{j,l}} = \frac{k_n}{k_0}\mathcal{T}_{\overrightarrow{j,l}} \\ \frac{k_n}{k_0}\left|\mathcal{T}_{\overrightarrow{j,l}}\right|^2 + \left|\mathcal{R}_{\overrightarrow{j,l}}\right|^2 = 1 \\ \left(\mathcal{T}_{\overleftarrow{j,l}}\right)^{\star}\mathcal{R}_{\overrightarrow{j,l}} + \mathcal{T}_{\overleftarrow{j,l}}\left(\mathcal{R}_{\overleftarrow{j,l}}\right)^{\star} = 0 \end{cases}. \tag{A.23}$$

The total number of conditions is five, two result from each of the first two equations and one from the third equation.

Case 2: $k_j = i\kappa_j$, $\kappa_j \in \mathbb{R}$ and $k_l \in \mathbb{R}$. The conservation laws for this situation can be derived in full analogy to Eq. (A.23). In this specific situation, we find

$$\begin{cases} \mathcal{T}_{\overleftarrow{j,l}} = \frac{i\kappa_j}{k_l}\mathcal{T}_{\overrightarrow{j,l}} \\ \mathcal{T}_{\overleftarrow{j,l}} + \left(\mathcal{T}_{\overleftarrow{j,l}}\right)^{\star}\mathcal{R}_{\overrightarrow{j,l}} = 0 \\ \left(\mathcal{R}_{\overleftarrow{j,l}}\right)^{\star} - \mathcal{R}_{\overleftarrow{j,l}} = \frac{k_l}{i\kappa_j}\left|\mathcal{T}_{\overleftarrow{j,l}}\right|^2 \end{cases}, \tag{A.24}$$

which also yields five conditions.

Case 3: $k_j \in \mathbb{R}$ and $k_l = i\kappa_l$, $\kappa_l \in \mathbb{R}$. In this situation, the relations (A.8) and (A.15) eventually produce

$$\begin{cases} \mathcal{T}_{\overleftarrow{j,l}} = \frac{k_j}{i\kappa_l}\mathcal{T}_{\overrightarrow{j,l}} \\ \left(\mathcal{T}_{\overleftarrow{j,l}}\right)^{\star}\mathcal{R}_{\overleftarrow{j,l}} = \mathcal{T}_{\overleftarrow{j,l}} \\ \left(\mathcal{R}_{\overrightarrow{j,l}}\right)^{\star} - \mathcal{R}_{\overrightarrow{j,l}} = \frac{k_j}{i\kappa_l}\left|\mathcal{T}_{\overrightarrow{j,l}}\right|^2 \end{cases}, \tag{A.25}$$

which again imposes five conditions on the transfer matrix.

Case 4: $k_j = i\kappa_j$, $\kappa_j \in \mathbb{R}$ and $k_l = i\kappa_l$, $\kappa_l \in \mathbb{R}$. If we combine Eqs. (A.15) and (A.8), we obtain the five independent conditions in the form

$$\begin{cases} \mathcal{T}^{\leftarrow}_{j,l} = \frac{k_n}{k_0}\mathcal{T}^{\rightarrow}_{j,l} \\ \mathcal{R}^{\rightarrow}_{j,l} = \left(\mathcal{R}^{\rightarrow}_{j,l}\right)^{\star} \\ \mathcal{R}^{\leftarrow}_{j,l} = \left(\mathcal{R}^{\leftarrow}_{j,l}\right)^{\star} \\ \left(\mathcal{T}^{\leftarrow}_{j,l}\right)^{\star} = \mathcal{T}^{\leftarrow}_{j,l} \end{cases}. \tag{A.26}$$

We now see that all these four cases yield a common condition

$$\mathcal{T}^{\leftarrow}_{j,l} = \frac{k_j}{k_l}\mathcal{T}^{\rightarrow}_{j,l}. \tag{A.27}$$

In general, the conservation laws are very useful in checking the correctness and numerical accuracy of the transfer matrix programs. They also provide analytical insight into the properties and orthogonality of mode functions, as seen in Chapter 4.

Further reading

This derivation is based on the discussion presented in:

M. Kira (1995). *Complementary quantum dynamics*, University of Helsinki Report Series in Theoretical Physics, HU-TFT-IR-95-3.

Index

N-band generalizations, 502
N-particle
  cluster, 310
  expectation value, 296, 311
  operator, 228, 296, 298
Γ-point, 243
δ-function, 14
$\hat{\mathbf{A}} \cdot \hat{\mathbf{p}}$ picture, 345
$\hat{\mathbf{A}} \cdot \mathbf{p}$ interaction, 325
$\hat{\mathbf{E}} \cdot \hat{\mathbf{x}}$ picture, 345
π pulse, 384
$\pi, 3\pi, \cdots$ pulse, 384
$g^{(2)}$ function, 511
n-th-order coherences, 511
p-polarization, 181
p-polarized plane waves, 181
s-polarization, 181
s-polarized plane waves, 181
$\mathbf{k} \cdot \mathbf{p}$ analysis, 243

absorption
  linear, 536
  spectra, 534
abstract notation, 307
acoustic phonons, 236, 266
action, 3
  electromagnetic field, 130
angle of incidence, 179
angular momentum, 7
anharmonic corrections, 12
anharmonic oscillation, 11
anharmonic oscillator, 19
annihilation operator, 204, 223
antibunching, 511
anticrossing, 434
associated Laguerre polynomial, 329
associated Legendre polynomial, 329
atom
  classical instability, 48
  stability, 65
Auger scattering, 271, 275
auxiliary space, 196
auxiliary space vs. real space, 197
average value, 54
axioms, 1, 26, 49

background susceptibility, 165
band gap, 242, 522
  renormalization, 522
  unrenormalized, 245
band structure, 240, 242
  calculations, 240
basis, 60
  vector, 90, 241
biexciton, 490
binary semiconductor compounds, 513
binding energy, 329
Bloch
  function, 241
  sphere, 380, 381
  theorem, 241
  vector, 380, 391, 393, 476
  wave function, 248
Bloch equations
  magnetic, 391
  Maxwell-optical, 395, 425, 500
  optical, 391
  self-consistent, 399
Bogoliubov–Born–Green–Kirkwood–Yvon hierarchy, 298
Bohr radius, 329
Boltzmann scattering, 559
Born's theorem, 94, 198, 458
Born–Oppenheimer approximation, 162
Bose–Einstein
  condensation, 620
  distribution, 236, 465
Boson, 211, 220
  commutation relation, 199
  exciton approximation, 555

bound state, 78
  solutions, 72
boundary
  classical–quantum, 52
box potential, 79
bra–ket notation, 89
Brewster angle, 184
bright excitons, 599
broadening
  inhomogeneous, 502
  radiative, 354, 536
build-up of correlations, 318
bunching, 511

canonical commutation relation, 59, 233
canonical momentum, 27, 124, 126
canonical quantization, 60, 193
canonical velocity, 361
carrier correlations
  photon assisted, 610
carrier destruction operator, 282
carrier energy
  average, 560
carrier quantum dynamics, 574
Casimir effect, 215
Cauchy principal value, 84
Cauchy's integral theorem, 268, 276
Cauchy–Schwarz inequality, 96, 99
cavity, 190
cavity mode, 427
cavity polariton, 433
center-of-mass coordinate, 325, 484
chain rule of differentiation, 7
chaotic evolution, 19
chaotic light, 465
chaotic trajectories, 13
charge density, 26
charge distribution, 27, 135
charged particles, 26
chirp, 44, 62
classical averaging, 21, 48, 466
classical chaos, 13
classical electromagnetism, 26
classical factorization, 305, 425, 445, 468
classical forbidden region, 8
classical limit for light, 366
classical Liouville equation, 16
classical phase-space distribution, 13
classical trajectory, 37
classically forbidden regions, 72
cluster expansion, 304, 425
  Jaynes–Cummings model, 467
cluster-expansion factorization, 468
cluster-expansion method, 304
cluster-expansion truncation, 311, 471
coherence, 472
  quantum optical, 511
coherence time, 512
coherent correlation, 472
coherent exciton, 550, 554, 556
coherent limit, 502, 505, 551, 562, 620
coherent quantity, 472, 499, 508, 509
coherent state, 446, 459, 511
  photon statistics, 448
  Wigner function, 461
coherent-limit wave function, 553, 555
collapse, 450
collective interaction, 593
collective phonon frequency, 235
collective photon operator, 525
commutation relation
  Bosonic, 199
  canonical, 59, 233
commutator, 58, 281
completeness relation, 39, 66, 88, 90
conditional probability, 483, 557
conduction band, 242, 485
confinement function, 247, 249, 266, 272, 274, 487, 573
conservation law, 7, 473, 545, 575, 577, 608
conservation of angular momentum, 8
conservation of energy, 7
contraction, 310
contrast of the interference, 73
conversion
  polarization-to-population, 559, 562, 601
convolution theorem, 333
coordinates
  relative and center-of-mass, 325, 484
Copenhagen interpretation, 95
correlated clusters, 311
correlated doublets, 468, 574
correlated electron states, 310
correlated electron–hole plasma, 565, 576, 586, 587
correlated plasma, 312, 586
correlation
  coherent, 472
correlation build-up, 318
correlation function
  electron–hole-pair, 556
correlations, 304
  higher-order, 312
  quantum optical, 524
  two-photon emission, 611
correspondence of atoms and exciton correlations, 603
correspondence principle, 60
Coulomb enhancement, 531
Coulomb force, 347
Coulomb gauge, 143, 231, 325
  generalized, 166
Coulomb Hamiltonian, 267
Coulomb interaction, 267, 268, 272, 274, 298
  generalized, 170

quantum dot, 275
Coulomb matrix element, 268, 270
effective, 528
Coulomb potential, 144, 151
Coulomb-induced dephasing, 543
counterpropagating pulses, 30
coupled-clusters approach, 304
creation operator, 204
critical angle, 182, 183
crystal momentum, 241
current
kinetic, 346
current density, 26
current distribution, 27, 135
current operator
macroscopic, 286
current-matrix element, 261, 270, 272, 274, 275
currents
macroscopic, 293

d'Alambert operator, 28
damped oscillator, 503
dark energy, 216
dark excitons, 599
decay constants, 408
deformation potential, 266
degeneracy, 66, 88
degeneracy removal, 433
degenerate bands, 248
degenerate states, 88
degree of degeneracy, 88
delta function, 14, 171
density distribution
microscopic, 483
density matrix, 116, 212, 413, 440
reduced, 441
two-level system, 406
density of β excitons, 564
density of states, 250, 534
density operator
macroscopic, 286, 490
density-functional theory, 300
density-matrix theories, 299
dephasing, 409
radiative, 397
dephasing approximation
triplets, 595
dephasing model
diffusive, 545
detuning, 370
diagrammatic rules, 262
dielectric constant, 26
difference operator, 294
diffraction, 37, 170
diffusive dephasing model, 545
dipole approximation, 342
dipole emission, 353
dipole matrix element, 423, 573
dipole selection rules, 338, 367
dipole self-energy, 358, 366
dipole source
planar, 354
dipole strength, 338
dipole transitions, 367
Dirac notation, 89
direct diagonalization, 616
direct transitions, 485
discretization
Helmholtz equation, 67
disorder, 230
dispersion relation, 30, 38, 40, 50, 51, 153, 240
displacement operator, 358, 447, 556
displacement parameter, 447
dissipative processes, 408
distributed Bragg reflector, 190
double-slit experiment, 48
doublet correlations, 574
doublets, 304
Drude response, 604
dyadic tensor, 155
dynamics-controlled truncation, DCT, 320

effective Coulomb matrix element, 528
effective field, 610
effective mass, 522
effective mass tensor, 245
effective mode function, 488
effective mode strength, 423
eigenvalue problem, 87
eigenvalue problem, matrix form, 91
eikonal approximation, 35
eikonal equation, 33, 50
electric field, 26, 347
effective, 389
longitudinal, 151
electrical charge, 26
electrical field, 291
pulsed, 28
electromagnetic fields, 27
electron distribution function, 486
electron–hole distribution, 552
electron–hole droplets, 490
electron–hole mass
reduced, 245
electron–hole pair-correlation function, 556
electron–hole-pair recombination, 572
electron–hole pairs, 550
electron–hole picture, 486
electron–hole plasma
correlated, 565, 576, 586, 587
electron–hole recombination
photon assisted, 575
electron–phonon interaction, 253
electronic vacancy, 245, 271, 486

elementary electron charge, 330
Elliott formula, 533
Elliott luminescence formula, 585, 587
energetic shift, 397
energy conservation, 7
energy density of states, 251
energy dynamics, 594
energy quantization, 78
energy renormalization, 522
entanglement, 433, 615
entropy
  thermal light, 465
envelope
  slowly varying, 33, 395
envelope-function approximation, 247
envelope-function matrix element, 257
equation hierarchy, 347
equation-of-motion technique, 299, 304
equator, 381
evanescent decay, 183
evanescent propagation, 37, 170, 294
evanescent wave, 35, 72
excitation-induced dephasing, 542, 543
excited states, 81
exciton, 330, 556
  coherent, 550, 554, 556
  radiative decay, 600
exciton basis, 529, 563, 576, 585
exciton binding energy
  two-dimensional, 331
exciton correlation, 557
exciton correlation function, 575
exciton distribution, 564, 601
  nonequilibrium, 601
exciton formation, 568
exciton ionization, 528
exciton operator, 554
exciton population, 527, 556
exciton resonance shift, 529
exciton states, 527
excitonic emission, 577
excitonic plasma luminescence, 594
excitonic resonances, 533
excitons, 527
  Fermionic substructure, 586
  optically bright, 599
  optically dark, 599
expansion coefficients, 89
expectation value, 54, 196
  momentum, 56
  normally ordered, 305, 313, 465
  symmetrized, 106
external fields, 165
external potentials, 165

Fabry Pérot cavity, 83
factorization
  classical, 305, 425, 445, 468
Fermi energy, 242
Fermi pressure, 528
Fermion, 219
Fermion anticommutation, 223
Fermion anticommutation relations, 228
Fermionic field operator, 306
Fermionic substructure, 555
Fermions
  field operator, 227
Feynman path integrals, 299
field operator, 227, 228, 483, 557
fluctuations, 95
  light field, 444
  photon number, 448, 465
Fock number states, 208
Fock space, 222
Fock state, 430, 460
  field expectation value, 214
  Wigner function, 462
force, 5, 6
four-point correlation, 492
four-wave mixing, 491
Fourier expansion, 265, 267
Fourier transformation, 33
fractal structure, 19
free-electron mass, 330
Fresnel coefficients, 70, 84, 182
functional, 126
functional dependence, 3
functional derivative, 126, 127
functional displacement operator, 556
fuzziness, 60

Göppert-Mayer transformation, 357
gain, 544
gamma function, 45, 208
gauge invariance, 125, 361
gauge transformation, 125, 143
Gauss theorem, 56
Gaussian, 14
Gaussian integration, 461, 466
Gaussian solutions, 29
Gaussian wave, 42
generalized Coulomb gauge, 166
generalized Poisson equation, 167
generalized wave equation, 38, 167
Glauber–Sudarshan function, 455, 463
Green's function approach, 91
Green's functions, 299
ground state, 48, 81, 555
  semiconductor, 485
guided mode, 189

Hamilton equations, 4, 5, 123, 170
Hamilton formalism, 2
Hamilton operator, 59

Hamilton–Jacobi theory, 34
Hamiltonian, 5, 60, 92
  electromagnetic, 144
  electromagnetic field, 129
  generalized Coulomb gauge, 169
  interaction picture, 422
  Jaynes–Cummings, 429
  quantum two-level system, 423
  semiclassical, 366
  semiclassical two-level system, 406
  total, 269
harmonic oscillator eigenstates, 210
harmonic oscillator Hamiltonian, 203
harmonic potential, 10
Hartree contribution, 494
Hartree–Fock approximation, 309
Hartree–Fock factorization, 494
Heaviside function, 68, 171, 505
Heisenberg equation, 98, 279
Heisenberg picture, 97, 218
Heisenberg uncertainty principle, 57
Heisenberg uncertainty relation, 96
Helmholtz equation, 32, 65, 291, 294
  discretization, 67
  generalized, 168
  numerical solution, 67
Hermite polynomials, 209
Hermitian adjoint, 87
Hermitian conjugation, 87
Hermitian operator, 87
hierarchy problem, 98, 297, 298, 304, 347, 467
  quantum optical, 425, 427
Hilbert space, 222, 225
hole, 245, 271, 486
hole burning, 601
hole distribution function, 486
hole energy, 327
hole mass, 327
hole operator, 326
holes, 486
homodyne detection, 459, 617
homogeneous excitation, 484
Hubbard-type models, 300
Husimi function, 463
hydrogen atom, 324
  eigenstates, 328
hydrogen problem, 83, 328

identity operator, 225
image charge, 173, 174
implicit notation, 312
incoherent doublet correlations, 574
incoherent exciton, 550, 603
incoherent exciton populations, 576
incoherent quantity, 472, 499, 508, 573
incoherent regime, 573, 581
incoherent region, 609
indistinguishable particles, 219
inhomogeneous broadening, 502
inner product, 90, 223, 307
intensity distribution, 30
interaction Hamiltonian, 254, 264
  light–matter, 261
interaction matrix element, 489
interaction picture, 375, 422, 469
interaction vertex, 269
interactions, 283
interband processes, 286
interband transition, 262, 264, 508
interference, 37, 73, 170
  phase space, 109, 110
  quantum mechanical, 198
interference contrast, 32
interference fringes, 615
interference pattern, 31
intraband processes, 286
intraband transitions, 261, 264, 508
intrinsic inhomogeneous broadening, 502
intrinsic semiconductor, 484
inversion, 380
inversion factor, 379, 390, 528
  dynamics, 391
inversion operator, 407
inversion point, 528
ionic motion, 232
ionic polarization, 137
ionic susceptibility, 136
isotropic approximation, 246

Jaynes–Cummings eigenstates, 433
Jaynes–Cummings Hamiltonian, 429
Jaynes–Cummings ladder, 434, 442, 621
Jaynes–Cummings model, 467
jellium model, 269

ket, 89
kinetic current, 346
kinetic energy
  renormalized, 496
kinetic momentum, 27, 126

labile equilibrium, 12
ladder
  Jaynes–Cummings, 434, 442, 621
Lagrange equation, 124
Lagrange theory, 124
Lagrangian, 3
  Maxwell's equations, 124
Laguerre polynomial, 462
Laplace expansion, 221, 309
laser cooling, 368
laser cooling effects, 390
lattice periodicity, 265
lattice potential, 240

lattice vector, 241
lattice vibrations, 232, 253
left-handed solution, 527
Legendre transformation, 124, 128
Leibniz formula, 220, 236
Lennard-Jones potential, 369
level crossing, 434
Levi–Civita symbol, 149, 181
light
  classical limit, 366
  polarized, 599
light modes
  generalized, 177
light quantization, 420
light quantization hypothesis, 121
light rays, 34
light–matter correlations, 467
light–matter coupling, 125
  effective, 431
light–matter coupling strength, 423
light–matter Hamiltonian, 256
light–matter interaction, 263
limited operator space, 470
Lindblad form, 416
Lindhard screening, 542
line width
  natural, 354, 397
linear potential, 9
linear superposition, 39
linear susceptibility, 28, 532
linear-response relation, 28
Liouville equation, 414
  classical, 16
  general solution, 17
Liouville operator, 115
long-wave-length approximation, 341
long-wave-length limit Hamiltonian, 360
longitudinal component, 150
longitudinal field, 148
Lorentz force, 129, 132, 345, 348, 362
Lorentz gauge, 143
Louis de Broglie, 48
luminescence
  Elliott formula, 585, 587
  time-resolved, 597
luminescence equation, 472, 593
luminescence spectrum, 582

macroscopic optical polarization, 524
macroscopic polarization, 293, 524, 532, 551
magnetic Bloch equations, 391
magnetic field, 26
magnetic quantum number, 328
main-sum terms, 559
many-body interaction, 268
marginal distribution, 449, 459, 460
  Wigner function, 103
marginal distributions, 21
Markov approximation, 320
master equation, 416
matrix element, 257
  envelope-function, 257
matrix form
  eigenvalue problem, 91
Maxwell equations, 26, 135
  Lagrangian, 124
Maxwell–Bloch equations
  quantum optical, 424
  quantum-optical corrections, 470
Maxwell–Lorentz equations, 135, 170
Maxwell-optical Bloch equations, 395
Maxwell-semiconductor Bloch equations, 523, 550
mean-field contribution, 579
mean-field theory, 310
mean-value theorem of integration, 372
measurement ensemble, 458
measurements
  accuracy, 95
  correlated, 95
memory effects, 416
mesoscopic, 249
  structures, 246
mesoscopic nanostructures, 246
microcavity, 618
microscopic dipole, 338
microscopic polarization, 483
microscopic scattering terms, 524
minimal substitution, 129, 164
minimum-uncertainty state, 97
mixed operator terms, 298
mode expansion, 39, 66, 481
mode function, 153
mode strength
  cavity, 427
modes, 29
momentum conservation, 7, 263, 575
momentum matrix element, 245, 260, 339, 517, 525
momentum operator, 55, 57
  k-space, 60
momentum spread, 57
momentum states, 231
momentum transfer
  optical, 369
momentum uncertainty, 22
Mott transition, 544
  excitonic, 528
multicycle pulses, 513

N-representability problem, 300
nanostructure, 229
natural line width, 354, 397
near-field contributions, 353
near-resonance condition, 370
negative values

Wigner function, 107
Newton's equation
charged particle, 129
Newton's first law, 5
Newton's laws, 5
Newton's second law, 5
Newton's third law, 6
Newtonian axioms, 1
nonclassical states, 215
nonequilibrium exciton distribution, 601
nonradiative process, 562
nonradiative recombination, 602
nonthermal exciton distribution, 601
normal order, 214, 227, 445
normal-mode coupling, 614, 618
normalizability, 53
normally ordered expectation value, 305, 313, 465
north pole, 381
notation
Dirac, 89
number states, 208

observables, 87
open system, 405, 416
operator
formal aspects, 86
momentum, 55, 57
quantum mechanical, 54
operator class, 298
operator classes, 296
operator function, 16
operator product, 58
operator representation, 60
operator, adjoint, 87
operator, self-adjoint, 87
operators
ordering, 59
optical Bloch equations, 391, 425, 500
generalized, 426
self-consistent, 399
optical gain, 544
optical momentum transfer, 369
optical phonon, 236
optical pulses, 513
optical transitions, 367
optically active excitons, 599
orbital quantum number, 328
order
normal, 214, 227, 445
order parameter, 554
orthogonality, 88
orthogonality relation, 39, 66, 89
oscillation frequency, 10
oscillator strength, 531

pair-correlation function, 483
particle aspects, 50
particle vacuum, 222
particle wave, 53
particle waves, 49
particle–wave duality, 48
path
classical, 3
path integral, 91
Pauli blocking, 242, 528
Pauli exclusion principle, 220, 307
Pauli matrices
commutation relations, 407
Pauli operator dynamics, 424
Pauli spin matrices, 406, 421
periodic motion, 10
perturbation theory, 243
phase, 30
arbitrary, 54
slowly varying, 51
phase difference, 392
phase space, 3
phase-space averages, 105
phase-space filling, 528, 544
phase-space interference, 109, 110
phase-space trajectory, 9
phonon contributions, 610
phonon destruction operator, 282
phonon displacement, 482
phonon energy, 266
phonon field, 273
collective, 288, 489
phonon frequency
collective, 235
phonon Hamiltonian, 235
phonon interaction, 266, 274
phonon matrix element, 275
phonon operators, 287
phonon sideband emission, 615
phonon–electron coupling, 266
phonon, 230, 234
photoluminescence, 579
photon, 211, 218, 261
photon antibunching, 616
photon assisted carrier correlations, 610
photon cutoff, 573
photon field
collective, 489
photon number distribution, 441
photon number dynamics, 573
photon operator
collective, 525
effective, 427, 581
photon operator dynamics, 573
photon statistics, 441, 460
coherent state, 448
photon–carrier interaction, 283
photon–phonon system, 289

photon-singlet dynamics, 573
photon-assisted density correlations, 610
photon-assisted electron–hole recombination, 575
photon-assisted polarization, 473, 574
photon-number fluctuations, 448, 465
photon-operator dynamics, 424
photons-on-demand, 616
planar excitations, 491
plane-wave solution, 33, 122
plasma
  correlated, 312, 586
plasma frequency, 294
plasma heating, 597
plasma luminescence
  excitonic, 594
plasma properties, 312
plasma term, 295
Poisson distribution, 448
Poisson equation, 143
  generalized, 167
polariton
  cavity, 433
polarizability, 165
polarizable matter, 138
polarization, 28, 380, 390, 472
  ionic, 137
  macroscopic, 293, 524, 532, 551
  microscopic, 483
  photon-assisted, 473, 574
  transversal, 358
polarization decay, 562
polarization direction, 181, 261
polarization dynamics, 391
polarization operator, 407
  macroscopic, 286
polarization-operator dynamics, 580
polarization-to-population conversion, 559, 562, 601
position
  average, 54
position spread, 22
position–momentum uncertainty, 97
potential
  harmonic, 10
potential momentum, 126
principal quantum number, 328
probabilistic interpretation, 53, 95
probability density, 14
probability distribution, 13, 53
propagation angle, 35
pulse area, 383
pulses
  optical, 513
Purcell effect, 615
pure state, 407

quadrature distribution, 445
quadrature fluctuations, 215
quadrature operators, 205, 213
quadratures
  single mode, 617
quadruplets, 304
quantization
  canonical, 60, 193
  energy, 78
  second, 198
quantization box, 154
quantization volume, 250, 332
quantized phonon field, 482
quantized Rabi frequency, 423, 443
quantum confinement, 246
quantum degeneracy, 220
quantum degenerate states, 619
quantum dots, 249, 333, 616
quantum dynamics, 279
  Jaynes–Cummings, 439
quantum features, 52
quantum fluctuations, 55, 81
quantum interference, 198
quantum light
  fundamental definition, 463
quantum number, 80, 88
  energy, 65
quantum Rabi flopping, 442, 474
quantum Rabi oscillations, 445
quantum statistics, 101, 222, 440, 458, 463
quantum well, 246, 480
quantum wire, 249, 330, 480
quantum-confined system, 334
quantum-dot system, 274
quantum-dynamical properties, 290
quantum-optical correlation, 524
quantum-optical effects
  strong, 427
quantum-optical spectroscopy, 459, 466, 512, 618, 621
quantum-statistical might, 117
quantum-statistical pluralism, 463
quantum-well systems, 262
quasi-momentum states, 231
quasiequilibrium situation, 602
quasiparticles, 218, 240
quasiprobability distribution, 107

Rabi cycle, 382
Rabi flopping, 382, 384, 442, 501, 617
Rabi frequency, 377, 395, 500
  generalized, 379, 390, 474
  quantized, 443
  renormalized, 496, 522, 551
  self-consistent, 397
Rabi oscillation, 382
radiative broadening, 354, 536
radiative coupling constant, 536
radiative decay, 397, 399, 401, 411

nonlinearity, 412
radiative shift, 397
radiative width, 354
random-phase approximation, 483
ray optics, 34
reduced density matrix, 441
reduced electron–hole mass, 245
reduced mass, 325, 584
reflected part, 36
reflection, 180, 535
reflection coefficient, 70, 536
reflection probability, 74
refraction, 170, 535
refractive index, 28, 137, 165
refractive index changes, 35
regions
classically forbidden, 72
relative coordinate, 325, 484
renormalized kinetic energy, 496
renormalized Rabi frequency, 496, 522, 551
renormalized transition energy, 522
representation, 61
representation, state, 90
reservoir coupling, 408, 412
resonance fluorescence, 616
resonant contribution, 514
resonant tunneling, 83, 85
retardation effects, 146
retardation time, 147
retarded time, 146
reversible spontaneous emission, 445
revival, 450
revival time, 452
right-handed solution, 527
rotating frame, 374, 377, 469
rotating-wave approximation, 375, 406, 423, 515
Rydberg series, 531

scalar potential, 124
Lorentz gauge, 145
point charges, 144
scaled susceptibility, 535
scaling
spontaneous emission sources, 579
scaling law, 502
scattering terms
microscopic, 524
scattering-level approximation, 320
Schrödinger equation, 219, 389
abstract form, 92
classical aspects, 50
stationary, 65
time dependent, 49
Schrödinger picture, 97
screening, 166
second Born approximation, 320, 542
second moment, 214
second quantization, 198
second-order coherence, 511
secondary emission, 617
selection rules, 485
dipole, 338, 367
selective spontaneous recombination, 579
self-consistency, 394
self-interaction, 144
semiclassical approximation, 366
semiconductor Bloch equations, 521
semiconductor ground state, 485
semiconductor luminescence equations, 582, 609
semiconductor nanostructures, 480
separation ansatz, 327, 369, 389
separation of length scales, 256
separation of time scales, 499
shot noise, 460, 465
shot-noise limit, 449, 451
SIAS
state-in-auxiliary-space concept, 212
similarity transformation, 378
single-cycle pulses, 512
single-mode approximation, 427, 429
single-particle operators, 298
single-photon source, 212
singlet contribution, 468
singlet factorization, 309
singlet source, 558
singlet–doublet analysis, 490
singlets, 304
Slater determinant, 220, 306, 485, 552
Snell's law, 36, 180, 182, 188
south pole, 381
spectroscopy
quantum optical, 459, 466, 512, 618, 621
speed of light, 26
spherical harmonics, 329
spin matrices, 406
spontaneous emission, 445, 575, 577, 608
spontaneous-emission source, 473, 587
exciton basis, 585
spreading
wave packet, 51, 62
spreading time, 51
square normalizability, 53
squeezing, 617
state representation, 90
statistical uncertainty, 21
stimulated contribution
luminescence, 473
strong confinement limit, 248
strong quantum-optical effects, 427
subband, 248
subdeterminant, 221
subdeterminant expansion, 221, 237
subsingle-cycle pulses, 513
sum rules, 340

summation convention, 200, 307
superposition principle, 29, 92
superposition state, 30
superposition-based averaging of waves, 29
superradiance, 615
susceptibility
  background, 165
  ionic, 136
  linear, 28, 532
symmetrized expectation values, 106
system
  quantum-confined, 334
system energy
  average, 594
system Hamiltonian, 169, 253, 269

Taylor expansion, 4
terahertz spectroscopy, 603
thermal state, 464
third peak, 618
three-particle correlations, 468
three-particle operators, 298
time order, 378
time retardation, 352
time scale separation, 515
torque, 8
total Hamiltonian
  quantum wells, 272
total mass, 6, 325
trajectory, 51
transfer matrix, 71
  recursive, 627
transfer-matrix approach, 70
transient correlations, 509
transition amplitude, 530
transition energy, 500
translational symmetry, 241
transmission, 180, 535
transmission coefficient, 70, 536
transmission probability, 74
transposition, 220
transversal component, 150
transversal current, 152, 153
transversal field, 148
transversal polarization, 358
transversal wave equation, 153
transversal–longitudinal
    decomposition, 148
triplet contributions
  scattering approximation, 582
  spontaneous emission, 581
triplet dynamics, 319
triplets, 304
tunneling, 72, 170, 183
  Gaussian wave packet, 76
tunneling depth, 76
tunneling phenomena, 65
tunneling solution, 75
two-band approximation, 244
two-band model, 485
two-level approximation, 372
two-level model, 373
two-level system (TLS), 373
two-particle contributions, 468
two-particle correlations, 425
  homogeneous systems, 492
two-particle operators, 298
two-photon emission correlations, 611
two-photon processes, 262

ultraviolet catastrophe, 121
uncertainty principle, 44, 57
uncertainty relation, 44
  Heisenberg, 96
uncorrelated electron states, 310
uncorrelated electron–hole plasma, 579
uncorrelated light–matter system, 467
unit cell, 232, 236, 241
unitary transformation, 200, 356

vacuum permeability, 26
vacuum Rabi splitting, 614
vacuum state, 204, 219
vacuum wave length, 30
vacuum-field amplitude, 199, 215, 423
vacuum-field fluctuations, 215, 473, 578
valence band, 242, 485
valence-band hole, 271
variance, 55
variational principle, 3
variations, 3
vector field decomposition, 148
vector potential, 124, 255, 274
  fully quantized, 290
vectorial differentiation, 6
velocity of sound, 266
vertex, 269

Wannier equation, 527
Wannier excitons, 557
wave aspects, 71
wave averaging, 29, 53, 79, 197
wave equation, 27
  generalized, 38, 167
  scalar, 28
  transversal, 153
wave fronts, 41
wave function, 53
  representation, 89
wave length, 29
wave packet, 61
  spreading, 51, 62
wave packet dynamics, 50
wave packets, 32

wave vector, 29, 36, 153
wave-averaging, 39
wave-front propagation, 38
wave–particle dualism, 48, 50, 193
wetting layer, 616
which-way experiments, 615
Wigner function, 103, 461
  dynamics, 115
  Fock state, 462
  negativity, 107
  phase-space averages, 104
Wronskian, 628

zero-point energy, 81, 215, 248